3rd International Conference on Materials and Manufacturing Engineering (ICMME-2018)

IOP Conference Series: Materials Science and Engineering
Volume 390

Tamilnadu, India
8 – 9 March 2018

Editors:

T. Rajmohan
A. Tamilarasan
D. Vijayan

ISBN: 978-1-5108-6821-2
ISSN: 1757-8981

Printed from e-media with permission by:

Curran Associates, Inc.
57 Morehouse Lane
Red Hook, NY 12571

Some format issues inherent in the e-media version may also appear in this print version.

Printed by Curran Associates, Inc. (2018)

For permission requests, please contact the Institute of Physics
at the address below.

Institute of Physics
Dirac House, Temple Back
Bristol BS1 6BE UK

Phone: 44 1 17 929 7481
Fax: 44 1 17 920 0979

techtracking@iop.org

Additional copies of this publication are available from:

Curran Associates, Inc.
57 Morehouse Lane
Red Hook, NY 12571 USA
Phone: 845-758-0400
Fax: 845-758-2633
Email: curran@proceedings.com
Web: www.proceedings.com

Table of contents

Volume 390

The 3rd International Conference on Materials and Manufacturing Engineering 2018

8–9 March 2018, Tamilnadu, India

Accepted papers received: 6 July 2018
Published online: 30 July 2018

Preface

Papers

PREFACE

The **Third International Conference on Materials and Manufacturing Engineering (ICMME-2018)** was organized by Department of Mechanical Engineering, SRI CHANDRASEKHARENDRA SARASWATHI VISWA MAHAVIDYALAYA(SCSVMV), Kancheepuram, India, on March 8 & 9, 2018. The conference goal was to provide a global platform for professors, researchers, scientists, business giants and engineers coming from academia and industry to present their research results and activities in the relevant research field. The conference was also focused on today's technical challenges, research updates and breakthrough innovations that are shaping the future of materials and manufacturing engineering. The Conference convenes engineers, scientists and technologists for the purposes of exploring solutions to global challenges and for the advancement of materials and manufacturing engineering worldwide.

The first and second International conferences were organized in April 2015 and April 2017 at SCSVMV (Deemed to be University), Enathur, Kanchipuram, India. The series of conferences were very successful events and articles were published in SCOPUS indexed Journals. To continue our successful story, this year ICMME 2018 was comprised with broader coverage of all disciplines engineering areas.

We received overwhelming responses from national and international researchers, and 116 papers were finalized (i.e. on basis of Innovations and cutting-edge technologies in relevant to research on a global scale) to be included in the IOP Conference Series: Materials Science and Engineering.

We would like to thank and acknowledge all people who were involved in ICMME 2018. Each individual committee member and institution's financial support was very meaningful for the success of this conference. Especially, we would like to thank the organizing committee for their valuable advice in the organization and helpful peer review of the papers. We also to thank our inspiring keynote speakers, scientific reviewers, presenters and participants.

Dr.T.Rajmohan,

Associate Professor & Head, SCSVMV.,

Convenor - ICMME 2018.

IOP Conf. Series: Materials Science and Engineering **390** (2018) 011001 doi:10.1088/1757-899X/390/1/011001

GUEST EDITORS

Dr.T.RAJMOHAN
Associate Professor & Head,
Department of Mechanical Engineering,
Sri Chandrasekharendra Saraswathi Viswa Mahavidyalaya(SCSVMV)
Kanchipuram,
Tamilnadu-631561, India.
E.Mail: rajmohanscsvmv@yahoo.com

Dr.A.TAMILARASAN
Assistant Professor,
Department of Mechanical Engineering,
Sri Chandrasekharendra Saraswathi Viswa Mahavidyalaya(SCSVMV)
Kanchipuram,
Tamilnadu-631561, India.
E.Mail: tamilrj2010@gmail.com

Dr.D.VIJAYAN
Assistant Professor,
Department of Mechanical Engineering,
Sri Chandrasekharendra Saraswathi Viswa Mahavidyalaya(SCSVMV)
Kanchipuram,
Tamilnadu-631561, India.
E.Mail: vijaiand2012@gmail.com

IOP Conf. Series: Materials Science and Engineering **390** (2018) 011001 doi:10.1088/1757-899X/390/1/011001

COMMITTEE

Chairman : Prof. Dr. G. Sriram, Dean, Engineering & Technology, SCSVMV

Convenor : Dr. T. Rajmohan, Associate Professor & Head, SCSVMV

Organizing Secretary : Dr.A.Tamilarasan & Dr.D.Vijayan, SCSVMV

Organizing Committee : Dr.M.Vetrivel, SCSVMV
Dr.S.Arumugam, SCSVMV
Dr.A.Arun premnath, SCSVMV
Dr.K. Venkatesh, SCSVMV
Dr.R. Vinayagamoorthy, SCSVMV

Technical Committee : Dr.J. Paulo Davim, University of Aveiro, Portugal
Dr. Nikhil Ranjan Dhar, Professor, BUET, Bangladesh
Dr. Immanuel Edinbarough, University of Texas, Brownville
Uday S Dixit. Professor, IIT Guwahati
Dr.N Ramachandran , Professor & Head, NIT Calicut
Dr. Vinod Yadava, Professor, MNNIT Allahabad
M. Chandrasekaran, NERIST Nirjuli, Arunachal Pradesh
Dr. S. Basavarajappa, Professor, IIIT Dharwad
Dr. G. Nagarajan, Professor, Anna University, Chennai
Dr. N. Krishnamohan, Professor, Annamalai University
Dr. K. Elangovan, Oman
Dr. P. Ramkumar, IIT Madras
Dr. I. A. Palani, IIT, Indore
Dr. J. Ramkumar, IIT Kanpur
Dr. P. Asokan, NIT, Trichy
Dr. J. Srinivas, NIT, Rourkela
Dr. L. Karunamoorthy, Anna University, Chennai
Dr. A. Azad, Anna University, Chennai
Dr. M. Pradeepkumar, Anna University, Chennai
Dr. Gaitentode, BVB college of Engineering, Hubli
Dr. J. Sornakumar, Thiagaraja College of Engg.,
Dr. G. Subba Rao, Geethanjali Inst. of Sci & Tech, Nellore
Dr. K. Rama Kotaiah, K L Deemed University, Vijayawada

IOP Conf. Series: Materials Science and Engineering **390** (2018) 011001 doi:10.1088/1757-899X/390/1/011001

KEYNOTE SPEAKERS

Kittichai Sojiphan, Ph.D.

Department of Welding Engineering Technology,

College of Industrial Technology,

King Mongkut's University of Technology North Bangkok,

Thailand.

Prof.B.V.R.Chowdari,

Senior Executive Director,

President's Office,

School of Materials Science & Engineering,

Nanyang Technological University,

Singapore.

DOCUMENTATION

INAUGURAL SESSION WITH KEY NOTE SPEAKERS AND OFFICIALS OF SCSVMV

IOP Conf. Series: Materials Science and Engineering **390** (2018) 011001 doi:10.1088/1757-899X/390/1/011001

INAUGURAL SESSION WITH PARTICIPANTS

IOP Conf. Series: Materials Science and Engineering **390** (2018) 011001 doi:10.1088/1757-899X/390/1/011001

HONOURING OF KEYNOTE SPEAKERS

HONOURING OF KEYNOTE SPEAKER

IOP Conf. Series: Materials Science and Engineering **390** (2018) 011001 doi:10.1088/1757-899X/390/1/011001

HONOURY WELCOME OF OUR KEYNOTE SPEAKERS

CERTIFICATE DISTRIBUTION BY KEYNOTE SPEAKER

Peer review statement

All papers published in this volume of *IOP Conference Series: Materials Science and Engineering* have been peer reviewed through processes administered by the proceedings Editors. Reviews were conducted by expert referees to the professional and scientific standards expected of a proceedings journal published by IOP Publishing.

IOP Conf. Series: Materials Science and Engineering **390** (2018) 012001 doi:10.1088/1757-899X/390/1/012001

Investigation of structural, optical and electronic properties of (Co, Ni) codoped CeO_2 nanoparticles

Saurabh Tiwari[a], Nivedha Balasubramanian[b], Sajal Biring[c], Somaditya Sen[a,*]

[a]Metallurgical Engineering and Material Sciences, Indian Institute of Technology, Indore, India

[b]Department of Physics, PSG College of Arts and Science, Coimbatore, India

[b]Electronic Engg., Ming Chi University of Technology, New Taipei City, Taiwan

E-mail: sens@iiti.ac.in

Abstract. Cerium oxide co-doped with cobalt and nickel, is synthesized using sol-gel method. The phase purity and crystal structure are confirmed using x-ray powder diffraction (XRD). Rietveld analysis of XRD data confirms cubic fluorite structure of undoped and (Co, Ni) co-doped CeO_2. Lattice strain increases with codoping which results in reduction of crystallite size, lattice disorder and defects. UV-vis absorption analysis shows that co-doping leads to huge reduction in bandgap from 3.05 eV to 2.31 eV; this is due to the presence of defects and disorder as confirmed by Urbach energy calculation. Raman and Photoluminescence (PL) analysis shows that oxygen vacancies related defects are increasing with codoping. These defects leads to lattice disorder and shifts the bandgap from UV region to visible region by modifying its electronic levels.

1. Introduction

Nanocrystalline materials with crystallite size less than 100 nm has attracted a lot of interest of researchers by its better mechanical, electrical, optical, catalytic and thermal properties that notably differ from their bulk form. In recent times, CeO_2 nanostructures have received significant attention because of its novel properties i.e. hardness, UV absorption, high temperature stability, ability for reacting [1]. These properties leads to its wide application in sunscreens [2], degradation of dyes [3], in solid oxide fuel cells [4] etc.. It has cubic fluorite structure with each cerium is surrounded by 8 oxygen sites and each oxygen with 4 cerium sites. Recently, lots of work has been done on transition element doped CeO_2 such as enhancement in its photocatalytic activity, ferromagnetism and anticancer properties [5–7]. Oxygen vacancies related defects play crucial role in versatile application of CeO_2 as mentioned above.

But very less work has been done on transition element co-doped CeO_2. In the present study undoped and (Co, Ni) co-doped CeO2 samples were synthesized by sol-gel method. We have applied different characterization techniques such as X-ray diffraction, Raman spectroscopy, UV-visible

IOP Conf. Series: Materials Science and Engineering **390** (2018) 012001 doi:10.1088/1757-899X/390/1/012001

spectroscopy and Photoluminescence spectroscopy for understanding the effect of (Co, Ni) co-doping on structural, optical and electronic properties of CeO_2.

2. Experimental

Starting material used for synthesizing CeO_2 and $Ce_{0.95}Co_{0.025}Ni_{0.025}O_2$, were $(Ce(NO_3)_3.6H_2O)$, $(Co(NO_3)_2.6H_2O)$ and $(Ni(NO_3)_2.6H_2O)$, all were obtained from Alfa Aesar. The precursors were dissolved in double distilled de-ionized water in separate beakers. These precursors were mixed together and final solution is stirred for 2 h. For formation of the chelating chains and burning of gel, citric acid and glycerol were used. The resulted solution was maintained at constant temperature (80°C) which in the end gives a black powder. For denitrification and decarburization, the samples were calcined at 450°C for 6 h. The calcined samples were used for further characterization and the different compositions were referred as C1 for undoped CeO_2 and C2 for $Ce_{0.95}Co_{0.025}Ni_{0.025}O_2$ respectively.

For identification of the crystal structure X-ray diffraction (XRD) having Cu K_{α}, wavelength 1.54 Å (Bruker D_2 Phaser) was used. UV-visible spectroscopy (Shimadzu (UV-2600)) has been used for determining the optical bandgap. Photoluminescence (PL) measurement was done using fluorescence spectrometer (Perkin Elmer LS55, 340nm). Raman spectra were recorded using LabRAM HR ((HORIBA Scientific, He-Ne laser (632.8 nm)) having spectral resolution of 0.4 cm^{-1}.

3. Result and discussion

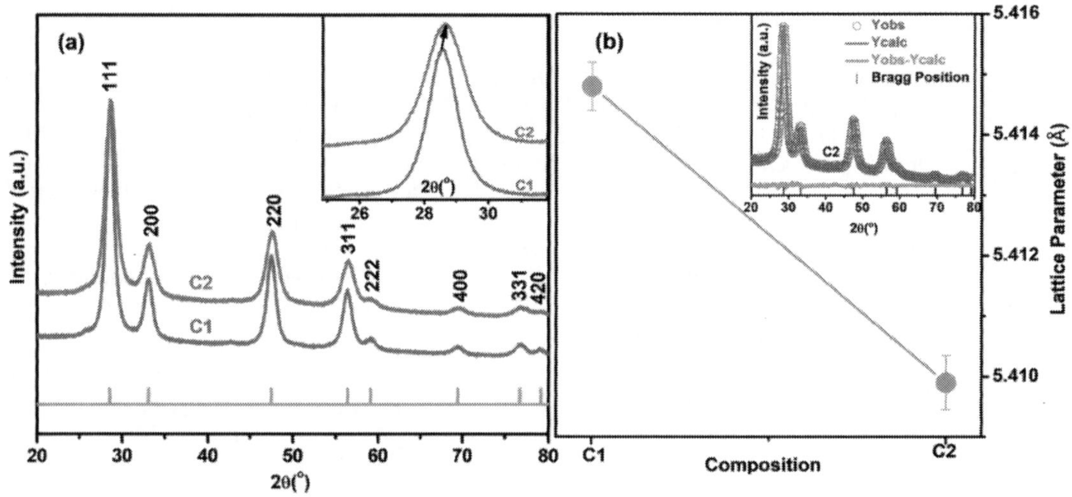

Fig.1. (a) XRD patterns of undoped and (Co, Ni) codoped CeO_2 (-inset shows peak shifting) (b) Variation of lattice parameter with cooping (-inset shows Rietveld refinement).

XRD patterns for CeO_2 (C1) and $Ce_{0.95}Co_{0.025}Ni_{0.025}O_2$ (C2) were shown in Fig. 1. (a), which shows the typical cubic fluorite structure having characteristic (111), (200), (220), (311), (222), (400), (331) and (420) planes (CIF file-4343161, Crystallography open database). No impurity phases were detected. The lattice parameter was determined using Rietveld analysis using Fullprof suite software, found decreasing from 5.4148 Å to 5.4099 Å [Fig. 1 (b)]. The lattice parameters were decreased because of lower ionic radii of $Co^{2+/3+}$ (0.75-0.9 Å), $Ni^{2+/3+}$ (0.7-0.83 Å) than Ce^{4+} (0.97 Å). This was inferred from Figure 1 (a-inset) peak which shift to higher angle that hints reduction in lattice parameter. The mean crystallite size (D) were evaluated from Scherer's formulae; $D = k\lambda/(\beta\cos\theta)$, using X-ray line broadening, where k (0.9) is a constant, θ is angle of diffraction, β is full width at half maxima FWHM) and λ is X-ray wavelength. The crystallite sizes was found to be 6.7 nm for undoped and 4.9 for (Co, Ni) codoped CeO_2 samples [Fig. 2]. Decrease in crystallite size and lattice parameter lead to lattice distortion and strain. Strain was evaluated by Williamson-Hall equation [8]; $\beta Cos\theta/\lambda = 1/D + \eta Sin\theta/\lambda$, where η is strain. Co-doping of (Co, Ni) increase the strain in CeO_2 lattice [Fig. 2].

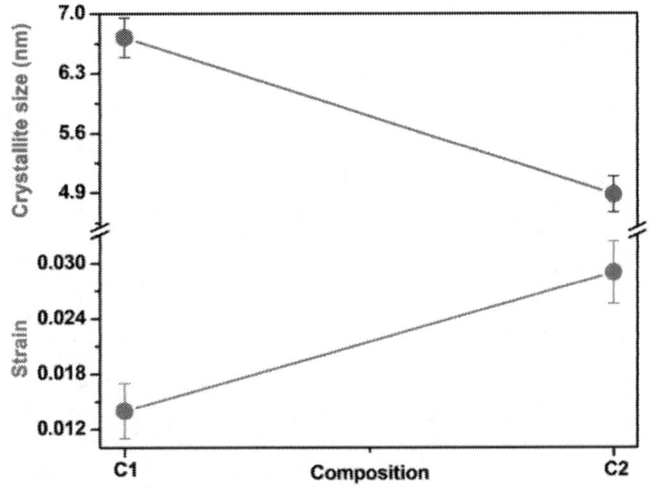

Fig. 2. Change in crystallite size and strain with composition.

The UV-visible absorption spectra of the C1 and C2 samples were shown in Figure 3 (a). An intense absorption below 400 nm was observed for both the samples, owing to charge transition from O 2p band to Ce 4f band [9]. For estimating bandgap, absorption data has been fitted with indirect transition relationship and by extrapolating linear part of the plot to zero absorption; $(\alpha h\nu)^{1/n} = A(h\nu - E_g)$, where A is a constant, $h\nu$ is photon energy, E_g is bandgap, α is absorption coefficient and n is a constant (n = 2 for indirect and n=1/2 for direct transition) [10].

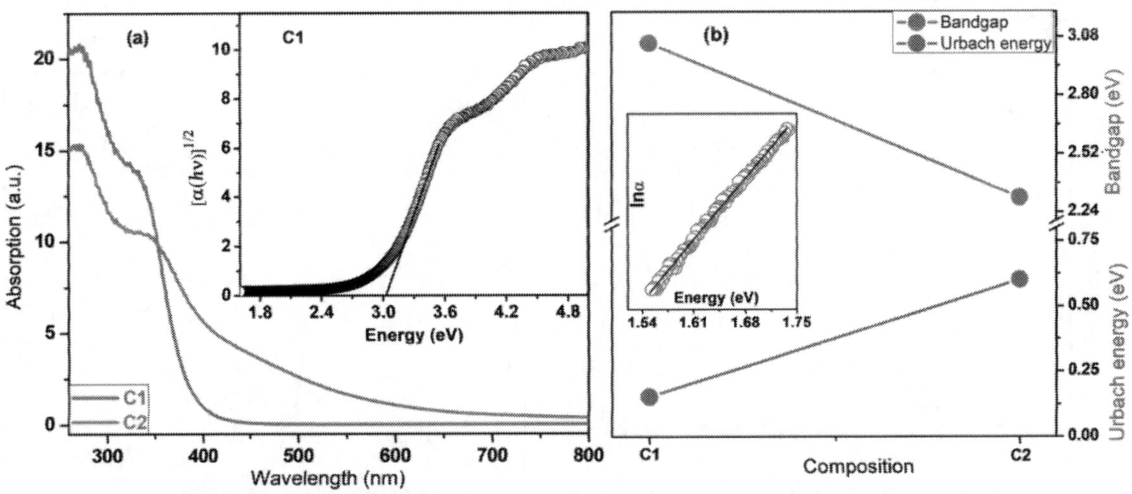

Fig. 3. (a) Optical absorption spectra of C1 and C2 (-inset shows plot of $(\alpha h\nu)^{1/2}$ vs energy); (b) Changes in bandgap and Urbach energy with composition (-inset shows Urbach fitting).

The bandgap of undoped CeO_2 is red shifted from UV region to visible region with (Co, Ni) co-doping from 3.05 eV to 2.31 eV [Fig. 3 (b)]. This red shifting of bandgap is due to the formation of some localized band between valance band (VB) and conduction band (CB) [11]. To confirm this Urbach energy is calculated ;(E_U): $\alpha = \alpha_0$*exp (E/Eu),where E_u is Urbach energy (eV) and E is photon energy.

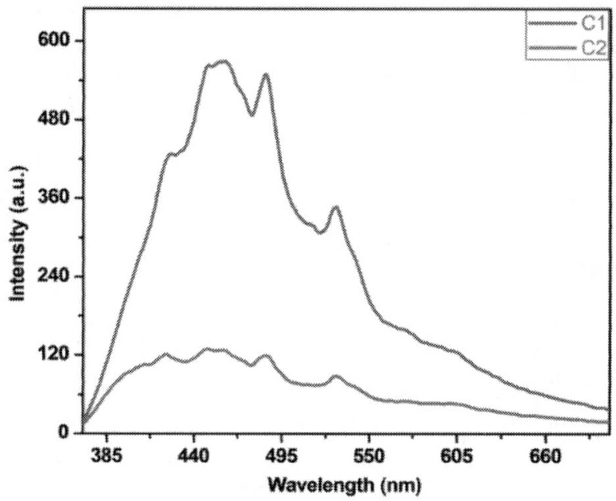

Fig. 4. Room-temperature PL spectra of C1 and C2 samples.

The plot between lnα and E has been plotted and the inverse of the slope gives Urbach energy [12]. Increase in Urbach energy with (Co, Ni) co-doping in CeO_2 [Fig. 3 (b)] pointing out increase in defect states between VB and CB. These defect states forms new band between VB and CB and thereby reduces its bandgap.

Photoluminescence spectra of undoped and (Co, Ni) co-doped samples were obtained at room temperature using excitation wavelength of 340 nm are shown in Figure 4. Cerium oxide (CeO_2) reversibly takes and release oxygen and form oxygen vacancies while maintaining the cubic fluorite structure [13].

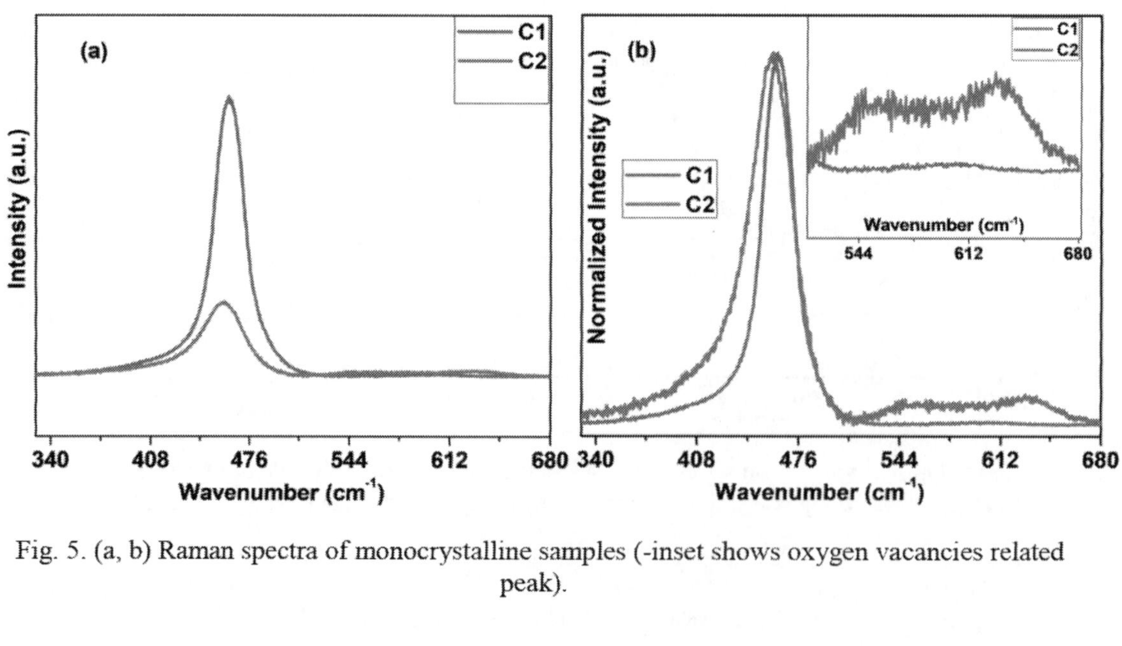

Fig. 5. (a, b) Raman spectra of monocrystalline samples (-inset shows oxygen vacancies related peak).

Oxygen vacancies related defects were present just below the Ce 4f band which is situated ~3 eV from VB. During electronic transition, excited electron goes to Ce 4f band and also to the bands that were related to oxygen vacancies related defects. The spectra of undoped and (Co, Ni) co-doped CeO_2 were almost similar. It has peaks at pale blue ~ 424 nm due to surface related defects, blue ~ 457 nm, blue-green ~485 nm due to dislocation or oxygen vacancies, and green ~ 528 nm due oxygen vacancies related defects [14]. Reduction in PL emission occurs due to increase in oxygen vacancies related defects which cause nonradiative recombination [15] of charge carriers.

Figure 5 (a, b) shows Raman spectra of undoped and (Co, Ni) co-doped CeO_2 nanoparticles. The cubic fluorite structure of prepared samples were confirmed by characteristic Raman active F_{2g} mode at 460.8 cm^{-1} [16]. It was a symmetrical stretching mode of Ce-O8 vibrational unit. The peak intensity of undoped CeO_2 decreases and become asymmetric with (Co, Ni) co-doping. This occurs due to increased strain and decreased phonon lifetime. The broad peak ~ 540-640 cm^{-1} has been attributed to oxygen vacancies [17]. Raman analysis confirmed the Urbach and PL study which hints the generation of defects related to oxygen which cause quenching of PL emission and reduction of bandgap.

4. Conclusion

Pure and cobalt, nickel co-doped CeO_2 nanoparticles were successfully synthesized using sol-gel method. Rietveld refinement of XRD data confirms cubic fluorite structure for CeO_2 (C1) and $Ce_{0.95}Co_{0.025}Ni_{0.025}O_2$ (C2) nanoparticles. Lattice parameter decreases owing to lesser ionic radii of cobalt and nickel than that of cerium. This lattice contraction increases the strain ~ 93% ,which has been reflected in reduced crystallite size. Lattice strain and disorder increases with cobalt, nickel co-doping.It were reflected in notable band tail and increase in Urbach energy (0.15 eV→0.60 eV). Co-doping lead to shifting of bandgap from UV region (3.05 eV; C1) to visible region (2.33 eV; C2) which enhances its utility for photocatalytic and other solar energy applications. The PL analysis shows that the emission intensity decreases with Co, Ni co-doping in CeO_2 due to increase in non-radiative centers which are oxygen vacancy related defects. The oxygen vacancy related defects were increasing and it has been confirmed by Raman analysis (~ 540-640 cm^{-1}).

Acknowledgements

The authors would like to thank IIT Indore for the funding, Dr. V. K. Jain (Amity University, Noida, India) for UV-Vis measurement facility, Dr. Manoj kumar (IISER, Bhopal, India) for Raman spectroscopy of the samples.

References

[1] Maensiri S, Masingboon C, Laokul P, Jareonboon W, Promarak V, Anderson P L and Seraphin S 2007 Egg White Synthesis and Photoluminescence of Platelike Clusters of CeO2 Nanoparticles *Cryst. Growth Des.* **7** 950–5

[2] Yabe S and Sato T 2003 Cerium oxide for sunscreen cosmetics *J. Solid State Chem.* **171** 7–11

[3] Channei D, Inceesungvorn B, Wetchakun N, Ukritnukun S, Nattestad A, Chen J and Phanichphant S 2014 Photocatalytic Degradation of Methyl Orange by CeO2 and Fe–doped CeO2 Films under Visible Light Irradiation *Sci. Rep.* **4** 5757

[4] Steele B C H 2000 Appraisal of Ce1−yGdyO2−y/2 electrolytes for IT-SOFC operation at 500°C *Solid State Ion.* **129** 95–110

[5] Mansingh S, Padhi D K and Parida K M 2016 Enhanced photocatalytic activity of nanostructured Fe doped CeO2 for hydrogen production under visible light irradiation *Int. J. Hydrog. Energy* **41** 14133–46

[6] Ranjith K S, Saravanan P, Chen S-H, Dong C-L, Chen C L, Chen S-Y, Asokan K and Rajendra Kumar R T 2014 Enhanced Room-Temperature Ferromagnetism on Co-Doped CeO2 Nanoparticles: Mechanism and Electronic and Optical Properties *J. Phys. Chem. C* **118** 27039–47

[7] Abbas F, Jan T, Iqbal J, Ahmad I, Naqvi M S H and Malik M 2015 Facile synthesis of ferromagnetic Ni doped CeO2 nanoparticles with enhanced anticancer activity *Appl. Surf. Sci.* **357** 931–6

[8] Kuriakose S, Satpati B and Mohapatra S 2014 Enhanced photocatalytic activity of Co doped ZnO nanodisks and nanorods prepared by a facile wet chemical method *Phys. Chem. Chem. Phys.* **16** 12741–9

[9] Soni S, Kumar S, Meena R S, Vats V S and Dalela S 2015 Interplay of structural, optical and magnetic properties in Gd doped CeO2 *AIP Conf. Proc.* **1665** 130029

[10] Khatun N, Rini E G, Shirage P, Rajput P, Jha S N and Sen S 2016 Effect of lattice distortion on bandgap decrement due to vanadium substitution in TiO2 nanoparticles *Mater. Sci. Semicond. Process.* **50** 7–13

[11] Veis M, Kucera M, Zahradnik M, Antos R, Mistrik J, Bi L, Kim H-S, Dionne G F and Ross C A 20140507 Optical and magneto-optical properties of Co-doped CeO2−δ films in the 0.5 to 4 eV range *J. Appl. Phys.*

[12] Tiwari S, Bajpai G, Srivastava T, Viswakarma S, Shirage P, Sen S and Biring S 2017 Effect of strain due to Ni substitution in CeO2 nanoparticles on optical and mechanical properties *Scr. Mater.* **129** 84–7

[13] Masui T, Fujiwara K, Machida K, Adachi G, Sakata T and Mori H 1997 Characterization of Cerium(IV) Oxide Ultrafine Particles Prepared Using Reversed Micelles *Chem. Mater.* **9** 2197–204

[14] Wang G, Mu Q, Chen T and Wang Y 2010 Synthesis, characterization and photoluminescence of CeO2 nanoparticles by a facile method at room temperature *J. Alloys Compd.* **493** 202–7

[15] Alla S K, Mandal R K and Prasad N K 2016 Optical and magnetic properties of Mg2+ doped CeO2 nanoparticles *RSC Adv.* **6** 103491–8

[16] Weber W H, Hass K C and McBride J R 1993 Raman study of ${\mathrm{CeO}}_{2}$: Second-order scattering, lattice dynamics, and particle-size effects *Phys. Rev. B* **48** 178–85

[17] Bernardi M I B, Mesquita A, Béron F, Pirota K R, Zevallos A O de, Doriguetto A C and Carvalho H B de 2015 The role of oxygen vacancies and their location in the magnetic properties of Ce1−xCuxO2−δ nanorods *Phys. Chem. Chem. Phys.* **17** 3072–80

IOP Conf. Series: Materials Science and Engineering **390** (2018) 012002 doi:10.1088/1757-899X/390/1/012002

Friction and Wear Behaviour of Tungsten Carbide and E Glass Fibre reinforced Al7075 based Hybrid composites

Santhosh Kumar.B.M[1], D.P.Girish[2]
[1]Department of Mechanical Engineering, JSSATE, Bangalore.
[2]Department of Mechanical Engineering, Government Engineering College, Ramanagaram.

Abstract

Al7075 based hybrid metal matrix composite reinforced with tungsten carbide and E glass fiber particles were fabricated by liquid metallurgy. The developed hybrid composites were subjected microstructure studies, hardness and dry friction and wear tests. Pin-on-disc setup was utilized to perform friction and wear tests over a load range of 20-100N and sliding speeds of 0.314-1.57m/s. Micro-structure affirms uniform scattering of reinforced particles. Hybrid composites show higher hardness, enhanced friction and wear characteristics in the presence of E Glass fiber and tungsten carbide particles when evaluated with the unreinforced alloy.

Keywords: Hybrid Composites, Al7075, E-Glass fiber, Tungsten Carbide, Friction, wear

1.0 Introduction

Metal matrix composites (MMCs) constitute an imperative class of design and weight-productive auxiliary materials that are empowering each circle of engineering applications.In recent years, metal matrix composites are gaining tremendous popularity over the conventional metals/alloys in various sectors such as automobile, aerospace and sports owing to the fact that properties can be tailored as per ones requirement- A uniqueness ofcomposites. The combinations of properties are obtained only by combining multiple materials and selecting the right type of reinforcements which can be developed by using different metal matrices such as copper, aluminum, magnesium, Titanium etc. Among various types of metal matrices, Aluminum and its alloys have magnificent physical properties combined with excellent corrosion resistance. Al7075 is a very well-known choice as an alloy to get ready metal matrix composites inferable from their better formability attributes and alternative of alteration of the strength of composites by embracing ideal heat treatment.In the recent years researchers have extensively focused on developing and characterization of aluminum based hybrid metal matrix composites[1-4].

In another work Basavarajappa et al [5] deliberated the effect of sliding speed on wear behaviour of Al2219/SiC/Gr hybrid composites. The wear rate for all the materials that is unreinforced alloy and hybrid composites was found to be same up to sliding velocity of 3 m/s. After that there was abrupt increase in the wear rate of unreinforced alloy when evaluated with that of hybrid composite. Rana *et al.* [6] have studied the friction co-efficient existing between Al-1.5% Mg reinforced with SiC and a tool steel counter face. The co-efficient of friction decreases with addition of SiCcontent. Roy *et al.* [7] have characterized friction behavior of aluminum composites reinforced with SiC, TiC and TiB_2. Results indicate that a maximum reduction of 30% was observed for composite when evaluated with unreinforced alloy. The type and size of the reinforcement phase were reported to have a negligible influence on the friction co-efficient.Various types of reinforcements used to produce composites such as SiC, TiO_2, Al_2O_3, TiB_2, TiC [2, 8-10].Tungsten Carbide (WC) is also used as reinforcement's owing to their highest hardness, low CTE, temperature stability and chemical inertness [11-12]. On the other hand, E-Glass is a fiber type inexpensive reinforcement which possess lower density, high hardness and fracture toughness [13-14]. Meager information available onthecombined effect of glass fiber and Tungsten carbide on tribological characteristics of Al7075 hybrid metal matrix composites. In view of the above facts, currentstudyemphases on study of tribological behaviour ofE glass fiber and tungsten carbide reinforced Aluminum 7075 metal matrix composite.

2.0 Materials and Methods

Aluminum 7075 alloy was used as matrix material. Table.1 demonstrates chemical composition of Al7075 alloy.

Table: 1 Al7075 alloy composition

Element	Cu	Cr	Mn	Mg	Si	Ti	Zn	Fe	Al
Wt. %	1.8	0.2	0.4	1.9	0.5	0.15	3.25	0.5	Balance

Chopped E-glass fiber and tungsten carbide particles were used as hybrid reinforcements in Al7075 matrix material. 0wt%, 1wt%, 3wt% and 5wt% E-glass fiber was incorporated in the matrix alloy for synthesizing composites whereas tungsten carbide particles were varied from zero to 6wt% in steps of 2wt%. Composites were fabricated by liquefying Al7075 alloy in an electrical resistance furnace utilizing clay graphite pot. In a batch, 3 Kgs of Al7075 ingots were used for composite preparation. The furnace temperature was maintained at 750 °Cfor melting of Al7075 alloy.Hexachloroethane degassing billets were added to remove entrapped gases. The mixing blade was rotated at 250 rpm and a vortex was made in the liquid aluminum. The slag was taken away from entering the molds while pouring through the spout. Hardness tests were conducted utilizing Brinell hardness analyzer. A load of 10 kg were applied into metalographically polished samples of Al7075 alloy and its hybrid composites. Five readings were considered and the average values have been ascertained for each sample.The specimen for friction and wear studies have been machined and examined in computerized pin on disc testing system. Tests were accompanied for specific loads and speeds at constant track radius (135mm) and fixed sliding distance(1KM). The mating surface of the disc and pin turned into cleaned with the aid of using acetone. The wear samples (8 mmØX15 mm) were tried in agreement with ASTM G99 standard procedure utilizing a pin-on-disc sliding wear testing machine. The test specimen was fixed and positioned against the turning EN24 steel plate.

3.0 Results and Discussions

3.1 Microstructure

Fig.3.1 (a-b) displays optical micrographs of Cast Al7075 and al7075+E-glass fiber+WC hybrid composites. It is observed that dispersion of reinforced phase is fairly uniform. Further, it is visible from the micrographs that there exists good bond between Al7075 matrix alloy and E-glass fibers and tungsten carbide particles. There are no visible defects or porosities associated with reinforced particles indicating good quality of composites.

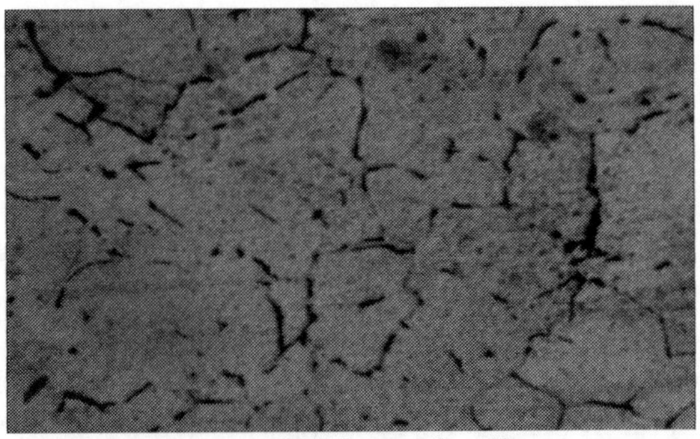

3.1a Al7075 matrix alloy

3.1b Al7075+2% E glass fibre+3% WC

3.2 Hardness

Fig.3.2 shows variation of hardness of Al7075+E-glass fiber+WC hybrid composite with increase in tungsten carbide particles at constant percentage of E-glass fiber. It is observed that, for a given percentage of E-glass fibers, there is a continuous increase in the hardness of the hybrid metal matrix composites with increase in tungsten carbide particle. In the absence of E-glass fiber a maximum hardness is noticed for 6wt% WC particles (Al7075+6%WC+0%E-glass fiber). In the presence of E-glass fiber, the maximum hardness recorded with increase in the percentage of tungsten carbide alone from 0 to 6wt%, a maximum improvement is observed, whereas with addition of E-glass fibers from 0% to 5% with corresponding tungsten carbide content a maximum hardness was recorded Al7075+4%WC+3%E-glass fiber hybrid composite. Among all the combinations of hybrid composites studied, the highest hardness was recorded for Al7075+4%WC+3% E-glass fiber hybrid metal matrix composites.

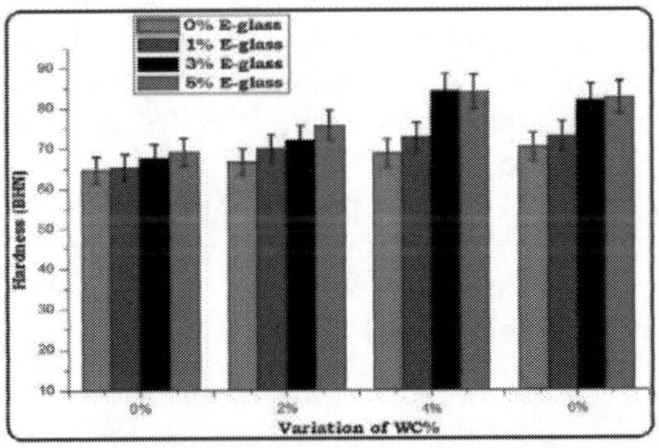

Fig.3.2 Variation of hardness with dual reinforcements

3.3 Coefficient of Friction

Fig.3.3 demonstrates the after effect of Tungsten carbide and Glass fiber content on the friction of AL7075 hybrid composites. It is seen from the graph that all the composites have shown lower friction values when assessed with to that of AL7075. The lowest friction values was obtained for AL7075-3%Tungsten carbide-3%E GF composite is almost 40% less that of AL7075. This is mainly because of presence of homogeneously distributed hybrid reinforcements in the Al7075 matrix material. The reinforcements are dispersed consistently because of virtue of optimum processing conditions adopted during stir casting and dispersing them uniformly. The presence of multiple reinforcements on the surface of the AL7075-Tungsten carbide-E GF composite acts as projections which protect the aluminum alloy from having direct contact with the steel surface. Owing to their non- sticking characteristics both the hybrid reinforcements minimizes the contact area and in turn reduces the adhesion of two mating surfaces.

Fig. 3.4 demonstrates the influence of load on frictional values of AL7075 and its hybrid composites. It can be seen from the graph that as the load is expanded from 20 N to 100 N, the friction coefficient is diminishing for both AL7075 and hybrid combinations. Most minimal friction coefficient was seen for AL7075-6%WC-5%GF composite. The abatement in the coefficient of friction is largely credited to slippage between the surfaces steel plate and composite pin. As saw in both AL7075 and composites, with the expansion in load the contact area between mating surfaces increases however this will offer raise to the temperature at the interface. This expansion in temperature at the contact surface will prompt softening of this surface which causes the slippage. Because of this slippage there is a diminishment in coefficient of friction in both Al7075 and its hybrid metal matrix composites.

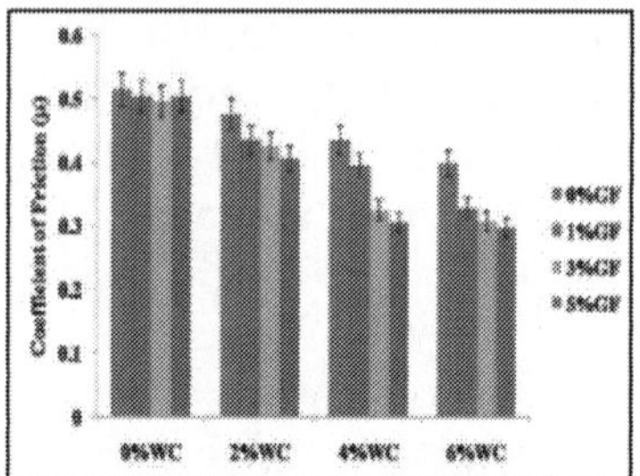

Fig. 3.3 Variation of Co-efficient of friction

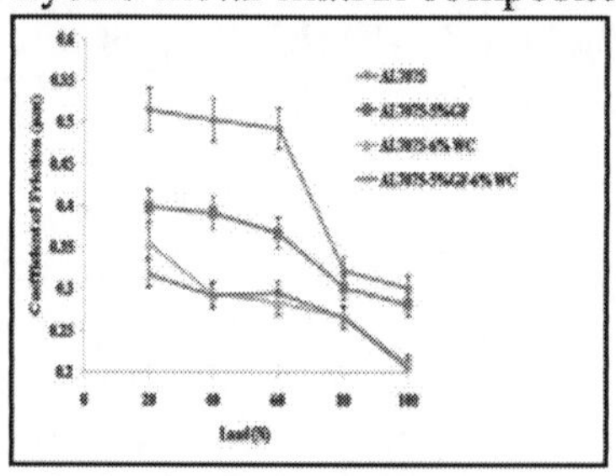

Fig.3.4 Coefficient of friction with respect to load

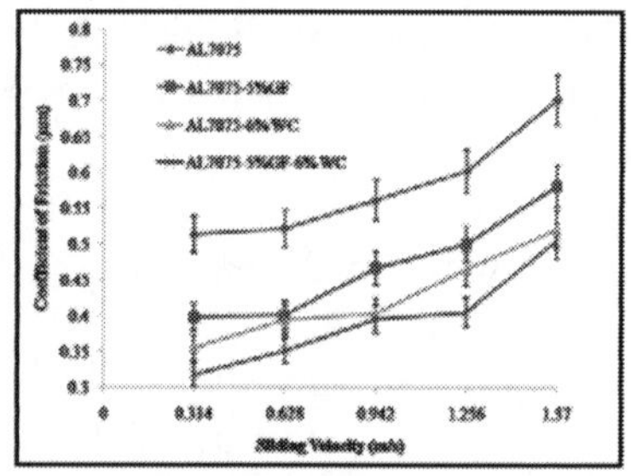

Fig.3.5 Coefficient of friction with respect to sliding velocity

From Fig. 3.5 it can be seen that, with the expansion in sliding speed from 0.314 to 1.570 m/s, there is an increment in frictional values. Highest friction coefficient wasrecorded for AL7075 composite especially at 1.570 m/s. most minimal coefficient of friction was seen for AL7075-6%WC-5%E GF for 0.314 m/s sliding speed. It is observed that, with the expansion in friction coefficient with the increase in sliding speed can be credited to development of plastic strains. This plastic strain will progresses the bond between the stick surfaces and counter surface which can prompt raised coefficient of friction. The plastic strain will increases with the expansion in sliding speed and the measure of localized adhesion of various reinforcements with the counter surface plate will be high. This is the reason we are observing higher friction coefficient at higher sliding speeds for all the combinations [15].

3.4 Adhesive Wear

Fig. 3.6displays the influence of hybrid reinforcements on the wear rate of AL7075 hybrid composites. The reduction in wear rate is almost 52% which elucidate imperative wear resistance by high weight percentage of Tungsten carbide and E glass fibre content. It can be noticed from Fig. 3.6 that AL7075-6%WC-5%E-GF hybrid composite has got highest hardness. As according to law of Archard's the wear loss of the material during wear test is inversely proportional to the hardness. So in present case as AL7075-6%WC-5%E-GF has got highest hardness the wear rate in this composite was also established low when evaluated to others. This is mainly because composites with highest hardness than the AL7075 alloy display tough confrontation to the sliding wear. Further, due to its high hardness the Tungsten carbide particles restrict the plastic flow of AL7075 matrix during sliding wear.Fig. 5.7demonstrates the influence of load on wear rate of AL7075 and hybrid composites with fixed sliding velocity of 0.314 m/s. With the increase in load from 20 N to 100 N, the wear rate of all the materials is increasing linearly for all the materials. Maximum wear rate was recorded for AL7075 for 100 N load while that of lowest wear rate was observed for AL7075- AL7075-6%WC-5%E-GF for 20 N load. At starting load of 20 N, the hard asperities of offset surface enter the composite pin surface which leads to ploughing and formation of grooves on the matrix surface. Further, the formation of grooves is mainly due to delamination of matrix surface at this load indicates mild and oxidative wear. According to Archard's law, if the load or pressure applied on the material is increased then the wear rate also increases linearly. With rise in applied load the asperities of hard counter surface penetrate the soft matrix surface. This can lead into higher material removal rate from the soft matrix surface. Overall when compared to that of AL7075 alloy, the hybrid composites displayed lower wear rate at all loading conditions. This is mainly due to high hardness of both the reinforcement particles which enables the composites to sustain higher loads and restrict the flow of AL7075 matrix during sliding. In addition to this the strong bonding of matrix with reinforcement is the one more reason for low wear rate of hybrid composites.

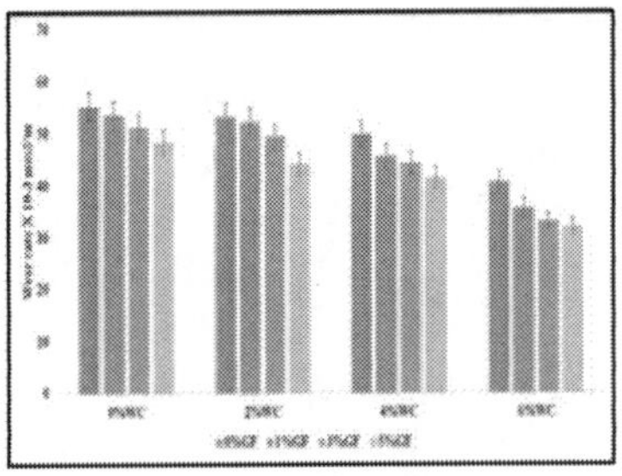

Fig.3.6 Effect of reinforcement content on wear rate

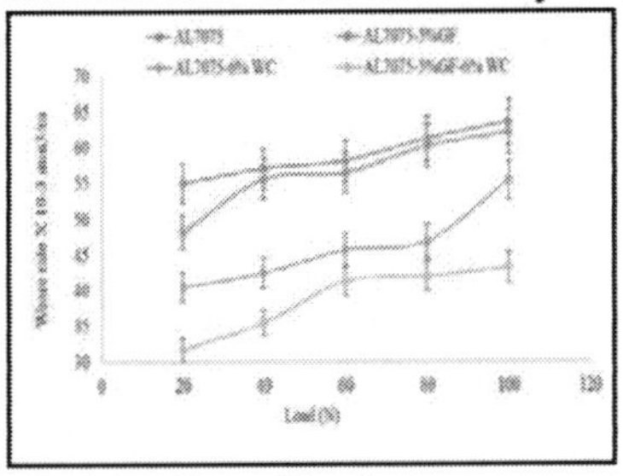

Fig.3.7 Effect of load on wear rate

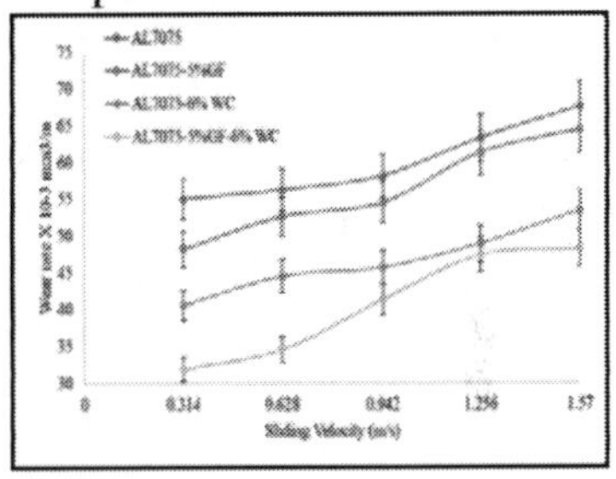

Fig.3.8 Effect of sliding velocity on wear rate

Fig. 3.8 shows the impact of velocity on the wear behaviour of Al7075 alloy and its hybrid combinations. The sliding speed was varied from 0.314 to 1.570 m/s by keeping load of 20 N constant for all the sliding tests. It was seen that with the enhanced sliding speed the wear rate were increased. The maximum wear rate was recorded for Al7075 alloy at sliding velocity of 1.570 m/s while lowest wear rate was obtained for hybrid composites with AL7075-6%WC-5%E-GF at 0.314 m/s. The increase in wear rate is almost 15% for increase in sliding velocity is observed for alloy. When we compare the wear rate of both alloy and hybrid composite with AL7075-6%WC-5%E-GF at 0.314 and 1.570 m/s, the increase in wear rate is about 40% and 53% respectively. The significant reduction in wear rate at all sliding velocities for composites is mainly due the presence of hybrid reinforcement's.

4.Conclusions

Friction coefficient of Al7075 alloy decreases with addition of tungsten carbide and E glass fiber. Coefficient of friction decreases with load and exhibits higher values with sliding speed. Wear rate of hybrid materialswere minimal when evaluated with aluminum. Addition of hard reinforcements stimulates the reduced wear rateunder all loads and sliding speeds.

References.

[1]. R.Keshavamurthy, MankariPavan, Mohammed Abdul TazamulHaq, DnyaneshwarZinge, Mohammed. (2105) "Hardness and Tensile behavior of forged Al7075-SiC composite"International Journal of Applied Engineering Research 10 (55), 3769-3772.

[2]. CS Ramesh, R Keshavamurthy, PG Koppad, KT Kashyap(2013), Role of particle stimulated nucleation in recrystallization of hot extruded Al 6061/SiCp composites, Transactions of Nonferrous Metals Society of China 23 (1), 53-58.

[3]. K V Shivananda Murthy, D P Girish, R.Keshavamurthy, Praveennath Koppad, TemelVarol, "Mechanical and Thermal Properties of AA7075/TiO_2/Fly ash Hybrid Composites Obtained by Hot Forging", Journal of Progress in Natural Science: Materials International, 2017, 27 (4), 474-481.

[4]. R Keshavamurthy, JM Sudhan, N Gowda, RA Krishna (2016), Effect of Thermo-Mechanical Processing and Heat Treatment on the Tribological Characteristics of Al Based MMC's", IOP Conference Series: Materials Science and Engineering 149 (1), 012118.

[5]. S. Basavarajappa, G. Chandramohan, A. Mahadevan, M. Thangavelu, R. Subramanian, P. Gopalakrishnan, Influence of sliding speed on the dry sliding wear behaviour and the subsurface deformation on hybrid metal matrix composite, Wear 262 (2007) 1007–1012.

[6]. F. Rana. and D.M. Stefanescu, "Friction properties of Al-1.5Pct Mg/SiC particulate Metal-Matrix Composites". Metallurgical Trans. A, (1989), 20A, 1564-1566.

[7]. Roy M., Venkataraman B., Bhanuprasad V.V., Mahajan Y.R. and Sundararajan G., "The effect of particulate reinforcement on the sliding wear behaviour of Aluminium Matrix Composites", Metallurgical Trans. A 23A (1992), 2833-2847.

[8]. K.V.Shivananda Murthy, D.P.Girish, R.Keshavamurthy, "Mechanical Characteristics of Hot Forged Al6061-Al_2O_3 Composite" Journal of Applied Mechanics and Materials, Vol. 787, pp. 598-601, 2015.

[9]. K.V.Shivananda Murthy, D.P.Girish and R.Keshavamurthy, "Investigation on mechanical Behavior of Hot Forged AL7075-TiO_2-flyash Hybrid Metal Matrix Composites", International Journal of Applied Engineering Research, Vol. 10 (2015), 55, pp.4105-4109.

[10]. GSP Kumar, PG Koppad, R Keshavamurthy, M AlipourMicrostructure and mechanical behaviour of in situ fabricated AA6061–TiC metal matrix composites Archives of Civil and Mechanical Engineering 17 (3), 535-544.

[11]. K. Ravikumar, K. Kiran, V.S. Sreebalaji "Characterization of mechanical properties of aluminium/tungsten carbide composites" Measurement 102 (2017) 142–149.

[12]. SJerryAndrewsFabian,B.Selvam,"DensificationbehaviourofAluminiumreinforcedwithTung sten Carbide particulate Metal Matrix Compositeprocessedby P/M",IOSRJournalof MechanicalandCivil Engineering (IOSR-JMCE),PP24-29.

[13]. J. W. Pinto, G. Sujaykumar, Sushiledra R. M "Effect of Heat Treatment on Mechanical and Wear Characterization of Coconut Shell Ash and E-glass Fiber Reinforced Aluminum Hybrid Composites" American Journal of Materials Science 2016, 6(4A): 15-19.

[14]. SiddharthPatela and R.S. Rana and Swadesh Kumar Singh, "Study on mechanical properties of environment friendly Aluminium E-waste Composite with Fly ash and E-glass fiber" Materials Today: Proceedings 4 (2017) 3441–3450.

[15]. R.N. Rao, S. Das, Mondal DP, Dixit G. Dry sliding wear behavior of cast high strength aluminium alloy (Al–Zn–Mg) and hard particle composites. Wear 267 (2009) 1688-95.

IOP Conf. Series: Materials Science and Engineering **390** (2018) 012003 doi:10.1088/1757-899X/390/1/012003

Investigation on thermal properties of Styrene Acrylonitrile (SAN) matrix with Polytetrafluroethylene (PTFE) particle reinforced composites

R.Mahesh Kumar, N.Rajini, K.Mayandi*, M.Gnanaguru, S.Haja Jasim Farid.

Department of Mechanical Engineering, Kalasalingam Academy of Research and Education,
Krishnankoil-626126, Tamilnadu, India

*E-Mail of the corresponding author: *k.mayandi@gmail.com*

Abstract— Temperature utilization of composites materials depends on its thermal conductivity and co-efficient of thermal expansion. Polymers are exceptionally combustible because of their structure and they can't fulfill a few applications. Thermal treatment of soften polymer because of the preparing setup by injection molding fabrication process able to control polymer structure (fitting) and in this manner deciding its thermal conduct. The fundamental point of this work is to contemplate the impact of Polytetrafluroethylene polymer (PTFE) as particle with Styrene Acrylonitrile polymer (SAN) as matrix to examine its thermal properties. Samples of 100% SAN and 10% PTFE by weight, 20% PTFE by weight blended specimens with SAN are prepared successfully by Injection molding process. Heat capacity, Thermal expansion, Thermal conductivity and Thermal degradation are the thermal properties examined. From the observations it was found that PTFE particles blended samples exhibits maximum thermal conductivity and good dimensional stability.

Keywords — Polytetrafluroethylene particle, Styrene Acrylonitrile matrix, Injection molding fabrication process, Thermal properties investigation.

1. Introduction

The utilization of Polymer-based materials is quickly expanding. The enterprises of today are progressively seeking techniques to supplant metal parts with polymer based materials as a result of light weight, simple support and modest costs [1]. The ceaselessly developing requirement for composite materials with new or enhanced properties brings about the preparation of different polymer mixes with various arrangements, morphologies and properties.

The change of the thermal properties of SAN has been the subject of various fascinating logical investigations. One of the effective techniques for this reason conceivably making utilization of added substances, for example, particles, enhances thermal conductivity and coefficient of thermal expansion [2]. The acknowledgment of natural fiber as fillers in the plastic business has additionally been developing widely as a result of their simple openness, biodegradability, moderateness, and ecological maintainability. Additionally, there is an abnormal state of enthusiasm for utilizing polymer particles for making polymer composites with enhanced thermal and combustion properties [3].

SAN is a broadly utilized designing thermoplastic inferable from its alluring properties, which incorporates great mechanical properties, synthetic protection and simple preparing attributes. It has numerous imperative applications, for example, fittings of portable industry and home appliances, instrument board, and different switches, and so on. In this article, we utilized PTFE particles as support of SAN matrix to set up the high thermal stability SAN/PTFE composites by melt shaping procedure [4].

PTFE is an excellent material with high insulation (i.e. better than SAN), finds many potential applications in thermal and medical fields [5-6]. This investigation deals with the improvement of thermal properties of SAN composites over pure SAN. In this regard, SAN/PTFE composites were prepared by means of melt, and the thermal properties of the readied composites were contemplated [7]. To the best of our insight, the thermal properties of SAN/PTFE composites have not been yet detailed. From this investigation it is expected that thermoplastic composites with high thermal properties can be created for use in various applications.

2. Materials and methods

Materials

Polytetrafluroethylene powder particles (PTFE) were used for reinforcement in this work [8]. Powder PTFE polymers were procured from M/s. Modoplast Company Pvt. Ltd, Kolkata, West Bengal, India. Styrene Acrylonitrile (SAN) resin was used as an effective matrix system in polymer reinforced composites to investigate its thermal properties. SAN polymer resin in pellet form is commercially available and procured from Sree Meenakshi Plastics, Madurai, Tamilnadu, India. The average diameter of SAN pellets was approximately 2mm. The particle size of PTFE powder was about 5 to 15 μm. For fabrication of test sample specimens size (200x140x3mm), polymer injection mold suitable for semi-automatic injection molding machine (MACFIELD, MAC60) were prepared by using EN8 mild steel plates and obtained from Mahesh Plastics, Madurai, Tamilnadu, India.

Sample Preparation

The samples used for this work are: Sample 1: 100% virgin SAN polymer resin, Sample 2: 90% by weight composition of SAN polymer resin + 10% by weight composition of PTFE powder particle and Sample 3: 80% by weight composition of SAN polymer resin + 20% by weight composition of PTFE powder particle. The materials procured, SAN pellets and PTFE powder were pre-dried to moisture free condition at 100°C in vacuum oven for 24 h separately. The materials as per requirement are taken in three batches, mentioned as samples and ready for injection molding. The three batches one by one were fed into the hopper of semi-automatic reciprocating screw injection molding machine DGP Windsor, 50T - **figure 1** to produce rectangular size (200x140x3mm) molded test plate samples [9-11]. The processing temperature was maintained at 220°C for SAN resin and 280°C for SAN/PTFE composites.

Fig. 1 Injection molding machine

Thermal conductivity and Coefficient of thermal expansion tests

The heat energy (Q) transfer across a plane of area (A), with a temperature gradient (ΔT/Δl) can be expressed as Q = kA (ΔT/Δl), where k is the thermal conductivity of the material. The unit for 'k' is W/m-K and for polymers its value in the order of 0.3 [7]. As per ASTM E1530 disc shaped specimens with diameter of 50mm and thickness of 3mm are used for measuring thermal conductivity under steady state condition [12].The experiment was conducted by heat flux method: Fox 50-110 where heat is applied from one side of the specimen.

Coefficient of linear thermal expansion (α) is the change in length per unit by the addition of heat energy given by α = ε / ΔT, where ε is the strain. The unit for thermal expansion co-efficient is $°C^{-1}$, and its values for polymers in the order of 50 to 400 x 10^{-6}. The materials having strong atomic bonds have high melting point and low thermal expansion coefficients [6]. Dilatometer instrument was used to measure the thermal expansion coefficient. According to ASTM D 696 rectangular plate specimens with 50 X 10 mm and thickness 3mm are used for coefficient of linear thermal expansion measurements.

Optical Microscopy

The PTFE distribution was analysed using Motic Optimal Microscope (Fig 2) that was equipped with moticam 2500 camera, digitally controlled by motic images plus 2.0 ML image processing software.

The PTFE distribution images were taken for pure SAN, 10% by wt. composition of particle reinforced PTFE, and 20% by wt. composition of particle reinforced PTFE particle filled composite materials.

Thermogravimetric analyzer test

The powder sample taken in a crucible can be heated in a steady rate in a gas environment (inert-Nitrogen gas) to obtain weight loss values for the samples at various time and temperature [13]. This test provides us a graph depicting the nature of the sample at various loading conditions. TGA measures the percent volatile, water, solvent by weight.

3. Results and discussion

Thermal conductivity analysis

The heat transfer rate of a material depends on its thermal conductivity [14]. The values measured in five samples are taken in each category and the average is plotted in **figure 2**. The experimental analysis reveals that sample 2 and Sample 3 exhibited maximum thermal conductivity. The composite sample has good thermal conductivity because PTFE powder enhances the heat transfer rate. It is observed that composites filled by 10% and 20% by wt. composition PTFE particle, also exhibited good thermal conductivity than pure SAN.

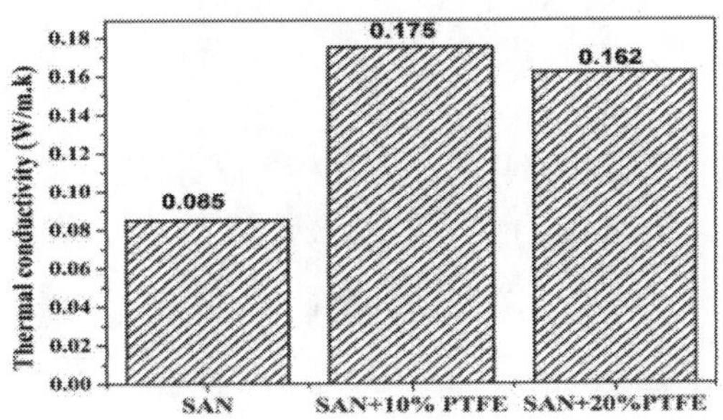

Fig. 2 Thermal conductivity for different composition of composite materials

Coefficient of linear thermal expansion analysis

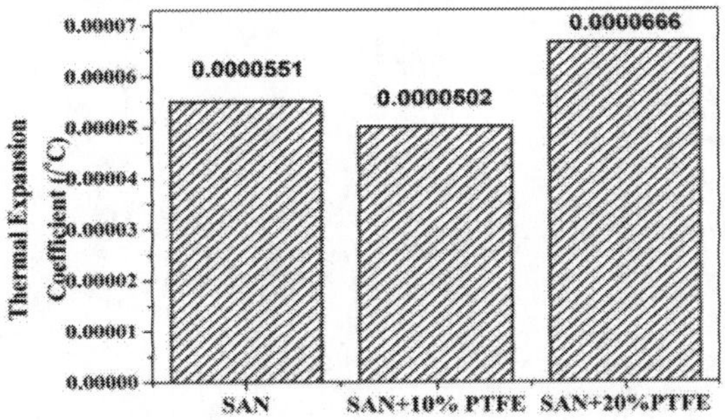

Fig. 3 Thermal expansion coefficient for different composition of composite materials

The thermal expansion coefficient is a measure of thermal property and lower values finds many applications. From literature survey, composite material with particle fillers has a lower thermal expansion coefficient [15]. From **figure 3** SAN Composites with PTFE particle fillers (10% by weight) has low thermal coefficient (5.02 x 10^{-5}/°C) when compared with other (20% by weight) composites. The measurement of the thermal conductivities of composite samples indicated that PTFE would be a better thermal barrier in the presence of SAN.

Optical Microscopy

The PTFE distribution images taken for 10% by wt. composition of PTFE particles reinforced composites and 20% by wt. composition of PTFE particles reinforced composites materials are given.

Fig. 4 SAN + 10% PTFE Composite

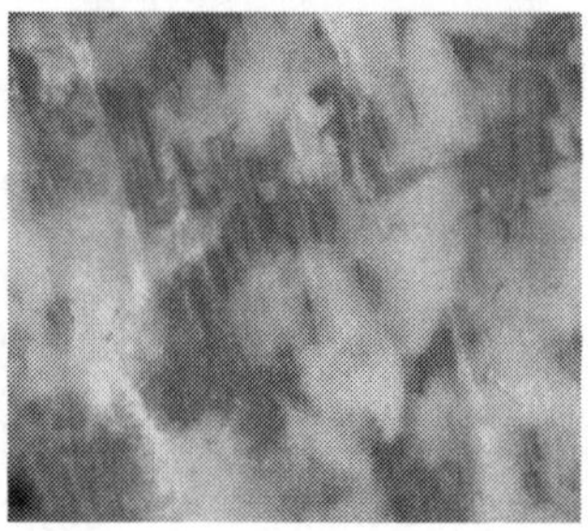

Fig. 5 SAN + 20% PTFE Composite

Thermogravimetric Analysis

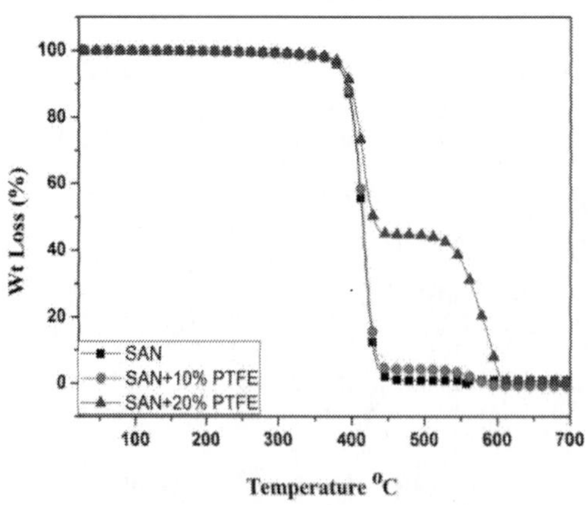

Fig. 6 TGA test report of SAN & SAN Composites

Thermogravimetric (TGA) analysis were completed utilizing a METTLER STAR SW15.00 thermo analyzer instrument from 25-700°C utilizing a continuous heating rate of 10°C/min under nitrogen stream. The nitrogen flow was 25ml/min. Powdered Samples for SAN and SAN/PTFE composites one by one were estimated in a crucible closed, with a lid on the hole with a mass of around 10mg [16]. From **figure 6** the decomposition temperature curves obtained from the three samples the radius of curvature is bigger at the beginning than at the end of the heating process. The decomposition temperature curve for 20% by wt. composition of PTFE particle reinforced composite has two steps. The result obtained explains that the thermal stability of this sample has remarkable potential value. The PTFE particle plays an essential role on thermal stability. The TGA curve shown in **figure 6** depicts the mass change in the process. The SDTA curve (measures exothermic effects in similar to DSC) gives the enhancement of thermal stability of SAN/PTFE composites. The thermal decomposition curves for SAN/PTFE particle reinforced composites withstand higher temperature range than SAN sample.

Table 1.The TGA and DTGA data SAN and SAN/PTFE particle composites

Samples	$T_{-5\,wt\%}$ degradation (°C)	$T_{-50\,wt\%}$ degradation (°C)	T_{max} (°C)
Sample 1 Pure SAN	381	413	439.3
Sample 2 10% PTFE	381.5	415	443
Sample 3 20% PTFE	386	429	666

Table 1 gives the degradation temperature values. From the values obtained SAN/PTFE particle reinforced composite has the maximum values. This enhancement is due to the physical protective barrier of the PTFE particles on the surface layer of the matrix. The PTFE particle behaves as thermal guard to the volatile materials generated during the thermal decomposition as a result of higher thermal stability.

4. Conclusions

- The PTFE particle reinforced SAN polymer composites fabricated successfully by injection molding process possess higher thermal conductivity and low thermal diffusivity than existing individual SAN copolymer.
- The thermal conductivity and thermal expansion co-efficient values of SAN, 10% by wt. composition of PTFE particle reinforced composite and 20% by wt. composition of PTFE particle reinforced composite are (0.085 W/m.K, 5.51×10^{-5} /°C); (0.175 W/m.K, 5.02×10^{-5} /°C); (0.162 W/m.K, 6.66×10^{-5} /°C) respectively.
- The characterization of the morphology of all three samples investigated by Optical Microscopy and TGA ensures improved morphology with PTFE particle based SAN has significant structure. By physical transitions, the adsorption of gaseous substances on samples such as active charcoal residue at 700° C is less for PTFE particle reinforced composite samples 2 and3.
- From overall analysis made in Thermal properties, Sample 2 and Sample 3 have better thermal properties.

References

[1] Srinivasulu,NV, Tejaswi,V, 2012, *Mechanical properties of polymer composite materials*, IJRET: International Journal of Research in Engineering and Technology, 2319-1163, **01**

[2] Chang.L, Zhang Z, Ye L and Friedrich.K, 2007, *Tribological properties of high temperature resistant polymer composites with fine particles*, Tribology international **40**, 1170-1178

[3] Devendra.K, Rangaswamy.T, 2012, *Evaluation of thermal properties of E-glass/ Epoxy composites filled by different filler materials,* International Journal of Computational Engineering Research **2, 5**

[4] Lih-Long Chu, Stephanie K. Anderson, Joseph D. Harris, Mark W. Beach and Alexander B. Morgan, 2004, *Styrene Acrylonitrile (SAN) layered silicate nanocomposites prepared by melt compounding*, Science direct, Polymer **45**, 4051-4061

[5] Kunal Singha1, Mrinal Singha, 2012, *Cardio Vascular Grafts: Existing Problems and Proposed Solutions*, International Journal of Biological Engineering, **2(2),** 1-8

[6] Arzu Tezvergil, Lippo VJ. Lasslla and Pekka K. Vallittu, 2003, *The effect of fiber orientation on the thermal expansion coefficients of fiber-reinforced composites*, Academy of Dental materials, **19, 6**, 471–477

[7] Yibing Cai, Yuan Hu, Junfeng Xiao, Lei Song, Weicheng Fan, Huaxia Deng, Xinglong Gong and Zuyao Chen, 2007, *Morphology, Thermal and Mechanical Properties of Poly (Styrene-Acrylonitrile) (SAN)/Clay Nanocomposites from Organic-Modified Montmorillonite*, Polymer-Plastics Technology and Engineering, **46**, 541–548

[8] Vail.J.R, Krick.B.A, Marchman.K.R, Gregory Sawyer.W, 2011, *Polytetrafluoroethylene (PTFE) fiber reinforced polyetheretherketone (PEEK) composites*, Journal ISSN 0043-1648, Wear **270**, 737–741

[9] Tang.S.H, Kong.Y.M, Sapuan.S.M, Razali Samin and Shamsuddin Sulaiman,2006, *Design and thermal analysis of plastic injection mould*, Journal of Materials Processing Technology, **171**, 259-267

[10] Bilewicz.M 2, Viana.J.C, Dobrzanki.L.A, 2008, *Polymer composite strengthening by developed injection moulding technique*, Archives of Material Science and Engineering, **30,** 69-72

[11] Liliane Pimenta de Melo, Gean Vitor Salmoria, Eduardo Alberto Fancello and Carlos Rodrigo Fancello, 2017, *Effect of Injection Moulding Melt Temperatures on PLGA Craniofacial Plate Properties during In Vitro Degradation*, International Journal of Biomaterials, **2017**, 1256537, 11

[12] Raju.G.U, Kumarappa.S, 2012, *Experimental study on Mechanical and Thermal properties of Epoxy Composites filled with agricultural residue,* Polymers from renewable resources, **3, 3**

[13] Xiaoping Zhang, Fang Wang and Leon M.. Keer, 2015, *Influence of Surface Modification on the Microstructure and Thermo-Mechanical Properties of Bamboo Fibers*, Materials, **8**, 6597–6608

[14] Geon-WoogLee, MinPark, JunkyungKim, Jae IkLee, Ho GyuYoon, 2006, *Enhanced thermal conductivity of polymer composites filled with hybrid filler,* Composites Part A: Applied Science and Manufacturing, **37, 5**, 727-734

[15] Wong.C.P, Raja S. Bollampally, 1999. *Thermal conductivity, elastic modulus and coefficient of thermal expansion of polymer composites filled with ceramic particles for electronic packaging,* Journal of Applied Polymer Science, **74, 14**, 3396-3403

[16] Marcos Vinicius FonsecaFerreira, Anna Carolina CerqueiraNeves, Caroline Gomes deOliveira, Felipe Perisse DuarteLopes, Frederico MuylaertMargem, Carlos Mauricio FontesVieira and Sergio NevesMonteiro, 2017, *Thermogravimetric characterization of polyester matrix composites reinforced with eucalyptus fibers,* Journal of Materials Research and Technology, **6, 4**, 396-400

IOP Conf. Series: Materials Science and Engineering **390** (2018) 012004 doi:10.1088/1757-899X/390/1/012004

Influence of grinding parameters on grinding force and temperature in grinding of AA6061-TiB_2/ZrB_2 in-situ composite

A Mahamani and S Jawahar
Department of Mechanical Engineering
Sri Venkateswara College of Engineering and Technology (Autonomous),
Chittoor- 517127, A.P., India.

Abstract: In-situcomposite material synthesis has been widely recognized by the researchers owing tofine particle size, clean interface, and smart wettability of the reinforcement with the matrix.Variation of wheel speed, work speed, depth of cut and feed rate on grindability of AA6061-5%TiB_2/ZrB_2 in-situ composite are investigated.The present work is aimed to investigate the Grinding force and grinding temperature of the AA6061-TiB_2/ZrB_2 in-situ composite under various grinding parameters and wheel materials. A hike of wheel speed decreases the grinding force an increase in temperature. Increase in work speed, depth of cut and feed rate increase the grinding force and grinding temperature. This experimental investigation helps to control the grinding force and grinding temperature at the various grinding condition.

1. Introduction

Composite materials are received the worldwide attention of material scientists owing to the potential for weight reduction, superior properties and tailorability [1]. These extreme properties motivate the aerospace and automotive industrialists to consider the composites to replace the conventional materials [2]. In-situ composites manufactured by the halide salt- aluminum melt reaction offers many advantages including low cost, easy to manufacture and single step processing [3]. The existence of micro size, coarse free and freshly generated reinforcements have enhanced the wettability and interfacial strength between the particles and matrix and thereby increase the mechanical properties of the composites [4]. Excessive addition of reinforcements into the matrix dilutes the mechanical properties and restricts the widespread application. In-situ composites may prevail over this problem by the way of obtaining the highest mechanical properties at lower reinforcement ratio. TiB_2 possess high melting temperature, incomparable corrosion resistance, and exceptional chemical stability, venerable creep resistance [5]. Adding an additional reinforcement into matrix helps to explore the benefits of both ceramics. The combination of ZrB_2 and TiB_2 ceramics reinforced into the aluminum matrix is expected to put forward a typical high-temperature structural application [6].In spite of countless miraculous properties, the machining of the composite is still challenging to industries [7]. In-spite of many near net shape manufacturing technology, the composite components require finishing operation to meet the desired dimensional tolerances, surface roughness and surface integrity [8].Cylindrical grinding is one of the processes to finish the surface of the machined components with the cylinder, hollow shaft and pipes. Grinding is a time-consuming machining process among the conventional machining category and it is associated with some challenge to attain the optimal surface finish with economic machining rate. The generated surface should have an ability to withstand the fatigue load, creep load, friction, temperature, and corrosion during the service condition. Quality of grounded surface is completely re-laid on the product performance [9]. The mechanism of the grinding of the composite is associated with the ceramic compound available with the grinding wheel makes physical contact with all regions of the composite namely: matrix, reinforcements and matrix-reinforcement interface, and facilitates material removal in form of small chips. Grinding process is allied with many parameters: wheel speed, work speed, feed rate, grinding depth, number of passes, wheel material, grit size, nature of grinding matrix material, type of reinforcement, size of reinforcement and fineness of the reinforcement also influence the grounded surface. The components of grindability include grinding force, temperature, surface roughness, subsurface material removal rate. Grindabilitty studies on ex-situ composites widely reported in the literature. Performed an experimental work by grinding of Al/SiC composites. The effect of cylindrical grinding parameters including speed of the wheel, the speed of workpiece, rate feed, grinding depth and SiC content on the grinding force, surface roughness, and grinding temperature are examined. Analysis of the experimental result depicts that, the superior surface finish and scratch free surfaces are found at high wheel speed and workpiece speed [11]. Reported the grindability study of the Al-Al_2O_3 composites. SiC verified and diamond resin bonded matrix grinding wheels used for the experimental work. Surface roughness, grinding force, and subsurface damage are evaluated under different grinding parameters. The experimental result reveals that the SiC wheel well for rough grinding and diamond wheel for finish Grinding [12]. Established the response of grinding operation on the surface smoothness in machining of particles fiber reinforced aluminum matrix composites. Size, orientation, and method of manufacturing are considered for investigation. Microscopic examination of the grounded surface reveals the subsurface damage due to particle or fiber pulling effect and crack formation [13].

2. Materials and Method

AA6061-TiB_2/ZrB_2 in-situ composites with 0%, 2.5%, 5% and 7.5% reinforcement ratio are used for experimental work. These composites are made by reacting K_2TiF_6, KBF_4, and K_2ZrF_6 salts. The microstructure and XRD pattern of the composites is presented in the Figures 1and2. Small size (1-2μm) reinforcement particles, the clean interface between particle and matrix, and uniform distribution of particle are seen in the microstructure. Al, TiB_2 and ZrB_2 phases are detected in the XRD pattern. The dimension of the workpiece is Φ30x300mm. Grinding experiments are conducted by using horizontal spindle cylindrical grinding machine (Heavy duty machine, Indian make). Al_2O_3, CBN, and Diamond grinding wheels are used for the experimental work. Grinding force values are measured by using Variable Frequency Drive and grinding temperature values are measured using Non- contact infrared thermometer.The experimental work is conducted with various grinding parameters as per Table 1.A parametric study is planned in such a way that influence of one parameter is assessed by keeping other parameters as constant.

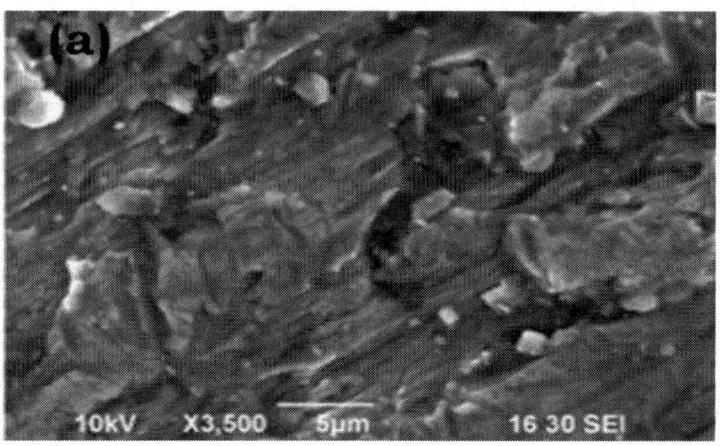

Figure 1. SEM image

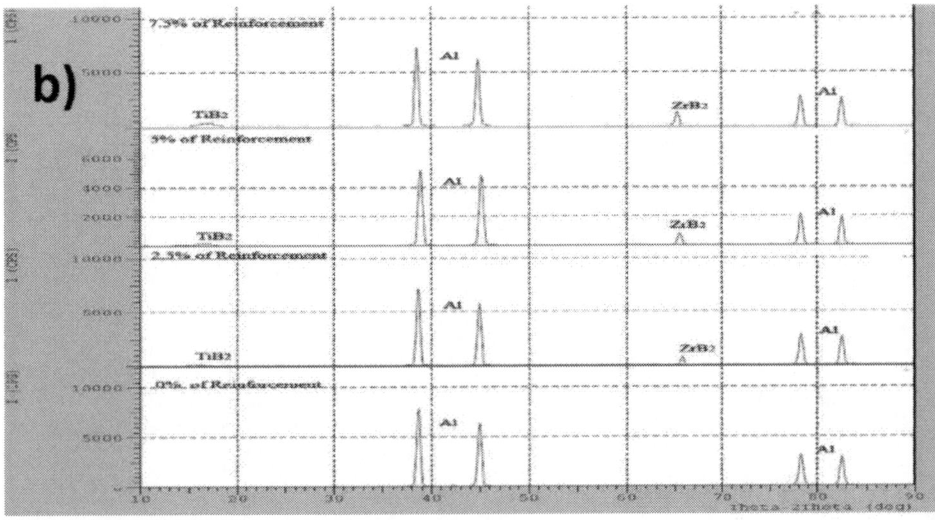

Figure 2. XRD pattern

Table 1. Parameters and their levels

Parameters	Units	Levels		
		Level 1	Level 2	Level 3
Wheel speed	m/s	23.5	33.7	43.9
Work speed	m/min	6	12	26
Depth of cut	µm	10	20	30
Feed rate	m/min	0.06	0.09	0.17

3. **Result & Discussion**

3.1. Grinding force

The effect of wheel speed on grinding force as a function of wheel materials are shown in Figure 3. This investigation is carried out by fixing the work speed as 12m/min, depth of cut 20µm, feed rate 0.09m/min and 5% reinforcement ratio. It is seen from the Figure 3 the diamond wheel performed better than other wheels in terms of generating very low grinding force. Al_2O_3 wheel generating higher grinding force for the given experimental conditions. It also is seen from the Figure 3 increase in wheel speed decrease in grinding force.The grinding force, that is the tangential one, has the strong influence on the removal modes of the reinforcements and the matrix of PTMCs during grinding. The grinding force increases with an increase in workpiece speed and depth of cut. Higher force ratio always corresponds to severe tool wear; therefore, low tool sharpness in the super-high-speed grinding process [14].Diamond gives lesser grinding force and better surface finish compared to SiC wheel. Also grinding of MMC reinforced with SiC, using SiC wheel usually result in clogging and attrition wear of SiC abrasive. Although it is known that electroplated bond is hard (prevents pull out of abrasive) compared to resin bonded wheel, during grinding, the resin bonded diamond wheel will enable self-sharpening of the diamond abrasive, enabling better grindability [10].

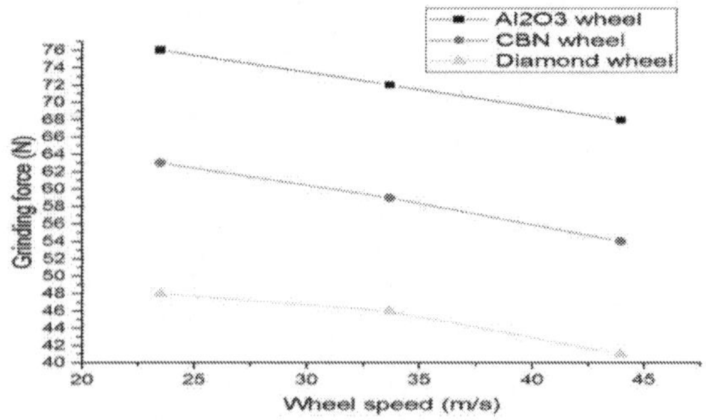

Figure 3. Wheel speed vs Grinding force

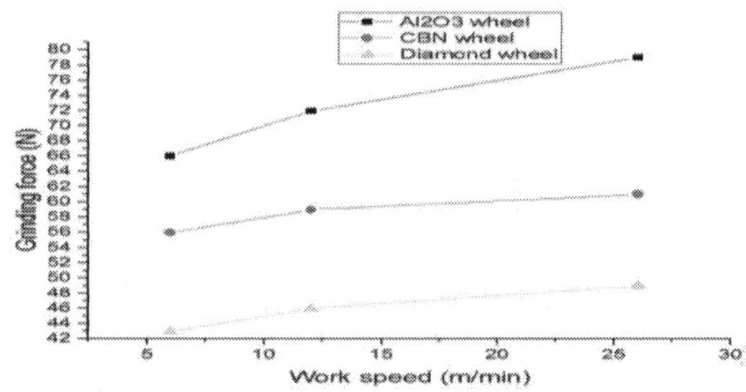

Figure 4. Work speed vs Grinding force

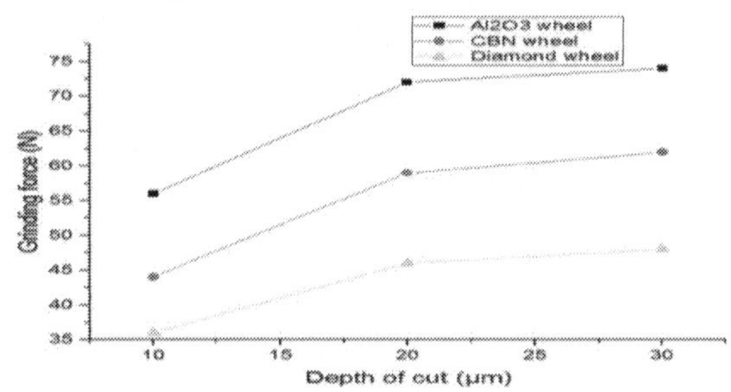

Figure 5. The depth of cut vs Grinding force

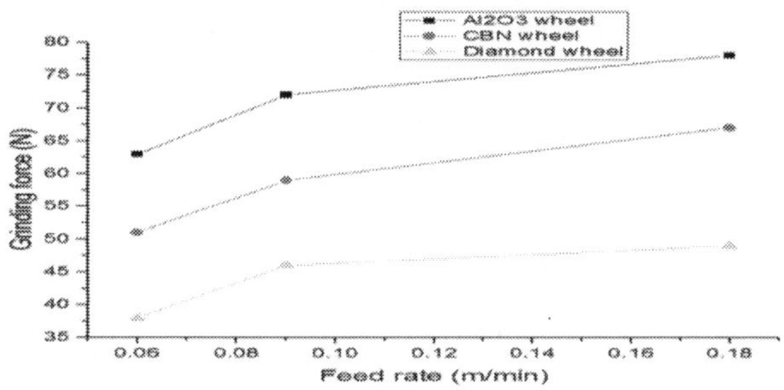

Figure 6. Feed rate vs Grinding force

The effect of work speed on grinding force as a function of wheel materials are shown in Figure 4. This investigation is carried out by fixing the wheel speed as 33.7m/s, depth of cut 20μm, feed rate 0.09m/min and 5% reinforcement ratio. It is seen from the Figure 4 the diamond wheel performed better than other wheels in terms of generating very low grinding force. Al_2O_3 wheel generating higher grinding force for the given experimental conditions. It also is seen from the Figure 4 increase in work speed increase in grinding force. As the increase of workpiece speed, the moving speed of grinding source is increasing. So the heat source works little time on the workpiece and a large amount of grinding heat transfer quickly into the atmosphere. Increasing workpiece speed and grinding depth result in the increase of the maximum undeformed chip thickness and grinding force [15]. Both the normal and the tangential components of the grinding force show moderate increases with increasing workpiece speed. Thus the grinding force is inversely related to the material hardness, topography after machining non-reinforced Al and Al-SiC composites [16].

The effect of depth of cut on grinding force as a function of wheel materials is shown in Figure 5. This investigation is carried out by fixing the wheel speed as 33.7m/s, work speed as 12m/min, feed rate 0.09m/min and 5% reinforcement ratio. It is seen from the Figure 5 the diamond wheel performed better than other wheels in terms of generating very low grinding force. Al_2O_3 wheel generating higher grinding force for the given experimental conditions. It also is seen from the Figure 5 increase in Depth of cut increase in grinding force. As the grinding wheel velocity increases, the heat generated in the deformation zone increases and thereby softening the aluminium matrix thus reducing the force required to remove the material, the surface roughness decreases with an increase in wheel velocity and workpiece velocity. This

IOP Conf. Series: Materials Science and Engineering **390** (2018) 012004 doi:10.1088/1757-899X/390/1/012004

is due to the increase in relative velocity between the wheel and workpiece and the reduction in contact time thereby reducing the chip thickness, tangential grinding force and surface roughness increase with an increase in feed and depth of cut, increase in material removal rate and the increase in chip thickness account for the increase of the tangential force (F_T), increase with an increase of percentage of SiC volume fraction [11].

The effect of feed rate on grinding force as a function of wheel materials are shown in Figure 6. This investigation is carried out by fixing the wheel speed as 33.7m/s, work speed as 12m/min, depth of cut 20μm, and 5% reinforcement ratio. It is seen from the Figure 6 the diamond wheel performed better than other wheels in terms of generating very low grinding force. Al_2O_3 wheel generating higher grinding force for the given experimental conditions. It also is seen from the Figure 6 increase in Feed rate increase in grinding force. The relative velocity between the wheel and the workpiece and the reduction in contact time reduces the chip thickness. When the feed and depth of cut are increased, the increase in material removal rate and the increase in chip thickness account for the increase of the Ft and Ra values [17].

3.2. Grinding Temperature

The effect of varying the wheel speed on grinding temperature as a function of wheel materials are shown in Figure 7. An experimental investigation is conducted through varying each parameter and keep remaining parameters as constant and 5% reinforcement ratio. It is seen from the Figure 7 the diamond wheel performed better than other wheels in terms of generating very low grinding temperature. Al_2O_3 wheel generating higher grinding temperature for the given experimental conditions. It also is seen from the Figure 7 increase in wheel speed increase in grinding temperature.Higher grinding temperature as a result of the rise of the energy needed to grind a unit volume of the material [18].

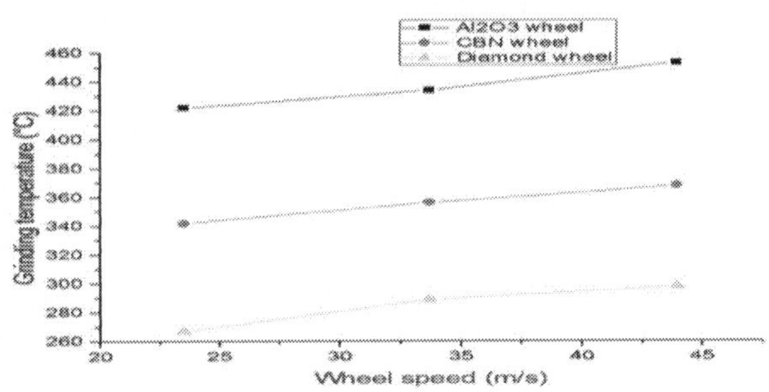

Figure 7. Wheel speed vs Grinding temperature

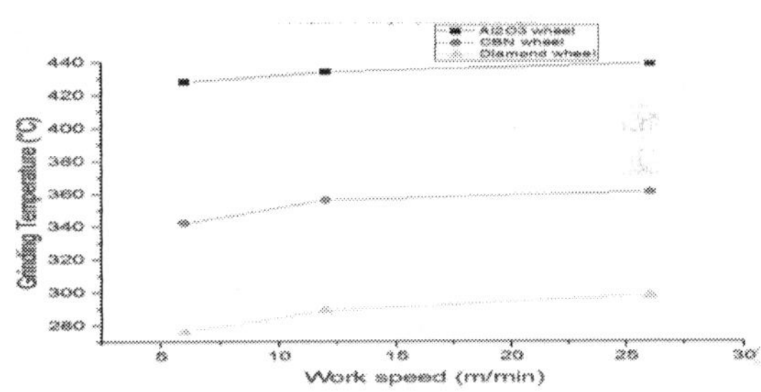

Figure 8. Work speed vs Grinding temperature

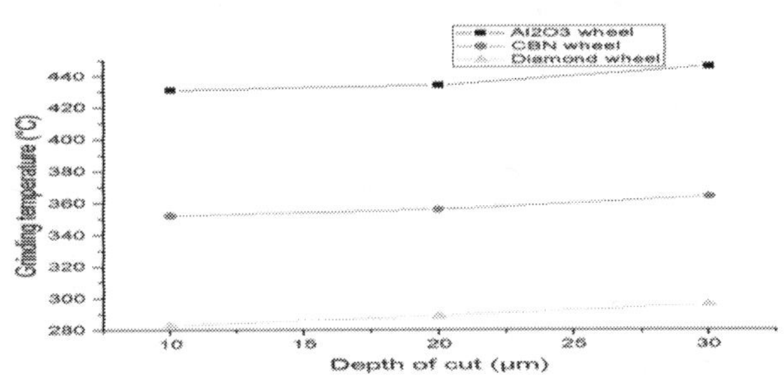

Figure 9. The depth of cut vs Grinding temperature

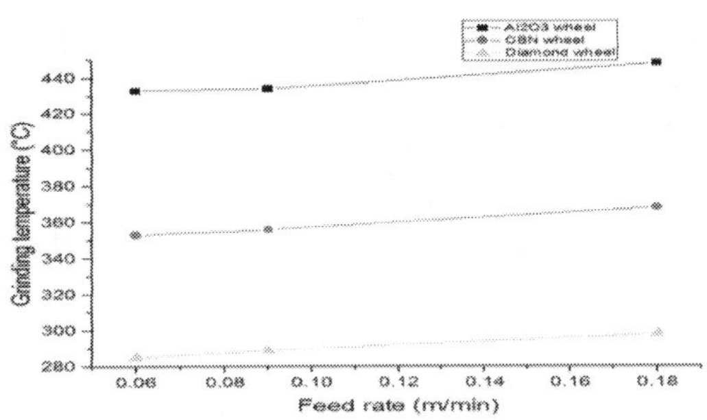

Figure 10. Fed rate vs Grinding temperature

The effect of varying the work speed on grinding temperature as a function of wheel materials are shown in Figure 8. An experimental investigation is conducted through varying each parameter and keep remaining parameters as constant and 5% reinforcement ratio. It is seen from the Figure 8 the diamond wheel performed better than other wheels in terms of generating very low grinding temperature. Al_2O_3 wheel generating higher grinding temperature for the given experimental conditions. It also is seen from the Figure 8 increase in work speed increase in grinding temperature.The greater depth of cut and the faster workpiece speed always result in the increase of the grinding temperature in the current investigation.Obviously, the grinding temperature in the super-high-speed grinding process of 120 m/s is always higher than that in the conventional speed grinding process of 20m/s [14].

The effect of varying the depth of cut on grinding temperature as a function of wheel materials are shown in Figure 9. An experimental investigation is conducted through varying each parameter and keep remaining parameters as constant and 5% reinforcement ratio. It is seen from the Figure 9 the diamond wheel performed better than other wheels in terms of generating very low grinding temperature. Al_2O_3 wheel generating higher grinding temperature for the given experimental conditions. It also is seen from the Figure 9 increase in Depth of cut increase in grinding temperature. The increase of wheel speed and grinding depth results in an increasing trend of grinding temperature. More grains enter the grinding zone in unit time quantities of grains plowing and sliding on the workpiece surface become greater. Therefore, more energy is consumed which makes the increasing temperature [19].

The effect of varying the feed rate on grinding temperature as a function of wheel materials are shown in Figure 10. An experimental investigation is conducted through varying each parameter and keep remaining parameters as constant and 5% reinforcement ratio. It is seen from the Figure 10 the diamond wheel performed better than other wheels in terms of generating very low grinding temperature. Al_2O_3 wheel generating higher grinding temperature for the given experimental conditions. It also is seen from the Figure 10 increase in wheel speed, work speed, Depth of cut and feed rate increase in grinding temperature. As the cutting speed, feed rate and or depth of cut increase, the depth of the affected layer increases. Hence, the increase in cutting temperatures, which leads to thermal deformation in the matrix material, could be one reason behind the hardening of machined sub-surface [20].

4. Conclusion

The Grinding force and grinding temperature of the AA6061-5 % TiB_2/ZrB_2 in-situ composite under different grinding parameters and wheel materials are experimentally evaluated. Increase in wheel speed decrease in grinding force due to the reduction in undeformed chip thickness. Increase in work speed, depth of cut and feed rate increase in grinding force. Increase in wheel speed, work speed, depth of cut and feed rate increase in grinding temperature due to excessive sliding action between grit and workpiece. The diamond wheel shows the lower grinding force and grinding temperature as compared to the other grinding wheels harder grits with the higher thermal conductivity of diamond wheel improves the grindability.

5. Acknowledgment

The Authors acknowledge the financial support from Science and Engineering Research Board, DST, New Delhi under Young Scientist Scheme (Ref. No. SB/FTP/ETA-0190/2014).

6. References

[1] Hossain D M, Kabir M S and Kurny A S W 2015 *J. Inst. Eng. India Ser. D* DOI 10.1007/s40033- 015-0066-8.
[2] Xu J and Mansori M E 2016*Int. J. Precis. Eng. Manuf.* 17 99-107.

IOP Conf. Series: Materials Science and Engineering **390** (2018) 012004 doi:10.1088/1757-899X/390/1/012004

[3] Zhao Y, Zhang S, Chen G and Cheng X 2007 *Mat. Sci. Eng.* **457** 156–161.
[4] Aikin R M Jr 1997 *JOM*49 35-39.
[5] Zhao G, Huang C, He N, Liu H and Zou B 2016 *Int. J. Refract. Met. Hard Mat.*61 13–21.
[6] Ping L, Shijie1 D, Anzhuo Y, Shixuan S, Zhixiong X, Zhong Z and Wei Y 2016 *Rare Met. Mat.E ng.*45 1381-1385.
[7] Yadav R N and Yadava V 2013 *Int.l J. Precis. Eng. Manuf.*14 1293-1299.
[8] Muthukrishnana N and Davimb J P 2009 *J. Mat. Proces.Tech.*2 0 9 225–232.
[9] Basheera A C, Dabadea U A, Joshia S S, Bhanuprasadc V V and Gadreb V M 2008*J. Mat. Proces s. Tech.*1 9 7 439–444.
[10] Ronald B A, Vijayaraghavan L and Krishnamurthy R 2009*Mat.Design*30 679–686.
[11] Thiagarajan C, Sivaramakrishnan R and Somasundaram S 2011 *APRN J. Eng. Appl.Sci.*6 1819-6608.
[12] Zhong Z and Hung P N 2002 Grinding of Alumina/Aluminium composites *Journals of Materials Processing Technology*123 13-17.
[13] Chandrasekaran H and Johansson J O 1997 *CIRP* **46** 493-496.
[14] Zhao B, Ding W F, Dai J B, Xi X X and Xu H J 2014 *Int. J. Adv. Manuf. Technol.*72 69-75.
[15] Lin K, Wang W, Jiang R, Xiong Y and Son G 2016 *Int. J. Adv. Manuf. Technol.*DOI: 10.1007/s00170-016-8841-8.
[16] Ilio D A, Paoletti T A, Tagliaferri T V and Veniali F 1996 *Int. J. Mach. Tool. Manuf.*36 673-685.
[17] Thiagarajan C, Sivaramakrishnan R and Somasundaram S 2012*ABCM* **34** 32-40.
[18] Thiagarajan C, Somasundaram S and Shankar P 2013 Int. *J. Eng. Sci.*12 58-66.
[19] Lin K, Wang W, Jiang R, Xiong Y and Son G 2016 *Int. J. Adv. Manuf. Tech.*DOI: 10.1007/s00170-016-8841-8.
[20] El-Gallab M and Sklad M 1998 *J. Mat. Process. Tech.* **83** 277–285.

Temperature Distribution and Stress Analysis of Aluminium Oxide Work-Piece during Laser Micro Machining Operation

Salila Ranjan Dixit[1], Prof. Debabrata Dhupal[2], And Prof. B.K Nanda[3]

[1]*PhD Research Scholar, Department of Production Engineering VSSUT, Burla, Sambalpur, Odisha, PIN: 768018.*

[2](Professor, Department of Production Engineering VSSUT, Burla, Sambalpur, Odisha, PIN: 768018.)

[3](Retired Professor, Department of Mechanical Engineering NIT, RKL, Odisha,PIN-769008

Corresponding Author E-Mail—salila.simple@gmail.com

Abstract. The current study investigates the temperature distribution, stress analysis and material removal of aluminum oxide work piece. Temperature distribution of Aluminum Oxide work-piece when a laser beam (of laser cutting machine) is focused on it to create a micro groove as a part of machining process. Present Demand of industry currently focus on ceramics because of its vast and potential use in many fields such as vehicle sector(automobile), aerospace and bio-medical engineering applications, etc. As part of machining process, a micro groove needs to be created on a crude design material. Aiming at this, a laser cutting machine is employed to create a micro groove which works on a principle that when a high power laser beam is focused on a surface of the body, material gets removed as a result of large amount of heat energy released by the laser beam. Achieving a desired groove of specific dimensions is possible by having control over process parameters of a laser cutting machine which in turn governs the amount of material removed. To tackle with this it is essential to study the temperature distribution (due to heat energy released by laser beam) in the body that is responsible for material removal. ANSYS FEA is used to simulate and analyze this temperature distribution. In this paper temperature graphs for 40W,50W , Material removal for 40W, 50W, Equivalent stress for 40W, 50W are shown.

Keywords: FEM, laser, Temperature, Thermal stress, ceramic material, Ansys

1. Introduction

Increased demand for advanced machine-intensive materials and the availability of high-power lasers have stimulated interest among researchers in the development of laser beam machining (LBM) processes. The LBM, which is a thermal energy-based machining process, is now widely applied to meet current needs for high flexibility and productivity, non-contact processing, finishing operations elimination, automation adaptability, increased material utilization, processing of materials regardless of electrical conductivity, minimum area affected by heat (HAZ) and green manufacture. In this process, the material is evacuated by (a) melting, (b) vaporization, and (c) chemical degradation where substance bonds break causing corruption of materials. At the point when a high energy density laser beam is centered around a work surface, the heat energy that is heated , assimilated and the work volume is changed into a liquid, vaporized or artificially changed express that can be effortlessly evacuated by the jet stream of High weight auxiliary gas. This procedure likewise does not suggest any mechanical power of cutting and wear of the device. July-August, 2015 Anita Pritam[9] In this paper, a test and numerical investigation on laser percussion drilling was done. A two-dimension (2D) axisymmetric finite element (FE) show for simulation of temperature field and continuing of entire arrangement amid percussion drilling was produced. The impacts of laser peak power, pulse frequency, pulse width on MRR were investigated by the developed models and experiments comes about were in great concurrence with the analyses. In light of the exploratory and numerical investigation, the procedure parameters were improved and a penetrated opening with low tapper and low scatter affidavit was gotten utilizing a 2.5kW CO2 laser. In this paper, CO2 laser percussion penetrating of thick-area (5-mm-thickness) 316 stainless steel was examined. It was discovered that the MRR increments with increment in pulse width, pulse frequency and peak power at a given thickness. A. M. Sifullah JAN 2016 [6]amid laser cutting of stainless steel-304 sheet, fleeting variety happens, while stresses are produced. Likewise grain refinement and carbide arrangement is occurred, which starts heat affected zone (HAZ) that need to distinguish. In the mean time, far reaching simulation of CO2 laser cutting procedure of stainless steel-304 sheet is perplexing as it includes thermo-mechanical issue. In this manner, in this investigation a coupled thermo-mechanical finite element method is created utilizing ANSYS to anticipate the worldly variety together with thermal stress and width of (HAZ). The reproduction results of warm examination demonstrate that the width of HAZ increments with an expansion in laser power and reductions with an expansion in cutting speed. Besides the temperature rises strongly close to the engaged laser beam. However, thermal gradient decreases sharply and gradually with the increase of distance from the cutting edge. From the structural analysis the outcomes of thermal stress was identified.. From the basic examination the results of thermal stress was recognized. It was discovered that maximum stress concentrations were observed at the cutting edge especially at the middle of the work piece which was supported by the analysis of optical microscope and SEM. C.H. Ful 2015 [7] In this investigation, design of experiment (DOE) based 3-dimensional (3D) finite element analysis was produced to reveal insight into process instruments of laser cutting NiTi. The impacts of cutting speed, peak pulse power, and pulse width on kerf width, temperature, stress, and HAZ were examined. A DFLUX client subroutine was created to demonstrate a moving volumetric (3D) heat flux of a pulsed laser utilized that consolidated super versatility and shape memory of NiTi. Expanding cutting speed decreasing kerf width, stress magnitude in the subsurface, and HAZ thickness, while the impact of pulse peak power is in actuality. pulse width does not have noteworthy effect on kerf width, temperature distribution, and HAZ thickness. Notwithstanding, it will influence the stress distribution. Pulse peak power rules stress arrangement and normal power rules kerf age. Among the different sorts of lasers utilized for machining in industries, the CO2 and Nd: YAG lasers are the most settled. In spite of the fact that CO2 lasers are broadly utilized as a part of business sheet metal cutting tasks, the advantages of the Nd: YAG laser make it a fascinating field of research. The exploratory outcomes demonstrate that the Nd: YAG laser has some one of kind qualities. Despite the fact that the normal beam power is moderately low, the beam intensity might be generally high because of a shorter pulse duration and better centering conduct. A littler cutting width, smaller micro holes, smaller HAZ and a superior cutting profile can be gotten in the Nd: YAG LBM process. The lower thermal load offered by the Nd: YAG laser permits the machining of some brittle materials, for example, SiC ceramics, which can't be machined by CO2 laser without splitting harm. As based on the demand in current industry laser micro machining operation paced a important role but to find a good and smooth dimensional groove, analysis of temperature distribution and analysis of thermal stress is required because for temperature distribution and thermal stress is played a vital role in laser micromachining process. The temperature and Thermal stress modelling is one amongst the complicated method that uses the weld parameters and material properties at higher temperatures. The temperature distribution is non-uniform like fusion zone with liquefied metal, heat affected zone and therefore the base metal zone. Stress analysis may be a general term accustomed describes analyses wherever the results quantities embody stresses and strains. In the present work Temperature dissemination, stress analysis and material removal of Aluminum Oxide work-piece when a laser beam is centered around it to make a micro groove as a piece of machining procedure.

2. ANSYS:

Based upon the today's demand of researcher ANSYS is best simulation software for simulation work. This crates vast field to simulate the problem very quickly and easily. in my research work I am continuously using Ansys for simulation work. This paper mostly focuses on simulation work using Ansys.

3. Geometry of Work Piece:

A dummy body is created to simulate the effect of Laser beam and is small enough to control the heat flow from this body into work piece. The dimensions of Work-Piece= 4mm X 25 mm X 25 mm. The Fig. 1 shows the geometry of work-piece

Fig.1 Geometry of work piece where laser like a spot move in work piece

4. Material properties.

Following material properties are the input in Ansys for simulation work.

TABLE: 1—MATERIAL PROPOTIES

Properties	Unit	Value
Density	Kj/Cm3	3.97
Specific Heat	J/Kg	0.900
Thermal conductivity	W/mk	35
Thermal Expansion Coefficient	10^{-6} 1/k	5.4
Vaporization Temperature	k	3250
Emissivity		60%
Volumetric resistivity		10^{16}
Relative permittivity	k'	10
Young's Modulus	Gpa	380
Shear Modulus	Gpa	158
Hard Ness		2000
Poisons Ratio		0.22
Dielectric Strength	Ac-) kv/mm (volts/mil	16.9
Compressive Strength		2600
Flexural Strength	Mpa	379
Bulk Modulus	Gpa	228
Elastic Modulus	Gpa	375
Dielectric Constant	@1MHz	9.8
Dissipation Factor	@1KHz	0.002
Convective Heat Transfer Coefficient	W/m ^{2} k	20

5. Boundary condition.

1. 40W and 50W of Heat is contribution as Heat Flux characterized on the base face of the fake laser body.
2. 25 mm of displacement is defined on the face of a dummy body to simulate the laser beam movement.
3. The base faces of the work-piece are constrained all over.
4. Convection of is described on each one of the qualities of the work-piece displayed to including temperature.
5. Surrounding Temperature is considered as 35 °C.
6. Intensity distribution of laser body 0.1mm

6. Results Analysis:

The Temperature distribution for 40W, 50W and are figured here. With this Material removal and equivalent stress charts also showed. As we know that the Aluminum oxide ceramic material have m melting temperature is 2050 ^{0}c. When the laser beam is applied on the material at that time it increases the temperature of that material. When the material temperature is increased to 2050 ^{0}c at the point where laser beam was focused the material starts melted. When the temperature again increased at that point above 3050^{0}c the material stars vaporizing then we find the required groove dimension. Simulation result it start melted at 2050 ^{0}c and which shown in deep yellow color and start vaporizing at 3050 ^{0}c which shows in red color in fig 2,Fig 3 of temperature distribution diagram of laser power 40w, 50w, respectively. Fig 2 And Fig 3 generally represents, when laser is applied to material at that time how many areas of that material affected by temperature known as Heat Affected Zone (HAZ) In fig 2 which is temperature distribution diagram of laser power 40w,the generation temperature is maximum approximately 10000^{0}c but it immediately fluctuates at very instant of time to 5000^{0}c which maintain at constantly till end of the process which is shown in fig 4. In fig 3 which is temperature distribution diagram of laser power 50w, the generation temperature of material will goes to maximum 12000^{0}c but it immediately fluctuate to 7000^{0}c which is maintain at constantly at end of process which is shown in fig 5.Fig 4 And Fig 5 represents the how temperature fluctuates with respect to time(s). So, in this research found that when we increase the laser power the generation temperature will increase so that the material removal process will good because it will achieve its specified groove dimension very easily and quickly due to generation temperature will more shown in fig 6, fig 7, for 40w,50w respectively. As we know that the thermal stress effect in material removal process has a good hold for achieving the desire result. In fig 8 the thermal stress effect more on red color will varies from 4.60e^{-14} MPa to 3.68e^{-5} MPa as the value is very less and it will fluctuate constantly till the end of process as shown in fig 9. Those have less effect in material removal process. Fig 8 and Fig 10 shows the effect of stress on material when laser beam is applied and Fig 9 and Fig 11 shows the Fluctuation of Stress with respect to time(s) when laser beam applied on material.

In fig 10 the thermal stress effect more on red color which varies from varies from 4.60e^{-14} MPa to 3.68e^{-5} MPa as the value is very less and it will fluctuate constantly till the end of process as shown in fig 11 that have less effect in material removal process.

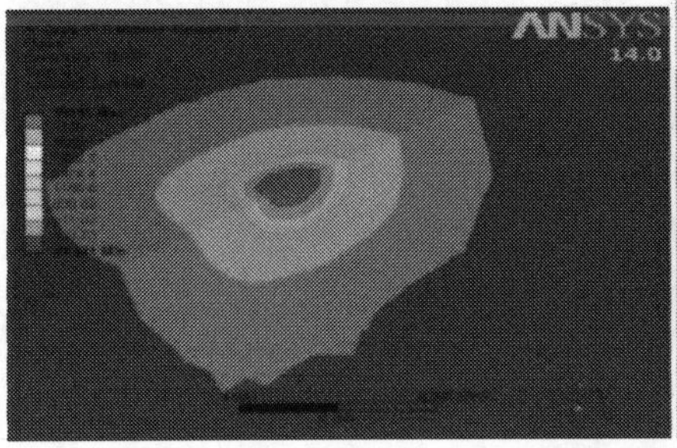

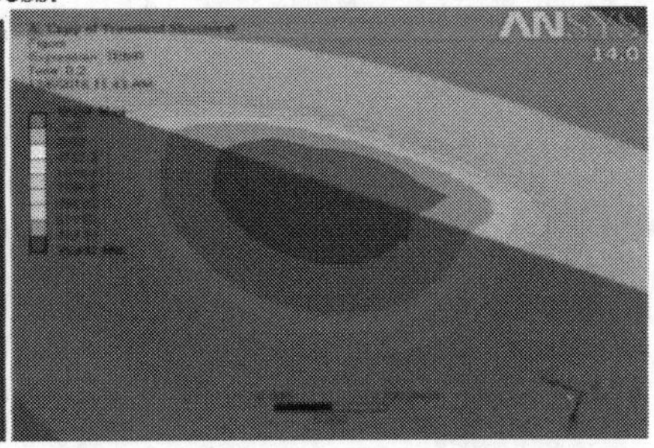

Fig. 2: Temperature Distribution Diagram of 40WFig. 3: Temperature Distribution Diagram 50 W

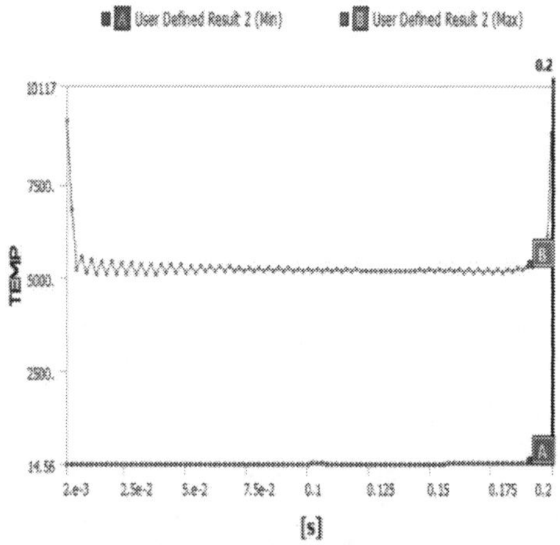

Fig. 4: Temperature Graph of 40W

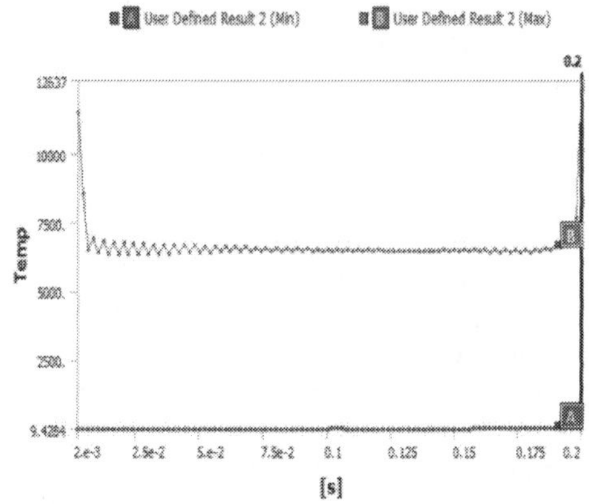

Fig. 5: Temperature Graph of 50W

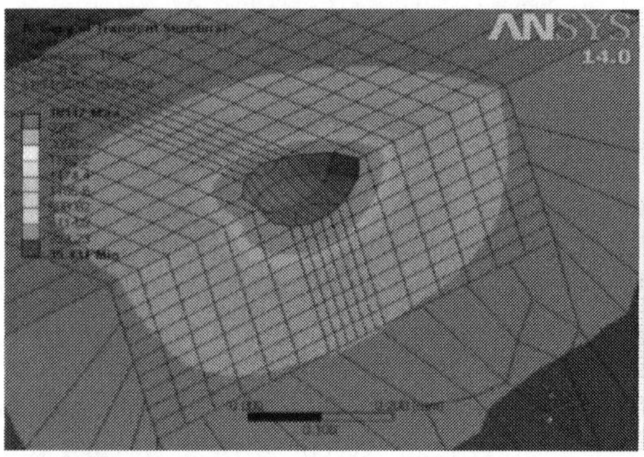

Fig. 6: Material Removal For 40 W

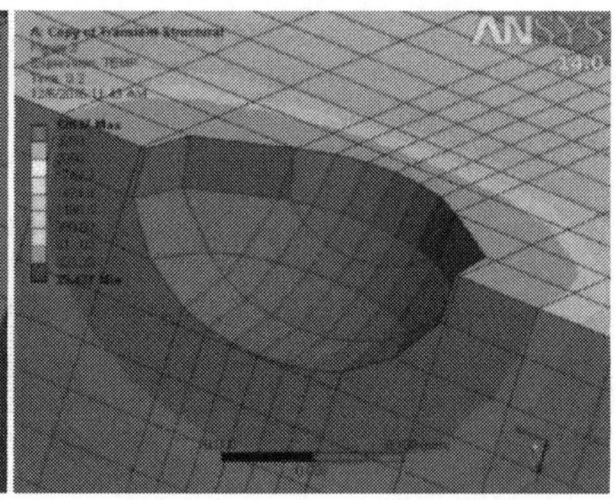

Fig. 7: Material Removal For 50 W

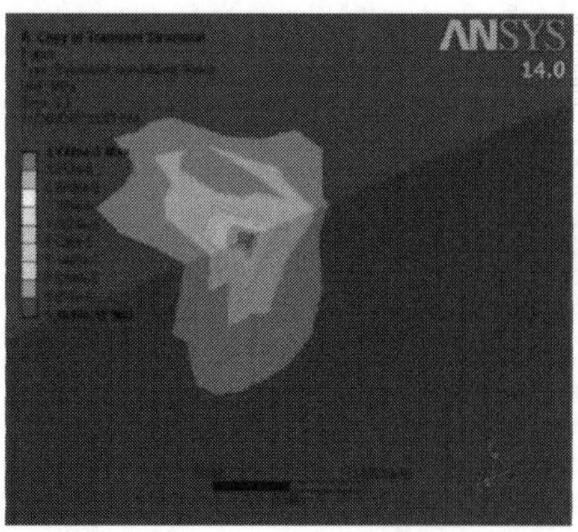

8: Equivalent Stress Diagram for 40 W

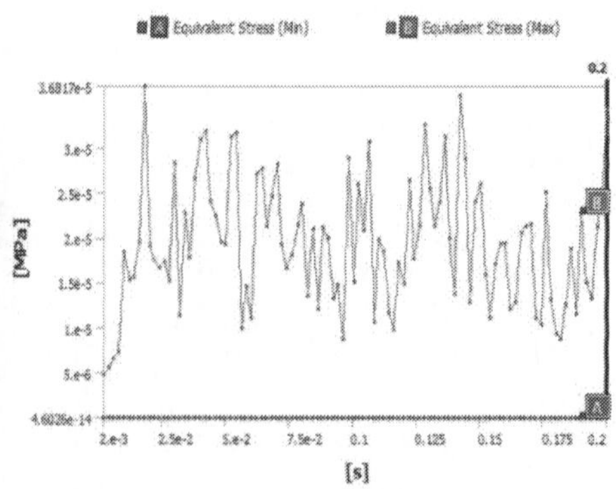

Fig. 9: Equivalent stress Graph 40W

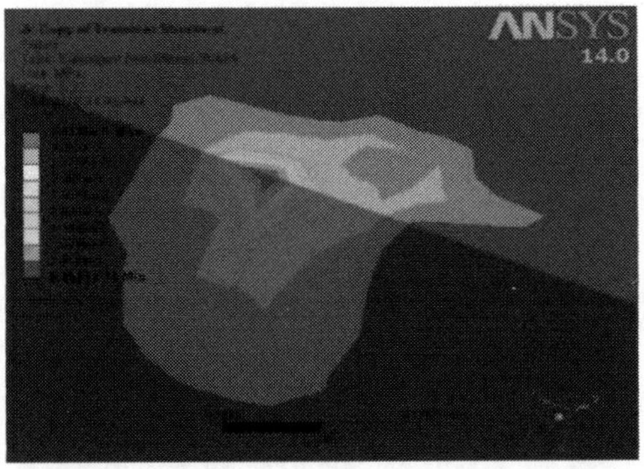

Fig. 10: Equivalent stress Diagram for50W

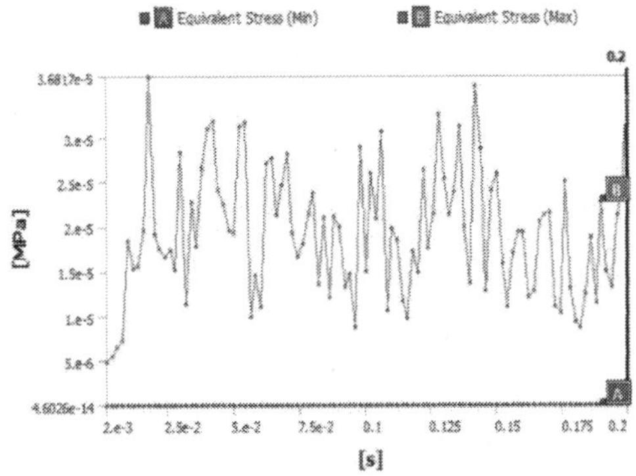

Fig. 11: Equivalent Stress Graph 50W

In this simulation result found that in 40W, 50W,the thermal stress variation is increased and decreased at every instant of time and give less effect on material removal process but the laser power increased above 50W the thermal stress is continuously increased and give more effect on material removal process. Also when increased the Laser power above 50W then the thermal stress value also higher.

As we know that the aluminum oxide ceramic material will start vaporizing at 3250ºc. In every laser power the material will vaporize at 3250ºc but when we increased the laser power the generation temperature will more and achieve good quality groove with specified dimension very easily and firstly because it takes very less time for vaporization as it is shown in fig for 40W, 50W .

7. Conclusions:

As the laser beam moves along the straight line, the heat flows into the work-piece through convection phenomenon ultimately rising the temperature of the work piece. Observation from the results is that, as the laser beam allows the heat flow into the work piece the temperature is raised at that instant. As the melting & vaporization temperature of aluminum oxide are 2050°C & 3250°C respectively, with respect to stress analysis. it was also found that when laser applied to the aluminum oxide material at a single point where it transfer heat through convection at that point and created a heat affected zone(HAZ)which is easy to simulate temperature distribution at that point. Because of creation of HAZ at that point it takes very less time for temperature generation at point beyond melting point temperature of that material again heat is continuously supplied and vaporize the material at that point and created a accurate dimensional groove which is shown in fig. The Advantage of that is no thermal damage, high machining quality, heat sensitive material machining unmatched accuracy. This is possible by increasing the heat flow of the laser beam (power) or the diameter of the laser beam. Further improvement in the temperature distribution result shall be possible through even smaller time step & mesh size. Because in this paper we are only showing the temperature distribution at a single point where laser is applied but when laser is moving through along a work piece it required more area to simulate ,However, this becomes computationally expensive. In simulation, playing around with the input heat of the laser beam may result in the desired dimensions of material removal. Temperature distribution, stress analysis and material removal of aluminum oxide work piece explained.

Acknowledgements

Authors would like to acknowledge the higher authorities of University to arrange the lab facilities.

References

[1] Chryssolouris G, Laser Machining: Theory and Practice, Springer, New York, 1991.

[2] Laser Institute of America, LIA Handbook of Laser Materials Processing, Magnolia Publishing Inc.,2001, ISBN: 0-941463-02-08.

[3] Z. Zhang, Numerical simulation of short-pulsed laser processing of materials, Numerical Heat Transfer, Part A 40 (2001) 497–509

[4] Kuar A.S ,Doloi B, Bhattacharyya B, Modeling and analysis of pulsed Nd:YAG laser machining characteristics during micro-drilling of zirconia(ZrO2), International Journal of Machine Tools and Manufacture 46 (2006) 1301–1310.

[5] Engin D,Kirby K W, Development of analytical model for the laser machining of ceramic and glass–ceramic materials, Journal of Applied Physics 80 (2) (1996).

[6] A. M. Sifullah, Y. Nukman, M. A. Hassan and A. Hossain, finite element analysis of fusion laser cutting on stainless steel-304, vol. 11, no. 1, january 2016 issn 1819-6608 arpn Journal of Engineering and Applied Sciences

[7] C.H. Fu, M.P. Sealy, Y.B. Guo1 , X.T. Wei, Finite Element Simulation and Experimental Validation of Pulsed Laser Cutting of Nitinol, 43rd Proceedings of the North American Manufacturing Research

[8] Black I, Livingstone S.A.J,Chua K.L, A laser beam machining (LBM) database for the cutting of ceramic tile, Journals of Materials Processing Technology 84 (1998) 47.

[9] Anita Pritam, , Dr. R.R. Dash, Dr. R.K. Mallik, An experimental and numerical study on laser percussion drilling of 316- Stainless Steel, International Journal of Engineering Research and General Science Volume 3, Issue 4, July-August, 2015

[10] Dhupal D, Doloi B, Bhattacharyya B, Pulsed Nd:YAG laser turning of micro-groove on aluminum oxide ceramic (Al2O3), International Journal of Machine Tools & Manufacture 48 (2008) 236–248.

Optimization of the electrochemical machining parameters in drilling of AA6061-TIB_2 in-situ composites produced by K_2TiF_6-KBF_4 reaction system

S.Chandrasekhar and N B V Prasad

Department of Mechanical Engineering, K L Deemed to be University,Guntur- 522 502 ,A.P, India

Abstract. Utilization of un conventional machining is increased exponentially due to its specific advantages like good machinabilty on the ceramic materials, ability to make complex profiles and supreme surface finish. Among the various un conventional machining, the electro chemical machining has unique identity including reasonably higher machining rate along with thermal and mechanical stress free machined surface. In-situ composites are made by reinforcing the micro size reinforcement by chemical reaction. The mechanical properties of these composites superior due its good interfacial bonding and smaller grain size of the matrix. The current paper deals the optimization of the electrochemical drilling parameters in machining of AA6061-5% TiB_2 in-situ composites produced by K_2TiF_6- KBF_4 reaction system. Current, concentration of the electrolyte and applied voltage are preferred as matrix whereas removal rate of the material and taper of the electrochemically drilled hole is identified to assess the machinbility. L_{27} orthogonal layout is followed for experimental work and Taguchi method deployed for optimization. Analysis of result and variance indicates that the preferred responses have significant influence on the responses.

1. Introduction

Aluminum matrix composite components are widely used aerospace, heavy vehicles and skeletal part of ships due its special property like higher strength at lowest density. High toughness, fatigue resistance, stiffness and corrosion resistance are notable properties of the composite material which attract the material designers [1].In-situ composites are widely preferred due to the provision for achieving the greater mechanical properties by reinforcing the relatively lower volume fraction of ceramics. Reacting the aluminum melt with halide salt is one of the popular and cost effective routes for producing the in-situ composites [2]. Presence of ceramic reinforcement into the composite give up the poor machinability. Conventional machining of the composite has associated with higher tool wear, poor surface finish and excessive residual stress formation on the machined surface [3].Un conventional machining process is suitable for machining hard materials, making complex shape and attaining high degree surface finish.

Among the various un conventional machining process, the electro chemical machining process is found advantageous like lack of thermal and mechanical stress on the machined surface. Electro chemical machining has potential advantages together with relatively higher machining rate, enhanced precision, controlled material removal, and also wider variety of materials that can be machined[4]. Hence, this machining can be applied successfully for making dies, turbine components and medical implants [5].Optimization is referred as process of doing the maximum things with least effort or practice of obtaining the most effectual or gratifying value [6]. Electro chemical machinabilty of the various materials is extensively reported in the literatures. Kao and Hocheng [7] carried out an experimental work to gauge the electro chemical machining performance of stainless steel in terms of surface roughness and passivation strength of electrolyte. Concentration of electrolyte, flow rate of electrolyte and voltage are optimized using grey relational analysis. This study results that the process parameters have momentous influence on the responses which are considered in this work. Jagannath and Bhattacharya [8] developed a mathematical models to foretell material removal rate and radial over cut by using response surface methodology based approach. Result of the investigation make known that the applied voltage is most influential on material removal rate and it was increased by increasing the applied voltage. It is also found that current density is most powerful parameters in generating the radial over cut and it was increased by enhancing the current densities. Bhattacharya and Sorkhel [9] developed an empirical relation to envisage the material removal rate with minimal radial overcut in electro chemical machining of EN-8 steel by linking the time of machining , concentration of electrolyte , voltage and standoff distance. It is observed from the investigation the removal rate of material and radial overcut can be controlled by choosing the appropriate parameters. Young et al, (2010) [10] aimed to make the complex structure by using electrochemical micro machining by controlling the machining voltage, pulse on time, pulse frequency, feed rate, electrode diameter, and shape of the electrode. An experimental result discloses that lesser machining voltage, slighter pulse on time and smaller diameter of the electrode helps to enhance the accuracy of the machining whereas electrode by way of arch working end is capable of perk up the quality of the machined.

Machinability assessment of aluminum based composites by electro chemical machining is also found in the literatures. Senthilkumar et al [11] developed a mathematical model to foretell the material removal rate and surface finish in machining of Al-SiC composites by electrochemical action using response surface methodology. This optimization was performed by preferring the voltage, concentration of electrolyte, flow rate of electrolyte and feed rate of the tool. The formed contour plots depicts that the material removal rate and roughness of surface is increased by enhancing all process parameters. They also discussed the effective deployment of the electro chemical machining in order to obtain the higher material removal rate and smoother finish on the surface. Senthilkumar et al [12] further destined a multi-characteristics optimization to appraise the machining performance of electro chemical machining of Al/15% SiC composites using non-dominated sorting genetic algorithm. The anticipated optimization scheme give up the combination of voltage, concentration of electrolyte, flow rate of electrolyte and feed rate of the tool to accomplish the lesser surface finish and upper removal rate of material. Result of authentication test designates that the optimization results are good agreement with experimental results. Rao ,S.R and Padmanabhan, G [13] reported an experimental investigation on electrochemical machining of the (Al-Si /B_4C) composites. The persuade of electro chemical machining parameters together with voltage, feed rate of electrode , concentration of electrolyte and weight percentage of B_4C on roughness of the surface roughness and radial over cut were studied. Optimization of process parameters also conducted using response surface methodology to minimize the roughness of the surface roughness and radial over cut. It is found from the above literature review the electro chemical machinability of the aluminum matrix composite made by in-situ route is not addressed by any author. Therefore, an effort has been made to deal with the electro chemical machinability issues of AA6061-5 % TiB_2 in-situ composite made by K_2TiF_6-KBF_4 reaction. The objective of the paper is to optimize the electro chemical drilling parameters in such a way that to enhance the removal rate of materials and decline the taper of the drilled hole by Taguchi method.

IOP Conf. Series: Materials Science and Engineering **390** (2018) 012006 doi:10.1088/1757-899X/390/1/012006

2. Materials and Methods

AA6061 alloy is selected as a matrix material. K_2TiF_6 and KBF_4 halide salts are used to generate the TiB_2 reinforcements. AA6061 melted in an induction furnace and brought the melt up to 850° C. The preheated K_2TiF_6 and KBF_4 are introduced into the melt. Halide salts react with molten aluminum and discharge about 1400°C temperature due to exothermic reaction At the end of reaction slag is formed and floated on the melt. The slag is warily detached from the melt. Slag free melt form the composite which shown in equation.

$$3K_2TiF_6 + 13\,Al \rightarrow 3TiAl_3 + 3KAlF_4 + K_3AlF_6$$

$$2KBF_4 + 3Al \rightarrow AlB_2 + 2KAlF_4$$

$$AlB_2 + TiAl_3 \rightarrow TiB_2 + 4Al$$

Energy-dispersive X-ray spectroscopy analysis is carried out for the produced composite and corresponding report is illustrated in the Figure 1. It is understandable from the Figure 1, the aluminum and titanium peaks are detected in the pattern. Quantitative elemental analysis report divulges the 4.71 weight percentage of TiB_2 phases is found. The proposed weight percentage of TiB_2 is 5. Deviation in weight percentage of TiB_2 may be attributed that the vaporization due to the exothermic reaction. Scanning electron microscopic image of the produced composite is exposed in the Figure 2. Approximately, 0.5 to 1 μm size of TiB_2 particles can be seen in the Figure 2.

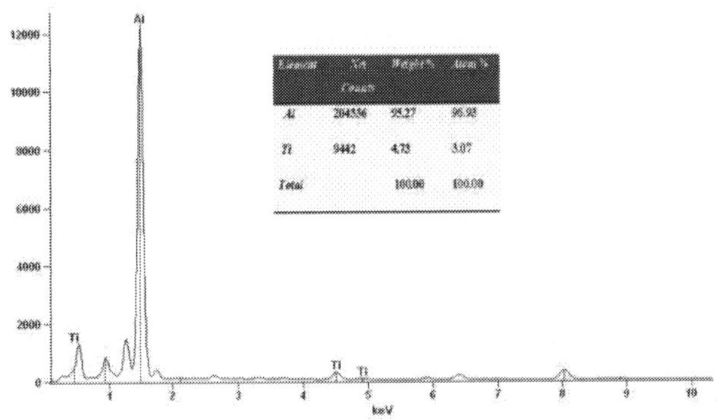

Figure 1. EDAX Pattern of the composite

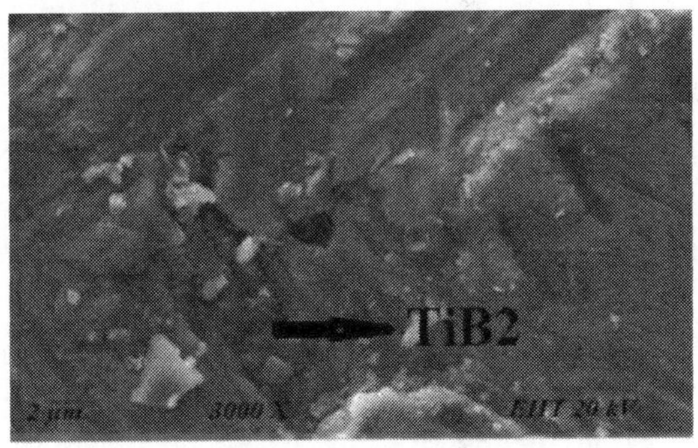

Figure 2. SEM image the composite

3. Experimental work

The produced AA6061-5 % TiB_2 in-situ composite is made into 0.5 mm sheet and used for the experimental investigation. A micro electrochemical machining setup is used for the experimental work. Concentration of electrolyte, voltage and current are preferred as machining parameters whereas material removal rate and taper are selected as response. Copper tool is selected and sodium nitrate electrolyte preferred for the current work. The other parameters specific power consumption is 7 W/mm3/min, inter electrode gap is 0.5 mm and flow rate of electrolyte is 10 Litter /min is fixed during the experimental work. Factors and their level of the experimental work are exemplified in the Table 1. L_{27} orthogonal array is followed for the experimentation. Material removal rate is calculated by the ratio of mass loss and time taken for machining and taper angle is calculated by the flowing relation and tabulated in Table 2.

Taper angle = $\tan^{-1}$ (Entry diameter-Exit diameter)/2*Thickness of the sheet -----------(1)

Table 1: Factors and levels

Factors	Units	Levels		
Electrolyte concentration	Moles	1	2	3
Voltage	Volts	14	16	18
Current	Amps	2	3	4

Table 2: L_{27} Orthogonal array for 3 factors

S.No	Electrolyte concentration	Voltage	Current	*MRR*	Taper	S.No	Electrolyte concentration	Voltage	Current	*MRR*	Taper
1	1	1	1	0.029	0.279	15	2	2	3	0.062	0.301
2	1	1	2	0.031	0.283	16	2	3	1	0.035	0.288
3	1	1	3	0.037	0.293	17	2	3	2	0.049	0.296
4	1	2	1	0.028	0.281	18	2	3	3	0.056	0.306
5	1	2	2	0.032	0.289	19	3	1	1	0.033	0.28
6	1	2	3	0.044	0.299	20	3	1	2	0.042	0.288
7	1	3	1	0.029	0.287	21	3	1	3	0.043	0.298
8	1	3	2	0.038	0.295	22	3	2	1	0.037	0.286
9	1	3	3	0.043	0.305	23	3	2	2	0.048	0.294
10	2	1	1	0.033	0.276	24	3	2	3	0.055	0.304
11	2	1	2	0.04	0.284	25	3	3	1	0.040	0.292
12	2	1	3	0.048	0.294	26	3	3	2	0.057	0.300
13	2	2	1	0.039	0.282	27	3	3	3	0.056	0.310
14	2	2	2	0.048	0.290						

4. Result and Discussion

L_{27} Orthogonal array is generated by using Minitab 18 and corresponding material removal rate values are entered. Response graph for the material removal rate is created and illustrated in the Figure 3. It is found from the Figure 3, 3 moles of electrolyte concentration, 18 volts applied voltage and 4 ampere current is optimal process parameters to attain the most advantageous removal rate of material with in selected levels of parameters. It is also observed from the response graph, the enhancement of electrolyte concentration, applied voltage and current is maximizing the removal rate of material. This is happening due to enhancement of electrical conductivity at higher level of concentration in turn maximize the machining current within the gap of the inter electrode [14]. Higher voltage allows maximum machining current within the gap of the inter electrode thus the removal rate of material is increased [15].As the current increases, the anodic metal dissolution thereby the removal rate of the material is increased. For the same L_{27} layout, the analysis of variance for the rate of removal of the material also performed and exhibited in the Table 3. Table 3 clearly brings out the percentage of the contribution of the each parameter on the material removal rate. Applied current plays a dominant role than the other parameters. The contribution of the current has 51.47 %, concentration of the electrolyte has 33.91 % and applied voltage has 13.40 % on the removal rate of the materials

The response graph produced for the taper of the electro chemically drilled hole is disclosed in the Figure 4. Figure 3 make known that 1 mole of electrolyte concentration, 14 volts applied voltage and 2 ampere current is optimal process parameters to conquer the minimal taper of the hole within selected level of parameters. It is clear from the Figure 3, the hike of the all parameters maximize the taper of the hole. Gap of inter electrode is maximum at entry diameter whereas minimum at exit diameter. This phenomenon maximize the taper at high levels all parameters. Smaller range of the parameters within the selected level gives minimal taper. Analysis of the variance for the taper of the hole is presented in the Table 3. The contribution of the current has 66.09 %, concentration of the electrolyte has 5.3 % and applied voltage has 28.5 % on the taper of the hole. Approximately 2 % error of removal rate of materials and taper indicates the statistical significance of the variance. The regression equations also generated by relating the process parameters and removal rate of materials and taper of the hole and presented in the following equations.

MRR= 0.041926 - 0.007370 Electrolyte concentration_1 + 0.003630 Electrolyte concentration_2
+ 0.003741 Electrolyte concentration_3 - 0.004593 Voltage_1 + 0.001741 Voltage_2
+ 0.002852 Voltage_3 - 0.008259 Current_1 + 0.000852 Current_2 + 0.007407 Current_3------ (2)

Taper = 0.291815 - 0.001704 Electrolyte concentration_1
- 0.001148 Electrolyte concentration_2 + 0.002852 Electrolyte concentration_3
- 0.005704 Voltage_1 - 0.000148 Voltage_2 + 0.005852 Voltage_3 - 0.008370 Current_1
- 0.000815 Current_2 + 0.009185 Current_3-- (3)

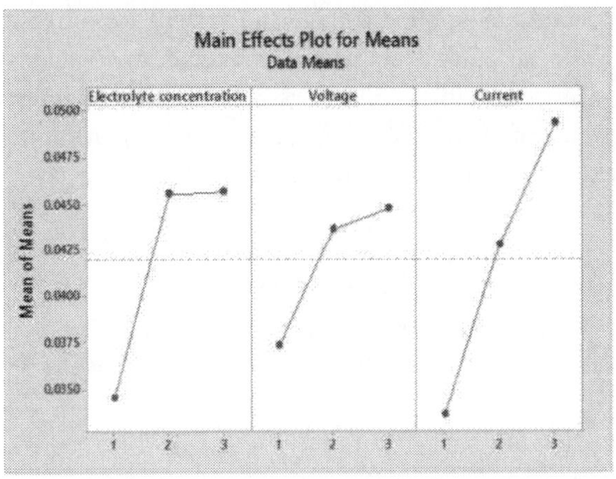

Figure 3: Response graph for MRR

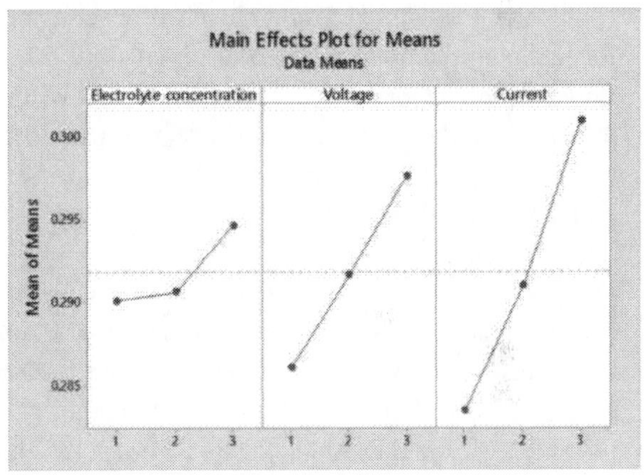

Figure 4: Response graph for Taper

Table 3: Analysis of variance for material removal rate and taper

Source	DF	Adj SS	Adj MS	% Of Contribution	DF	Adj SS	Adj MS	% Of Contribution
	Material removal rate				Taper			
Electrolyte concentration	2	0.000734	0.000367	33.918	2	0.0001	0.000056	5.30303
Voltage	2	0.00029	0.000145	13.401	2	0.0006	0.000301	28.50379
Current	2	0.001114	0.000557	51.478	2	0.0013	0.000698	66.09848
Error	20	0.000026	0.000013	1.2014	20	0.0000	0.000001	0.094697
Total	26	0.002164	0.001082	100	26	0.0021	0.001056	100

5. Conclusion

AA6061-5 % TiB_2 composite is successfully produced by K_2TiF_6-KBF_4 halide salt reaction. TiB2 phase and its volume fraction and confirmed by Energy-dispersive X-ray spectroscopy and scanning electron microscopy analysis. Optimization of the electro chemical drilling parameters in machining the in-situ composite is discussed. Higher levels of electrolyte concentration, applied voltage and current are suitable for attaining the optimal material removal rate. Increase in all parameters maximizes the both response. Lower levels of parameters are found good to minimizing the taper. Current is offering the strongest influence on the material removal rate and taper. Inter electrode gap plays significant role in generating both responses. The generated regression equation helps to predict the material removal rate and taper with considerable accuracy.

References

[1] Das D.K, Mishra P.C, Singh S. and Thakur R.K 2014 *Int J Mech Mater Eng* **9** 12.

[2] Mahamani A and Anantha Chakravarthy, V.V 2017 *Mater and Manuf* **32:15** 1700-1706

[3] Pramanik A., Arsecularatne J and Zhang L 2008 Machining of Particulate-Reinforced Metal Matrix Composites. In: Machining. *Springer*, London

[4] Shibli M.A and R. Manu 2008 , *J Mater Sci* **43** 4282–4289.

[5] Kozak, J., Dabrowski, L., Lubkowski, K., Rozenek, M and Slawinski, R 2000 *J Mater Process. Technol* **107** 293–299

[6] Kelly TR. 2010, *Technology Teacher*.**18** 23
https://digitalcommons.usu.edu/cgi/viewcontent.cgi?article=1031&context=ncete_publications

[7] Kao P. S. and Hocheng,H. 2003 *J Mater Proc Techno* **140** 255-259

[8] Jagannath M and Bhattacharyya B 2008 *Int J Adv Manuf Techno* **35** 821-832.

[9] Bhattacharyya and Sorkhel S.K. 1998 *J Mater Proc Techno* **86** 200-207.

[10] Young L, Di Z, Zong Y.B, Huang S and Yu H.B 2010 *Chinese J Aero* **23** 578-584.

[11] Senthil Kumar C, Ganesan G and Karthikeyan R 2009 *Inter J Adv Manuf Techno* **43** 256-263**12|**

[12] Senthil Kumar C, Ganesan G and Karthikeyan R 20**11** *Trans Non ferrous Met Soc China* 21 2294-2300

[13] Rao S.R and Padmanabhan,G 2013 *Int_J_ManufMater_Mech Eng* **3** 42-56

[14] Bhattacharya B and Sorkhel SK 1997 *Proc of the 17th All India machine tool design and research conf* **1** 307–311.

[15] Neto JCDS, Silva EMD and Silva MBD 2006 *J Mater Process Technol* **179** 92–96

IOP Conf. Series: Materials Science and Engineering **390** (2018) 012007 doi:10.1088/1757-899X/390/1/012007

Important factors of self-efficacy and its relationship with life satisfaction and self-esteem – with reference to gen y and gen z individuals.

M Kishore Kumar[1,2] and Dr.R G Priyadarshini[1,3]

[1] Amrita School of Business, Coimbatore, Amrita Vishwa Vidyapeetham, India.
[2] kishore8274@gmail.com
[3] rg_priyadarshini@cb.amrita.edu

Abstract. Internet applications, social media and sedentary lifestyle followed by individuals put a lot of stress on Gen Y and Gen Z individuals. This study seeks to find out what these individuals think they are capable of which is an individual's self-efficacy, with respect to gender, age and educational qualification, using subjective evaluation questionnaire which would enable the respondents to evaluate the questions posed and respond according to their perception. This study was done on students who about to become a part of the workforce in the near future. The study was aimed at finding meaningful relationships between students' overall quality of life and self-esteem with self-efficacy and it was found to be positive. The other factors-age, gender and educational qualification that were considered in the study failed to show considerable significance with self-efficacy, as self-efficacy was found to have negligible variation with respect to those factors.

1. Introduction

In today's world, people are living in the era of internet and instant gratification. This is enabled by social media and internet applications that are presenting a sedentary life style among individuals (Alejandro Silva Cortés et al, 2016), alongside which students are made to deal with high pressure environments which has adverse effects on them. Today's youth (Gen Y and Gen Z in this case) live in a highly dynamic environment which is demanding more from them and the society puts a lot of pressure on them. This reduces life satisfaction and causes stress in them. Thus there was a need identified to study the relationship between (i) an individual's perceived ability to accomplish ore succeed at certain tasks. (Bandura 1977, 1982, 1986) which is identified as self-efficacy (ii) the general sense of an individual's own value or worth as perceived by the individual himself/herself (Makikangas and Kinnunen 2003) which is identified as self-esteem and (iii) the overall life quality of a person (Diener, 1995) which is identified as life satisfaction.

Early work about life satisfaction was mainly concentrated on adults. Only recently, researchers have begun to include children and adolescents in these types of studies. It has been found that there is strong association between life satisfaction and self-esteem in countries that are oriented towards individualism (Heubner, 2004). It was also found that high youth life satisfaction facilitates better self-esteem and self-efficacy (Gilman et al 2006; Gilman et al. 2000, MacDonald et al. 2005, Suldo et al. 2006). Thus a need, has been identified, to study the life satisfaction in today's young generation that is Gen Z and Gen Y, who have just started to go for jobs or who are going to join the work force soon.

2. Literature review

Until now, very few evidence is available to show whether self-efficacy has an association with life satisfaction over a period of time and if self-efficacy definitely reduces the negative effects of caused on young people's life satisfaction by perceived stress (Burger et al. 2016). Though stress is not considered in this study, self-efficacy's relationship with life satisfaction is being considered in this study, which can give an insight into the lives of today's youth and their wellbeing, as it was found that higher scores in Life Satisfaction was associated with less depression and less social stress. (Proctor et al. 2009).

2.1. Self-Efficacy and Self Esteem

Self-efficacy is an individual's perceived ability to accomplish ore succeed at certain tasks. (Bandura 1977, 1982, 1986) and Self-esteem is the general sense of an individual's own value or worth as perceived by the individual himself/herself (Makikangas and Kinnunen 2003).

2.2. Life Satisfaction

Many researchers conceptualize SWB Subjective Wellbeing (SWB) as made up of three interrelated, distinguishable components such as positive affect, global life satisfaction and negative affect (Lucas and Diener, 2000). This study is taking this Global Life Satisfaction for assessing a part of the wellbeing of individuals and the author refers to Global Life Satisfaction while mentioning the phrase "Life Satisfaction". Life Satisfaction refers to subjective evaluation of overall life quality of a person (Diener, 1995). It refers to a cognitive appraisal of the overall life quality of a person based on subjective evaluation which means the respondent can assume standards while answering the questionnaire (Diener et al., 1999).

In large-scale international research, India is typically among the lower-scoring countries on indicators of life satisfaction and happiness (Biswas-Diener et al. 2012; Diener et al. 1995; Schyns 1998)

Life satisfaction is indeed a construct worthy of special attention because it gives a clear understanding of well-being through a global index generated through criteria determined by respondents themselves rather than researchers defining the standards. (Diener, 2000) as the questionnaire is subjective evaluation questionnaire.

2.3. Relationship between Self-Efficacy, Life Satisfaction and Self-Esteem,

The association between individuals' Life Satisfaction and self-efficacy, or between Self-Efficacy and Self-esteem are not causal. Self-esteem is considered to predict satisfaction with life (Diener and Diener, 1995) also, higher life satisfaction is found to bring about feelings better self-esteem (Proctor et al. 2009). Higher levels of self-efficacy is found to reduce negative effects on life satisfaction by factors like stress (Burger et al, 2016), thus individuals scoring high on self-efficacy tend to have high Life satisfaction scores. This study seeks to find meaningful relationships between the three parameters of Life Satisfaction, Self Esteem and Self Efficacy.

3. Factors

A number of demographic factors were identified to have significant relationship with life satisfaction which is the overall life quality of a person. As higher self-efficacy is found to improve the quality life (Burger et al, 2016), the effects of the following factors on self-efficacy is studied.

3.1. Gender

The overall quality of life varies between men and women of the same age group. (Daig, et al. 2009) and also gender is likely to cause differences in level and kind of satisfaction in academic career (Thoreson, 1990)

However, in children of ages 8 to 12 years, the overall quality of life is unrelated to gender (Heubner 2004). None of the surveyed individuals were below 12 years of age which enables us to consider gender as an influencing factor for the whole sample.

3.2. Educational Background

Overall quality of a person's life varies with education level. Those with higher education qualifications usually rend towards having higher life satisfaction (Feldman, 2010) .Studies show modest relationships with Life Satisfaction and demographic variables. (Heubner, 2004). Age and educational level predicted life satisfaction (Chengedzai, et al. 2016). This study's respondents are students and the sample contains students from two different educational backgrounds such as ones who are currently pursuing their Under Graduation degrees and ones who are currently pursuing their Post-Graduation degrees.

3.3. Age

The overall quality of life varies with different age groups. (Daig, et al. 2009) In a study on people above 60 years of age, it was found that the quality of life decreased as age advanced, and it is related to their life experiences in their early adulthood. (Chaonan Chen, 2000). In this study, respondents are grouped into Generation Y and Generation Z based on their age at the time of the survey. Individuals belonging to the same generation are bound by the kinship within their generation people as they encounter same events which changed everything around them when they were growing up. They would have witnessed the same economic and political changes around them which would set their attitudes in a similar way. What is more important about that to this study is that they share similar expectations for life (Marcie Merriman 2015).

4. Gen Y and Gen Z

Generation Y are individuals who are born between 1981 and 1996 and Generation Z are individuals born after 1997. Though there is no consensus among different publications, on the end year of Generation Z, there are a few evidences for the end year being 2003 and others suggest later years. Since the respondents of this study are Undergraduate and Post Graduate students, none of the respondent was born after 2003 thus all respondents of this study qualify as either Generation Y or Generation Z, based on their age at the time of the survey. (Marcie Merriman 2015).

The novelty of this study is preserved by the fact that this study is done on students who belong to the Generations Y and Z, wherein Generation Z is only now coming of age. Moreover at the time of survey, they were a few years away from entering the workforce. Through this study, those individuals' ability to accomplish various tasks in their lives can be understood which can be used at the workplace to get the best out of these individuals, by making means to get what they can give at their best.

5. Research Questions

- What are the important factors that affect Self-efficacy in individuals?
- How does Life satisfaction and Self-esteem affect Self efficacy in individuals?
- Does Self-efficacy in individuals differ pertaining to various generation types?
- Does Self-efficacy in individuals differ pertaining to various age groups?
- Does Self-efficacy in individuals differ pertaining to various educational qualification backgrounds?

6. Hypotheses

H1: Life satisfaction and Self-esteem does not have a significant association on Self efficacy in individuals.

H2: There is no significant variance in scores of self-efficacy in individuals pertaining to various generation types.
H3: There is no significant variance in scores of self-efficacy in individuals pertaining to various age groups.
H4: There is no significant variance in scores of self-efficacy in individuals pertaining to various educational backgrounds.

7. Conceptual Model

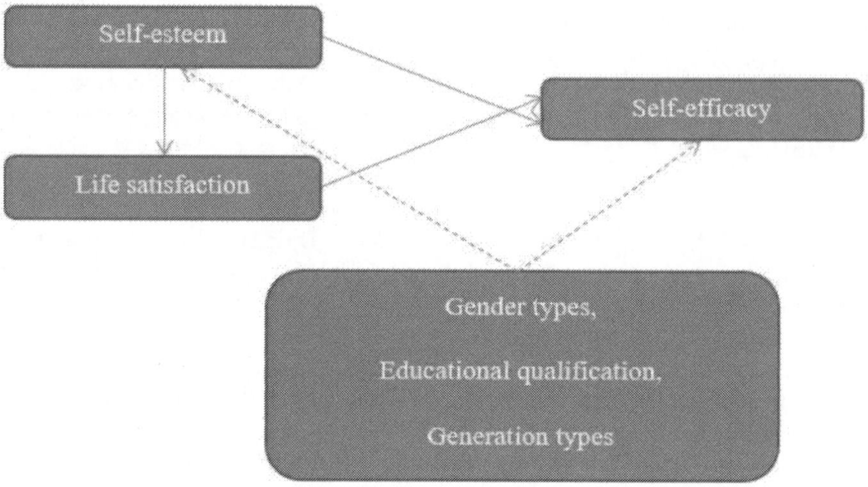

Figure 1. Conceptual model showing the relationship between parameters

8. Methodology

A survey was conducted on Gen Y and Gen Z individuals using a survey questionnaire through online platforms.
The survey questionnaire contained questions in the following pattern.

- Demographics (Age, Gender, Education, Home town)
- BMSLSS: Brief Multidimensional Students' Life Satisfaction Scale (Seligson et al. 2003) – 6 items
- Rosenberg Self Esteem Scale (Rosenberg, 1965) – 10 items
- General Self Efficacy Scale (Schwarzer et al. 1995) – 10 items.

These questions were subjective evaluation questions which means the response to any question in the questionnaire (except from the 1st section), is based on the respondents' perceptions of the questions. All items in section 2 were answered in 5-Point Likert Scale, items in section 3 and 4 were answered on 4-Point Likert Scale. Scoring is straight forward, for most items except for five items which were reverse-keyed.
The survey finally yielded a sample size of 145 out of which 79 respondents belong to Generation Y and 66 belong to Generation Z when surveyed. 76 of the respondents were males and 69 were females. When the survey was conducted, 67 students were pursuing their Under Graduation degree and the remaining 78 were pursuing their Post-Graduation.

9. Analysis

The data was analyzed using SPSS and the analysis is as follows:

9.1. Association of Life satisfaction and Self-esteem on Self Efficacy

Self-Efficacy has a significant association (R-=0.658) with self-esteem and life satisfaction as alpha value is less than 0.05 as shown in 'table 1'.
Self-Efficacy = 0.468 (Life satisfaction) + 0.574 (Self Esteem) -0.18

9.2. Does variation in age groups (Generations) affect Self efficacy?
There is no significant variation in generation types with respect to Self-efficacy since alpha value if greater than 0.05.

Table 1. Coefficients[a].

Model		Separation energies		Standardized Coefficients		
		B	Std. Error	Beta	t	Sig
1	(Constant)	-.018	2.854		.006	.995
	Life Satisfaction	.468	.092	.344	5.079	.000
	Self-Esteem	.574	.086	.450	6.643	.000

[a] Dependent Variable: Self-Efficacy

9.3. Does variation in gender types affect Self efficacy?
There is no significant variation in gender types affect Self efficacy since alpha value if greater than 0.05

9.4. Does variation in educational qualification affect Self efficacy?
There is no significant variation in educational qualification affect Self efficacy since alpha value if greater than 0.05

10. Results and Discussion
The following are the findings of the analysis:

Self-Efficacy has a significant association (R-=0.658) with self-esteem and life satisfaction as alpha value is less than 0.05. The regression equation is Self-Efficacy = 0.468 (Life satisfaction) + 0.574 (Self Esteem) -0.18. Thus H1 is rejected.

There is no significant variation in generation types with respect to Self-efficacy since alpha value if greater than 0.05, therefore H2 is not rejected. There is no significant variation in gender types affect Self efficacy since alpha value if greater than 0.05, therefore H3 is not rejected. There is no significant variation in educational qualification affect Self efficacy since alpha value if greater than 0.05, therefore H4 is not rejected.

The results indicate that life satisfaction and self-esteem are positively related with Self efficacy for both Gen Y and Gen Z individuals. Similarly, both gender and educational qualification does not have a significant impact on self-efficacy with regard to the sample of 145 students.

11. References
[1] Alejandro Silva Cortésa, Ana María Correa-Díazb, Martha Luz Benjumea-Ariasc Alejandro Valencia-Ariasa and Lemy Bran-Piedrahitaa 2016 Motivational factors and effects associated with physical-sport practice in undergraduate students Procedia- Social and Behavioral Sciences vol. **237** pp 811-815

[2] Anna Zajacova, Scott M Lynch and Thomas J Espenshad, 2005 Self-efficacy, Stress, And Academic Success in College *Research in Higher Education* vol. **46(6)** pp 677-706

[3] Bandura A 1977 Self-efficacy: Toward a unifying theory of behavioral change *Psychological Review* **84(2)** pp 191-215

[4] Carmel L Proctor, Alex Linley P and John Maltby 2009 Youth Life Satisfaction: A review of literature Journal of Happiness Studies vol. **10(5)** pp 583-630

[5] Carmel Proctor, Alex Linley P and John Maltby 2009 Very Happy Youths: Benefits of Very High Life Satisfaction among Adolescents Social Indicators Research vol. **98** pp519-532
[6] Chaonan Chen 2001 Aging and Life Satisfaction *Social Indicators Research* vol **54** pp 57-79
[7] Chengedzai Mafini and Daniel Meyer 2016 Life Satisfaction by Demography in Low Income Residential Areas: Evidence from South Africa *Studia Universitatis Babeş-bolyai Oeconomica*, vol. **61(2)** pp 5-20
[8] Chiungjung Huan 2012 Gender differences in academic self-efficacy: a meta-analysis *European Journal of Psychology of Education* vol. **28(1)** pp 1-35
[9] Feng Kong, Jingjing Zhao, and Xuqun You 2012 Self-Esteem as Mediator and Moderator of the Relationship between Social Support and Subjective Well-Being among Chinese University Students Social Indicators Research vol. **112(1)** pp 151-161
[10] Fred Feldman 2010 *What is this thing called happiness?* (New York: Oxford)
[11] Isolde Daig, Peter Herschbach, Anja Lehmann, Nina Knoll and Oliver Decker 2009 Gender and Age Differences in Domain-Specific Life Satisfaction and the Impact of Depressive and Anxiety Symptoms: A General Population Survey from Germany Quality of Life Research vol. **18(6)** pp 669-678
[12] Kate Ann Levin, Lorenza Dallago and Candace Currie 2012 The Association Between Adolescent Life Satisfaction, Family Structure, Family Affluence and Gender Differences in Parent–Child Communication *Social Indicators Research* vol. **106(2)** pp 287-305
[13] Rich Gilman and Scott Huebner E 2006, Characteristics of Adolescents Who Report Very High Life Satisfaction *Journal of Youth and Adolescence* vol. **35 (3)** pp 293-301
[14] Rich Gilman and Scott Huebner E Characteristics of adolescents who report very high life satisfaction *Journal of Youth and Adolescence* vol. **35(3)** pp 311-319
[15] Richard W. Thoreson, CarolAnne M. Kardash, David A. Leuthold and Kelly A.Morrow, (1990), Gender Differences in academic career *Research in Higher Education* vol. **31(2)** pp 193-209
[16] Rosenberg M 1965 Society and the adolescent self-image (Princeton, NJ: Princeton University Press)
[17] Seligson, Huebner and Valois 2003 BMSLSS: Brief Multidimensional Students' Life Satisfaction Scale *Social Indicators Research* vol **61** pp 121-145
[18] Schwarzer R. & Jerusalem M 1995 Generalized Self-Efficacy scale *Measures in health psychology: A user's portfolio. Causal and control beliefs* ed Weinman J Wright S & Johnston M (Windsor, UK: NFER-NELSON) pp 35-37
[19] Scott Huebner E 2004 Research on Assessment of Life Satisfaction of Children and Adolescents (Quality of Life Research on Children and Adolescents) *Social Indicators Research* vol. **66(1)** pp. 3-33
[20] Susan Antaramian, Shanmukh V Kamble and Scott Huebner E 2015 Life Satisfaction and Coping in Hindu Adolescents in India. Journal of Happiness Studies vol. 17(4) pp 1703-17
[21] Tammy D Gilligan and Scott Huebner 2007 Initial Development and Validation of the Multidimensional Students' Life Satisfaction Scale –Adolescent Version *Applied Research in Quality of Life* vol. **2(1)** pp 1-16
[22] Yongquan Huo and Feng Kong 2013 Moderating Effects of Gender and Loneliness on the Relationship between Self-esteem and Life Satisfaction in Chinese University Students *Social Indicators Research* vol. **118** pp 305–314
[23] Marcie Merriman 2015 What id the next big disruptor isn't a what but a who? *Ernst and Young Global Limited*

IOP Conf. Series: Materials Science and Engineering **390** (2018) 012008 doi:10.1088/1757-899X/390/1/012008

Effect of Hot Extrusion through Mathematically Contoured Die on Mechanical and Tribological Characteristics of the AMC Prepared through P/M Route

Sambit Kumar Mohapatra[1]*, Kalipada Maity[2], Sataya Brata Ghadei[3], Akhtar Khan[4]

[1]School of Mechanical Engineering, KIIT University, Bhubaneswar, Odisha 751024, India
[2, 4] Department of Mechanical Engineering, NIT Rourkela, Odisha 769008, India
[3]Mahindra & Mahindra Ltd., Tamil Nadu 603004, India

*E-mail : sambit.mohapatrafme@kiit.ac.in

Abstract. Thermo mechanical treatment of the components prepared by powder metallurgy route instigates the extensive research, due to its ameliorated impacts on the properties. In this research aluminium matrix composites (AMCs) {Al (92) - Mg (5) + Gr (1) + Zn (2)} have been prepared by following powder metallurgy route with dual axial cold compaction and subsequent sintering. Hot extrusion was considered as the secondary processing operation for the AMCs. Mathematical contoured die was utilised for the extrusion to avoid severe surface defects. Comparative analysis of different mechanical as well as tribological properties of the specimen before and after extrusion was focused. Extrusion engenders significant amount of improvements of the properties those are attributed to excellent bond strength and uniform density distribution due to high compressive stress. To furnish the suitable explanation scanning electron microscopies have been performed for the wear surfaces.

Keywords: AMC, Cosine Die, Extrusion, Powder Metallurgy, Tribology.

1. Introduction

Materials having high specific modulus, specific stiffness and good tribological properties have the significant implications in the automotive, aerospace, hand hold tools, defense and sports sectors. Aluminium matrix composites (AMCs) are the aforementioned segment of materials, have(P/M) been encroached the manufacturing industry to fulfill the current demands [1]. Powder metallurgy route of composite manufacturing is one of the suitable techniques adopted. This route of manufacturing provides the facility of all proportions of the compositions and unlimited combination of the co8nstituents [2, 3].

Aluminium base products are utilized in different thrust areas due to its enhanced and tailorable properties which could be achieved by composing it with different ceramic reinforcements [4, 5]. For improving the properties of the AMCs the bond formation between ceramics with matrix element along with its distribution need to be taken care of. The bond strength depends on the wettability of the matrix materials which is improved by the implication of pressure at different temperature environments as a secondary treatment like extrusion[6], forging, rolling etc.. Secondary thermo-mechanical treatments helps to enhance mechanical and tribological characteristics of the product [5, 7]. Homogeneously distributed solid lubricant like graphite helps to improve tribological property with conciliating hardness and flexural strength [8-10]. The proportions of the reinforcements always carry the major deciding factor for achieving a specific property.

IOP Conf. Series: Materials Science and Engineering **390** (2018) 012008 doi:10.1088/1757-899X/390/1/012008

The P/M process of manufacturing the components instigates the investigators to tailor the proportions of the compositions, different variables of process parameters, implications of different secondary treatments for analyzing the correspondent change in the product behaviour [11]. The main concern in AMC manufacturing process through P/M route is formation of oxide layer around the particles. Removal of the layer for a better bond formation can be achieved by inducing shearing acionby the implementation of a secondary thermo-mechanical treatment [12]. For extrusion of P/M specimen, shear faced dies does not perform well in concern to flow characteristics and surface properties. Hence a mathematical contoured die is used for the extrusion operation.

Here in this research an AMC composition has been selected for preparing the cylindrical specimen by P/M route along with subsequent thermo-mechanical treatment: hot extrusion. To ameliorate the process, the specimens have been extruded through a mathematical contoured cosine die. The melioration of different mechanical and tribological characteristics due to secondary hot extrusion operation has been investigated by comparative analysis.

2. Experimental procedure

2. 1. Specimen manufacturing

The composition with weight proportions of Al (92) + Mg (5) + Gr (1) + Zn (2) was selected for the specimen preparation. A centrifugal blender at 200 RPM was utilized for preparing homogeneous blend of the composition. Different densities of the blend were checked prior to cold compaction. The double action compaction process was followed for preparing the green sample. For achieving a good flow of the powder, zinc stearate was used as lubricant. With slow rate of rise the blend was subjected to a maximum pressure limit of 275 MPa with ten minutes of dwell period. Green specimens were sintered in an argon atmospheric tubular furnace. Dwell periods of 20, 30, 90 minutes at 110°C to remove water vapour, 450°C to remove zinc stearate lubricant, 590°C to form the metallic bond were maintained respectively.

2. 2. Thermo-Mechanical treatment

The sintered specimens were subjected to a thermo-mechanical treatment: extrusion. The cylindrical specimens were extruded to a square shape through designed contoured cosine die. The cosine contour for round to square die in one quadrant is shown in Figure 1. The specimens were extruded at the temperature range of 400-450°C. For metal matrix composite extrusion flow characteristics plays an important role for the properties hence shear faced dies are not suitable [13]. To provide homogeneous flow characteristics, a specially designed cosine contoured die was utilised for minimising the surface defects. Figure 1 depicts the cross sectional variation of the profile generated by MATLAB. The profile is engraved indigenously on a tool material (H-13) in three dimensions.

2. 3. Physical, mechanical as well as tribological depiction

Apparent, tap, green, sintered, theoretical and extruded densities were checked at different stages during the process. Both sintered and extruded specimen densities were assessed by adopting Archimedes' principle.

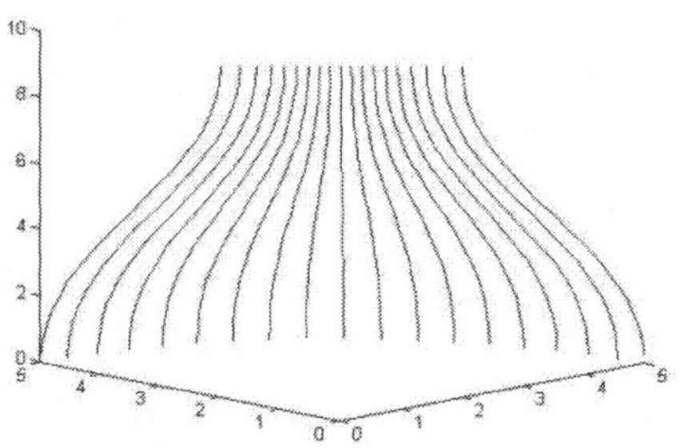

Figure 1. Cosine profile for round to square die in one quadrant

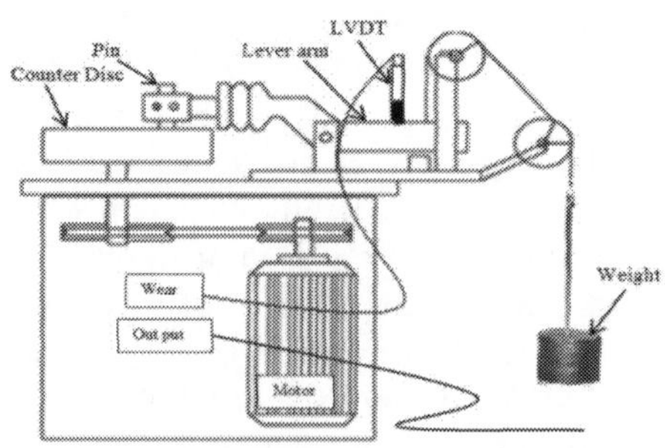

Figure 2. Schematic of two-body wear testing apparatus

Transverse rupture strength (TRS) of both types of specimens like sintered as well as extruded was found by 3-point bend test. The test was conducted with the ram travel rate of 2 mm/min by considering 30 mm span specimen in the universal testing machine (Instron -5979)

Pin-on-disc type two-body wear testing apparatus was employed to assess the wear characteristics of both the P/M as well as extruded matrix alloy. The cylindrical pins of dimension Ø 10*25 mm were prepared for the sintered specimens and (6.3 * 6.3 *25) square bars with flat surface contact were considered for the extruded type. The EN-31 steel of hardness Rc 60 and the average surface roughness of 02 μm (Ra) was employed as the counter object. The schematic of the wear testing apparatus is shown in Figure 2. Three variable parameters were considered for the analysis and the values chosen for the study are tabulated in Table 1. Each experiment is continued for ten minutes duration. The wear measurement in terms of weight loss was performed precisely with an average accuracy of 0.1 mg.

Table 1 Variable parameters opted for test

Variable parameters	Level-1	Level-2	Level-3
Normal load (L), (in N)	40	60	80
RPM of the counter body (N),	200	400	600
Track diameter (D), (in mm)	50	70	90

3. Results and discussion

3. 1. *Physical depiction*

Physical properties like shape and size of the powder have a great influence on flow characteristics during cold compaction, hence density of the specimen get affected and consequent effects diminish mechanical and tribological properties. Among different suitable techniques like LASER diffraction, sieve analysis and microscopy along with image analysis, to measure the shape and size of the particles, latter one is used in this work. The shape of the powder is clear by the scanning electron microscopy (SEM) images depicted in Figure 3. The size and shape of the composition are listed in Table 2.

Table 2 Physical characteristics of the selected powders

Selected Powder	Supplier of the powder	Average particle size	Overall Particle shape	Purity / Assay of the powder
Magnesium	Loba Chemie	140 μm	Flakey type	99.0%
Aluminium		45 μm	Spherical, elliptical and sub-rounded	98.0%
Graphite		20 μm	Rounded and Flakey type	98.0%
Zinc		22 μm	Spherical as well as sub-rounded	98.0%

Apparent or bulk and tap densities were computed. To improve the compcatability, negligible weight percentage of zinc stearate was added with the blend. 30-35% enhancement of density occurs due to tapping, depicted in the Figure 4 which indicates a good flow characteristics of the blend. A proper consolidation of particles in green compact was observed at the compaction pressure of 275 MPa. A non-dimensional densification parameter ascertains the effect of sintering in terms of shrinkage or growth mentioned in Equation 1 [14].

$$\text{Densification parameter} = \frac{(\text{sintered density} - \text{green density})}{(\text{theoretical density} - \text{green density})} \quad (1)$$

Shrinkage and swelling or growth were affirmed by Positive and negative densification parameter respectively. The green specimens possess swelling behavior in this case, which has an adversarial impact on the properties. To overcome the effect as well as improve the properties a thermo-mechanical intervention i.e (extrusion) was considered as a secondary operation. The improvement of density due to the secondary treatment is calculated. 50% reduction of cross sectional area by extrusion causes 15-20% improvement in density. All densities are presented in

Figure 4 where theoretical density occupied the highest among all. The calculated data of densification factor, different densities, porosities are listed in Table 3.

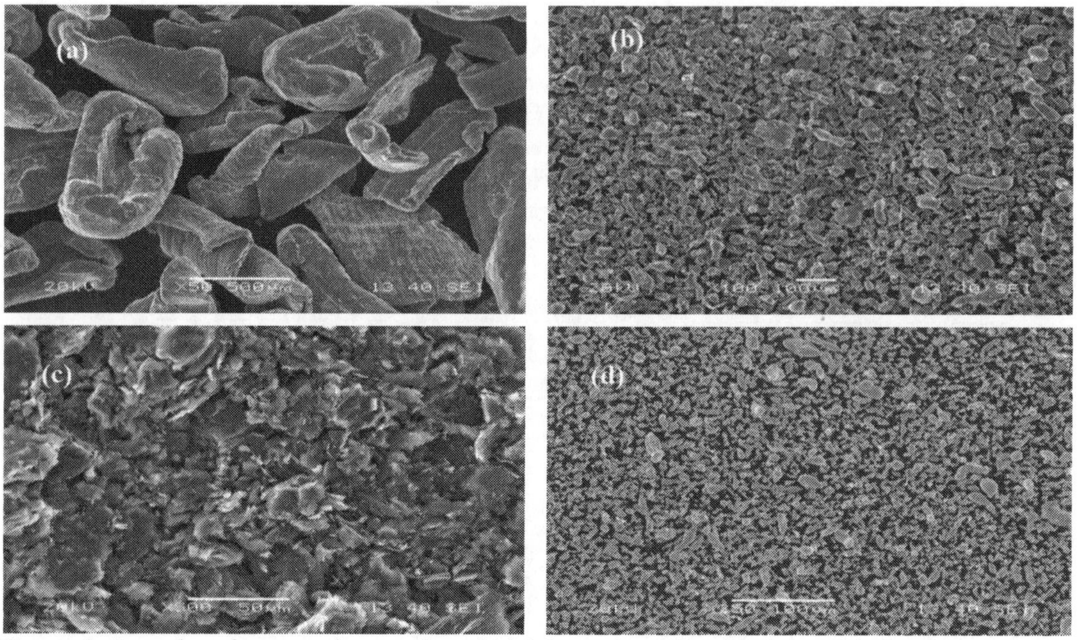

Figure 3. Scanning electron micrography of (a) Magnesium (b) Aluminium (c) Graphite and (d) Zinc powder

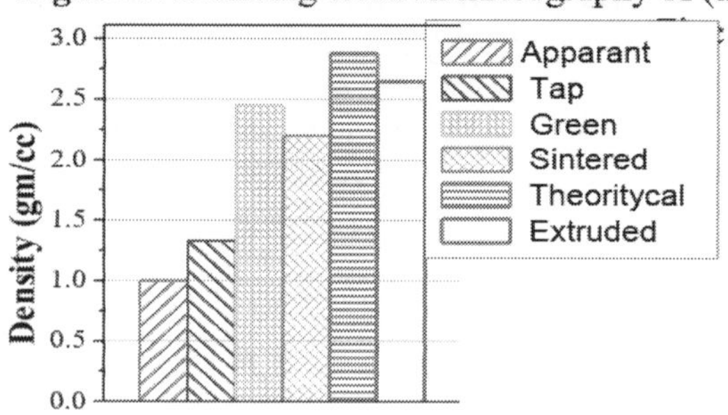

Figure 4. Density plot

Table 3 Porosity analysis

Parameters	Data
Densification factor	-0.742
Improvement in densification due to extrusion	17.35 (%)
Porosity of the AMC before extrusion	22 (%)
Porosity of the AMCafter extrusion	8.53 (%)

3. 2. Mechanical depiction

Ram travel rate of 3 mm/min was considered for compression test of sintered specimens. The compression test was performed in the Instron-setec series UTM at room temperature. The ultimate stress in average found is 409MPa, depicted from the stress versus strain plot in Figure 5. For determining hardness, ten readings were taken at different locations and the average was calculated. The same procedure was followed for both sintered as well as thermo-mechanically treated specimen. The hardness in average found was 82.3 HRV and 115.7 HRV for sintered and extruded specimen respectively. The thermo-mechanical treatment induces large thermal and internal stresses, which establishes improved properties.

The Transverse rupture strength (TRS) in MPa was estimated for both sintered circular section and square extruded section. TRS of the sintered P/M specimen is comparatively very low as shown in Figure 6, which could be attributed to the less bond strength as well as presence of porosity. Bond strength improves due to the uniform distribution of liquid Zinc by the application of high pressure on the matrix at high temperature condition during secondary treatments.

3. 3. Wear morphology

Two-body dry sliding wear test was conducted for both types of specimens. By considering three variables in three levels as in Table 4, an L_9 orthogonal array was formed to analyse the wear rate in terms of mass loss. The mass loss is converted to volume loss per unit sliding distance and the same is plotted in Figure 7 for two types of specimen comparatively for all the nine runs. It was observed wear rate is reduced by the implication of secondary treatments to the P/M specimens. The improved bond strength avoids erosion of the pin material and consequent three body wear [10].

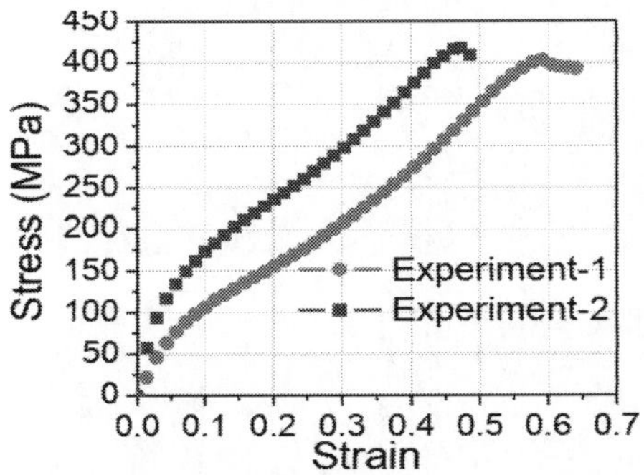

Figure 5. Stress versus strain

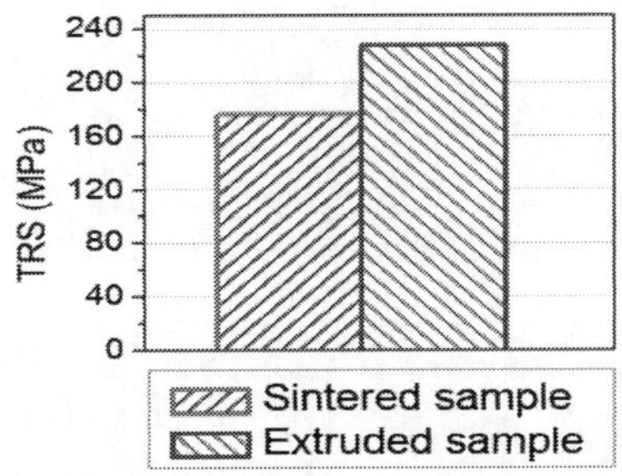

Figure 6. Transverse rupture strength

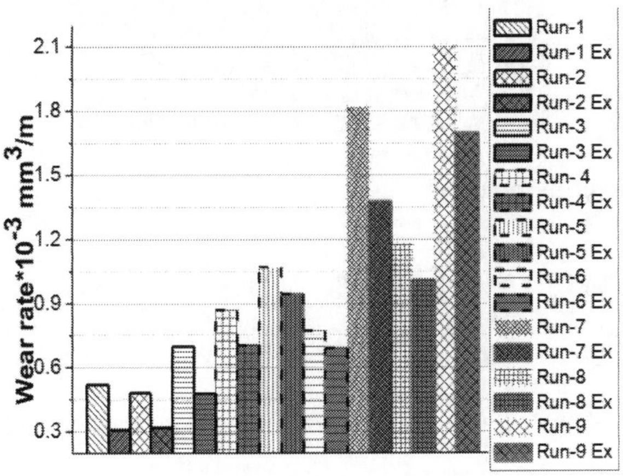

Figure 7. Wear rate for L_9 array

Table 4 L_9 array adopted for wear characterisation

Run	Normal Load (N)	Track Diameter (mm)	RPM of the counter body
1	20	50	200
2	20	70	400
3	20	90	600
4	40	50	400
5	40	70	600
6	40	90	200
7	60	50	600
8	60	70	200
9	60	90	400

Randomly selected weared surfaces were captured for FESEM images. Specimen surfaces for run-2 and 7 are shown in Figure 8. It was observed delamination as well as combination of three i.e adhesion, delamination and abrasion is the predominating mechanism of wear for high sliding velocity with heavy loading conditions. Presence of porosity before the accomplishment of secondary treatments causes severe erosion, crack formation and fracture of the specimen. Shear deformation of the surface occurs by the continuous sliding over the crack. Presence of graphite at the tribosurface works as a solid lubricant, which minimizes friction and wear. Due to low melting point of zinc at adversarial conditions thermal softening and adhesion also come into picture.

4. Conclusion

Effect of the secondary thermo-mechanical intervention i.e extrusion, considering a three dimensionally designed mathematically contoured die on the P/M specimen was investigated successfully. It was observed, the liquid phase sintering and its distribution due to hot extrusion causes a significant improvement of mechanical as well as tribological properties. Minimal amount of surface defects and smooth flow characteristics obtained by zero entry and exit angled die extrusion made the extrudate stronger. Wear rate of extrudade composites for every test conditions were lower in comparision with the P/M specimen without thermo-mechanical treatment. Presence of graphite ameliorates the wear resistance property by compromising with a little strength. A mixed type of wear (oxidative, adhesive, abrasion and delamination) pattern is observed in AMC surfaces.

IOP Conf. Series: Materials Science and Engineering **390** (2018) 012008 doi:10.1088/1757-899X/390/1/012008

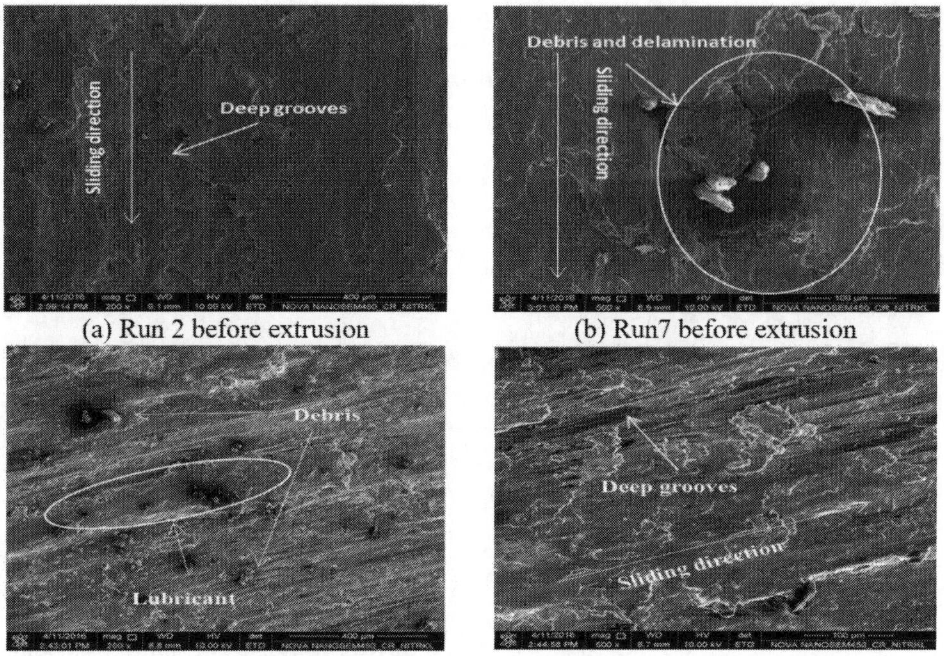

(a) Run 2 before extrusion (b) Run7 before extrusion

(c) Run-2 after extrusion (d) Run-7 after extrusion

Figure 8. Wear morphology

5. References

[1] Fogagnolo JB, Velasco F, Robert MH and Torralba JM 2003 Materials Science and Engineering: A 342(1):131-43.

[2] Torralba J, Da Costa C and Velasco F 2003 Journal of Materials Processing Technology 133(1):203-06.

[3] Min KH, Kang SP, Kim D-G and Do Kim Y 2005 Journal of alloys and compounds 400(1):150-53.

[4] El-Kady O and Fathy A 2014 Materials & Design (1980-2015) 54:348-53.

[5] Taleghani MJ, Navas ER and Torralba JM 2014 Materials & Design 55:674-82.

[6] Adamiak M, Fogagnolo JB, Ruiz-Navas EM, Dobrzański LA and Torralba JM 2004 Journal of Materials Processing Technology 155-156(Supplement C):2002-06.

[7] Abdollahi A, Alizadeh A and Baharvandi HR 2014 Materials & Design 55:471-81.

[8] Ravindran P, Manisekar K, Narayanasamy P, Selvakumar N and Narayanasamy R 2012 Materials & Design 39:42-54.

[9] Ravindran P, Manisekar K, Narayanasamy R and Narayanasamy P 2013 Ceramics International 39(2):1169-82.

[10] Baradeswaran A and Perumal AE 2014 Composites Part B: Engineering 56:472-76.

[11] Yigezu BS, Jha PK and Mahapatra MM 2013 Materials and Manufacturing Processes 28(9):969-79.

[12] Bykov YV, Rybakov K and Semenov V 2001 Journal of Physics D: Applied Physics 34(13):R55.

[13] Prasad VB, Bhat B, Mahajan Y and Ramakrishnan P 2001

[14] Padmavathi C, Upadhyaya A and Agrawal D 2011 Materials chemistry and physics 130(1):449-57.

IOP Conf. Series: Materials Science and Engineering **390** (2018) 012009 doi:10.1088/1757-899X/390/1/012009

A Comprehensive Review on Machinability Aspects in Hard Turning of AISI 4340 Steel

S Roy, R Kumar*, R K Das and A K Sahoo

School of Mechanical Engineering, Kalinga Institute of Industrial Technology Deemed to be University, Bhubaneswar, Odisha, India

E-mail: ramanujkumar22@gmail.com

Abstract. AISI 4340 steel is an alloy steel of nickel, molybdenum, and chromium with medium carbon content. It is known for its high toughness and the ability of further enhancement in strength by applying heat treatment condition. It is used in aircraft landing gear, oil and gas drilling, automobile industry and general engineering industries to manufacture heavy duty gears, shafts, pins, spindles, couplings, chucks etc. As it is a hard to machine material so many researchers are taking keen interest to study the effects of different machining environments and machining aspects on the machining performance of 4340-grade steel. The present paper presents a comprehensive review on important aspects associated to machining of 4340 grade steel using different cooling environments and cutting tools. Experimental and theoretical observations for tool wear, surface roughness, tool life, residual stress, machining force and cutting temperature using different types of cutting inserts under distinguish cooling environments are also been discussed.

1. Introduction

The AISI 4340 is a highly commercially utilized medium carbon low alloy steel which falls into the category of hard to cut material. In recent decades hard turning operation is mostly preferred for machining of AISI 4340 steel due to its high hardness value (45-60 HRC). It projects good balance of strength, wear resistance and toughness properties due to its alloying element for which it outscores the normal steel. [1]. It primarily treasures its application in manufacturing of aircraft landing gear, power transmission gears, shafts along with automobile industries to manufacture shafts, gears, bearings and cams which demands tight geometric tolerance, advanced surface quality and longer service life [2]. To implement AISI 4340 steel in above fields dry turning operation has been introduced which provides improved results regarding surface quality and maximum tool life [3]. The most commonly used coated tools are CVD coated carbide, PVD coated carbide, ceramic, CBN and PCBN tools. Using coated tool provides a longer life of the tool, better surface finish and lower machining temperature as compared to uncoated tool [4]. Machining under dry condition generates huge amount of heat at the tool-work specimen interface due to friction between tool and work material surface which highly affects the surface quality of work material and tool life [5]. So for the reduction of frictional force and competent removal of heat from the machining zone the use of coolant plays a vital role. During MQL technique the coolant is directly applied over the machining zone which provides easy removal of heat from the tool-workpiece interface. MQL technique not only removes the heat generated during machining but it also provides better machining performances than dry machining condition [6]. Due to heat generation at the machining zone, the surface topography of the workpiece is altered. To prevent these phenomena cryogenic cooling is being introduced. As using cryogenic cooling the heat generation at the machining zone is very less, the change in microstructure is negligible as compared to other cooling environments. Using cryogenic cooling technique with MQL environment provided the best surface finish and longer tool life among other cooling environments [7].

In the present paper, efforts have been made to present a brief literature review on turning operation of 4340-grade steel using different coated insert under different cooling environments (dry, cryogenic and MQL). Figure 1 presents the aspects of study covered in the present review work and Table 1 describes the chemical constituent of grade 4340 steel. This literature review also describes the relationship between different machining parameters with different machining performances as surface finish, wear of the tool and tool life.

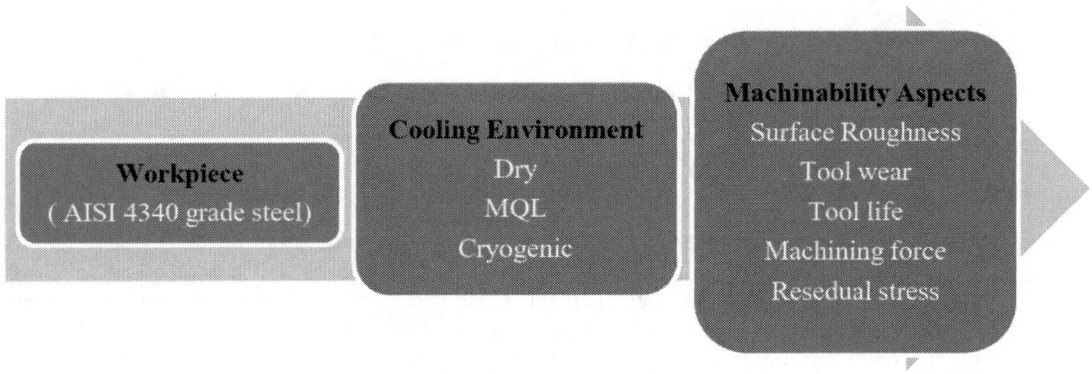

Figure 1. Schematic view of machining aspects

Table 1. Chemical constitution of 4340 grade steel [7]

Ni	Mn	S	Si	Cr	P	C	Fe
1.8	0.70	0.04	0.25	0.8	0.035	0.41	Balance

2. Cutting performance analysis in various cooling environments

In present days, due to the growing demand for an eco-friendly and cost effective machining process, green machining technique came into huge existence. Green machining technique mostly involves dry machining, MQL, spray cooling, compressed air cooling and cryogenic cooling machining environments. Many researchers investigated the role of dry, MQL and cryogenic cooling environment during turning operation of grade 4340 steel.

2.1. Dry machining environment

Due to the growing demand for machining of hard and tough steels, dry turning operation came into existence because of its better machining performances while turning difficult to machine materials. Dry turning operation is used to machine harder materials using mostly coated insert without application of lubricants or coolants [3]. Satish Chinchanikar and S.K. Choudhury observed that life of the tool was mostly affected by the hardness of the work specimen. CVD coated tool displayed longer tool life as compared to PVD coated tool. Adhesion and abrasion at the tool work specimen interface were found to be most predominant wear mechanism factor followed by notch wear and chipping of the coated layer from the tool surface [4]. Varun Sharma and P.M. Pandey studied the machining parameters which affect the residual stress generated during turning operation of grade 4340 steel. It was found that residual stress generation was only dependent on machining speed and feed rate whereas the depth of cut had the negligible effect on hoop residual stress [8]. R. Suresh et al. found that the best possible combination of low depth of cut, low feed rate and high machining speed resulted in lesser machining force and higher feed rate minimized the specific machining force. The increment in machining tool wear and machining force were found linear with increase in feed rate and machining speed. During every machining condition the abrasion was the prime wear mechanism on the tool surface [9]. Ashok Kumar Sahoo and Bidyadhar Sahoo experimentally investigated machining under dry condition and concluded that machining speed directly controls the wear and life of the tool. Longer tool life and lower wear of the tool was achieved at lower machining speed. Best parametric combination for machining was obtained at machining speed (150m/minute), depth of cut (0.4 millimeter) and feed (0.15m/minute) [10]. R. Suresh et al. ascertain that increase in machining force was directly proportional with increase in depth of cut and feed rate during turning operation. Best machining performance was achieved at a combination of low depth of cut, low feed rate and short machining time along with high machining speed [11]. Reginaldo T. Coelho et al. observed that by using TiAlN nano coating over PCBN tool can extend the life of the tool by 38%. Machining force and surface roughness were found maximum using the uncoated tool during turning operation of grade steel [12]. Virginia Garcı´a Navas studied that as the feed increases the residual stress lean to become more tensile and the surface quality deteriorates with increase in machining feed. The residual stress was found to be less tensile at a medium machining speed range of 200 to 300 m/minutes [13].

2.2. Minimum quantity lubrication

A huge amount of heat is being generated during dry turning operation at the machining zone due to the friction between tool and work specimen surface. So to achieve improved surface quality and minimize the heat generation, the use of coolant becomes very necessary. Thus MQL technique is used to fulfill all the demands which were not achieved during dry turning operation [5,6]. N.R. Dhar et al. experimentally found that the machining performance using MQL was best as compared to dry and flooded condition during turning operation of 4340-grade steel. Using MQL environment also resulted in better surface finish, longer life of the tool and lower wear of the tool surface [14]. According to Kedare et al. MQL technique can become the best alternative for traditional flooded cooling method during milling operation of grade steel because of it consumes lesser coolant. MQL technique provided both economic and ecological benefits as the coolants are eco friendly and its consumption during machining was found only 900 ml/hour which is very minimal when compared with conventional cooling methods [15]. Sanjeev Kumar et al. compared the machining performance during turning operation of AISI 4340 grade steel using different machining environments (dry, wet and MQL). Using MQL environment resulted in around 7%-10% increase in surface finish and greatly reduced the wear of the tool tip [16]. D.V. Lohar and C.R. Nanavaty tried to evaluate the machining performance of MQL condition during turning operation of AISI 4340 grade steel. Experimental results showed that while using MQL environment, the machining force and machining temperature got reduced by around 40% and 36% respectively as compared to dry machining condition. Surface finish was also found to be better as compared to other machining conditions [17]. Nikhil Ranjan Dhar et al. found that using MQL environment during turning operation reduces the wear of the tool, enhances the surface quality of the workpiece and increases the life of the tool. MQL also provided benefits like reduction of machining temperature and maintaining the sharpness of machining edges [18]. L. R. Silva et al. found that using MQL technique significantly reduced the diametric wear and surface roughness value. As compared to other traditional cooling methods, reduction in tangential machining force during machining of grade steel was maximum using MQL environment [19].

2.3. Cryogenic environment

Cryogenic environment further decreases the heat generated at the machining zone and the work specimen microstructure is hardly affected during the machining operation. So using cryogenic environment allows high speed machining even using higher machining parameters [7]. Mirghani I. Ahmed researched to increase the effectiveness of cryogenic cooling environment by modifying the tool holder. It was observed that using customized tool holder during cryogenic environment was extra effective when the coolant out flow was directed away from the machining edge of the cutting insert.

Table 2. Summary of types of coolant and its effects on machining performances

Ref.	Year	Cooling environment	Process parameters			Response
			Cutting Speed (m/min)	Feed (mm/rpm)	Depth of cut (mm)	
4	2013	Dry	100,300	0.1,0.2,0.3	1.5	Wear mechanism, machining force, surface roughness, tool life
8	2016	Dry	10 - 90	0.1 - 0.3	0.1 - 0.5	Residual stress, hoop residual stress
9	2012	Dry	140,200,260	0.10, 0.18, 0.26	0.6, 0.8, 1	Surface roughness and Tool wear
10	2012	Dry	150	0.15	0.4	Wear mechanism, machining force
11	2012	Dry	80,140,200 260	0.10,0.18,0.26	0.8,1,1.2	Machining force, tool wear, surface roughness
12	2007	Dry	150	0.07	0.2	Machining force, surface roughness, tool life
13	2012	Dry	200,255,300	0.075,0.1,0.125	0.4	Residual stress, thermal stress, surface roughness
14	2006	MQL	110	0.16	1.5	Tool wear,surface roughness
16	2017	MQL	75,100,125,	0.1,0.125,0.15	0.2,0.4,0.8	Surface quality
17	2013	MQL	40,80,120	0.05,0.075, 0.10	0.5,1.0	Cutting force, cutting temp, surface finish
18	2007	MQL	63, 80, 95, 110,128	0.1, 0.13,0.16, 0.2	1, 1.5	Tool wear, surface roughness
19	2005	MQL	20	1	0.1	Surface roughness, tool wear
21	2017	Cryogenic	50,150,250	0.045,0.90.135	0.2,0.4,0.6	Surface roughness

The freshly modified tool holder resulted in better surface quality, reduction in coolant consumption and longer life of the tool even at higher machining speed [20]. Sanchit Kumar Khare and Sanjay Agarwal found that the depth of cut and machining speed were the most influencing factors on the surface roughness. Experimental results proved that cryogenic environment generated better surface quality than dry machining environment [21]. Khalid A Al-Ghamdi et al. studied the machinability of AISI 4340 steel under cryogenic cooling environment using CO_2 snow as coolant. It was observed that machining speed and feed rate were the prime factors affecting the machining performances and longer life of the tool was only ensured at a lower combination of machining parameters (machining speed and feed rate) [6]. The different types of the coolant used by the various researchers and its effect on the machining performance are mentioned in Table 2.

3. Performance analysis of different cutting tools

Modern machining operation demands for machining tools with enhanced properties, especially to achieve improved surface quality. The relationship between physical, chemical and mechanical properties of machining operation acts as the key factors for both uses and the machining tool manufacturers. Many researchers studied the performance of ceramic, CBN and coated carbide tools during machining of 4340-grade steel [9]. Satish Chinchanikar and S.K. Choudhury came to a conclusion that the life of the tool was mostly affected by workpiece hardness, depth of cut and machining speed. The longer service period of the tool was obtained using CVD coated tools nose wear and flank wear was found to be the prime wear mechanism for wear of the tool surface [4]. Ashok Kumar Sahoo and Bidyadhar Sahoo found that TiN coated tool have longer service period as compared to ZrCN and uncoated carbide tool during turning operation. From the experimental results it was proposed that hard turning of grade 4340 steel using ZrCN and TiN can be done ever at a higher machining speed of 150 m/minutes [10]. Many researchers investigated on the wear mechanism and surface roughness of the workpiece during turning operation using different types of coated tools. They came with a conclusion that machining parameters and the nature of the coated material over the tool surface mainly influences the surface quality of the workpiece [21,22,23,24]. Anshuman Das et al. experimentally found that using coated cermet tool during turning operation helped in reducing wear of the tool, machining temperature and machining force as compared to uncoated carbide tool [25].
Summary of investigation on machining tools during turning operation is presented in Table 3.

Table 3. Summary of various machining tools during turning operation

Ref	Year	Tool	Workpiece hardness (HRC)	Response	Findings
4	2013	CVD coated carbide, PVD coated carbide	35 and 45	Wear mechanism, machining force, surface roughness, tool life	CVD coated tool performed better at lower speed, PVD coated tool performed better at higher speed
9	2012	(TiN/TiCN/Al2O3) CVD coated cemented carbide	48	Surface roughness and Tool wear	Machining speed and feed rate highly influence machining force
10	2012	Uncoated carbide, CVD coated carbide	47±1	Wear mechanism, machining force	Longer tool life and better surface quality using coated tool,
14	2006	SNMM 120408 carbide	_	Tool wear, surface roughness	Reduction in tool wear, longer tool life, better surface finish using MQL

16	2017	K5625 CBN tool	45	Surface quality	Surface finish enhanced by 7-10%
17	2013	TNMA160404 Grade PB250 CBN	45 ~ 55	Cutting force, cutting temp, surface finish	Improvement in Ra, Decrease in machining force and temperature
18	2007	SNMM 120408 carbide	_	Tool wear, surface roughness	Improvement in tool life, tool wear and surface finish
21	2017	HSS T-42, S-400	_	Surface roughness	Surface roughness was better in cryogenic
23	2013	MT KB 5625 CBN	45	Machining force, Surface finish	Machining environment and depth of cut highly effect the machining force
31	2006	TiN (CVD) coated, PCBN compact	53	Surface roughness, tool wear, life of the tool	Machining performance of PCBN was found better than TiN (CVD) coated tool
32	2014	TiC mixed alumina ceramic	35, 45, 55	Surface roughness, machining temperature, machining force	Machining force increases with increase in depth of cut, feed and specimen hardness
33	2008	CBN grades 7020 and 7050	56	Wear of the tool, life of the tool	Tool life was longer in interrupted machining as compared to continuous machining
34	2009	PCBN, silicon carbide reinforced ceramic	56	Life of the tool, wear of the tool	Surface finish was better using PCBN tool

4. Analysis of Surface roughness

Surface roughness is an important index for machinability of any cutting operation. As in every manufacturing industry surface finish is given maximum priory, many researchers investigated on machining parameters which affect the surface roughness of the work specimen [14]. Sanchit Kumar Khare and Sanjay Agarwal investigated on machining parameters which affect the surface roughness during turning operation of 4340 grade steel using cryogenic cooling environment. Experimental results suggested that surface roughness was mostly dependent on the machining speed and depth of cut [21]. Sanjeev Kumar et al. experimentally found that using MQL environment during turning operation resulted in 7-10% enhancement in surface finish as compared to other cooling environments. The reduction in machining temperature and lower wear of the tool were the prime factors for improvement in the surface quality of the work specimen [16]. Using MQL environment resulted in thinner chips formation during machining operation which reduces the frictional force produced between the tool-chip interface, which added further enhancement in surface quality of the workpiece [17, 23]. Bailey et al. observed that initially the surface roughness increases continuously with increase in machining speed for some extent and then it became independent of machining speed with further increment in machining speed. Sudden damage on the machined surface arises due to discontinuous chip formation during machining operation [24]. Nilrudra Mandal et al. studied the surface roughness produced during turning operation of 4340-grade steel using coated alumina tool. During experiment best surface finish was obtained at a machining parameter combination of machining speed 420 m/minute, depth of cut 0.5 millimetre and feed rate 0.24 mm/revolution [26].

Machining insert nose geometry along with feed rate was found to strongly affect the surface roughness values in the turning operation [27]. Most of the researchers found that during turning operation of 4340 grade, the main machining parameters which alter the surface roughness were machining speed, depth of cut and feed rate [28,29,30].

5. Conclusion

In this review paper efforts are being made to present a brief description on turning operation of AISI 4340-grade steel using different machining tools in different cooling environment (Dry, MQL and Cryogenic). This paper also depicts the affect of different machining parameters and cooling environments on surface roughness of the work specimen. Due to the growing demand for eco-friendly and cost efficient machining process, many researchers are taking a keen interest in new cooling environments so the comparison between different cooling conditions are presented in this review paper.

- Best possible combination of low depth of cut, low feed rate and high machining speed resulted in smaller machining force and higher feed rate minimized the specific machining force. The increment in machining tool wear and machining force were found linear with increase in feed rate and machining speed. For every machining condition the abrasion was the major wear mechanism on the tool surface.

- During turning operation with TiN and ZrCN coated carbide tool, longer tool life and lower wear of the tool was achieved at lower machining speed. Best parametric combination for machining was obtained at machining speed (150 m/min), depth of cut (0.4 mm) and feed (0.15 m/min).
- CVD coated insert displayed longer tool life as compared to PVD coated insert. Adhesion and abrasion on the tool work specimen interface were found to be most predominant wear mechanism factor followed by notch wear and chipping of coated layer from the tool surface.
- Using MQL cooling condition resulted in the better surface finish, lower wear of the tool and longer tool service life as compared to dry machining condition. Cryogenic cooling environment displayed maximum heat removal rate from the machining zone during turning operation and best cooling condition was achieved under hybrid (Cryo + MQL) environment for short machining time.
- Machining environment highly controls the quality of the machined surface and it was found to be mostly dependent on machining speed followed by feed and depth of cut.

6. References

[1] Rashid W B, Goel S, Davim J P and Joshi S N 2016 Parametric design optimization of hard turning of AISI 4340 steel (69 HRC) *International Journal of Advanced Manufacturing Technology* **82** pp 451-462

[2] Rashid W B, Goel S, Luo X and Ritchie J M 2012 An experimental investigation for the improvement of attainable surface roughness during hard turning process *Proc IMechE Part B: Journal of Engineering Manufacture* pp 1-5

[3] Liew P J, Shaaroni A, Sidik N A C and Yan J 2017 An overview of current status of cutting fluids and cooling techniques of turning hard steel *International Journal of Heat and Mass Transfer* **114** pp 380-394

[4] Chinchanikar S and Choudhury S K 2013 Investigations on machinability aspects of hardened AISI 4340 steel at different levels of hardness using coated carbide tools *International Journal of Refractory Metals and Hard Materials* **38** pp 124-133

[5] Gaur APS and Agarwal S 2010 Improvement in surface quality with molybdenum disulphide as solid lubricant in turning AISI 4340 steel *Proceedings of the ASME 2010 International Mechanical Engineering Congress & Exposition IMECE2010* pp 1-6

[6] K.A.A Ghamdi, A Iqbal and G Hussain 2014 Machinability comparison of AISI 4340 and Ti-6Al-4V under cryogenic and hybrid cooling environments: A knowledge engineering approach *Proceedings of the Institution of Mechanical Engineers, Part B: Journal of Engineering Manufacture* pp 1-21

[7] Zhirafar S, Rezaeian A and M. Pugh 2007 Effect of cryogenic treatment on the mechanical properties of 4340 steel *Journal of Materials Processing Technology* **186** pp 298- 303

[8] Sharma V and Pandey P M 2016 Optimization of machining and vibration parameters for residual stresses minimization in ultrasonic assisted turning of 4340 hardened steel *Ultrasonics* **70** pp 172-182

[9] Suresh R, Basavarajappa S and Samuel G L 2012 Some studies on hard turning of AISI 4340 steel using multilayer coated carbide tool *Measurement* **45** pp 1872-1884

[10] Sahoo A K and Sahoo B 2012 Experimental investigations on machinability aspects in finish hard turning of AISI 4340 steel using uncoated and multilayer coated carbide inserts *Measurement* **45** pp 2153-2165

[11] Suresh R, Basavarajappa S, Gaitonde V N and Samuel G L 2012 Machinability investigations on hardened AISI 4340 steel using coated carbide insert *International Journal of Refractory Metals and Hard Materials* **33** pp 75-86

[12] Coelho R T, Ng E G, Elbestawi M A 2007 Tool wear when turning hardened AISI 4340 with coated PCBN tools using finishing cutting conditions *International Journal of Machine Tools & Manufacture* **47** pp 263-272

[13] Navas V G, Gonzalo O, Bengoetxea I 2012 Effect of cutting parameters in the surface residual stresses generated by turning in AISI 4340 steel *International Journal of Machine Tools & Manufacture* **61** pp 48-57

[14] Dhar N R, Kamruzzaman M and Ahmed M 2006 Effect of minimum quantity lubrication (MQL) on tool wear and surface roughness in turning AISI-4340 steel *Journal of Materials Processing Technology* **172** pp 299-304

[15] Kedare S B, Borse D R and Shahane P T 2014 Effect of Minimum Quantity Lubrication (MQL) on Surface Roughness of Mild Steel of 15HRC on Universal Milling Machine *Procedia Materials Science* **6** pp 150-153

[16] Kumar S, Singh D and Kalsi N S 2017 Analysis of Surface Roughness during Machining of Hardened AISI 4340 Steel using Minimum Quantity lubrication *Materials Today: Proceedings* **4** pp 3627-3635

[17] Lohar D V and Nanavaty C R 2013 Performance Evaluation of Minimum Quantity Lubrication (MQL) using CBN Tool during Hard Turning of AISI 4340 and its Comparison with Dry and Wet Turning *Bonfring International Journal of Industrial Engineering and Management Science* **3** pp 102-106

[18] Dhara N R, Islama S and Kamruzzaman M 2007 Effect of Minimum Quantity Lubrication (MQL) on Tool Wear, Surface Roughness and Dimensional Deviation in Turning AISI-4340 Steel *G.U. Journal of Science* **20(2)** pp 23-32

[19] Silva L R, Bianchi E C, Catai R E, Fusse R Y, França T V and Aguiar P R 2005 Study on the behavior of the minimum quantity lubricant - MQL technique under different lubricating and cooling conditions when grinding ABNT 4340 steel *Journal of the brazilian society of mechanical sciences and engineering* **27** pp 1-15

[20] Ahmed M I, Ismail A F, Abakr Y A and Nurul A A K M Effectiveness of cryogenic machining with modified tool holder *Journal of Materials Processing Technology* **185** pp 91-96

[21] Khare S K and Agarwal S 2017 Optimization of Machining Parameters in Turning of AISI 4340 Steel under Cryogenic Condition Using Taguchi Technique *Procedia CIRP* **63** pp 610-614

[22] Kamble P D, Waghmare A C, Askhedkar R D and Sahare S B 2015 Multi objective optimization of turning AISI 4340 steel considering spindle vibration using Taguchi-

Fuzzy Inference system *Materials Today: Proceedings* **2** pp 3318-3326

[23] Naigade D M, Patil D H and Sadaiah M 2013 Some investigations in hard turning of AISI 4340 alloy steel in different cu tting environments by CBN insert *International Journal of Machining and Machinability of Materials* **14 2** pp 165-192

[24] Bailey J A, Jeelani S and Becker S E 1976 Surface integrity in machining AISI 4340 steel *Journal of Engineering for industry* pp 999-1006

[25] Das A, Mukhopadhyay A, Patel S K and Biswal B B 2016 Comparative Assessment on Machinability Aspects of AISI 4340 Alloy Steel Using Uncoated Carbide and Coated Cermet Inserts During Hard Turning *Arab Journal Science Eng* **41** pp 4531-4552

[26] Mandal N, Doloi B and Mondal B *2013* Predictive modeling of surface roughness in high speed machining of AISI 4340 steel using yttria stabilized zirconia toughened alumina turning insert, Interantional Journal of Refractory *Metals and Hard Materials* **38** pp 40-46

[27] Godoy V AA d and Diniz A E 2011 Turning of interrupted and continuous hardened steel surfaces using ceramic and CBN cutting tools *Journal of Materials Processing Technology* **211** pp 1014-1025

[28] Chakraborty P, Asfour S, Cho S, Onar A and Lynn M 2008 Modeling tool wear progression by using mixed effects modeling technique when end-milling AISI 4340 steel *Journal of materials processing technology* **205** pp 190-202

[29] Agrawal A, Goel S, Rashid W B and Price M 2015 Prediction of surface roughness during hard turning of AISI 4340 steel (69 HRC) *Applied Soft Computing* **30** pp 279-286

[30] Kountanya R K 2008 Optimizing PCBN cutting tool performance in hard turning *Processing IMechE Vol. 222 Part B: J. Engineering Manufacture* pp 969-979

[31] More A S, Jiang W, Brown W D and Malshe A P 2006 Tool wear and machining performance of cBN–TiN coated carbide inserts and PCBN compact inserts in turning AISI 4340 hardened steel *Journal of Materials Processing Technology* **180** pp 253-262

[32] Pal A, Choudhury S K and Chinchanikar S 2014 Machinability Assessment through Experimental Investigation during Hard and Soft Turning of Hardened Steel *Procedia Materials Science* **6** pp 80-91

[33] Diniz A E and Oliveira A.J.d 2008 Hard turning of interrupted surfaces using CBN tools *Journal of materials processing technology* **195** 275-281

[34] Oliveira A.J.d, Diniz A.E and Ursolino D J 2009 Hard turning in continuous and interrupted cut with PCBN and whisker-reinforced cutting tools *Journal of Materials Processing Technology* **209** pp 5262-5270

IOP Conf. Series: Materials Science and Engineering **390** (2018) 012010 doi:10.1088/1757-899X/390/1/012010

Modeling the Performance of Vortex Tube using Response Surface Methodology and Artificial Neural Networks

G Suresh Kumar[1*], B Veerabhadra Reddy[1], G Sankaraiah[1], P Venkateshwar Reddy[1]

[1]Department of Mechanical Engineering, G Pulla Reddy Engineering College (Autonomous), Kurnool – 518 007, A.P., India
Abstract. Vortex tube gives hot and cold streams of gas taking pressurized gas as input. Though this device is mainly used for spot cooling purposes, hot temperature of outlet gas as a response variable, is also of concern for experimentation. The present study deals with modelling and analyzing the effect of five input controllable parameters, each considered with two levels (Max. and Min) viz., internal Diameter of the hot tube in mm, D_t, Length of the hot tube in mm, L, inlet Pressure of air, Kgf / cm^2, P, nozzle Diameter, in mm, D_n, and Diameter of the orifice, in mm, D_o, on the responses - cold and hot temperatures of outlet air streams (T_c & T_h) in °C. Box Behnken design is used for experimentation. The interrelationship between responses and input variables is modelled using Response Surface Methodology (RSM) and Artificial Neural Networks (ANN) separately. Regression equations have been developed for the responses using RSM. ANN modelling is observed giving better prediction results compared to RSM modelling as is evidence that through R-square and Average Absolute Deviation (AAD) for both the models.

E-mail: gudimettask@yahoo.com

1. Introduction

Compressed or pressurized gas / air are the input to the vortex tube. (Counter flow) Vortex tube separates the inlet gas / air into hot and cold streams exhausting from either ends of the vortex tube. Streams separation of hot and cold air is depicted in Fig. 1.

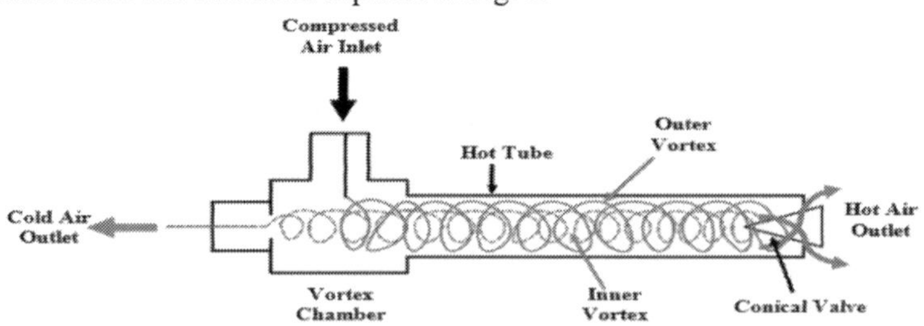

Figure 1. Working of Counter flow Vortex tube

Understanding of heat transfer phenomena, influence of various design parameters, have been the subject matters of various research studies on Vortex Tube. Employing Taguchi method, Pinar *et al.* [1], attempted to maximize the temperature gradient between the two ends of the Vortex Tubes. Statistical significance of the three factors and response has been established through this study. Prabakaran *et al.* [2] used Response Surface Methodology (RSM), and found that the selected parameters had significant effect on temperature difference. Cooling the vortex tube externally better the performance of vortex tube was the conclusion [3, 4]. Using Box- Behnken design, five controllable parameters taken at two levels (max. and min.), and experimentation is carried out and results for cold and hot outlet air temperatures (responses) for all the 46 trial conditions have been recorded. Modelling of the performance of the vortex tube is attempted using both RSM and Artificial

Neural Networks (ANN). The objective of the present study is to develop a prediction system that predicts the values of the responses at different combinations of five input parameters.

2. Methodology

2.1 Response Surface Methodology

Design layouts can be either Central Composite Designs (CCD) or Box Behnken Designs and the appropriate design can be chosen depending on the no. of variables. In the present work, Box Behnken Design is used for experimentation. For the five parameters, each at two levels, viz., internal Diameter of the hot tube (13 mm and 15 mm), D_t, Length of the hot tube (150 mm and 180 mm), L, inlet Pressure of air (1 Kgf / cm^2 , 3 Kgf / cm^2), P, nozzle Diameter, (9 mm and 12 mm), D_n, and Diameter of the orifice, (6 mm and 8 mm), D_o, 46 experimental trials / test points were required to be conducted in all. The design layout, with the uncoded level values of the parameters and the two responses values noted after experimentation (cold and hot temperatures of outlet air streams (T_c & T_h) in ° C), is presented in Table 1.

Table 1. Box Behnken Design and Experimental Results

	Factor 1	**Factor 2**	**Factor 3**	**Factor 4**	**Factor 5**	**Response 1**	**Response 2**
Run	L	Dt	Dn	Do	P	Tc	Th
1	165	15	10.5	8	2	17	38
2	180	13	10.5	7	2	14	45
3	165	14	12	8	2	19	41
4	165	14	9	8	2	17	40
5	180	14	10.5	7	3	12	47
6	165	14	10.5	7	2	18	40
7	165	14	9	7	1	19	35
8	150	14	9	7	2	15	38
9	165	15	10.5	6	2	17	39
10	165	15	10.5	7	3	13	41
11	165	14	10.5	7	2	18	40
12	165	13	9	7	2	14	40
13	165	14	10.5	7	2	18	40
14	150	14	10.5	7	1	19	43
15	180	14	10.5	7	1	20	41
16	165	13	10.5	6	2	15	44
17	165	15	12	7	2	16	37
18	165	14	10.5	7	2	17	40
19	165	14	12	7	1	20	39
20	165	14	12	6	2	16	45
21	150	15	10.5	7	2	16	44
22	180	14	12	7	2	15	41

23	150	13	10.5	7	2	15	50
24	165	14	10.5	8	3	14	42
25	165	14	10.5	7	2	17	40
26	180	14	10.5	8	2	18	42
27	150	14	10.5	7	3	13	50
28	165	15	10.5	7	1	20	34
29	165	14	10.5	6	3	13	46
30	180	14	9	7	2	16	47
31	165	15	9	7	2	16	39
32	165	14	10.5	8	1	21	35
33	180	14	10.5	6	2	16	45
34	150	14	10.5	8	2	17	44
35	180	15	10.5	7	2	17	40
36	165	14	10.5	6	1	22	38
37	165	14	9	7	3	13	42
38	165	14	10.5	7	2	17	40
39	150	14	12	7	2	16	55
40	165	13	10.5	7	1	20	39
41	165	13	10.5	8	2	18	41
42	165	13	10.5	7	3	12	47
43	150	14	10.5	6	2	15	48
44	165	14	9	6	2	16	40
45	165	13	12	7	2	16	48
46	165	14	12	7	3	13	48

2.2 Artificial Neural Networks

In a neural network, the first important stage is the training step. In the training step, an input is introduced to the network accompanied by the desired output. Initially, the weights were set randomly. Since the output may not be what is expected, the weights may need to be altered [5, 6]. During the training phase, random weights are changed by the back-propagation algorithm to produce a satisfactory level of performance. Back Propagation algorithm is a learning technique that adjusts weights in neural network by propagating weight changes backward from the output to the input neurons. After training, the weights contain meaningful information. When a satisfactory level of the performance is reached, the training will stop. Then the network uses these weights to make decisions. In this paper, to evaluate model performance, absolute fraction of variance (R-Squared (R^2)) was computed from the results produced by the ANN model. R-Squared measures the proportion of the variation around the mean. R-square is 1 if the model fits perfectly.

3. Results and Discussion

The results obtained after experimentation are modelled using RSM and ANN.

3.1 Modelling using RSM

The mathematical relationships for the correlation of the output responses like Cold Temperature (T_c) and Hot Temperature (T_h) are obtained from the MiniTab 17 software using RSM. Eqs 1 & 2 are the regression equations for Cold Temperature (T_c) and Hot Temperature (T_h) respectively.

$$T_c = -413.1 + 1.522\ L + 30.67\ Dt + 14.90\ Dn + 2.90\ Do - 2.12\ P - 0.005093\ L*L - 0.979\ Dt*Dt - 0.3981\ Dn*Dn + 0.271\ Do*Do - 0.313\ P*P + 0.0333\ L*Dt - 0.0222\ L*Dn - 0.0000\ L*Do - 0.0333\ L*P - 0.333\ Dt*Dn - 0.750\ Dt*Do + 0.250\ Dt*P + 0.333\ Dn*Do - 0.167\ Dn*P + 0.500\ Do*P \quad (1)$$

$$T_h = 84 - 4.517\ L + 5.63\ Dt + 62.71\ Dn - 7.71\ Do + 7.35\ P + 0.02056\ L*L + 0.125\ Dt*Dt + 0.389\ Dn*Dn + 0.292\ Do*Do + 0.208\ P*P + 0.0167\ L*Dt - 0.2556\ L*Dn + 0.0167\ L*Do - 0.0167\ L*P - 1.667\ Dt*Dn + 0.500\ Dt*Do - 0.250\ Dt*P - 0.667\ Dn*Do + 0.333\ Dn*P - 0.250\ Do*P \quad (2)$$

3.2 Modelling using ANN

For training the neural network model, the inputs and outputs must be between 0 and 1 so the values are normalized. The mathematical models developed by RSM and ANN are evaluated for the prediction capability of the outputs for the cold and hot temperatures in vortex tube. The coefficient of determination (R^2) and AAD (absolute average deviation) are determined by Eq. (3) and Eq. (4), respectively [7, 8]; and their values are calculated in order to compare the best model.

$$R^2 = 1 - \frac{\sum_{i=1}^{n}(Y_{i\,Cal} - Y_{i\,exp})^2}{\sum_{i=1}^{n}(Y_{avg,exp} - Y_{i\,exp})^2} \quad (3)$$

$$AAD = \{[\sum(\frac{|y_{i,exp} - y_{i,cal}|}{y_{i,exp}})]\frac{1}{n}\}x100 \quad (4)$$

Table 2. Comparison of ANN Predicted values and Experimental Values along with Errors

Test No	Experimental Results		ANN Predicted		Error	
	Cold Temperature	Hot Temperature	Cold Temperature	Hot Temperature	Cold Temperature	Hot Temperature
1	17	38	16.99744	37.99893	0.002558	1.07E-03
2	14	45	14.00052	44.99836	-0.00052	1.64E-03
3	19	41	19.00091	41.0013	-0.00091	-1.30E-03
4	17	40	16.99955	39.99845	0.000451	1.55E-03
5	12	47	11.99945	47.00023	0.000546	-2.30E-04
6	18	40	18.00027	40.00039	-0.00027	-3.85E-04
7	19	35	18.99895	34.99828	0.001046	1.72E-03
8	15	38	15.00184	38.00374	-0.00184	-3.74E-03
9	17	39	16.99849	38.99811	0.001513	1.89E-03
10	13	41	13.00183	41.00279	-0.00183	-2.79E-03
11	18	40	18.00027	40.00039	-0.00027	-3.85E-04
12	14	40	14.00025	40.0021	-0.00025	-2.10E-03
13	18	40	18.00027	40.00039	-0.00027	-3.85E-04
14	19	43	18.99885	42.99669	0.001145	3.31E-03
15	20	41	19.99855	41.00025	0.001452	-2.54E-04
16	15	44	14.99993	44.00057	6.51E-05	-5.66E-04
17	16	37	15.9998	37.00326	0.000203	-3.26E-03
18	17	40	16.99955	39.99845	0.000451	1.55E-03
19	20	39	20.00098	39.00172	-0.00098	-1.72E-03
20	16	45	16.00052	45.00225	-0.00052	-2.25E-03
21	16	44	16.00044	44.00236	-0.00044	-2.36E-03
22	15	41	14.99818	40.99844	0.001821	1.56E-03
23	15	50	14.99244	49.98928	0.007557	1.07E-02

24	14	42	13.99844	41.99819	0.001556	1.81E-03
25	17	40	16.99955	39.99845	0.000451	1.55E-03
26	18	42	18.00165	42.00081	-0.00165	-8.11E-04
27	13	50	13.00057	50.00068	-0.00057	-6.81E-04
28	20	34	20.00055	33.99794	-0.00055	2.06E-03
29	13	46	13.00013	45.99787	-0.00013	2.13E-03
30	16	47	15.9987	46.99772	0.001303	2.28E-03
31	16	39	15.99865	38.99775	0.001353	2.25E-03
32	21	35	20.99979	35.00281	0.000207	-2.81E-03
33	16	45	16.00052	45.00225	-0.00052	-2.25E-03
34	17	44	17.00058	44.00194	-0.00058	-1.94E-03
35	17	40	16.99955	39.99845	0.000451	1.55E-03
36	22	38	22.00012	37.99931	-0.00012	6.94E-04
37	13	42	13.00032	41.99896	-0.00032	1.04E-03
38	17	40	16.99955	39.99845	0.000451	1.55E-03
39	16	55	15.92466	54.88134	0.075342	1.19E-01
40	20	39	20.00098	39.00172	-0.00098	-1.72E-03
41	18	41	18.00132	41.00069	-0.00132	-6.94E-04
42	12	47	11.99945	47.00023	0.000546	-2.30E-04
43	15	48	14.99959	47.99756	0.000412	2.44E-03
44	16	40	15.99872	39.99765	0.001285	2.35E-03
45	16	48	15.99615	47.99281	0.003845	7.19E-03
46	13	48	12.99988	48.00176	0.000116	-1.76E-03

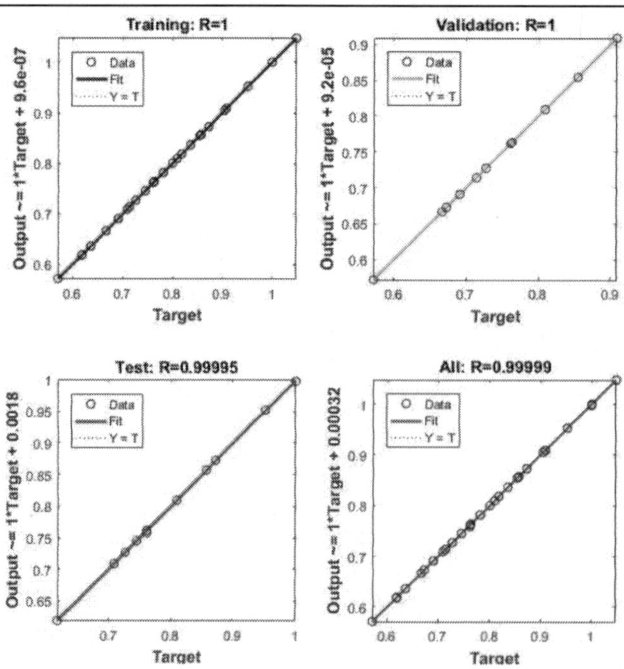

Figure 2. Regression Curve of the ANN model

Table 3. Comparison of R^2 and AAD values for the outcomes

	Cold Temperature (T_c)		Hot Temperature (T_h)	
	RSM	ANN	RSM	ANN
R square	0.9899	0.9999	0.9609	0.9999
Average Absolute Deviation	0.819	0.573	0.462	0.255

The coefficient of determination (R^2) and average absolute deviation are calculated for the responses and found that R^2 values for both the values are near to 1 in both the methods. But to elaborate more the R^2 values of ANN is very much closure to 1 when compared to RSM values, and also the AAD values of ANN shows lesser deviation than the RSM values.

Conclusions

Experimentation has been performed under various input conditions using Box-Benkhen design with 46 experimental trials. Using the experimental results, the performance (cold and hot temperatures of outlet air) of the Vortex tube is modeled using both Artificial Neural Networks and Response Surface Methodology. It has been evaluated that the ANN model is predicting better than the RSM model as is evident from the R- Square(99.99, 99.99 for ANN (T_c and T_h); 98.99, 96.09 for RSM (T_c and T_h)) and AAD values(0.573,0.255 for ANN (T_c and T_h); 0.819,0.462 for RSM (T_c and T_h)). Compared to the hitherto works, this work, considering five input controllable parameters, is comprehensive in modeling the performance of Vortex tube.

References

[1] Ahmet Murat Pinar, Onuralp Uluer, Volkan Kirmaci, (2009), "Optimization of counter flow Ranque-Hilsch Vortex Tube performance using Taguchi method" *International Journal of Refrigeration*, vol. 32, 1487-1494 .

[2] J., Prabakaran,, S., Vaidyanathan, D., Kanagarajan, (2012) "Establishing empirical relation to predict temperature difference of vortex tube using Response Surface Methodology", *Journal of Engineering Science and Technology*, vol. 7, no.6, 722–731

[3] S., Eiamsa-ard, K., Wongcharee, P., Promvonge, (2010) "Experimental investigation on energy separation in a counter-flow Ranque-Hilsch vortex tube: Effect of cooling a hot tube", *International Communications in Heat and Mass Transfer*, vol. 37, 156-162.

[4] Kirmaci, V. (2017). Experimental Investigation of Cooling - Heating Performance of Counter Flow Ranque-Hilsch Vortex Tubes Having Different Length Diameter Ratio. *Cumhuriyet Science Journal, 38*(4), 813-821

[5] Dincer, K., Tasdemir, S., Baskaya, S., & Uysal, B. (2008). Modeling of the effects of length to diameter ratio and nozzle number on the performance of counterflow Ranque–Hilsch vortex tubes using artificial neural networks. *Applied Thermal Engineering, 28*(17-18), 2380-2390.

[6] Korkmaz, M. E., Gümüşel, L., & Markal, B. (2012). Using artificial neural network for predicting performance of the Ranque–Hilsch vortex tube. *International Journal of Refrigeration, 35*(6), 1690-1696.

[7] Uluer, O., Kırmacı, V., & Ataş, Ş. (2009). Using the artificial neural network model for modeling the performance of the counter flow vortex tube. *Expert Systems with Applications, 36*(10), 12256-12263.

[8] Pouraria, H., Kia, S., Park, W., & Mehdizadeh, B. (2016). Modeling the cooling performance of vortex tube using a genetic algorithm-based artificial neural network. *Thermal Science, 20*(1), 53-65.

Evaluating Tensile Properties of Animal and Hybrid Fiber Reinforced Polyester Composites

Gunti Rajesh[1*], B Hemanth Nadh[2], Raja Chandra Chowdary Guduru[3]

[123]Department of Mechanical Engineering,DVR & Dr.HS MIC College of Technology, Kanchikacherla, India-521180.

**Email: rajesh.mtech@gmail.com*

Abstract

Partially degradable composites were made with short Goat hair at different weight percentages in polyester matrix and also hybrid composite with Madar fiber and Goat hair at different weight ratios in polyester matrix using hand layup technique. The tensile properties of goat hair-polyester composite and hybrid fiber composites were evaluated. From the results, it is observed that the tensile strength of goat hair-polyester composite was increased up to a fiber content of 7.5% (24.7 MPa) and then reduced with further fiber addition in the composite. The strength of hybrid composite is increased up to 12.5% (27 MPa) madar fiber and goat hair loading and then reduced with further hybrid fiber loading. The strength of hybrid composite is 9.31% higher than the goat hair fiber composite. The tensile modulus is increased in both the composites with increase in fiber loading. The tensile modulus of goat hair fiber reinforced composite at 15% of fiber loading is high and the value is 367 MPa and for hybrid fiber it is 444.13 MPa at 15% of hybrid fiber loading which is 21.01% higher than the goat hair fiber composite. The %Elongation at break of goat hair fiber composite gradually decreased and higher value of elongation is obtained at 2.5% of the fiber loading. For the hybrid fiber it is first increased and then decreased to 4.4% at 15% of fiber loading.

Key words: Polyester, Madar Fiber, Goat Hair, Hybrid Composite

1. Introduction

Bio fibers like agave, sisal, coir, hemp, flax, jute, banana etc. are now considered to be best alternatives to synthetic fibers for reinforcing polymer composites. Natural fibers have numerous advantages over established reinforcing synthetic fibers such as carbon fiber, glass fiber etc. For applications such as the automotive, construction or packaging industries they posses high strength and toughness, light weight, non-corrosive nature, good thermal properties, low cost, low density, biodegradability and renewability [1].

The chief chemical constituents of the plant fiber are cellulose portion, hemicellulose portion and lignin. The amount of these constituents will alter from one plant to another and also from various parts of same plant based on age, species. These basic components partially determine the physical properties of the fiber. Table 1 shows an average of chemical composition of few plant fibers [2-7].

Animal fibers like wool, feathers, hair, silk etc., are said to be next most significant resource of natural fibers after plant based fibers for reinforcing composites. Even in terms of availability the animal fibers are next most important natural fiber after plant fibers. Many sources from each type of animal fibers are wool, silk, hair, and feather [8].

Table.1. Chemical composition of few natural fibers

Fiber	Cellulose (wt%)	Lignin (wt%)	Hemicellulose (wt%)	Wax (wt%)
Bast				
Flax	71.0	2.2	18.6–20.6	1.7
Hemp	70.2–74.4	3.7–5.7	17.9–22.4	0.8
Jute	61.0–71.5	12.0–13.0	13.6–20.4	0.5
Kenaf	31.0–39.0	15.0–19.0	21.5	—
Ramie	68.6–76.2	0.6–0.7	13.1–16.7	0.3
Leaf				
Abaca	56.0–63.0	7.0–9.0	20.0–25.0	3.0
Curaua	73.6	7.5	9.9	—
Henequen	77.6	13.1	4.0–8.0	—
Pineapple	70.0–82.0	5.0–12.0	—	—
Sisal	67.0–78.0	8.0–11.0	10.0–14.2	2.0
Seed/Fruit				
Coir	36.0–43.0	41.0–45.0	0.15–0.25	-
Cotton	82.7	—	5.7	0.6
Oil Palm	65.0	—	29.0	—
Grass				
Bagasse	55.2	25.3	16.8	—
Bamboo	26.0–43.0	21.0–31.0	30.0	—
Straw				
Rice	41.0–57.0	8.0–19.0	33.0	8.0–38.0
Wheat	39.0–45.0	13.0–20.0	15.0–31.0	—
Others				
Rice Husk	35.0–45.0	20.0	19.0–25.0	14.0–17.0

Onuegbu et al. examined the consequences of animal hair fiber on properties of polyester resin. The animal hair fiber was taken from the Hausa – specie she-goat [9]. The hair fiber was added to the polyester resin which is cured with accelerator and catalyst, stirred and cast into Teflon mould. The composite specimen were cured at normal room temperature for 24 hours and then heat treated in an oven at 80^0C for a period of one hour to get dimensional stability of the composite. The results depicted an increase in the tensile strength of the composite from 3% - 9% with a quick drop in 12% volume ratio. Also the tensile strength of the plain sample was higher than that of reinforced composites.

Madar fiber reinforced polymer composites were fabricated and mechanical properties were investigated [10]. In this work, newly identified madar fiber is used for making composite material the fiber is extracted from the stem by manual process. The fiber is separated into small parts and then it is mixed with polyester resin .Five plates were going to be made by 20%, 30%, 40%, and 50% by mass of the fiber. The mechanical properties of the plates such as tensile strength, compressive strength, impact strength and compressive strength of each plate were tested. Initially the tensile strength values increase from 20% fiber to 40% fiber plate and finally decreased at 50% fiber plate. The result shows that the tensile strength value is better at 60% resin and 40% fiber plate. The flexural strength value for the madar fiber increase initially from 20% fiber plate to 40% fiber plate and suddenly decreased at 50% fiber and 50% resin this implies flexural strength value is better at 60% resin and 40%fiber.

A mixture of two or more fibers in a polymer matrix will lead to hybrid composites. The probable mixtures of hybrid composites may include artificial–artificial, natural–natural fiber and natural–artificial types. They possess high strength to weight ratio, low cost and ease of manufacturing and have wide range of applications in engineering [11]. Also, these hybrid composites possess stiffness, ductility and strength that cannot be obtained by single fiber reinforced composites. They also have improved fatigue life, improved fracture toughness and lesser notch sensitivity when compared to that of single fiber reinforced composites [12].

With the combination of natural–natural and artificial–natural fibers, hybrid composites based on natural fiber may be obtained. The natural fiber itself can be considered as composite which make them tough when compared with synthetic fibers, but a rightly configured high quality natural fiber reinforced hybrid composite can possess good strength and stiffness competitive to glass fiber reinforced composites [13].

In this work an effort is made to use hybrid natural fiber reinforcement one is from plant and the other is from animal in polyester matrix. The plant fiber used is Madar and goat hair is another fiber to form hybrid fiber polyester composite. Tensile properties of these hybrid composites were evaluated at different weight ratios of both plant and animal fibers in polyester matrix.

2. Experimental Work

2.1. Materials Used

2.1.1. Polyester

Polyester is a category of polymers that contain the ester functional in their main chain. They are exceptionally important polymers. Their well-known applications include clothing, plastic water, food packaging and carbonated soft drinks bottles. It is often known as polyethylene terephthalate (PET). polyester is purchased from Bindu Agencies, Vijayawada.

2.1.2. Catalyst

The catalyst is the agent that is mixed directly in to the polyester in a composition of 1.2ml for 100gm of resin. The job of is that it starts the curing action of the polyester composite to get set hard. The name of the catalyst used for curing of polyester is Methyl Ethyl Ketone Peroxide [MEKP].

2.1.3. Accelerator

Accelerator is also directly mixed with the polyester in combination with the catalyst MEPK. The job of the accelerator is to make the process faster. The name of the accelerator used is cobalt octate ($C_{16}H_{30}CoO_4$). When cobalt octate is mixed with MEKP a chemical reaction takes place and heat is evolved and it is a good and preferred combination for polyester.

2.1.4. Goat hair

Goat hair is said to be a natural fiber extracted from the skin of goat. It is Eco- friendly, Bio-degradable, elastic, soft, smooth, low-cost and is abundantly available. Disposal of goat hair in tannery process is a big headache for them in waste management. Hence the goat hair can be a good choice for reinforcing polymer in the green composites.

2.1.5. Madar fiber

Madar fiber is a natural fiber extracted from the stem of Calotropis gigantean plant. Calotropis gigantean is a medium sized bush that grows up to a length of 4m with usually waxy exterior and abundant milky sap. The stem is ash colored, smooth, branching almost from the base of the plant. The leaves appear grey-green, opposite, waxy, thick and rounded-ovate.

2.2. Methodology

2.2.1. Extraction of fibers

Goat hair is obtained from local Kabela point in Vijayawada, India. The Madar fibers were extracted by retting process. The stem is cut into certain small length pieces and soaked in water for nearly 20 days and later they were beaten gently with mallet to extract the fibers from the stem.

Prior to the development of the composites, both the fibers were washed thoroughly and cut to 5mm length.

2.2.2. Preparation of mould

The moulds were prepared to fabricate the composite using hand layup technique. The moulds were prepared with rubber sheet of 3mm thickness which is cut to a size that can accommodate 5 specimens as per the ASTM D638 standard for tensile test (figure 1).

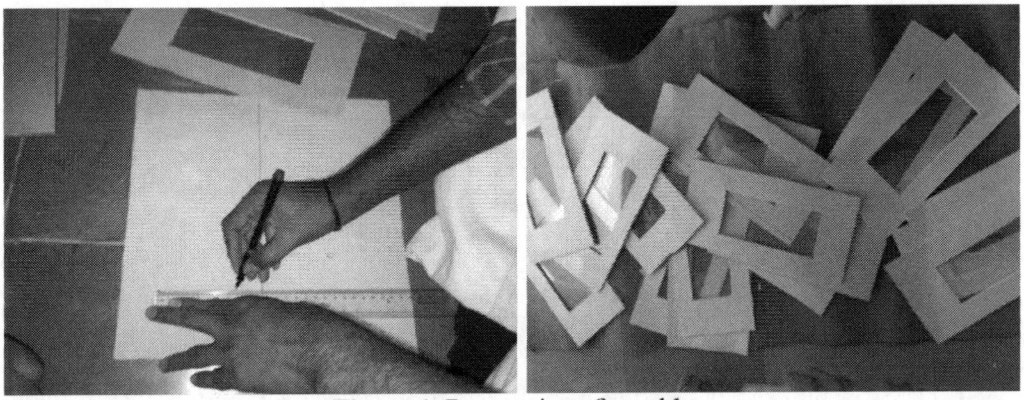

Figure 1. Preparation of moulds

2.2.3. Preparation of samples

The samples were prepared using hand lay-up technique. The goat hair fiber reinforced polyester composite and goat hair- madar fiber hybrid reinforced polyester composite samples were prepared for tensile test as per ASTM standard in six different compositions from 2.5% to 15% as mentioned in the tables 2 and 3.

Table 2. Compositions of goat hair fiber reinforced polyester composites

Composition	W%	W%	W%	W%	W%	W%
Goat Hair fiber (GHF)	2.5	5	7.5	10	12.5	15
Polyester	97.5	95	92.5	90	87.5	85

Table 3. Compositions of goat hair, madar fiber hybrid reinforced polyester composite

Composition	W%	W%	W%	W%	W%	W%
Goat Hair fiber (GHF)	1.25	2.5	3.75	5	6.75	7.5
Madar Fiber (MF)	1.25	2.5	3.75	5	6.75	7.5
Polyester	97.5	95	92.5	90	87.5	85

2.2.4. Testing of composites

Diameter testing of fiber: Tool maker microscope is used for finding the diameter of fiber

Density testing of fiber

Fiber density is determined by using Archimedes' principle which states that the volume of an object which is fully immersed in a fluid or to that fraction of the volume below the surface for an object partially submerged in a liquid is equivalent to volume of displaced fluid.

Tensile testing

These samples are tested for tensile property on a universal testing machine and the results are analyzed as per ASTM standards (ASTM D638). The samples are loaded on the machine and the maximum load is set to 200 kg and the load is applied and the samples were tested at a crosshead speed of 0.5 mm/min and the strain was measured with an extensometer of the machine. The elongation and their respective loads are noted

3. Results and Discussion

3.1. Calculations

The diameter and densities of madar and goat hair fibers were depicted in the table 4. The density of madar fiber is about 50%higher than goat hair considered in the study.

Table 4. Physical properties of fibers

	Goat hair	**Madar fiber**
Diameter, Microns	76-107	150-237
Density, gm/cc	0.73	1.45

From the figure 2, it is observed that the tensile strength of goat hair reinforced polyester composite is increased up to a fiber loading of 7.5% and then reduced with further fiber loading. The tensile strength of hybrid composite is increased up to 12.5% fiber loading and then reduced with further hybrid fiber loading. The highest tensile strength of 24.7 MPa is obtained for goat hair fiber reinforced composite at 7.5% fiber loading and 27 MPa for the hybrid fiber at 12.5% of fiber loading which is 9.31% higher than that of goat hair fiber polyester composite.

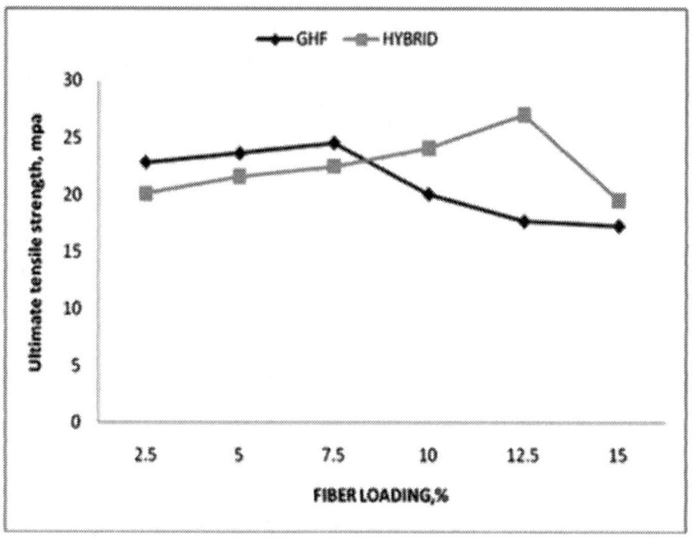

Figure 2. Variation of Ultimate tensile strength of goat hair fiber and hybrid fiber reinforced polyester composite with fiber loading

Form figure 3, it is observed that the tensile modulus is increased in both the composites with increase in fiber loading. The tensile modulus of goat hair fiber reinforced composite at 15% of fiber loading is high and the value is 367 MPa and for hybrid fiber it is 444.13 MPa at 15% of hybrid fiber loading which is 21.01% higher than the goat hair fiber composite.

The %Elongation at break of goat hair fiber composite gradually decreased and higher value of elongation is obtained at 2.5% of the fiber loading and the value is 8.27. For the hybrid fiber it is first increased and then decreased to 4.4% at 15% of fiber loading (figure 4).

As the fiber is much stronger than that of the matrix material, and because there is an adequate fiber-matrix interfacial bond due to the existence of hybrid fiber, it is clear that an increase in fiber content leads to increase in the composite strength. However, the trend is only up to definite quantity of fiber, after which there is insufficient polymer in the composite to provide adequate wetting of the fibers. The increase in fiber content beyond certain limit also resulted in reduction of fiber dispersion in the composite and leading to fiber-fiber interaction, which may have further contribution to the observed decrease in composite strength.

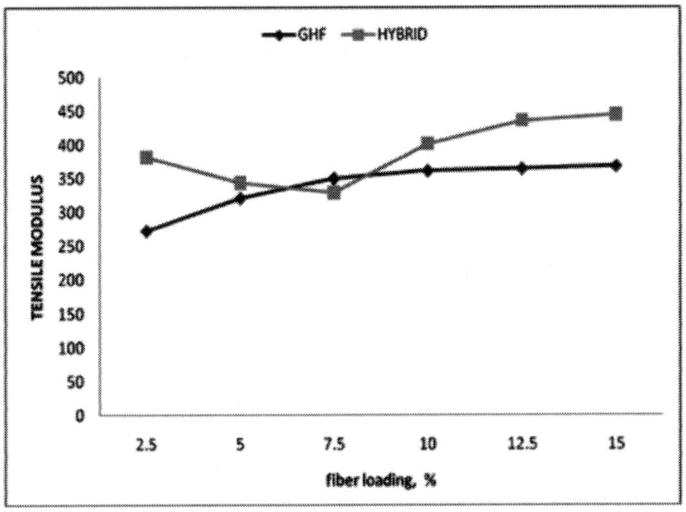

Figure 3. Variation of Tensile Modulus of goat hair fiber and hybrid fiber reinforced polyester composite with fiber loading.

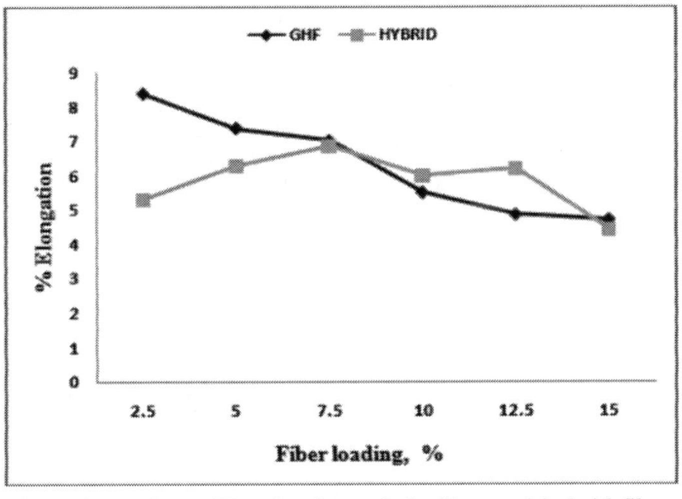

Figure 4. Variation of % Elongation of Break of goat hair fiber and hybrid fiber reinforced polyester composite.

The results from figures 2 and 3 have shown that an increase in fiber content can greatly improve the tensile strength and tensile modulus of the composites. It is also observed that the mixing of hybrid composite increase the interface adhesion between the fiber and the matrix. Therefore, hybrid composites may be taken into consideration to modify the properties of natural fibers.

4. Conclusions

With the results obtained from the experimental procedure, the following conclusions are made.

- The tensile strength of goat hair reinforced polyester composite is increased up to a fiber loading of 7.5% and then reduced with further fiber loading.
- The tensile strength of hybrid composite is increased up to 12.5% fiber loading and then reduced with further hybrid fiber loading.
- The highest tensile strength of 24.7 MPa is obtained for goat hair fiber reinforced composite at 7.5% fiber loading and 27 MPa for the hybrid fiber at 12.5% of fiber loading which is 9.31% higher than the goat hair fiber composite.
- The tensile modulus is increased in both the composites with increase in fiber loading. The tensile modulus of goat hair fiber reinforced composite at 15% of fiber loading is high and the value is 367 MPa and for hybrid fiber it is 444.13 MPa at 15% of hybrid fiber loading which is 21.01% higher than the goat hair fiber composite.
- The %Elongation at break of goat hair fiber composite gradually decreased and higher value of elongation is obtained at 2.5% of the fiber loading and the value is 8.27.
- For the hybrid fiber it is first increased and then decreased to 4.4% at 15% of fiber loading.

References

[1] A.S.Singha, Vijay Kumar Thakur 2008 *Bul. Mater. Sci.* **31** 791.
[2] O.Faruk, A.K. Bledzki, H.P. Fink, M. Sain 2012 *Prog. Polym. Sci.* **37** 1552.
[3] A.K.Mohanty, M. Misra, G. Hinrichsen 2000 *Macromol. Mater. Eng.* **276–277** 1.
[4] A.K.Bledzki, S. Reihmane, J.J. Gassan 1996 *Appl. Polym. Sci.* **59** 1329.
[5] D.N.S.Hon 1992 *Polym. News.* **17** 102.
[6] S.C.O.Ugbolue 1990 *Tekst. Prom.* **20** 1.
[7] M.S.Kucharyov 1993 *Tekst. Prom.* 23 8.
[8] J. Mussig, *Wiley: NJ.* 2010.
[9] T.U.Onuegbu, A.J.J.Azih 2015 *Natur. Sci. Resear.* **5** 42.
[10] C.Kandeepan, K Raja, P Ganeshan 2016 *Proceed. Int. Conf. Cur. Resear. Eng. Sci. Technol.* **2016** 110.
[11] M.N.Gururaja, A.N. Hari Rao 2012 *Int. J. Soft Comp. Eng.* **1** 2231.
[12] A.Shahzad 2011 *J Rein. Plast. Compos.* **30** 1389.
[13] D.N.Goswami, M.F.Ansari, A.Day 2008 *Ind. J. Chem. Technol.* **15** 325.

Effect on microstructures, hardness and SDAS of primary Mg_2Si/Al-Si eutectic phases of centrifugally cast functionally graded Al-(Mg_2Si)p in-situ composites

S C Ram*, K Chattopadhyay, I Chakrabarty

Department of Metallurgical Engineering, Indian Institute of Technology (Banaras Hindu University), Varanasi-221005, Uttar Pradesh, India
*Corresponding Author :(Email:subh.mtech07@yahoo.com), Mob.9919740549

Abstract. In the present study, the centrifugal casting route was employed to synthesize the Al-(Mg_2Si)p functionally graded in-situ composites at 1200 rpm rotating speed of mold. The formation of primary Mg_2Si particles/Al-Si eutectic phases by the in-situ chemical reaction has been investigated. In the Al-Mg-Si alloys system, in-situ formed Mg_2Si particles are lately introduced a new aspect, because of due to lighter density (1.99 gm/cm^3) than Al-matrix (2.27 gm/cm^3) the Mg_2Si particles are more capable of migrating into inner zone of cast tubes due to centrifugal force. The X-Ray diffraction (XRD) technique was carried out to confirm the presence of in-situ formed Mg_2Si/Si particles in the cast composites. Optical microscopy analysis was carried out to reveals distribution, size and volume fraction of Mg_2Si particles of cast cylindrical tubes. Scanning electron microscopy (SEM) analysis was also carried out to reveal the morphological characteristics and some intermetallics (π & β phases) present into the cast FGM tubes. The Vickers hardness test was carried out along the radial direction of tubes. Resultant, the hardness value is remarkably improved by 20% to 30% at the inner zone due to high volume percent of reinforced primary Mg_2Si hard particles. The objective of this investigation was to observe the effect on microstructures, hardness and secondary dendrites arm spacing (SDAS) of Mg_2Si particles/Al-Si eutectic phases of fabricated FGMs tube and analyzed their gradient properties.

Key words: Mg_2Si and Al-Si eutectic phase, FGMs, SDAS, Hardness, and Microstructures.

1. Introduction

Functionally graded materials can be considered as new class of advanced composite materials. Recently, the various researchers have been developed the composite structures depending on the design prospects and requirements [1]. Now days, the Al-Si-Mg casting alloys are increasingly being used in the fabrication of engine components in the automotive industry as an alternative material for cast iron. The in-situ Al-Mg_2Si composites have great potential applications in the automotive and aerospace industries due to their excellent properties, e.g., high hardness, low thermal expansion coefficient; low density, high elastic modulus, high strength and superior wear resistance properties.

Functionally graded materials (FGM) are of practical interest because of the gradation of physical, mechanical and chemical properties can be achieved across volume depending on design.

In Al-base alloys the several reinforcing phases e.g.Al_2O_3, TiB_2, SiC, Al_3Ti, etc. can be used but the segregation of Si and Mg_2Si primary crystal particles seems to be of great interest in FGMs. Moreover, functionally graded materials (FGM) are a group of advanced materials in which composition and microstructures, as well as specific properties, vary continuously along the thickness. Thus, in such types of materials the reasonable features such as mechanical, physical and wear resistance properties can be achieved. Besides that, in FGMs, the thermal stability may also obtain, so a wide range of engineering applications especially in automotive engine components can be used where thermal fluctuation may cause damage the cylinder liners and pistons [2-6]. There are several fabrication techniques of in-situ FGMs composite include control mold, chemical vapor deposition(CVD), pressure vapor deposition(PVD), the plasma spray technique, and powder metallurgy techniques can be used, but centrifugal casting method is more economical for fabrication of cylindrical components(engine block liners). However, among these fabrication processes, centrifugal castings have been lately introduced a new concept to produce the in-situ metal matrix composites (PMCs) tubes with gradient properties [7-8].

In Al-base alloys the several reinforcing phases e.g.Al_2O_3, TiB_2, SiC, and Al_3Ti, etc. can be used but due to light density (1.99 g/cm^3) and high strength of Mg_2Si primary particles seems to be of great interest in FGMs. Moreover, most of the authors are fabricated the PMMCs with several particulate reinforcements have larger densities (SiC, TiB_2, etc.) than the Al matrix. Thus, with few PMMC systems, the reinforcements having lighter densities may be used with a concentration towards the inside area of the tube through centrifugal casting [3, 9]. In general, as a major reinforcing phase (e.g., SiC, TiB_2, ZrB_2, Mg_2Si and other ceramic phases), to the reduction in weight, the Mg_2Si is the most effective reinforcing phase for engine block applications. In situ formed Mg_2Si phase in the matrix are the new aspect, because of it is more capable of migrating into inner surfaces of cylindrical cast tubes due to their lighter density (1.99 gm/cm^3) than other phases and Al-matrix (2.27 gm/cm^3) using centrifugal casting. Furthermore, there is very limited information about the centrifugal casting for producing the FGMs, especially as in-situ composites, by centrifugal casting. The present work especially shows a practical way to fabricate an FGM with Mg_2Si reinforcements. Centrifugal casting can be used to fabricate both (i) reinforced towards inside and (ii) reinforced towards the outside of the tube specimen depending on the densities of reinforcements. Farahany et al. [10] have investigated some intermetallics formation during solidification of Al- 13 Mg-7Si-2Cu in-situ composite by gravity casting. They have also reported, the formation of primary Mg_2Si begins during solidification followed by eutectic Al- Mg_2Si, Al_5FeSi and simultaneous precipitation of $Al_5Cu_2Mg_8Si_6$ and Al_2Cu complex intermetallic phases. Shabastari et al. [11] studied the effect on the morphological changes in Mg_2Si particles of Al- 25%Mg_2Si composites processed by slope casting and semi-solid techniques. They have also observed the particle size refinement due to increased nucleation rate and fragmentation of particles by fluid shearing stress in slope casting. The spheroidization of α-Al and Mg_2Si particles were observed in semi-solid processing.

E.Georgatis, et al. [12] have adopted the modified investment casting technique and studied the effect of microstructure, heat treatment and consequent mechanical properties of Al- 11% Mg_2Si-Si in-situ composite. They have been found that fine dendritic arm spacing to improve the aging performance.

However, the effect of particles distribution into the core with varying processing conditions, and compositions are not reported yet, which are very crucial for automotive engine blocks and liners.

In the present investigation, a stringent work is made to correlate secondary dendrites arm spacing (SDAS) and its effect on microstructures, hardness properties and grain refinements mechanism of fabricated Al-Mg_2Si in-situ composites by centrifugal casting. Generally, at the inner surface of cylinder liners of an automotive engine is required more strength and wear resistance properties. So in this context, the hard Mg_2Si particles with light density than matrix have more capable of migrating into the inner region of tubes due to centrifugal force. However, due to segregation of Mg_2Si particles along the radial direction of the components are achieved the gradient properties and have the great potential for automotive engine blocks and liners.

2. Experimental procedure

2.1 Materials and method

Industrially pure commercial A356 (Al-0.3Mg-7.3Si) alloy was used as matrix materials to produce the Al-Mg_2Sip functionally graded in-situ composites. The required amount of Mg-turnings about 20 % of additional pure Mg turning (wrapped in the Al foil) was added to the molten matrix materials to fabricate the master alloys ingots. Here, the 20%Mg turning was used to recover the melting loss of Mg. In this work, centrifugal casting technique has been adopted for manufacturing of axi-symmetric cylindrical tubes with the gradient structure. At first, three ingots of A356+0.30%Mg, A356+10%Mg, and A356+20%Mg master alloy were prepared with the addition of pure Mg tunings in required proportions. After that, the pure A356 ingots along with master alloy were melted at 800°C in the electrical resistance furnace with $MgCl_2.6H_2O$ and KCl salts as a flux. By the in-situ reaction, the Mg_2Si particles are formed in the molten alloy, and the molten composite was then transferred to a preheated stir casting melting furnace attached to vertical centrifugal casting machine (Figure 1) and was stirred for 10 min with a graphite stirrer for homogeneous mixing.

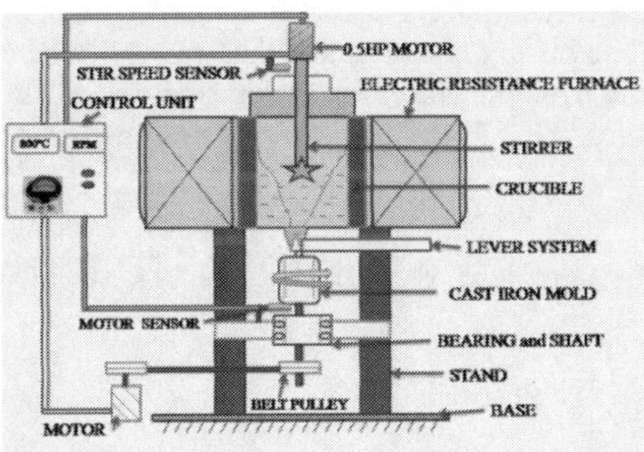

Figure 1. Centrifugal casting machine Set-Up.

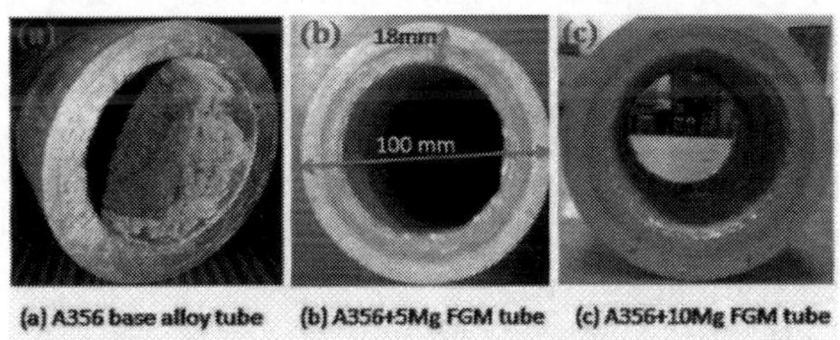

Figure 2. Centrifugally cast alloy and FGM tubes.

The melt was then poured the bottom into a rotating cast iron mold at a constant rotational speed of 1200 rpm. The mold was preheated to 250°C before pouring. The composite is poured into rotating cast iron mold at 1200 rpm. Finally, the three cylindrical casting tubes were made as (a) A356 base alloy (b) A356+Mg (c) A356+10Mg with gradient properties. The tubes of dimensions 100 mm in the outer diameter, 18 mm in thickness and 150 mm in length were centrifugally cast as shown in Figure 2. The chemical composition of the base alloy and composites (Table 1) were analyzed with an Optical Emission Spectrometer (Foundry Master).

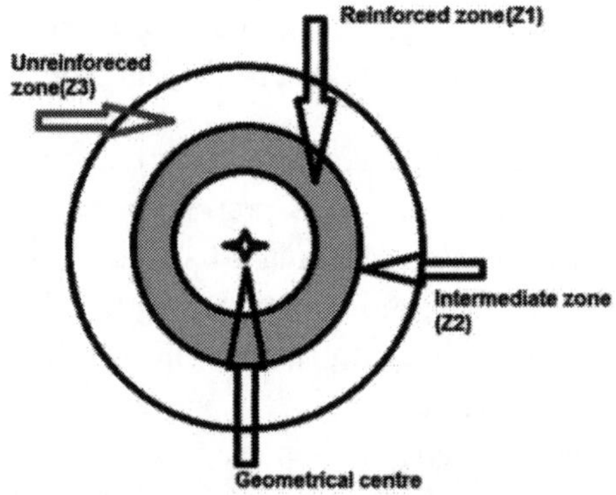

Figure.3 Cross-section of cast tube showing different zone: inner (z_1), middle (z_2) and outer (z_3).

For the identification of Mg_2Si/Si, and α-Al phases in the casting, the XRD analysis was carried out using CuKα radiation of wavelength 1.541836Å with Ni filter. Optical micrographs were taken to reveals the microstructures of the centrifugally cast A356 alloy and of FGMs, in the different zones at inner (z_1), intermediate (z_2) and outer (z_3). The hardness of the base alloy and composites was estimated by Vickers Hardness Tester (Leica) at 5 Kgf loads along three zone z_1, z_2 and z_3 for each composite as shown in Figure 3.

Table.1 Chemical composition of as-cast base alloy and FGMs

Alloy and composites	Si	Mg	Cu	Fe	Al
A356 Alloy	6.50	0.30	0.60	1.40	bal.
A356+5.0Mg composite	5.48	5.13	0.63	0.81	bal.
A356+10.0Mg Composite	5.54	10.50	0.36	0.44	bal.

2.2 Determination of secondary dendrite arm spacing (SDAS) and solidification time

The image analyzer was used to evaluate the SDAS by identifying and measuring aligned groups of each dendrite cells on the screen. The SDAS is then calculated using equation as SDAS=L/nM, where L is the length of the line, M is the magnification, and n is the number of dendrite cells. The solidification times (t_E) have been calculated after knowing the SDAS from the given equation as follows:

$$t_E = \{SDAS/A\}^3$$

Where 'A' is a material constant, for the type Al- Si7-Mg aluminum alloy, the value of 'A' as 11.7 [13].

3. Results and Discussion

3.1 Microstructural and Hardness analysis

The microstructures of A356 base alloy, 356-5Mg and A356+10Mg FGMs tube composite using the vertical centrifugal machine is shown in Figures 4-6(a), (b) and (c) at zone z_1, z_2 and z_3 respectively. Graded distribution of particles have seen in A356+5Mg and A356+10Mg FGMs tubes at three different locations from the inner(z_1) (a), middle(z_2) (b) and (c) periphery(z_3) of 18 mm thick ring of 100 outer diameters casting as Figure.5-6. In contrast, for A356 alloy eutectic Al-Si, α-Al dendrites cells and grains structures have been observed refer Figure.4 (a-c). However, in FGM the outer periphery of the casting showed Figure. 5(c) has the low concentration of Mg_2Sip than the interior of the casting with a sharp interface separating the particle-rich and particle free zone and the inner area

IOP Conf. Series: Materials Science and Engineering **390** (2018) 012012 doi:10.1088/1757-899X/390/1/012012

with the higher concentration of particles Figure. 5(a). Similarly, as Figure.6 a-c shows the particles segregation from inner to periphery of cast tube for A356+10Mg FGMs.

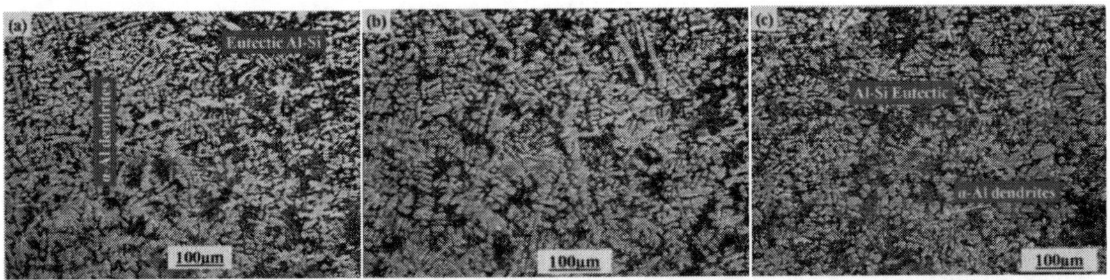

Figure.4: (a-c) Microstructures of A356 base alloy tube at (a) inner zone z_1, (b) middle zone z_2 and (c) outer zone z_3 respectively.

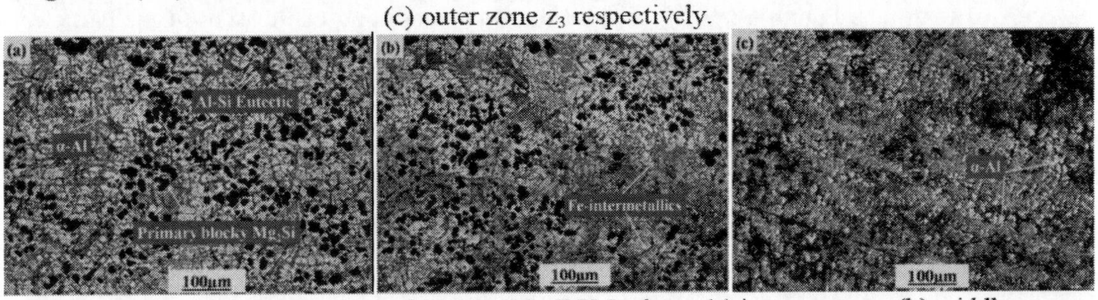

Figure 5: (a-c) Microstructures of A356+5Mg FGM tube at (a) inner zone z_1, (b) middle zone z_2 and (c) outer zone z_3 respectively.

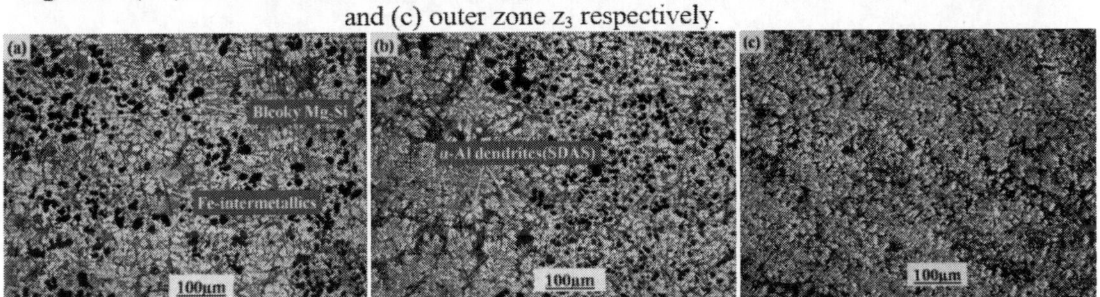

Figure. 6: (a-c) Microstructures of A356+10Mg FGM tube at (a) inner zone z_1, (b) middle zone z_2 and (c) outer zone z_3 respectively.

The segregation of particles was seen because of action of centrifugal force and depending on the density of particles. The lighter dense particles are moving toward inner zone and Al-grains (high density) toward the periphery due to centrifugal forces. The size distribution and volume percentages of Mg_2Sip evaluated using the image-J analyzer software shows (Figure 8. a-c) for A356+5Mg FGM and A356+10Mg FGMs. The outer periphery of the cylinder contained less vol. % Mg_2Sip gradually increases towards the core of tubes respectively. The Mg_2Si particles (blocky shape), Al-Si eutectic phase and intermetallic phase (Needle like shape) have been observed in the microstructural examination Figures5-6(a& b). Intermetallics phases were not detected in XRD patterns as shown in Figures 7 because of very low vol.% of the π & β phases were observed under SEM analysis.

The Vickers hardness values evaluated along the radial directions of the composite FGM tubes fabricated are shown in Figure 10. The same trends were observed in each alloy and composites; the maximum hardness at the inner zone and gradually decreasing, then increasing to some extent near the outer zone. In the base alloy, the graphs show the reverse trend, the value of hardness increasing towards the outer zone due to fast chilling effect and finer SDAS. Although, in inner areas, the SDAS are coarser which lower the hardness due to the high volume fraction of Mg_2Si particles hardness values are increases. These effects on hardness could explain and possible by observing the SDAS and solidification time as shown in Figure 9. However, in the base alloy at inner zone hardness values are decreases due to coarser SDAS but towards outer periphery the hardness of base alloy increases due to finer SDAS.

3.2XRD Analysis and phase distribution observation

The XRD analysis was carried out to confirm the phases formed in alloy and FGM composites as shown in Figure 7. So clearly we can see the peak intensity of Mg_2Si particles and eutectic Si as well as α-Al at 2θ angles. When Magnesium amount increases the peak intensity was higher in case of A356+10Mg FGM and lower with A356+5Mg FGM composites. The microstructure of centrifugal cast A356+10%Mg tube Figure 6 (a)-(c)) showed the very interesting distribution of the particles. A small portion of Mg_2Sip is segregated in the core z_1 of the tube and gradually decreases toward the periphery (z_3) along which shows the gradient distribution of particles. Due to gradient distribution of particles along the thickness of the tube, the hardness values increases and gradually varies along the thickness this may depend on coarsening and de-coarsening of secondary dendrite arm spacing.

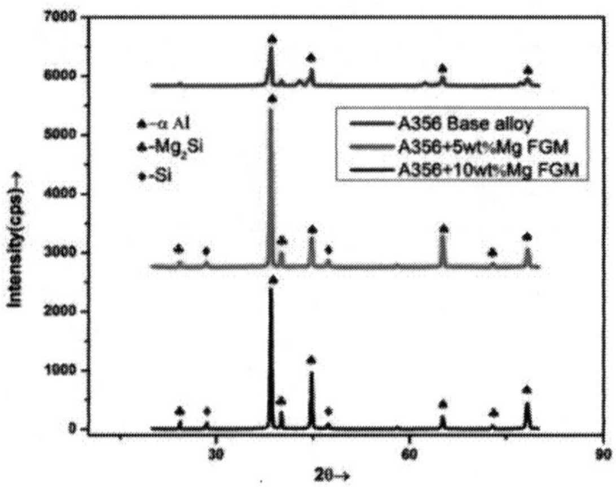

Figure.7 XRD analysis graphs of base alloy and FGMs showing the phases formed.

Chirita et al.[14] have found the same trend and valid with this observation in the case of composites, they have also explained due to rapid chilling on the surface causes some primary Mg_2Si particles are entrapped in the surfaces of tubes and due to finer SDAS the hardness increase near the outer zone. These increments in hardness values are more prominent in A356+5Mg and A356+10Mg FGMs. During the solidification processes, most of the primaries Mg_2Si are formed, and the majority of these particles are moved to the inner zone due to centrifugal force. After that, at the time of semi-solid state the particles movement become slow by centrifugal force.

For evidence, refer to Figure 8 due to the high volume fraction of blocky Mg_2Si particles are showed maxima hardness values around 125-130HV in case of A356+10Mg FGM than A356+5Mg FGMs and gradually decreased to some extent toward periphery as shown in Figure 10.But in case of base alloy due to the absence of Mg_2Si particles, Vickers hardness values are minimum at inner 80 HV and gradually increases toward periphery around 90 HV to 100 HV due the finer dendrites cell. These finer dendrites cells are formed due to rapid chilling effect near the mold wall.

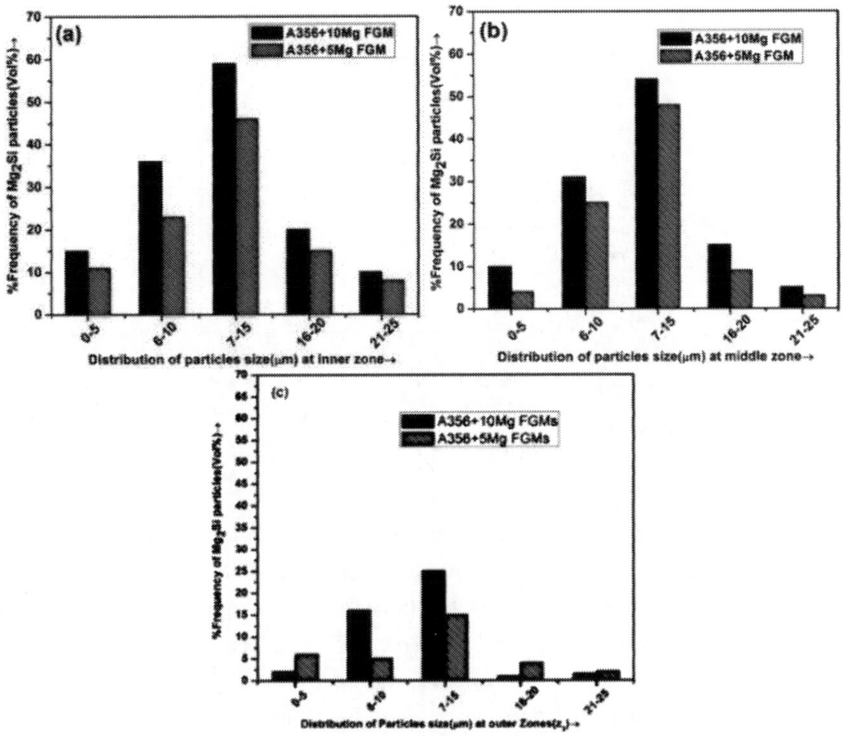

Figure.8. %Frequency and size distribution of Mg_2Si particles in three zones of A356+5 %Mg and A356+10Mg FGM composite at (a) inner (z_1), (b) middle (z_2) and (c) outer(z_3) zones.

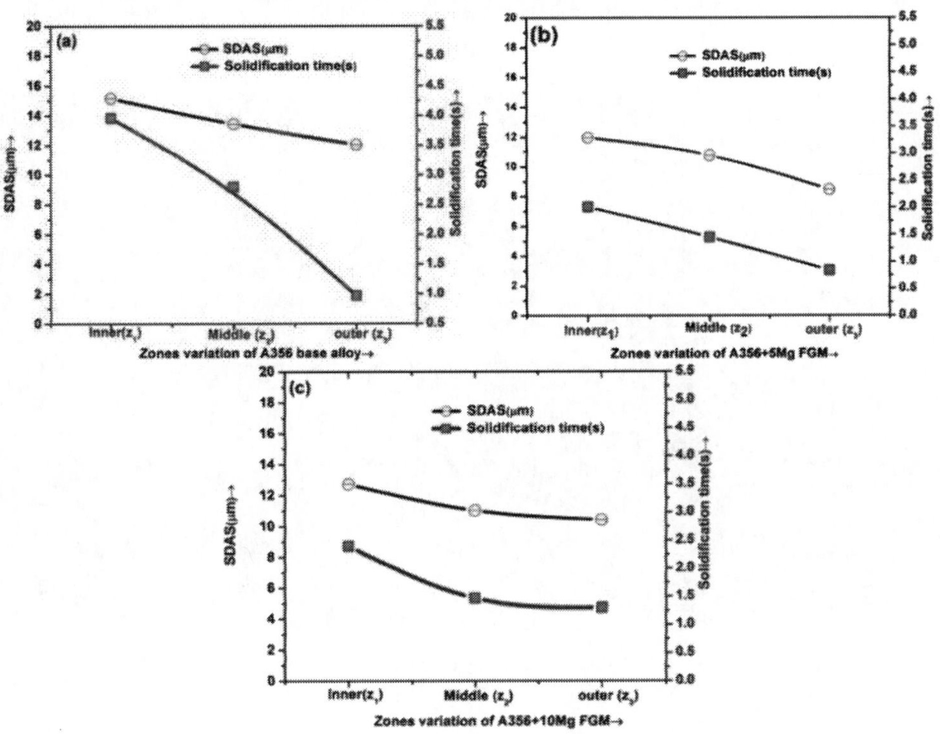

Figure.9. SDAS Vs Solidification time graphs for different zones of base alloy and FG composites (a) A356 alloy (b) A356+5%Mg and (c) A356+10%Mg respectively.

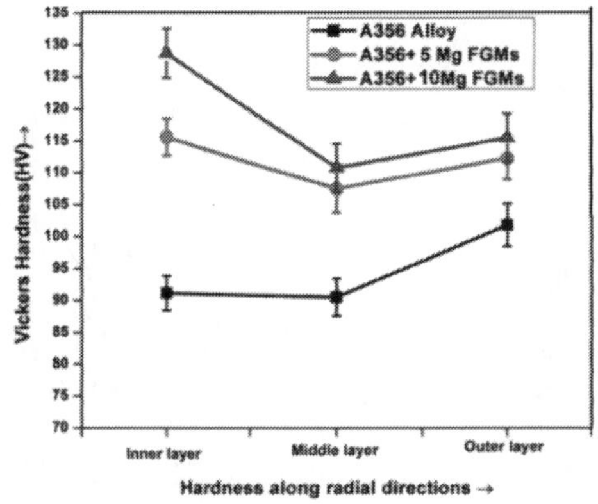

Figure.10. Vickers hardness graphs for different zones of base alloy and FG composites; A356 alloy, A356+5%Mg and A356+10%Mg respectively.

SEM micrographs as shown in Figurev11 for FG-composites of A356+5Mg and A356+10Mg shows the primary Mg_2Si particles, and other phases form during solidification like α-Al, β & π phases and Chinese script like morphologies. These constituents of intermetallic phases, the strength and SDAS affected and finally, the microstructures and graded behavior of composites have changed. Sometimes these phases acted as the obstacle to dislocation movements and enhanced the mechanical strength. Thus, the Al-Mg_2Si functionally graded composite cylinders liner prepared for automotive engine block replacing the existing cast iron liners have been successfully agreed to the benefit of weight saving as well as high hardness and strength. By using the Al-Mg_2Si in-situ functionally graded composite provide the significant saving in the weight and new aspect in case of automotive industries is possible. The effect of different Mg content on the in-situ A356- Mg_2Si composite functionally graded material has been synthesized by in-situ centrifugal casting.

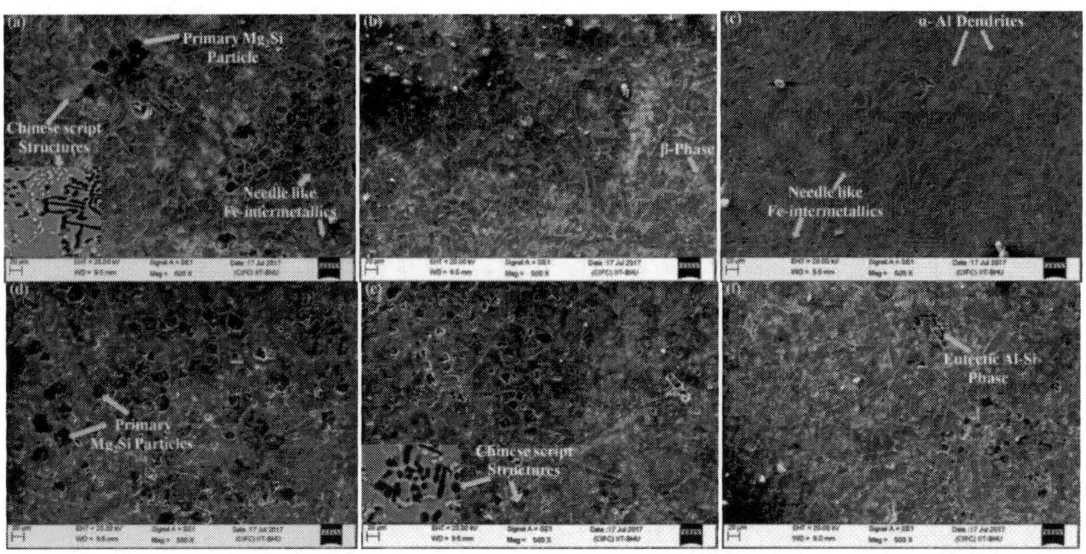

Figure 11. SEM graphs for different zones o f FG composites ;(a),(b) &(c) for A356+5%Mg and (d),e) &(f) for A356+10%Mg respectively.

4. Conclusions

The effect of solidification time and SDAS along the radial direction of centrifugally cast tubes on microstructure and hardness properties of Al-Mg_2Si in-situ composite was investigated. From the observation of results, the following conclusions can be made:

The microstructure of centrifugally cast Al-Mg_2Si FGM was analyzed and found that, the most of the particles are migrated into inner zone of tubes and some of the Mg_2Si particles are stick to the surface and middle zone which formed the gradient structure of casting. The Mg_2Si volume fraction and size were also evaluated from the surface to core of FGMs. In both A356+5Mg and A356+10Mg FGM tubes, the volume fraction was maximum in the inner zone about 47 to 57% and size range about 20-30 μm was observed. In the SEM analysis, the Mg_2Si particles (blocky shape), Al-Si eutectic phase and Fe-intermetallics phase (Needle like shape) and some π & β phases were observed which affected the gradient properties of the composites. Intermetallic phases were not detected in XRD patterns because of very low vol.% of the are present. As expected, the hardness is maximum in the inner layer of cast tubes which varies from 110 HV to 120 HV with 5wt.% Mg alloys. When Mg content is increased to 10 Wt% Mg then hardness value reaches up to 130 HV in the inner core of the cast FGM tubes. Hence hardness of in-situ FGMs increased by 25% to 30% as compared to base alloys.

Apart from these, the hardness values have correlated with secondary dendrites arm spacing (SDAS) cells from inner to the outer periphery of cast base alloy and FGM tubes. Resultant, because of rapid chilling effect to the surface of tubes, the finer SDAS was observed hence the hardness of base alloy increases at the outer zone but at intermediate layer, hardness was decreased due to coarser dendrites cells. Although, in the case of composites coarse dendrite can not affect the hardness values due to the presence of hard Mg_2Si particles. Hence, the hardness will increase at the inner zone and gradually decreases towards surfaces. Further in respect of scope and applications, such types of functionally graded materials are most capable in the casting industries especially in automotive manufacturing sectors for engine blocks and liners.

Highlights:

- The Al-Mg_2Si in-situ composites were developed by centrifugal casting method.
- Secondary dendrites arm spacing (SDAS) effect was correlated with solidification time.
- In the case of A356 base alloy casting, the hardness values are affected due to coarser dendrites cells into inner zone of tubes.
- In the case of FGM tubes, the SDAS effect has seen on the size and volume fraction of Mg_2Si particles.

References

[1] Ulukoy, Arzum, Muzaffer Topcu, and Suleyman Tasgetiren. (2016).An Experimental Crack Propagation Analysis of Aluminum Matrix Functionally Graded Material. *Advances in Functionally Graded Materials and Structures*. InTech.
[2] Watanabe, Yoshimi, and Hisashi Sato.(2011)Review fabrication of functionally graded materialsunder a centrifugal force.Nanocomposites with Unique Properties and Applications in MedicIne and Industry. InTech.
[3] Zhang, J., et al. (2000).Microstructural development of Al-15wt.% Mg_2Si in situ composite with

mischmetal addition. Journal of Materials Science and Engineering: A 281.1 104-112.

[4] Zhai, Yanbo, Xiuteng Ma, and Zhen Mei. (2013).Structure and mechanical properties of Al-based gradient composites reinforced with primary Si and Mg_2Si particles through centrifugal casting.Journal of Wuhan University of Technology-Mater. Sci. Ed. 28.4 813-818.

[5] Xie, Yong, et al. (2009).Centrifugal casting processes of manufacturing in situ functionally gradient composite materials of Al-19Si-5Mg alloy. Rare metals 28.4 405-411.

[6] Song, Chang-Jiang, Zhen-Ming Xu, and Jian-Guo Li. (2007). Fabrication of in situ Al/Mg_2Si functionally graded materials by electromagnetic separation method." Composites Part A: Applied Science and Manufacturing 38.2 427-433.

[7] Bo, L. I., et al. (2013).Effects of temperature on fracture behavior of Al-based in-situ composites reinforced with Mg_2Si and Si particles fabricated by centrifugal casting. Journal of Transactions of Nonferrous Metals Society of China 23.4 923-930.

[8] Rajan, T. P. D., R. M. Pillai, and B. C. Pai.(2010).Characterization of centrifugal cast functionally graded aluminum-silicon carbide metal matrix composites.Journal of Materials characterization 61.10 923-928.

[9] Arsha, A. G., et al. (2015). Design and fabrication of functionally graded in-situ aluminium composites for automotive pistons." Materials & Design 88 1201-1209.

[10] Saeed Farahany, Nur Azmah Nordin, Ali Ourdjini, TutyAsma Abu Bakar, Esah Hamzah, Mohd Hasbullah Idris, Alireza Hekmat-Ardakan.(2014). The sequence of intermetallic formation and solidification pathway of an Al-13Mg-7Si-2Cu in-situ composite, Mater. Character 98 119-129.

[11] S.G. Shabestari, H. Saghafian, F. Sahihi, M.H. Ghoncheh (2015).Investigation on microstructure of Al-25 wt-%Mg_2Si composite produced by slope casting and semi-solid forming, Inter. J. Cast. Met. Res. 28 (3) 158-166.

[12] E. Georgatis, A. Lekatou, A.E. Karantzalis, H. Petropoulos, S. Katsamakis, A. Poulia(2013). Development of a cast Al-Mg_2Si-Si in situ composite: microstructure, heat treatment, and mechanical properties, J. Mater. Eng. Perform. 22 (3) 729-741.

[13] www.giessereilexikon.com/en/foundry-lexicon.

[14] G. Chirita, D. Soares, F.S. Silva.(2008) . Advantages of the centrifugal casting technique for the production of structural components with Al-Si alloys, Mater. Des. 29 20-27.

IOP Conf. Series: Materials Science and Engineering **390** (2018) 012013 doi:10.1088/1757-899X/390/1/012013

Studies on friction stir processing parameters on microstructure and micro hardness of Silicon carbide (SiC) particulate reinforced Magnesium(Mg) surface composites

T Rajmohan[1], K Gokul Prasad[1], S Jeyavignesh[1], K Kamesh[1], S Karthick[1], S Duraimurugan[1].

[1] Department of Mechanical Engineering, Sri Chandrasekharendra Viswa Mahavidyalaya, Kanchipuram -631561, Tamilnadu, India.

Abstract. Friction stir processing (FSP) is promising as a capable method for producing surface composites in aluminum and magnesium-based alloys. In FSP a specifically intended rotating cylindrical tool that includes of a pin and shoulder that have dimensions relative to the plate thickness. The pin of the rotating tool is forced into the material and the shoulder comes into contact with the surface of the sheet, and then pass through in the desired direction. The present work focused to fabricate Silicon carbide (SiC) particulate reinforced Magnesium (Mg) surface composites. The effect of FSP parameters such as tool rotational speed, processing speed and wt % SiC on microstructure and micro hardness was investigated. A groove was contrived on the 6mm thick Mg plates and packed with SiC particles. FSP was carried out using Taguchi's L9 orthogonal array. Scanning electron microscopy and XRD were employed to study the microstructure of the fabricated surface composites. The results indicated that the selected FSP parameters significantly influenced the area of surface composite, distribution of SiC particles and microhardness of the surface composites.

1. Introduction

The study of magnesium and its alloys as one of the capable research topics for their load bearing capacity and good mechanical properties [1]. Surface composites are apposite materials for engineering applications bumping into surface contacts [2]. Friction stir processing (FSP) is one of the procedure for making surface composites and varying microstructural characteristics [3]. Many research have established that rigorous plastic deformation (SPD) is a useful method of producing ultrafine-grained materials [4-5]. Mishra et al. [6] effectively made Al matrix SiC reinforced surface composite using FSP. The majority of the accounted work has been on craming the microstructural evolution, hardness, corrosion and wear properties [7–8].

In FSP process variables are divided into three categories of machine variables, tool design variables and material properties. Main machine variables are tool rotating rate and tool traverse speed. Tool rotational and transverse speeds decide the quantity of heat generated in the work piece [9]. Tool geometry mostly comprises shoulder diameter, shoulder feature, probe shape, probe size and probe feature. Stream of plasticized material in processed zone is influenced by tool geometry as well as traverse and rotational motion of the tool [10]. Tool geometry is an important aspect of FSP as it affects heat generation, material flow and ensuing microstructure [11]. Nano-sized grains can be accomplished by proper cooling arrangements. Cooling during FSP also provides the extra function of plummeting tool wear [12]. Su et al. [13] examined the ensuing microstructure of friction stir processed 7075 Al using TEM. Mahoney et al. [14] proved that the FSP microstructures of NiAl Bronze alloy have considerably greater mechanical properties compared to the as-cast microstructure. Ma et al. [15] investigated the super plasticity in friction stir processed cast A356 at different temperature and strain rate.

There have been limited published studies concerned with the effect of FSP parameters on microstructure and micro hardness SiC reinforced Mg matrix composites. Experiments were conducted using Taguchi L9 orthogonal array. Microstructure of prepared composites were examined by SEM.

2. Experimental

2.1 Materials and Methods

The principal resources used for synthesis were Mg billet acquired from Micro Fine chemicals, India. and SiC particles from US Research Nanomaterials Inc, USA. Mg is used as the matrix material. The Mg matrix was reinforced with 5-15 wt% of SiC particles with the particle size varying from 100 μm to 150 μm

2.2 Experimental design

The FSP are intended using a Taguchi method which involves sinking the deviation in a process through strong design of experiments. The experimental design proposed by Taguchi involves using orthogonal arrays to organize the parameters affecting the FSP and the levels at which they should be varied. From the literature and trials done by the authors in the field, the independently controllable predominant FSP parameters, which influences the performance of MMC is identified and presented Table 1. This experimentation has 4 variables at 3 diverse settings. Taguchi experiment with a L9 (34) orthogonal array was implemented to FSP of MMCs.

Table 1. Control parameters and their levels

Control parameters	**Symbol**	**Level**		
		1	**2**	**3**
Speed rpm	*S*	600	700	800
Transverse speed mm/min	*f*	30	40	50
Axial depth mm	*d*	1.5	2	2.5
wt % SiC	*w*	5	10	15

2.3 Experimental work

A specifically intended tool with a straight cylindrical pin of 6-mm diameter, 5.7mm length, and a shoulder of 18-mm diameter was used. FSP was carried out using a tool which was mounting in a vertical CNC milling machine. Figure 1 shows the tool and the complete experimental setup. Before the beginning of the process, the samples were fastened firmly using ridged backing and holding plates. By using a drilling tool equipped with a CNC machine, a small hole with a diameter of that of the pin was drilled. The pin of the tool going ahead revolving and at the same time it was strained into the sample. The shoulder was then in get in touch with the surface of the specimen. Finally, the tool transmitted along the required length.

Nine FSP tests were conducted using different combinations as per Taguchi's experimental design. Table 2 shows an example of effectively friction stir processed composites. Vickers micro-hardness was measured using a QV-1000DM digital micro-hardness tester with a load of 1,000 gram and dwell time of ten seconds. Samples were taken from the crossways sections at the middle of the pass. Then, the micro-hardness was measured at three different positions within the thickness of the composites. Three to five readings were taken and averaged for each position and presented in Table 2.

IOP Conf. Series: Materials Science and Engineering **390** (2018) 012013 doi:10.1088/1757-899X/390/1/012013

Figure 1. Experimental setup

Table 2. Experimental results

Sample no	Speed rpm	Transverse feed	Axial depth mm	% SiC	Micro hardness VHN	Prepared FSP composites
1	600	30	1.5	5	65.2	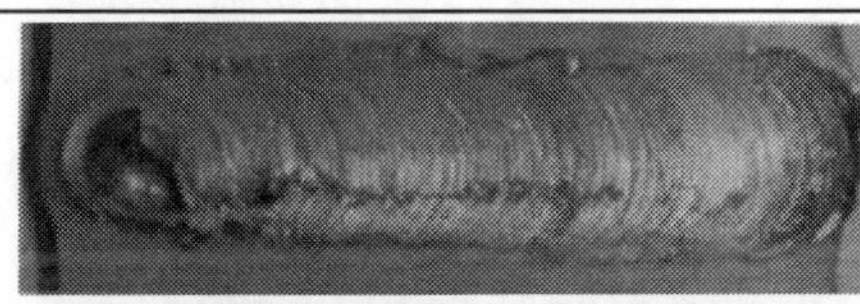
2	600	40	2	10	70	
3	600	50	2.5	15	71.2	
4	700	30	2	15	66.2	

5	700	40	2.5	5	65.6	
6	700	50	1.5	10	71.4	
7	800	30	2.5	10	80	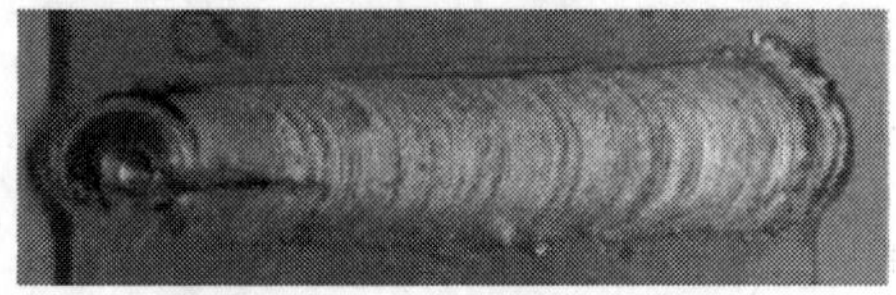
8	800	50	1.5	15	73.6	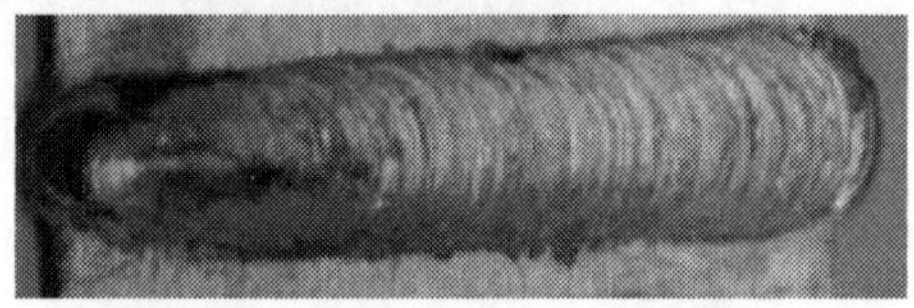
9	800	50	2	5	69.4	

3. Results and Discussion

3.1 SEM analysis

The micro structural and surface morphology of Mg metal matrix and SiC particles reinforced composites were done using scanning electron microscope with EDX. EDX was used to evaluate the dispersion of SiC in magnesium matrix composites. Figure 2 shows the SEM micrograph of Mg metal matrix reinforced with SiC and attachment of silicon carbide particles can be established over the surface of Mg. In order to confirm the presence of SiC particle, EDX analysis was done over a particular region. EDX analysis of the 10% SiC reinforced Mg matrix composite is presented in Figure 3. It is clear that the peaks for magnesium, silicon and carbon were attained.

FSP is a novel procedure for fabrication of surface composites and it is beneficial in ease of dispersion and elimination of clusters of reinforcement particles. The uniform distribution of reinforcement particles in Mg matrix is shown in **Figure 2.** Dispersion and distribution of reinforcement particles in the surface composite can be successfully achieved after multi-pass FSP. Also, a consistent particle distribution is attained. This can afford outstanding combinations of strength and ductility in surface composites

Figure 2. SEM micrograph for Friction stir processed Mg matrix reinforced with 10% SiC

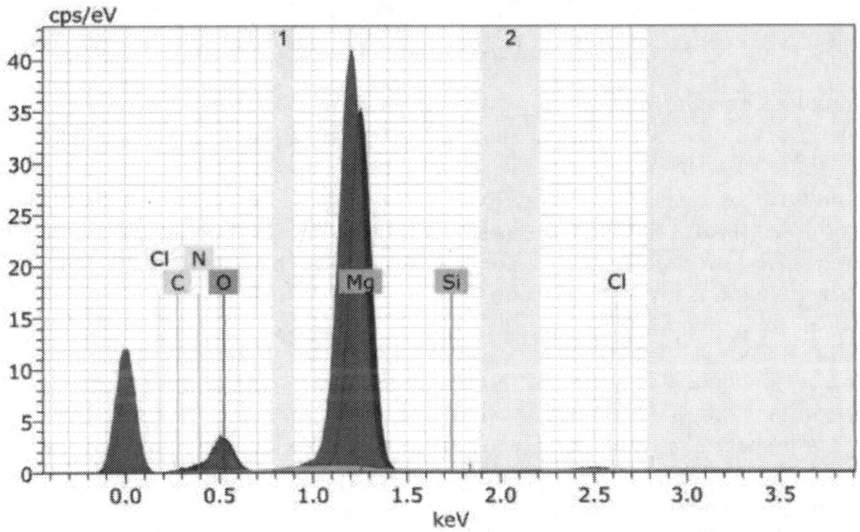

Figure 3. EDAX for Friction stir processed Mg matrix reinforced with 10%SiC

3.2 XRD analysis

Diffraction pattern was composed using BRUKER D8 FOCUS (9 kW) at approximately 2 min intervals and all XRD analysis were matched with the JCPDS database available in the operating software to determine all phases present. The XRD pattern of the Friction stir processed Mg-SiC composites are shown in Figure 4. As Shown in figure the diffraction peaks at $2\theta = 33°,34°,65,80°$ can be attributed to the reflection of Mg (JCPDS card no 36-1451), $2\theta = 69°,72°,81°$ correspond to those of SiC (JCPDS file 29-1129).

IOP Conf. Series: Materials Science and Engineering **390** (2018) 012013 doi:10.1088/1757-899X/390/1/012013

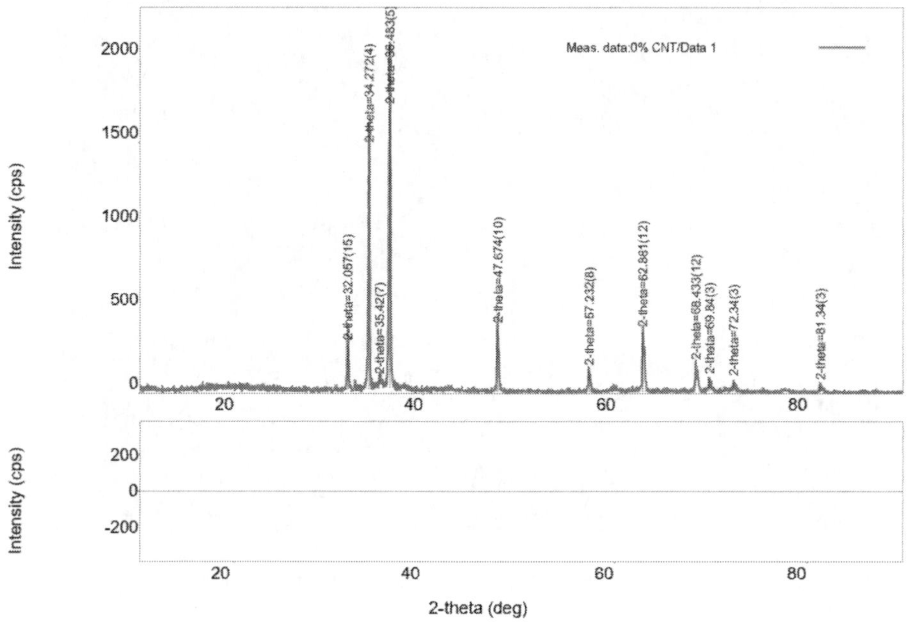

Figure 4. XRD pattern of friction stir processed Mg matrix reinforced with 10% SiC

3.3 Optimization of FSP parameters

The analysis of variance was used to set up statistically considerable FSP parameters and the percent contribution of these parameters on the micro hardness. The intention of Taguchi's quality loss function is quantitative assessment of quality loss due to practical variation. A quality characteristic is the objective of interest of a product or process. It is called as fundamental characteristic. The difference between the functional value and objective value is emphasized and identified as the loss function. The loss function is given in equation 1

$$L(Y) = \frac{L(m)(y-m)2}{2!} = K(y-m)^2 = K(MSD) \quad (1)$$

where, $L(y)$ is loss function, y is value of the quantity characteristics, m is target value of Y, K is proportionality constant and *MSD* is mean square deviation.

The category the-higher -the-better is always preferred to calculate the S/N ratio for quality characteristics of micro hardness. The equation for calculation S/N ratio for smaller the better characteristic is given in equation 3 (in decibels)

$$S/N = -10*\log\left(\Sigma(1/Y^2)/n\right) \quad (2)$$

Where Y = responses for the given factor level combination and n = number of responses in the factor level combination.

Despite of type of the performance characteristics, a greater S/N ratio value communicates to improved performance. Therefore, the optimal level of the machining parameters is the level with the greatest S/N ratio value. By applying the Eqs. 1 and 2, the S/N ratio values for each experiment of L_9 (Table 3) were calculated. Figure 5 shows mean S/n ratio for micro hardness. Based on the analysis of S/N ratio, the optimal machining performance for the Micro hardness was obtained at 800 rpm speed (level 3), 50 mm/min transverse feed (level 3), 5 mm of axial depth (level 3) and 10% of SiC (level 2). The comparative significance among the FSP parameters for the micro hardness was investigated by using the ANOVA method so that optimal combinations of the cutting parameter levels can be determined more accurately. The results of ANOVA are shown in Table 4. Larger sum of square indicates that the variation of the process parameter makes a big change on the performance characteristics. Percent contribution indicates the relative power of a factor to reduce variation. For a factor with a high percent contribution, a small variation will have a great influence on the micro hardness. It is clear that the most effect on the micro hardness is wt % of SiC followed by Speed and axial depth.

IOP Conf. Series: Materials Science and Engineering **390** (2018) 012013 doi:10.1088/1757-899X/390/1/012013

Table 3. S/N ratio for experimental data

Ex no	S/N ratio
1	36.28495
2	36.90196
3	37.0496
4	36.41716
5	36.33808
6	37.07396
7	38.0618
8	37.33756
9	36.82719

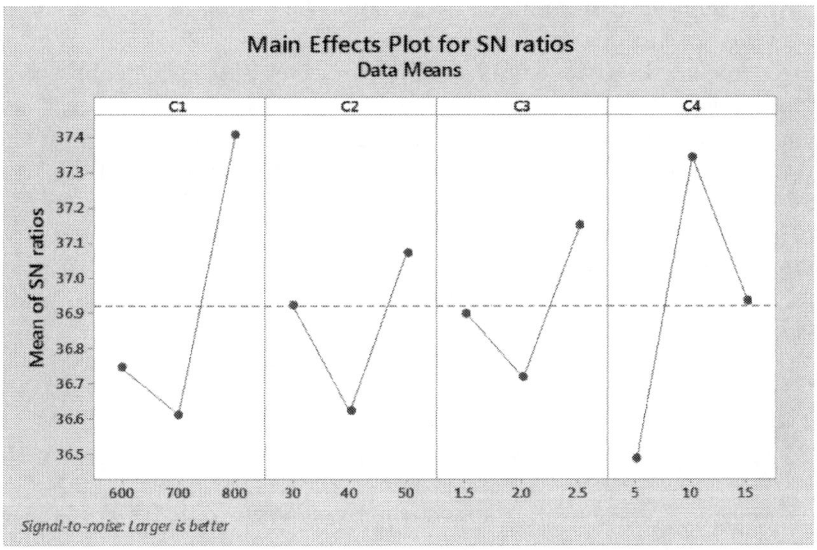

Figure 5. Mean S/N ratio for micro hardness

Table 4. ANOVA for S/N ratio

Term	DOF	Sum Square	Mean Square	% Contribution
Speed	2	1.09705	0.55	43.49482776
Transverse speed	2	0.023207	0.012	0.920083474
Axial depth	2	0.285328	0.14	11.31242496
wt % of SiC	2	1.116669	0.56	44.2726638

3.4 Confirmation experiments

The estimated/predicted S/N ratio, using the optimal level of the FSP parameters, can be calculated from following equation

$$\eta predicted = \eta m + (\eta 0 - \eta m)$$

where η_m is the total mean of the S/N ratio, $\eta 0$ is the mean S/N ratio at optimal level and N is the number of main design parameters that affect the performance characteristics. Table 5 shows the comparisons of predicted and actual FSP performance for micro hardness using their optimal cutting parameters. The analysis of result (mean, from Fig. 3) indicates that the optimal solution is obtained at $S_3f_3d_3w_2$ (Micro hardness=37.01VHN).

Table 5 Comparison between FSP performance using the initial and optimal level

	Initial parameter	Optimal FSP parameter	
		Prediction	Experimental
Setting level	$S_1f_1d_1w_1$	$S_3f_3d_3w_2$	$S_3f_3d_3w_2$
Micro hardness VHN	65.2	71.2	72.5
S/n ratio	36.28	37.01	37.25

4. Conclusions

The results from the experimental investigation of FSP of SiC reinforced Mg matrix surface composites can be summarized as follows

- FSP of magnesium-SiC composites was accomplished using different combinations of tool rotational, transverse feed, axial depth and wt% SiC.
- Taguchi model was used to optimize FSP parameters by considering micro-hardness as performance indicator.
- An optimum parameter combination for the MAXIMUM micro hardness is obtained by using the analysis of S/N ratio. Based on the analysis of S/N ratio, the optimal FSP performance for the micro hardness is obtained between 800 rpm speed (level 3), 50 mm/min transverse feed (level 3), 5 mm of axial depth (level 3) and 10% of SiC (level 2).
- Maximum Micro hardness was obtained at ANOVA analysis was carried out to find importance of FSP parameters. wt % of SiC followed by Speed and axial depth on micro hardness.

5.Reference

[1] Song G L 2007 *Corros Sci.* **49** 1696–701.
[2] Mishra, R S, Mahoney M W, McFadden S X, Mara N A and Mukherjee A K 1999 *Scripta Materialia* **42** 163-168
[3] Valiev R Z, Korznikov A V and Mulyukov R R 1993 *Mater. Sci. Eng. A* **168** 141-148.
[4] Ma Z.Y 2008 *Met. Mat. Trans. A* **39** 642-658.
[5] Sabirov I, Murashkin M Y and Valiev R Z 2013 *Mater. Sci. Eng. A* **560** 1-24.
[6] Mishra R S, Ma Z Y and Charit I 2003 *Mater Sci Eng A* **341** 307–10.
[7] Morisada Y, Fujii H, Nagaoka T, Nogi K and Fukusumi M 2007 *Composites A.* **38** 2097–2101.
[8] Lee W B, Lee C Y, Kim M K, Yoon J I, Kim Y J, Yoen Y M and Jung S B 2006 *Compos Sci Technol.* **66** 1513–20.
[9] Dolatkhah A, Golbabaei P, Givi M K B and Molaiekiya F 2012 *Mater. Design* **37** 458-464.
[10] Palanivel R., Koshy Mathews P, Murugan N and Dinaharan I 2012 *Mater. Design* **40** 7-16.
[11] Mishra R S and Ma Z Y 2005 *Mater. Sci. Eng. R: Reports* **50** 1-78.
[12] Xue P, Xiao B L and Ma Z Y 2014 *Mater. Design* **56** 848-851.
[13] Su T, Nelson and Sterling C 2005 *Scripta Materialia* **52** 135-140.
[14] Mahoney W, Bingel S, Sharma and Mishra R 2003 Mater Sci. Forum **426-432** 2843-2848.
[15] Mishra R and Mahoney M 2004 *Scripta Materialia* **50** 931-935.

Performance Evaluation of Indian Education Sector using Interpretive Structural Modelling

Samrat Majumdar and Sushanta Tripathy
School of Mechanical Engineering,
KIIT University, Patia, Bhubaneswar, Odisha, 751024, India
E-mail: samratmajumdar26@gmail.com E-mail: sushant.tripathy@gmail.com

Abstract: In twenty first Century, the performance of the service sector is much prominent than that of the other sectors comprising to the GDP of any country. The productivity of the service sectors of any country is dependent on type of the sectors, their culture, venture capital and availability of the skilled workers. The research paper highlights the different factors affecting the service productivity in different countries with different sectors. The paper also emphasizes the various techniques used to measure the performance of different service sectors. Finally, the paper attempts to propose a unique measurement methodology of a target sector for a specific country with some meaningful extracted key factors.

Keywords: Service sector productivity, different countries, key factors affecting, performance measurement methods.

1. **Introduction:** In this 21st Century, service sector is highly dominating part of gross domestic product (G.D.P.) for any country. It has remained the distinctive sector when it comes to the contributions to national and state incomes. Its notable contributions are also found in employment, foreign direct investment (F.D.I.) inflows and also trade flows. In an economic survey carried out during 2015-16 it was observed that 66.1 percent of the Indian G.D.P. growth obtained was from the Indian service sector. It was also interpreted that service sector made up 28 percent of the total employment for that year and 51 percent of total F.D.I inflows in India. Thus, it was evaluated from the observations that service sector has a high potential as a net foreign exchange earner and is very attractive for F.D.I. The present growth rate of the service sector in India is about 10 percent and also India has ranked among the top ten service exporters securing the number 8 spot. The service sector is considered to be the tertiary sector of Indian economy while the primary sector is associated with agriculture and fisheries and the second sector mainly highlighting the manufacturing domain. The services provided are not really physical in nature and consists of hospitality, banking and insurance services, retail, logistics, education, healthcare, community services, real estate, broadcasting services, defense, business services, tourism,

information technology, consultancy services, research and development services and shipping services.

2. Literature Review:

Table 2.1: Tabular representation of the studied papers

Author	Country	Sector	Model
Lee and Worthington (2016)	Australia	Education	DEA
Prajogo and McDermott (2011)	Australia	Retail	IPA matrix
Gomes and Andrade (2012)	Brazil	HR	AHP
Zareinejad (2014)	Brazil	Education	SERVQUAL
Zsidóa et al. (2015)	Hungary	Agriculture	DEA
Islam & Rasad (2006)	Hawaii	HR	AHP
Zhang et al. (2013)	China	Logistics	AHP
Shaverdia et al. (2014)	Iran	OIL	AHP
Yarmohammadian et al. (2011)	Iran	Education	SERVQUAL
Jowkar et al. (2015)	Iran	Supply Chain	ISM
Naidoo et al. 2010	Mauritius	Public Service	SERVQUAL
Gabriel et al. (2013)	Kenya	Hr	Regression Analysis
Wadongo et al. (2010)	Kenya	Hospitality	Custom Factorial Univariate Analysis of Variance
Ukko et al. (2012)	Finland	HR	REVIEW PAPER
Lozano (2016)	Spain	Banking	DEA
Kazan and Bayder (2013)	Turkey	Banking	DEA
Elango and Wieland (2015)	USA	All Sectors	AHP
Shaout and Yousif (2014)	USA	HR	Fuzzy, TOPSIS, FAHP, MCDM
Tyagi and Gupta (2013)	USA	IT	Balanced scorecard
Keebler and Plank (2009)	USA	Retail	Triangulation approach
Dey et al. (2008)	United Kingdom	Healthcare	AHP
Longbottom et al. (2011)	United Kingdom	All Sectors	SERVQUAL
McAdam et al. (2008)	United Kingdom	All Sectors	Performance measurement & lead benchmarking
Wilderom et al. (2012)	Netherlands	Public Service	PMS
Nitirath and Bangchokdee(2011)	Thailand	Finance	Path Analysis
Hamali et al. (2011)	Thailand	Banking	SERVQUAL
Shang et al. (2008)	Taiwan	Hospitality	DEA
Karimi et al. (2011)	Pakistan	HR	Regression Analysis
George et al. (2008)	India	Railways	DEA
Bhanot and Singh (2012)	India	Logistics	DEA
Masson et al. (2016)	India	All Sectors	DEA
Kumar and Gulati (2009)	India	Banking	DEA
Jothimani and Sarmah (2013)	India	Retail	KPI, SCOR model, FAHP and TOPSIS
V. and Manohar (2015)	India	HR	Regression Analysis

Tripathi et al. (2011)	India	HR	Regression Analysis
Katiyar et al. (2015)	India	Retail	Fuzzy cross-impact matrix multiplication
Satapathy (2014)	India	Retail	Structural equation modelling
Saad and Patel (2006)	India	Retail	Factor Analysis
Ranjan et al. (2016)	India	Railways	DEMATEL and VIKOR methods
Kiran Banga Chhokar (2010)	India	Education	Review Paper
Chakraborty et al. (2015)	India	Tea Industry	ISM
Abdullah (2005)	India	Education	HEdPERF, SERVPERF
Aich & Tripathy (2014)	India	Supply Chain	ISM
Arambewela & Hall (2009)	India	Education	ANOVA
Shankar et al. (2012)	India	Hospital	ISM
Dev et al. (2012)	India	Supply Chain	ISM
Gunasekaran et al. (2012)	India	Logistics	ISM
Dubey et al. (2016)	India	Utility	ISM
Kannan et al. (2012)	India	Logistics	ISM
Jayalakshmi and V. (2013)	India	IT	ISM
Sharma et al. (2013)	India	All Sectors	ISM
Tripathy et al. (2012)	India	R&D	ISM
Prashant et al. (2012)	India	Supply Chain	ISM
Joshi (2012)	India	Education	Review Paper
Khan and Haleem (2015)	India	HR	ISM and fuzzy-Micmac modelling
Nangia et al. (2016)	India	Education	ISM
Sohani and Sohani (2012)	India	Education	ISM
Ravi and Shankar (2017)	India	Supply Chain	ISM
Karunes et al. (2016)	India	Education	TQM
Lokachari et al. (2010)	India	Education	TQM
Arulraj and Senthilkumar (2011)	India	Education	SQM
Mahapatra and Sreekumar (2011)	India	Education	DEA
Thakkar et al. (2007)	India	Supply Chain	ISM
Thakkar et al. (2008)	India	IT	ISM
Tripathy et al. (2013)	India	R&D	ISM
Umashankar and Dutta (2007)	India	Education	Balanced Scorecard
Yeravdekar and Behl (2017)	India	Education	ISM

Identified factors: From studying the above-mentioned papers, the factors affecting any educational institution are discussed below.

- *Modern Facilities (MF):* Modern facilities are an integral part of every educational institutes now days it aims for better convenience of the students and also helps them develop themselves. Modern facilities often include digital library, Wi-Fi, classroom accessories, etc.

- *Physical facilities and infrastructure (PFL):* Physical facilities and infrastructure includes a well-equipped lab, lay grounds, recreational areas, clubs and associations, modern canteens, etc. It is a vital factor which can directly influence the brand value of an institute.
- *Pre-placement training (PPT):* Now a day it is very crucial to receive a pre-placement training course before the commencement of campusing. This may consist of training sessions as well as corporate talks and visits. It helps the student prepare and have a greater oppurtunity of getting placed as well as it can simulate the environment that goes on during the tests and interview. This can motivate the student leading to a higher place ability.
- *Accommodation (A):* The quality of accommodation as well the facilities available with the accommodation plays a crucial part in the performance of education institutes and can indirectly affect the brand name of the institute.
- *Safety and security (SS):* In today's world safety and security is a very important aspect for the students. It can directly create effects on the brand name of the institute. The tighter the safety and security of an institution the higher will be the brand value and therefore the performance.
- *Brand name (BN):* It is a major factor that denotes the performance of the educational institutes on the basis of other direct and indirect factors like infrastructure, safety, etc.
- *Quality enhancement (QE):* Quality enhancement plays a major role in improving the brand name of the institute and also in improving the brand name. It can be done by optimizing various factors like placement, curriculum, infrastructure, etc.
- *Management vision (MV):* The accuracy and aim of management vision results in better efficiency achievement, better brand name, better alumni interaction, better employee motivation and also improving the efficiency of the students.
- *Relevant curriculum (RC):* A relevant curriculum helps the students become more efficient and they can performance more with more knowledge. It can directly affect the performance of the faculty too apart from influencing the quantitative and qualitative results and at the end industry readiness.
- *Staff motivation and development (SMD):* Staff motivation and development can be linked with providing welfare facilities, a better infrastructure, PBAS and encouraging professional development.
- *Quality lecture (QL):* Quality lecture can also be related to relevant curriculum and quantitative and qualitative results apart from influencing the performance of students and the brand name of the institute.
- *Industry interface (II):* Industry interface aids in improving the placements of the institute and also it helps in the self-development of the students by providing various courses that can enhance the knowledge and industry readiness.
- *Alumni focus (AF):* Alumnis are considered to be assets of an institute if the connection between educational institutes and alumnis are well established they can directly affect the placement of the institute along with various other developments that enhances the brand name.

3. **Proposed methodology: Interpretive Structural Modelling (ISM)** Interpretive structural modelling (ISM) is an established methodology to identify the interdependency among selected factors among higher technical education. Chakraborty et al. (2015) used ISM to evaluate the Indian tea industry with critical factors like wages, health & safety issues, housing issues, discrimination, labour casualisation, child labour, worker representation, etc. While Aich and Tripathy (2014), Dev et al. (2012), Prasant et al. (2012), Ravi and Shankar (2017), Thakkar et al. (2007) decided to use ISM as a tool for the scrutiny of the Indian supply chain sector where the key factors were green supply chain management, green procurement, environmental accounting,

green design, green manufacturing, green logistics, economic performance, environmental performance, management vision and direction, customer support, operational performance, supplier management, trust, adaptation, commitment, dependence, interdependence, co-operation, distributive justice, procedural justice, relational behaviour, satisfaction, mutuality, transparency, etc. Shankar et al. (2012) used ISM for a model generation between different critical factors like ERP, small incision cataract surgery, extracapsular cataract extraction, intrascapular cataract extraction, endophthalmitis, high pressure in the eye, drooping eyelid, detachment of the retina, choroidal haemorrhage, etc. Gunasekaran et al. (2012) and Kannan et al. (2012) focussed on the logistic sector of India and used ISM with factors like transitional processing , and delivery , reclaim , recycle , remanufacture , reuse and disposal , effective communication , service improvement, cost saving, and overall working relations , customer satisfaction , frequent updating , profitability , and employee morale , quality , cost , time, flexibility , customer satisfaction, and service , warehouse management , order management , supply chain planning , lack of application and knowledge of advanced information technology, lack of qualification for employees in TPL, environmental issues, etc. Dubey et al. (2016) focussed on India's utility area using ISM considering factors like poor waste management planning, wrong selection of waste to energy technology, wrong selection of location, poor contract management, no benchmarking to assess efficiency of services, incomplete legislation and insufficient enforcement, etc. The Indian education sector was discussed using ISM by Yeravdekar and Behl (2017), Sohani and Sohani (2012), Nangia et al. (2016), Debnath and Shankar (2017) using factors like strategy orientation of regulatory body, curriculum design and development, academic flexibility, teaching learning process, teacher quality, evaluation process and reformation, research and promotion, extension activities and collaboration, physical facility and infrastructure, library and learning, student support and progression, vision and leadership, finance and resource management, quality enhancement, stake holder relationship, competitiveness of entrance exam, fees of the course, diversity of courses, extra co-curricular activities, diversity in background of students, advertisements, etc.

4. Research Methodology:

4.1*Questionnaire design:* Questionnaire development being the most vital part of the present research work, utmost care was taken to design the same after thorough discussion with the quality experts. For the ease of filling by the target, the template was kept as simple as possible with close-ended questions only. The questions were clearly laid out which in turn made the respondents feel comfortable to be a part of the survey and give their precious views on the critical factors affecting the performance of their institutions. Form the questionnaires reviews we got some factors as inputs and outputs of education sector and on the basis of that a model is generated indicating the factors critically affecting the performance of education sector in India by applying ISM mathematical models.

4.2. *Structural self-interaction matrix development:* The contextual relationship among the 13 critical factors is developed with the help of 6 experts from the field of education. To express the relationship between key success factors, mainly four symbols are used for denotation of the direction of relationship between the factors *i* and *j* where it is assumed $i< j$. *The symbols are as follows:*

A: parameter *j* will lead to *I*, V: parameter *i* will lead to *j*, X: parameters *i* and *j* will lead to each other, O: parameters *i* and *j* are unrelated. Considering the contextual relationships above, a structural self-interaction matrix (SSIM) is developed in the table below.

Table 4.1. Structural Self-Interaction Matrix

	BN	QE	MF	SS	QL	A	RC	II	PPT	SMD	PFL	AL	MV
BN	1	A	A	A	A	A	A	A	A	A	A	A	A
QE		1	A	A	A	A	A	A	A	A	A	A	A
MF			1	O	O	O	O	O	O	O	A	O	A
SS				1	O	O	O	O	O	O	O	A	A
QL					1	X	O	O	O	A	O	O	A
A						1	O	O	O	A	O	O	A
RC							1	X	X	O	O	A	A
II								1	X	O	O	A	A
PPT									1	O	O	A	A
SMD										1	O	A	A
PFL											1	O	A
AL												1	A
MV													1

4.3. *Reachability matrix construction:* From the contextual relationships, the SSIM is constructed. Following this, conversion of SSIM into a binary matrix which is also termed as the initial reachability matrix by replacing A, V, X,0 by 1 and 0. The substitution is done on the basis of the following rules:

1. In the event that the (i, j) value in SSIM ends up being V, at that point the (I, j) in the reachability matrix is substituted by 1 and the (j, I) is substituted by 0.
2. In the event that the (I, j) in SSIM ends up being A, then at that point the (I, j) in the reachability matrix is substituted by 0 and the (j, I) is substituted by 1.
3. In the event that the (I, j) in SSIM ends up being X, at that point the (I, j) in the reachability matrix is substituted by 1 and the (j, I) is substituted by 0.
4. In the event that the (I, j) value in SSIM ends up being O, at that point the (I, j) in the reachability network is substituted by 0 and the (j, i) is substituted by 0.

Table 4.2. Reachability matrix

	BN	QE	MF	SS	QL	A	RC	II	PPT	SMD	PFL	AL	MV	Driving power
BN	1	0	0	0	0	0	0	0	0	0	0	0	0	1
QE	1	1	0	0	0	0	0	0	0	0	0	0	0	2
MF	1	1	1	0	0	0	0	0	0	0	0	0	0	3
SS	1	1	0	1	0	0	0	0	0	0	0	0	0	3
QL	1	1	0	0	1	1	0	0	0	0	0	0	0	4
A	1	1	0	0	1	1	0	0	0	0	0	0	0	4
RC	1	1	0	0	0	0	1	1	1	0	0	0	0	5
II	1	1	0	0	0	0	1	1	1	0	0	0	0	5
PPT	1	1	0	0	0	0	1	1	1	0	0	0	0	5
SMD	1	1	0	0	1	1	0	0	0	1	0	0	0	5
PFL	1	1	1	0	0	0	0	0	0	0	1	0	0	4
AL	1	1	0	1	0	0	1	1	1	0	0	1	0	7
MV	1	1	1	1	1	1	1	1	1	1	1	1	1	13
Dependency	13	12	3	3	3	4	5	5	5	2	2	2	1	

Reachability matrix is developed for the critical factors by using the above rules. For both dependent and independent of each variable the driving power is also visualized. The driving power for each variable is the total number of variables, which it may help to achieve. Dependence is therefore the total number of variables which may help in achieving it. The driving power and dependency can be used to classify the variables into four groups namely autonomous, dependent, linkage and independent driver variables

5.Result and discussion: In this research work an attempt is being made to identify the various critical factors affecting performance of higher technical education through questionnaire survey. Among the cluster of factors to establish interrelationship, to understand the driver and dependence a systematic approach "Hierarchy based ISM" is proposed. In this model, the 13 operational factors which are base of improvement in performance of education sectors are well structured to interrelate & categorized them into Autonomous, Driver, Dependent, & Linkage variables. Finally, from the ISM model, it can be concluded that management vision drive factors like alumni focus, infrastructure and facilities, staff motivation. Again, alumni focus can control factors like pre-placement training, relevant curriculum and industry interface which are interdependent and influence each other directly. In case of infrastructure and facilities, the factors controlled are modern facilities and safety and security which are not inter dependent and not affect each other. Staff motivation can drive factors like quality lecture, etc. This all contribute to the overall quality enhancement and there by directly influencing the brand performance or the brand name of the institute.

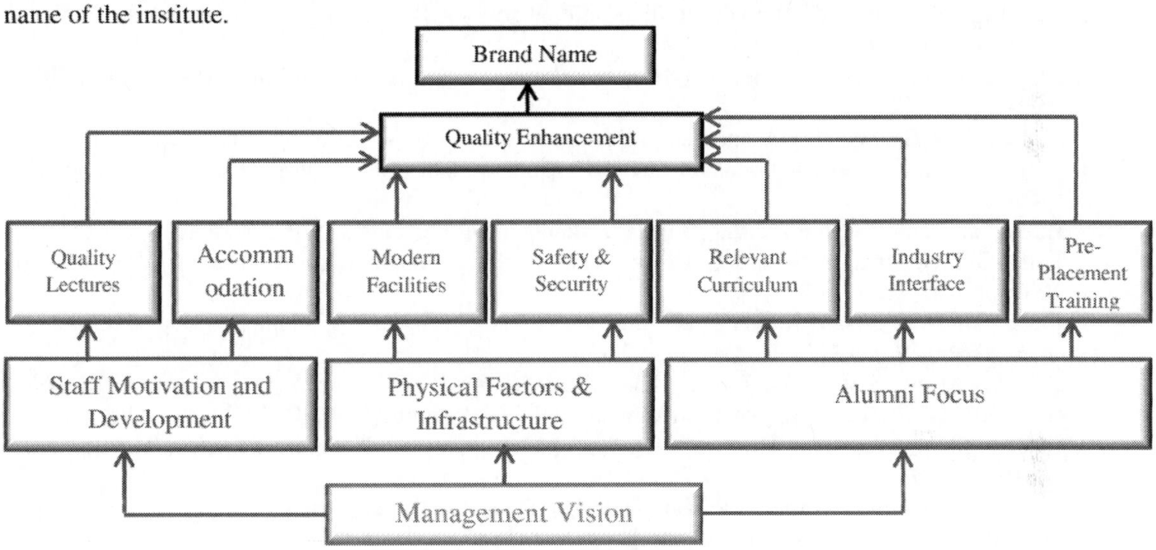

Figure 1. Interpretive Structural Model

6.Conclusion: This research was done taking data from a few numbers of colleges with limited number of operational factors. If more number of colleges are focused on in this study along with further incorporation of strategic factors, a better ISM model can be proposed with a clear understanding of the relationships among the factors and how they affect the performance. Along with the technical side of performance evaluation management side can be incorporated also giving this study a new dimension.

7.References

1. Adnan Shaout and Mohamed Khalid Yousif,"International Journal of Computer Sciences & Information Technology (IJCSIT), Vol 6 No 4, 2014.
2. Ahmed Ali Foroughi Abaria, Mohammad Hossein Yarmohammadian and Mina Estekic, Procedia Social and Behavioral Sciences, Vol 15, 2011, pp. 2299-2304.
3. Ajit Kumar Nigam, Sonia Nongmaithem, Sudeep Sharma, Nachiketa Tripathi, Journal of Indian Business Research, Vol 3, Iss 3, 2011,pp 148-167
4. Aki Jääskeläinen, Harri Laihonen, Antii Lönnqvist, Miikka Palvalin, Virpi Sillanpää, Sanna Pekkola, Juhani Ukko, Measuring Business Excellence, Vol 16 Iss 1,2012, pp. 43-52.
5. Ali Diabat Abdallah Khreishah Govindan Kannan Vinay Panikar and Angappa Gunasekaran ,Benchmarking: An International Journal, Vol. 20 Iss 6 , 2013,pp. 805 – 824.

6. Begum Sayeda, Chandrasekharan Rajendran and Prakash Sai Lokachari, Benchmarking: An International Journal, Vol. 17 Iss 5, 2010 pp. 728 – 767.
7. Bianca A.C. Groen, Mirthe van de Belt, Celeste P.M. Wilderom, International Journal of Productivity and Performance Management, Vol 61, 2012, Iss, pp. 839-862.
8. Billy Wadongo, Edwin Odhuno, Oscar Kambona, Lucas Othuon, Benchmarking: An International Journal, Vol 17 Iss 6, 2010, pp. 858-875.
9. Boon L. Lee and Andrew Worthington, Omega: The International Journal of Management Science, Vol 60, 2016, pp. 26-33.
10. B. Elango and Jamie R. Weiland, Journal of Service Management, Vol 26 Iss 4, 2015, pp. 588-607.
11. Daniel I. Prajogo Peggy McDermott, Managing Service Quality: An International Journal, Vol. 21 Iss 5 , 2011,pp. 465 – 483.
12. David Longbottom and Julie Hilton, International Journal of Quality and Service Sciences, Vol 3 Iss 1, 2011, pp. 39-59.
13. Dhanya Jothimani and S.P. Sarmah, Benchmarking: An International Journal, Vol 21 Iss 6, 2014, pp.944-963.
14. Firdaus Abdullah, Quality Assurance in Education, Vol 13 Iss 4, 2005, pp. 305 – 328.
15. Gautam Rajkhowa, International Journal of Organizational Analysis, Vol. 21 Iss 3 , 2013, pp. 471 – 484.
16. Halim Kazan and Mevlut Baydar, Business Management Dynamics, Vol.3 No.5,2013, pp. 37-50.
17. James S. Keebler and Richard E. Plank, Benchmarking: An International Journal, Vol 16 Iss 6, 2009, pp. 785-798.
18. Jitesh Thakkar, Arun Kanda and S.G. Deshmukh, Journal of Manufacturing Technology Management, Vol. 19 No. 1, 2007, pp. 92-124.
19. Jocelyn Mary Seruya Omusebe, Dr. Kimanichege Gabriel and Dr. Musiega Douglas, (International Journal of Innovative Research and Development, Vol 2 Iss 9, 2013, pp. 51-57.
20. Jui-Kou Shang, Wei-Ting Hung and Fei-Ching Wang, Applied Economics Letters, Vol 15, 2008 pp. 1053-1057.
21. Kannan Govindan, Murugesan Palaniappan, Qinghua Zhu, Devika Kannan , International Journal of Production Economics, vol. 140, 2012, pp 204-211.
22. Kiran Banga Chhokar, International Journal of Sustainability in Higher Education, Vol. 11 Iss 2, 2010, pp. 141 – 152.
23. Lewis Hawke, International Journal of Productivity and Performance Management, Vol 61 Iss 3, 2012, pp. 310-328
24. Luiz Flávio Autran Monteiro Gomes and Renata Marques de Andrade, Pesqui. Oper, Vol 32 No 1,2012, pp. 31-54
25. Meysam Shaverdia, Mohammad Rasoul Heshmatib and Iman Ramezanic, Procedia Computer Science, Vol 31, 2014, pp. 995-1004.
26. Mohammed Saad and Bhaskar Patel, Benchmarking: An International Journal, Vol 13 Iss 1, 2006, pp. 36-53
27. Mohsen Zareinejad, Iberoamerican Journal of Industrial Engineering, Vol 6 No. 11, 2014, pp. 271-293
28. N. Senthilkumar A. Arulraj, (2011),"SQM-HEI – determination of service quality measurement of higher education in India", Journal of Modelling in Management, Vol. 6 Iss 1 pp. 60 – 78.
29. Navin K. Dev Ravi Shankar Prasanta Kumar Dey, Benchmarking: An International Journal, Vol. 21 Iss 3, 2011, pp. 386 – 411.

IOP Conf. Series: Materials Science and Engineering **390** (2018) 012014 doi:10.1088/1757-899X/390/1/012014

30. Navin K. Dev, Ravi Shankar and Kamal Arvind, International Journal of Health Care Quality Assurance. 26 No. 2, 2013, pp. 93-108
31. Neena Sohani and Nagendra Sohani, Journal of Engineering. Sc. Management Education, Vol-5 Iss 2, 2012, pp.495–501
32. Neeraj Bhanot and Harwinder Singh, Benchmarking: An International Journal, Vol 21 Iss 1,2014, pp. 101-120.
33. P. Shahabadkar, S.S. Hebbal & S. Prashant, International Journal of Industrial Engineering & Production Research, Volume 23, Number 3, 2012, pp. 195—205.
34. Poornima V. and Dr. S. John Manohar , International Journal of Science and Research (IJSR), Vol 4 Iss 3,2015, pp.1169-1174.
35. Prabha Ramseook-Munhurrun, Soolakshna D., Lukea-Bhiwajee and Perunjodi Naidoo, International Journal of Management and Marketing Research, Vol 3 No 1, 2010, pp. 37-50.
36. Prasanta K. Dey, Seetharaman Hariharan and Ozren Despic , Benchmarking: An International Journal, Iss. 15 No. 4, 2008, pp. 444-468.
37. Rafikul Islam and Shuib bin Mohd Rasad, Asia Pacific Management Review, Vol 11 No. 3, 2006, pp. 163-176.
38. Rajeev Ranjan, Prasenjit Chatterjee, Shankar Chakraborty, Benchmarking: An International Journal, Vol. 23 Iss 1, 2016, pp.78-95.
39. Rajesh Katiyar, Mukesh Kumar Barua, Purushottam and L. Meena, Benchmarking: An International Journal, Vol 22 Iss 4, 2015, pp. 665-696.
40. Roma Mitra Debnath and Ravi Shankar, Quality Assurance in Education, Vol 20 No.4, 2012, pp 387-407.
41. S. Sreekumar S.S. Mahapatra, Benchmarking: An International Journal, Vol. 18 Iss 2, 2011, pp. 221 – 239.
42. Sajeev Abraham George and Narayan Rangaraj, Benchmarking: An International Journal, Vol 15 Iss 5, 2008, pp. 599-617.
43. Sandra Moffett, Karen Anderson-Gillespie and Rodney Mc Adam, Benchmarking: An International Journal, Vol 15 Iss 4, 2008, pp. 368-381.
44. Satyabrata Aich and Sushanta Tripathy , International Journal Procurement Management, Vol. 7, No. 3, 2014, pp. 239-256.
45. Siddhant Masson, Rachit Jain, Narendra Mani, Ganesh Sajeev, Abraham George, Benchmarking: An International Journal, Vol 23 Iss 4,2016, pp. 893-915.
46. Soham Chakraborty, K. Das, S. Satapathy , International Journal of Advanced Technology in Engineering and Science, Vol 03, Special Issue No. 01, 2015, pp. 1377-1391.
47. Suchismita Satapathy, Benchmarking: An International Journal, Vol 21 Iss 6, 2014, pp. 964-986.
48. Sunil Kumar and Rachita Gulati, International Journal of Productivity and Performance Management, Vol 59 Iss 1, 2009, pp. 51-74.
49. Sushanta Tripathy Sadananda Sahu Pradip Kumar Ray, Journal of Modelling in Management, Vol 8 Iss 2 , 2013, pp. 212 – 240.
50. Sushanta Tripathy, Sadananda Sahu and Pradip Kumar Ray, Journal of Modelling in Management, Vol. 8 No. 2, 2012, pp. 212-240.
51. Suwit Srimai Nitirath Damsaman Sirilak Bangchokdee, Measuring Business Excellence, Vol 15 Iss 2,2011, pp. 57-69.
52. Swati Yeravdekar Abhishek Behl, Benchmarking: An International Journal, Vol. 24 Iss 3, 2017.
53. Urfi Khan Abid Haleem, Journal of Manufacturing Technology Management, Vol. 26 Iss 6, 2015.
54. V. Ravi, Ravi Shankar, Journal of Modelling in Management, Vol. 12 Iss 1, 2017.

IOP Conf. Series: Materials Science and Engineering **390** (2018) 012014 doi:10.1088/1757-899X/390/1/012014

55. Veronika Fenyvesa, Tibor Tarnóczia and Kinga Zsidóa, Procedia Economics and Finance, Issue 32, 2015, pp. 423-431.
56. Wang X., Chen Y. and Zhang L., Journal of Industrial Engineering and management, Vol 6 No. 4, 2013, pp. 1084-1104.

Engineering careers to enhance organizational commitment among employees in the IT industry

Arjun Ananthanarayanan and Dr.R G Priyadarshini

Amrita School of Business, Amrita Vishwa Vidyapeetham, Coimbatore, India

E-mail: arjun_ananthanarayanan@yahoo.com; rg_priyadarshini@cb.amrita.edu

Abstract. Organizations are facing tremendous pressure to enhance employee commitment and retain them in the organization. Among various factors that affect Organizational commitment, career anchors are seen as the prominent factor. It is then seen pertinent to identify the dominant and supporting career anchor that affects their commitment in organization which would help design relevant interventions to support the same. The objective of the study is to explore the relation between career anchors (as measured by Career Orientation Inventory) and Organizational commitment (as measured by Organizational Commitment Questionnaire). A survey was conducted among 100 employees working at information technology companies in India with variations in age, hierarchical levels and gender. Results of the study can be used to design appropriate work place interventions to enhance career orientation of employees and thereby increase their commitment towards the organization.

1. INTRODUCTION

In today's growing era of technology employees gain a lot of job opportunities in the market and hence it becomes difficult for the IT (Information Technology) organizations to manage and retain employees. Supporting to this there has been substantial research that has been going on for the past 20 years [1], [2], [3]. IT organizations invest heavily on their human resources in the form of recruitment, training & development and employee engagement but organizations fail to understand that the career needs of an individual employee may vary and it can influence on their career decision and their willingness to work for the organization[4], [5]. Career anchors are the key determinants of the individual's career choices that they make inside or outside the organization [4]. Career anchors help in understanding the long term needs of the individuals and measure the achievements or goals that they accomplish to fulfil their career choices.

Study on career anchors can also help organization understand the influence of career design over the commitment level of an employee [6]. In this study we try to find the influence of career anchors of the individuals over their organizational commitment. The research made by Meyer & Allen [7] reveals that organizational commitment can be measured through 3 dimensions of commitment; they are affective commitment, continuance commitment and normative commitment. According to Meyer, Allen & Smith [8] the commitment shown by an individual varies for individuals and it's associated with combination of 3 dimensions of commitment.

2. CAREER ACNHORS

According to Schien [1], [9], [10] career anchors can be seen as a self-perceived talents and values that influences a person to take career related decision in their life. Schein in 1978 in his seminal work on career anchors suggested that there are five career anchors that strongly influence the individual's career choice. They are (1) *security and stability* (2) *autonomy and independence* (3) *technical/functional competence* (4) *managerial competence* (5) *entrepreneurial creativity*. In his further research in 1990 he added three more career anchors along with the existing five anchors, namely (6) *service or dedication to a cause* (7) *pure challenge* and (8) *lifestyle*. Schien also proposed that an individual will strongly possess only one among these anchors which strongly influences their career decisions. In relevance to Schien's [11] findings Feldman and Bolino [5] reframed the 8 career anchors into 3 groups; Talent based anchors, Need based anchors and Value based anchors. Schein [4] in his further research found that an individual career anchor would differ for the same individual in his lifetime. An individual possessing a strong Autonomy/Independence career anchor might evolve into Security/Stability once the same individual gets older. Ituma [12] found that highly skilled labour in IT industry of Nigeria are motivated by a variety of career anchors Quesenberry and Trauth [13] found that an individual might possess more than one career anchors at a particular point in their lifetime. Further research by Quesenberry [14] found that Women working in IT profession show a varied variety of the career anchors which were proposed by Schein [11] and hence recommended that organizations must provide multiple career opportunities for them to have a satisfied career in their life.

3. ORGANIZATIONAL COMMITMENT

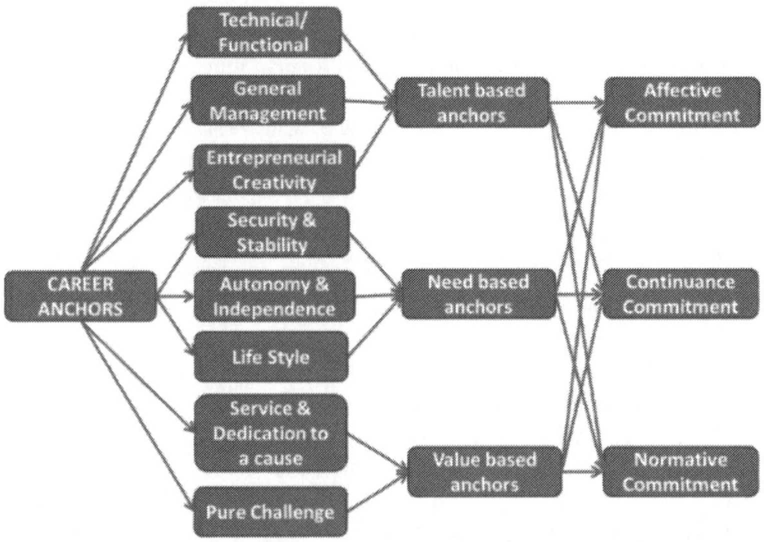

Figure 1. Relation model of career anchor and organizational commitment

Commitment can be defined as the individual's involvement in the job that they perform in the organization. Commitment can be studied in attitudinal, behavioural or motivational perspective [15]. In our study we will look at organizational commitment in attitude perspective. It can be seen in an attitude perspective because organizational commitment can be seen as a stable mind set of individuals towards the organization [16]. Meyer & Allen's [7] 3 dimensional model of organizational commitment is used to measure the organizational commitment in this study. They are (1) a*ffective commitment* (2) *continuance commitment* (3) *normative commitment*. . Normative Commitment is seen in employee when they have a high sense of obligation towards the organization. It is the behaviour of the employee to stick on to the current organization even if they feel they are not happy with the current job. Continuance Commitment can be seen when the employee analyses the pros and cons of leaving the organization. It is the behaviour of the employee to weigh the benefits or salary between the current and the new organization. Affective commitment can be defined as a strong emotional attachment of the employee towards the organization and the work they do. This type of commitment can reduce the rate of attrition in the organization.

4. RESEARCH PROPOSITION

Research proposition examines the influence of career anchors over organizational commitment among Indian IT employees.

Q1: *Does the career anchors impact the organizational commitment among employees in the Indian IT industry?*

Prior research has been done to understand the strongest career anchors influencing commitment using Schein [11] model. In this paper we will be studying the influence of Career anchors as grouped by Feldman and Bolino [5] were the career anchors are grouped into talent based anchors, need based anchors and value based anchors and to find out which among these strongly influence the organization commitment. Yousefi, Behboudi and Zarghamifard [17] found that organizational structure influences the career anchors of the employee and hence the organizations must reduce the structural formation and formalities present within the organization. With that finding in this research we will try to find the influence of the organizational structure over the career anchors

Q2: *Does the relation between career anchors and organizational commitment vary pertaining to various employee age groups in the IT industry?*

Q3: *Does the relation between career anchors and organizational commitment vary pertaining to various hierarchical levels among employees in the Indian IT industry?*

Q4: *Does the relation between career anchors and organizational commitment vary pertaining to various hierarchical levels among employees in the Indian IT industry?*

5. RESEARCH DESIGN

This is a descriptive study. Survey method is used to get the primary data. A set of questionnaire was developed based on the previous studies which is relevant to this study.
For measuring Career anchors the seminal work of Edgar Schein [11]; Career Orientation Inventory (COI) was used to collect data. It consists of 40 questions and it measures all 8 career anchors. A 6 point Likert-type scale was used with “6” representing “Always True” and “1” representing “Never True”.

For measuring organizational commitment, the Organizational Commitment Questionnaire (OCQ) designed by Jaros [18] is used. Jaros redesigned Allen & Meyer [16] questionnaire of organizational commitment. It consists of 18 questions and a 7-point scale ranging from Strongly Disagree to Strongly Agree is used.

6. SAMPLING FRAMEWORK

The research and data collection is done in the IT organizations in India.

7. ORGANIZATION PROFILE

The organization has been classified based on their existence in the industry and number of employees working in the current organization

8. RESPONDENTS PROFILE

The Respondents profile has been classified based on Gender, Age and Work Experience.

9. ANALYSIS AND FINDINGS

The ranking of the career anchor as per the average rating is seen in the Table 1. Lifestyle is seen as the dominant career anchor influencing the IT professionals. Technical Function is seen as a supportive career anchor. Table 2 shows the ranking of the commitment types as per the average rating given by the respondents .Normative commitment is seen as the dominant commitment type followed by continuance commitment.

Table 1.Rank wise representation of Career anchors

Career Anchors	Average Rating
Life Style	4.75
Technical Function	4.61
Pure Challenge	4.59
Service Dedication	4.54
Security/ Stability	4.45
Entrepreneurial Creativity	4.43
Autonomy	4.39
General Management	3.88

Table 2.Rank wise representation of Commitment types

Rank	Commitment Types	Average Rating
1	Normative	4.43
2	Continuance	4.12
3	Affective	4.08

9.1. Gender Wise

Table 3 shows the correlation of career anchors and commitment types for Men and Women. From a total of 50 Men respondents it is seen that security and stability has the highest correlation with affective commitment (0.3678) .General Managerial career anchor has high correlation with continuance commitment (0.6257) and normative commitment (0.6338). From a total of 50 Women respondents it is seen that security and stability has the highest correlation with affective commitment (0.3715) and service & dedication to a cause career anchor have high correlation with continuance commitment (0.3360) and normative commitment (0.5495).

Table 3.Correlation of Career anchors with Commitment based on gender types

Gender	Number of responses	Commitment Variable	Career Anchor	Correlation value
Men	**50**	AC	SE	0.3678
		CC	GM	0.6257
		NC	GM	0.6338
Women	**50**	AC	SE	0.3715
		CC	SV	0.3360
		NC	SV	0.5495

9.2. *Organizational Hierarchy*

Table 4 shows the correlation of Career anchors with Commitment type for junior level and middle level employees in the IT organization. From a total of 43 junior level employees it can be seen that

Security and Stability career anchor is highly correlated to Affective Commitment (0.3954). General Management career anchor is seen highly correlated to Continuance Commitment (0.3174) and Normative Commitment (0.6367). From a total of 55 middle level employees it can be seen that Autonomy is highly correlated with the Affective Commitment (0.3235), General management career anchor is seen highly correlated with Continuance Commitment (0.5002) and service & dedication to a cause career anchor is highly correlated with continuance commitment (0.5086)

Table 4.Correlation of Career anchors with Commitment based on hierarchy levels

Role	Number of responses	Commitment Variable	Career Anchor	Correlation value
Junior Level	45	AC	SE	0.3954
		CC	GM	0.3174
		NC	GM	0.6367
Middle	55	AC	AU	0.3715
		CC	GM	0.3360
		NC	SV	0.5495

9.3. Organizational Size

Table 5 shows the correlation between career anchors and commitment types based on the number of employees in the organization. For small organization were employee strength is less than 50,000. From a total of 31 respondents it can be seen that Pure Challenge career anchors is highly correlated to Affective Commitment (0.4399), Entrepreneurial Creativity is highly correlated to Continuance Commitment (0.6321) and General Management is highly correlated with Normative Commitment (0.6858). From 28 respondents in Type 2 organization where the employee strength is between 50,000 to 2 lakh it can be seen that Security and Stability career anchors is highly correlated to Affective Commitment (0.6037), Entrepreneurial Creativity is highly correlated to Continuance Commitment (0.1547) and General Management is highly correlated with Normative Commitment (0.4456).

Table 5.Correlation of Career anchors with Commitment based on employee's strength

Employee Strength	Number of responses	Commitment Variable	Career Anchor	Correlation value
<50,000	31	AC	CH	0.4399
		CC	EC	0.6321
		NC	GM	0.6858
50,000 to 2 lakh	28	AC	SE	0.6037
		CC	EC	0.1547
		NC	GM	0.4456
above 2 lakh	41	AC	SE	0.2367
		CC	GM	0.6358
		NC	GM	0.5536

From 41 respondents in Type 3 organizations where the employee strength is above 2 lakh it can be seen that Security and Stability career anchors is highly correlated to Affective Commitment (0.2367) and General Management in highly correlated with Continuance Commitment (0.6358) and Normative Commitment (0.5536)

10. DISCUSSION

Lifestyle, Technical Function and Pure Challenge are the highest rated career anchors by the gen Y IT professionals in India. That may be evident from the study done by Martin [19] who defines Generation Y IT professionals as employees who are willing to take up challenging task in their current role and also seeking to find a work life balance. Taking the gender into consideration, the gen Y women working in IT organizations who perceive to have a security and stability in their career, tend to show more emotional belonging to the organization. Apart from this, gen Y women who perceive Service and Dedication to cause career anchor tend to project higher amount of continuance and normative commitment. This is because they look forward in adding value to the work and it creates a sense of obligation within the organization. Commitment among gen Y Men could be seen influenced by their General Managerial career anchors. Research explains that many factors (dominant number of male, managing complex task, gender statistics at executive level) contribute to the sense of belonging within an IT organization [20]. It can be also seen that all the career anchors of women are positively correlated with all commitment types; this is evident from the study done by Quesenberry and Traut [13]; who found that women exhibit a combination of career anchors and it varies based on the temporal characteristics.

Taking hierarchy into consideration, junior level Gen Y IT employee who perceives to have security and stability as their dominant career anchors tend to exhibit high degree of emotional attachment towards the organization. Also junior level gen Y IT employee who perceive general managerial career anchors look to have more loyalty toward the organization (high on continuance and normative commitment). In the middle level gen Y IT employees it can be seen that people perceiving autonomy tend to have more emotional connect and thereby have high affective commitment. Continuance commitment can be seen high among gen Y employees perceiving general managerial career anchors. This clearly shows that gen Y employees have a high degree of fear in losing their job and hence look forward for opportunities to move up in the hierarchy and remain attached to the same organization. Gen Y employees with service dedication to a cause as their perceived anchors tend to exhibit more of normative commitment. They tend to be committed to the organization because they feel that their work in the organization is valuable [21].

Taking the employee strength in the organization we can see that across the three categories; people with a general managerial competency tend to be more loyal to the organization and try to stay back and grow up within the organization. Similarly we can see that continuance commitment is exhibited highly by the GenY IT employees with entrepreneurial creativity career anchor. These people tend to weight the pros and cons of leaving the organization. They can be seen as more productive if entrepreneurial opportunities are given within the organization.

11. CONCLUSION

From the study done by Holt, Marques and Way [22] it is found that the gen y workforce look forward for Challenge, Personal growth, making positive impact, success, financial stability and ability to help in their workplace. This can be seen from our study that in Gen Y IT employees Pure Challenge, Technical function and Lifestyle are the highest rated anchors. Research found that job security and lifetime employment are replaced by flexibility, creativity and innovation [23]. It can be also seen that HRM practices done within the IT organization in India have a positive correlation with the organizational commitment of the software professionals [24]. Hence the major implication from our study is that generation Y IT employees have a specific career expectation which has to be recognized by the HR partners and design a job accordingly to increase the commitment level in the employees. In the world of opportunities if you are not able to bind the job with the expectation of the people then it becomes a challenge for the organization. It is recommended that organization must use generational diversity as an opportunity to design and tailor the retention strategies [25]. In today's context it can be seen that attrition is a major challenge for the IT organizations. Hence this research suggests that HR partners design work based on the generational expectations around the job which will positively influence the commitment level among the IT employees and thereby reduce the rate of attrition in IT organizations.

12. REFERENCE

[1] Gqubule T 2006 The real retainer *Financial Mail 187*(1) p 1
[2] Ramakrishna H and Potosky D 2002 Structural shifts in career anchors of information systems personnel: A preliminary empirical analysis *Journal of Computer Information Systems 42*(2) pp 83-89
[3] Van der Merwe W 2008 Determining employees' perceptions of non-financial recognition initiatives (*Unpublished masters dissertation, University of Pretoria. Pretori).*
[4] Schein E H 1996 Career anchors revisited: Implications for career development in the 21st century *The Academy of Management Executive 10*(4) pp 80-88
[5] Feldman D C and Bolino M C 1996 Careers within careers: Reconceptualizing the nature of career anchors and their consequences *Human Resource Management Review 6*(2) pp 89-112
[6] Smit C W 1992 Using personal orientation and career anchors to predict commitment and performance in life insurance salespeople (Doctoral dissertation, University of Cape Town).
[7] Meyer J P and Allen N J 1991 A three-component conceptualization of organizational commitment *Human resource management review 1*(1) pp 61-89
[8] Meyer J P, Allen N J and Smith C A1993 Commitment to organizations and occupations: Extension and test of a three-component conceptualization *Journal of applied psychology 78*(4) p 538
[9] Schein E H 1974 Career anchors and career paths: A panel study of management school graduates
[10] Schein E H 1975 How career anchors hold executives to their career paths *Personnel 52*(3) pp 11-24
[11] Schein E H 1990 Career anchors and job/role planning: the links between career pathing and career development
[12] Ituma A 2006 The internal career: an explorative study of the career anchors of information technology workers in Nigeria *Proceedings of the 2006 ACM SIGMIS CPR conference on computer personnel research: Forty four years of computer personnel research: achievements, challenges & the future* (pp 205-212) ACM
[13] Quesenberry J L and Trauth E M 2007 What do women want?: an investigation of career anchors among women in the IT workforce *Proceedings of the 2007 ACM SIGMIS CPR conference on Computer personnel research: The global information technology workforce* (pp 122-127) ACM
[14] Quesenberry J L 2012 Re-Examining the Career Anchor Model: An Investigation of Career Values and Motivation among Women in the Information Technology Profession *Globalization, Technology Diffusion and Gender Disparity: Social Impacts of ICTs: Social Impacts of ICTs* p 169
[15] Mester C, Visser D, Roodt G and Kellerman R 2003 Leadership style and its relation to employee attitudes and behaviour *SA Journal of Industrial Psychology 29*(2) pp 72-82

[16] Allen N J and Meyer J P 1990 The measurement and antecedents of affective, continuance and normative commitment to the organization *Journal of occupational and organizational psychology 63*(1) pp 1-18

[17] Yousefi M, Behboudi M R and Zarghamifard M 2016 Examining the Impact of Organizational Structure on Career Anchors of Employees in Public and Private Organizations *International Business Management 10*(11) pp 2177-2184

[18] Jaros S 2007 Meyer and Allen model of organizational commitment: Measurement issues *The Icfai Journal of Organizational Behavior 6*(4) pp 7-25

[19] Martin C A 2005 From high maintenance to high productivity: What managers need to know about Generation Y *Industrial and commercial training 37*(1) pp 39-44

[20] Perrons D 2002 Gendered divisions in the new economy: Risks and opportunities *GeoJournal 56*(4) pp 271-280

[21] Cameron K and Dutton J eds 2003 *Positive organizational scholarship: Foundations of a new discipline* Berrett-Koehler Publishers

[22] Holt S, Marques J and Way D 2012 Bracing for the millennial workforce: Looking for ways to inspire Generation Y *Journal of Leadership, Accountability and Ethics 9*(6) p 81

[23] Naim M F and Lenkla U 2016 Knowledge sharing as an intervention for Gen Y employees' intention to stay *Industrial and Commercial Training 48*(3) pp 142-148

[24] Paul A K and Anantharaman R N 2004 Influence of HRM practices on organizational commitment: A study among software professionals in India *Human Resource Development Quarterly 15*(1) pp 77-88

[25] Bansal N and Srivastava D K 2017 Motivation & attitude of generation y in India: an exploratory study *Indian Journal of Industrial Relations 53*(1) pp 102-115

[26] Rathi N and Lee K 2015 Retaining talent by enhancing organizational prestige: An HRM strategy for employees working in the retail sector *Personnel Review 44*(4) pp 454-469

IOP Conf. Series: Materials Science and Engineering **390** (2018) 012016 doi:10.1088/1757-899X/390/1/012016

Effect of alkali treatment on tensile properties of short madar fibre reinforced polyester composites

Gunti Rajesh[1], Divya Dasari [2] and Arava Venkata Sai Kiran[3]

[123]Department of Mechanical Engineering, DVR & Dr. HS MIC College of Technology, Kanchikacherla, India

Email: rajesh.mtech@gmail.com

Abstract

Natural fibre reinforced polymer composites were fabricated with madar fibre as reinforcement and polyester as matrix. The madar fibres were treated with 10% NaOH solution (Mercerization) for surface modification. The composite samples were made with different weight proportions (2.5% to 15%) of both untreated and treated fibre in polyester according to ASTM standards using hand layup technique and tested for tensile properties. From the results, it was seen that there was improvement in the tensile strength of untreated madar fibre reinforced polyester composite to a value of 23.33MPa with increased fibre content up to 10% and with further addition of fibres there was drop in the strength. There was a significant outcome of alkali treatment on tensile strength of composite. The strength of alkali treated madar fibre reinforced composite was increased to a value of 26.23MPa with increase in the treated fibre content up to 15% which is 12.43% higher than that of untreated fibre reinforced polyester composite. There is increase in the modulus of untreated madar fibre reinforced polyester composite to a value of 493.23MPa with increased fibre loading up to 10% and with further fibre loading the value is reduced. It was also depicted that the modulus of treated fibre reinforced composite increased to a value of 444.57MPa with increased fibre loading up to 15% but the value is 9.86% less compared to that of untreated fibre composite. The % elongation of untreated fibre reinforced polyester composite decreased with increase in fibre content and was same with the treated fibre reinforced polyester composite also.

Key words: Madar fibre; Alkali Treatment; Polyester; Tensile Strength;

1. Introduction

From the past three decades the natural fibre reinforced composites have become promising substitutes to glass fiber reinforced polymer composites in several applications. It is experimentally proved that adding of natural fibres to the polymer improves strength, stiffness, thermal and acoustic insulating properties of that polymer composite and it is also found that these composites have high resistance to fracture and biodegradability[1-2]. Several natural fibres such as sisal, coir, bamboo, flax, jute, hemp etc. are used as reinforcing materials for thermosetting and thermo plastic composites [3-5].

The cellulose of natural fibre contains large quantity of hydroxyl group and gives hydrophilic property to natural fibre when utilised to reinforce hydrophobic matrices. Hence the effect is a poor resistance to moisture absorption and very poor interface leading to reduction in strength of composite [6]. To overcome such difficulties and to increase the bonding strength of fibre, chemical treatments are used. Among various treatments, treatment with alkali solution is one of the best among various treatments on natural fibres [7-9]. The significant modification done by alkali treatment is the interruption of hydrogen bonding in the network structure of fibre to increase the surface roughness [10]. It is also depicted that the treatment of natural fibre with NaOH will lead to the ionization of the hydroxyl group to alkoxide [11].

$$\mathbf{Fiber - OH + NaOH \rightarrow Fiber - O - Na + H_2O}$$

Chemical treatment on natural fibre at different curing temperatures and times were examined [12]. Jute fibres were treated with 5% NaOH solution for 2 to 8 h at 30°C. The results depicted an increase in the modulus by 79% after 8 hours of treatment respectively. Also there was fibre pullout when cured between 0 and 4 h, whereas there was transverse fracture with minimum fibre pullout during 6 and 8 hours of treatment.

Another important parameter for improving strength of composite is fibre length. The composites were made with fibres of different lengths (5, 10, 15, and 20 mm) and weight ratios (8, 12, 16, and 20%) in the epoxy resin and tensile strength was evaluated [13]. The results showed improvement in tensile strength and modulus with increase in fibre length up to 15mm and weight ration up to 12%. It was also observed that the maximum tensile strength and modulus for the fibre length of 5mm and fibre content of 12%. In another study, the composites were fabricated using kenaf fibres in unsaturated polyester resin [14]. The resin was mixed with 2.5mm and 5mm kenaf fibres with varying fibre weight fraction (2.5 and 5%). The results showed that 5 mm length of fibre in the composites has higher strength than 2.5 mm length of fibre.

Present work is aimed at fabrication of composite with untreated and alkali treated madar fibre as a reinforcement in polyester matrix. The composites were fabricated with varied weight fractions of treated and untreated fibres by means of hand layup technique, and tested for tensile properties. The effect of chemical treatment on the fibre and fibre weight fraction on the tensile properties were depicted.

2. Experimental work

2.1. Materials Used

2.1.1. Polyester

Polyester is a type of polymer which contains the ester functional in their main chain. Their well-known applications include clothing, plastic water, food packaging and carbonated soft drinks bottles. It is often known as polyethylene terephthalate (PET). polyester is purchased from Bindu Agencies, Vijayawada.

2.1.2. Catalyst

The catalyst is the agent that is mixed directly in to the polyester in a composition of 1.2ml for 100gm of resin. The name of the catalyst used for curing of polyester is Methyl Ethyl Ketone Peroxide [MEKP].

2.1.3. Accelerator

Accelerator is also directly mixed with the polyester in combination with the catalyst MEPK. The job of the accelerator is to make the process faster. The name of the accelerator used is cobalt octate ($C_{16}H_{30}CoO_4$).

2.1.4. Madar fibre

Madar fibre is a natural fibre extracted from the stem of Calotropis gigantean plant (figure 1). Calotropis gigantean is a medium sized bush that grows up to a length of 4m with usually waxy

IOP Conf. Series: Materials Science and Engineering **390** (2018) 012016 doi:10.1088/1757-899X/390/1/012016

exterior and abundant milky sap. The stem is ash colored, smooth, branching almost from the base of the plant. The leaves appear grey-green, opposite, waxy, thick and rounded-ovate.

Figure 1. Madar Plant (Calotropis Gigantea)

2.2. Methodology

2.2.1. Extraction of fibres

The Madar fibres were extracted by retting process. The stem is cut into certain small length pieces and soaked in water for nearly 20 days and later they were beaten gently with mallet to extract the fibres from the stem. Prior to the development of the composites, the fibres were washed thoroughly and cut to 5mm length.

2.2.2. Chemical treatment

The obtained fibres were soaked in 10% of aqueous sodium hydroxide solution for 6 hours. Then the fibres were washed for number of times with distilled water to eliminate any alkali adhering on their surface, then neutralized with dilute acetic acid and then finally washed again with distilled water, followed by drying at 70°C for 48 hours in hot air oven.

2.2.3. Mould preparation

The moulds were prepared with rubber sheet of 3mm thickness which is cut to a size that can accommodate 5 specimens as per the ASTM D638 standard for tensile test (figure 2).

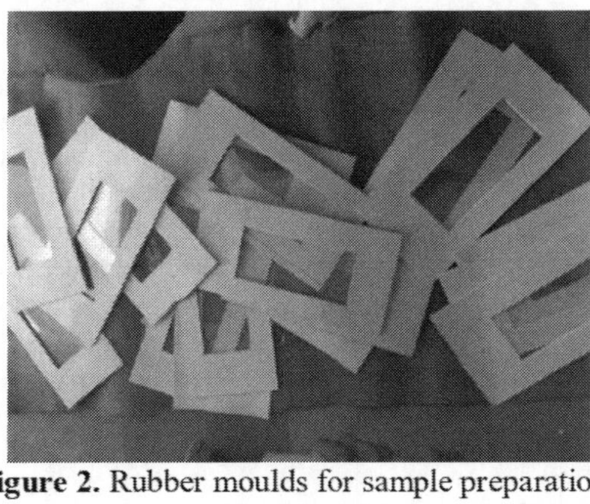

Figure 2. Rubber moulds for sample preparation

2.2.4. Specimen preparation

The samples were prepared in the moulds with untreated and alkali treated short madar fibres at different weight proportions (2.5%, 5%, 7.5%, 10%, 12.5% and 15%) in the polyester using hand layup technique. Catalyst and accelerators were also added for faster curing process. 5 samples of each weight proportions were fabricated.

2.3. Testing of composites

2.3.1. Diameter and density of fibre

Tool maker's microscope is used to measure the diameter of both treated and untreated fibres. Density of fibre is determined based on Archimedes principle which states that the volume of fluid displaced is equal to the volume of any object which is fully immersed in that fluid.

2.3.2.Tensile test

These samples are tested for tensile property on a universal testing machine as per ASTM standards (ASTM D638). The samples are loaded on the machine and the maximum load is set to 200 kg and the load is applied and the samples were tested with a crosshead speed of 0.5mm/min. The elongation and their respective loads are noted.

3. Results and discussion

3.1. Diameter and density of fibre

The physical properties such as diameter and density of untreated and alkali treated fibres were evaluated and depicted in the table 1. From the results it is seen that the alkali treatment on the fibre has reduced the diameter and density of fibres which is an indication of alkali effect of removal of hemi cellulose and lignin from the fibre surface.

Table 1. Physical properties of Madar fibre

	Diameter, mm	Density, gm/cc
Untreated Madar fibre	0.237	1.45
Alkali treated Madar fibre	0.204	0.54

From figure 3, it was depicted that the tensile strength of untreated madar fiber reinforced polyester composite was increased with increased fibre loading up to 10% and with further addition of fibres there was decrease in the tensile strength. The maximum value attained for 10% fibre loading was 23.33MPa. Also it was depicted that the strength of treated madar fibre reinforced composite was increased with increase in the fibre loading even up to 15%. The strength of treated fibre composite was 26.23MPa which is 12.43% higher compared to that of untreated fibre reinforced polyester composite.

From the past research is fact that fibres are stronger than that of the matrix material, and as there is a sufficient fibre-matrix adhesion due to the presence of modified fibre (chemical treated), it is clear that with an increase in fibre content the composite strength can be improved. Also it was observed from the literature that alkali respond on the cementing materials of the fibre mainly hemicellulose which leads to the splitting of the fibre into finer filament leading to increase in the surface area of contact for bonding with matrix leading to increased strength. However, in case of untreated fiber composite the trend of increase in strength is only up to certain volume of fibre, and after that there is deficient polymer component in the composite for providing sufficient wetting of fibres. The increase in the fibre content beyond certain limit also led to reduction in fibre dispersion and fiber-fiber interaction than fiber-matrix interaction in the composites that may have further contribution to the decrease in the strength of composite.

From figure 4, it is observed that there is increase in the modulus of untreated madar fibre reinforced polyester composite to a value of 493.23MPa with increased fibre loading up to 10% and with further fibre loading the value is reduced. Moreover the tensile modulus of treated fibre composite was increased to a value of 444.57MPa with increased fibre loading up to 15% but the value is 9.86% less compared to that of untreated fibre polyester composite.

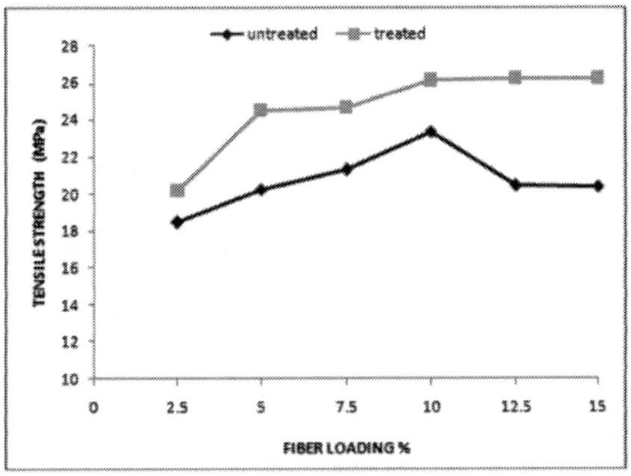

Figure 3. Variation of tensile strength of madar fibre reinforced polyester composite with fibre loading.

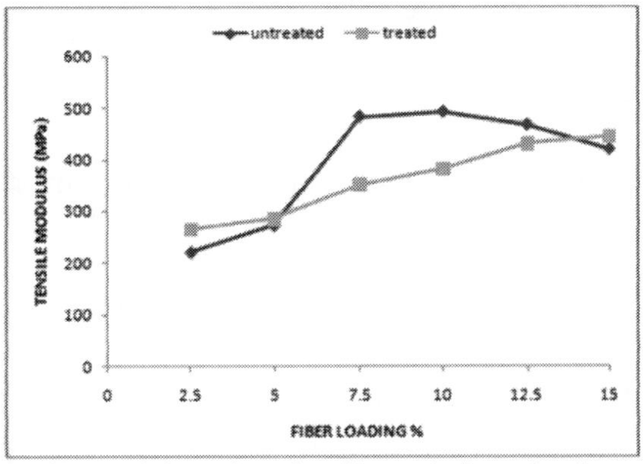

Figure 4. Variation of tensile modulus of madar fibre reinforced polyester composite with fibre loading

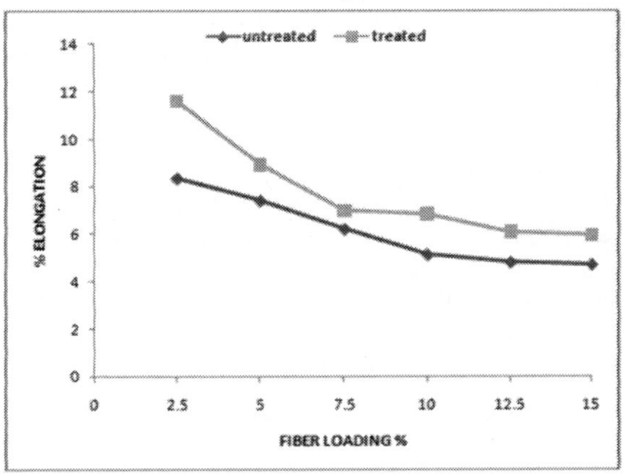

Figure 5. Variation of % elongation at break of madar fibre reinforced polyester composite with fibre loading

It is observed from figure 5 that the % Elongation at break of untreated fibres and treated fibres decreases as the fibre weight fraction increases. There is reduction in % elongation at break to 4.7% for untreated fibre reinforced composite with increase in the fibre content and is further reduced to 5.96% with addition of treated fibre in the polyester.

4. Conclusions

With the results obtained from the experimental procedure, the following conclusions are made.

- The tensile strength of untreated madar fibre reinforced polyester composite was increased with increased fibre loading up to 10% and with further addition of fibres there was decrease in the tensile strength. The maximum value attained for 10% fibre loading was 23.33MPa.

- It is also depicted that the tensile strength of treated madar fibre reinforced composite was increased to a value of 26.23MPa with increase in the fibre loading even up to 15% of treated fibre loading and the value is 12.43% high compared to that of untreated fibre reinforced polyester composite.

- There is increase in the modulus of untreated madar fibre reinforced polyester composite to a value of 493.23MPa with increased fibre loading up to 10% and with further fibre loading the value is reduced.

- Moreover the tensile modulus of treated fibre polyester composite has increased to a value of 444.57MPa with increased fibre loading up to 15% but the value is 9.86% less when compared with that of untreated fibre polyester composite.

- The % elongation of untreated fibre reinforced polyester composite decreases with increase in weight ratio and is same with the treated fibre reinforced polyester composite.

- Finally, it is concluded that, chemical treatments on fibre has produced the stronger and stiffer fibre with a good level of fibre separation leading to more surface area of contact and hence better interface with matrix. This has lead to better strength of the composite.

5. References

[1] A V Ratna Prasad and K Mohana Rao 2011 *Mater. Des.* **32** 4658
[2] C W Nguong, S N B Lee and D Sujan 2013 *Int. J. Chem. Mol. Nucl. Mater. Metall. Eng.***7** 33
[3] Thakur V K, Thakur M K and Gupta R K 2014 *Int. J. Polym. Anal. Charact.* **19** 256
[4] Cao Y and Wu Y 2008 *J. Cent. South Univ. Technol.* **15** 564
[5] Lee B H, Kim H J and Yu W R. 2009 *Fib. Polym.* **10** 83
[7] Xue Li, Æ Lope G, Tabil Æ and Satyanarayan Panigrahi 2007 *J. Polym. Environ.* **15** 25
[8] Rajesh Gunti, A V Ratna Prasad and A V S S K S Gupta 2016 *Polym. Compos.* **37** 2160
[9] Rajesh Gunti and A V Ratna Prasad 2014 *Proc. Mater. Sci.* **5** 2188
[10] Mohanty A K, Misra M and Drzal LT 2001 *Compos. Interf.* **8** 313
[11] Agrawal R, Saxena N S, Sharma K B, Thomas S and Sreekala M S 2000 *Mater. Sci. Eng: A* **277** 77
[12] Dipa Ray, B K Sarkar,A K Rana and N R Bose 2001 *Bull. Mater. Sci.* **24** 129
[13] N Venkateshwaran, A Elaya Perumal and M S Jagatheeshwaran 2011 *J Rein. Plast. Compos.* **30** 1621
[14] Gholamhossein Nodeh Farahani,Ishak Ahmad and Zeinab Mosadeghzad 2012 *Polym. Plast. Technol. Eng.***51** 634
[15] Laly A Pothan, Sabu Thomas and N R Neelakantan 1997 *J Rein. Plast. Compos.* **16** 744

Influence of morphology of anti-wear nano additives on Tribological behavior of Chemically Modified Rapeseed Oil

S Arumugam[1*], S Baskar[a], S Sankaranarayanan[b], S H Athreya[c], N Lakshmi Narayanan[d], and Shyam Sundar Devi Prasad[e]

[1, b, c, d & e] Department of Mechanical Engineering, Sri Chandrasekharendra Saraswathi Viswa Mahavidyalaya, Enathur, Kanchipuram – 631561, Tamilnadu, India.

[a]Department of Mechanical Engineering, S.A. Engineering College, Chennai-600 077.

[*1]Corresponding Author: aru_amace@yahoo.co.in

Abstract. This study compared the improvement of tribological characteristics of chemically modified rapeseed oil with two different anti-wear nano additives such as titania (TiO_2) and alumina wire (Al_2O_3) as potential bio-degradable automotive lubricant. The tribological investigations were performed using a pin-on-disc tribometer. The commercial TiO_2 nano particle of anatase and rutile phase, commercial Al_2O_3 nano wire of α and γ phase have been characterized by X-Ray diffraction (XRD), Scanning Electron Microscopy (SEM), Transmission Electron Microscopy (TEM), and Selected Area Electron Diffraction (SAED) pattern. The chemically modified rapeseed oil with fibrous nano alumina wire showed higher friction coefficient than one with spherical titania nano particles. The SEM images of the worn pin surfaces tested using spherical titania based nano lubricant is found smoother compared to that of nano alumina wire-based lubricant.

1. Introduction

In recent years, deliberate and accidental lubricant losses to the environment by means including evaporation, leakages and spills have led to major concerns regarding pollution and environmental health. Further, the depletion of the world's crude oil reserve, increasing crude oil prices, and issues related to conservation have brought about renewed interest in the use of bio-based materials as reported by Horner [1]. The present emphasis is on the use of vegetable oils for non-edible purposes. Attention has been focused by Aluyor et al. [2] and Metzger [3] on technologies that incorporate vegetable oil as biofuels and industrial lubricants due to the fact that they are renewable and bio-degradable. Vegetable oil-based lubricants and derivatives have excellent lubricity when compared to mineral oils, for which they are being investigated as a base stock for automotive lubricants. However, their widespread use in formulating lubricants is restricted owing to insufficient thermal and oxidative stability, poor low temperature fluidity and hydrolytic instability. The genetic and chemical modification of vegetable oils can overcome these limitations. Although plenty of information is available on synthesis of polymeric materials from rapeseed oil, there are few literatures on the synthesis of automotive lubricants from rapeseed oil. Unquestionably, the chemical modifications like epoxidation, hydroxylation and, esterification greatly improves thermo-oxidative stability and cold flow behaviour of rapeseed oil. Xuedong Wu et al. described the application of epoxidized rapeseed oil as a potential biodegradable lubricant [4]. Hwang and Erhan produced chemically modified soybean oil with improved low-temperature and oxidation stability by a ring-opening reaction of epoxidized fatty acid esters with alcohols followed by esterification process. They showed that the epoxidation treatment has no adverse effect on the biodegradability of the base stock. However, the anti-wear behavior of chemically modified vegetable oil is still inferior [5].

Nano technology is a revolutionary concept introduced in the 21st century that has the potential to improve the characteristics of materials in many fields. In recent years, nano-sized materials have emerged as a new additive alternative because nano materials possess many special properties, such as the quantum size effect, the small size effect, and surface and interface effects; the application of nano particles in tribology has received considerable attention. A variety of mechanisms have been proposed to explain the lubrication enhancement of the nano particle suspended lubricating oil (i.e., nano-oil), including the ball bearing effect by Wu et al; mending effect by Liu et al; and polishing effect by Tao et al [6-8].

Titanium dioxide or titania (TiO_2) is obtained from a variety of ores. The bulk material of TiO_2 is widely nominated for three main phases of rutile, anatase and brookite. Among them, the TiO_2 exists mostly as rutile and anatase phases which both of them have the tetragonal structures as reported by Kim, et al. [9]. Perry and Wu et al. have reported alumina is a hard, highly resistant towards bases and acids, allows very high temperature applications and has excellent wear resistance characteristics [10, 11]. Xian Chen et al. demonstrated that nano-TiO_2 is an effective heterogeneous catalyst for the ring-opening of epoxide with aromatic amines to afford β-amino alcohols in good to excellent yields at room temperature under solvent-free conditions [12]. Hu and Dong have reported less than 1% of TiO_2 and 0.05% of DDP-Pbs by Chen et al., nano particles is sufficient to improve tribological properties of vegetable oil [13, 14]. Yujin Hwang et al. investigated the carbon-based nano particles dispersed in mineral oil to systematically examine the effect of the size and shape of nano particles on the properties of frictional behavior [15]. Wu et al. demonstrated nano particles such as CuO, TiO_2 and nano diamond used as additives in API-SF engine oil and mineral oil exhibit good friction reduction and anti-wear behavior [16]. Recently Manu Varghese Thottackkad et al. studied the use of copper oxide (CuO) nano particles as anti-wear additives in coconut oil as possible automotive lubricants [17].

Many researchers have tried to improve the tribological characteristics of mineral and synthetic lubricants to minimize the friction and wear. However, the studies devoted to tribological improvement of vegetable oil-based lubricants with nano additives are limited. In the present investigation, the base lubricant used was chemically modified rapeseed oil, and spherical titania nano particle of particle of size ~20nm and alumina nano wire of size ~10nm were used to formulate the nano lubricants focusing on the influence of nano additives on tribological behavior of chemically modified rapeseed oil.

2. Materials & Methods

2.1. Synthesis of Nano based bio lubricant

In the present investigation, bio lubricant was formulated through chemical modification process such as epoxidation, hydroxylation and esterification process. The detailed procedure for the chemical modification of rapeseed oil was tracked from study by Arumugam et al. **[18]**. The commercially available Degussa spherical TiO_2 nano particle of size ~20nm and fibrous Al_2O_3 nano wire of size ~10nm supplied by M/s Sigma-Aldrich Ltd., Bangalore, was used to improve anti-wear characteristics of chemically modified rapeseed oil. One of the major hurdles in introducing the nano additive into oil is agglomeration, inhibiting ideal homogeneity and dispersion. An ultrasonic sonicator–OSCAR®-PR-1000, with 750W maximum power output and 20kHz operating frequency has been used to ensure homogeneous mixing and dispersion of the nano additives into chemically modified vegetable oil without agglomeration.

The size and distribution of the nano particles was verified using XRD, SEM and TEM. These nano additives were added to the chemically modified rapeseed oil by 0.05% (mass basis) to prepare the required sample and 1% of glycol was used as dispersant. Two different nano lubricants have been prepared. 1. Chemically modified rapeseed oil with 0.05% spherical TiO_2 nano particles. 2. Chemically modified rapeseed oil with 0.05% fibrous Al_2O_3 nano wire. Each lubricant sample comprised of 90% chemically modified rapeseed oil and 10% of nano additives (0.05% TiO_2 or Al_2O_3 + 9.95% glycol). The mixture was then agitated using ultrasonic sonicator at room temperature to ensure uniform dispersion and good suspension stability **[16]**. Table 1 shows the properties of chemically modified rapeseed oil with and without nano additives.

Table 1. Tested properties of lubricant

Properties	Chemically modified rapeseed oil	TiO_2 based nano lubricant	Al_2O_3 based nano lubricant
Biodegradability (OECD 301F), %	>95	>85	>85
Viscosity index (ASTM D-2270)	160	155	148
Flash point (ASTM D93), °C	240	236	225
Pour point (ASTM D97), °C	<-15	<-15	<-15
Viscosity @40°C (ASTM D445), cSt	140.20	137.2	133.8
Viscosity @100°C (ASTM D445), cSt	16.52	15.6	14.5

2.2 Tribological Investigation

The friction coefficient is an important parameter in evaluating the characteristics of nano lubricants. To establish tribological characteristics of nano-lubricant, a pin-on-disc (TR-201C) tribometer was used. A pin was made out of piston ring material and the disc was made out of cylinder liner material with hardness of 65 HRC as per the standard dimensions specified by the manufacturer of pin-on-disc tribometer. The detailed descriptions of experimental preparation, conditions have been reported in our earlier study **[19]**. All the tests were performed in accordance with ASTM G99 standard at different sliding velocities., i.e., 1-4 m/sec. The initial lubricant temperature was maintained at 35°C and load applied at the pin and disc interface was 100N.

3.Results and discussion

3.1.Morphological and structural characterization

The phase composition of the titania nano particle and alumina nano wire were determined using X-ray diffraction (Rigaku 6000, Japan) using Cu- Kα radiation at 50kV and 20mA. The peak positions and relatively intensities of the powder pattern were identified by comparison with powder diffraction file (PDF) data. The calculation on the size of nano particle/wire was based on Scherrer's equation:

$$D= \{0.9\lambda/ \beta\cos \theta\}$$

Where D is the size of the particles, λ the wavelength of the X-ray and β is the width of the base of diffraction line at half its maximum intensity. The XRD pattern of nano-TiO_2 in anatase and rutile phases are as shown in Fig. 1 (a). The XRD pattern exhibited strong diffraction peaks at 25°, 48° indicating nano TiO_2 in the anatase phase, on the other hand 27°, 36° and 55° indicating nano TiO_2 in the rutile phase. All peaks were in good agreement with the standard spectrum of JCPDS no.:88-1175. These results suggested that the nano TiO_2 is composed of amorphous structure. The amorphous revealed a broad pattern with low intensity; the effect of amorphous materials on the broadening of the XRD pattern of nano sized TiO_2 is found. Fig. 1(b) shows the XRD pattern of fibrous alumina nano wire. It can be seen from the Fig. 1 (b) that both α-phase and γ-phase Al_2O_3 were found in lubricant sample. The XRD

IOP Conf. Series: Materials Science and Engineering **390** (2018) 012017 doi:10.1088/1757-899X/390/1/012017

pattern exhibited strong diffraction peaks at 38°, 44° indicating γ-phase alumina, while the other peaks indicating alumina at α-phase. Fig. 2 and 3 shows the Scanning Electron Microscope (SEM) image of spherical TiO_2 and fibrous Al_2O_3 nano wire. From the SEM image, the distributions of both the nano particles present in bio-lubricant (chemically modified rapeseed oil) are found more uniform.

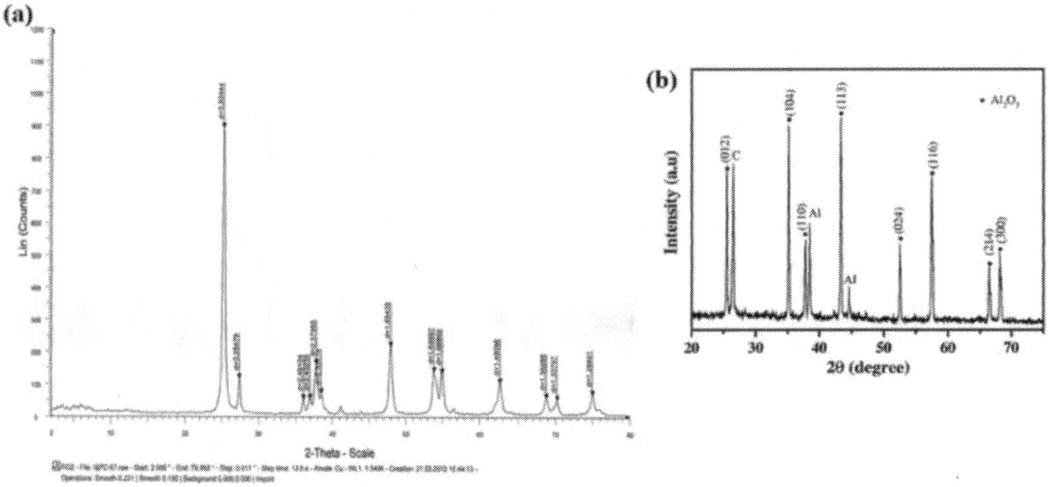

Figure 1. (a) XRD pattern of TiO_2 nano particles (b)XRD pattern of nano Al_2O_3 wire.

Figure 2. SEM image of nano TiO_2.

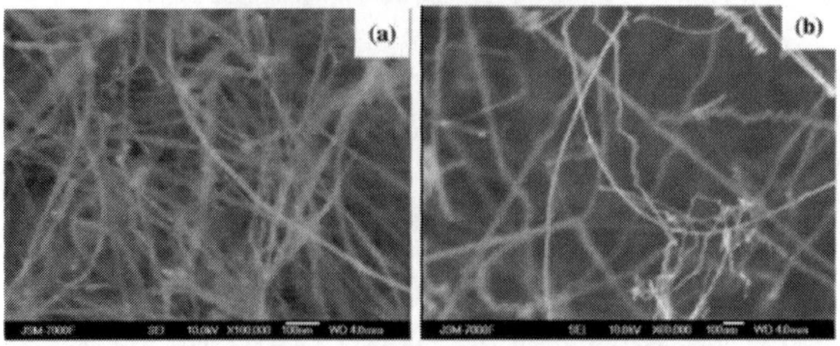

Figure 3. (a) & (b) - SEM images of nano Al_2O_3 wire.

JEOL JSM-1200 EX-Transmission Electron Microscope (TEM) was employed to examine the particle size and morphology of nano lubricants. TEM bright field images of TiO_2 nano particles in anatase and rutile phases are shown in Fig. 4. In general, the TiO_2 nano particles in rutile phase consist of both spherical and rod shapes but the anatase phase is of spherical morphology. From the TEM image, it is clearly observed that the TiO_2 particles present in chemically modified rapeseed oil is in anatase phase. Similarly, Fig. 5 (a) shows the bright field TEM image of alumina nano wire presents in chemically modified rapeseed oil. Fig. 5 (b) shows the bright field TEM image of single alumina nano wire and corresponding inner view shows the selected area electron diffraction (SAED) pattern. It reveals that the alumina nano wires are uniform and neat, with the wire diameter of about ~10nm. Also, by recording the SAED pattern of nano wires, it is evident that each individual α-Al_2O_3 nano wire is a single crystal.

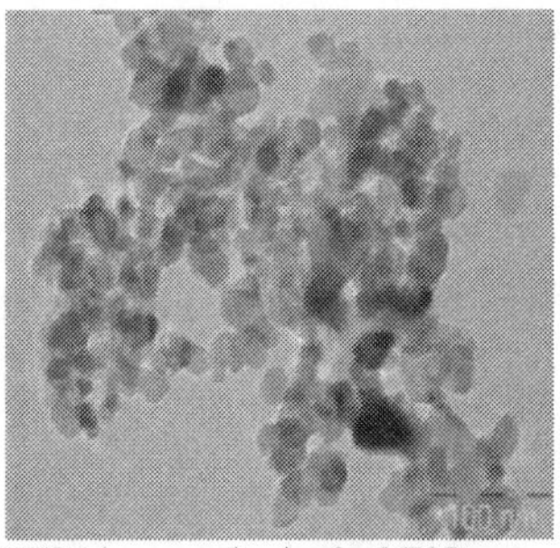

Figure 4. TEM image of spherical TiO_2 nano particles.

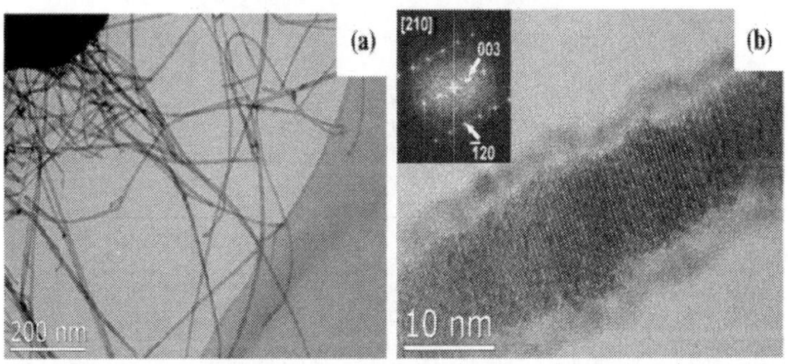

Figure 5. TEM image of nano wire a) general view b) view of one wire with SAED pattern (inner view).

3.2. Tribological characterization

Fig. 6 shows the tribological test results as a function of sliding velocity. It is observed that the friction coefficient of chemically modified rapeseed oil and nano particle included lubricants decreased gradually with an increase in sliding velocity. At higher sliding velocity, the friction coefficient was significantly decreased by 32% and 11% respectively for spherical titania based nano lubricant and fibrous alumina wire based nano lubricant compared to that of chemically modified rapeseed oil without nano additives. These results indicate that the presence of spherical titania nano particles in chemically modified oil resulted in less metal contact as compared to that of fibrous nano alumina wire. This is mainly attributed to the higher aspect ratio of fibrous nano alumina wire, resulting in more contact between the frictional surfaces. This result is also in good agreement with addition of spherical copper oxide (CuO) nano particles with coconut oil-based lubricant [17].

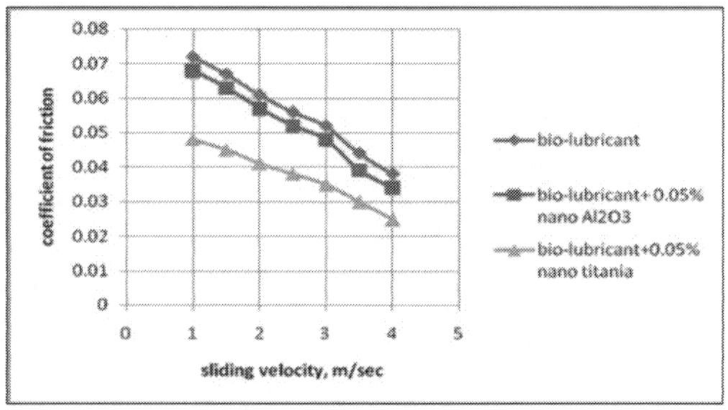

Figure 6. Variation of co-efficient of friction.

Fig. 7 shows the average temperature of the pin surface as a function of sliding velocity. The temperature is increased upto 120°C when using chemically modified rapeseed oil as bio-lubricant, whereas the average temperature of pin surface of the titania and alumina based nano lubricants were lower than that of bio-lubricant. In the case of nano based bio-lubricant, the lower friction between the disc-pin contact surfaces decreased the lubricant temperature, and also the presence of nano particles decreases the surface contact.

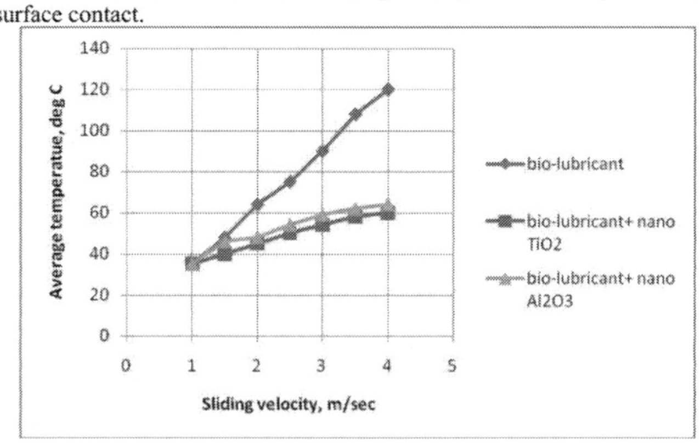

Figure 7. Variation of lubricant temperature.

Morphology of frictional surfaces lubricated by raw and nano lubricant were investigated by SEM and it is displayed in Fig. 8. Since the hardness of the pin surface is lower than the disc surface, only the morphology of pin was taken into account for the SEM analysis. Fig. 8(a) shows the SEM morphology of specimen corresponds to chemically modified rapeseed oil without nano additive. From the Fig. 8(b) it is clear that wear scar obtained is comparatively smoother while using spherical TiO_2 nano particles with chemically modified lubricant than with fibrous Al_2O_3 nano wire as shown in Fig. 8(c). This may be attributed to the fact that the spherical TiO_2 particles in lubricant at nano level develop a third body rolling effect between the sliding surfaces causes a lower wear, whereas the fibrous nano particles causes higher agglomeration, and agglomerated fibrous nano particles may not play the effect of roll or ball bearings and thus wear was somewhat higher than the spherical shaped nano particles.

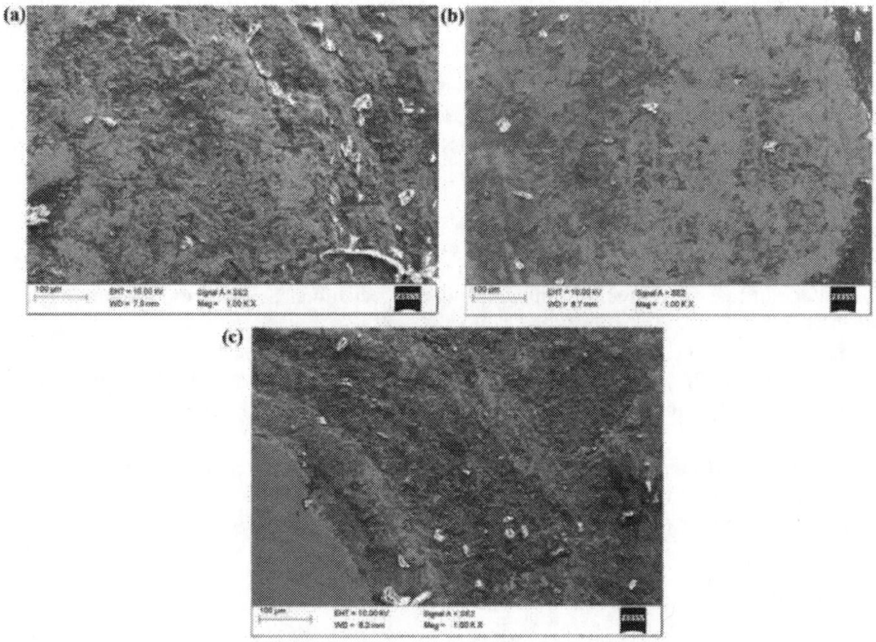

Figure 8. SEM image of the pin wear surface a) Chemically modified rapeseed oil b) titania based nano lubricant c) alumina based nano lubricant.

4.Conclusions

- The addition of spherical titania nano particles to the chemically modified rapeseed oil enhanced the lubrication characteristics as compared to fibrous nano alumina wire. This is mainly attributed to spherical titania nano particles possess zero-dimensional structure have fewer contact points than one-dimensional structured fibrous alumina nano wire.

- SEM images of the worn surfaces tested with spherical titania based nano lubricant is found smoother as compared to that of fibrous alumina nano wire based nano lubricant.

- The flash point of lubricant does not get changed with the addition of nano particles, due to the non-flammable characteristics of titania/alumina particles. Further, the addition of nano particles to the chemically modified rapeseed oil does not affect its oxidation stability and cold flow behavior.

- To sum up, the spherical shaped nano TiO_2 is found best suited as anti-wear nano additive to chemically modified rapeseed oil to improve its anti-wear and friction reduction behavior.

5.References

[1]. Horner,J. Syn. Lubr., 2002, 18, 327-347.
[2]. Aluyor, E. O., Obahiagbon, K. O., and Ori-jesu, M. Sci. Res. Essay 2009; 4: 543-548.
[3]. Metzger, J. O. European Journal of Lipid Science and Technology 2009; 111: 865-876.
[4]. Xuedong Wu., Xingang Zhang., Shengrong Yang., Haigang Chen and Dapu Wang. Chemical Society 2000; 77: 561-563.
[5]. Hwang, H., and Erhan, S. Z. Journal of American Oil Chemical Society 2001; 78: 1179-1184.
[6]. Wu, Y.Y., Tsui, W. C., and Liu, T. C. Wear 2007; 262: 819-825.

[7]. Liu, G., Li, X., Qin, B., Xing, Y., Guo, Y., and Fan, R. Tribology Letter 2004; 17(4): 961-966.
[8]. Tao, X., Jiazheng, Z., and Kang, X. Journal of Physics D Applied Physics 1996; 29: 2932-2937.
[9]. Kim, T. K., Lee, M. N., Lee, S. H., Park, Y. C., Jung, C. K., and Boo, J. H. Thin Solid Films 2005; 475: 71-177.
[10]. Perry, R. H. Chemical Engineers Handbook, 6th ed., McGraw-Hill, New York, 1984; 23.
[11]. Wu. Y. Q., Zhang, Y., Huang, X., and Guo, J. Ceramics International 2001; 27: 265-268.
[12]. Xi'an Chen., Huayue Wu., Shun Wang., and Shaoming Huang. An International Journal for Rapid Communication of Synthetic Organic Chemistry 2011.
[13]. Hu, Z. S., and Dong, J. X. Wear 1998; 216: 92-96.
[14]. Chen, S., and Liu, W. Material Research Bulletin 2001; 36: 137-143.
[15]. Yujin Hwang., Changgun Lee., Youngmin Choi., Seongir Cheong., Doohyun Kim., Kwangho Lee., Jaekeun Lee., and Hyung Kim. Journal of Mechanical Science and Technology 2011; 25: 2853-2857.
[16]. Wu, Y. Y., Tsui, W. C., and Liu, T. C. Wear 2007; 262: 819-825.
[17]. Manu Varghese Thottackkad., Rajendrakumar Krishnan Perikinalil., and Prabhakaran Nair Kumarapillai. International Journal of Precision Engineering and Manufacturing 2012; 13: 111-116.
[18]. Arumugam, S., Sriram, G., and Santhanam, V. IEEE-International Conference on Advances in Engineering, Science and Management, 2012: 25-30.
[19]. Arumugam, S., and Sriram, G. Tribology Transaction 2012; 55: 438-445.

MECHANICAL AND WEAR BEHAVIOR OF ALUMINIUM-ZIRCON SAND-FLYASH METAL MATRIX COMPOSITE

B.Vijaya Ramnath*, C.Elanchezhian, E.Naveen, P.Nagarajakrishnan, T.Chezhian, Saleem.A, and Srivalsan.M Department of Mechanical Engineering
Sri Sairam Engineering College, West Tambaram, Chennai-44.

***Corresponding author mail ID: vijayaramnathb@gmail.com**

Abstract. All Metal Matrix Composites are new materials which is a combination of molten metal called matrix and strengthening particles called reinforcement. This paper reveals the reinforcement of Zircon sand and fly ash with Aluminium alloy (LM6) with a specific ratio using stir casting technique to fabricate MMC. Here, Mechanical property namely tensile strength, hardness and Flexural strength are evaluated along with the Wear Analysis in Pin-on-Disc Apparatus under Abrasive Wear condition. The result shows that sample 4 has highest tensile strength of 170 MPa with hardness value of 57 HB. Also, the sample exhibited more wear resistance than other samples.

1.0 Introduction

Singla et al [1] found that ethanol solution at 50^0C is added while preheating the ash for the surface treatment, the addition of Mg increases the wettability which decreases the surface tension of the fly-ash. To remove the slag from the molten metal, hexachloro- Ethane tablets were added. Mohana Rao et al. [2] stated that by blowing the Argon gas at 2cc/min can minimize the high temperature oxidation problems. Also, to achieve sufficient turbulence and prevent deposition of Fly ash clusters on the crucible wall, the slurry was added at the rate of 4 mm/sec. Babu et al [3] studied that by preheating the Fly ash at 800^0C for two hours, it changes its colour from dark grey to brownish due to unburnt carbons deposition, which consists largely Silica (SiO_2), Alumina (Al_2O_3), Mullite ($3Al_2O_3.2SiO_2$) and small amount of hematite (Fe_2O_3). Okafor et al [4] studied that by increasing $ZrSi_4$ content in Aluminium during squeezing casting make the casting lesser porous and also improves the wettability and interfacial bond. By adding Zircon sand in the matrix alloy results in with elastic and plastic incompatibility due to difference in the co efficient of thermal expansion. They also observed that by adding 5-25% of Zircon sand in matrix material hardness value increases. Sunil Sharma et al [5] experimented by adding ZrO_2 in Al 6061 matrix, the percentage elongation will be reduced because of its brittleness. Kaur et al [6] stated that the reinforced composites will show less wear than the unreinforced alloy. Suchitharan et al [7] investigated the wear behavior of Al6063 and Zircon sand which was tested by pin on disc method. They found that by increasing Zircon sand percentage the wear rate was reduced and 8wt% $ZrSiO_4$ showed a high wear resistance. Sanjeev Kumar Yadav et al [8] stated that Zircon sand provides micro hardness and good internal bonding strength. By adding 50% TiC and 50% Zircon sand in ADC-12 alloy, better compressive strength was observed than the other combinations. Radhika et al [9] stated that the incorporation of graphite as primary reinforcement forms a protective layer between the pin and the counter face such that an increase in the wear resistance was observed. By adding alumina as a secondary reinforcement the significant changes in wear behavior of composite was observed. Singla et al [10] found that the sample which had 25% SiC content found to have higher hardness and toughness of 45.5 BHN and 36 N-m respectively. Beyond this weight fraction, the hardness will got decreased because of the interaction of SiC particles resulted in cluster formation. Vijaya Ramnath et al [11,12,14] investigated mechanical behavior of Aluminium metal matrix composites with Alumina-Boron Carbide, SiC and Boron Carbide as reinforcement. They found that addition of reinforcement increases mechanical properties of composites up to certain level beyond that the strength decreases. Elanchezhian et al [13,15] studied the compression and chemical Properties of Aluminium Nano Composites. It is also found that the wear volume loss and coefficient of friction were influenced by % Volume of Reinforcement addition [16]. Other than graphite, SiC reinforced composite showed more positive results on wear resistance when compared with fly ash reinforcements [17]. But in the case of impact properties comparative study between fly ash and S-Glass, a monotonically increase of impact strength were obsrved when fly-ash was reinforced in Aluminium matrix [18]. At dynamic and quasi-static states of Aluminium composite with Nano metricsized B_4C, strain softening and elastic near plastic behaviour was noticed [19].Being a light weight material, Aluminium finds most of its application in automotive industries especially the SiC reinforced Aluminium due to SiC's high ceramic hardness, high modulus of about 410 GPa and low CTE of 1 ppm/C [20]. Since, no work has been carried out using Zircon sand-fly ash as reinforcement with Aluminium, this paper aimed to fabricate metal matrix composites using Zircon sand-Fly ash with Aluminium.

2.0 Materials

Aluminium alloy (LM6):

This alloy has excellent corrosion resistance which is suitable for marine applications. This can be strengthened further by anodic treatment; however this changes the color from grey to a dark brown. Density of this material is 2.65g/cc.

Zircon sand and Fly ash:

Zircon sand is a naturally available material, which melts at 2100-2300^{0}C and has hardness of 7.0-7.5 Mohs.

3.0 Fabrication of composite

Raw Aluminium LM6 ingots are taken in the required amount in graphite crucible was placed for 2 hours inside the furnace maintained around 850-900^{0}c. At the same time, the Zircon sand and fly ash are preheated. When the Aluminium ingots melt completely, the Degasser was added to the molten metal which removed all the gasses and impurities presents in the molten metal. Then the cover all powder is added, which forms a layer on top of the molten metal since it retains heat inside the crucible. The preheated Zircon sand and fly ash were added 1% and 2.5% weight in the molten metal as shown in Table 1. The stirring process was carried out for 10 minutes and the stirring temperature was maintained around 650-700^{0}C. To avoid casting defects the die was be preheated to 500^{0}C, at which the molten metal poured in to the die (100x100x10mm) and later it was allowed to solidify. Table 1 shows composition of MMC fabricated in this work.

Table 1- Material Composition

Name of sample	Percentage of Aluminium alloy(LM6)	Percentage of Zircon	Percentage of fly ash
SAMPLE 1	100	-	-
SAMPLE 2	98	1	1
SAMPLE 3	95	2.5	2.5
SAMPLE 4	90	5	5

4.0 Testing of material

4.1 Tensile test: The composite sample was prepared as per ASTM: D638. This test was carried out in Universal Testing Machine, and done in normal atmospheric conditions (35^{0}C) in controlled relative humidity.

4.2 Hardness test: The hardness test was carried on Brinell Hardness Testing Machine, Avery make 3000 kgf with ASTM E10 standard.

4.3 Flexural Test: The Flexural (Compression) test was performed using the Universal Testing Machine [UTN 40 Model]. The test specimen was prepared as per ASTM E8 standard.

4.4 Wear Test: The Abrasion Wear test was conducted on Pin-on-Disc apparatus. The size of the specimen that was subjected to wear was 10mm diameter and 30 mm length as per ASTM G 99 standard. 60 Grade SiC abrasive sheet was used as Counter-Disc material. The rotational frequency was set to 24±1 rpm and the equivalent revolution is 84 times the normal revolution, and the load applied was 10N.

5.0 Results and discussions

5.1 Tensile Properties

The result of tensile test is shown in figure 1, from which it was observed that the sample-2 has higher strength than the other two compositions. This was due to the presence of Zircon sand and the ash in minimum percentage as compared to the other samples. Also, it is concluded that increase in reinforcement increases brittleness of materials which in turn reduces tensile behaviour.

IOP Conf. Series: Materials Science and Engineering **390** (2018) 012018 doi:10.1088/1757-899X/390/1/012018

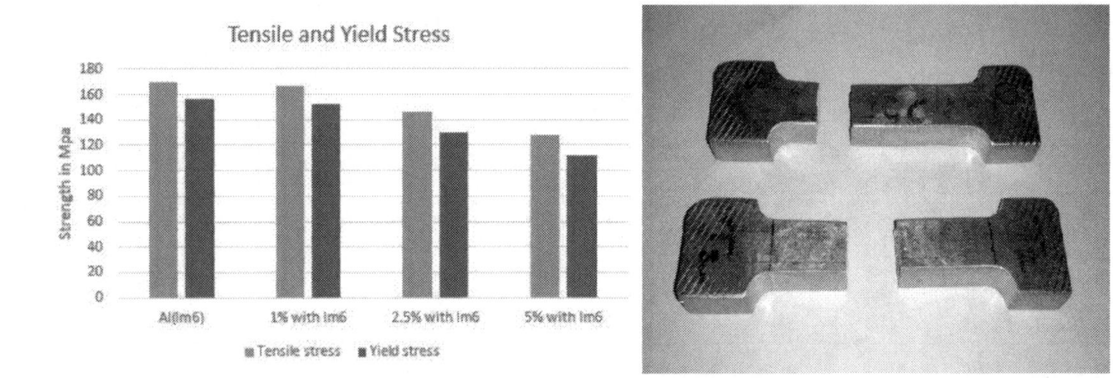

Figure1. Result of Tensile test

5.2 Hardness

Figure 2 shows the result of hardness test, from which it is found that sample 2 which has 1% of Zircon and Fly ash has more hardness than the other 2 samples. This was due to the presence of reinforcement with minimum percentage

Figure2. Result of Hardness test

5.3 Flexural Properties

Figure3 shows the result of flexural test, from which was identified that, the sample which contained 2.5% of Zircon and fly ash had better flexural strength than the other samples. This was due to the presence of Zircon sand and fly ash, which withstood more bending load on the composite as compared to the other samples.

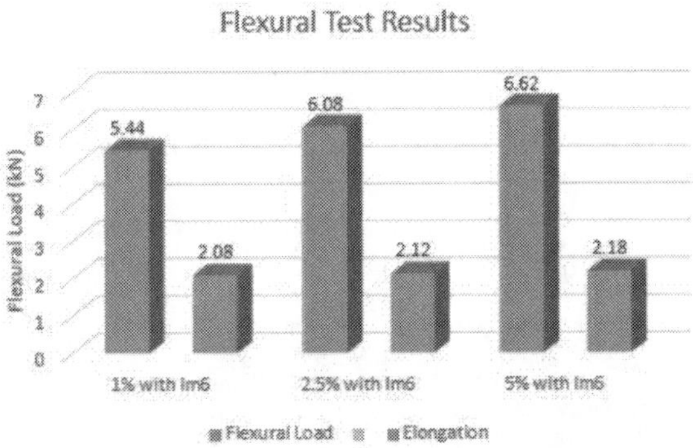

Figure3. Result of Flexural test

5.4 Wear Properties

The result of the wear test is given in Table 2. From the two samples, sample 4 had high wear resistance than the other samples due to the presence of high amount of reinforcement. The tested samples are shown in Fig 4.

Table2. Result of wear test

Sample No	Initial Weight (g)	Final Weight (g)	Abrasion Loss (g)	Loss (%)
2	5.4316	5.3077	0.1239	2.28
3	5.1068	4.9308	0.1760	3.45
4	5.1467	5.0394	0.1073	2.13

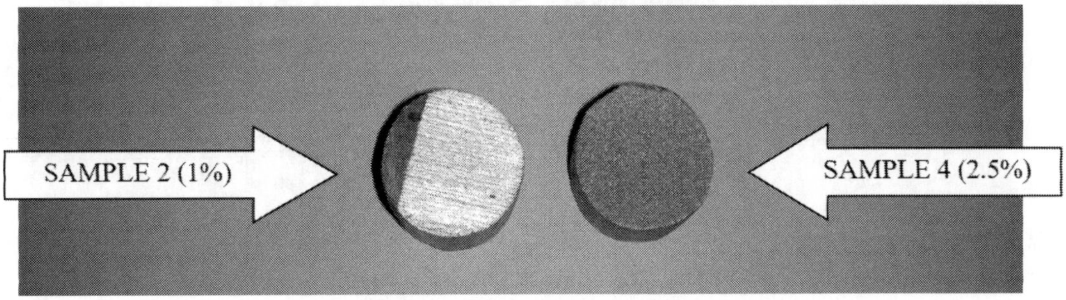

Figure 4. Wear test samples

5.5 Morphological Analysis

This analysis was carried out to find the internal surface defect in the composite fabricated like blow holes, crack propagation and accumulation of reinforcement. Figure 5 shows fly ash accumulation which was due to improper stirring of reinforcements. Crack initiation was observed the sample as shown in figure 6. Also the accumulation of Zircon sand is shown in figure 7.

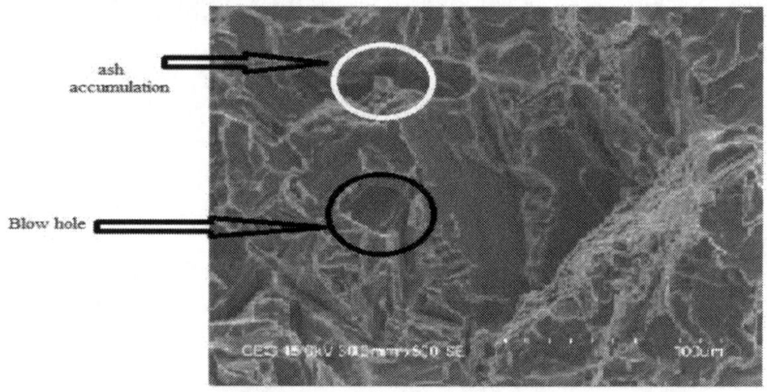

Fig.5 Accumulation of Fly ash

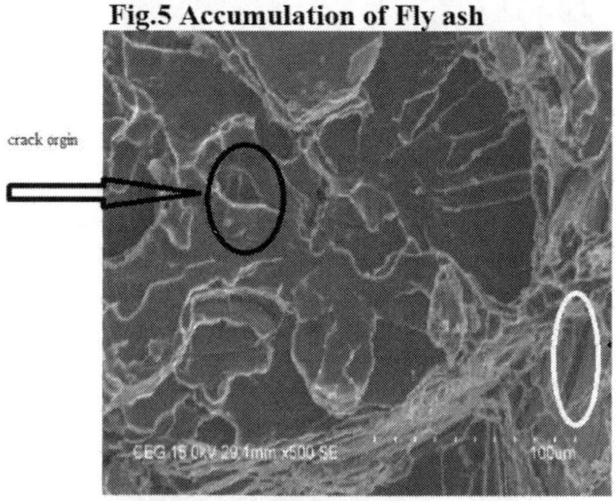

Fig.6 Crack Formation

Fig.7 Accumulation of Zircon

6.0 Conclusion

In this work the tensile, flexural and hardness of Al metal matrix composites reinforced with Zircon sand and Fly ash was investigated.

- It was found that samples which contain 1% of Zircon sand and fly ash has more tensile and hardness strength.
- In Flexural (bending) strength the sample which contains the 1% is less than the 2.5% of Zircon sand and fly ash.
- In Wear Resistant test the sample which contains 1% Zircon sand and fly ash has higher resistance than the other samples
- In figure 5-7 (Morphological Analysis) the accumulation of Fly ash and crack origin was reported

7.0 References

1. Babu Rao. J, D. Venkata Rao and N.R.M.R Bhargava Development of light weight ALFA composites. International Journal of Engineering Science and Technology. 2010,Vol. 2, No. 11, pp. 50-59

2. Deepak Singla and S.R. Mediratta Evaluation of mechanical properties of al 7075-fly ash composite material. IJIRSET. 2013 vol2, issue4,

3. Dharmesh M Patoliya and Sunil Sharma Preparation and Characterization of Zirconium Dioxide Reinforced Aluminium Metal Matrix Composites. IJIRSET. 2015 Vol. 4, Issue 5.

4. Don Hashiguchi, David Tricker, Andrew Tarrant, Jeffrey Campbell and Charles Pokross," Discontinuously reinforced aluminum MMC extrusions", Metal Powder Report, Volume 00, Number 00,February 2016.

5. Kaur. K and O. P. Pandey Wear and microstructural characteristics of spray atomized Zircon sand reinforced LM13 alloy. Mat.-wiss. u.Werkstofftech. 2010, 41, No. 7.

6. Mallikarjuna B.E, H.K. Shivanand, Hemachandra Reddy, M. Hanif Ahamadsab," Evaluation of Impact Properties of Fly Ash and S-Glass Reinforced Al-4046 Hybrid Metal Matrix Composites", Materials Today: Proceedings 4 (2017) 12285–12290.

7. Manoj Singla, D. Deepak Dwivedi, Lakhvir Singh and Vikas Chawla Development of Aluminium Based Silicon Carbide Particulate Metal Matrix Composite. Journal of Minerals & Materials Characterization & Engineering. Vol. 8, No.6, pp 455-467, 2009.

8. Mohammad Mohsin Khan, Gajendra Dixit," Comparative Study on Erosive Wear Response of SiC Reinforced and Fly Ash Reinforced Aluminium Based Metal Matrix Composite", Materials Today: Proceedings 4 (2017) 10093–10098.

9. Mohana Rao. D and Bapi Raju Bandam M.E. Preparation and Characterization of AL-Fly Ash Metal Matrix Composite by Stir Casting Method. IJISME. 2014. Vol3, Issue1.

10. Nathan Behm, Hanry Yang, Jianghua Shen, Kaka Ma, Laszlo J. Kecskes, Enrique J. Lavernia, Julie M. Schoenung, Qiuming Wei,"Quasi-static and high-rate mechanical behavior of aluminum-based
MMC reinforced with boron carbide of various length scales", Materials Science & Engineering AS0921-5093(15)30524- 4.

11. Okafor E. G. and V. S. Aigbodion Effect of Zircon Silicate Reinforcements on the Microstructure and Properties of as Cast Al-4.5Cu Matrix Particulate Composites Synthesized via Squeeze Cast Route. Tribology in industry, 2010 Vol.32, No.2.

12. Radhika. N, R. Subramanian, S. Venkat Prasat Tribological Behaviour of Aluminium/Alumina/Graphite Hybrid Metal Matrix Composite Using Taguchi's Techniques. Journal of Minerals & Materials Characterization & Engineering. , 2011. Vol. 10, No.5, pp.427-
443

13. Rajesh. S, A. Gopala Krishan, P. Rama Murty Raju, M. Duraiselvam," Statistical Analysis of Dry Sliding Wear Behavior of Graphite Reinforced Aluminum MMCs", Procedia Materials Science 6 (2014) 1110 – 1120.

14. Sanjeev Kumar Yadav, Surendra Kumar Patel, Raman Nateriya Microstructure, Micro hardness and compressive behaviour of dual reinforced particles Aluminium alloy composites. International Journal of Innovative Science Engineering & Technology, 2015, Vol. 2 Issue 4.

15. Siddique Ahmed Ghias B.Vijaya Ramnath, Investigation of Tensile Property of Aluminium SiC Metal Matrix. International Journal Applied Mechanics and Materials. 2015,Volume 766-767, Pages 252-256.

16. Subramaniya Reddy. P, R.Kesavan, and B.Vijaya Ramnath, "Investigation of Mechanical Properties of Aluminium6061-Silicon Carbide, Boron Carbide Metal Matrix Composite ", Silicon, Pages1-8, 2016.

17. Suchitharan. K.S, P.SenthilKumar, D.Shivalingappa, J.Jenix Rino Wear Behaviour of Al6063-Zircon Sand Metal Matrix Composite. IOSR Journal of Engineering (IOSRJEN). , 2013 Vol. 3, Issue 2, V3, 24-28.

18. Vijaya Ramnath. B, C. Elanchezhian, M.Jaivignesh, S. Rajesh, C. Parswajinan, A. Siddique Ahmed Ghias, Evaluation of Mechanical Properties of Aluminium Alloy-Alumina- Boron Carbide Metal matrix Composites. Materials and design. June 2014, Volume 58 , Pages 332-338.

19. Vijaya Ramnath. B, C. Elanchezhian, RM. Annamalai, S.Aravind, T. Sri Ananda Atreya, V. Vignesh, C.Subramanian, "Aluminium metal matrix composites – a review", Reviews on advanced material science, Volume 38, Pages 55-60.

20. Vijaya Ramnath. B, C. Parswajinan, C. Elanchezhian, S. V. Pragadeesh, C. Kavin,P. ,R. Ramkishore, V. Sabarish, Experimental Investigation of Compression and Chemical Properties of Aluminium Nano Composite. International Journal Applied Mechanics and Materials. July 2014, Volume 680, Page 7-10.

IOP Conf. Series: Materials Science and Engineering **390** (2018) 012019 doi:10.1088/1757-899X/390/1/012019

Mechanical properties of Al-SiC Nano Composites fabricated by Friction Stir Processing

A Arun Premnath[1*], P. Likhil Varma[a], P. Sathya sai RamaKrishna[b], P. Dheeraj[c], N and P.karthik[d]

[a, b, c, & d] Department of Mechanical Engineering, Sri Chandrasekharendra Saraswathi Viswa Mahavidyalaya, Enathur, Kanchipuram – 631561, Tamilnadu, India.

[*1]Corresponding Author: arun_premnath@yahoo.co.in

Abstract. Aluminium metal matrix composites finds its application in aerospace, automobile, and energy industry owing to their excellent mechanical properties such as low density high strength, light weight, high strength-to-weight ratio and fatigue and corrosion resistances. Further the addition of nano size particles improves mechanical properties of the composites. The present work is focused on the mechanical performances of aluminium SiC nano composites. Instead of making bulk composites, the surface layers on the substrate improve the mechanical properties. For the fabrication of such surface composite layer, the new modification technique, friction stir processing (FSP) is used. FSP technique is used to produce defect-free surface composite with a good bonding and uniform distribution of reinforcement particles within an aluminium matrix. Various process parameters such as rotational speed, transverse feed, and volume fraction of nano SiC, on mechanical properties is investigated. The mechanical properties such as micro hardness and flexural strength were measured using micro hardness tester and universal tensile testing machine respectively.. In addition to that the morphology of the specimen is examined through scanning electron microscope to study them in detail.

1. Introduction

In the modern era, a material with high strength is a basic requirement for various engineering applications. In order to satisfy these needs, composite materials have found a wide range of applications in various industrial components and structures because of their improved mechanical properties. Among various types of composites materials available, metal matrix composites plays a vital role in terms of high strength or stiffness, low density, good corrosion as well as fatigue resistance. For further improving the mechanical properties of such composites, the widely practiced method is the incorporation of either micron/Nano-sized inorganic particle in the metal matrix. Most of the properties improvement by incorporation of such nano particles in aluminium metal matrix composites fabricated by friction stir processing route is not fully explored yet.

Most of the research in composites materials start early in 1960, but addition of nano particles to increase its mechanical and wear properties comes to picture in the beginning of 21st century. Both mechanical and tribological properties are reported in the literature. Some of the work done in the last decade is summarized. In another work, Zhao et al (2015) studied the effect of friction stir processing with B4C particles on the microstructure and mechanical properties of 6061 aluminium alloy. In their research, Optical microscopy, scanning election microscopy, and energy-dispersive X-ray analysis were used to study the microstructure and the distribution of B4Cp. Shahi et al (2014) studied the influence of the rotational and traverse speed of the tool as well as the number of FSP passes on the in situ formation of Al3Ni in aluminium matrix. From the research they found that six passes of FSP showed the highest hardness, which was almost twice of that of the base metal. Liu et al. (2013) fabricated aluminium matrix composites reinforced by multi-walled carbon nanotubes via FSP. They reported that the maximum ultimate tensile of composite was improved about two times compared with as-received aluminium. Dolatkhah et al. (2012) produced metal matrix composite on surface of 5052 aluminium sheets by means of 5-μm and 50-nm SiC particles using FSP. They noticed that hardness value of fabricated composite was improved up to 55% and wear rate was reduced about 9.7 times comparing with as-received 5052 aluminium. Barmouz et al. (2011) used FSP to fabricate copper- SiC composites by varying the volume fraction and size of the SiC particle. They studied the microstructure and mechanical properties of the composites and concluded that Cu/SiC composite will be very attractive to surface composites manufacturers. Asadi et al. (2010) used FSP for modifying the mechanical properties and microstructure of AZ91/SiC composite. They considered the rotational and traverse speeds, tool penetration depth and tilt angle on the formation of defects such as cracks, tunnelling cavity and also on sticking of matrix material to the tool. From the result they concluded that FSP is an effective process to fabricate SiC/AZ91 composite layer with uniform distribution of SiC particles, good interfacial integrity and significant grain refinement. Puviyarasan et al. (2015) used FSP to produce composites reinforced with B4C particles. The result indicates that the composites fabricated using optimal process parameters exhibit a higher tensile strength (174 MPa) and microhardness (183 HV). After a thorough literature survey it was found that out friction stir processing route alter the microstructure which enhances the mechanical properties of the composites. This makes this composites one of the most important composite materials used in many industries, such as aerospace, construction, transportation, as well as for medical and military applications. Taking this into consideration, the present work is focussed on studying the mechanical properties such as micro hardness and flexural properties of Al-SiC nano composites.

2. Materials &Methods

2.1. Fabrication of Composites

Aluminium 6061 is used as the base metal. Nano SiC of size 60-90nm is used as the reinforcement which is purchased from U.S. Nanoresearch. The friction Stir Processing Set-up is shown in the figure1. A Hardened H13 steel with and without pin is used as tool for the FSP. Firstly a groove is made on the al 6061 alloy of 5mm width and 4 mm depth. The groove is filled with Nano SiC particles. Secondly a pin-less tool is allowed to go-through the groove to avoid scattering of the Nano SiC particles. Then a pin tool of shoulder 15mm, pin of 5mm diameter and 5 mm long is pressed along the groove for fabrication of the surface composites. The experiment is done for a rotational speed of 1000rpm and transverse feed of 30mm/min. The experiments were done by varying the number of passes from 1 to 3.

Table 1. Composition of Al6061

COMPONENT	Al	Cr	Cu	Fe	Mg	Mn	Si	Ti	Zn
WEIGHT %	98.6	0.35	0.4	0.7	1.2	0.15	0.8	0.15	0.25

Figure 1. Friction Stir Processing Set-up

2.2 Mechanical Testing

The composite material thus fabricated will be cut into required dimension for mechanical testing. Mechanical properties such as flexural test and micro hardness test are conducted as per ASTM standards. In order to perform the flexural test the different composite specimen is tested in the universal testing. Microhardness of the specimens was measured on the cross section of the specimen perpendicular to the processing direction using Vickers micro hardness tester. The effects of nano SiC on the mechanical properties are studied in terms of its microstructure, the nature of failure, the increase in strength of the composites.

3.Results and discussion

3.1.Morphological and structural characterization

Figure 1-3 shows the microstructure of the fabrication composites using scanning electron microscope. From the figure 2a, it is found that the Nano SiC particles were distributed evenly throughout the matrix for three passes which is not so in one and two passes. In figure 2b and 2c, it is noticed that the particles are agglomerated at one particular places and proper even distribution is not seen.

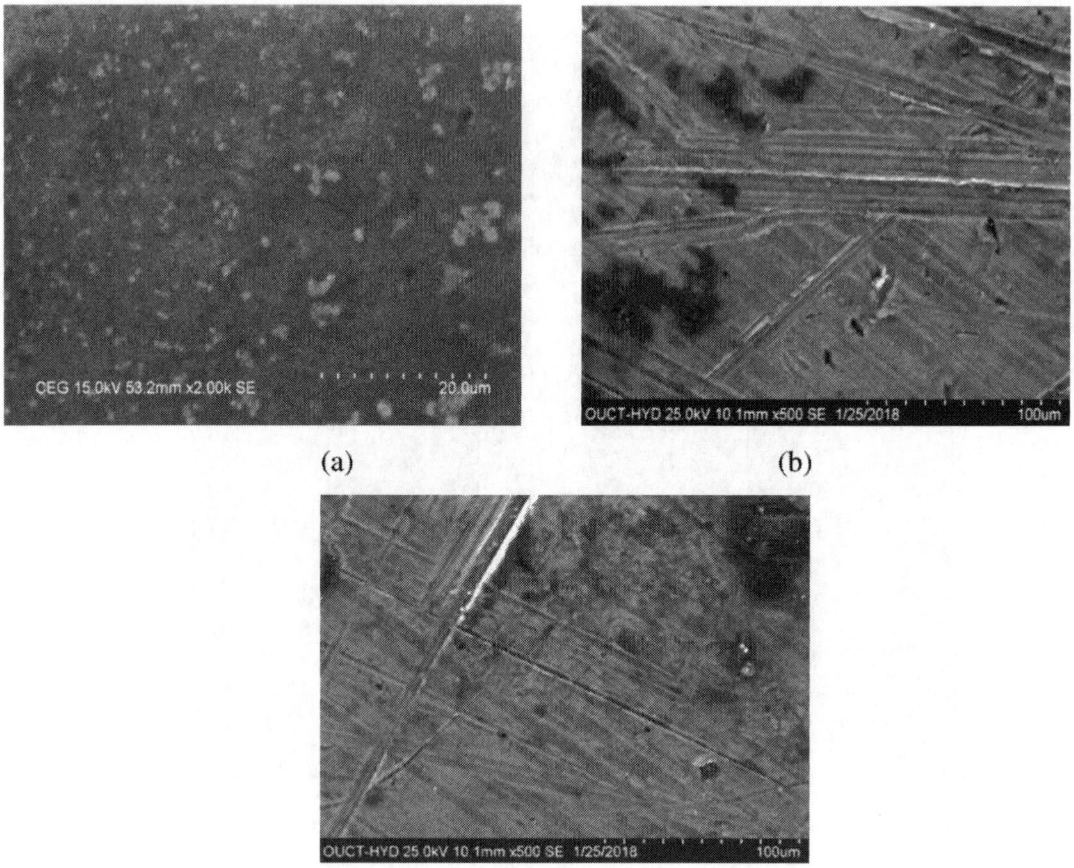

(a) (b)

(c)

Figure 2.SEM image FSPed specimen (a) Three Pass (b) Two Pass (c) One Pass.

3.2. Microhardness

Figure 3 shows the effect of number of pass on microhardness of FSPed specimen. From the figure, it is inferred that three number of pass results in higher microhardness and the microhardness decreases with decreasing the number of passes. At three pass, grain refinement, better bonding and even distribution of particles occurs which results in the enhancement of microhardness as compared to two and single pass. Also the increase in microhardness for all specimens follows the same trend which is similar to the results obtained by Yuvaraj et al. (2017).

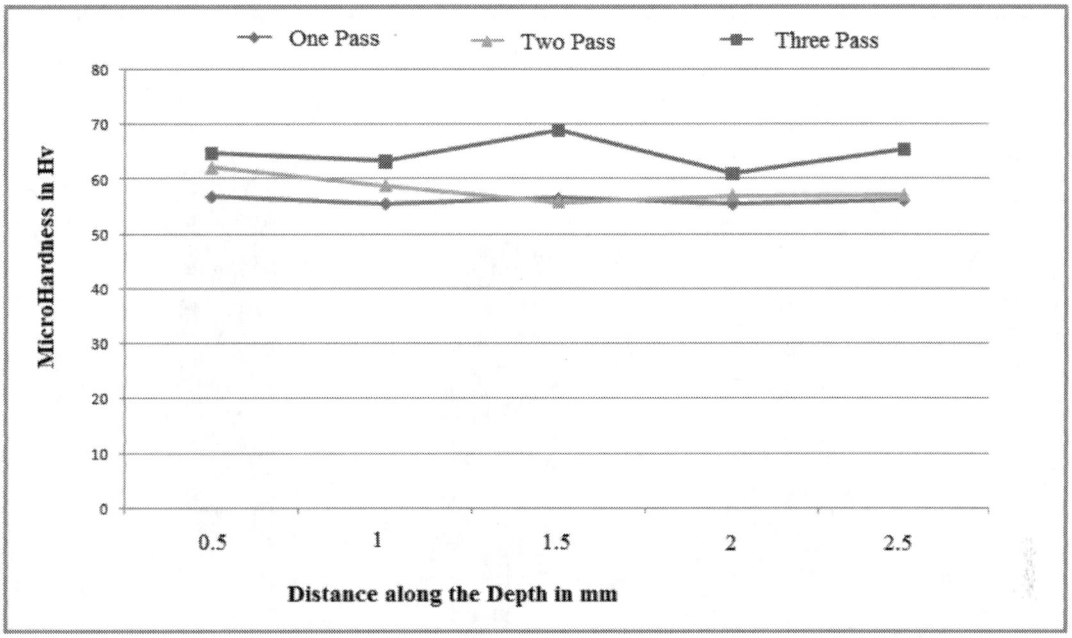

Figure 3. Microhardness values of the FSPed Specimen

3.3 Flexural Strength

The increase in flexural strength of the composites improves the crack expansion and also limits the deformation of the composites. The effect of number of passes on flexural strength of FSPed specimen is shown in the figure 2. It is depicted from the figure tat the flexural strength of all FSPed specimen increases with respect to the base alloy. Further the increase in flexural strength is found to high for three pass specimen compared to the other which indicates a better bonding and grain refinement during the FSP operation.

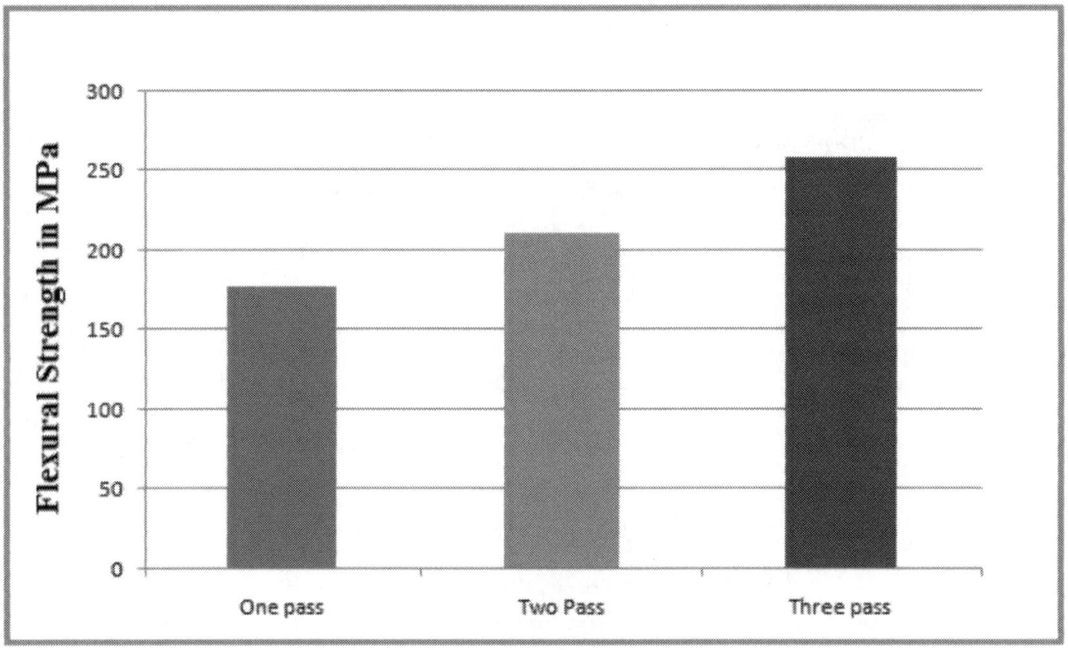

Figure 4. Flexural Strength values of the FSPed Specimen

4.Conclusions

From the study the following conclusion are arrived.

- The composites are successfully fabricated by FSP operation
- The microstructure of the composites shows even distributed nano SiC particles at three pass.
- The hardness and flexural strength of the composites increases for all fabricated composite specimen further three numbers of passes of FSPed specimen shows a better microhardenss and flexural strength.

5.References

[1].Zhao, Y., Huang, X., Li, Q., Huang, J., & Yan, K. IJAMT, 2015; 78(9-12), 1437-1443.

[2].Shahi, A., Sohi, M. H., Ahmadkhaniha, D., & Ghambari, M. IJAMT 2014; 75(9- 12), 1331-1337.

[3].Liu, Q., Ke, L., Liu, F., Huang, C., & Xing, L. Materials & Design, 2013;45, 343-348.

[4].Dolatkhah, A., Golbabaei, P., Givi, M. B., & Molaiekiya, F. Materials & Design, (2012); 37, 458-464.

[5].Barmouz, M., Asadi, P., BesharatiGivi, M. K., Taherishargh, M. Mater Sci Eng A. 2011; 528:1740–1749.

[6]. Sadi, P., Faraji, G., & Besharati, M. K. .IJAMT 2010; 51(1-4), 247-260
[7]. Puviyarasan M, Senthil Kumar VS Arab J Sci Eng 2045; 40(6):1733–1741.
[8]. Yuvaraj, N., & Aravindan, S. Transactions of the Indian Institute of Metals,2017; 70(4), 1111-1129.

IOP Conf. Series: Materials Science and Engineering **390** (2018) 012020 doi:10.1088/1757-899X/390/1/012020

Development of 3D Printed Brain Tumour for Surgical Biopsy

Y Sandeep Kumar[1], Dr S Ramalinga Reddy[2], Y Naveen Kumar[3]
H V Saran Gopal[4] and B Sai Charan[5]

[1]Assistant Professor, Department of Mechanical Engineering, Aditya college of Engineering, Madanapalli – 517325, India
[2]Professor and Principal, Aditya college of Engineering, Madanapalli – 517325, India
[3] U.G Student, Department of Mechanical Engineering, Aditya college of Engineering, Madanapalli – 517325, India
[4] U.G Student, Department of Mechanical Engineering, Aditya college of Engineering, Madanapalli – 517325, India
[5] U.G Student, Department of Mechanical Engineering, Aditya college of Engineering, Madanapalli – 517325, India

Email ID : ssandeep539@gmail.com

Abstract The traditional treatment for brain cancer includes surgery, radiotherapy and chemotherapy. The neurosurgeon usually plans and performs the surgery with the help of two dimensional MRI images. If the size and the exact location of tumour is not clear it is difficult to perform a safe surgery and sometimes may cause permanent damage to the normal tissues. Such problems could be minimized by having first hand information regarding the tumour and hence an attempt is made in the present research to create 3D printed models of brain tumours using a medical image processing software based on MRI datasets. The MRI data saved as DICOM (Digital Imaging and Communications in Medicine) would be integrated with 3D slicer. 3D printer produces a physical 3D models using STL file generated from 3D slicer. Thus the 3D printed model of Brain Tumour could be served as a tool for preoperative surgical planning.

1. INTRODUCTION

A Brain Tumour forms when cells grow abnormally within the brain. In India, every year 40,000 to 50.000 people are diagnosed with brain tumour, out of which 20% are children. The symptoms include new or increasingly strong headache, blurred vision, and loss of balance, confusion and seizures. The treatment depends on the stage of tumor and it includes surgery, radiation therapy and chemotherapy. In some situations tumors can't be removed, so doctors may refer it for biopsy. In biopsy a piece of tumor is collected and will be sent for pathology examination, where it is tested under a microscope and decides the type of cells it contains. Biopsy is performed with the help of a needle and the CT/MRI scans uses to pinpoint the exact location of the tumor. Doctors get only 2D cross sectional images for viewing the tumor in MRI scan. Hence it always takes a skilled neurologist to detect the tumor in brain anatomy.

3D Printing Technology has become a boom in recent years for the creation of customized prosthetics, implants, and surgical tools. 3D printed brain models were found convenient in understanding the anatomy of brain and tumour, surgical simulation [1]. The applicability of 3D printing technology is very limited in brain tumors analysis. Ivan Lau et al., presented the preliminary experience of creating patient specific 3D printed model of brain tumour in a pediatric patient and demonstrated the feasibility of using 3D printing [2]. Vakharia et al. conducted a systematic review of 3D printing on cranial neurosurgery simulation. This analysis focused on the benefits of 3D printed models in preoperative simulation and training [3]. A computer based image processing software for preoperative surgical planning was used by kikins et al in 1996. They used the MRI images on 14 cases to analyze the brain tumors [4]. The authors reviewed the important concepts and described the techniques of interactive 3D visualization from routine 3D T_1 – weighted, MR angiography and brain suite software [5]. Mirchael Kuhlmann et al., attempted to construct the model using conjoint CT and

MRI images on a 4 month old male patient and succeeded in creating a realistic 3D model of a pineal gland tumours, which was used on a trial basis in neuroendoscopic training and simulation [6]. Sayalilopes et al. presented a method for automatic tumour detection with an added feature of reconstructing its 3D image. This involves the implementation in various steps of detecting and exacting the tumour with 2D slices of MRI brain images by region growing technique [7].

2. METHODOLOGY

Based on the tumor location and type of tissue involved, the brain tumours are categorized as primary and metastatic. Primary brain tumours begin to form within the brain. A metastatic tumour is formed when cancer cells located elsewhere in the body break away and travel to the brain.

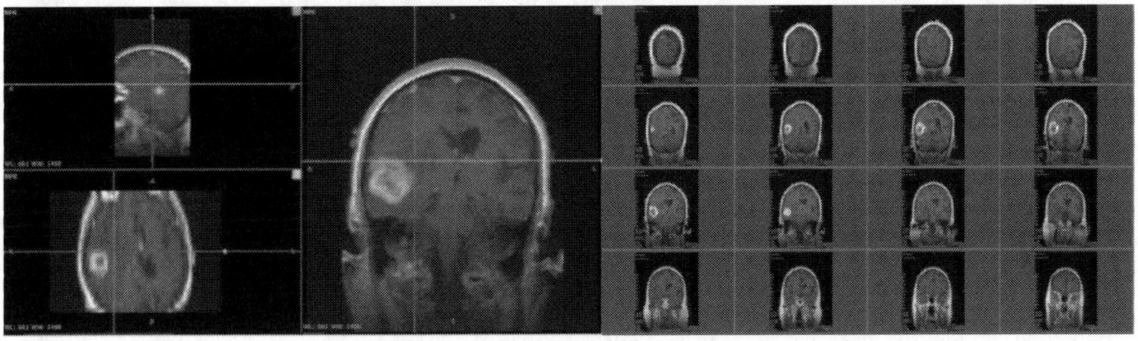

Figure 1 MRI data slices.

2.1 TREATMENT OPTIONS

A variety of therapies have been used to treat brain tumours. The type of treatment recommended depends on the size and type of the tumour, its growth rate, location of tumour, and the general health of the patient. Treatment options include surgery, radiation therapy, chemotherapy, targeted biological agents, or a combination of these. Surgery is generally the first step when the tumour is accessible and the vital structure does not get disturbed. Radiation is used to stop the tumour's growth and causes it to shrink. Chemotherapy destroys tumour cell that may remain after surgery, but it is not very effective for gliomas and at large level, because it is difficult for chemotherapeutic drugs to reach the brain.

3. CASE STUDY

Nine different cases have been considered for the study of size and position of the Brain tumour for pre surgical planning. Different stages that are included in the generation of 3D printed STL files are Image Acquisition, Segmentation, Creating ROI and Model Making. Medical Image Data could be obtained from CT or MRI data. CT scan images are to be used for orthopedic, dental applications and in the situation of high radiation exposure, where as the use of MRI images have the advantage of not exposing to radiation for image acquisition. It is used for medical imaging on soft tissues such as brain.

IOP Conf. Series: Materials Science and Engineering **390** (2018) 012020 doi:10.1088/1757-899X/390/1/012020

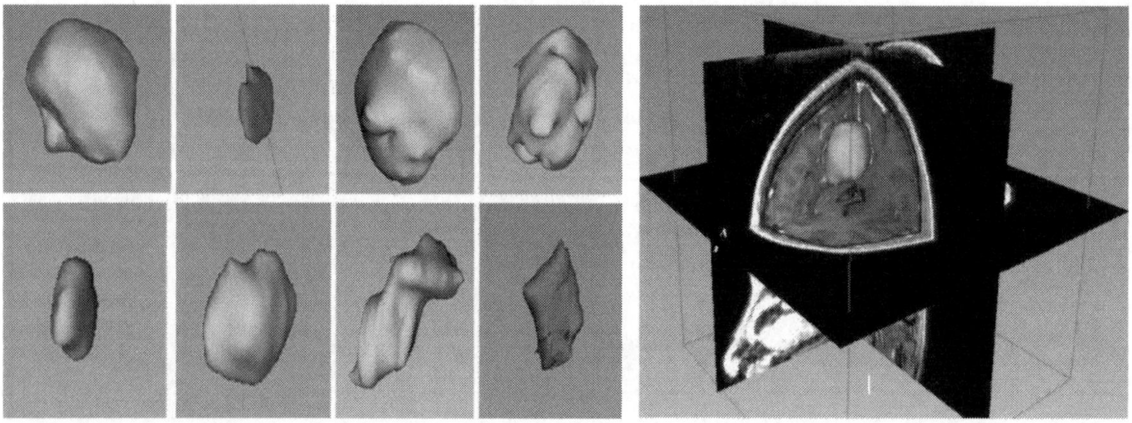

Figure 2(a) Tumours of different sizes (b) Position of tumour in the brain tissue.

3.1 IMAGE SEGMENTATION

Image segmentation is the crucial step used in many clinical applications in the image analysis. Brain data acquisition from MRI scan and the image segmentation are used for measuring and visualizing the brain's anatomical structure, tumour growth and size for surgical planning. Computerized methods for MR image segmentation and visualization have been extensively used to assist doctors to carryout qualitative diagnosis. The images saved in DICOM format would be imported into 3D slicer which is open access software for medical image processing for the purpose of image segmentation and 3D modelling.

3.2 CREATING ROI

The selected ROI (Region of Interest) could be painted using 3D slicer, preferred region could be segmented by opting the effect of grow cut. The cerebellum and brain tumour are considered separately while creating 3D printed model. The model maker function is used to create a computerized model of the ROI and the Laplacian filter is applied to optimize the model. The brain tumours in form of STL files have been exported for 3D printing using global 3D pramaan – 200 3D printer.

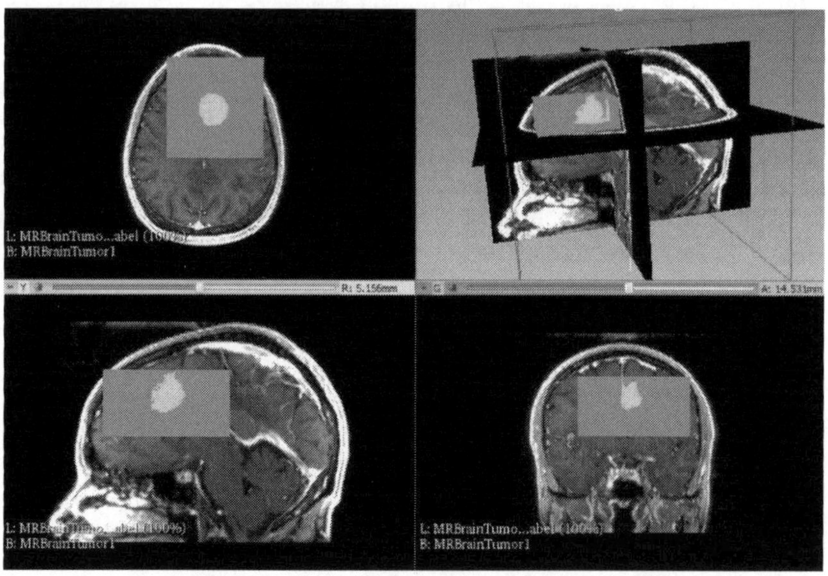

Figure 3 Grow cut effect and selection of ROI.

3.3 3D PRINTING

All 3D printers are encoded in STL files to build 3D models to deposition of materials layer by layer. Major considerations for the selection of 3D Printing Technology includes the parameters like time required to complete printing, cost of printer, and the choice of material and accuracy of model. The Radiologist can select a 3D printer which should be customized for surgical planning. Accurate printing of 3D model would be possible by minimizing the errors generated in each stage of the parameters like Image Acquisition, post processing and 3D printing.

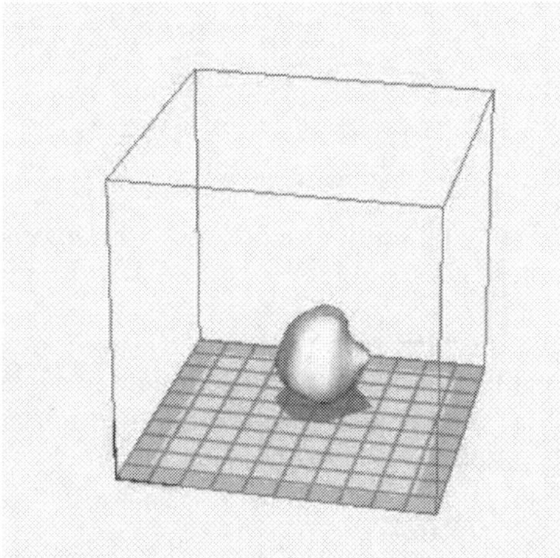

Figure 4 Brain tumour STL model.

In the present study, the tumour models are produced by controlling on the parameters like height of layer, amount of volume infill and time required to print the model. The STL file is imported into CURA open access software. Since the height of layer or thickness effects the accuracy of model, hence the radiologist should ensure for the desired accuracy for better treatment. The amount of volume infill can't effect on surgical planning, so that the radiologist can plan to generate 3D model with low volume infill.

Table 1 Effect of parameters on printing time of tumour model.

Layer Height (mm)	Volume Infill (%)	Time required for Printing	Layer Height (mm)	Volume Infill (%)	Time required for Printing	Layer Height (mm)	Volume Infill (%)	Time required for Printing
0.06	10	1 hr 13 min	0.1	10	40 min	0.15	10	22 min
	20	1 hr 27min		20	50 min		20	27 min
	30	1 hr 39 min		30	59 min		30	31 min
	40	1 hr 52 min		40	1 hr 09min		40	35 min
	50	2 hr 04 min		50	1 hr 18 min		50	39 min
	60	2 hr 16 min		60	1 hr 28 min		60	43 min
	70	2 hr 28 min		70	1 hr 37 min		70	47 min
	80	2 hr 40 min		80	1 hr 37 min		80	51 min
	90	2 hr 52 min		90	1 hr 56 min		90	55 min
	100	10 hr 47 min		100	3 hr 29 min		100	2 hr 16 min
0.2	10	20 min	0.4	10	10 min	0.6	10	08 min
	20	25 min		20	12 min		20	10 min
	30	30 min		30	15 min		30	11 min
	40	35 min		40	17 min		40	13 min
	50	39 min		50	20 min		50	14 min
	60	44 min		60	22 min		60	16 min
	70	49 min		70	24 min		70	17 min
	80	53 min		80	27 min		80	19 min
	90	58 min		90	29 min		90	20 min
	100	1 hr 44 min		100	52 min		100	35 min

IOP Conf. Series: Materials Science and Engineering **390** (2018) 012020 doi:10.1088/1757-899X/390/1/012020

4. CONCLUSION

External Radiation Therapy commonly results hair loss, the loss of skin on the scalp and sometimes kills the healthy brain tissue. So in such situations it is very convenient to use image processing and 3D printing to fix the location and the size of tumour. It is concluded that the use of image Processing, Preoperative 3D printed tumour models have led to the reduction in surgical time, decreased time under anesthesia and dose of radiation. Based upon the results in table, it is concluded that with the increasing in layer height and the volume infill, the time required for printing increases.

Acknowledgments
We would like to thank **Dr Ch Srinivasa Rao** Dean ME & CE, **Mr VVN Bhaskar** HOD, Aditya College of Engineering for their continues support in carrying out this research work. We express our sincere thanks to chair ICMME – 2018 for consideration for further level.

REFERENCES

[1] Mullar A, Krishna KG, Uhl E "The Applications of Rapid Prototyping techniques in cranial reconstruction and preoperative planning in neurosurgery", *J craniofacial surgery – **2003***, pg no 899 – 914.
[2] Iran Lau, Andrew squwlch, Yung wan, "Patient specific 3D printed model in delineating brain glioma and surrounding structure in a pediatric patient", published in *wolten kluwer – **2017**.*
[3] Vakharia VN, Vakharia NN, Hill CS " Review of 3D printing on cranial neurosurgery simulation training", *World neuro surgery – **2016***, pg no 188 – 198.
[4] Kikins R, Gleoson PL, Moriatry TM, "Computer Assisted Interactive 3D planning for neurosurgical procedures", *neurosurgery – **1996.***
[5] Zafar Neyar, Rajendra V phadke, vivek singh, chaitanya godbole, " Three Dimensional Visualization of Intracranial Tumors with Cortical Surface and Vasculature from Routine MR Sequences", Published in *Neurology India – **2017***, **volume 65**, issue – 2, pg no 330 – 340.
[6] Michael kuhlmann, James Drake, "Development of 3D patient specific brain model with a pineal region tumor and hydrocephalus for neuroendoscopic simulation and training," published in *RCSI* - 2013, **volume – 6**, number – 1, pg no 68 – 71.
[7] Syali Loper, Deepak Jayaswal "A Methodical Approach for Detecting and 3D Reconstruction of Brain Tumor in MRI", Published in *International Journal of Computer Applications* – 2015, **volume – 118**, pg no 37 – 43.

BIO-GRAPHY:

Mr Y Sandeep Kumar Assistant Professor in Department of Mechanical Engineering Aditya College of Engineering Madanapalle. He received the Bachelor's degree in Mechanical Engineering and Master's degree in CAD/CAM from JNTU and Pursuing PhD from Visweswaraya Technological University Belagavi. He has 7 years of teaching experience. His research interests include Bio materials, Biomechanical Analysis and Finite Element Analysis. He published more than 10 papers in International and National Journals and conferences.

Dr S Ramalinga Reddy Professor and principal in Aditya College of Engineering Madanapalle. He received the Bachelor's degree in Mechanical Engineering from SV University and Master's degree from Manipal University and PhD from SK University Anantapur.. He has 20 years of teaching experience. He published more than 30 papers in International and national journals and conferences.

Mr Y Naveen Kumar pursuing B.Tech from Mechanical Engineering in Aditya college of Engineering, Madanapalle, Andhra Pradesh.

Mr H V Saran Gopal pursuing B.Tech from Mechanical Engineering in Aditya college of Engineering, Madanapalle, Andhra Pradesh.

Mr B Sai Charan pursuing B.Tech from Mechanical Engineering in Aditya college of Engineering, Madanapalle, Andhra Pradesh.

Glass ceiling factors affecting women career advancement in IT industry in India

Nilufer Azeez P V[1] and R G Priyadarshini[2]

[1] Amrita School of Business, Coimbatore, Amrita Vishwa Vidyapeetham, India, nilufer.nilu@gmail.com
[2] Amrita School of Business, Coimbatore, Amrita Vishwa Vidyapeetham, India, rg_priyadarshini@gmail.com

Abstract. Women in IT industry is concentrated more in the lower management level and encounter a glass ceiling which is a hindrance as they climb up the career ladder to top management level. This study presents about the glass ceiling factors that affect the career advancement of women in IT industry in India. This study attempts to identify factors which contribute to the glass ceiling which affect women career advancement. Several factors were identified while reviewing the literature; they are personal factors, organizational factors and societal factors. It also aims at finding out which factor from the identified factors contribute more hindrance to the career advancement opportunity for women. Conceptual model is designed based on the information from literature review; whole study is structured based on this conceptual model. Data is collected by floating a questionnaire to the women employees working in IT companies in India. Sample consists of 57 female employees from different IT companies who are in middle level management and in senior level management. From the analysis of the data collected, it is seen that societal factors contribute more hindrance to the women in advancing their careers, women perception about themselves and management perception about women reaching higher position also plays an important role.
Key Words: Glass Ceiling, Personal factors, Organizational factors, Societal factors, Career advancement.

1. Introduction

There exists a prevalent issue of gender bias with respect to career advancement of women in big corporations (Sharma and Seharwat, 2014). In India women are still seen weak, emotional and unaggressive to hold higher level management positions. It is observed that gender based discrimination is still subsisting to scale up to senior level. (Bhosle and Barik, 2014). Corporate career typically have a span of 40 to 45 years (Lynch and Katherine, 1998) and even in their 40-45 years of service, they are not able to reach the C-level. Out of 1000 fortune companies, only 5% of the women are serving as senior managers (Federal glass commission report,1995). Many women are working in the IT sector, but less women on top management of the corporation.

Achievements of women are devalued and it is represented as luck rather than their ability or skill (Deaux, 1976; Heilman, 1983), this effect them as it reduces the chance to advance in their career. It is also seen that the difference is not only in the career development, but also in the compensation they receive. (Blau& Ferber, 1992). Developmental opportunities for women are less when compared to men which indicate there is gender bias is significant for developmental opportunities. Job characteristics associated with higher level positions are absent in women(Ohlott, Ruderman, & McCauley, 1994). Men are chosen for job which involve risk, less chances are there to chose women for such assignments. Primary importance for any organization is the effectiveness of the team(Pavithra et.al.,2017)

2. Literature Review

According to Wall street journal,1983 "Glass Ceiling" is defined as the barriers that prevent women from ascending to senior management positions in corporations. Glass ceiling is not one ceiling or wall, they are varied and pervasive forms of gender bias that occur frequently in both overt and covert ways. (Chamaru and Bambuvela, 2013).

Corporate argues glass ceiling doesn't exist (Rai and Srivastava ,2008). According to them women are not promoted to higher positions as they have a tendency to leave the jobs in middle, lesser working hours, doing low risk jobs. Number of opportunities are there for women in advancing their career as they are present at organizations that operate globally. Women are hardly at the higher level of managerial jobs(Kelkar et.al 2002, Upadhya 2006). Individual are judged according to the congruence between their characteristics. Organizations prefer female leadership when the organization performance is poor (Bruckmüller et.al ,2010) and Ryan et.al,2011).Studies reveal that to perform successfully in such situation leader requires feminine attributes such as being sympathetic, relationship-orientation, service oriented and caring.

In India limited research is done in the area of barrier women face while climbing up the corporate career. (Gupta, 1976). Men presume they have more managerial skills than women, they believe "think manager- think men". Many of the jobs in IT, specially leadership roles are considered as masculine than feminine (Gutek, 2001). Competitiveness, self-confidence, objectiveness, aggressiveness, and ambition are seen as agentic attributes and they are seen as prerequisites for success in the IT sector in general and especially for leadership roles in IT organizations.

Career development is directly linked with the goals and objectives set by an individual. Career development is starting with self actualisation and self assessment. The interest of the individuals is taken into consideration and matched them with the available options. Employee need to train himself/herself to acquire the skills which is absent in them to excel in their career. After acquiring required competencies employee have to perform and achieve the targets and goals set by them.

Number of studies now revealed that there exists "Glass ceiling" or feminization in IT industry, because of which more women are at the lower level of job in the IT industry (Ghosh and Bhattacharya,2012). At the managerial position very few women are working, few of them working middle management as Project managers, Team leads, Lead Consultants whereas a majority of the employees are working in the lower level as programmer analyst, support analyst, specialist and testers. A Study conducted by Shankar (2008) among IT Companies in Bangalore revealed that female employees are mostly concentrated at the lower management roles, than at the middle and top management roles, which creates glass ceiling. Some of the factors affecting glass ceiling is identified and some of them are:

1) Personal factors
2) Organizational factor
3) Societal factors
4) Demographic features

Personal Factors: Under personal factors, ability to work, willingness to do the assigned job, self-perception about themselves, family work balance have been cited as factors determining career development. This shows the extents to which individual factors come as barriers for women's career development Family work balance shows how relation of the female employees affect their performance.

Organizational factors: Organization policy, organization culture, perception of the management towards the advancement of the women in higher roles are some of the factors under organizational factors. Organizations are reluctant to invest in women employees, as they perceive women may quit the job. Women have fewer promotion opportunity, despite having high coordinating ability.

Societal Factors: Beliefs and stereotype are the factors identified under societal factors. This refers to the extent to which the beliefs, traditions and influence the employee development.

3. Objectives of the Study

There is a significant rise in the women participation in the workforce over past few years. But it is commonly seen career advancement opportunities are less for women when compared to men. Reasons of this bias are clearly associated with the idea of a glass ceiling. (Afza and Nawaz,2008).
Some of the objectives of the study are:

1. To identify which factors contributes to glass ceiling.
2. To find out how glass ceiling effect the career advancement of women employees
3. To rank the glass ceiling factors that exists for women in IT sector.
4. To examine how management's perception affects the female employee's promotion to senior level

4. Research Gap

Literature have spoken about the glass ceiling factors and and there is a dearth of work about which factor (personal, organizational and societal) contribute more to the glass ceiling and career development, what measures are taken by the organization to help women, breaking the glass ceiling. Societal factors influencing the organization in framing the organizational policy are limited in previous studies, which are studied here.

5. Conceptual Model

The main idea highlighted by this conceptual model is, how Glass Ceiling affects career advancement of women. Factors contributing to glass ceiling is considered as the independent variable. Career advancement of women as the dependent variable(CA). Societal factors(SF), Personal factors (PF), Organizational factors (OF) as independent variables in this model.
SF (stereotype and belief), PF (ability to do work, willingness to do work, self perception and family work balance), OF (organizational policy, organization culture and management perception).

The model using in the study is as follows:

CA = function (Glass Ceiling)

CA = function (PF, OF, SF)

Figure 1. Conceptual Model

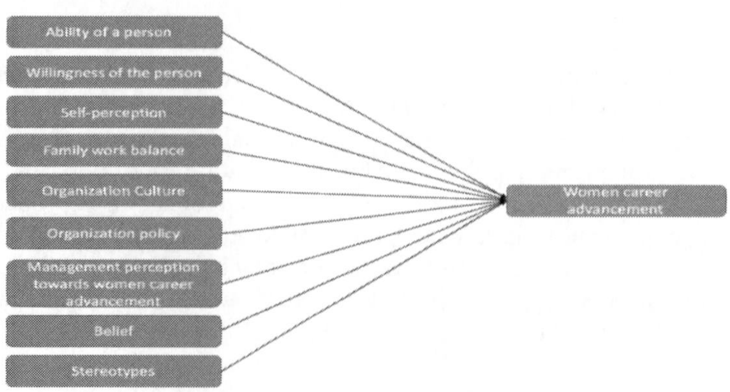

Figure 2. Detailed conceptual model

6. Methodology

Relevant data for the study is collected by survey method. Primary data were collected to diagnose the problem answers the research questions and achieve the objectives. Thus the survey method is regarded as most suitable one. Survey instrument is the questionnaire. The study was conducted among women from different IT companies, sample is exclusively comprised of women employees as it the research is on women oriented issue. Women who are working in the middle level and senior level were considered. A sample of 57 female employees was studied using a questionnaire, based on five point Likert scale ranging from 5 to 1, they are asked to mark their response on based on the agreement to the sentence, from to great extent to no extent.54.4%(31) of the respondents are in the mid level and the rest 45.6%(26) is in senior level position. Glass ceiling factors were included in the questionnaire namely, personal factors, organizational factors and cultural factors. Demographics of the employees are also taken into consideration.

7. Hypothesis Creation

Maheshwari (2012) spoke about glass ceiling barriers which can be broadly classified into three categories namely, individual, societal and organization related barriers. Based on the literature reviewed, this study take into consideration three factors which effects the women's career development, namely personal factors, organizational factors and societal factors and formulated following alternative hypotheses,

Hypothesis-H1: Individual Factors does not effect significantly on career development of women.

Hypothesis-H2: Organizational Factors does not effect significantly on career development of women.

Hypothesis-H3: Societal Factors does not effect significantly on career development of women.

8. Analysis and findings

The data obtained from the response of the questionnaire is analyzed and found out which glass ceiling factor contribute more to the career advancement of the women in organization. According to Marshall and Rossman(1999), data analysis is the process of bring the collected data in an order, structuring it and interpreting the same. The data collected from questionnaire were systematically organized in order to facilitate the analysis. In data analysis the data is cleaned and coded. It is coded in Microsoft excel which is compatible to the tool WEKA. The data is then loaded to the tool and regression is done. One of the objective of the study is to find which glass ceiling factor contribute more to the advancement of the career, so portions of regression is done. Linear Regression of the factors gave the result how glass ceiling factors the advancement of women in the career. Ranking among the factors is done. The result of the Analysis is following in the next session.

9. Results and discussion

In order to find out which factor contribute more to the career advancement, regression is done in portions and it is seen that societal factor contribute more the career advancement of the women. Societal factor is highly correlated to the career advancement, by 77.77%. The next factor which is having significant effect is the organizational factors by having a correlation coefficient of 75.39% and then the personal factors which is having a correlation coefficient of 75.24%.

While performing the linear regression on the personal factors and the career advancement, it is seen that out of Ability, Willingness, Self perception and Family work balance the one which is having significant impact is willingness and family work conflict. In case of the organizational factors which include organizational policy, organizational culture and the perception of the management, the significant one is organization policy. While considering the societal factors, which are societal belief and stereotypes. It is seen that both have significant effect in the advancement of the career.

While performing a linear regression with all the factors, it is seen that the correlation of all these factors with career advancement is 90.29%. The significant factors are ability, family work balance, organization policy, organization culture, perception of the management, belief and stereotype. Table 1 summarizes the result.

Table 1. Result Summary

X-Variable	Y- variable	Correlation coefficient
Personal factors	Career Advancement	75.24%
Organizational Factors	Career Advancement	75.39%
Societal Factors	Career Advancement	77.77%
Glass ceiling factors	Career Advancement	90.29%

All the factors are significant, as the correlation coefficient of all the factors are above 75%. Even though from the three the highest value is seen for societal factors. All the hypotheses are rejected as it is seen from the result that, there is significant impact of all the factors (Personal, Organizational and Societal) on women career advancement. In order to rank the factors, the data is converted from numeric to nominal, and the ranking is done using WEKA. The factors which comes first is the Self perception, stereotype and belief in second and third position respectively. Rank of the factors is mentioned in table 2.

Table 2. Rank of the factors

Rank	Factors
1	Self Perception about the ability
2	Stereotyping in society
3	Societal belief
4	Management perception in employing women in senior level
5	Willingness
6	Organization Culture
7	Family work balance

8	Ability
9	Organization policy

When considering the equal opportunity for men and women while recruitment, 91.2% of the respondents said in their organization there is equal opportunity considered while recruiting a candidate. After their recruitment, the next check is on the equal opportunity in their deployment, 87.7% of the respondents tells that they have equal opportunity in deploying the employees.

While checking on the number of employees in different levels of management the study result shows that women are majorly in the lower management and very less in the top management. In the middle level of management even men are more but not very high in number as top management.It seen from the study that, if the perception of the management is positive towards the progression of women to higher position, women face less barrier while promotion, less discrimination in the performance appraisal and task assignment. 49.1% of the respondents believe that discrimination of women in appointments affect their progression to top management.

10. Conclusion

This paper discusses about the glass ceiling factors, how these factors affect women in their advancement of the career. The study was done in particular to the IT industry. Prominent factors of glass ceiling factors identified through literature review are personal, organization and societal factors. The factors were identified, validated in the study, how they affect career advancement is studied. From the data analysis it is found that societal factors are the one which hinder the women from climbing their career ladder. Other factors like personal and organizational factors contribute almost in equal amount to the hindrance in climbing up the career ladder. Better condusive environments is attained in an organization through providing the employees good work life balance (Ambily et.al,2015). Social support from the colleagues and family plays an important role in career advancement (Rathi. N et.al,2013). The level of mobility, decision making power of women is now changing and more women are into entrepreneurship.

IT industry is the fifth largest industry in India, it contributes to almost 8 percent of the country's GDP (India Brand Equity Foundation report,2017). From the result we could understand that it is the societal belief and stereotypes need to be changed to reach in heights. Despite being this big industry, women are less in top management, organization can help them in reaching the higher position by reducing the gender bias in the organization and by having positive attitude toward the women promotion to higher positions, by this organization can be trend setters to other organization. The one next to be changed is the way women perceive about themselves, from the analysis we could understand that women are capable of doing things and they are willing too but the way they perceive about themselves hinder themselves from reaching heights n their career.

11. Limitations

Limitations of the study are, detailed discussion can be done with the managers and the respondents. Face to face interview of at least five people can be taken into account to understand much broader perspective of the issue. Here the study considered only one country, India global perspective can be taken to have a broader picture.

References

[1] Afza, S., &Newaz, M. (2008). Factors determining the presence of GC and influencing women's career advancement in Bangladesh. *Bbrac University Journal*, 5(1), 85-92.

[2] A.S.Ambily and Suresh,V.N (2015).,"Quality of Work Life of Non-managerial Employees in Manufacturing Industries",*OPUS:HR Journal*,**4**

[3] Ashby, J., Ryan, M. K., and Haslam, S. A. (2007) Legal work and the glass cliff: Evidence that women are preferentially selected to lead problematic cases. *Journal of Women and the Law*, **13**, pp.775-794.

[4] Bakan, D. (1966), The duality of human existence: An essay on psychology and religion. Chicago: Rand McNally

[5] Bhosle and Barik(2014), Time to crack glass ceiling: India context, *International Journal of research in commerce and management*. **5**, Issue no . 07

[6] Blau, E D., & Ferber, M. A. (1992). *The economics of women, men, and work* (2nd ed.). Englewood Cliffs, NJ: Prentice Hall.

[7] Bruckmüller, S., and Branscombe, N. R. (2010) The glass cliff: When and why women are selected as leaders in crisis contexts. *British Journal of Social Psychology*, **49**, pp.433-451. doi:10.1348/014466609X466594.

[8] Chamaru and Bombuwela (2013), Effects of Glass Ceiling on Women Career Development in Private Sector Organizations, *Journal of Competitiveness*, **5**, Issue 2, pp. 3-19, June

[9] Dimovski et.al (2009), Is There a 'Glass Ceiling' for Mid-Level Female Managers?,*International Journal of Management and Innovation*,**1** Issue 1

[10] Eagly, A. H., and Johnson, B.T. (1990), Gender and leadership style: A meta-analysis. Psychological Bulletin, **108**, 233-256. doi: 10.1037/0033- 2909.108.2.233

[11] Eagly, A. H., and Karau, S. J. (2002), Role congruity theory of prejudice toward female leaders. Psychological Review, **109**, 573-598. doi:10.1037/0033-295X.109.3.573

[12] Fischer and Vianen(2002), Illuminating the glass ceiling: The role of organizational culture preferences, *Journal of Occupational and Organizational Psychology*/ (2002). **75**, 31S-337

[13] Garcia-Retamero, R. and López-Zafra, E. (2006), Prejudice against women in male-congenial environments: Perceptions of gender role congruity in leadership. Sex Roles, **55,** 51-61. doi: 10.1007/s11199-006-9068-1

[14] G.G,Jose,E.M., and John,r.(2014), Women Empowerment Through Entrepreneurship,*New Century Pub*, **3**

[15] Glass Ceiling Commission, (1995). Good for business: Making full use of the nation's human capital. Washington, DC: Glass Ceiling Commission.

[16] Greenhaus, J. H., &Parasuraman, S. (1993). Job performance attributions and career advancement prospects: An examination of gender and race effects. *Organizational Behavior and Human Decision Processes, 55,* 273-297.

[17] Gupta A, Koshal M and KoshalR J (1998), Women Managers in India: Challenge and Opportunities, *Equal Opportunities International*, **17**, No. 8, pp. 4-18.

[18] Gutek, B. A. (2001), Women and paid work. Psychology of Women Quarterly, 25, 379–393. doi:10.1111/1471-6402.00036

[19] Heilman, M. E. (2001), Description and prescription: How gender stereotypes prevent women's ascent up the organizational ladder. *Journal of Social Issues*, **57**, 657-674. doi:10.1111/0022-4537.00234

[20] Heilman, M. E. (1997), Gender discrimination and the affirmative action remedy: The role of gender stereotypes. *Journal of Business Ethics*, **16**, 877-889. doi: 10.1023/A:1017927002761

[21] Hoobler and Wayne, (2009), Bosses' Perceptions of family work conflict and women's promotability: Glass ceiling effects, *Academy of Management Journal* ,**52**, No. 5, 939–957.

[22] Korn/Ferry International & UCLA Anderson Graduate School of Management. (1993). Decade of the Executive Woman. New York: Korn/Ferry International.

[23] McCauley, C. D., Ruderman, M.N., Ohlott, P. J., & Morrow, J.E. (1994). Assessing the developmental components of managerial jobs. *Journal of Applied Psychology,* **79**, 544- 560.

[24] Nieva, V. E, &Gutek, B. A. (1980). Sex effects on evaluation. *Academy of Management Review,* **5**, 267-276.

[25] N.Rathi and Bharath,M. (2013), "Work-family conflict and job and family satisfaction:Moderating effect of social support among police personnel",*Equality, Diversity and Inclusion*,**32,**pp.438-454.

[26] Oakley J G (2000, October), Gender-based Barriers to Senior Management Positions: Understanding the Scarcity of Female CEOs, *Journal of Business Ethics*, **27**, No. 4, pp. 321-335.

[27] Pravitha N.R and Indrajith(2017), "A study on factors influencing team effectiveness in a resort environment", *International Journal of Scientific Research,* **6**

[28] Ryan, M., &Haslam, A. (2005). The glass cliff: Evidence that women are overrepresented in precarious leadership positions. *British Journal of Management*, **16(2)**,81-90.

[29] SarivothaRath et.al (2015), Career Advancement of Women Bank Managers in India: A Study in State Bank of India, Vilakshan, *XIMB Journal of Management*, **12 (1)**

[30] Sczesny, S. (2003), A closer look beneath the surface: Various facets of the think-manager–think male stereotype. Sex Roles, 49, 353-363. doi:10.1023/A:1025112204526

[31] Sharma and Sehrawat(2014), Glass ceiling for women, does it exists in Modern India?, *Journal of Organisation& Human Behaviour*,**3**, Issue, 2 & 3

[32] Thomson and Lyness , (1997) , Above the Glass Ceiling? A Comparison of Matched Samples of Female and Male Executives, *Journal of Applied Psychology*. 1997, **82**, No. 3, 359-375

[33] Wood, R. G., Corcoran, M. E., & Courant, P. N. (1993). Pay differences among the highly paid: The male-female earnings gap in lawyers' salaries. *Journal of Labor Economics,* **11**, 417-441.

[34] Woody, B., & Weiss, C. (1994). Barriers to workplace advancement experienced by White women workers (Paper prepared for the Glass Ceiling Commission). Boston: University of Massachusetts.

IOP Conf. Series: Materials Science and Engineering **390** (2018) 012022 doi:10.1088/1757-899X/390/1/012022

Strength analysis of single lap dissimilar materials bolted joint

Dr. Sudev L J[1*] and Mr. Gajendra S[1]

[1] Department of Mechanical Engineering, Vidyavardhaka College of Engineering, Mysuru, India

Email: sudevlj@vvce.ac.in, gajendra407@gmail.com

Abstract

Presently, lot of emphasis is given on joining of dissimilar materials. Around 60-70% of the structures are failing at the joints. Therefore, it is critical to investigate the failure load and failure mode of dissimilar materials joint. The present work investigates the failure of single lap single bolted joints mainly on shear and bending strength. Two material combination joints such as Mild Steel - Aluminum and Mild Steel - Glass fiber reinforced polyesters bolted joints are analyzed using experimental and finite element methods. Present investigation shows that the Mild Steel - Glass fiber reinforced polyesters has better shear strength and bending strength than the Mild Steel - Aluminum material joint.

1 Introduction

Mechanically fastened joints are often used to join composite materials in aircraft structures. It has an advantage over adhesive joints in regard to joining of thick structures which are difficult to bond. They can be detached without damaging the structural members, making repair possible. However, mechanical joints have disadvantages, the main one being that bolt holes cause stress concentrations which reduce the resistance of the joint construction to applied loads, and thus its efficiency[1]. Loading conditions, such as type of loading and load configuration, are one of the most important factors in designing a joint in a structure. Most published papers deal with the influence of geometry parameters, material system, and joint configuration on the strength and fatigue performance of composite joints [2]. Presently, in case of upper class vehicle doors, a large drawing depth is required for the inner panels. Using standard stamping procedures, the consistent realization of such high drawing depths presents some problems with the available standard car body aluminum alloys in mass production. The fabrication of aluminum inner door panels with larger drawing depths is possible, but asks for more complex and closely controlled stamping procedures, which increases the cost.

Ricardo de Medeiros et, al [2] has investigated on the mechanical behavior of single lap single bolted joints for titanium and composite structure using experimental and finite element method. They have concluded that the new experimental method can help the engineers to design similar metal-composite joints and also possible to study the different design parameters such as strength, limit stress and modulus of elasticity for metal-composite joints. Esmaeil Ghanbari et. al [3] have investigated strength of bolted joints for different tightening torque in tension load and four point bending load. They have

concluded that the both tensile and bending strength of joint increases with increase in E/D ratio and tightening torque value. Hilton Ahmad et.al [4] have investigated the effect of stress distribution around the hole for different materials and plate thickness, such as steel-CFRP and CFRP-CFRP materials joints. They have concluded that, due to secondary bending, the stress distribution through the CFRP woven fabric system single-lap joint changes significantly and the stress distribution increases with increase in thickness.
The present study aims to determine the shear and bending strength of the metal to metal and metal to composite bolted materials joint using experimental and finite element method. Finally, the results obtained from the two methods are compared.

2 Methodology

The present design and material data are obtained or collected from various journals. The specimens were manufactured as per the ASTM standards for experimental analysis. From the test ultimate load for different joints were obtained and same load was used to determine the shear and bending strength using empirical formulae. The same experimental data are used to generate a 3D model for finite element analysis using Cero3.0 and simulation is carried out using Ansys workbench17.0. The same ultimate load obtained from test has been used in simulation and strength was analyzed. Finally, the results are compared with each other.

The existing car door is made using Mild Steel with Aluminum bolted materials joints. On account of the ductility behavior aluminum is preferred as interior panel or bumper materials for the automobile vehicle. In the present competitive market, fuel efficiency and high strength to weight ratio place an important role. By considering these points existing materials fails to meet the requirement. Therefore, the other structural based material has been selected and comparative study of different materials joint is carried out. Here, a randomly oriented chopped glass fiber with polyester is selected in place of aluminum due to high strength to weight ratio and randomly oriented fibers exhibit isotropic property in planar condition.

3 Experimental analyses

3.1 Shear test: In most bolted connection, the bolts are subjected to shear and bolts can fail in shear or in tension. So it is necessary to know the strength of the joint under the shear conditions. As there are no such specific standard methods for determining the shear strength of the dissimilar materials single bolted joint, ASTM D5961 [5] provides a methodology to determine the shear strength of the composite-composite material joint. Figure 1 shows the image of test specimen used to determine the shear strength of the joint.

Figure 1. Image of MS-GFRP and MS-AL for shear test

The test was performed in the universal testing machine as per ASTM D5961 standard for both the Mild Steel-Aluminum and Mild Steel-GFRP joints for two samples each. The test has been performed till the samples break. The Figure 2 shows experimental load value at different cross head travel for MS-AL and MS-GFRP materials joint in shear test. The ultimate load of the MS-AL and MS-GFRP is 6950N and 7450N respectively.

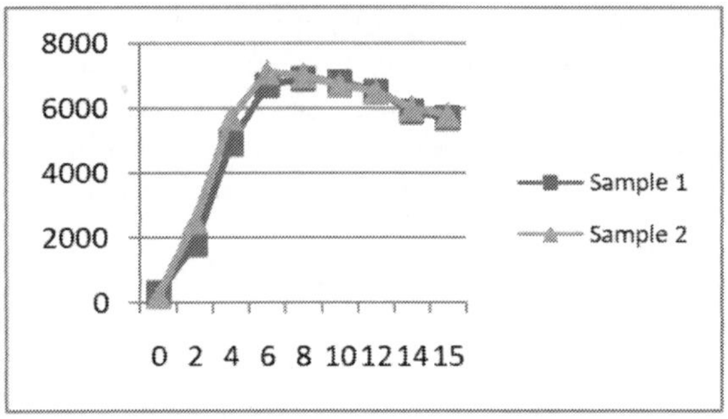

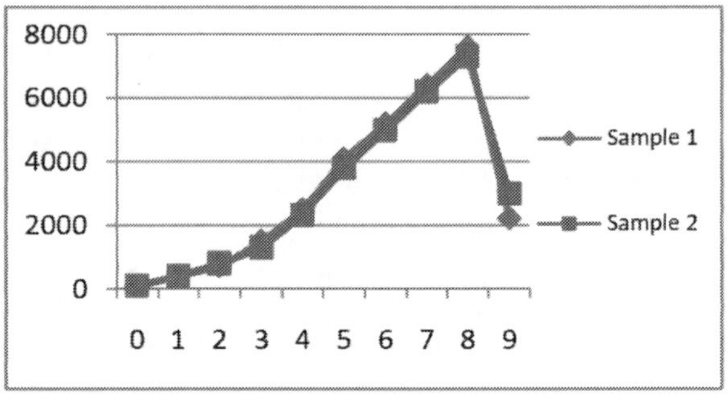

Figure 2. Loads (N) vs. cross head travel (mm) of MS-AL and MS-GFRP joint in shear

The Shear Strength of the joint is determined by using the formula, **S = F / A (MPa)**

Where, F= Load (N),

A= Area of hole (mm^2) = $(3.141/4) * d^2 = (3.141/4) * 6^2 =$ **28.3 mm^2**

Ultimate shear strength of MS-AL joint, S_{ult}= (6950/28.3) = **245.93 MPa.**

And, Ultimate shear strength of MS-GFRP joint, S_{ul}= (7450/28.3) = **263.6 MPa.**

From the analysis, it shows that the shear strength of the MS-GFRP joint is better than the MS-AL joint. The Figure 3 shows the image of the failed specimen. The failure mode of MS-AL is Shear out and of MS-GFRP is net tension failure.

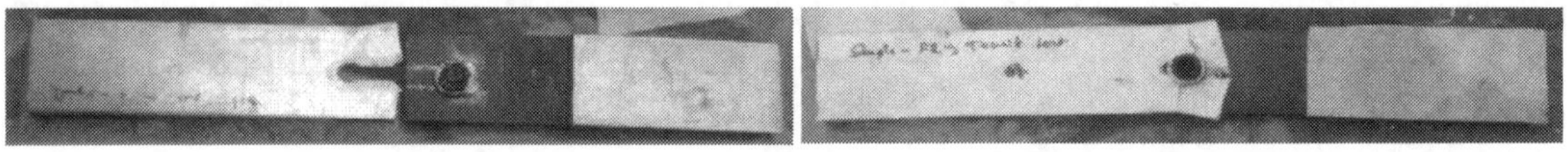

Figure 3. Failed Specimen of MS-AL and MS-GFRP in Shear test

3.2 Four-point bending test

The doors are subjected to bending load. Due to the closing of the doors, there will be generation of bending load on the door. So it is necessary to find the bending strength of the door. In this study, since we have concentrated at the joint, it is essential to find the bend strength of the joint. There is no standard procedure to find the bending strength of the dissimilar joint, ASTM E-855[6] standard is used to determine the bending strength of the joint. Figure 4 shows the image the bending test specimen.

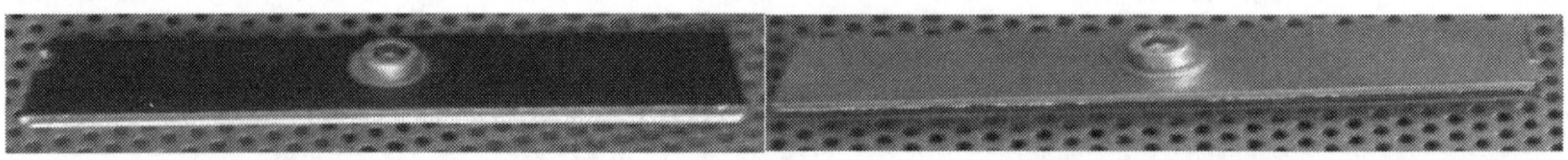

Figure 4. Image of Bending test specimen for MS-GFRP and MS-AL

The test has been carried out on universal testing machine using four bending fixtures for both MS-AL and MS-GFRP joints and the test as been performed for two specimens of each combination. Figure 5 show experimental load value at different cross head travel for MS-AL and MS-GFRP joint in bending test. The ultimate bending load obtained from the test is 7669N and of MS-AL and MS-GFRP joints.

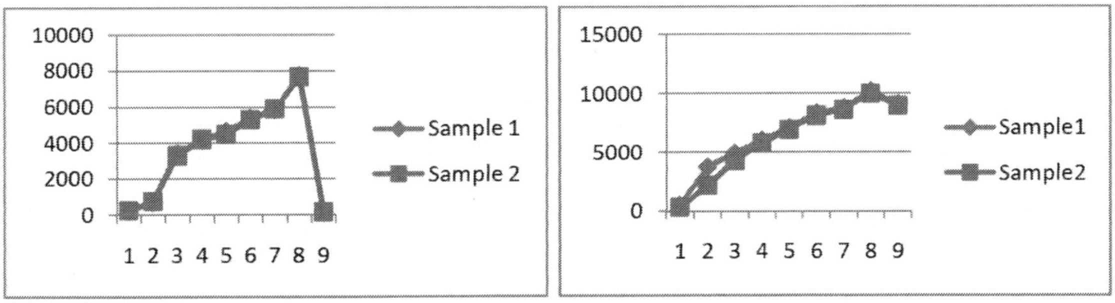

Figure 5. Loads (N) vs. cross head travel (mm) of MS-AL joint in bending

Bending Strength of the joint is determined by using the formula, $\mathbf{S_b = 3Wa / bh^2}$ (MPa).

Where, W= Load (N), b= Width of the specimen (mm), h= Thickness of the specimen (mm), a= distance from the support to the load applicator (mm).

Ultimate Bending strength of MS-AL, S_{bult}= [(3*7669*16) / (24*6^2)] = **426.1 MPa.**

And, Ultimate Bending strength of MS-GFRP, S_{ult}= [(3*10083*16) / (24*6^2)] = **560.2 MPa.**

From this analysis, it shows that the bending strength of the MS-GFRP joint is better than the MS-AL joint. The Figure 6 shows the image of the specimens failed under bending.

Figure 6. MS-AL and MS-GFRP combination specimen failed in bending test

4 Finite element analyses

The finite element analysis is carried out using average load from the experimental data and the bolted joint specimen is designed as per standards used in experimental method for different analysis. The simulation is carried using the analysis software Ansys 17.0.

4.1 Shear strength analysis

In meshing, a quadrilateral type of mesh is used for both types of material joint. The number of elements and nodes obtained are 3121 and 12612 respectively. The boundary conditions are applied as per experimental condition. Here, the Mild Steel end of joint model is fixed in all the three direction and the tensile load is applied at the AL or GFRP end of the joint model, which has free displacement in x direction. The Figure 7 shows the boundary conditions for materials joints under shear.

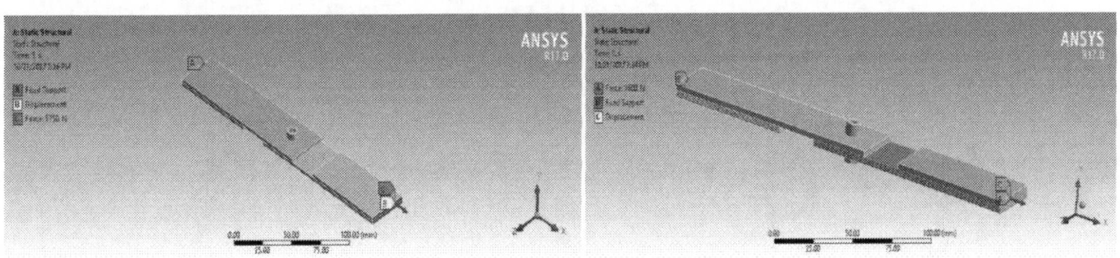

Figure 7. Boundary condition of MS-AL and MS-GFRP joint for shear

The finite element analysis has been carried out for ultimate load which is obtained from experimental verification for both types of joint. The Figure 8 shows the shear strength value for ultimate load of

6950N and 7450N for MS-AL and MS-GFRP materials joint respectively. The results obtained are 245.9 and 426.1Mpa for MS-AL and MS-GFRP materials joint respectively.

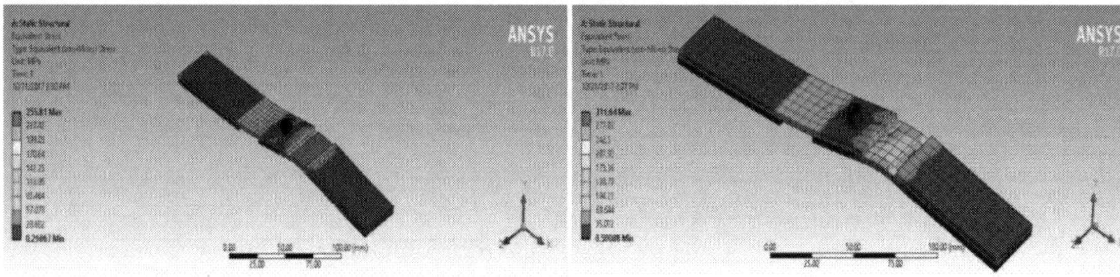

Figure 8. Shear stress of MS-AL and MS-GFRP materials joint for ultimate load

From this analysis, it shows that the Shear Strength of MS-AL joint as higher strength than the MS-GFRP joint.

4.2 FOUR POINT BEND ANALYSIS:

In meshing, a quadrilateral type of mesh is used for both the types material joint. The number of elements and nodes are 1690 and 7906 respectively. The Figure 9 shows the boundary condition for MS-AL and MS-GFRP materials joint. In this, the load is applied on Mild Steel upper face or on the loader, and support is fixed in all the direction. The specimen is fixed in x and z direction and free displacement in y direction. The boundary conditions are applied on the basis of experimental setup.

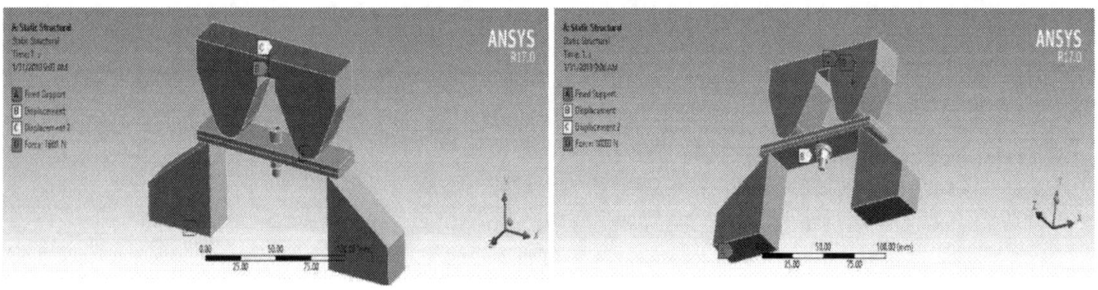

Figure 9. Boundary condition of MS – AL & MS-GFRP joints for Bending

The finite element analysis has been carried out for ultimate load which is obtained from experimental verification for both types of joint. The Figure shows the shear strength value for ultimate load of 7669N and 10083N for MS-AL and MS-GFRP materials joint respectively. The results obtained are 311.6 and 637 Mpa for MS-AL and MS-GFRP materials joint respectively

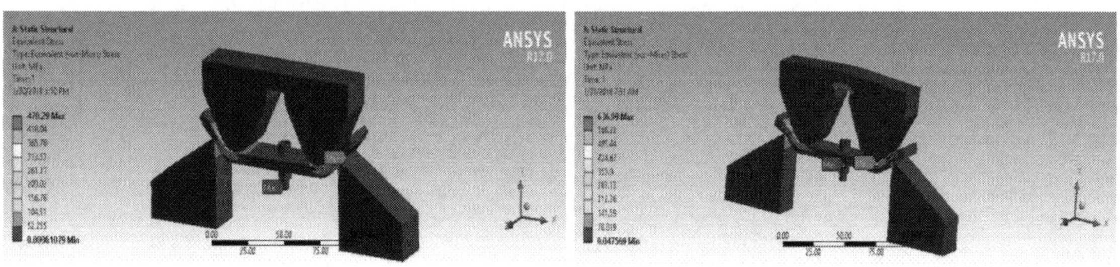

Figure 10. Equivalent stresses for MS – AL and MS-GFRP joints for ultimate load for Bending

From this analysis, it shows that the Bending strength of MS-AL joint as lower strength than the MS-GFRP joint.

The table 1 shows the comparative study between FEA and Experimental analysis for ultimate load under shear and bending test.

Table 1. Comparison between experimental and FE method

Types of materials joint	Ultimate shear Strength (Mpa)		Ultimate Bending Strength (Mpa)	
	Experimental	FEA	Experimental	FEA
MS-AL	245.9	255.8	426.1	470
MS-GFRP	263.9	311.6	560.2	637

5 Conclusion

In this present work, the MS-AL and MS-GFRP materials single bolted single lap joint has been analyzed. The shear and bending strength of both the joints have been analyzed using experimental and finite element method. In experimental analysis, a two set sample were prepared as per ASTM standards and carried out the shear and bending test were carried out. The average load obtained from the test has been used in empirical formulae to determine the shear and bending strength. In case of FE analysis, the models are generated as per ASTM standards which are used in experimental work using Creo3.0 and the simulation was carried out using Ansys package. In simulation, average load obtained from the experimental work has been applied on the model to verify the shear and bending strength. Finally the results of joints of dissimilar materials were compared and concluded as follows:

- MS-GFRP materials single bolted joint has better strength compared to MS-AL materials single bolted joint under Shear and Bending loading.
- Experimental analysis has good agreement with Finite Element analysis with minimal difference.
- Finally, MS-GFRP joint has better properties compare to MS-AL joint.

Reference:

1. Paul Kah, Raimo Suoranta, Jukka Martikainen, Carl Magnus "Techniques for joining dissimilar material joint: Metals and polymers", Rev. Advance material science, Vol. 36, 2014 pp.152-164.
2. Ricardo de Medeiros, Marcelo Leite Ribeiro and Volnei Tita, "Experimental Methodology for Testing Metal-Composite Bolted Joints", Journal of Mineral Metal and Material Engineering, E-ISSN: 2414-2115, Vol. 2, 2016 pp. 11-22.
3. Esmaeil Ghanbari, Onur Sayman, Mustafa Ozen, Yusuf Arman, "Failure Load of Composite Single-Lap Bolting Joint Under Traction Force and Bending Moment", Australian Journal of Basic and Applied Sciences, Vol. 6, ISSN: 1991-8178, 2012 pp. 683-692.
4. Hilton Ahmed, "Stress distribution of secondary bending in single lap bolted joints with dissimilar joining plates and plates type", Hilton Ahmad / Jurnal Teknologi (Sciences & Engineering) ISSN: 78:5–5, 2015, pp. 1–7.
5. American Society for Testing and Materials. D5961/D5961M-05, Standard test method for bearing response of polymer matrix composite laminates. West Conshohocken: ASTM, 2007.
6. American Society for Testing and Materials. E 855-90, Standard test method for Bend Testing of Metallic Flat Materials for Spring Applications Involving Static Loading. West Conshohocken: ASTM, 2007.
7. United States Department of Defense Military Handbook, MIL-HDBK-5J Metallic Materials Properties Development and Standardization. US Department of Defense: Philadelphia, 2003.

IOP Conf. Series: Materials Science and Engineering **390** (2018) 012023 doi:10.1088/1757-899X/390/1/012023

Multi response optimization of drilling performance of MWCNT filled banana-glass fibre reinforced composite

T. Rajmohan[1], G.M. Dinesh[1], Gangamalla Vinay Raghavendra[1], Gayatri.S[1], K. R. Charan reddy[1], Kovuru Yogeswar[1]

[1]Department of Mechanical Engineering, Sri Chandrasekharendra Viswa Mahavidyalaya, Kanchipuram -631561, Tamilnadu, India.

Abstract. Natural fibres reinforced composites (NFRC) with their extended account of portion mankind are extremely significant in a broad variety of applications, and they battle and co-exist in the twenty-first century with man-made fibres, particularly as far as excellence, sustainability and cost-cutting measure of production are troubled. Numerous researchers have synthesized NFRCs using various natural fibres as well as matrix materials. Though, real-world applications of NFRCs involve some machining operations in order to complete the assembly of the parts. In this work, a technique based on the factorial design-based desirability analysis is used to determine the desired optimum machining parameters for minimum delamination, thrust force and surface roughness in drilling of MWCNT filled banana-glass natural fibres reinforced composites. The speed, federate, wt% MWCNT and point angle of drill are the most important parameters that characterize the machining operation and have been selected for investigation. Results indicated that point angle is playing major role on the responses followed by feed rate.

1. Introduction

Natural Fibre Reinforced Composites (NFRC) have turn into very significant materials in the recent years [1]. They have been used in numerous industries such as in automotive, aerospace, building and construction, and furniture industries [2]. NFRCs are inexpensive to produce, Eco friendly, safe to health, light in weight, High stiffness to strength ratio. Even though NFRC are lack in thermal stability, strength degradation, water absorption and poor impact properties [3]. To solve this problem, material engineers are investigating the use of hybridization and different nano particles mixed in different polymers to improve mechanical and wear properties of the polymer. The hybridization may be Natural – Natural, Natural – Artificial. It provides an alternative to achieve a mix together properties such as, high stiffness, Increased fatigue life, high toughness, high impact resistance [4]. The banana fiber was used as reinforcing agent in cement and polymer-based composites by various researchers [5-8]. Due to low density, high tensile strength, high tensile modulus, and low elongation at break of banana fibers, composites based on these fibers have very good potential use in the various sectors like construction, automotive, machinery, etc. [9]

Drilling building is one of the most important machining operations to make easy the assembly operations of components formed from NFRCs. Drilling of composite materials is considered to be a critical operation owing to their tendency to delaminate when subjected to cutting forces. In order to defeat the issues such as delamination, poor surface roughness and etc., it is necessary to extend a appropriate method and choose suitable cutting parameters [10].

Durão et al. [11] assured that the decrease of delamination can be attained through the correct range of machining situation such as feed and cutting speed as well as tool material and tool geometry. Delamination mechanism takes place previous to the plates are entirely pierced by the drill, and it mostly depends on the feed rate and drill diameter [12]. Ramesh et al. [13] performed drilling by various drill geometries on hybrid glass-sisal,and glass-jute fiber reinforced composites. They concluded that the brad and spur drill created less induced break about the hole surface and produced lesser thrust force than the other two drills the delamination at hole exit increases with the increase in the feed rate and decreases with the increase in the speed and point angle [14]. Jayabal and Natarajan [15-16] showed that minimum drill forces are attained when using smaller diameter drill bits and functioning at higher spindle speed and lower feed rate in drilling of of coir fiber reinforced composites.

Patel et al. [14] found that the thrust force increases with the increase in the point angle and feed rate in drilling of banana fiber reinforced composites. Balaji et al. [17] found that treated woven coir mat exhibited low delamination when compared to non-woven coir reinforced polyester composites. Sakthivel et al. [18] found that the optimized parameter to control thrust force in basalt/sisal fiber. Venkateshwaran and ElayaPerumal [19] studied analysis of delamination at the entrance and exit of the composite plates as a function of the drilling process parameters. They used machine vision technique to analyse the quality of the hole and also the ultrasonic C-scan imaging method was carried out to find the amount of delamination.

Most of the published research work has been approved in machining of synthetic fibre reinforced composites. Not so much work has been reported on machining of NFRCs. Hence the present paper discusses the effect of drilling parameters on thrust force, surface roughness and delamination. Results revealed that point angle of drill are the significant factor followed by feed rate.

2. Materials and Methods

2.1Materials used.

Epoxy resin Araldite LY556 and hardener Aradur HY951 is acquired from Huntsman group and MWCNT manufactured by M/S US Research Nano materials Inc, USA were used as components of the starting materials. Glass and banana fibres are acquired from M/s Suntech Fibers Chennai, India were used as a reinforcing agent. An ultrasonic bath sonicator, was used to make sure consistent dispersal of MWCNT particles into resin with no agglomeration. The present NFRCs includes of five layers in total, in which the glass fiber layers are set in the top, middle and bottom of the NFRC, whereas the second and fourth layers are filled by banana natural fibre. Compression moulding procedure is used to make the NFRCs for the present investigation. The composition of NFRC is presented Table 1

Table 1 Composition of Banana -glass fibre reinforced composites

Specimen	wt %Epoxy	wt % Glass	Wt % of MWCNT	wt % banana
A1	60	25	0	15
A2	60	25	1	15

2.2 Experimental design

Design of experiments (DoE) is an organized precise method in engineering problem cracking that applies principles and methods at the data compilation stage so as to make sure the generation of suitable, justifiable, and manageable engineering conclusions of manufacturing processes. To solve this problem a design in which every setting of every factor appears with every setting of every other factor is a full factorial design. An ordinary experimental design is one with all input factors set at two levels each. These levels are called `high' and `low' or `+1' and `-1', in that order. A design with all probable high/low combinations of all the input factors is called a full factorial design in two levels. From the author's earlier available research and numeral of tests conducted in the laboratory for polymer composites, four major control parameters were selected for the experiment (Rajmohan et al. 2015). The parameters selected for the present investigation are: 1. Spindle speed 2. Feed rate 3. wt% of filler 4. Drill point angle. In this research 4 factors 2 level full factorial 2^4 experimental design is considered for conducting experiments are shown in Table 2.

Table 2 Drilling parameters and their levels

Sno	Spindle speed rpm	Feed rate mm/rev	Point angle deg	Wt of MWCNT %
Level 1	500	0.04	118	0
Level 2	1250	0.08	135	1

2.3 Experimental procedure

Drilling engages the taking away the material from work piece such that a hole is obtained. Drilling is a difficult process; it needs methodical and quantitative swots on all the major controlling parameters on machining performance in order to structure the rule base under a wide range of work conditions. In the present investigation the drilling experiments were carried out in radial drilling machine. The machining samples were prepared in the form of 100 mm × 100 mm × 10 mm blocks for each material. Solid carbide Drills with diameter of 10 mm, a helix angle of 30°, and a point angle of 90o and 118° were used. The schematic arrangement of the experimental setup is shown in Figure 1.

Figure 1 Experimental Setup

IOP Conf. Series: Materials Science and Engineering **390** (2018) 012023 doi:10.1088/1757-899X/390/1/012023

The thrust force is measured directly from the strain gauge dynamometer. The surface roughness of the workpiece is measured with a Mitutoyo portable Surftest SJ-201 P/M contact profilometer at 2.4 mm cut-off value. Delamination is a damage phenomenon that occurs due to the anisotropic and brittleness of composite materials. The delamination was measured by using a coordinate measuring machine (CMM). The delamination factor is determined by the ratio of the maximum diameter Dmax of the delamination zone to the hole diameter D. The observed response values for all of the 16 experiments are listed in Table 3.

Table 3. Experimental results

	wt % of MWCNT	Feed rate mm/rev	Point angle deg	Spindle speed rpm	Thrust force in N	Surface roughness um	Delamination factor
1	0	0.04	118	500	3	1.813	1.001
2	0	0.08	118	500	3	3.091	1.001
3	0	0.04	118	1250	3	1.963	1.002
4	0	0.08	118	1250	5	1.533	1.002
5	0	0.04	135	1250	5	2.178	1.012
6	0	0.08	135	1250	7	1.619	1.009
7	0	0.04	135	500	7	1.327	1.012
8	0	0.08	135	500	8	1.985	1.011
9	5	0.08	135	500	3	12.016	1.012
10	5	0.04	135	500	5	10.509	1.01
11	5	0.04	135	1250	3	12.939	1.009
12	5	0.08	135	1250	3	5.155	1.011
13	5	0.04	118	1250	3	12.016	1.001
14	5	0.08	118	1250	5	10.849	0.999
15	5	0.04	118	500	6	9.722	1.001
16	5	0.08	118	500	6	5.059	0.998

3. Results and discussion

The machinability in this work was assessed by surface roughness (Ra) of the drilled surface of the work piece, thrust force and delamination. The Factorial design of experiments and desirability function analysis are applied in this project work for the identification of best levels of drilling parameters, significance, and optimisation of parameters.

- The individual desirability (di) is calculated for all the responses depending upon the type of quality characteristics. Since all the responses are possessing minimisation objective, the equation corresponding to smaller the better type is selected. The computed individual desirability for each quality characteristics using following equation

$$d_i = \begin{cases} 1, & \hat{y} \leq y_{min} \\ \left(\dfrac{\hat{y} - y_{max}}{y_{min} - y_{max}}\right)^r, & y_{min} \leq \hat{y} \leq y_{max}, r \geq 0 \\ 0, & \hat{y} \geq y_{max} \end{cases}$$

where the y_{max} and y_{min} represent the upper/lower tolerance limits of by and s and t represent the weights.

- The composite desirability values (dG) are calculated using following Equation

$$d_G = \sqrt[w]{\left(d_1^{w_1} * d_2^{w_2} \dots\dots\dots * d_i^{w_i}\right)}$$

where di is the individual desirability of the property Yi, wi is the weight of the property "Yi" in the composite desirability and w is the sum of the individual weights.The weightage for responses are based on assumed weightage of 1:2:3 for thrust force, delamination and surface roughness, respectively. Finally, these values are considered for optimising the multi-response parameter design problem. The results are shown in the Table 4.

- From the value of composite desirability in Table 4, the parameter effect and the optimal level are estimated and tabulated in Table 5 Considering the maximisation of composite desirability value, the optimal parameter condition is wt % MWCNT at level 1, spindle speed at level 2 feed rate at level 2 and point angle at level 1.
- Using the composite desirability value, ANOVA is formulated for identifying the significant parameters. The results of ANOVA are given in the Table 6. Prediction of optimum condition: By using the identified optimal parameter condition, the quality characteristics are verified by conducting confirmation experiments

Table 4. Individual and composite desirability for responses

S.No	$D_{Thrust\ force}$	$D_{Surface\ roughness}$	$D_{delamination}$	Composite desirability
1	0.958147	0.785714	1	0.903242
2	0.848088	0.785714	1	0.849784
3	0.945229	0.714286	1	0.869079
4	0.98226	0.714286	0.6	0.813634
5	0.926714	0	0.6	0
6	0.974854	0.214286	0.2	0.451827
7	1	0	0.2	0
8	0.943334	0.071429	0	0
9	0.079487	0	1	0
10	0.209266	0.142857	0.6	0.219622
11	0	0.214286	1	0
12	0.670341	0.071429	1	0.339708
13	0.079487	0.785714	1	0.260157
14	0.179986	0.928571	0.6	0.380116
15	0.277041	0.785714	0.4	0.416905
16	0.678608	1	0.4	0.707109

Table 5. Mean composite desirability for responses

Level	wt % of MWCNT	Feed rate mm/rev	Point angle deg	Speed
1	**0.485**	0.333	**0.65**	0.387
2	0.29	**0.442**	0.126	**0.402**
Optimal level	W_1	$\mathbf{f_2}$	P_1	$\mathbf{S_2}$

Table 6. ANOVA for composite desirability

Source	Sum of Squares	df	Mean Square	F Value	p-value Prob > F
Model	0.458733	4	0.114683	0.942037	0.4755
wt % MWCNT	0.107652	1	0.107652	0.884284	0.3672
Feed rate	0.000417	1	0.000417	0.003426	0.9544
Point angle	0.197793	1	0.197793	1.624718	0.2287
Spindle speed	0.152871	1	0.152871	1.255721	0.2863

3.1 Confirmation test

The intention of the confirmation test is to corroborate conclusions strained through the analysis phase. Once the optimum level of the process parameters is selected, the final step is to predict and verify the improvement of the performance characteristics using the optimum level of the process parameters. The confirmation experiment is conducted at the optimum test conditions to prove the quality characteristics for drilling of NFRC. The summary of the results obtained from the confirmation tests are shown in table 7. The predicted values are very close to the experimental results, and hence, the developed model is suitable for predicting the performance in drilling of NFRCs.

Table 7. Confirmation Experiments

Optimum setting	Predicted values			Experimental values		
	Roughness	Delamination	Thrust force	Roughness	Delamination	Thrust force
$W_1f_2p_1s_2$	1.62	1.002	5.25	1.67	1.004	5.5

4. Conclusions

Drilling experiments were conducted based on Factorial design technique for banana-glass fibre reinforced composites using carbide drill. The experimentally composed data were subjected to desirability function analysis for optimisation of machining parameters. From this analysis, the following conclusions are drawn for thrust force, delamination and surface roughness.

- Desirability function in the factorial design for the optimisation of multi-response problems is a very useful tool for predicting surface roughness, thrust force and delamination in drilling of NFRCs.
- The optimal parameter condition is wt % MWCNT at level 1, spindle speed at level 2 feed rate at level 2 and point angle at level 1to improve the performance in drilling of NFRCs.
- Point angle is the significant parameter followed by spindle speed and feed rate for drilling of NFRCs.

5. References

[1] Zimniewska M, Myalski J, Koziol M, Mankowski J and Bogacz E 2012 *J. Nat. Fibers* **9** 229–239.
[2] John MJ, Varughese KT and Thomas S 2008 *J. Nat. Fibers* **5** 47–60.
[3] Mohan K and Rajmohan T *J. Nat. Fibers* 1-11.
[4] Rajmohan T, Mohan K and Palanikumar K *Appl. Mech. Materials* 766.
[5] Sgriccia N, Hawley M C and Mishra M 2008 *Comp.: Part A*, **39** 1632-37.
[6] Ju stiz-Smith N G, Virgo G J and Buchanan V E 2008 *J. Mat. Characterization* **59** 1273-78.
[7] Geethama V G, Reethama J and Thomas S 1995 *J. Appl. Poly. Science* **55** 583-94.
[8] Venkateshwaran N and Elayaperumal A 2010 *J. Reinf. Plast. Composites* **29** 2387.
[9] Abrão AM, Faria PE, Rubio JCC, Reis P and Davim JP 2007 *J Mater Process Technol* **186** 1–7.
[10] Durão LMP, Gonçalves DJS, Tavares JMRS, de Albuquerque VHC, Panzera TH, Silva LJ, Vieira AA and Baptista APM 2013 *Mater Sci Forum* **730–732** 301–306.
[11] Chandramohan D and Marimuthu K 2011 *Int J Adv Eng Res Stud* **1** 134–145.

[12] Ramesh M, Palanikumar K and Reddy KH 2014 *26th All India Manuf. Tech. Desi. Res. Conf.* December 2–7
[13] Patel K, Gohil PP, Chaudhary V and Patel K 2015 *Int. Conf. Civil Mat. Env. Sciences* **1** 201–205.
[14] Jayabal S and Natarajan U 2011 *Bull Mater Sci* **34** 1563–67.
[15] Jayabal S and Natarajan U 2010 *Int J Adv Manuf Technol* **51** 371–81.
[16] Balaji NS, Jayabal S, Kalyana Sundaram S, Rajamuneeswaran S and Suresh P 2014 *Adv Mater Res* **984–985** 185–193.
[17] Sakthivel M, Vijayakumar S and Prasad NK 2015 *Appl Math Sci* **9** 3285–3290.
[18] Venkateshwaran N and ElayaPerumal A 2013 *J Reinf Plast Compos* **32(16)** 1188–1197.

A Critical Evaluation of Impact of Human Error: In Rail System Failure with a View to Enhance Safety

Sudarson Nayak[1]and Prof. S Tripathy[2]

[1]Research Scholar, School of Mechanical Engineering, KIIT University, Bhubaneswar, Odisha, India

[2]Professor Scholar, School of Mechanical Engineering, KIIT UniversityBhubaneswar, Odisha, India

[1]Snayak03@gmail.com

Abstract. Modern Transportation is a 24x7 industry, fraught with the risk of occasional accidents/incidents. This makes every mode of transport a risky business. With perennial pressure for becoming faster, safer & better with every passing day, greatest challenge for the industry is to eliminate & manage all possible risks to ensure an incident/accident free operation to meet the growing demand of the society. Modern society cannot be conceived without an efficient & safe transport system. Amongst all modes of transport, rail transportation occupies a unique place being the bulk carrier of people & cargo at an affordable cost with least adverse impact on environment.

Despite massive modernization & induction of state of art technology, the safety of rail-road transportation depends largely on the operator. A railway accident always makes a big news across societies & has a particular impact on the psyche of public. There is always a cause most of the time trivial.

The bright side of railway accidents that all misfortune has been turned to profit with evolution of technology based on the lessons learnt. In fact, the modernization of railways so far has greatly depended & progressed on the accident investigation knowledge. Railways growth since its evolution almost two centuries ago is nothing short of a revolutionary technical adventure unique in human history.

In the paper a comprehensive study of all human factor issues involved in working of employees at sharp end of safety has been attempted with critical analysis of more than 1200 incidents/accidents over a period of decades in Indian Railways.

The study revealed many hidden areas & contributing factors in lowering the human performance limit, leading to unsafe acts & unpredicted tragic accidents/incidents. Many findings have belied the traditional knowledge, belief & logic.

The knowledge emerging from this study may help develop counter measures against risks in all transport sectors which may accrue immense societal benefit. Further study could be undertaken by researchers, academician, administrators and regulators in greater public interest.

Keywords: Human Error, Human Factor, Crew. Station Master, Rail-road crossing, Front line staff, Risk management.

1. Introduction

Rail transport accidents are sad example of 'Human Error' in full view, sometimes exposing the organization to a shameful situation. Systems are not basically safe. It is therefore wrong to assume absolute system safety. It is the people who create safety while negotiating multiple layer of system goals with time and resource constraints. What one think should have/have not been done, cannot explain people's behavior. The unsafe act is an observable error such as choice of wrong procedure, task prioritization, violation of rules, violation of methods etc.

New technologies do not remove 'human error'. It only changes it. Human error is not the conclusion of an investigation it is the starting point. Why he did what he did that made sense to him then is the big question.

There is always a mismatch between rules and practices. Do then one should write more rules as the practices have little option for change.

There is a constant 'tug of war'. If one concludes 'human error' then it is worthless to have spent money on investigation. One of the problems with safety is that one can do things in an unsafe manner and still get away with it, but not always. Railways suffer from this complacency syndrome.

Sub optimal human performance often result in unsafe work. To understand conditions that yield error it is necessary to evolve beyond limited scope of tracing the erroneous action of individuals to its root. For every detected error in the system, there are countless undetected errors which might have already taken place or waiting to take place in some form or other to breach the system safety.

Seeing the invisible hidden error trap is the skill which the front-line staff to be motivated and trained to acquire. This perhaps can be strongest antidote to 'human error'. Each human failing is an opportunity for learning. Human error is a function of (I-Mechanical Error). Symbolic express can be

$HE = \int (I - ME)$.

With a view to assess all the possible human error that has potential to lead to unsafe situation/accident are identified hereunder with a detailed analysis of accidents statistics of Indian Railway from 1980 to 2010 particularly the accidents attributable to human error. The generic errors committed by front line staff are identified as under.

2. Identification of Error

Working of four categories of front line staff and their vulnerability of committing error have been analyzed based on available data.

2.1 Crew:

Crew is one of the most critical safety staff. The various type of error committed by them are primarily for the following reasons.

Signal Passed at Danger - HFCR1, Failure to observe speed as per signal aspect – HFCR2, Failure to check Auto signal – HFCR3, Failure to clear Fouling mark - HFCR4, Unauthorized backing of train - HFCR5, Excess speed – HFCR6, Poor risk assessment during abnormal work – HFCR7, Psychological depression – HFCR8, Excessive speed under abnormal working – HFCR9, No tail lamp – HFCR10, Failure to switch on flasher – HFCR11, Poor response to danger signal – HFCR12, Failure to protect train – HFCR13, Non-reporting of abnormal riding – HFCR14, Failure to observe rule – HFCR15, Non-securing of train leading to Rolling back of train – HFCR16, Taking unauthorized people in cab/footboard – HFCR17, Poor knowledge of brake system – HFCR18, Not observing speed/over speeding – HFCR19, Rolling down of train after pantry – HFCR20, Over speeding after passing auto signal/IB signal danger – HFCR21, New assets acquaintance – HFCR22, Passing semi-automatic signal at danger – HFCR23, Over speeding on turn out. – HFCR24, working without BPC – HFCR25, Violation of rules to pass auto signal at danger – HFCR26, Alcoholism – HFCR27, Fatigue related failure – HFCR28, Lack of alertness – HFCR29, Delayed action in brake application – HFCR30.

These causes have been accordingly illustrated in the diagram.

2.2 *Station Master*

Station Master role is very critical in train running. Generic error committed by them are primarily for the following reasons.

Wrong reception of train – HFSM1, Shunting in face of approaching train – HFSM2, Wrong operation/inattention – HFSM3, Wrong reception of train (no track circuit) – HFSM4, Wrong procedure followed which issuing PLC – HFSM5, Failure to observe rules – HFSM6, Wrong procedure in reception of train in two-line station with no isolation – HFSM7, Gradient of station yard and seeing of stabled train – HFSM8, Non-reversal of point in rear after receipt of train – HFSM9, Wrong system of simultaneous reception – HFSM10, Parted goods train rolled down and collided with stabled train – HFSM11, SWR incomplete – HFSM12, Reception without ensuring clearance of line – HFSM13, Not changing point to normal in rear after reception of train – HFSM14, Starter advance starter was occupied - HFSM15, Violation of rules during NI work – HFSM16, Unauthorized taking off of signal – HFSM17, Use of defective block instruments – HFSM18, Communication failure – HFSM19, Signal put back – HFSM20, Cancellation of route – HFSM21, Issuing of wrong authority – HFSM22

2.3 *L.C Gate Keeper*

LC Gate keepers work in a very hostile working environment. They are vulnerable to commit error under pressure of work. The type of error identified are as follows.

Failed to close – HFLC1, Unauthorized opening of gate - HFLC2, Failure to lock the Gate- HFLC3, Unauthorized opening of gate- HFLC4, Gate getting closed safe- HFLC5, Failure in communication-HFLC6, Exchanging PN before gate closing- HFLC7, Violation of rules – HFLC8, Lack of knowledge of Gate keeper- HFLC9

2.4 *Electrical Signal Maintainer*

Electrical signal maintainers work under pressure to minimize delay to train movement. As they work under tremendous pressure against time, they are vulnerable to commit error. The main errors are due to the following reasons

Wrongfully energizing relay – HFSIG1, blanking of signal – HFSIG2, Signal lamp fuse – HFSIG3, Attending failure without traffic block. – HFSIG4, Short cut method to rectify fault – HFSIG5, Poor appreciation of risk – HFSIG6, Defective signal – HFSIG7, Interference when interlock failed – HFSIG8, Shortcut method adopted to give signal – HFSIG9, blanking off signal – HFSIG10, no light due to failure of power supply – HFSIG11, LC obtained with defective block instrument – HFSIG12.

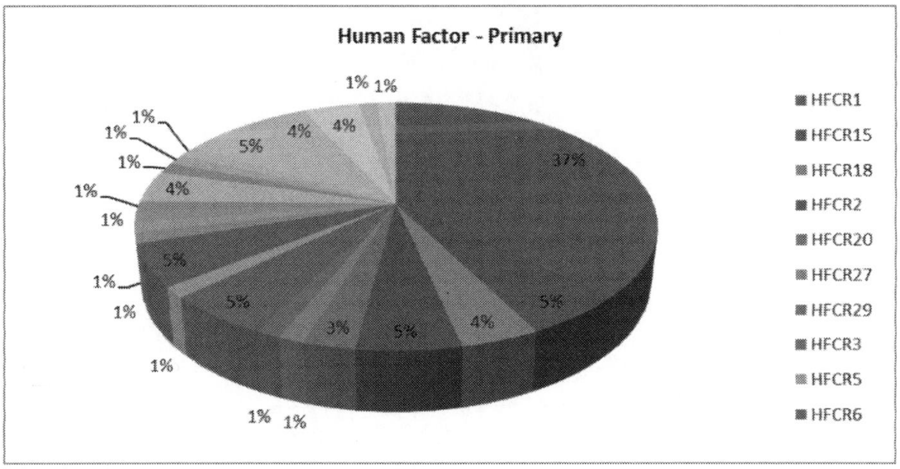

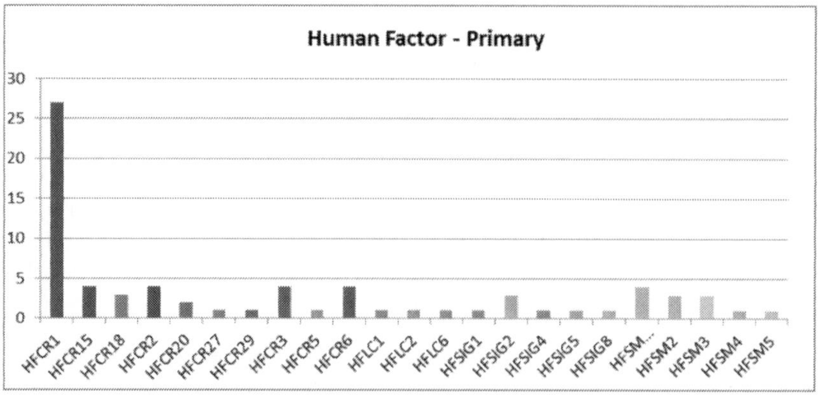

From the above revealed that the accidents attributable to human error (Primary) are broadly as under:

i. Signal Passed at Danger – Crew Fault (HFCR1) - 37%
ii. Failure to observe speed as per signal aspect (HFCR2)– 5%
iii. Failure to observe rule (HFCR15) – 5 %
iv. Reception without ensuring clearance of line (HFSM13) – 5%
v. Excess speed (HFCR6) – 6%
vi. Failure to check Auto signal (HFCR3) – 5%
vii. Poor knowledge of brake system (HFCR18) - 4%
viii. Blanking of signal (HFSIG2) – 4%
ix. Wrong operation/inattention (HFSM3) – 4%

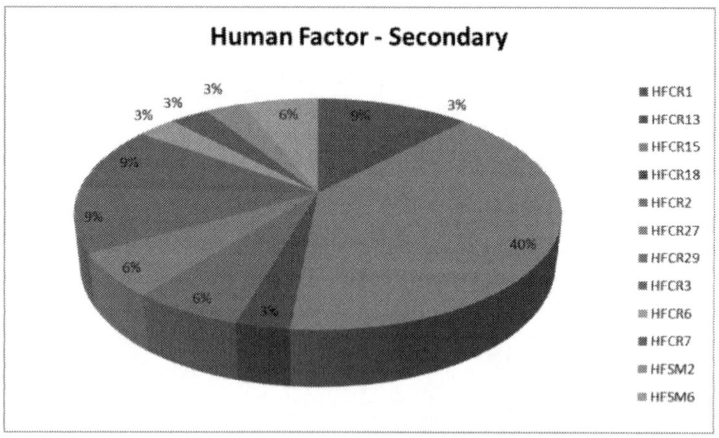

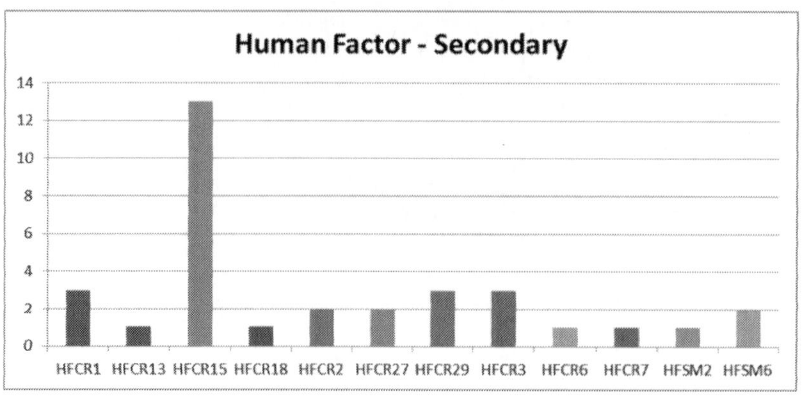

IOP Conf. Series: Materials Science and Engineering **390** (2018) 012024 doi:10.1088/1757-899X/390/1/012024

From the above it is revealed that the accidents attributable to human error (Secondary) are broadly as under:

i. Failure to observe rule (HFCR15) - 40%
ii. Lack of alertness (HFCR29) – 9%
iii. Failure to check Auto signal (HFCR3) – 9%
iv. Signal Passed At danger (Signal was blank) – 9%
v. Failure to observe rules (HFSM6) – 6%
vi. Failure to observe speed as per signal aspect (HFCR2) – 6%
vii. Alcoholism – 6%

2.5 Error Tolerant Rail Transport System

By addressing the susceptibility of frontline staff to the identified human error as mentioned above, it may be possible to reduce accidents and also develop an error tolerant rail transport system. The above could be aimed with following strategy:

Human error identification (HEI), Human error construction, Human Error Model, Strategies for error tolerant design

Skill, Rule, Knowledge based human behavior, Opportunity for human malfunction, Error Antidotes, infrastructure that is adopted to the limitation of human capacity

Pre-condition for unsafe Act: Latent, Operational factor, Environment factor, Personal factor, Unsafe Acts, Technology and Automation to monitor and control human action on the system. To bridge the gaps due to omission/commission, automation is less used to support human beings in their action/decision. Little regard has been paid to the fact that human flexibility and human capabilities can make an impact and important contribution when technology breakdown.

Human to be kept in battle ready condition all the time for ensuring safety. Slow decay of safe working practices after long spells of safe operation being checked. The technological advancement aimed at reducing/eliminations of human error in many cases actually increase it with great risk and severe adverse consequence which needs constant evaluation.

3 CONCLUSIONS

Human error is all pervasive and ubiquitous in any human activity especially in transport sector. The percentage of accidents and incidents caused due to human error as indicated above is as high as 70% to 80%. Hence, management of 'human error' is the key to sustainable rail transport system. Technology how advance may be when conducted by human beings is prone to human conditions. It is not possible to design man out of the system particularly in transport system. There is clear limitation to the scope of automation in rail transport. Hence, critical study of human factors providing in sight to possible human error remains the key to develop an error tolerant rail transport system through human error identification, human error model, identification of skill/rule/knowledge/best human knowledge, best human behavior, opportunity for human mal functioning and providing error antidotes with the objective of making the infrastructure and system compatible to the limitation of human capacity.

4Literature Review

1. Understanding human factor a must for Modern Rail Industry - *RSSB (Rail Standard and Safety Board/UK, Report 2008*
2. Human factors Development of Safety - *Critical Railway System – Simon Zhang – IRSC 2012*
3. People and Rail System Human factors at the Heart of the Railways - *Edited by John R. Wilson*
4. Learning from Accident – *Alle Handler.*
5. Beyond the Midnight oil Drew Dawson, Adam Flectcher – *Frank Husser*
6. Centre for Sleep Research

7. University of South Australia.
8. *Parliament of Australia House Committee or Communication, Transport and Arts report.*
9. Accident Investigation Technique – *Jeffer S. Oakley. University of Houston – American Society of safety Engineer.*
10. Accident Investigation – *The driver, method and one come prepared by Human Reliability Associates for Health and Safety Executive.*
11. Method for Accident Investigation by Snorre *SK Jet, Norwegian University of Science and Technology.*

12. Field guide to understanding Human Error – *Sidney Dekker*

13. Human Error – *J.T. Reason*
14. *Accident Data of Indian Railway (Year 1980 – 2010)*

Tribological investigation of a compressor liner-ring material under various bio-based lubricants using HFRR tribometer

P. Chenga Reddy[1 *], S. Arumugam [2], P. Nithin Sai Krishna[3], N. Jaya Sainath[4]

1, 2 Department of Mechanical Engineering, Sri Chandrasekharendra Saraswathi Viswa Mahavidyalaya, Enathur, Kanchipuram - 631561, Tamil Nadu, India.

Chengareddy.p@kanchiuniv.ac.in

Abstract. The main aim of this paper is to investigate the tribological performance of vegetable – based bio-lubricant as reciprocating air compressor oil. The tribological test at different operating conditions like frequency, stroke, load, temperature, speed and operating time was performed by the help of HFRR (High Frequency Reciprocating Rig) tribometer. The cylinder liner and piston ring material of reciprocating air compressor was taken for this tribological test under five different lubricants namely SAE30, PE, PE25, PE50 and PE75. The tribological results were shown that the experimental COF and Wear depth was reduced with PE 50 and PE75 as compared to other lubricants i.e., PE, PE25 and SAE 30. Also PE75 bio-lubricant was shown that the smoother worn surfaces of the tested specimen, which was confirmed by SEM and EDAX analysis.

1. Introduction

In all types of applications, namely industrial, automotive and other applications are fully depended on petroleum based fuels and lubricants because of its excellence in performance. But, on the other side, there is a severe problem with these fuels and oils in terms of pollution and degradation. Also, there is a depletion problem with crude oil, which relates to hiking in price. Therefore, it is necessary to produce an environmentally friendly oil and fuels from vegetable oils. All the vegetable oils have excellent lubricity and biodegredability, non-toxic and safety. But they have insufficient oxidation and thermal stability [1]. To overcome these drawbacks from raw vegetable oils, many researchers have done many works like transesterification, epoxydation etc. [2-3]. All process (transesterification, epoxydation etc.) was carried out using conventional method (Mechanical stirring), which consumes more reaction time and also obtained low yield percentage. Therefore, researchers are chosen ultrasonication technic to carry over the transesterification process for the production of biodiesel, which reduces the reaction time and obtained high yield [4-5] of the product by considering various parameters like frequency, temperature, alcohol ratio and catalyst concentration [6-7]. Chenga Reddy et al [8] was produced bio-lubricant from raw rapeseed oil using ultrasonication followed by transesterification process. Reported that the ultrasonication technic is an efficient one to produce bio-lubricant and also thermo-oxidative stability was improved by transesterification process, which were confirmed by TGA and DSC analysis. Investigated the physical and chemical properties of the modified vegetable based bio-lubricant, which impacts the vegetable oil applications in lubrication [9].

Bio-lubricants are not enough to prove their physico-chemical properties and thermo-oxidative stability of lubricity applications, also it have been proven their tribological behaviour in terms of wear and friction coefficient under lubrication regimes. With this connection Arumugam et al [10-12] was completed the experimental work to find the tribological behaviour of the liner-ring material under various lubricants using a pin-on-disk, four ball and high frequency reciprocating rig tribometers. Experimental results of the chemically modified bio-lubricant shows better performance of wear and friction co-efficient than that of mineral based lubricants. Tiong chiong ing [13] proved that the palm oil based lubricant had performed better in wear and friction reduction when compared with the paraffinic mineral oil using four ball tribometer under the conditions of speed, load, varying temperatures and at 1 hr running time. Sevim z. Erhan [14] claimed that the chemically modified vegetable oil improves thermo-oxidative stability and reduces the friction and wear using four ball and ball-on-disk tribometer. Siniawski et al [15] studied friction and wear characteristics of the sunflower and soyabean oils as alternative to petroleum based lubricants using ball-on-disc tribometer. Reported that the friction and abrasion rate were lower than that of mineral oil.

From the literature study, many researchers are studying about bio based lubricants and its tribological behaviour using a pin-on-disk, four ball, ball-on-disk, HFRR tribometer etc. with this connection, main aim of this investigation is to synthesis of pentaerythryl ester (PE) from raw rapeseed oil as reciprocating air compressor oil and study its tribological behavior in terms of wear depth and friction co-efficient using high frequency reciprocating rig tribometer. The surface of the tested specimens was also discussed by scanning electron microscopy (SEM) and energy-dispersive x-ray spectroscopy (EDAX).

2. Experimental Work

2.1 Synthesis of transesterification process

A pentaerythryl ester was prepared by a two-step transesterification process. In the first step, synthesized the methyl ester of rapeseed oil via conventional methods under atmospheric pressure. Ultrasonication was used for the conversion of methyl ester to pentaerythryl ester under nitrogen atmospheric pressure, as a second step. The various materials were taken for the process and the procedure of synthesis of pentaerythryl ester was followed by Chengareddy and Arumugam [8].

The prepared pentaerythryl ester was blended with the synthetic compressor oil (SAE 30) on a percentage basis of 25%, 50% and 75% by volume. These blends were prepared using mechanical stirring process at 300 rpm by maintaining the temperature of 80^{0}C for the purpose of homogeneous mixing.

2.2 Properties of Pentaerythryl Ester

Kinematic viscosity at 100^{0}C and viscosity index measurements were found, according to ASTM D 445 and ASTM D 2270 for PE, PE25, PE50, PE 75 and SAE 30 taken as base oil for comparison. The comparative properties of oils were shown in Table 1.

Table 1. Properties of PE and its Blends

Properties		
Lubricants	Kinematic Viscosity, cSt @ 100^{0}C (ASTM D 445)	Viscosity Index (ASTM 2270)
SAE 30	9.16	95
PE	9.02	124
PE 25	8.23	98
PE 50	8.42	100
PE 75	8.66	103

2.3 HFRR Tribometer

This method is developed to characterize the tribological performance of materials and lubricants by varying the operating conditions in a small volume of lubricant as shown in Fig. 1. For this purpose, a ring material was fixed in a steel plate with a fixed force and moved with a bracket back

IOP Conf. Series: Materials Science and Engineering **390** (2018) 012025 doi:10.1088/1757-899X/390/1/012025

and forth (reciprocating). The ring specimen slides in a liquid bath from the lubricating oil to be tested on the liner material back and forth. The specimens of liner-ring material before and after the tribological tests are shown in Fig. 2

By following the standard method ASTM D 6079, the lubricity performance was used HFRR with a 10 N load; 10 mm of sliding stroke; a frequency of 15 Hz; 50^{0}C of lubricity temperature and operating time of 1 hr for each sample separately. The liner-ring (specimen) material of reciprocating air compressor was chosen for tribological tests. The specimens are shown in Fig. 2, which were cut by using wire cut EDM and the dimensions of the specimens were followed by Arumugam [12].

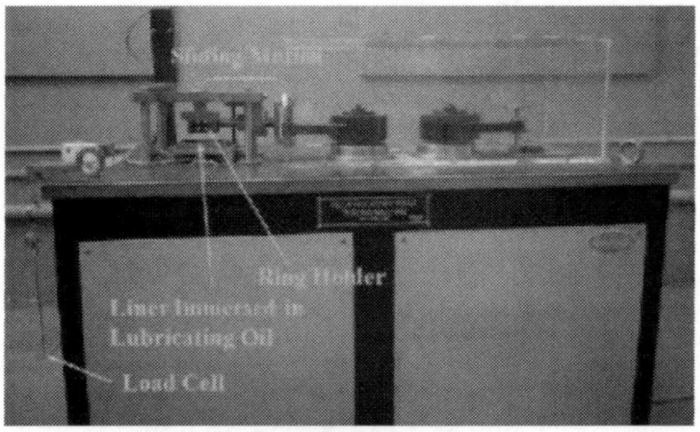

Fig. 1. HFRR Tribometer

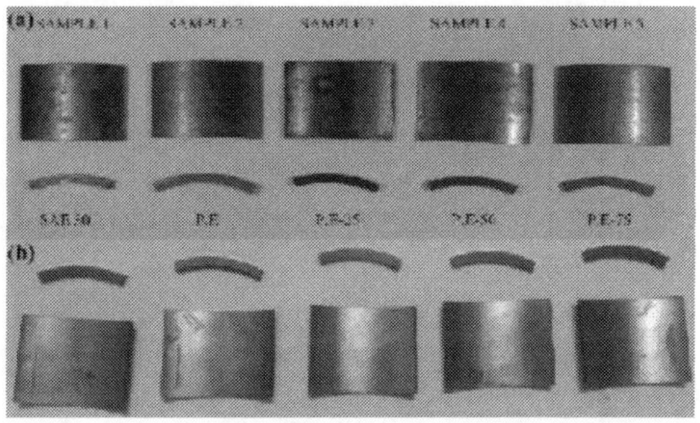

Fig.2. Liner-Ring Specimens (a) before and (b) after

3 Results and Discussions

3.1 Properties of Pentaerythryl Ester

All the four samples (i.e. PE, PE25, PE50 and PE 75) of kinematic viscosity range is 8-9 cSt, which were closer to the viscosity of SAE 30 synthetic grade compressor oil. From the Table 1. shows that the viscosity index of pentaerythryl ester had highest about 124, which indicates that have viscosity stability even at high temperatures.

3.2 HFRR Tribometer Testing Analysis

3.2.1 Co-efficient of Friction

The tribological properties of vegetable based PE as bio lubricant was investigated using HFRR tribometer under various test conditions. Fig 3. Shows the coefficient of friction of the liner material, lubricated with various blends of PE with SAE 30 and also with pure SAE 30 synthetic reciprocating air compressor oil as base oil for comparison purpose. During the starting of operation there is a rapid increase in coefficient of friction due to the sudden contact between the liner and ring specimens. However, within a few seconds the coefficient of friction was stabilized and there are some slight variations till the end of operations. Among all the five lubricants, PE 75 lubricant was shown the lowest coefficient of friction, which avoids the metal-to-metal contact because of the formation of an oil film between the specimens.

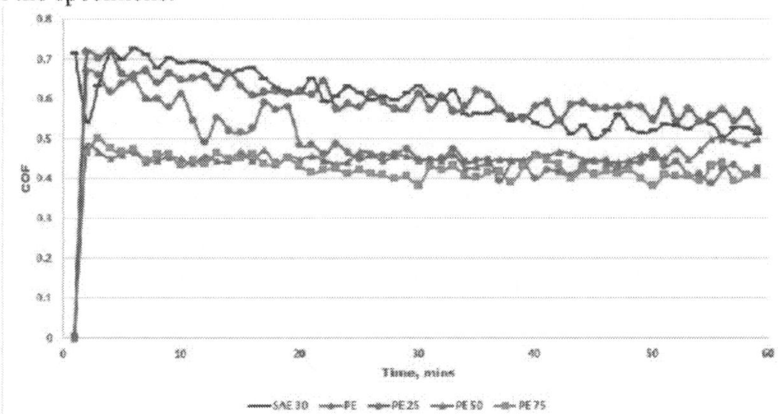

Fig. 3. Co-efficient of Friction

3.2.2 Wear Characterization

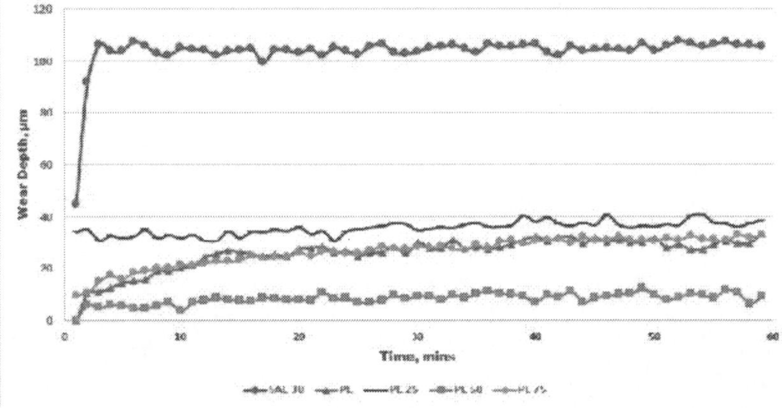

Fig. 4. Wear Depth

Fig 4. Shows the linear wear plotted against the sliding time for various vegetable based pentaerythryl ester under the load of 10 N and frequency of 50 Hz with a 1 hr operating time. It is observed that the highest wear depth was occurred on SAE 30 lubricated tested specimen. Whereas the lowest wear was observed from all the four blended lubricants (i.e. PE, PE25, PE50 and PE 75) than that of SAE 30. Out of which PE 50 shows the least wear, because of it may be the presence of oil film thickness on the liner material throughout the running time (about 1 hr) and it showed better lubricity in reducing the metal to metal contact.

3.3 SEM and EDAX Analysis

Reduction of lubricant film thickness leads to the surfaces to come closer to each other and can higher wear. Fig. 5 shows the SEM and EDAX images of the wear scars on each liner specimen treated by lubricants under the test conditions. It was observed that the ploughings are there in all the wear scar images due to its sliding motion during the test of 1 hr. However, there is some other scars were also observed like deep grooves, cavities, pits, ruptures etc. from the micrographs of specimen. The deep grooves, cavities and pits were observed from SAE 30, PE and PE 25 lubricated specimens, which indicates that a chance of oil film breakdown between the specimens due to the long running time of operation. From the SEM images observed that the PE 50 and PE 75 shows the minimum wear scars than that of the SAE 30, PE and PE 25. Therefore, it concluded that the wear scar can be reduced by the increase of PE in volume basis with SAE 30.

From the EDAX (Fig. 5 right side) spectrum, indicating that the percentage of carbon present in the PE 50 and PE 75 is little higher than that of other lubricants. Therefore, an increase in hardness and tensile strength of material due to the presence of carbon, should be an increase of wear resistance. Various elements are appearing in the specimens from the analysis are carbon (C), Oxygen (O_2), Chromium (Cr), Aluminium (Al), Silicon (Si) and Chromium (in SAE 30 only), as shown in Table 2.

Table 2. Elemental Analysis of PE and its Blends

Lubricants					
Elements	SAE 30	PE	PE 25	PE 50	PE 75
C	9.41	9.68	9.32	11.03	12.06
O_2	1.82	1.39	1.65	1.32	1.56
Al	0.23	0.45	0.22	0.16	0.17
Si	1.22	1.42	1.24	1.34	1.31
Fe	84.55	87.05	85.86	87.87	87.56
Cr	0.11	----	----	----	----

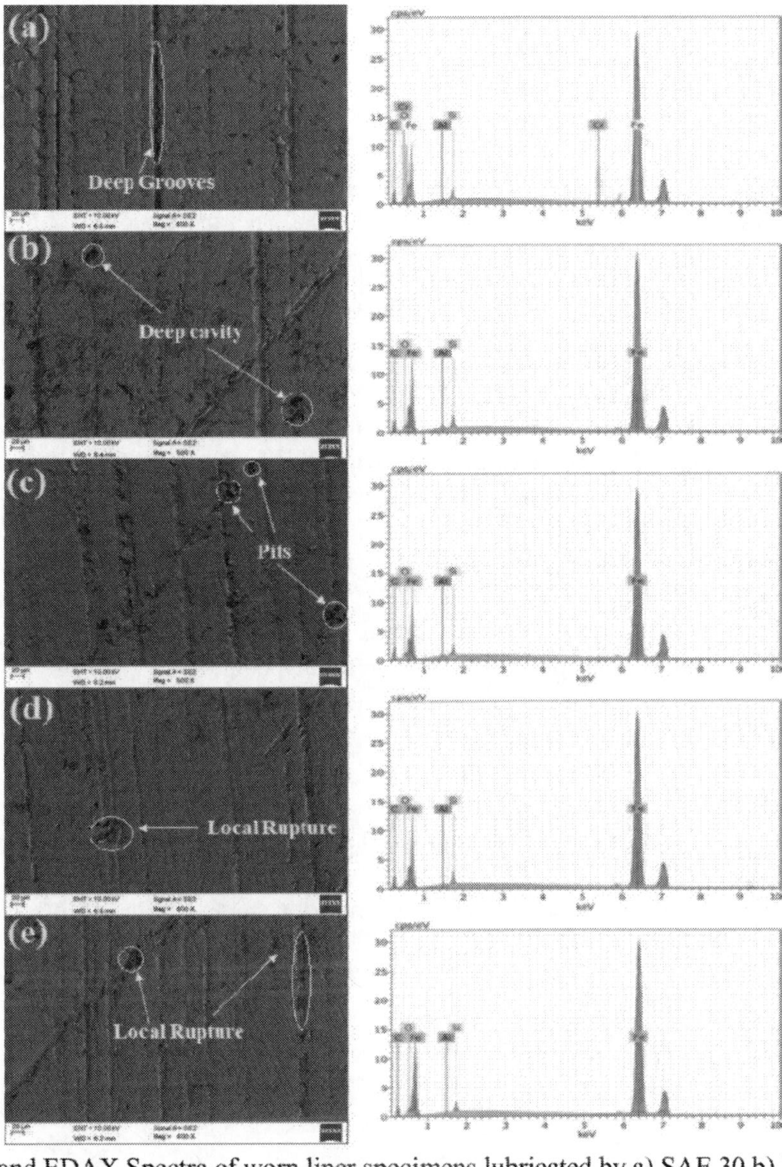

Fig. 5. SEM and EDAX Spectra of worn liner specimens lubricated by a) SAE 30 b) PE c) PE 25 d) PE 50 e) PE 75

Conclusions

The tribological behaviour of various PE based blended lubricants was evaluated using High Frequency Reciprocating Rig (HFRR) tribometer under various conditions and the results were compared with the SAE 30 grade compressor oil. The following conclusions were made from the experimental study:

- PE was produced successfully from raw rapeseed oil by the transesterification process using ultrasonication technic.
- The properties of the produced PE and its blends were closer its viscosity and viscosity index with the SAE 30 grade compressor oil.
- During the 1 hr running time, the COF (0.41) of PE 75 bio-based lubricant was shown better result than that of other four lubricants.
- Among all the five lubricants PE 50 shows the lower value of wear depth (10μm) compared to SAE 30, PE, PE 25 and PE 75.

- The addition of PE 75 and PE 50 with the synthetic lubricant SAE 30 grade compressor oil shows the smoother worn surfaces, which means an increase of PE percentage could be increase of lubricant performance.

References

[1] N.J.FoxG. W.Stachowiak 2007, Tribol. Int. 40 pp 1035-1046.
[2] Zoran S. Petrovic, Alisa Zlatanic, Charlene, C. Lavaa, Snezana Sinadinovic-Fiser 2002. Eur. J Lipid. Sci. Tech. 104. pp 293–299.
[3] Chandu S. Madankar, Subhalaxmi Pradhan, S.N. Naik 2013, Ind. Crops Prod. 43 pp 283– 290.
[4] Dharmendra Kumar Gajendra Kumar Poonam C.P. Singh 2010. Ultrason. Sonochem. 17 pp 555-559.
[5] K.G.Georgogianni M.G.Kontominas P.J.Pomonis D.Avlonitis V.Gergis 2008, Fuel Process Technol. 89 pp 503-509.
[6] Veera Gnaneswar Gude, Georgene Elizabeth Grant 2013, Appl. Energ. 109 pp 135-144.
[7] Priyanka Chand, Venkat Reddy Chintareddy, John G. Verkade, and David Grewell 2010, Energy Fuels 24 pp 2010–2015.
[8] P. Chenga Reddy, S. Arumugam, G. Sriram, and R. Parthasarathy 2018, Advanced Science Engineering and Medicine 10 pp 1-5, doi:10.1166/asem.2018.2171.
[9] A.Adhvaryu Z.Liu S.Z.Erhan 2005, Ind. Crops Prod. 21 113-119.
[10] S. Arumugam and G. Sriram 2012, Tribol. Trans. 55 pp 438-445.
[11] S Arumugam and G Sriram 2012, J Engg. Tribol. 227 pp 3-15.
[12] S. Baskar a, G. Sriram a, S. Arumugam 2015, Tribology in Industrial 37 pp 449-454.
[13] Tiong Chiong Ing, A. K. Mohammed Rafiq, Y. Azli and S. Syahrullail 2012, Tribol. Trans. 55 pp 539-548.
[14] Sevim Z. Erhan, Brajendra K. Sharma, Zengshe Liu, and Atanu Adhvaryu 2008, J. Agr. Food Chem. 56 pp 8919-8925.
[15] Matthew T, Siniawski, nader Saniei, Bigyan Adhikari and Lambert A 2007. Journal of synthetic lubrication 24 pp 101-110.

Practically Feasible Optimal Assembly Sequence Planning with Tool Accessibility

MVA Raju Bahubalendruni*[1], Gulivindala Anil Kumar[1], K Madhu sudana sankar[2]

[1]GMR Institute Technology, Rajam, Andhra Pradesh, India.
[2]Tata Lockheed Martin aero structure Ltd., Hyderabad, Telangana, India.

E-mail: bahubalindruni@gmail.com, anilgulivindala@gmail.com, Sankarmadu@yahoo.co.in

Abstract: Practically Feasible Optimal Assembly Sequence Plan (PFOASP) achievement is most important for an industrial engineer to meet the demand of producing the final product in less time and cost. A PFOASP directs the industrial engineer to generate a complete product with less number of manipulations from individual parts. An assembly sequence is said to be optimal from its nature of performing parallel operations to form stable subassemblies which directly reduces the overall assembly time for handling large scale products. Most the researchers followed to ignore tool accessibility in their approaches to avoid the computational complexities. However, the generated solution lacks completeness in practice because tools play the significant role while assembling operation. PFOASP determination is most challenging due to validation of multiple feasibilities qualifying criterion. In this paper, an attempt is made to explain the complexity and the importance of tool accessibility consideration in PFOASP with suitable illustrations. A tool based bounding box method is proposed to solve geometrical feasibility in PFOASP.

Key words: Assembly sequence planning, Liaison, Geometrical feasibility, Stability, Mechanical feasibility, Tool accessibility, Bounding box

1. Introduction:

Manufacturing industries are facing challenges in meeting the demands of producing new products in shortest time and less cost with their fabrication techniques. An industrial engineer is always paying effort to reduce the overall production cost and time of products to enter early into the market[1-2]. Researchers opinioned assembling a product was the significant time-consuming segment of the entire production process.

Assembling a product not only involves in bringing the individual parts together through a collision-free path but also maintaining its contact once positioned with the mating part and must allow for further joining operation[3-5]. So, a proposed assembly sequence must be optimal and practically feasible to save the time and cost in joining the individual parts in finished product generation. But PFOASP determination is most challenging due to validation of multiple assembly feasibility checking criteria. The criteria involved in the investigation of qualifying necessary assembly predicates such as liaison, stability, geometrical feasibility and mechanical feasibility. Several researchers investigated the importance of assembly predicate consideration for various assembly configurations [6].

Unfortunately, most of the researchers ignored tool accessibility consideration in their approaches to avoid the complexities of their implementation. Not only tool accessibility but also stability, mechanical feasibility has given less importance in assembly sequence generation problem. However, the generated solution lacks completeness in practice because tool accessibility plays the significant role in assembling operation. The next section composed to ascertain the background in PFOASP and importance of tool accessibility in assembly sequence generation.

2. Another section of your paper Back Ground of PFOASP:

Product data is extracted either in the matrix or graphical format to represent relations between the parts of an assembly. Assembly predicates such as liaison, geometrical feasibility, stability and mechanical feasibility to validate an assembly sequence for its execution in physical environment. But conventionally and recently most of the researchers supplementing few assembly predicates (stability and mechanical feasibility) in assembly sequence generation due to complexity in predicate consideration.

Smith made an attempt to represent part stability during assembly operation with any pairing part later Kumar and Bahubalendruni extended the concept and also proven assembly stability as a one of essential assembly predicate to result a practical possible solution. However, these concepts did not consider the involvement of assembly tool geometry and accessibility for establishing the assembly stability and mechanical feasibility in real time environment [7].

Unfortunately, researchers worked on DFA concepts and exploded view generation did not consider the involvement of assembly tool geometry and accessibility; though it is a key factor influence the cost and practicability [8-11].

Table. 1 presents assembly predicate consideration and tool accessibility consideration by numerous researchers in their approaches for ASP problem.

Table 1. Assembly predicates consideration from cited research literature

S. No.	*Liaison*	*Geometric feasibility*	*Stability/ connection*	*Mechanical feasibility*	*Tool accessibility*	*Ref.*
1	✓	✓	×	×	×	[12]
2	✓	✓	✓	×	×	[13]
3	✓	✓	✓	✓	×	[14]
4	✓	✓	✓	✓	×	[15]
5	✓	✓	✓	✓	×	[16]
6	✓	✓	✓	×	×	[17]
7	✓	✓	✓	×	×	[18]
8	✓	✓	✓	×	×	[19]
9	×	✓	×	×	✓	[20]
10	✓	✓	×	×	×	[21]
11	✓	✓	×	×	×	[22]
12	✓	✓	×	×	×	[23]
13	✓	✓	×	×	×	[24]
14	✓	✓	×	×	×	[25]
15	✓	✓	×	×	×	[26]
16	✓	✓	✓	×	×	[27]
17	✓	✓	✓	×	×	[28]
18	✓	✓	✓	×	×	[29]

19	✓	✓	✓	×	×	[30]
20	✓	✓	✓	×	×	[31]

Though all predicates are considered in ASG, still the solution is lacking for completeness. Because a tool is essential to bring a component into contact and to establish a connection between two or multiple parts to develop a complete product. It is crucial to consider tool accessibility in ASG to make solution exist in practically feasible in the physical environment.

2.1. Assembly tools:

Assembly tools classified into three main types based on their application in assembly development and they are namely Pre tools, In tools and Post tools .

Pre tools: These are the tools used to bring parts together before the assembling operations. Often assembly fixture to hold the components are treated as Pre-tools.

Ex: Grippers

In tools: These are the tools used to move parts relative to each other and for simple snapping and insertions.

Ex: spanner

Post tools: These are the tools used operate the physical connectors after positioning the primarily parts.

Ex: Wrench, screw driver

On varying the combinations of tools addressed above, assembly operations are performed. Some configurations need only pre and post tools for assembling, CAD model of vertical axis wind turbine in fig.4.

3. Methodology:

Tool accessibility holds for verifying the collision-free path to the tool and appending part with other parts of the assembly. Tool accessibility is nothing but enriched geometrical feasibility testing which considers tool volume/tool swept volume and part volume instead of part volume only.

3.1. Bounding Box

Bounding box technique is the most popular method which is successfully employing for collision detection. A bounding box is a minimum cuboid receptacle for a part/assembly generated with edges parallel to the principal axes. The bounding box represented by two extreme diagonal points of the cuboid. Co-ordinates of the part bounding box are used to evaluate the distance to be progressed by a part of a specified orientation to complete an assembly operation. The difference between upper and lower limit elements along a principal axis direction is considered to examine the geometric feasibility of along it.

Tool volume and part volume consideration is enough to perform Pre and Post assembly operations.Fig.1 depicts the TA for pre and post assembly operations.

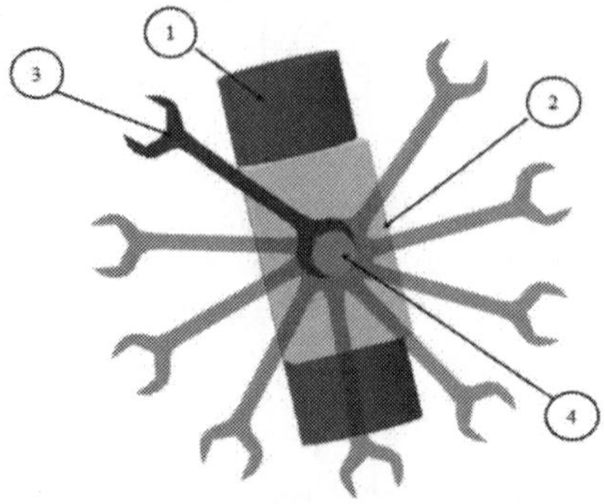

Fig.1. Tool accessibility for Pre and Post assembly operations

Tool volume, Part volume and the Swept volume of the tool during operation has to consider for inner assembly operations. Fig.2 depicts TA for Inner assembly operations.

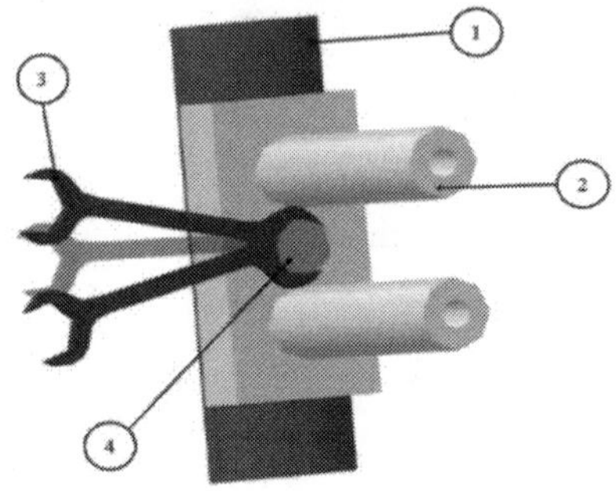

Fig.2. Tool accessibility for Inner assembly operations

3.2. Implementation

Tool accessibility predicate is implemented on two CAD models namely vertical axis wind turbine and switch board assembly represented in fig.3 and fig.4. Assembly sequences are generated with TA consideration and without consideration.

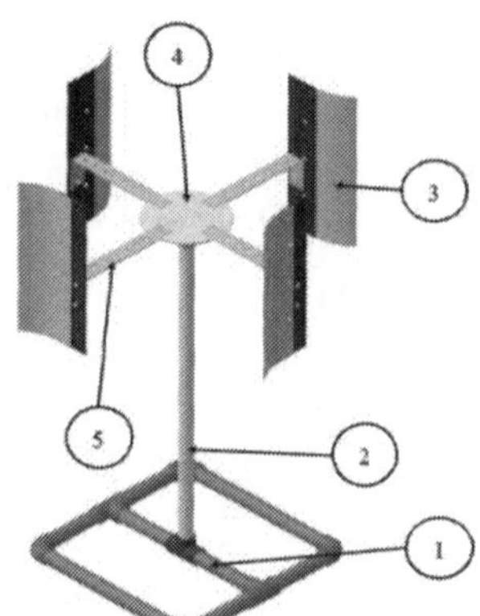

Fig.3. CAD model of Vertical axis wind turbine

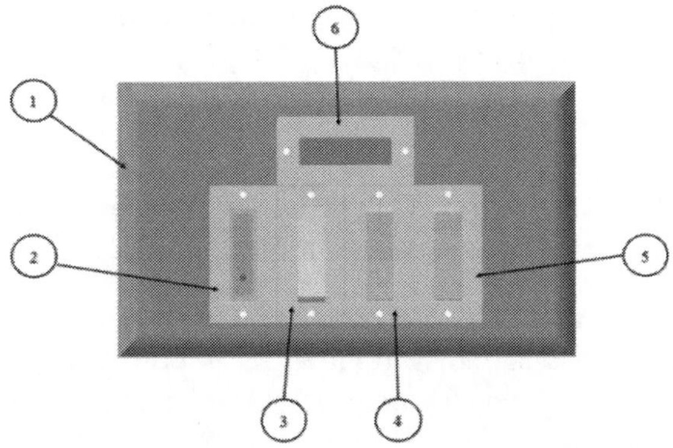

Fig.4. Switch board assembly

After implementing TA on preferred assembly model, results are presented in table 2 which explains the practical feasibility of the solution with and without TA.

Table.2. Influence of Tool Accessibility on PFOASP

Name of the assembly	*Assembly sequence without TA*	*Assembly sequence with TA*	*Practical feasibility of sequence without TA*
Wind mill	1-2-4-5-3	1-2-4-5-3	valid
Switch assembly	1-5-4-3-2-6	Not possible	Not valid

4. Discussions

Conventionally, researchers followed the consideration of connectors as parts of the assembly. But computational complexity issue raises with the increase in part count. Connection matrixes are used to represent type/presence of connection in an assembly instead of considering connectors considering directly/separately. However, these connection matrixes failed at explaining accessibility to install a connector between assembly parts. Later, Mechanical feasibility matrix used to represent the feasibility in establishing a connection between two different parts of an assembly among the presence of other parts. But Mechanical feasibility not conceded tool accessibility for placing a connector and assembly part.

Tool accessibility plays a crucial role in the case of restricted degrees of freedom between subassemblies of an assembly. Assembly sequence generation for Switch assembly is such a case which directly explains the error in ASG without considering tool accessibility predicate.

5. Conclusion

In this article, a novel tool based bounding box method is proposed to examine tool accessibility to perform assembly operations. It is observed that without considering tool assembly one may not achieve practical solution for all assembly configurations. Consideration of this assembly predicate may increase complexity and reducing solution space but practicality is achieved. Tool accessibility predicate consideration enriches the quality of solution enormously in PFOASP generation. Still, the efficiency of solution is limited because testing of assembly predicated confined to principal axis only. Space for assembling a product is more by performing in oblique directions too. Future scope of this research is to cover the identified issues in PFOASP.

Acknowledgement: The research presented in the article is funded by Science and Engineering Research Board-DST under File No. SERB/ECR/2017/000341

References

1. Bahubalendruni MR, Biswal BB, Upadhyaya V Assembly sequence generation and automation In *International Conference on Design, Manufacturing and Mechatronics* 2014 185-192
2. Bahubalendruni MR, Biswal BB A review on assembly sequence generation and its automation *Proceedings of the Institution of Mechanical Engineers, Part C: Journal of Mechanical Engineering Science* 230(5) 824-38.
3. Shiang-Fong Smith S, Smith G and Liao X 2001 Automatic stable assembly sequence generation and evaluation *Journal of Manufacturing Systems* 20 225-235
4. Bahubalendruni MV, Biswal BB Computer aid for stability testing between parts towards automatic assembly sequence generation *Journal of Computer Technology & Applications* 7(1) 22-6
5. Kumar M, Bahubalendruni M, Biswal B and Nayak R 2016 Identification of Stable Configurations between Constituent Parts of an Assembly *Applied Mechanics and Materials* 852 595-601
6. Bahubalendruni M, Biswal B, Kumar M and Nayak R 2015 Influence of assembly predicate consideration on optimal assembly sequence generation *Assembly Automation* 35 309-316
7. Bahubalendruni MR, Biswal BB, Kumar M, Deepak BB. A note on mechanical feasibility predicate for robotic assembly sequence generation *In CAD/CAM, Robotics and Factories of the Future* 397-404
8. Murali GB, Deepak BB, Bahubalendruni MR, Biswal BB. Optimal Assembly Sequence Planning Towards Design for Assembly Using Simulated Annealing Technique. *In International Conference on Research into Design* Jan 397-407
9. Sudhakar U, Bahubalendruni MR, Biswal BB, Deepak BB. DFA Model Through Assembly Contact Data And Geometrical Feasibility Testing *International Journal of Mechanical Engineering and Technology* 8(2) 131
10. Bahubalendruni M 2018 An efficient method for exploded view generation through assembly coherence data and precedence relations *World Journal of Engineering* 00-00
11. Sharma G 2017 An Automated Computer Aided Procedure for Exploded View Generation *International Journal of Performability Engineering*
12. RAJU BAHUBALENDRUNI M and BISWAL B 2016 Liaison concatenation – A method to obtain feasible assembly sequences from 3D-CAD product *Sadhana* 41 67-74

13. Bahubalendruni M, Biswal B and Deepak B 2016 Optimal Robotic Assembly Sequence Generation Using Particle Swarm Optimization *Journal of Automation and Control Engineering* 4 89-95
14. Bahubalendruni M and Biswal B 2016 An intelligent approach towards optimal assembly sequence generation *Proceedings of the Institution of Mechanical Engineers, Part C: Journal of Mechanical Engineering Science* 232 531-541
15. Bahubalendruni M, Deepak B and Biswal B 2016 An advanced immune based strategy to obtain an optimal feasible assembly sequence *Assembly Automation* 36 127-137
16. Bahubalendruni M and Biswal B 2015 A novel concatenation method for generating optimal robotic assembly sequences *Proceedings of the Institution of Mechanical Engineers, Part C: Journal of Mechanical Engineering Science* 231 1966-1977
17. Murali G, Deepak B, Bahubalendruni M and Biswal B 2017 Optimal Assembly Sequence Planning Using Hybridized Immune-Simulated Annealing Technique *Materials Today: Proceedings* 4 8313-8322
18. Gunji A, Deepak B, Bahubalendruni C and Biswal D 2018 An Optimal Robotic Assembly Sequence Planning by Assembly Subsets Detection Method Using Teaching Learning-Based Optimization Algorithm *IEEE Transactions on Automation Science and Engineering*1-17
19. Gunji B, Deepak B, Raju Bahubalendruni M and Biswal B 2017 Hybridized genetic-immune based strategy to obtain optimal feasible assembly sequences *International Journal of Industrial Engineering Computations* 333-346
20. Wilson R 1998 Geometric reasoning about assembly tools *Artificial Intelligence* 98 237-279
21. Lee H and Gemmill D 2001 Improved methods of assembly sequence determination for automatic assembly systems *European Journal of Operational Research* 131 611-621
22. Wang J, Liu J and Zhong Y 2004 A novel ant colony algorithm for assembly sequence planning *The International Journal of Advanced Manufacturing Technology* 25 1137-1143
23. Choi Y, Lee D and Cho Y 2008 An approach to multi-criteria assembly sequence planning using genetic algorithms *The International Journal of Advanced Manufacturing Technology* 42 180-188
24. Shan H, Zhou S and Sun Z 2009 Research on assembly sequence planning based on genetic simulated annealing algorithm and ant colony optimization algorithm *Assembly Automation* 29 249-256
25. Tseng H, Chang C, Lee S and Huang Y 2018 A Block-based genetic algorithm for disassembly sequence planning *Expert Systems with Applications* 96 492-505

26. Kim H, Park C and Lee D 2018 Selective disassembly sequencing with random operation times in parallel disassembly environment *International Journal of Production Research* 1-15

27. Moradi H, Goldberg K, Lee S. Geometry-based part grouping for assembly planning. InAssembly and Task Planning *IEEE International Symposium on* 281-286

28. Chang C, Tseng H and Meng L 2009 Artificial immune systems for assembly sequence planning exploration *Engineering Applications of Artificial Intelligence* 22 1218-1232

29. Li X, Qin K, Zeng B, Gao L and Su J 2015 Assembly sequence planning based on an improved harmony search algorithm *The International Journal of Advanced Manufacturing Technology* 84 2367-2380

30. Tao F, Bi L, Zuo Y and Nee A 2017 Partial/Parallel Disassembly Sequence Planning for Complex Products *Journal of Manufacturing Science and Engineering* 140 011016

31. Nayak R, Bahubalendruni MR, Biswal BB, Kumar M. Comparison of liaison concatenation method with simulated annealing for assembly sequence generation problems. *InNext Generation Computing Technologies (NGCT)* 531-535

A study on the effect of workplace negativity factors on employee engagement mediated by emotional exhaustion

Archana S Menon[1] and R G Priyadarshini[2]

[1] Amrita School of Business, Amrita Vishwa Vidyapeetham, Coimbatore, India, s.archanamenon@gmail.com
[2] Amrita School of Business, Amrita Vishwa Vidyapeetham, Coimbatore India, rg_priyadarshini@gmail.com

Abstract. Business development executives across industries face negativity in their work due to multiple reasons. The negative actions of one person and the impact caused may spread among the employees in an organization and can even affect their performance and the organization itself. This paper attempts to study the effect of negative factors like abusive supervision by managers and negative gossips that prevail in the organizations on employee engagement. It also aims at finding out the mediating effect of emotional exhaustion on employee engagement. The conceptual model for the study is supported by the earlier research conducted in this area and the whole study is based on this model. Data is collected from business development executives of both service based and product-based organizations in India. The sample consists of responses of 91 business development executives across India. From the analysis of data collected, abusive supervision and negative gossips have a positive effect on emotional exhaustion which in turn has a negative relationship with employee engagement.

1. Introduction

Research has proven that negative actions spread from person to person in the same way as how happiness spread. A small act of being rude may spread from person to person and affect people whom we may not interact directly. A survey which was conducted in the HR division of a huge international manufacturing organization came up with the conclusion that 10 percent of the employees who were prone to negative interactions had 30 percent less overall performance scores compared to other employees. Negative or de-energizing experiences in an organization can lead to job dissatisfaction, lack of motivation and increased attrition. Another study has shown that high performers who experience significant negativity in their working environment tend to leave their organization at an alarming rate of 13 times more than other employees [13]. There are also studies that show that workplace incivility causes a higher rate of absenteeism among employees [23]. Thus, a positive work environment leads to positive emotional states of employees which can eventually help them be more engaged in work, growing and productive [11]. This study concentrates on the extent of impact on employee engagement of negativity factors like abusive supervision and negative gossips. The effect of these factors on emotional exhaustion is studied which in turn affects the employee engagement in the organization.

2. Literature Review

2.1. Employee engagement

An employee is the most important and valuable asset of an organization, if engaged and managed in the right way. Hence employee engagement is a factor that can best measure the vigor of the company. Employee engagement is the level of involvement and commitment of an employee towards the values and norms of his organization. An engaged employee is responsible to fulfill his individual goals and encourages his colleagues to fulfill their goals for the success of the organization. According to Kahn (1990, p.694), for an employee to be engaged, three psychological conditions are inevitable which are meaningfulness, safety and availability. Another significant definition for employee engagement was put forward by Hewitt Associates LLC (2004) as how individuals are intellectually and emotionally committed to their group or organization which is measured with the following three behaviors: say, stay and strive. (Anita, 2014). According to the study conducted by (Anita, 2014), there are 7 determinants of employee engagement namely work environment, leadership, team and co-worker, training and career development, Compensation, Organizational policies and workplace well-being. This study shows that there is a significant relationship between employee engagement and performance of employees in the organization. It also concludes that the strongest factors that affect the employee engagement is the working environment and the team and co-worker relationship [1].

According to Topchik (2000), the negativity at workplace begins with worried or unhappy employees. Once a few employees get affected by this, it spreads like a virus to the entire organization affecting the productivity and people start leaving the place as it is not anymore, a fun place to work. The Bureau of Labor Statistics states that, about 3 billion dollars of loss is experienced by U.S companies due to the effects of negativity in workplaces. A negative workplace can lead to increased turnover, lateness and absences among employees [19].

2.2. Abusive supervision

Abusive supervision is defined as the perception of subordinates of the extent to which the supervisors show hostile nonverbal and verbal behaviors. This excludes physical contact. The perception of a superior's behavior could be different in different contexts. Also, different subordinates will have different perceptions about their superiors. It is also studied in the literature that subordinates whose supervisors are highly abusive had higher turnover and unfavorable attitude towards their job and life [17]. An employee who perceive more abuses from supervisors gets emotionally exhausted and ends up abusing their coworkers [21].

2.3. Negative Gossips

Gossips are common in the social world and it occurs both in public forums as well as face to face interactions. The definition of gossip used in this study is negative information exchanged about a third party in his/her absence [10]. From earlier studies it has been proved that employees do spread negative and positive gossips about colleagues from their own teams and negative gossips are directed towards those employees who have fewer friends [8]. There are also studies that show gossiping about others acts as a stimulus to relax and can help to bring the employees closer in an organization [2]. Gossips between female pairs are tend to be more negative than between males or cross-gender pairs. Also, women were found to encourage gossips in same gender friendships to make social comparisons [12]

2.4. Emotional exhaustion

The definition of burnout was first derived from Maslach's theory. According to this theory burnout involves three components namely, emotional exhaustion, diminished personal accomplishment and depersonalization out of which emotional exhaustion has the primary influence on an individual. It also states that there are several physical as well as mental consequences due to burnout in an individual [22]. Emotional exhaustion at workplace is defined as a state which is caused by emotional and psychological demands made on employees. It is also studied that work-family conflicts and work-role conflicts are positively related to emotional exhaustion [5]. Also, social support from co-workers is said to have a significant moderating effect between work-to-family and family-to-work conflict with satisfaction of employees [14]. The various factors that affect the exhaustion levels of employees are job related tension, job satisfaction, age and tenure on the job. Exhaustion can also lead to absenteeism, job changes and intention to leave [15]. Employees who are surrounded by optimal job demands are tend to have less exhaustion and maximum vigor compared to those employees who has many job demands [20]. A study conducted on the emotional exhaustion levels of college coaches showed that female coaches had higher exhaustion levels than male coaches. Hence the gender difference is significant when it comes to emotional exhaustion [6]. Emotional exhaustion is said to have positive mediating roles in studies that relate abusive supervision and performance and dedication of employees towards work [3]. Employees who experience burnout in the early stages of their career were found to be less likely to switch jobs and were more flexible [7].

3. Objective of the study

The objective of this study is to understand the effect of workplace negativity factors on emotional exhaustion. It also aims at finding out the most prominent factor that affect the exhaustion levels of employees and tries to find how the exhaustion levels of individuals has an effect on their engagement in job. The study justifies these effects with gender, age and type of organization as moderating variables.

4. Conceptual Model

The conceptual model is derived from previous literature. The most common workplace negativity factors are considered for the study. The direct effect of abusive supervision on emotional exhaustion which is a main constituent of burnout is studied along with its effect on employee engagement. Also, the level by which negative gossips in an organization affect the emotional exhaustion levels of employees is studied. The combined effect of these factors on emotional exhaustion and the final effect of exhaustion of employees in an organization on their job engagement is studied. Exhaustion is the mediating variable that enhances the relation between abusive supervision, negative gossips and employee engagement in work. The study also concentrates on the effect of these factors on engagement levels of male and female employees individually. The experience levels of individuals are also considered for the study.

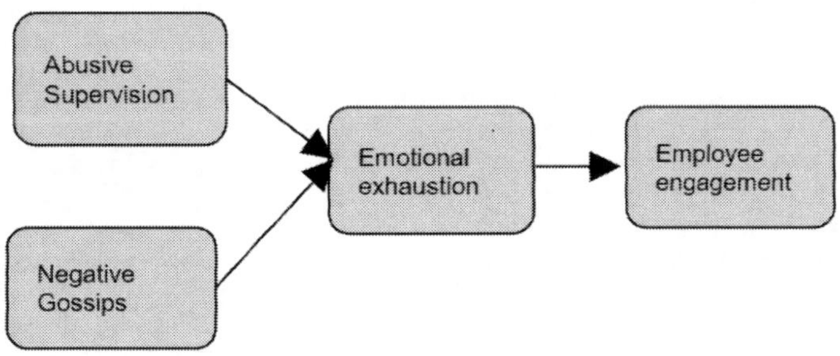

Figure 1. Conceptual model of the study

5. Hypotheses

From previous studies, it is found that abusive supervision has a negative effect on the work engagement levels of employees [4]. Also, negative gossips are negatively associated to job engagement specific to dedication and vigor [9]. Another study shows that engagement and burnout are negatively related to each other [16]. The moderating effect of emotional exhaustion on the effect of these negativity factors on employee engagement is not mentioned in any studies. These insights were used to develop the following null hypotheses.

Hypothesis 1: Abusive supervision has no effect on emotional exhaustion

Hypothesis 2: Negative gossips has no effect on emotional exhaustion

Hypothesis 3: Emotional exhaustion does not mediate the relation between negativity factors and employee engagement

6. Methodology

The study was conducted on a group of business development executives across India. The sample was randomly chosen from service as well as product-based organizations. A sample of 91 business development executives who were directly involved in sales and other marketing activities were chosen. The method of data collection was a questionnaire with 49 questions gathered from the literature. The questions were floated among the business development executives to understand the relation between abusive supervision, negative gossips, emotional exhaustion and employee engagement. The participants consisted of both 56 males and 35 female business development executives with different experience levels. The respondents belonged to either product based or service-based organization type.

The abusive supervision questionnaire consisted of 9 items measured using a 5-point Likert scale. A few items of the questionnaire are "My immediate supervisor is rude to me", "My immediate supervisor tells me I am an incompetent employee" etc. The questionnaire to measure the effect of negative gossips consisted of 11 items and the scale used was the 5-point Likert scale ranging from Never to Always. Emotional exhaustion was measured using a 9-item questionnaire which includes "I feel emotionally drained from my work", I feel frustrated from my job" etc. The scale used in this questionnaire was 5-point Likert scale which extended from Strongly Disagree to Strongly Agree. The employee engagement was measured using a 20 items scale from Aon Hewitt model which also used the above mentioned 5-point Likert scale measurement.

7. Analysis and Findings

Source	SS	df	MS		
				Number of obs	= 91
				F(2, 88)	= 72.33
Model	3791.33123	2	1895.66562	Prob > F	= 0.0000
Residual	2306.20723	88	26.2069003	R-squared	= 0.6218
				Adj R-squared	= 0.6132
Total	6097.53846	90	67.7504274	Root MSE	= 5.1193

emotionalex	Coef.	Std. Err.	t	P>|t|	[95% Conf.	Interval]
abussup	.3735608	.0630229	5.93	0.000	.2483161	.4988055
neggos	.3789323	.0717684	5.28	0.000	.2363076	.5215569
_cons	5.167267	1.913439	2.70	0.008	1.364709	8.969825

Figure 2. This figure shows the result of regression analysis of workplace negativity factors and emotional exhaustion performed in STATA. The R-squared value and p-values for overall significance (f-test) of the model is mentioned along with individual p-values (t-test).

IOP Conf. Series: Materials Science and Engineering **390** (2018) 012027 doi:10.1088/1757-899X/390/1/012027

The effect of abusive supervision and negative gossip on emotional exhaustion is studied with the help of multiple linear regression analysis. The result of the predicted model is given in figure 1. As the p- value of the model connecting abusive supervision and negative gossips to emotional exhaustion is significantly less than 0.05 (F-test), the overall significance of the model is satisfied. Also, the individual p values of the factors abusive supervision and negative gossips are less than 0.05. Hence these factors can make a significant change in emotional exhaustion which is the dependent variable. At the same time this does not explain causality that any change in the dependent variable is only because of the changes in the independent variables. The R square value is 62.18 percent. This value which is greater than 50 percent explains the validity of the model and proves a right fit for the relationships. Both the

Source	SS	df	MS		
				Number of obs	= 91
				F(1, 89)	= 90.48
Model	12559.3873	1	12559.3873	Prob > F	= 0.0000
Residual	12353.9094	89	138.80797	R-squared	= 0.5041
				Adj R-squared	= 0.4986
Total	24913.2967	90	276.814408	Root MSE	= 11.782

empeng	Coef.	Std. Err.	t	P>|t|	[95% Conf.	Interval]
emotionalex	-1.435182	.1508794	-9.51	0.000	-1.734976	-1.135388
_cons	100.2396	3.913926	25.61	0.000	92.46275	108.0165

Figure 3. This figure shows the result of regression analysis of emotional exhaustion and employee engagement performed in STATA. The R-squared value and p-values for overall significance (f-test) of the model is mentioned along with individual p-values (t-test).

coefficients of the model corresponding to abusive supervision and negative slopes proving the theory pertaining to both these attributes having a positive relationship with burnout which means that an increase in either negative gossip or abusive supervision results in an increase in emotional exhaustion.

Similarly, the model connecting burnout and employee engagement, as seen in figure 2, also has a p value significantly less than 0.05 suggesting an overall significance of the model. Further scrutinizing of the result revealed that the individual significance is also maintained as the corresponding p values of the dependent variable involved are also significantly less than 0.05. An R squared value of 50.41 percent was obtained for this relationship model which explains the proposed relationship. Burnout also was found having a negative slope in the regression equation suggesting a negative relationship between burnout and employee engagement. Further analysis revealed a stronger dependence between the dependent variable emotional exhaustion and the independent variables abusive supervision and negative gossips, in female candidates than male candidates as the R squared value is greater for female candidates. Similar trend was found in the case of the second relationship pertaining to exhaustion and employee engagement.

Moreover, a strong relationship was found between the independent variables, negative gossips and abusive supervision and the dependent variable, burnout, for employees working in product-based organizations than the ones in service-based organization. A similar trend was found in the relationship between emotional exhaustion and employee engagement. The model fits better for the age group 20 to 25 as the R squared value for both the relationships are higher than that of the age group 26 to 30.

IOP Conf. Series: Materials Science and Engineering **390** (2018) 012027 doi:10.1088/1757-899X/390/1/012027

8. Results and Discussion

The analysis clearly explains the conceptual model. That is, in an organization if there is a presence of abusive supervision or negative gossips among its employees, then the emotional exhaustion element of the burnout factor is affected positively increasing with an increase in the magnitude of abusive supervision and negative gossips. This rejects the null hypotheses H1 and H2. Emotional exhaustion element has a negative relationship with employee engagement resulting in the decrease of employee

Table 1. The dependent and independent variables in each regression analysis performed is mentioned along with their R-squared values. Gender and type of organization is considered as moderating variables to arrive at conclusions

	R squared values	
Dependent variable	***Independent variable***	***R-squared value (%)***
Emotional exhaustion	Abusive supervision, Negative gossip	62.18
Employee engagement	Emotional exhaustion	50.41
Emotional exhaustion (Male)	Abusive supervision, Negative gossip	55.84
Employee engagement (Male)	Emotional exhaustion	32.31
Emotional exhaustion (Female)	Abusive supervision, Negative gossip	69.66
Employee engagement (Female)	Emotional exhaustion	74.14
Emotional exhaustion (Product)	Abusive supervision, Negative gossip	79.61
Employee engagement (Product)	Emotional exhaustion	59.47
Emotional exhaustion (Service)	Abusive supervision, Negative gossip	59.73
Employee engagement (Service)	Emotional exhaustion	44.82

engagement with an increase in emotional exhaustion. Thus, it is proved that emotional exhaustion plays a mediating role between the negativity factors (abusive supervision and negative gossips) and employee engagement thus rejecting the null hypothesis H3.

As the obtained R squared values are not significantly high, the model does not imply causality among the variables involved. Changes in employee engagement or emotional exhaustion on all occasions cannot be completely explained by the changes in the independent variables of the model, abusive supervision and negative gossips. As the R squared values are sufficiently high we can say that the studied factors have a significant effect on employee engagement.

9. Conclusion

The study concentrates on finding out the effect of workplace negativity factors, abusive supervision and negative gossips on employee engagement and the mediating role of emotional exhaustion. Recent studies show that negative factors like abusive supervision has significant effect on employee well-being, performance, absenteeism as well as turnover [18]. After the analysis conducted on a sample of 91 business development executives, it was found that there is significant positive relationship between the workplace negativity factors and emotional exhaustion. Moreover, emotional exhaustion mediates the effect of workplace negativity factors on employee engagement. It was found that an increase in emotional exhaustion due to negativity factors has a negative effect on employee engagement. Female employees were found to have a greater effect of abusive supervision and negative gossips on their exhaustion levels and job engagement levels. Similarly, product-based companies were found to have a greater impact compared to service-based organizations.

10. Limitations

Current study does not focus on various other factors which causes either emotional exhaustion or employee engagement. Moreover, only one particular element of burnout, emotional exhaustion was taken as it was found to have primary influence in the work environment. This study was conducted without controlling various other factors which affect employee engagement. Future scope of the study would be to analyze the same by controlling these factors to get a more efficient model to explain the same.

11. References

[1] Anitha J 2014 Determinants of employee engagement and their impact on employee performance. *International journal of productivity and performance management* **63** 308

[2] Alshehre R A M 2017 Positive Effects of Gossiping at Work *Open Journal of Medical Psychology* **6** 126

[3] Aryee, S, Sun L Y, Chen Z X G and Debrah Y A 2008 Abusive supervision and contextual performance: The mediating role of emotional exhaustion and the moderating role of work unit structure *Management and Organization Review* **4** 393-411

[4] Barnes C M, Lucianetti L, Bhave D P and Christian M S 2015 You wouldn't like me when I'm sleepy: Leaders' sleep, daily abusive supervision, and work unit engagement *Academy of Management Journal* **58** 1419-1437

[5] Boles J S, Johnston M W and Hair Jr J F 1997 Role stress, work-family conflict and emotional exhaustion: Inter-relationships and effects on some work-related consequences *Journal of Personal Selling & Sales Management* **17** 17-28

[6] Caccese T M and Mayerberg C K 1984 Gender differences in perceived burnout of college coaches *Journal of sport psychology* **6** 279-288

[7] Cherniss C 1992 Long-term consequences of burnout: An exploratory study *Journal of Organizational behavior* **13** 1-11

[8] Ellwardt L, Labianca G J and Wittek R 2012 Who are the objects of positive and negative gossip at work?: A social network perspective on workplace gossip *Social Networks* **34** 193-205

[9] Georganta K, Panagopoulou E and Montgomery A 2014 Talking behind their backs: negative gossip and burnout in hospitals *Burnout Research* **1** 76-81

[10] Grosser T J, Lopez-Kidwell V and Labianca G 2010 A social network analysis of positive and negative gossip in organizational life *Group & Organization Management* **35** 177-212

[11] Harter J K, Schmidt F L and Keyes C L 2003 Well-being in the workplace and its relationship to business outcomes: A review of the Gallup studies *Flourishing: Positive psychology and the life well-lived* **2** 205-224

[12] Leaperand C and Holliday H 1995 Gossip in same-gender and cross-gender friends' conversations. *Personal Relationships* **2** 237-246
[13] Porath C 2016 Isolating toxic employees to reduce their negative effects. *Harvard business review*
[14] Rathi Neerpal and Barath M 2013 Work-family conflict and job and family satisfaction: Moderating effect of social support among police personnel *Equality, Diversity and Inclusion: An International Journal* **4** 438-454
[15] Saxton M J, Phillips J S and Blakeney R N 1991 Antecedents and consequences of emotional exhaustion in the airline reservations service sector *Human relations* **44** 583-595
[16] Schaufeli W B and Bakker A B 2004 Job demands, job resources, and their relationship with burnout and engagement: A multi-sample study *Journal of organizational Behavior* **25** 293-315
[17] Tepper B J 2000 Consequences of abusive supervision *Academy of management journal* **43** 178-190
[18] Tepper B J, Simon L and Park H M 2017 Abusive supervision *Annual Review of Organizational Psychology and Organizational Behavior* **4** 123-152
[19] Topchik G S 2000 Managing workplace negativity *AMACOM Div. American Mgmt. Assn.*
[20] Van den Broeck A, Vansteenkiste M, De Witte H and Lens W 2008 Explaining the relationships between job characteristics, burnout, and engagement: The role of basic psychological need satisfaction *Work & stress* **22** 277-294
[21] Wheeler A R, Halbesleben J R and Whitman M V 2013 The interactive effects of abusive supervision and entitlement on emotional exhaustion and co-worker abuse. *Journal of Occupational and Organizational Psychology* **86** 477-496
[22] Wittmer J L and Martin J E 2010 Emotional exhaustion among employees without social or client contact: The key role of nonstandard work schedules *Journal of Business and Psychology* **25** 607-623
[23] Zia-ud-Din M, Arif A and Shabbir M A 2017 The Impact of Workplace Incivility on Employee Absenteeism and Organization Commitment *International Journal of Academic Research in Business and Social Sciences* **7** 205-221

SELECTION OF OPTIMUM FUEL BLEND IN AN IC ENGINE USING FAHP-PROMETHEE

Warke Vivek Ramesh[1], G. Sakthivel[2]
School of mechanical and building sciences, Vellore Institute of Technology, Chennai

Emil ID: vivekwarke231@gmail.com: sakthivel.g@vit.ac.in

Abstract:
Energy is important criteria in determining the world's economy. The increasing demand on energy due to industrialisation, population growth and rise of living standards have led to considerable use of fossil fuels. Biodiesel proves to be a good alternative for fossil fuels. But the feasibility of biodiesel is the major factor for determining it as an alternate fuel for CI engines.The selection of suitable biodiesel with the appropriate blend for the IC engine plays a vital role in the energy sector. The objective of this study to identify the apt fuel blend using FAHP-PROMETHHE. FAHP method is applied to find the relative weights of the criteria, while PROMETHEE has been used to identify the best alternatives. NOx, Smoke, HC, CO, CO2, BTE, EGT, ID, CD and MRPR were considered as the assessment criteria. From the results, it is observed that B20 is the best blend.
Keywords: Energy, Fish oil, Diesel, Engine, Emission, MCDM, FAHP

1. Introduction

With the rapid development of the global economy, energy requirements have increased remarkably. Energy is central to achieve the inter related goals of modern societies to meet human needs. The realization that fossil fuel resources required for the generation of energy are becoming scarce. The average global energy consumption grows at the rate of 1.6 % p.a. This indiscriminate consumption of fossil fuels will lead to depletion of petroleum reserves between 2050 and 2075 [1]. It is mandatory to shift the energy supply system from the fossil fuel to the renewable alternative fuel. Biofuels are the convincing alternative for the fossil fuels since they are less pollutant, ecological and 100% natural fuel with the similar properties to dieselwhich enhances the energy security. A few researches are going around the world in search of alternative renewable fuels for diesel engines. The important advantage in using biodiesel as an alternate fuel is that it can be used in diesel engine without any modification.

Biodiesel can be produced from renewable resources by combining vegetable oils and animal fats with an alcohol to form alkyl esters which enhances the energy security and economy independence [2]. The cultivation of crops for biodiesel production poses a threat to food security and also contributes to drop in soil richness [3].In the meantime, the animal fat present in the waste parts of fish attends to be a best alternative for biodiesel.It is assessed that, consistently, a surplus quantity of fish parts is disposed by different fish products manufacturers at every year. More than one lakh tonnes of shrimp were produced as industrial fish waste as stated by Central Institute of Fisheries Technology (CIFT). The global fish oil production was 1.01 million tonnes as stated by International Fishmeal and Fish Oil Organization (IFFO) and it will be increased tenfold by next 5 years[4]. Hence, fish oil has increasing attention to be a good source of biodiesel for diesel fuel subsequently reducing the ecological pollutants and energy crisis. Jayasinghe and Hawboldtanalysed the fish oil properties and identified the possibility of using fish oil as a biodiesel [5]. Considerable researchers examined the engine performance characteristics operated with fish oil as a biodiesel [6-17]. The test results showed that the engines operated normally with reduction in emissions and increase in efficiencywith no apparent adverse operational or maintenance impacts. Hence an effort has been made to analyse the engine performancecharacteristicsusing ethyl ester of fish oil as a biodiesel. The performance, emission and combustion parameters of the engine are considered at different load conditions for choosing the optimum blend which is a difficult task [18-21]. To overwhelm the shortcomings of the existing research, all the performance, emission and combustion characteristics are considered in this research.

MCDM provides sophisticated methodological tools that are oriented towards the support of the decision makers in facing complex real-world decisions. The use of MCDM in the automotive has been progressively increasing in the past couple of decades. Poh and Ang (1999) applied an AHP technique to identify and evaluate the best alternative fuel for land transportation in Singapore [22]. Goumas and Lygerou (2000) ranked the alternative energy projects using PROMETHEE method in fuzzy environment [23]. Yedla et al (2003) have developed a multi-criteria decision making model for the selection of alternative options for environmentally sustainable transport system in Delhi [24]. Tsita and Pilavachi (2012) performed the evaluation of alternative fuels for the transport sector using FAHP [25]. Etghani et al. (2013) investigated the performance of a diesel engine by considering environmental criteria [26]. Sakthivel et al. (2013) applied PROMETHEE for evaluation of an automobile purchase model [27]. Chand (2017) proposed PROMETHEE to identify a sedan car by considering various criteria [28].

The aforementioned literature sections confirmed the significance of MCDM methods in the blend selection. It has also identified that there is no trace of research that deals with MCDM technique PROMETHEE with FAHP for the selection of suitable fuel blend based on the performance, combustion and emission characteristics. Hence this paper proposed a novel hybrid MCDM technique for evaluating optimum blend to achieve maximum engine performance and minimising environmental pollution.

2. Experimental Procedure

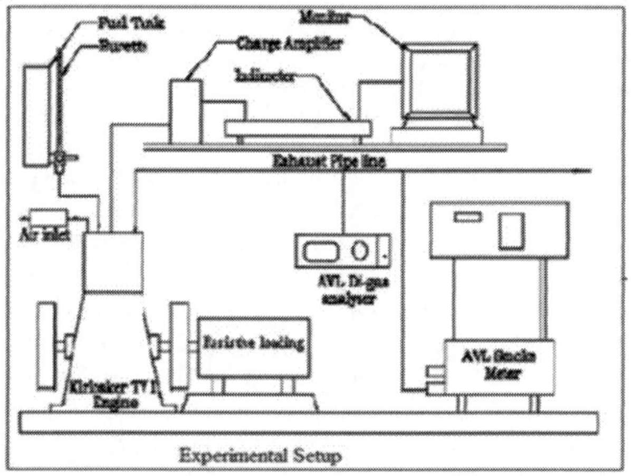

Figure 1. Schematic diagram of the engine setup

A four stroke single cylinder, direct injection, air cooled, constant speed, compression ignition engine was used in the test. The schematic diagram of the engine setup is shown in Figure 1. The engine was coupled to an electrical dynamometer with a control system to provide the brake load. The fuel consumption was measured by a burette and a stop watch. A provision was made to mount a piezoelectric pressure transducer on the cylinder head surface in order to measure the in-cylinder pressure. The pressure transducer was connected to a charge amplifier to condition the signals. A series of tests were carried out at a constant speed by varying the concentrations of the blends such as diesel and biodiesel with 20% interval. Each test was repeated three times to ensure the reproducibility of data and shown in Table 5. The AVL 437 smoke meter was used to measure the intensity of smoke and AVL 444 di gas analyser to measure the levels of carbon monoxide (CO), carbon dioxide (CO_2), oxides of nitrogen (NO_x) and hydrocarbon (HC) emissions. Technical specifications of the test engine are given in the Table 1.From the measured values, the performance parameters brake power, specific fuel consumption and brake thermal efficiency were calculated.

3. Literature Review

3.1 FAHP method

AHP method is proposed by Saaty (1980) [29]. In AHP, the decision problem is structured hierarchically at different levels with each level consisting of a finite number of elements (Khajeeh, 2010) [30]. A fuzzy set is characterized by a membership function which assigns a grade of membership ranging between zero and one to each object of the class (Zadeh, 1965) [31]. Fuzzy logic is integrated with AHP and proposed

as FAHP [32]. The shape of the membership functions can be either trapezoidal or triangular according to the situation. However, the triangular fuzzification method was most often used as it is the simplest and widely used method [33-34]. The reason for using a triangular fuzzy number is that it is instinctively easy for the decision-makers to use. Modelling using triangular fuzzy approach enables the experts to deal better with the vague decision problems [35]. According to the nature of TFN, it can be defined as triple points (p,q,r). Here P and R represent the fuzzy probability between the lesser and higher boundaries of evaluation. The triangular fuzzy number is shown in Figure 2. From the literature, it is observed that FAHP has been extensively applied in many complex decision making problems.Sakthivel et al. (2016) incorporate this technique to identify the criteria weights of the engine performance parameters [36]. Shen et al. (2010) incorporate this technique for exploiting renewable energy sources [37].Afsordegan et al (2016) applied this technique for assessing the weights to criteria to find the best energy alterative [38]. Sivarajaet al. (2017) used this technique for assessed renewable energy development in the engines [39].
The procedural steps of FAHPare shown below:
Step 1.The hierarchy is structured with respect to the criteria for a complex decision making problem. It has three levels in which objective of the study is at top, multi criteria attributes defines in the middle section and the alternatives at the bottom [56]. The objective of hierarchy is ultimately to find out the alternatives.
Step 2. The crisp pair wise comparison matrix is fuzzified using TFN M= (p, q, r) where p and are indicateshigher and lower range respectively that may exist in the preference criteria of decision maker's criteria. M1, M2, M3, M4, M5, M6, M7, M8, M9 are the membership function to indicate the assessment. The membership function of triangular fuzzy number for FAHP are expressed on five scales and also listed in Table 2.
Let Q = $\{Q_j, j = 1, 2..., z\}$ be a set of criteria. The result of the pair-wise comparison on 'z' criteria can be summarized in an (z × z) evaluation matrix T in which every element t_{ij} (i, j = 1, 2..., z) is the quotient of weights of the criteria, as shown:

$$T=\begin{bmatrix} t_{11} & t_{12}\cdots & \cdots t_{1z} \\ t_{21} & t_{22}\cdots & \cdots t_{2z} \\ t_{z1} & t_{z2}\cdots & \cdots t_{zm} \end{bmatrix} t_{ii} = 1, t_{ji} = \frac{1}{t_{ij}}, t_{ij} \neq 0. \quad (1)$$

Step 3. The mathematical process is normalized to identify the relative weights of decision criteria which is in each matrix. The relative weights are given by using right Eigen vector (Z) which relates to the maximum (largest) Eigen value λ max as

$$T_z = \lambda\text{max } z \quad (2)$$

The consistency of the pair-wise comparison matrix is defined by the relation between entries of T ; $t_{ij} \times t_{jk} = t_{ik}$.
And hence the consistency index is

$$CI = \frac{\lambda\text{max} - z}{z-1}. \quad (3)$$

Step 4. The pair wise matrix is normalized and the weights are prioritized. The values in this priority are summing equal to 1. The consistency ratio (CR) is the ratio of consistency index (CI) with random consistency index (RCI) using pair wise comparison matrix and it is indicated below

$$CR = \frac{CI}{RCI}. \quad (4)$$

Table 3 lists the value of the RCI for matrices of order 1 to 10 obtained by approximating random indices using a sample size of 500.

The acceptable CR range varies as per the matrix size and for 3 x 3 matrixis 0.05, for 4 x 4 matrix is 0.08 and for all greater matrices is 0.1. The value of CR is equal to, or less than the range shows that the evaluation matrix is acceptable which implies the consistency level is good. Ifit is more than the value, there is an inconsistency and therefore evaluations have to be reconsidered or reviewed until it improves.

3.2 PROMETHEE method

PROMETHEE is MCDM method for ranking a solution of any alternatives with respect to its related criteria which developed by Brans et al. [41]. Since it follows transparent computations and easily understandable it is best known method for outranking. PROMETHEE introduces preference functions to describe the decision maker's preferences along each criterion which is characterized by ease of use and decreased complexity. It also identifies the difference level between alternatives while calculating the ranking priority. Mladineo et al. (1987) proposed the technique for alternative location ranking [42]. Tzeng and Shiau (1987) applied PROMETHEE to analyse the strategies of energy conservation strategies in urban transportation [43].Turcksin et al. (2011) identified a multi-instrumentality policy to reduce environmental externalities by selecting the most appropriate energy policy scenario using this technique [44].Alsayed et al. (2014) used this technique for design of hybrid power generation systems [45].Wei et al. (2016) proposed this technique for study of grid energy storage systems [46]. The procedure for PROMETHEE is explained below.

Step 1 Normalization of the decision matrix

$$Y = \frac{z_{ij}}{\sqrt{\sum z_{ij}^2}} \quad (5)$$

Step 2 Computation of the evaluative difference of kth alternative with other alternatives

The differences in each criteria value is calculated for different alternatives in a pair wise manner.

Step 3 Compute the preference function

The preference function requires the definition of some preferred parameters such as the parameter which is to be preferred or indifference. But in real world application it is difficult to find such parameter for decision makers. Hence the following simple preference function is employed.

$$P_l(k,k') = 0 \; if \; Y_{kl} < Y_{k'l} \quad (6)$$

$$P_l(k,k') = (Y_{kl} - Y_{k'l}) \; if \; Y_{kl} > Y_{k'l} \quad (7)$$

Step 4 Calculate the aggregated preference function

The aggregated preference function isbe calculated by considering the weights for each criterion using the below equation

$$S(k,k') = \frac{\sum_{l=1}^{m} v_l \times P_l(k,k')}{\sum_{l=1}^{m} v_l} \quad (8)$$

Where v_l is the relative importance or weight for the lth criterion.

Step 5 Determination of leaving flow and entering flow

The leaving flow or positive flow of kth alternative is calculated as,

$$\varphi^+(k) = \frac{1}{n-1} \sum_{k'=1}^{n} S(k,k') \qquad (k \neq k') \quad (9)$$

The entering or negative flow for kth alternative is calculated as,

$$\varphi^-(k) = \frac{1}{n-1} \sum_{k'=1}^{n} S(k',k) \qquad (k \neq k') \quad (10)$$

Where n is the number of alternatives.

In this application for each alternative there are (n-1) other alternatives.

The entering flow determines how much a one alternative is conquered by other alternative and the leaving flow identifies how much one particular alternative governs the other alternatives. PROMETHEE I provide partial outranking for the alternatives based on the leaving and entering flows.

Step 6 Calculating net outranking flow

$$\varphi(k) = \varphi^+(k) - \varphi^-(k) \quad (11)$$

Step 7 Determination of the outranking values

The ranking of the alternatives can be determined based on the net outranking flow. Higher the value of net outranking flow is the better alternative.

4. THE PROPOSED MODEL

The proposed model contains of three basic phases to evaluate the best blend namely, exploratoryobservations, criteria weights computations, applying PROMETHEE to identify the suitable blend. Exploratoryanalysis was carried out in a four stroke single cylinder, direct injection, air cooled, constant speed and variable load, compression ignition engine to observe the performance and emission characteristics. In the second stage, the decision hierarchy is structured using the assessed criteria and the alternative blends. Third,the relative weights of the identified criteria are calculated using FAHP.Based on the percentage of relative weights of each criterion, the impact on the ranking may be changed. Thecalculated weights are given as the input for PROMETHEE methodto evaluate the best alternative.

IOP Conf. Series: Materials Science and Engineering **390** (2018) 012028 doi:10.1088/1757-899X/390/1/012028

5. COMPUTATIONS

In this investigation, the authors have identified the evaluation criteria from the literature to select the best blend. The criteria are prioritized in the decision making process as per the collective feedback from the experts. Oxides of Nitrogen, Smoke, Brake thermal efficiency, Carbon dioxide, Carbon monoxide, Hydrocarbon, Exhaust gas temperature, Ignition delay, Combustion duration, Maximum rate of pressure rise are the criteria considered.

5.1. FAHP-PROMETHEE

5.1.1. Identification of weights using FAHP

The decision hierarchy is framed with the use of identified criteria and the alternatives. Here, a feedback is collected from the engine experts and the manufactures to identify the relative importance of the criteria by comparing one with other using questionnaire design. Experts ranked the criteria to meet the better efficiency and minimum emissions. The relative weights are calculated using pair wise comparison matrix with satty's scale. The pair wise decision matrix with respect to the ten criteria is framed using equation 1. The relative weights and consistency ratio of the criteria are also computed from the pair wise comparison analysis using equations 2-4. From the relative analysis, it is identified that Oxides of Nitrogen and smoke is the significant emission factor by stands first and second with the relative weight. The identified criteria with the relative weights using pair wise comparison process are detailed in Table 4.

5.1.2 Determination of best alternative using PROMETTHEE

The computational steps of the PROMETHEE are demonstrated below for 50% load to identify the suitable blend with the observed engine performance parameters.

Step 1: Normalize the experimental performance and emission parameters of engine by equation 5

Step 2: The preference (P) functions are calculated using equations 6 and 7 is given below. for each criterion at various loads and shown in Table 6.

Step 3: The weighted preference function is calculated using FAHP weights for each criteria and also shown in Table 7.

Step 4: The resulting aggregated (S) preference function is calculated using equation 8 for all the blends with different loads and detailed in Table 8.

Step 5: The leaving and negative flow is then computed using equations 9 and 10 and shown in Table 9.

Step 6: The net outranking flow values is computed using equation 11 and tabulated in Table 10.Based on the net outranking values, rank for each alternative were obtained.

A similar computationis carried out for no load, 25%, 75% and 100% load to identify the best blend.

6. RESULTS AND DISCUSSION

The results of the proposed method FAHP and PROMETHEE isdetailed in Table 10. From the observed outranking values, it is identified that B20 blend stands first at 50%, 75% and 100% load and second at no load and 25% load. It is observed that B20 is the suitable blend among the all alternatives to minimise the emissions by improving the engine efficiency. Generally, the influence of performance, emission and combustion characteristics of the engine with respect to the load and blend differs and it is challenging for the researchers to identify the suitable alternative among the different alternatives. The fuel characteristics are also closer which creates inconsistency to satisfy the emission norms and meet the fuel economy. To overwhelm the above problem, the computational model of FAHP with PROMETHEE was proposed to identify the suitable fuel blend.

CONCLUSION

The selection of optimum blend will provide a vehement support to enhance the biodiesel usage in IC engines. The suggested hybrid decision method will assist the engine experts and researchers to identify the suitable blend to improve the energy competence in the engines. In multi criteria group decision making process, the decision makers may not express their opinions exactly in the numerical values. So, fuzzy sets are implemented to eliminate the uncertainties arising with the decision maker's opinions. The proposed hybrid MCDM method is perfect and effective tool for the experts to determine the apt blend between different alternatives.

Table 1 Engine specifications

Items	Specification
Make	Kirloskar
Cylinder number	1
Type	Four-stroke, stationary, constant speed, direct injection, air cooled, diesel engine
Bore x stroke	80mm x 110mm
Displacement	661cc
Compression ratio	17.5 : 1
Max. power/speed	4.4 kW/ 1500rpm
Injection timing	24° bTDC
Injection pressure	210bar

Table 2 Pair wise relative importance

Degree of Preference	Definition	Explanation
(1 1 1)	Equally Preferred(M_1)	Two activities contribute equally to the objective
(2 3 4)	Moderately Preferred (M_3)	Experience & judgment slightly favor one activity over another
(4 5 6)	Strongly Preferred (M_5)	Experience & judgment strongly or essentially favor one activity over another
(6 7 8)	Very strongly Preferred (M_7)	An activity is strongly favored over another and its dominance demonstrated in practice
(8 9 9)	Extremely Preferred (M_9)	The evidence favoring one activity over another is of the highest possible order of affirmation

Table 3 Random Consistency Index (RCI)

No	1	2	3	4	5	6	7	8	9	10
RCI	0	0	0.52	0.89	1.11	1.25	1.35	1.40	1.45	1.49

Table 4 Results obtained with FAHP

Criteria	Weights	λ_{max},CI, RCI	CR
NO_X	0.2748	λ_{max} = 11.714 CI = 0.05 RCI = 1.49	0.034
SMOKE	0.1933		
BTE	0.1698		
CO_2	0.1132		
CO	0.0851		
HC	0.0558		
EGT	0.0351		
ID	0.0340		
CD	0.0252		
MRPR	0.0137		

Table 5. Experimental performance and emission readings observed from engine for various alternative blends

CRITERIA / LOAD	BLENDS	NO_x (ppm)	SMOKE (%)	BTE (%)	CO_2 (%vol)	CO (%vol)	HC (ppm)	EGT (^{0}C)	ID (^{0}CA)	CD (^{0}CA)	MRPR (bar/^{0}CA)
0 %	Diesel	239	5	0	2.2	0.07	29	157	16	46.28	4.55
	B 20	221	10.1	0	2	0.06	27	164	15.9	45.16	4.01
	B 40	219	12.4	0	2	0.06	26	165	15.82	43.08	4.01
	B 60	201	14.2	0	2	0.05	25	168	15.44	44.44	3.48
	B 80	191	11.8	0	2	0.05	24	166	15.42	42.68	3.48
	B 100	178	20.5	0	2	0.04	23	165	15.34	41.135	3.09
25 %	Diesel	519	15.2	17.31	3.2	0.08	35	204	15.16	49.3	5.95
	B 20	522	16	18.03	3.2	0.07	31	205	14.64	47.26	5.61
	B 40	520	16.3	16.32	3.3	0.07	27	211	14.48	44.78	4.81
	B 60	518	19.4	16.3	3.2	0.06	26	212	14.48	45.95	4.08
	B 80	508	20.3	15.69	3.3	0.06	26	204	14.28	44.74	4.01
	B 100	492	25.7	14.54	3.5	0.05	24	215	14.02	41.44	3.38
50 %	Diesel	986	18.5	26.17	4.8	0.09	38	266	14.22	52.14	6.42
	B 20	987	18	27.05	4.5	0.08	34	266	13.64	50.52	5.61
	B 40	964	18.6	24.49	4.5	0.08	29	270	13.52	47.38	4.81
	B 60	944	23.6	23.99	4.7	0.07	31	270	13.42	48.46	4.4
	B 80	935	23.4	21.73	4.7	0.07	29	262	13.24	45.24	4.28
	B 100	904	31.4	22.55	4.9	0.06	28	272	12.84	47.12	3.8
75 %	Diesel	1357	25.8	30.71	6.2	0.09	37	321	13.72	53.28	6.42
	B 20	1358	22.5	30.91	5.8	0.09	35	325	13.52	52.66	5.68
	B 40	1351	23.9	30.41	6	0.08	33	330	13.46	48.28	5.35
	B 60	1346	29.8	27.68	6.2	0.08	31	327	13.42	49.73	5.08
	B 80	1340	28.8	27.16	6.2	0.07	29	332	12.66	51.76	4.55
	B 100	1336	37.4	26.71	6.4	0.07	31	340	11.14	50.34	3.7
100 %	Diesel	1700	32	32.63	8.2	0.11	40	398	13.42	54.81	6.68
	B 20	1689	33.3	34.78	7.4	0.09	37	392	13.38	53.42	6.15
	B 40	1666	36	33.91	7.7	0.09	35	403	13.3	52.22	5.61
	B 60	1642	41.3	31.99	7.7	0.08	32	398	13.04	50.55	5.35
	B 80	1642	41.6	30.17	7.7	0.08	33	397	11.68	49.34	4.55
	B 100	1606	50.9	29.68	8.2	0.07	32	399	10.92	48.94	3.9

Table 6: Preference Function matrix for 50% load

LOAD	BLEND	CRITERA	NO_x	SMOKE	BTE	CO_2	CO	HC	EGT	ID	CD	MRPP
50%	DIESEL	**(p1,p2)**	0.000	0.009	0.000	0.026	0.054	0.052	0.000	0.018	0.014	0.067
		(p1,p3)	0.009	0.000	0.030	0.026	0.054	0.116	0.000	0.021	0.040	0.132
		(p1,p4)	0.018	0.000	0.039	0.009	0.108	0.090	0.000	0.024	0.031	0.166
		(p1,p5)	0.022	0.000	0.080	0.009	0.108	0.116	0.006	0.030	0.058	0.176
		(p1,p6)	0.035	0.000	0.065	0.000	0.162	0.129	0.000	0.042	0.042	0.215
	B20	**(p2,p1)**	0.000	0.000	0.016	0.000	0.000	0.000	0.000	0.000	0.000	0.000
		(p2,p3)	0.010	0.000	0.046	0.000	0.000	0.064	0.000	0.004	0.026	0.066
		(p2,p4)	0.018	0.000	0.055	0.000	0.054	0.039	0.000	0.007	0.017	0.099
		(p2,p5)	0.022	0.000	0.096	0.000	0.054	0.064	0.006	0.012	0.044	0.109
		(p2,p6)	0.036	0.000	0.081	0.000	0.108	0.077	0.000	0.024	0.029	0.149
	B40	**(p3,p1)**	0.000	0.002	0.000	0.000	0.000	0.000	0.006	0.000	0.000	0.000
		(p3,p2)	0.000	0.011	0.000	0.000	0.000	0.000	0.006	0.000	0.000	0.000
		(P3,P4)	0.020	0.000	0.030	0.000	0.100	0.000	0.000	0.003	0.000	0.040
		(P3,P5)	0.012	0.000	0.050	0.000	0.054	0.000	0.012	0.008	0.018	0.044
		(P3,P6)	0.026	0.000	0.035	0.000	0.108	0.013	0.000	0.021	0.002	0.083
	B60	**(P4,P1)**	0.000	0.092	0.000	0.000	0.000	0.000	0.006	0.000	0.000	0.000
		(P4,P2)	0.000	0.101	0.000	0.017	0.000	0.000	0.006	0.000	0.000	0.000
		(P4,P3)	0.000	0.090	0.000	0.017	0.000	0.026	0.000	0.000	0.009	0.000
		(P4,P5)	0.004	0.004	0.041	0.000	0.000	0.026	0.012	0.005	0.027	0.010
		(P4,P6)	0.017	0.000	0.026	0.000	0.054	0.039	0.000	0.018	0.011	0.049
	B80	**(P5,P1)**	0.000	0.088	0.000	0.000	0.000	0.000	0.000	0.000	0.000	0.000
		(P5,P2)	0.000	0.097	0.000	0.017	0.000	0.000	0.000	0.000	0.000	0.000
		(P5,P3)	0.000	0.086	0.000	0.017	0.000	0.000	0.000	0.000	0.000	0.000
		(P5,P4)	0.000	0.000	0.000	0.000	0.000	0.000	0.000	0.000	0.000	0.000
		(P5,P6)	0.013	0.000	0.000	0.000	0.054	0.013	0.000	0.012	0.000	0.039
	B100	**(P6,P1)**	0.000	0.232	0.000	0.009	0.000	0.000	0.009	0.000	0.000	0.000
		(P6,P2)	0.000	0.241	0.000	0.035	0.000	0.000	0.009	0.000	0.000	0.000
		(P6,P3)	0.000	0.230	0.000	0.035	0.000	0.000	0.003	0.000	0.000	0.000
		(P6,P4)	0.000	0.140	0.000	0.017	0.000	0.000	0.003	0.000	0.000	0.000
		(P6,P5)	0.000	0.144	0.015	0.017	0.000	0.000	0.015	0.000	0.016	0.000

IOP Conf. Series: Materials Science and Engineering **390** (2018) 012028 doi:10.1088/1757-899X/390/1/012028

Table 7: Weighted Preference function matrix for 50% load

LOAD	BLEND	CRITERA	NO_x	SMOKE	BTE	CO_2	CO	HC	EGT	ID	CD	MRPP
	WEIGHTS		**0.275**	**0.193**	**0.170**	**0.113**	**0.085**	**0.056**	**0.035**	**0.034**	**0.025**	**0.014**
		(p1,p2)	0.000	0.002	0.000	0.003	0.005	0.003	0.000	0.001	0.000	0.001
		(p1,p3)	0.003	0.000	0.005	0.003	0.005	0.006	0.000	0.001	0.001	0.002
	Diesel	**(p1,p4)**	0.005	0.000	0.007	0.001	0.009	0.005	0.000	0.001	0.001	0.002
		(p1,p5)	0.006	0.000	0.014	0.001	0.009	0.006	0.000	0.001	0.001	0.002
		(p1,p6)	0.010	0.000	0.011	0.000	0.014	0.007	0.000	0.001	0.001	0.003
		(p2,p1)	0.000	0.000	0.003	0.000	0.000	0.000	0.000	0.000	0.000	0.000
		(p2,p3)	0.003	0.000	0.008	0.000	0.000	0.004	0.000	0.000	0.001	0.001
	B 20	**(p2,p4)**	0.005	0.000	0.009	0.000	0.005	0.002	0.000	0.000	0.000	0.001
		(p2,p5)	0.006	0.000	0.016	0.000	0.005	0.004	0.000	0.000	0.001	0.001
		(p2,p6)	0.010	0.000	0.014	0.000	0.009	0.004	0.000	0.001	0.001	0.002
		(p3,p1)	0.000	0.000	0.000	0.000	0.000	0.000	0.000	0.000	0.000	0.000
		(p3,p2)	0.000	0.002	0.000	0.000	0.000	0.000	0.000	0.000	0.000	0.000
	B 40	**(P3,P4)**	0.005	0.000	0.005	0.000	0.009	0.000	0.000	0.000	0.000	0.001
		(P3,P5)	0.003	0.000	0.008	0.000	0.005	0.000	0.000	0.000	0.000	0.001
50%		**(P3,P6)**	0.007	0.000	0.006	0.000	0.009	0.001	0.000	0.001	0.000	0.001
		(P4,P1)	0.000	0.018	0.000	0.000	0.000	0.000	0.000	0.000	0.000	0.000
		(P4,P2)	0.000	0.019	0.000	0.002	0.000	0.000	0.000	0.000	0.000	0.000
	B 60	**(P4,P3)**	0.000	0.017	0.000	0.002	0.000	0.001	0.000	0.000	0.000	0.000
		(P4,P5)	0.001	0.001	0.007	0.000	0.000	0.001	0.000	0.000	0.001	0.000
		(P4,P6)	0.005	0.000	0.004	0.000	0.005	0.002	0.000	0.001	0.000	0.001
		(P5,P1)	0.000	0.017	0.000	0.000	0.000	0.000	0.000	0.000	0.000	0.000
		(P5,P2)	0.000	0.019	0.000	0.002	0.000	0.000	0.000	0.000	0.000	0.000
	B 80	**(P5,P3)**	0.000	0.017	0.000	0.002	0.000	0.000	0.000	0.000	0.000	0.000
		(P5,P4)	0.000	0.000	0.000	0.000	0.000	0.000	0.000	0.000	0.000	0.000
		(P5,P6)	0.004	0.000	0.000	0.000	0.005	0.001	0.000	0.000	0.000	0.001
		(P6,P1)	0.000	0.045	0.000	0.001	0.000	0.000	0.000	0.000	0.000	0.000
		(P6,P2)	0.000	0.046	0.000	0.004	0.000	0.000	0.000	0.000	0.000	0.000
	B 100	**(P6,P3)**	0.000	0.044	0.000	0.004	0.000	0.000	0.000	0.000	0.000	0.000
		(P6,P4)	0.000	0.027	0.000	0.002	0.000	0.000	0.000	0.000	0.000	0.000
		(P6,P5)	0.000	0.028	0.002	0.002	0.000	0.000	0.001	0.000	0.000	0.000

Table 8: Aggregated Preference Function matrix

LOAD	BLEND	Diesel	B 20	B 40	B 60	B 80	B 100
0%	Diesel	-	0.09	0.06	0.05	0.04	0.09
	B 20	0.05	-	0.10	0.05	0.03	0.04
	B 40	0.02	0.03	-	0.05	0.05	0.07
	B 60	0.01	0.02	0.01	-	0.05	0.08
	B 80	0.05	0.05	0.05	0.02	-	0.02
	B 100	0.06	0.02	0.01	0.05	0.02	-
25%	Diesel	-	0.05	0.04	0.04	0.03	0.03
	B 20	0.01	-	0.02	0.03	0.03	0.05
	B 40	0.04	0.01	-	0.01	0.05	0.02
	B 60	0.02	0.01	0.02	-	0.02	0.04
	B 80	0.02	0.02	0.02	0.01	-	0.01
	B 100	0.01	0.00	0.01	0.01	0.03	-
50%	Diesel	-	0.05	0.05	0.05	0.03	0.03
	B 20	0.05	-	0.04	0.06	0.04	0.05
	B 40	0.02	0.02	-	0.02	0.01	0.02
	B 60	0.00	0.02	0.02	-	0.03	0.04
	B 80	0.02	0.02	0.02	0.00	-	0.01
	B 100	0.00	0.00	0.01	0.02	0.02	-
75%	Diesel	-	0.01	0.02	0.02	0.03	0.03
	B 20	0.03	-	0.05	0.04	0.02	0.03
	B 40	0.01	0.02	-	0.03	0.03	0.04
	B 60	0.01	0.02	0.02	-	0.00	0.00
	B 80	0.00	0.01	0.02	0.01	-	0.03
	B 100	0.00	0.01	0.01	0.02	0.02	-
100%	Diesel	-	0.02	0.04	0.03	0.02	0.02
	B 20	0.02	-	0.04	0.04	0.03	0.04
	B 40	0.01	0.01	-	0.01	0.02	0.03
	B 60	0.02	0.02	0.01	-	0.01	0.01
	B 80	0.01	0.01	0.02	0.01	-	0.02
	B 100	0.02	0.02	0.01	0.00	0.01	-

Table 9: Leaving and Entering flow matrix

LOAD	0%		25%		50%		75%		100%	
BLENDS	Ψ+	Ψ-	Ψ+	Ψ-	Ψ+	Ψ-	Ψ+	Ψ-	Ψ+	Ψ-
DIESEL	0.07	0.04	0.04	0.02	0.04	0.02	0.02	0.01	0.03	0.01
B 20	0.05	0.04	0.03	0.02	0.05	0.02	0.03	0.01	0.03	0.01
B 40	0.04	0.05	0.03	0.02	0.02	0.03	0.03	0.02	0.01	0.02
B 60	0.03	0.04	0.02	0.02	0.02	0.03	0.01	0.02	0.01	0.02
B 80	0.04	0.04	0.02	0.03	0.01	0.03	0.01	0.02	0.01	0.02
B 100	0.03	0.06	0.01	0.03	0.01	0.03	0.01	0.03	0.01	0.02

Table 10: Net Outranking Flow matrix

LOAD	0%		25%		50%		75%		100%	
BLENDS	Ψ	RANK	Ψ	RANK	Ψ	RANK	Ψ	RANK	Ψ	RANK
DIESEL	0.03	1	0.02	1	0.02	2	0.01	2	0.01	2
B 20	0.01	2	0.01	2	0.03	1	0.02	1	0.02	1
B 40	0.00	3	0.00	4	-0.01	4	0.01	3	-0.01	3
B 60	-0.01	5	0.00	3	-0.01	3	-0.01	4	-0.01	4
B 80	0.00	4	-0.02	5	-0.01	5	-0.01	5	-0.01	5
B 100	-0.03	6	-0.02	6	-0.02	6	-0.02	6	-0.01	6

References

[1] Kumar N, Varun, Kumar A. 2010 Biodiesel as an alternative fuel for CI engines: Environmental effect. *Indian Journal of Science and Technology*; **3** pp. 602- 606.

[2] Vilela L, Mata TM, Caetano NS 2010. Biodiesel production from fish oil with high acidity. *In: Third international symposium on energy from biomass and waste.*(Italy) pp. 1-14.

[3] Boyd M, Murray HA, Schaddelee K. 2004. Biodiesel in British Columbia-feasibility study report. Canada: WISE Energy Co-op/Eco-Literacy. pp. 1-126.

[4] Senthamaraikannan P, Sivaprakasam S, Saravanan CG. 2010. Combustion and emission analysis on DI diesel engine run on animal fat oil. *8th Asia Pacific Conference on Combustion (ASPACC 2010)*, (India)pp. 1005-09.

[5] Jayasinghe P and Hawboldt K. 2016. A review of bio-oils from waste biomass: Focus on fish processing waste. *Renewable and sustainable energy reviews*. **16.** pp. 798-821.

[6] Reyes JF, Sepulveda MA. 2006. PM-10 emissions and power of a diesel engine fueled with crude and refined biodiesel from salmon oil, *Fuel*. **85**. pp. 1714-19.

[7] Preto F, Zhang F, Wang J. 2007. Fish oil as an alternative fuel for conventional combustors. *The Arctic Energy Summit, Anchorage*. (Alaska) pp. 1-7.

[8] Aryee ANA, Van de VoortFR,Simpson BK. 2009. FTIR determination of free fatty acids in fish oils intended for biodiesel production. *Process Biochemistry*. **44** pp. 401- 405.

[9] S, Murthy CS, Reddy R. 2010. Performance and emission characteristics of a Kirloskar HA394 diesel engine operated on fish oil methyl esters. *Renew Energy*.**35**. pp. 355-59.

[10] Lin CY, and Li RJ. .2009a. Fuel properties of biodiesel produced from the crude fish oil from the soapstock of marine fish. *Fuel Process. Technol***90**. pp. 130-36.

[11] Lin CY, and Li RJ. 2009b. Engine performance and emission characteristics of marine fish-oil biodiesel produced from the discarded parts of marine fish. *Fuel Process Technol*. **90**. pp 883-888.

[12] Kato S, Kunisawa K, Kojima T, Murakami, S. 2004. Evaluation of ozone treated fish waste oil as a fuel for transportation. *Journal of Chemical Engineering of Japan*. **37**. Pp 863-870

[13] Steigers JA. 2003. Demonstrating the use of fish oil as fuel in a large stationary diesel engine. *Alaska energy authority*. pp 1-14.

[14] Cherng YL, Rong JL. 2009. Fuel properties of biodiesel produced from the crude fish oil from the soap stock of marine fish. *Fuel Processing Technology*. **90**. pp 130-36.

[15] Godiganur S, Murthy CS, Reddy R. 2010. Performance and emission characteristics of a Kirloskar HA394 diesel engine operated on fish oil methyl esters. *Renew Energy*.**35**.pp 355-359.

[16] Sakthivel G, Nagarajan G, Ilangkumaran M, Gaikwad AB. 2014. Comparative analysis of performance, emission and combustion parameters of diesel engine fuelled with ethyl ester of fish oil and its diesel blends. *Fuel*.**132**. pp 116-124.

[17] Behcet R. 2011. Performance and emission study of waste anchovy fish biodiesel in a diesel engine. *Fuel Process. Technol*. **92**. pp 1187-94

[18] Sakthivel G, Nagarajan G. 2011. Experimental studies on the performance and emission characteristics of a diesel engine fuelled with ethyl ester of waste fish oil and its diesel blends. *SAE International Journal of Fuels and Lubricants*. pp 1-6.

[19] Vedaraman N, Puhan S, Nagarajan G, Velappan KC. 2011. Preparation of palm oil biodiesel and effect of various additives on NOx emission reduction in B20: An experimental study. *International Journal of Green Energy*. **8**. pp 383-397.

[20] Shanmugam P, Sivakumar V, Murugesan A, Umarani C. 2011. Experimental study on diesel engine using hybrid fuel blends. *International Journal of Green Energy*.**8**. pp 655-668.

[21] Sakthivel G. 2016. Prediction of CI engine performance, emission and combustion characteristics using fish oil as a biodiesel at different injection timing using fuzzy logic *Fuel***183**. pp 214-229.

[22] Poh KL, and Ang BW. 1999. Transportation fuels and policy for Singapore: an AHP planning approach. *Computers and Industrial Engineering*. **37**. pp 507-525.

[23] Goumas M, Lygerou V. 2000. An extension of the PROMETHEE method for decision making in fuzzy environment: Ranking of alternative energy exploitation projects. *European Journal of Operational Research*. **123**. pp 606-13.

[24] Yedla S, and Shrestha RM. 2003. Multi-Criteria approach for the selection of alternative option for environmentally sustainable transport system in Delhi. *Transportation Research Part A*.**37**. pp 717-729.

[25] Tsita KG, and Pilavachi PA. 2012. Evaluation of alternative fuels for the Greek road transport sector using the analytic hierarchy process. *Energy policy*.**48**.pp 677-686.

[26] Etghani MM, Shojaeefard MH, Khalkhali A, and Akbari M. 2013. A hybrid method of modified NSGA-II and TOPSIS to optimize performance and emissions of a diesel engine using biodiesel. *Applied Thermal Engineering*. **59**. pp 309-315.

[27] Sakthivel G, Ilangkumaran M, Nagarajan G, Raja A, Ragunadhan PM, and Prakash J. 2013. A hybrid MCDM approach for evaluating an automobile purchase model. *International journal of information and decision sciences*. **5**. pp 50-85.

[28] Chand M, Hatwal D, Singh S, Mundepi V, Raturi V, RashmiAvikal S. 2017. An Approach for purchasing a sedan car from Indian car market under fuzzy environment. *Proceedings of sixth international conference on soft computing for problem solving*. pp 239-244.

[29] Saaty TL. 1980. The Analytic Hierarchy Process. *McGraw-Hill, New York*

[30] Khajeeh M. 2010. Water conservation in Kuwait: a fuzzy analysis approach. International *Journal of Industrial Engineering*. **6**. pp 90–105.

[31] Zadeh LA. 1965. Fuzzy sets. *Information and control*. **8**. pp 338-358.

[32] Laarhoven V, PJM, and Pedrycz W. 1983. A fuzzy extension of Saaty's priority theory. *Fuzzy Sets and Systems*. **11**. pp 229-241.

[33] Buyukozkan G, and Guleryuz S. 2016. Fuzzy multi criteria decision making approach for evaluating sustainable energy technology alternatives. *International journal of renewable energy sources*. **1**. pp 1-6.

[34] Ding JF, and Liang GS. 2005. Using fuzzy MCDM to select partners of strategic alliances for liner shipping. *Information Sciences*. **173**. pp 197-225.

[35] Wang TC, and Chang TH. 2007. Application of TOPSIS in evaluating initial training aircraft under a fuzzy environment. *Expert Systems with Application*. **33**. pp 870-80.

[36] Sakthivel G, Ilangkumaran M, Ikua BW. 2016. Selection of optimum fish oil fuel blend to reduce the greenhouse gas emissions in an IC engine-A hybrid multiple criteria decision aid approach. *International journal of green energy*. **13**. pp 1517-1533.

[37] Shen YC, Lin GT, Li KP, and Yuan BJ. 2010. An assessment of exploiting renewable energy sources with concerns of policy and technology. *Energy Policy*.**38**. pp 4604-4616.

[38] Afsordegan A, Sanchez M, Agell N, Zahedi S, Cremades LV. 2016. Decision making under uncertainty using a qualitative TOPSIS method for selecting sustainable energy alternatives. *International Journal of Environmental Science and Technology*.**13**. pp 1419-1432.

[39] Sivaraja CM, Sakthivel G. 2017. Compression ignition engine performance modelling using hybrid MCDM techniques for the selection of optimum fish oil biodiesel blend at different injection timings. *Energy*.**139**.pp 118-141.

[40] Albayrak E, and Erensal YC. 2004. Using analytic hierarchy process (AHP) to improve human performance: An application of multiple criteria decision making problem. *Journal of Intelligent Manufacturing*. **15**. pp 491-503.

[41] Brans JP, and Vincke PH. 1985. A preference ranking organization method. *Management Science*. **31**. pp 647-656.

[42] Mladineo N, Margeta J, Brans JP, and Mareschal B. 1987. Multi criteria ranking of alternative locations for small scale hydro plants. *European Journal of Operational Research*.**31**. pp 215-22.

[43] Tzeng GH, and Shiau TA. 1987. Energy conservation strategies in urban transportation, application of multi-criteria decision making. *Energy Systems and Policy*. **11**. pp 1-19.

[44] Turcksin L, Bernardini A, and Macharis C. 2011. A combined AHP-PROMETHEE approach for selecting the most appropriate policy scenario to stimulate a clean vehicle fleet. *Proceedia social and behavioral Sciences*. **20**. pp 954-965.

[45] Alsayed M, Cacciato M, Scarcella G, Scelba G. 2014. Design of hybrid power generation systems based on multi criteria decision analysis. *Solar Energy*. **105**. pp 548-60.

[46] Wei L, Hou J, Qin T, Yuan Z, and Yan Y. 2016. Evaluation of grid energy storage system based on AHP-PROMETHEE-GAIA. *IEEE Chinese Control conference*. pp 1-8.

IOP Conf. Series: Materials Science and Engineering **390** (2018) 012029 doi:10.1088/1757-899X/390/1/012029

Challenges on the synthesis, characterization and machining of green fiber plastics: A review

R. Vinayagamoorthy[1*], I. V. Manoj[2], G. Narendra Kumar[3], I. Sai Chand[4], G. V. Sai Charan Kumar[5], K. Suneel Kumar[6]

[1-6] Department of Mechanical Engineering, Sri Chandrasekharendra Saraswathi Viswa Mahavidyalaya, Kancheepuram, Tamilnadu - 631561, India.

Email: vin802002@gmail.com

Abstract. The need of composite material in industries is increasing day by day due to its competent characteristics with the traditional materials. The plastic composite materials made of polymers are used as structural elements in different fields. The present day polymeric composites are made by mainly focussing the natural fibers as reinforcements because of their usability, properties and cost. While preparing these new composites, utmost care must be taken for synthesizing the natural fibers and polymers before fabrication. Also, during the fabrication of a new composite the process and parameter selection plays a vital role. Once the composites are fabricated, it needs several machining operations as secondary processing. This paper gives a detailed review of different challenges during the synthesis, fabrication, characterization and machining of natural fiber reinforced plastic composites.

1. Introduction

A composite material is prepared by adding two different elements together in the form of a solid shape. The two elements are the matrix which provides the shape and the reinforcements which gives the characteristics [1, 2]. During the initial stage artificial reinforcements are used but later, due to the usability, characteristics and cost green fibers obtained various natural sources are utilized for reinforcements and are said to be green fiber plastics (GFP) [3-4]. The green fibers cannot be used in the original form. It requires some kind of synthesis in order to exhaust the useless substances possessed by it and this may be achieved by using suitable pre-treatments [5-9]. Polymer composites are widely utilised as structural and supporting components in automobile, railway, aircraft and marine industries. They are majorly used in making bodies for boats and radial fan blades [10, 11]. On the other hand, in view of the weight reduction of automobile parts, the GFP is superior to conventional material and therefore, the fuel consumption and efficiency may also be enhanced [12].

Fabrication of GFP is also a big challenge as the operating conditions majorly influences the properties of the composite. Among the several methods, wet layup is a mostly used fabrication technique to prepare the material in the form of slabs [13, 14]. Filament winding process, pultrusion and resin transfer moulding are other techniques to make the composites in different shapes. Once a GFP is fabricated, it requires some secondary operations to bring it in the exact size and shape [15, 16]. At the time of machining, the GFP is prone to damage in many ways and these damages affect the characteristics of the composite. Considering the machinability, a GFP must adapt itself to easily

IOP Conf. Series: Materials Science and Engineering **390** (2018) 012029 doi:10.1088/1757-899X/390/1/012029

machine and also with lesser damages. Hence, this review presents the outcome of synthesis, characterization and machining of GFP carried out by the researchers.

2. Green fibers and their properties

Natural fibers are taken from three bases. They are plant, animal and mineral. Fibers from plants are obtained from its stem, leaves, fruits, seeds and roots. In the same way the fibers from animals are extracted from hair, leather, internal organs etc. Mineral fibers are derived from the earth and are pre-treated before usage. Some important natural fibers with their properties and chemical compositions are listed in Table 1.

Table 1. Properties of some important natural fibers [17-21]

Fiber	Density (g/cm^3)	Diameter (μm)	Tensile Strength (MPa)	Young's Modulus (GPa)	Failure strain (%)	Chemical Composition (%)			
						Cellulose	Lignin	Hemi-Cellulose	Pectin
Sisal	1.5	50-80	511-635	9.4-22	2.0-2.5	84.93	7-11	10-24	0.8
Nettle	1.5	20-80	650	38	1.7	53-82.6	0.5	-	0.9-4.8
Banana	1.35	50-250	500-550	12-20	5-6	63-64	65.2	18.6	-
Coir	1.2	10-16	175	4-6	30	43	45	1.7	3-4
Cotton	1.5-1.6	20	287-597	5-13	7-8	92-95	-	5.7	1.2
Flax	1.5	50-100	345-1035	50-70	2.7-3.2	62-71	2-2.5	16-18	1.8-2
Hemp	1.1	120	389-900	35	1.6	77.07	3.7-13	14-22.4	0.8
Jute	1.3	260	393-773	23-27	1.4	63.24	12-26	13.6-21	0.2-4.4
Kenaf	1.31	106	427-519	23-27	1.8	65.7	15-19	21.5-23	-
Pineapple	0.8-1.6	25-34	400-627	14.5	1.44	71.6	10.5	80.5	2-4
Linen	1.4	12-60	800	50-70	2.7-3.5	82	4	2	-
Abaca	1.5	10-30	400	12	3-10	56-63	7-9	15-17	0.5
Henequen	1.49	180	430-580	10.1-	1.6	77.6	13.1	4-8	-
Ramie	1.5	34	560	24.5	2.5	91	0.7	5-16.7	1.9-2.1
Oil Palm	0.7-1.55	150-250	248	25	3.2	65	19	-	-
Isora	1.2-1.3	10	500-600	-	5-6	74	23	-	-

3. Synthesis of green fibers and fabrication of composites

In order to enhance the surface properties of a natural fiber, it is subjected to a series of pre-treatment. These pre-treatments include, cleaning of fibers with distilled water, soaking of fibers in chemicals followed by heat treatment. A research work studied the influence of three different chemical namely sodium hydroxide, hydrogen peroxide and benzoyl chloride on vetiver fibers and proved that all chemicals have the capacity to improve the properties of the natural fibers [22]. Silane treatment is also a treatment method which helps to improve the tensile strength of the natural fibers and in turn the GFP [23]. Each chemical treatment has a different effect and as a result, it will enhance the characteristics of the reinforcement. Treating with acetic anhydride is a good way to improve the fiber-matrix bonding strength and thereby enhance the tensile strength and thermal stability of the GFP [24]. Several research works are carried out by giving different chemical treatments to green fibers and some of them are listed in Table 2.

A GFP may be fabricated in different ways. Hand Layup or wet layup is the oldest and easiest technique for fabricating parts like bath tubs, bodies of simple boats and furniture. This method has been used by many researchers due to its simplicity [25-27]. Compression moulding is the next level to the hand layup method in which, the load is applied by the ram after layering the composite

IOP Conf. Series: Materials Science and Engineering **390** (2018) 012029 doi:10.1088/1757-899X/390/1/012029

constituents in the mould. This method has got several advantages over the hand layup method like, uniform resin coverage, good bonding and compaction [28]. Automobile components are manufactured using press moulding technique in which the vacuum pressure level decides the quality of component. A nominal range of pressure creates products with high strengths whereas high pressure decreases the strengths of the composite [29]. Vacuum bag moulding is another important method which suits well for preparation of hybrid composites. It has been proved that, this method promotes the cross linking of fiber and resin when a setting time of more than 70 hours is maintained [30]. Composite tubes and rods are manufactured by using filament winding method and resin transfer moulding is highly suitable for producing GFP [31].

Table 2. Chemical treatments and their influence in research

S. No	Green fiber	Chemical for treatment	Properties achieved	Ref
1.	Vetiver	Sodium hydroxide	Tensile, flexural, compression and impact strengths	3
2.	Luffa	Sodium hydroxide	Tensile, flexural, compression and impact strengths	7, 14
3.	Kenaf	Sodium hydroxide	Tensile, flexural and impact strengths	32
4.	Date palm	Sodium hydroxide	Tensile strength	33
5.	Kapok	Sodium hydroxide	Tensile strength and chemical resistance	34
6.	Banana	Acrylic acid	Flexural and impact strengths and water absorption	35
7.	Kenaf	Maleic anhydride	Tensile and impact strength	36
8.	Hemp	Sodium hydroxide	Bonding strength	37
9.	Basalt	Silane	Tensile strength	38

4. Characterization of GFP

Green fibers are used majorly with artificial fibers in order to enhance the characteristics of the composite material. Thus they work effectively in hybrid form and this has been proved by several researchers. It has been proved that, the composition of all fibers must be equal in order to get a maximum strength of the hybrid composite material [7]. As each fiber has its own best property, the use of multiple fibers improves majority of the composite properties. Hence, the advantage of hybrid composite is that, a design engineer may decide well in advance about the property of the composite by properly selecting the fiber [9]. In an experimental study, it has been reported that the use of sisal fiber along with glass increased the impact strength and tensile strength of the composite. Also, it is concluded that, a combined use of green fiber and artificial fiber is more advantageous than the green fiber and green fiber combination [39].

Table 3. Optimum composition of green fiber arrived in research

S. No	Green fiber	Optimum composition (%)	Properties achieved	Ref
1.	Flax	40	Flexural and impact strengths	44
2.	Menkuang	20	Tensile strength	45
3.	Aloe vera	30	Tensile, flexural, compression and impact strengths	46
4.	Bagasse	30	Tensile strength	47
5.	Bamboo	50	Impact strength	48
6.	Coir	15	Tensile, flexural, compression and impact strengths	49
7.	Sisal	40	Tensile and flexural strength	50
8.	Hemp	20	Tensile and compressive strengths	51
9.	Kenaf	70	Flexural strength	52
10.	Oil palm	20	Tensile, flexural and impact strengths	53
11.	Jute	30	Tensile, flexural and impact strengths	54
12.	Basalt	20	Compressive strength and Hardness	55
13.	Pine needle	30	Tensile, flexural, compression and wear resistance	56

The number of laminas used in a GFP has influence on its characteristics. A study reported that, a double fiber layered composite produces a maximum bending and impact strength than a triple fiber layered composite. Also, it is concluded that in order to increase the strength of composite, an appropriate fiber must be used as top and bottom layers in a composite [40]. The use of artificial fiber as top and bottom layers in a composite and a green fiber as a central layer gives highest strength to the hybrid GFP [41]. The fiber and matrix composition in a composite decides majority of the characteristics of the composite material. Many research works have been performed to study the influence of fiber loading on the characteristics of the composite material. A research showed that, 30 % okra fiber is the optimum composition for achieving highest bending and tensile strength [42] whereas in another research, it is proved that 20 % composition sisal fiber is the apt proportion to acquire maximum mechanical properties [43]. The optimum composition of fiber arrived in some important research works are shown in Table 3.

5. Machinability of GFP

As a GFP after fabrication require few secondary operations to make it suitable for assembly. Therefore machining is mandatory and the ease of machining a GFP is to be studied [57]. The ease of machining a GFP depends upon the property of green fiber present in the composite as well the property of the matrix used. Thermoset matrices are always harder than the thermoplastic matrices thereby hard matrices produces chips in the form of particles and soft matrices produces chips in the form of continuous threads [58]. A study on the vetiver reinforced vinyl ester composite reported that the presence of green fiber reduces the drilling induced damages to a maximum extent [59, 60]. In another study, optimum conditions have been arrived for machining jute reinforced composite and reported that, a high level of spindle speed, medium level of spindle feed and a medium level of drill size as the optimum values for drilling. In order to arrive at the optimum conditions, any one of the tools of design of experiment is used by all researchers working on the machinability [61]. It is also proved that the green fiber present in a composite has two capacities; one is to improve the mechanical characteristics and other if to improve the machinability of the composite [62-64]. Many research studies have used green fibers along with an artificial fiber and some research has also been worked out with a combination of a green fiber and a metal fiber. It has been proven that, the green works well in respect to both properties as well as in terms of machinability. A research utilised steel fibers as reinforcements along with woven jute and made machining studies. The research reported that, only the factors that can be controlled by an operator of the machine affect the damage of hole to a maximum effect whereas the tool diameter does not influences the damage [65, 66]. A recent research on drilling of hybrid composites concluded that a spindle speed of 1450 rpm, a spindle feed of 0.2 mm/rev, a point angle of 60^0 and a tool diameter of 10 mm are the optimum combinations for drilling [67].

6. Conclusions

This paper has presented a comprehensive review on the literature pertaining to the green fiber reinforced plastic composites. The challenges faced during synthesis of a green fiber, fabrication of green fiber plastics and the ease of machining have been elaborately addressed. Some of the major conclusions obtained are as follows:

- A green fiber may not be used in its native form. A series of pre-treatment methods are to be followed in order to exhaust the unwanted substances present in it and also to make it suitable for usage as reinforcement in composite.
- Among the several pre-treatment methods, alkali treatment and followed by heat treatment is majorly done by many researchers and they proved that alkali treatment is the best method of chemical treatment for green fibers.
- During composite fabrication, a suitable technique must be chosen in order to adapt the size and shape of the component to be manufactured. Hand layup is one the best and easiest method which has been used by many researchers for fabrication of green fiber plastics.

- During characterization of GFP, the influence of fiber loading on the properties has been reported by many researchers. It has been shown that, a fiber loading between 25 % and 35 % will be an optimum value to arrive best properties.
- Machining associated studies have been conducted by several researchers and showed that the presence of green fiber in a GFP improves the ease of machining and reduces the damages in all forms. Working under optimum conditions will be a suitable technique considering the machinability of GFP.

References

[1] Mahmoud Nassar M A, Ramanathan Arunachalam and Khalid Alzebdeh I 2017 *Int. J. Adv. Manuf. Technol.* **88** 2985.
[2] Vinayagamoorthy R 2017 *J. Reinf. Plast. Compos.* **36** 1577.
[3] Vinayagamoorthy R and Rajeswari N 2014 *J. Reinf. Plast. Compos.* **33** 81.
[4] Michael Fuqua A, Shanshan Huo and Chad Ulven A 2012 *Polym. Rev.* **52** 259.
[5] Vinayagamoorthy R, Sivanarasimha,S, Vijay Padmanabhan, Vedula Ganesh and Karthikeyan S 2015 *Int. J. Appl. Eng. Res.* **10** 663.
[6] Sunil Kumar Ramamoorthy, Mikael Skrifvars and Anders Persson 2014 *Polym. Rev.* **55** 107.
[7] Vinayagamoorthy R, Sivanarasimha S, Vinay Kumar K R and Vijay Padmanabhan 2015 *Appl. Mech. Mater.* **813** 14.
[8] Rajani Samal K and Giti Sudha Giri 1992 *J. Macromol. Sci. Part C: Polym. Rev.* **32** 55.
[9] Vinayagamoorthy R, Karthikeyan S, Prem Bhargav R S and Rajivalochan T V 2015 *Appl. Mech. Mater.* **813** 101.
[10] Costa C S M F, Fonseca A, Serra A C and Coelho J F J 2016 *Polym. Rev.* **56** 362.
[11] Vinayagamoorthy R and Madhavan S 2011 *Int. J. Bus. Rev. Manuf. Manage.* **1** 5.
[12] Vinayagamoorthy R, Mothilal T and Madhavan S 2009 *Int. J. Des. Manuf. Technol.* **3** 63.
[13] Marissa Paglicawan A, Susana Cabillion Ma, Rosito Cerbito P and Elizabeth Santos O 2005 *Philli. J. Sci.* **134** 113.
[14] Vinayagamoorthy R, Saswath Koundinya, S L, Mani Teja G L S N and Kovuru Adithya 2016 *Indian J. Sci. Technol.* **9** 1.
15] Davim P J, Silva L R, Festas A and Abrão A 2009 *Mater. Des.* **30** 228.
[16] Vinayagamoorthy R 2017 *J. Reinf. Plast. Compos.* **37** 49.
[17] Osorio L, Trujillo E, Van Vuure AW and Verpoest I 2011 *J. Reinf. Plast. Compos.* **30** 396.
[18] Thiruchitrambalam M and Shamnugam D 2012 *J. Reinf. Plast. Compos.* **30** 1400.
[19] Yupaporn Ruksakulpiwat, Nitinat Suppakarn, Wimonlak Sutapun and Wandee Thomthong 2007 *Compos. Part A: Appl. Sci. Manuf.* **38** 590.
[20] Zainudin E S, Sapuan S M, Abdan K and Mohamad MTM 2009 *Polym. Polym. Compos.* **17** 55.
[21] Yuvaraj G, Hemanth Kumar and Saravanan G 2017 *Poly. Polym. Compos.***25** 221.
[22] Vinayagamoorthy R 2017 *J. Nat. Fib.* doi: 10.1080/15440478.2017.1410513.
[23] Sunija A J, Siva Ilango S and Vinod Kumar K P 2015 *J. Nat. Fib.* **12** 481.
[24] Md. Minhaz-Ul Haque, Maniruzzaman M and Md. Shahmim Reza 2016 *J. Nat. Fib.* **13** 125.
[25] Vinayagamoorthy R, Rajeswari N, Vijayshankar S and Balasubramanian K 2014 *Int. Rev. Mech. Eng.* **8** 952.
[26] Murali K, Vigneshwaran T, Luke Justin Johnson G, Karthikeyan G and Suthan C 2016 *Polym. Polym. Compos.* **24** 461.
[27] Vinayagamoorthy R, Rajeswari N and Balasubramanian K 2014 *Jordan J. Mech. Ind. Eng.* **8** 385.
[28] Mei-po Ho, Hao Wang, Joong-Hee Lee, Chun-kit Ho, Kin-tak Lau, Jinsong Leng and David Hui 2012 *Compos. Part B: Eng.* **43** 3549.
[29] Luisa Medina, Ralf Schledjewski and Alois Schlarb K 2009 *Compos. Sci. Technol.* **69** 1404.
[30] Antoine Le Duigou, Jean-Marc Dux, Peter Davies and Christophe Baley 2011 *Appl. Compos. Mater.* **18** 421.

[31] Sreekumar P A, Joseph K and Unnikrishnan G 2007 *Compos. Sci. Technol.* **67** 453.
[32] Nor Azowa Ibrahim, Wan Md Zin Wan Yunus, Maizatul Othman and Khalina Abdan 2011 *J. Reinf. Plast. Compos.* **30** 381.
[33] Abdalla Abdal-hay, Ngakan Putu Gede Suardana, Do Yeon Jung, Kwang-Seog Choi and Jae Kyoo Lim 2012 *Int. J. Precis. Eng. Manuf.* **13** 1199.
[34] Venkata Reddy G, Shobana Rani T, Khanam P N, John K and Venkata Naidu 2009 *J. Reinf. Plast. Compos.* **28** 175.
[35] Jannah M, Mariatti M, Abu Bakar A and Abdul Khalil 2009 *J. Reinf. Plast. Compos.* 28 1519.
[36] Kwang-Jea Kim 2015 *Compos. Interfaces.* **22** 433.
[37] Sangappaa, Lakshmeesha Raoa B, Ashaa S, Madhu Kumara R and Somashekar R 2014 *Compos. Interfaces.* **21** 153.
[38] Denni Kurniawanab, Byung Sun Kimc, Ho Yong Leeb and Joong Yeon Lim 2015 *Compos. Interfaces.* **22** 553.
[39] Sathish S, Kumaresan M, Karthi N and Dilip Kumar T 2014 *Int. J. Mod. Eng. Res.* **4** 9.
[40] Maries Idicula, Kuruvilla Joseph and Sabu Thomas 2010 *J. Reinf. Plast. Compos.* **2** 12.
[41] Amico S C, Angrizani, C C and Drummond M L 2010 *J. Reinf. Plast. Compos.* **29** 179.
[42] Arifuzzaman Khan G M, Abdullah-Al-Mamun M, Ashraful Abedin S M, Jonayed Choudhury M, Gafur M A and Shamsul Alam M 2014 *Res. Rev: J. Mater. Sci.* **2** 32.
[43] Gunti Rajesh, Ratna Prasad A V and Gupta AVSSKS. 2015 *J. Reinf. Plast. Compos.* **34** 951.
[44] Sathish S, Kumaresan K, Prabhu L and Vigneshkumar N 2017 *Poly. Polym. Compos.* **25** 229.
[45] Mohd Razi Mat Piah, Azizah Bahrum and Ibrahim Abdullah. 2016 *Polym. Polym. Compos.* **24** 767.
[46] Saurabh Chaitanya and Inderdeep Singh 2016 *J. Reinf. Plast. Compos.***35** 1411.
[47] Sivarao Subramonian, Aidy Ali, Mohammad Amran, Sivakumar L D, Shukor Salleh and Rajaizam A 2016 *Adv. Mech. Eng.* **8** 1.
[48] Wong K J, Low K O and Israr H A 2015 *Proc. IMechE: J. Mater. Des. Appl.* **231** 683.
[49] Geetanjali Das and Sandhayarani Biswas 2016 *J. Reinf. Plast. Compos.***35** 628.
[50] Venkatesh Gupta N S, Akash, Sreenivasa Rao K V and Arun Kumar D S 2016 *IOP Conf. Ser.: Mater. Sci. Eng.* **149** 012093.
[51] Somasekar S M, Manjunath V, Gowtham M J and Balasubramaniam N S 2016 *Int. J. Mech. Eng. Technol.* **7** 182.
[52] Chandra Mohan D and Murali B 2014 *J. Chem. Pharma. Res.* **6** 626.
[53] Obasi H C, Iheaturu N C, Onuoha F N, Chike-Onyegbula C O, Akanbi M N, Ezeh VO 2014 *Am. J. Eng. Res.* **3** 117.
[54] Md Rezaur Rahman, Mahbub Hasan, Md Monimul Huque and Nazrul Islam 2010 *J. Reinf. Plast. Compos.* **29** 445.
[55] Shoufan Cao, Hongtao Liu, Shirong Ge and Gaofeng Wu 2011 *J. Reinf. Plast. Compos.* **30** 347.
[56] Singha A S and Vijay Kumar Thakur 2010 *J. Reinf. Plast. Compos.* **29** 700.
[57] Vinayagamoorthy R and Rajeswari N 2012 *Int. J. Compos. Mater. Manuf.* **2** 15.
[58] Kishore Debnath, Inderdeep Singh and Akshay Dvivedi. 2014 *Mater. Manuf. Processes.* **29** 1401.
[59] Vinayagamoorthy R, Rajeswari N, Vijayshankar S, Vivekanandan M, Sri Rama Murthy Bellala and Venkata Subramaniam KR. *Procedia Mater Sci.* **5** 2075.
[60] Jayabal S, Natarajan U and Sekar U 2011 *Int. J. Adv. Manuf. Technol.* **55** 263.
[61] Kannan P, Balasubramanian K and Vinayagamoorthy R 2015 *Int. Rev. Mech. Eng.* **9** 536.
[62] Velumani S, Navaneethakrishnan P, Jayabal S and Ramprasath R 2013 *Int. J. Mach. Mach. Mater.* **14** 1.
[63] Vinayagamoorthy R, Rajeswari N and Karthikeyan S 2015 *Appl. Mech. Mater.* **766-767** 812.
[64] Vinayagamoorthy R, Rajeswari N, Sivanarasimha S, Balasubramanian K 2015 *Appl. Mech. Mater.* **787** 265.
[65] Vinayagamoorthy R 2017 *Mater. Manuf. Processes.* **32** 645.
[66] Vinayagamoorthy R 2016 *Indian. J. Sci. Technol.***9** 1.

[67] Vinayagamoorthy R, Manoj I V, Narendra Kumar G, Sai Chand I, Sai Charan Kumar G V and Suneel Kumar K 2018 *J. Mech. Sci. Technol.* **32** 2011.

IOP Conf. Series: Materials Science and Engineering **390** (2018) 012030 doi:10.1088/1757-899X/390/1/012030

Tribological behaviour on aluminium –nano Titanium Di Oxide with addition of solid lubricant nano Gr powder metallurgy composite.

D Balaji [1], M Vetrivel [1*], M Ekambaram [1], S Vijayaraghavan [2] and K D Gowtham[2].

[1] Department of Mechanical Engineering, SCSVMV, Enathur 631 561,Tamil Nadu, India.

[2] Department of mechanical engineering, Shri Andal Alagar College of Engineering, Mamandur, Tami Nadu, India.

balajibala81@gmail.com, km_vetrivel@yahoo.com , e4ekam@gmail.com

Abstract. The aluminium-nano titanium di oxide hybrid composite with self lubricant nano graphite is processed by powder metallurgy. The wear behaviour was experimentally investigated on the aluminium combined with titanium di oxide 5wt% and graphite 5wt% at nano scale in reinforcement with different density com at different sliding velocity of 1.5m/s, 2m/s, 2.5m/s along with the different loads of 5N, 7N, 10N. The wear test was conducted on the pin on disc apparatus. The equal weight percentage of TiO_2 and Gr in Al MMCs changing the hardness and density of the hybrid composites due to the occupancy of volume fraction in the Al-TiO_2-Gr hybrid composites. The TiO_2 provide strength of the hybrid composites and the graphite exhibited both excellent wear resistance and friction coefficients.

1. Introduction

The modern engineering applications require materials that are stronger, lighter, wear resistance and less expensive [1,2]. This paper is concentrated on the special metal matrix composite composed of aluminium, titanium di-oxide in nano scale and graphite [3-8]. In recent trends, MMCs are fast replacing conventional metallic alloys in wide range of applications as their use have been emerging aerospace, manufacturing and automobile to defense, marine, sports and machine tool industries [5]. These MMC‘s are materials in which one constituent is necessarily a metal and other constituent embedded metal matrix usually reinforcement [7]. These reinforcements are characterized by their chemical composition, shape, dimension and mainly their mechanical and physical properties [8].

In this study we consider Aluminium Titanium di-oxide Graphite metal matrix composite for their enhanced wear and hardness properties that are obtained from the experimental research work [9]. This Al-TiO_2-Gr MMC's consists of the constituent materials that are in nano scale there by it can be called as Al-TiO_2-Gr nano metal matrix composites [10,11].

2. Materials and Experiment

2.1 Material

The detail of materials used in present research work are listed in the Table 1

Table 1 Materials detail

Detail	Aluminium	Titanium di Oxide	Graphite
Symbol	Al	TiO_2	Gr
Particle size	130 – 180μm	5 nm	400nm
Density (g/cm^3)	2.69	3.78	2.66
Purity (%)	98.00	99.89	98.00
Supplier	Loba Chemi, India	US Research Nano Materials Inc, USA	M K nano Impex Cop,Canada

2.2 Specimen preparation

The hybrid Al-TiO_2-Gr metal matrix composites prepared through powder metallurgy method. The blending process is done by high energy planetary ball milling machine to mixing of Al powder with equal weight percentage of nano TiO_2 and nano solid lubricant partial Gr as per the details given in the Table 2 and milling was done for 3 hour at a speed of 300 r.p.m

The Measured amount of Al -TiO_2 –Gr powder is poured into the die cavity diameter of 10 mm and pressure is applied. The compaction was done in a Universal Testing Machine to attain 30mm height with a pressure of 8.5 MPa .Sintering is carried at a temperature of 580°C for a time of 2 hours to increases strength, ductility, thermal conductivity and thermal conductivity.

Table 2 Composition detail of various samples

Specimen	Composition by weight in percentage			Sample weight in grams
	Aluminium (Al)	Titanium di-oxide (TiO_2)	Graphite (Gr)	
A1	90	5	5	4
A2	90	5	5	5
A3	90	5	5	6

After preparation of the pin samples the average hardness was tested with micro hardness tester and average density of the each set of specimens are calculated and listed in the Table 3.

Table 3 Average Vicker's Hardness and density

Specimen	Average Hardness (VH)	Average density (g /cm^3)
A1	22.83	1.739
A2	85.67	2.045
A3	58.60	2.495

3. Experimental method

The following experimental testing were made on the prepared specimen,

3.1 Tribological testing

Tribometer is used measure tribological quantities, such as coefficient of friction, friction force, wear volume between 2 surfaces in contact. A pin on disc tribometer consists of a stationary pin under an applied load in contact with a rotating disc as per ASTM G99 based on the design variable mention in Table 4.

Table 4. Process parameters of tribological study.

Parameter	**A1**	**A2**	**A3**
Normal Load (N)	5	7	10
Sliding velocity (m/s)	1.5	2	2.5
Sliding distance (m)	1000	1000	1000

The tribological test was conducted with the above combinations on a pin on disk apparatus. The specific wear rate was calculated wear volume per unit and unit distance and the coefficient of friction is determined by the ratio of the frictional force to the loading force on the pin. The calculated Specific wear rate (SWR) and Coefficient of friction (COF) tabulated Table 5 and table 6 respectively.

Table 5. Specific wear rate for deferent sample

Sample	**Load (N)**	**Sliding distance (m)**	**Sliding velocity (m/s)**	**Time (min)**	**Specific wear rate (mm^3/Nm) In terms of 10^{-6}**		
					A1	A2	A3
1	5	1000	1.5	11.0	5.156	2.048	4.640
2	5	1000	2.0	8.27	6.948	2.780	6.040
3	5	1000	2.5	7.01	12.878	6.040	7.256
4	7	1000	1.5	11.0	10.284	5.148	8.217
5	7	1000	2.0	8.27	32.741	5.583	12.382
6	7	1000	2.5	7.01	37.018	7.876	19.510
7	10	1000	1.5	11.0	41.246	3.723	18.006
8	10	1000	2.0	8.27	47.096	11.560	22.170
9	10	1000	2.5	7.01	57.471	20.050	47.740

Table 6. Coefficient of friction for deferent sample

Sample	Load (N)	Sliding distance (m)	Sliding velocity (m/s)	Time (min)	Coefficient of friction		
					A1	A2	A3
1	5	1000	1.5	11.0	0.498	0.994	0.454
2	5	1000	2.0	8.27	0.194	0.588	0.468
3	5	1000	2.5	7.01	0.230	1.139	0.438
4	7	1000	1.5	11.0	0.532	0.677	0.504
5	7	1000	2.0	8.27	0.361	0.464	0.51
6	7	1000	2.5	7.01	0.342	0.705	0.481
7	10	1000	1.5	11.0	0.497	0.326	0.517
8	10	1000	2.0	8.27	0.413	0.266	0.513
9	10	1000	2.5	7.01	0.411	0.336	0.514

4. Results analysis

4.1 Specific wear rate versus sliding velocity

The specific wear rate for Al- TiO_2 -Gr shows in the Figure 1 to Figure 2, versus sliding velocity at different applying load 5N, 7N, 10N respectively.

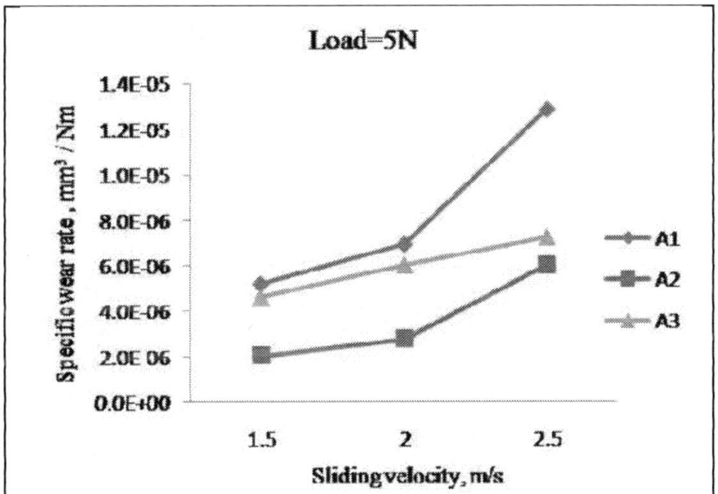

Figure 1. Specific wear rate versus sliding velocity at 5N Load condition.

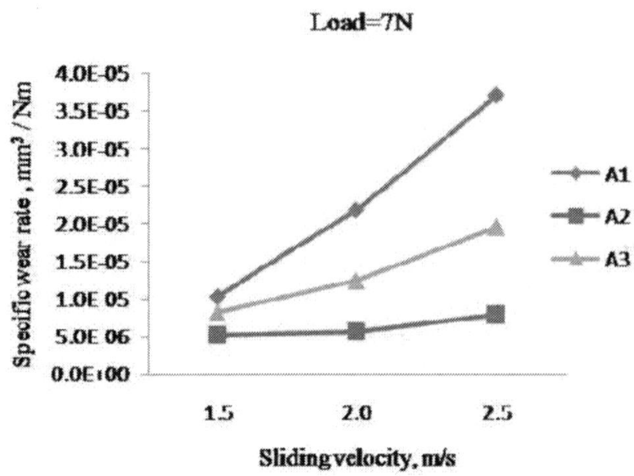

Figure 2. Specific wear rate versus sliding velocity at 7N Load condition.

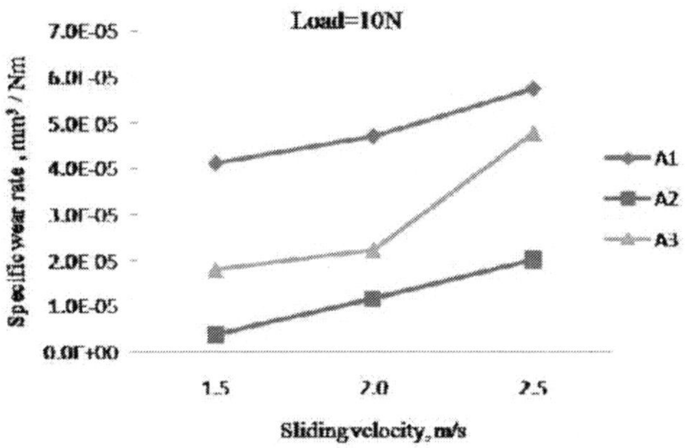

Figure 3. Specific wear rate versus sliding velocity at 10N Load condition.

From the above graphs the specific wear rate increase with respect to increasing load and sliding velocity but specimen A2 gives the minimum specific wear rate compared to the other specimen due to high hardness value of 85.67(VH) .

IOP Conf. Series: Materials Science and Engineering **390** (2018) 012030 doi:10.1088/1757-899X/390/1/012030

4.2 Coefficient of friction versus sliding velocity

The coefficient of friction for Al- TiO_2 -Gr shows in the Figure 4 to Figure 6, versus sliding velocity at different applying load 5N, 7N, 10N respectively.

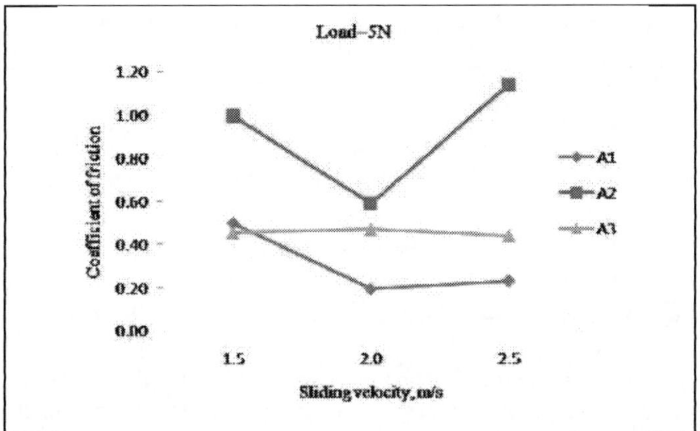

Figure 4. Co efficient of friction versus sliding velocity at 5N Load condition.

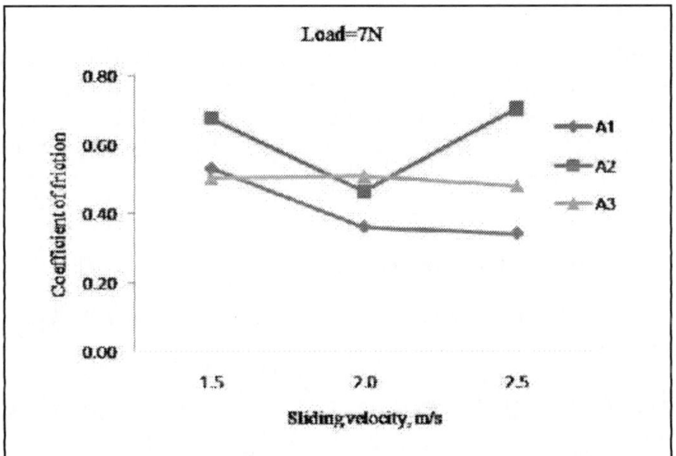

Figure 5. Co efficient of friction versus sliding velocity at 7N Load condition.

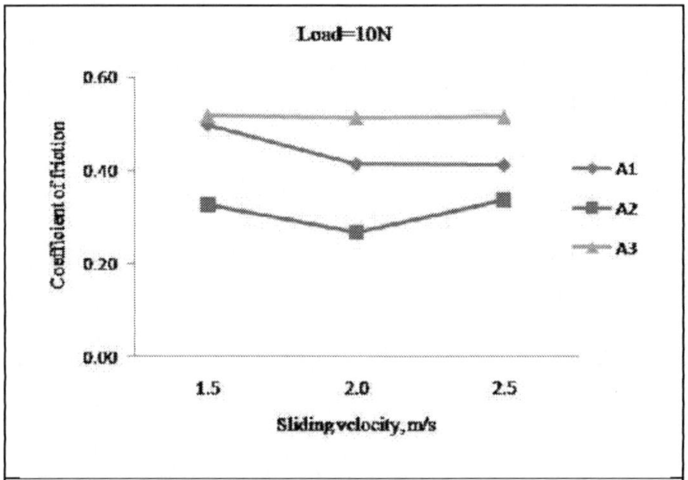

Figure 6. Co efficient of friction versus sliding velocity at 10N Load condition.

From the above graphs the coefficient of friction provides uniform COF for specimen A3 may be due to more volume fraction of graphite to presence of lubrication effect. For the specimen A1, COF decreases from the sliding velocity 1.5 m/s to 2.00 m/s and then maintaining almost same to the sliding velocity of 2.5 m/s. The minimum COF recorded at highest load condition of 10 N at all sliding velocity conditions due to the high hardness and medium density of specimen A2.

5. Conclusion

In this investigation, Al-TiO_2-Gr hybrid composites were successfully fabricated using the powder metallurgy route and the tribological wear properties are investigated. The equal weight percentage of TiO2 and Gr in Al MMCs varying hardness and density of the hybrid composites due to the occupancy of volume in the Al-TiO_2-Gr hybrid composites. The graphite exhibited both excellent wear resistance and friction coefficients acted as lubricant.

References

[1] Kanthavel K, Sumesh K R and Saravankumar P *Alexandria Engineering Journal.*2016 **55** 13–17
[2] Ravindran P, Manisekar K, Narayanasamy R and Narayanasamy P *Ceramics International.* 2013 **39** 1169–1182
[3] Suresha and Sridhara B K *Materials and Design.* 2010 **31** 1804–1812
[4] Padmavathi K R and Ramakrishnan R *International Journal of Engineering and Technology.* **04** 3368-3373
[5] Rousses Papagiannakis G *Rev.Tec Ing univ Zulia.* 2014 **2** 33-36
[6] Soorya prakash K, Gopal P M, Anburose D and Kavi mani V *Ain Shams Engineering Journal.* 2016 2090-4479
[7] Maninder Singh, Khushdeep Goyal and Deepak Kumap Goyal *Universal Journal of Mechanical Engineering.* 2015 **3**(4) 142-146
[8] Moslah shirazi S, Akhlaghiand F and Li D Y *Tribology International.* 2016 **103** 620-628
[9] Mina Bastwros M H, Amal Esavia M K and Abdalla wifi *Wear.* 2013 **307** 164-173
[10] Afsaneh Dorri Moghadam, Emad Omrani, Pradeep Menezes C and Pradeep Rohatgi K *Composite.* 2015 **B77** 402-420
[11] Ch Kaushik N , R N Rao *Tribology International.* 2016 **96** 184-190

IOP Conf. Series: Materials Science and Engineering **390** (2018) 012031 doi:10.1088/1757-899X/390/1/012031

Enhancing recruitment effectiveness in IT industry

R G Priyadarshini
Amrita School of Business, Coimbatore
Amrita Vishwa Vidyapeetham, India
rg_priyadarshini@cb.amrita.edu

Abstract. The Major Decision making challenge for any HR Manager starts with Recruitment. Recruiting Metrics directly impact the Return on Investment in any Business. The traditional ways of measuring HR metrics has taken a seismic shift. Previously the focus of recruiting managers was only on efficiency, but now effectiveness has gained a lot of relevance. This study on organizational recruitment is an attempt to understand the choice of recruitment source and its impact on the cost per hire, quality of hire and time to hire aspects. The measures for this assessment include both pre-hire and post hire recruitment outcomes that further elucidates how the recruitment source and the recruitment variables contribute to candidates' success as well as organization success.

1. Introduction

There is a lot of emphasis on effective sources of recruitment in the staffing world. The current context presents various sources of recruitment to choose from for laterals and new entrants. Literature showcases studies on Recruitment effectiveness but there is paucity of research with respect to industry wise comparison and level wise comparison. It is seen as to how how HR managers perceive recruitment effectiveness with respect to various sources of recruitment [1]. The avenues of recruitment they prefer for entry and lateral level jobs and their perceptions on cost, quality and time aspects with respect to entry and lateral level jobs. Entry level jobs in this context are the technical jobs and the lateral level jobs are the managerial level jobs.

The paper draws reference from the work done by renowned HR organizations like SHRM (Society of Human Resource Management) and American National Standards for Human Resource Management. The various heads under which the metrics are classified are: Recruitment cost (sourcing cost, pre-screening cost, Background check expenses etc) Quality of Hire and time to hire. Quality of hire theoretically aims at ensuring whether it is a right hire or not. The methods to ensure the quality of a hire were very subjective earlier but now due to the growing popularity for Analytics, people started quantifying it.[2][3][4]The Quality of hire is ensured both in performance attributes –pre hire and post hire and as well as on the credibility of the information provided by the candidates, others being experience in the previous organization, most importantly managerial satisfaction and applicant satisfaction throughout the process of Recruitment.[10] The time to hire is an efficiency metric which focuses on reaching a large talent pool in a short span of time, ensuring a period to check new hires performance. With the growing emphasis on quality rather than volumes, HR managers have changed their perspectives towards the definition of time to hire as well.[5][6]

Another source of recruitment which has steered the focus of HR Managers to reach a large and diverse pool of candidates is Social media. The recent shift of candidates' presence in digital platforms, employers also started finding ways to reach to this diverse talent pool via social media. There are tendencies to prefer social media over others due to its inexpensiveness and impact but there are also apprehensions towards using it due to some misconceptions about security.

2. Background for the study

This study on recruitment draws majorly from the recruitment framework designed by Breaugh and Starke [7] which emphasizes on sources of recruitment and its impact on pre-hire and post-hire metrics. The metrics selected for the study have been further validated by practicing managers of various industries. The basic metrics like Number of vacancies, time to fill, offer letters issued to acceptance ratio have become very obvious. In order to add value strategically, industry experts have tried out some very interesting ideas that worked out well. One such is that used in Saratoga institute acquired by Price Waterhouse Coopers. The firm has developed a score card using matrix approach to observe the source of Successful hires. The metrics that had a strategic edge are : Turnover, Pay increases and performance rating. This data has transformed the way the hiring process is looked at and the whole idea is to know where we are getting our ROI –Return on Investment from. This data has driven companies to improve the quality of hires and to retain new hires. (Jim Hatch, Managing partner of Saratoga institute, HR Magazine, SHRM-The next generation of Hiring Metrics, Garvey, charlotte)

One size does not fit for all. Every organization is unique in its own way. Designing specific metrics internally is another way of looking at things. Wachovia Corporation re-examined its hiring process and focused on linking it with business strategy. Their core competency in hiring is to ensure workforce diversity. So this encouraged their recruiters to align to the business strategy by developing a recruiter card for them. They involved the compensation and incentive compensation related metrics in the card. Some of the newly added metrics to the score card are: Managerial satisfaction & new hire satisfaction) says George –CEO of Wachovia (Wachovia corporation, a consumer banking and financial services company based in charlotte)

Research on Recruitment is broadly classified into 3 key research areas namely, recruiter effects, administrative recruitment policies and procedure effects and recruitment source effects.

The first component of the study focuses on Recruitment source effects, with special reference to assessing cost, quality and speed aspects of recruitment with special reference to entry and lateral level jobs. The Second component assesses if there are any differences in the hiring preferences of HR Managers of IT & Non IT Companies towards various avenues of recruitment while hiring for entry level jobs and lateral level jobs. It is important that recruitment places emphasis on competencies required for being effective on the job, especially with respect to managerial level jobs [8]. In lateral hires, competency based training need assessment can be performed so that appropriately qualified candidates get into the positions, and hence gaps if any in the competencies could be bridged [9].

3. Recruitment Efficiency

Efficiency is concerned with speed and cost. For example hiring more people in less time would be one of objectives to achieve efficiency. To assess the efficiency of a recruitment process, the following metrics are involved in the checklist.

- Count of resumes submitted
- Count of screened resumes
- Avg. Days between resume submission and resume screening
- Count of approved resumes
- Avg. Days between resume screening and approval

IOP Conf. Series: Materials Science and Engineering **390** (2018) 012031 doi:10.1088/1757-899X/390/1/012031

4. Recruitment Effectiveness

Effectiveness is a measure of quality. The quality of hire is any day the most sought after metric for any business. Since recruitment carries a cost and the only tool which decides whether it is a successful recruitment or not is the quality of hire. The quality of hire involves a set of metrics like post hire performance, managerial satisfaction, applicant satisfaction.
Hiring effectively is often more important than hiring efficiently to gain more return-on-investment (ROI). The American National Standard for Human Resource Management defines Cost per hire metric in three ways. The first one is Cost per hire internal which is restricted to only that particular organization. The second one is cost per hire comparable, in which we compare the CPH metric across different organization and set a benchmark. The third one is Recruiting cost Ratio, which implies the ratio of total cost of hiring to the total compensation paid in the first year of employment period.(Cost per hire metric –SHRM report) (ANSI/SHRM 06001.2012)

5. Objectives

1. To assess the perceptions of HR Managers on effectiveness of recruitment source across entry level and lateral level jobs.
2. To assess the perceptions of HR Managers on Cost, Quality and time considerations across various sources of recruitment within entry and lateral level jobs.

The important variables considered for the study are:

Recruitment Costs	**Quality of Hire**	**Speed**
Sourcing Costs	Pre-employment Screening test	Lead time to source
Pre - screening Costs	Competency Profiling	Lead time to staff
Administrative Expenses	Retention rate of new hires	Lead time to hire
Background Check costs	On the job Performance	Lead time to ensure proper background Checks
Advertising and Marketing expenses	Managerial Satisfaction	Lead time to ensure Productivity
	Applicant Satisfaction	

6. Hypothesis

- Hiring preferences of HR managers of IT & Non IT companies would not vary their considerations towards **recruitment cost, quality of hire and recruitment time for entry level jobs.**
- Hiring preferences of HR managers of IT & Non IT Companies would not affect their considerations towards **recruitment cost, quality of hire and recruitment time lateral level jobs.**

7. Research Methodology

The sampling Frame consists of HR Managers of IT and Non-IT Companies that are in existence for more than one decade. Some of the Non IT Companies are e commerce, Public Sector Units, pharmaceutical companies etc. The sampling method used is convenience sampling. As identified in Literature Review the earlier research was primarily focused on recruitment strategy and sources, but not on the variables which effect recruitment effectiveness. Further these outcomes were not explained with respect to Cost, Quality and Time. The only qualification for Applicants is that they should be in the field of HR in the organization. There was no bar on Age or Education. In total, 40 HR managers participated, of which 22 were HR managers from non-IT companies and 18 were from IT companies.

8. Results and Discussion

The HR Managers of IT Companies do not vary their perceptions towards entry level jobs in terms of recruitment cost as per the significance value. But the mean of Entry level jobs is more than that of Lateral level jobs, so it is inferred that the IT Companies perceive cost as a crucial factor while hiring candidates for entry level jobs. There is a variation in the quality considerations as per the significance value .From the mean values we can infer that IT Companies consider quality of hire as more crucial factor in hiring candidates for entry level jobs. There is no variation in the considerations towards recruitment time but IT Companies perceive Recruitment time as a crucial factor in hiring candidates for lateral level jobs than that of entry level jobs.

The HR Managers of Non- IT companies vary their perceptions towards entry level jobs in terms of recruitment cost as per the significance value. But the mean of Entry level jobs is more than that of Lateral level jobs, so it is inferred that the Non- IT Companies perceive cost as a crucial factor while hiring candidates for entry level jobs. There is no variation in the quality considerations as per the significance value .From the mean values we can infer that Non- IT Companies consider quality of hire as more crucial factor in hiring candidates for entry level jobs. There is a variation in the considerations towards recruitment time across entry and lateral level jobs, but Non- IT companies perceive Recruitment time as a crucial factor in hiring candidates for lateral level jobs than that of entry level jobs.

8.1. Hiring Considerations of HR Managers of IT& Non IT Companies with respect to Entry Level Jobs

The Hiring Considerations of HR Mangers of IT and Non IT Companies were analyzed and it was found that the hiring preferences towards recruitment cost do not vary for Entry level jobs .The Significance value is 0.621.This means that recruitment cost for entry level jobs is perceived as equally important across IT & Non IT Companies. The preferences towards Quality of hire would vary across IT & Non IT Companies for entry level jobs. The HR Managers of IT Companies perceive quality as more important factor when compared to Non IT Companies for entry level jobs. The significance value is 0.014. But it is seen that there is no variation in time considerations for entry level jobs across IT & Non IT Companies. The Significance value is 0.055.

8.2. Recruitment effectiveness with respect to entry level jobs

Cost considerations: For Entry level jobs, HR managers of IT and Non IT have the same perceptions towards cost, since the sources of recruitment are similar and the skill sets are more or less on the same level in terms of complexity. So the perceptions on recruitment costs do not vary for entry level jobs across IT & Non IT(alpha value is 0.631).
Quality Considerations: The mean value towards quality considerations of HR managers of IT Companies is higher than Non IT. It was inferred from the discussion with HR managers of IT Companies that they are more concerned about Client satisfaction and managerial satisfaction. The reason is attributed to IT Companies taking a lot of pre-requisites to fulfill the competencies required by the client side and the managers. But for Non-IT Companies there is not much interaction with the client and hence usually the selection process is designed as per the company's requirement. The alpha value (0.014) shows that there is a significant variation in the quality considerations of IT HR Managers and Non IT HR Managers.
Time considerations: There is no variation in the time to hire considerations across IT & Non IT Companies for entry level jobs. HR Managers of both IT & Non IT Companies perceive entry level jobs as equally important and incur same amount of time. The significance value is 0.55.As discussed earlier, the sources of recruitment are more or less similar for entry level jobs so it justifies that the time considerations do not defer across IT & Non IT Companies.

8.3. Recruitment Effectiveness for Lateral level jobs

Cost Considerations: Variation exists among IT & Non IT HR managers. The mean of preferences of HR Managers of IT companies towards recruitment costs is more than that if Non IT. It is attributed to the reason of HR managers of IT Companies incur more cost in third party consultancies in hiring laterals since they perceive that laterals add more strategic value to their business because they have prior experience. The significance value is 0.003.It shows that there is a significant variation in the cost considerations of IT and Non IT Managers.

Quality Considerations: No variations exist among IT and Non IT HR managers' considerations towards quality of hire. This is attributed to HR managers of both IT & Non IT Companies perceiving quality as an equally important metric and there are no variations in their perceptions. Further when Lateral level jobs are the managerial jobs which add more strategic value to any business, both IT & Non IT HR Managers perceive Quality as a very crucial factor. The significance value is 0.185.Since it is greater than 0.05, the null hypothesis is accepted .Hence there is no variation in the perceptions of IT & Non IT HR Managers considerations towards quality.
Time Considerations: Variation exists among IT and non IT HR managers' consideration towards recruitment time. It is seen that the mean of HR managers' preferences towards recruitment time for IT companies is more than that of Non IT for lateral level jobs. The HR managers of IT Companies spend more time in hiring laterals and they mostly prefer third party consultants. The significance value is 0.003.

9. Conclusion
This study has assessed the perceptions of HR managers on certain very important metrics –Cost, quality and time which define recruitment effectiveness. There are two broad hypothesis framed from the gaps in the literature review. The first one is the perceptions of HR managers of IT & Non IT on the cost, quality of hire and time to hire aspects for entry and lateral level jobs and the second hypothesis is based on the HR Managers preferences for the various Avenues of Recruitment. Both the hypothesis gave interesting results. The perceptions of HR managers of IT & Non -IT towards recruitment cost does not vary for entry level jobs but varies for lateral level jobs. But the perceptions of HR Managers of IT & Non IT Companies towards quality varies for entry level jobs but does not vary for lateral level jobs. And the perceptions of HR Managers of IT & Non IT companies towards time to hire does not vary for entry level jobs but varies for Lateral level jobs. The sources of recruitment preferred for entry level jobs by IT and Non- IT companies is campus hiring and for lateral level jobs it is Third party consultants. The preference for social media is more prevalent in IT companies than Non IT. There is a lot of scope to go in-depth of each aspect i.e cost, quality and time and compare them across IT & Non IT companies and also for entry and lateral level jobs.

References

[1] Karl, K., Peluchette, J & Schlaegel, C. (2010). Who's posting Facebook faux pas -A cross-cultural examination of personality differences, *International Journal of Selection & Assessment* **18(2)** 174–186
[2] Sohel Ahmad & Roger G. Schroeder. (2002) "The importance of recruitment and selection process for sustainability of total quality management", *International Journal of Quality & Reliability Management* **19(5)** 540-550
[3] John , A & Geisinger, Kurt F. (1989). The relationship between recruiting source, applicant quality, and hire performance: an analysis by sex, ethnicity, and age, *Personnel Psychology*, **42**
[4] Michael J. Curran, Peter Draus Michael, Schrage Steve Zappala. (2014). College students and HR professionals-Conflicting views on information available on Face book, *Human Resource management Journal* **24(4)**
[5] Sinha, Vinita & Thafy, Priya. (2013). A review on changing trend of recruitment practice to enhance the quality of hiring in global organizations", *Journal of Contemporary Management Issues*, **18(2)**
[6] Collins, C. (2006). The interactive effects of recruitment practices and product awareness on job seekers' employer knowledge and application behaviors Ithaca, NY: Center for Advanced Human Resources
[7] Breaugh, A. J & Starke, M. (2000). Research on employee recruitment, *Journal of Management*", **26 (3)** 405-434
[8] Murale, V.,& Preetha, R. (2011).A Study among Indian Middle Level Managers on Their Competencies and Role Effectiveness, *SCMS Journal of Indian Management* **8(1)** 38-50

[9] Priyadarshini R G and Dave, D., (2012) "Competency-based Training Needs Assessment Model", *Management and Labour Studies* **37** 195–207
[10] Jepsen, Denise, Knox-Haly, Martha, Townsend, Daniel 2015 Recruitment practices in Australia: A review and comparative research agenda, *Employment Relations Record* **15(2)** 5-19

Investigation of Mechanical properties of Aluminium reinforced Fly ash with CNT

C. Parswajinan[1, a*], B. Vijaya Ramnath[2,b], I. Anish Hilary[3,c], N. Chenbhaga Ram[4,d], A. Mayandi[5,e]

[1,a,*]Assistant Professor, Department of Mechanical Engineering, Sri Sai Ram Engineering College, Chennai, Tamil Nadu, India.
[2]Professor, Department of Mechanical Engineering, Sri Sai Ram Engineering College, Chennai, India.
[3,4,5] Department of Mechanical Engineering, Sri Sai Ram Engineering College, Chennai, India.

[a*]parswajinan@gmail.com, [b]vijayaramnathb@gmail.com

Abstract. Metal matrix composites are gaining enormous acceptance in the recent times. This is due to their remarkable characteristic of enhancement in their mechanical properties without any compromise in their weight. This paper deals with the investigation of mechanical properties of Aluminium 6061 reinforced fly ash with CNT. Here samples of four different compositions were fabricated using stir casting method and subjected to tensile, flexural, hardness and impact tests. The result indicated no change in the impact value of the samples. On addition of 2% flyash, the hardness value increased to a value of 59.2BHN when compared with pure Aluminium and the tensile strength increased to about 147.76 MPa on addition of about 3% fly ash.

1. Introduction

Carbon nano tubes (CNTs) are being used as the most common reinforcement material in metal matrix composites (MMCs) due to its unexceptional physical and chemical properties. Many MMC have been prepared using CNT as reinforcing material. Aluminium is one of the widely reinforced metal with CNT for the preparation of the composite. This is because of the collective properties such as low density, low weight, high strength, superior malleability, excellent corrosion resistant, high thermal and electrical conductivity of the aluminium. B. Vijayaramnath et al. [1] concluded that addition of aluminium increases the tensile strength of the sample. Jinzhi Liao and Ming-Jen Tan [2] observed increase in tensile strength and hardness of the composite by the addition of CNTs (0.5 wt%) to the pure matrix. C. Parswajinan et al. [3] concluded that the hardness and tensile strength of iron increased when reinforced with CNT. A.M.K. Esawi et al. (2009) [4] determined that there was uniform dispersion at the tensile testing fracture surface of the specimen. It is also found that the CNTs act as nucleation sites during testing. D.-S. Lim et al. (2005) [5] stated that when the CNTs are added up to 12 wt.%, the wear loss decreased significantly and the friction coefficients were maintained same. H.J. Choi et al. (2010) [5] observed that, as the grain size is reduced to 150 nm and MWCNT volume increases up to 4.5 vol.%, yield stress notably increases and hence wear resistance is significantly enhanced in the specimen. P. Naresh Narayanan [6] paper illustrates Carbon nanotubes reinforced pure Al (CNT/Al) composites and fly ash reinforced pure Al (FA/Al) composites produced by ball-milling and sintering it was indicated that the CNTs and fly ash were uniformly dispersed into the Al matrix as ball-milling time increased with increase in hardness. Oluwagbenga B. Fatile et.al [7] results show that the hardness, ultimate tensile strength, and percent elongation of the hybrid composites decrease with increase in fly ash content. B. Vijayaramnath et al. [8] concluded that the compressive strength increases with increase in CNT. Shadakshari. R et.al [9] concluded that on reinforcing CNT with metal matrix composite at about 400° C, high damping capabilities were achieved. C. Parswajinan et al. [10] concluded that addition of CNT did not have any impact on the hardness of the material, but there was an increase in compressive strength on addition of CNT. A. M. K. Esawi et al. (2010) [11] concluded in their studies that the mechanical properties were found to increase more than the expected values but fell short at 5 wt%. B. Vijaya Ramnath et al. [12] reviewed the properties of Aluminium and Magnesium matrix composites with CNT and suggested an increase in mechanical properties. C. Parswajinan et al. [13] studied the addition of CNT with iron and recommended an increase in strength of sample. Additon of reinforcements to the metal matrix increases the Mechanical Property [14,15,16,17]

2. Material selection

Aluminium (6061), CNT and fly ash are the materials used in this experiment. The aluminium powders used are of 99.5% purity with a mesh size of 200.The melting point and density of aluminium powders are 660°C and 2.70 g/cm^3, respectively. The properties of CNTs used has a purity greater than 95% on trace metal basis, melting point of 3652-3697°C, density of 2 g/cm3 and dimensions of 10-30 nm of outer diameter, 2-6 nm of inner diameter and 15-30 μm of length. As per TEM analysis, the impurities present in the CNTs used are amorphous carbon of about $<$ 3%. Fly ash, also known pulverised fuel ash is one of the combustion products of coal, composed of the fine particles that are driven out of the boiler with the flue gases. Ash that falls in the bottom of the boiler is called bottom ash. Fly ash is generally captured by electrostatic precipitators or other particle filtration equipment before the flue gases reach the chimneys.

2.1. Specimen preparation

The material is prepared using Stir Casting technique, in which the Aluminium 6061-CNT-fly ash mixture is combined in various compositions. The mixture is heated inside the furnace at a temperature of 800^{0}C for about 2 hours. The molten mixture is stirred using electric stirrer for the even distribution of all the elements throughout the material. It is again heated inside the furnace for about half an hour. The molten mixture is then poured into the die and allowed to cool in the ambient temperature for few hours.

Table 1. Composition of samples.

Samples	Composition
Sample 1	Al+2% Fly Ash
Sample 2	Al+2%Fly Ash+0.2%CNT
Sample 3	Al+3%Fly Ash+0.2%CNT
Sample 4	Al+4%Fly Ash+0.2%CNT

3. Testing conducted on the material

3.1. Tensile test

The ability of the material to resist breaking under the action of longitudinal force is defined as tensile strength of the material. From table 2 it could be inferred that the tensile strength reduces with addition of CNT but reaches a maximum strength of 147.76MPa at 3% of the fly ash. The graphical representation of tensile strength of various samples is shown in figure 1. Stress-Strain chart has been illustrated in figure 2 for the sample 3 while performing the tensile strength test. The experimented sample on which the tensile test had been performed is shown in figure 3.

Table 2. Tensile strength of samples.

Composition	Tensile Strength [MPa]
Sample 1	132.15
Sample 2	78.76
Sample 3	147.76
Sample 4	60.83

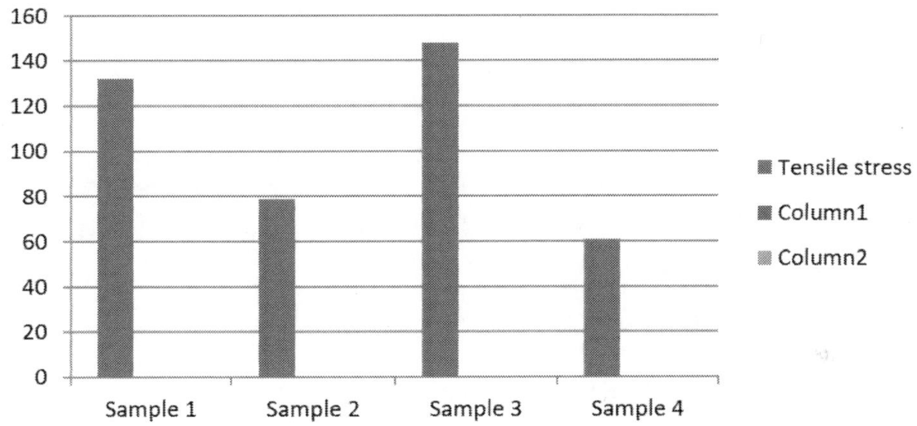

Figure 1. Graphical representation of Tensile test results.

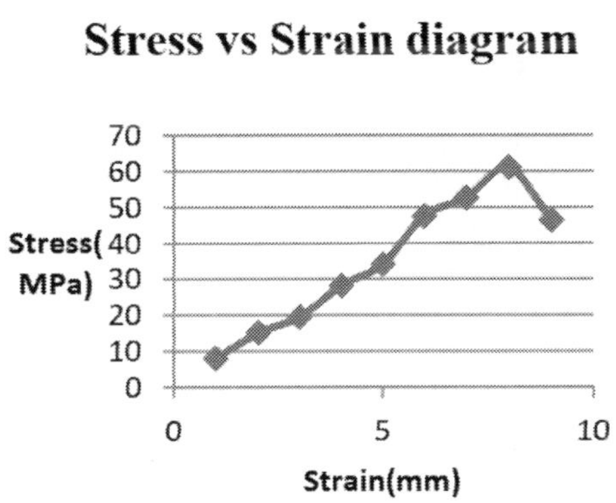

Figure 2. Stress strain diagram of sample 3.

Figure 3. Tensile sample after testing.

3.2. Flexural test

Flexural tests are generally used to determine the flexural modulus or flexural strength of a material. A flexural test is more affordable than a tensile test with a slight variation in values. The material is laid horizontally over two points of contact (lower support span) and then a force is applied to the top of the material through either one or two points of contact (upper loading span) until the sample fails as shown in figure 4. Variations of the flexural load values for different samples are tabulated in table 3. The maximum recorded force is the flexural strength of that particular sample which is 7.51 kN. The variation of the flexural strength of various samples is shown in figure 5.

Table 3. Flexural test results.

Composition	Flexural load in kN
Sample 1	5.55
Sample 2	6.76
Sample 3	5.41
Sample 4	7.51

Figure 4. Flexure sample after testing.

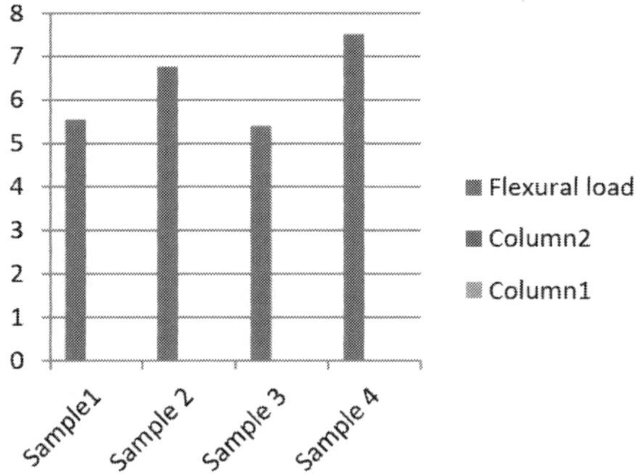

Figure 5. Graphical representation of flexure test results.

3.3. *Impact test*

The impact test is a method for evaluating the toughness and notch sensitivity of engineering materials. It is usually used to test the toughness of metals to determine its ability of being strong enough to withstand adverse conditions or rough handling. This test conducted on various samples with different composition gave a constant result. The tested sample is shown in figure 6.

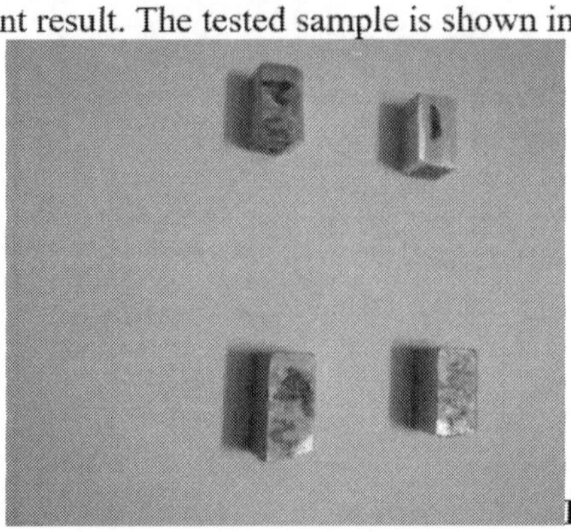

Figure 6. Impact sample after testing.

3.4. Hardness test

The measure of resistance to indentation is termed as hardness of the material. The hardness information facilitates to effectively use the material in the place of requirement. Hardness of these specimens are measured using Brinell method in which tungsten carbide indenter is used to produce the indentation with a force of 3000kgf which is then tabulated in the sequence of the samples in table 4. It is noticed that the hardness of the specimen with fly ash content of 2% along with 0.2% of CNT is much more than the other specimens. The tested sample is shown in figure 7. The variation of hardness values with various composition have been graphically represented in figure 8.

Table 4. Hardness of samples.

Composition	Hardness [HBW]
Sample 1	53
Sample 2	59.2
Sample 3	55.3
Sample 4	49.8

Figure 7. Hardness sample after testing

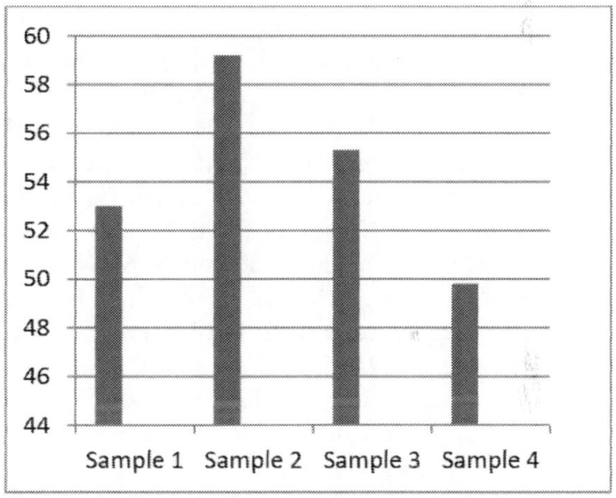

Figure 8. Graphical representation of Hardness test results.

4. Conclusion

The work done could be summarized as, Aluminium 6061, Fly ash, CNT are combined together performing stir casting technique and the results of various compositions are compared along with the effects on pure Aluminium. The results are as follows:

1) The tensile strength of the material has the highest value at 3% of flyash with 147.76 kN.
2) The flexural ability of the composite specimen seems to increase to a highest value on addition of 3% of fly ash and 0.2% of CNT.
3) The impact strength of the specimen remains the same without any variation.
4) The hardness of the material increases with decrease in the percentage of fly ash.

References

[1] B Vijaya Ramnath , C Elanchezhian , M Jaivignesh , S Rajesh , C Parswajinan and A. Siddique Ahmed Ghias 2014 Evaluation of mechanical properties of aluminium alloy–alumina–

IOP Conf. Series: Materials Science and Engineering **390** (2018) 012032 doi:10.1088/1757-899X/390/1/012032

boroncarbide metal matrix composites *Materials and Design* **58** 332–338.

[2] JinzhiLiao and Ming-Jen Tan 2011 Mixing of carbon nanotubes (CNTs) and aluminum powder for powder metallurgy use *Powder Technology* **208** 42–48.

[3] C Parswajinan, B Vijaya Ramnath, C Elanchezhian, S V Pragadeesh, P R Ramkishore and V Sabarish 2014 Investigation on Mechanical Properties of Nano Ferrous Composite *Procedia Engineering* **97** 513 – 521.

[4] A M K Esawi, K Morsi, A Sayed, M Taher and S Lanka 2009 Effect of carbon nanotube (CNT) content on the mechanical properties of CNT-reinforced aluminium composites *Composites Science and Technology* **70** 2237–2241.

[5] D S Lim, D H You, H J Choi, S H Lim and H Jang 2005 Effect of CNT distribution on tribological behavior of alumina–CNT composites *Wear* **259** 539–544.

[6] P Ramesh Narayanan and S V S Narayana Murty 2015 Experimental Investigation and Micro-Structural Evolution Analysis of CNT & Fly Ash Reinforced Aluminium Matrix Composites *Materials Science Forum* **830-831** 429-432.

[7] Oluwagbenga B Fatile, Felix U Idu, Olajide and T Sanya 2015 Fabrication Characteristics and Mechanical Behaviour of Fly Ash-Alumina Reinforced Zn-27Al Alloy MatrixHybrid Composite Using Stir-Casting Technique *World Academy of Science, Engineering and Technology International Journal of Materials and Metallurgical Engineering* **9** 10.

[8] B Vijaya Ramnath, C Parswajinan, C Elanchezhian, S V Pragadeesh, C Kavin, P R Ramkishore and V Sabarish 2014 Experimental Investigation on Compression and Chemical Properties of Aluminium Nano Composite *Applied Mechanics and Materials* **680** 7-10.

[9] Shadakshari R, Mahesha K, Dr Niranjan H B 2015 Carbon Nanotube Reinforced Aluminium Matrix Composites *Materials today: Proceedings* **2** 2886-2895.

[10] C Parswajinan, B Vijaya Ramnath, M Vetrivel C Elanchezhian, K Loganathan, R Sarvesh, C Rohit Prasanna and R N Karthik Babu 2015 Experimental Investigation of Mechanical and Chemical Properties of Aluminium reinforced with MWCNT *Applied Mechanics and Materials* **766-767** 287-292.

[11] A M K Esawi, K Morsi, A Sayed, M Taher and S Lanka, Effect of carbon nanotube (CNT) content on the mechanical properties of CNT-reinforced aluminium composites *Composite Science and Technology* **70** 2237–2241.

[12] B Vijaya Ramnath, C Parswajinan, C Elanchezhian, S V Pragadeesh, P R Ramkishore and V Sabarish 2014 A Review on CNT Reinforced Aluminium and Magnesium Matrix Composites *Applied Mechanics and Materials* **591** 120-123.

[13] C Parswajinan, B Vijaya Ramnath, M Vetrivel, P Ramanarayanan, S Bharath , T Ajay and R Raghav Chander 2015 A Review on Composite Materials with Ferrous, CNT and Powder Metallurgy *Applied Mechanics and Materials* **813-814** 9-13.

[14] Siddique Ahmed Ghias B.Vijaya Ramnath, 'Investigation of Tensile Property of Aluminium SiC Metal Matrix ', *International Journal Applied Mechanics and Materials*, **766-767** 252-256.

[15] C.Elanchezhian, B.Vijaya Ramnath, G. Ramakrishnan, K.N.SripadaRaghavendra, Mithun Muralidharan, V.Kishore, Review on metal matrix composites for marine applications, *Elsevier – Materials Today Proceedings*, **5(1),** 1211-1218.

[16] C. Parswajinan, B.Vijaya Ramnath, N. Sai Prasath, S. Kuralamudhan and S. Balaji, Mechanical investigation of Aluminum hybrid Composite Reinforced with CNT and Aluminum oxide (Al2O3), *Advanced Science, Engineering and Medicine*, **10**, 1–3.

[17] C. Parswajinan, B. Vijaya Ramnath, M. Vetrivel, M. Aswin, and B. Mohana Vignesh, Mechanical Behavior of Aluminium Metal Matrix Composite, *Advanced Science, Engineering and Medicine*, **10**, 1–3.

IOP Conf. Series: Materials Science and Engineering **390** (2018) 012033 doi:10.1088/1757-899X/390/1/012033

Grinding of MMC using MQL based vegetable oil - Review

A.Nandakumar[1] *, T.Rajmohan [2]

1, 2 Department of Mechanical Engineering, Sri Chandrasekharendra Saraswathi Viswa Mahavidyalaya, Enathur, Kanchipuram - 631561, Tamil Nadu, India.

Nandakumar.a@kanchiuniv.ac.in

Abstract. Grinding plays an essential role in obtaining high dimensional accuracy and surface finish. Therefore, a strategy to improve the grinding rate and surface finish is important for the economic production. Most important environmental hazards in machining operations are due to the use of cutting fluids. Nowadays to reduce the harmful effects coupled with cutting fluids, researchers have developed environmental cognisant machining such as dry cutting, machining with minimum quantity lubricant and Machining with vegetable based cutting fluids including other types of cutting fluids. In this review, the applicability of MQL based vegetable oil in grinding of MMC has been undertaken. The performances of with respect to the cutting force, surface finish of work piece and temperature of the cutting zone have been investigated. It has been reported in various literature that metalworking fluids, which are vegetable oil-based, could be an environmentally friendly mode of machining with similar performance attained using mineral oil-based metalworking fluids.

1.Introduction

In many developed countries and industries are continuing a research interest in MMCs for last 25 years due to its unique physical and mechanical properties. The Metal Matrix Composite is used in aircraft, automobile components, structural, Electrical, electronic applications and military industries. [1-3] Though MMCs are superior properties, but not widely applied due to their higher manufacturing cost and poor machinability. [4,5] Aluminium metal matrix composites are superior to other MMCs due to their low cost.

The Main difficulties such as fabrication and machining have to be overcome while applying composites in different applications. While machining of MMCs the subsurface damages are caused due to conventional and unconventional processes, to rectify the effect the finishing processes such as grinding is used to improve the surface integrity of the MMCs [6-8]. In conventional machining processes, limited scientific efforts directed towards the study of the grinding characteristics of MMCs. [9-13]

Grinding is used to finish workpieces to maintain high surface quality and high accuracy of shape and dimension. The grinding accuracy is 0.000025 mm; in most applications, finishing operation is very essential and removes material about 0.25 to 0.50 mm depth. The grinding wheels are made of tempered hardened steel, silicon carbide, aluminium oxide. This grinding operation is used to improve the surface finish of the workpiece by removing a small amount of material. But grinding of silicon carbide is difficult because of its low fracture toughness, making it very sensitive to cracking. The investigations and little study has been carried out on machining advanced composite materials. The studies on grinding of composites exhibit an improved grindability with respect to non-reinforced aluminium alloy for the better surface finish and the lower tendency to clog the wheel. [14-16].

The cutting fluids have many technological advantages in grinding operation, they have serious ecological and economic problems. On the other side in dry grinding, there exist thermal damages on the workpiece surface, to rectify this effect the efficient and effective alternative method is minimum quantity lubrication (MQL) technique is shown in Fig 1. MQL technique uses a spray of small oil droplets in a compressed air jet. The lubricant is sprayed directly into the cutting zone. Therefore, it provides efficient lubrication and improves cutting performance without using huge flow of fluids. [17]

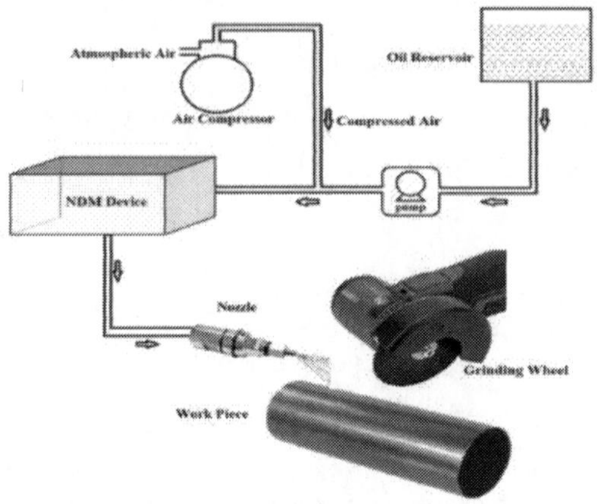

Fig 1 - Minimum Quantity Lubrication (MQL) Technique [36]

Nowadays, higher metal removal rate is used to get higher efficiency in cutting process. But the metal removal rate increase in results the tool life reduces due to increase in friction and heat generation at the cutting zone. This requires further improvement of MQL technique to meet more cooling and lubricating in high-efficiency machining. One way of enhancing the cooling and lubricating effect of MQL is to develop and employ fluids with high cooling and lubricating abilities. [18]

Recent studies have indicated that compared with conventional fluids, nanofluids and vegetable oils have high heat transfer rates and friction reduction properties. The characteristics of MQL fluids are shown in Table 1. These features make to improve the cooling and lubricating effect of MQL. The MQL with nanofluids and vegetable oils showed the benefits of reducing the grinding force and improving surface roughness and the G - ratio could be improved with high concentration nanofluids. [34,35]

Table 1 - MQL Fluids Characteristics [36]

SYNTHETIC ESTERS (VEGETABLE OILS)	FATTY ALCOHOLS (MINERAL OILS)
• Toxicologically harmless • low level of hazard • good biodegradability • Very good lubrication property • Good corrosion resistance • Inferior cooling properties • Vaporizes with residuals • High flash and boiling point with low viscosity	• Poor lubrication properties • Better heat removal due to evaporation heat • Little residuals • Low flash and boiling point, comparatively high viscosity

2. Grinding of MMC

Muhammad *et al.*. Studied the grindability of aluminium metal matrix composite. In this the Aluminium metal matrix composite 25 volume percent of fine (2-3 μm) embedded in an AA–2124 matrix alloy, was ground using a high performance polycrystalline diamond based grinding wheel. He analyzed the surface finish of the ground Al-MMC is mainly a function of the grinding wheel. He found that the surface and subsurface microstructure revealed that the damage was confined to 0-10 μm top-layer in terms of surface cracks, loosely adhered particles and surface pits/holes. [19]

Thiagarajan *et al.*. Investigated cylindrical grinding of sic particles reinforced aluminium metal matrix composites. In this study The Al/SiC composite specimens are produced aluminium alloys reinforced with 13 μm SiC particles of the size of 30 X 200 mm and the grinding experiments are carried out on a cylindrical grinding machine. He analyzed the various grinding parameters to affect the surface integrity of Al/SiC composites during cylindrical grinding. He found a better surface finish and damage free surfaces are produced due to low grinding force at high wheel and workpiece velocities. [20]

Anand Ronald *et al.*. Deals with Temperature measurement during grinding of metal matrix composites. He observed that although the temperature measured by the pyrometer is less owing to factors like emissivity and specific heat capacity, the actual temperature prevalent of the wheel-work interface could have been higher, as confirmed from the SEM of the machined surface. [21]

Li Zheng *et al.*. investicated to develop the high-efficiency and precision machining technique of TiCp/Ti–6Al–4V particulate reinforced titanium matrix composites. In this high-speed grinding experiments were conducted using the single-layer electroplated cubic boron nitride (CBN) wheel and brazed CBN wheel. The grinding performance was analyses in terms of grinding force and grinding temperature. He found that the grinding temperature and grinding forces for CBN wheel are always less than electroplated CBN wheel. [22]

Anand Ronald *et al.*. Studied the grindability of metal matrix composites on the influence of grinding wheel bond material. He analysed the performance indicators, grinding force and Surface finish using with resin bonded and plated diamond abrasive wheels. He conclude that the resin bonded wheel performed better surface finish better compared to the electroplated wheel. [23]

Suna *et al.*. studied the grinding characteristics of SiC-particle-reinforced aluminum-based metal matrix composite In this the grinding force, grinding temperature, surface roughness are analyzed by the various advanced measuring method. He conducted Grinding experiments in different grinding conditions to obtain good surface integrity. He found that the results are of great significance for high efficiency and precision grinding of MMCs. [24]

Antoniomaria *et al.*. studied a grinding performance of SiC-aluminium composites using conventional abrasives and superabrasives. In this the grindability of metal matrix composites were analysed with different types of grinding wheels, made with both conventional abrasives and superabrasives. He found that the conventional abrasives have given better performances than superabrasives in terms of low clogging, low grinding forces and better surface finish. [25]

3. Grinding of Hybrid MMC

Jaswinder Singh *et al.*. Deals with the various aspects regarding the fabrication and wear properties of Al/SiC/Gr hybrid composites. He found that the presence of Gr particles significantly enhances the wear properties under optimum conditions and the hybrid reinforcements can be successfully incorporated into the Al matrix. Also found that the morphology and size of wear debris are controlled by the amount of Gr particles in the composite. [26]

Jaswinder Singh *et al.*. Investigated the Effect of Machining Parameters on MRR and Surface Roughness in Internal Grinding using EN8, EN31 Steel. He analyzed the characteristics of material removal rate (MRR), surface roughness and tool life for various machining parameters such as cutting force, cutting speed, depth of cut. He concluded that Comparatively, EN31material produces minimum surface roughness and also increases the tool life. [27]

Kannayiram Ponappa *et al.*. Investigated the Grinding of magnesium metal matrix composites. In this work the Grinding performance of magnesium composite was carried out by varying process parameters. The forces were monitored using piezoelectric dynamometer and the performance of the grinding was analyzed and compared with the grinding forces. He concluded that the grinding forces were decreasing with the increase in hardness. [28]

Alakesh Manna *et al.*. Deals with the surface roughness Ra during nano finishing of Hybrid Al/(Al2O3+ZrO2) MMC in grinding operation. A metal bonded diamond grinding wheel is used to for finishing internal surfaces of a cylindrical workpiece. He proved that the surface roughness height is reduced and it reveals an improvement in the surface finish of Hybrid MMC. [29]

4. Grinding of MMC using MQL

Adibi *et al.*. Studied on minimum quantity lubrication (MQL) in grinding of carbon fiber-reinforced SiC matrix composites. In this the MQL grinding performance was compared with conventional fluid and dry grinding. Also the effects of main grinding parameters, such as cutting speed, feed speed, and depth of cut, on the grinding forces, specific grinding energy, wheel wear, surface roughness, and integrity were analyzed. He found that MQL grinding reduced the grinding forces and specific grinding energy, wheel wear was lowered and G-ratio was increased. [30]

Mohsen Emami *et al.*. Studied Minimum Quantity Lubrication in Grinding Process of Zirconia (ZrO_2) Engineering Ceramic. In this work the investication is carried out on MQL grinding of zirconia (ZrO_2)engineering ceramic and the performance were compared with MQL and conventional wet grinding. He found that in MQL grinding, the grinding force and surface roughness are decreased and increase in surface quality. [31]

Dinesh Setti *et al.*. Investigated the application of nano cutting fluid under MQL Technique to Improve Grinding of Ti-6Al-4V Alloy. He compared the results with conventional coolant and pure water, the grinding forces reduced when nano cutting fluid was used even at low concentration of the nano particles and surface finish has been found to improve with higher concentration of the nano particles. [32]

Rahman *et al.*. Investigated material removal rate and surface roughness on grinding of ductile cast iron using minimum quantity lubrication. In this the experiment was analysed according to the design of experiment principle and prepared based on central composite design. He conclude that the highest grinding ratio to be achieved in single-pass MQL and the surface quality of workpiece is mainly influenced by the depth of cut and table speed. [33]

As obvious from the review that the most of the available research work has been approved out in grinding of MMCs. Not so much work has been reported on MQL based grinding of MMCs. Many useful MMCs have not been tried as work material on Grinding using MQL process.

5. Conclusions

Sustainable manufacturing through environmentally-friendly machining begins from the steps that must be taken to apply ecological machining methods in order to create these technologies reliable, environmentally friendly and cost competent. The review centres on the performance and environmental puff of environmentally-friendly grinding for MMC. The salient features are as follows:

- MQL is an ecologically friendly and cost effective process of fluid release that can be used as an well-organized option in grinding process of MMC.
- MQL reduces the tangential and normal forces in grinding process of MMC.
- Surface roughness in MQL grinding was somewhat better than fluid grinding.
- MQL grinding generated a surface with somewhat narrower grazes and homogeneous scenery compared to fluid grinding.

6. References

[1] Shorowordi K M, Laoui T, Haseeb A S M A, Celis J P and Froyen L 2003 *Journal of Materials Processing Technology* **142** 738-43.

[2] Quilgley O, Monaghan J and Reilly P 1994 *Journal of Materials Processing Technology* **43** 21-36.

[3] Sedat Ozden, Recep Ekici and Fehmi Nair 2007 *Composites: Part A* **38** 484-94.

[4] Kwak J.S and Kim Y.S 2008 *Journal of Materials Processing Technology* **20** 596-600.

[5] Adem Onat, Hatem Akbulut and Fevzi Yilmaz 2007 *Journal of Alloys and Compounds* **436** 375-82.

[6] Hung N P, Boey F Y C, Khor K A , Phua Y S and Lee H F 1996 Journal *of Materials Processing Technology* **56** 966-77.

[7] Manna A and Bhattacharayya B 2003 *Journal of Materials Processing and Technology* **140** 711-16.

[8] Zhong Z W 2003 *International Journal of Advanced Manufacturing Technology* **21** 79-83.

[9] Zhong Z W 2003 *Int. J. Adv. Manuf. Technology* **21** 79-83.

[10] Sun F H, Li X K, Wang Y and Chen M 2006 *Key Engineering Materials* **304 – 305** 261-65.

[11] Antoniomaria D Ilio and Alfonso Paoletti 2000 *International Journal of Machine Tools & Manufacture* **40** 173-84.

[12] Zhong Z and Hung N P 2000 *Materials and Manufacturing Processes* **15/6** 853-865.

[13] Chandrasekaran H and Johansson J 1993 *Machining of Advanced Materials* 55-70.

[14] Sung hoon oh 2012 *Int. Journal of Control and Automation* **5(3)** 287-92.

[15] S. Nallusamy 2016 *International Journal of Performability Engineering* **12(2)** 143-54.

[16] Surender Kumar 2014 *International Journal of Engineering and Management Research* **4(4)** 162-70.

[17] Rabiei F, Rahimi A R,. Hadad M J and Ashrafijou M 2014 *J. of Cleaner Production*1-14

[18] Yu Su, Le Gong, Bi Li, Zhiqiang Liu & Dandan Chen 2016 *International Journal of Advanced Manufacturing Technology* **83** 2083–89.

[19] Muhammad A. Shahzad and Jan C Aurich *Institute for manufacturing technology and production systems, university of kaiserslautern, germany.*

[20] Thiagarajan C, Sivaramakrishnan R and Somasundaram S 2011 *ARPN Journal of Engineering and Applied Sciences* **6** 1819-6608.

[21] Anand Ronald B 2013 *Int. J. Machining and Machinability of Materials* **13** 2/3.

[22] Li Zheng, Ding Wenfeng, Shen Long, Xi Xinxin and Fu Yucan 2016 *Chinese Journal of Aeronautics* **29(5)** 1414–1424.

[23] Anand Ronald B, Vijayaraghavan L and Krishnamurthy R 2009 *Materials and Design* 679–86.

[24] Suna F H, Lib X K, Wangc Y and Chend M 2006 *Key Engineering Materials* **304-305** 261-65.

[25] Antoniomaria Di Ilio and Alfonso Paoletti 2000 *International Journal of Machine Tools & Manufacture* **40** 173–84.

[26] Jaswinder Singh 2016 *University Institute of Engineering and Technology Panjab University SSG Regional Centre Hoshiarpur Punjab india.*

[27] Jaswinder Singh 2017 *International Journal of Applied Engineering Research* **12** 2963-68.

[28] Kannayiram Ponappa, Sivanandam Aravindan and Venkateswara Rao 2012 P *Journal of Engineering Manufacture* **226** 1675–83.

[29] Alakesh Manna and Molla K Z 2014 *All India Manufacturing Technology, IIT guwahati, Assam, India.*

[30] Adibi H, Esmaeili H and Rezaeil S M 2017 *The International Journal of Advanced Manufacturing Technology.*

[31] Mohsen Emami, Mohammad Hossein Sadeghi and Ahmed Sarhan 2013 *International Journal of Mining, Metallurgy & Mechanical Engineering* 2320-4052.

[32] Dinesh Setti, Sudarasan Ghosh and Venkateswara Rao P 2012 *International Journal of Mechanical and Mechatronics Engineering* **6** 10.

[33] Rahman M M and Kadirgama K 2015 *International Journal of Automotive and Mechanical Engineering* 2471-2483.

[34] Rabiei F, Rahimi A R, Hadad M J and Ashrafijou M 2014 *Journal of Cleaner Production* 1-14.

[35] Bin Shen and Albert Shih 2007 *Proceedings of ASME/STLE International Joint Tribology Conference IJTC.*

[36] Paulo Davim J 2013 *Springer Heidelberg New York Dordrecht London.*

IOP Conf. Series: Materials Science and Engineering **390** (2018) 012034 doi:10.1088/1757-899X/390/1/012034

Application of crow search algorithm for the optimization of abrasive water jet cutting process parameters

A Tamilarasan[1*], A Renugambal[2], Manikanta D[1], G B C Sekhar Reddy[1], K Sravankumar[1], B Sreekar[1] and G V Prasadreddy[1]

[1]Department of Mechanical Engineering, Sri Chandrasekharendra Saraswathi Viswa Mahavidyalaya, Enathur, Kanchipuram – 631561, Tamilnadu, India.

[2]Department of Mathematics, University College of Engineering Kancheepuram, Kanchipuram (A Constituent College of Anna University, Chennai).

[1*]Corresponding Author: tamilrj2010@gmail.com

Abstract. The present work deals with the prediction of optimal parametric data-set to yield the minimum surface roughness in Abrasive Water Jet (AWJ) cutting of ceramic Tile. By means of a Box–Behnken experiment design technique, an experiment matrix with three factors and three levels was designed. Quadratic model for surface roughness was developed to fit with experimental data. Then, the improved optimal combination of the process parameters is evaluated by proposed methodology of most efficient crow search algorithm. Further, experimentally validation test has been conducted for the optimal cutting conditions suggested by crow search algorithm.

1. Introduction

Abrasive water jet cutting is a newly emerging non-traditional machining process which is used to cut tough materials[1]. In this process, high speed water jet accelerated hard abrasive particles. Then, high speed water mixed with the abrasive material impact to the workpiece, which removes the material. Because of the high velocity, the cutting could be done quickly[2]. Abrasive water jet machine cuts almost all types of materials (conductive and nonconductive). Where, the traditional machine unable to perform effectively especially on brittle and ductile materials. It is widely applied in the cutting of hard or complex-to-cut materials, such as titanium, stainless steel, aluminium alloys, etc., and it has been used in the aircraft and automobile industries[3]. Abrasive water jet machining does not use the traditional tool, so according to the material composition, structure, hardness, and the physical properties of different, abrasive water jet can be suitable process parameters selection for the processing of various materials[2]. Compared with other processing methods, the AWJ machining method has the advantages of flexible, rapid, extremely low machining force, does not have contact with tools and does not produce heat. Therefore, the considerable research and development effort has been made in recent years to develop new techniques to enhance the cutting performance of the process. Recently, many computational techniques are evolved for optimizing the machining process parameters such as Response surface Methodology, Genetic Algorithm , Particle Swarm Optimization etc. Lima et al[4] studied the influence of the main process parameters such as traverse speed and abrasive mass flow rate on the surface finish of agates machined by AWJ cutting process. It was found that the thickness of the sample is not a predominant factor for the surface finish, when the experiments were conducted for agate plates of 5 and 10 mm thickness. It also stated that , for the economic point of view, a combination of high traverse speeds and low abrasive mass flow rates leads to reducing the process-associated costs, making AWJ an attractive machining process. Babu and Muthukrishnan[5] focused on optimising the process parameters in abrasive water jet machining with the objective of minimising surface roughness (R_a) in brass-360 material. Abrasive flow rate, pump pressure, stand-off distance and feed rate were considered as process parameters. A mathematical model developed using ANOVA in this study was found to be satisfactory which gives R^2 value of 94.91%. It has been found that the Pump pressure is the most influential factor related to surface roughness, followed by abrasive flow rate. And, the surface roughness value was improved by 33% despite an increase in Pump pressure of 25% and a decrease in abrasive flow rate of 40%.Babu and Muthukrishnan[6] further discussed on the optimization of an abrasive water jet machining process with multiple characteristics, using the Taguchi orthogonal array and grey relational analysis (GRA) in cutting of Corian aurora (tile) .The performance characteristics of surface roughness and the kerf angle were optimized through the machining process variables, such as mesh size, nozzle diameter, abrasive flow rate, water pressure, stand-off distance, and feed rate. And, it is stated that the optimized parameters may used in manufacturing industries, for improving the non-conventional machining performance in AWJ cutting. Santhanakumar et al [7] examined the effect of AWJ cutting parameters like abrasive grain size, abrasive flow rate, nozzle–workpiece standoff, water pressure and jet traverse rate on the surface roughness and taper angle of cut produced with ceramic tiles. A new combined technique of grey-based response surface methodology (g-RSM) was disclosed for obtaining the optimal level of AWJ cutting parameters. The optimal parameter setting was validated by conducting a confirmation test. The cut surfaces were also examined using field emission scanning electron microscope images, P-profile plots and atomic force microscope images. Aich et al[8] conducted on cutting of borosilicate glass by AWJ cutting process. Optimum condition of control parameters setting was searched through particle swarm optimization (PSO). Abdullah et al[9] examined the effect of cutting parameters, namely standoff distance, nozzle traverse speed , abrasive flow rate , and material type on cutting performance for two types of marble workpieces, Carrara white and Indian green. Statistical analysis was assessed to understand the influence of the cutting parameters on the process performances in terms of surface roughness, surface waviness, and Kerf taper ratio. The results showed that the traverse speed and material type were the most significant factors that affected surface roughness and Kerf taper ratio. Hajdarevic et al[3] studied the effects of material thickness, traverse speed and abrasive mass flow rate during abrasive water jet cutting of aluminum workpiece. Surface roughness was measured across of depth of cut. The experimental results show that traverse speed has great effect on the surface roughness at the bottom of the cut.

Through the literature review, abrasive water jet cutting having many capabilities regarding process efficiency and effectiveness are being investigated through theory, understanding using experimentation and process modelling techniques. but many aspects of abrasive water jet cutting technology are still under development. Therefore , the present study to explore the optimum combinations of the abrasive water jet cutting process parameters in cutting of ceramic tile using a new type of crow search algorithm. The experiments were designed on the basis of the response surface methodology (RSM) based box-behnken design (BBD) technique. Through RSM[10], a mathematical model for surface roughness is developed. Further, a proposed crow search algorithm is used to obtain the optimum parametric setting for abrasive water jet cutting process parameters for yielding the minimum surface roughness.

2. Experimental work

Figure 1. CNC controlled AWJ Machining Centre

The CNC controlled OMAX 2652 AWJ machining centre, which is imported from the Germany, is used for carrying out the experimental work. The CNC controller based abrasive eater jet cutting machine includes the following basic components: high pressure pump, abrasive water jet cutting head, abrasive delivery system, abrasive material and water catcher, and x-y positioning table. There is a large number of process parameters associated with the AWJM process but it is practically impossible to include all these parameters during experimentation. The experiments were planned using response surface methodology based box-behnken design method to obtain the independent, interactive, and higher order effects of different process variables on surface roughness.

Table 1. Process Parameters and Levels`

Process parameters	Reference symbol	Units	Low(-1)	Centre(0)	High(+1)
Water jet pressure	X_1	bar	2000	3000	4000
Jet traverse speed	X_2	mm/min	150	200	250
Abrasive flow rate	X_3	g/min	300	400	500
Standoff distance	X_4	mm	1	2	3

Experimental range of process parameters and the levels of the independent variables are presented in Table 1, where -1 corresponds to the lower(minimum) value and +1 to the higher (maximum) value of each process variable. Four input process parameters namely the water jet pressure, jet traverse speed, abrasive feed rate and standoff distance, were considered. The values of X_1 (Water Jet Pressure)varied between 2000bar and 4000bar, X_2 (Jet traverse speed) between 150 mm/min and 250mm/min, X_3 (Abrasive flow rate) between 300 g/min and 500 g/min, and X_4 (Standoff Distance) between 1mm and 3mm. This BBD design scheme required 29 experimental runs . The other parameters that were kept constant during the tests included the nominal jet impact angle (90°), orifice diameter (0.3 mm), mixing tube or nozzle diameter (1.2 mm) and abrasive material (80 mesh garnet). A typical ceramic tile with dimension of the material is 300 × 50 ×10 mm thick plate is used for experimental work as shown in Figure 2. Because , ceramics are a large group of structural materials characterized by many useful properties. The most important are the high heat resistance, resistance to chemical agents, good mechanical properties and good dielectric and insulation properties[11]. The specimen was cut into 50 mm × 10mm slot through depth of 10 mm. The cutting zone of AWJ process as illustrated in Figure 3.All the experiments were conducted randomly to minimize the effects of unexplained variability in the observed responses because of externally influencing factors.

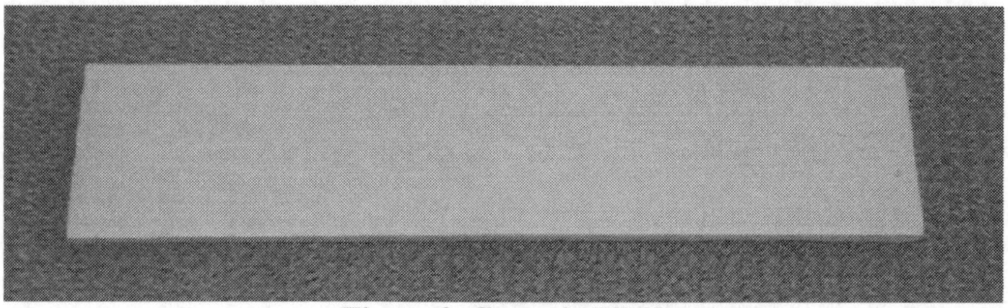

Figure 2. Ceramic Tile Specimen

Figure 3. AWJ cutting zone

Average arithmetic surface roughness, measured at 10 different locations on striation free zone of cut surface (R_a) – to be measured in microns using Maher Perth meter M1. The measurements were repeated twice and their average values used. In parts manufacturing, accuracy of the shape and dimension, and surface finish are the primary quality objectives. Surface roughness is the irregularity in the cut materials surface texture. Roughness becomes important when abrasive water jet machining produces parts requiring high quality. Therefore , in this study, the arithmetic average of the roughness was measured and used for further analysis.

IOP Conf. Series: Materials Science and Engineering **390** (2018) 012034 doi:10.1088/1757-899X/390/1/012034

Table 2. Experimental plan and collected responses

Exp.No	Process parameter				Response
	Water Jet Pressure	Jet traverse speed	Abrasive flow rate	Standoff Distance	Surface roughness
	bar	mm/min	g/min	mm	Micron
1	2000	150	400	2	2.173
2	4000	150	400	2	1.471
3	2000	250	400	2	2.407
4	4000	250	400	2	1.906
5	3000	200	300	1	2.641
6	3000	200	500	1	2.541
7	3000	200	300	3	3.277
8	3000	200	500	3	3.009
9	2000	200	400	1	1.906
10	4000	200	400	1	1.571
11	2000	200	400	3	2.792
12	4000	200	400	3	2.001
13	3000	150	300	2	2.641
14	3000	250	300	2	3.126
15	3000	150	500	2	2.675
16	3000	250	500	2	2.742
17	2000	200	300	2	2.474
18	4000	200	300	2	1.906
19	2000	200	500	2	2.34
20	4000	200	500	2	1.672
21	3000	150	400	1	2.34
22	3000	250	400	1	2.441
23	3000	150	400	3	2.842
24	3000	250	400	3	3.210
25	3000	200	400	2	3.109
26	3000	200	400	2	3.206
27	3000	200	400	2	3.226
28	3000	200	400	2	3.205
29	3000	200	400	2	3.207

3. Model development and ANOVA analysis

A box-behnken design is constructed using the software Design Expert (8.0.6 version) and was used to analyze the results of collected surface roughness as indicated in Table 2. The final quadratic model for surface roughness is represented as follows (uncoded units)

$$\text{Surfaceroughness}=-13.66927+0.0053634X_1+0.048620X_2+0.014139X_3+1.16228\ X_4+0.000001005X_1X_2-0.000114X_1X_4-0.0000209\ X_2X_3+0.001335\ X_2X_4-0.000000938925X_1^2-0.00010782\ X_2^2-0.00001358X_3^2-0.19493\ X_4^2 \quad (1)$$

Table 3. ANOVA Table for surface roughness

Source	Sum of Squares	df	Mean Square	F value	Prob > F
Model	8.461774	14	0.604412462	332.5497	< 0.0001
X_1	1.059099	1	1.059099188	582.7198	< 0.0001
X_2	0.238008	1	0.238008333	130.9529	< 0.0001
X_3	0.098282	1	0.098282118	54.07513	< 0.0001
X_4	1.13529	1	1.135290083	624.6402	< 0.0001
X_1X_2	0.0101	1	0.01010025	5.55719	0.0335
X_1X_4	0.051984	1	0.051984	28.60176	0.0001
X_2X_3	0.043681	1	0.043681	24.03343	0.0002
X_2X_4	0.017822	1	0.01782225	9.805859	0.0074
X_1	5.718353	1	5.71835282	3146.256	< 0.0001
X_2	0.471291	1	0.471290672	259.3056	< 0.0001
X_3	0.119621	1	0.119620733	65.81571	< 0.0001
X_4	0.246459	1	0.246459469	135.6028	< 0.0001
Residual	0.025445	14	0.00181751		
Lack of Fit	0.01682	10	0.001681995	0.780037	0.6599
Pure Error	0.008625	4	0.0021563		
Cor Total	8.48722	28			
Std. Dev.	0.042632		R^2	0.997	
Mean	2.55369		Adj R^2	0.994	
C.V. %	1.669438		Pred R^2	0.986	
PRESS	0.11036		Adeq Precision	58.902	

Analysis of variance (ANOVA) [12] was conducted to fit the mathematical model (equation (1)) to examine the statistical significance of the model terms. The adequacy of the models was determined using model analysis, lack-of-fit tests, coefficient of determination (R^2) , Adj R^2 , Pred R^2 and Adequate Precision. The calculated values of the determination coefficient (R^2) and adjusted determination coefficient (Adj. R^2) are more than 95%, which indicates a high significance of the model. Pred. R^2 is the correlation coefficient in prediction, which is more desirable when approaching unity. Therefore, the fitted quadratic model is reliable, and can be employed in optimization of the test output. The ANOVA of the quadratic regression model indicated that the model was highly significant, as the *F* value for the model was 332.5497. There was only a 0.01% chance that the "model *F* value" this large could occur because of noise. The *p* value Prob>*F* value of the model was <0.0001, which also confirmed that the model was highly significant. Lack-of-fit test was also carried out, which measures the failure of a model to represent the data in the experimental domain at points which are not included in the regression. A lack of-fit value of 0.780037 implies that the lack of fit is not significant relative to the pure error when *p* (0.6599 >0.05) also supports the fitness of the model. A normal probability plot of the residuals is depicted in Figure 5, which reveals that the residuals generally fall on a least-square line which is used to estimate the cumulative distribution function for the population. As evident from the figure, the errors are normally distributed and there are almost no serious violations of the assumptions that underlie the analysis[13].

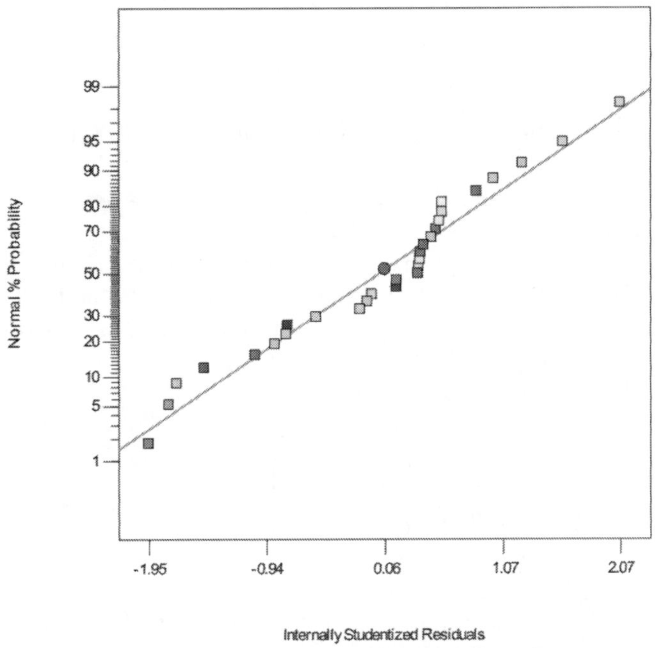

Figure 4. Normal probability plot

4. Optimization by crow search algorithm

The step by step procedure for the implementation of crow search algorithm[14] as follows:

Step 1: Initialize problem and adjustable parameters

The optimization problem, decision variables and constraints are defined. Then, the adjustable parameters of CSA(flock size(*N*), maximum number of iterations($iter_{max}$), flight length(*fl*) and awareness probability(*AP*)) are valued.

Step 2: Initialize position and memory of crows

N crows are randomly positioned in a d-dimensional search space as the members of the flock. Each crow denotes a feasible solution of the problem and d is the number of decision variables.

$$crows = \begin{bmatrix} x_1^1 & x_2^1 & \cdots & x_d^1 \\ x_1^2 & x_2^2 & \cdots & x_d^2 \\ \vdots & \vdots & \vdots & \vdots \\ x_1^N & x_2^N & \cdots & x_d^N \end{bmatrix} \tag{2}$$

The memory of each crow is initialized. Since at the initial iteration, the crows have no experiences, it is assumed that they have hidden their foods at their initial positions.

$$Memory = \begin{bmatrix} m_1^1 & m_2^1 & \cdots & m_d^1 \\ m_1^2 & m_2^2 & \cdots & m_d^2 \\ \vdots & \vdots & \vdots & \vdots \\ m_1^N & m_2^N & \cdots & m_d^N \end{bmatrix} \tag{3}$$

Step 3: Evaluate fitness(objective) function

For each crow, the quality of its position is computed by inserting the decision variable values into the objective function.

Step 4: Generate new position

Crows generate new position in the search space as follows: suppose crow i wants to generate a new position. For this aim, this crow randomly selects one of the flock crows (for example crow j) and follows it to discover the position of the foods hidden by this crow(m^j). The new position of crow i is obtained by the equation as below:

$$x^{i,iter+1} = \begin{cases} x^{i,iter} + r_i \times (m^{j,iter} - x^{i,iter}) & r_j \geq AP^{j,iter} \\ randomposition & otherwise \end{cases} \tag{4}$$

This process is repeated for all the crows.

Step 5: Check the feasibility of new positions

The feasibility of the new position of each crow is checked. If the new position of a crow is feasible, the crow updates its position. Otherwise, the crow stays in the current position and does not move to the generated new position.

Step 6: Evaluate fitness function of new positions

The fitness function value for the new position of each crow is computed.

Step 7: Update memory

The crows update their memory as follows:

$$m^{i,iter+1} = \begin{cases} x^{i,iter+1} & f(x^{i,iter+1}) is better than f(m^{i,iter}) \\ m^{i,iter} & otherwise \end{cases} \tag{5}$$

Where $f(.)$ denotes the objective function value.

It is seen that if the fitness function value of the new position of a crow is better than the fitness function value of the memorized position, the crow updates its memory by the new position.

Step 8: Check termination criterion

Steps 4-7 are repeated until $iter_{max}$ is reached. When the termination criterion is met, the best position of the memory in terms of the objective function value is reported as the solution of the optimization problem.

According to the above crow search algorithm, at first, mathematical models of polynomial type are developed to correlate the AWJ machining parameters and performance measures. Once the process model of surface roughness is constructed, an appropriate objective function and constraints are developed with the process models The optimization process of crow search algorithm is implemented Matlab software in order to find the optimal solution of surface roughness. The crow search is expected to give the minimum value of surface roughness for abrasive water jet cutting process. The convergence plot of surface roughness value over max number of generations using crow search algorithm is shown in Figure 5, which illustrates the algorithm, converges up to 200 generations.

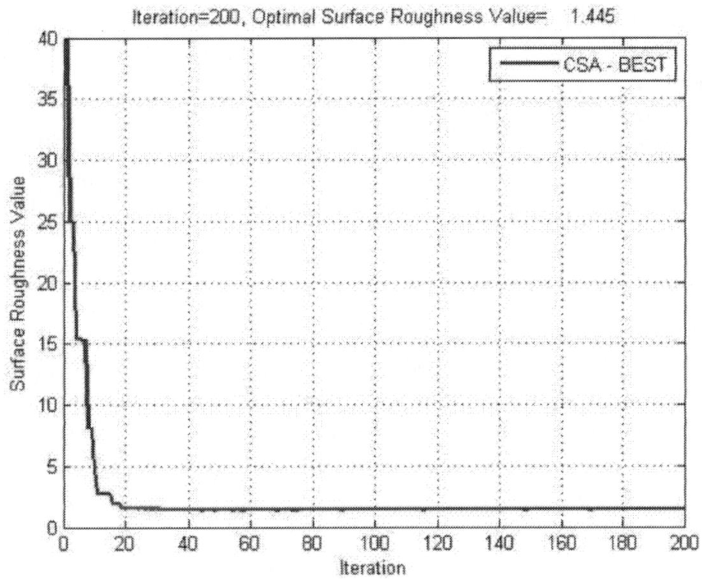

Figure 5. Convergence plot

Table 4. Comparison between the initial experimental condition and optimal condition achieved by CS algorithm

Model Summary		Initial condition	Optimal condition	% of improvement
	Water Jet Pressure(bar)	2000	3997.19	
Process parameters	Jet traverse speed(mm/min)	150	152.05	
	Abrasive flow rate(g/min)	400	381.08	
	Standoff distance(mm)	2	1.62	
Corresponding response parameters		2.173	1.445	33.5

Table 5. Experimental validation of optimal parameter settings

	Water Jet Pressure(bar)	Jet traverse speed(mm/min)	Abrasive flow rate(g/min)	Standoff Distance(mm)	Surface roughness(Micron)
Predicted by CS algorithm	3997.19	152.05	381.08	1.62	1.445
Experimental	3997.19	152.05	381.08	1.62	1.456
% Error					0.7612

On the other hand, the percentage of improvement of fitness value to corresponding optimal condition is achieved by 33.5%(Table 4) , and it is clear that the predictive capability of the crow search algorithm. After the selection of AWJ optimal process parameters, further experiments were carried out to verify the corresponding surface roughness predicted by algorithm. Table.5 shows that the percentage of error between the predicted and experimented values. From this analysis, it is observed that the calculated error(i.e. 0.7612) is very small which confirms the excellent reproducibility of the experimental conditions.

5. Conclusions

In this paper, a response surface methodology and PSO algorithm has been introduced for abrasive water jet cutting of ceramic tile. The present approach comprises two stages. Mathematical model is developed to predict the surface roughness in stage 1. Stage 2 involves the optimization of process parameters using the crow search algorithm. From the proposed approach, the following conclusions are drawn.

1. The Box-Behnken design approach is used for experimental design. The water jet pressure, Jet traverse speed, abrasive flow rate and standoff distance were chosen as process parameters or factors , and three levels of each factor were considered in the experimental design. As response variable of surface roughness was taken into consideration.

2. The predicted values match the experimental values reasonably well, with R^2 of 0.997 for surface roughness. Thus the developed model is more suitable for further analysis.

3. The social behaviour of crows is an effective model for constructing the crow search algorithm.

4. The optimal performance characteristics are observed to have minimum surface roughness when the process parameters are water jet pressure(3997.19bar), Jet traverse speed (152.05mm/min), abrasive flow rate (381.08 g/min) and standoff distance (1.62mm).

5. A significant improvement in surface roughness characteristics is observed at the optimum parameter settings in comparison to the initial settings.

References

[1] Chen L, Siores E and Wong W 1998 Optimising abrasive waterjet cutting of ceramic materials *Journal of Materials Processing Technology* 74 251-254.

[2] Kovacevic R, Hashish M, Mohan R, Ramulu M, Kim T and Geskin E 1997 State of the Art of Research and Development in Abrasive Waterjet Machining *Journal of Manufacturing Science and Engineering*119 776-785.

[3] Begic-Hajdarevic D, Cekic A, Mehmedovic M and Djelmic A 2015 Experimental Study on Surface Roughness in Abrasive Water Jet Cutting *Procedia Engineering* 100 394-399.

[4] de Abreu e Lima C, Lebrón R, de Souza A, Ferreira N and Neis P 2015 Study of influence of traverse speed and abrasive mass flowrate in abrasive water jet machining of gemstones *The International Journal of Advanced Manufacturing Technology* 83 77-87.

[5] Naresh Babu M and Muthukrishnan N 2018 Investigation on Surface Roughness in Abrasive Water-Jet Machining by the Response Surface Method.

[6] Naresh Babu M and Muthukrishnan N 2018 Investigation of multiple process parameters in abrasive water jet machining of tiles.

[7] Santhanakumar M, Adalarasan R and Rajmohan M 2015 Experimental Modelling and Analysis in Abrasive Waterjet Cutting of Ceramic Tiles Using Grey-Based Response Surface Methodology *Arabian Journal for Science and Engineering* 40 3299-3311.

[8] Aich U, Banerjee S, Bandyopadhyay A and Das P 2014 Abrasive Water Jet Cutting of Borosilicate Glass *Procedia Materials Science* 6 775-785.

[9] Abdullah R, Mahrous A and Barakat A 2018 Surface quality of marble machined by abrasive water jet

[10] Santhanakumar M, Adalarasan R and Rajmohan M 2017 Analysis on abrasive waterjet cutting of glazed vitreous tiles using Taguchi-based RSM *International Journal of Manufacturing Technology and Management* 31 472-489.

[11] Gudimetla P, Wang J and Wong W 2002 Kerf formation analysis in the abrasive waterjet cutting of industrial ceramics *Journal of Materials Processing Technology* 128 123-129.

[12] Tamilarasan A and Marimuthu K 2014 Multi-response optimisation of hard milling process parameters based on integrated Box-Behnken design with desirability function approach *International Journal of Machining and Machinability of Materials* 15 300.

[13] Tamilarasan A and Rajamani D 2017 Multi-response optimization of Nd:YAG laser cutting parameters of Ti-6Al-4V superalloy sheet *Journal of Mechanical Science and Technology* 31 813-821.

[14] Askarzadeh A 2016 A novel metaheuristic method for solving constrained engineering optimization problems: Crow search algorithm *Computers & Structures* 169 1-12.

Mechanical and Microstructural Characterization of Diffusion Bonded Super Martensitic Stainless Steel Joints

C B Maheswaran, Seshadri Sridharan, Anuj A and Arun kumar Srirangan*

School of Mechanical Engineering, SASTRA Deemed to be University

Thanjavur-613401, Tamilnadu, India.

*Corresponding author: arun1988@gmail.com; Tel: +91-9943989906

Abstract. Super Martensitic Stainless Steels (SMSS) are low carbon high chromium steels. They possess high strength and resistant to corrosion for fluids like CO_2 and H_2S. Diffusion bonding is a welding technique capable of joining similar or dissimilar metals wherein the solid materials intersperse themselves with time. In this present study, the SMSS specimens were bonded at temperatures ranging from 800-1050^{0}C in the non-inert atmosphere and their microstructure and mechanical properties were analyzed.

Keywords: Diffusion bonding; super martensitic stainless steel; high strength; corrosion resistance; non- inert atmosphere.

1. Introduction

Development of SMSS happened to reduce brittle chromium carbide precipitation on welding of traditional martensitic stainless steels. They are currently produced for the applications of Oil Country Tubular Goods (OCTG). They are based on Fe-Cr-Ni-Mo System with 4-6 % of Ni, 0.5-2.5% of Mo and low carbon contents (≤0.02%) [1]. To obtain martensitic lath microstructure SMSS are air-cooled in solution heat treatment temperature at the austenitic field due to their adequate hardenability [2]. Diffusion bonding was an efficient way to manufacture the complex structures with less weight and significant cost savings. The establishment of diffusion bonding process was to manufacture components like honeycomb structures, aircraft engine parts, and hollow engine blades. Diffusion bonding creates permanent joints among materials by establishing interatomic bonds between work pieces through the action of local or general heating or plastic deformation or both [3]. The bonding pressure or pressing load causes the edge to move within the range of atomic force. If there is a molten layer then the decisive factor is the pressing load, which expels the interlayers from the joint. To squeeze out all of the interlayers the pressing load has to be raised. The most stable bonding system was formed by bringing atoms to the distance of minimum energy or equilibrium. Any increase or decrease in inter atomic distance would increase the energy of atomic interaction. Diffusion bonding process was carried out in a bonding chamber in which vacuum is maintained [6]. The vacuum in the chamber reduces the detrimental effect of faying surfaces. The materials to be bonded are to be placed in the chamber with their mating surfaces touching each other and heated to a specific temperature [4-7]. A pressing load was then applied for a specific amount of time. After the joining was complete, the weldment was cooled in vacuum or quenched depending on the properties of materials involved. The mechanism of diffusion bonding begins with initial point contact of metals, then the contact area increases as the local stress reach the limit of yield stress, then the grain boundary rearrangement happens and the final stage is the volume diffusion of grain boundaries thus the process completes [8].

2. Literature Review

Ayush Barmola.et.al., [8] investigated the effect of recrystallization temperature on super martensitic stainless steel, reported that SMSS on tempering contains tempered martensite and retained austenite with a little amount of α –ferrite present. **Chellappan.et.al.,** [9] discussed the effect of heat input on mechanical and metallurgical properties of SMSS on A-TIG welding found that amount of δ-ferrite increases with the heat input and the tensile strength decreases as the heat input increases and toughness is directly proportional to heat input. **Sanjeev Kumar.et.al.,** [10] discussed the effect of preheat temperature on weldability of martensitic stainless steels reported that the ultimate tensile strength gets increased for the specimen welded at the temperature of 1000^0 C and preheated at the temperature of 121^0C. **Surinder Singh.et.al.,** [11] discussed the effects of alloying and heat treatment on properties of SMSS found that While following the heat treatment route, most of the researchers have taken different austenitizing/ heat treatment temperatures (with a wide gap) on which experiments were performed. The temperatures selected either been based on an earlier study or selected randomly without providing the supporting logic. Similarly, some authors conducted heating at different heating rates, with the same limitations as discussed for temperature and very limited work was reported about the magnetic properties of these steels. When reported, it has been stated that austenite is a paramagnetic phase. Thus, the paramagnetic properties were found in the steel until the martensite started to form. However, in some SMSS's that ferromagnetism was observed above M_S temperature (340-230 ^{0}C). The underlying reason behind this unusual behavior was not adequately explained. **Emel Taban.et.al.,** [12] on his discussion about the weldability of super martensitic stainless steels recommends a post weld heat treatment of 620^0 C-650^0C for 5 minutes to prevent intergranular stress corrosion cracking. **Kang.et.al.,** [13] on his discussion about the passivity of SMSS found that the passivity combinations of SMSS in both acidic and alkaline environments are Fe oxides and Cr oxides. **Tarun Nanda.et.al.,** [14] on his discussion about the effect of isothermal annealing on microstructure of SMSS reveals that the thermo-mechanical processing route resulted from information of fine austenite grains due to recrystallization of the cold rolled martensitic matrix. The fine austenite grains were much stable at room temperature than the coarse grains. This austenite retention resulted in ductility enhancement. Further, the formation of strain- free grains due to recrystallization resulted in the annihilation of dislocations, which also improved ductility without significantly deteriorating the strength.

A.Dethelfs.et.al., [4] discussed the effect of hybrid diffusion bonding on aluminium tube to sheet connections in the coil wound heat exchangers, found that the weld microstructure, as observed by optical microscopy, consisted of three zones: the TMAZ, the HAZ, and non-affected BM. The TMAZ undergoes significant mechanical deformation under temperature and was characterized by fine equiaxed grains. The heat affected zone was characterized by a gradual grain growth and an ill-defined boundary around the unaffected base metal. **Zhang Guoge.et.al.,** [14] on his discussion about the solid -state diffusion bonding of Inconel 718 to 17-PH stainless steel reported that a continuous intermetallic film forms along the joint interface when bonding Inconel alloy 718 with 17-4 PH stainless steel at 1000^0 C for 60 min, with pressure ranging from 16 MPa to 48 MPa. The formation of such phases was caused by the agglomeration of Nb and Ti. In all joints, failure occurred at the interface, nearly without plastic deformation. The maximum tensile strength of the joint reached 774 MPa, more than 70% of that of 17-4 PH stainless steel (1007 MPa). The influence of pressure on joint tensile strength was more significant than that in area reduction. **N.L.Loh.et.al.,** [15] discussed the effect of diffusion bonding of ceramics to metals reported that comparing diffusion bonding to Hot isostatic pressing (HIP) process was considered to be more effective and they are further researching on diffusion bonding using the hot isostatic pressing technique. **M.Balasubramaniam** [16] while discussing the application of box benkhen design in diffusion bonding of Ti-6Al-4V with 304 stainless steel states that the bonding strength increases with increase in bonding temperature. The increase in bonding strength can be attributed to the formation of intermetallics like Ti-Ag at the interface. **M. Mazar Etabakei.et.al.,** [17] on his discussion about diffusion bonding of Al/Mg_2Si metal matrix composite with AZ91D magnesium alloy at different heating rates state that the average shear strength of the bonded joints showed higher value at lower heating rate. **Kemal Aydin.et.al.,** [18] on his discussion about diffusion bonding of copper to titanium substantiates that suitable weld parameters should be selected carefully. Otherwise, the material with low yield strength would be subjected to a deformation due to the high temperature and pressure. The high temperature lead to grain growth.

From the literature survey it was found that, there was no sufficient literature on diffusion bonding of super martensitic stainless steel, and the novelty of this particular study was the efforts, which had been taken to bond SMSS specimens in non- inert atmosphere in variation to the actual process.

3. Experimental Procedure

The SMSS work pieces to be diffusion bonded are brought as rectangular bars of dimension 195x20x6mm and then cut into pieces of dimension 20x20x6 mm and they were polished by following sequence of steps like surface grinding of the work piece and polishing them using various grades of emery sheets and disc polishing. The hardness of the work piece was measured using Brinell's hardness scale, which showed the value of 66.3 BHN. The chemical composition of SMSS specimen are given in Table 1

Table 1 Chemical composition of SMSS

ELEMENT	Cr	Ni	Fe
COMPOSITION (%)	13	6	81

Tool steel was used as the die material because their suitability comes from the distinctive hardness, resistance to abrasion, deformation and their ability to hold their cutting edges at elevated temperatures. They have the carbon content between 0.5%-1.5% and their four major alloying elements are tungsten, chromium, vanadium, and molybdenum. The modeling and design of the die were done in AutoCAD and CREO parametric. To obtain the desired accuracy of the die was machined using Electric Discharge Machining (EDM). The specimens were placed inside the die and the die was placed inside the muffle furnace and the experiments were carried out at temperatures ranging from 800-1050^{0}C.. Figure (1.1) and Figure (1.2) shows the images of the die and the experimental setup.

Figure 1.1. Die with the screws

Figure 1.2. Experimental set up

4. Results and Discussions

4.1. Bonding Process

Table 2 shows the experimental table for bonding the SMSS specimens. The main parameters taken were bonding time and bonding temperature, which influences the bonding strength of the specimen .

Table 2. Process table

S.NO	Time(min)	Temperature (^{0} C)	Result	Reason
1	45	800	No bonding	Oxidation
2	45	950	Bonding achieved	-
3	45	1050	Bonding achieved	-

Figure 2(a-c) shows the images of the specimen at the trials, while bonding the specimens. Figure 2 (a) shows the failure in the bonding of the specimen due to oxide formation, since the environment was considered as a major factor in determining the strength of the bonded joints the interface contact was optimized by the treatment of the surface to be bonded by mechanical polishing, etching, and cleaning. At 950 ^{0}C and 1050^0 C perfect bonding was achieved which can be seen from figure 2(b) and 2(c)

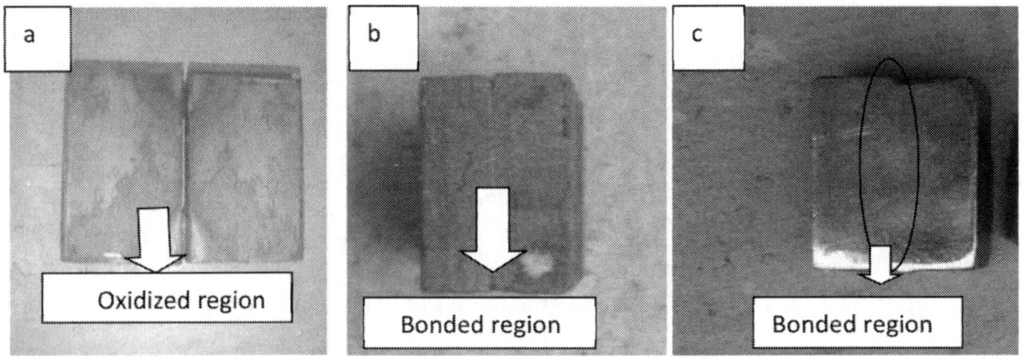

Figure 2. Experimental trials at a) 800^0 C b) 950^0C c) 1050^0 C

4.2 Microstructural Characterization

The microstructure of the base metal, Figure 3(a), shows rolled, panacke structure with varied grain size. The phase mapping structure of SMSS is presented in Figure 3(b).Figure 3(a) shows clearly a well -distinct grain boundary area. The base material has elongated grains of δ–ferrite and streaks (like plates) of δ- ferrite and also randomly dispersed globular particles of Cr_2C_3[2]. Figure 3(b) shows that the delta ferrite phase in the matrix of martensite is 47% and the alpha ferrite phase, look like to be elongated bands, is 43% [9].

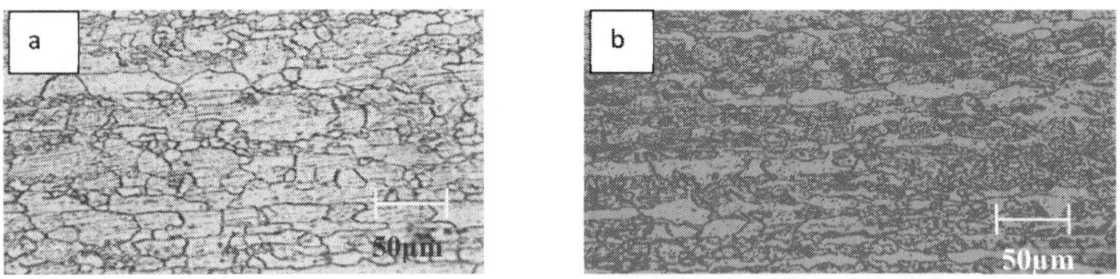

Figure 3. a) Microstructure of base metal b) Phase mapping image of the base metal

Figure 4(a-c) shows the weld metal microstructure of SMSS specimens at different bonding temperatures. Figure 4(a) shows tree - like structure dark etching martensite in a ferrite matrix along with globular particles are Cr_2 carbides. Figure 4(b) presents, Precipitated particles are Cr_2 carbides which are evenly distributed in a matrix .Figure 4(c) shows martensite in a ferrite matrix, feather type of martensite is observed, but Cr_2C_3 evenly distributed in the matrix [9].

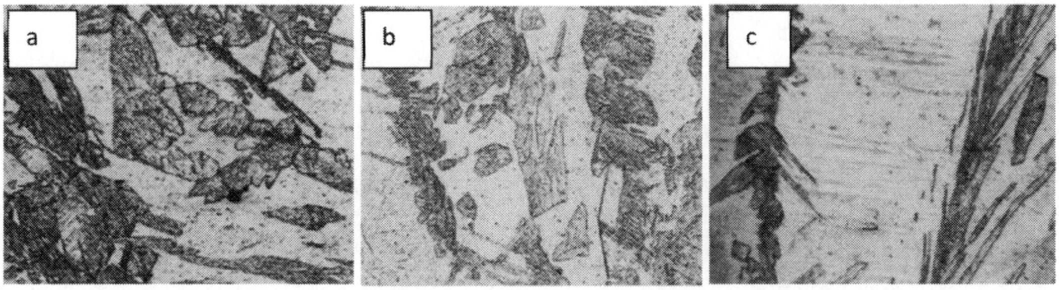

Figure 4. Weld metal microstructure of specimen at a) 800^{0}C b) 950^{0}C c) 1050^{0}C

Figure 5(a&b) shows the microstructure of bonded region. From figure 5(a) it is seen that the perfect bonding has occurred due to diffusion of larger number of atoms to form an uniform structure, as the plastic flow of materials increases. This is highly attributed to increase in heat input because as the temperature increases the grain size becomes smaller and the grain boundaries participating in the diffusion process becomes higher resulting in proper diffusion. From figure 5(b) it is seen that the irregular grain size has resulted in the improper bonding [20].

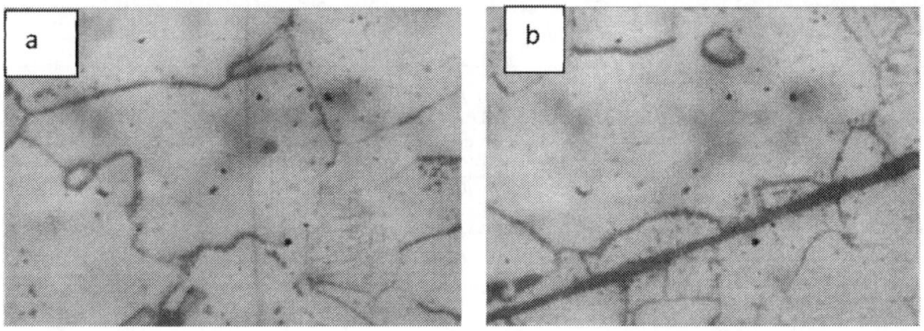

Figure 5. Microstructure of bonded region at a) 1050^{0}C b) 950^0 C

Table 3 shows the areal fraction of phases present in the weld metal. The analysis was done with the help of phase analysis software and the ferrite content was measured using feritscope. From the table, it is clear that the higher heat input at 1050^0 C reduces the formation of martensite and the higher cooling rate at 900^0 C increases the formation of martensite in the weld region [9].

Table 3. Areal fraction of phases present in weld metal

TEMPERATURE(^{0} C)	% OF MARTENSITE	% OF FERRITE
1050	54.61	45.39
950	55.60	43.4
800	67.58	32.42

4.2 Hardness Measurement

The hardness measurements were carried out at different bonding temperatures in weld metal. In each temperature, five measurements were taken and their average values are taken as the result. Table 4 shows the hardness measurements of weld metal. From the table, it is clear that the weld metal hardness value is higher at 800 ^{0}C when compared to the 950 and 1050 ^{0}C. The difference in micro-hardness at 800 and 1050 ^{0} C can be attributed to the differences in the grain size and also major constituents of the martensite phases present at 800^0 C [2].

Table 4. Hardness measurements on weld metal

TEMPERATURE (^{0} C)	HARDNESS (HV)
1050	224
950	243
800	257

5. Conclusions

Thus the diffusion bonding of super martensitic stainless steels at temperatures ranging from800^0 C to 1050^0 C were done at different time periods and the mechanical and the microstructural characterization were done and the following results were obtained.

- Presence of martensite at lower heat inputs prevents proper bonding of specimens.
- This is contributed to the grain structure of martensite which reduces the diffusion rate and interlocking of grains
- At higher heat input, martensite gets converted into austenite and the grain size becomes finer allowing proper bonding
- Microhardness values decreases at higher heat input (i.e.) at the higher temperature.

IOP Conf. Series: Materials Science and Engineering **390** (2018) 012035 doi:10.1088/1757-899X/390/1/012035

6. References

[1] Ma X P Wang L Liu C M & Subramanian S V 2012 Microstructure and properties of 13Cr5Ni1Mo0.025Nb0.09V0.06N super martensitic stainless steel *Materials Science and Engineering.* ***539***, 271–279

[2] Muthusamy C Karuppiah L Paulraj S Kandasami D & Kandhasamy R 2016 Effect of heat input on mechanical and metallurgical properties of gas tungsten arc welded lean super martensitic stainless steel *Materials Research.* ***19*****(3)**, 1–12

[3] Kumar S Chaudhari G P Nath S K & Basu B 2012. Effect of preheat temperature on weldability of martensitic stainless steel *Materials and Manufacturing Processes.* ***27*****(12)**, 1382–1386.

[4] Dethlefs A Roos A dos Santos J F & Wimmer G 2014 Hybrid friction diffusion bonding of aluminium tube-to-tube-sheet connections in coil-wound heat exchangers *Materials and Design*, ***60***, 7–12.

[5] Z Jiang S Zhang K Lu Z Li B & Zhang D 2016 The structural design and superplastic forming/diffusion bonding of Ti2AlNb based alloy for four-layer structure *Materials & Design.* ***104***, 242–250.

[6] Guoge Z Chandel R S & Seow H P 2001 Solid-state diffusion bonding of Inconel alloy 718 to 17-4 PH stainless steel *Materials and Manufacturing Processes.* ***16*****(2)**, 265–279.

[7] Loh N L & Wu Y L 1993 Diffusion bonding of ceramics to metals *Materials and Manufacturing Processes.* ***8*****(2)**, 159–181

[8] Barmola A (2015) Establishing Recrystallization Temperature of Super martensitic Stainless Steel *Materials and Manufacturing process.* ***9(5)***, 234-240.

[9] Singh S & Nanda T 2013 Effect of Alloying and Heat Treatment on the Properties of Super Martensitic Stainless Steels *J. Materials and design.* ***10(2)***, 6–9.

[10] Taban E Dhooge A Kaluç E 2008 Plasma Arc Welding of Modified 12%Cr Stainless Steel *Materials and Manufacturing Processes.* **24**, 649–656.

[11] Liu Y R Ye, D Yong Q L Su J Zhao K Y & Jiang W 2011 Effect of heat treatment on microstructure and property of Cr13 super martensitic stainless steel *Journal of Iron and Steel Research International.* ***18*****(11)**, 60–66.

[12] Nanda T & Kumar B R 2014 Effect of Isothermal Annealing on Microstructural Morphology of Martensite in a Super Martensitic Stainless Steel Subjected to Different Prior Conditions *Materials and Manufacturing process.* **10(2)**, 37–41

[13] Taban E Kaluc E & Ojo O 2016 Properties, weldability, and corrosion behaviour of super martensitic stainless steels for on- and offshore applications. *Materials and Manufacturing process.* ***11(1)***, 32- 35

[14] Liu Y R Ye D Yong Q L Su J Zhao K Y & Jiang W 2011 Effect of heat treatment on microstructure and property of Cr13 super martensitic stainless steel *Journal of Iron and Steel Research International.* ***18*****(11)**, 60–66..

[15] Atabaki M M & Idris J 2012 Low-temperature partial transient liquid phase diffusion bonding of Al / Mg 2 Si metal matrix composite to AZ91D using Al-based interlayer *Materials and Design.* ***34*****,** 832–841.

[16] Balasubramanian M 2015 Application of Box – Behnken design for fabrication of titanium alloy and 304 stainless steel joints with silver interlayer by diffusion bonding *J. materials & design.* ***77***, 161–169.

[17] Aydın K Kaya Y & Kahraman N 2012 Experimental study of diffusion welding /bonding of titanium to copper *J. materials &design.* ***37*****,** 356–368.

[18] Zhang X Miyamoto G Toji Y Nambu S Koseki T & Furuhara T 2018 Orientation of austenite reverted from martensite in Fe-2Mn-15Si- *Acta Materialia.* ***144***, 601–612.

IOP Conf. Series: Materials Science and Engineering **390** (2018) 012036 doi:10.1088/1757-899X/390/1/012036

Analysis of Hole Taper, Recast Layer and Heat Affected zone in pulsed O_2 and N_2 laser drilling of Difficult-to-cut alloy Nimonic C-263

C.Ezhilarasan[1], A.Velayudham[2], M.Nagaraj[3], R.Anburaj4
[1]Professor, Department of Mechanical Engineering, SMK Fomra Institute of Technology, Kelambakkam, Chennai-602103.
[2]Scientist 'F', CVRDE, Avadi, Chennai-600054.
[3]Assistant Professor, Department of Mechanical Engineering, MNM Jain engineering College, Chennai-603103, India.
[4]IITM, Chennai-600032
Corresponding Author:ezhilshriram.c@gmail.com

Abstract. The quality and performance of the drilled component is of prime importance in the development of aero-engine components and are directly related to the quality of HAZ, Recast Layer and Hole taper. In order to maintain and/or improve the reliability of aerospace components, it is first essential to be aware of the possible damages that can be imparted to a material when it is drilled. Therefore, an attempt has been taken to investigate the effect of laser drilling parameters such as Gas Type: O2, N2; Power: 3200Watts, 3900Watts and Pressure: 4 bars, 5 bars, while laser drilling the Nimonic C-263 super alloy. Laser drilling trails have been conducted with different combinations of drilling parameters. The quality of the drilled holes such as Hole Tapper, HAZ and Recast Layer were evaluated quantitatively and analyzed through scanning electron microscope (SEM).
Key Words: Laser Drilling, N2 –Nitrogen, O2- Oxygen, Nimonic C-263, Hole Taper, Recast Layer and Heat Affected Zone (HAZ).

1.Introduction

Nickel based super alloys are finding wider applications in the development of critical components for high-performance aircraft gas turbine engines due to their superior mechanical properties that are maintained to elevated temperatures and the reliability is also most important in the manufacture of aerospace components and therefore, aerospace manufacturers need to maintain high-quality on a consistent basis. [1].The microstructure also affects the machinability of alloys in two ways: The presence of graphite or sulphide phases greatly improves machinability. Hard phases, such as carbides, nitrides, carbonitrides, oxides, silicates and possibly the γ' phase (Ni3 (Al, Ti)), are abrasive causing rapid tool wear. The super alloy or high-performance alloy exhibits excellent mechanical strength and creep resistance at high temperatures, good surface stability, and corrosion and oxidation resistance. Super alloys typically have a matrix with an austenitic face-centred cubic crystal structure.Their high temperature performance is distinguished in their tensile strength, hot hardness, creep resistance, and corrosion resistance at elevated temperatures are the important mechanical properties of interest [2].

Among the varieties of heat resistant materials, nickel based Nimonic C-263 super alloy finds wider use in the combustion portion of the gas turbine, heat exchangers, power generators and it is becoming important in the manufacturing of components such as casings, rings, seals etc. and some components are shown in the Fig.1, in which the drilling operations are used to make holes.

Laser drilling technique is widely used in the aerospace industry for manufacturing effusion-cooling holes on aerospace components as this technique is able to drill super alloys at acute angles [3]. Sanasam Sunderlal Singh et al [4] stated that the Laser- Induced Micro machining (LIMM) produced better machining efficiency than other non-conventional machining processes in terms of the machining rate, metal removal, surface morphology and capability to machine wide range of materials irrespective of its hardness and electrical conductivity. The micro scale mechanical drilling is difficult in the parts made of difficult to cut materials and also number of researchers said that the chisel edge contributes to the total thrust force [5]. Ahn et al [6] investigated the effects of machining parameters on thermal characteristics, kerfwidth and the heat transfer phenomenon in laser cutting of Inconel 718 alloy sheet using a CW Nd: YAG laser and the temperature distribution in the vicinity of the cut section has also been quantitatively evaluated. Venkata Rao et al [7] reviewed the optimization aspects of various parameters of the modern machining processes such as laser beam machining, electric discharge machining, abrasive jet machining, ultrasonic machining, electrochemical machining, , micro-machining, nano-finishing and they also stated that these machining processes are useful to do precise machining operations.

R. Biswas et al [8] said that the taper formation and production of non-circular holes are the characteristics of the laser drilling and the laser machining depends on the interaction of a laser beam with inherent focusing characteristics and they further said that the modeling and optimization of the process parameters are needed to control the laser drilling characteristics. Habib Alavi et al [9] experimentally investigated ultrasonic vibrations-assisted continuous wave laser drilling of materials. They stated that the drilling rate and the quality of laser drilled holes depend on the material removal mechanism. Sanjay Kumar et al [10] observed that the laser trepan drilling (LTD) produces better quality holes compared to laser percussion drilling (LPD) and they optimized multiple quality characteristics using computer-aided genetic algorithm-based multi objective optimization (CGAMO). Tatjana V. Sibalija et al [11] presented a hybrid design such as Quality loss function Grey relational analysis and Neural network Genetic algorithm for the determination of the optimum laser drilling parameters which simultaneously meets the requirements of quality characteristics of the holes produced during pulsed Nd:YAG laser drilling of a thin sheet of nickel-based super alloy Nimonic 263.

Bandyopadhyay et al [12] conducted an experiment to study the effect of the process variables such as pulse energy, pulse repetition rate, pulse duration, focal position, nozzle standoff, type of gas and gas pressure on the quality of the hole based on the Taguchi design and these process variables were optimized and they reported that the low level of pulse duration and pulse energy are the ideal conditions to obtain minimum taper-drilled hole. Ghoreishi et al [13] investigated the interaction effect of laser peak power, laser pulse width, pulse frequency, number of pulses, assist gas pressure and focal plane position on the hole taper and circularity in laser percussion drilling on stainless steel and mild steel based on the central composite design (CCD). They reported that the pulse frequency has significant on the hole entrance diameter and hole circularity in drilling stainless steel.

Sanjay Mishra and Vinod Yadava [14] developed a prediction model for laser beam percussion drilling (LBPD) using coupled methodology of FEM, ANN. They developed FEM based thermal models for LBPD by considering the temperature-dependent properties, optical properties and phase change of aluminum and the results obtained through FEM is validated and compared with experimental results for hole taper. Further, they developed ANN model to predict the hole taper and material removal rates by considering the process variables such as pulse width, pulse frequency, in which they found out that pulse width affects significantly the responses where as the pulse frequency affects significantly the HAZ. Naifei Ren et al [15] investigated the influence of laser parameters such as laser energy and pulse number on the quality of the drilled holes and they observed that the crater depth and the diameter are increased by increasing the number of laser pulses and laser energy.

Based on the literature, it was found there is no comprehensive study on the influence of laser drilling parameters in laser drilling of Nimonic C-263 alloy. In the present investigation, the laser drilling of Nimonic C-263 is performed by using O_2 and N_2 gases and the effects of process parameters such as power, pressure and type of gases on the responses such as Hole taper, Recast layer and Heat Affected Zone (HAZ) have been studied and analysed.

2. EXPERIMENTAL DETAILS AND PROCEDURE

Nimonic C-263 alloy of 75 mm diameter and 5mm thickness was used as work material for all the experiments. The chemical composition of the Nimonic C-263 alloy is such as:Ni52.49%, Si0.19%, Mn0.46%, Cr20%, Mo6.29%, Cu0.07%, Fe1.0%, co16.7%, Ti1.94%, Al0.48%,W0.15%,Va0.02%, C0.02%, S0.001% and Tan 0.007%.The microstructure of Nimonic C-263 alloy is shown in figure.1. In addition, the Nimonic C-263 alloy was identified by the Electro Dispersive Spectroscopy (EDS) analysis to contain Ni (higher content), Cr, Co, Ti, Mo, Al and V elements as shown figure.2. From the microstructure of this alloy, significant amounts of MC type carbides (enriched with Ti) within the grains are seen (as black dots) in the solution treated specimen of the nimonic C-263 alloy. The mechanical properties of the Nimonic C-263 alloy are given in Table 1. The experimental study was carried out by using TRU Laser 3040 machine with the following specification: Laser capacity of 4000Watts, Main supply of 400V/50HZ, connected load of 88 KVA, Control voltage of 24V, Preliminary fuse of 3x/160A, Air connection of 7-10 bar, Nitrogen supply of 30 bar, Oxygen supply of 20 bar and Machine No AO24OAO536. The experimental set-up is shown in figure. 3.

Laser drilling tests were carried by varying the machining parameters such as: Power; 3200Watts, and 3900Watts; Pressure: 4 bars, 5 bars and Gas Type: N2 and O2. The laser drilling performance was evaluated by the following responses such as Hole Taper angle, Heat Affected Zone (HAZ) and Recast layer. The laser drilled work piece at different level machining parameters is shown in figure .4.

Table 1 Mechanical Properties of Nimonic C-263 alloy

S.No	Properties	Units	Values
1	Yield strength	MPa	689
2	0.2 % Proof Stress	MPa	692
3	Ultimate strength	MPa	1060
4	Elongation	%	35
5	Reduction in area	%	28
5	Hardness	HV	320

3. RESULTS AND DISCUSSION

3.1 Influence of the machining parameters on the Hole Taper, HAZ and Recast Cast Layer

The hole taper angle of a drilled hole is one of the main characteristics of its quality since it affects geometrical characteristics and ultimately, the reliability of the product. The hole taper angle can be calculated from the Eq. (1).

Taper (deg) = $(d_{max}-d_{min})/2t \times 180/\pi$ ------------------------------------ (1)

Where, d_{max} is the entrance diameter of the hole, d_{min} is the exit diameter of the hole, t is the material thickness [13]. Hole Taper angles in laser drilling of Nimonic C-263 alloy at different level of power and pressure while O2 and N2 gases are illustrated in figure .5. (a-b).From the illustration it is observed, that, as the power is increased, the hole taper angle reduced at higher pressure level of 5 bar while using O2 gas, it is due to that, the effect of high pressure at hole exit diameter is higher than the hole entrance diameter and high power produces large entrance diameter. [3], however the Hole Taper angle is increased while drilling using N2 gases.

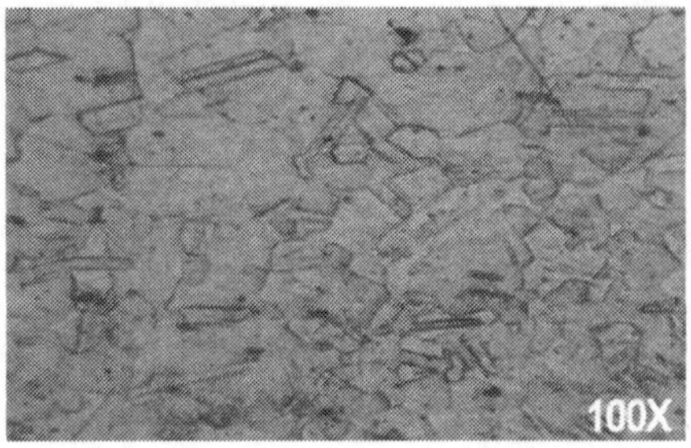

Figure .1.Microstructure of Nimonic C-263 Alloy

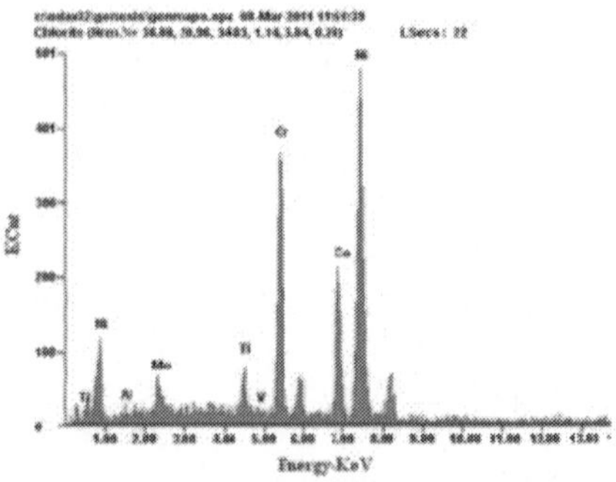

Figure.2.EDS Analysis of Nimonic C-263 Alloy

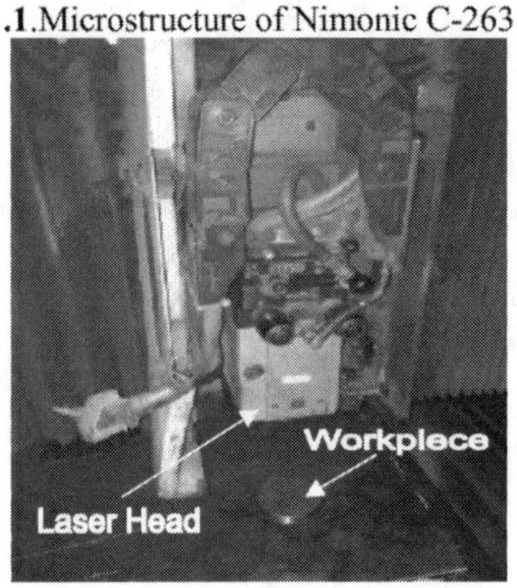

Figure. 3 .Experimental set up

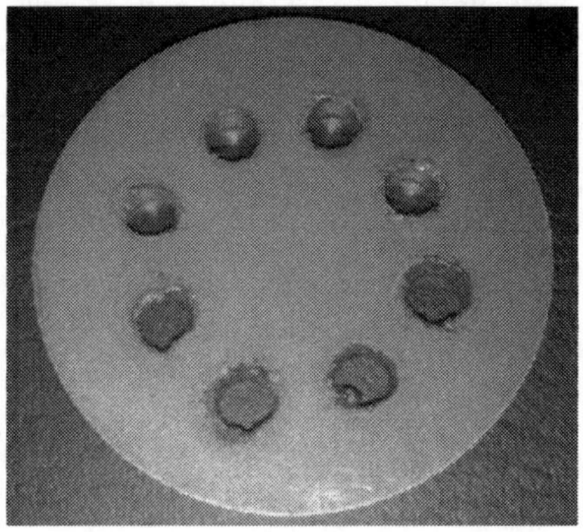

Figure.4. Laser Drilled Work piece at different level of Machining Parameters

Hole taper angle is increased at high power and low pressure in laser drilling of Nimonic C-263 alloy while using O2 and N2 gases. Hole taper angle observed at 3900 watts, 4 bars during laser drilling while using O2 gas is 1.57 times higher than the Hole Taper observed at 3200 watts, 4 bars by using N2 gas, which is shown in figure.5.(b).The hole taper angle observed at 3900 watts, 5 bars in laser drilling of Nimonic C-263 alloy while using N2 is 1.064 times higher than the hole taper angle observed at 3900 watts, 5 bars while using O2 gas and the hole taper angle observed at 3200 watts, 5 bars while using O2 gas is 1.14 times higher than the hole taper angle observed at 3200 watts, 5 bars while using N2 gas, which is shown in figure.5. (a).From the Fig.6 (b), it is further observed that the hole taper angle at 3900 watts, 4 bars during laser drilling of Nimonic C-263 alloy while using O2 is 1.50 times higher than the hole taper angle observed at 3900 watts, 4 bars while using N2 gas and the hole taper angle observed at 3200 watts, 4 bars while using O2 gas is 1.47 times higher than the hole taper angle observed at 3200 watts, 4 bars while using N2 gas.

From figure.5. (a-b), it is revealed that the parameters such as power, pressure and gas type significantly affect the Hole Taper angle. The quality and the performance of a drilled component are also directly related to the quality of HAZ. The hole quality was examined by analyzing the SEM image of the hole entrance and exit, after laser drilling at different levels of machining parameters and the affected zone were also measured quantitatively.

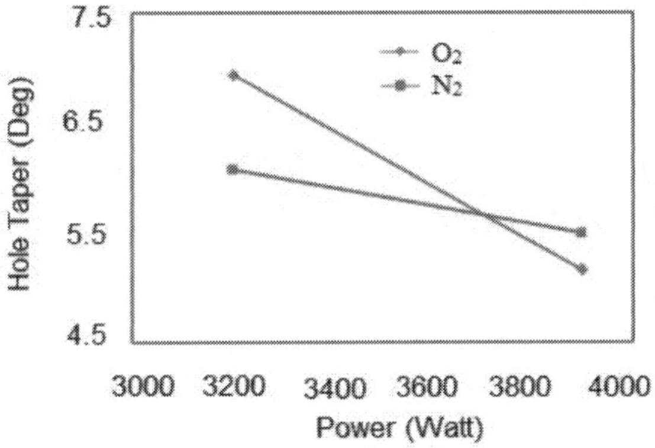

Figure.5. (a) Hole Taper for different level of Power at a Pressure of 5 bars.

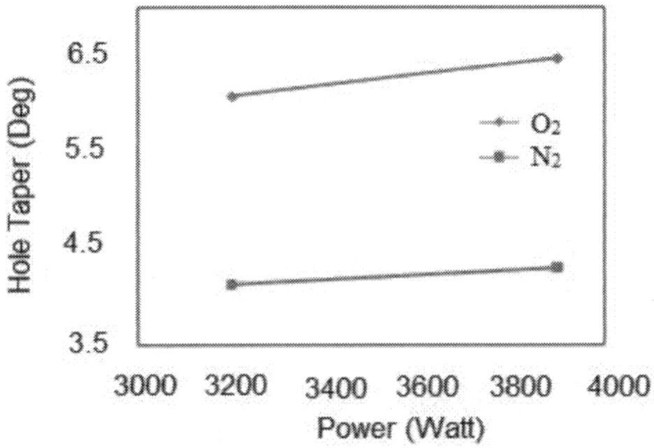

Figure.5. (b) Hole Taper for different level of Power at a Pressure of 4 bars.

Figure.5. (a-b) Hole Taper at different power and pressure While using O_2 and N_2 Gas

The Heat Affected Zone (HAZ) of laser assisted drilled hole is defined as a thin region on hole walls in which microstructure and properties are altered by laser drilling and HAZ determines the region with which the micro structure properties of the work piece is altered because of the thermal energy of the laser beam. [3]. The defects such as HAZ, Recast layer and spatter limit, the application of laser drilling and elimination of these defects are important to carry out precision drilling of variety of components. [16].HAZ in laser drilling of Nimonic C-263 alloy at different levels of power and pressure using O2 and N2 gases is illustrated in figure.6. (a-b).From the illustration it is observed that as the power is increased, the HAZ reduced at higher pressure level of 5 bars when O2 gases is used, however the HAZ is increased while laser drilling by using N2 gases. The HAZ is reduced when the power is increased from 3200 Watts to 3900 Watts, while drilling using O2 gas at 4 bars, whereas the HAZ is increased when the power is increased from 3200 Watts to 3900 Watts, while laser drilling by using N2 gas at 4 bars, which is 1.0923 times higher than the HAZ observed at 3900 watts, 4 bars while using O2 gas, which is shown in figure.6. (b), it is due to hot molten metal flow [13], it is eventually shown in the Fig.9. (h), in which larger HAZ is observed in the drilled hole at exit. From figure.6. (a-b) revealed that the parameters such as power, pressure and gas type significantly affect the HAZ.

HAZ observed at 3900 watts, 5 bars while laser drilling of Nimonic C-263 alloy using N2 is 1.20 times higher than HAZ observed at 3900 watts, 5 bars while using O2 gas and the HAZ observed at 3200 watts, 5 bars while using N2 gas is 1.071 times higher than the HAZ observed at 3200 watts, 5 bars while using O2 gas, which is shown in figure.6. (a). From the figure.6 (b), It is observed that the HAZ at 3900 watts, 4 bars while laser drilling of Nimonic C-263 alloy using N2 is 1.15 times higher than the HAZ observed at 3900 watts, 4 bars while using O2 gas and the HAZ observed at 3200 watts, 4 bars while using O2 gas is 1.06 times higher than the HAZ observed at 3200 watts, 4 bars while using N2 gas. From figure.6. (a-b), it is revealed that the parameters such as power, pressure and gas type significantly affect the HAZ.

The hole quality is also related to the formation of Recast layer in the drilled hole and is examined by analyzing the SEM image of the hole entrance and exit after laser drilling at different levels of machining parameters and the Recast layer were also measured quantitately.The Recast layer in laser drilling of Nimonic C-263 alloy at different levels of power and pressure using O2 and N2 gases is illustrated in figure.7. (a-b).From the illustration it is observed, that as the power is increased, the Recast layer is increased at higher pressure level of 5 bars when N2 and O2 gases are used, however higher Recast layer is observed at 3900watts, 4 bars while drilling using N2 gas which 1.68 times higher than the Recast Layer observed at 3200 watt, 4 bars using N2 gas.

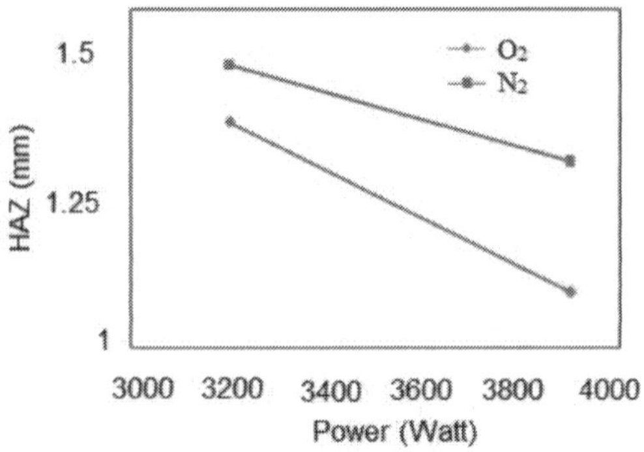

Figure.6. (a) HAZ for different level of Power at a Pressure of 5 bars.

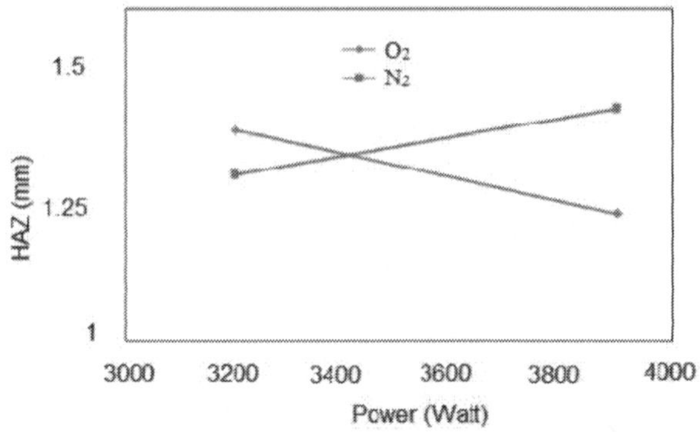

Figure.6. (b) HAZ for different level of Power at a Pressure of 4 bars.

Figure.6. (a-b) HAZ at different power and pressure While using O_2 and N_2 Gas

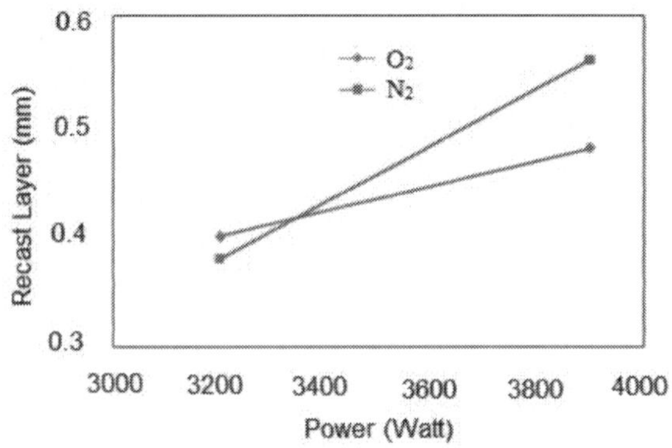

Figure.7. (a) Recast Layer for different level of Power at a Pressure of 5 bars.

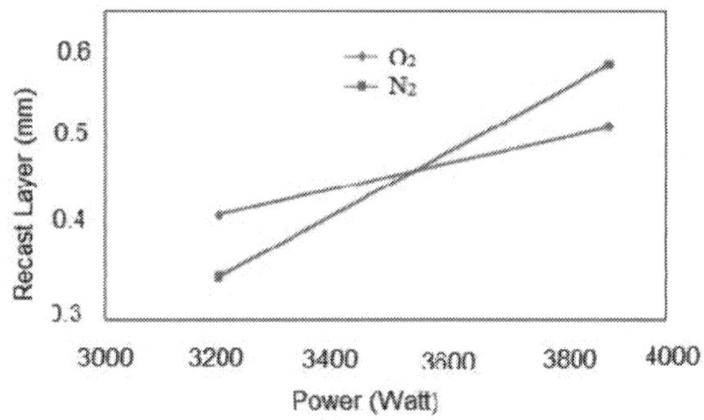

Figure.7. (b) Recast Layer for different level of Power at a Pressure of 4 bars.

Figure.7. (a-b) Recast Layer at different power and pressure While Using O_2 and N_2 Gas

The Recast layer observed at 3900 watts, 5 bars while laser drilling of Nimonic C-263 alloy by using N2 is 1.164 times higher than the Recast layer observed at 3900 watts, 5 bars while using O2 gas and the Recast layer observed at 3200 watts, 5 bars while using O2 gas is 1.05 times higher than the Recast layer observed at 3200 watts, 5 bars while using N2 gas, which is shown in figure.7. (a).From the figure.7 (b), it is observed that the Recast layer at 3900 watts, 4 bars while laser drilling of Nimonic C-263 alloy by using N2 is 1.134 times higher than the Recast layer observed at 3900 watts, 4 bars while using O2 gas and the Recast layer observed at 3200 watts, 4 bars while using O2 gas is 1.2 times higher than the Recast layer observed at 3200 watts, 4 bars while using N2 gas. From figure.7. (a-b), it is revealed that the parameters such as power, pressure and gas type significantly affect the Recast layer.

Figure.8. (a-p) shows the SEM images of laser drilled holes while using O2 and N2 gases at different power and pressure. The damages such as recast layer, HAZ, oxide layer and spatter are all observed. Figure.8. (a-p) revealed that the HAZ is observed mostly at the drilled hole at exit, however the HAZ is also observed at the hole entry which is shown in Figure 3.1.4 (k,m,o) and these HAZ are lower than the HAZ observed at exit drilled hole. The Recast layer is observed in the laser drilled hole at both entry and exit of the drilled hole for all cutting conditions, which is eventually shown in figure.8. (a-p), however the higher Recast layer is observed in figure.8. (a-h).The oxide layer is observed in the drilled hole at exit only while drilling this Nimonic C-263 alloy which is eventually shown in the figure.8 (f, j, l, n, p),however higher oxide layer is observed at 3200 watts while drilling at 4 and 5 bars with respect to O2 and N2 gases , which is eventually shown in figure.8. (J, l, n) and oxide layer is not observed at hole entry. The distinct oxide inclusions were also observed at 3200 watts, 5 bars while using O2 gas which is shown in figure.8.(j). It can also be observed that the laser assisted drilling exhibit spatter is shown in figure.8.(f, g, n).

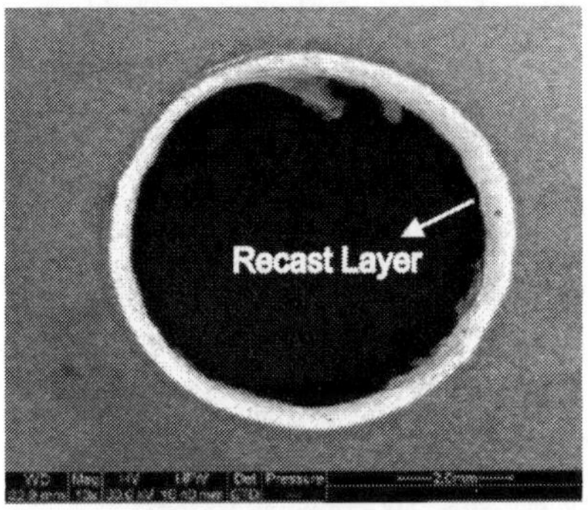

a. Laser drilled hole at entry at 3900 Watt,5 bars using O_2

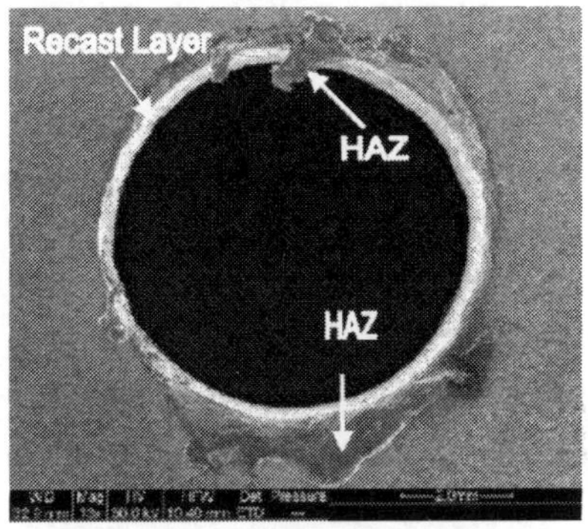

b. Laser drilled hole at exit at 3900 Watt, 5 bars using O_2

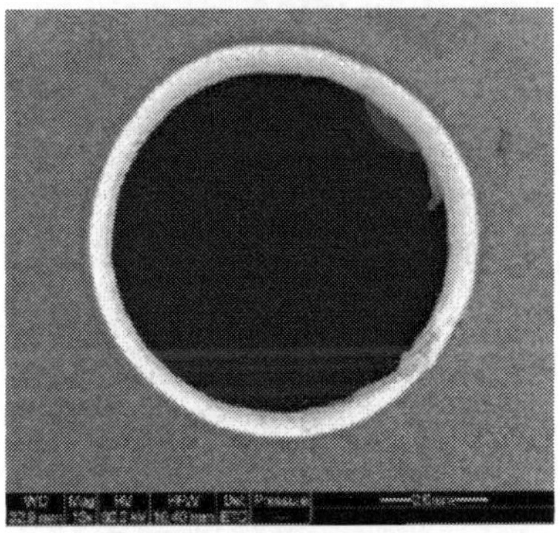

c. Laser drilled hole at entry at 3900 Watt, 4 bars using O_2 gas

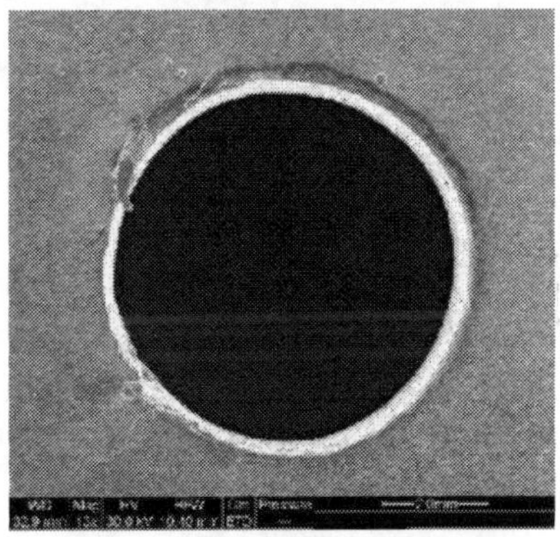

d. Laser drilled hole at exit at 3900 Watt, 4 bars using O_2 gas

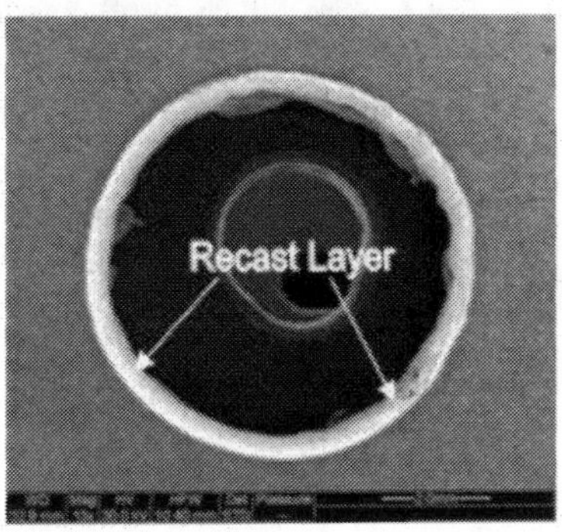

e. Laser drilled hole at entry at 3900 Watt, 5 bars using N2

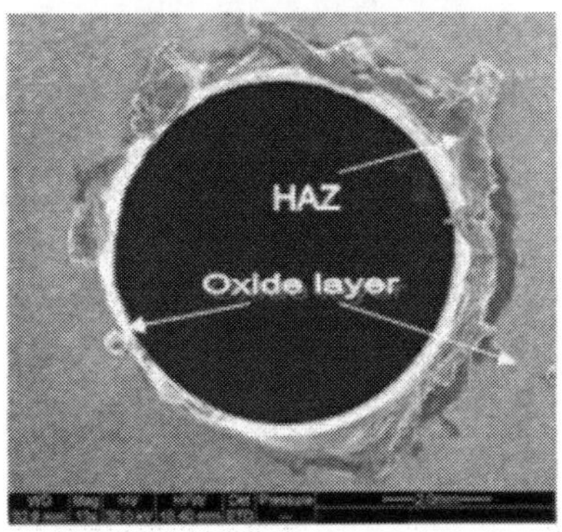

f. Laser drilled hole at exit at 3900 Watt, 5 bars using N2

IOP Conf. Series: Materials Science and Engineering **390** (2018) 012036 doi:10.1088/1757-899X/390/1/012036

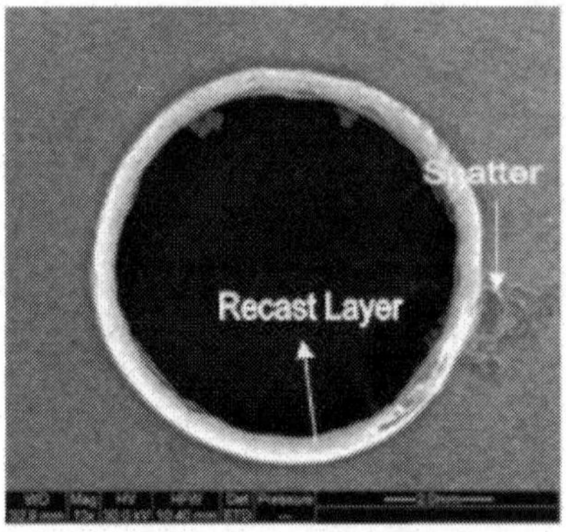

g. Laser drilled hole at entry at 3900 Watt, 4 bars using N2

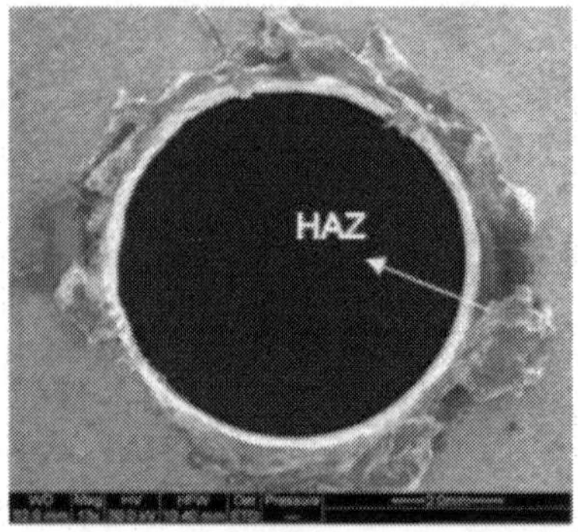

h. Laser drilled hole at exit at 3900 Watt, 4 bars using N2

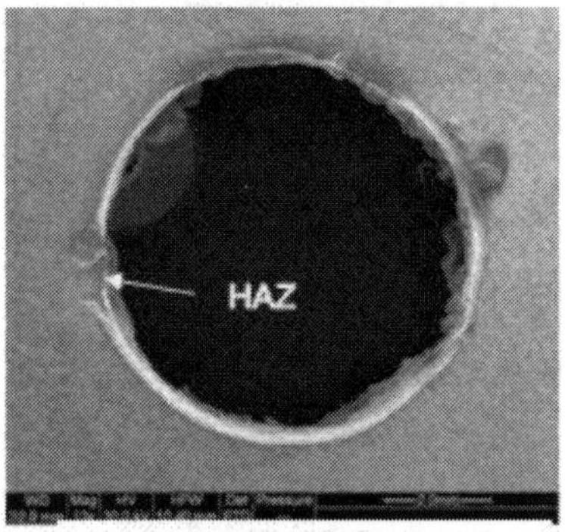

m. Laser drilled hole at entry at 3200 Watt, 5 bars using N2

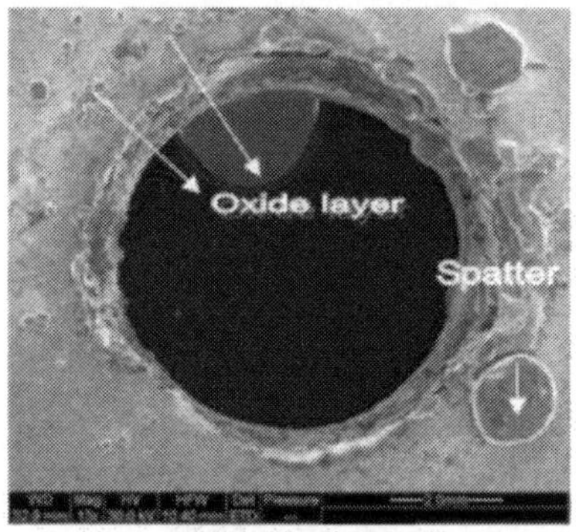

n. Laser drilled hole at exit at 3200 Watt, 5 bars using N2

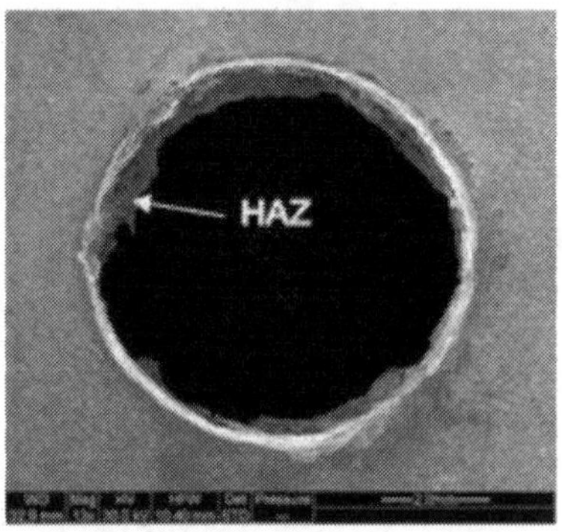

o. Laser drilled hole at entry at 3200 Watt, 4 bars using N2

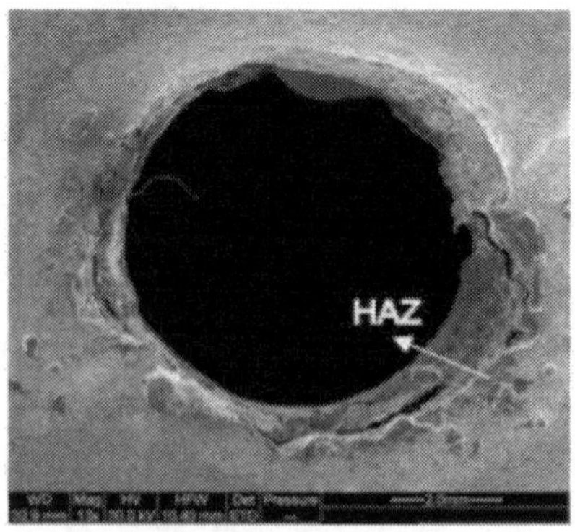

p. Laser drilled hole at exit at 3200 Watt, 4 bars using N2

Figure.8. (a-p) Typical SEM Images of Laser drilled hole at different power and pressure using N_2 and O_2 gases

Figure.9. shows the experimental results of HAZ, Recast layer of the drilled hole at different cutting conditions with respect to N2 and O2 gases and the corresponding typical SEM images of the drilled holes are also shown. It is observed from these figure.9, that the thickness of the HAZ and Recast layer varied when cutting conditions varied. An increase in HAZ and Recast layer occurs while drilling Nimonic C-263 when the power is reduced from 3900watts to 3200 watts.

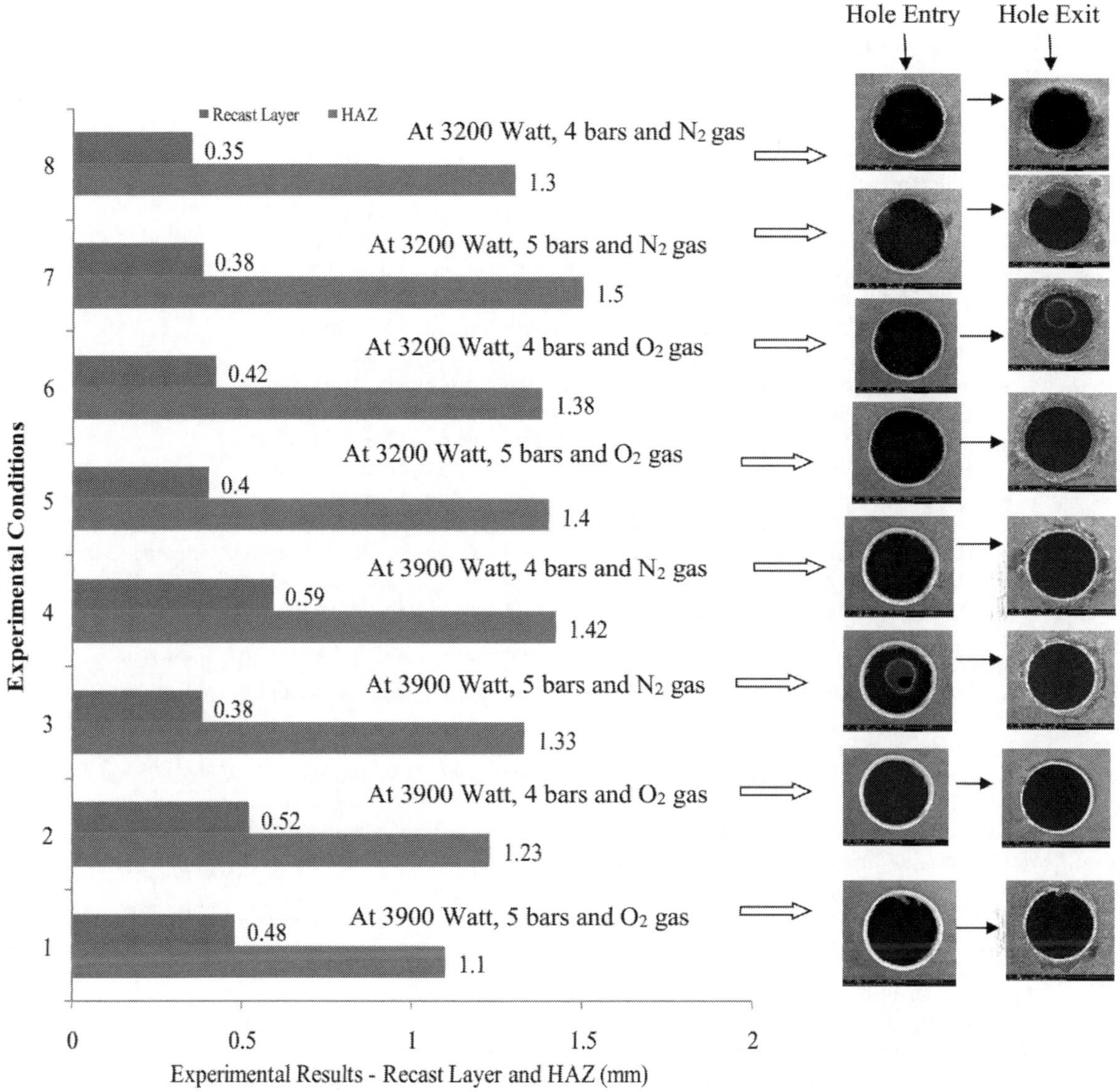

Figure.9. Experimental Results of HAZ, Recast Layer and Typical SEM Images of the drilled hole at different machining conditions.

4. Conclusion

The laser drilling of a difficult-to-cut alloy Nimonic C-263 using O2 and N2 gases has been performed and the effect of input parameters such as Pressure, power on Tapper angle, HAZ, and Recast layer were determined and the following conclusions were drawn.

- Higher Hole Tapper angle is observed at 3900 watt, 4 bars while drilling using O2 gas, which is 1.57 times higher than the Hole Tapper angle observed at 3200 watt, 4 bars using N2 gas.
- More cylindrical hole with less Hole Tapper angle of 4.12 is observed at 3200 Watt and 4 bars while drilling using N2 gas.
- Higher HAZ is observed at 3900 Watt and 4 bars while drilling using N2 gas, which is 1.0923 times higher than the HAZ observed at 3900 watt, 4 bars using O2 gas.
- Higher Recast layer is observed at 3900 watt, 4 bars while drilling using N2 gas which 1.68 times higher than the Recast Layer observed at 3200 watt, 4 bars using N2 gas.
- The effect of laser parameters on the responses are observed as follows:

- When the power reduces from 3900 watts to 3200 watts at 5 bars and O2:
  Hole Tapper angle – Increases, HAZ-Increases, Recast Layer-Reduces
- When the power reduces from 3900 watts to 3200 watts at 5 bars and N2:
  Hole Tapper angle – Increases, HAZ-Increases, Recast Layer-Reduces.
- When the power reduces from 3900 watts to 3200 watts at 4 bars and O2:
  Hole Tapper angle – Reduces, HAZ-Increases, Recast Layer-Reduces.
- When the power reduces from 3900 watts to 3200 watts at 4 bars and N2:
  Hole Tapper angle – Reduces, HAZ- Reduces, Recast Layer-Reduces.
- When the power increases from 3200 watts to 3900 watts at 5 bars and O2:
  Hole Tapper angle – Reduces, HAZ-Increases, Recast Layer- Increases
- When the power increases from 3200 watts to 3900 watts at 5 bars and N2:
  Hole Tapper angle – Reduces, HAZ- Reduces, Recast Layer- Increases.
- When the power increases from 3200 watts to 3900 watts at 4 bars and O2:
  Hole Tapper angle – Increases, HAZ- Reduces, Recast Layer- Increases.
- When the power increases from 3200 watts to 3900 watts at 4 bars and N2:
  Hole Tapper angle – Increases, HAZ- Increases, Recast Layer- Increases.

➢ The oxide layer is observed in the drilled hole at exit only while drilling this Nimonic C-263 alloy.

➢ The damages such as Recast Layer, HAZ, Oxide Layer and Spatter were observed.

References

[1].A. Thakur, S. Gangopadhyay, International Journal of Machine Tools and Manufacture, 100(2016) 25–54,

[2] Ezilarasan, C., Senthil Kumar, V.S. and Velayudham, A,Measurement, 46 (2013)185-199,

[3]. Sezer.H.K, Li.L, Schmidt.M, Pinkerton.A.J, Anderson.B, Williams.P, International Journal of Machine Tools & Manufacture.46 (2006)1972-1982.

[4] Sanasam Sunderlal Singh,S. N. Joshi , Alika Khare, Laser Based Manufacturing. (2015)179-200.

[5]. Okasha.M.M, Mativenga.P.T, Li.L,Journal of Manufacturing Science and Engineering. 133(2011) 011008-8.

[6]. D.G. Ahn, K.W. Byun, M.C. Kan,. Journal of Materials Science & Technology 26(2010) 362-366.

[7]. Venkata Rao .R, Kalyankar.V.D,International Journal of Advanced Manufacturing Technology 73 (2014)1159-1188.

[8]. Biswas.R, Kuar.A.S, Sarkar.S, Mitra.S,Optics & laser Technology 42 (2010) 23-31.

[9]. Habib Alavi.S, SandiP, Harimkar.P, Research Letters 4(2015)1-5.

[10].Sanjay Kumar, Avanish Kumar Dubey , Arun Kumar Pandey,.International Journal of Advanced Manufacturing Technology 14 (2013)1119-1125.

[11]. Tatjana V. Sibalija , Sanja Z. Petronic, Vidosav D. Majstorovic, Radica Prokic-Cvetkovic, Andjelka Milosavljevic,Journal of Advanced Manufacturing Technology.54 (2011)537-552.

[12].Bandyopadhyay.S,Hina Gokhale, Sarin Sundar.J.K, Sundarjan.G, Joshi.S.V,Optics and lasers in Engineering. 43(2005)163-182.

[13]. Ghoreishi.M, Low.D.K.Y, Li.L. International Journal of Machine Tools & Manufacture .42(2002)985-995.

[14]. Sanjay Mishra, Vinod Yadava,Optics & laser Technology. 48(2013)461-474.

[15]. Naifei Ren ,Leilei Jiang,Dan Liu,Liu Lv,Quan Wang,International Journal of Advanced Manufacturing Technology.76 (2015)735-743.

[16]. Sumanta Panda, Debadutta Mishra , Bibhuti B. Biswal, International Journal of Advanced Manufacturing Technology.54 (2011) 957-967.

IOP Conf. Series: Materials Science and Engineering **390** (2018) 012037 doi:10.1088/1757-899X/390/1/012037

Effect of circular web openings on dynamic behaviour of cantilever steel thin-walled tapered I - beams

K. S. Vivek[1*], M. Jugal Kishore[2], K.V.S. Manoj[3] and K.S.V.S Pujitha[4]
[1]Asst. Professor, Department of Civil Engineering, VVIT, Guntur-522508, A.P., India
[2] Asst. Professor, Department of Civil Engineering, KHIT, Guntur-522019, A.P., India
[3] P.G. Student, Department of Civil Engineering, VVIT, Guntur-522508, A.P., India
[4] U.G. Student, Department of Civil Engineering, KHIT, Guntur-522019, A.P., India

*Corresponding Author, Email ID: vivek3393@yahoo.com

Abstract. In this study, free and harmonic vibration analyses of cantilever steel thin-walled tapered beams without and with web openings are performed. The analyses are carried out based on finite element method with the help of ANSYS. SHELL 63 element is used for the analyses. Tapered I - beams with symmetric and un-symmetric cross-sections are considered. Initially, free vibration analysis is performed for tapered beams without and with circular web openings to find out the effect of web openings on natural frequencies. The free vibration analysis is then followed by the harmonic analysis. The peak amplitudes and corresponding frequency ranges are presented for the considered cases. Web tapered beam of symmetric cross-section with circular web openings resulted in maximum fundamental natural frequency on performing free vibration (modal) analysis and resulted in peak amplitude at higher frequency on performing harmonic analysis among the considered cases of beams with and without web openings. Thus web openings and type of cross – section affect the dynamic behaviour.

1. Introduction

Steel tapered beams are commonly used in modern era of steel structures. They are economical when compared with the prismatic members. Recent advancement in engineering design enabled the structural designer to provide web openings in the tapered beams for further increase in economy and to meet the functional requirements. These beams are quite commonly used for roofing in stadiums and other metal structures. The presence of web openings may influence the dynamic behaviour and is worth to be studied.

Lot of research on dynamic behaviour of non – prismatic / non – uniform beams was carried out by previous researchers [1-6]. Improved formulation for free vibration of thin – walled tapered beams was proposed by researchers Kim and Kim [7]. Free vibration analysis of tapered I-beams for various tapered ratios and boundary conditions using power series method and ANSYS was performed and the results were presented by Soltani and et al.[8]. From the brief literature review, it is evident that the effect of web openings on dynamic behaviour of tapered beams is worth to be studied. In this paper, the effect of circular web openings on free and harmonic vibration of tapered beams is discussed.

IOP Conf. Series: Materials Science and Engineering **390** (2018) 012037 doi:10.1088/1757-899X/390/1/012037

2. Validation

Free vibration analyses of tapered steel beams without web openings were previously performed [8]. For validation, three cases of cantilever tapered beams are modelled and again free vibration analysis is performed with the help of ANSYS. The results obtained from present analysis are presented in the Table 1, along with results of the previous study [8]. SHELL 63 element (four noded element) is used for analysis. The material is modelled as linearly elastic isotropic. Young's modulus (E) = 210 GPa; Poisson's ratio (υ) = 0.3; Density = 7800 kg/m^3. Block Lanczos method is used for finding out the natural frequencies. The obtained results from present study are in good agreement with the previous values of previous study [8].

Table 1. Natural frequencies ω (rad/sec)

Data case [8]		Mode	Previous study[8]	Present study
1	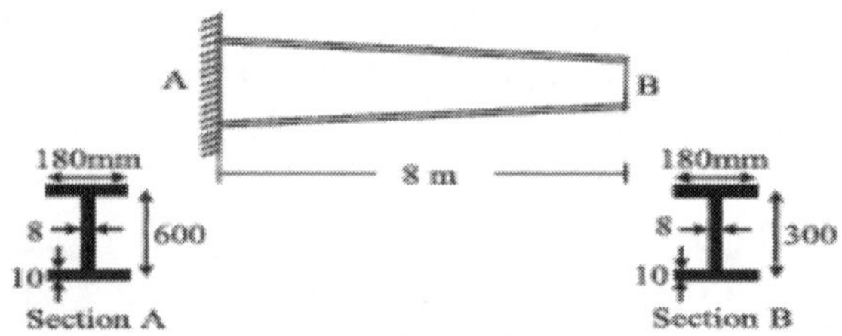	1	11.06	11.0654
		2	33.98	33.5240
		3	65.001	66.6143
2	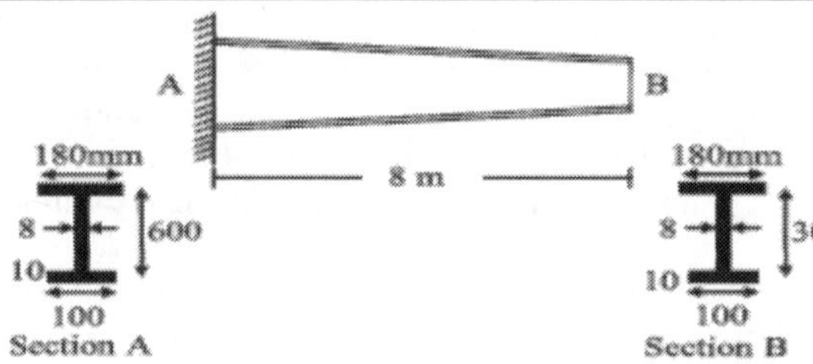	1	8.86	8.8499
		2	29.41	29.0742
		3	50.45	50.1769
3	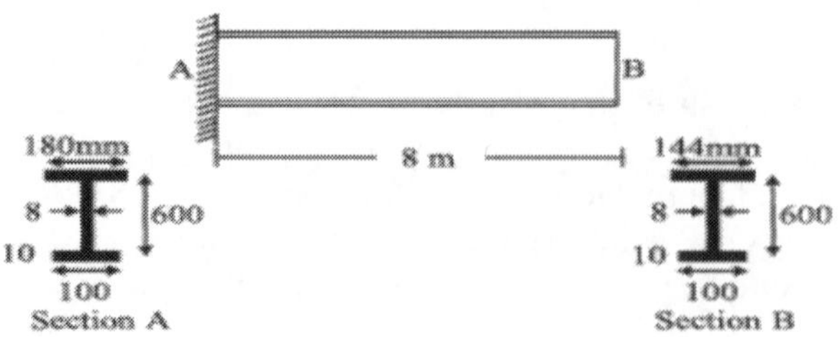	1	7.14	7.1264
		2	19.93	19.8165
		3	37.31	38.5876

3. Methodology

Circular openings are provided for the beams considered in above cases. The openings are provided such that (i) Web opening is away from the support by twice the beam depth, (ii) Clear spacing between openings greater than depth of beam and (iii) The diameter of circular openings is limited to 0.5 times the depth of beam [9].

As one end of the beam is free (cantilever beam), shear force and moment are zero at the free end for uniformly distributed / concentrated load. Hence near the free end, a circular opening is provided. From supported end (fixed end at A, Figure 1), the distance of first opening is 1.2 m for beam. Diameter of circular opening is taken as 0.125m and spacing between the (centre to centre distance) adjacent holes is taken as 0.8m. For 8 m length beam, a total number of nine web openings are provided. Modelled image of case 1 beam is shown in Figure 1. Likewise case 2 and case 3 beams are modelled. Linear elastic material behaviour of steel is considered. Modulus of Elasticity (E) is taken as 210 GPa. Poisson's ratio (υ) = 0.3. Density is assumed to be 7800 kg/m^3. SHELL 181 which is a four noded isoparametric element is considered for proper meshing near the holes. The element has six degrees of freedom (translations along X, Y and Z axes along with rotations about X, Y and Z axes at each node). The element is well suited for both linear and non-linear analyses. Fine meshing is provided with a total of around 705 elements for analysing the tapered steel beam with circular web openings. Figure 1 also depicts the image of meshing. Beams with and without circular cut outs corresponding to data cases 1 – 3 (Table 1) are analysed.

Block Lanczos method is used for free vibration analysis to find out the natural frequencies. For performing harmonic analysis, a vertical point load of -10 kN is applied at top tip of web and flange near free end as shown in Figure 2. Sparse solver method is chosen for solving the dynamic equations. The harmonic frequency is taken as 1 to 800 Hz. The load is divided into 50 sub steps. Stepped condition is assumed. The beams without and with circular web openings are analyzed.

Figure 1. Model of case 1 beam in ANSYS

Figure 2. Application of point load 10 kN at free end

4. Results and Discussion

4.1 Free Vibration Analysis

Natural frequencies of for the first six modes are presented in the Table 2 for beams of cases 1-3 with and without circular cut out. The mode shapes are same for beams of all cases considered. Mode shapes 1-6 are depicted in Figures 3 – 8. Flexural-torsional vibration of the beams occurred for mode shapes 2, 3, 5 and 6. Flexural vibration of the beam occurred for mode shapes 1 and 4.

Table 2. Natural Frequencies ω (rad/sec)

Data case	Mode	Without opening	With opening
1	1	11.0654	11.4612
	2	33.5240	33.0759
	3	66.6143	68.7820
	4	66.8405	68.9203
	5	109.7987	108.7808
	6	180.9369	186.3530
2	1	8.8499	9.2086
	2	29.0742	28.7783
	3	50.1769	51.5083
	4	64.8299	67.0981
	5	94.9326	94.9955
	6	121.4288	123.2447
3	1	7.1264	7.3520
	2	19.8165	19.4848
	3	38.5876	39.5231
	4	64.7105	66.4824
	5	77.4340	77.0633
	6	96.9190	98.8282

From Table 2 and Figures 3 – 8, the following inferences may be drawn

- Fundamental natural frequency of the web tapered beam of symmetric cross – section (case 1) along with circular web openings is greater than the fundamental natural frequency of the web tapered beam of symmetric cross – section without circular web openings by 3.45%.
- Fundamental frequency of the web tapered beam of un- symmetric cross – section (case 2) along with circular web openings is greater than fundamental frequency of the web tapered beam of un- symmetric cross – section without circular web openings by 4.05%.
- Fundamental frequency of the flange tapered beam of un-symmetric cross –section (case 3) with circular web openings is greater than fundamental frequency of the flange tapered beam of un-symmetric cross –section without circular web openings by 3.166%.
- Web tapered beam of symmetric cross-section (case 1) resulted in higher fundamental natural frequency when compared with web tapered beam of un-symmetric cross-section and flange tapered beam of un-symmetric cross-section either with / without web openings.

Thus, presence of circular web openings resulted in increase of fundamental natural frequencies for web / flange tapered beam with symmetric / un-symmetric cross sections and also among the considered cases, web tapered beam of symmetric cross-section resulted in higher fundamental frequency either for with / without circular web openings. Hence, type of cross-section also affects the dynamic behaviour.

Figure 3. Mode shape 1

Figure 4. Mode shape 2

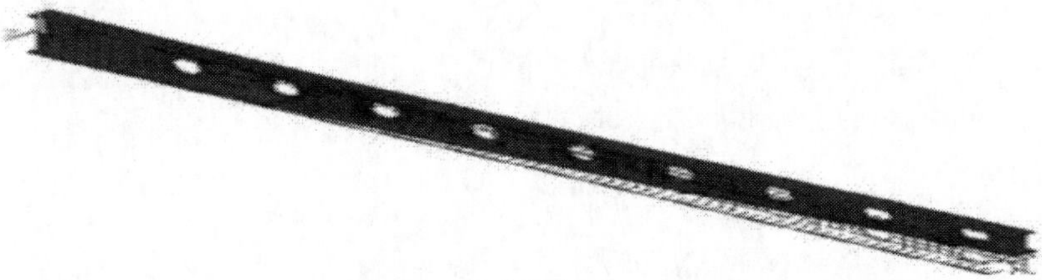

Figure 5. Mode shape 3

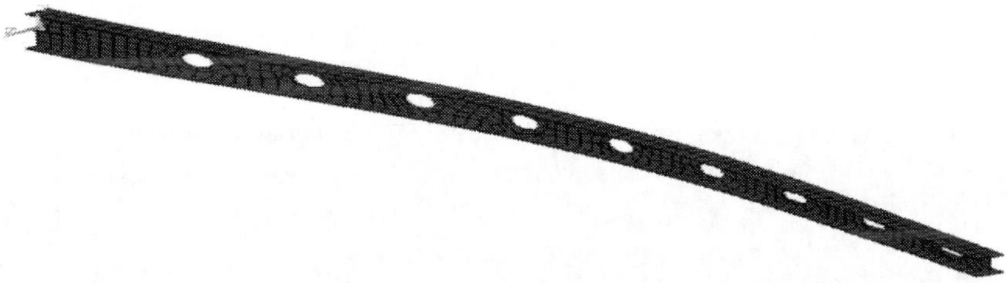

Figure 6. Mode shape 4

Figure 7. Mode shape 5

Figure 8. Mode shape 6

4.2. Harmonic Vibration Analysis

Plot between harmonic frequency and amplitude – UY (VALU) at the top junction of web and flange is plotted for data cases 1-3 with web openings as shown in Figures 9 - 11.

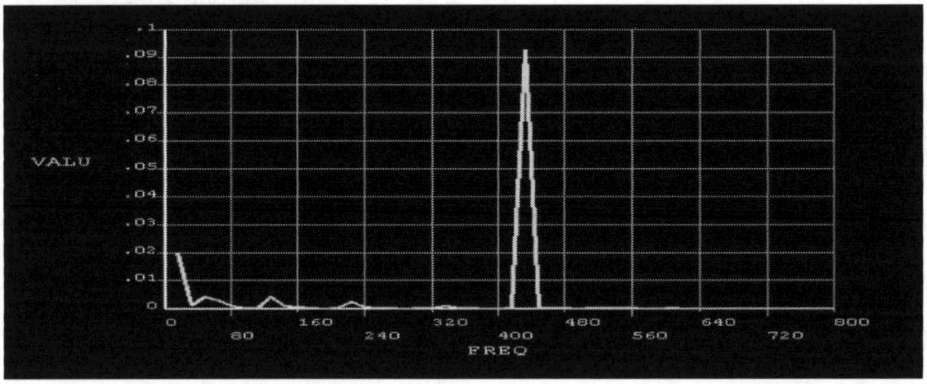

Figure 9. Harmonic frequency (FREQ) vs Amplitude UY for case 1 beam with web opening

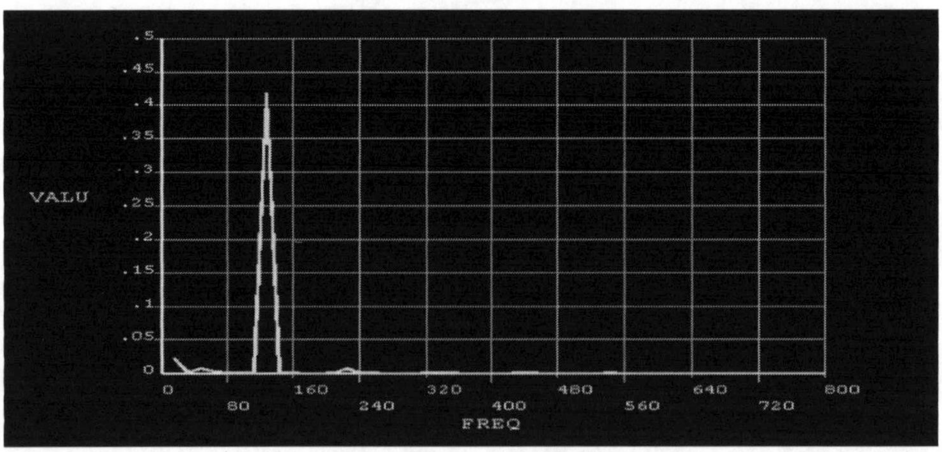

Figure 10. Harmonic frequency (FREQ) vs Amplitude UY for case 2 beam with web opening

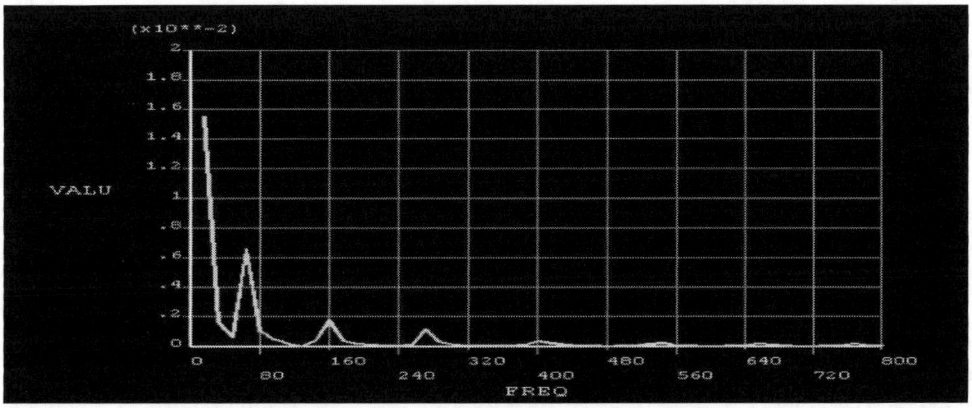

Figure 11. Harmonic frequency (FREQ) vs Amplitude UY for case 2 beam with web opening

Table 3. Peak Amplitude UY vs Frequency Range

Data Case	Peak Amplitude UY (m)		Frequency Range (Hz)	
	Without Web Opening	With Web Opening	Without Web Opening	With Web Opening
1	$3.4x10^{-2}$ (avg.)	0.095 (avg.)	320-400	400-480
2	$1.9x10^{-2}$ (avg.)	0.425 (avg.)	0-80	80-160
3	$1.4x10^{-2}$	$1.45x10^{-2}$ (avg.)	0-80	0-80

From Figures 9-11 and Table 3, it can be observed that beams with web openings resulted in higher amplitudes at greater frequency ranges for beams of cases 1 and 2. For case 3 beam there is not much difference in the dynamic behvaiour when subjected to harmonic loading.

5. Conclusions

Dynamic analyses of cantilever steel tapered I beam with and without circular web openings are performed with the help of ANSYS based on finite element method. SHELL 181 (4 noded isoparametric element) is considered for meshing. Linear – Elastic material behaviour of steel is assumed. Both symmetric and un-symmetric sections are considered. The obtained natural frequencies and mode shapes are presented. From this study it is evident that the provision of circular web openings resulted in increase of fundamental natural frequencies. The main inferences from the study are as follows

1) Presence of circular web openings resulted in increase of fundamental natural frequencies for web / flange tapered beam with symmetric / un-symmetric cross sections.
2) Web tapered beam of symmetric cross-section (Case 1) resulted in higher fundamental frequency for either with / without circular web openings when compared with the web tapered beam of un-symmetric cross-section (case 2) and flange tapered un-symmetric cross-section (case 3) .
3) Web tapered beam of symmetric cross-section (Case 1) resulted in higher amplitude at greater harmonic frequency for either with / without circular web openings when compared with the web tapered beam of un-symmetric cross-section (case 2) and flange tapered un-symmetric cross-section (case 3).

References

[1] Klein, L. (1974). Transverse vibration of non-uniform beams. *Journal of Sound and Vibration,* 37(4), 491- 505.

[2] Goel, R.P. (1976). Transverse vibrations of tapered beams. *Journal of Sound and Vibration,* 47(1), 1-7.

[3] Downs, B. (1978). Reference frequencies for the validation of numerical solutions of transverse vibrations of non-uniform beams. *Journal of Sound and Vibration,* 61(1), 71-78.

[4] Laura, P. A. A. and Valerga De Greco, B. (1988). Numerical experiments on free and forced vibrations of beams of non-uniform cross section. *Journal of Sound and Vibration,* 120(3), 587-596.

[5] Lee, S. Y., Ke, H. Y. and Kuo, Y.H. (1990). Analysis of non-uniform beam vibration. *Journal of Sound and Vibration,* 142 (1), 15-29.

[6] Zhou,D. and Cheung, Y.K. (2000). The free vibration of a type of tapered beams. *Comput. Methods Appl. Mech. Engrg.* 188, 203-216.

[7] Kim, S.B. and Kim, M.Y.(2000). Improved formulation for spatial stability and free vibration of thin-walled tapered beams and space frames. *Engineering Structures*, 22, 446–458.

[8] Soltani, M., Asgarian.B. and Mohri.F. (2014). Elastic instability and free vibration analyses of tapered thin-walled beams by power series method. *Journal of constructional steel research*, 96, 106-126.

[9] Subramanian, N. (2016). *Design of steel structures*. Oxford university press- India.

A study on Thermodynamic, Acoustic and Elastic properties of aqueous solutions of Hydroxyethyl Cellulose

R Padmanaban, A Gayathri and K Venkatramanan

Faculty of Science, Sri Chandrasekharendra Saraswathi Viswa Mahavidyalaya, Enathur, Kanchipuram – 631561, India

E-mail :kv.scsvmv@gmail.com

Abstract. Hydroxyethyl cellulose is a thickening and gelling derived from cellulose. It is broadly used in household products, cosmetics and cleaning solutions. Hydroxyethyl cellulose is frequently used with hydrophobic drugs in capsule formulations, to improve the drugs dissolution in the gastrointestinal fluids. Ultrasonic technique is one of the basic non-destructive methods for evaluation of materials and structures. A significant part of every ultrasonic inspection is the way in which the ultrasonic energy is transferred between the transducer and the tested object. In the present investigation, we have tried to compute the activation energy, ultrasound investigations, elastic property (bulk modulus) and molecular interaction properties of aqueous solutions of Hydroxyethyl cellulose at different concentrations (0.05%, 0.10%, 0.15%, 0.20%, 0.25%, 0.30%, 0.35%, 0.40%, 0.45% and 5%) for different temperatures (303K, 308K, 313K, 318K) by determining the relative viscosity.

1. Introduction

Ultrasonic technique has been engaged as commendable tool for non-destructive testing and imaging. Although the relationship between material properties and acoustic parameters has been studied for a long period of time [1-5].In the recent years, the usage of ultrasound sound waves are escalated, because of the advantage it offers to synthetic chemistry. This includes time consumption in reaction, saving energy, higher yield, milder conditions and higher purity of the product. For analyzing the molecular interaction properties, ultrasonic method has the added advantage of being less costly with efficiency comparable to other methods [6-8]. The molecular interaction parameters such as free volume, internal pressure, attenuation, compressibility, acoustic impedance and scattering provide us a clear picture of structural changes in cellulose derivate polymers and biological substances [9].When ultrasound waves are propagated in polymeric materials, acoustic waves are influenced by the polymer structure and molecular relaxation process [10]. Activation energy is an important thermodynamic property and used to describe the minimum energy that must be available for initiating a chemical process. For analyzing the properties and interactions taking place in polymer-polymer systems, knowledge of rheological properties is essential and viscometry is one of the efficient techniques for studying the properties [11]. Most commonly, the elastic property of the polymer solution particles has been accessed using atomic force microscopy (AFM). In this approach, it is calculated by theoretical method. In the present investigation, we have tried to compute the activation energy of aqueous solutions of Hydroxyethyl cellulose at different concentrations (0.05%, 0.10%, 0.15%, 0.20%, 0.25%, 0.30%, 0.35%, 0.40%, 0.45% and 5%) for different temperatures (303K, 308K, 313K, 318K) by determining the relative viscosity and ultrasound velocity and density studies were taken for the above concentrations at 303K for aqueous solutions of Hydroxyethyl cellulose. From these experimental values, the acoustical parameters such as adiabatic compressibility, intermolecular free length and internal pressure etc were calculated and effect of solute solvent interaction were analysed. From adiabatic compressibility values, the elastic property (bulk modulus) was calculated for the above concentrations. The important reason for choosing this material is it has variety of applications in household cleaning products, soaps, and shampoos and the detailed literatures showed that no measurements have been previously reported for the mixtures reported in this paper.

2. Experimental Details

The cellulose derived polymer Hydroxyethyl cellulose (HEC – SPECTRA Grade) was purchased from Sri Ganapathy Trading Company, Kanchipuram. The polymer HEC is dissolved in double distilled water and the polymer solutions are prepared for various concentrations (0.05%, 0.10%, 0.15%, 0.20%, 0.25%, 0.30%, 0.35%, 0.40%, 0.45% and 5%). A magnetic stirrer (REMI make) operating at a rate of 1000 rpm is used for preparing the polymer solutions. BROOKFIELD Viscometer (USA) is used to determine the relative viscosity of the prepared polymer solutions at different concentrations at 303K, 308K, 313K & 318K. The viscometer was connected to an electronically controlled thermostat having a thermal stability of ± 0.05K for taking measurements at different temperatures. Pyknometric method is employed to determine the density of the solutions. A K-ROY electronic balance having an accuracy of ±0.001gm is used to measure the mass of the liquid. A digital ultrasonic velocity meter (VCT-70A model, VI MICROSYSTEMS) is used to measure the ultrasonic velocity. The measurements were taken at a frequency of 2 MHz (accuracy ± 0.1m/s). The ultrasound velocity measurements are taken at 303K by circulating water from a thermostat having a thermal stability of ± 0.05 K.

3. Results and Discussions

3.1. Determination of Activation Energy

Viscosity studies are performed for aqueous solutions of Hydroxyethyl cellulose for different concentrations at different temperatures as mentioned above. Fig. 1 shows the variation of relative viscosity against concentration for aqueous solutions of Hydroxyethyl cellulose at different temperatures.

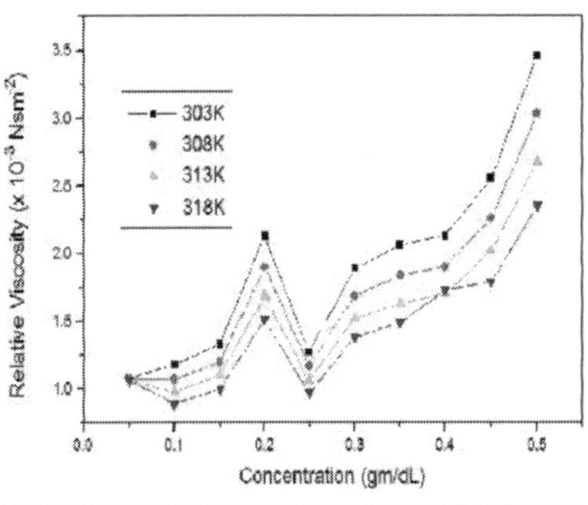

Figure 1. Variation of Relative Viscosity against concentration of aqueous solutions of HEC

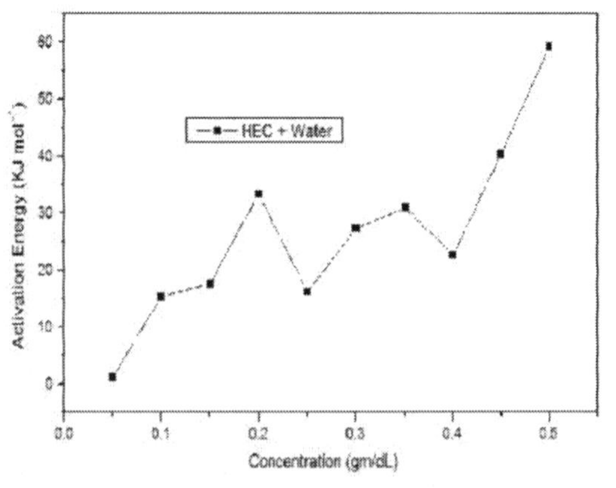

Figure 2. Variation of Activation Energy against concentration of aqueous solutions of HEC

In this system, the relative viscosity shows a non-linear variation with increase in concentration and temperature. The nonlinear variation of viscosity with concentration of HEC indicates presence of intermolecular interaction within the system. The decreasing trend of viscosity indicates that water is a poor solvent for HEC. Arrhenius expression is valid for pure solvents and also for dilute polymer solutions. According to Venkatramanan et al. [12],

$$\eta = A\, e^{Q/R}$$

Where Q is the apparent activation energy of flow and A is the pre-exponential term with an activation entropy significance.

The logarithm of this equation leads to a straight-line type given by

$$\ln \eta = \ln A + (Q / R)\frac{1}{T}$$

The graph is drawn between lnη against 1/ T. The slope of these straight lines gives the value for activation energy. The variation of activation energy (Q) with concentration (C) of aqueous solutions of HEC is shown in Fig. 2.

In this system, it is observed that there is a sudden decrease in the activation energy for 0.2% and 0.35% concentration. In most cases, activation energy for the viscous flow of polymer solution increases continuously with increasing polymer concentration, except in particular cases of critical mixtures [13]. The reason being the activation energy for the viscous flow of polymer solutions in poor solvent decreases with polymer concentration at high dilution and this is ascribed to the temperature dependence of the limiting viscosity number [14, 15].

The experimentally observed values of density and ultrasonic velocity of the above systems are presented in Table 1. It is observed that the density and ultrasonic velocity values are non-linear. The pronounced increase or decrease in these parameters with composition of mixtures indicates the presence of interactions between the components of molecules [16].

Table 1. Values of density and ultrasonic velocity of aqueous solutions of HEC

Concentration (w/v)	Density (kg/m^3)	Ultrasonic Velocity (m/s)
0.05	995.6609	1493.675
0.10	995.8627	1493.762
0.15	995.9636	1493.159
0.20	996.3673	1494.364
0.25	996.0645	1493.762
0.30	996.8718	1494.967
0.35	996.2663	1494.967
0.40	996.5691	1494.967
0.45	996.7709	1493.762
0.50	997.1745	1494.967

3.2. Determination of Molecular Interaction Parameters

Normally a decrease in adiabatic compressibility indicates closed packing and decreased ionic repulsion. The variation of adiabatic compressibility for aqueous solutions of HEC with concentration is shown [Fig. 3].It is observed that there is a non-linear variation with increase in concentration of HEC. The actual deviation depends on the resultant effect. The observed decrease or increase in adiabatic compressibility with composition is an evidence of significant interaction between the component molecules in the binary mixtures. This shows the complex formation and molecular dissociation. It also indicates weakening of hydrogen bond at this molar concentration [17].Intermolecular free length shows a similar behavior as reflected by adiabatic compressibility. Free length varies nonlinearly with increasing concentration of HEC at lower concentration which suggests that there is a significant interaction between solute and solvent molecules due to which structural arrangement is also affected [Fig. 4].Free volume refers to the void space between the molecules, i.e., the volume present as holes because of irregular packing of the molecules. It may be defined as the average volume in which the central molecule can move inside the hypothetical cell freely without affected by the repulsion of surrounding molecules. The free volume plot follows reverse trend of viscosity. Therefore viscosity rather than velocity determines the free volume of our system. The free volume reduces when the internal pressure increases [6]. In the present case, free volume for aqueous solutions of HEC is maximum at 0.05% and minimum at 0.50% [Fig. 5] and this variation is found to be nonlinear. The reverse trend of free volume with internal pressure is clearly shown in Fig.5.

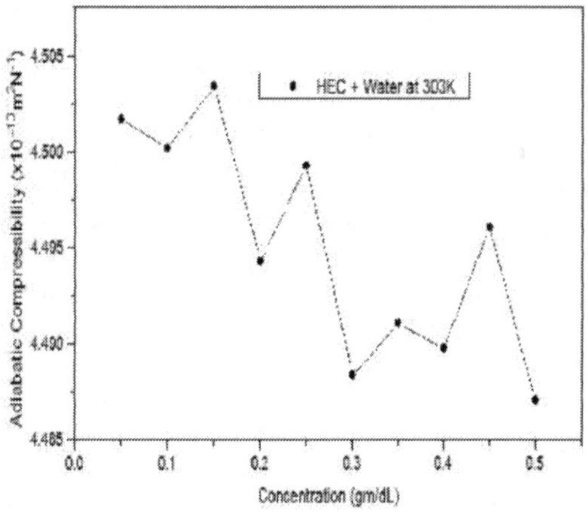

Figure 3. Variation of Adiabatic compressibility against concentration of aqueous solutions of HEC

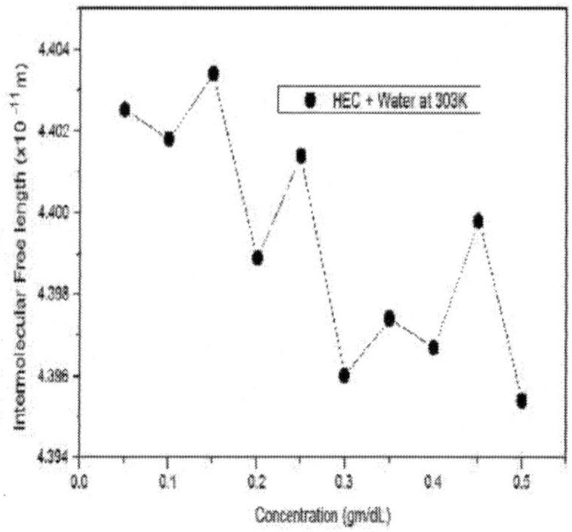

Figure 4. Variation of Intermolecular free length against concentration of aqueous solutions of HEC

The internal pressure of a liquid reflects the molecular interaction. Internal pressure studies show a nonlinear variation with increase in concentration of HEC and thus it is the evidence for the existence of significant interaction between the solvent – polymer molecules. Internal pressure shows dip at 0.25% concentration, whereas the free volume shows peak at corresponding concentration. This shows the complex formation and molecular dissociation. It also indicates weakening of hydrogen bond at this molar concentration.In the present study the classical absorption coefficient has been calculated for Aqueous HEC at different concentrations of 0.05%, 0.10%, 0.15%, 0.20% 0.25%, 0.30%, 0.35%, 0.40%, 0.45% and 0.50% at 303 K and it is shown in Fig 6.A non-linear variation is observed. The values of viscous relaxation time, specific acoustical impedance, relative association, Rao's constant and Wada's constant are reported in Table 2.

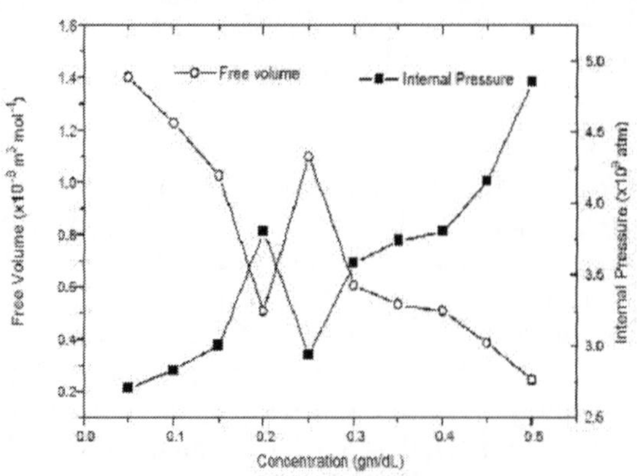

Figure 5. Variation of Free volume and Internal pressure against concentration of aqueous solutions of HEC

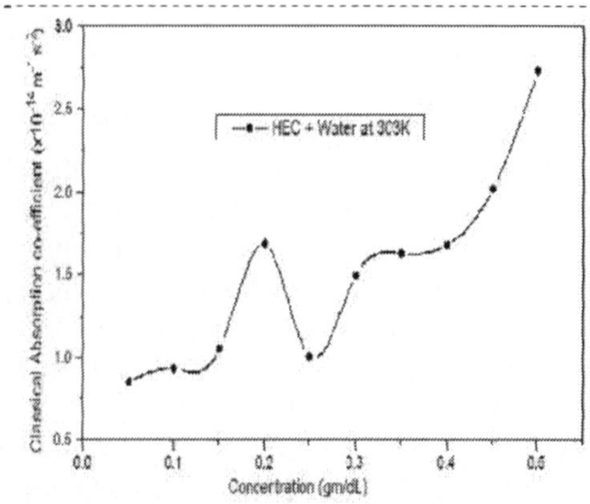

Figure 6. Variation of Classical absorption co-efficient against concentration of aqueous solutions of HEC

Table 2. Values of Molecular Interaction Parameters

Concentration (w/v)	Specific Acoustical Impedance (x 10^6 kg m^{-2} s^{-1})	Relaxation Time (x10^{-12} sec)	Relative Association	Rao's Constant (X 10^{-4} [m^3mol-1][ms^{-1}]$^{1/3}$)	Wada's Constant (X 10^2 m^3 mol^{-1} [Nm^{-2}]$^{1/7}$)
0.05	1.4871	0.6482	0.9996	2.0168	3.8817
0.10	1.4875	0.7080	0.9999	2.0164	3.8826
0.15	1.4871	0.7986	1.0001	2.0160	3.8826
0.20	1.4889	1.2764	1.0002	2.0157	3.8853
0.25	1.4878	0.7618	1.0001	2.0160	3.8835
0.30	1.4902	1.1310	1.0006	2.0149	3.8880
0.35	1.4893	1.2335	1.0000	2.0162	3.8853
0.40	1.4889	1.2751	1.0003	2.0156	3.8866
0.45	1.4889	1.5286	1.0008	2.0146	3.8866
0.50	1.4907	2.0700	1.0009	2.0143	3.8893

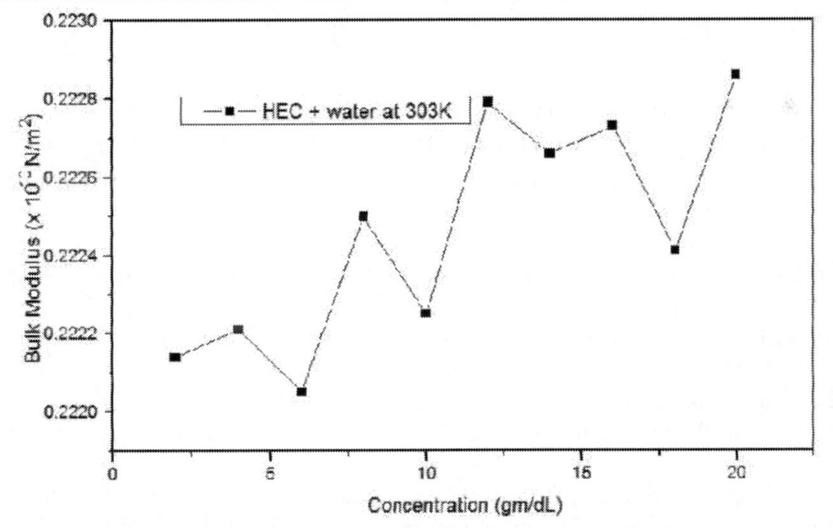

Figure 7. Variation of Bulk modulus against concentration of aqueous solutions of HEC

The relaxation time depends upon the size and shape of the rotating molecular entities in the solution. From Table 2, it is observed that the viscous relaxation time shows non-linear variation with increase in concentration throughout the system. This nonlinear variation with increase in concentration is due to the existence of significant molecular interaction between the solute and solvent molecules [18]. Specific acoustic impedance is defined as impedance offered to the sound wave by the components of mixture. Mathematically it is directly proportional to ultrasonic velocity and inversely proportional to adiabatic compressibility. The non-linear decrease of impedance with concentration shows that the molecular interaction of aqueous solutions of HEC is dissociative [17]. Relative association is influenced by two factors (i) breaking up of the associated solvent molecules on addition of solute and (ii) the solvation of solute molecules. The former leads to decrease and latter leads to increase in relative association. In the present system, the Relative Association is minimum at 0.05% concentration and maximum at 0.50% concentration. Rao's constant and Wada's constant also show a non-linear variation with increase in concentration.

3.3 Determination of Bulk modulus

The variation of bulk modulus with different concentration of HEC in binary mixture with water at room temperature is plotted in Fig. 7, which shows the non-linear variation with increase in concentration of HEC. Bulk modulus determines how compressible a fluid is. In other words, it tells us how the density of a liquid changes when the liquid is subjected to pressure. Higher the values of Bulk modulus indicate that it is difficult to compress the fluid [19].

4. Conclusion

The nonlinear variation of ultrasonic velocity and other thermo acoustical parameters with increase in concentration of HEC in water provides useful information about the nature of intermolecular forces existing in the mixture. The observed complex formation in the binary liquid mixture may be due to the formation of hydrogen bonding and the tendency of solute-solvent interaction. This complex formation is very much important in solid pesticides in water-based sprays for agriculture formulations.

Acknowledgments

The authors acknowledge the financial support of SCSVMV University, Enathur, Kanchipuram for carrying out this research work.

References

[1] SumanGahalyan, Manju Rani, and SanjeevMaken (2016)*J. Mol. Liq.***219** 1107
[2] Kushare S K, Terdale S S, Dagade D H and Patil K J (2007) *J. Chem. Thermodyn.***39** 1125
[3] Amalendu Pal, Harsh Kumar, RituMaan and Harish Kumar Sharma (2014)*Thermochim. Acta* **590** 127
[4] Yadava S S andNeetuYadav (2010) *J. Mol. Liq.***157** 6
[5] Krzysztof Klimaszewski, EwaStronka-Lewkowska, KatarzynaAbramczyk and Adam Bald (2015) *J. Chem. Thermodyn.***89** 212
[6] Venkatramanan K, Padmanaban R and Arumugam V (2015)*Phys. Procedia***70** 1052
[7] BushraNaseem, Mahreen Khan and Muhammad Asghar Jamal (2016) *J. Mol. Liq.* 220 581
[8] Kulkarni S S and Khadke U V (2016) *Indian J. Mater. Sci.* **2016** 1
[9] Punitha S, Uvarani R, Panneerselvam A and Nithiyanantham S (2014) *J. Saudi Chem. Soc.***18** 657
[10] Kannappan V, Santhi B S and Reuban Jonathan D (2014) *Indian J. Chem.***53A** 688
[11] Al-Furhood J A, Alsewailem F D and Almutabaqani L A (2014) *Eur. Chem. Bull.***3** 93
[12] Venkatramanan K and Arumugam V (2006) *Int. J. Thermophys.***27** 66
[13] Debye P, Chu B and Woermann D (1963) *J. Polym. Sci.A Polym. Chem.***1** 249
[14] Padmanaban R and Venkatramanan K (2015) *AIP Conf. Proc.***1675** 020019
[15] Reisuke Okada and Hiroshi Tanzawa (1965) *J. Polym.Sci. A***3** 4294
[16] Ravichandran S (2011) *Res. J. Chem. Sci.***1** 12
[17] PriyankaShamkuwar and Chimankar O P (2014) *Int. J. Sci. Res. Phy. Appl. Sci.***1** 14
[18] Venkatramanan K, Padmanaban R and Kavitha B (2016) *Adv. Sci. Lett.***22** 3948
[19] Monupal and Suthar B (2016) *AIP Conf. Proc.***1728** 020003

IOP Conf. Series: Materials Science and Engineering **390** (2018) 012039 doi:10.1088/1757-899X/390/1/012039

Gamification Elements used in Employee Retention and Enhancing Employee Productivity

S Shyam Nivedhan[1] and R G Priyadarshini[2]

[1] Amrita School of Business, Coimbatore, Amrita Vishwa Vidyapeetham, India, shyamnivedhans@gmail.com
[2] Amrita School of Business, Coimbatore, Amrita Vishwa Vidyapeetham, India, rg_priyadarshini@cb.amrita.edu

Abstract- Nowadays in IT industries retention of employees is a crucial task. Companies try hard to have win-win situation so that both the employees as well as the companies get mutual benefit. Most of the employees tend to change companies regularly as there is no growth in their careers, non-availability of challenging roles and lack of recognition. Another Important problem is to increase the employee's productivity. Most of the roles are saturated as employees keep doing the same work daily and no new challenges have been introduced in their work, to motivate their employees. This paper helps to understand and compare what all elements the IT companies are using to retain their employees and increase the productivity of the company. This study consider some of the reputed companies in the IT industry, comparing their Gamification elements with each other and finding the effective ones.

1. Introduction

IT Companies have been spending a lot in research and money to control the attrition rate of employees which in-turn increases the productivity of the organization. In sports, athletes has been self-motivated because of the passion for the games and they perform extremely well to achieve their goals and rewards. In the case of sports there is no need of any extra motivation given to the people, they are 100% committed towards achieving their rewards. So the IT organization use this strategy of gamification to motivate their employees at work, which reduces the attrition rate of the organization and which in turn increases the productivity of the Organization. The gamification elements used differs from different organizations. The gamification elements used are Learning and development, Rewards and recognitions, Employee Engagement, One on One with every employees

2. Literature Review

The gamification elements are used to understand the process well and obtain maximum efforts from people to achieve the goals." In this gamification elements are used in real time scenarios and confirmed that it motivates people to achieve their goals" (**Michael Sailer**, 2017).The word gamification has been used in many of the organizations. "The gamification elements such us leader

boards, rewards, promotions enhances intrinsic motivation and also productivity of the organization" (**Elisa D Mekler**, 2017).
The effect of spiritual values of employees plays an important role in organizational commitment." When employees are spiritually experience at work, they feel more commited to their organization, experience a sense of responsibility, loyality towards them and feel less materially committed" (**Morteza Raei Dehaghi**, 2012). In today's real life scenario organizations need talented people to improve the productivity. "Improving in Talent management like Strategic Leadership, Enhancing employee quality, Performance evaluation, Evaluation of expectations leads to organizational commitment"(**Yalcin Vural**, 2012).
In recent years job satisfaction and organizational commitment is becoming more important in working life." If a job done gives a satisfaction for the talented employees they ensure that they will stay committed to the organization" (**Mehmet Altinoz**, 2012).In this technological era, organizations find it difficult to satisfy employees in order to cope up the evolving environment. "In order to increase productivity, effectiveness and job commitment of the employees, organization must satisfy the needs of employees by providing good working conditions" (**Abdul Raziq**, 2015). In this not only satisfaction but emotions also play a major role in organizational commitment." Negative emotions in employees tends the employees to leave the organizations" (**Oya Erdil**, 2014).To overcome the process of Retention of employees in small and medium industries where the attrition rate is high some of the HR practices must be made available for the employees." The major HR practices that must be given to the employees are compensation and benefit, performance management, training and employee relations" (**Choi Sang Long**, 2014).
Learning and development is a process where employee could learn and develop their skills so that they can be efficient for futuristic roles. "The employees can be retained if the organization can provide the wants, needs and the expectations of the employee" (**Roya Anvari**, 2013).
In every organization productivity play an important role." If the Customer information System has been made easy in usage, content and format it helps the employees to increase the productivity of the organization"(**Norfazlina G**, 2016). In an organization the productivity of the each and every employees must be equally captured so that appraisal will be fair. "Employee based monitoring devices are used to monitor the employee's performance, productivity, work related behavior, learning and development and mainly it is transparent with the employees" (**David L Tomczak**, 2017). This provides fair appraisal which increases productivity and retention of employees.

3. Conceptual Model

In this model we could see what all the factors that affect the employee productivity and retention of employees. The factors affecting the model are Learning & Development, Rewards and Recognition, Employee Engagement and Taking Feedbacks. We have considered the important variables that are responsible for Employee Retention and enhancing Employee productivity they are learning and development, Rewards and Recognitions, Employee Engagement and Taking Feedback.

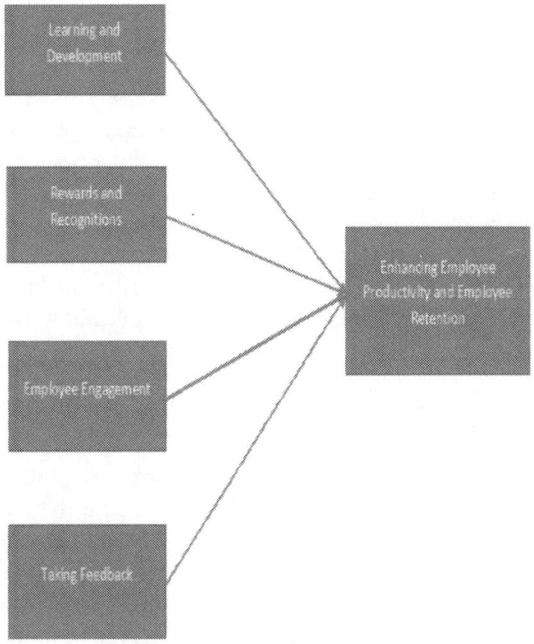

Figure 1. Conceptual model of Gamification

4. Methodology

In this an in-depth interview has been conducted with Assistant Managers/ Managers of Human Resources of different IT sectors in the Industry. After the discussing about what all problems in the IT sectors and what all the solution that can be made feasible in the real time scenario are noted down. A pilot study has been conducted in enhancing productivity and retention of employees. Then Questionnaires has been modified and in-depth interview has been conducted to identify the feasible solutions.

5. Questions

1. What are the Gamification elements used by the company?
2. Is the Gamification elements used monthly, quarterly or yearly?
3. After introduction of Gamification elements have employee absenteeism and boredom reduced and by what percentage?
4. What percentage retention and productivity has been increased or decreased after Gamification elements has been introduced?
5. Is there any Drawbacks of Gamification so that it can be resolved in the future?
6. Whether Gamification provide equal opportunities for all the employees in the organization?
7. Whether Introduction of Gamification reduces conflicts between employees and Mangers?

6. Analysis

To increase productivity and in retention of employees following elements play an important role:-

6.1. Optimized Hiring

Many Organizations hire people who are mismatch for the organization which in turn result in increase in attrition rate for the organization. To avoid this, many organization provide a process called employee referrals, where the employees within the organization could refer his/her colleagues by providing the information of the organization, work culture and what all they expect from new employees. This process decreases the recruitment cost and employees are rewarded with incentives for the each new recruits. This process is mainly done for experience candidates to make sure they needed less training programs. Incentives are used as a gamification element to engage the employee to find suitable candidates to improve productivity, to retain employees and to reduce the cost of the Organization.

6.2. Effective Induction program

Most of the organizations have taken induction seriously and they are doing the best to educate the employees about the organization and what all they expect from their employees in return. Some companies have scheduled first 1 or 2 months of induction program for the employees to understand well about the company. In the Induction program the employees will be given exposure to what all roles available in the project and will make sure employees are aware about the projects and their functional roles. Some organizations even cross train their employees in the induction program and when they have vacancies in the future they use the employees to make sure the project have delivered in time. Employees are made to work in challenging roles and reduces the stag mentation of work in their life. This motivates the employees to work for the organization and stay committed.

6.3. Learning and Development

IT companies have investing heavily to make sure all the employees has been trained to the current application in which they are working and also being cross trained to other tools that are available. Many companies have added Learning and development as a part of their KRA to every employees and ratings are given accordingly. In some companies it has made mandatory to complete 2 to 3 learning and development courses in order to be eligible for the appraisal.
In many IT industries, learning and development has been made available in internal portals for all the employee of the organizations for free of cost. The courses differ from different positions according to their positions the courses will be made available for the employees. Employees can utilize this courses from anywhere, at any point of time in order to increase their knowledge in their own domain and also can work on their futuristic roles. This decreases the boredom in organization and which in turn increases the productivity of employees.

6.4. Maintaining good relationship with employees

Team work is the essential part in an IT organizational, employees can't work alone and complete the projects, it can be only achieved with help of team work. Organizations started providing HR games where employees are put into different groups and the employees must work together as one to resolve the issues. Organizations also provide team outing, team dinner where employee get to know each other better and also relationship between employees improves. In this way there will be less conflict between employees and top level management.

6.5. Improve the work conditions

Organization must make sure the working condition are made available for every employees. Most of the Organizations have changed their computers into laptops and each employee is given a laptop to work on, employee can work on a place where they are comfortable. Some of the Organizations help employees by providing work from home policies especially for woman employees for working at night shifts, when they are pregnant. It is also applicable for employees who have broken legs or unable to travel to office for work. And they are also given flexible work schedules for the employees to help promote a work-life balance.

6.6. Rewards and Recognitions

Employees have been rewarded for their performance quarterly and yearly to make sure their productivity has been maintained, it also motivates other employees to improve their productivity. The rewards will be in types of incentives, gift vouchers and international tours. Some companies provide flash points which in turn can be used for online shopping. Some of the rewards have been given to the whole team as a whole for the successful competition of project by having team dinner or incentives for all the employees in that team.
Employees have been recognized by providing momentum and also internal promotion according to the performance of the employees. These recognitions has been mentioned in the leader board of the organization to motivate other employees to work towards recognitions.

6.7. Employee Engagement

Organizations have started to concentrate more on this part by celebrating all the National festivals in office and making sure employees doesn't feel they have been ill-treated. This reduces employee absenteeism and also provide relaxation for the employees.
They also have started introducing Tech Talks which improves the imagination of employees to come up with new ideas and it has been taken into to account by the organization. If it's a genuine solution it has been applied in the organization and employees are rewarded for the extra effort they have made.

6.8. One on One with every Employees

In this process the organization tries to understand the needs/ wants of the employees such as trainings, futuristic roles, higher education, whether there is any conflict with managers, new ideas for the project etc, have been accomplished. This helps the organization to know more about the employees and could easily help the employees in the area of weakness. Organizations have started following the process and it's mandatory that all employees have gone through the process. This increases the trust of employees towards the organization.

6.9. Taking Feedback

All the IT companies have started accepting feedback from each and every employees in the organization to make sure that they really grow together as one. Organization as a whole will find it difficult to identify the problems in a team or in a project, but employees could easy able to identify the problems in this kind of scenario but won't be able to resolve the issues alone. Employees use this feedback as a platform to showcase their views about the company, manager, project on which he/she is working, wants and needs of the company, conflict with team members/ managers. This has been made mandatory to every employees and they must fill the feedback every 6 months or 1 year. Separate team is allocated to understand the views of the employees and within a limited period of time the problems will be resolved.

7. Limitations

1. Most the employees are not aware of the gamification elements and its benefits.
2. Lack of Transparency between the organizations and their employees

3. This may increase the conflict between managers and employees as the managers are skeptical in nature.

8. Conclusion

By using the Gamification elements like Optimized Hiring, Effective Induction program, Learning and development, Maintaining good relationship with employees, Improve the work conditions, Rewards and Recognitions, Employee Engagement, One on One with Every employees and Taking Feedback the absenteeism and boredom has been reduced to 30 to 35%

The Productivity and retention of employees has been increased to 10 to 15%.

9. References

[1] Norfazlina, G, Sharidatul Akma A S, Nurul Adrina S & Noorizan, M M (2016) *Procedia Economics and Finance* **37**, 7 – 12
[2] Jalal Hanaysha (2016) *Procedia - Social and Behavioral Sciences* **229**, 298 – 306
[3] Roya Anvari, Nur Naha Abu Mansorb (2014*) Procedia - Social and Behavioral Sciences* **129**, 172 – 177
[4] Choi Sang Longa, Musibau Akintunde Ajagbeb, (2014) *Procedia - Social and Behavioral Sciences* **129**, 99 – 104
[5] Hely Innanem, Asko Tolvanen, Katariina Salmela-Aro (2014) *Burnout Research* **1**, 38–49
[6] Abdul Raziqa, Raheela Maulabakhsha (2015) *Procedia Economics and Finance* **23**, 717 – 725
[7] Tuna Cenkci, Ayşe Begüm Ötkenb (2014) *Procedia - Social and Behavioral Sciences* **150**, 404 – 412
[8] Jalal Hanaysha (2016) *Procedia - Social and Behavioral Sciences* **229** , 289 – 297
[9] Mehmet Altinoz, Demet Cakiroglu, Serdar Cop (2012*) Procedia - Social and Behavioral Sciences* **58**, 322 – 330
[10] Morteza Raei Dehaghi, Masoud Goodarzi , Zahra Karimi Arazi (2012) *Procedia - Social and Behavioral Sciences* **62**, 159 – 166
[11] Oya Erdila , Büşra Müceldilib (2014) *Procedia - Social and Behavioral Sciences* **150**, 447 – 454
[12] Yalcin Vural, Pelin VARDARLIER, Abdullah AYKIR (2012) *Procedia - Social and Behavioral Sciences* **58** , 340 – 349
[13] Iveta Ozolina-Ozolaa (2014) *Procedia - Social and Behavioral Sciences* **156**, 223 – 226
[14] Xhavit Islamia,, Enis Mulollia, Naim Mustafab (2018) *Future Business Journal* **4** ,94–108
[15] Elisa D Mekler , Florian Brühlmann, Alexandre N Tuch, Klaus Opwis (2017) *Computers in Human Behavior* **71** ,525-534
[16] Richard N Landers , Kristina N Bauer , Rachel C Callan (2017) *Computers in Human Behavior* **71** , 508-515
[17] Boopathi, K., Sreejith, S., & Bithin, A. (2015). Learning cyber security through gamification. *Indian Journal of Science and Technology*, *8*(7), 642-649.
[18] Rathi, N., & Lee, K. (2015). Retaining talent by enhancing organizational prestige: An HRM strategy for employees working in the retail sector. *Personnel Review*, *44*(4), 454-469.

IOP Conf. Series: Materials Science and Engineering **390** (2018) 012040 doi:10.1088/1757-899X/390/1/012040

Free vibration analysis of stiffened laminated composite plate in a thermal environment

Sudhansu S Patro[1*], Dibakar Sutradhar[1], Ranjan K Behera[1], Nitin Sharma[1]

[1]School of mechanical engineering, Kalinga Institute of Industrial Technology, Deemed to be University, Bhubaneswar, Odisha-751024, India

*sudhansu.patrofme@kiit.ac.in

Abstract. In this paper, the free vibration responses of stiffened composite plates in thermal environment are investigated. The flat panel geometry is modelled by using ANSYS parametric design language (APDL) code based on first order shear deformation theory (FSDT)midplane kinematics. Firstly, the validity of the present model is established along with the convergence. Further, several numerical examples have been solved and discussed in detail to bring out the effect of change in temperature, modular ratio and the coefficient of thermal expansion on the free vibration responses of stiffened laminated composite flat panels. It is observed that the values of natural frequencies decrease with temperature.

1. Introduction

Now-a-days for better performance, longevity and increase in number of life cycles before fatigue fracture engineering structures have to meet the requirements such as high strength to weight ratio, high resistance to failure due to dynamic, shock loading etc. So, various studies have been performed to study the free vibration responses of rectangular plates with various prestress environment and other factors such as temperature, support conditions, material properties, lamination schemes. The plates are modelled with stiffeners to have better strength and stiffness without increasing significant weight. The free vibration of laminated composite and sandwich plate based on a higher-order refined theory was studied by T. Kant [1] analytically and heat induced vibration of a rectangular plate was studied by N. D. Jadeja [2] and Theodore R Tauchert [3]. Also various studies have done on symmetric cross-ply by A. A. Khdeir [4]. The buckling analysis and vibro acoustic analysis in clamped composite laminated plate is done by Xiangyang Li [5]. Buckling, free vibration and acoustic analysis in isotropic plate with and without uniform and non-uniform thermal prestress is studied by P. Jeyaraj [6-7]. thefree vibration in stiffened plates is studied by Thinh and Khoa [8] and Guanghui Qing[9]. Also free vibration and buckling analysis of composite stiffened shells and plates was studied by R Rikards [10] and eccentric stiffeners was studied by Chao and Lee [11].

From this brief review of literature, it is clear that no study investigating the free vibration of different composite plates with stiffeners in a prestressed thermal environment has been reported till date. So, the present work aims at analysing the vibration of laminated composite rectangular shell platewithstiffeners with thermal prestressusing ANSYS 15.0 APDL. The free vibration responses (natural frequency) of the plate in a thermal environment are obtained and validated with the published literature.

2. Formulation of equations

The composite laminated plate with stiffeners, considered for the present investigation is shown in the Figure 1. Plates and stiffeners have orthotropic properties. Three different lamination scheme are used in this investigation.

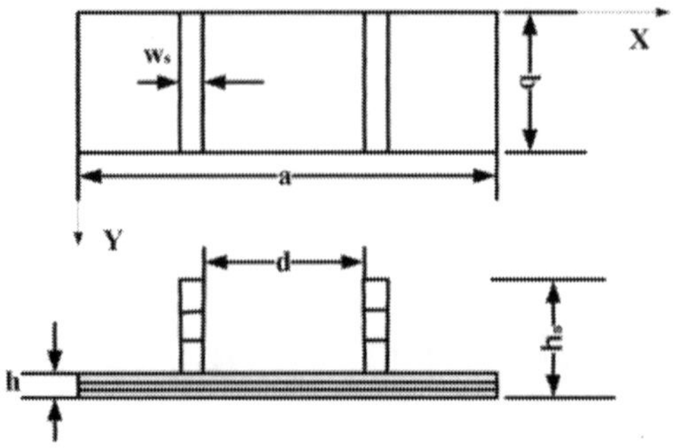

Figure 1. Diagram of stiffened composite plate

In the present study, a laminated flat composite plate is considered. The plate consists of a number of orthotropic layers of equal thickness. The lamination schemes, boundary conditions are varied for different cases. The extension induced shear and the coupling effect of bending and twisting are neglected throughout the present investigation. The stiffeners are defined as to consists of reinforcing rib and a strip of rib width in the plating. Shear due to extension and twisting due to bending are neglected. $[p\,q\,r]' = [p_0 + z\ \ \theta_x q_0 + z\theta_y\ \ r_0 + z\theta_z]'$ (1)

where, p, q and $r$are the displacement vector of any point in the layer at time t along x, y and z direction respectively. Modal analysis is used for this structure and its modal parameters are calculated by solving the eigen-value equations: $[(K_L - K_G) - \omega^2 M]\{\varphi\} = 0$ (2)

whereK_L, K_G, M, ω and $\{\varphi\}$ are the linear global stiffness matrix, global geometry stiffness matrix, mass matrix, natural frequency of vibration and corresponding mode shape vector, respectively.

3. Results and Discussions

The dynamic responses of eccentrically stiffened (single and double) laminated composites flat plates in thermal environment have been investigated using ANSYS 15. The flat panel as well as the stiffeners have been modelled using an eight noded serendipity shell element. First, the values of buckling temperature are computed for different lamination schemes and boundary conditions in order to find the suitable temperature range to conduct the modal analysis in prestressed condition. The values are then compared with the published values. The test of convergence and the validation with the previously published results are carried out to verify the reliability and accuracy of the present analysis. The influence of change of temperature, modular ratio and ratio of coefficient of thermal expansions on free vibration responses are investigated for different boundary conditions and lamination schemes. The different boundary conditions used in the present analysis are CCCC (clamped at all edges), CHCH (clamped at two edges and hinged at others), SCSC (simply supported at two edges and clamped at others). A flat laminated composite rectangular plate of dimension 600x400x5 mm^3 with single and double eccentric stiffeners is considered throughout the present study. The following geometrical as well as material properties, reported by Li et al[5], are considered for the analysis unless mentioned otherwise: E_1=132GPa, E_2=E_3=10.3GPa, G_{12}=G_{13}=6.5GPa, G_{23}=3.91GPa, v_{12}=.3, ρ=1570kg/m3, α_1=1.2×10^{-6}/°C, α_2=2.4×10^{-5}/°C.

3.1. Study of convergence and validation of natural frequencies

In the present section, the critical buckling temperatures are calculated for different mesh sizes to verify the convergence behavior of present analysis. For convergence study, a clamped flat laminated composite rectangular plate [0°/90°/0°/90°/0°] with the dimensions mentioned earlier with single and double stiffeners is considered. The variation of critical buckling temperature for different modes of frequencies with mesh sizes are presented in Figure 2. It can be observed that, the values of critical buckling temperatures are nicely converging for all modes with refinement of mesh. The validation study is made in two stages. First, the results (critical buckling temperature and natural frequency) computed for laminated plate without stiffeners in thermal environment are compared with the results published by Li et al. [5].

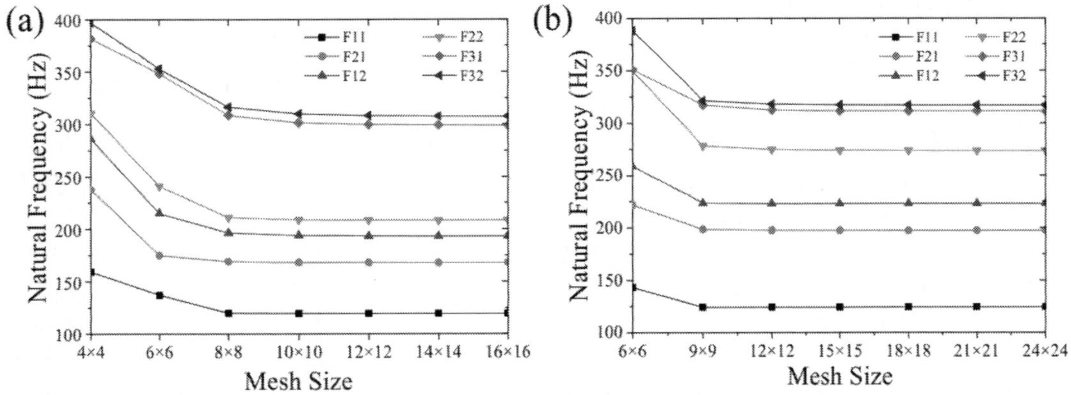

Figure 2. Convergence of critical buckling temperature for eccentrically stiffened plate (a) Single Stiffener, (b) Double Stiffeners

The validation study is made in two stages. First, the results (critical buckling temperature and natural frequency) computed for laminated plate without stiffeners in thermal environment are compared with the results published by Li et al. [5]. The comparison is presented in the table no 1. Then, natural frequencies (at ambient temperature) obtained for laminated composites with single and double eccentric stiffeners are compared with the results published by Chao et al. [11]. The geometrical and material properties are kept same as reported by Chao et al.[11].The results are presented in Table no 2. In both of the validation studies it is found that the present values of natural frequencies are very close to the reference values.

Table 1. Validation of natural frequency in thermal environment

ΔT = 0°C			ΔT= 20°C			ΔT= 40°C		
Li et al. [5]	Present	Difference(%)	Li et al. [5]	Present	Difference (%)	Li et al. [5]	Present	Difference (%)
200.21	199.97	0.120	175.81	176.85	-0.588	149.74	149.82	-0.055
370.13	371.34	-0.325	348.26	349.30	-0.297	324.71	325.60	-0.274
445.00	447.12	-0.475	416.40	417.93	-0.366	383.59	386.37	-0.723
562.77	564.01	-0.219	533.33	534.58	-0.234	502.21	503.28	-0.213
665.40	664.30	0.166	644.37	643.56	0.126	624.18	622.06	0.340
806.73	807.92	-0.147	778.96	780.50	-0.196	752.05	752.01	0.005

3.2. Effect of temperature on natural frequency

The effect of change in temperature on natural frequency of a five layered of equal thickness cross-ply [0/90/0/90/0] laminated composite flat plates with single and double eccentric stiffeners is investigated in this example. The dimensions of the plate are taken as a=0.4 m, b=0.3m, h=0.0034m andh_s=0.008+hm,w_s=0.003m. The boundary condition used for this example is CCCC (all edges are clamped). First the critical buckling temperatures(T_{cr}) are computed for each case. The value of buckling temperature for the plate with single stiffener is 119.52°C and for plate with double stiffener is 124.43°C. The temperature is increasedwith a step of 20. As the temperature approaches critical buckling temperature the increasing step is reduced to visualise the variation of frequencies at temperatures near the critical temperature. The variation of natural frequencies for different modes with temperature is presented in the Figure 3 (a) and (b). From the figure it is clear that the values of frequencies for all modes are decreasing with temperature. Also, for F11 mode, the values of frequencies at temperatures near the critical temperature are decreasing rapidly as the value of frequencies approaches zero. The observations are similar for both the cases (single and double stiffeners). It can also be observed that as the number of stiffeners increases the value of frequencies have increased for all modes of frequency except F32 mode.

Table 2. Validation of natural frequency of plates with stiffeners

		Single Stiffener		Double Stiffeners			
Eccentricity	Modes	F11	F12	F11	F21	F12	F22
1	Chao [11]	56.86	198.20	58.90	95.38	200.51	225.18
	Present	55.94	194.27	56.52	94.28	196.83	222.56
	Difference (%)	1.64	2.02	4.21	1.16	1.87	1.18
2	Chao [11]	59.15	205.89	62.25	96.58	213.52	239.26
	Present	58.42	202.79	60.66	95.63	212.64	230.33
	Difference (%)	1.25	1.53	2.63	0.99	0.41	3.88
3	Chao [11]	63.60	215.74	66.68	99.94	237.25	259.78
	Present	62.90	212.97	67.99	98.29	233.94	240.74
	Difference (%)	1.11	1.30	-1.93	1.67	1.42	7.91

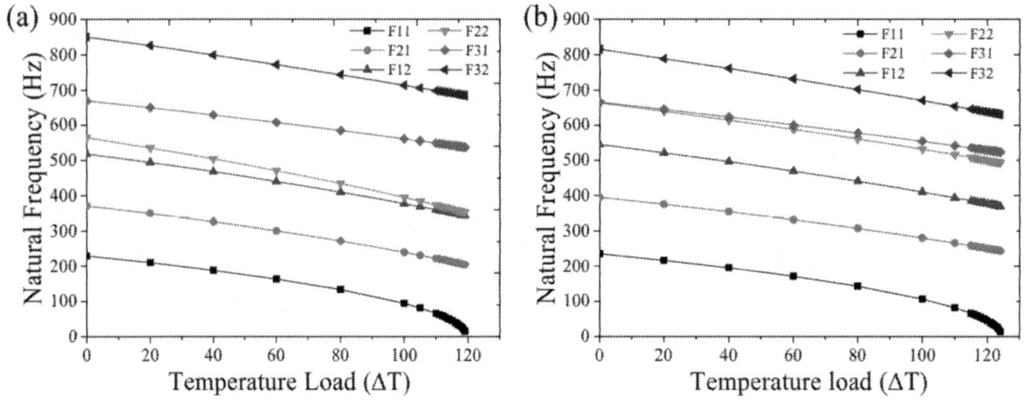

Figure 3.Effect of temperature on eccentrically stiffened plate (a) Single Stiffener, (b) Double Stiffeners

3.3. Influence of modular ratio on natural frequencies at different temperatures

In this example, the influence of modular ratio on natural frequencies of a four layered of equal thickness angle-ply $[45/-45]_s$laminated composite panel with single and double stiffeners is investigated at different temperatures. The plate dimensions are kept same as the plate used in previous example. A CHCH type boundary condition is used in this study. The values of natural frequencies are computed for modular ratio (E_1/E_2)= 5, 10, 20, 30, 40 at three different temperature change, i.e., ΔT=0°C, 40°C, 60°C. The variation of natural frequencies with modular ratio is presented in the Figure 4. It can be observed that the values of frequencies are increasing with modular ratio for all temperatures. As the value of E_1 increases, the stiffness of the plates increases hence the value of frequencies increase. The variation is higher at the lower values of modular ratio.

3.4. Influence of ratio of coefficient of thermal expansion on natural frequencies

The influence of ratio of coefficient of thermal expansion (α_1/α_2) on natural frequency of a four layered cross-ply symmetric $[0/90]_s$ laminated composite flat panel with single and double eccentric stiffeners is investigated at different temperatures in this section. The support condition used for this example is SCSC. The values of natural frequencies are calculated for α_1/α_2= 0.05, 0.1, 0.15, 0.2, 0.25, 0.3 at two different temperatures change, i.e., ΔT=20°C, 40°C. The variation of natural frequency with $\alpha_1/\alpha_2$is shown in the Figure 5. The values of frequencies for all modes decreases with α_1/α_2 at all temperatures and the same can be observed from the figure. At ΔT=20°C the amount of variation in the values of frequencies is not significant when compared to the same at ΔT=40°C. it can also be observed at ΔT=40°C, the variation in the first mode (F11) of frequency is large when compared to other modes of frequency.

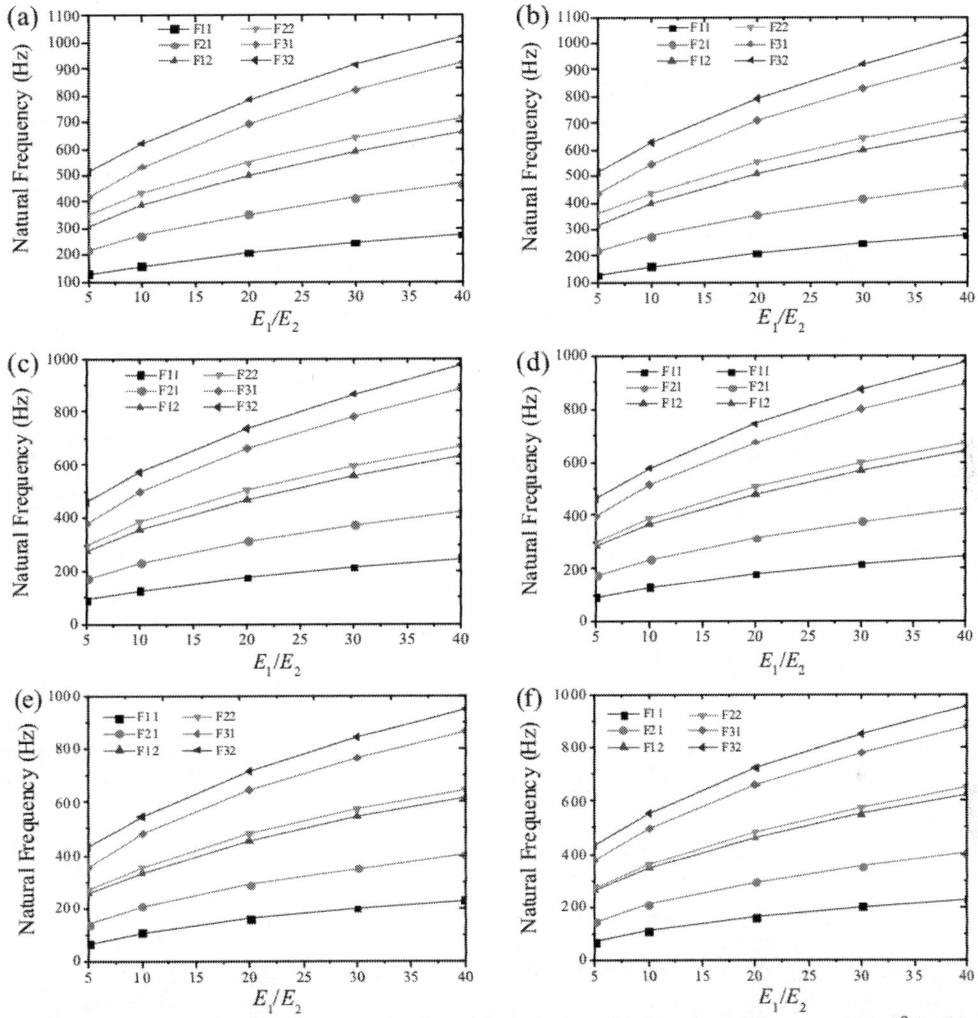

Figure 4. Effect of modular ratio on eccentrically stiffened plate (a) Single Stiffener at ΔT=0°C, (b) Double Stiffeners at ΔT=0°C, (c) Single Stiffener at ΔT=40°C, (d) Double Stiffeners at ΔT=40°C, (e) Single Stiffener at ΔT=60°C (f) Double Stiffeners at ΔT=60°C

4. Conclusion

The dynamic responses of laminated composite flat panel with eccentric stiffeners in thermal environment is analyzed in ANSYS using APDL code. To establish accuracy and reliability of model, first, the vibration characteristics computed are compared and validated with previously reported results. After that, the values of natural frequencies are computed at different temperatures along with critical buckling temperature. The frequency values are also calculated for different values of modular ratio and ratio of coefficient of thermal expansion at various temperatures. It is observed that the values of frequencies for all modes are decreasing with temperature. The amount of variation in the values of frequency is more as the temperature approaches critical buckling temperature. It is also observed that the values of natural frequencies are increasing with modular ratio for all modes of frequencies. The variation in the values of frequencies is large at lower value of modular ratio. It is also evident from the analysis that, the values of natural frequencies are decreasing with ratio of coefficient of thermal expansion (α_1/α_2). It is found that the variation is larger at ΔT=40°C.

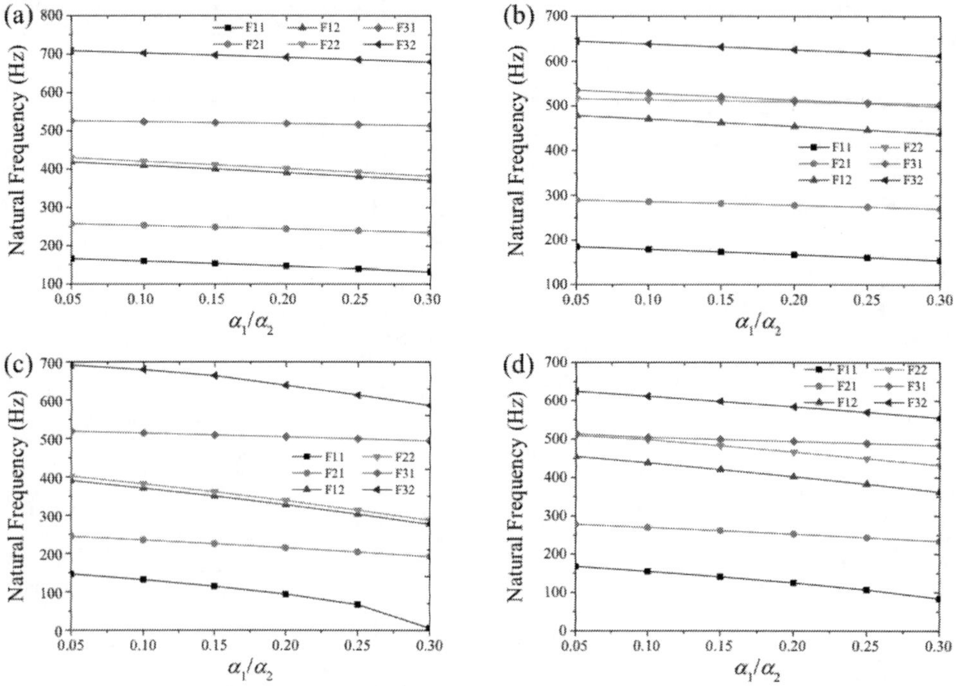

Figure 5. Effect of α_1/α_2 for eccentrically stiffened plate (a) Single Stiffener at ΔT=20°C, (b) Double Stiffeners at ΔT=20°C, (c) Single Stiffener at ΔT=40°C, (d) Double Stiffeners at ΔT=40°C

References

[1] Kant T, Swaminathan K, Analytical Solution for Free Vibration of Laminated Composite and Sandwich Plate Based on a Higher-Order Refined Theory, Composite Structures 53 (2001) 73-85

[2] Jadeja N. D, Loo Ta-Cheng, Heat Induced Vibration of a Rectangular Plate; Aug 1974/1015, Journal of engineering for industry, ASME

[3] Tauchert T. R, Thermally induced flexure, Buckling and Vibration of Plates, ASME book no.-AMR099; Appl Mech Rev vol 44, no 8, Aug 1991

[4] Khdeir A. A, Librescu L, Analysis of Symmetric Cross-ply Laminated Elastic Plates Using a Higher-Order Theory: Part II – Buckling and Free Vibration; Composite Structures 9 (1988) 259-277

[5] Li Xiangyang, Yu Kaiping, Buckling and Vibro-acoustic Response of the Clamped Composite Laminated Plate in Thermal Environment, (2016); 119:10.021

[6] Jeyaraj P, Buckling and Free Vibration Behaviour of an Isotropic Plate under Non-Uniform Thermal load, International journal of structural stability and dynamics, Vol. 13, No.3 (2013) 1250071

[7] Jeyaraj P, Ganesan N, Vibration and Acoustic Response of an Isotropic Plate in a Thermal Environment, (Oct,2008); 130/051005-1; Journal of vibration and acoustics, ASME

[8] Thinh T. I, Ngo Nhu Khoa, Free vibration analysis of a stiffened laminated plate using a new stiffened element, Vietnam journal of mechanics, Vol. 29, No. 4, (2007)

[9] Qing G, Qiu J, Liu Yanhong, Free Vibration analysis of stiffened laminated plates,International journal of solids and structures, Vol. 43, Issue 6, (Mar, 2006), 1357/1371

[10] Rikards R, Chate A, OzolinsO. H, Analysis for buckling and vibration of composite stiffened shells and plates, 51, 361-370

[11] Chao C.C, Lee J.C, Vibration of eccentrically stiffened laminates, Journal of composite materials, Vol. 14, (Jul, 1980), p-233

Synthesis and tribological investigation of compressor's liner-tribo pair material under the influence of biodegradable compressor oil

P. chenga reddy[1]*, S. Arumugam [2], M. Lalith Babu[3], V.V.K. Krishna Teja[4]

[1, 2] Department of Mechanical Engineering, Sri Chandrasekharendra Saraswathi Viswa Mahavidyalaya, Enathur, Kanchipuram - 631561, Tamil Nadu, India.

Chengareddy.p@kanchiuniv.ac.in

Abstract. The present investigation is intended to synthesis biodegradable compressor oil for reciprocating air compressors and analyses the tribological behavior of liner-ring tribo pair material under the influence of new formulations. A pin-on-disc tribometer was used to perform the tribological investigations. EN 31 steel and EN-JL 1020 was selected as disc and pin material, respectively. An ultrasonication based transesterification process was used to formulate a biodegradable compressor oil. TGA and DSC analysis was performed to examine the thermos-oxidative stability of compressor oil. The newly formulated bio compressor oil showed better tribological parameters in terms of coefficient of friction, wear and specific wear rate as compared with SAE 30 grade compressor oil. SEM and EDAX analysis of testing pin surfaces proved that the worn surfaces were smooth with bio compressor oil.

1 Introduction

At present, demand for crude oil and petroleum products are projected to continue climbing, further increasing the price of products and also rapid depletion of crude oil reserves in the world due to their usages in many applications like transport (63.7%), industrial (8.5%) and non-energy use and others (27.8%). On the other side, environmental issues with the mineral oils due to its non-degradability, air pollution and toxic problems [1]. This situation creates greater opportunity to produce environmentally friendly lubricating oil from vegetable oil for air compressors, because of its biodegradability, high viscosity index, good anti wear characteristics, low toxicity and high flash point in nature [2]. However, the main problem with the vegetable oil is relatively poor oxidation stability. To overcome this drawback, vegetable oil was formulated by chemically modified via epoxidation, hydroxylation followed by esterification process, which improves the oxidation stability and cold flow properties of the oil [3]. Waste cooking oil methyl ester (WCOME) was prepared using *In-situ* epoxidation process for industrial applications, for this reaction acidic ion-exchange resin was used as a heterogeneous catalyst. The modified WCOME had improved thermo-oxidative stability and also viscosity, which was confirmed by NMR, oxirane analysis by HBr titration method and FTIR reports [4]. Investigations were also made to improve the oxidative stability of vegetable oil by transesterification process. Hashem et al [5] was converted some vegetable oils into bio lubricants via two successive transesterification process using conventional method and the converted oils were shown good thermal and oxidation stability.

By conventional method, the synthesis of vegetable oil is consuming more reaction time with a low yield percentage of oil which leads to high processing cost. In this connection, researchers are making the attempts to reduce the production cost and faster reaction time by the use of ultrasound energy. Ponappa et al [6] was employed to optimize the process parameters for transesterification of kapok seed oil to produce biodiesel using ultrasonic energy. From the results found that the highest biodiesel yield was obtained by some of the important parameters like methanol/oil molar ratio, catalyst concentration, reaction time, ultrasonic frequency and temperature [7-9].

Many researchers have reported that the liner-ring Tribo pair is the important tribological component in reciprocating internal combustion engine, same as in reciprocating air compressors. Arumugam et al [10] was conducted several tests on cylinder liner and piston ring combination using a pin-on-disc tribometer under different lubricants. From the experimental results found that the newly formulated bio-lubricant and biodiesel were shown better performance in terms of friction, frictional force and wear, also its impact was shown positively on engine component life. Yashvir Singh et al [11-12] was developed bio based blended lubricants in the ratio of 15, 30, and 50% by volume with pongamia pinnata and jatropha oils for tribological investigations. It was found that the addition of pongamia and jatropha oil to the base lubricants (SAE 40 and SAE 20W40) resulted better reduction in terms of friction and wear. Trivedi et al [13] was investigated and compared the tribological behavior of the realistic engine oils (SAE 10W30, SAE 20W40 and SAE 20W50) with constant load and incremental load of 20 N to 140 N for 105 min run using a pin-on-disc Tribo tester and observed that the effect of viscosity of oil and variation of load plays an important role in wear and friction characteristics. Nuraliza et al [14] have studied the effect of vegetable oil as bio-lubricant on the friction and wear of piston ring and cylinder liner materials using a pin-on-disk tribometer under variable loads (50N and 100N) and varying sliding speeds (1, 2, 3, 4, 5 m/s), base oil as hydraulic and engine oil used for the comparison purpose. It is observed that the new alternative lubricant has shown good results in terms of wear, coefficient of friction and also shows the smother worn surface of the specimen, confirmed by SEM analysis.

In the present investigation ultrasonication energy is used to develop bio-lubricant from rapeseed oil by the transesterification process for the application of reciprocating air compressors and its tribological tests were conducted on liner-ring tribo pair material using a pin-on-disc tribometer. The worn surfaces of the tested liner-ring tribo pair was examined by using scanning electron microscopy (SEM) and EDAX analysis.

2 Experimental Work

2.1 Synthesis of bio-lubricant

Raw rapeseed oil was synthesized following the two-step reaction. First the rapeseed oil was converted into methyl ester. In this process, the products of raw rapeseed oil (1000 ml), methanol (178.4ml) as alcohol, and sodium hydroxide (8.92 grams) as catalyst were taken in a 2000 ml capacity of three-necked round bottomed flask. The products were stirred at 200 rpm with 55^0C for the time of 3 hr. After the reaction, the contents were well washed with warm water for the purpose of removing the dissolved excess alcohol. The final content was formed as the mixture of methyl ester and glycerol as by product. Then the methyl ester was separated from glycerol by using a separate funnel without any disturbance and heated up the reactants at 55^0C to evaporate the water residue.

In the second step, methyl ester was converted into pentaerythritol ester (bio-lubricant) with the influence of low frequency ultrasound energy (20kHz). For this reaction, the obtained Methyl ester (640ml), pentaerythritol (149.366 grams), p-TSA (1 gram) as a catalyst and xylene (20 ml) as solvent were taken in a three-necked round bottomed flask and this was introduced into the ultrasound chamber in the transesterification process. Then the reactants were heated up on a heating mantle under a nitrogen atmosphere at 150^0C for 3hr reaction time. After the completion of the reaction, the pentaerythritol ester was well washed with warm water to remove the excess alcohol and heated up the reactants at 70^0C to evaporate the water residue.

2.2 Lubrication properties test

The physico-chemical properties of biodegradable compressor oil were measured in Ita labs, Chennai, Tamil Nadu, India, according to the ASTM (American Society for Testing and Materials) standards. The obtained results were shown in table 1.

IOP Conf. Series: Materials Science and Engineering **390** (2018) 012041 doi:10.1088/1757-899X/390/1/012041

Table 1. Properties of Biodegradable compressor oil

Parameters	Testing Standards	SAE 30	Biodegradable compressor oil
Kinematic Viscosity at 100^{0}C	ASTM D7279	10.5	9.02
Flash Point (^{0}C)	ASTM D93	250	206
Pour Point (^{0}C)	ASTM D97	-40	<-10
Viscosity Index	ASTM D7279	145	124

2.3 Wear and friction Test method

The Pin-on-disc tribometer is used to investigate the effect of PE on wear and friction properties of the liner-ring tribo pair. ASTM G99 standard test method was used to evaluate the wear preventive characteristics of lubricant. The test conditions are 50 N load, 50 mm track diameter, 1800 m of sliding distance, 5 m/s of sliding velocity, operational speed of 1061 rpm and 60 min of operational time.

2.4 Preparation of specimen

The specimens were prepared from EN 31 steel and grey cast iron EN JL 1020 material. EN 31 steel was used as disc specimen and EN JL 1020 material is used for the pin specimen. Before to conduct the tests, it was mandatory to clean the specimens properly to free from the foreign particles and debris with alcohol.

3 Results and Discussion

3.1 TGA (Thermogravimetric Analysis)

Thermal stability is one of the most important property of the lubricant and it is essential to determine for the oil and this analysis was carried out using TGA machine. The approximately sample weight was taken as 14.061 mg for the analysis. It can be seen that the biodegradable oil was stable up to 230^{0}C (onset temperature). From the TGA curve (Fig. 1), the weight loss occurred between 230^{0}C to 455^{0}C due to the combustion of organic matter present in the biodegradable oil. The maximum weight loss occurred at the decomposition temperature of 365^{0}C and 455^{0}C.

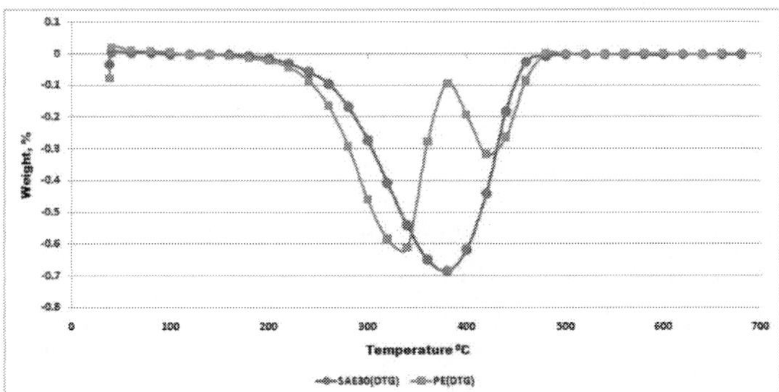

Fig 1. TGA Analysis

3.2 DSC (Differential Scanning Calorimetry)

It is a technique of thermoanalytical, which determines the temperature and heat flow difference between a sample and reference material. This technique is mainly used for the measurement of oxidative stability which is very important for the lubricating oil. The sample of oil was evaluated at a rated temperature of 10^{0}C/min with the sample weight of 14.723 mg under the air of 150 psi. From the DSC curve in Fig 2. The thermic peak represents the melting point between 230^{0}C and 263^{0}C.

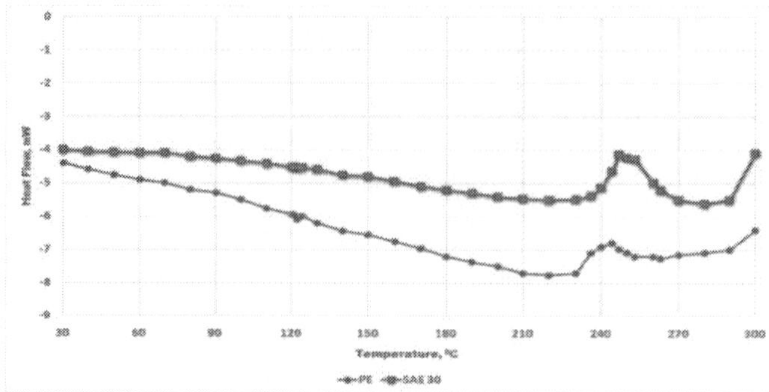

Fig 2. DSC Analysis

3.3 COF (Coefficient of friction)

Pin-on-disc tribometer was used to examine the wear and friction characteristics of the liner-ring material under biodegradable compressor oil and SAE 30 compressor oil. The testing oil was filled up in a tank and a pump, which is used to monitor the oil continuously between the contact area of rotating disc and a pin. The coefficient friction was measured for the constant load (50N) condition and working speed of 1061 rpm for the duration of 3600s under lubrication. From the Fig 3. observed that the coefficient of friction at starting of the test of biodegradable oil was lower than that of SAE 30 grade compressor oil. At one stage, the COF value of the lubricants were same. This is due to viscosity of the lubricants were closer each other, that forms the thinner film between the sliding surfaces in contact.

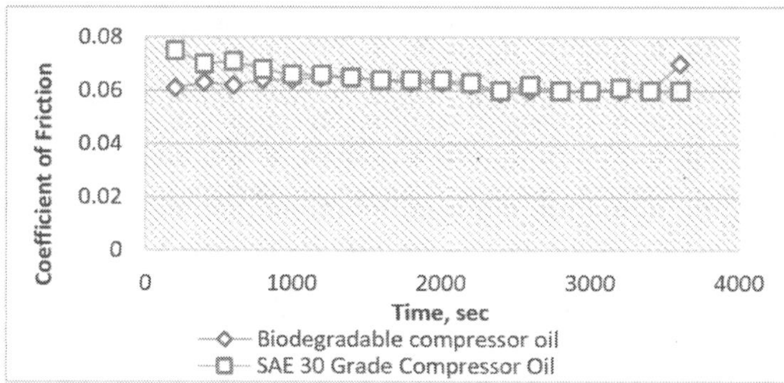

Fig 3. Coefficient of Friction

3.4 Wear characteristics

For the same load conditions, wear experiments were conducted with same lubricants. The variations in wear with the load for one-hour duration is shown in Fig 4. In general wear behavior increases with the increase of time and always sliding surface plays a vital role to increase the wear.Also, the irregularities of the surface increase the area of contact between the sliding surfaces, resulted maximum wear. In this study, Biodegradable oil showed minimum wear as compared to the synthetic air compressor oil due to maintain its film thickness between the contact areas. The wear characteristics of SAE 30 grade compressor oil was increased with the increase of running time.

The specific wear rate of biodegradable compressor oil and SAE 30 compressor oil were calculated by the equation given below. The values are 3.397E-06 mm^3/N-m for bio lubricant and 7.078E-06 mm^3/N-m for SAE grade compressor oil.The specific wear rate of biodegradable compressor oil was lower than the SAE 30 grade compressor oil. Where V is volume loss in cubic millimeter, F is applied load in Newton and D is Sliding distance in meter.

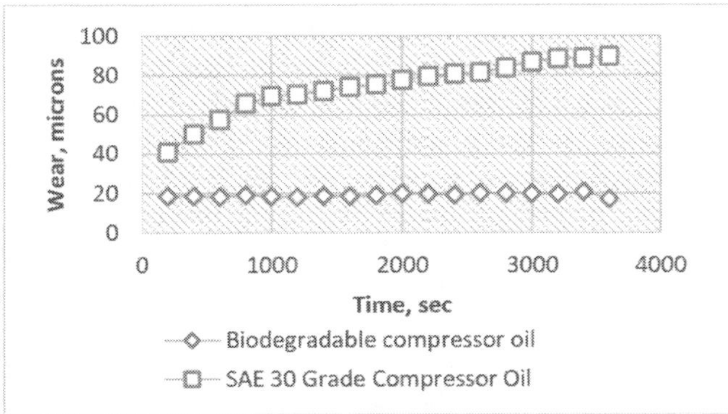

Fig 4. Wear Characteristics

$$\text{Specific Wear Rate } SWR = \frac{V}{F \times D} \text{ mm}^3\text{/N-m}$$

3.5 SEM Analysis

The rubbed surfaces of the pin were examined by the SEM (Scanning Electron Microscope) analysis. The obtained worn surfaces of the samples were shown in Fig 5 (a) and 5 (b) for the synthetic oil and biodegradable oil. It can be easily found that after running a 3600s of time, Fig 5 (a) represents the presence of cavities and pits under synthetic oil. From Fig 5 (b) observed that, with the biodegradable air compressor oil, appears cavities and grooves due to the metal-metal contact between liner-ring surfaces.

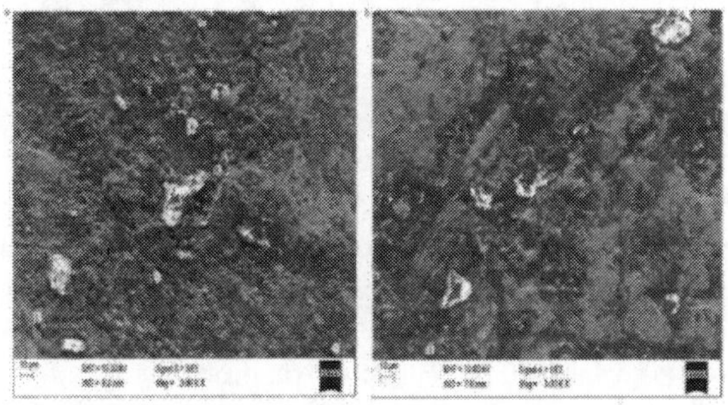

Fig 5. SEM image of the cylinder liner wear surface: a) Synthetic lubricant b) Bio-lubricant.

3.6 EDAX Analysis

Energy Dispersive Analysis of X-rays (EDAX) of the pin worn surface is shown in figure 6 (a) and 6 (b). From the figure 6 (a) it is clear that C, O, Fe and Si peaks are observed in the energy dispersive analysis X-ray lubricated by SAE 30 grade compressor oil. Figure 6 (b) shows worn surface under bio-lubricant, the elements like Al, Ca, C, Na, Fe, Mg and Si peaks were observed.

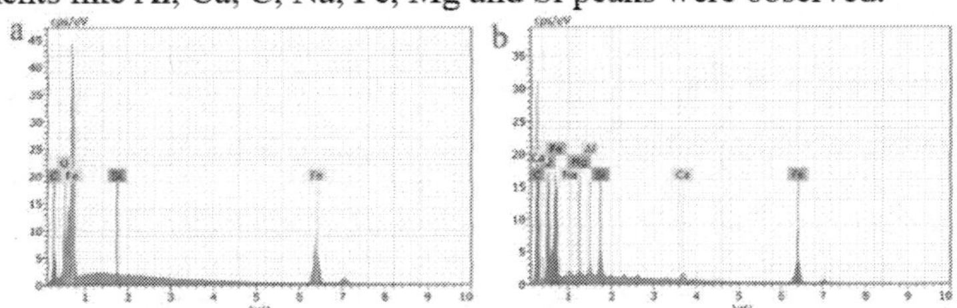

Fig 6. EDAX images of the cylinder liner wear surface: a) Synthetic lubricant b) Bio-lubricant.

4 Conclusions

In this study coefficient of friction, wear and specific wear rate of liner-ring material at constant load were carried out using a pin-on-disc tribometer under lubricants. The following conclusions were made from the obtained results:

- The ultrasonic energy is found to be suitable for transesterification process to produce biodegradable compressor oil by reducing the reaction time with high yields.
- Physico-chemical properties of the bio-lubricant meets the ASTM standards.
- Improved in thermo-oxidative stability and heat flow properties of the bio-lubricant by transesterification process, which was analyzed by TGA, DSC and DTA.
- The biodegradable oil reduced the coefficient of friction, wear and specific wear rate compared with SAE 30 grade compressor oil. With this proved that the bio-lubricant had excellent lubricity as compared to the synthetic lubricant.
- From the observation of the worn surfaces, biodegradable compressor oil showed smoother surface than the SAE 30 grade compressor oil.
- EDAX technique was used to show the chemical composition of the specimens, the various peaks of the elements were shown clearly in figure 6.

References

[1] N.J. Fox, G.W. Stachowiak 2007, Vegetable oil-based lubricants—A review of oxidation (Tribology International, vol. 40) pp 1035-1046.

[2] Sevim Z. Eehan, SvajusAsadauskas 2000, Lubricant base stocks from vegetable oils (Industrial Crops and Products, vol.11) pp 277-282.

[3] S Arumugam and G Sriram 2012, Synthesis and characterisation of rapeseed oil bio-lubricant – its effect on wear and frictional behaviour of piston ring–cylinder liner combination (Journal of engineering Tribology, vol. 227(1)) pp 3-15.

[4] VenuBabuBorugadda 2015, Improved thermo-oxidative stability of structurally modified waste cooking oil methyl esters for bio-lubricant application (Journal of Cleaner Production) pp 1-10.

[5] A. I. Hashem, W. S. I. AbouElmagd, A. E. Salem, M. El-Kasaby& A. M. El-Nahas 2013, Conversion of Some Vegetable Oils into Synthetic Lubricants via Two Successive Transesterifications (Energy Sources, vol. 35) pp 909-912.

[6] K. Ponappa, V.Velmurugan, P. Arul Franco, T.R. Kannan and R. Ragurajan 2016, Optimization of biodiesel production from ceiba pentandra (kapok seed oi) using response methodology assisted by ultrasonic energy method (International Journal of Chem Tech Research, vol. 5) pp 794-803.

[7] Xiaohu Fan, Xi Wang, and Feng Che 2010, Ultrasonically assisted production biodiesel from crude cottonseed oil (International Journal of Green Energy, vol. 7) pp 117-127.

[8] Francisco F.P. Santos, Jackson Q. Malveria, Marcia, Marcia G.A. Cruz, Fabiano A.N. Fernandes 2010, Productio of biodiesel by Ultrasound assisted esterification of Oreochromis niloticus oil (Fuel, vol. 89) pp 275-279.

[9] Priyanka Chand, Venkat Reddy Chintareddy, John G. Verkade, and David Grewell 2010, Enhancing Biodiesel Production from Soybean Oil Using Ultrasonics. (Energy and fuels, vol. 24) pp 2010-2015.

[10] S Arumugam, G Sriram andSubhadhra L 2012, Synthesis, chemical modification and tribological evaluation of plant oil as bio-degradable low temperature lubricant (Procedia Engineering, vol. 38) pp 1508-1517.

[11] Yashvir Singh, Amneesh Singla, Anshul Kumar Singh, and Avani Kumar Upadhyay 2016, Tribological characterization of pongamia pinnata oil blended bio-lubricant. (Biofuels,vol. 38) pp 1-8.

[12] Yashvir Singh, rajnish Garg and Suresh Kumar 2016, Effect of load on friction and wear characteristics of jatropha oil bio-lubricants (Biofuels, vol. 8) pp 125-133.

[13] H.K. Trivedi, D.V. Bhatt 2017, Effect of lubricating oil on tribological behavior in pin on disc test rig (Tribology in Industry, vol. 39) pp 90-99.

[14] N. Nuraliza& S. Syahrullail: Tribological effects of vegetable oil as alternative lubricant 2016, a pin-on-disk tribometer and wear study (Tribology Transactions, vol. 59) pp 831-837.

IOP Conf. Series: Materials Science and Engineering **390** (2018) 012042 doi:10.1088/1757-899X/390/1/012042

Sensitivity analysis of process parameters on tensile properties in Plasma Arc Welding of AA8011-H24 aluminium alloys and Ti3Al2.5V Titanium alloys using Response Surface Methodology

K Palani[1], C Elanchezhian[2] and M Raju[3]

[1] Department of Mechanical Engineering, Sri Chandrasekharendra Saraswathi Viswa Mahavidyalaya (Deemed to be University), Kanchipuram, 631561, Tamilnadu, India

[2] Department of Mechanical Engineering, Sri Sairam Engineering College, Chennai, 600 044, Tamil Nadu, India

[3] Department of Mechanical Engineering, Eswari Engineering College, Chennai, 600 089, Tamil Nadu, India

E-Mail: palani.k@kanchiuniv.ac.in

Abstract. The quality of a welded joint is strongly decided by the input parameters of the plasma arc welding process. The mathematical models used to predict the optimum tensile properties based on the input process parameters of the plasma arc welding. In this investigation, the response surface methodology is applied to optimize the tensile properties of the Plasma arc welding of AA8011-H24 aluminium alloys and Ti3Al2.5V Titanium alloys by proper selection of process parameters such as welding speed, welding current and gas flow rate. The three factors, five levels central composite design was applied to conduct the experiments. The impact of the process parameters on tensile properties of the plasma arc welding process was investigated by using the sensitivity analysis of the mathematical models. The welding speed was the more sensitive parameter on the tensile properties of the plasma arc welding compared to the other parameters followed by the welding current and gas flow rate.

1. Introduction

Owing to high strength-to-weight ratio, good corrosion resistance, low density and low cost of lightweight hybrid materials, both aluminum alloy and titanium alloy are having more demand in the fields of aerospace, marine, automotive and transportation industries [1-3] and it reduces the effort of the fuel consumption. Even though, the formation of brittle intermetallic structures (IMCs) and residual stress were found which degrade the strength properties at the weld centre of the dissimilar aluminium alloys with the titanium alloys due to their different chemical properties, viz., melting temperature and physical properties, viz., Coefficient of linear expansion [4, 5]. Umanath K et al. [6] made an investigation on the austenitic stainless steel rods using friction welding process and studied the effect of individual parameters of friction welded joints. They concluded that the least influencing parameters of forging pressure and friction pressure followed by the rotational speed of the rods on the tensile strength of the friction welded joints. Tahsini SM et al. [7] made an investigation on the microstructure of the similar friction welded (FW) joints of AA 7005 aluminum rods with the parameters, viz., friction pressure and friction time. The fine equiaxed grains were appeared at the rotating side due to higher cooling rate. The parameters selection of the welding process was the key factor in deciding the quality of the welded joints discussed by Palani et al. [8, 9].

The responses of the welding process, viz., bead geometry, tensile properties, hardness and grain size were optimized by using the different optimization techniques, such as Taguchi method and response surface methodology reported by Montgomery DC [10]. The enhancement of mechanical properties was decided by investigating the microstructures at the different zones of the welded joints studied by Palani et al. [11]. Some of the previous literature was made on the effects of process parameters on the microstructures and mechanical properties of the welded joints for the different welding processes. Shi Y et al. [12] studied the effect of underwater wet flux cored arc welding (FCAW) parameters on weld bead geometry using Taguchi L_{16} orthogonal array design and they predicted the responses of the bead geometry in welding compared with the experimental responses. Raja Dhas JE et al. [13] attempted the investigation on the bead geometry of the low alloy steel using submerged arc welding process (SAW) with the parameters of arc voltage, welding speed and welding current and also applied the sensitivity analysis on the SAW process parameters of the welded joints. They reported that the more sensitiveness of the dilution was found for arc voltage and welding current compared to the other bead geometry of the joints. Yeni C et al. [14] made an investigation on the microstructure and mechanical properties of the AA 7075 aluminium alloy joints using both friction stir welding process and conventional fusion welding process, viz., Metal inert gas welding (MIG) and tungsten inert gas welding (TIG) process. They observed that the coarse grains for both fusion welding process were found in weld centre of the joints, whereas the fine grains appeared in stir zone of the joints.

Dongxia Y et al. [15] successfully made a laser welding of dissimilar non-ferrous Al-Mg alloys with the proper selection of laser welding process parameters, such as welding laser power, welding speed, wire feed rate to produce the good quality of the welded joints. They optimized that the process parameters of 2.4 kW laser power, 50 mm/sec welding speed and 35 mm wire feed rate using the Taguchi method. They reported that the welding speed was the dominant significant parameter compared to other parameters on the bead geometry of the joints. The lightweight nonferrous alloys were focused in aerospace, transportation and automobile industries due to its properties such as high strength-to-weight ratio and corrosion resistance using the friction stir welding process by Elanchezhian C et al. [16]. They used the Taguchi L_9 orthogonal array design to optimize the tensile strength of the welded joints with the optimum parameters of tool rotational speed of 1400 RPM, transverse speed of 75mm/min, plunge force of 7 kN, 15.54 mm diameter of the tool, 5.13 mm diameter of the tool pin with tool hardness of 600 HV. Arun Kumar N et al. [17] applied the friction stir welding process on the AA 7075-T651 aluminium alloy 6.35 mm thick plates and optimized the process parameters viz., rotational speed, transverse speed and plunge force on the impact strength of the welded joints using Taguchi L_9 orthogonal array design. The most contributing parameters of transverse speed 35.81%, axial force 22.23% and rotational speed 7.49% was found using the analysis of variance and also they reported the maximum impact strength 43.89 for the optimum process settings of rotational speed 1200r/min, axial force 2 kN and transverse speed 62 mm/min for the FSW joints. The sensitivity of the responses based on small changes in process parameters of the welding process was studied using the sensitivity analysis. It is used to enhance the output responses of the welding process by selecting the suitable process parameters of the joints. Rajakumar S et al. [18] successfully friction stir welding on AA 6061-T6 aluminium alloys and developed the mathematical models to study the nature of the sensitiveness of the parameters on the welded joints. Also they reported that the optimum parameter settings of rotational speed of 1150 r/min, axial force of 7.16 kN, transverse speed of 84.3 mm/min, shoulder and pin diameter of the tool of 15.5 mm and 5.2 mm with 45 HRC tool hardness exhibited the maximum strength (225 Mpa) of the welded joints.

The transverse speed showed that as more sensitive than the other parameters on the strength of the joints using sensitivity analysis. Karthikeyan R et al. [19] focused on the lightweight materials (AA 7075-T6) in aerospace applications using the friction stir spot welding process with the suitable parameters of plunge depth, rotational speed, plunge rate and dwell period and optimized these parameters using the response surface methodology. They reported that the initial plunge depth involvement was more in joining of the alloys, further increment of the plunge depth, there were no changes occurred, whereas the strength of the joints at certain limit increased by increasing the other parameter then it was decreased further increment of the other parameters. Saravanan V et al. [20, 21] successfully investigated the effect of the rotational speed of the FSW tool on the dissimilar AA 2014-T6 and AA 7075-T6 aluminum alloys joints. They reported that the enhancement of strength properties owing to the formation of fine equiaxed grain regions in the stir zone with proper material mixing and plastic deformation of the alloy materials by microstructural analysis. The very limited work was carried out earlier literature on the dissimilar nonferrous alloys using the plasma arc welding (PAW) process and the discussion was made only on the bead properties of the welded joints. Ther is no work was carried out on the sensitiveness of the tensile properties of the PAW process. This work is focused to investigate the sensitivity of the PAW process parameters on the tensile properties of the joints.

2. Experimental Procedure

In this investigation, the AA 8011 aluminium alloys [22] and Ti3Al2.5V Titanium alloys (purchased from Kataria steel distributors, Mumbai, Maharashtra) with the dimensions of 100 mm x 100 mm x 6 mm were selected to make the plasma arc welded (PAW) butt joints. The chemical composition (wt. %) and mechanical properties of the AA 8011 aluminium alloys and Ti3Al2.5V Titanium alloys are presented in Table 1 and Table 2 respectively. The plates were initially cut by using the power hacksaw and the cleaning process was done with Ultrasonic bath sonicator and acetylene.

Table 1 Chemical composition (wt. %) and Mechanical properties of the AA8011-H24 Aluminium alloy

Chemical composition (wt. %)								
Fe	Si	Mn	Cu	Mg	Cr	Ti	Zn	Al
0.74	0.56	0.10	0.13	0.08	0.10	0.005	0.08	Balance

Mechanical properties		
Yield strength (Mpa)	Ultimate tensile strength (Mpa)	Elongation (%)
110	140	5.6

Table 2 Chemical composition (wt. %) and Mechanical properties of the Ti3Al2.5V Titanium alloy

Chemical composition (wt. %)								
Si	Fe	Al	V	O	C	H	N	Ti
0.0.14	0.30	3.0	2.5	0.12	0.05	0.02	0.02	Balance

Mechanical properties		
Yield strength (Mpa)	Ultimate tensile strength (Mpa)	Elongation (%)
518	620	15

The alloy plates were placed in a horizontal position with 45° chamber angle and maintained 0.5 mm gap between the plates. The standard ER4043 filler rod with 2 mm diameter and argon gas as shielding gas was used to join the alloy materials using the plasma arc welding machine is shown in Figure 1 (fabricated from Sri Multi Weld Systems, Chennai, Tamil Nadu) and the welding conditions are shown in Table 3. The selection of proper parameters and clamping fixtures decide the good quality of the welded joints.

In this investigation, the welding speed, welding current and gas flow rate were selected as the welding process parameters of the PAW process based on the previous literature on the non-ferrous alloy materials. The three factors, five levels central composite design of response surface methodology [23] was introduced to conduct the effective experimental trails by minimizing the effort and cost through avoiding the unnecessary experiments. The range of parameters and their levels are presented in Table 4. The twenty different combination of experiments were conducted on these aluminium alloys and titanium alloys using plasma arc welding process.

Figure 1 Plasma Arc Welding Machine

Table 3 Plasma Arc Welding process conditions

PAW Parameters	Details
Welding Speed, (mm/min)	20 - 100
Welding Current, A	120 - 300
Gas flowrate, (lpm)	10 - 30
Shielding gas	Argon
Electrode rod and its diameter	ER4043, 2 mm
Nozzle to plate distance, mm	1-6
Torch angle, deg	75-100

Table 4 Welding process parameters and their levels for Plasma Arc Welding (PAW)

Factors	Units	Notation	Levels				
			-1.672	-1	0	+1	+1.672
Welding Speed	mm/min	S	23	30	40	50	57
Welding Current	Amp	I	156	180	215	250	274
Gas Flow rate	l/min	Q	16	18	21	24	26

The tensile specimens were prepared to perform the tensile behaviour of the PAW butt joints according to the ASME SEC IX standards [24] using the Tensile testing machine (Model: FIE, UTN 40) and the tensile testing was conducted in Omega inspection and analytical laboratory, Chennai. The tensile test response of the PAW butt joints, viz., the tensile strength was shown in Table 5 along with the design matrix.

3. Results and Discussion

3.1 Developing a mathematical model for PAW process

The statistical method was used to analyze the responses using the input parameters by response surface methodology through the collective information of the process shown in equation (1),

$$Y = f(x_1, x_{2,} ......, x_k) \pm \varepsilon \qquad (1)$$

Table 5 Design matrix and Experimental results for Plasma Arc Welding (PAW)

Exp. No.	PAW process parameters			Tensile strength (Mpa)
	Welding Speed (mm/min)	Welding Current (Amp)	Gas Flow rate (l/min)	
1	-1	-1	-1	82
2	1	-1	-1	85.6
3	-1	1	-1	82.8
4	1	1	-1	90.7
5	-1	-1	1	93.2
6	1	-1	1	88.8
7	-1	1	1	66.6
8	1	1	1	68.5
9	-1.682	0	0	88.2
10	1.682	0	0	95
11	0	-1.682	0	72.8
12	0	1.682	0	57.6
13	0	0	-1.682	94
14	0	0	1.682	86.4
15	0	0	0	88.2
16	0	0	0	86
17	0	0	0	85.8
18	0	0	0	85.2
19	0	0	0	86.6
20	0	0	0	85.4

Where Y is the response variable and x1, x2 and xk are the input parameter variables of the process. The present work deals with the developed regression models shown in equation (2) which gives the relationship between the PAW parameters and the tensile properties of the butt joints. These second order models used to predict the tensile properties of the PAW process with acceptable values of the experimental responses [],

$$Y=b_o+\sum b_i x_i+\sum b_{ii} x_i^2+\sum b_{ij} x_i x_j+\varepsilon \tag{2}$$

Where b_o is the mean value of the responses and b_i, $b_{ii,}$ b_{ij} is the parameter coefficients of the main and interaction effects. The three factors, five levels central composite design was applied on the second order quadratic models of the welded joints using Design-Expert 8.1 software [25]. The mathematical models of the response, viz., the tensile strength of the PAW process are represented in equations (3)

$$\begin{aligned} TS(MPa)=&86.19+1.50(S)-4.87(I)-2.69(Q)+1.33(SI)-1.75(SQ)-6.60(IQ) \\ &+1.95(S^2)-7.39(I^2)+1.45(Q^2) \end{aligned} \tag{3}$$

Table 6 Analysis of variance for Tensile strength

Source	Sum of Squares	df	Mean Square	F Value	p-value (Prob > F)
Model	1787.50	9	198.61	161.51	< 0.0001
S-Welding Speed*	30.58	1	30.58	24.87	0.0005
I-Welding Current*	324.43	1	324.43	263.82	< 0.0001
Q-Gas Flowrate*	99.06	1	99.06	80.56	< 0.0001
SI*	14.05	1	14.05	11.42	0.0070
SQ*	24.50	1	24.50	19.92	0.0012
IQ*	348.48	1	348.48	283.38	< 0.0001
S^2*	54.60	1	54.60	44.40	< 0.0001
I^2*	786.48	1	786.48	639.55	< 0.0001
Q^2*	30.36	1	30.36	24.69	0.0006
Residual	12.30	10	1.23		
Lack of Fit	6.30	5	1.26	1.05	0.4795
Pure Error	6.00	5	1.20		
Core Total	1799.80	19			
R-Squared		0.9932	Adj R-Squared		0.9870
Adeq Precision		48.112	Pred R-Squared		0.9682

* Significant Terms

3.2 Checking for adequacy of the model

In this work, the analysis of variance (ANOVA) was used to check the adequacy of the developed regression model for tensile strength using the Design-Expert 8.1 software. The model F-value of the tensile strength was found as 161.51 which indicates the model is significant at 95% confidence level shown in Table 6. The main effects of S, I, Q and interaction effects of SI, SQ, IQ, S^2, I^2 and Q^2 were studied as a significant model for the tensile strength, whereas there is 47.95% chance of lack of fit found for the tensile strength which could be more due to noise effects. Furthermore, the regression coefficient (R^2 value) is closer to one which has a good reasonable agreement with the responses [26]. The predicted R^2 value of 0.9682 is in good agreement with the adjacent R^2 value of 0.9870 for the tensile strength of the PAW butt joints.

3.3 Sensitivity analysis on the tensile strength of the PAW process

The Sensitivity analysis is to study the sensitiveness of the objective function by the variation of input parameters of the process which is used to predict the quality of the welded joints. The tendency of increment or decrement of the objective function based on the critical input process parameters and priority of these parameters by their contribution were identified to optimize the parameters on the responses [27]. The partial derivatives were applied on the objective function with respect to the design parameters to identify the sensitivity of the parameters on the responses [28].

$$\frac{\partial TS}{\partial S} = 1.5 + 1.33(I) - 1.75(Q) + 3.9(S) \quad (4)$$

$$\frac{\partial TS}{\partial I} = -4.87 + 1.33(\mathrm{S}) - 6.60(Q) - 14.78(\mathrm{I}) \quad (5)$$

$$\frac{\partial TS}{\partial Q} = -2.69 - 1.75(S) - 6.60(I) + 2.9(Q) \tag{6}$$

In this work, the sensitivity equations are attained by partial differentiating the mathematical model of tensile strength with respect to the input parameters, viz., welding speed, welding current and gas flow rate of the PAW joints and are presented in the equations (4), (5) and (6) respectively.

Table 7 Sensitivity analysis on tensile strength (Welding Current = 250 A)

Welding Speed (mm/min)	Gas Flow rate (lpm)	Tensile Strength (Mpa)	Sensitivity of Tensile strength (Mpa)		
			∂TS/∂N	∂TS/∂I	∂TS/∂Q
20	16	93.5	-1.47	-9.11	-6.21
	18	83.3	-3.22	-15.71	-3.31
	21	76.1	-4.97	-22.31	-0.41
	24	71.7	-6.72	-28.91	2.49
	27	70.3	-8.47	-35.51	5.39
30	16	93.9	2.43	-7.78	-7.96
	18	82.1	0.68	-14.38	-5.06
	21	73.1	-1.07	-20.98	-2.16
	24	66.9	-2.82	-27.58	0.74
	26	63.8	-4.57	-34.18	3.64
40	16	98.3	6.33	-6.45	-9.71
	18	84.7	4.58	-13.05	-6.81
	21	73.9	2.83	-19.65	-3.91
	24	66.1	1.08	-26.25	-1.01
	26	61.2	-0.67	-32.85	1.89
50	16	106.6	10.23	-5.12	-11.46
	18	91.2	8.48	-11.72	-8.56
	21	78.7	6.73	-18.32	-5.66
	24	69.1	4.98	-24.92	-2.76
	26	62.4	3.23	-31.52	0.14
60	16	118.2	14.13	-3.79	-13.21
	18	101.6	12.38	-10.39	-10.31
	21	87.4	10.63	-16.99	-7.41
	24	76.1	8.88	-23.59	-4.51
	26	67.6	7.13	-30.19	-1.61

In this analysis, the increment of the objective function by a small change in parameter showed as the positive sensitive behaviour, while the decrement of the objective function showed the negative sensitive behaviour in the variation of a small amount of parameters of the welded joints [29]. The partial derivatives of the tensile strength with respect to the input PAW parameters were given in Table 7. The sensitivity analysis of the tensile strength of the PAW butt joints with respect to a small change in input process parameters, such as welding speed, welding current and gas flow rate is presented in Figure 2.

The figure 2(a) showed that the sensitivity of the tensile strength of the PAW process with the small variations of the welding speed. It can be seen that the tensile strength of the joints is decreased at the lower welding speed and at constant welding current (250A) whereas tensile strength of the welded joints gradually increases further increment of welding speed due to proper diffusion of the alloys [30]. The lower tensile strength was observed at the higher welding speed and the constant current shown in figure 2(b) which is due to improper material reaction at the weld zone and it produces the solidification defects. The figure 2(c) shows the decrement of the tensile strength of the joints by the small changes in welding speed and gas flow rate. The combinations of lower gas flow rate and welding speed produced the higher tensile strength compared to other flowrates while it is lower than the measured tensile strength.

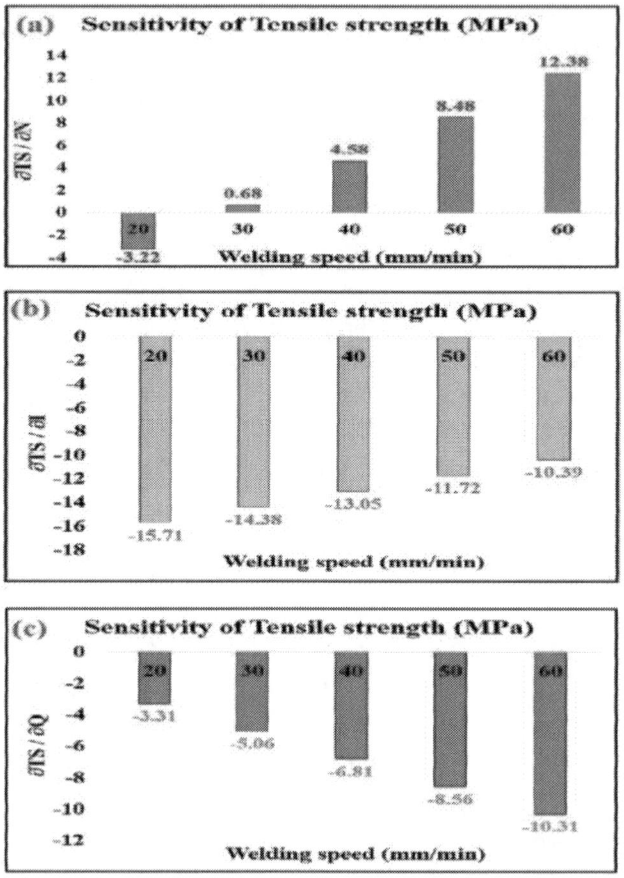

Figure 2 Sensitivity analysis on the tensile strength of the PAW process, (a) Welding Speed, (b) Welding Current and (c) Gas Flowrate

It can be seen that the maximum tensile strength was obtained at the higher welding speed with a constant welding current (250A) due to proper shielding gas nearby and at the welded region which controls the environmental oxidation process is presented in figure 3. The lower strength was achieved at the higher gas flow rate with constant welding current (250A) due to the oxidation process was occurring at the welded region [31] which degrades the strength of the welded joints. The results were revealed from the figure 2, the welding speed was the more sensitive than the other parameters followed by the welding current and gas flow rate.

IOP Conf. Series: Materials Science and Engineering **390** (2018) 012042 doi:10.1088/1757-899X/390/1/012042

3.4 Microstructural Analysis

The microstructural analysis of the plasma arc welded joints was discussed to decide the behaviour of the welded joints on the strength properties. The dissimilar plasma arc welded joints of AA8011-H24 aluminium alloys and Ti3Al2.5V Titanium alloys was presented in Figure 3. In this work, the microstructures were examined at different zones of the welded joints. Figure 3(a) showed that the coarse grains appeared in the base AA8011-H24 of aluminium alloys whereas the grain growth was found at the heat affected zone (HAZ) in Figure 3(b).

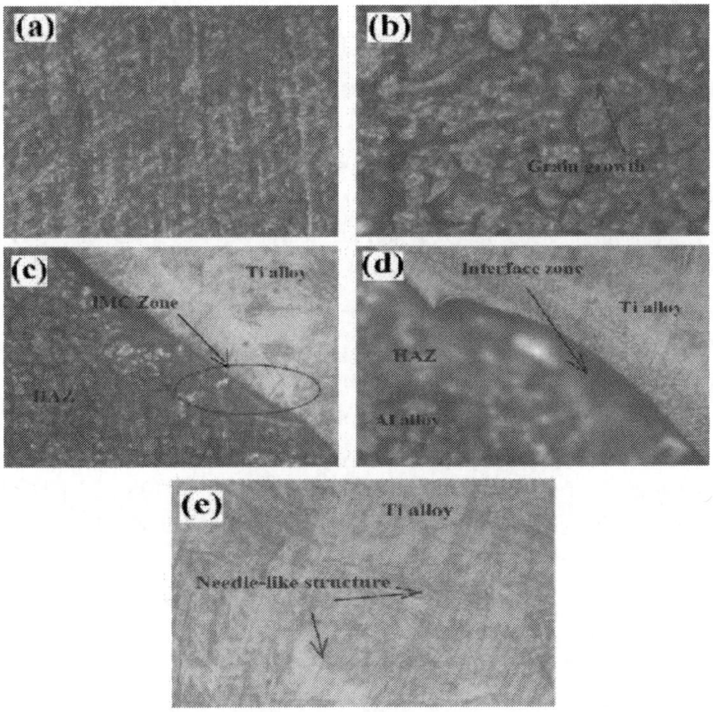

Figure 3 Microstructural Analysis of the PAW butt joints

The different crystalline structure was formed at the weld interface between the aluminium alloys and titanium alloys in Figure 3(c) which degrades the strength of the welded joints. The clear interface zone was found in Figure 3(d) nearby the titanium alloys which degrades the strength of the joints and induced the grain growth due to the heat input [32]. The needle-like structure appeared in Figure 3(e) which was shown in the titanium material.

4 Conclusions

The optimization of the PAW process parameters on the tensile properties of the AA8011 aluminium alloys and Ti3Al2.5V Titanium alloys were investigated using the Response surface methodology. From this investigation, the following important conclusions are listed below:

- The mathematical relationship was obtained to predict the tensile properties of the plasma arc welded dissimilar AA8011 aluminium alloys and Ti3Al2.5V Titanium alloys joints by incorporating the input parameters at 95% confidence level.
- The sensitivity analysis was applied to investigate the sensitiveness of the response, tensile strength of the welded joints with small variations of process parameters of the dissimilar joints.

- The welding speed was the more sensitive parameter compared to the other PAW parameters followed by the welding current and gas flow rate.
- The intermetallic region was observed at the interface which was nearby the titanium alloys which reduces the strength of the joints and induced the grain growth due to the heat input.

5 References

[1] Kreimeyer M, Wagner F and Vollertsen F 2005 *Opt. Las. Engg.* **43** 1021-35
[2] Chen YC and Nakata K 2009 *Mater. Des.* **30** 469-74
[3] Palani K, Elanchezhian C, Vijaya Ramnath V, Bhaskar GB, Sai Jagadeesh J and Kumar GM 2017 *Int. J. Res. Appl. Sci. Engg. Technol.* **5** 437-42
[4] Tomashchuk I, Sallamand P, Méasson A, Cicala E, Duband M and Peyre P 2017 *J. Mater. Proc. Tech.* **245** 24-36
[5] Peyre P, Berthe L, Dal M, Pouzet S, Sallamand P and Tomashchuk I 2014 *J. Mater. Proc. Tech.* **214** 1946-53
[6] Umanath K, Palanikumar K, Balasubramanian V and Selvamani ST 2015 *Appl. Mecha. Mater.* **766** 757-64
[7] Tahsini SM, Halvaee A and Khosravi H 2016 *J. Cent. South Univ.* **23** 1839-46
[8] Palani K and Elanchezhian C 2015 *App. Mech. Mat.* **814** 451-55
[9] Palani K and Elanchezhian C 2015 *App. Mech. Mat.* **766** 921-27
[10] Montgomery DC 2017 9th ed. Wiley J & Sons, New York
[11] Palani K, Elanchezhian C, Bhaskar GB and Ramnath BV 2018 *Adv. Sci. Engg. Med.* **10** 1-6
[12] Shi Y, Zheng Z and Huang J 2013 *Trans. Nonfer. Met. Soc. China* **23** 1977-84
[13] Raja Dhas JE and Satheesh M 2013 *Ind. J. Eng. Mater. Sci* **20** 425-34
[14] Yeni C, Sayer S and Pakdil M 2009 *Kovov. Mater.* **47** 341-47
[15] Dongxia Y, Xiaoyan L, Dingyong H, Zuoren N and Hui H 2012 *Opt. Las. Tech.* **44** 2020-25
[16] Elanchezhian C, Ramnath BV, Venkatesan P, Sathish S, Vignesh T, Siddharth RV, Vinay B and Gopinath K 2014 *Proc. Engg.* **97** 775-82
[17] Arun Kumar N, Elanchezhian C and Ramnath BV 2014 *Int. J. Engg. Res.* **3** (1) 148-50
[18] Rajakumar S, Muralidharan C and Balasubramanian V 2011 *Int. J. Manuf. Res.* **6** (3) 215-35
[19] Karthikeyan R and Balasubramanian V 2012 *Int. J. Manuf. Res.* **7** (3) 257-72
[20] Saravanan V, Rajakumar S and Muruganandam A 2016 *J. Adv. Mech. Engg. Sci.* **2** (4) 19-24
[21] Saravanan V, Banerjee N, Amuthakkannan R and Rajakumar S 2015 *Appl. Mech. Mater.* **787** 350-54
[22] Palani K, Elanchezhian C 2015 *Appl. Mecha. Mater.* **813** 446-450
[23] Palanivel R and Koshy Mathews P 2012 *J. Cent. South Univ.* **19** 1-8
[24] Juang SC and Tang YS 2005 *J. Mater. Proc. Tech.* **160** 128-37
[25] Vijayan D and Rao VS 2016 *J. Cent. South Univ.* **23** 1847-57
[26] Benyounis KY, Olabi AG and Hashmi MSJ 2005 *J. Mater. Proc. Tech.* **164** 978-85
[27] Karthikeyan R and Balasubramanian V 2013 *Exp. Tech.* **37** (2) 6-15
[28] Kim IS, Son KJ, Yang YS and Yaragada PKDV 2003 *Int. J. Mach. Tools Manuf.* **43** 763-69
29] Karaoglu S and Secgin A 2008 *J. Mater. Proc. Tech.* **202** 500-07
[30] Tian X, Liu Y, Deng W and Liu G 2015 *Int. J. Adv. Manuf. Tech.* **78** (1-4) 481-92
[31] Taban E and Kaluc E 2007 *Kovov. Mater.* **45** 241-48
[32] Balasubramanian M, Jayabalan V and Balasubramanian V 2007 *Int. J. Manu. Res.* **2** (4) 403-13

IOP Conf. Series: Materials Science and Engineering **390** (2018) 012043 doi:10.1088/1757-899X/390/1/012043

Eigen-frequency analysis of spherical shell laminated composite plates with and without central cutouts using finite elements

Ranjan K. Behera[1,*], Kushagra Garg[1], Sudhansu S. Patro[1], Nitin Sharma[1]

[1]School of Mechanical Engineering, Kalinga Institute of Industrial Technology, Deemed to be University, Bhubaneswar, Odisha-751024, India

*Email: ranjan.beherafme@kiit.ac.in; ranjancet@gmail.com

Abstract. This article presents the free vibration responses of spherical shell laminated composite panels (SSLCP) without and with central cutouts (square, circular and rectangular) using finite elements. The SSLCP midplane kinematics has been modeled using ABAQUS commercial finite element tool in the framework of first order shear deformation theory. Firstly, the validity of the present model is established by comparing the present natural frequencies values with the available benchmark results. Subsequently, the influence of both varying and constant aspect ratio for different central cutouts on the natural frequencies of SSLCP are investigated by solving appropriate numerical examples and findings are discussed in detail. The results are given in the form of tables and figures.

1. Introduction

Now a day's spherical shell laminated composite panels (SSLCP) are extensively used in various engineering applications such as ships, aircrafts, missiles, automobiles, mining equipments, railway wagon and civil structures etc. SSLCP are widely used due to weight sensitive, increase strength and stability, corrosion resistance, easy for fabrication and low cost.

Most of the researchers have used modal analysis for finding modal parameters by using different techniques. For dynamic analysis finite element analysis (FEA) is the best tool among the different numerical techniques. So many theories such as classical laminated plate/shell theory, first order deformation theory, higher order deformation theory etc. have been used for modal analysis of SSLCP.

Different computational models for laminated composite shell/plates were proposed by Reddy [1], Noor et al [2], Reddy and Chandrashekhar [3]. Chakravorty et al [4] solved many numerical problems of doubly curved laminated composite panels. Chakravorty et al [5] analyzed both free and force vibration on laminated shells with and without cutouts. Sahoo [6] solved numerical problems of laminated composite stiffened spherical panels with and without cutouts. Tornabene [7] formulated higher order layer wise theory for free vibration analysis of doubly-curved laminated composite panels. Singh [8] solved many numerical problems both static and free vibration behavior of laminated composite spherical shell panel. Fazzolari [9] formulated higher order shear deformation theory (HSDT) for free vibration analysis of cross-ply laminated composite both cylindrical and spherical shells. Chao et al [10] calculated ultimate strength of curved stiffened pates experimentally. Tornabene et al [11] formulated 2D higher order layer theory for free vibration of thin and thick doubly curved laminated composite panels. Lee and Han [12] computed forced vibration of laminated composite plates subjected to arbitrary loading.

Thus, it is seen that free vibration of SSLCP with and without cutouts requires in depth for proper understanding of the mechanical behavior. The objective of this work is to model and simulate the free vibration behavior of spherical shell laminated composite panels (SSLCP) with and without cutouts having different boundary conditions. Firstly, the SSLCP is modeled using commercial FEA tool (ABAQUS) in the framework of first order shear deformation theory (FSDT). An eight-nodded serendipity element having six degrees of freedom per node is utilized to discretize the panel. Subsequently, the validity of the present model is established along with the convergence. Finally, several numerical examples have been solved to study the influence of the variation of size of cutouts like square, rectangular and circular on the natural frequencies of the SSLCP and discussed in detail.

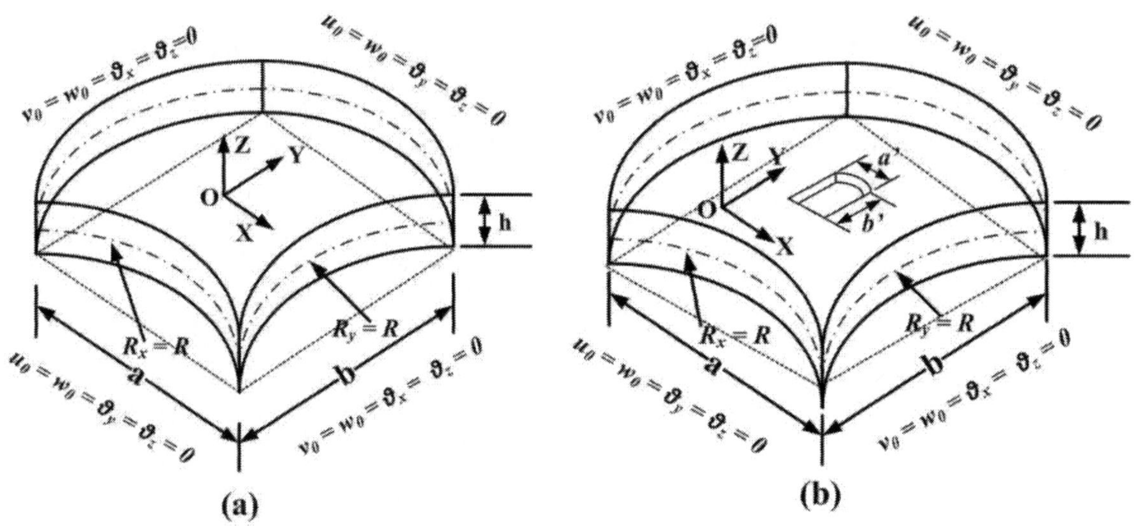

Figure 1. Simply supported (a) Spherical shell panel without cutout, (b) Spherical shell panel with central cutout

Nomenclature

a,b	length and width of SSLCP in plan
a',b'	length and width of cutout in plan
r	radius of circular cutout in plan
h	thickness of panel
E_{11}, E_{22}	elastic moduli
G_{12}, G_{13}	shear moduli
v_{12}	poison's ratio
R_{xx},R_{yy}	radii of curvature panel
ρ	density of material
ω_n	natural frequency
ϖ	non-dimensional natural frequency $\left[\varpi = \omega_n a^2 \sqrt{\rho / E_{22} h^2}\right]$

2. Vibration analysis of SSLCP

The SSLCP made of orthotropic layers having equal thickness and placed symmetrically about its middle surface, which subjected to the simply supported boundary conditions as shown in Figure 1 (a) & (b) without and with central cutouts respectively. Neglect coupling effect due to bending-twisting and neglect shear extension. The middle plane dynamics for plate or shell structure is calculated based on the FSDT and given as:

$$\begin{aligned} u(x,y,z,t) &= u_0(x,y) + z\theta_x(x,y) \\ v(x,y,z,t) &= v_0(x,y) + z\theta_y(x,y) \\ w(x,y,z,t) &= w_0(x,y) + z\theta_z(x,y) \end{aligned} \quad (1)$$

where, u, v and w are the displacement vector of any point in the layer at time t along x, y and z direction respectively. Modal analysis is used for this structure and its modal parameters are calculated by solving the eigen value equations:

$$([K]) - \omega^2[M]\{\varphi\} = 0 \quad (2)$$

where $[K]$, $[M]$, ω and $\{\varphi\}$ are the stiffness matrix, mass matrix, natural frequency of vibration and corresponding mode shape vector, respectively.

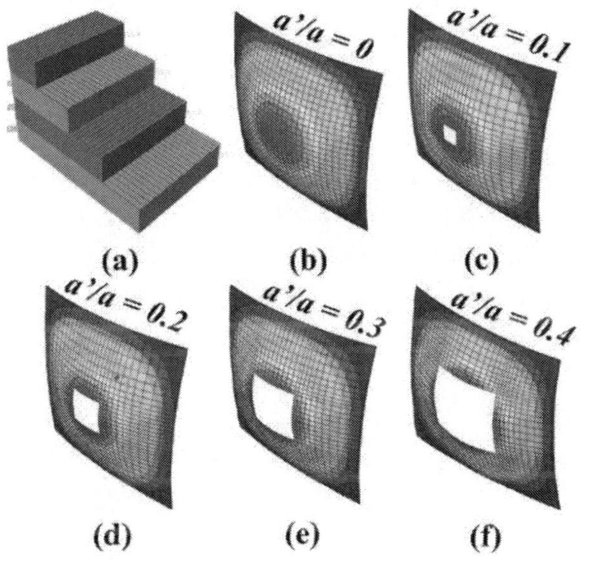

Figure 2. (a) Ply orientation ($0^0/90^0/0^0/90^0$) of laminated panel, (b) - (f) First mode shapes of SSLCP for different sizes of the central square cutouts

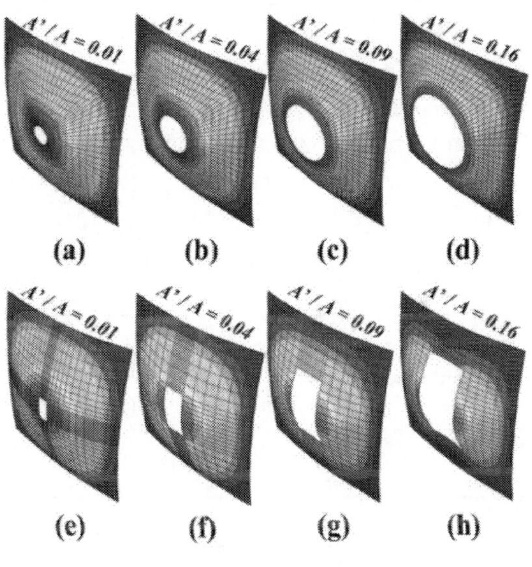

Figure 3. (a)-(d) First mode shapes of SSLCP for different sizes of the central circular cutouts, (e)-(h) First mode shapes of SSLCP for different sizes of the central rectangular cutouts

3. Results and discussions

Modal analysis of SSLCP is carried out using ABAQUS to calculate modal parameters and compared with Reddy [1], Chakravorty [4] [5] and Sahoo [6]. The material properties [4] are taken for modal analysis as given below:

$$E_{11}/E_{22} = 25,\ G_{12} = G_{13} = 0.5E_{22},\ v_{12} = 0.25.$$

Element type S8R (from the ABAQUS element library) having eight nodes and six degrees of freedom per node has been used for discretization purpose. The first mode non-dimensional modal frequencies (ϖ) are taken into account for analysis. Figure 2 (a) represent ply orientation of laminated composite panel and Figure 2 (b) shows SSLCP without cutouts.

In the present section, the non-dimensional modal frequencies (ϖ) are calculated for different mesh sizes to verify the convergence of present analysis. For convergence study, the simply supported spherical shells laminated composite panel ($0^0/90^0$) is considered. The variation of non-dimensional modal frequencies (ϖ) with mesh sizes are shown in Table 1. It can be observed that, the values non-dimensional modal frequencies (ϖ) are nicely converging with refinement of mesh. Beyond 14 × 14 mesh size is not improved the results. The result of 14 × 14 mesh is good agreement with Reddy [1]and Chakravorty [4].

Table 1. Non-dimensional natural frequencies ($\left[\varpi = \omega_n a^2 \sqrt{\rho / E_{22} h^2}\right]$) for simply supported SSLCP ($0^0/90^0$)

Mesh Size	4 × 4	6 × 6	8 × 8	10 × 10	12 × 12	14 × 14	16 × 16	Reddy [1]	Chakravorty [2]
ϖ	47.862	46.841	46.594	46.358	46.129	45.940	45.799	46.002	45.801

[$h/R_{yy} = h/R_{xx} = 1/300$, $a/b = 1$, $a/h = 100$, $E_{11}/E_{22} = 25$, $G_{12} = G_{13} = 0.5E_{22}$, $v_{12} = 0.25$]

3.1. Free vibration analysis of simply supported spherical shells laminated composite panel (SSLCP) without cutout

Non-dimensional natural frequency (ϖ) of SSLCP having different ply orientations ($0^0/90^0$) and ($0^0/90^0/0^0$) with different h/R_{xx} ratios are presented in Table 2 along with exact results obtained by Reddy [1] and Chakravorty [4]. In both cases the present results exhibit good agreement with others. With increase the R_{xx} or decrease the h/R_{xx}, the non-dimensional natural frequency (ϖ) decreases.

Table 2. Non-dimensional natural frequencies (ϖ) for simply supported SSLCP without cutout

	($0^0/90^0$)			($0^0/90^0/0^0$)		
h/R_{xx}	Reddy [1]	Chakravorty [4]	Present	Reddy [1]	Chakravorty [4]	Present
Plate	9.6873	9.6893	9.7500	15.183	15.192	15.181
1/300	46.002	45.801	45.799	47.265	47.035	47.972
1/400	35.228	35.126	36.078	36.971	36.89	37.412
1/500	28.825	28.778	29.216	30.993	30.963	31.305

[$R_{xx}/R_{yy} = 1$, $a/b = 1$, $a/h = 100$, $E_{11}/E_{22} = 25$, $G_{12} = G_{13} = 0.5E_{22}$, $v_{12} = 0.25$]

3.2. Free vibration analysis of spherical shells laminated composite panel (SSLCP) with central square cutout

Non-dimensional natural frequency (ϖ) of SSLCP ($0^0/90^0/0^0/90^0$) with central square cutouts in two boundary conditions, simply supported (SSSS) and clamped (CCCC), with different *a'/a* ratios are presented in Table 3 along with exact results obtained by Sahoo [6] and Chakravorty [5]. In Figure 2 (b)-(f) shows first mode non-dimensional natural frequency with varying aspect ratios (*a'/a*). In SSSS boundary conditions, with increase the aspect ratios (a'/a) the non-dimensional natural frequency (ϖ) increases. But this type of results are not seen in CCCC boundary conditions.

Table 3. Non-dimensional natural frequencies (ϖ) for SSLCP ($0^0/90^0/0^0/90^0$) with central square cutouts as shown in Figure 2 (b)-(f) in different boundary conditions

	(SSSS)			(CCCC)		
a'/a	Sahoo [6]	Chakravorty [5]	Present	Sahoo [6]	Chakravorty [5]	Present
0.0	47.100	47.109	47.866	117.621	118.197	118.205
0.1	47.114	47.524	48.522	104.251	104.274	104.578
0.2	48.801	48.823	49.763	97.488	98.299	99.015
0.3	50.920	50.925	51.789	113.226	113.766	114.236
0.4	53.788	53.789	54.503	110.094	110.601	110.986

[$h/R_{yy} = h/R_{xx} = 1/300$, $a/b = a'/b' = 1$, $a/h = 100$, $E_{11}/E_{22} = 25$, $G_{12} = G_{13} = 0.5E_{22}$, $v_{12} = 0.25$]

3.3. *Free vibration analysis of simply supported spherical shells laminated composite panel (SSLCP) with central square, circular and rectangular cutouts*

Non-dimensional natural frequency (ϖ) of simply supported SSLCP ($0^0/90^0/0^0/90^0$) with central square, circular and rectangular cutouts with different A'/A ratios are presented in Table 4. Here we take $a'/b' = 2$ for rectangular cutout. Increase the A'/A ratio increase the non-dimensional natural frequency (ϖ) for both square and circular central cutouts. But in case of rectangular cutouts, the non-dimensional natural frequencies (ϖ) are more very at higher A'/A ratio. For constant A'/A ratio, the square cutouts give more natural frequencies followed by circular cutouts and then rectangular cutouts. But at higher A'/A ratio, the natural frequencies of circular cutouts give more value than square cutouts.

Table 4. Non-dimensional natural frequencies (ϖ) for simply supported SSLCP ($0^0/90^0/0^0/90^0$) with central square, circular and rectangular cutouts as shown respectively

A'/A	Square cutout	Circular cutout	Rectangular cutout
0.0	47.866	47.866	47.866
0.01	48.522	48.217	48.396
0.04	49.763	49.518	48.965
0.09	51.789	51.704	48.161
0.16	54.503	54.823	42.928

[$h/R_{yy} = h/R_{xx} = 1/300$, $a/b = 1$, $A = ab$, $A' = a'b' = \pi r^2$, *Square cutout* ($a'/b' = 1$), *Rectangular cutout* ($a'/b' = 2$), *Circular cutout* ($r = \sqrt{a'b'/\pi}$, $a/h = 100$, $E_{11}/E_{22} = 25$, $G_{12} = G_{13} = 0.5E_{22}$, $v_{12} = 0.25$]

4. Conclusion

The following conclusions are drawn from the present study:

(a) Among simply supported and clamped boundary conditions, clamped boundary condition is the more desirable for SSLCP with or without cutouts.

(b) With increase the R_{xx} or decrease the h/R_{xx}, the non-dimensional natural frequency (ϖ) decreases in case SSLCP without cutouts.

(c) In SSSS boundary conditions, with increase the aspect (a'/a) ratios the non-dimensional natural frequency (ϖ) increases. This type of results are not seen in CCCC boundary conditions. But its values are more desirable than former.

(d) Increase the A'/A ratio increase the non-dimensional natural frequency (ϖ) in square, circular and rectangular central cutouts. But in case of rectangular cutouts, the non-dimensional natural frequencies (ϖ) are not so desirable as compared to other two cutouts. At higher A'/A ratio, the rectangular cutouts give very less desirable outputs. For constant A'/A ratio, the square cutouts give more natural frequencies followed by circular cutouts and then rectangular cutouts. But at higher A'/A ratio, the natural frequencies of circular cutouts give more value than square cutouts.

References

[1] Reddy J N 1984 Exact Solutions of Moderately Thick Laminated Shells *J. Eng. Mech.* **110** 794–809

[2] Noor A K, Burton W S and Peters J M 1991 Assessment of computational models for multilayered composite cylinders *Int. J. Solids Struct.* **27** 1269–86

[3] Reddy J N and Chandrashekhara K 1985 Geometrically non-linear transient analysis of laminated, doubly curved shells *Int. J. Non. Linear. Mech.* **20** 79–90

[4] Chakravorty D, Bandyopadhyay J N and Sinha P K 1996 Finite element free vibration analysis of doubly curved laminated composite shells *J. Sound Vib.* **191** 491–504

[5] Chakravorty B D and Bandyopadhyay N 1998 Applications of Fem on Free and Forced Vibration of Laminated Shells *J. Eng. Mech.* **124** 1–8

[6] Sahoo S 2014 Laminated composite stiffened shallow spherical panels with cutouts under free vibration – A finite element approach *Eng. Sci. Technol. an Int. J.* **17** 247–59

[7] Tornabene F 2016 General higher-order layer-wise theory for free vibrations of doubly-curved laminated composite shells and panels *Mech. Adv. Mater. Struct.* **23** 1046–67

[8] Singh V K 2015 LINEAR STATIC AND FREE VIBRATION ANALYSES OF LAMINATED COMPOSITE *ASME 2013 Gas Turbine India Conf. GTINDIA 2013* 1–7

[9] Fazzolari F A 2014 A refined dynamic stiffness element for free vibration analysis of cross-ply laminated composite cylindrical and spherical shallow shells *Compos. Part B Eng.* **62** 143–58

[10] Cho S-R, Park H-Z, Kim H-S and Seo J-S 2007 Experimental and numerical investigations on the ultimate strength of curved stiffened plates *10th Int. Symp. Pract. Des. Ships other Float. Struct. PRADS 2007* **1**

[11] Tornabene F, Viola E and Fantuzzi N 2013 General higher-order equivalent single layer theory for free vibrations of doubly-curved laminated composite shells and panels *Compos. Struct.* **104** 94–117

[12] Lee W H and Han S C 2006 Free and forced vibration analysis of laminated composite plates and shells using a 9-node assumed strain shell element *Comput. Mech.* **39** 41–58

IOP Conf. Series: Materials Science and Engineering **390** (2018) 012044 doi:10.1088/1757-899X/390/1/012044

Design and Development of Peristaltic Soft Robot Using Shape Memory Alloy Actuators with different control strategies

Rohit Patil[1*], Nandini Patra[2], Abhinav Sharma[2], Kavitha P[1] and I APalani[2*]

[1]Control and Automation Department, VIT University, Vellore, Tamil Nadu, India
[2]Metallurgy Engineering and Materials Science, IIT Indore, Indore, India

E-mail:*patilrohit.manohar2016@gmail.com, palaniia@iiti.ac.in

Abstract: This paper presents design, development and different optimal control strategies for a spiral, shape memory alloy (SMA) actuators for robotics applications. The actuator material used is thermally activated Ni-Ti SMA which exhibits high torque capacity. SMA based actuators are widely used in robotics applications because of their large energy density and super-elasticity. Here describe the multi-direction movement of soft robot based on SMA actuators. To controlling precise position in minimum time for SMA actuator suitable controller require. Here we investigated the different ideal controlling method for design soft robot model. The Control strategies for soft robot developed using Matlab Simulink; here we investigate P, PI, PD and PID types of controller. For real-time controller implementation we compare the controller responses with respect to the desired set point of operating robot. Thus, we can identify the PID controller is suitable for real-time designing the soft robot model depending upon different controlling responses and least 3.47 integrate absolute error (IAE) and 2.119 integrate square error (ISE).

1. Introduction:

Humans have often looked at biological systems for that answers are engineering problems throughout our existence. Self-assembly, Self-healing abilities, resistance, hydrophobicity, and environmental exposure tolerance etc, this type of engineering problem solving inspired by nature. This ability and knowledge of designing and building bio motivated apparatus are known as biomimicry or biomimetic [1]. The inspiration for designing of a soft robot has come from the snake and earthworm creatures. Here the designing of the soft robot is based on PET (Polyethylene terephthalate) material and SMA actuator for providing the flexible movement of robot body [2].The mechanisms of traditional rigid-body consist of stiff links coupled at movable joints, but in a dynamic environment, they may not survive. Thus, flexible mechanism helps to change the shape and geometry of robot according to nature. However, the rigid mechanism also transfers motion, force, and energy. But the major difference is that the flexible mechanism is more advantageous than rigid one in respect to the mobility point of view. Most of the flexible mechanisms are a single body and does not encounter any frictional losses. Limbless crawling is a primary form of biological locomotion adopts by a wide selection of species, including the earthworm, snails as well as the snake. To achieve a successful peristaltic locomotion certain elements are required. Polyethylene terephthalate material used as a body of the soft robot, it also useful for longitudinal movement of the robot. At the initial stage, PET material was a hard type of material. Therefore, the training of PET segment is essential [3]. Thus, the training of PET tube is done by MS rod in a furnace. We have fixed the PET tube around the 5cm MS steel rod and put inside the furnace at 225°Ctemperatures up to 15 minutes. Therefore, the PET tube is flexible towards the longitudinal movement of the soft robot. Moreover, for radial muscle movement, the combination of normal bias spring and SMA spring are used. We have specifically used the SMA spring material; SMA shows unique properties of shape memory effect and super elasticity with the application heat energy. SMA is a material which remembers its original shape or properties. Thus, when the SMA material deformed due to the external force, it returns to its pre-defined shape due to the application of heat above its actuation temperature. Moreover, when the electric supply provided to the SMA spring, the compression of SMA spring was observed. Further, the force generated from the bias connected normal spring helped to bring the SMA to its original position [4, 5 &6]. This paper described the fabrication of peristaltic soft robot and developed the different type controller to control the directions of the soft robot by using Simulink. Here different types of controller used like proportional, proportional integral, proportional derivative, cascade PID. Each controller has their own application and advantages like Proportional (P) controller used for stabilized the gain of system response but produces steady state error because of that integral action required to reduce offset error. The Proportional integral (PI) controller used to reduce steady-state error as well as stabilize the gain of the system, but system response gets slow. To improve system response derivative action is required but in proportional derivative (PD) steady state error still present. Cascade PID controller take all advantage of proportional, integral, derivate and operated on the system, PID also type intelligent controller. To design the ideal controller for soft robot help to reach at the précised position in the least time as well as least losses.

2. Working principle:

To achieve the robotic movements, some important elements are required such as longitudinal muscle, Radial muscle and Coelom. The Table1 details the bio inspired robotics part for earthworm robot;

Table1. Bio inspired soft robotics part mechanism

Sr. No.	Peristaltic Body Part	Robot Body Part
1	Main Body	Braided mesh tube of PET material
2	Radial Muscles	NiTi actuators with bias spring
3	Longitudinal Muscles	Braided mesh tube made of PET or NiTi actuators
4	Coelom	Longitudinal, flexible, tendons
5	Brain	Arduino and Relay
6	Intelligence	Intelligent control algorithm

This investigation describes the fabrication of robot where different parts of the soft robot are detailed in above table1. Figure 1 shows the working principle of the peristaltic soft robot where the robot made by four number of segment and each segment containing two radial actuators. The electric supply provided to the first segment of a radial actuator. Thus, when the first segment compressed, the diameter gets decreased and thereafter expand in a horizontal direction [2]. Longitudinal muscle of PET tube helps to bring back the compressed segment in original position. Similarly, the operating procedure is applied to all the segments sequentially. Thus, the robot moves in forward and backward directions due to the sequential segment movement. Depend upon requirements of application we can arrange the three SMA actuator on the body of the soft robot for achieving the right, left and upward directional movement. Increases in a number of segments improve the flexibility and accuracy in the form of displacement. Figure3 shows the different way of fabrication of real time Peristaltic soft robot. The SMA spring actuator placed inside the body of the robot according to the required movement.

3. Mathematical Model:

The mathematical representation of peristaltic soft robot model [7] with all operating applicable parameter of a mechanical model as shown in below figure 4;
Where,
M_1=M_2= mass of actuator (half of SMA spring and half of PET mesh tube weight) =9.5×10^{-3}kg
K_1 and K=spring constant for 7 no. of turns and 20 no. of turns respectively=10kg/m

F= force in the form 3 Ampere current and a 5Volt voltage applied for movement
B= Friction in terms of the effect of environmental condition, kinematic friction & power losses etc.

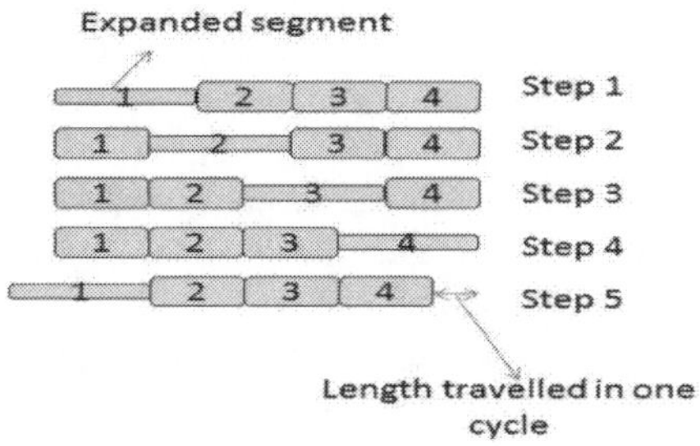

Figure1. Working principle of movement the peristaltic Soft Robot.

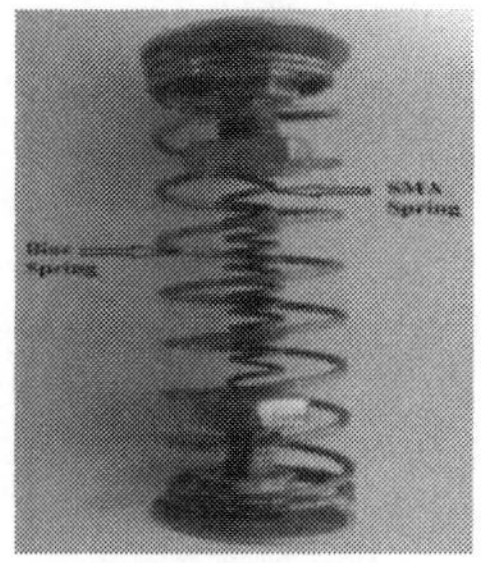

Figure2. Configuration of SMA and bias spring actuator for radial muscles Robot.

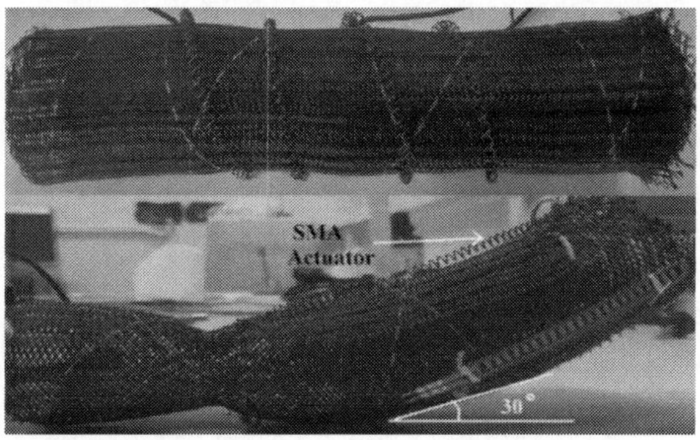

Figure3. Fabrication of real time Peristaltic soft robot model

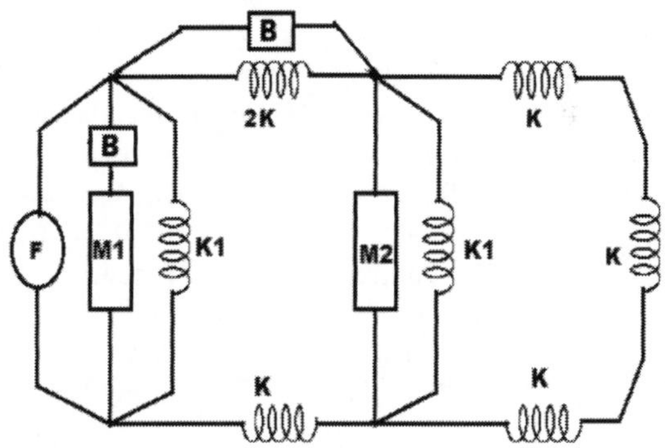

Figure4. Mechanical model of peristaltic soft robot segment

The mathematical representation above mechanical model in the form of linear differential equation and transfer function of s- domain are shown in below equations;

$$BM_1\frac{d^3x_1}{dt^3}+B\frac{d}{dt}(x_1-x_2)+k_1x_1+3k(x_1-x_2)=f \tag{1}$$

$$M_2\frac{d^2x_2}{dt^2}+B\frac{d}{dt}(x_2-x_1)+k_1x_2+3kx_2+3k(x_2-x_1)=0 \tag{2}$$

$$M_1\frac{d^2x_1}{dt^2}+B\frac{d}{dt}(x_1-x_2)+k_1x_1+2k(x_1-x_2)=f \tag{3}$$

$$M_2\frac{d^2x_2}{dt^2}+B\frac{d}{dt}(x_2-x_1)+k_1x_2+2kx_2+2k(x_2-x_1)=0 \tag{4}$$

$$T_1(s)=\frac{S^2+17S+42.5}{S^5+4S^4+16S^3+55S^2+50S+36} \tag{5}$$

$$T_2(s)=\frac{S^2+18S+40.5}{S^4+55S^3+30S^2+63S+47} \tag{6}$$

The number of segment are connected into the series because of that the overall transfer function of peristaltic soft robot model is,

$$T(S)=\frac{S^4+31S^3+261S^2+1063S+1620}{S^9+8S^8+65S^7+282S^6+1031S^5+2531S^4+4644S^3+5832S^2+4419S+1620} \quad (7)$$

4. Controller Designing:

Designed soft robot model is nonlinear type system. Thus, we required to developing a nonlinear controller which will control the model. The designed controllers will subsequently reduce the delay time in process monitoring and command in the mesh worm to drive in the desired trajectory and it works earlier it will be highly effective in terms of energy saving capabilities [8]. Here we implement a different controller for soft robot model in matlab Simulink, response of soft robot model which help to identify the better controller to implement in real-time system with hardware of soft robot model [9].

4.1. PID controller:

Each controller has its own application and advantages. The P controller used for stabilized the gain of system response but produces steady-state error because of that integral action required to reduce offset error. The PI controller used to reduce steady-state error as well as stabilize the gain of the system. However, the system response gets slow. To improve the system response, a derivative action is required. In the cascade PID controller, each controller responses are performs separately, because of that we get all controller advantages. Moreover, an ideal controller designing for the soft robot we have to compare the controller output responses, Integral absolute errors (IAE), and integral square errors (ISE). The basic block diagram of PID controller block diagram is shown in figure 4. The parameters of PID controller are calculated with the help of Ziegler- Nichol's method [10 &11]. The block diagram of cascade PID controller shown in below Figure 5 and its basic differential equation is,

$$u(t)=k_p e(t)+k_i\int_0^t e(\tau)d\tau+k_d\frac{d}{dt}e(t) \quad (8)$$

By using the Z-N method the standard parameter of PID controller, from below table2 and a table3 parameter of PID used to find out different controller responses of the system.

Table2. Standard value of controller by Z-N method

Controller	$K_p = 0.6K_{cr}$	$K_i = {}^1/_{T_i} = {}^1/_{0.5P_{cr}}$	$K_d = T_d = 0.125P_{cr}$
P	1	0	0
PI	1	0.9242	0
PD	1	0	0.72
PID	1	0.9242	0.72

Table3. Tuned value of controller by Z-N method [12]

Controller	K_p	K_i	K_d
P	1.1621	0	0
PI	0.01825	0.2134	0
PD	1.04421	0	0
PID	0.2477	0.3081	0

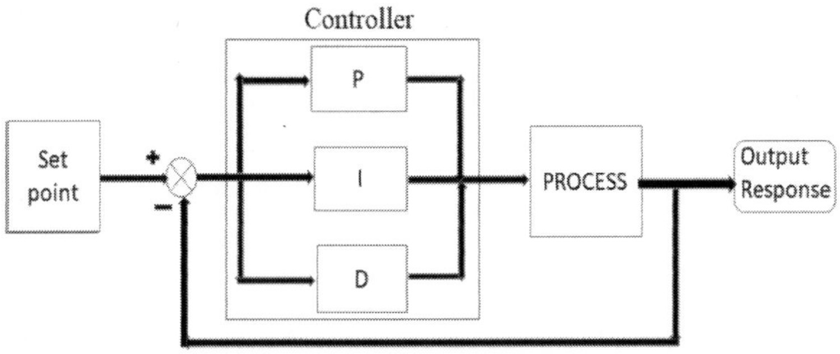

Figure5. Block diagram of cascade PID controller

5. Result and discussion:

Based on the mathematical model of the soft robot, we have investigated the controller which will perform better than the other controllers. By tuning the parameters of PID controller we have calculated different type error in the form area under the response like an ISE and IAE etc. Table 4 and table 5 show the ISE and IAE for standard PID Z-N parameter and tuned PID parameter. Based on the observation from both table, tuned PID parameters gives the best performance. It is also observed that cascade PID controller encounter less error which provides a great support towards the controller design in real time. Based on the analysis of the data (table 4 and table 5), we can conclude that cascade PID controller with tuned parameter generates less error and provides the value near to desired set point. Also, we observed that standard Z-N integral parameter is not suitable for the system because it covers a large amount of error area [1].

Table4. IAE and ISE of standard parameter

Controller	P	PI	PD	PID	Without Controller
IAE	10.62	24.64	10.29	3.826	12.62
ISE	6.552	39.98	7.19	2.474	7.556

Table5. IAE and ISE of tuned parameter

Controller	P	PI	PD	PID	Without Controller
IAE	9.878	3.474	10.29	3.47	12.62
ISE	6.36	2.73	6.364	2.119	7.556

Figure 6 shows the comparisons of different controller responses to the desired set point. Here we have assumed that the value of the desired set point is 1. Accordingly, all controllers try to get settled into the system at a desired point. Therefore, we can control the real-time system at a required desired location for movement of the soft robot. We already mentioned that depending upon the different movement the SMA actuator fabricated or embedded on a soft robot body. Figure 3 shows the real-time fabrication, which inclined 30 degrees from the ground with the help of top mounted SMA actuator on a tube. However, we can check the performance of robot according to the requirement through simulation before implementation or interfacing with hardware. Similarly, figure 7 shows the comparisons of different controller responses for tuner PID parameter. We can observe from table 5 and figure 7, those tuned PID parameters show improved system control comparing to the standard PID. Figure 7 shows all the responses are trying to reach closer to the desired set point in less error area. Such type of accuracy in controller help to performed desired operation. It is observed that cascade PID controller provide better control response in comparison to the other controller responses.

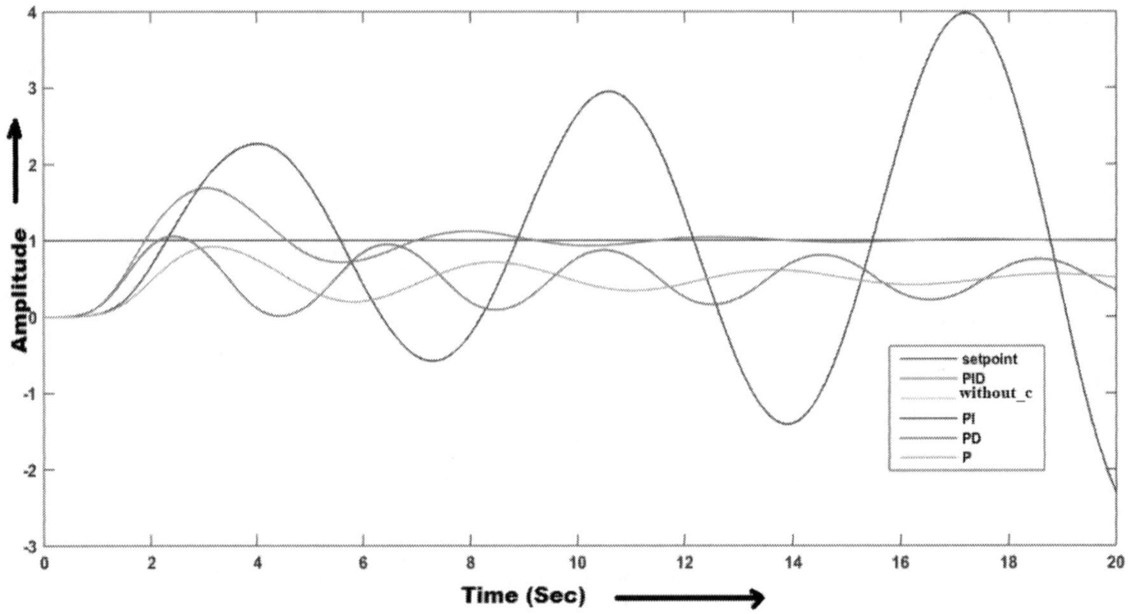

Figure6. Comparisons step response of different controller for standard parameter

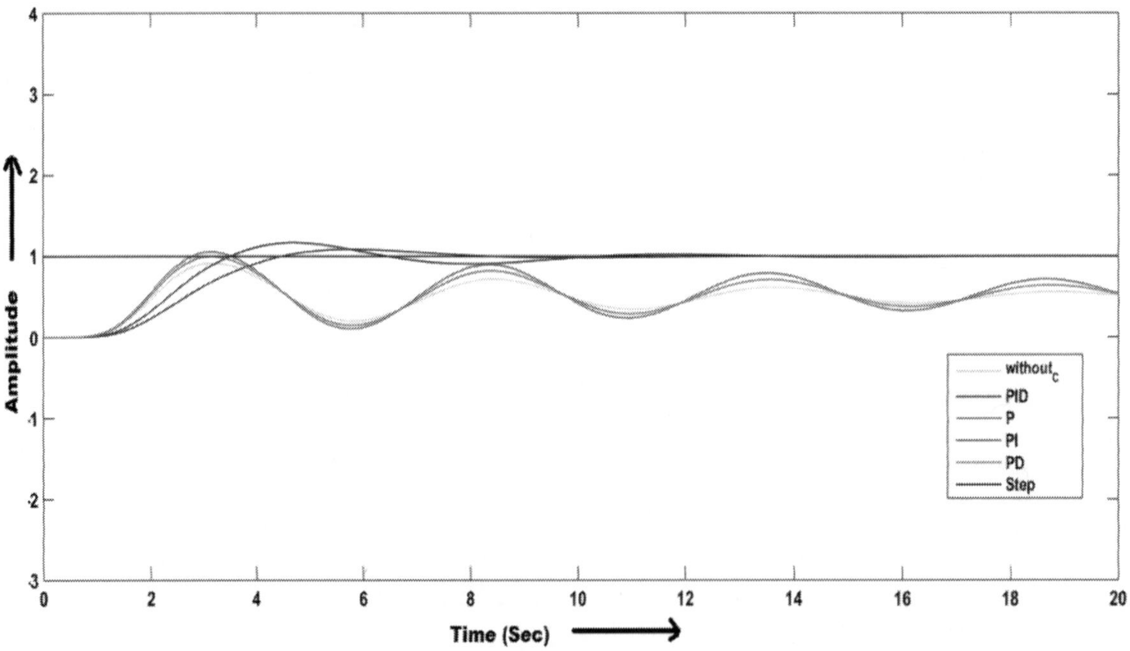

Figure7. Comparisons step response of different controller for tuned parameter

6. Conclusion:

This paper described the fabrication and development of different type PET material soft robot model with SMA actuator and bias spring actuator. We have also investigated the control strategies with the help of real-time mathematical model in Matlab Simulink. From the different controlling response, we identified cascade PID controller response provide better control towards the desired set point of the soft robot. Accordingly, depend upon the requirement of the multi-directional movement of soft robot we can easily change the desired position. The comparison of different controller response at a desired position for standard and tuned PID parameter was observed in figure 6 and figure 7. In cascade PID controller, we get steady-state response in minimum time and reduce the overshoot as well as an error of the system at desired position comparing to P, PI, PD and without a controller of the original system for a tuned parameter. From table 4 and table 5, the error (like IAE and ISE) comparisons between standard and tuned PID controller is clearly visible. Cascade PID controller provides less 3.47 IAE, 2.119 ISE and better controlled at desired position comparing to other controllers for tuned parameter [1]. With the help of this controller strategy, we can easily and accurately move the soft robot in different directions, like an example 20° left, 30° right and 30° upward etc according to operator requirement.

Reference:

[1] Seok S, Onal CD, Wood R, Rus Dand Kim S 2010 May. Peristaltic locomotion with antagonistic actuators in soft robotics. In *Robotics and Automation (ICRA) 2010 IEEE International Conference on* pp. 1228-33

[2] Seok S, Onal CD, Cho KJ, Wood RJ, Rus D and Kim S 2013 Meshworm: a peristaltic soft robot with antagonistic nickel titanium coil actuators. *IEEE/ASME Transactions on mechatronics* ***18*(5)** pp.1485-97

[3] Zhang Y, Su M, Li M, Xie R, Zhu H and Guan Y 2017 August A spatial soft module actuated by SMA coilIn *Mechatronics and Automation (ICMA) 2017 IEEE International Conference on* pp. 677-82

[4] Koh JS, Lee DY and Cho KJ 2012 November Design of the shape memory alloy coil spring actuator for the soft deformable wheel robot In *Ubiquitous Robots and Ambient Intelligence (URAI) 2012 9th International Conference on* pp. 641-42

[5] Vergata T, Politecnico V 2010 *Shape Memory Alloys*

[6] Souri M 2014 *Finite element modeling and fabrication of an SMA-SMP shape memory composite actuator* University of Kentucky

[7] Torres WL, Araujo IBQ, Menezes Filho JB and Junior AGC 2017 Mathematical Modeling and PID Controller Parameter Tuning in a Didactic Thermal Plant. *IEEE Latin America Transactions* ***15*(7)** pp.1250-56

[8] Karray F, Gueaieb W and Al-Sharhan S 2002 The hierarchical expert tuning of PID controllers using tools of soft computing *IEEE Transactions on Systems, Man, and Cybernetics Part B (Cybernetics)* ***32*(1)** pp.77-90

[9] Salem FA 2014 Modeling, Simulation and Control Issues for a Robot ARM Education and Research (III) *International Journal of Intelligent Systems and Applications* **6(4)** p.26

[10] Wang XB, Song SB and Li H 2016 Vehicle control strategies analysis based on PID and fuzzy logic control *Procedia engineering* ***137*** pp.234-43

[11] Karthick S and Palani IA 2016 July Fractional order PID control of an underactuated balancing system. In *Power Electronics Intelligent Control and Energy Systems (ICPEICES) IEEE International Conference on* pp. 1-5

[12] Heikkinen J, Minav T and Stotckaia AD 2017 May Self-tuning parameter fuzzy PID controller for autonomous differential drive mobile robot In *Soft Computing and Measurements (SCM) 2017 XX IEEE International Conference on* pp. 382-85

IOP Conf. Series: Materials Science and Engineering **390** (2018) 012045 doi:10.1088/1757-899X/390/1/012045

Study to measure the impact of social media usage on work-life balance

K P Aravinda Kumar[1,2] and Dr.R G Priyadarshini[1,3]

[1] Amrita School of Business, Amrita Vishwa Vidyapeetham, Coimbatore, India
[2] kparavind94@gmail.com
[3] rg_priyadarshini@cb.amrita.edu

Abstract: In today's world both individuals and the organisations have tough time in bringing work-life balance. In organizational perspective, a high standard of work–life balance is essential to continue attracting and retaining its employees; while at the individuals perspective it has a positive impact on attitude, behaviour, and commitment towards both professional and personal life. Work-life balance has become an on-going challenge in contemporary times. Though work-life balance is discussed for various decades and measured with different dimensions, this is high time to measure work-life balance in accordance with present lifestyle. Here comes the role of internet which is twisted with every human being in day to day activities. Internet without social media is unimaginable. Usage of social media results in both productive and unproductive behaviour. Therefore this is a study to measure the usage of social media in personal life and professional life on work-life balance. It is found that social media usage in professional life has more impact on work-life balance which caused due to pre-occupation with social media in workplace.

1. Introduction

Everyone have a number of roles to hold throughout life. It has become a great challenge when time demands us by our many roles become conflicting with one another. Work-life balance is termed as the general the idea of managing a job and also spending quality time to family. It is defined as the bifurcation of one's time for personal and professional activities without affecting each other. Everyone wants to lead a happy personal life and also to succeed in their career life.

In today's internet era, there is a major constraint which affects this work-life balance. It is found that most of the internet users spend their time in social media in the form of websites or applications. Social networking in social media is done by allowing its users to create and share content or to participate in the same. It is an online platform that allows people to build virtual relationship with other people including strangers who share similar professional or career interests, activities, backgrounds or real-life connections. LinkedIn is a social networking site designed specifically for the business community that helps business people and individuals to network. Other social networking sites include Facebook, twitter, YouTube, Pinterest, Google+ etc.

The usage of social media is exponentially increasing in day to day life. This can be broadly classified into personal and professional life usage. Personal life usage includes sharing of information, and other activities to a bounded group of friends, relatives etc., for non-work purpose. Similarly social media usage for professional life involves only for work related activities and create a network within the workplace for sharing data or for communication purposes. Both kinds of usage in social media have high impact towards the work-life balance. So this paper is all about measuring the impact of social media usage on work-life balance which has some serious effect on the individual, group & at the organisation level.

2. Literature Review

For a long period of time, various studies are proposed to measure work-life balance. It can be measured in various dimensions such as Personal Life Spillover in Work, work Spillover in Personal Life, Work/Life Behavioural Constrainers and Work/Life Behavioural Enhancers (Smita Singh, 2014). Studies also suggest that work-life balance has effect on organisational commitment (Pradhan, Jena, Kumari, 2016). It was also found that Generation Y employees have a positive attitude towards varying work timings, and also willing to participate in e-commuting compared to other generations. Generally, Generation Y is expected to give importance an ideal work/life balance (Uba, Dark, Megan, Yendt, Gnatek, 2012).

On the other side internet with social media is twisted with our day to day activities. Social media and the major functional blocks which contribute to social media such as groups, identity, conversations, reputations, sharing, presence and relationships (Jan, Hermkens, McCarthy, 2011). Social media sites and applications attract people based on common language or shared racial, sexual, religious, or nationality based identities and also connects strangers with communicating tools (Ellison, Boyd, 2008). Research are also done on how social media usage varies across different generations of people were Gen-Y contribute more to social media usage (Bolton, A. Parasuraman, Ankie, Gruber, Yuliya, Loureiro, Solnet, 2013). The study on social media is extended on the various consequences of its usage describing how it affects the day to day activities in work (Zoonen, Verhoeven, Vliegenthart, 2017). Technology and media use affect stress (cortisol) and inflammation (interleukinIL-6) in dual earning parents and their adolescents (Tamara, Nicole, Kathryn, Micelle, 2017)

Eventually the usage of social media ends up in consuming time which effects the work-life balance. It was found that increase in personal social media usage led to lower performance on the task, and lower happiness (Brooks, 2015). Contradicting the previous finding, it is also found that use of social networks at work helped employees improve their performance. Thereby suggesting the organizations to comprehend while setting appropriate policies at workplace for social media usage (Kishokumar, 2016).

3. Objective

The objective of the study is to measure the usage of social media in personal & professional life on work-life balance.

4. Research Question

- Does social media usage at professional life or personal life affect the work-life balance?
- Does social media usage affect work-life balance differ with respect to various types of gender?
- Does social media usage affect work-life balance differ with respect to various levels of work experience?

IOP Conf. Series: Materials Science and Engineering **390** (2018) 012045 doi:10.1088/1757-899X/390/1/012045

5. Null Hypothesis

- There is no significant association between social media usage at professional life and work life balance
- There is no significant association between social media usage at personal life and work life balance

6. Sample Characteristics & Methodology

The study is conducted in India and the sample is taken from various working professionals in the sectors namely, IT & services. The Respondents profile has been classified based on Gender, Age, total work experience and experience in the current organisation. The sample is collected from different age groups to eliminate the bias while majority of the samples fall under Gen-Y employees. A total of 110 respondents was received.

7. Conceptual Model

The following conceptual model is adopted for the study to measure social media usage in work-life balance. Social media usage in professional life is influenced by four factors as Intensity of social media use, task-oriented social media behaviour, relationship building social media behaviour, and deviant social media behaviour, pre-occupation with social media.

Intensity of social media use depends on how often an individual connects to a social media which may be due to a positive outcome of continuous social media usage. Productive behaviours in workplace such as using social media to get a work done or to build a good relationship with other colleagues in the workplace include task-oriented and relationship building social media behaviour. This contribute to the betterment of the individual and the organisation.

Unproductive behaviours in workplace include where employees are addicted with social media at work which is not for the work purpose and because of this they are more deviated from doing their work. This includes pre-occupation with social media and deviant social media behaviour

In this model Work-life balance is also determined with various factors such as work Spillover in Personal Life where family activities is changed because of work demands, Personal Life Spillover in Work where work activities is changed because of family demands, Work/Life Behavioural Enhancers that makes one effective at work also help in being a better family member. The decision making and problem solving approaches used at work is equally effective in resolving personal problems and Work/Life Behavioural Constrainers the work life expects to be more different than in personal life.

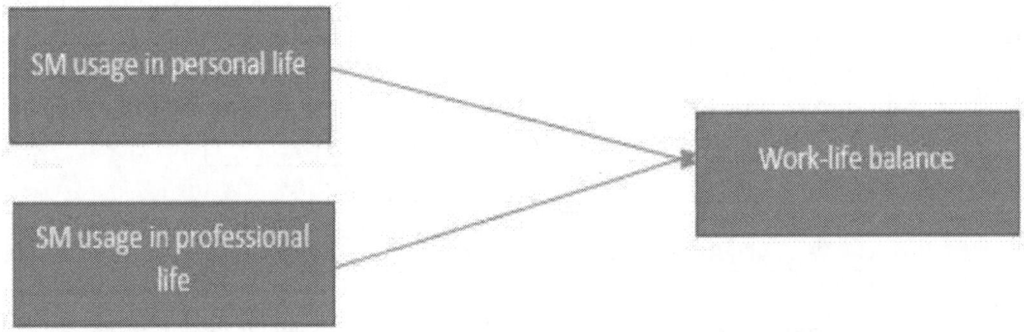

Figure 1. Conceptual model

8. Research Design

This is a quantitative study. Primary data is collected by the survey method. A set of questionnaire was developed based on the previous studies which is relevant to this study.

For measuring social media usage in personal life a set of seven questions is designed based on the paper "Social media? Get serious! Understanding the functional building blocks of social media" by Kietzmann, J. H., Hermkens, K., McCarthy, I. P., & Silvestre, B. S. (2011). Those functional blocks include groups, identity, conversations, reputations, sharing, presence and relationships.

For measuring social media usage in professional life based on the intensity of social media use, task-oriented social media behavior, relationship building social media behavior, deviant social media behavior, preoccupation with social media the paper "Social Media Use in the Workplace: A Study of Dual Effects" by Carlson, Zivnuska, Harris, Kenneth J. Harris, Dawn S. (2016) is used.

For measuring work-life balance in various dimensions such as work Spillover in Personal Life, Personal Life Spillover in Work, Work/Life Behavioural Enhancers and Work/Life Behavioural Constrainers the work of, Measuring work-life balance in India by Smita Singh (2014) is used which contains a set of 23 questions.

9. Analysis and Findings

9.1 Effect on work-life balance based on personal life and professional life social media usage

Table 1 represents the correlation between the social media usage in personal and professional life in accordance with work-life balance. 110 working professionals responded. From the below table it can be said that work-life balance is more positively correlated towards social media usage in professional life. This states that employees tend to use social media in both personal and professional life which has a positive effect on work-life balance

Table 1.WLB on professional & personal life SM usage

	Professional life SM usage	Personal life SM usage
Work-life Balance	0.390201	0.387412

9.2 Effect on work-life balance by personal life and professional life social media usage based on gender

Table 2 represents the correlation between the social media usage in personal and professional life towards work-life balance with respect to gender. The sample comprises of 77 male and 33 female. It is inferred from the table that work-life balance is more correlated (0.486496) by professional life social media usage for male respondents, while work-life balance is more correlated (0.223017) by personal life social media usage for female respondents.

Table 2. WLB by personal life and professional life SM usage based on gender

	Work-life balance	
	Male	Female
Professional life SM usage	0.486496	0.088578
Personal life SM usage	0.439339	0.223017

9.3 Effect on work-life balance by personal life and professional life social media usage based on work experience

Table 3 represents the correlation between the social media usage in personal and professional life towards work-life balance with respect to total work-experience. The analysis is compared between respondents of work-experience two years and below (62), and above 2 years (50). It is inferred from the table that work-life balance is more correlated (0.500062) by professional life social media usage for respondents with work-experience above 2 years, while work-life balance is more correlated (0.401921) by personal life social media usage for respondents with experience 2years and below.

Table 3. WLB by personal life and professional life SM usage based on work experience

	Work-life balance	
	2 years & below	Above 2 years
Professional life SM usage	0.377732	0.500062
Personal life SM usage	0.401921	0.446397

9.4 Factors influencing professional life social media usage

As social media usage in professional life is highly influenced towards work-life balance from the figures in table1, the below table4 explains that pre-occupation with social media has high correlation with WLB when compared with other variables of SM usage in professional life.

Table 4. Factors influencing professional life social media usage

	Professional life SM usage				
	Intensity of SM	*Task-oriented SM behaviour*	*Relationship building SM behaviour*	*Deviant SM behaviour*	*Pre-occupation with SM*
WLB	0.077146	0.283868	0.156323	0.269653	0.409361

9.5 Factors influencing Work-life balance

Personal life spill over work is more correlated with work-life balance when compared to other variables such as work spill over personal life, work-life balance enhancers, and work-life balance constrainers. This can be stated personal life demands interferes with responsibilities at work.

Table 5. Factors influencing Work-life balance

	WSPL	PLSW	WLBE	WLBC
WLB	0.616006	0.787208	0.628096	0.512821

10. Results and Discussion

The following are the results of the analysis. Work-life balance has a positive co-relation with social media usage at both personal and professional life where professional life social media usage is more correlated (0.390201). The results also indicate that work-life balance has a positive influence (0.486496) on professional life social media usage for men, while work-life balance is more influenced (0.223017) by personal life social media usage for women. The results highlight that work-life balance is affected only by personal life social media usage for employees with work experience two years and below (0.500062) while work-life balance is affected only by professional life social media usage for employees with work experience above two years. This may be due to the usage of social media increase as experience of individuals increase with experience in worklife

11. Conclusion

It can be concluded that irrespective of gender, age and work experience: work-life balance has a positive correlation the usage of social media. The study on social media is extended on the various consequences of its usage describing how it affects the day to day activities in work (Zoonen, Verhoeven, Vliegenthart, 2017). It is also found that social media usage in professional life has more impact on work-life balance which is caused due to pre-occupation (unproductive behaviour) with social media in workplace when compared to other productive variables like task-oriented social media behaviour or relationship building social media behaviour. As the respondents of this study majorly comprises of Gen-Y employees, future study can be done across different generations.

12. References

[1] Singh S 2014 Measuring Work-life Balance in India *International Journal 2*(5)
[2] Pradhan R K, Jena L K and Kumari I G 2016 Effect of Work–Life Balance on Organizational Citizenship Behaviour: Role of Organizational Commitment *Global Business Review 17*(3_suppl) pp.15S-29S
[3] Kietzmann J H, Hermkens K, McCarthy I P & Silvestre B S 2011 Social media? Get serious! Understanding the functional building blocks of social media. Business horizons 54(3) 241-251
[4] Ellison N B 2007 Social network sites: Definition, history, and scholarship Journal of computer-mediated Communication 13(1) 210-230
[5] Bolton R N, Parasuraman A, Hoefnagels A, Migchels N, Kabadayi S, Gruber T & Solnet D 2013 Understanding Generation Y and their use of social media: a review and research agenda Journal of service management 24(3) 245-267
[6] Van Zoonen W, Verhoeven J W & Vliegenthart R 2017 Understanding the consequences of public social media use for work. European Management Journal 35(5) 595-605
[7] Brooks S 2015 Does personal social media usage affect efficiency and well-being? *Computers in Human Behavior 46* 26-37
[8] Carlson J R, Zivnuska S, Harris R B, Harris K J & Carlson D S 2016 Social Media Use in the Workplace: A Study of Dual Effects. *Journal of Organizational and End User Computing JOEUC 28*(1) 15-31
[9] Munene A G & Nyaribo Y M 2013 Effect of social media pertication in the workplace on employee productivity *International Journal of Advances in Management and Economics 2*(2) 141-150
[10] Tajudeen F P, Jaafar N I & Ainin S 2017 Understanding the impact of social media usage among organizations *Information & Management*
[11] Deal J J, Altman D G & Rogelberg S G 2010 Millennials at work: What we know and what we need to do *Journal of Business and Psychology 25*(2) 191-199
[12] Adame-Sánchez C, González-Cruz T F & Martínez-Fuentes C 2016 Do firms implement work–life balance policies to benefit their workers or themselves? *Journal of Business Research 69*(11) 5519-5523
[13] Armour S 2005 Generation Y: They've arrived at work with a new attitude *USA Today 6* 2005
[14] Kishokumar M R Influence of Social Networking in the Work place on Individual Job Performance: Special reference to the Financial Sector in Batticaloa District

[15] Ashraf N & Javed T 2014 Impact of social networking on employee performance *Business Management and Strategy 5*(2) 139-150

[16] Afifi T D, Zamanzadeh N, Harrison K & Callejas M A 2018 WIRED: The impact of media and technology use on stress (cortisol) and inflammation (interleukin IL-6) in fast paced families *Computers in Human Behavior 81* 265-273

[17] Rathi N & Barath M 2013 Work-family conflict and job and family satisfaction: Moderating effect of social support among police personnel *Equality, Diversity and Inclusion: An International Journal 32*(4) 438-454

Comparative study on Cu, Al and Cu-Al alloy nanoparticles synthesized through underwater laser ablation technique

N Patra[1], R Patil[2], A Sharma[1], V Singh[1] and I A Palani[1,*]

[1]Metallurgy Engineering and Materials Science, Indian Institute of Technology Indore, Indore, Madhya Pradesh, India
[2]Control and Automation Department, VIT University, Vellore, Tamil Nadu, India

E-mail: phd1401281003@iiti.ac.in, *palaniia@iiti.ac.in

Abstract: In this work, we have done a comparative study on Cu, Al and Cu-Al alloy nanoparticles synthesized through underwater pulsed laser ablation technique. Second harmonic of Nd: YAG nanosecond pulsed laser with 50 J/cm² fluence was used to ablate the Cu (99.99%), Al (99.99%) and Cu-45% Al alloy targets. However, the laser synthesis of Cu-Al alloy nanoparticles is not yet investigated best of our knowledge. Thus, it is the first time we have synthesized alloy nanoparticles through underwater laser ablation in a chemical free environment. Moreover, we have used abundant earth metals to synthesize the nanoparticles. The particles morphology, shape and size were analysed through SEM and DLS analysis where we have observed that the average size of alloy nanoparticle was much smaller than its monometallic counterparts. The crystallinity and elemental analysis of synthesized nanoparticles were studied through XRD and EDS respectively. The alloy formation of nanoparticles was evident from such structural investigations. Further a comparative study on optical properties of alloy and monometallic particles were investigated through UV-Vis absorption spectroscopy. Alloy particle shows a remarkable improvement in optical absorption in comparison to its parent materials.

1. Introduction:

Laser ablation in liquid environment is a promising top-down approach to generate nanoparticles[1]. This is a favorable technique where laser energy is used to synthesize the particles in chemical free environment from variety of materials [2]. Laser parameters like laser wavelength, laser fluence and pulse duration have significant effect towards the particle shape, size and concentration of colloidal solution [3]. Number of studies have been done on noble metals like Au and Ag, synthesized through underwater ablation technique. However, the availability and cost effectiveness should be addressed while analyzing with noble metals. Hence, the copper and aluminum nanoparticles have attracted great attention due to their low cost, high conductivity and easy availability compared to the noble metals [4, 5]. Such metal nanoparticles are mostly used for optical, electronic and catalytic applications due to their unique properties. Moreover such nanoparticles show good surface Plasmonic property in UV and visible range of spectrum [6, 7]. Number of literatures are there on Cu and Al nanoparticles synthesized through laser ablation technique. Miranda et al. have generated copper nanoparticles through laser ablation in liquid and investigated the surface enhance Raman scattering of nanoparticles [8]. Whereas, Kumar et al. have investigated the pulsed laser synthesis of copper nanoparticles in ambient air and water [9]. There are reports on optical properties of Cu nanoparticles where Cu NPs are synthesized through laser ablation in different synthesis medium [10, 11]. Moreover, Oguchi et al. have done little different investigation on underwater laser ablation of Cu where they have studied the effect of pulse duration on plasma plume formation [12]. Besides that, number of studies are there on Al nanoparticles where Shaheen et al. have shown the effect of spot size on formation of Al alloy nanoparticles through laser ablation in air [13]. Hu et al. have done a relative study on femtosecond laser ablation of aluminium in ambient air and water [14]. Furthermore, Dudoitis et al. have done a comparative study on aluminium nanoparticles in different ablation environment (in ambient air and argon gas) [15]. However the studies on optical absorption of Cu and Al nanoparticles are very limited. It was reported that the extinction spectra of both of the metal particles mostly lie in between UV-Vis range of spectrum [16]. But, the literatures on synthesis of Cu-Al alloy nanoparticles are rarely being discussed [17]. Moreover the optical properties of such bimetallic nanoparticles have not yet been discussed before.

Here, we have synthesized Cu, Al and CuAl alloy nanoparticles with the help of underwater pulsed lased ablation technique where second harmonic (532 nm) of pulsed laser at 50 J/cm^2 fluence were used to generate the nanoparticles. We have done a comparative study towards the size and optical properties of nanoparticles synthesized from different metal targets. It was observed that the size of bimetallic (Cu-Al) particles are comparatively smaller (~ 90 nm) than the monometallic one. Moreover, we have noticed that the Cu-Al alloy nanoparticles show improved optical absorption than its monometallic counterparts. Thus, we can claim that we have used abundant earth metals (Cu and Al) to generate nanoparticles and show improved optical absorption.

2. Experimental Setup:

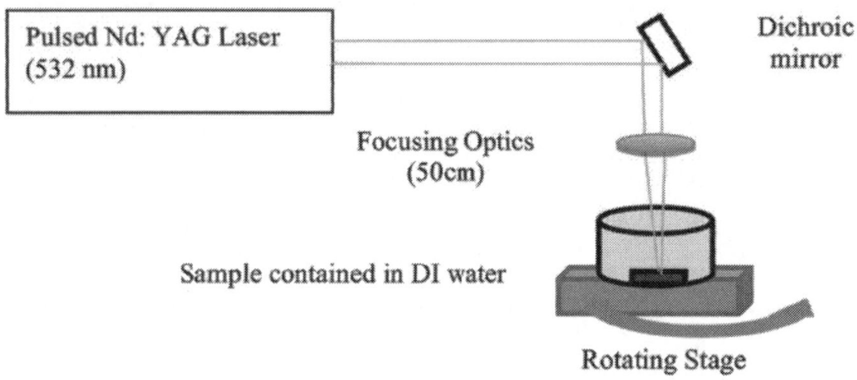

Figure 1. Experimental setup of nanoparticle generation

Figure 1 shows the schematic diagram of underwater nanoparticle generation. Pulsed nanosecond Nd:YAG laser at 532 nm wavelength with 50 J/cm^2 fluence were used to synthesize Cu-Al alloy, Cu and Al nanoparticles from three different metal targets. The pulse duration of laser pulse was 9 ns with 10 Hz of repetition rate. The laser beam was diverted with the help of dichroic mirror and focused on the target (Diameter: 30 mm, thickness: 0.5 mm) surface with a converging lens (focal point: 50 cm). The target materials (Cu-Al, Cu and Al) were placed inside a glass container filled with 20 ml of deionized water which corresponds to the liquid height of ~8 mm above the target. The glass container along with the target material were rotated in a clockwise direction with a constant speed of 10 RPM by a motorized rotating stage. The ablation time was set to 45 mins for individual laser irradiation. Generated samples were characterized through SEM, EDS, DLS, XRD and Uv-Vis spectrum analysis. The surface morphology and shape of particles were characterized by SEM images whereas the size distribution was studied through DLS analysis. The elemental analysis and crystalline formation were investigated through EDS and XRD respectively. The absorption spectrum of monometallic and bimetallic nanoparticles were detected from UV-Vis spectrum analysis.

3. Results and discussions:

3.1 Investigations on Cu-Al alloy nanoparticles synthesized at 532 nm laser wavelength with 50 J/cm^2 laser fluence:

Figure 2 shows the underwater laser ablated Cu-Al alloy nanoparticles at 532 nm laser wavelength with 50 J/cm^2 laser fluence. The surface morphology of such alloy nanoparticles were observed from SEM image where spherical alloy nanoparticles with average diameter ~90 nm was detected. The particle size distribution range from 55-145 nm was analyzed from DLS characterization. The crystallinity was examined through XRD analysis whereas the main elements were observed in EDS characterization. The XRD shows the different phases of Cu-Al alloy nanoparticles with their oxidation behavior. The oxidation of nanoparticles is visible from such structural analysis due to the high oxidation potential of Cu and Al. Further, the EDS analysis shows the presence of main elements like Cu, Al, O in the samples. Thus, EDS analysis validates the XRD results which confirms the alloy formation of Cu-Al alloy nanoparticles.

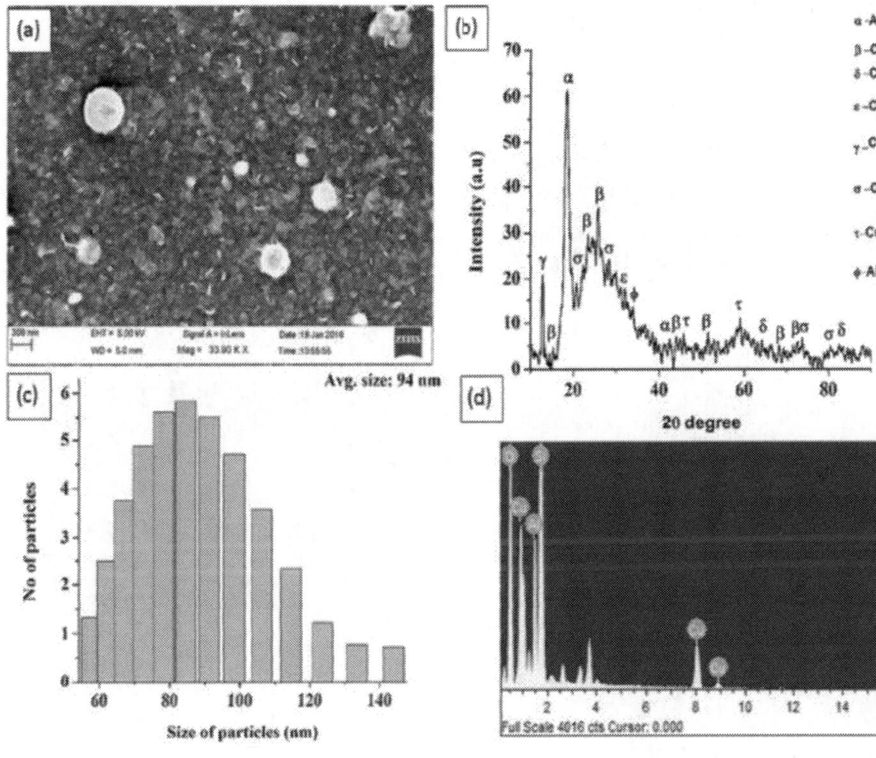

Figure 2. SEM (a), XRD (b), DLS (c), and EDS (d) analysis of laser ablated CuAl nanoparticles

3.2 Investigations on Al nanoparticles synthesized at 532 nm laser wavelength with 50 J/cm^2 laser fluence:

Figure 3 shows the underwater laser ablation of Al nanoparticles synthesized at 532 nm laser wavelength with 50 J/cm^2 fluence. SEM image (figure 3a) has shown the size, shape and surface morphologies of as synthesized Al nanoparticles where particles were detected in spherical shape. Further, particle size distribution was investigated through DLS analysis where average particle size of 98 nm was observed. XRD and EDS characterization shows the crystalline nature of Al nanoparticles. Clear oxide formation due to high oxidation potential of Al nanoparticles was evident from XRD image.

Table1. Size distribution analysis of different metal nanoparticles

Material	Average Size (nm)	Size distribution range (nm)
Cu-Al	94	55-145
Al	98	65-165
Cu	134	85-190

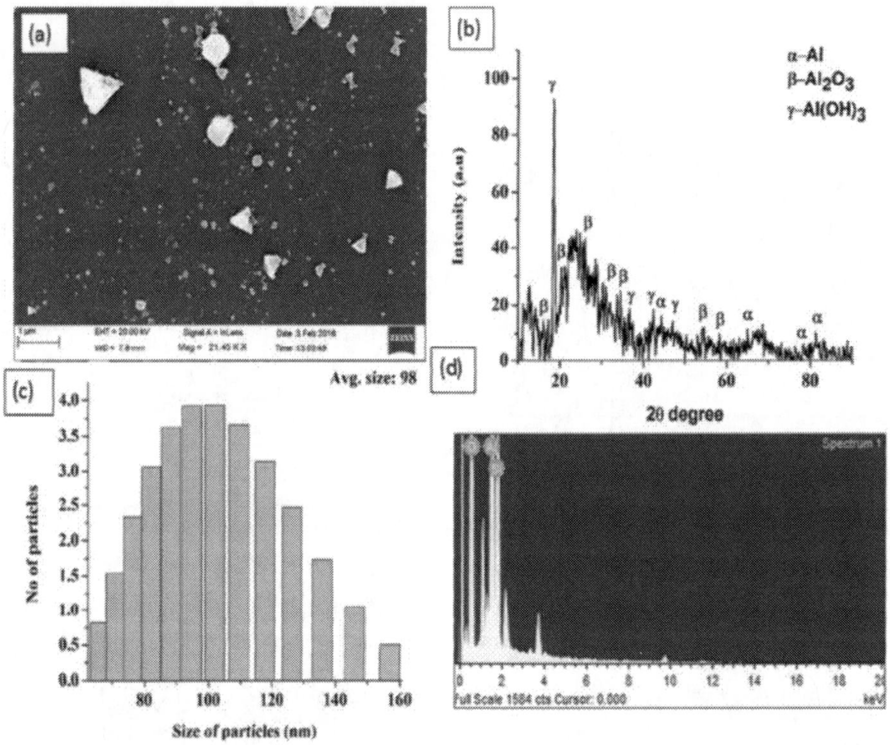

Figure 3. SEM (a), XRD (b), DLS (c), and EDS (d) analysis of laser ablated Al nanoparticles

3.3 Investigations on Cu nanoparticles synthesized at 532 nm laser wavelength with 50 J/cm² laser fluences:

Laser ablated Cu nanoparticles at 532 nm wavelength with 50 J/cm^2 fluence is shown in figure 4. SEM image shows the spherical shape, size and surface morphology of as synthesized Cu nanoparticles where particle size distribution was investigated through DLS analysis (figure 4(c)). Average particle size of 134 nm was calculated from DLS analysis. Crystalline formation of Cu nanoparticles with their oxidation behaviour were investigated from XRD and EDS characterizations respectively.

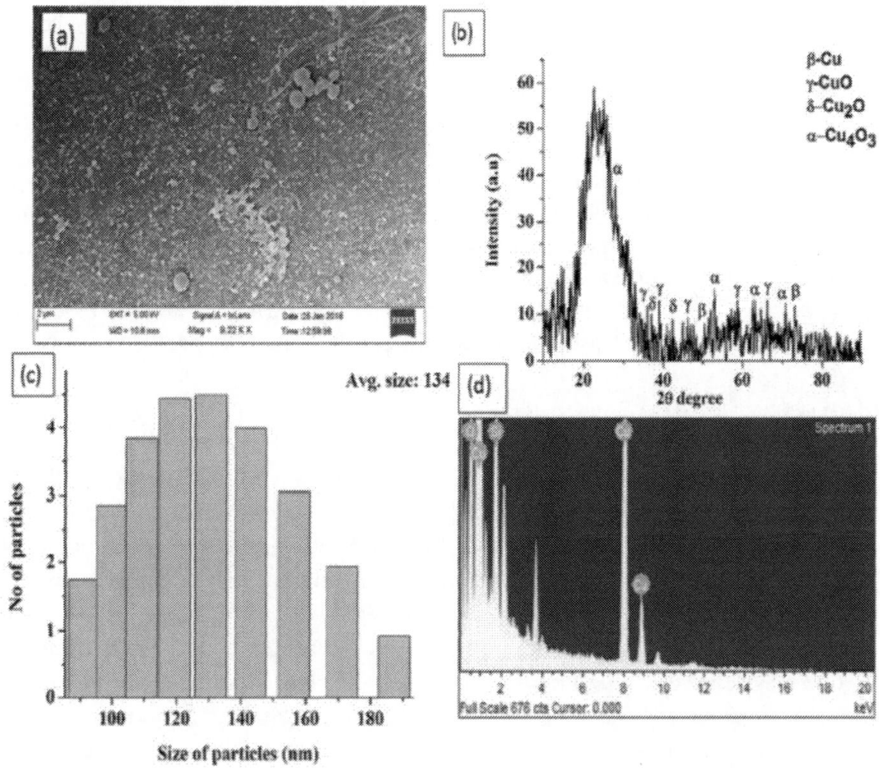

Figure 4. SEM (a), XRD (b), DLS (c), and EDS (d) analysis of laser ablated Cu nanoparticles

Thus, it has been scrutinized from three different DLS graphs, that the particle size is smaller in case of alloy nanoparticle than its monometallic counterparts. The probable reason might be the growth rate of crystallite is slower in case of alloy particles than the pure one [18]. The crystallinity and alloy formation of particles were confirmed from XRD and EDS analysis. Moreover, the oxidation behaviour was observed for all of the metal particles (Cu, Al, and Cu-Al) due to the high oxidation potential.

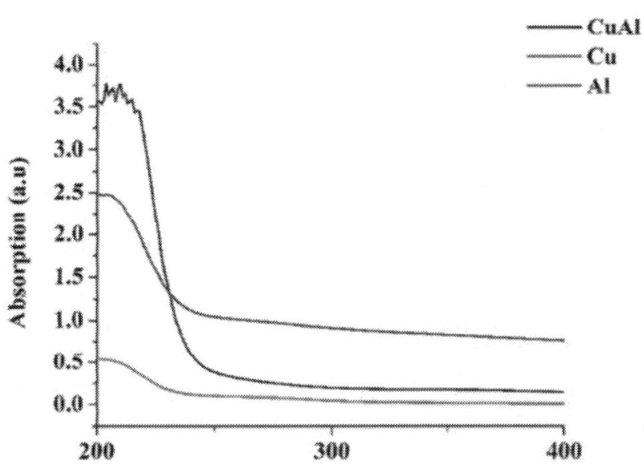

Figure 5. Absorption spectra of Cu-Al, Cu and Al nanoparticles

Furthermore, the optical absorption spectra of three different metal nanoparticles are shown in figure 5. It was evident from spectra that the absorption peak of metal nanoparticles mostly lie in the UV region of spectrum [11, 19]. Though Cu nanoparticle typically shows its absorption in the visible region [10, 11] but the absorption peak of such nanoparticles lies in the UV range owing to their oxidation behaviour. It was also observed that the absorption intensity of alloy nanoparticle is significantly higher than Cu and Al nanoparticles. Hereafter, such high intensity congruent to the nanoparticle concentration in solution. Thus, we can conclude that the ablation rate of alloy nanoparticles was more than the parent materials. Further, we have calculated the bandgaps of different metal nanoparticles from their absorption spectra. Bandgaps around 5.2 eV, 4.8 eV and 4.5 eV were calculated for Cu-Al, Cu and Al nanoparticles respectively. It was evident from the absorption spectra that the band gap of alloy particle is larger than the pure particles.

4. Conclusion:

In summary, we have used earth abundant material (Cu and Al) to generate Cu, Al and CuAl alloy nanoparticles through underwater laser ablation technique. Importantly, the laser synthesis of CuAl alloy nanoparticles have been investigated for the first time in this report. It was observed that the size of alloy particle is much smaller (~90 nm) in comparison to its monometallic particles. Crystalline formations of monometallic and bimetallic particles were verified through XRD whereas the presence of dominating elements was examined through EDS analysis. Hence, the alloy formation of nanoparticles is confirmed from such structural analysis. Herein we can claim that CuAl alloy nanoparticles can be synthesized through underwater laser ablation technique. Moreover, such binary alloy particles show improved optical absorption than monometallic particles. Thus it can be concluded that, alloy nanoparticles with improved optical absorptivity can be achieved through underwater laser ablation technique.

Acknowledgements:

Authors are grateful to the sophisticated instrumentation center, IIT Indore to carry out the above investigation.

References:

[1] Barcikowski S and Compagnini G 2013 Advanced nanoparticle generation and excitation by lasers in liquids *Phys. Chem. Chem. Phys.* **15,9** pp 3022–26

[2] Hahn A 2008 Influences on nanoparticle production during pulsed laser ablation *J. Laser Micro/Nanoengineering* **3,2** pp 73–77

[3] Barcikowski S, Hann A, Kabashin A V and Chichkov B N 2007 Properties of nanoparticles generated during femtosecond laser machining *Appl. Phy. A: Matr. Sci. Process.* **87** pp 47–55

[4] Musa A, Ahmad M B, Hussein M Z, Izham S M, Shameli K and Sani H A 2016 Synthesis of nanocrystalline cellulose stabilized copper nanoparticles *J. Nanomater.* **2016** p 1-7

[5] Ghosh S, Das R, Chowdhury I H, Bhanja P and Naskar M K 2015 Hierarchical copper nanoassembly and its use as a reusable catalyst for 4-nitrophenol reduction *RSC Adv.* **5,25** pp 101519–24

[6] Knight M W, King N S, Liu L, Everitt H O and Nordlander P 2014 Aluminum for plasmonics *ACS Nano* **8,1** pp 834–40

[7] Chan G H, Zhao J, Hicks E M, Schatz G C and Duyne R P 2007 Plasmonic properties of copper nanoparticles fabricated by nanosphere lithography *Nano Lett.* **7** pp 1947-52

[8] Miranda M, Gellini C and Giorgetti E 2011 Surface-enhanced raman scattering from copper nanoparticles obtained by laser ablation *J. Phys. Chem. C* **115,**

12 pp 5021–27

[9] Kumar B and Thareja R K 2013 Laser ablated copper plasmas in liquid and gas ambient *Phys. Plasmas* **20** p 53503

[10] Moniri S, Ghoranneviss M, Hantehzadeh M R and Asadabad M A 2017 Synthesis and optical characterization of copper nanoparticles prepared by laser ablation *Bull. Mater. Sci.* **40,1** pp 37–43

[11] Sadrolhosseini A R, Noor A S B M, Shameli K, Mamdoohi G, Moksin M M and Adzir Mahdi M 2013 Laser ablation synthesis and optical properties of copper nanoparticles *J. Mater. Res.***28,18** pp 2629–2636

[12] Oguchi H, Sakka T and Ogata Y H 2007 Effects of pulse duration upon the plume formation by the laser ablation of Cu in water *J. Appl. Phys.* **102,2**

[13] Shaheen M E, Gagnon J E and Fryer B J 2016 Excimer laser ablation of aluminum: Influence of spot size on ablation rate *Laser Phys.* **26,11** p 116102

[14] Hu H, Liu T and Zhai H 2015 Comparison of femtosecond laser ablation of aluminum in water and in air by time-resolved optical diagnosis *Opt. Express* **23,2** pp 237–41

[15] Dudoitis V, Ulevičius V, Račiukaitis G, Špirkauskaitė N, Plauškaitė K 2011 Generation of metal nanoparticles by laser ablation *Lith. J. Phys.* **51,3** pp 248–259

[16] Temple T L and Bagnall D M 2011 Optical properties of gold and aluminium nanoparticles for silicon solar cell applications *J. Appl. Phys.* **109,8**

[17] Svarovskaya N V, Berenda A V, Bakina O V, Glazkova E A, Lozhkomoev A S, Khorobraya E G, Domashenko V V, Lerner M I and Fomenko A N 2015 Chemica behaviour of Al/Cu nanoparticles in water *Prog. Nat. Sci. Mater. Int.* **25,1** pp 1–5

[18] Chen M and Schmidt L D 1979 Morphology and composition of PtPd alloy crystallites on SiO_2 in reactive atmospheres *J. Catal.* **56, 2** pp 198–218

[19] Mahdieh M H and Mozaffari H 2017 Characteristics of colloidal aluminum nanoparticles prepared by nanosecond pulsed laser ablation in deionized water in presence of parallel external electric field *Phys. Lett. Sect. A Gen. At. Solid State Phys.* **381,38** pp 3314–3323

Assessment on Tribological Characteristics of Waste Ayurvedic Oil Biodiesel Blends using High-Frequency Reciprocating Rig Tribometer

R. Balakumar [a,*], G. Sriram [b], S. Arumugam [c]
[a, b, c] Department of Mechanical Engineering,
Sri Chandrasekharendra Saraswathi Viswa Mahavidyalaya,
Kanchipuram-631561. Tamilnadu, India
Corresponding author : [a,]*balavhd@gmail.com, [b]drg.sriram@gmail.com,
[c]aru_amace@yahoo.co.in

Abstract. In general the components of the engine create more wear and friction in metal to metal contact surface, when the engine runs with petroleum based fossil fuel. This may damage the mating parts especially cylinder liner and piston ring, this may lead engine life reduction, increase of service cost. The biodiesel will give good lubrication property than the petroleum based fossil fuels. Basically, the researchers are mainly focusing on waste management system and its processes in the way of reduction in the virgin feed stocks used as a alternate fuel. This study dealt to investigate the tribological property of Waste Ayurvedic Oil (WAO) biodiesel blends by comparing with diesel fuel to improve fuel lubricity by using High Frequency Reciprocating Test Rig. The experimental investigation was conducted with load: 10N, frequency: 50Hz, stroke length: 15mm, time: 75min, temperature: 60°C. From the test results it is concluded that the range of blending of biodiesel from 10%, 15% and 20% of WAO with diesel fuel will results remarkable amount of improvement in the lubricity. Coefficient of friction and wear has been decreased with the increases percentages of biodiesel in blends. The WAO20 gives better reduction values in terms of co-efficient of friction, wear depth and frictional force than other WAO blended fuel and diesel fuel. The tested surfaces of cylinder liner were investigated using Scanning Electron Microscope (SEM) and atomic force microscopic analysis (AFM). The image shows a better impact and less wear scars for the WAO blended fuels than diesel fuel.
Keywords: Waste Ayurvedic Oil Biodiesel, HFRR, wear, friction.

1. Introduction

The drastic falling down of petroleum based fossil fuel reservoirs and fastest increasing pollution is increasing every day. Many researchers are working to bring out the suitable alternative fuels as a substitute for petroleum based fossil fuels [1]. The biodiesels are mainly called as fatty acid methyl esters (FAME) which is of long-chain derived from many sources such as vegetable oils, animal fats and various waste oil sources. Basically the biodiesel will have better lubrication property which used as an additive which used in petroleum based fossil fuels [2, 3]. The most and critical component of internal combustion engine is cylinder liner and piston ring tribo pair. The clear study about the wear and friction of liner rig pair will reduce the wear and friction of engine simultaneously it will lead the increase of engine life [4, 5]. There are more number of researches were undertaken to investigate the feasible method of production, performance in the diesel engines and tribological studies of biodiesel from various waste oils and animal fats [6]. The reduction in frictional force, frictional coefficient and wear of biodiesel blends over diesel fuel indicates that the rapeseed oil biodiesel blended fuel will produce an excellent lubricity characteristic. From the physical observation on the wear surfaces of the tested specimens, the rapeseed oil biodiesel contaminated lubrication oil appears a smoother surface than diesel contaminated lubrication oil [7].

The fuel quality is makes a major difference in the lubricity level. The important parameter in the engine is wear and friction between cylinder liner and piston ring tribopair material. Many researchers have analysed the biodiesel in terms of tribology perspective from vegetable oils, waste frying oil, and various fats from animals [8-14]. Goodrum and geller [8] investigated the lubricity of castor and lesquerella oil esters over diesel fuel in HFRR test. The high concentrations of methyl esters will give better lubricity for the mating parts and which lead the reduction in wear. Bhatnagar [9] investigated the lubricity and wear scar diameter of different biodiesels from nonedible oils like Jatropha Pongamia glabra, curcas, Madhuca indica, and Salvadora oleoides using HFRR. From the analysis, the addition of biodiesels to diesel fuel improves the lubricity and reduced the wear scar diameter. From the test results the biodiesel blend range varies from 5% to 20%.

Taravus et al. [10] investigated the physical properties of sunflower oil and beef tallow methyl ester blends. Even though the beef tallow gives more viscosity and poor physical properties but, the main advantage is to reduce the raw source cost which consist of 70% of total cost of biodiesel. Fevzi Yasar [11] studied about the combination of canola oil and inedible animal tallow in the volume basis of 0% to 100% (25% interval). The HFRR results experiences that the kinematic viscosity increased for biodiesels especially because of animal tallow which gives better wear prevention. Sulek et al. [12] analysed about the rapeseed oil methyl ester along with diesel using HFRR (ball on flat) tribometer. The blended rapeseed oil biodiesel reduces 5% of frictional coefficient and 20% reduction in wear than that of diesel fuel. The stable file thickness leads the reduction in wear and friction.

Arumugam et al. [13] studied about environmentally lubricant i.e. rapeseed oil bio lubricant using HFRR by adding the copper oxide (CuO) nano particle with synthetic lubricant and bio lubricant. SEM and AFM results proved that the surface morphology of cylinder liner tested with CMRO addition with 0.5 wt% of CuO explicit less friction. Xianguo et al. [14] investigated the waste cooking oil ethyl ester (WCOEE) using four ball tribometer and compared with diesel fuel. The investigation reveals that the improvement in the wear resistance of WCOEE blend fuel resulted from free fatty acids in biodiesel. The scanning electron microscopy (SEM) and atomic force microscopic analysis (AFM) was used to investigate the image and texture of the tested surface.

From the above survey it is very clear that no one touched through in the area of waste ayurvedic oil (WAO).The main theme of this investigation is to analyze the effectiveness of waste ayurvedic oil methyl ester blends as a better lubricity enhancer for diesel fuel. The tribological study was conducted using high frequency reciprocating rig as per the ASTM standards. In this investigation the commercially available cylinder liner and piston ring was used as a test specimen. After the tribological study, the tested liner and ring was undergone image morphology and surface texture analysis.

2. **Materials and Experimental methods**

2.1. Synthesis of Waste Ayurvedic Oil Biodiesel

The waste ayurvedic oil (after treatment) was collected from Sri Jayendra Saraswathi Ayurveda College and Hospital, Chennai, India. The chemicals used for transesterification like sodium hydroxide, methanol, hydrochloric acid and anhydrous sodium sulphate was purchased from M/S Ganapathy trading company, Chennai. The collected WAO was filtered to remove the dust and unwanted particles present in it. Transesterification process was carried out to convert the waste ayurvedic oil into waste ayurvedic methyl ester (WAOME). Finally, the WAO biodiesel blends were prepared by mixing with diesel fuel via. 90% of diesel fuel with 10% WAOME referred as WAO10, 85% of diesel fuel with 15% WAOME referred as WAO15 and 80% of diesel fuel with 20% WAOME referred as WAO20. The properties of biodiesel and diesel were considered from balakumar et al. [17].

2.2. Preparation of cylinder liner and piston ring test samples

The lubricity and frictional coefficient of WAO biodiesel blends along with diesel fuel was analysed by using HFRR tribometer. The test samples was cut using wire cut EDM process from the real engine cylinder liner (AISI-316) and piston ring (AISI-304) segment as per the manufacturer at Sree Krishna Enterprises, Chennai. The figure 1 shows the photographic view of cylinder liner and piston ring segment. Separate cylinder liner and piston ring segment were used for all the tests.

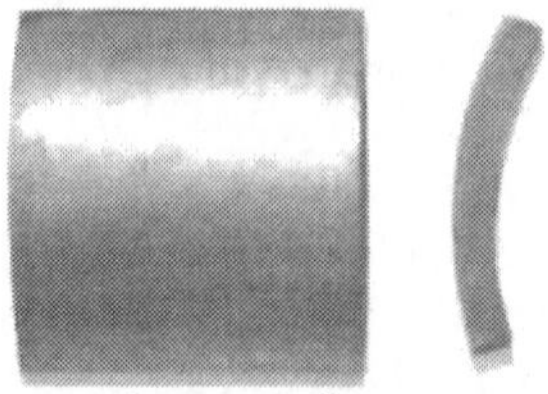

Figure 1 Photographic view of cylinder liner and piston ring segment

2.3. High frequency tribometer setup

The tribological study of engine liner ring segment under various biodiesel blends were analysed using HFRR tribometer (Model – Ducom TR-282) as per standard ASTM G181. The HFRR machine setup is shown in figure 2. The cylinder liner is fixed on the bottom of the oil tray and which is filled with the test fuel samples and the piston ring was fixed in the fixture holder which reciprocates over the liner. The operating conditions of HFRR are listed in the table 1.

Figure 2 Photographic view of HFRR tribometer

Table 1 HFRR test conditions

Parameter	Value
Load	10N
Frequency	50Hz
Temperature	60°C
Time	75 min
Stroke length	15mm

3. Result and discussion

The HFRR testing results on the blends of WAO biodiesel blends and the diesel fuel via coefficient off friction, wear depth along with time are shown in figure 3 (a, b).

3.1. Coefficient of friction

In this present analysis the coefficient of friction for all WAO biodiesel blends were tested and compared with diesel fuel. The frictional value is high at the beginning of the test. After few minutes the values become stable fluctuations. Markable results were absorbed for all the WAO biodiesel blends in terms of friction, wear depth and frictional coefficient than the diesel fuel. The coefficient of friction (COF) for WAO biodiesel blend was considerably lesser than the diesel fuel. The average COF for WAO20 is 56.85% better than that of diesel fuel. Similarly, for WAO10 and WAO15 is 4.54% and 27.66% better than diesel fuel. From the COF value results, the viscosity values of all the biodiesel are gives better scuffing protection behavior which leads a reduction in frictional values for the applied test conditions. For all the WAO biodiesel blends it is analysed that the addition of fatty acid methyl esters gives beneficial result on the lubricity of diesel fuel. The experimental results were consistent with Zulkifli et al. [15].

3.2. Wear depth

The wear depth is depends on the fuels used. The wear depth was less for the increasing concentrations of WAO biodiesel than the diesel. At the initial testing the wear depth value is fluctuates very high for all the WAO biodiesel blends and diesel fuel after that there is a stable results were marked. The average wear depth for diesel is 65.67% more than WAO20. Similarly for the other biodiesel blends like WAO10 and WAO15 which is 22.41% and 53.47% better than the diesel fuel. The increase in WAO biodiesel concentration reduces the wear depth than the diesel because of better scuffing protection behavior of biodiesel properties and close binding of molecules.

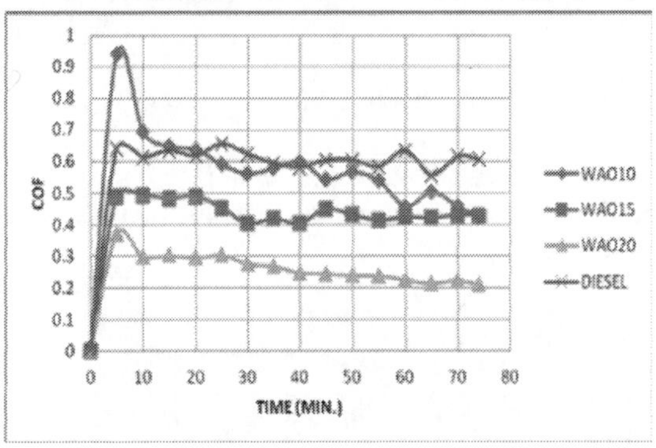

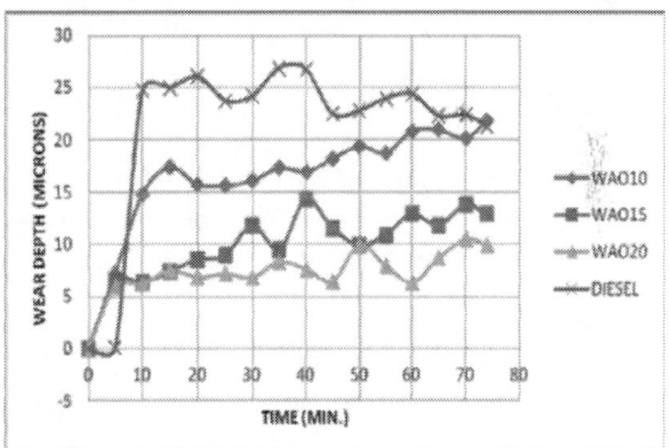

Figure 3 a) Time vs COF b) Time vs wear depth

3.3. Surface morphology analysis of tested cylinder liner surfaces using SEM

The tested surface was cleaned properly and the cylinder liner segment was investigated using scanning electron microscope (SEM) to find out the type of wear mechanism. The figure 4 (a-d) shows the SEM image of wear surfaces of liner with WAO biodiesel blends and diesel fuel. The cylinder liner surface deformation is quite more and severe for diesel fuel. The rubbed surface lubricated by the diesel fuel has more number of wide and deep furrows. Furthermore the severity of wear scar, cavities, deformation surface and grooves were decreased with increase of WAO biodiesel concentrations because of more free fatty acids present in the biodiesel. The closest analysis of tested surface of engine cylinder liner segment for diesel fuel shows that the depth longitudinal grooves and more pits which is the result of adhesive wear. But, abrasive wear were found with the WAO biodiesel blended fuels because of pull out materials form the cylinder liner and piston ring segment was less with biodiesel fuels. Especially, comparing with all the biodiesel blends WAO20 gives less wear depth, furrows and grooves than the other WAO blended fuels and diesel fuels. The test results consistent with the results from Fazal et al. [16].

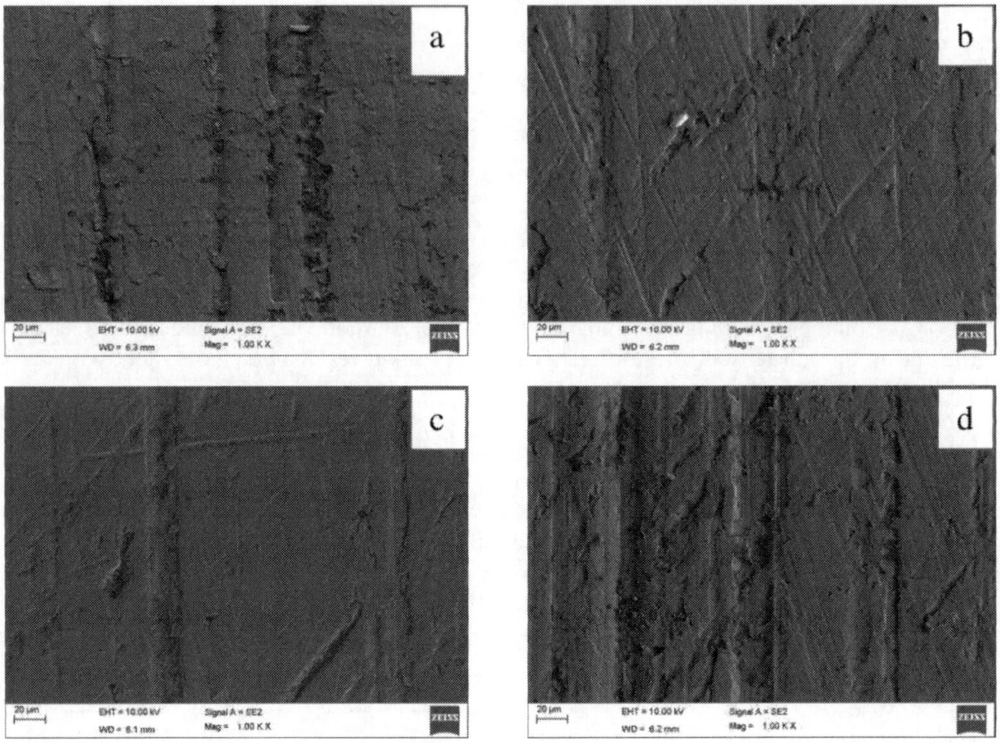

Figure 4 SEM images of tested liners for different fuels (a) WAO10 (b) WAO15 (c) WAO20 (d) diesel

3.4. Surface morphology analysis of tested cylinder liner surfaces using AFM (2D and 3D)

The three dimensional (3D) and two dimensional (2D) view of a tested cylinder liner surface by Atomic Force Microscopic (AFM) analysis were shown in figure 5 (a-d)

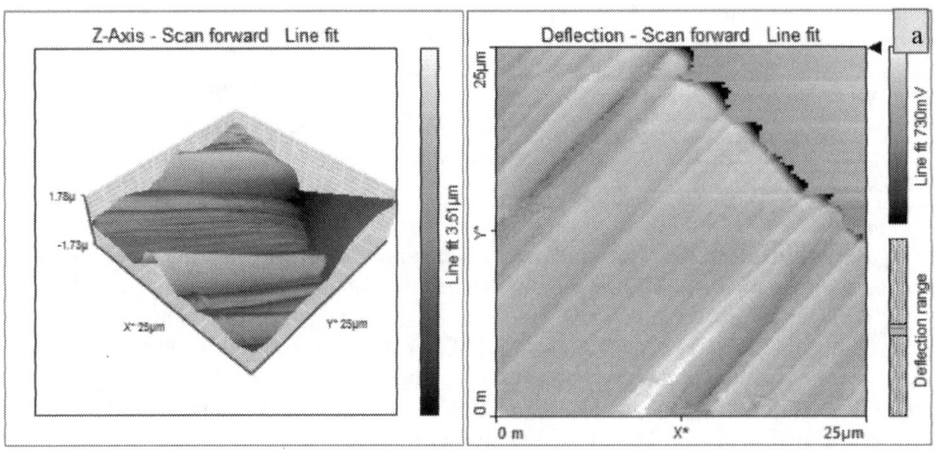

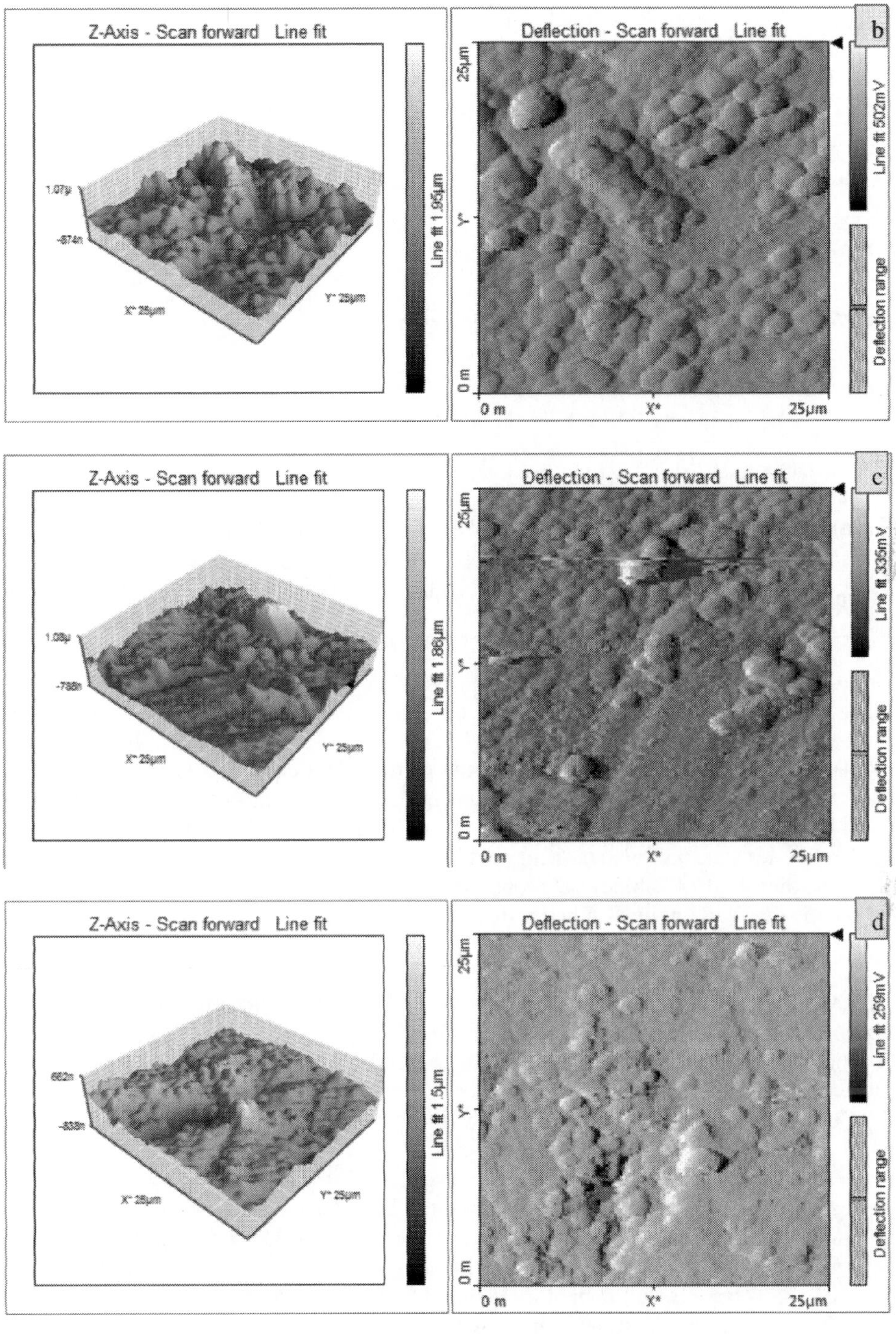

Figure 5 AFM images of tested liners for different fuels (a) WAO10 (b) WAO15 (c) WAO20 (d) diesel

The AFM surface profile investigation represents the roughness of the engine cylinder liner surface. The various roughness parameters of engine cylinder liner analysed by AFM were Sa (average roughness), Sq (root mean square value), Sy (peak-valley height), Sp (peak height), Sv (valley depth), Sm (mean value) listed in the table 2.

Table 2 Roughness parameters of tested engine cylinder liner for different fuels

Parameters	B10	B15	B20	Diesel
Sa (nm)	257.1	196.7	107.41	119.5
Sq (nm)	319.2	260.17	163.88	362.72
Sy (nm)	2508.7	2157.6	1955.3	2810.1
Sp (nm)	1479.2	1262.4	1081.7	1672.52
Sv (nm)	-1729.7	-873.6	-895.68	-837.62
Sm (pm)	-10.024	-10.024	-10.028	-10.025

From the above table and AFM images it is very clear that all the WAO biodiesel blends gives less roughness values than the diesel fuel. From both 2D and 3D images more number of red regions, valleys and peaks can be seen. The average roughness value (Sa) for WAO20 gives lower roughness value of 107.41nm which is very lesser than the diesel fuel. The other WAO blended fuels also gives lesser average roughness value than diesel fuel and quite higher than WAO20. The root mean square value (Sq) also gives less value i.e., 163.88 nm for WAO20 than other blended fuels and diesel fuels. The roughness values are less because of fatty acid contents present in the biodiesel. From all the above AFM images it is found that the WAO biodiesel blends represents less average surface roughness. Among all the biodiesel blends the WAO20 gives a better markable result than the diesel fuel.

Conclusions

The results obtained from the current investigation with HFRR tribometer, the following conclusions can be summarized.

- The frictional coefficient and wear depth were decreased with increase of WAO biodiesel concentrations due to the availability of ester content, unsaturated molecules etc.
- The COF values for WAO biodiesel fuels vary between 5% to 56% which is efficient than diesel fuel.
- The wear depth for WAO20 is 65% better than the diesel fuel. Similarly other biodiesel blends produces less wear depth than diesel but more the B20. This is because of
- From the SEM image it is concluded that less wear scar, cavities and grooves were observed in all the WAO biodiesel concentrations than the diesel fuel. Among all the SEM analysis images WAO20 gives a better result than other blended fuels and diesel fuel.
- The AFM results expose that the average roughness value, peaks and valleys were more for diesel fuel than WAO biodiesel blends. Comparatively with all the biodiesel concentrations WAOB20 gives marginal better result than the diesel fuel.

  From the above findings it is suggested that the WAO20 explores fabulous lubricating performance in the way of wear, friction and it can be used as a partial substitute for diesel engine.

Reference

[1] Restrepo-Florez, J.-M., Bassi, A., Rehmann, L., Thompson, M.R., (2014). Investigation of bio film formation on polyethylene in a diesel/biodiesel fuel storage environment. *Fuel* 128,240–247.

[2] Stoldt SH, Dave H, Heights H. (1998) Esters derived from vegetable oils used as additives for fuels. *United States Patent, US* 5730029.

[3] Bhushan, B. (2013). Introduction to Tribology. *John Wiley & Sons, New York,* DOI:10.1002/9781118403259.

[4] Hosni M. (1984) Reduction friction losses in automobile engines. *Tribol Int*; 17: 185–189.

[5] Ting LL. (1993) Development of reciprocating test rig for tribological studies of piston ring engine moving components-Part I: Rig design and piston ring friction coefficient measuring method. *Tribology of engines and engine oils. SAE International*: SP-959, 930685, pp.1–8.

[6] R. Balakumar, G. Sriram, S. Arumugam. (2018) Waste Ayurvedic Oil as a Engine Fuel-A Review. *Adv. Sci. Eng. Med.* 10, 249–254

[7] Arumugam S, Sriram G, Tulika Jha, (2012) Simulation of Wear and Frictional behavior of Cylinder liner - Piston ring combination with Diesel and bio-diesel contaminated lubricant, *National Conference on Research Trends in Mechanical Engineering – NCRTME.*

[8] J.W. Goodrum, D.P. Geller, (2005) Influence of fatty acid methyl esters from hydroxylated vegetable oils on diesel fuel lubricity, *Bioresource Technology* 96: 851–855.

[9] A. K. Bhatnagar, Savita Kaul, V. K. Chhibber, A. K. Gupta, (2006) HFRR Studies on Methyl Esters of Nonedible Vegetable Oils. *Energy & Fuels*, 20, 1341-1344.

[10] Taravus S, Temur H, Yartasi A. (2009) Alkali-catalyzed biodiesel production from mixtures of sunflower oil and beef tallow. *Energ Fuels*;23:4112–4115.

[11] Fevzi Yasar, Sehmus Altun, Hamit Adin. (2011) Fuel properties of biodiesels produced from blends of canola oil and animal tallow. *Energy Education Science and Technology Part A: Energy Science and Research* Volume (Issue) 27(1): 199-208.

[12] M. W. Sulek, A. Kulczycki, A. Malysa, (2010). Assessment of lubricity of compositions of fuel oil with bio components derived from rape-seed, *Wear* 268, 104.

[13] S Arumugam and G Sriram. (2014) Synthesis and characterization of rapeseed oil bio-lubricant dispersed with nano copper oxide: Its effect on wear and frictional behavior of piston ring–cylinder liner combination. *Proc IMechE Part J: J Engineering Tribology* 0(0) 1–11.

[14] Xianguo Hu, Xiangyang Wang, Jinsi Chen, Yufu Xu, and Enzhu Hu, (2016) Tribological behavior of diesel/biodiesel blend from waste cooking oil and ethanol, *Energy Sources, Part A: Recovery, Utilization, and Environmental Effects,* 38:8, 1062-1067.

[15] Zulkifli, N., Kalam, M., Masjuki, H., Shahabuddin, M., Yunus, R., (2013) Wear prevention characteristics of a palm oil-based TMP (trimethylolpropane) ester as an engine lubricant. *Energy* 54: 167-173.

[16] M.A. Fazal, A.S.M.A. Haseeb, H.H. Masjuki. (2013) Investigation of friction and wear characteristics of palm biodiesel. *Energy Conversion and Management* 67: 251–256.

[17] R. Balakumar, G. Sriram, S. Arumugam. (2018) Investigation of Friction and Wear Characteristics of Waste Ayurvedic Oil Biodiesel Blends. *Adv.* Sci. Eng. Med, 10, 472–475.

Modeling of Plasma Arc Welding of Inconel 617 Super Alloy Plates using RSM

Kadivendi Srinivas[1], Pandu R Vundavalli[2] and M Manzoor Hussain[3]
[1]Department of Mechanical Engineering, DVR & Dr. HS MIC College of Technology, Kanchikacherla, AP, India.
[2] School of Mechanical Sciences, IIT Bhubaneswar, Odisha, India
[3] Department of Mech. Engineering, JNTUH College of Engineering, Hyderabad, Telangana, India.
[1]kadivendi@gmail.com, [2]pandu@iitbbs.ac.in, [3]manzoorjntu@gmail.com

Abstract:

In the present work, an attempt is given to study the outcomes of input process parameters, namely weld current, weld speed and plasma gas flow rate on the bead geometry, that is bead width (BW) and bead height (BH) of Plasma Arc Welded Inconel 617 plates. In the present study, Bead on Plate (BoP) welding was done on 2 mm thick plates with the help of experiments designed using central composite design of experiments (CCD). The non-linear regression equations developed during this study were validated through experimental test cases. Further, it was depicted that the developed regression models were found to forecast the bead geometry with a reasonably good accuracy.

Key words: Inconel 617 super alloy, plasma arc welding, bead geometry, RSM, regression models.

1. Introduction

Plasma arc welding is an advancement of Gas Tungsten Arc Welding Process (GTAW) with a better control due to its constricted Nozzle and gas stream. It is seen that the constriction increased the heat per unit volume of the plasma arc [1]. Due to this, high temperatures in the order of 11000°c will be developed, and is suitable for welding high temperature resistant alloys like Ni based super alloys. In addition to the above fact, it is important to note that the weld distortion will also be very less, which makes the Inconel 617 to be used for elevated temperature applications, like gas turbines, nuclear reactors and boilers. Central composite design (CCD) of experiments is a methodology by which one can conduct experiments to establish the non-linear regression models, which are extensively used to model various manufacturing processes. This technique was also applied for welding processes to determine the relation between input and output parameters of arc welding process. Gunaraj et al. [2] used CCRD with five-level four-factor to develop the mathematical models that were utilized to forecast the weld-bead geometry in the SAW. Further, they analyzed the outcome of OC voltage, plate-nozzle distance, weld feed rate and welding speed on output parameters, like bead width, penetration, reinforcement and dilution using RSM technique. Elangovan et al. [3] developed a mathematical model using CCD. In their research, they considered welding parameters, namely rotational speed of tool, axial force, welding speed and tool pin profile as inputs and tensile strength of AA6061 alloy as output parameter. Further, Sivaprasad et al. [4] used response surface method (RSM) based on CCD of experimental procedure to carryout experiments on Inconel 625 material with Pulsed current Micro plasma arc welding. Nanda et al. [5] developed second-order response surface model to predict the depth of penetration of duplex steel alloy 2205 weld bead due to variation of current, torch speed and arc gap. Moreover, Babu et al. [6] reported the effect of axial force, welding speed and rotational speed to predict the maximum tensile strength of the joint on aluminum alloy 5059 using central composite design. The results were analyzed with the help of response surface graphs. Xu et al. [7] developed statistical models to forecast and enhance the geometry of weld bead after utilizing CCD based on RSM of experimentation in an automated all-position GMA welding. Rahimzadeh et al. [8] established a mathematical model to identify the hardness and grain size of joints of AA 7020 aluminum alloy which were friction stir-welded using RSM in conjunction with CCRD. Further, Ragavendran et al. [9] used CCD matrix to conduct the welding experiments. The multiple regression equations associating the process factors with the responses were established for the hybrid laser – TIG welding of 316LN steel. Benyounis et al.[10] designed experiments based on 3-level Box–Behnken method, to study the interactions between focal point position, welding speed and laser power on HAZ, penetration and width of the weld in laser welding on medium carbon steels. Tarang et al [11] have done the experimentation in GTAW by incorporating taguchi method to forecast the geometry of weld bead behavior with change in process parameters. Ren et al. [12] made a comparative study of fusion area associated with the fiber laser welding and CO_2 welding of Inconel 617.

From the literature, it has been observed that many researchers had used RSM technique to found the associations among the process factors and responses in different types of welding methods. They extensively used central composite design for experiments to perform welding on variety of advanced metals. The present work demonstrate the attempts made for modeling the plasma arc welded Inconel 617 plates using RSM technique by conducting bead on plate experiments using central composite design. Welding parameters, such as gas flow rate, weld current and welding speed are considered as inputs, and bead width and bead height treated as outputs. The non-linear models developed are validated with the assistance of real test scenarios.

2. Experimental Details

The investigational particulars related to the welding of Inconel 617 plates using transferred Plasma Arc Welding in melt-in mode is given below.

2.1. Specimen preparation

In the present Research, Inconel 617 plates of size 150 mm x 100 mm x 2 mm are utilized as work piece. Before welding, the surface of the specimen is cleaned with the help of wire brush, and acetone is used for degreasing and to remove the effects of oxidation. Autogenously plasma arc welding machine is used to conduct the experimentation. Special care has been taken to prevent the distortion of the weld plates by providing suitable clamping.

2.2. Design of Experiments

Central component design of experiments is used to conduct the experiments. Process factors, such as plasma gas flow rate (G), welding speed (N) and welding current (I) are considered to vary within the limits provided in Table 1. Once the experiments are conducted, the responses, namely bead width (BW) and bead height (BH) are measured for the BoP welds after taking their macro graphs with the help of optical microscope.

Table 1. Process factors and their varieties

Process factor	Units	Minimum value	Maximum value
Current(I)	amp	80	105
Speed(N)	mm/min	250	300
Gas flow rate(G)	LPM	2	2.5

IOP Conf. Series: Materials Science and Engineering **390** (2018) 012048 doi:10.1088/1757-899X/390/1/012048

2.3 Experimental Results of Bead on plate Trials:

The bead on plate weld is sectioned in transverse direction by using wire cut electric discharge machine. The samples are mounted on the base for proper support for the weld bead. The sample are etched and polished with the suitable etchant. Computer connected metallurgical microscope is used to check the height and width of bead profile for the specimens.

3. Results and Discussion

The MINITAB 17 software is used to analyze the experimental data. Response surface methodology (RSM) is used to obtain non-linear regression equations. Obtained non-linear equations are analyzed with the help of Analysis of variance (ANOVA) for both the responses, namely bead width (BW) and bead height (BH).

3.1 Responses- Bead width& Bead height

The non-linear regression equations obtained for bead width and bead height of BoP trials are given in Eqns. (1) and (2), respectively and are given below.

$$BW = 45.76 + 0.1263\ I - 0.3217\ N - 1.82\ G + 0.00186\ I*I + 0.000626\ N*N + 1.294\ G*G - 0.00072\ I*N - 0.10044\ I*G + 0.01635\ N*G \quad (1)$$

$$BH = -22.76 - 0.6486\ I + 0.5473\ N - 17.70\ G + 0.001970\ I*I - 0.000953\ N*N + 4.767\ G*G + 0.000507\ I*N + 0.06872\ I*G - 0.03507\ N*G \quad (2)$$

Once the non-linear regression equations that represent the relationship between the input process parameters and bead geometry of PAW of Inconel 617 plates is obtained, its statistical validity has been tested with the help of ANOVA and surface plots. The ANOVA table for the response bead width of BoP trail is given in Table 2.

Table 2. ANOVA table for the response – bead width of BoP

Source	DF	Adj SS	Adj MS	F-Value	P-Value
Model		7.59038	0.84338	230.82	0.000
Linear	9	4.06965	1.35655	371.27	0.000
I	3	3.26167	3.26167	892.67	0.000
N	1	0.41209	0.41209	112.78	0.000
G	1	0.39589	0.39589	108.35	0.000
Square	1	2.23675	0.74558	204.05	0.000
I*I	3	0.23367	0.23367	63.95	0.000
N*N	1	0.42031	0.42031	115.03	0.000
G*G	1	0.01800	0.01800	4.93	0.051
2-Way	1	1.28398	0.42799	117.14	0.000
Interaction	3	0.41228	0.41228	112.83	0.000
I*N	1	0.78820	0.78820	215.72	0.000
I*G	1	0.08350	0.08350	22.85	0.001
N*G	1	0.03654	0.00365		
Error	10	0.03190	0.00638	6.88	0.027
Lack-of-Fit	5	0.00464	0.00093		
Pure Error	5	7.62691			
Total	19				

IOP Conf. Series: Materials Science and Engineering **390** (2018) 012048 doi:10.1088/1757-899X/390/1/012048

From Table 2, it was depicted that apart from the square term of plasma gas flow rate, all other terms such as interaction, linear and square terms have played significant role on the response, bead width. Similar study has also been conducted on the bead height of the BoP trails. It has been observed that the effect of linear and square terms of flow rate of plasma gas was insignificant when compared with other terms, because their p terms obtained are more than 0.05. Further, figure 1 and 2 shows the variation of responses such as bead width and bead height.

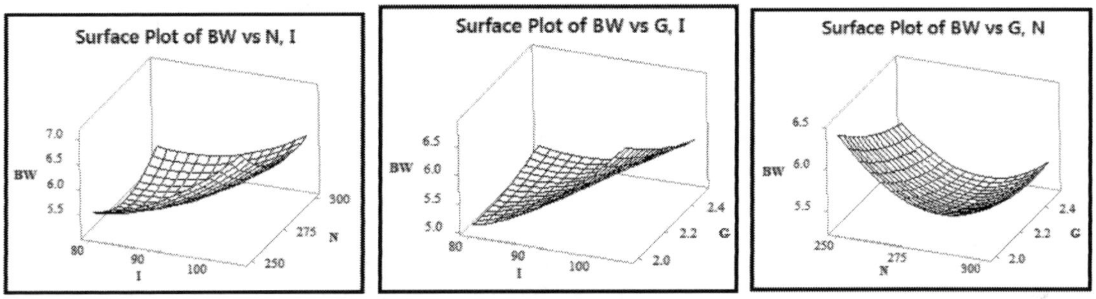

(a) (b) (c)

Fig. 1: Response plots showing the relationship between bead width with (a) welding speed and current, (b) gas flow rate and welding current, (c) gas flow rate and welding speed.

From the surface plots it has been found that the bead width is seen to increase with an increase in the combination of welding speed and current. Further bead width is found to be more at lower and high speed values and minimum at medium speed values. From Fig. (3), it has been observed that the bead height is found to be more for low and high values of current and at medium speed bead height is high compared to low at low and high speed values. The effect of gas flow rate on width and height of bead was observed to be very less compared to other input parameters, namely welding current and speed.

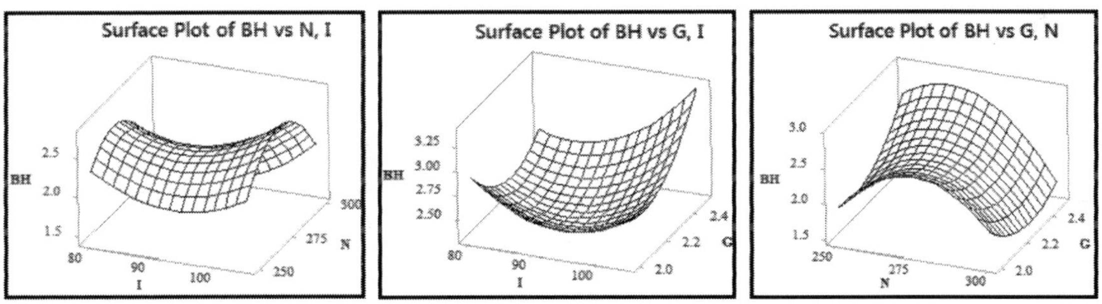

(a) (b) (c)

Fig. 2: Response plots showing the relationship between bead height with (a) welding speed and current, (b) gas flow rate and welding current, (c) gas flow rate and welding speed.

Further, the coefficient correlation values obtained for BW and BH were 0.98 and 0.99, respectively. From the results of ANOVA, surface plots and coefficients of correlation values, it can be concluded that, the non-linear regression equations established are statistically tolerable and can be utilized for performing predictions. The table showing the experimental and predicted values of bead width and bead height of PAW of Inconel 617 plates are given in Table 3. It has been observed that the predicted values obtained from the non-linear regression equations for BW and BH of the bead geometry are very close to the experimental values.

Table 3. The experimental and predicted data obtained for BW and BH of PAW welding

S. No	Weld current (I)	Weld speed (N)	Gas flow rate (G)	Expt. BW	Expt. BH	Predicted BW	Predicted BH
1	85	260	2.4	5.2510	2.4291	5.4064	2.5352
2	82	270	2.3	5.4040	2.5506	5.2359	2.5129
3	90	250	2.5	5.9365	2.6754	5.7638	2.5801
4	95	290	2.1	5.8976	2.2435	5.7553	2.1442
5	100	280	2.2	6.1234	2.3685	5.9581	2.4523
6	105	260	2.3	6.6956	2.6750	6.6302	2.7040
7	90	285	2.2	5.3456	2.0987	5.4053	2.1505
8	98	265	2.1	6.0234	2.4327	6.2273	2.3525

Further, the percentage error obtained for bead width and bead height of PAW was shown in Figure 3. It has been observed that the range of error percentage prediction of the responses such as BW and BH were {-3.38 to +3.11} and {-4.37 to +4.43} %, respectively.

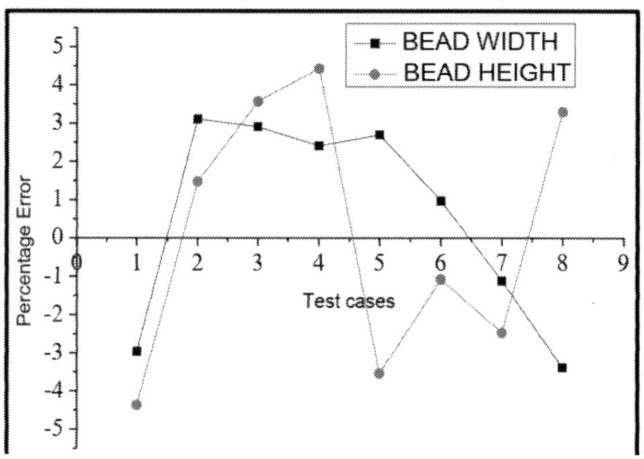

Fig. 3: Graph showing the percentage error for bead width and bead height of the PAW.

This shows that the non-linear regression equations have performed the predictions with a reasonably good accuracy.

4. Conclusion

The current research has made an attempt to establish the non-linear regression models for the bead width and bead height of BoP plasma arc welding of 2 mm thick Inconel 617 plates. The results of ANOVA and response surface plots show that the models are statistical adequate. Further, it has also been observed that the welding speed and welding current are having significant impact on the bead geometry that is bead height and bead width, when compared with the other process parameter – plasma gas flow rate. Moreover, the prediction accuracy of the models show that the non-linear models are capable of predicting the responses without conducting the real experiments, which are expensive and time consuming.

References

[1] C.S. Wu, L. Wang, W.J. Ren, X.Y. Zhang. Plasma arc welding: Process, sensing, control and modeling. *Journal of Manufacturing Processes*. 2013, **16**(1):74-85.

[2] V. Gunaraj, N. Murugan. Application of response surface methodology for predicting weld bead quality in submerged arc welding of pipes. *Journal of Materials Processing Technology*, 1999, **88**:266-275.

[3] K. Elangovan, V. Balasubramanian, S. Babu. Predicting tensile strength of friction stir welded AA6061 aluminium alloy joints. *Materials and Design*, 2009, **30**:188-193.

[4] K S Prasad & S Rao Ch & N Rao D Application of grey relational analysis for optimizing weldbead geometry parameters of pulsed current micro plasma arcwelded inconel 625 sheets. *The International Journal of Advanced Manufacturing Technology*, 2015, **78**(1-4):625-632.

[5] N N Korra, M Vasudevan KR Balasubramanian.Optimization of A-TIG welding of duplex stainless steel alloy 2205 based on response surfacemethodology and experimental validation *.Journal of material design and applications*.2015, **230**(4): 837-846

[6] N. Babu, N. Karunakaran V. Balasubramanian. A study to estimate the tensile strength of friction stir welded AA5059 aluminium alloy joints, *The International Journal of Advanced Manufacturing Technology*,2017, **93**(1-4): 1–9.

[7] W. H. Xu & S. B. Lin & C. L. Fan & C. L. Yang, Prediction and optimization of weld bead geometry in oscillating arc narrow gap all-position GMA welding, *The International Journal of Advanced Manufacturing Technology*,2015,**79**(1-4): 183–196.

[8] A. Rahimzadeh I, R. Soufi, G. Hussain, R. V Barenji, A. Heidarzadeh, Establishing Mathematical Models to Predict Grain Size and Hardness of the Friction Stir-Welded AA 7020 Aluminum Alloy Joints, *Metallurgical and Materials Transactions B*,2014,**46**(1) :357-365.

[9] M. Ragavendran, N.Chandrasekhar, R.R kumar, Rajesh S, M.Vasudevan,A.K. Bhaduri, Optimization of hybrid laser – TIG welding of 316LN steel using response surface methodology (RSM), *Optics and Lasers in Engineering*, 2017,**94**:27-36.

[10] K.Y. Benyounis, A.G. Olabi, M.S.J. Hashmi.Effect of laser welding parameters on the heat input and weld-bead profile. *Journal of Materials Processing Technology*, 2005. **164–165**: 978–985.

[11] Y S Tarng, W H Yang.Optimization of the weld bad geometry in gas tungsten arc welding by Taguchi method.*Journal of Advanced Manufacturing Technology*.,1998,**14** :549-554.

[12] Wenjie Ren, Fenggui Lu, Renjie Yang, Xia Liu, Zhuguo Li, Seyed Reza ElmiHosseini. A comparative study on fiber laser and CO_2 laser welding of Inconel 617. *Marerials & Design*, 2015, **76**:207-214.

Multi-Response Optimization Of Machining Parameters In CNC Turning Of AISI 316L Stainless Steel Using MQL Nano fluids

S D Sathishkumar[1] and T Rajmohan[2]

[1,2]Department of Mechanical Engineering, Sri Chandrasekharendra Saraswathi Viswa Maha Vidyalaya Deemed to be University, Enathur, Kanchipuram-631561, India

E-mail: [1]sathishsd.2@gmail.com and [2]rajmohanscsvmv@yahoo.com

Abstract. In this research work, optimization of machining parameters in plain turning operation to minimize surface roughness and cutting temperature by applying Taguchi based Desirability function analysis and ANOVA method under Nano cutting fluid environment is performed. Experiments have been performed by plain turning of AISI 316L stainless steel on CNC lathe under MQL setup with Multi-walled carbon Nanotube (MWCNT) inclusions using carbide inserts. Carbon Nano tubes are mixed in a specific ratio (0.20% by wt.) with the cutting fluid. The impact of input parameters on the response variables such as surface roughness and cutting temperature are analysed by using desirability function based approach. Experimental results reveal that feed rate is said to have more effect on the response variables followed by depth of cut, type of cutting fluid and spindle speed.

Keywords: CNC lathe, AISI 316L stainless steel, Surface roughness, Cutting temperature, Nano fluids, Taguchi, Optimization.

1. Introduction

Sharma et al. [1] made an experimental work to optimize cutting variables to reduce surface roughness and to maximize material removal rate in plain turning of EN-31 alloy steel by using coated carbide tool. They have developed a mathematical model using regression technique and optimized the cutting conditions by using Box-Behnken of response surface methodology. The experimental results reveal that the spindle speed is the most significant factor affecting the responses followed by depth of cut and feed rate. Liew et al. [2] made an experimental investigation to optimize cutting variables such as cutting speed, feed rate and type of coolant to minimize surface roughness and flank wear in plain turning of AISI D2 steel. They have selected Carbon Nano fiber Nano fluid and Deionized water as their coolants. The experimental results reveal that cutting speed is influencing more for tool wear and feed rate is influencing more for surface roughness. They have found that 100m/min cutting speed, 0.2mm/rev feed rate and CNF coolant as optimal parameters for flank wear by using Taguchi method. They have found that 200m/min cutting speed, 0.1mm/rev feed rate and CNF coolant as optimum conditions for surface roughness by Taguchi method. They have found that 144.58m/min cutting speed, 0.14mm/rev feed rate and CNF coolant as optimum parameters for minimizing tool wear and surface roughness according to RSM. Ramana et al. [3] made a research work to optimize machining parameters to minimize flank wear in plain turning of Titanium alloy by using Taguchi Design methodology. The ANOVA results indicate that cutting speed is contributing more to minimize tool wear followed by tool material, depth of cut, feed rate and coolant condition. Sonu Ram and Raj Kumar Yadav [4] made an experimental work to find optimum machining conditions in CNC turning of EN-31 alloy steel. The experimental results reveal that spindle speed is influencing more for surface roughness and depth of cut is influencing more for material removal rate. Shamnad and Thulaseedharan Raman [5] made an experimental investigation to study the effect of process parameters in plain turning of AISI 304 steel under different machining conditions. They have used Taguchi's L18 orthogonal array to design the experiment. They have made a comparison between machining environment like mineral oil MQL, vegetable and mineral oil flooded and dry cutting. The experimental results reveal that flow rate is the most significant parameter for cutting temperature and frequency of pulse is the most significant parameter for surface roughness. They have found that MQL provides good performance than other conventional fluid application while measuring cutting temperature and surface finish. Katle et al. [6] made an experimental investigation to study the effects of different cutting parameters on material removal rate in straight turning operation of EN31 steel under different environmental conditions. They have used Taguchi L9 orthogonal array to analyse the effect of cutting parameters on material removal rate. The experimental results reveal that depth of cut is most significant factor for material removal rate in the turning operation. Shrikant Borade and Kadam [7] made an experimental work to study the effect of Minimum Quantity Lubrication and Al_2O_3 Nano fluids on the machinability characteristics of EN353 steel in CNC turning operation by using CBN cutting tool inserts. The experimental results reveal that Nano cutting fluids reduced surface roughness and temperature significantly as compared to MQL vegetable oil. They have concluded that depth of cut has significant effect on responses under MQL Nano fluids conditions. Ramana et al. [8] made an experimental work to investigate the effects of process parameters in plain turning of Titanium alloy under different lubricant conditions by using Taguchi's robust design methodology. The ANOVA results indicate that feed rate has major effect in optimizing the surface roughness. They have found that MQL exhibits better performance and improvement to reduce surface roughness as compared to dry and flooded lubricant conditions. Mamundi azaath et al. [9] in their research work, Taguchi's L9 orthogonal array to optimize the process parameters in plain turning of EN24 steel under different cutting environment. They have concluded that Nano MQL machining gives the better results as compared to dry and MQL machining. They have found that Nano MQL machining gives 70% betterment in surface finish of the work piece. Vishnu et al. [10] have employed Taguchi's L9 orthogonal array to optimize the machining parameters in plain turning of EN353 steel under different cutting environment. They have found that 1500rpm spindle speed, 0.8mm/min feed rate, 2.5mm depth of cut and Gold coated carbide tip as tool type are the optimum parameters for maximizing material removal rate under MQL conditions. They have concluded that MQL is a better alternate cooling technique than flood cooling. Patole and Kulkarni [11] made an experimental investigation to study the impact of process parameters in CNC turning of AISI 4340 alloy steel under MQL mode with Nano fluid. The ANOVA results reveal that feed rate is influencing more for surface roughness and depth of cut is influencing more for cutting force in the turning operation. They have found that 0.04mm/rev feed rate, 75m/min cutting speed, 1mm depth of cut, 0.8mm tool nose radius are the optimal conditions for lowering the surface roughness. They have found that 0.04mm/rev feed rate, 90m/min cutting speed, 0.5mm depth of cut, 0.4mm tool nose radius are the optimum conditions for lowering the cutting force. From the literature survey, the machinability characteristics of hard materials can be improved by introducing minimum quantity lubrication Nano cutting fluids. Therefore, the present work deals with the process of improving the machining characteristics of AISI 316L stainless steel by using MQL Nano cutting fluids.

2. Materials and methods

2.1. Materials used

The stainless steel type AISI 316L of 60mm diameter and 150mm length was used for all the experiments. The chemical analysis has been carried out using the spark emission spectrometer. The chemical composition of AISI 316L stainless steel is given in Table 1. In this investigation, coated carbide insert was chosen as the tool material and Multi-walled carbon Nano tube (MWCNT) dispersed in SAE20W40 engine oil as lubricant for conducting turning trials. The Nano fluids were prepared by dispersing 10g of MWCNTs into the SAE20W40 oil base fluid (4000ml) using Esterification process. The properties of Nano fluids are tested in Delta lab, Chennai, India and presented in Table 2.

Table 1. Chemical composition of AISI 316L Stainless steel

Element	C	Si	Mn	S	P	Ni	Cr	Mo	N	Fe
Wt. %	0.018	0.350	1.664	0.014	0.043	10.176	16.115	2.197	0.10	Rest

Table 2. Properties of Nano fluids

Thermal property	**Without MWCNTs**	**With MWCNTs**
Flash Point (°C)	242	251
Fire Point (°C)	262	255
Pour Point (Deg. C)	-17	-15
Viscosity index (cST)	117.54	120.37
Kinematic Viscosity @ 40°C (cST)	13.7	15.1
Thermal Conductivity (W/mK)	0.137	0.176

2.2 Experimental design

The experiments for this work were planned using Taguchi's design of experiments (DoE). For the experimental plan, the Taguchi's mixed level design was selected as it was decided to keep two levels of cutting fluid type. The rest three parameters were kept at three levels. The machining parameters used in this work and their levels are shown in Table 3.

Table 3. Machining parameter and levels

Control parameters	Symbol	Level		
		1	2	3
Type of cutting fluid	A	SAE20W40	SAE20W40 + 0.2% MWCNT	-
Spindle speed rpm	B	500	750	1000
Feed rate mm/rev	C	0.1	0.15	0.2
Depth of cut mm	D	0.4	0.7	1.0

2.3 Experimental Procedure

The experimental study was carried out on DX160 JYOTI CNC Lathe machine with the following specifications: spindle power 5.5/7kw, working temperature 10°C to 50°C, spindle speed range 50 to 4000rpm, swing over bed 500mm and maximum turning length 500mm. The experiment is conducted for MQL (with cutting fluid) of using AISI 316L stainless steel as work material and coated carbide as tool material on CNC Lathe machine. The output responses to be measured are surface roughness and cutting temperature. The surface roughness of the turned surface was measured using a Mitutoyo Surfest SJ-201 contact profile meter as shown in Figure 1. The instrument was set to a cut off length of 0.8mm with a transverse length of 5mm. The cutting temperature at the tool chip interface was measured using a non-contact infrared thermometer. Turning operation is carried out on work piece with 85mm length. The experimental set up is presented in Figure 2. Based on the Taguchi's L18 Orthogonal array design, eighteen turning experiments are conducted as shown in Table 4.

Figure 1. Surface roughness measurement setup

Figure 2. Experimental set up

Table 4. Experimental results

Order	Run	Type of cutting fluid	Spindle speed rpm	Feed rate mm/rev	Depth of cut mm	Surface Roughness µm	Cutting Temperature °C
1	6	SAE20W40	500	0.10	0.40	0.28	39.9
2	4	SAE20W40	500	0.15	0.7	0.705	39.8
3	5	SAE20W40	500	0.20	1.0	1.1	39.7
4	7	SAE20W40	750	0.10	0.40	0.575	36.2
5	16	SAE20W40	750	0.15	0.7	0.78	42.4
6	2	SAE20W40	750	0.20	1.0	1.065	46.8
7	10	SAE20W40	1000	0.10	0.7	0.615	41.7
8	9	SAE20W40	1000	0.15	1.0	0.775	42.7
9	17	SAE20W40	1000	0.20	0.40	1.07	47.7
10	18	SAE20W40+MWCNT	500	0.10	1.0	0.635	43.4
11	12	SAE20W40+MWCNT	500	0.15	0.40	1.07	42.7
12	1	SAE20W40+MWCNT	500	0.20	0.7	1.54	45.2
13	8	SAE20W40+MWCNT	750	0.10	0.7	0.50	43.5
14	11	SAE20W40+MWCNT	750	0.15	1.0	0.88	42.2
15	15	SAE20W40+MWCNT	750	0.20	0.40	1.34	40.9
16	13	SAE20W40+MWCNT	1000	0.10	1.0	0.535	41.6
17	14	SAE20W40+MWCNT	1000	0.15	0.40	0.87	40.5
18	3	SAE20W40+MWCNT	1000	0.20	0.7	1.32	41.5

2.4 Desirability Function Analysis

Desirability function analysis is used to convert the multi response parameters into single response parameters. By using DFA, multi-response optimization of machining parameters can be converted into single-response optimization called composite desirability. The responses such as surface roughness and cutting temperature are combined as composite desirability.

The individual desirability (D_i) is calculated for surface roughness and cutting temperature by using the formula given in equation 1.

$$D_i = [y-y_{max}]/[y_{min}-y_{max}] \quad \text{----------------------------(1)}$$

Where y_{min} represents the minimum value of the response variable and y_{max} represents the maximum value of the response variable.

In this research work, the smaller-the-better desirability function is used to find the individual desirability for the response variables.

To compute the composite desirability (D_G), the following equation 2 is employed.

$$D_G = (D_1^{W_1} \times D_1^{W_2} \times \ldots\ldots \; \ldots\ldots \; \ldots\ldots \times D_i^{W_i})^{1/W} \quad \text{------------------------------(2)}$$

Where, D_i is the individual desirability of the response variable Y_i, W_i is the weight of the response variable "Y_i" in the composite desirability. W is the sum of the individual weights of the response variables.

IOP Conf. Series: Materials Science and Engineering **390** (2018) 012049 doi:10.1088/1757-899X/390/1/012049

3. Results and Discussions

3.1 Results from Desirability Function Analysis (DFA)

The multiple performances characteristics of the experimental results are evaluated using Desirability Function Analysis (DFA). After finding the individual desirability values of the response variables, then the composite desirability is calculated by using the formula given in equation 2. The resulting values are tabulated as shown in Table 5.

Table 5. Individual Desirability and Composite Desirability Values of Responses

Order	Run	Type of cutting fluid	Spindle speed rpm	Feed rate mm/rev	Depth of cut mm	Individual Desirability (D_i)		Composite Desirability (D_G)
						Surface Roughness µm	Cutting Temperature °C	
1	6	SAE20W40	500	0.10	0.40	1	0.6509434	0.77196541
2	4	SAE20W40	500	0.15	0.7	0.662698413	0.66037736	0.67877339
3	5	SAE20W40	500	0.20	1.0	0.349206349	0.66981132	0.55286888
4	7	SAE20W40	750	0.10	0.40	0.765873016	1	0.9149252
5	16	SAE20W40	750	0.15	0.7	0.603174603	0.41509434	0.50411896
6	2	SAE20W40	750	0.20	1.0	0.376984127	0	0.13217139
7	10	SAE20W40	1000	0.10	0.7	0.734126984	0.48113208	0.58464558
8	9	SAE20W40	1000	0.15	1.0	0.607142857	0.38679245	0.48597255
9	17	SAE20W40	1000	0.20	0.40	0.373015873	-0.08490566	0
10	18	SAE20W40+ MWCNT	500	0.10	1.0	0.718253968	0.32075472	0.46480746
11	12	SAE20W40+ MWCNT	500	0.15	0.40	0.373015873	0.38679245	0.41313363
12	1	SAE20W40+ MWCNT	500	0.20	0.7	0	0.1509434	0
13	8	SAE20W40+ MWCNT	750	0.10	0.7	0.825396825	0.31132075	0.47927918
14	11	SAE20W40+ MWCNT	750	0.15	1.0	0.523809524	0.43396226	0.49298575
15	15	SAE20W40+ MWCNT	750	0.20	0.40	0.158730159	0.55660377	0.38144085
16	13	SAE20W40+ MWCNT	1000	0.10	1.0	0.797619048	0.49056604	0.60769617
17	14	SAE20W40+ MWCNT	1000	0.15	0.40	0.531746032	0.59433962	0.5929133
18	3	SAE20W40+ MWCNT	1000	0.20	0.7	0.174603175	0.5	0.37023711

3.2 Effect of input parameters on Composite Desirability (D_G):

Table 6 shows the mean response values for each level of machining parameters on composite desirability. Figure 3 shows the response graph of means for composite desirability.

Table 6. Response table for the composite desirability

Control factors	Average composite desirability			Max-Min	Rank
	Level-1	Level-2	Level-3		
Type of cutting fluid, A	0.51393	0.42249		0.09144	3
Spindle speed, B	0.48025	0.48415	0.44024	0.04391	4
Feed rate, C	0.63721	0.52798	0.23945	0.39776	1
Depth of cut, D	0.53098	0.43475	0.43749	0.09623	2
Total mean of the composite desirability = 0.46808					

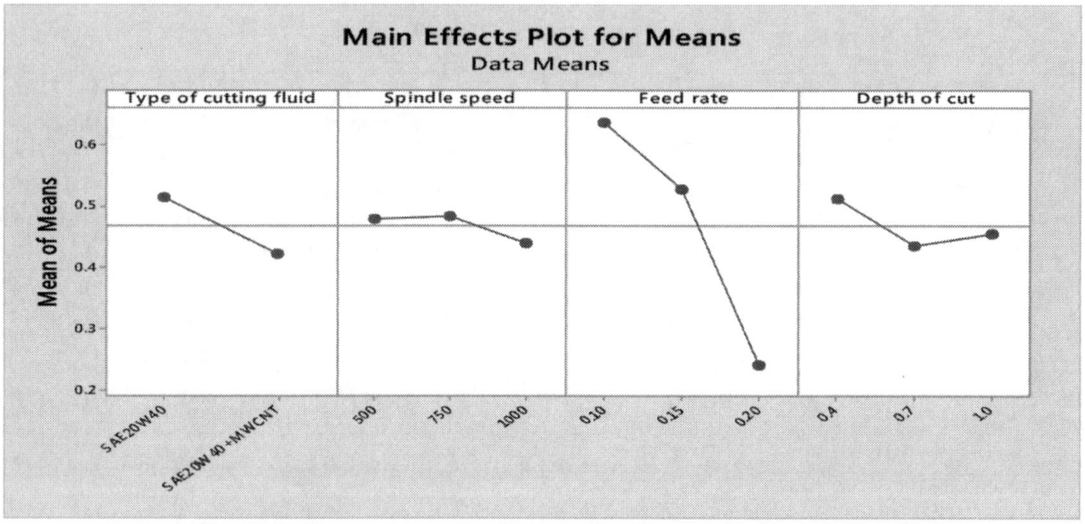

Figure 3. Response graph of means for Composite Desirability (D_G)

Basically, when the composite desirability is higher, the better is the output response variables. The optimal machining parameters for the surface roughness and cutting temperature are obtained as A1B2C1D1 as shown in table 7.

Table 7. Optimal machining parameters

Exp. No.	**Type of cutting fluid**	**Spindle speed**	**Feed rate**	**Depth of cut**
1	SAE20W40	750rpm	0.10mm/rev	0.40mm

3.3 ANOVA for composite desirability

The main objective of the statistical ANOVA is to find the significant machining parameters which affect the performance characteristics. From the ANOVA Table 8, it is observed that the feed rate is the most influencing cutting parameter for affecting the output responses. Simultaneously, the interaction between the type of cutting fluid and spindle speed affects the surface roughness and cutting temperature.

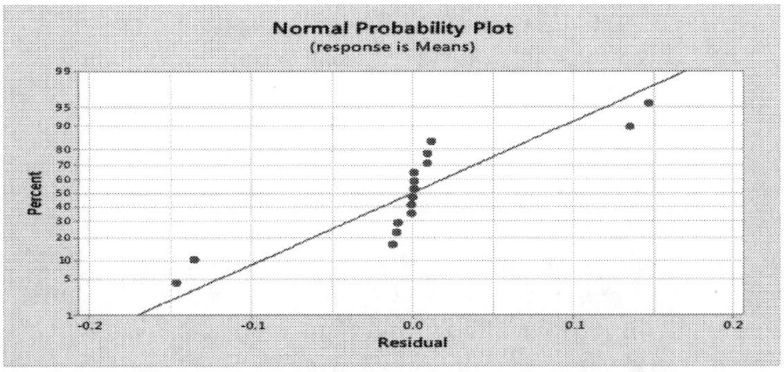

Figure 4. Residual plot for Composite desirability

Table 8. Analysis of Variance for the composite desirability (D_G)

Source	DOF	Seq SS	Adj SS	Adj MS	F - test	P (%)
A	1	0.04233	0.024190	0.024190	1.21	8.92
B	2	0.01935	0.003699	0.001850	0.09	4.09
C	2	0.17378	0.220139	0.110070	5.49	36.77
D	2	0.05937	0.013807	0.006904	0.34	12.56
A×B	2	0.07725	0.086746	0.043373	2.16	16.37
A×C	2	0.02024	0.020236	0.010118	0.50	4.32
Residual Error	4	0.08020	0.080197	0.020049		16.97
Total	15	0.47250				100

From the figure 4, it is found that all the points fall close to the straight line in the normal probability plot. This indicates that the developed models are significant.

3.4 Confirmation Experiments

The composite desirability value is predicted for the optimal combination of parameters (A1B2C1D1) and its value is 0.76203. Finally, Confirmation test was conducted using the optimum combination of parameters (A1B2C1D1). Table 9 shows the comparison of predicted composite desirability value with the actual one.

Table 9. Results of Confirmation Experiment

	Optimal machining parameters	
	Experimental	Prediction
Setting level	A1B2C1D1	A1B2C1D1
Surface roughness	0.575	
Cutting temperature	36.2	
Composite desirability value	0.91495	0.76203

4. Conclusions

In this study, the L_{18} taguchi orthogonal array design is used for conducting the experimental runs. The desirability function analysis method is applied to optimize the responses for the machining parameters during CNC turning of AISI 316L stainless steel with carbide tool inserts. It is shown that the multiple performance characteristics during machining of AISI 316L stainless steel are improved by using the desirability function analysis method in this study. The optimal parametric setting obtained using DFA is A1B2C1D1. From the ANOVA and response tables, it is concluded that feed rate is the most influencing cutting variable on the response variables.

References

[1] Mahesh K Sharma, Sanjay Singh and Rakesh Kumar 2016 *International Journal of Engineering Trends and Technology* **36**(4) 166-173

[2] Pay Jun Liew, Ainusyafiqah Shaaroni, Jeefferie Abd Razak, Mohd Shahir Kasim and Mohd Amri Sulaiman 2017 *International Journal of Applied Engineering Research* **12**(10) 2243-2252

[3] Venkata Ramana M, Krishna Mohan Rao G and Hanumantha Rao 2014 *International Journal of Materials, Mechanics and Manufacturing* **2**(4) 272-277

[4] Sonu Ram and Raj Kumar Yadav 2016 *International Journal of Advanced Technology in Engineering and Science* **4**(6) 275-287

[5] Shamnad M and Thulaseedharan Raman 2016 *International Journal of Advances in Mechanical and Automobile Engineering* **3**(1) 18-23

[6] Karan A Katle, Abhishek Rehpade and Faruk Qureshi 2017 *International Research Journal of Engineering and Technology* **4**(3) 1680-1685

[7] Shrikant Borade and Kadam M S 2016 *International Journal of Mechanical Engineering and Technology* **7**(1) 203-213

[8] Venkata Ramana M, Venkata Vishnu A, Krishna Mohan Rao G and Hanumantha Rao D 2012 *IOSR Journal of Engineering* **2**(1) 86-101

[9] Mamundi azaath L, Natarajan U and Arivanantham K 2017 *SSRG International Journal of Mechanical Engineering* Special Issue 15-22

[10] Venkata Vishnu A, Akhil J, Akhil A S, Raju B, Praveen Ch and Pavan A 2017 *International Journal of Advance Research and Innovative Ideas in Education* **2**(2) 20-27

[11] Patole P B and Kulkarni V V 2018 *Materials Today: Proceedings* **5** 104-112

IOP Conf. Series: Materials Science and Engineering **390** (2018) 012050 doi:10.1088/1757-899X/390/1/012050

Suitability and adaptability of Lean manufacturing in Indian pharmaceutical sector

Priya Harikumar[1] and P.G. Saleeshya[2]

[1]Department of Electronics & Communication Engineering, Amrita School of Engineering, Coimbatore, Amrita Vishwa Vidyapeetham, India

[2]Department of Mechanical Engineering, Amrita School of Engineering, Coimbatore, Amrita Vishwa Vidyapeetham, India

Email: h_priya@cb.amrita.edu

Abstract. Today's manufacturing companies are measured by how fast they meet the customer and user requirements, how responsive they are to the changing market demands and how capable they are in producing good quality products and supply them at optimal cost. Lean manufacturing is a manufacturing concept which has demonstrated results of successful implementation especially in discrete manufacturing companies. This paper presents a case study of an Indian pharmaceutical firm to outline the scope of a successful lean based manufacturing strategy implementation in this sector. Analytic Network Process (ANP) is applied to measure and rate the present leanness of a pharmaceutical firm. A multi-level model along with the questionnaires developed by the authors was used for the initial assessment. The approach helped to assess the current leanness level. Since inhibitors for lean interventions were identified the authors studied them and established the correlation among different parameters using Interpretive structural modelling (ISM) approach. The model was also validated through MICMAC analysis. The results of ANP, ISM and MICMAC showed the root causes for the unsuccessful implementation of lean in the firm. They were the lack of a responsive supply chain and the excessive Quality regulations and enforcements by the Authorities. This study gives guidelines to companies planning for lean implementation to deploy methods to remove the inhibitors of lean at the initial stages. Identifying the barriers can guide managers in developing strategies to eliminate them at the inception. This paper recommends supplier qualification programmes which managers can use to get the suppliers be part of their extended lean family. Through this supplier's involvement can be improved with an aim of improving the product lead time.

1. Introduction

The philosophy of Lean management was pioneered by Toyota Motor Company in the 1950's. Competing with US automakers they realized that to be profitable they should manufacture high quality products with very short lead times. This management philosophy focuses on waste elimination in all business processes. The authors Womack *et al.*, (1990) claim that lean methods make the companies specify value, make a best sequence of value creation activities, run these activities with no interruptions whenever there is a request and perform them continuously aiming for perfection. The primary interest in doing this research is to investigate whether pharmaceutical industries feel that Lean is a useful technique for them. A detailed study of pharmaceutical sector becomes indispensable to understand the scope of lean implementation. From literature review we realize that process industry which includes chemical, pharmaceutical, food and beverage, rubber, paper, steel, plastics, textile, and cement industries have not utilized lean techniques to the extent possible. [1] Applied lean tools in a steel plant to bring the lead time down by 50 % and production cost by 8 %. [3] in a case study in a chemical plant showed lead time reduction by 25 % and inventory reduction through JIT In textile industry. [2] implemented SMED and could bring down the setup time. Kanban pull production and levelling to be used once the product becomes discrete during the last stages of production in food and chemical industry [3]. According to [4] for lean to be implemented in a cGMP environment it should be done in an equal partnership [5] explained how lean sigma can be successfully implemented in pharmaceutical sector. They mention that structured change management programmes are essential for sustained results of lean sigma implementation. [6] have showed positive results like reduced lead time, throughput time, WIP and walkthrough time. in a pharmaceutical company.

The pharmaceutical firm selected as a case for our study is located in South India. It is a WHO-GMP certified medicine manufacturing facility with around 30 years of experience with emphasis on quality... The data for our study was collected from twelve senior level Managers with immense experience in the firm, working in Production, Quality and Logistics department. All the forty products manufactured go through four major processes which are Dispensing, Blending, Filling, and Packing. Based on the extensive literature review we developed a framework which includes all the elements of lean pillars and lean tools. The model developed should help to identify the strong lean practices followed and also should hint about the neglected ones which company should address so as to increase the leanness level of the firm.

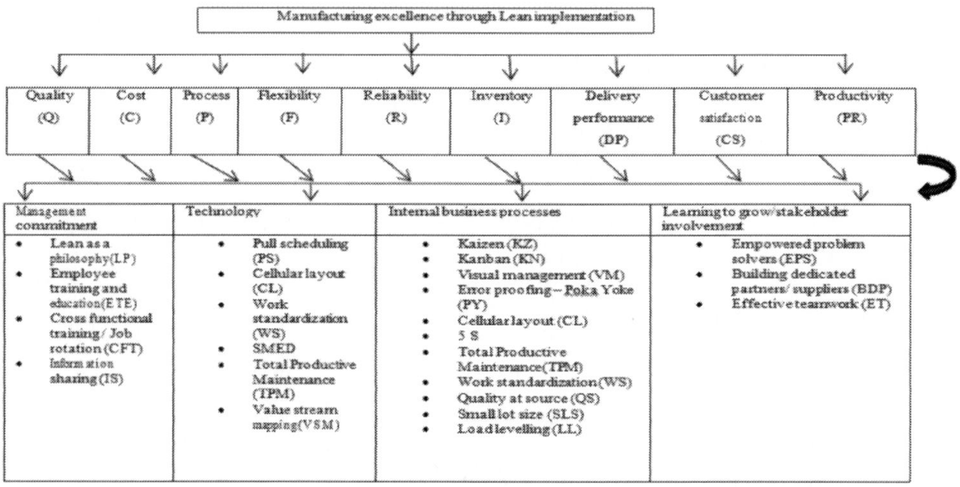

Figure 1. Model developed to assess the leanness of the firm

The model, the selection criteria, main clusters and their attributes were discussed with the managers. It was noticed that many factors in the model were interdependent. For example, improving the product quality can yield better customer satisfaction. Investment in process equipment can better the process. This in turn will reduce defects and improve quality. Reduced inventory can result in reduced manufacturing cost. Similarly delivery performance affects the customer satisfaction. This is at the top level of the model. Similar inter dependencies can be seen at the bottom levels as well. For example, effective teamwork in the cluster "Learning to grow" can happen only with a proper committed management which is the first cluster. Team work is vital in the second cluster as well. To practice Kanban (Third cluster) pull scheduling (Second cluster) is a required criteria.

Since managerial judgments, experience and knowledge are directly modelled, a tool which can solve strategic decision making problems is required. It should also account for the interdependencies among the main objective, nine selection criteria, four main clusters and twenty four criteria in all the four clusters. For this purpose we use Analytic Network Process (ANP) method for understanding the inter relationships and also to get the priorities for each of the clusters in relation to the top most objective, at the present functional level of the firm.

2. ANP methodology and results

ANP, developed by [6] considers the interdependencies among and between the various levels of top objective, criteria, clusters and its various elements. This method involves representing the elements in a hierarchical structure but allows for the dependencies among all the levels regardless of this hierarchy. ANP is advanced AHP which was developed by [6]. [7] used ANP to select an appropriate concept in attaining product sustainability. Lean manufacturing system was selected as the best strategy to implement in the operations department of a valve manufacturing firm by [9] and the method they used was ANP.

The method of ANP is applied to the conceptual model (Figure 1) developed to understand the level of leanness of the firm and its measurement. The overall organizational objective is to excel in manufacturing their products. There are nine selection criteria considered significant while trying to do so. However the list can be modified to suit any organization based on their requirements. The lower level in the hierarchy is the main four categories which are the areas where leanness is measured. Each cluster has own elements which are the lean enablers. It is clear from the model that these criteria are interdependent. This multifaceted dimension necessitates the use of a multi decision modelling technique for which Analytic Network Process (ANP) modelling method is preferred.

Construction of ANP model

1) Identify the organizational objective which needs to be catered to by being lean
2) Identify the criteria for selection
3) Develop a hierarchical structure with the main objective at the top, selection criteria at the second, main clusters where leanness is measured at third level and at the next level include the sub-attributes for each cluster - the lean enablers.

Calculation of rankings

- Make pair-wise comparison matrices to compare within each hierarchical level, to find the contribution of each criterion to the immediate upper level, to find the relative contribution of clusters to the higher levels. This was done by the experts in the firm once a consensus was reached among the managers on the weightages to be assigned to each factor when compared with another factor.
- Calculate the local priorities for selection criteria and the clusters
- Using those find the overall priorities for each cluster
- Use these priorities to rank the clusters.
- Choose the cluster with the highest priority.

ANP methodology calculations are quite tedious for this complex structure with one objective, nine criteria, four clusters and twenty four attributes as it involves determining interrelationships between all the factors in the model. Hence modelling the elements in the hierarchical structure, the pairwise comparison results given by the experts in the questionnaire survey and its result synthesis was done using Super decisions software. The final result analysis (Table 2) gives the current leanness level with respect to each of the category and makes the management aware of the scope for improvement.

2.1. Data collection using Contact survey
Questionnaires were distributed to the managers to get their input in a pair wise comparison mode. Since ANP methodology requires input in a matrix based comparison mode, the questionnaires were designed to gather inputs on a comparison scale (Table 1) and the comparison was made among

- Each of the nine selection criteria with respect to the main top level objective
- Each main four clusters with respect to the main top level objective
- The four main clusters with respect to each of the nine selection criteria
- The sub criteria factors in each of the four clusters with respect to the main cluster.
- Sub criteria factors in each main four categories with respect to each of the selection criteria factor. (For example pairwise comparison of Lean as philosophy , Employee training, cross functional training and information sharing with respect to each selection criteria Quality, cost, process etc.)

The pairwise comparison scale of 1 – 9 was used to rate the importance of each element over another. The table 1 below explains how the numbers were used. Here i correspond to the row element and j the column element. When comparing i to j, if i is 3 compared to j, then j is 1/3 when compared to i. (Reciprocals)

Table 1. Explanation of Numeric Values in Scale

Intensity of relative importance	Definition
1	Two elements in i and j contribute equally
3	i is favoured over j
5	i is strongly favoured over j
7	i is strongly favoured over j and dominance is demonstrated in practice
9	i favoured over j with the maximum possible confirmation
2,4, 6,8	Number in the middle of two adjacent judgments

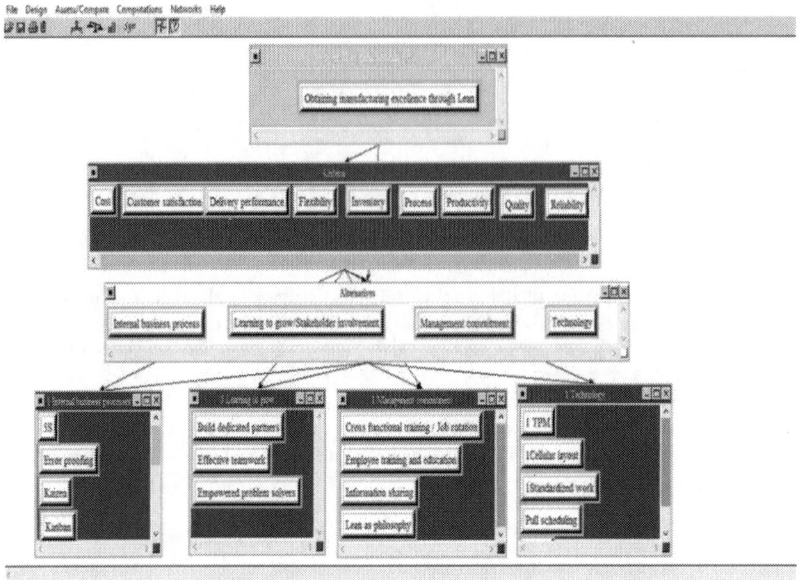

Figure 2. Model developed using Super decisions software

Table 2. Results obtained from Super decisions software

Cluster name	Normalized value	Rank
Internal business process	0.322840	1
Management commitment	0.250749	2
Technology	0.240343	3
Learning to grow / stakeholder involvement	0.186068	4

3. Interpretation of the ANP results and suggestions

The ANP results show the overall leanness of the firm is low which implies several factors exist which prevent lean to be fully implemented. Looking at the results cluster wise:

The leanness of the cluster Internal business process is 32.28%, management commitment as 25.07%, technology which is very close at 24.03% and results show the least lean area 18.60% for the cluster learning to grow/ stakeholder involvement. Knowing that the lean implementation is improper, we tried to identify the main barriers for the implementation. Our literature survey has showed several positive results of lean implementation in pharmaceutical sector. Hence we extended our study to the adaptability of lean tools to this environment.

4. Identification of barriers of lean implementation

Following the expert discussions in the firm, we wanted to perform a root cause analysis study on the Inhibitors of Lean in the firm. This is based specifically on the case under study. Hence cannot be generalized. But majority of the factors do represent the actual production related problems faced by a pharmaceutical firm in India. The main problems in lean implementation - "inhibitors" were identified and shown in Table 3.

Table 3. Lean inhibitors

	Lean Inhibitors	Causes for the inhibitors
1	Pull scheduling is difficult to attain.	The process is a continuous flow type. This required batching products along the assembly line. Single piece flow is almost impractical

2	Small lot sizing is not economical	Large setup time needed. Managers were apprehensive about minimum capacity utilization in case if they
3	Load leveling (Heijunka) could not be done	Large set up time and high inventory maintained.
4	Set up time reduction not possible	Long hours of machine cleaning needed and this was enforced by the
5	Very high quality regulations strictly followed (time consuming) and standards	Required as per WHO mandates. Hence difficult to reduce the lead time below a certain minimum
6	Lead time reduction not possible	Several point quality checks, some of which are out sourced due to non-availability of testing facilities which
7	Work standards already enforced	SOP's followed which are regulated by the health authorities and hence standardization through continuous
8	Supply chain responsiveness extremely low.	Suppliers located far away. Very few suppliers and hence autonomous
9	Minimum inventory impractical	All the above reasons

4.1. Interpretive Structural Modeling (ISM)
The result of ANP showed that lean is poorly implemented in all four clusters of the model. We wanted to understand the relation among the inhibitors (nine variables as inhibitors) if it exists and also make an informed judgement about the existence of a root cause for the poor implementation using ISM methodology. [12 & 13] have studied Lean in service sector and use of ISM in knowing about the contributing factors.

4.1.1. ISM applied to the case under study. Developing Structural self-interaction matrix (SSIM) of the lean inhibitors - SSIM provides the direction of relationships among the variables (Table 4). These are developed based on the interactions with experts from industry as well as from academics. Four symbols are used to develop the relationships between variables to form the SSIM. i represents row and j represents column.

V – Factor i will influence factor j
A – Factor j will influence factor i
X – Factors i and j help to attain each other
O – Factors i and j are unrelated

Table 4. Structural self – Interaction matrix of lean Inhibitors

Lean Inhibitors	*9*	*8*	*7*	*6*	*5*	*4*	*3*	*2*	*1*
1	*A*	*A*	*0*	*A*	*A*	*A*	*A*	*A*	*1*
2	*X*	*A*	*0*	*A*	*A*	*A*	*A*	*1*	
3	*X*	*A*	*V*	*0*	*A*	*A*	*1*		
4	*V*	*0*	*A*	*V*	*A*	*1*			
5	*V*	*0*	*V*	*V*	*1*				
6	*V*	*A*	*0*	*1*					
7	*V*	*0*	*1*						
8	*V*	*1*							
9	*1*								

*4.1.2. Initial reachability matrix (IRM) of lean inhibitors:*Here SSIM is converted to a matrix of binary representation called IRM (Table 5). For this conversion following rules are used.

- SSIM is V for (i, j) then replace IRM by 1 and (j, i) by 0.
- SSIM is A for (i, j) then replace IRM by 0 and (j, i) by 1.
- SSIM is X for (i, j) then replace IRM by 1 and (j, i) by 1.
- SSIM is 0for (i, j) then replace IRM by 0 and (j, i) by 0.

Table 5. Initial Reachability matrix

Lean Inhibitors	*1*	*2*	*3*	*4*	*5*	*6*	*7*	*8*	*9*
1	*1*	*0*	*0*	*0*	*0*	*0*	*0*	*0*	*0*
2	*1*	*1*	*0*	*0*	*0*	*0*	*0*	*0*	*1*
3	*1*	*1*	*1*	*0*	*0*	*0*	*1*	*0*	*1*
4	*1*	*1*	*1*	*1*	*0*	*1*	*0*	*0*	*1*
5	*1*	*1*	*1*	*1*	*1*	*1*	*1*	*0*	*1*
6	*1*	*1*	*0*	*0*	*0*	*1*	*0*	*0*	*1*
7	*0*	*0*	*0*	*1*	*0*	*0*	*1*	*0*	*1*
8	*1*	*1*	*1*	*0*	*0*	*1*	*0*	*1*	*1*
9	*1*	*1*	*1*	*0*	*0*	*0*	*0*	*0*	*1*

4.1.3. Final reachability matrix .Transitivity is checked while forming the final reachability matrix (Table 6). According to the transitivity rule, if variable 1 leads to variable 2 and variable 2 leads to variable 3, then variable 1 leads to 3. Following this rule, a final reachability matrix is formed. For instance, value in row 3 (variable 2) and column 4(variable 3) is changed to 1. This is because variable 2 is related to variable 9 (implied by value 1) and variable 9 is related to variable 3 (implied by value 1 in the initial reachability matrix) which imply variable 2 is related to variable 3. Hence changed the value in (3, 4) from 0 to 1. Similarly checked for transitivity for each cell value. (Table 7)

Table 6. Final reachability matrix

Lean Inhibitors	*1*	*2*	*3*	*4*	*5*	*6*	*7*	*8*	*9*
1	*1*	*0*	*0*	*0*	*0*	*0*	*0*	*0*	*0*
2	*1*	*1*	*0**	*0*	*0*	*0*	*0*	*0*	*1*
3	*1*	*1*	*1*	*0**	*0*	*0*	*1*	*0*	*1*
4	*1*	*1*	*1*	*1*	*0*	*1*	*0**	*0*	*1*
5	*1*	*1*	*1*	*1*	*1*	*1*	*1*	*0*	*1*
6	*1*	*1*	*0**	*0*	*0*	*1*	*0*	*0*	*1*
7	*0**	*0**	*0**	*1*	*0*	*0**	*1*	*0*	*1*
8	*1*	*1*	*1*	*0*	*0*	*1*	*0**	*1*	*1*
9	*1*	*1*	*1*	*0*	*0*	*0*	*0**	*0*	*1*

Table 7. Final reachability matrix with transitivity's included

Lean Inhibitors	*1*	*2*	*3*	*4*	*5*	*6*	*7*	*8*	*9*	***Driving power***
1	*1*	*0*	*0*	*0*	*0*	*0*	*0*	*0*	*0*	*1*
2	*1*	*1*	*1*	*0*	*0*	*0*	*0*	*0*	*1*	*4*
3	*1*	*1*	*1*	*1*	*0*	*0*	*1*	*0*	*1*	*6*
4	*1*	*1*	*1*	*1*	*0*	*1*	*1*	*0*	*1*	*7*
5	*1*	*1*	*1*	*1*	*1*	*1*	*1*	*0*	*1*	*8*
6	*1*	*1*	*1*	*0*	*0*	*1*	*0*	*0*	*1*	*5*
7	*1*	*1*	*1*	*1*	*0*	*1*	*1*	*0*	*1*	*7*
8	*1*	*1*	*1*	*0*	*0*	*1*	*1*	*1*	*1*	*7*
9	*1*	*1*	*1*	*0*	*0*	*0*	*1*	*0*	*1*	*5*
Dependence power	*9*	*8*	*8*	*4*	*1*	*5*	*6*	*1*	*8*	

4.1.4. Level partition. Levels are partitioned based on Reachability matrix. The reachability set of each inhibitor is obtained by adding the number of factors that have value one in the rows corresponding to that inhibitor. Likewise the antecedent set of each inhibitor is obtained by collecting the elements that have value one in the column corresponding to that element. A diagraph (Figure 4) is constructed using the reachability matrix. The level partitions in Table 10 are used to form the final ISM model (Figure 5) after the transitivity's are removed.

Table 8. Iteration 1

Lean Inhibitors	*Reachability set*	*Antecedent set*	*Intersection set*	*Level*
1	*1*	*1,2,3,4,5,6,7,8,9*	*1*	*I*
2	*1,2,3, 9*	*2,3,4,5,6,7,8,9*	*2,3,9*	
3	*1,2,3,4,7,9*	*2,3,4,5,6,7,8,9*	*2,3,4,7,9*	
4	*1,2,3,4,6,7,9*	*3,4,5, 7*	*3,4,7*	
5	*1,2,3,4,5,6,7,9*	*5*	*5*	
6	*1,2,3,6,9*	*4,5,6, 7,8*	*6*	
7	*1,2,3,4,6,7,9*	*3,4,5,7,8,9*	*3,4,7, 9*	
8	*1,2,3,6,7,8,9*	*8*	*8*	
9	*1,2,3,7,9*	*2,3,4,5,6,7,8,9,*	*2,3,7, 9*	

Table 9. Iteration 2

2	*2,3,9*	*2,3,4,5,6,7,8,9*	*2,3,9*	*II*
3	*2,3,4,7,9*	*2,3,4,5,6,7,8,9*	*2,3,4,7,9*	*II*
4	*2,3,4,6,7,9*	*3,4,5,7*	*3,4,7*	
5	*2,3,4,5,6,7 ,9*	*5*	*5*	
6	*2,3,6,9*	*4,5,6,7,8*	*6*	
7	*2,3,4,6,7,9*	*3,4,5,7,8,9*	*3,4,7,9*	
8	*2,3,6,7,8,9*	*8*	*8*	
9	*2,3,7,9*	*2,3,4,5,6,7,8,9*	*2,3,7,9*	*II*

Further iterations showed final level partitions as shown in Table10.

Table 10. Final level partitions

No	Elements	Level
1	**1**	**I**
2	**2,3,9**	**II**
3	**6**	**III**
4	**4,7**	**IV**
5	**5,8**	**V**

5.0 Barrier classification using MICMAC Analysis

The ISM model developed is validated using MICMAC Analysis (Figure 3). The idea of using MICMAC is to identify the driving power and dependence of all inhibitors and classify them into autonomous, dependent, linkage and independent factors. This is obtained from the final reachability matrix in ISM model. Table 7 above shows the driving power and dependence power of each inhibitor.

5.1. Driving and dependence power of variables

The driving power of each inhibitor is the entire number of elements it may assist to achieve, including itself. The dependence power is the total number of variables including itself which may assist to achieve it. Driving power of an inhibitor is the total number of ones in the equivalent rows of the final reachability matrix and dependence power is the total number of ones in the corresponding columns for each inhibitor. Higher dependence value for an inhibitor means that to eliminate this barrier several other inhibitors would have to be removed. Higher driving power suggests that if this inhibitor is removed many others can be eliminated.

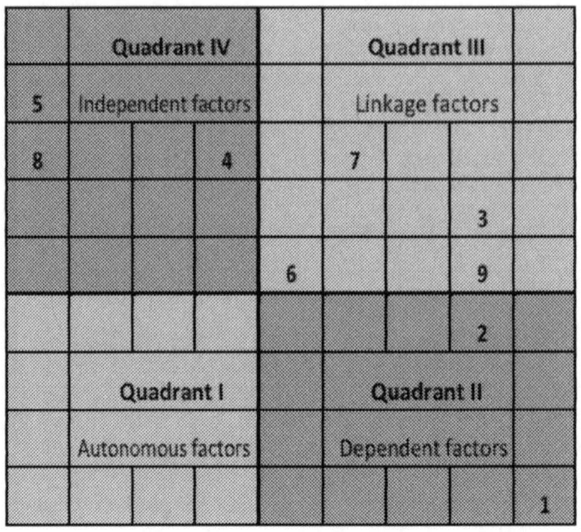

Figure 3. MICMAC Analysis

Autonomous factors (Quadrant I): The factors in this quadrant have weak driving power and weak dependence power. They are aloof from the system. They neither affect nor get influenced by the system. Our study did not find any barrier of this sort in this segment. This means that all the nine variables or inhibitors in the system are significant and all needs attention from the managers in the organization.

Dependent factors (Quadrant II): The factors in this quadrant have weak driving power and strong dependence power. In our analysis factors 1 and 2 belong to this cluster. Pull scheduling (1) has driving power 1 and dependence of 9. Similarly small lot sizing (2) has driving power 4 and dependence of 8. These values are plotted in the MICMAC graph. This falls in the Quadrant II. It means these factors do not drive the other inhibitors but many factors (9 for inhibitor 1 and 8 for inhibitor 2) may assist in achieving them. Hence to remove these inhibitors several other factors would have to be removed earlier.

Linkage factors (Quadrant III): The factors here have high driving and dependence power. Hence actions taken on these inhibitors will have high influence on all other inhibitors. Management should focus on these factors as entire system improves if steps are taken to prevent them from occurring. Linkage factors in our model are Load levelling (3), Lead time reduction (6), work standardization (7), and minimum inventory (9).

Independent factors (Quadrant IV): The factors have very high driving power and low dependence power. These factors are Setup time reduction (4), very strict quality standards (5), and very less responsive supply chain (8). These factors drive all other inhibitors and are not dependent on them. These factors are usually the ones which need proper attention. Hence termed as "key factors" in our analysis. These constitute the root causes for improper lean implementation and management focus should be on removing these so that it won't drive more inhibitors into the system. If these inhibitors can be eliminated several others can be easily removed.

6. Results of ANP, ISM and MICMAC analysis

- Our study captures the operations in a pharmaceutical company. The ANP analysis showed that business processes in the firm are comparatively leaner than the other areas. This means the company practices maximum lean in their processes. This is because of the enforcement of Quality throughout. But the leanness value is still very low at 32.28%.

- Management commitment is found high at 25.07% since employee training and employee involvement, job rotation, information sharing are practiced.

- Technologically lean is lagging behind at 24.03%. This was clearly visible as Pull scheduling, cellular layout, SMED and work standardization by means of continuous improvement was lacking in their operations.

- The lowest leanness value for the stakeholder involvement which is 18.60% is due to the least priority given to supplier partnership.

Based on the final ISM model, strictly enforced quality regulations and irresponsive supply chain are identified as the root cause and the worst barriers for effective implementation of lean practices in the pharmaceutical sector. These factors are driving all the other barriers and making lean implementation difficult. MICMAC refers to the inhibitors such as irresponsive supply chain, inability to reduce set up times and excessive quality regulations as independent drivers of all other barriers.

6.1. Conclusions from the study

cGMP quality checks makes less scope for work standards to be updated as they follow routine procedures without a continuous improvement approach. These result in inability to reduce the set up time. This further leads to difficulty in shortening the product lead time. The fact that lead time cannot be reduced makes it impossible to manufacture in small lot sizes. Hence big lots result in high inventory and Load levelling becomes impractical. All these lead to the top level inhibitor which is difficulty to maintain pull scheduling which is the basic requirement for lean implementation. Hence for lean to be implemented in a cGMP environment, both have to be made adaptable to each other if results are to sustain in the long run.

7.0 Managerial and research implications

Governmental / health authority enforced rules and policies also act as the major hindrance.

- Since this industry demands quality as highest priority and with all regulatory checks it is indeed a hindrance to improving the Lead time of the product. Hence options should be weighed as to how lead time can be reduced. Our recommendation is supplier partnership which is detailed in next section.

- If the Quality procedures are little relaxed without compromising on quality it can be a main contributor of lean since improved quality in processes will make the system defect free which means elimination of waste.

Major problems due to suppliers can be eliminated by installing a supplier qualification process as necessitated by FDA (Food and Drug Administration, 2006). This begins with identification and specification of type of materials, service and product needed from them. Proper audits, contractual obligations defined within a quality agreement and supply agreement are critical aspects. The compliance of the selected suppliers with the requirements and user requirement specification should be demonstrated and covered in the audit process. The compliance should be periodically evaluated. Any change if made by the supplier, which could have an impact on the GMP status or the production, have to be mutually agreed to before any such changes are implemented. A supplier must also notify the manufacturer immediately upon discovery of any deviation or non-conformance that may have impact on the services provided and corrective actions need to be taken immediately.

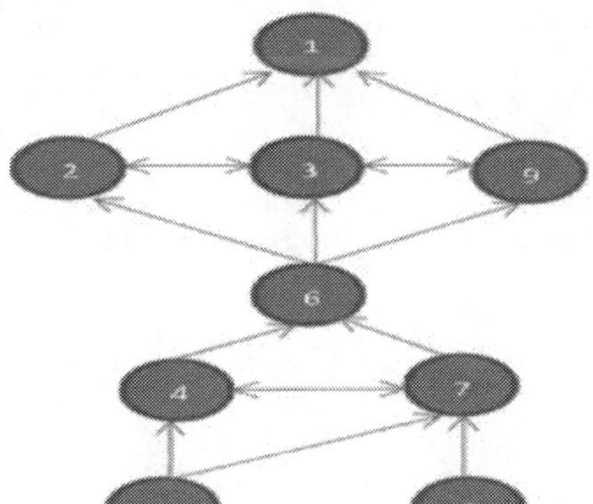

Figure 4. Development of Diagraph

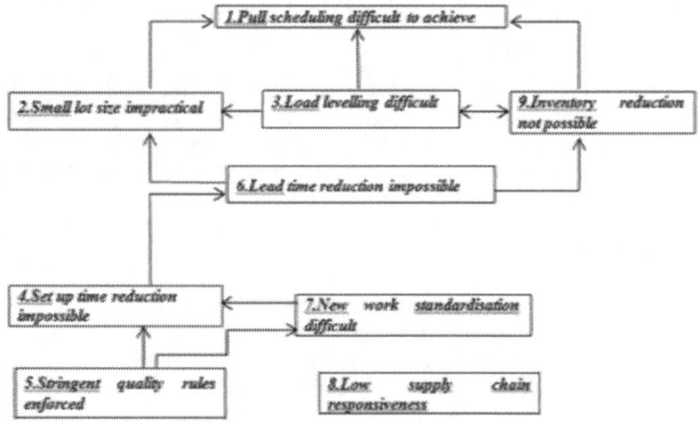

Figure 5. Final ISM developed

References

[1] Dhandapani V, Potter A and Naim M 2004 Applying lean thinking a case study of an Indian steel plant *International Journal of Logistics Research and Applications* **7,3** pp 239-250

[2] Cook R C and Rogowski R A 1996 Applying JIT principles to continuous process manufacturing supply chains. *Production and Inventory Management Journal* **37,1** pp 12–16

[3] Moxham C and Greatbanks R 2001 Prerequisites for the implementation of the SMED methodology:a study in a textile processing environment. *Journal of Quality and Reliability Measurement* **18,4** pp 404–414

[4] Greene A and Rourke O D 2006 Lean manufacturing practice in a cGMP environment *Pharmaceutical Technology Europe* **18,10** p 33

[5] Carleysmith S W, Dufton A M and Altria K D 2009 Implementing Lean Sigma in pharmaceutical research and development: a review by practitioners. *R&D Management*. **39,1** pp 95-106

[6] Saaty T 1996 Decision making with dependence and feedback: the analytic network process RWS

[7] Jayakrishna K., Vimal K. and Sekar V 2012 ANP based sustainable concept selection *Journal of Modelling in Management* **10,1** pp 118 – 136

[8] Anand G and Rambabu K 2009 Selection of lean manufacturing systems using the analytic network process – a case study Journal of Manufacturing Technology Management **20,2** pp 258-289

[9] Saleeshya P G, Babu A S and Vishnu A S 2011 A model to assess the agility of manufacturing organizations: systems approach and applications *International Journal of Productivity and Quality Management* **8,3** pp 247-264

[10] Saleeshya P G, Karthick S. and Raghuram P 2012 A combined AHP and ISM based model to assess the agility of supply chain – a case study *International Journal of integrated supply Management* **7,1** pp 167-191

[11] SuperDecision for Decision-Making 2011 Available at http://www.superdecisions.com

[12] Vignesh V and Suresh M 2016 Factors influencing lean practises in super market services using Interpretive Structural Modelling *IEEE International conference onComputational intelligence and computing research* ICCIC 7919662

[13] Vignesh V, Suresh M and Aramvalarthan S 2016 Lean in service industries, A literature review *IOP Conference series, Materials science and Engineering* **149,1**

IOP Conf. Series: Materials Science and Engineering **390** (2018) 012051 doi:10.1088/1757-899X/390/1/012051

Review of WEDM studies on metal matrix composites

S Vijayabhaskar[1*], T Rajmohan[1],TVPranay Sisir[1], JVS Phani Abishek[1], D Sharukh khan[1]and RMohankrishna Reddy[1]

[1]Department ofMechanical Engineering, SriChandrasekharendraViswa Mahavidyalaya, Kanchipuram -631561, Tamilnadu, India.

*mailvijayabhaskar@gmail.com

Abstract. At present, the growing needs and design requirements with significant weight savings as well as high strength-to-weight ratio as compared to conventional materials raised a growing interest towards composite materials. Metal matrix composites (MMCs) are extensively used in fields like aerospace, automobiles and army-defence applications due to their high specific strength, low coefficient of thermal expansion, wear resistance, high durability, excellent physical and mechanical properties. The filler reinforcements like SiC, TiC, Al_2O_3 improve the stiffness, specific strength, creep properties of MMCs, but the presence of such hard reinforcements makes MMCs difficult to machine through conventional machining processes. Wirecut electrical discharge machining (WEDM) is highly precise thermoelectrical process and is considered ideal for machining very hard materials, intricate geometries and tooling. This paper reviews the researches in WEDM of metal matrix composites and the relation between different process parameters like pulse on time, pulse off time, wire feed rate,voltage, wire tension, dielectric flow rate on the machining performances such as material removal rate (MRR), Kerf widthand surface roughness (SR). The future trends in WEDM researches were also recommended in this paper.

1. Introduction

Metal matrix composites are materials with two or more constituent parts, one of them being a metal composition called matrix and the other hard materials like ceramic or organic compound known as reinforcement. MMC's are obtained by dispersing the reinforcing material into the metal matrix. In structural applications, lighter metal such as aluminium, magnesium, or titanium are employed for impartingamenable medium for the reinforcement. In high temperature applications, cobalt and cobalt–nickel alloy matrices are common. Reinforcement generally improves rigidity and greatly obstructs crack propagation. The composite's strength usually depends on the diameter of the particles, the inter-particle spacing and the volume fraction of the reinforcement. Some of the commonly used reinforcements are Al_2O_3, SiC, TiC, B_4C, etc., impart high hardness to the matrix in the composites. The presence of these hard and abrasive reinforcements makes the production of complex shapes difficult through traditional machining processes.

Among different unconventional process WEDM plays a prominent role in different manufacturing industries and it is mostly used in the aerospace, defence and army applications. WEDM is a typical thermal machining techniquethat can accurately machine parts with varying hardness and intricate shapes. It is a spark erosion process which can able to produce complex 2D and 3D work pieces in which the wire does not touch the work piece resulting in reduction of the physical pressure imparted on the work piece. When compared with the other conventional machining processes like the grinding and milling cutters where serious tool wear occurs due the presence of reinforcement, WEDM is considered as an effective non-contact machining process. While other non-conventional methods like abrasive jet, electron beam machining and laser cutting are restricted to linear cutting, WEDM exhibits advanced capability for cutting complex 3D shapes with high accuracy.

Despite of these advantages, WEDM process is limited in terms of material removal rate because of poor surface finish and danger of wire breakage. This calls for the study of WEDM process parameters to obtain maximum material removal rate with better surface finish. This paper primarily reviews the major WEDM research activities on MMCs which include the different process parameters such as pulse on time, pulse off time, servo voltage,peak current, wire feed rate, dielectric flow rate and wire speed. These process parameters have a major effect on the performance characteristics such as material removal rate, surface finish, kerf width, dimensional deviation and wire wear rate.

2. Working principle of WEDM

Wire cut electrical discharge machining is a thermal based process in which spark is generated between work piece and tool material. Here the wok piece and electrode are placed in a dielectric medium which are connected through electronic circuit and generates DC pulses from the wire to the work piece. In this WEDM process material removal takes places through continuous repetition of spark discharge between the tool and work piece. By using servo-controlled mechanism the gap of 0.025 – 0.075 is continuously maintained between the tool and work piece. In this process de ionized water is used as coolant. This de ionized water continuously passes between the work piece and the wire. In this process wire feed should be continuous.

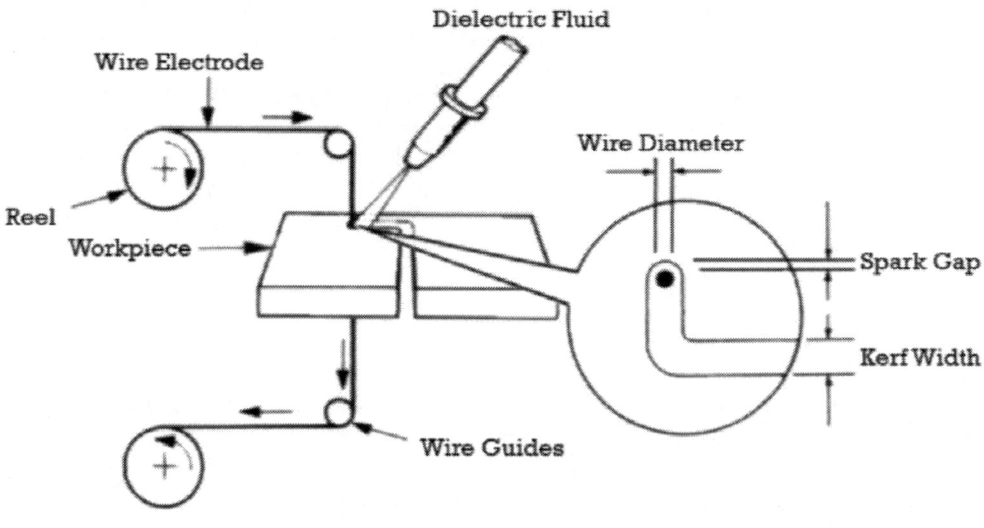

Figure 1. Schematic diagram of WEDM process

During this process spark is generated which generate huge amount of heat. This much amount of heat is sufficient to melt the material along with the work piece. This results in removing of molten mass from the work piece. Then the tool profile is transferred to the work piece. Figure 1 shows the schematic diagram of WEDM process. Usually, WEDM is used for machining hard materials such as High-Speed Steel, hardened steel, Metal Matrix Composites etc. WEDM enables machining of different types such as blind, inclined, micro holes and complicated profiles. This is the only machining process that can machine hard materials with high accuracy and good surface finish.

3. WEDM process parameters

The performance WEDM is observed in terms of output responses like material removal rate, kerf width, surface roughness, etc. These responses were mainly affected by input process parameters such as pulse on time, pulse off time, voltage, feed rate, dielectric flow rate, wire speed etc. Figure 2 shows the cause and effect diagram for various performance measures in WEDM process.

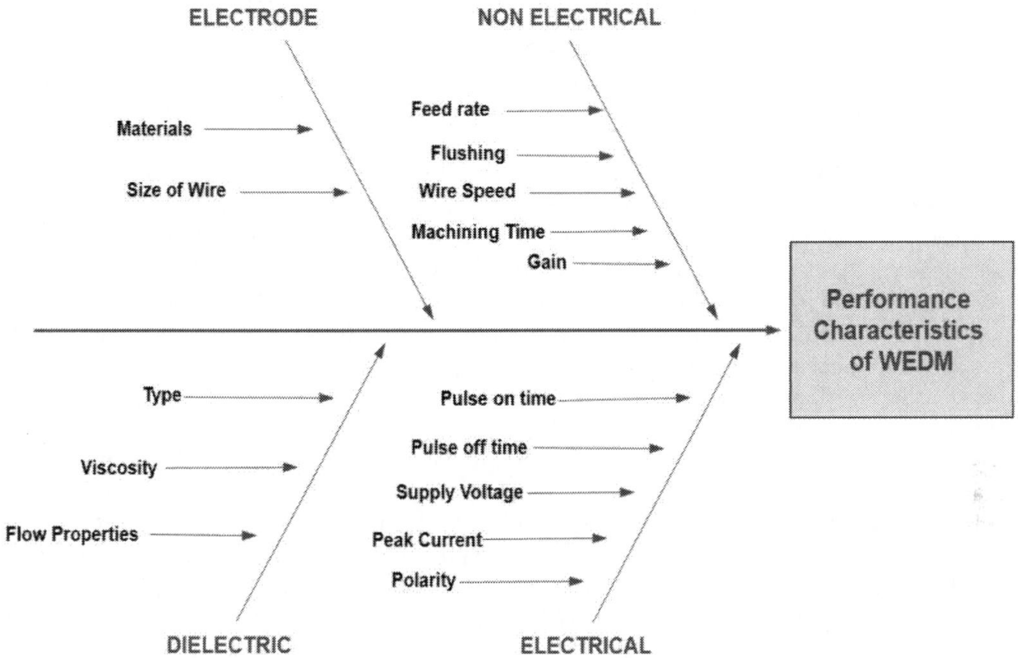

Figure 2. Cause and effect diagram for WEDM process

Pulse on time and pulse off time: In wire cut electrical discharge machining, the material removal occurs and stops alternately during pulse on time and pulse off time. During pulse on time, the voltage is applied to the gap between the electrode and the inserted work piece, resulting in electric discharges between them. If there is a long duration of electrical discharge then the greatest value of pulse on time is taken, to avoid wire breakage pulse off time is inserted between the pulse on time.

Peak Current: Peak current is the amount of power utilized in WEDM process. During pulseon time, the current goes on increasing till it reaches the pre-set level which is expressed as peak current. In case of roughing operations and large surface areas, higher peak current values are used.

*Servo Voltage and feed rate:*The movement of wire is controlled using servo voltage while the table is controlled by the servo feed rate. If the value of servo voltage is higher, the gap between the work piece and electrode is higher. In the case of higher servo voltage, the electric sparks and machining rate is low. Whereas lower servo voltage results electric sparks and machining rate is high.

Dielectric flow rate: Dielectric fluid is essential for obtaining stable electric discharge as a result of efficient cooling and chip removal. In WEDM process, de-ionized water is used as dielectric fluid because it is environmental friendly nature. High flushing pressure will result in better material removal rate.

Wire tension and speed:
The wire remains straight if the wire tension high, otherwise it may get bend and results in breakage. Increased wire speed increases the wire consumption while wire breakage occurs with low wire speed.

4. WEDM performance characteristics

4.1. Effect of process parameters on WEDM Performances

The performance characteristics of WEDM such as materials removal rate, surface roughness, kerf width, etc., were influenced by various parameters such as pulse on time, pulse off time, current,voltage and dielectric flushing pressure. Scott et al. [1] give a methodology to study optimal combination of parameters such as pulse duration, pulse frequency, wire speed and dielectric flow rate in WEDM of D2 tool steel. In this process, the methods used are enumeration and dynamic programming. From the result they conclude that pulse duration and pulse frequency are significant factors for material removal rate and surface finish, whereas dielectric flow rate and wire tension are insignificant.Dhar et al. [2] founded to study the effects of machining parameters like pulse on time and gap voltage on the response parameters like material removal rate, tool wear rate and radial over cut in WEDM of Al4 Cu-6Si alloy reinforced with 10% SiC particles. From the results, they developed the mathematical model to organize bonding among the parameters.The material removal rate and surface roughness were increased by increasing the current [3]. The regulation in pulse off time based on the sparking frequency affects the material removal rate. But, the sudden increase in sparking frequency leads to wire breakage [4].

The other parameters relating the electrode wire and workpiece also play a vital role in WEDM process. The open circuit voltage increases when there is an increase in wire speed and dielectric fluid pressure [5]. However, wire feed rate and wire tension had no effect on MRR. MRR found to be increased when there is an increase in pulse on time and get decreases with the increasing pulse off time [6]. The materials which are used for fabrication of wire electrodes should be characterized by high melting and evaporation temperatures. Metals like Mg, alkaline metals are easily machined by using of copper, brass and steel wires which can increase cutting efficiency of WEDM [7]. The size of craters formed and the MRR are increased with increase in relative speed. Also, high relative speed can produce higher melting efficiencies [8].Some of properties like electrode wear ratio and surface roughness were better on the conductive materials than insulating materials while MRR is better in insulating materials than conductive materials.

4.2. WEDM on metal matrix composites

WEDM is considered to be one of the ideal methods for machining metal matrix composites with hard reinforcements and varying hardness. Patel and Brahmankar [9] carried out experimental studies on WEDM of the aluminium metal matrix reinforced with alumina particle and found that increased percentage of ceramic particulates reinforcement in the MMC decreases MRR. The pulse on-time and thermo-physical properties like melting point temperature, thermal diffusivity and coefficient of thermal expansion areimportant parameters influencing MRR.Ahamed et al. [10] investigated EDM performances of hybrid Al-SiC composites in addition with B_4C and glass reinforcements. The presence of these ceramic particulate reinforcements acts as a barrier to the machining of hybrid Al-SiC_p Composites. It is observed the long sparking duration and high dielectric flushing pressureare required to remove the tightly packed reinforcement like B_4C and SiC. The reinforcements with better wettability result in increased pulse on time and reduced flushing pressure.

Lal et al. [11] investigated the influence of parameters such as discharge period, pulse interval, wire drum speed on kerf width during WEDM of Al7075/SiC/Al_2O_3 hybrid composite obtained by the inert gas assisted electromagnetic stir casting process. The discharge current has the most significant effect on kerf width followed discharge duration and wire drum speed, while the effect of pulse duration is insignificant on kerf width.Rozenek et al [12] studied the WEDM characteristics of metal matrix composites such as AlSi7Mg/SiC and AlSi7Mg/Al2O3. They found that the cutting speed while machining AlSi7Mg/SiC and AlSi7Mg/Al2O3 through WEDM is 3 times and 6.5 times slower than the aluminium alloy respectively. The wire feed rate during WEDM of these composites also depends on the type of reinforcement used. Material removal rate in WEDM and EDM of MMCs is lower than that of the alloy due to decreased the electrical conductivity and thermal conductivity.Yan et al [4] indicated that the surface roughness, kerf width and cutting speed during WEDM of alumina reinforced 6061 aluminium matrix composites were significantly depend on volume fraction of reinforced alumina particles. It is also revealed from the test results that a slight wire tension, a larger dielectric flushing rate and a high wire speed are important to avoid wire breakage

4.3. Process modelling and optimization.

The works done until now have not been associated with any modelling andoptimization techniques. Therefore, Patil and Brahmankar [9] developed two different models of MRR in WEDM of MMCs. The first model is anempirical model employing response surface methodology (RSM) and the second model was a semi-empirical using dimensional analysis based on physical, electrical and thermal properties of the work material in addition to the machining parameters such as pulse-on time and voltage. The predictability of the both the models found to be more than 99%.Liao et al. [13] identified some of machining parameters like voltage, resistance, pulse- generating circuit and capacitance as the significant parameters affecting the surface roughness of WEDM in finishing process with the help of Taguchi quality design, ANOVAand F-test.Rao et al. [14] performed parametric analysis of Wire EDM parameters on surface roughness and material removal rate using Taguchi method. Hybrid genetic algorithm is used to develop linear regression model to optimize the performance measures such as surface roughness and material removal rate simultaneously. The results obtained are in a good agreement with experimental values.

Puri et al. [15]developed mathematical modeling of white layer to correlate the significant input parameters of WEDM with rough cut followed by trim cut through response surface methodology. The second order rotatable central composite design have been used for experimental plan to carry out the investigation employing four variable such as pulse on time in rough cutting, pulse on time in trim cutting, offset and cutting speed with five levels.Iqbal et al [16] used response surface methodology to investigate the relationships and parametric interactions between the performance measures such as MRR, electrode wear rate and surface roughness in EDM milling. The model coefficients of the factors were estimated using central composite experimental design and ANOVA is performed to obtain significant coefficients.

Datta et al. [17] derived the quadratic mathematical models to represent parametric influence on the process responses of WEDM such as material removal rate, surface roughness and kerf. Response surface method is used for prediction of process responses with different combinations of factor settings. Predicted data obtained by Taguchi's L27 orthogonal array is used for achieving an optimal combination of parameters to attain maximum MRR, minimum surface roughness and maximum dimensional accuracy. This multi-objective criterion is convert into an equivalent single objective function using grey relational analysis. Overall grey relational gradehas been optimized using Taguchi technique and verified through confirmatory test. Kumar et al. [18] investigated WEDMof pure titanium by modelling the response such as surface roughness, machining rate, dimensional accuracy and wire wear ratio using response surface methodology. The experiments were carried by changing the parameters such as pulse on time, pulse off time, gap voltage, peak current, wire tension and feed rate based on Box–Behnken design and the responses were optimized through desirability analysis, while the significance of the model is verified by ANOVA.

Shah et al [19] optimized the process parameters during WEDM of Inconnel-600 using response surface methodology (RSM). The experimental plan is proposed by Taguchi's robust design and the experimentation has been carried out using Taguchi's Mixed L18 orthogonal array. The input parameter such as pulse on time, pulse off time, peak current and wire feed rate were considered for optimizing the response in terms of material removal rate. ANOVA is used to study the significance of the process parameters on the response of WEDM also the response surface model has been developed for the performance characteristic such as MRR. Sharma et al. [20] investigated the effect of parameters such as pulse on time, pulse off time, servo voltage, peak current and wire tension on the responses such as metal removal rate and surface roughness during WEDM of HSLA. The central composite rotatable design (CCRD) is employed in conducting the experiments and response surface methodology is applied in formulating mathematical model to optimize the process parameters.

Lakshmanan et al [21] performed electrical discharge machining of EN 31 tool steel to correlate the process parameters with the responses such as MRR. The performances of the process were modelled using response surface methodology and model adequacy checking is validated using analysis of variance.Majhi et al [22] presented a hybrid optimization method for the obtaining of the optimal process parameters such as Pulse on time, pulse off time and pulse current for maximizing material removal rate and minimizing surface roughness and tool wear rate. The plannedexperimental results are applied in the grey relational analysis and the weight of the quality characteristics are obtained through entropy measurement method.The influence of the process parameters on the output responses were predicted by response surface methodology based on the results from optimization.

5. Conclusions

From the above studies on WEDM of metal matrix composites it can be concluded that:

- WEDM is found to be effective and efficient process for the machining of metal matrix composites with varying hardness, intricate shapes and presence of hard reinforcement which can cause serious tool wear when machining through conventional process.
- Most of the studies were carried out mainly on WEDM of tool materials like die steels, hard metals like titanium alloys, MMCs such as aluminium matrix composites reinforced with SiC, Al_20_3, B_4C, etc.,
- Some of the WEDM process parameters such as pulse on time, pulse off time, voltage, peak current, wire tension and wire feed rate were considered for optimising in many investigations to improve the performance such as material removal rate and surface roughness.
- The effects of other non-electrical parameters such as dielectric flushing pressure, volume % of the reinforcements, workpiece material properties, etc., can also be studied in improving the machining performances of WEDM.
- In most of the investigations, the experimental plan is devised by Taguchi method, mathematical models were derived by response surface methodology to optimize the process parameters using desirability analysis and finally the significance of the different models were tested using ANOVA.
- From the review of various studies carried out in WEDM of MMCs, it is clear that there is a future scope for WEDM of the light materials like magnesium matrix composites reinforced with nano particles such as carbon nanotubes, nano SiC, etc.,

6. References

[1] DanS, Boyina S andRajurkar KP 1991 *Int. J. Prod. Research* **29**2189-207
[2] Dhar S, Purohit R, Saini N,Sharma A and Kumar H G 2007 *J. Mat. Pro.Tech***194:**24–29
[3] Karthikeyan R, Raju S and Nagavazen R.S 2001 *J. Mat. Sci. Tech* **17** 57-60
[4] Yan B, Chung H, Huang FY, and Lee LC 2005 *Int. Mach. & Manuf.* **43** 251 – 259.
[5] Nihat T and Cogun C 2002 *J. Mat. Sci. Tech***6471** 1-6.
[6] Singh H and Garg R *2009 J. Ach. Mat. &. Manf. Engg***32** 70-74
[7] Prohaszka J, Mamalis AG, and Vaxevaxevanidis NM 1997 *J. Mat Tech***69** 233-238
[8] Weingartner E, Wegener K and Custer F 2013 *Procedia CIRP***6** 238-243
[9] Patil N G and Brahmankar P K 2010 *Int. J. Adv. Manuf. Tech.* **48** 537-555
[10] Ahamed A, Asokan P and Aravindan S 2009 *Int. J. Adv. Manuf. Technol.***44** 520–528
[11] Lal S, Kumar S, Khan ZA, and Siddiquee A.N 2014 *J Braz. Soc. Mech. Sci. Eng***36** 335-346
[12] Rozenek M, Kozak J, Dąbrowski L and Łubkowski K 2001 *J. Mat. Tech***109** 367–37015.
[13] Liao Y S, Huang J T and Chen Y H 2004 *J. Mat. Proc. Tech.***149** 165-171.
[14] RaoS P, Ramji K and Satyanarayana B 2014 *Procedia Material Science* **5** 2197-2206.
[15] Puri A and Bhattacharya B 2005 *Int. J. Adv. Manuf. Tech.*301-307.
[16] Iqbal A AKM and KhanA A2010 *American J. Eng. & Appl. Sci.3(4)* 611-619
[17] Saurav D and Sankar M2010 *Int. J. Eng. Sci. & Tech.***2(5)** 162-183.
[18] Kumar A, Kumar V and Kumar J 2012 *Int. J. Eng. & Tech.***2(4)**708-719
[19] Shah, C D, Mevada J R and Khatri BC2013 Int. J. Tech. & Adv. Eng. **3(4)**260-267.
[20] SharmaN, Khanna R and Gupta R 2013 *Int. Conf. Manuf.***64** 710-719
[21] Lakshmanan S and Kumar M2013 *Int. j. Eng. Sci. & Tech.* **2(5)** 64-71.
[22] Majhi SK, Pradhan M K and SoniH 2013 *Int. J. App. Mech. Eng.* **3(1)** 82-87.

IOP Conf. Series: Materials Science and Engineering **390** (2018) 012052 doi:10.1088/1757-899X/390/1/012052

Experimental Investigation on Tribological Characteristics of Biodegradable Refrigeration Oil using Pin on Disc Tribometer

P. Vithya[1*], G. Sriram[2], S. Arumugam[3], Ankur Sharma[4], Sistila.Venkata sai sathyanarayana sastry[5]

[1-5] Department of Mechanical Engineering, Sri Chandrasekharendra Saraswathi Viswa Mahavidyalaya, Kanchipuram, Tamilnadu - 631561, India.

Email: vithyagautham@gmail.com

Abstract. The present study addresses the issue of the tribological behaviour of cylinder liner and piston ring material of a hermetically sealed reciprocating compressor of Vapour Compression Refrigeration system under the influence of pentaerythritol ester based biodegradable refrigeration oil and compare with synthetic refrigeration oil (Polyol ester-ISO 68grade) using the pin on disc tribometer. Synthesis of pentaerythritol ester based biodegradable refrigeration oil for household domestic refrigeration compressors was carried out via successive transesterification of rapeseed oil. The biodegradable refrigeration oil was blended equally with synthetic refrigeration oil to formulate the best biolubricant. Further Energy Dispersive X-ray (EDX) technique was adopted for the chemical characterisation of the pin material surfaces. The wear behaviour of the tested pin materials were evaluated using scanning electron microscope. The results show that under same investigational conditions 50% blend of pentaerythritol ester biodegradable refrigeration oil (PBRO 50) with synthetic refrigeration oil demonstrated superior friction reduction performance and inferior wear reduction performance as compared to synthetic refrigeration oil.

1. Introduction

The compressors are considered to be an influential component of vapour compression refrigeration system. The major cause of loss of energy in a hermetically sealed reciprocating refrigeration compressor is due to friction and wear, which should be minimised by lubrication. A secured and sustainable operation of the vapour compression refrigeration system has accomplished by the lubrication system. The compressors consume approximately 80% of the whole power consumption of the vapour compression refrigeration system. The energy conservation is becoming an indispensable factor in the aspect of environmental and global issue. Thus the coefficient of performance escalation of the vapour compression refrigeration system can be enacted by the reduction in the compression work, which in turn reduces the energy consumption.

The environmental concerns give great significance of using bio-based lubricants for the commercial purpose which leads to the major transformational growth in the lubrication market [1]. The main limitations of vegetable oil are its poor low temperature behaviour, oxidation and thermal stability and gumming effect [2,3]. Menkiti et al [4] prepared the polyol ester from sesame seed oil as biodegradable base stock through chemical transesterification of sesame methyl ester with trimethylolpropane (TMP) using calcium hydroxide catalyst. They found that temperature and mole ratio have the major effects on the synthesis of transesterification process. Journal bearing tribological behaviour specifically for automobile utilization under the impact of a synthetic lubricant such as SAE20W40 and biolubricant from rapeseed oil (CMRO) dispersed with TiO_2, WS_2 and CuO nano particles utilized concerning illustration anti-wear additives analysed by Baskar et al [5]. They concluded that maximal friction coefficient and wear was settling up in concern with biolubricant (CMRO) than the synthetic lubricant, while the addition of nano CuO to biolubricant (CMRO) reduce the friction coefficient by 27 % and wear by about 47 % as compared with a synthetic lubricant. Habibullah et al [6] evaluated tribological characteristics of Calophyllum inophyllum (CI) based trimethylolpropane (TMP) ester as biodegradable lubricant using four ball tribometer and describe the resemblances with commercial lubricant and paraffin mineral oil on energy saving aspects. The outcomes illustrate that biodegradable lubricant had the improved friction behaviour with lower energy consumption. However, surface morphology analysis shows that CI TMP ester comprise higher wear match up to paraffin mineral oil and lower wear to commercial lubricant.

Zulkifli et al [7] conducted a detailed investigation to access the wear prevention characteristics of palm oil based TMP ester as engine lubricant through a four-ball tribo tester for various lubrication regimes. They observed that 30% reduction of friction coefficient and higher amount of load (220 kg) carrying capacity in inclusion of 3% of palm oil based TMP ester with commercial lubricant for boundary lubrications, 50% reduction of friction coefficient in inclusion of 7% of palm oil based TMP ester with conventional commercial lubricant for hydrodynamic lubrication, thus suggested that can be a surrogate as automotive engine lubricant. Nuraliza and Syahrullail [8] examined the friction and wear aspects of double fractionated palm oil (DFPO) as biolubricant using pin-on-disk tribo-tester under loads of 50N and 100N, with rotating speeds of 1-5m/s for testing time of 3600 sec and referred with hydraulic oil and SAE 40 engine oil. On the view of coefficient of friction, wear and rise in sliding speed on whole process time, the DFPO lubricant demonstrated most excellent performance, can be suggested as a base lubricant for mechanical applications.

A lot of research works are conceded using bio lubricant as replacement to synthetic lubricant for engine oil application. Likewise, plenty of investigation work is focused on the performance enhancement of vapour compression refrigeration system by using synthetic lubricant with nano particle additives. Conversely, lesser studies have been carried out with bio lubricant as refrigeration oil. From the literature, it is identified that currently no refrigeration oil is available in the market with vegetable oil as base oil for the formulation of refrigeration oil. The present study formulated the bio degradable refrigeration oil from rapeseed oil using successive transesterification with pentaerythritol through chemical modification to perceive the lubrication characteristics appropriateness as refrigeration oil for domestic compressors. Moreover, tribological behaviour of the bio degradable refrigeration oil has been accessed using pin-on-disc tribometer for examining the compatibility for using as refrigeration oil.

2. Experimental

2.1 Materials

The raw rapeseed oils were collected from the local vendor. Methanol%, NaOH, Pentaerythritol, Para-toluene sulphonic acid and xylene were purchased from Ganesh Enterprises, Kanchipuram.

2.2. Synthesis of pentaerythritol ester based Biodegradable Refrigeration oil

Rapeseed oil of 1 litre was placed in the 5 litre double necked round bottom flask. Potassium Hydroxide of 16g and methanol of 200 ml was added to the contents and were mixed by means of a stirrer and it was maintained at atmospheric temperature. The stirrer was kept at a stirring speed of 300 rpm and temperature is then raised slowly to 80- 85°C by keeping the setup as shown in Fig. 1and maintained at same temperature. The raw rapeseed oil and mixture of methanol and potassium hydroxide are mixed, heated and stirred for 2 hours. The product of the reaction was cooled and decanted to effect a separation of the methyl ester from water phase. The obtained methyl ester is mixed with three times of water and then heated up to 100°C to remove the presence of methanol. The blend preparation of pentaerythritol ester and ISO 68 was carried out in batch type glass reactor consisting of a three necked round bottom flask of 2 litre capacity. A motor driven speed regulator stirrer inserted in the reactor through the central neck while the other neck was used for inserting thermometer. The third neck is used for dropping the raw materials into the reactor via dropping funnel. The reactor was heated by an electric heating mantle having an arrangement for accurate control of the temperature within ±1°C of the desired temperature.

2.3. Lubricant blend sample preparation

The blend preparation of biodegradable refrigeration oil of 500 ml and synthetic refrigeration oil of 500 ml was carried out in batch type glass reactor consisting of a three necked round bottom flask of 2liters capacity. A motor driven speed regulator stirrer inserted in the reactor through the central neck while the other neck was used for inserting thermometer. The third neck is used for dropping the raw materials into the reactor via dropping funnel. The reactor was heated by an electric heating mantle having an arrangement for accurate control of the temperature within ±1°C of the desired temperature. Further the contents were well mixed and were performed at a stirring speed of 300 rpm. The temperature is then raised slowly to 55-60°C and finally blend was obtained after the completion of 60 minutes. Table.1 shows the percentages of lubricant blend in each sample.

Table.1. Volume percentages of lubricant blend in each sample.

Lubricant	Biodegradable Refrigeration Oil	Synthetic Refrigeration Oil
POE 68	0	100
PBRO100	100	0
PBRO50	50	50

2.4. Tribological investigation

The pin on disc tribometer [supplied by DUCOM Instruments, TR-20LE-PHM400] was used to confirm the compatibility between the formulated biodegradable refrigeration oil and synthetic refrigeration oil and cylinder liner-piston ring tribo pair materials. The tribological tests were conducted as per ASTM G99 standards. A photographic view of the pin on disc tribometer is illustrated in Fig1. The compressor cylinder liner was taken as the disc material SAE J431 grade grey cast iron heat treated hardened up to 60HRC and piston ring material of the compressor as FG 150 grey cast iron of hardness 210 BHN.

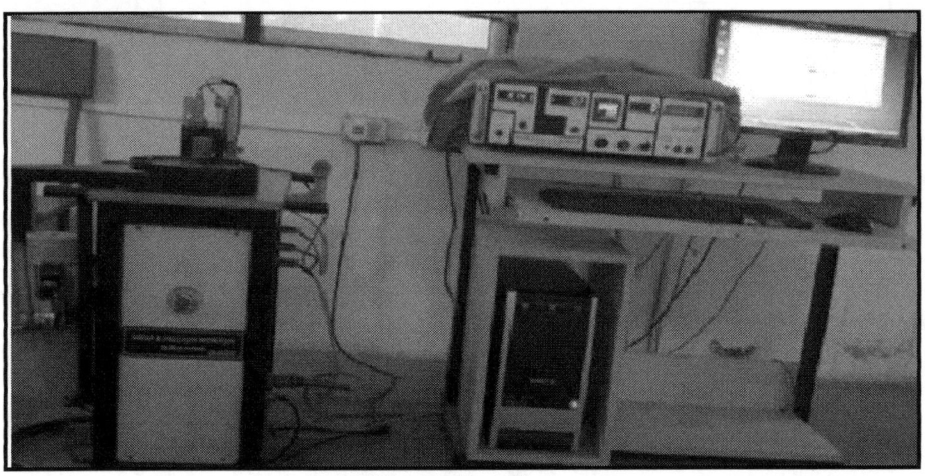

Fig.1 Pin on disc tribometer

Based on the load applied the pin is pressed against the disc. All the test section components collet, pin, disc, tube, oil tub were cleaned thoroughly using acetone. The pin will be stationary, where as the disc will be rotating in the bottom. The sliding distance and rotational speed of the disc can be set according to the requirement as 1000m and 1200 rpm. The load of 10 kg that is to be applied is done using the lever mechanism on the pin. The combination of the rotating motion and the normal load produces frictional force indicated by a digital meter connected to the tribometer. Scanning Electron Microscope (SEM) was used to evaluate the wear characteristics of the pin surfaces under high magnification (500X).
The specific wear rate was calculated by using the following equation:

$$\text{specific wear rate} = \frac{\text{volume loss}}{\text{applied load x sliding distance}}$$

$$\text{volume loss} = \frac{\text{weight loss} * 1000}{7.2}$$

$$\text{sliding distance} = \frac{\Pi * D * N}{60000}$$

Where,
D=track diameter
N=speed of disc in rpm
Density of the material= 7.2 kg/m^3

3. Results and Discussion

3.1 Thermo-physical properties of Biodegradable Refrigeration oil

The viscosity of the bio degradable refrigeration oil and blend PBRO50 at 100^0 C was 11cSt and 9.16cSt. The higher viscosity index (VI) tends to the lesser change in lubricant viscosity with respect to change in temperature. Table.2 shows the thermo physical properties of synthetic refrigeration oil, bio degradable refrigeration oil and blends. The viscosity index of the bio degradable refrigeration oil is superior to the synthetic refrigeration oil for the reason due to chemical nature and higher molecular weight. The pour point is the lowest temperature of the lubricant that remains in a fluid state. The lower pour point is the critical feature of the lubricant under cold working conditions. The pour point of bio degradable refrigeration oil and blend PBRO50 were -15^{0}C and -7^{0}C. The lower pour point of the bio degradable refrigeration oil is because of complex chain formation and branched oils [9]. The total acid number (TAN) is to determine the quantity of degradation of the lubrication oil. The TAN value of bio degradable refrigeration oil and PBRO50 were 1.2 mg KOH/g and 0.79 mg KOH/g respectively. The higher TAN value for bio degradable refrigeration oil indicates the moisture contamination of the vegetable oil.

Table.2. Thermo-Physical Properties of POE 68, PBRO100 and PBRO50

Test parameters	Test method	Units	POE 68	PBRO100	PBRO50
Viscosity at	ASTMD445	cSt	68	67	55.3
Viscosity at 100°C	ASTMD445	cSt	9.3	11	9.16
Viscosity index	ASTMD445	-	114	156	147
Flash point	ASTM D92	^{0}C	220	242	220
Pour point	ASTM D97	^{0}C	-39	-5	-7
Total Acid Number	ASTMD974	mgkoh/g	0.30	1.2	0.79

3.2. Tribological investigation of biodegradable refrigeration oil

3.2.1 Friction Behaviour

The tribological studies clearly show the assessment of the lubrication characteristics of the biodegradable refrigeration oil from the base stocks of rapeseed oil.

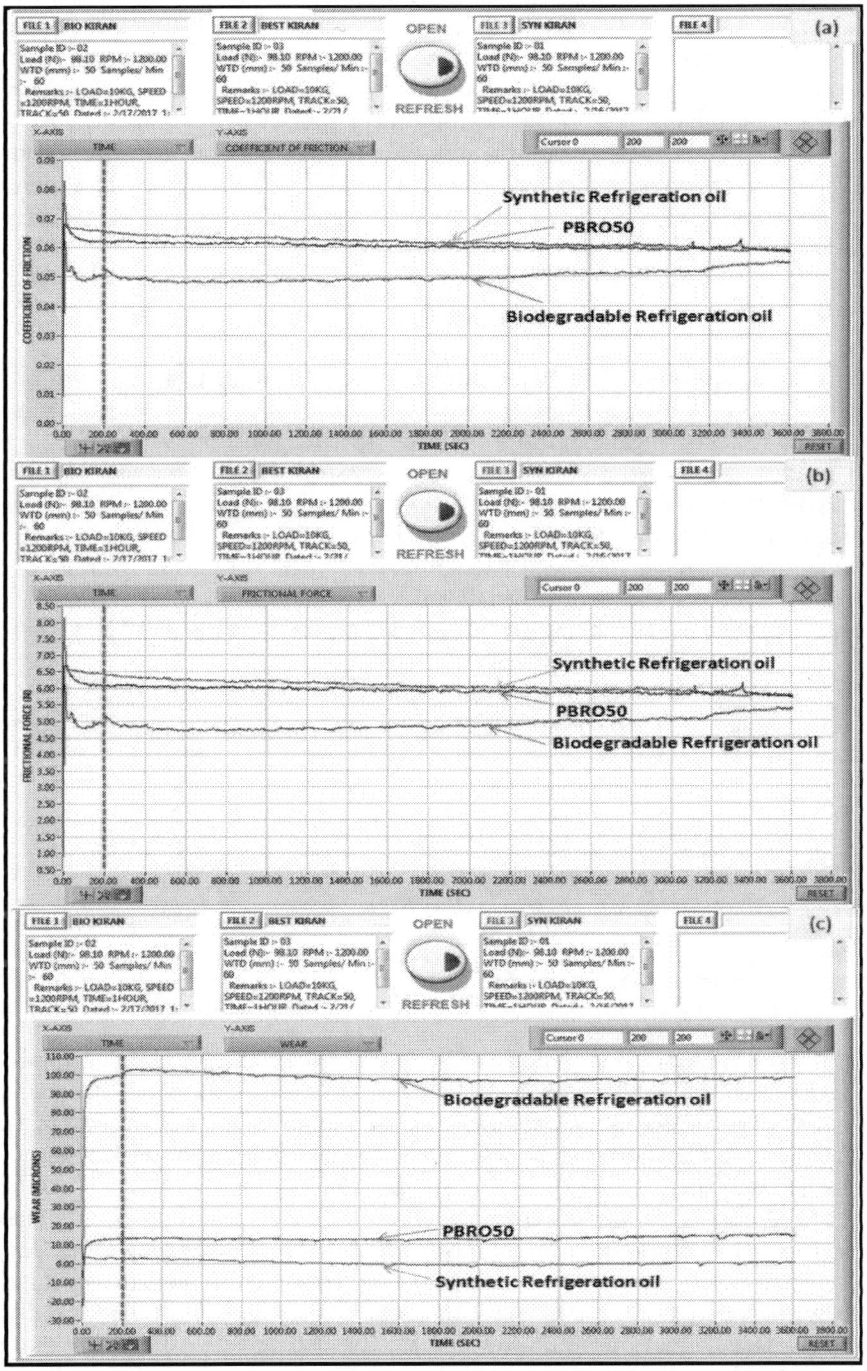

Fig.2 (a) (b) (c) Variation in the Coefficient of Friction, Friction Force and Wear with respect to sliding distance

Fig. 2 (a) (b) and (c) show the variation in the coefficient of friction, friction forces and wear with respect to sliding distance for different percentages of biodegradable refrigeration oil at constant load of 10 kg. The magnitude level (μ value lies between 0.001 and 0.2) of the COF demonstrates that boundary lubrication abrasion take place in the abrasion zone. The biodegradable refrigeration oil has a lower coefficient of friction when correlated with synthetic refrigeration oil. The results reveal that the coefficient of friction of bio degradable refrigeration oil (PBRO100) was noticeably 23% and PBRO50 was 10% lower than that of synthetic lubricant. The pointed out reduced friction coefficient is owing to the presence of higher polar functional groups of carboxylic acids and ester and functions as a protective agent on pin/disc that decreases the direct contact of rubbing surface[10].Free fatty acids in vegetable oils permit monolayer film formation on the surface that offers sliding surfaces prevention of direct metal to metal contact and preservation the surface from damage. In the examination of frictional force, the mean value for bio degradable refrigeration oil was 4.8 N was smaller than the average frictional force of the synthetic lubricant. The results depicts that the friction force of bio degradable refrigeration oil (PBRO100) was noticeably 22% and PBRO50 was 9% lower than that of synthetic lubricant shown in Fig.2(b).

3.2.2 Effect of biodegradable refrigeration oil on surface roughness

Surface roughness is a term relates to wear resistance and coefficient of friction of mating surfaces. The lower surface roughness value indicates that the surface is smoother, easier for the materials slide together and the material loss is at the lowest. The value of surface roughness of the pin and disk lubricated with PBRO50 was lowest. However, the value of surface roughness for PBRO100 was higher when compared to the synthetic refrigeration oil (POE 68). According to the concept of adhesive theory, at first the tougher asperities adhere mutually and the plastic shearing of the junction produced will remove the softer asperities leaving them sticking to the hard surface. Therefore, the wear will be augmented as long as rubbing action between the disc and the pin [8].

Table.3. Surface Roughness of POE 68, PBRO100 and PBRO50

Lubricant	Load (kg)	Duration (min)	Initial Weight of pin (gm)	Final Weight of pin (gm	Roughness on raw material Surface (μm)	Roughness on tested Surface (μm)	Specific Wear Rate (mm^3/N-m)
POE 68	10	60	16.7738	16.7100	0.27	0.45	7.990×10^{-6}
PBRO100	10	60	17.0676	16.9990	0.39	0.75	8.591×10^{-6}
PBRO50	10	60	16.8785	16.8140	0.37	0.54	8.078×10^{-6}

3.3 Wear Behaviour

Fig.3 shows the surface morphology of the pin surface under the influence of the synthetic refrigeration oil, biodegradable refrigeration oil and PBRO50 blend. Scanning Electron Microscope (SEM) was used to evaluate the wear characteristics of the pin surfaces under high magnification (500X). The abrasive and adhesive type of wear mechanism was observed on the boundary lubrication regime. From the Fig. 3 (a), it is clear that the synthetic refrigeration acquire smooth surface with superior wear behaviour because of improved surface adsorption rate on metal surface. The abrasive lines on the wear surface on the pin lubricated with biodegradable refrigeration oil are confirmation of abrasive wear action as shown in Fig.3 (b). This will originate the shearing of physical bonding and chemical bonding due to aggressive mechanical shearing and straight metal-to-metal [11, 12]. The higher surface fatigue, deep furrows and wear delamination found in the case of bio degradable refrigeration oil.

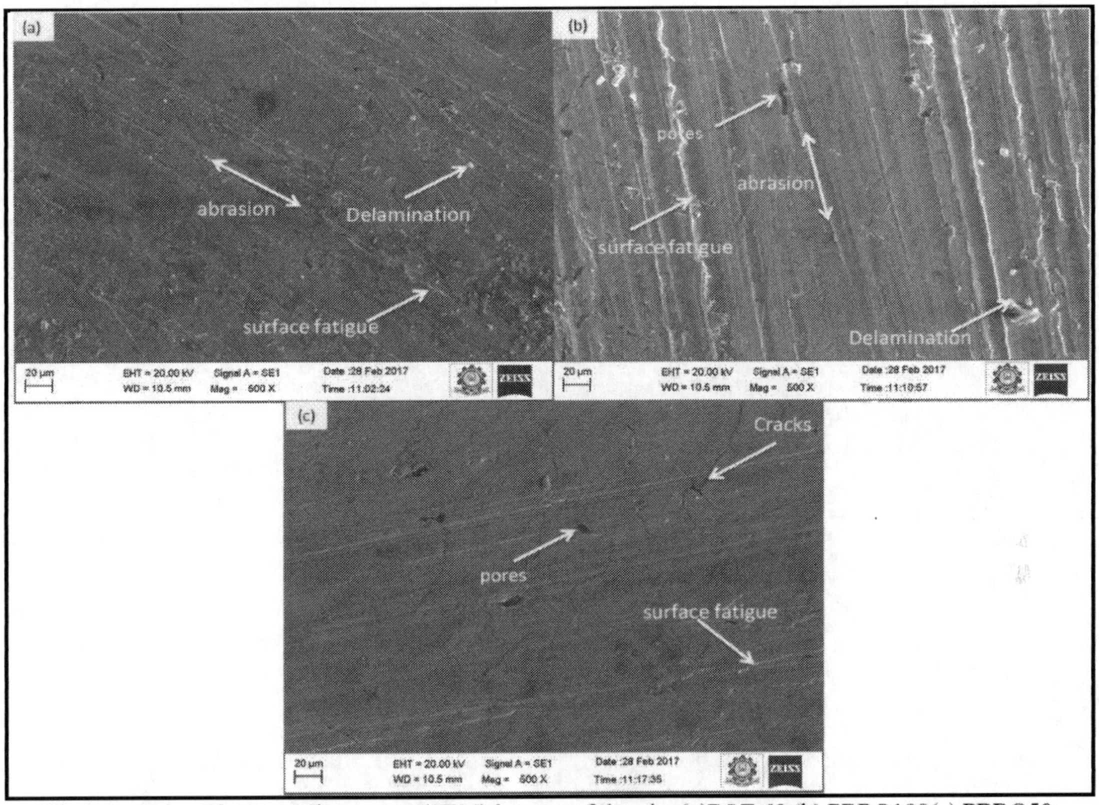

Fig.3.Scanning Electron Microscopy (SEM) images of the pin (a)POE 68 (b) PBRO100(c) PBRO50

The PBRO50 exhibit the superior tribological properties in terms of wear when evaluate the biodegradable refrigeration oil shown in Fig.3 (c). This owes to that the fact; synthetic refrigeration oil was prepared with anti wear, anti-detergent and anti-foam agents. Further, the addition of anti-wear additives like Zinc dialkyl-dithiophosphate will increase the characteristics related to the wear behaviour.

3.4 Elemental Analysis

The intend of the elemental investigation by using a Energy Dispersive X-Ray Spectroscopy is to observe the amount of metals containing in the surfaces of the pin material lubricated with the synthetic refrigeration oil, biodegradable refrigeration oil and their blends. Fig. 4(a-c) depicts the SEM images of the abrasion surfaces lubricated by synthetic refrigeration oil, biodegradable refrigeration oil and their blends with the consequent EDS analysis. From Fig. 4(a, b), it can be the found the worn surface lubricated by synthetic refrigeration oil have micro voids and pits, meanwhile the surface lubricated by biodegradable refrigeration oil was smoothest and no sharp furrows were present on the wear surfaces. The elevated abrasive wear and intensive scratches have found in synthetic refrigeration oil. The acids and oxides developed during oxidation reduces and produce the negative effect in lubricity [13].The C element content on the metal surface for biodegradable refrigeration oil is 19.74wt% on the EDX examination was due to the reason of main component present in the steel ball material undergoes carbonizations on lubrication and the ferrous content were increased to 63.54% because of material loss in the ball surface as shown in the Fig 4(b). The result from the PBRO50 indicates that the superior anti-wear and friction reduction properties which were confirmed by the SEM and EDS investigation.

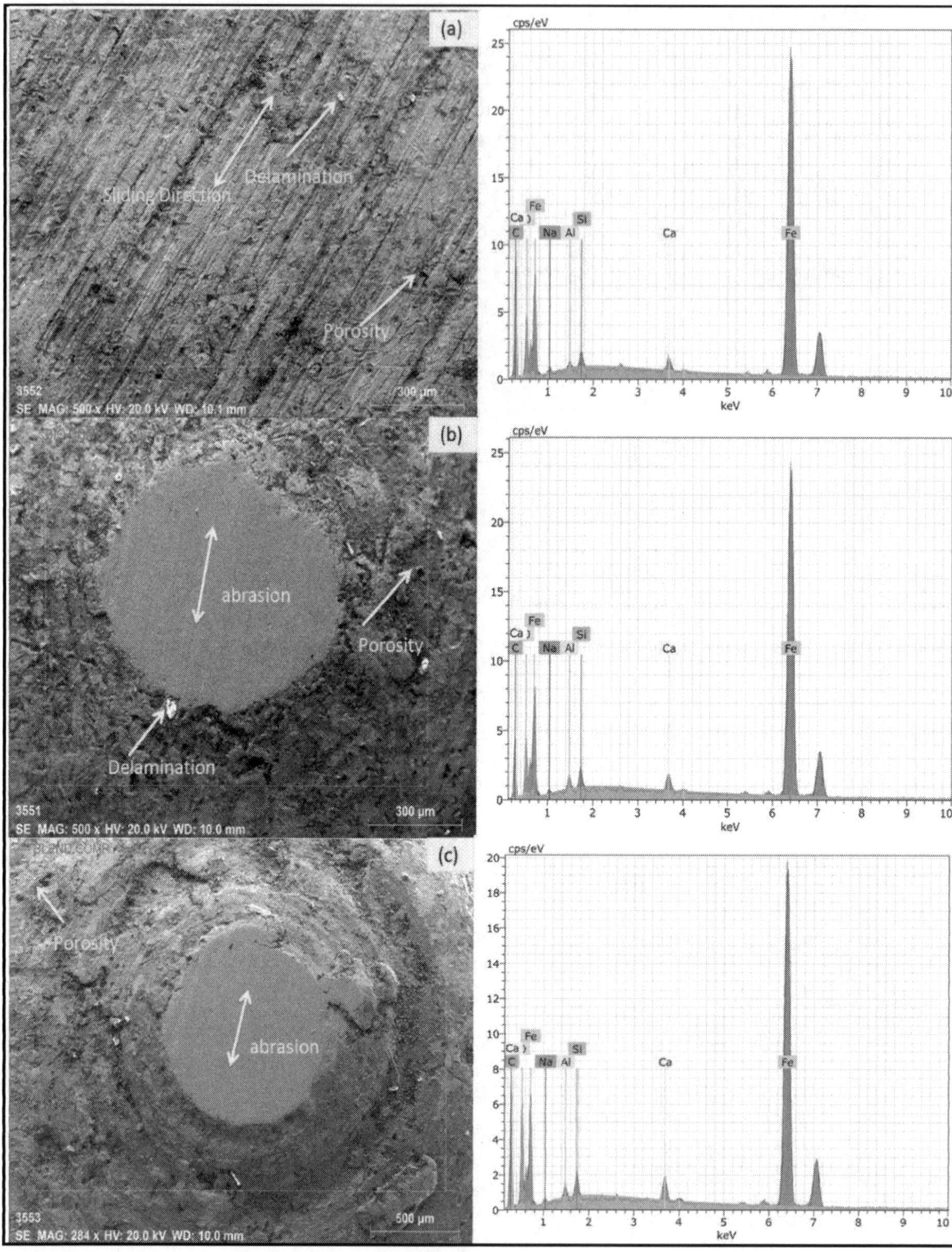

Fig.4 Scanning Electron Microscopy(SEM) images and corresponding (EDX) spectra of the pin worn surface (a)POE 68 (b) PBRO100(c) PBRO50

The rubbing exterior of PBRO50 enclose even surfaces and content of C element was 25.16 wt%, which was 33% higher than biodegradable refrigeration oil shown in Fig 5 (c). The higher percentage of oxygen in the PBRO50 suggests that more oxide layer is retained on the surface in reducing the metal to metal contact. The higher percentage of oxygen in the SBO specimen than the PO counterpart suggests that more oxide layer is retained on the surface in reducing the metal to metal contact thus lower mass loss is produced in the SBO specimen. A similar reason could be applied to the specimen lubricated with MO in which a higher oxygen element is detected, suggesting that more oxide layer exists in preventing the metal to metal contact found by Mayank Anand et al [14]. Table.4 illustrates the summary of SEM/EDX analysis results of the elements present in the pin material surface. These results demonstrate that PBRO50 blend forming a greater continuous protective film layer and to prevent the abrasion surfaces from direct contact and improve the tribological performance of the lubricant.

Table.4. Summary of EDX analysis results

Lubricant	Weight percentages (%)						
	C	O	Na	Al	Si	Ca	Fe
POE 68	18.55	8.20	0.48	0.78	0.76	1.11	70.12
PBRO100	28.00	9.22	0.43	0.30	0.48	0.64	60.93
PBRO50	25.75	13.53	0.37	0.50	0.69	1.27	57.89

4. Conclusions

In this study the tribological performance of biodegradable refrigeration oil in the lubricating contact pair FG 150 grey cast iron sliding against SAE J431 grade grey cast iron was investigated and following conclusions can be deduced from the experiment results.

- The biodegradable refrigeration oil (PBRO100) and blend PBRO50 exhibited good physicochemical properties and could be favorably used as lubricant feedstock in refrigeration applications.
- The results from the tribological study reveal that the coefficient of friction of bio degradable refrigeration oil (PBRO100) was noticeably 23% and PBRO50 was 10% lower than that of synthetic lubricant.
- The surface roughness of the pin and disk lubricated with PBRO50 was lowest and found to be 0.54.
- The higher surface fatigue, deep furrows and wear delamination found in the case of bio degradable refrigeration oil. The PBRO50 exhibit the superior tribological properties in terms of wear when evaluate the biodegradable refrigeration oil.
- From the elemental analysis, Fe and Al were increased due to the loss of material from the pin and the disc, other element like Si, Ca and Na were decreased by oxidizing and other chemical interaction. The higher percentage of oxygen in the PBRO suggests that more oxide layer is retained on the surface in reducing the metal to metal contact.

References

[1] Tirth M Panchal, Ankit Patel, Chauhan D D, Merlin Thomas, Jigar V Patel, Renewable and Sustainable Energy Reviews, 70 (2017) 65–70.
[2] Ponnekanti N, Kaul S, Renewable and Sustainable Energy Reviews, Vol.16, 1, (2012) 764-775.
[3] Mofijur M, Masjuki H H, Kalam MA Energy Procedia,14,(2012) 1748-1753.
[4] Matthew Chukwudi Menkiti, Ocheje Ocholi, Kechinyere Bibian Oyoh, Okechukwu Dominic Onukwuli, Journal of the Chinese Advanced Materials Society, Vol.3,2, (2015) 71-88.

[5] Baskar S, Sriram G, Arumugam S, Tribology Transactions Vol.60,3,(2017) 428-436.
[6] Habibullah M, Masjuki H H, Kalam M A, Lubricants, Vol. 58,6, (2015) 1002-1011.
[7] Zulkifli N W M, Kalam M A, Masjuki H H, Shahabuddin M, Yunus R, Energy 54 (2013) 167-173
[8] Nuraliza N, Syahrullail S, Tribology Transactions, 59, 5 (2016), 831-837.
[9] Cermak S C, Isbell T A, Improved oxidative stability of estolide esters, Ind. Crops and Products 18, (2003) 183-196.
[10] Sharma B K, Adhvaryu A, Erhan S Z , Wear 42,(2009) 353–358.
[11] Shahabuddin M, Masjuki H H, Kalam M A,Procedia Engineering 56 (2013) 597 – 606.
[12] Iman Golshokouh, Syahrullail S, Farid Nasir Ani, Masjuki H H, Journal of Palm Research, Vol 26,1, (2014) 25-36.
[13] Bhale P V, Deshpande N, Thombre S, Soc. Automot. Eng.(2008) 2008-28-0107.
[14] Mayank Anand ,Mark Hadfield, Jose-Luis Viesca ,Ben Thomas, Ruben Gonzalez ,Rob Cantrill, Antolin Hernandez Battez , Lubricants 4, 17,(2016) 1-14.

IOP Conf. Series: Materials Science and Engineering **390** (2018) 012053 doi:10.1088/1757-899X/390/1/012053

Formability studies on Ni-Ti shape memory alloy using laser forming technology

Abhinav Sharma[1], S.S.Mani Prabu[1], I.A.Palani[1*], S.S.Hosmani[1], Rohit Patil[2]

[1]Metallurgy Engineering and Materials Science, Indian Institute of Technology Indore, Indore-453552
[2]Control and Automation Department, VIT University, Vellore-632014

E-mail: mt1602181001@iiti.ac.in, *palaniia@iiti.ac.in

Abstract. Laser forming is an advanced manufacturing process which is extensively used in automotive and manufacturing industry by deforming metallic and non-metallic sheets. In this process, a continuous fiber laser is irradiated over Ni-Ti sheet. As the laser beam imbibed over the sheet, a step temperature thermal gradient got induced across the thickness of the sheet. The bending moment arises due to uneven thermal stresses leads to the deformation of the sheet without using any external forces. This paper shows the experiments performed on Ni-Ti sheets of dimension 60mm×50mm with a maximum power of 50W. The process parameters which were varied are number of passes (10 to 20), power (30W to 50W) to get the maximum bending angle with a uniform spot diameter of 2mm. The study between variations of the bending angle with process parameters is done. After this, the characterization of samples was done using differential scanning calorimeter (DSC) and the changes in phase transformation of Ni-Ti SMA with respect to change in different powers and number of passes is studied.

1. Introduction

Shape memory alloy (SMA) is a class of smart material which is known for two distinct properties such as superelasticity (SE) and thermal shape memory effect (SME). SMAs have the ability to achieve high deformation and return to a certain predefined shape upon unloading or after heating above a certain temperature. SME refers to the phenomena where the SMA will automatically come to its original position after heating and SE is the effect where SMAs may undergo huge non-linear deformation and comes back to its original shape upon unloading. Due to these amusing properties, SMA's have found various applications in the region of biomedical field as well as in industrial sector such as for making fasteners, coupling, actuators, etc.[1-3]

SME can be seen in various alloys such as Cu-Zn-Al, Cu-Al-Ni, Nb-Ti, Cu-Sn, Ni-Ti, Ti-Nb, Cu-Al-Be. Among the various options, Ni-Ti alloys are being widely used because of its ability to retain SME and practicability for longer time [4-5]. They are light in weight, corrosion resistance and have excellent fatigue resistance with low stiffness due to which they show high performance and thus are widely used in MEMS field.

Laser forming is an advanced manufacturing technique which has been developed during recent years. It is an advanced unconventional forming technique used for deforming metallic and non-metallic sheets. The forming of sheets generally occurs due to the induction of non-uniform thermal stresses. When a high power beam is irradiated on the surface of sheet, it leads to the generation of thermal compressive stresses peculiarly near to the irradiated area. The heated area of the workpiece will get expanded but because of the resistance from the surrounding material, localized plastic deformation occurs. The surrounded area of the heated region will remain in the elastic state and upon cooling this region under tension, bending of the workpiece takes place [6-8]. The laser line beam's power, spot diameter, marking speed and number of passes will determine the final pattern of the workpiece.

This technique is more advantageous than other unconventional techniques as it requires no external force, due to which spring back can be negligible [9]. It is a non-contact process by which complex parts can be easily formed. Also, this technique is mostly used nowadays for the fabrication of very small complex parts as this method involves less cycle time and moreover it eliminates the need for expensive dies.

This work focuses on the laser forming of Ni-Ti SMA sheet. A detailed investigation has been carried out to find out the corollary of laser parameters such as power and number of passes on the bending angle of Ni-Ti sheet. In addition, the outcome of forming parameters over the phase transformation behaviour of the laser irradiated area of the Ni-Ti sheets has been analyzed using differential scanning calorimeter (DSC) [10].

2. Method of experiments

2.1 Experimental setup

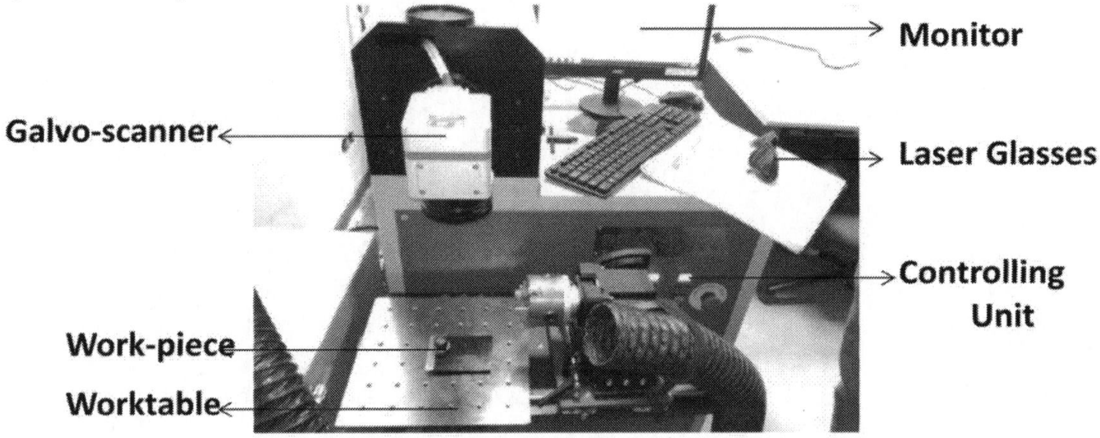

Figure1 – Experimental setup used for laser forming

A fiber laser (Scantech laser Pvt. Ltd) doped with rare earth elements like erbium, ytterbium, neodymium, etc is used in our case. It has a rated capacity of 50 W with a galvo scanner which deflects the beam in X and Y direction with a focal length of 160mm. The Ni-Ti sheet of dimension 60×50×0.25 mm is used in our experiments. The sheet is clamped at one end of the worktable as shown in Figure1.

2.2 Experimental procedure

Firstly the parameters of the laser such as power, spot diameter, mark speed and number of passes were fixed. The laser line beam is then marked at a distance of 20 mm from the extreme end with the help of scanmark software. Now the laser line beam is irradiated on the surface of the Ni-Ti sheet as shown in figure2. The forward and the backward stroke are considered as one single pass. The sheets were scanned for the different number of passes in order to get the maximum bending of sheets. The different laser and process parameters affect the bending angle. The parameters which were varied are Power (30W to 50W), Number of passes (10 to 30). There is no time lapse given between each pass. A total of 6 Ni-Ti samples were bent having a thickness of 0.25mm as shown in figure3. Here the spot diameter is kept 2mm for all the samples.

Figure2 - Irradiation of laser on the Ni-Ti sheet.

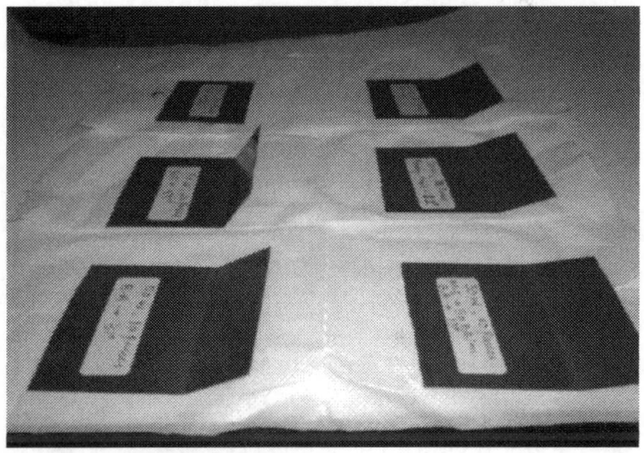

Figure3 – Bended Ni-Ti samples after laser irradiation

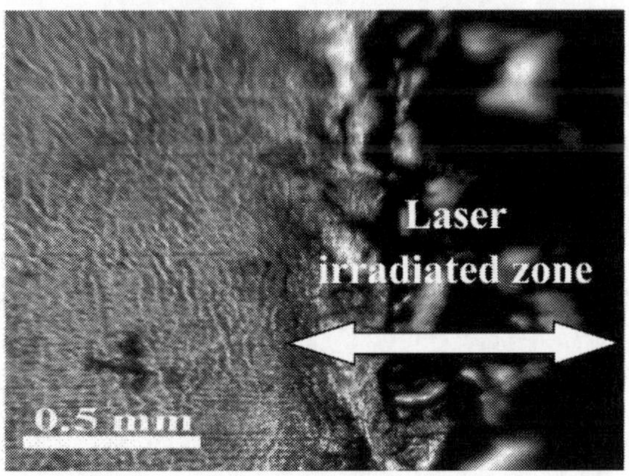

Figure4 - Microscopic image of Formed Sheet

After the laser processing of all the samples, their bending angles were measured manually. Also, the samples were sectioned for the DSC analysis so as investigate the phase transformation of Nitinol (Ni-Ti) with respect to the variation in power and number of passes.

IOP Conf. Series: Materials Science and Engineering **390** (2018) 012053 doi:10.1088/1757-899X/390/1/012053

3. Results and discussion

The deviation of bending angle with respect to power and number of passes are discussed in this chapter. Also, the phase transformation behaviour of the Ni-Ti samples with respect to the change in process parameters is also analyzed.

3.1 Parametric Analysis

A non-linear relationship can be seen between the bending angle and power. As the power is increased, the bending angle also gets increased due to the generation of more thermal compressive stresses on the irradiated area. But after a certain time, the increment in the angle is not so steeper because of the decrement of the thermal extension and plastic deformation of the irradiated surface. It indicates that there is an optimal temperature after which there is no such large variation in the increment of the bending angle.

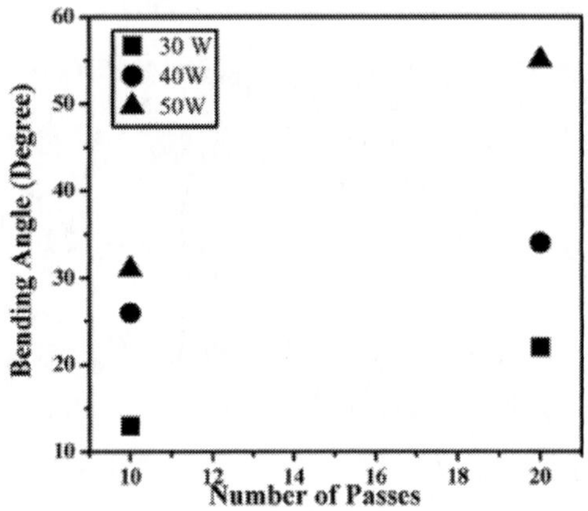

Figure5- Bending angle variation with number of passes

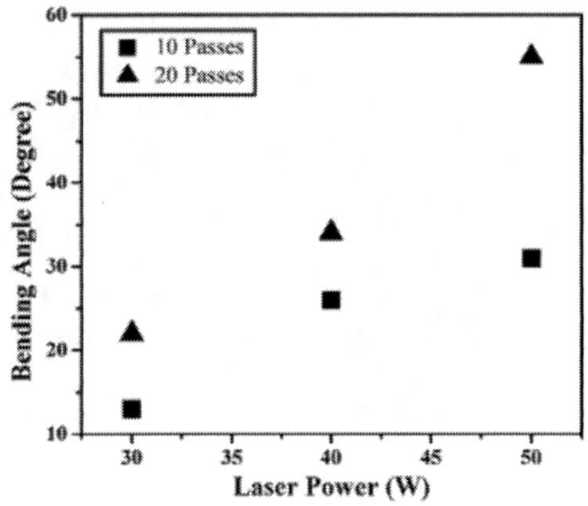

Figure6- Bending angle variations with power

Figure5 shows the variations in the bending angle with reference to the number of passes for three different laser powers 30W, 40W and 50W.Whereas figure6 shows the change in the bending angle with reference to power at two different number of passes i.e. 10 and 20 passes. It can be noticed that power and number of passes are the two important parameters which are affecting the bending angle of the sheet. With the increase of laser power and the number of passes the thermal compressive stresses will increase and leads to thermo-mechanical effects over the sheets due to which the bending angle increases. The increase in the compressive stresses leads to more plastification of the irradiated surface and thermal expansion to the opposite side of the sheet leading to more forming of the desired sheet.

3.2 DSC analysis

Figure7 shows the differential scanning caliometry curves of Ni-Ti sheet for 30W power at two different passes i.e. 10 and 20. It can be clearly seen from the DSC curves that there is a significant single stage transformation seen in both the heating and cooling curves. The Austenite start temperatures are (A_s) = -52.5°C and -50.9°C whereas the austenite finish temperatures are (A_f) = -32.6°C and -33.2°C. Also in the cooling curve the martensite start are (M_s) = -71°C and -70.7°C whereas the martensite finish temperatures are (M_f) = -100.9°C and -94.4°C. During heating, martensite will transform into austenite and while cooling austenite will transform into martensite.

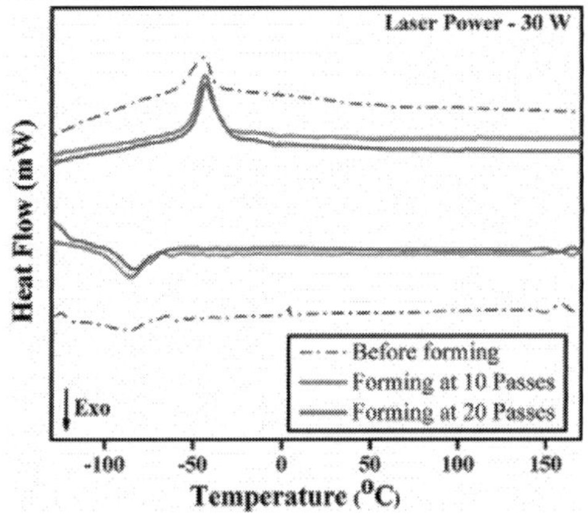

Figure7- DSC curves at 30 W power

Figure8 shows the DSC curves of the Ni-Ti sheet at 40W for 10 and 20 numbers of passes. Here the Austenite start temperatures are (A_s) = -52°C and -52.1°C whereas the austenite finish temperatures are (A_f) = -35.8°C and -19.4°C. Also in the cooling curve the martensite start are (M_s) = -74.8°C and -50.2°C whereas the martensite finish temperatures are (M_f) = -97.8°C and -101.1°C. The curves in both the heating and cooling curves are clearly noticeable. Here, there is a single stage transformation in the Ni-Ti SMA only for 10 passes while for 20 passes there is a two-stage transformation in the Ni-Ti due to the generation of R-phase in the material [7].

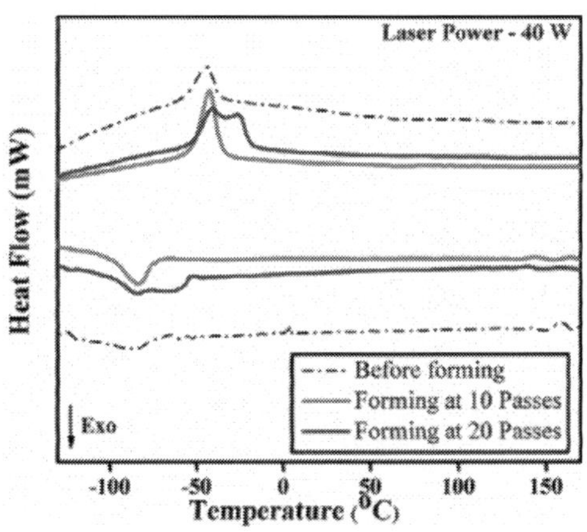

Figure8- DSC curves at 40 W power

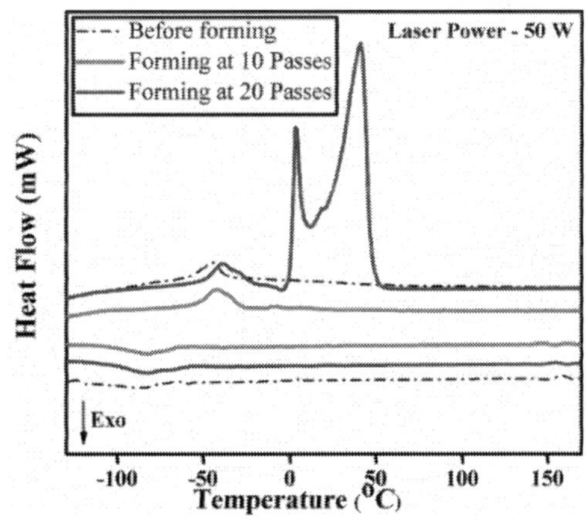

Figure9- DSC curves at 50 W power

Figure9 shows the DSC curves of Ni-Ti sheet for 50W power at two different passes i.e. 10 and 20. The austenite start(A_s) temperatures for 10 and 20 passes are -53.7°C and 0.4°C whereas austenite finish(A_f) temperatures are -28.9°C and 47°C, likewise martensite start(M_s) temperatures are -65.3°C and -58.7°C and martensite finish(M_f) temperatures are -103°C and -103.9°C.Here also we see can that for 10 number of passes there is only a single stage transformation in the Ni-Ti SMA while for 20 number of passes there is a two-stage transformation due to the generation of R-phase which entail that with the increase in laser power and number of passes, the forward transformation has changed from one stage to two stage.

4. Conclusion

In current work laser forming on Ni-Ti sheets of 0.25 mm were performed. The variation in the bending angles with respect to the process parameters were calculated and it is found that there is a certain non-linear increment in the bending angle with an increment of power and number of passes. The DSC analysis done for the samples also gives the idea that there is only a single stage transformation in the Ni-Ti SMA for 30W laser power in 10 and 20 passes while for 40W and 50W laser power there is a two stage phase transformation occurs due to the generation of R-phase.

Acknowledgement

We are very thankful to the sophisticated instrumentation centre of IIT Indore for their continuous help and support.

References

[1] Birnbaum A and Yao Y 2010 *Journal of Manufacturing Science and Engineering* **132** p 041002
[2] Wu H and Schetky L 2000 *International Conference on Shape Memory and Superelastic Technolgies* **182** pp 171-82
[3] Lagoudas D 2010 *Shape Memory Alloys: Modeling and Engineering Applications* (Texas: Springer)
[4] Shiva S, Palani I, Mishra S, Paul C and Kukreja L 2015 *Optics and Laser Technology* **69** pp 44-51
[5] Wang W, Chen L and Zhao X 2008 *3rd Pacific International Conference on Applications of Lasers and Optics, PICALO 2008* **31** pp 162-86
[6] Singh S, Gautam S and Dixit U 2014 AIMTDR 2014
[7] Paunoiu V, Squeo E, Quadrini F, Gheorghies C and Nicoara D 2008 *International Journal of Material Forming* **1** pp 1371-74
[8] D'Souza A, Palani I and Naikwad S 2015 RACE 2015
[9] Shi Y, Yao Z, Shen H and Hu J 2006 Int. J. Mach. Tools Manuf. **46** pp 1689-97
[10] Duerig T and Bhattacharya K 2015 *Shap. Mem. Superelasticity* **1** pp 153-61

IOP Conf. Series: Materials Science and Engineering **390** (2018) 012054 doi:10.1088/1757-899X/390/1/012054

A PRELIMINARY INVESTIGATIONS ON SECOND GENERATION NANO COMPOSITE SUPER NITRIDE COATINGS ON ASTM A681 TOOL STEELS

[1*]D. Vijayan, [2]Y.G. Manoj kumar
[1]Assistant Professor, Department of Mechanical Engineering, Sri Chandrasekharendra Saraswathi Viswa Mahavidyalaya, Enathur, Kanchipuram – 631561, Tamilnadu.
[2]Student, Department of Mechanical Engineering, Sri Chandrasekharendra Saraswathi Viswa Mahavidyalaya, Enathur, Kanchipuram – 631561, Tamilnadu.

Corresponding author: [1*]vijaiand2012@gmail.com

Abstract

It is essential to sustain ever increasing challenges looked by cutting tool producers in enhancing the performance of the tools amid machining of hard to-cut materials which have been created as of late. Though coated tools have discovered wide application in industries, there still remains an extensive scope of change of the properties of coatings with a view to accomplish friendly cutting environment. Surface treatment has been recognized as one of the conceivable avenues which can possibly expand properties and performance of coated tools in machining. The current research work has utilized a second generation AlTiN super nitride nano composite coating commercially called Hyperlox® was deposited on ASTM A681 tool steels by PVD magnetron sputtering technique. The friction and wear properties were investigated using pin on disc test, has been correlated to microstructural and micromechanical characteristics. The outcomes demonstrated that the Hyperlox® coatings introduced lower friction coefficient and lower wear rate among all loading conditions.

1. Introduction

Medium carbon steels are widely used in fabrication of components in engineering, construction and mining industries due to its strength and high hardness. Many sliding materials were subjected to severe wear, abrasion, high stresses, corrosive or aggressive environments due to its robust construction in nature (1). Hence, wearable materials are much attentive topic to the specific researchers. Usually medium carbon steels are considerably modified by thermo mechanical processing and heat treatment for withstanding such destructive forces not only to the tool/work piece interface, but also to a substantial depth below that interface. On the other hand, tremendous research activities around the world dealt with new types of protective nano-coatings and their corresponding developments, which showed excellent wear resistant, hardness, thermal stability, fracture toughness enable them to improve tool life greatly (2, 3). These tendencies gained a massive support to the respective researchers that allowed to study micro and nano structure of coating materials. Many commercial surface modification techniques available in the market for steels namely surface implantation, plating or coating, thermo chemical or thermal treatments. Surface implantation change surface chemistry similar to thermo chemical treatments; however, surface layer implantation is considered as a separate surface treatment. Whereas in plating or coating, a thin layer of hard particles of nitrides, carbides are deposited on the surface that exhibit a little change in surface chemistry as a result tool life can be increased. A historical review of the advancement of CVD and PVD processes shows a steady difference in viable coating materials through changes in the manufacturing quality and coating strategy. Especially nano structured PVD coatings made sense of how to develop themselves accessible surprisingly quickly. Hence, manufacturing industries often searching new coating materials and their corresponding techniques involved for their own standard / customized tools, especially where high stresses, temperature or wear induced directly and indirectly, in particular, cutting tools for rapid manufacturing processes (4-7). Two different processing approaches are used in coatings to deposit a thin film layer of nitrides or carbide's namely physical vapor deposition (PVD) and chemical vapor deposition (CVD). PVD is a plasma coating process involves a physical process of depositing thin solid layers of another material onto a surface. PVD does not require any chemical reaction to be happen to deposit a durable particle. In CVD process, required many inorganic chemicals and the surface to be well cleaned, sometime it takes more time and is not environmental friendly. Nevertheless, the main limitation of PVD process is the rate at which the deposition is relatively very slow when coating undercuts and obtaining similar surface features. Especially, if the product has cylindrical or curved features therefore complex machines, and skilled operators are required. Quality of PVD coating on a tool steels are depending on the following: chemical inertness, a good adhesion, a high hardness and high wear resistance. Attaining these facts by a monolayer of nitride or carbide coating increases the hardness, brittleness and lowers the yield strength. Therefore, a good solution to the above discrepancies is binary / ternary carbide or nitride coatings (8, 9). In recent year's combination of aluminum titanium nitride (Al-Ti-N) coatings have been deposited as binary and ternary layer systems such as TiAlV (10), TiAlN (11-14), TiCN (2), CrAlN (15, 16). Several studies confirmed the effects of coatings have much superior in enhancing the mechanical, corrosion, tribological properties and to extend tool life (17-20). However, due to their implementing difficulties they were not drawn much attention by the researchers when comparing to commercial tribological coatings (21). The major limitations of binary / ternary nitride and carbide coatings are extremely difficult, consumes more time and numerous trials to be required to obtain good surface integrity. Hence, the present investigation mainly focuses on bringing surface integrity of ASTM A681 Super nitride coating on tool steel.

2. Experimental procedures

2.1. Specimen preparation

A second-generation AlTiN supernitride coating layers was deposited onto machined premium grade ASTM A681 tool steel substrate via DC reactive PVD magnetron sputtering is used in an Ar/N_2 gas mixture. The ASTM A681 tool steel samples were cut into small size and subsequently austenized at 1030°C for 30 min, quenched in circulating air and finally tempered at 620°C for 2 h yielding a hardness of 434±20 HV. With a specific end goal to minimize the impact of the substrate surface severity, samples were mechanically cleaned utilizing precious diamond slurry in addition to lubricant to a normal roughness (Ra) smaller than ~0.01 m before they were nitrided to a case depth of 5±2microns. A vacuum chamber was evacuated to a base pressure of 3×10^{-4} Pa and the operating pressure consisting of pure 99.99% of Ar and N_2 gases with a constant flow was set at 0.3 Pa during all supernitride coatings. A Cemecon DC sputtering with booster technology coating system (Model: Cemecon-CC800®) is used for depositing second generation super nitride coatings on all ASTM A681 tool steel samples. The nominal composition of ASTM A681 tool steel is 0.32% - 0.40% C, 5.13% - 5.25% Cr, 1.33% - 1.4% Mo, 1.0% Si, 1.0% V. All the A681 samples were rotated at 10 rpm throughout the deposition process in order to attain a homogeneous coating therefore each sample were located at a distance of approximately 160mm from the target. A 10mm holder is used to hold the samples of ASTM A681 tool steel samples for each coating at a pulsed DC bias voltage of 25V was applied during the deposition process at a temperature of 250°C. A Zeiss Sigma VP -Schottky thermal field emission electron gun scanning electron microscope (FEG - SEM) with an Oxford INCA model energy dispersive X-ray analysis (EDS) was employed to analyze the topography on both coated and uncoated surfaces of A681 tool steel. The phase structure of coated surface layers was analyzed using PANalytical X'Pert Pro MPD X-ray Diffractometer (XRD) equipped with Cu Kα radiation for qualitative and quantitative identification of crystalline phases. In order to conduct a thorough investigation, diffractometer was operated with a scan rate of 0.02° $s^{-1}$was used with a grazing incidence of 0.5°.

3. Results and Discussion

3.1. Surface morphology

Figure. 1(a) and Figure 1(b) show the high magnification SEM micrograph of heat treated sample of ASTM A681 samples. The samples are relatively cleaned and typically coarser, grooved with small pores as shown in Figure 1(a). On few samples, it can be observed a macro crack having adhered debris particles in plane of machined direction as presented in Figure 1(b). Figure 1(d) show the XRD result confirm no contaminations on the surface to be coated and the presence of great amount of chromium oxide, iron oxide on the heat-treated samples. The surface morphology of coated and heat-treated ASTM A681 tool steel was characterized by using Nanosurf© - compact Atomic Force Microscopy (AFM) (Model: NaioAFM). AFM is best suitable for measuring the thin film coatings for structural analysis on the nano scale also used to study the mechanical properties of the coatings with a variation of loads.

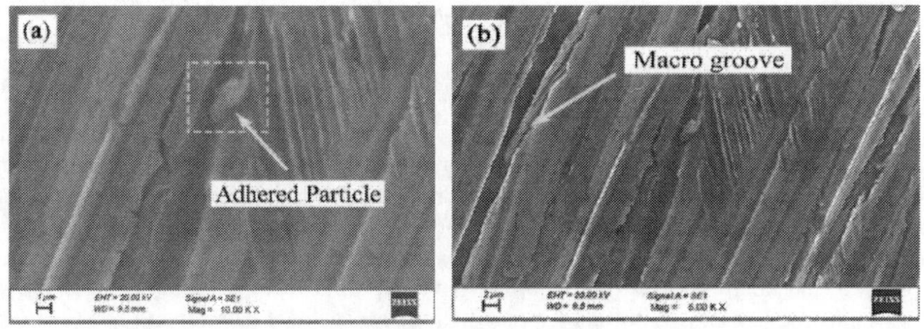

Figure 1. Heat treated ASTM A681 tool steel samples (a). Grooved, (b). macro cracked

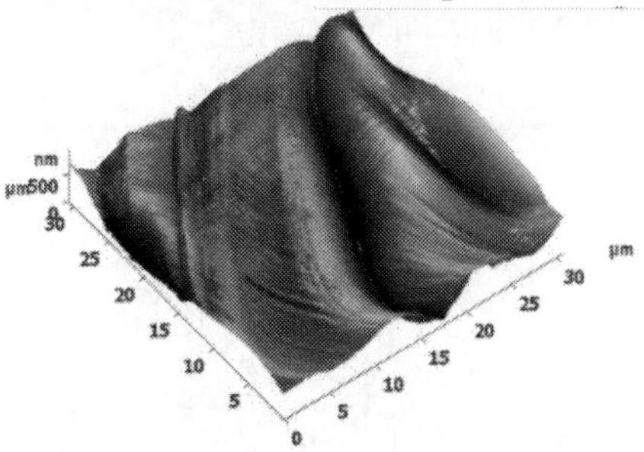

Figure 2. AFM image of heat treated A681 sample

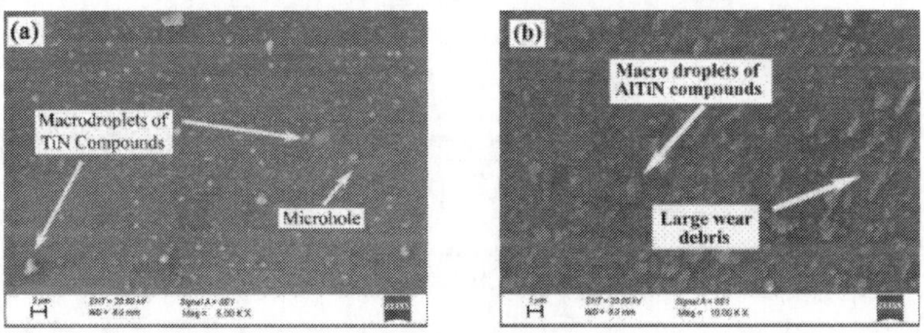

Figure 3. Coated ASTM A681 tool steel samples (a). Macro droplets of TiN Compounds, (b). Macro droplets of AlTiN compounds

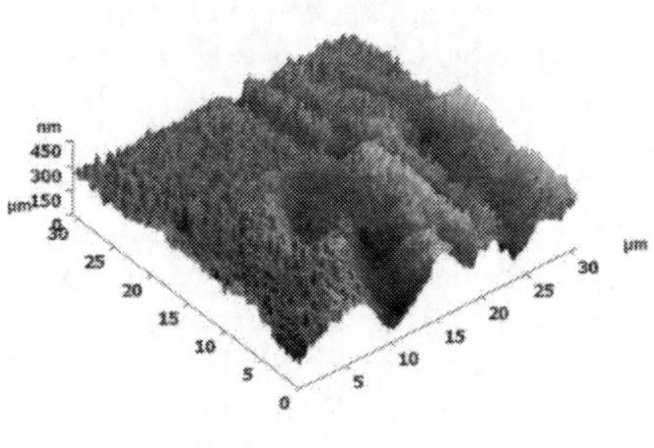

Figure 4. AFM morphology of AlTiN Coated sample

Figure 2 show AFM topography of heat treated ASTM A681 steel that seems machined wafer surface include narrow machining streaks consequences a coarse surface finish. The surface roughness of heat treated samples was in the range between 270 μm to 290 μm. In addition to that, it can be observed that the coating was inconsistent due to low agglomeration as a result a dotted wafer coated surface exhibited on the coating surfaces. The magnitude of change in color of different regions is due to variation in signal with depth which corresponds to the height difference and surface roughness of each sample. therefore, an average surface roughness was considered to understand the effective coating roughness in each coating. The coating surface has less uniformity, micro holes and macro droplets were enclosed by TiN intermetallic compounds as shown in Figure 3(a) and Figure 3(b). Consequently, the surface roughness of 208 μm was observed irrespective of all samples when fabricated by AlTiN coatings. At last, usually wear resistant coatings onto tool steels consequences substantial increase of titanium, oxides of iron, silicon and chromium substrates. Therefore, XRD analysis was performed on the all layers coating samples, for instance, AlTiN coated and heat-treated sample results were graphically presented in Figure 5. The obtained results confirm the presence of addition of titanium, oxides of iron, silicon and chromium substrates. The coating surfaces display heterogeneities covered by debris are in the form of elongated micro-particles which are originated probably due to splattering on the substrate surface during coating process. With this it was concluded that, the adhered nano composite particles (Ti,Al,N) have significantly increasing micro hardness over the surface layer and contributing to decrease in wear intensity of ASTM A681 tool steel.

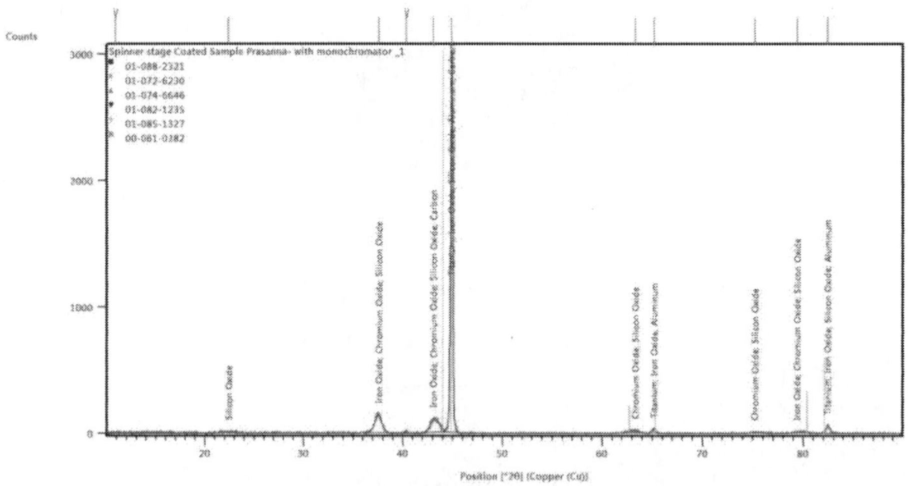

Figure 5. XRD pattern of coated AlTiN sample

3.2. Coating analysis

The hardness (H), elastic modulus (E) and indentation load (P) is a displacement function of coated and heat-treated samples of A681 tool steel. For measuring these parameters Berkovich tip indenter showing a face of 78.9 deg is used. A graphical representation of the elasto-plastic indentation of heat-treated and coated samples are presented in Figure 6(a) to Figure 6(b). h_c is the contact depth, h_f is the final depth of the contact impression after unloading, P is the indentation load, and h_{max} is the indenter displacement at peak load. The area A_1 (OBC) represents the plastic work done in the viscoelastic-plastic segment and the area A_2 (CBC_1) represents the elastic work recovered during the unloading. If the material has completely plastic, the unloading curve is a straight line (BD) and $h=h_{max}$ therefore $A_2= 0$. S is the contact compliance and P_{max} is the peak load. It can be seen that, the area of A_1 is less when compared to area D_1 which reveal the coated sample relatively low displacement compared to heat treated sample. The hardness of the heat treated and coated was determined using nano indenter at several points. Six different loads were applied while testing: 5.91N, 15.8N, 19.7N, 30N, 35N and 39.7N. The average hardness values were calculated and plotted against load for all coatings and heat-treated samples.

6(a)

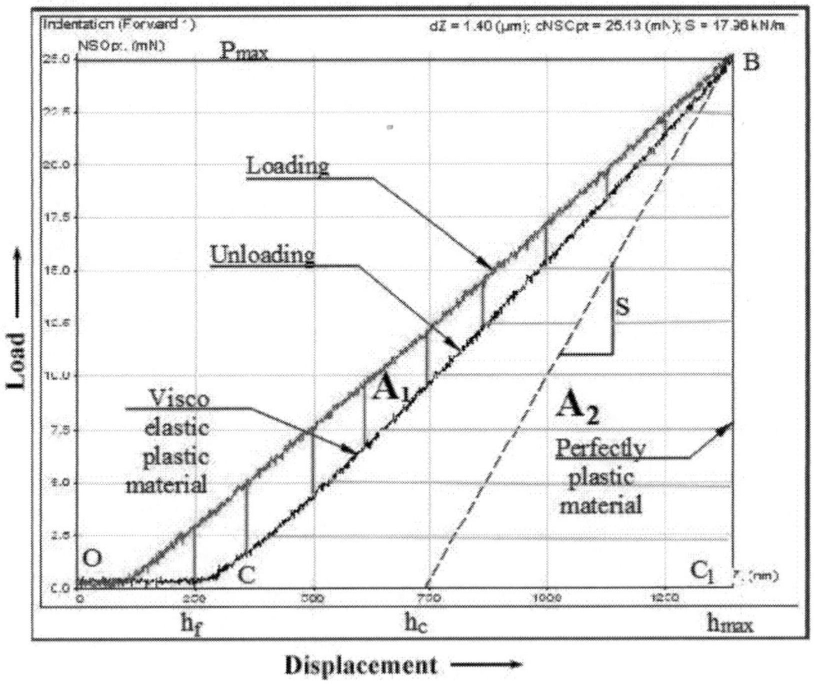

6(b)

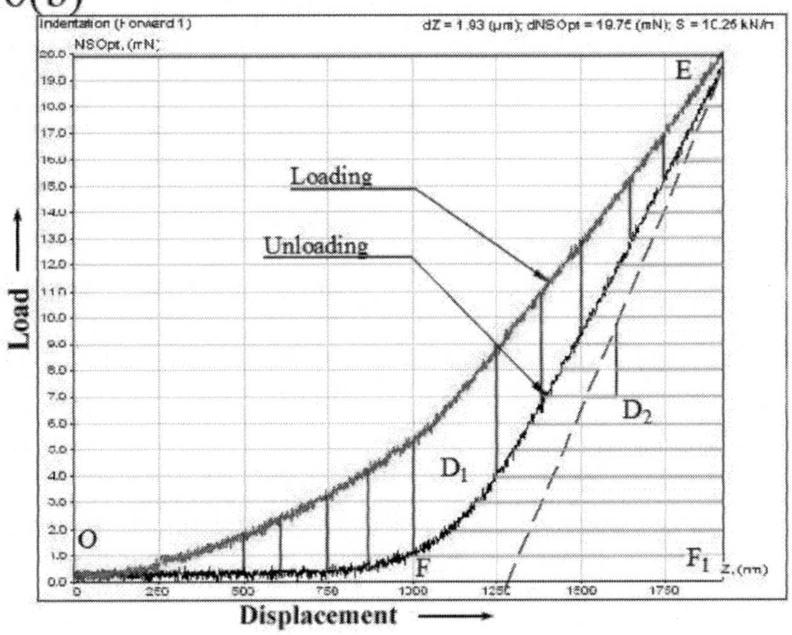

Figure 6 (a) and Figure 6(b) elasto-plastic indentation of heat-treated and coated samples

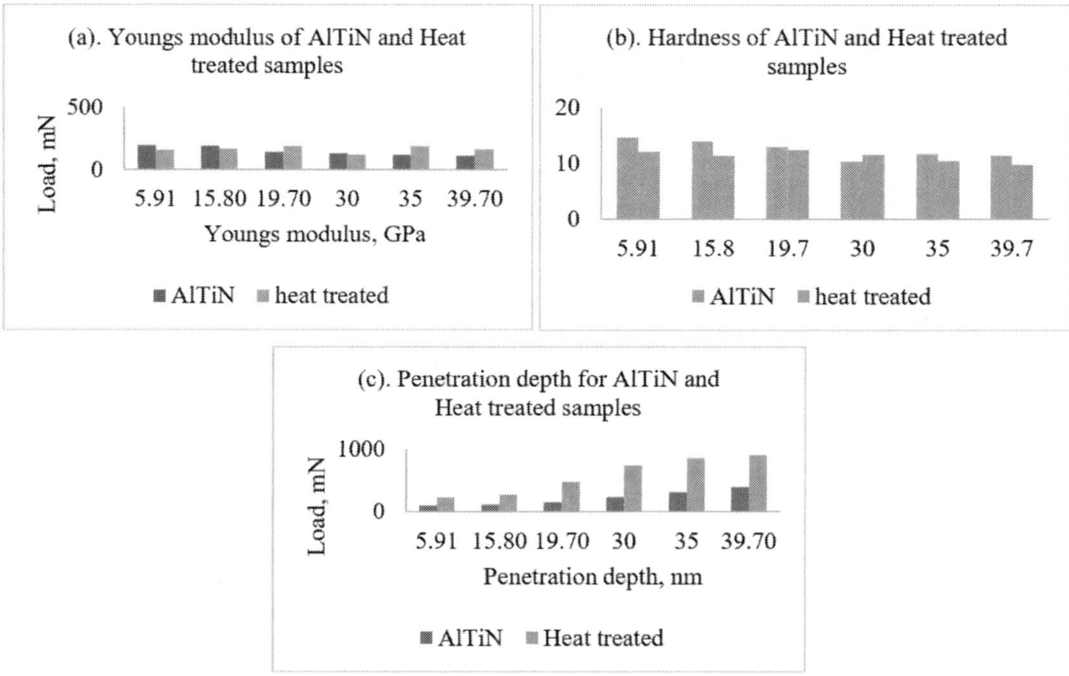

Figure 7 (a). Young's modulus, (b). Hardness, (c). Penetration depth of AlTiN coated and heat-treated samples

The young's moduli of heat treated and coated samples were presented graphically in Figure 7(a). The average young's modulus of 200 GPa was observed in AlTiN coatings whereas 170 GPa was observed in heat treated samples. It is observed that, the young's moduli of all coatings were high when low indentation load whereas it has low at high indentation load. It is due to the variation in the coating substrate thickness around the periphery of each sample. Figure 7(b) show a plot of average hardness of all coatings and heat-treated sample. It is observed that the samples coated with AlTiN exhibited high average hardness value of 13.47 GPa with an average load of 24 mN whereas 10.21 GPa with an average load of 16 mN was observed in heat treated samples. Similarly, the maximum penetration depth of all coatings and heat-treated samples is presented in Figure 7(c). It is observed that an average penetration depth of 422.70 nm and 480.833 nm on AlTiN coatings and heat-treated samples respectively. The lowest penetration depth of 201nm was observed in AlTiN sample. In general, the penetration depth is a function of applied load of the coated sample. Therefore, as the load increases to 5.91N to 19.7N, penetration depth is increases with decrease in hardness of the all coatings irrespective of type of coated substrate. In addition to that, based on the hardness readings observed, the hardness of the all coated samples more or less same and no significant hardness improvement can be seen in any samples when increasing load. The increase in hardness was related to utilizing of multilayered materials with nanometrics thickness kept over mechanical steel substrate. The nanometric multilayer's of AlTiN have a microstructure subject to the measure of the bilayer period, which decides the improvement of the mechanical properties. Similar observation was observed by Noorakma et al (22) in hydroxyapatite coated magnesium based bio-degradable alloys. The hardness increased from substrate to outer surface of the coated samples.

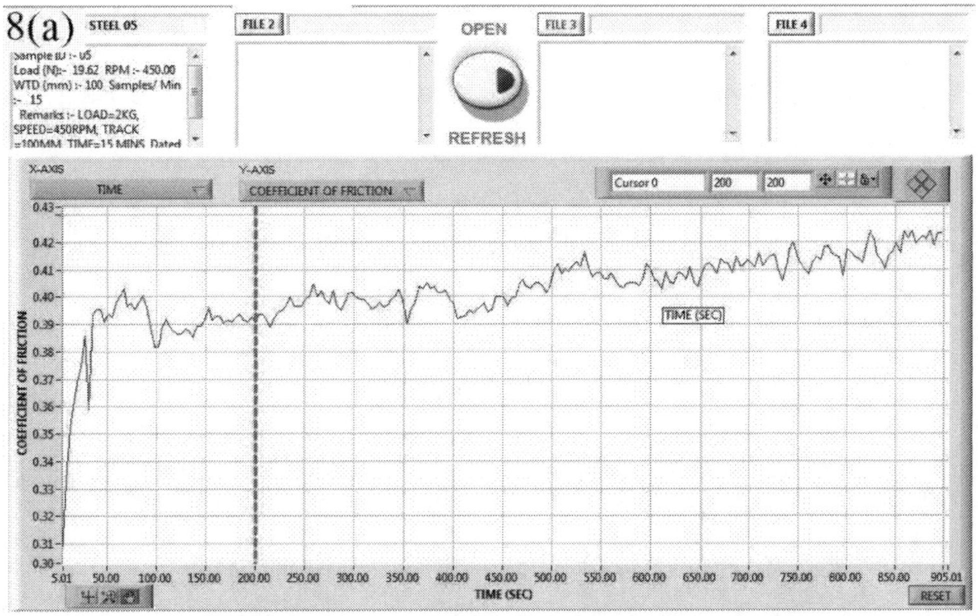

Figure 8(a) Result of coefficient of friction of AlTiN coated sample

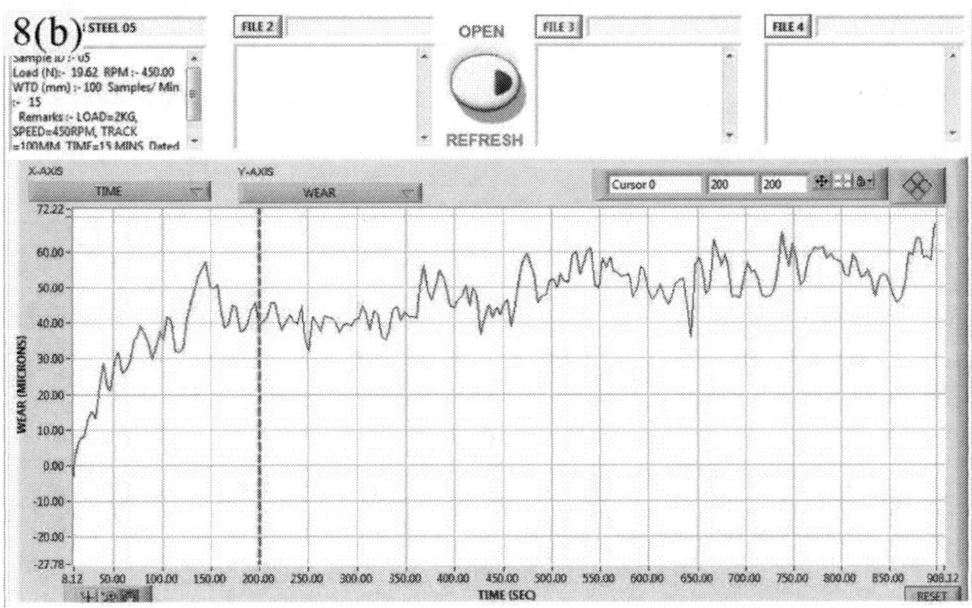

Figure 8(b) Wear result of AlTiN coated sample

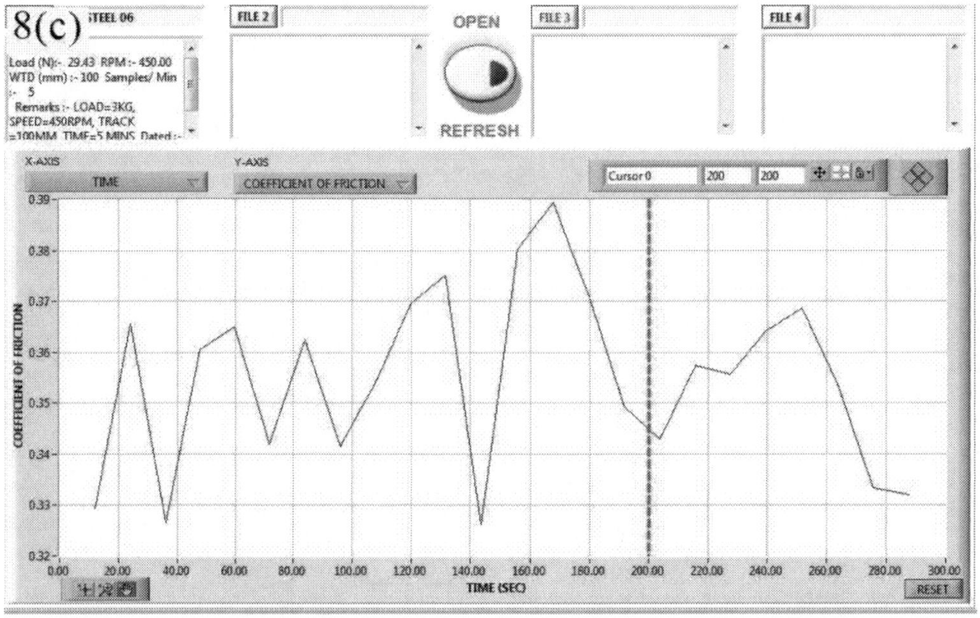

Figure 8(c) Result of coefficient of friction of heat treated sample

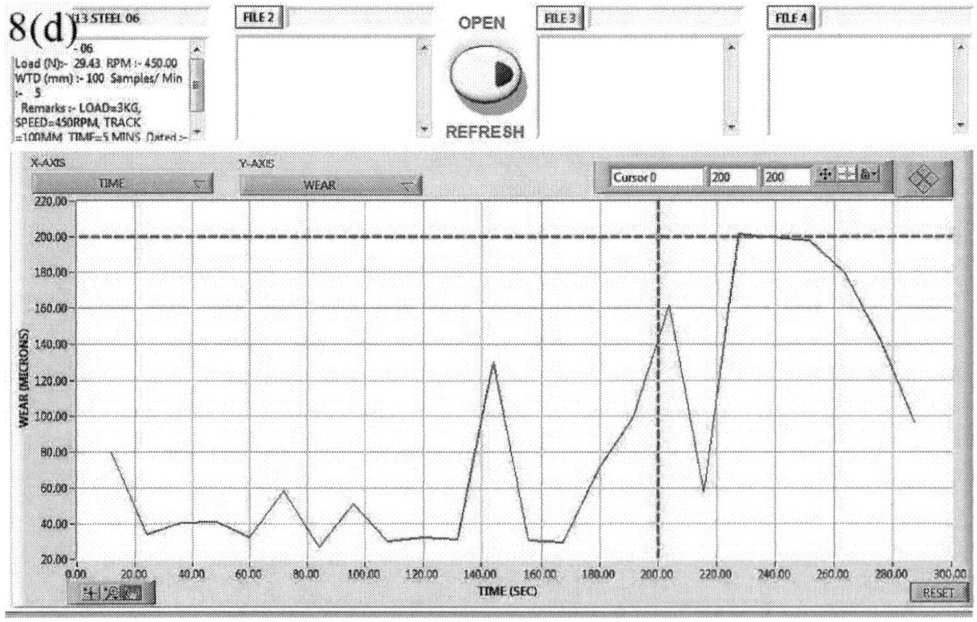

Figure 8(d) Wear result of heat treated sample

3.3. *Tribological properties*

Tribological performance of the coatings were carried out using pin on disc method on the Ducom tribometer at room temperature. All coated samples were mechanically polished after deposition of coating using a comparable level. The substrates of coated sample of all multi-layer systems were tested at room temperature against steel disc and the obtained coefficient of friction and wear rate results of coated and heat-treated samples were presented in Figure 8(a) – Figure 8(d). Test load and sliding speed 2N and 0.1m/s were given in order to establish the initial contact between pin and disc. The initial Coefficient of Friction (CoF) is begun at low level 0.3 was observed in each coated and heat-treated surface. The dependency of friction coefficient was evaluated on the number of cycles with the standard number of cycles 12000 for AlTiN coatings and heat-treated samples. The heat-treated samples have characterized run with rapid increase of the CoF value, while the AlTiN coating samples comparatively steady wear regime after about 2200 cycles with noticeable fluctuation. The rapid increase of wear rate of the pin, wear debris around the pin stimulated until the friction reached a steady-state. Using wear track profiles, the wear rate of the coatings was calculated. Each coating wear rate was estimated at room temperature using the formula, $w = V / (F \times S)$, where V is wear volume in mm^3, which was evaluated according to the cross-sectional area of the wear track, F is load in N and S is the sliding distance in m. Volume of the material removed on the wear track during the tests were measured to estimate wear rate of each coating. In addition to that, built up edge is also to be considered for estimating the wear of the coatings, since, built-up edge is the growth of material around or in the wear track (23). However, no significant built-up edge can be seen around the wear track. The estimated wear rate is presented graphically in Figure 9. The highest wear rate can be observed in heat treated samples. Wear rates of coating and heat-treated samples such as AlTiN and heat-treated samples are 180 and 453 m3/N/m $\times 10^{-17}$.

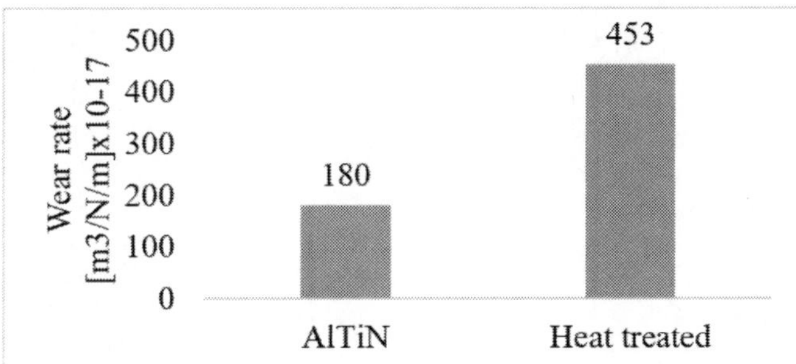

Figure 9. Wear rate of AlTiN and heat-treated sample

4. Conclusion

A Second generation AlTiN based coatings characterize the state-of-the-art for PVD hard coatings used in industrial machining applications. However, new difficulties primarily in the field of metal cutting in stamping and framing applications require constant enhancements. Other than mechanical and physical properties, oxidation protection and the affinity between the workpiece material and the cutting tools are essential for further advancements. These properties can be affected from one viewpoint by different coating parameters and coating innovations yet additionally by a specific chemical composition of the elements utilized for the coatings. As this coating is quite new in the cutting tool market further machining tests have to be done to define the potential for this type of coating.

References

1. Wang X, Kwon PY, Schrock D, Kim D. Friction coefficient and sliding wear of AlTiN coating under various lubrication conditions. Wear. 2013;304(1):67-76.
2. Yang W, Xiong J, Guo Z, Du H, Yang T, Tang J, et al. Structure and properties of PVD TiAlN and TiAlN/CrAlN coated Ti(C, N)-based cermets. Ceramics International. 2017;43(2):1911-5.
3. Aihua L, Jianxin D, Haibing C, Yangyang C, Jun Z. Friction and wear properties of TiN, TiAlN, AlTiN and CrAlN PVD nitride coatings. International Journal of Refractory Metals and Hard Materials. 2012;31(Supplement C):82-8.
4. Roberts GA, Kennedy R, Krauss G. Tool Steels, 5th Edition: ASM International; 1998.

5. Long Y, Zeng J, Shanghua W. Cutting performance and wear mechanism of Ti–Al–N/Al–Cr–O coated silicon nitride ceramic cutting inserts. Ceramics International. 2014;40(7, Part A):9615-20.
6. Ipaz L, Caicedo JC, Esteve J, Espinoza-Beltran FJ, Zambrano G. Improvement of mechanical and tribological properties in steel surfaces by using titanium–aluminum/titanium–aluminum nitride multilayered system. Applied Surface Science. 2012;258(8):3805-14.
7. Bobzin K. High-performance coatings for cutting tools. CIRP Journal of Manufacturing Science and Technology. 2017;18:1-9.
8. 0Hovsepian PE, Lewis DB, Münz WD. Recent progress in large scale manufacturing of multilayer/superlattice hard coatings. Surface and Coatings Technology. 2000;133–134:166-75.
9. Warcholinski B, Gilewicz A. Multilayer coatings on tools for woodworking. Wear. 2011;271(11–12):2812-20.
10. N S, R MRP. Microstructure, surface topography and sliding wear behaviour of titanium based coating on AISI 1040 steel by magnetron sputtering. Archives of Civil and Mechanical Engineering. 2017;17(2):281-92.
11. Chen L, Chang KK, Du Y, Li JR, Wu MJ. A comparative research on magnetron sputtering and arc evaporation deposition of Ti–Al–N coatings. Thin Solid Films. 2011;519(11):3762-7.
12. Chen L, Yang Y, Wu MJ, Xu YX, Du Y, Kolozsvári S. Correlation between arc evaporation of Ti–Al–N coatings and corresponding Ti0.50Al0.50 target types. Surface and Coatings Technology. 2015;275:309-15.
13. Elmkhah H, Zhang TF, Abdollah-zadeh A, Kim KH, Mahboubi F. Surface characteristics for the TiAlN coatings deposited by high power impulse magnetron sputtering technique at the different bias voltages. Journal of Alloys and Compounds. 2016;688, Part A:820-7.
14. Glatz SA, Koller CM, Bolvardi H, Kolozsvári S, Riedl H, Mayrhofer PH. Influence of Mo on the structure and the tribomechanical properties of arc evaporated Ti-Al-N. Surface and Coatings Technology. 2017;311:330-6.
15. Yang B, Chen L, Chang KK, Pan W, Peng YB, Du Y, et al. Thermal and thermo-mechanical properties of Ti–Al–N and Cr–Al–N coatings. International Journal of Refractory Metals and Hard Materials. 2012;35:235-40.
16. Birol Y. Sliding wear of CrN, AlCrN and AlTiN coated AISI H13 hot work tool steels in aluminium extrusion. Tribology International. 2013;57:101-6.
17. Wei Y, Gong C. Effects of pulsed bias duty ratio on microstructure and mechanical properties of TiN/TiAlN multilayer coatings. Applied Surface Science. 2011;257(17):7881-6.
18. Bodnarchuk VI, Petrov P, Kozlenko DP, Dechev D, Ivanov N, Martev I, et al. Study into the properties of TiN/WN multilayer nanocoatings prepared via magnetron sputtering. Journal of Surface Investigation: X-ray, Synchrotron and Neutron Techniques. 2017;11(1):186-9.
19. Faisal NH, Prathuru AK, Goel S, Ahmed R, Droubi MG, Beake BD, et al. Cyclic Nanoindentation and Nano-Impact Fatigue Mechanisms of Functionally Graded TiN/TiNi Film. Shape Memory and Superelasticity. 2017:1-19.
20. Vannan RRRM, Moorthy TV, Hariharan P, Prabhu P. Effect of physical vapor deposited bilayer (AlCrN + TiAlN) coating on high-speed steel single point cutting tool. Materials and Manufacturing Processes. 2017;32(4):373-6.
21. Al-Bukhaiti MA, Al-hatab KA, Tillmann W, Hoffmann F, Sprute T. Tribological and mechanical properties of Ti/TiAlN/TiAlCN nanoscale multilayer PVD coatings deposited on AISI H11 hot work tool steel. Applied Surface Science. 2014;318(Supplement C):180-90.
22. Noorakma AC, Zuhailawati H, Aishvarya V, Dhindaw B. Hydroxyapatite-coated magnesium-based biodegradable alloy: cold spray deposition and simulated body fluid studies. Journal of materials engineering and performance. 2013;22(10):2997-3004.
23. Bolelli G, Cannillo V, Lusvarghi L, Manfredini T. Wear behaviour of thermally sprayed ceramic oxide coatings. Wear. 2006;261(11-12):1298-315.

Optimization of Wear Phenomenon of Al6061/Gr MMCs using Non-Traditional Optimization Methods

P. Gangadhara Rao[1], A. Gopala Krishna[2], Pandu R Vundavilli[3],
[1]Department of Mechanical Engineering, KL University, Vaddeswaram, AP-522502.
[2]Department of Mechanical Engineering, JNT University, Kakinada, AP-533002.
[3]School of Mechanical Sciences, IIT Bhubaneswar, Odisha-752050.
[1]audibalav@gmail.com, [2]dr.a.gopalakrishna@gmail.com, [3]pandu@iitbbs.ac.in

Abstract.
The tribological properties like wear rate/weight loss plays a significant role when the newly developed metal matrix composites are put in to the service condition. Therefore, deciding the optimal parameters for wear is an important research area. In the present manuscript, two non-traditional optimization algorithms, namely invasive weed optimization (IWO) and particle swarm optimization (PSO) algorithms are used to optimize the said process. The non-linear regression equations developed for as cast and heat-treated Al6061/Gr MMCs using response surface methodology is used for the said purpose. The four independent process parameters, namely sliding velocity, percentage of reinforcement, load and sliding distance are considered to optimize the weight loss during wear test. The performance of the developed optimization algorithms is compared in terms of their ability to produce optimal solution.

Keywords: Weight loss, Al6061/Gr MMCs, PSO, IWO.

1. Introduction

The aluminum-based metal matrix composites (MMCs) are most commonly used in several engineering applications, namely aerospace, automobile and military applications. Researchers had tested several materials such as Al_2O_3, SiC, TiC, B_4C, TiB_2 and ZrC etc. as reinforcement materials and examined for their wear characteristics [1]. In [2], the authors' studied the effect of particle size of Al_2O_3 abrasive particles on wear. It was observed that the infiltration of SiC increased the resistance to wear of Al alloys. Sakip et al. [3] used Taguchi's method to experimentally determine the influence of various factors on dry sliding wear phenomenon of AlB_2/Al composites. Later on, Rajesh et al. used grey-Taguchi method to optimize the wear properties such as wear volume rste and coefficient of friction of Al/SiC [4] and Al/Gr [5] metal matrix composites. The optimized values obtained were verified experimentally. Rao et al. [6] also evaluated the tribological and mechanical properties of Al7075/TiC MMCs with as cast and heat-treated conditions. It was observed that the heat condition had shown an improvement in the properties. Further, Praveen et al. [7] response surface methodology (RSM) to model the wear behavior of LM13/SiC Aluminum MMCs. The process parameters, namely normal load, sliding distance, sliding velocity and silicon carbide were used as the input process parameters and wear rate was used as the response. Not much work is reported on the optimization of wear phenomenon for aluminum metal matrix composites. In [8], genetic algorithm (GA) was used to model and optimize the mechanical properties of Al nanocomposites. Chandrasekaran and Santosh [9] used GA and desirability analysis to optimize the multi response optimization of Al-SiC_p composites. The optimal Pareto optimal fronts related to the higher material rate and desired surface roughness were obtained. In addition to the above approaches, Vinoth et al. [10] used GA to optimize the nonlinear regression equation of wear rate obtained from RSM approach. They considered sliding speed, percentage of silicon and applied load as the input process parameters and wear rate as response. Moreover, Ali et al. [11] used particle swarm optimization (PSO) algorithm to optimize the stir casting process that produce more uniform composites. Further, wear test was conducted on those samples. In the present paper an attempt is made to optimize the wear rate of as cast and heat-treated Al6061/Gr MMCs using PSO and IWO algorithms. The nonlinear regression equation developed from RSM is utilized for the said task.

2 Mathematical Model of Al6061/Gr Wear Process

In the present research, the as cast and heat-treated Al6061/Gr MMCs (ref. to Fig. 1) produced using stir casting process has been considered to conduct the experiments on the pin-on-disc apparatus (ref. to Fig. 2). The independent parameters, namely applied load, sliding distance, percentage of reinforcement and sliding velocity are considered as input process parameters and the weight loss (WL) is considered as the response. The range of input process parameters considered in this study are given in Table 1.

Fig.1 Schematic diagram showing plates of Al6061/Gr composite produced using stir casting route.

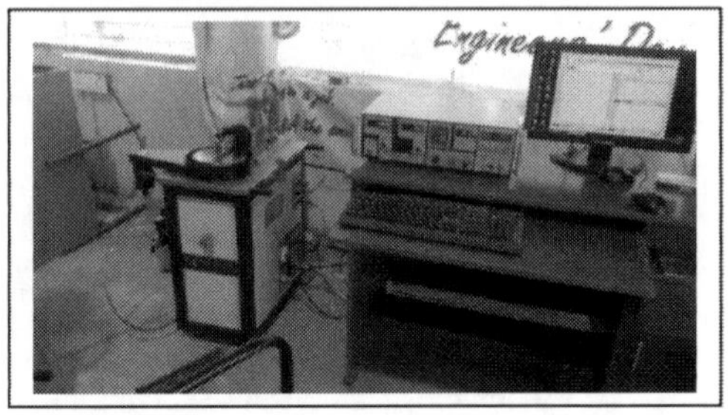

Fig. 2 Schematic diagram showing the pin-on-disc wear apparatus.

Table 1: Table showing the ranges of the variables used in the study.

S. No	Factors	Symbol	Ranges	
			Low (-1)	High (+1)
1	Reinforcement (%)	A	3	9
2	Load (N)	B	10	30
3	Sliding distance (m)	C	500	2500
4	Sliding velocity (m/s)	D	1	3

The nonlinear regression equations are developed after conducting the experiments as specified by the central composite design of experiments. The response equations that represent the weight loss for as cast and heat treated Al6061/Gr composites are given in Eqns (1) and (2), respectively.

$$WL_{as\ cast} = 0.01815 + 0.00069*A + 0.000649*B - 0.000003*C - 0.01452*D - 0.000193*A*A + 0.000008*B*B + 0.00626*D*D + 0.000019*A*B - 0.000146*A*D + 0.000006*B*D + 0.000010*C*D \quad (1)$$

$$WL_{heat\ treated} = 0.04900 + 0.00456*A + 0.002475*B - 0.000016*C - 0.00779*D - 0.000939*A*A + 0.000006*B*B + 0.00555*D*D + 0.000048*A*B + 0.000104*A*D - 0.000000*B*C + 0.000019*B*D \quad (2)$$

The statistical adequacy of the above models is tested and it was observed that the coefficient of correlation for weight loss of heat treated and as cast Al6061/Gr MMCs are found to be equal to 0.9978 and 0.9984, respectively.

3 Formulation as an Optimization Problem

The weight loss during wear test of as cast and heat-treated Al6061/Gr MMCs is formulated as an optimization problem. The as cast weight loss model that represent the formulation is given below.

$$Minimize\ (f) = WL \tag{3}$$

Subject to:

$$3 \le A \le 9$$
$$10 \le B \le 30$$
$$500 \le C \le 2500$$
$$1 \le C \le 3$$

where the variables *A, B, C* and *D* are having their usual meaning. Similar approach is used to optimize the wear loss for heat-treated AMMCs also.

4 Optimization Methods Used

In the present study, two non-traditional optimization methods, namely PSO and IWO are used to optimize the wear process of Al6061/Gr MMCs. The explanation for the algorithms is given in the subsequent subsections.

4.1 Particle Swarm Optimization

PSO is a population based stochastic optimization algorithm developed by Eberhart and Kennedy in 1995 [12]. The operation of this algorithm is mainly inspired by the social behavior of organisms such birds flocking or fish schooling. In order to determine the food source each particle have some velocity based on their personal experience as well as getting the information from the other members of the swarm. Initially, the particles are generated randomly in a D-dimensional solution search space. Each particle is a solution to the problem and it is moving in the multidimensional search space of the problem with a specific velocity and position for searching the optimal solution. In the present paper the variables of the problem, namely % reinforcement, load, sliding velocity and sliding distance are considered as the particles that explore the solution space and weight loss is treated as the objective function in the present problem. The particle velocity V_{id}^{t+1} and position X_{id}^{t+1} are two main characteristics of the particle while moving in a multidimensional solution space, and are determined by using the following two equations.

$$V_{id}^{t+1} = W * V_{id}^{t} + C_1 * rand_1 * \left(P_{id}^{t} - X_{id}^{t}\right) + C_2 * rand_2 * \left(G_{id}^{t} - X_{id}^{t}\right) \tag{4}$$

$$X_{id}^{t+1} = X_{id}^{t} + V_{id}^{t+1} \qquad i = 1, 2, \ldots\ldots\ldots\ldots\ldots\ldots n \tag{5}$$

where t is the iteration number, W is the inertia weight, V_{id}^{t} and V_{id}^{t+1} denotes the velocities and X_{id}^{t} and X_{id}^{t+1} represents the position of the particles i in t and $t+1$ generations, respectively, P_{id}^{t} and G_{id}^{t} are the local best (*p-best*) and global best (*g-best*), $rand_1$ and $rand_2$ are two random numbers varying in the range [0, 1] and C_1 and C_2 are two positive acceleration constants, respectively.

4.2 Invasive Weed Optimization

IWO is a stochastic optimization algorithm that is constructed using the concept of weed biology and ecology developed by Mehrabian and Lucus in 2006 [13]. In a cropping field, weeds are dispersed randomly and they grow to become flowering weeds and produce new weeds by consuming unused resources. The fitness/qualification of the flowering weed is going to determine the no. of weeds generated by every flowering weed. The weeds with a good adaptation in the location will consume additional unexploited resources from the field and grows faster and harvest additional new seeds. Further, the newly formed seeds dispersed arbitrarily throughout the field and grows to a flowering weed. The above process will continue to reach the maximum no. of weeds in the field by using the limited resources. Finally, the weeds having better fitness/qualification could continue and harvest new weeds. The step by step procedure of the algorithm is explained below.

Initialization: Initially the number of weeds (i.e variables) is dispersed randomly in a D-dimensional solution space and assigns the maximum and minimum value of each weed.
Reproduction: In this stage each weed produces number of seeds based on the fitness/qualification value of the flowering weed in a cluster. The number of seeds reproduced from each weed is calculated based on the fitness value and it increases in linear manner from lower to higher fitness value. The higher fitness of the flowering weed produces more seeds and lower fitness of the flowering weed produces less seeds. The number of seeds (S) evaluated by each flowering weed is decided by using the resulting equation.

$$S = Floor\left[S_{\min} + \frac{f - f_{\min}}{f_{\max} - f_{\min}} \times S_{\max} \right] \tag{6}$$

Where f_{min} and f_{max} represents the minimum and maximum fitness value in the colony, respectively and S_{max} and S_{min} denotes the maximum and minimum production by each pant respectively.
Spatial dispersal: The seeds produced by each flowering weed are dispersed throughout the exploration space with mean equal to the position of the plants and varying standard diviation. Based on the value of variation of variance the new seeds are distributed around the parent weed. At the end of each generation the standard deviation of the present problem is calculated by using the following equation.

$$\sigma_{Gen} = \frac{(Gen_{\max} - Gen)^n}{(Gen_{\max})^n}(\sigma_{initial} - \sigma_{final}) + \sigma_{final} \tag{7}$$

Where Gen_{max} represent the maximum number of generations, n indicates the modulation index, and $\sigma_{initial}$ and σ_{final} denotes the initial and final standard deviation, respectively.

5 Results & Discussions

In the present research work, two optimization algorithms that is, PSO and IWO are used to optimize the wear phenomenon of Al6061/Gr MMCs. Initially, a parametric study has been conducted to identify the optimal parameters of the said algorithms. For conducting this study the parameters of PSO, namely number of generations, maximum population size, acceleration constants C_1 and C_2, and Inertia weight are kept equal to 30, 50, 2.5, 2.5 and 0.99, respectively. Fig. 3 shows the results of parametric study conducted to PSO for both the as cast and heat treated Al6061/Gr MMCs. The optimal parameters for population size and maximum number of generations for as cast and heat treated AMMCs are found to be equal to {45, 90} and {50, 70}, respectively.

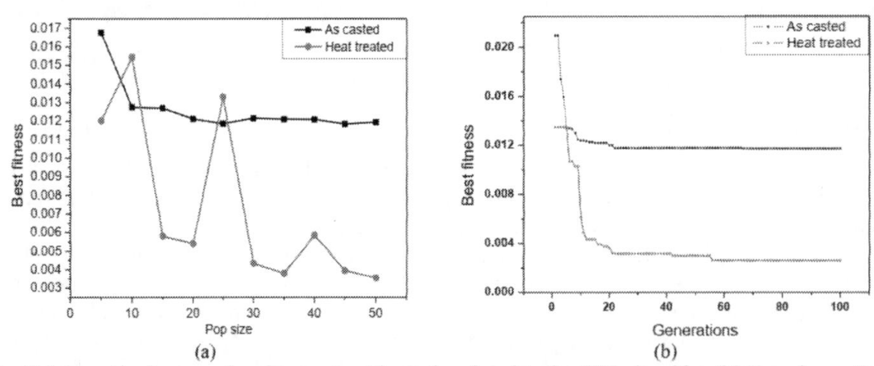

Fig. 3 Schematic diagram showing parametric study related to the PSO algorithm (a) Pop size vs Best fitness (b) Generations vs Best fitness

In the case of IWO algorithm also, a similar study has been conducted to evolve the optimal parameters that resulted in a reduced weight loss during the wear test of Al6061/Gr MMCs. During parametric study, the parameters related to the IWO algorithm are set at the following levels: sigma final = 0.001, exponent =5, minimum no. of seeds = 1, maximum no. of seeds = 3, initial population size = 10, maximum population size = 20, no. of generations = 30, respectively. The results of parametric study related to IWO algorithm for as cast and heat treated AMMCs are given in Fig. 4. The optimal parameters, namely sigma initial, exponent, maximum number of seeds, maximum population size and maximum number of generations for as cast and heat treated AMMCs are found to be equal to {8%, 4, 8, 50, 60} and {8%, 8, 6, 30, 80}, respectively. Once the parameters related to the algorithm are optimized, the optimal values of the parameters that are responsible for the better performance of the response for Al6061/Gr MMCs are given below.

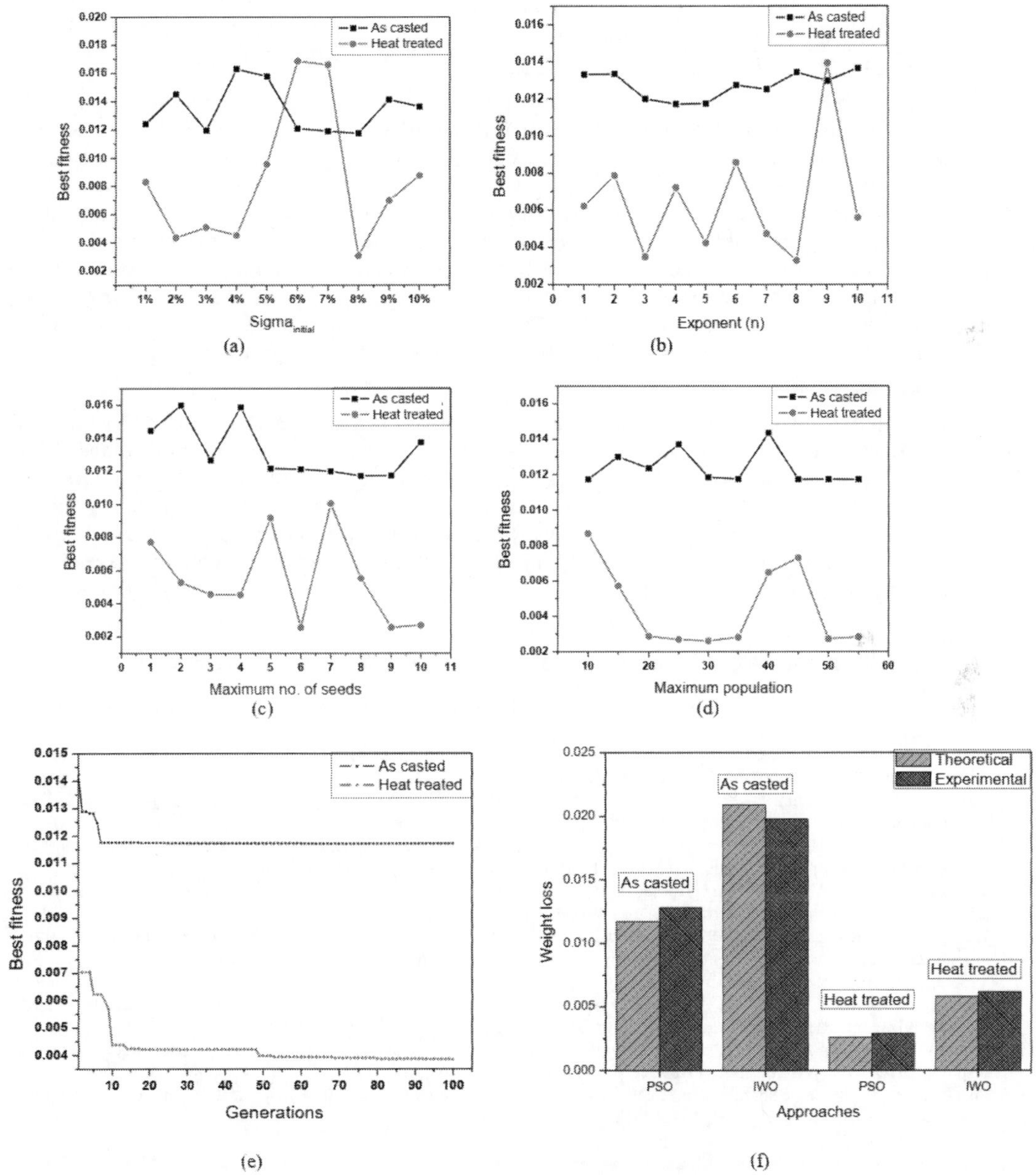

Fig. 4 Schematic diagram showing parametric study related to the IWO algorithm (a) Sigma_initial vs Best fitness, (b) Exponent vs Best fitness, (c) Maximum no. of seeds vs Best fitness, (d) Maximum population vs Best fitness (e) Generations vs Best fitness, (f) showing the comparison of experimental and theoretical weight loss during wear test.

Table 2: Comparison of optimal parameters and their corresponding response values.

	%reinforcement	load	sliding distance	sliding velocity	wear loss
As casted					
PSO	8.9995	10.0024	500.0024	1.0024	**0.0117**
IWO	9.0000	16.0000	927.0000	1.0000	0.0209
Heat treated					
PSO	8.9955	10.0025	2500.0000	1.0025	**0.0026**
IWO	9.0000	10.0000	2297.5000	1.0000	0.0058

From Table 2, it has been observed that PSO algorithm is found to perform better than simple IWO algorithm. It may be due to the reason PSO is a less sensitive, limited parameters, derivative free technique, stable convergence and shorter calculation time and stable convergence characteristics compared to other heuristic optimization algorithms. It is also important to note that the confirmation experiments are conducted for the optimal sets of parameters obtained with PSO and IWO algorithms for both the as cast and heat treated AMMCs. The results of comparison for the wear test are shown in Fig. 4 (f). From Fig. 4(f), it has been observed that the experimental results obtained for wear test are very close to the optimal values obtained from various algorithms. The percentage error between the experimental values and optimal values for PSO and IWO algorithms for as cast and heat treated AMMCs are seen to be equal to {8.59%, -5.55%} and {10.34%, 6.45%}, respectively.

6 Conclusions

In the present research an attempt is made to optimize the wear phenomenon of as cast and heat treated Al6061/Gr MMCs with the help of two non-traditional optimization algorithms, namely PSO and IWO. The weight loss equation obtained using the response surface methodology is utilized as the objective/cost function to optimize the said process. It has been observed that PSO algorithm is found to perform better than IWO algorithm in terms of weight loss during wear test. It has also been observed that the experimental verification of optimal parameters obtained using different algorithms have shown good agreement with the experimental results.

References

1. Zhu H G, Wang H Z, Ge L Q, Xu W J, Yuan Y Z: (2010), Dry sliding wear behavior of Al-based composites fabricated by exothermic dispersion reaction in an Al-ZrO_2-C system, Wear, 268, 1465-1471.
2. Ahlatci H, Candan E and Cimenoglu H: (2004), Abrasive wear behavior and mechanical properties of Al-Si-SiC composites, Wear, 257(5-6), 625-632.
3. Sakip K, Ferit F, Ramazan K and Omer S: (2012), Experimental optimization of dry sliding wear behavior of in situ AlB_2/Al composite based on Taguchi's method, Materials and Design, 42, 124-130.
4. Rajesh S, Gopala Krishna A, Rama Murthy Raju P and Muthukannan D: Application of grey-Taguchi method for optimization of dry sliding wear properties of aluminum MMCs, Front. Mech. Eng., 7(3), (2012), 279-287.
5. Rajesh S, Gopala Krishna A, Rama Murthy Raju P and Muthukannan D: (2014), Statistical analysis of dry sliding wear behavior of graphite reinforcedaluminum MMCs, Procedia Materials Science, 6, 1110-1120.
6. Ramakoteswara Rao V, Ramanaiah N, Sarkar MMM: (2016), Mechanical and tribological properties of AA7075-TiC metal matrix composites under heat treated (T_6) and cast conditions, J. Mater Res Technol., 5(4), 377-383.
7. Praveen J E D, Robinson Smart DS, Babu R and Vijay Prabhu A: (2017), Investigation on wear behavior of LM13/SiC aluminum metal matrix composites by response surface methodology, International Journal of Mechanical Engineering and Technology, 8(6), 530-543.
8. Ali M and Mohsen OS: (2012), Assistance of novel artificial intelligence in optimization of aluminum matrix nanocomposite by genetic algorithm, Tran. Of Metallurgical and Materials, 43A, 5279-5285.
9. Chandrasekaran M and Santosh T: Desirability analysis and genetic algorithm approaches to optimize single and multi-response characteristics in machining Al-SiC_p, 5th International and 26th AIMTDR, 12th – 14th Dec. 2014, 653-1-6.
10. Somasundara Vinoth K, Subramanian R, Dharmalingam S and Anandavel B: (2014), Optimization of dry sliding wear conditions for AlSi10Mg/SiC_p composites using response surface: genetic algorithm approach, Industrial Lubrication and Tribology, 66/5 593-600.
11. Ali M, Mohsen OS and Ahmed E: (2014), Searching for the superior solution to the population-based optimization problem: Processing of the wear resistant commercial AA6061 AMCs, International Journal of Damage Mechanics, 23(7), 899-916.
12. Kennedy J,Eberhart R. Particle Swarm Optimiza-tion', Proceedings of IEEE International Conference on Neural Networks, Perth, 27, pp. 1942-1948, 1995.
13. Mehrabian A. R, C. Lucas. 2006, A novel numerical optimization algorithm inspired from weed colonization, Ecological Informatics, Vol. 1, No. 4, pp. 355–366.

Factors affecting Supplier Relationship Management: An AHP Approach

A Dash[1] , L.K.Pothal[2] , S Tripathy[3]

[1] Industrial & System Engineering, IIT Kharagpur, West Bengal, India

[2&3] School of Mechanical Engineering, KIIT University, Bhubaneswar, Odisha, India

[1] aishigit93@gmail.com

Abstract. Supplier Relationship Management has been attached paramount importance with a view to maintaining a superior performance of the supply chain in a manufacturing industry. Both academic and corporate are striving hard to establish a long term understanding with the manufacturer and supplier as this will pave the way for a greater success. This paper provides an insight to understand the Supplier Relationship Management and the significant factors affecting the relationship. An attempt has been made to prioritize or rank the Key factors and subsequently Analytical Hierarchy Process (AHP) has been resorted as a tool to rank the different sub factors of the Key factors responsible for the afore-said supplier Relationship management Indubitable, this paper will guide the budding managers of the manufacturing industries as to how to create the buyer supplier bond which will be fulfilling and will provide a win- win situation.

Keywords: Supplier Relationship management (SRM), Analytical Hierarchy Process (AHP), Buyer supplier relationship

1.Introduction

Paradigm of modern business era has changed completely that competition is no longer between companies rather it is between supply chains. Supplier relationship management (SRM) playing key role and become relevant management area for sustainability in this globalised competition. Reason is Global competition forced organizations to focus on innovation to add flexibility to the existing process to be highly customer responsive and to manufacture customized products but this can't be achieved through one or two area of excellence like-production or distribution. Complex, customized & unpredictable demand, shortened life cycle of product can only be fulfilled from innovative components and services those are through procurement. Academic and corporate interest in buyer-supplier relationship area increased remarkably only to build strong and long term relationship as SRM is one of the emerging areas of emphasis of today's procurement practices. Focus of SRM has changed from cost minimization to value creation and promise delivery Hong and Kwon (2012) [17]. Suppliers are not defined as who supplies goods rather as strategic partners of firm because of their important role in value chain require good understanding of customer's need. SRM is emerged as global management network that involves skills & knowledge only to enhance supply chain performance Lintukangas (2011) [16].To sustain in the global competition the relationship has to be strong, long term and developed continuously. But the developed relationship should be not be competitive rather cooperative Loppacher et al (2011) [15]. Short term and long term goals of suppliers to fulfil needs of buyers, short terms are delivery, order cycles time and quality & long terms are strengthening supplier's managerial, product development and process competences. To develop SRM suppliers should be proactive and it is based upon the mutual interest of both suppliers and buyers. So Aim of the paper is to study the factors affecting mutual interest of both suppliers and manufacturers to maintain long term relationship. Paper is organized as follows. Section 2 reviewed past literature and extract factors, Section 3 represents methodology to be used, Section 4 entails discussion and section 5 concludes the paper.

2. Literature Review

Good amount of research has been reported in this area. In electronic database we have searched for papers using keywords like- supplier relationship management, Supplier Manufacturer Coordination. Numerous papers have been reviewed to obtain the factors affecting SRM. Reviewed papers from 200 to 2017 are represented in tabular form showing focus and findings of eminent authors in different journals.

Table 1

Author	Findings
Eskandari et al. (2010)	A model based on linear demand was introduced. Simulation optimization is used to develop optimum decision variables so that above coordination problem effectively analyzed and solved.
Hematyar and Chaharsooghi (2014)	Due to high demand uncertainty and return of product it is difficult for retailer to determine order quantity. Hence Insurance contract was proposed to solve coordination problem. It is also compatible for both the manufacturer and retailer &solution is better than other contract.
Alaei et al (2014)	Although RFM (Retail fixed mark up) policy can't coordinate the channel but can bring considerable improvements over the channel. Each member's profit can be improved through this about 96%.
Giannakis, M. (2012)	Importance of procurement in supplier relationship management Four fundamental dimensions mainly focused are Trust, Power, Commitment and Involvement
Sucky and Durst (2013)	Focus on importance of supplier development programme. Balance of cost incurred and benefit gained from supplier skill and capability enhancement
Gan and Sethi (2004)	Three specific cases were considered where first Pareto-optimal sharing rule was considered and then combined with external actions like- Order quantity, production volume to obtain Pareto-optimal solutions.
Hematyar and Chaharsooghi (2014)	Due to high demand uncertainty and return of product it is difficult for retailer to determine order quantity. Hence Insurance contract was proposed to solve coordination problem. It is also compatible for both the manufacturer and retailer &solution is better than other contract.
Alaei et al (2014)	Although RFM (Retail fixed mark up) policy can't coordinate the channel but can bring considerable improvements over the channel. Each member's profit can be improved through this about 96%.
Gan and Sethi (2004)	Three specific cases were considered where first Pareto-optimal sharing rule was considered and then combined with external actions like- Order quantity, production volume to obtain Pareto-optimal solutions.
Habib M. (2014)	Study showed the detailed SCM evolution in product Industry. A Tertiary Educational supply chain management model is discussed using Structural equation modeling & its future implication for the betterment of society also highlighted.
Moyaux et al (2007)	Order is assumed to be a 2D vector & two principles are proposed based on each dimension. Firstly ordering policy should be lot for lot, secondly during variation of market consumption company should react once. Each time company should specify the reason of under or over order to the suppliers through information centralization

Amoako and Boye (1998)	Modern Practices like TQM ,JIT practitioners have lower unit manufacturing cost,higher quality and hence competitive. Supplier Development is a long term resource consuming business activity which requires commitment from both manufacturing firms and supplier.
Yao et al(2008)	A model is developed to find the impact of price dependant and stochastic demand on return policy.
Prahinski and Benton (2004)	communication strategies play key role to improve supplier Performance
Saranga.H (2009)	Working capital is important .High Cost capital and advance for Raw Material affect Lower tier smaller firms dealing with simple parts as they have to wait long for payment.
Kahkonen, A.K. (2011)	From Finnish food industry it is revealed that power relation strongly influence nature of relationship. Further Actions, decisions, Strategies to be used in the supply chain are determined from the nature of relationship.

Detailed literature review shed light on different factors affecting SRM. Due to different demand for products, skill of labor, venture capital, nature of market these factors varies from country to country. Each factor represents its unique contribution towards long term relationship. Different factors and sub factors are enlisted below.

Objective to maintain long-term relationship

Long term relationship can only be maintained through alignment of local objective, communication, trust, commitment, value addition and satisfaction between buyer and manufacturer. In long term relationship simple supplier becomes strategic partner. So sub factors with notations which affecting the long term relationship are

Open communication (LTR1)
Mutual trust (LTR2)
Role & Responsibility (LTR3)
Joint action for value creation (LTR4)
Common goal (LTR5)

Continuous improvement orientation

Continuous improvement is must. Today's market will soon get extinct by the overpowering innovations and aggressive strategies of competitors. For continuous improvement the sub factors to be focused are

Supplier Development program (CI1)
Benefit Measurement (CI2)
Supplier evaluation (CI3)

Supplier Competency

Competency refers to the ability of supplier to do the assigned job accurately and precisely. Ability of a supplier to deal with supply and demand efficiently by incorporating the principles of industrial management such as lead time and various models to minimize cost and increase output. Various sub factors are

Supplier quantity discount (SC1)
Supply lead time (SC2)
Supply Uncertainty (SC3)
Quality Issue due to delivery (SC4)

Management Vision & Mission

Managers are the pivotal point in the supply chain as they are the ones who are responsible for implementing their prowess to the smooth maintenance of supplier relations. The various factors included in this section are as stated below.

Good Management Practice (MVM1)

Established Mutual Relationship (MVM2)

Top Management Support (MVM3)

3. Methodology

Extraction of factors and sub factors is not enough. For decision making or to set a managerial guideline ranking of factors is must. Several Methods are there to rank the sub factors under main factor such as- Nominal Group Technique (NGT), Descriptive Analysis. But AHP (Analytical Hierarchy Process) is a multi criteria decision making preferred technique to rank the sub factors. AHP makes pair wise comparison of sub factors. So experts are asked to make pair wise comparison of sub factors on 1 to 9 scales.

Table 2 Scale and Rating used in AHP

Scale	Rating	Scale	Rating
Equally Preferred	1	From Equally Moderately Preferred	2
Moderately Preferred	3	From Moderately to Strongly Preferred	4
Strongly Preferred	5	From Strongly to Very Strongly Preferred	6
Very Strongly Preferred	7	Very Strongly to Extremely Preferred	8
Extremely Preferred	9		

Table 3 to 6 representing Comparison Normalised Matrix with weight obtained from AHP analysis

Table 3 Comparison Normalised Matrix with weight for sub factors of Objective to maintain long-term relationship factor

	Comparison Matrix					Normalised Matrix					
	LTR1	LTR2	LTR3	LTR4	LTR5	LTR1	LTR2	LTR3	LTR4	LTR5	Weight
LTR1	1	0.142	0.333	0.2	0.5	0.055	0.084	0.058	0.015	0.025	0.047
LTR2	7	1	4	6	8	0.388	0.593	0.704	0.478	0.41	0.514
LTR3	3	0.25	1	5	7	0.166	0.148	0.176	0.398	0.358	0.249
LTR4	5	0.166	0.2	1	3	0.277	0.098	0.035	0.079	0.153	0.128
LTR5	2	0.125	0.142	0.333	1	0.111	0.074	0.025	0.026	0.051	0.057

Table 4 Comparison Normalised Matrix with weight for sub factors of Continuous improvement orientation factor

	Comparison Matrix			Normalised Matrix			
	CI1	CI2	CI3	CI1	CI2	CI3	Weight
CI1	1	0.2	3	0.157	0.149	0.272	0.192
CI2	5	1	7	0.789	0.745	0.636	0.723
CI3	0.333	0.142	1	0.052	0.106	0.090	0.082

Table 5 Comparison Normalised Matrix with weight for sub factors of Supplier Competency factor

	Comparison Matrix				Normalised Matrix				
	SC1	SC2	SC3	SC4	SC1	SC2	SC3	SC4	Weight
SC1	1	0.25	0.333	2	0.117	0.038	0.201	0.117	0.118
SC2	4	1	0.2	6	0.470	0.155	0.120	0.352	0.274
SC3	3	5	1	8	0.352	0.779	0.603	0.470	0.551
SC4	0.5	0.166	0.125	1	0.058	0.025	0.075	0.058	0.054

Table 6 Comparison Normalised Matrix with weight for sub factors of Management Vision & Mission factor

	Comparison Matrix			Normalised Matrix			
	MVM1	MVM2	MVM3	MVM1	MVM2	MVM3	Weight
MVM1	1	5	2	0.588	0.555	0.600	0.581
MVM2	0.2	1	0.333	0.117	0.111	0.100	0.109
MVM3	0.5	3	1	0.294	0.333	0.300	0.309

4. Discussion

From AHP analysis it is revealed that Mutual trust is most important sub factor with a weight of 0.514 to accomplish the objective of maintaining long term relationship between manufacturer and supplier. For strategic partnership trust is must it leads to more transparent and seamless flow of information, risk free investment and higher satisfaction in building such relationship. Again Benefit measurement with a weight of 0.723 proved to be a sub factor through which continuous improvement of relationship possible because it is challenging to sustain in the keen competition without measuring improvement of skill & capabilities of suppliers. Further supplier uncertainty ranked highest with a weight of 0.551 to measure supplier competency in supplier selection. Finally Good management practice with a weight of 0.581 is found to be the driving factor in setting the management vision to build relation.

5. Conclusion

Supplier relationship management is the backbone of supply chain management. Now a day's Suppliers are not defined as who supplies goods rather as strategic partners of firm because of their important role in value chain in understanding of customer's need. SRM is emerged as global management network that involves skills & knowledge to enhance supply chain performance & to meet the short term and long term needs. So to improve performance of whole supply chain it is high time to focus on the high ranked sub factors under each extracted factor obtained from AHP analysis. An attempt is being made in this paper to represent significance of factors and its sub factors that affecting supplier relationship. This paper will be a managerial guideline to build and maintain buyer supplier relationship.

6. Reference

[1] Eskandari H, Darayi M and Geige 2010 *Proceedings of the 2010 Winter Simulation*.
[2] Hematyar S and Chaharsooghi K 2014 *International Journal of Materials, Mechanics and Manufacturing* **2** 202-209.
[3] Alaei S, Behravesh M and Karegar N 2014 *Prague Economics Papers* **2** 198-216.
[4] Giannakis M 2012 *Int. J. Procurement Management* **5** 368-408.
[5] Chen Y M 2010 *Journals of computing* **2** 42-47.
[6] Gan X, Sethi S P and Ya H 2004 **13** 135-142.
[7] Habib M 2014 *Open Journal of Social Sciences* **2** 238-246.
[8] Moyaux T and Draa B C 2007 *IEEE transactions on system, man & Cybernetics* **37** 396-409.
[9] Amoako-Gyampah K and Boye S S 1998 *International Journal of Production Economics* **54** 143–161.
[10] Yao Z, Leung S C H and Lai K K 2008 *European Journal of operation research* **187** 275-282.
[11] Prahinski C and Benton W C 2004 *Journal of Operation management* **22** 39-62.
[12] Saranga H 2009 *European Journal of Operational Research* **196** 707–718.
[13] Kahkonen A K 2011 *International Journal of Procurement Management* **4** 386-401.
[14] Sucky E and Durst S M 2013 *International Journal of Procurement Management* **6** 92-127.
[15] Loppacher J S, Caglliano R and Spina G 2011 *Int. J. Procurement Management* **4** 156-180.
[16] Lintukangas K 2011 *Int. J. Procurement Management* **4** 1-19.
[17] Hong P and Kwon H B 2012 *International Journal of Procurement management* **5** 452-469.

The Intention towards Social Entrepreneurship among Students and its link with Big 5 Model

Preethi CM[1] and R G Priyadarshini[2]

[1] Amrita School of Business, Coimbatore, Amrita Vishwa Vidyapeetham, India, preethichandran7@gmail.com
[2] Amrita School of Business, Coimbatore, Amrita Vishwa Vidyapeetham, India, rg_priyadarshini@gmail.com

Abstract. In the recent decade there is a developing interest of people getting attracted towards social entrepreneurship. So, what is social entrepreneurship? It is the process of applying innovation and business expertise in a non-profit sector by having a social objective and creating a social value. While on the other side there is commercial entrepreneurship which is vastly found, it is the process of applying various techniques to increase profit and wealth of an individual. This paper is to measure, analyze and interpret the intention of the budding generation towards social entrepreneurship as the earth have attained a position where the environment, society and the survival of fellow humans is at stake rather than considering profit and economic growth as important. There is also a trait based analysis, considering the big 5 personality traits such as openness, extraversion, agreeableness, neuroticism and conscientiousness and it natural affect on their inclination towards this.

1. Introduction

1.1. Social Entrepreneurship

It is the process of planning and executing a mission that has a social value and this can be achieved by continuous learning and adaptation of new strategies which are sustainable. The innovative usage of available resources in a way to pursue all the opportunities so as to bring out a social change and/or social needs is the process of Social Entrepreneurship (Mair and Marti, 2006).Also a social entrepreneur recognizes a social issue and uses entrepreneurial principle to make a change in the society by creating and managing a venture. These entrepreneur's never expect profit as returns and their aim is to make the world a better place to live (Verma and Vohra, 2014). They focus on the social issues which are vast and diverse, it is not particular to a community or a region and sometimes they may take up an issue which is having a global presence such as poverty, inadequate education facility, health problems, environmental issues, unemployment, etc,. It is expected that the future research could focus on the difference in social entrepreneurship and innovation ecosystems in developed countries and in developing countries. How social entrepreneurship can be brought in at various levels of the firms, economies and industries, how it can diffuse innovations across global level and whether social entrepreneurship has got the potential to alter the technological trajectories (Surie and Groen, 2017).

1.2. Personality traits

The personality is the individual differences between the people in the way in which they think, feel and behave. There are many ways to measure personality and the best among them is the Big 5 Personality Trait method consisting of openness, conscientiousness, extroversion, agreeableness, neuroticism. The personality of a person can be explained as reflection of factors affecting the emotions of that person and it can be internal or external factors, thus if personality of that particular individual is being evaluated then it not reflects his own traits but also the trait of the society and group of such individual and also all human kind in certain level (Irengun and Arikboga, 2015). In this study we have looked at the students who did professional courses and try to understand and analyze about their inclination and also the influence of big 5 personality towards social entrepreneurship.

2. Literature Review

2.1. Social Entrepreneurs
The one, who takes up an initiative to solve important social issues that are happening in their surroundings are called Social Entrepreneurs, it can be an individual or any private organization. They also develop new programs, services and solutions for any particular problem that affects certain community (Korosec and Berman, 2006). In social entrepreneurship these people may end up in creating a start up and do activities and processes that enhances the social wealth by discovering and using the available opportunities (Zahra, Gedajlovic, Neubaumand and Shulman, 2009).

2.2. Importance of Social Entrepreneurship
The socio-economic inequalities are being looked by the combined effort of the government and Non Governmental Organizations. These social entrepreneurs are always concerned about serving the un-served needs of underserved populations who could not pay for the things that they need desperately. One of the prominent examples of an enterprise that functions for a social cause is Aravind Eye Hospital, as they provide free and low cost eye care for the people who could not afford it and substantiate it from one's who can afford to pay (Bornstein and Davis, 2010; Dacin, Dacin and Matear, 2010). The impact of social entrepreneurship is that they initiate social change in the emerging markets under the conditions such as institutional complexity, which means the environmental governance and policies. In that the aspects that enables institutional complexity and the possibilities for contradicting institutional logics in order to provide opportunity for social entrepreneurship (Cherrier, Paromita and Ray, 2017).

2.3. Different fields of Social Entrepreneurship
Social Entrepreneurship is into various sectors, thus the innovations in developing countries will require mechanisms that allow new market, industries and an ecosystem to evolve and integrate with global systems. The paper speaks about two contributions; they are developing a framework mechanism for forming an ecosystem and new markets through social entrepreneurship and providing support framework for renewable energy sector in rural India. There is another study which analyses the relevant factors in social enterprises that affect the job creation (Surie, 2017, Marti, Domingoand Sanchez-Garcia, 2015).

2.4. Education on Social Entrepreneurship
The main objective consists of evaluation of the impact and assessment in relation with the course than is being undertaken by the entrepreneurship students and also to bring forward knowledge about assessment and impact outside the college, it contexts the humanities with a focus on the subarea of social entrepreneurship and their attention towards social value and non-profit dimensions (Jensen, 2014). There is another study which analyses the way in which social entrepreneurship can be integrated into management studies through experiential learning. It helps in identifying the benefits of incorporating experiential learning in MBA programs and also to integrate some components that are already in the program (Awaysheh and Bonfiglio, 2017).

2.5. Relationship between Personality Trait and Social Entrepreneurship
There are certain performance measures to measure the relationship of Big Five personality factors to entrepreneurial performance.The incubator, such as Venture Labhas been used to measure entrepreneurial learning. To find the entrepreneurial performance based on education, a multiple regression analysis has been used to test the different relationships, from the results it is noted that extraversion has a significant positive effect on entrepreneurial performance, but the entrepreneurial performance after the amount of courses they have undergone gave a negative effect on it (Beek, 2017).

It depends on the perspective of students towards the concept of entrepreneurship, the various factors that influences them and the role of the institution in cultivating that spirit. It is also revealed from the findings that these factors constitute a major determinant of the student's entrepreneurial perspectives (Obembe, Otesile and Ukpong, 2014). The study is the comparison of the personality traits based on the attitudes of university students towards entrepreneurship. The result states that the students who are innovative, high need for achievement and greater internal locus of control had a good intention towards entrepreneurship than others (Çolakoglua and Gozukara, 2016).

3. Objectives of the study

3.1 To identify the prominent Big 5 factors that affects the intent of social entrepreneurship
3.2 To identify the prominent personality factors that affects the factors affecting the intent of social entrepreneurship

4. Conceptual Model

The conceptual model defines the link between Education background and big 5 personality traits to social entrepreneurship. Social Entrepreneurship intent of a particular person is analyzed based on the big 5 personality traits which are usually the trait of people who are present and also the combined influence of society on that person's behaviour. The education background of that particular person was also being examined so that the difference in impact based on their education background can be measured. Thus the effects and outcome of these components on social entrepreneurship is measured and analyzed.

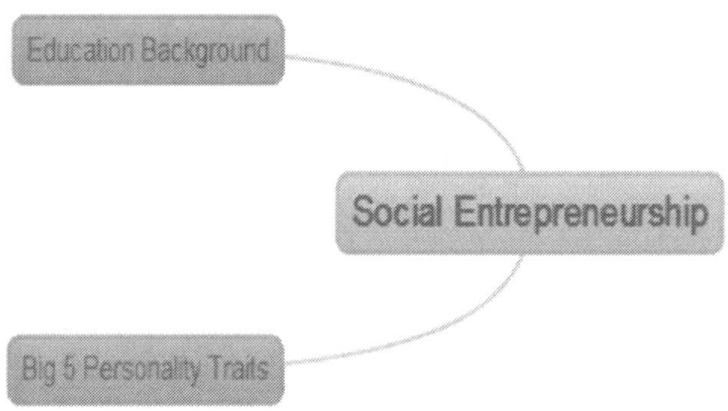

Figure 1. Conceptual Model of the Study

5. Hypotheses

5.1 There is no significant association between personality traits and their orientation towards social entrepreneurship
5.2 There is no significant difference in the orientation towards social entrepreneurship with respect to individuals of various professional backgrounds

6. Methodology

The study was conducted with the sampling frame consisting students who are undergoing undergraduate and postgraduate program in various colleges in Tamil Nadu region. The undergraduate students are engineers and post graduates are students doing their management studies. A questionnaire was floated with 56 questions which consisted of 5 dimensions for big 5 personality traits such as openness, extraversion, agreeableness, neuroticism and conscientiousness and also 5 dimensions for measuring social entrepreneurship intent such as social vision, financial returns, innovation, sustainability and social networks. The sample size consisted of 100 students and their responses were collected in 5-point likert scale measurements.

7. Analysis and Findings

In this study the social entrepreneurship is taken as the dependant variable where as the independent variables are the education background of that person and their big 5 personality trait. The analysis in Figure 2 is done using path analysis considering the variables big 5personality traits and Social Entrepreneurship.

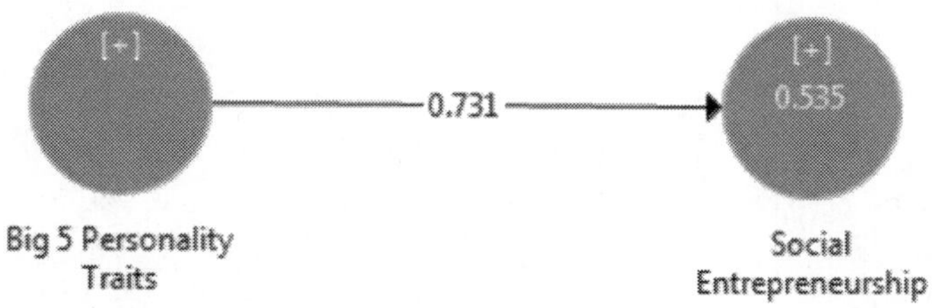

Figure 2. This figure shows the result of the analysis done using SmartPLS and the value of significance for all the variables are found out

After running the analysis the standardized regression rates and the R square value are founded to be significant and the value is 0.535. Thus by seeing the results it is found that the Big 5 personality trait has a greater impact on the social entrepreneurship intention as it has a higher standardized regression value. The Big 5 personality trait is said to have 73% significance.

Table 1. Analysis and calculation of mean using SPSS

	Educational Qualification	N	Mean	Std. Deviation	Std. Error Mean
Social Vision	UG	67	3.79	.576	.070
	PG	32	3.89	.707	.125
Financial returns	UG	67	3.57	.699	.085
	PG	32	3.97	.677	.120
Innovation	UG	67	3.88	.585	.071
	PG	32	3.99	.598	.106
Sustainability	UG	67	3.90	1.156	.141
	PG	32	3.74	1.065	.188
Social networking	UG	67	3.23	.440	.054
	PG	32	3.52	.427	.076

A group statistics is being done between the education backgrounds of the students to the various dimensions of social entrepreneurship. From the table1 it is found that the mean value of the post graduate students in various dimensions of social entrepreneurship have no a significance compared to the mean values of the undergraduates except Social Vision which is relatively having good significance than others (refer table2).

IOP Conf. Series: Materials Science and Engineering **390** (2018) 012057 doi:10.1088/1757-899X/390/1/012057

Table 2. Independent Sample T Test

		Levene's Test for Equality of Variances					95% Confidence Interval of the Difference	
		F	*Sig.*	*t*	*df*	*Sig. (2-tailed)*	*Lower*	*Upper*
Social Vision	*Equal variances assumed*	3.804	.054	-.689	97	.492	-.357	.173
	Equal variances not assumed			-.641	51.297	.524	-.380	.196
Financial returns	*Equal variances assumed*	.248	.619	-2.717	97	.008	-.700	-.109
	Equal variances not assumed			-2.748	62.934	.008	-.698	-.110
Innovation	*Equal variances assumed*	.002	.969	-.866	97	.389	-.361	.142
	Equal variances not assumed			-.859	59.933	.394	-.365	.146
Sustainability	*Equal variances assumed*	.497	.482	.643	97	.522	-.325	.637
	Equal variances not assumed			.662	65.925	.510	-.314	.626
Social networking	*Equal variances assumed*	.168	.683	-3.094	97	.003	-.476	-.104
	Equal variances not assumed			-3.127	62.834	.003	-.475	-.105

An independent sample t test is done to find the significance and from the values it shows that there may not be a great difference as the value is least significant and it will vary 90% of the times. Thus this defines that the intent of social entrepreneurship does not have a greater impact based on the education background. They may be postgraduates or undergraduates their values won't show a significant difference.

8. Result and Discussion

Thus from the analysis it is found the big 5 personality traits have a great impact on the intention towards entrepreneurship whereas the educational background of that particular person does not affect to that extend. So the various dimensions of big 5 such as openness, extraversion, agreeableness, neuroticism and conscientiousness whereas the dimensions of social entrepreneurship such as social vision, financial returns, innovation, sustainability and social networks returns were correlated and the corresponding results were found

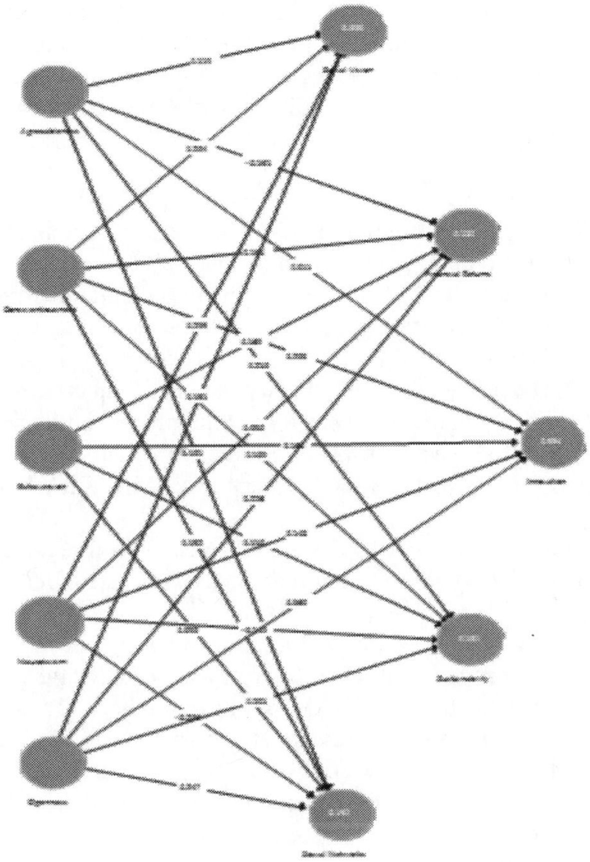

Figure 3. The factor analysis between the dependent and independent variables

In the correlation between various variables, some variables show positively very good impact like the correlation between openness and innovation, the people who are open to experience are always ready to learn and think about new ideas. Neuroticism people have an average to moody character and they have a very high anxiety which makes them to have a greater impact on Social Vision, the personality trait Openness has a great impact towards Financial Returns these people are often have a great level of acceptance. Agreeableness is a trait that makes them warm and kind towards the society which in turn results in a good impact on sustainability. People with openness are always ready to explore new things and hence will make them to reach out a wide range in Social Networking.

9. Limitations

The study is not done on an extensive geographic location, hence the results may vary depending upon the geographic area taken into consideration. When considering education background the people with no education background are not taken into account. The big 5 personality test and its correlation may differ according to the way in which the samples are collected and also the total number of samples. Also there are chances where results may differ according to cultural and demographic differences.

10. Conclusion

The study tries to focus mainly on the big 5 personality traits of a particular person and his/her education background and its impact on their intention towards social entrepreneurship. And from the study it is found that the education background has the least impact, while the big 5 personality traits of a particular person greatly influence the various dimensions of social entrepreneurship. The analysis also proves that some of the dimensions of the variable have a reasonable direct impact, while others do not have that much positive impact towards the main dimensions. The factors also depends on the study as in some other study the personality traits that were taken into account are locus of control, need for achievement, risk of tolerance, entrepreneurial alertness and from this only locus of control was the only factor that was found to have an effect on entrepreneurial intentions (Karabulut, 2016). Hence it is found that different personality traits have their own impact on various social entrepreneurship dimensions.

11. References

[1] Austin J, Stevenson H and WeiSkillern J 2006 Social and commercial entrepreneurship: same, different, or both? *Entrepreneurship theory and practice*, **30**(1), 1-22.

[2] Awaysheh A and Bonfiglio D 2017 Leveraging experiential learning to incorporate social entrepreneurship in MBA programs: A case study *The International Journal of Management Education*, **15**(2), 332-349.

[3] Beek R C 2017 The influence of personality on entrepreneurial performance: An investigation of the effects of the big five personality factors and entrepreneurial learning on the subjective performance of start-up entrepreneurs *Master's thesis, University of Twente*.

[4] Bornstein D and Davis S 2010. Social Entrepreneurship: What Everyone Needs to Know® *Oxford University Press*.

[5] Çolakoğlu N and Gözükara İ 2016 A comparison study on personality traits based on the attitudes of university students toward entrepreneurship *Procedia-Social and Behavioral Sciences*, **229**, 133-140.

[6] Dacin P A, Dacin M T and Matear M 2010 Social entrepreneurship: Why we don't need a new theory and how we move forward from here *The academy of management perspectives*, **24**(3), 37-57.

[7] İrengün O and Arıkboğa Ş 2015 The effect of personality traits on social entrepreneurship intentions: a field research *Procedia-Social and Behavioral Sciences*, **195**, 1186-1195.

[8] Jensen T L 2014 A holistic person perspective in measuring entrepreneurship education impact– Social entrepreneurship education at the Humanities *The International Journal of Management Education*, **12**(3), 349-364.

[9] Korosec R L and Berman E M 2006 Municipal support for social entrepreneurship *Public administration review*, **66**(3), 448-462.

[10] Karabulut A T 2016 Personality traits on entrepreneurial intention *Procedia-Social and Behavioral Sciences*, **229**, 12-21.

[11] Mair J and Marti I 2006 Social entrepreneurship research: A source of explanation, prediction, and delight *Journal of world business*, **41**(1), 36-44.

[12] Nga J K H and Shamuganathan G 2010 The influence of personality traits and demographic factors on social entrepreneurship start up intentions *Journal of business ethics*, **95**(2), 259-282.

[13] Obembe E, Otesile O and Ukpong I 2014 Understanding the students' perspectives towards entrepreneurship *Procedia-Social and Behavioral Sciences*, **145**, 5-11.

[14] Rey-Martí A, Ribeiro-Soriano D and Sánchez-García J L 2016 Giving back to society: Job creation through social entrepreneurship *Journal of Business Research*, **69**(6), 2067-2072.

[15] Surie G 2017 Creating the innovation ecosystem for renewable energy via social entrepreneurship: Insights from India *Technological Forecasting and Social Change*, **121**, 184-195.

[16] Thampi P P, Jyotishi A and Bishu R 2015 Cultural characteristics of small business entrepreneurs in India: examining the adequacy of Hofstede's framework *International Journal of Business and Globalisation*, **15**(4), 475-495.

[17] Zahra S A, Gedajlovic E, Neubaum D O and Shulman J M 2009 A typology of social entrepreneurs: Motives, search processes and ethical challenges *Journal of business venturing*, **24**(5), 519-532.

IOP Conf. Series: Materials Science and Engineering **390** (2018) 012058 doi:10.1088/1757-899X/390/1/012058

Correlation between Micro hardness and Microstructure of CMT Welded AA 7075 Al Alloy

S.T.Selvamani[1,a,*], P.Govindarajan[2,b], M.Ajaymohan[3,c], S.J.Hariharan[3,d], M.Vigneshwar[1,e,f]

[a] Associate Professor, Department of Mechanical Engineering, Vel Tech Multi Tech Engineering College, Avadi, Chennai – 600062; Email id: *selva_balu@rediffmail.com

[b] UG Student, Department of Mechanical Engineering, Sri Sai Ram Institute of Technology, Tambaram, Chennai – 600044; Email id: govindarajanmech007@gmail.com

[c] UG Student, Department of Mechanical Engineering, Sri Sai Ram Institute of Technology, Tambaram, Chennai – 600044; Email id: ajaymohanraj@gmail.com

[d] UG Student, Department of Mechanical Engineering, Institute of Road and Transport Technology, Erode - 638316; Email id: harijosun@gmail.com

[e] Junior Research Fellow (JRF), Department of Mechanical Engineering, Vel Tech Multi Tech Engineering College, Avadi, Chennai – 600062.

[f] PG Scholar, Metallurgy and Materials Science Engineering, Vel Tech Dr.RR Dr.SR R&D Institute of Science and Technology, Avadi, Chennai – 600062;
Email id: vigneshwar@gmail.com

Abstract. The Cold Metal transfer process have recently gained significant importance in the manufacturing sectors due to its excellent droplet transfer of consumable in the molten pool with no spatter. The 7xxx series of Al alloys are the most preferable material to dynamic applications. Hence, in this research work, the AA7075 Al alloys were joined using an advanced TPS 400i Cold Metal Transfer (TCMT) welding machine in order to understand the homogeneity of the weldment with respect to its microhardness and the correlation between the microhardness and microstructure of the joints were analyzed. Finally, it was found that the hardness was decreased in the Weld Zone (WZ) compared to Heat Affected Zone (HAZ) and the Base Metal (BM) and also the metallurgical characterization reveals that the joint is free of defects and fine grains were observed in the WZ.

1. Introduction

The current, demands reduction in vehicle emissions to meet stringent pollution norms. Among several approaches like hybrid vehicle and electric vehicle, the prominent is using lightweight vehicles without compromising on safety. To develop light weight vehicles, manufacturers rely on dissimilar combinations of alloys like aluminium, magnesium and advanced high strength steels where the aluminum alloys were the prime choice for the fabricators due to its availability and also economically. There are variety of welding approaches for joining Al alloys but most of them result in severe defects in its own form. All of the solid state and fusion state welding techniques were ideal under certain conditions. This made the research society to develop a new technology to exhibit better properties in dynamic conditions compared to the existing processes [1-3]. Finally, in 2004, **FRONIUS** came up with an idea of joining the aluminum alloys with a renowned MIG welding process known as Cold Metal Transfer (CMT) welding process which has the same basic principle as that of conventionally used MIG except the fact it uses a droplet transfer mechanism to the electrode tip whose actions are controlled by a servomotor during the process [4].

A novel joining route for Aluminum alloys using CMT welding process was explained by **Pickin** et al [5] in which the mechanical properties of the joints were found to be better compared to the conventionally used MIG/MAG process. The control of mass and heat during cold metal transfer process was initially found to be very difficult which was substantiated by **Mezrag** [6] by means of reducing the heat transfer modes during the process which contributes for the homogeneity in the weld zone. The investigation of droplet transfer behaviors in cold metal transfer process on welding Ti-6Al-4V alloy was performed by **Zhe Sun** [7] and the innovative use of low heat input in CMT welding process was explained by **Lorenzin** [8] from which the droplet transfer modes and the subsequent solidification were well understood. The dissimilar joining of steel to aluminum was carried by **Shang** [9] et al to identify the feasible working limits for joining the dissimilar materials and also to find the weldability of steel to aluminum alloys. Though there are many recent developments in the CMT process for joining variety of materials yet there is no sufficient technical reports regarding the mechanical properties and the microstructural variations of the joints. Hence, in this research work, an advanced TPS 400i CMT (TCMT) welding process have been utilized for joining the aluminum alloy and the corresponding mechanical, microstructural properties were identified using advanced facilities and their properties were correlated to understand the droplet transfer mode during the advanced CMT process.

2. Experimental work

The AA 7075 Aluminum alloy (5mm thick) were used in this research work due to its wider applications and availability. The **Zn** (5.83%) and **Mg** (2.3%) were found to be the major alloying elements in this aluminum alloy along with **Cr** (0.072), **Cu** (1.54) and **Fe** (0.57) which was identified using a vacuum spectrometer. The mechanical properties of the base Al alloy are displayed in Table 1. The metal was then joined using an advanced CMT welding machine (TPS 400i) wherein the heat input was controlled while the electrode tip makes contact with the molten pool by a droplet mechanism. This short circuit transfer creates no spatter during welding and also the wave cruciform will be synchronous and uniform throughout the process. The figure 1 (a-b) shows the Advanced CMT welding machine and the corresponding joints were fabricated under the welding conditions of current 90A, voltage 12.5 V and welding speed 275 mm/min.

Table1. Mechanical properties of the Base Al alloy

Yield Strength (MPa)	Tensile Strength (MPa)	Fatigue Strength (MPa)	Hardness (Hv)	Elongation (%)
508	542	173	160	12

The welded joints were drawn using a wire cut EDM and the microhardness and microstructural specimens were prepared as per ASTM specifications. The hardness of the joints were evaluated using a Vicker's microhardness tester wherein the hardness values were displayed digitally and the results were recorded. The optical microscope was used to visualize the grain orientation at different zones in the joints and the corresponding grain formation were observed to understand the nature of the joints.

IOP Conf. Series: Materials Science and Engineering **390** (2018) 012058 doi:10.1088/1757-899X/390/1/012058

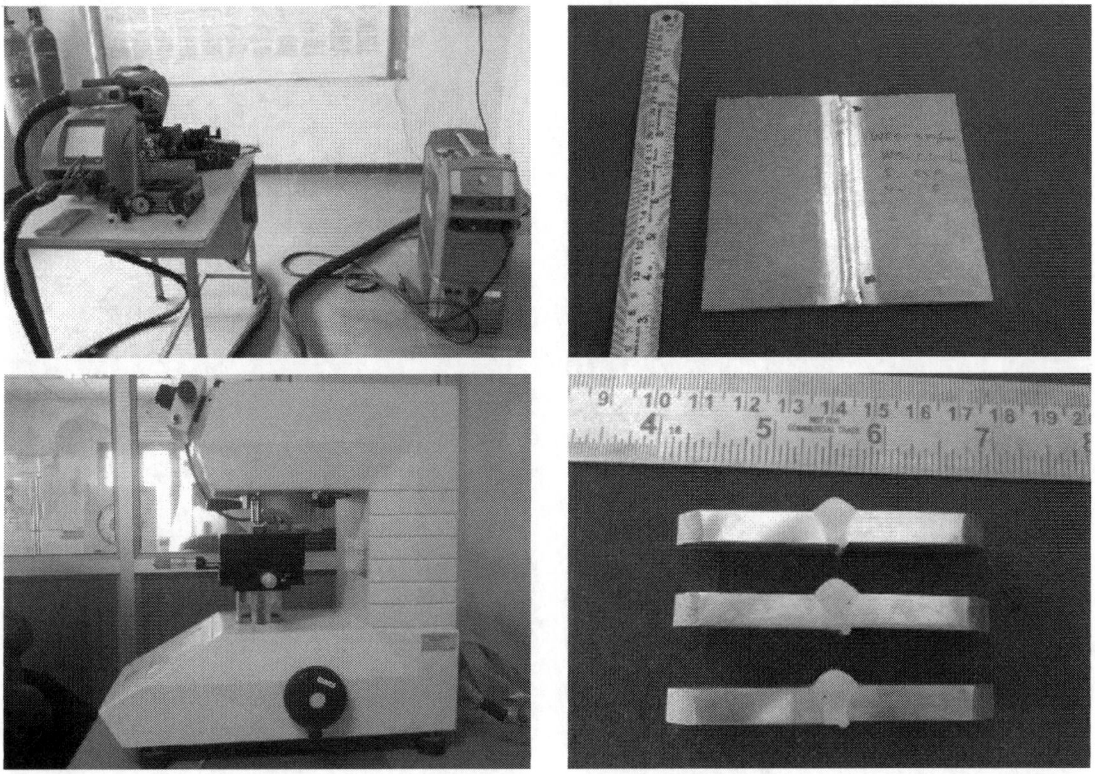

Figure 1. The Experimental work (a) CMT welding machine, (b) CMT joints, (c) Vicker's microhardness testing machine, (d) Microhardness specimen

3. Results and Discussion

The microhardness variation across the weldments were recorded and the average of the results were taken. It was found that the hardness was decreased in the Weld Zone (WZ) compared to the Heat Affected Zone (HAZ) and the Base Metal (BM) which is achieved due to the short circuit transfer between the electrode tip and the molten pool. The average microhardness across the weld zone was found to be 123.2 Hv and on the other hand, the HAZ shows an average microhardness value of 136 Hv which is an increase of 10.38 % compared to the WZ which was found to be better than the conventional fusion welding techniques which was also highlighted by Elrefaey [10].

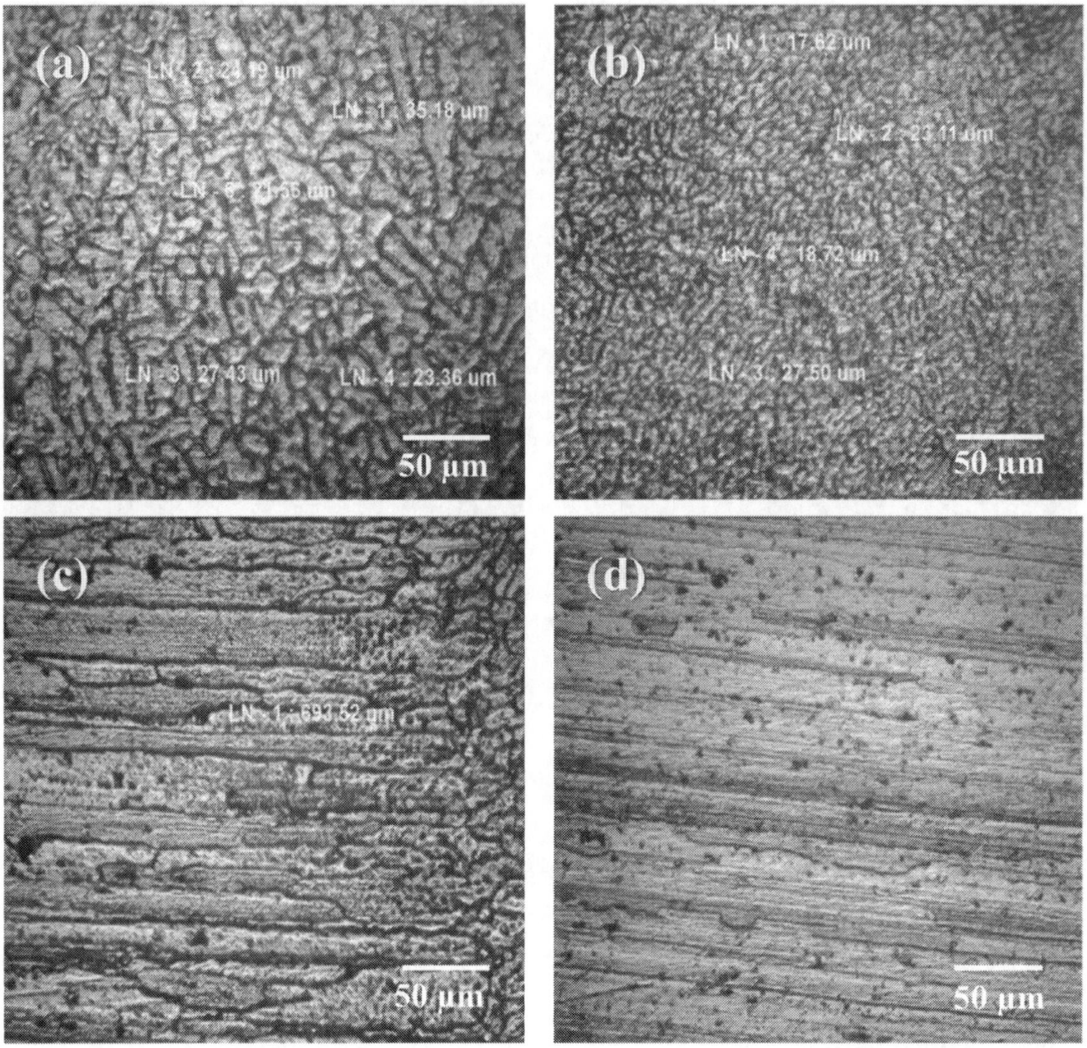

Figure 2. The Microstructural Results (a) Micrograph of WZ at 400 X, (b) Micrograph of the WZ at 200X, (c) Micrograph of the HAZ at 200X, (d) Micrograph of the BM at 200X

The microstructure of the joints analyzed by optical microscope shows no defects throughout the weldment and also the grains were uniformly distributed with only minute pores been observed. The figure 2 (a-d) depicts the microstructural results of the various zones of the CMT welded joints. The fine grains were attained in the weld zone with no solidification defects and the HAZ was found to be very narrow which is possible due to the controlled heat input during the welding process and rapid solidification of the molten pool [11-12].

Also a relationship between microhardness and microstructure has been established to understand the effect of microhardness attributing to the grain size. Hence, the hardness of various zones and the corresponding grain size of the joints acquired from the experimentations were related and expressed in figure 3. Using a straight line, the experimental data points were fitted which was governed by the corresponding regression equation:

$$\text{Hardness} = --1.1017\ (\text{Grain size}) + 169.97\ \text{Hv} \qquad (1)$$

From the above regression equation, the slope (−1.1017) is found to be negative, which indicates that if the grain size increases, the hardness of fully deformed zone decreases. From the Fig. 3, it can be inferred that the hardness is inversely proportional to the grain size of the CMT welded AA7075 aluminum alloy joints.

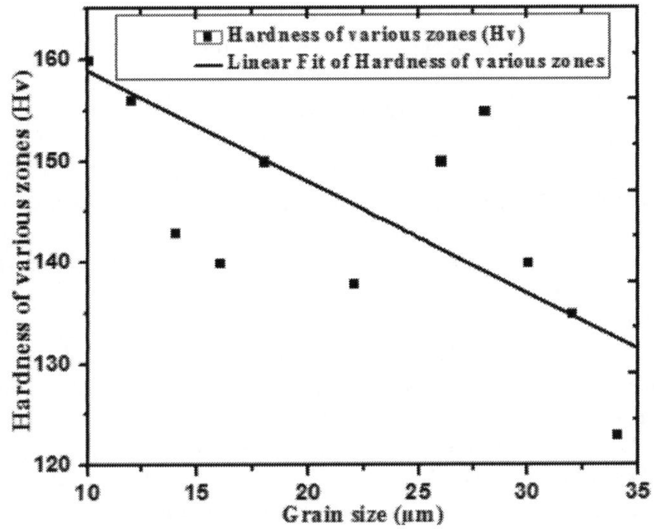

Figure 3. The relation between hardness of various zones and grain size

Therefore, from the mechanical properties and the microstructural characteristics attained it can be concluded that the weldment doesn't undergo any phase transition because of the innovative metal transfer (hot-cold-hot) in this process which provides a limited heat input across the weld zone due to the retraction of the wire which provide better weld aesthetics.

4. Conclusion

- The AA7075 Al alloy were successfully joined using advanced TPS 400i CMT welding machine
- The microhardness of the joint was found to be 123.2 Hv across the weld zone and the heat affected zone shows a microhardness of 136 Hv which is an increase of 10.8 % compared to the weld zone
- The grains were homogenous and very fine grains were observed in the weld zone with no solidification defects
- The microhardness exhibits inversely proportional relationship with the grain size of various zones
- The rapid heating and rapid cooling in the process facilitates an enhanced mechanical and metallurgical properties of the joints

Acknowledgement

The authors would like to express their sincere gratitude to **DST-SERB** for providing financial support to carry out this research work under EEQ scheme **(F.No.EEQ/2016/000114)**. Also we would like to thank **Dr.V.Balasubramanian**, Director (CEMAJOR) for providing facility to undergo characterization part of this research work. Also, we would extend our gratitude to our Chairman **Dr.R.Rangarajan**, Vel Tech Dr.RR Dr.SR Institutions for providing the consistent support to carry out the research work.

References

[1] Totten G E, Mackenzie S 2003. *Handbook of aluminum. vol 1. Physical metallurgy and processes* (Marcel Dekker Inc: USA)

[2] Mathers G 2002 *The welding of aluminium and its alloys.Cambridge* (UK Woodhead, publishing Limited)

[3] David S A, Babu S S, and Vitek J M 2003 Welding: Solidification and microstructure *JOM* **55** 14-20

[4] Robert Talalaev, Renno Veinthal, Andres Laansoo and Martinš Sarkans 2012 Cold metal transfer (CMT) welding of thin sheet metal products *Estonian Journal of Engineering* **18** 243-50

[5] Pickin C G, Williams S W and Lunt M 2011 Characterisation of the cold metal transfer (CMT) process and its application for low dilution cladding *Journal of Materials Processing Technology* **211** 496–502

[6] Mezrag B, Deschaux-Beaume F and M. Benachour 2015 Control of mass and heat transfer for steel, aluminium joining using cold metal transfer process *Science and Technology of Welding and Joining* **20** 189-98

[7] Zhe Sun, Yaohui Lv, Binshi Xu, Yuxin Liu, Jianjun Lin and Kaibo Wang 2015 Investigation of droplet transfer behaviours in cold metal transfer process on welding Ti-6Al-4V alloy *The International Journal of Advanced Manufacturing Technology*, **80**, 2007-14

[8] Lorenzin G and Rutili G 2009 The innovative use of low heat input in welding- experiences on 'cladding' and brazing using the CMT process *Welding International* **23** 622-32

[9] Yang S, Yang J, Zhang J and Lian Lei Y 2013 Welding of aluminium alloy to zinc coated steel by cold metal transfer *Mater. Des* **49** 602-12

[10] Elrefaey A Effectiveness of cold metal transfer process for welding 7075 aluminium alloys *Science and Technology of Welding and Joining* **20** 280-85

[11] Fengyuan Shu, Yaohui Lv, Yuxin Liu, Fujia Xu, Zhe Sun, Peng He and Binshi Xu 2014 Residual stress modeling of narrow gap welded joint of aluminum alloy by cold metal transferring procedure *Construction and Building Materials* **54** 224–35

[12] Jair Carlos Dutra, Régis Henrique Gonçalves e Silva, Bruna Martinello Savi, Cleber Marques and Orestes Estevam Alarcon 2015 Metallurgical characterization of the 5083H116 aluminum alloy welded with the cold metal transfer process and two different wire-electrodes (5183 and 5087) *Weld world* **59** 797-807

Dynamical analysis of Nano filled - Sisal fiber hybrid reinforced composites

K. Mohan[1*], T. Rajmohan[2], J. Sankar raman[3] and V. Ramanujam[4]

[1-4]Department of Mechanical Engineering,
Sri Chandrasekharendra Saraswathi Viswa Mahavidyalaya,
Kancheepuram, Tamilnadu - 631561, India.

Email: mohank84@gmail.com@gmail.com

Abstract. In this present work, the nano filled (Multi walled carbon nano tube) glass and sisal fiber reinforced hybrid composites were fabricated with varying the wt. % of MWCNT's by 0%, 0.5% and 1% respectively. Compression molding technique is used to fabricate the composites and are subjected to dynamic analysis over a range of temperature (30-190℃) at the frequency rate of 5Hz. The effect of temperature and nano filler on storage modulus (E′) loss modulus (E ″) and damping efficiency (tan δ) are determined by using dynamic mechanical analyser (DMA). The shear and loss modulus obtained maximum values for 0.5wt.% MWCNT composites. Damping value is higher for 1wt.% MWCNT incorporated composites. By using ultrasonic probe sonicator, nano fillers are mixed in to the epoxy resin to prepare a modified resin. The dynamic properties are evaluated as per as ASTM standards.

1. Introduction

In the recent years many researchers are focused in natural fiber reinforced hybrid composites due their beneficial properties like recycling, biodegradable and environmental friendly. Reubmann, T. Et.al [1] investigated on flax fiber reinforced polypropylene composites, and reported that higher fiber content leads to higher elastic shear modulus (E′) and with a tan(δ) peak is lower. When increasing the temperature from 23 to 80°C, the E′of 30% FFRP was decreased into 41.8%. Kaiser, J.et.al [2] concluded, when the volume fraction of a filler is increased from 0 to 50% in the epoxy matrix, the storage shear modulus (E′) is increased upto 176% and significant improvement have been observed in the damping co efficient (tan δ).
Yuan et al [3] studied on the effect of plasma treatment on enhancing the performance of wood fiber polypropylene composites. It states that dynamic mechanical analysis test of the composites have shown the improvement in the storage modulus after chemical treatment. Significant changes in the results have been observed by varying the frequency during DMA analysis. Jacob et.al [4], Mohamed et.al [5] investigated especially on natural fibers like flax, jute, sisal, banana, etc. They found that reinforced-foamed materials have especially important significance for reduction of density in automobile construction components due to its higher specific stiffness and specific tensile strength. Yuanming X et.al [6] studied on the dynamic mechanical properties of epoxy resin are dependent on the resin system with different monomial components and manufacturing technology.

Martins and Mattoso [7] reported on the dynamic mechanical behaviour of sisal fiber reinforced with tyre rubber. It states that in certain cases, when the fiber was chemically modified and subjected to DMA analysis its indicating the enhancements in the results. Amash et.al [8] have been revealed that outputs of the DMA will vary by changing the input frequencies. They have reported that effectiveness of cellulose fibre is increasing in stiffness and decreasing in damping coefficient. Thermal properties of sisal/glass hybrid composites investigated and concluded that, the addition of glass fibers will improve thermal properties of the glass/sisal comp of alkali treated jute fiber composite. The storage modulus and damping factor were manipulated by the alkali treatment of jute fiber. By swelling reaction it can change the native crystalline structure of the fiber. It can provide the improved interfacial bonding between fiber and matrix frequently. Andrej et al[12]'s investigated and states that the shear modulus increases when the fiber volume fraction upto 42 vol% and the damping co efficient values has been decreased. Further increasing the in the fiber content may leads to decrease the shear modulus.

Mou'ad. A et al [13] results stated that the storage modulus (E') and the loss modulus (E'') has been increased due to the addition of hybrid nano fillers. The OMMT and MWNTs incorporated hybrid composites were compared and concluded that the higher Tg values has been observed in MWCNT incorporated composites rather than OMMT composites.

2. Materials and methods

2.1.Materials

Sisal fibers obtained from Eco Craft, Hosur. Glass fibers, Epoxy resin Araldite LY556 and hardener HY951obtained from M/s. Suntech fibers Chennai, From M/S US Research Nano materials Inc, USA, the MWCNTs has procured and used as constituents for fabricating the composite.

2.2. *Preparation of modified Epoxy resin*

MWCNT particles were initially taken by the Wt% of 0%, 0.5% &1%. Using ultrasonic probe sonicator the MWCNT were dispersed into the epoxy resin without agglomeration. Then the blends are allowed to vibrate for the period of 2hrs using a rotary shaker for homogeneous mixing of MWCNT particles into resin. [14-16].

2.3. Preparation of MWCNT filled composites

In the present investigation, the composites specimens were prepared using compression molding technique. Sisal and glass fibers of 30 X 30 cm size used to fabricate the specimen. The specimens having five layers in which the glass fibers were used in extreme layers in-between sisal and glass fiber have arranged alternatively. The figure (1-3) shows the fabrication of composites using compression molding machine and dynamic mechanical analysis (DMA) of the composites. Figure 4 shows the prepared specimens for DMA testing.

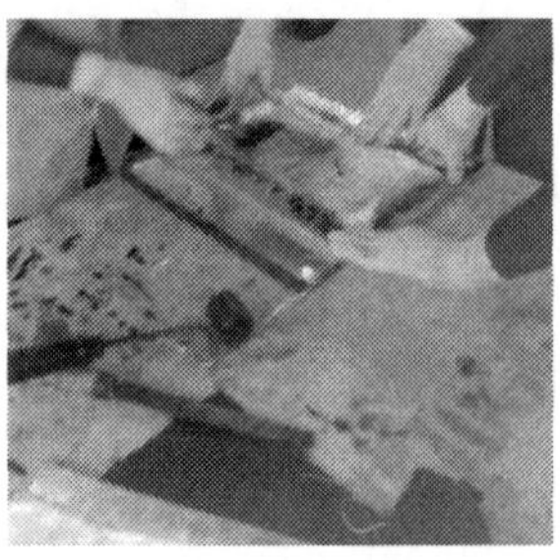

Fig 1. Preparation of composites

Fig 2. Compression molding

Fig 3. Dynamic mechanical analyser

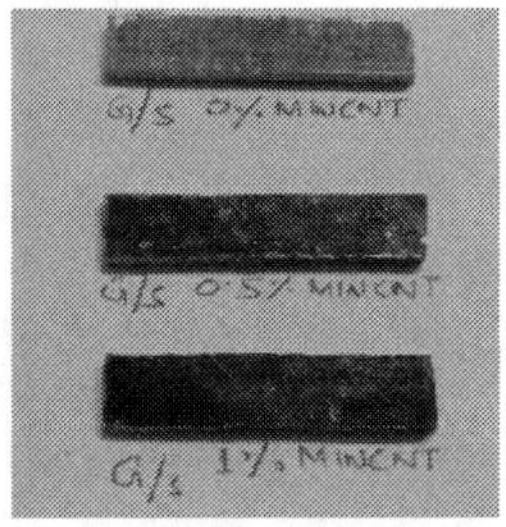

Fig 4. Composite Specimens

Natural fibers are taken from three bases. They are plant, animal and mineral. Fibers from plants are obtained from its stem, leaves, fruits, seeds and roots. In the same way the fibers from animals are extracted from hair, leather, internal organs etc. Mineral fibers are derived from the earth and are pre-treated before usage. Some important natural fibers with their properties and chemical compositions are listed in Table 1.

3. Result and Discussion

Dynamic mechanical analysis is one of the important technique to analyse the modulus of the material as well as structure of the material. The effect of temperature on shear modulus, loss modulus and damping co efficient at the frequency rate of 5Hz for various wt% of MWCNT hybrid composites are discussed below.

3.1.Shear Modulus

The fabricated composites predominantly cutted into a size as per ASTM standard and subjected to dynamic mechanic analysis. The fig (5) shows the effect of temperature on shear modulus for various wt% of MWCNT filled sisal glass fiber reinforced hybrid composites.

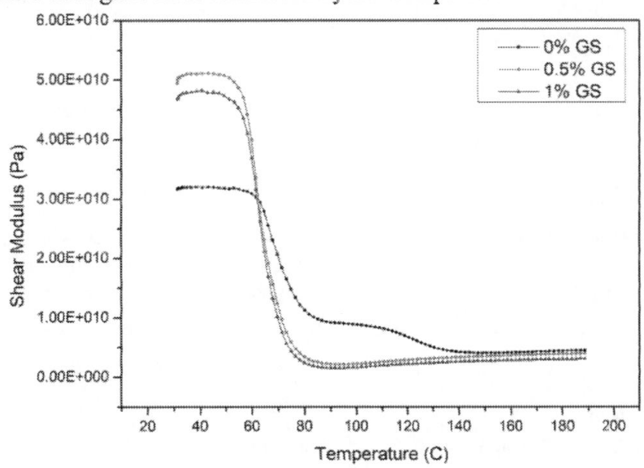

Fig 5. Effect of temperature on the shear modulus of different hybrid samples (frequency 5 Hz).

The shear modulus for 0.5% MWCNT incorporated composites shows the maximum value followed by 1wt% and 0wt% composites. The peak value of each respective specimens is maintained up to 55°c thereafter it's reduced dramatically when the temperature increased up to 90°c due to the molecule motions and interfacial strength bonding between the fibers and matrix. Then the shear modulus is increased slightly while increasing the temperature after 90°c. At minimum temperature the shear modulus for each specimen indicating maximum values due to their stress distribution between the two fibers.

3.2.*Loss modulus*

The loss modulus curve as shown in the figure (6) and it is the influence of the adhesive materials in the polymer composites. The loss modulus peaks are found to high for 0.5% incorporated MWCNT rather than the 1% and 0% respectively. The prompt rise in the loss modulus indicates an increasing the structural agility of the polymer composites. The maximum peak value of loss modulus indicates the maximum heat dissipation occurred at the temperature which shows the T_g of the system. Also observed incorporation of MWCNT into the matrix, the T_g is shifted to the higher temperature region. Higher wt % of nano fillers agglomerate in the composite may be the reason for reduction in loss modulus. Similar results have been reported by pothan et al.

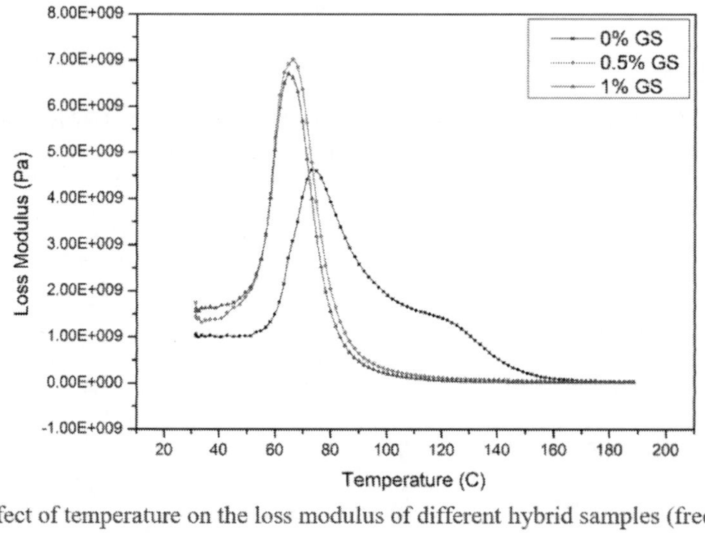

Fig 6. Effect of temperature on the loss modulus of different hybrid samples (frequency 5 Hz).

3.3.*Damping coefficient*

The ratio of the energy dissipation potential and absorption of the viscous elastic material is represented by tan D. From the figure 7, comparing the various wt% of MWCNT's incorporated composites, 1wt% composites shows the maximum tan D peak value than the other two composites. The peak value of the 1wt% GS is obtained at 0.73 in 72°C which indicates the material has more energy dissipation potential compared with remaining composites. Also it's steadily decreases up to 120°C and it maintains constantly till 190°C. Due to the strong interactions between the fibre and matrix in which it tends to decrease the mobility of the molecular chains results in diminishing of damping. The peak values might have been lowered because the amount of polymer due to fibre merging is not fair enough.

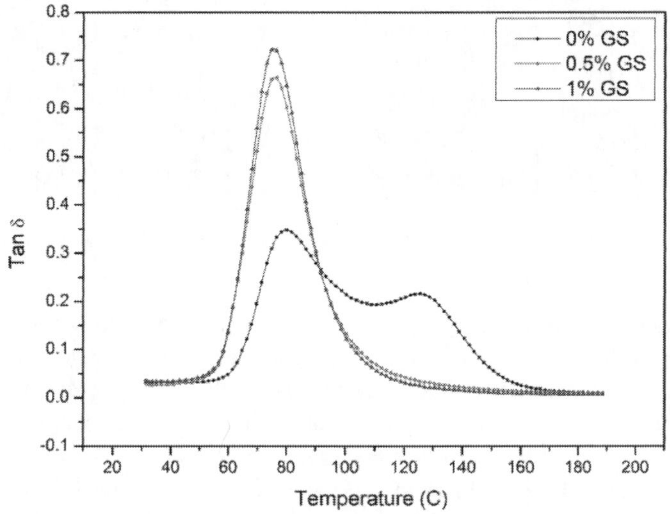

Fig 7. Effect of temperature on the tan D curve of different hybrid samples (frequency 5 Hz).

4. Microstructural analysis

Microstructure of the developed composites incorporated with 1wt.% and 0.5wt.% as shown in the figure 8 and 9. From the SEM images it is clearly observed that uniform dispersion of MWCNT into the epoxy resin. Hence the ultrasonic probe sonicator process is suitable for dispersing the nano fillers into the matrix .Also well interfacial bonding between the matrix and fibers are confirmed from the SEM images. Pulled out fibers and fiber breakages were identified during the deformation.

Fig 8. SEM micrographs for 1 Wt.% MWCNT incorporated composites

IOP Conf. Series: Materials Science and Engineering **390** (2018) 012059 doi:10.1088/1757-899X/390/1/012059

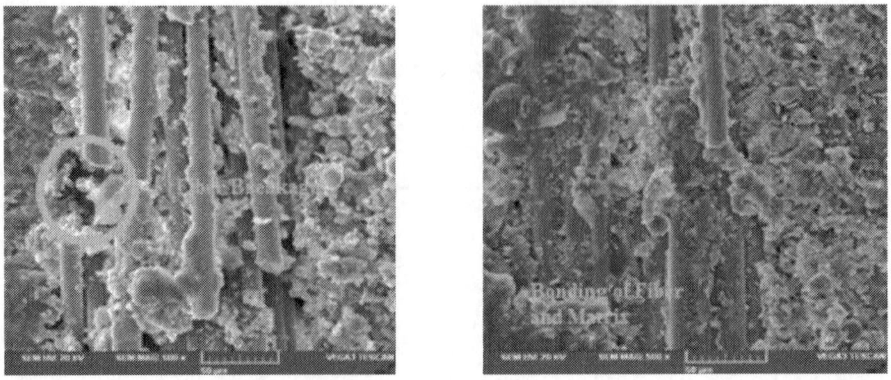

Fig 9. SEM micrographs for 0.5 Wt. % MWCNT incorporated composites

5. Conclusions

- The addition of nano fillers increase the storage modulus up to 0.5wt.% MWCNT. Further increasing with the nano fillers leads to reduce the storage modulus. It's may be due to agglomeration of nano fillers in the composites. In this work, storage modulus of 0.5wt% G/S composites shows vast improvement comparing other composites. Due to the agglomeration of MWCNT's leads to decrease in properties.
- The loss modulus for 0.5wt% MWCNT incorporated composites shows the highest value due to the maximum heat released at the peak value of Tg for further addition of the nano fillers leads the Tg values negative which indicates lower in the loss modulus.
- Higher damping value is obtained for 1%MWCNT incorporated composite rather than other Composites. Increasing the nano filler content improves the damping which indicates the damping characteristics of the nano filled polymer system.
- The micro structural analysis clearly shows that, presence of the MWCNT's in the composites which is dispersed in to the epoxy resin. Also the bonding between fibers and matrix and pulled out fibers during deformation were identified in the composites.

References

[1] Reußmann T, Mieck T P and Herz H P. *Tagung Polymer werkstoffe* in Merseburg 1998. **74** 82.
[2] Kaiser J 1989*J. Kunststoffe*.**79** 1064.
[3] Yuan X, Jayaraman K and Bhattacharyya 2004 D *Compos*.**35** 1363.
[4] Jacob M, Jose J, Jose S, Varughese KT and Thomas S 2006. *J. Appl. Polym. Sci.***117** 614.
[5] Mohamad SQ, Rani WS and Wenchao W 2010. *Compos Structure*.**93** 14.
[6] Yuanming X and Yang W J 2007. *Compos*.**35** 31.
[7] Martins MA and Mattoso LHC 2004*J. Appl. Polym. Sci*.**91** 670.
[8] Amash A and Zugenmaier P *2000 Polym. Rev*.**41** 96.
[9] Nayak SK and Mohanty S 2010 *J. Reinf. Plast. Compos*.**29** 68.
[10] Samal SK, Mohanty S and Nayak SK 2009 *J. Reinf. Plast. Compos*.**28** 47.
[11] Dipa Ray, Sarkar BK, Das S and Rana AK 2002 *Compos. Sci. Technol*.**62** 917.
[12] Andrzej K. Bledzki and Wenyang Zhang 2001 *J. Reinf. Plast. Compos*.**20** 1263.
[13] Mou'ad A. Tarawneh and Sahrim H 2011 *J. Reinf. Plast. Compos*.**30** 1745.
[14] Rajmohan T, Mohan K and Palanikumar K 2015 *J. Appl. Mech. Mater. **766** 767.*
[15] Mohan K and Rajmohan T 2017 *J. Nat. Fib*. **10** 1080.
[16] Mohan K and Rajmohan T 2018 *Mat. Today proceedings*. **5** 11628.

Developing Empirical Relationship to Predict Maximum Tensile Strength on AA 7075 CMT Welded Al Alloy

S.T.Selvamani[1,a,*], P.Govindarajan[2,b], M.Ajaymohan[3,c], S.J.Hariharan[3,d], M.Vigneshwar[1,e,f]

[a] Associate Professor, Department of Mechanical Engineering, Vel Tech Multi Tech Engineering College, Avadi, Chennai – 600062; Email id: *selva_balu@rediffmail.com

[b] UG Student, Department of Mechanical Engineering, Sri Sai Ram Institute of Technology, Tambaram, Chennai – 600044; Email id: govindarajanmech007@gmail.com

[c] UG Student, Department of Mechanical Engineering, Sri Sai Ram Institute of Technology, Tambaram, Chennai – 600044; Email id: ajaymohanraj@gmail.com

[d] UG Student, Department of Mechanical Engineering, Institute of Road and Transport Technology, Erode - 638316; Email id: harijosun@gmail.com

[e] Junior Research Fellow (JRF), Department of Mechanical Engineering, Vel Tech Multi Tech Engineering College, Avadi, Chennai – 600062.

[f] PG Scholar, Metallurgy and Materials Science Engineering, Vel Tech Dr.RR Dr.SR R&D Institute of Science and Technology, Avadi, Chennai – 600062;
Email id: vigneshwarmr@gmail.com

Abstract. Aluminum alloys are predominantly utilized by transportation industries globally to reduce the vehicle mass which in turn reduces fuel consumptions and greenhouse emissions. In case of joining these alloys into a successful product, a suitable welding technique is required. Therefore, in this research work, the Al alloys were joined using an Advanced Cold Metal Transfer (ACMT) welding machine to understand the integrity of the joints. Also to predict the maximum strength of the ACMT joints, a relationship has been developed utilizing Design of Expert (DOE).

1. Introduction

The aluminum alloys provide considerable amount of intrinsic and extrinsic properties which made this material the prime choice for manufacturers. In general, majority of the products are produced from components where welding plays a key role for manufacturers [1-2]. Many joining technologies are available in the market but it is not reliable in rapid conditions due to the dynamic changes caused by the operation. The fusion welding technologies more often produces many defects in the joints due to the high amount of heat generated during the process which resulted in poor mechanical properties that drives interest of many researchers to develop a reliable fusion welding technology to address the scenario. Finally, in 2004, FRONIUS came up with a process known as Cold Metal Transfer (CMT) welding process which is an advancement of gas metal arc welding process by controlling wire feed motion which turns to be a major breakthrough by providing enhancement in physical properties of the weldment compared to conventional fusion joining techniques [3]. The metal transfer behavior and arc characteristics of CMT process during joining was investigated by **Zhang** [4] wherein the significance of transfer modes during joining was highlighted.

Amin [5] developed a model for the cold metal transfer welding process which proclaims that this technology can be used for joining of dissimilar alloys and will produce better properties compared to the other fusion welding techniques. The dissimilar joining of Al to Mg [6] and Al to steel [7] were carried out by using CMT welding technique and sound joints were attained with better strength in the weldment. The influence of process parameters on AA6061 aluminum alloy was carried out by **Pavan kumar** [8] to determine the vulnerability in the weldment area caused by parameter attributes. The impact of heat input on the weldment area and the porosities obtained in Al-Cu alloys were investigated by **Cong Baoqiang** [9] to determine the suitable working conditions for joining the materials.

From the overview technical report under CMT welding process, it can be moralized that major research works are focused on joining the various materials and researchers are focusing on property enhancement but are not concentrating to develop a suitable working constraints for a particular material. Therefore, in this research work, the AA 7075 aluminum alloy was joined using an Advanced Cold Metal Transfer (ACMT) welding process and an empirical relationship was developed using Response Surface Methodology incorporating ANOVA design matrix to identify the maximum strength attained in the CMT joints.

2. Experimental work

A high strength light weight AA 7075 grade Al alloys comprising of zinc and magnesium as the principle alloying element was chosen as the parent metal in this investigation. The Al-Zn-Cu alloys were preferably the widely used aluminum alloys in the material processing industry due to its excellent mechanical and corrosion resistant properties. The material was joined using an advanced TPS 400i cold metal transfer welding machine which provides low thermal input to the molten pool by controlling the wire deposition rate. Also, in CMT welding process, the current drops to non-zero during the metal transfer which results in no spatter formation. The suitable working constraints for the CMT joints are presented in Table 1 and the corresponding response (output) was displayed in Table 2 respectively. The CMT welding machine and the CMT welded joints were shown in Fig. 1 (a-b) respectively.

The mechanical testing such as microhardness and tensile testing was carried out using advanced E-UTM and Vicker's hardness machine. The tensile tested specimens are displayed in Fig. 1 (c). In order to govern the tensile strength of the fabricated joints, the microstructure and macrostructure of the advanced cold metal transfer welded joints were examined to correlate it with the optimized joint. The microstructure of the ACMT weldments are examined using light optical metallurgical microscope with various magnification (Make: MEJI, Japan; Model: MIL-7100). Shown in figure1 (d).

Table 1. Suitable working constraints for CMT welding parameters

Process Parameter	Representation	Unit	Level				
			-1.68	(-1)	(0)	(+1)	+1.68
Current	D	A	81.59	85	90	95	98.40
Voltage	S	V	11.65	12	12.5	13	13.34
Welding Speed	T	mm/min	266.59	270	275	280	283.40

Table 2. The design matrix and the output response

Std. Order	Run Order	Coded Value			Tensile Strength (MPa)
		D (A)	S (V)	T (mm/min)	
1	15	-1	-1	-1	321
2	20	1	-1	-1	335
3	14	-1	1	-1	330
4	13	1	1	-1	335
5	16	-1	-1	1	346
6	19	1	-1	1	342
7	12	-1	1	1	353
8	2	1	1	1	340
9	5	-1.68	0	0	324
10	8	1.68	0	0	327
11	17	0	-1.68	0	335
12	3	0	1.68	0	341
13	10	0	0	-1.68	348
14	6	0	0	1.68	376
15	4	0	0	0	396
16	1	0	0	0	401
17	9	0	0	0	400
18	11	0	0	0	399
19	18	0	0	0	402
20	7	0	0	0	402

3. Results and Discussion

In the current work, the RSM was utilized to develop the empirical relationship to decide the quality characteristics for the ACMT joints. The responses of strength of the CMT joints are dependent of the process parameters that includes current (D) A, voltage (S) V, welding speed (T) mm/min as expressed in equation (1)

$$TS = f\{D, S, T\} \qquad (1)$$

The second-order polynomial equation was utilized to represent the tensile strength of the joints by formulating an empirical relationships and the developed equations are represented in equation (2).

$$TS = [399.019 + 0.51 * D + 1.76 * S + 7.84 * T - 2.25 * D * S - 4.5 * D * T - 0.5 * S * T - 26.11 * D^2 - 21.69 * S^2 - 13.21 * T^2]\ Mpa \qquad (2)$$

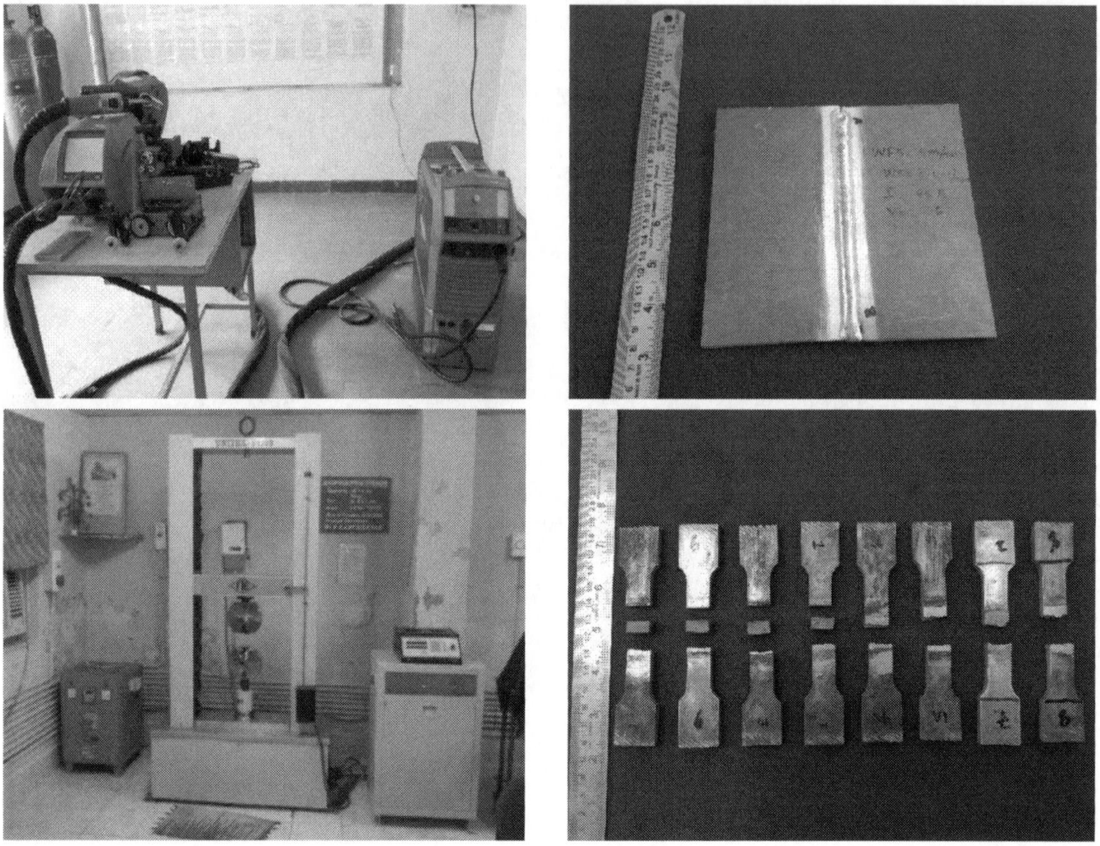

Figure 1. (a) The CMT welding apparatus (b) The CMT Joints, (c) E-UTM (INSTRON), (d) Tensile specimens (After test)

The ANOVA method is followed to predict the appropriateness of the established mathematical relationship. The level of confidence was found to be 94%. From the ANOVA design matrix, it was found that the developed model is found to be adequate and also calculated 'F' and 'R' values doesn't exceed beyond the tabulated 'F' and 'R' values. Also the residuals are falling over the straight line which can be significantly concluded that the errors are distributed normally [10].

An optical microscopic view for the CMT joints at different zones in optimized condition were shown in figure 2 (a-d). The micrographs reveals that the weld zone displays fine-equiaxed grains and also the heat affected zone is similar to 80% compared to the base metal structure. Fine grains were observed in this process due to continuous contraction (heating) and retraction (cooling) of the filler wire in the weld pool because of the wire feed mechanism. Also no defects were encountered in the joints which is because in this welding technique, the short circuit transfer takes place during the deposition of filler without the assistance of electromagnetic force and the same was observed by Zhang [4]. On the other hand, the spatter and heat input can be minimized significantly which is the main reason for achievement of enhanced properties in this process compared to other fusion welding techniques and on the contrary, it can be concluded that the strength achieved in this process is in close relation with that of the solid state friction stir welding technology [13].

IOP Conf. Series: Materials Science and Engineering **390** (2018) 012060 doi:10.1088/1757-899X/390/1/012060

Figure 2. The Micrographs of various zones (a) Macrostructure, (b) WZ, (c) HAZ, (d) Base

4. Conclusion

The detailed report of this research work is summarized as follows:

- The advanced cold metal transfer joining was successfully carried out for the AA7075 aluminum alloys
- The mathematical models are established to calculate the maximum strength for the CMT welded joints
- Very fine-equiaxed grains were observed in the fusion zone; no defects were observed throughout the weldment area
- The maximum strength was predicted to be 402 Mpa which is achieved under the welding condition of current of 90 A, voltage of 12.5 V and welding speed of 275 mm/min in the CMT welding process

Acknowledgements

The authors are grateful for the **DST-SERB** for providing funding under EEQ (F.No.EEQ/2016/000114) and also wish to thank the Chairman, **Col.R.Rangarajan**, Vel Tech Institutions for providing support and we wish to express our gratitude to **Dr.V.Balasubramanian**, Director (CEMAJOR), for providing characterization testing facilities to carry out this investigation.

References

[1] Linnert, G. E. (1965) 'Chapter 8, Welding Metallurgy', Vol 1, Aws, pp.367-396.
[2] Olson, D.L., Edwards, G.R., and Marya, S.K. (1992) 'The Physical and Chemical Behavior Associated with Slag Detachability during Welding', Ferrous Alloy Weldments, Vol 67-70, Trans Tech, pp.253-268
[3] Yang, X.R., 2006. CMT: Cold Metal Transfer MIG/MAG dip-transfer process for auto- mated applications. Arc Welding Machine 36, 5–7.
[4] Zhang, H.T., Feng, J.C., He, P., Zhang, B.B., Chen, J.M., Wang, L., 2009. The arc char- acteristics and metal transfer behavior of cold metal transfer and its use in joining aluminum to zinc-coated steel. Materials Science and Engineering A 499, 111–113.
[5] Amin S. Azar, A heat source model for cold metal transfer (CMT) welding, Journal of Thermal Analysis and Calorimetry, 2015, 122:2, 741-746.
[6] S. Madhavan, M. Kamaraj and L. Vijayaraghavan, Cold metal transfer welding of aluminium to magnesium: microstructure and mechanical properties, Science and Technology of Welding and Joining, 2016, 21:4, 310-316.
[7] Shanglu Yang, Jing Zhang, Jin Lian, Yongpin Lei, Welding of aluminum alloy to zinc coated steel by cold metal transfer, Materials and Design, 49 (2013), 602-612.
[8] N. Pavan Kumar, S. Arungalai Vendan and N. Siva Shanmugam, Investigations on the parametric effects of cold metal transfer process on the microstructural aspects in AA6061, Journal of Alloys and Compounds, 2016, 658, 255-264.
[9] Cong Baoqiang, Ouyang Ruijie, Qi Bojin and Ding Jialuo, Influence of Cold Metal Transfer Process and Its Heat Input on Weld Bead Geometry and Porosity of Aluminum-Copper alloy welds, Rare Metal Materials and Engineering, 2016, 45:3, 606-611.
[10] R. Cao, Gang Yu, J.H. Chen, Pei-Chung Wang, Cold metal transfer joining aluminum alloys-to-galvanized mild steel, Journal of Materials Processing Technology 213 (2013) 1753– 1763
[11] Shang J, Wang K, Zhou Q, Zhang DK, Huang J, Li GL. Microstructure characteristics and mechanical properties of cold metal transfer welding of Mg/Al dissimilar metals. Mater Des 2012; 34:559–65.
[12] Zhang, H.T., Liu, J.K., 2011. Microstructure characteristics and mechani- cal property of aluminum alloy/stainless steel lap joints fabricated by MIG welding–brazing process. Materials Science and Engineering A 528, 6179–6185.
[13] Beytullah Gungor, Erdinc Kaluc, Emel Taban, Aydin SIK SS, Mechanical and microstructural properties of robotic Cold Metal Transfer (CMT) welded 5083-H111 and 6082-T651 aluminum alloys, Materials and Design 54 (2014) 207–211.

Multi response Optimization of Friction stir welding process parameters in dissimilar alloys using Grey relational analysis

K Palani[1] and C Elanchezhian[2]

[1] Department of Mechanical Engineering, Sri Chandrasekharendra Saraswathi Viswa Mahavidyalaya (Deemed to be University), Kanchipuram, 631561, Tamilnadu, India
[2] Department of Mechanical Engineering, Sri Sairam Engineering College, Chennai, 600 044, Tamilnadu, India

E-mail: palani.k@kanchiuniv.ac.in

Abstract. A multi response optimization of friction stir welding parameters in dissimilar alloys using Taguchi based Grey relational analysis was studied in this investigation. The dissimilar nonferrous AA8011-H24 aluminium alloys and Ti3Al2.5V titanium alloys were friction stir welded under different process parameters, viz., rotational speed, tool pin profiles and welding speed on tensile strength, tensile elongation and joint efficiency have been optimized by applying multi response analysis. The grey relational grade is attained which correlates between the friction stir welding parameters and responses and the optimum level of welding parameters have been determined by grey analysis. The rotational speed is the most significant contributing parameter to decide the good quality of the welded joints followed welding speed and the tool pin profiles was found using analysis of variance. The confirmation test results showed that the grey relational grade was improved 0.059 with the predicted responses at the optimal conditions.

1. Introduction

The focus on dissimilar joining of aluminum and titanium alloys is considered in the potential application between low and high melting alloys in the field of aerospace and automobile industry, which has the unique properties of high strength to weight ratio, corrosion resistance, reduction of weight and cost (due to Al alloy) and improve strength (due to Ti alloy). The great challenge of the joining of aluminum and titanium alloys due to the different properties mainly, melting temperature and coefficient of heat transfer coefficient is considered in material joining industries [1]. Due to the temperature variation in melting and solidifying the metals in the fusion welding process, the brittle intermetallic compounds are formed at the interface of the dissimilar Aluminium and Titanium joints which deteriorates the strength of the welded joints [2, 3]. The conventional fusion welding process such as metal inert gas welding [4], tungsten inert gas welding [5] and laser welding [6] is carried out on the dissimilar materials. The other solid state joining processes viz., diffusion bonding [7, 8] and friction welding [9] for joining of titanium and aluminium alloys have been investigated. The newly invented solid state joining technique, The Friction stir welding (FSW) process is applied in joining of dissimilar materials to overcome the problems such as solidification shrinkage, hot cracking, improper fusion, porosity and loss of work hardening which occurred in the fusion welding process [10]. This process was initially applied to the similar aluminium alloys and developed by The Welding Institute (TWI), Cambridge in 1991 [11].

The micro structural evolution was examined in the friction stir welding (FSW) of dissimilar materials with the parameters of rotational speed, transverse speed, plunge depth, tool tilt angle, diameter ration of tool shoulder and tool pin and various pin shapes [12] effectively. Thereby the selection of FSW parameters is very critical to decide the quality of the welded joints for the dissimilar materials. The multi objective optimization techniques are used to optimize the responses of the FSW process with suitable process variables of dissimilar materials. For instance, Shanmuga Sundaram N et al. [13] successfully welded the hot worked AA2024 aluminium alloy with the cold worked AA5083 aluminium alloy with five different tool pin profiles using the friction stir welding. They developed the mathematical models which give the good correlation between the process parameters and the responses of the friction stir butt welded joints. They examined that the highest tensile elongation and tensile strength of the joints was found in tapered hexagon pin profile due to its higher pulsating effect and smooth material flow, whereas the straight cylinder pin profile gave the lower strength and elongation of the joints. Palani et al [14] discussed the effect of the friction stir welding process parameters of the AA 8011 aluminium alloys and successfully optimized the parameters using the response surface methodology. They examined that the higher relative efficiency decides the quality of the welded joints. Ghosh M et al. [15] successfully optimized the friction stir welding process parameters of the dissimilar aluminium alloys and found that due to the zone of Si rich particles in the retreating side, the interface microstructure occurred within the stir zone of the welded joints. They reported that the lower tool rotational speed and traverse speed gave the fine equiaxed grains near the AA6061 alloy interface due to proper material intermixing and dynamic crystallization of the materials.

Vijayan et al. [16] optimized the FSW process parameters of AA5083 aluminum alloy with multiple responses with grey analysis. They found that the maximum tensile strength and minimum power of the FSW joints based on the optimum levels of the process parameters welded joints. Palani et al [17] successfully applied the Taguchi method for the optimization of multiple responses of the friction stir welded AA8011 aluminium alloys. They reported the highest signal to noise ratio was the key factor to decide the quality of the welded joints. Kasman [18] optimized the multiple responses for the dissimilar FSW process of AA6082-AA5754 aluminum alloys successfully by Taguchi based grey analysis. In the present investigation, the attempt has been carried out on the multi response optimization of friction stir welding process parameters in dissimilar non-ferrous alloys of Ti3Al2.5V titanium alloy and AA8011 aluminium alloy using Taguchi based grey relational analysis.

2. Grey relational analysis

The quality of the welded joints is decided by the proper selection of parameters of the friction stir welding process. The optimum settings of process parameters are mainly focused to enhance the properties of the joints using different statistical approaches. In this work, the Taguchi method was applied to conduct the experiments based on L_{27} orthogonal array design and to improve the fabricated joint properties of the welded joints using friction stir welding process on two dissimilar non-ferrous alloys. The objective of the work is to optimize the friction stir welding parameters on the multiple responses such as, tensile strength, tensile elongation and joint efficiency. The multiple responses of the dissimilar friction stir welding process have considered as the higher-the-better- performance characteristics using the Taguchi method. The Grey relational analysis was applied to improve the all the response characteristics without degrading the performance of the joints [19]. The following steps are considered to find the optimum responses based on the grey relational grade using Taguchi based grey relational analysis.

2.1 Normalization of the responses

In the grey relational analysis, data preprocessing was initially made in order to normalize the experimental data and the linear normalization of the signal to noise ratio was attained in the range between 0 and 1 for further analysis [20]. In this study, the larger-the-better characteristic was considered for the responses of the dissimilar friction stir welding, such as tensile strength, tensile elongation and joint efficiency and these can be expressed as follows:

$$x_{ij} = \frac{\eta_{ij} - \min_j \eta_{ij}}{\max_j \eta_{ij} - \min_j \eta_{ij}} \tag{1}$$

Where, η_{ij} is the original sequence for the ith response in the jth experiment, and x_{ij} is the normalized value of the ith response in the jth experiment. The normalized response is closer to one which gives the best experimental results according to Deng grey rule [21].

IOP Conf. Series: Materials Science and Engineering **390** (2018) 012061 doi:10.1088/1757-899X/390/1/012061

2.2 Calculation of Grey relational Coefficient

After normalizing the responses, the grey relational coefficient was determined to give the relationship between the best and actual normalized responses and can be expressed as:

$$\xi_{ij} = \frac{\Delta \min_j + \zeta \Delta \max_j}{\Delta_{oij} + \zeta \Delta \max_j} \tag{2}$$

Where, Δ_{oij} is the normalized response to the ith response in the jth experiment and ζ is the distinguishing coefficient, which is considered as 0.5 for the analysis. The table 3 shows the grey relational coefficient for each response characteristic.

2.3 Determination of Grey relational grade

The average grey relational grade was computed for the each response of the friction stir welding process and expressed as:

$$\alpha_j = \frac{1}{n} \sum_{i=1}^{n} \xi_{ij} \tag{3}$$

Where, n is the number of responses and α_j is the grey relational grade for all the responses of the welded joints. The experiment 7 shows the highest grey relational grade, which is closer to the ideally normalized response.

3. Experimental work

The titanium alloy Ti3Al2.5V alloy and AA8011 aluminium alloy [22] (Procured from Kataria Steels, Mumbai) were used in this investigation with the dimensions of 100 mm x 100 mm x 6 mm and used for butt joint configuration using friction stir welding process. The chemical composition of the AA8011 aluminium alloy and Ti3Al2.5V titanium alloy are presented in Table 1. The plates for investigation initially washed with an ultrasonic bath sonicator and the butt surfaces were cleaned with acetone. The dissimilar butt welding was carried out using Computerized Friction stir welding machine (Fabricated by Universal Industries, Model AMS 400S, Kanchipuram) shown in Figure 1 and the welding operations was performed by clamping the plates properly with the fixture.

Table 1. Chemical composition of the AA8011-H24 Aluminium alloy and Ti3Al2.5V Titanium alloy

Chemical composition (Wt.%) of Aluminium alloy (AA8011-H24)				
Fe – 0.74%	Si – 0.56%	Mn – 0.10	Mg – 0.08	Cu – 0.13
Zn – 0.08	Cr – 0.10%	Ti – 0.05	Remaining Al	
Chemical composition (Wt.%) of Titanium alloy (Ti3Al2.5V)				
Fe – 0.30%	Si – 0.14%	Al – 2.9%	V – 2.5%	O – 0.15%
C – 0.05%	H – 0.02%	N – 0.02%	Remaining Ti	

The newly fabricated friction stir welding tools made of M2 tool steel (Manufactured by SRS Diamond tools, Chennai) with the dimensions of 18 mm shoulder diameter, 6 mm pin diameter and 5.7 mm pin length were used to produce the friction butt welded joints. The three different tool pin profiles, namely Straight Square, Straight Pentagon and Straight Hexagon pin profiles were used to make the butt welded joints of dissimilar alloys shown in Figure 2. The quality of the weld was decided by proper selection of process parameters of the friction stir welding process.

Table 2. Friction stir welding parameters and their levels

Control process parameters	**Unit**	**Symbol**	**Levels**		
			1	**2**	**3**
Rotational Speed	r/min	A	300	350	400
Tool pin profile	---	B	1 (Square)	2 (Pentagon	3 (Hexagon)
Welding Speed	mm/min	C	30	35	40

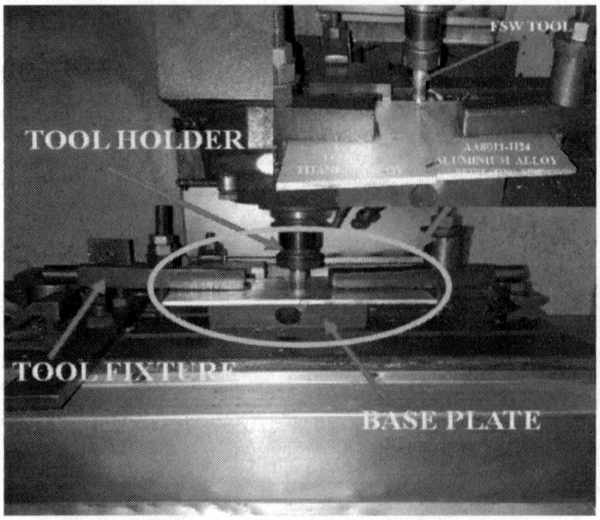

Figure 1. Friction stir welding (FSW) machine

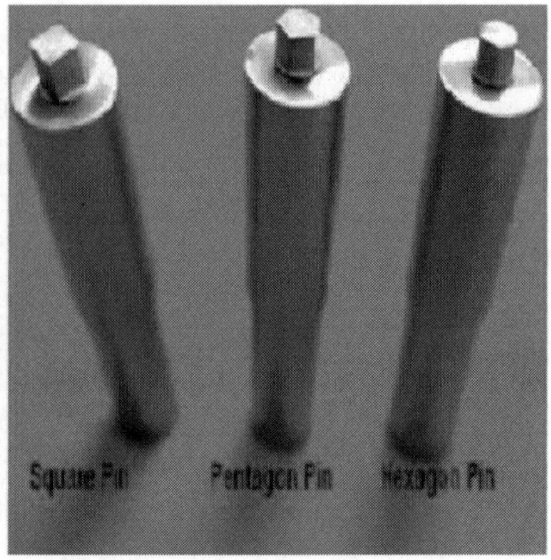

Figure 2. FSW tool pin profiles

The FSW parameters of tool rotational speed, tool pin profiles and welding speed were used in these experiments based on the trial experiments [23] and study made in the previous literature work. Table 2 shows the FSW parameters and their levels, which are used to decide the design matrix for the experiments. The three factors, three levels of the L27 orthogonal array was applied to conduct the experiments based on the Taguchi method. The tensile specimens were sliced perpendicular to the welded direction using a power hacksaw and machined as per ASTM-E08 standards and which were tested in the Omega Inspection and Analytical Laboratory, Chennai).

4. Result and Discussions

The analysis of Variance (ANOVA) for the responses of tensile strength, tensile elongation and joint efficiency in dissimilar friction stir welding of titanium alloy Ti3Al2.5V alloy and AA8011 aluminium alloy are represented in Table 3 and it illustrates the most dominant contributing factors of tool rotational speed affecting the tensile strength (85.10%) and joint efficiency (85.15%), whereas for tensile elongation, the welding speed (38.72%) is the most dominant factor followed by rotational speed (20.31%) and tool pin profiles (16.04%) of the welded joints. These results conflict with tensile strength, tensile elongation and joint efficiency in friction stir welding of dissimilar non-ferrous alloys. The multiple response analysis is very important to decide the good quality of welded joints based on the proper selection parameter in friction stir welding process.

The optimization of the parameters for the tensile strength, tensile elongation and joint efficiency was done by using the Taguchi based grey relational analysis. The normalized responses, grey relational coefficients of the responses, average grey relational grade, and its rank for each experiment are shown in Table 4. The maximum average grey relational grade gives the better multiple performance in dissimilar friction stir welding process which is closer to the ideal responses.

Table 3. Analysis of Variance (ANOVA) for the responses

Tensile Strength (Mpa)					
FSW Parameters	**Degrees of Freedom**	**Sum of squares**	**Mean squares**	**F-Value**	**Contribution (%)**
Rotational speed	2	1191.35	595.68	154.13	85.1
Tool pin profile	2	56.15	28.08	7.26	4.01
Welding speed	2	75.11	37.55	9.72	5.37
Error	20	77.29	3.86		5.52
Total	26	1399.91			100

Tensile Elongation (%)					
FSW Parameters	**Degrees of Freedom**	**Sum of squares**	**Mean squares**	**F-Value**	**Contribution (%)**
Rotational speed	2	3.916	1.958	8.14	20.31
Tool pin profile	2	3.092	1.546	6.43	16.04
Welding speed	2	7.465	3.733	15.52	38.72
Error	20	4.808	0.2404		24.93
Total	26	19.281			100

Joint Efficiency (%)					
FSW Parameters	**Degrees of Freedom**	**Sum of squares**	**Mean squares**	**F-Value**	**Contribution (%)**
Rotational speed	2	30.992	15.496	153.15	85.15
Tool pin profile	2	1.452	0.726	7.17	3.99
Welding speed	2	1.928	0.964	9.54	5.30
Error	20	2.024	0.101		5.56
Total	26	36.396			100

Table 4. Data preprocessing, Grey Relational coefficient and Grey relational Grade with Order

Exp. No	Tensile Strength (TS) (MPa)	Tensile Elongation (TE) (%)	Joint Efficiency (JE) (%)	Grey Relational Coefficient			Grey Relational Grade	Rank
				TS	TE	JE		
1	0.877	0.064	0.876	0.802	0.348	0.802	0.651	6
2	0.585	0.000	0.583	0.546	0.333	0.545	0.475	20
3	0.854	0.329	0.852	0.774	0.427	0.772	0.658	5
4	0.777	0.521	0.776	0.691	0.511	0.691	0.631	9
5	0.705	0.493	0.707	0.629	0.496	0.631	0.585	11
6	0.604	0.764	0.605	0.558	0.680	0.559	0.599	10
7	1.000	0.671	1.000	1.000	0.603	1.000	0.868	1
8	0.685	0.193	0.683	0.613	0.383	0.612	0.536	15
9	0.804	0.493	0.802	0.718	0.496	0.717	0.644	7
10	0.615	0.632	0.614	0.565	0.576	0.565	0.569	12
11	0.478	0.296	0.486	0.489	0.415	0.493	0.466	21
12	0.688	0.936	0.688	0.616	0.886	0.616	0.706	2
13	0.458	0.618	0.457	0.480	0.567	0.479	0.509	19
14	0.346	0.421	0.348	0.433	0.464	0.434	0.444	23
15	0.488	1.000	0.488	0.494	1.000	0.494	0.663	4
16	0.681	0.764	0.681	0.610	0.680	0.610	0.633	8
17	0.615	0.582	0.614	0.565	0.545	0.565	0.558	13
18	0.612	0.993	0.612	0.563	0.986	0.563	0.704	3
19	0.177	0.493	0.179	0.378	0.496	0.378	0.418	25
20	0.000	0.343	0.000	0.333	0.432	0.333	0.366	27
21	0.112	0.971	0.112	0.360	0.946	0.360	0.555	14
22	0.219	0.857	0.219	0.390	0.778	0.390	0.520	18
23	0.092	0.754	0.093	0.355	0.670	0.355	0.460	22
24	0.115	0.907	0.117	0.361	0.843	0.361	0.522	17
25	0.292	0.843	0.293	0.414	0.761	0.414	0.530	16
26	0.177	0.179	0.179	0.378	0.378	0.378	0.378	26
27	0.162	0.643	0.162	0.374	0.583	0.374	0.444	24

The best multiple performance was achieved for the experiment 7 which has the highest grey relational grade compared to the other experimental responses. Figure 3 shows the Pareto chart for the average grey relational grade for the welded joints. It is clearly shown that the combined effort of the rotational speed and welding speed decide the 97% of the quality of the butt welded joints and similar work has been done by Palanikumar K et al. [19]. In addition, the total mean of the Grey relational grade is also calculated and given in Table 5.

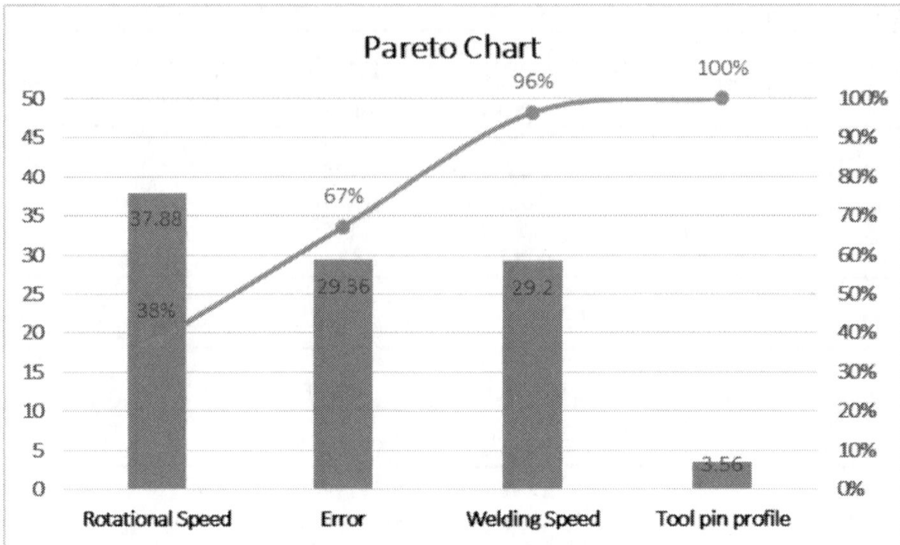

Figure 3. Pareto Chart for Grey relational Grade

The most dominant parameters based on the grey relational grade and its rank are shown in Table 5 in dissimilar friction stir welding of AA8011-H24 Aluminium alloy and Ti3Al2.5V Titanium alloy joints and Figure 4 shows the grey relational grade graph for each parameter of friction stir welding at different levels respectively. It shows that the low rotational speed, higher welding speed with the pentagon pin profile are the optimum friction stir welding parameters of the dissimilar butt joints and the average grey relational grade for all the experimental responses are shown as the dotted line in the graph and similar report as Rajyalakshmi G et al. [24].

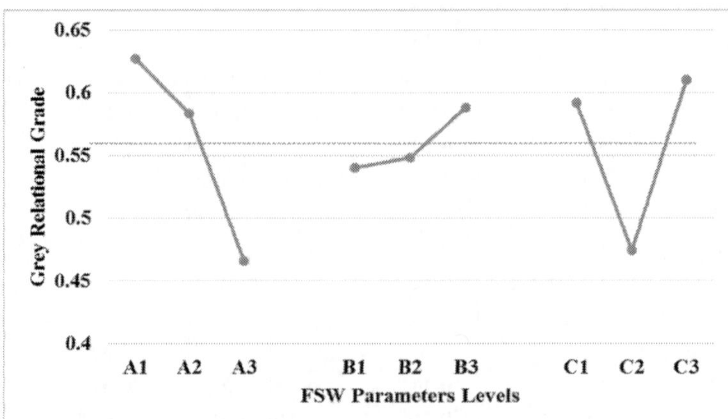

Figure 4. Response graph of grey relational grade

The confirmatory test was conducted for the optimal levels of the parameters of the welding process to improve the multiple responses of the welded joints.

$$\alpha_{Predicted} = \alpha_m + \sum_{i=1}^{N} (\alpha_0 - \alpha_m) \quad (4)$$

Where, α_m is the average grey relational grade for all the responses, N is the number of responses and α_0 is the average grey relational grade at the optimal level.

Table 5 Results of friction stir welding responses using the initial and optimal levels

	Initial FSW Parameters	**Optimal FSW Parameters**	
		Experimental	**Predicted**
Optimum FSW Setting Levels	$A_1B_1C_1$	$A_1B_3C_3$	$A_1B_3C_3$
Rotational Speed (r/min)	300	300	-
Tool Pin Profile	3	3	-
Welding Speed (mm/min)	30	40	-
Grey Relational Grade	0.651	0.710	0.7082

Improvement in grey relational grade: 0.059

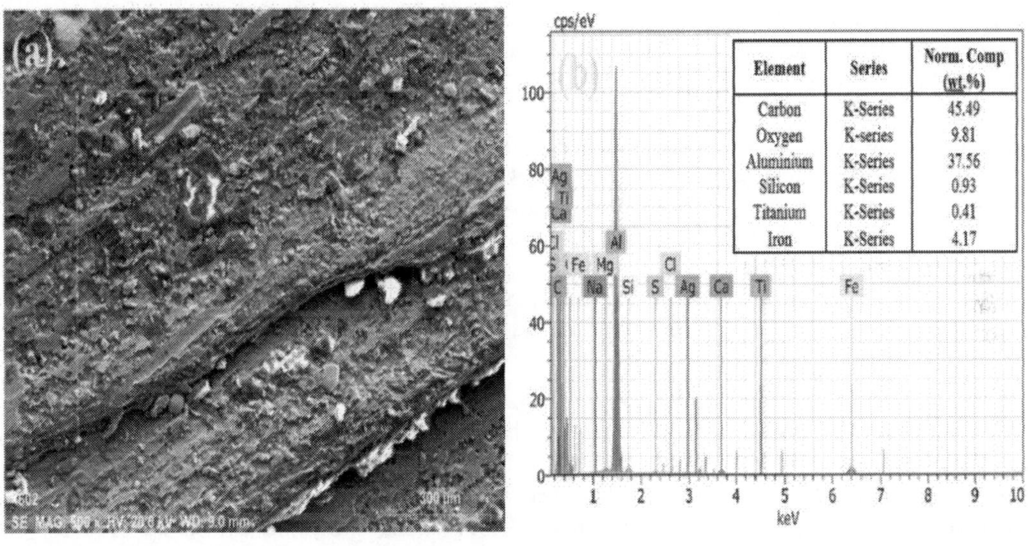

Figure 5. SEM micrograph and EDS image of the optimum process settings at the weld Interface

The comparisons of predicted and actual welding responses for multiple performance at optimal levels are shown in Table 5. There is an improvement of 0.059 of the predicted responses at optimal levels ($A_1B_3C_3$) compared with the initial levels ($A_1B_1C_1$) and the experimental response at the optimum level is much closer to the predicted responses at the optimum level was found and similar comparison done by Datta S et al. [25]. It is clearly evident from the multiple performance characteristics in the dissimilar friction stir welding process are improved through the Taguchi grey relational analysis. The microstructure of dissimilar friction stir welded joints of aluminium alloy and titanium alloy in the weld interface at the optimum process condition shown in figure 5 and showed that that the proper mixing and material flow of the material which initiates the plastic deformation between the these two nonferrous alloys using friction stir welding process.

5. Conclusions

- The Taguchi based grey relational analysis was successfully applied to optimize the multiple responses in the dissimilar friction stir welding of AA8011-H24 Aluminium alloy and Ti3Al2.5V Titanium alloy joints
- There is a significant improvement of 0.059 was found based on the grey relational grade at the optimal levels of the parameters of the dissimilar friction stir welding process
- The tool rotational speed is the dominant factor to decide the good quality of the welded joints followed welding speed and tool pin profile based on the analysis of variance on the grey relational grade.
- The microstructure of the friction stir welded joint at the weld interface for the optimum process setting showed that the material mixing was properly done due to the friction between the tool and butt surfaces of the base materials.

6. References

[1] Yu-hua C, Quan NI and Li-ming K 2012 *Trans. Nonferrous Met. Soc. China 22* (2) 299–304
[2] Bang K, Lee K, Bang H and Bang H 2011 *Mater. Trans* **52** (5) 974-78
[3] Aonuma M and Nakata K 2011 *Mater. Trans* **52** (5) 948-52
[4] Kim T, Lee J, Min B, Yoo D and Kim C 2007 *J. Elect. Eng. Tech* **2** (3) 366-72
[5] Lakshminarayanan AK and Balasubramanian V 2011 *J. Mater. Eng. Perf* **20** 82-89
[6] Vaidya WV, Horstmann M, Ventzke V, Petrovski B, Koc M, Kocik R and Tempus G 2010 *J. Mater. Sci* **45** 6242-54
[7] Wei Y, Aiping W, Guisheng Z and Jialie R 2008 *Mater. Let* **62** 2836-39
[8] Ilangovan M, Boopathy S R and Balasubramanian V 2015 *Def. Tech* **11** (2) 174–184
[9] Kimura M, Nakamura S, Kusaka M, Seo K and Fuji A 2005 *Sci. Tech. Weld. Join* **10** (6) 666-73
[10] Chen Y, Yu L and Ni Q 2012 *Adv. Mater. Res* **413** 439-43
[11] Rajakumar S and Balasubramanian V 2015 *Int. J. Adv. Manuf. Tech* **86** (1-4) 1095-1112
[12] Ma Z, Wang C, Yu H, Yan J and Shen H 2013 *Mater. Des* **45** 72–79
[13] Sundaram NS and Murugan N 2010 *Mater. Des 31* (9), 4184-93
[14] Palani K and Elanchezhian C 2015 *Appl. Mech. Mater* **813** 451-55
[15] Ghosh M, Kumar K, Kailas SV and Ray AK 2010 *Mater. Des* **31** (6), 3033-37
[16] Vijayan S, Raju R and Rao SRK *Mater. Manuf. Proc* **25** 1206-12
[17] Palani K and Elanchezhian C 2015 *Appl. Mech. Mater* **766** 921-27
[18] Şefika Kasman 2013 Int. J. Adv. Manuf. Tech **68** 795–804
[19] Palanikumar K, Karunamoorthy L and Karthikeyan R 2006 *Mat. Manuf. Proc* **21** (8) 846-52
[20] Ranganathan S and Senthilvelan T 2011 *Int. J. Adv. Manuf. Technol.* (2011) **56** 455-62
[21] Gagliardi F, Ciancio C and Ambrogio G 2017 *Int. J. Adv. Manuf. Technol.* **94** (1-4) 719-28
[22] Palani K, Elanchezhian C, Ramnath BV, Bhaskar GB Jagadeesh JS and Kumar GM 2017 *Int. J. Res. App. Sci. Eng. Tech* **5** 437–442
[23] Palani K and Elanchezhian C 2015 *Appl. Mech. Mater* **813** 446-450
[24] Rajyalakshmi G and Venkata Ramaiah G 2013 *Int. J. Adv. Manuf. Technol.* **69** (5-8) 1249-62
[25] Datta S, Bandyopadhyay A and Pal PK 2008 *Int. J. Adv. Manuf. Technol.* **39** (11-12) 1136-43

IOP Conf. Series: Materials Science and Engineering **390** (2018) 012062 doi:10.1088/1757-899X/390/1/012062

REVIEW ON NATURAL FIBER COMPOSITES

B.VIJAYA RAMNATH*, R.ARVIND, I.DINESH, M. HARI PRASADH
Department of Mechanical Engineering
Sri Sairam Engineering College, West Tambaram, Chennai-44.

***Corresponding author mail ID: vijayaramnathb@gmail.com**

Abstract. In recent years more research is carried out on natural fiber composites due to their low weight and reusable property and improved mechanical characteristics. The natural fiber based composites are economical and can be fabricated easily in places fibers are abundantly available. Such composites are recyclable and are nontoxic in nature. They are replacing rapidly synthetic fiber based composites in various engineering applications.. This paper just reviews the recent works carried out on natural fiber based composites.

1.0 Review on natural fiber composites

A composite material made from two or more constituent materials with significantly different physical or chemical properties that when combined produces a material with characteristics different from individual component.

2.0 Reviews on kenaf fiber

Majid Niaz Akhtara et al. [1] studied two different types of fibers such as chemically treated kenaf fiber and untreated kenaf fibers. Polypropylene was reinforced with treated and untreated kenaf fibers. 10%, 20%,30%,40%,50% weight of kenaf composites were prepared during fabrication. The process used for fabrication of kenaf/pp reinforced composite was Injection molding process. Testing includes X-ray diffraction, Differential scanning calorimetric, Thermo gravimetric analysis, Fourier transform infrared spectroscopy, Scanning electron microscopy. It is found that chemically treated kenaf/pp composites with 40% wt loading showed the improved mechanical properties(tensile strength, flexural strength) when compared to untreated kenaf/pp reinforced composite and other loadings such as 10% , 20 % , 30 %, 50% wt of kenaf/pp reinforced composite. In testing process, it was found that under 50% wt loading burning of kenaf fibers occurred and there was no proper mixing of kenaf and polypropylene.

Neng Sri Suhartya et al. [2] studied two different reinforcements such as primary and secondary were made. Here, recycled polypropylene (RPP) was used as a matrix, Divinyl benzene (DVB) as a cross linker, PP-g-AA as a bonding agent as a primary reinforcement. Whenever these materials were bonded together, it was found that the tensile strength and flexural strength percentage was increased gradually. Clay or Kenaf fiber was used as secondary reinforcement. When 20% kenaf fiber was added as a secondary reinforcement to the RPP/DVB/PP-g-AA/hall composite it was found that tensile strength was once again improved gradually approximately to 20% and flexural strength was increased by 30% and impact toughness by 28%. Synthesis process was carried by mixing RPP/DVB/PP-g-AA/hall composite with boiled Xylene solvent, mechanical stirrer and nitrogen gas for 1 hour. Tensile testing and flexural testing were carried out using universal testing machine and Izod impact test machine. It was found that addition of kenaf fiber with RPP/DVB/PP-g-AA/hall composite increased the tensile, flexural strength and impact toughness.

2.1.Hybrid composites Vijaya Ramnath et al. [3] dealt the paper on Hybrid Composites consisting of Flax and Abaca and Glass fiber Reinforce Polymer in which Flax and Abaca is a matrix GFRP is a reinforcement. Among other fibers Flax proved to have improved Mechanical properties when combined with glass. It is found that Abaca+Flax+GFRP found to have improved Flexural properties and mechanical properties. Satish et al. [4] studied the Properties of Hybrid composites varying the mechanical properties and thermal properties when orientation of fibers. Here Banana Kenaf reinforced with GFRP is studied. Reinforcement agent is epoxy resin and hardener. Fabrication is done using Hand layup method. It was found that fibers having 45degree orientation produced higher mechanical and thermal properties.SEM is used to analyses internal structure.

2.1.1 Composites and polymers Vijaya Ramnath et al. [5] deals with selection of fibers play a major role in day to day implementation. Epoxy resin is used as a matrix and manila as a reinforcing agent which is placed between glass fibers. Manila is placed alternatively by using Hand layup technique. The bonding between fiber and glass plays a major role on tensile strength. It is performed in three samples and average is noted down. Mohamed Hafiz Zemra et al. [6] used Kenaf fiber was used as reinforcement and unsaturated polyester resin was used as a matrix. The process used for fabrication is thermoset pultrusion process in which Kenaf fiber was first impregnated on unsaturated polyester resin. Then the Kenaf fiber impregnated with resin was pulled out at particular speed from steel die pulling machine. Thus the fiber pulled was cured at a temperature of 120°. Then, Pultruded kenaf fiber reinforced composite was parted into different sizes of 1400,2200,3300. Compression and flexural properties were studied. Compression and flexural strength were increased with the decrease in Tex number and it was found that tensile strength increased with increase in tex number.

2.1.2. Properties of natural fiber in Acoustics Lima et al. [7] used Kenaf fiber is used to absorb the sound energy. When the amount of fiber is more in the specimen, the absorption coefficient also increases according to it. It was tested using Natural incidence sound absorption where the specimen was fitted to the one end of the impedance tube and random incidence sound absorption where kenaf fibers were made into sheets of size 0.43m*0.43m. In natural incidence sound absorption method, the fiber was made into required thickness and diameter by hot compression method. In random incidence sound absorption the sheets were placed in the frame by providing the air gaps of three distances 10mm, 20mm, 30mm. It was found that there was an increase in absorption coefficient whenever there was a increase in thickness and bulk density. It was found that when the air gap was increased, the absorption was increased at lower frequencies.

2.1.3. Short fibers and long fibers Reza Mahjoub et al. [8] studied the influence of short fibers and long fibers as reinforcements. Polyethylene and polypropylene were used as matrix. Kenaf fiber was used as a reinforcement and maleic anhydride polyethylene (MAPE) and maleic anhydride polypropylene e(MAPP) were used as coupling agents. When short fibers were treated with different alkaline composition such as 2%, 3%, 6% and 9% of NaOH, it was found that tensile strength and modulus of elasticity increased at 6% NaOH alkaline treatment when compared with untreated kenaf fiber. When alkaline treated kenaf fiber was bonded with MAPP, it was found that composite can able to withstand the higher loads (impact loads) when compared with Kenaf fiber reinforced with MAPE.

Shifeng Zhang et al. [9] chopped Kenaf into total of 100 gms in 50mm length. Kenaf fiber was treated with NaOH solution (5%-10%). This retted fiber was treated with ALCL3 solution which involved heating to a particular temperature, cooling process and drying for more than 2 hours. Fabrication process was carried out by forming the bond between treated and untreated kenaf fibers with unsaturated polyester and t-BP catalyst using Vacuum Assisted Resin Transfer Molding (VARTM). It was found that thickness swelling percentage varied with time. Tensile test was carried out using Universal testing machine. Mechanical properties like modulus of rigidity and modulus of elasticity were tested as per standards. Such properties increased by certain percentage when aluminium was added by (2-4%). Hence Treated kenaf fibers can replace the glass fibers in automotive applications.

2.1.4. Natural fiber composites in construction Mohd Suhard Meona et al. [10] used Natural fibers in construction of buildings where the concrete was wrapped up with Kenaf fiber reinforced polymers. Glass fiber reinforced polymer had higher tensile strength compared to the KFRP and has high modulus of rigidity and ductility. Initially kenaf was chemically treated and the polyester resin was reinforced with kenaf fiber. Improvements on Kenaf fibers could be made by increasing the amount of fibers and thickness and amount of wrappings in the concrete. Axial load was applied on the concrete wrapping. Testing on unwrapped samples, samples wrapped with GFRP and samples wrapped with KFRP were made and improvements were studied. Researches started to focus on hybrid composites.

2.1.5. Study of Mechanical and Thermal properties of natural fiber composites

Jamal Mirbagheri et al. [11] used hybrid composite consisted of wood flour/ kenaf fiber reinforced with polypropylene matrix. Bonding agents such as maleic anhydride and cumylin peroxide were used. There were two reinforcement and the ratio was carried out as fixed ratio which is reinforcement1 and reinforcement 2. The process carried out for fabrication of sample was injection molding process. Tensile test was carried out by using Universal testing machine as per standards. Rule of mixtures equation was used to find out the elastic modulus of composites. They identified that addition of kenaf fiber to wood flour/pp matrix composite increased the property of tensile strength and tensile modulus. Edynoor bin osman et al. [12] studied about morphological and electrical properties of Composites that are used in electrical fields. Kenaf fiber was treated with (5-10%) of NaOH solution. Indium/zinc solution was prepared by dissolving zinc acetate dehydrate in ethanol and stirred mechanically for number of hours and then finally Indium solution was added to the mixture. Indium/zinc oxide solution was coated over the layer of kenaf fiber. The process carried out was dip coating method. After that annealing process was carried out at 150-200°. Scanning Electron Microscope was used for testing process in order to ensure whether the indium/zinc oxide solution was properly coated entirely. Electrical conductivity of this composite is entirely high and it is used in textile industries, and for electrostatic coating protection. Dunne et al. [13] discussed about Material charecterization of natural fiber based Composites are environmental friendly. Some of them include Kenaf and Sisal in which kenaf is chosen because it has been used in automotive applications and used in industries. Sisal is chosen because it has been used in making coir's and twines. Hybrid composite was fabricated by using kenaf/sisal as the reinforcement and Acrylonitrile Butadiene Styrene matrix (ABS) immersed in acetone solution which acted a bonding agent. The bonding was so strong that it increased the mechanical properties such as tensile strength and elastic modulus. Testing such as tensile testing, air flow resistivity, and impact strength with different densities were measured. There was an increase in tensile strength, impact strength and found automotive applications.

Srinivasan et al. [14] discussed about increasing mechanical and thermal properties based on types of fibers such as banana and flax reinforced with glassfiber.Epoxy resin of particular quantity such as certain percent increases the property. Hand layup is used to fabricate the fibers reinforced with epoxy resin glass fiber. It is found that thermal stability has been increased. Testing such as SEM is used. Chaithanyan et al. [15] investigated about tensile behaviour of Fibers such as Sisal is mixed with vinyl ester resin which is fabricated using hand layup method.GFRP is used with particular weight is used in this process.Different composition of Fiber and resin is mixed and different results are generated. Testing is carried out using universal testing machine. As a result it is found that it has high specific strength, good stiffness and light weight. Vijaya Ramnath et al. [16] investigated about the mechanical behaviour of Twisted Natural Fiber Hybrid Composite and Fabrication is done by using Vacuum assisted Compression moulding technique. For the improvement of mechanical properties fibres are also alkalized. To enhance the strength and stiffness of laminates two fibres are sandwiched between the layers of glass fibres It shows that due to the presence of twisted fibres there is an significant improvement in its mechanical properties.

3.0 A Review on sisal fiber

Felipe dos santos et al. [17] Studied about the sisal fibre when it was mixed to the cementitious composite along with Ethylene Vinyl Acetate (EVA) and piassava fibre was mixed to cementitous composite along with EVA and then the mechanical properties of both were compared. Workability, tensile strength and compression test were made. Specimens were prepared by adding required % of EVA and fibre (sisal or piassava) to the cementitious material. Fibers were cut into the required length. The tensile strength was tested using a Universal servo controlled machine. Compressive strength was tested using hydraulic press. Workability was tested in a flow table device. The sisal fiber reinforcement in light weight cementitious composite with EVA was more efficient than piassava fiber. Fláviode Andrade Silva et al. [18] studied the fatigue behaviour of sisal fiber when it was used as reinforcement to cementitious material. The composite was fabricated using the molding process. The sisal fiber was cut into the required length and then the fiber is mixed to the cementitious material. Fatigue test was carried out and the micro structural investigation was done. The composite did not undergo fatigue when the composite was subjected to maximum stress. Ashishkumre et al. [19] fabricated a composite by using sisal as reinforcement and epoxy as matrix. The fabrication of composite was done using hand layup method. Tensile test was done using Universal testing machine. Flexural test was also done using Universal testing machine. Impact test was done using impact machine. They found that by the addition of glass to sisal fibre-epoxy polymer composite, the mechanical property increased. Tensile strength of epoxy remained same by the reinforcement of sisal but the tensile modulus, flexural and impact properties increased.

Mohan et al. [20] studied the chemical treatment of sisal fibre. The composite (sisal-pp composite) was fabricated using the extrusion method. The three types of specimen were made, one was made using the untreated sisal fibre and second one was using chemically (Na OH) treated sisal fiber and the third one was using chemically-clay (Na OH-clay) treated sisal fiber. Moisture absorption testing for the three specimens was conducted. Fibre pullout test for all the three specimens were done as per the standards. Tensile test was conducted in universal testing machine. The researchers found that the chemically treated sisal fibre had greater thermal, physical and mechanical properties than the untreated sisal fibre. Whereas the chemically-clay treated sisal fibre had much greater thermal, physical and mechanical properties than chemically treated sisal fiber. Easwar Prasad et al. [21] tested the Impact strength of both treated (4-10% Na OH) and untreated sisal polyester composites. Sisal fibres were cut into 10mm length. 2mm, 3mm,4mm,5mm and 6mm thick specimens with fiber volume of 10% ,15%, 20%, 25% and 30%. The process used for making specimen was manually operated hot compression moulding. Izod impact strength of each specimen was tested according to ASTM standard D-256. After the testing it was found that the impact strength of both treated and untreated sisal fibre embedded polyester matrix composite increase with increase in thickness and volume fraction of the fibre. Yuvaraj et al. [22] investigated flexural and water absorption behaviour of epoxy hybrid composites in which Epoxy is used as a resin and hand layup method was used to make up the composites. This results in improved mechanical properties for glass fiber- sisal fiber construction than dispersed construction. .It seems that there is a slight increase in flexural property of composites with increase in fibre loading and also there is an decrease in water absorption property with increase in glass fibre composition in composites. Yuvaraj et al. [23] studied mechanical behaviour of Sisal epoxy Hybrid composites which shows that the breaking load of double shear test, hardness, inter delamination tests are increased It seems that the fiber is fabricated by hand layup process. The results prove that the composite having equal proportion of glass and sisal enhanced the mechanical properties when compared with composite having minimum sisal fibre proportion.

4.0 Conclusion

In this paper the review has been done based purely on natural fiber composite such as Kenaf and sisal. It is observed that fibers may enhance the physical and mechanical properties by chemical treatment. It shall certainly be applied in various fields such as automotive, construction and for domestic applications.

References

[1] Majid Niaz Akhtara, Abu Bakar Sulongb, M.K. Fadzly Radzib, N.F. Ismailb, M.R. Razab, Norhamidi Muhamadb, Muhammad Azhar Khane, Influence of alkaline treatment and fiber loading on the physical and mechanical properties of kenaf/polypropylene composites for variety of applications, *Progress in Natural Science: Materials International*, 26 (2016), 657-664 .

[2]Neng Sri Suhartya, Hanafi Ismailb, Kuncoro Diharjoc, Desi Suci Handayania, and Maulidan Firdausa ,Effect of Kenaf Fiber as a Reinforcement on the Tensile, Flexural Strength and Impact Toughness Properties of Recycled Polypropylene/Halloysite Composites, *Procedia Chemistry*, 19(2016), 253 – 258.

[3]B.VijayaRamnath,C.Elanchezian,C.VJayakumar,Manickavasagam,S.Ashwin,H.Eashwar,P.Kavirajan,D.Murugan,Investigation of flexural characteristics of flax and abaca hybrid epoxy composites,*AppliedMechanics and Materials*,766-767(2015),183-186

[4]P.Sathish,R.Kesavan,B.Vijaya Ramnath,C.Vishal,Effect of fiber orientation and stacking sequence on mechanical and thermal characteristics of banana kenaf hybrid epoxy composites,*Silicon ISSN* 1876-990X Silicon.

[5] Vijaya Ramnath B, Manickavasagam V.M Rajendra Prasad A , Elanchezhian C , Kavin C Karthik Subramanian Rahul V , Investigation of Tensile Properties of Manila Fibre Reinforced Composite , *Applied Mechanics and Materials* ,766-767 (2015) ,96-99.

[6]Mohd Hafiz Zamria, HazizanMdAkila,b, ZainalAriffinMohdIshaka,b, Pultruded Kenaf Fibre Reinforced Composites: Effect of Different Kenaf Fibre Yarn Tex , *Procedia Chemistry*, 19 (2016) 577 – 585

[7]Z.Y. Lima, A. Putraa,⁎, M.J.M. Nora, M.Y. Yaakobb ,Sound absorption performance of natural kenaf fibres ,applied acoustics, 130 (2018), 107-114.

[8]RezaMahjoub,Jamaludin MohamadYatim[b], Abdul RahmanMohd Sam, Sayed HamidHashemi ,Tensile properties of kenaf fiber due to various conditions of chemical fiber surface modifications, *construction and building materials*, 55 (2014), 103-113.

[9] ShifengZhang, SheldoN,Q.Shi, LipingCai, JonathanHuang, Property enhancement of kenaf fiber reinforced composites by in situ aluminum hydroxide impregnation, *Industrial Crops and Products* ,79 (2016), 131-136.

[10]Mohd Suhairil Meona, , Muhamad Fauzi Othmana, Hazran Husaina, Muhammad Fairuz Remelia, Mohd Syahar Mohd Syawala ,Improving tensile properties of kenaf fibers treated with sodium hydroxide, *Procedia Engineering* , 41 (2012) ,1587 – 1592.

[11]Jamal Mirbagheri, Mehdi Tajvidi, John C. Hermanson, Ismaeil Ghasemi, Tensile Properties of Wood Flour/Kenaf Fiber Polypropylene Hybrid Composites, *Wiley Periodicals, Inc. J Appl Polym Sci* , 105 (2007), 3054–3059

[12]dynoor bin Osman, Toshihiro Moriga,Kei-ichiroMurai, aMohd Warikh bin AbdRashid ,Study of morphology and electrical properties of indium zinc oxide-modified kenaf fiber, *Industrial Crops and Products* ,100 (2017), 171-175.

[13]R.Dunne ,D.Desai[a],R.Sadiku ,Material characterization of blended sisal-kenaf composites with an ABS matrix, *Applied Acoustics* ,12 (2017) ,184-193

[14]V.S.Srinivasan S.RajendraBoopathy D.Sangeetha B.Vijaya Ramnath ,Evalvuation of mechanical and thermal properties of banana-Flax based natural fibre composite, *materials and design*,60(2014),620-627

[15]C.Chaithanyan, T.Pannerselvam S.Raghuraman B.Vijaya Ramnath, Investigation of tensile behaviour of sisal and coir Reinforced Hybrid Composites using venyl Ester Resin,*Applied mechanics and materials*,591(2014),146-149

[16]B.Vijaya Ramnath S.Rajesh C.Elanchezian A.santosh Shankar S.Pithchai Pandian S.Vickneshwaran R.Sundar Rajan, Investigation on Mechanical Behaviour of Twisted Natural Fiber Hybrid Composite Fabricated by Vacuum Assisted Compression Molding Technique, *Fibers and Polymers* 2016, Vol.17, No.1, 80-87

[17]Felipe M.R. dos Santos, Thiago F. de Souza, Danilo M. Barquete, Franco D.R. ,Comparative analysis of the sisal and piassava fibers as reinforcements in lightweight cementitious composites with EVA waste, *Construction and Building Materials* ,128 (2016) 315–323.

[18]Fláviode Andrade Silva, BarzinMobasher, Romildo Dias ToledoFilho, Fatigue behavior of sisal fiber reinforced cement composite, *Materials Science and Engineering*, 527 (2010) 5507-5513

[19]Ashishkumre,R.S.Rana, Rajesh Purohit, A Review on mechanical property of sisal glass fiber reinforced polymer composites, *Materials today*, 4 (2017), 3466-3476.

[20]T.P. Mohan, K. Kanny, Chemical treatment of sisal fiber using alkali and clay method, *Composites Research Group*, 43 (2012) 1989–1998.

[21]Easwara Prasad G L, Keerthi Gowda B Sb, Velmurugan R, Comparative Study of Impact Strength Characteristics of Treated and Untreated Sisal Polyester Composites, *Procedia Engineering* ,173 (2017) 778 – 785.

[22] G. Yuvaraj, B. Vijaya Ramnath, G. Hemanth ragavendiran, R. Karthikraja, Investigation Of Flexural And Water absorption behaviour of epoxy hybrid composites, *Advances in Natural and Applied Sciences*, 2016 Special10(7): pages 34-39

[23] G. Yuvaraj, B.Vijaya Ramnath, A.Abinash, B.Srinivasan, R.Vikas Nair, Investigation of Mechanical Behaviour of Sisal Epoxy Hybrid Composites, *Indian Journal of Science and Technology*, Vol 9(34), DOI: 10.17485/ijst/2016/v9i34/10

Optimization of roller burnishing process parameters using lion optimization algorithm

A Tamilarasan[1*], A Renugambal[2], T Mohan[1], Akshay R.Iyer[1], V Pramoth Kumar Krish[1], A Rajkumar[1] and Himanshu N Sakriya[1]

[1]Department of Mechanical Engineering, Sri Chandrasekharendra Saraswathi Viswa Mahavidyalaya, Enathur, Kanchipuram – 631561, Tamilnadu, India.

[2]Department of Mathematics, University College of Engineering Kancheepuram, Kanchipuram (A Constituent College of Anna University, Chennai).

[1*]Corresponding Author: tamilrj2010@gmail.com

Abstract. The aim of this work is to optimize the roller burnishing process parameters for minimizing the surface roughness using response surface methodology(RSM) and lion optimization algorithm. A RSM based box-behnken design is utilized for experimentations. The empirical model of surface roughness is developed for showing the relation between inputs and output of the burnishing process. The analysis of variance (ANOVA) method was used to check the accuracy of the developed mathematical model. With the developed mathematical model, the roller burnishing process parameters were then optimized to minimize the surface roughness by a new type of lion optimization algorithm. Further, validation test has been conducted for the optimal conditions suggested by lion optimization algorithm. The experimental value agreed with those predicted value with 3.225 of relative error percentage.

1. Introduction

Burnishing is a plastic deformation process in which force is applied on to a surface through a hard roller or ball[1].It is a cold working process that can be used to improve surface finish and surface hardness of workpieces [2]. Burnishing offers many advantages; improved size control, densely compacted and fine surface finishes and also increases in the surface hardness[3]. This kind of burnishing process may find many uses in manufacturing components and machine tools since it is applicable for a wide range of industrial finishing operations. Roller burnishing generally uses tools with a hydrostatically supported ball as the rolling element[4]. This is mounted in a suitable holder and supported by a hydraulic piston, which, in turn, may be moved axially in a sleeve mounted on the tool shank and which ensures a constant rolling force corresponding to the hydraulic pressure. This process belongs to a group of manufacturing technologies, which are used for the mechanical strain hardening of the surface layer[5]. Esme [6] investigated the multi-response optimization of burnishing process for an optimal parametric combination to yield favourable surface roughness and micro hardness using the Grey relational analysis and Taguchi method. The objective functions have been selected in relation of burnishing parameters; burnishing force, number of passes, feed rate and burnishing speed.

The results showed that the application feasibility of the Grey relation analysis in combination with Taguchi technique for continuous improvement in product quality in manufacturing industry. Sagbas[7] used an optimization strategy of desirability function approach together with response surface methodology in ball burnishing process of 7178 aluminium alloy. The process model variables are burnishing force, number of passes, feed rate and burnishing speed were considered.Babu et al[3] studied the effects of various burnishing parameters on the surface characteristics, surface microstructure, micro hardness in roller burnishing of EN Series steels (EN 8, EN 24 and EN 31), Aluminum alloy (AA6061) and Alpha-beta brass. The following burnishing parameters were considered as burnishing speed, burnishing force, burnishing feed and number of passes. Taguchi technique is employed in the present investigation to identify the most influencing parameters on surface roughness. The optimum levels were found for minimum roughness values by theoretical Taguchi method. Basak and Goktas[8] conducted the burnishing experiments on Aluminum alloy (Al 7075 T6) by considering the different burnishing parameters (number of revolution, feed, number of passes, and pressure force)Using the experimental results a fuzzy logic model has been used to achieve the best parameters for the burnishing process. The surface roughness is directly affected negatively if the applied force is increased was found. The root mean square errors are obtained around in 2.55% and 0.62% for surface roughness and surface hardness respectively. It was found that the fuzzy logic is a suitable technique that may be efficiently used to optimize the burnishing process. Tayeb and Brevern [9] investigated the impact of burnishing speed, burnishing force and burnishing tool dimension on the surface qualities and tribological properties on Aluminium 6061 under different parameters and different burnishing orientations. Gharbi et al[10] identified the effect of burnishing parameters (i.e., burnishing speed, burnishing force, and feed rate) on surface roughness, surface hardness, and microstructure of burnished surfaces in burnishing of AISI 1010 steel using the Taguchi technique and response surface methodology. It was found that the burnishing force has the most influential effect on the surface roughness and hardness, followed by the burnishing speed, and least influence by the feed rate. In addition, microstructural examinations of the burnished surface indicate that burnishing force more than 400 N causes flaking of the burnished surfaces. The optimal burnishing parameters obtained through Taguchi method. Using the optimal parameters, the mean surface roughness has been improved from Ra = 2.48 to 1.75 micron, while the hardness increases from 59 to 65.5 HRB. John and Vinayagam[11] dealt with optimization of burnishing process in CNC machining centre using response surface methodology. The ball burnishing tool is used in a CNC machining centre for finishing after an end milling process. The tool and work piece materials are Tungsten carbide and Tool steel.The input parameters are burnishing force, feed, speed and number of passes. The output parameters are surface roughness and surface hardness. The output parameters are modelled and optimized using response surface methodology. When the burnishing force is increased above the yield stress of the material, the deformation takes place at the surface. Hence the surface roughness is minimized. When the number of passes is increased, the surface roughness is gradually increased because more contact force is generated on the surface which produces more heat and softens the top layer of the work piece. Hence, the surface roughness is increased and surface hardness is decreased. At lower feed rates, the more contact time exists between tool and work piece. Hence, the surface roughness is minimized at lower feed compared to higher feed rate. At medium speed, more contact force is generated on the surface compared to at the other speeds. Hence, the surface roughness is minimized at average speed. Taweel and Axir [12] employed the Taguchi technique to identify the effect of burnishing parameters, i.e., burnishing speed, burnishing feed, burnishing force and number of passes, on surface roughness, surface micro-hardness, improvement ratio of surface roughness, and improvement ratio of surface micro hardness. The results of analysis showed that the burnishing force with a contribution percent of 39.87% for surface roughness and 42.85% for surface micro-hardness had the dominant effect on both surface roughness and micro-hardness followed by burnishing feed, burnishing speed and then by number of passes. The published literature indicates that few studies have been reported for the optimization of process parameters in roller burnishing of brass. Therefore, the present paper aims at the optimization of best input process parameters for high performance burnishing process to yield the good quality surface roughness. A RSM-based box-behnken design matrix[13] has been used for conducting the experiments. The experiments was carried out using three different burnishing process parameters (Burnishing speed, feed rate and burnishing depth) to develop a mathematical model on the outcome of surface roughness. Then, a lion optimization algorithm was utilized to optimize roller burnishing process parameters.

2. Experimental work

The experimental work was carried out on an all gear lathe machine with a maximum speed of 2000 RPM and a 8kW drive motor. A special custom-made burnishing tool was used for roller burnishing. When the roller was pressed against the surface of the metallic specimen, a precalibrated spring became compressed. This spring was used mainly to reduce the possible sticking of the tool onto the surface of the specimen. The shank of the burnishing tool was designed in such a manner that it could be simply mounted or fixed onto the tool holder of a machine tool. The roller burnishing experimental setup as shown in Figure 1.

Figure 1. Experimental setup

Table 1. Parameters and Levels for BBD

Process parameters	Reference symbol	Units	Low(-1)	Centre(0)	High(+1)
Burnishing speed	X_1	rpm	350	450	550
Burnishing feed rate	X_2	mm/rev	0.06	0.09	0.12
Burnishing Depth	X_3	mm	0.1	0.15	0.2

The brass as a workpiece material was selected and widely used in manufacturing industry. The general composition of brass is:0.831% S, 2.21% Pb, 36.37% Zn, 0.216% P, 0.293% Fe, 0.442% Al, <59.23% Cu, and 0.237% Ni. The bars are received from Narendra steels, Mumbai, as 50 mm diameter and depicted in Figure 2. The external roller burnishing tests were performed under

IOP Conf. Series: Materials Science and Engineering **390** (2018) 012063 doi:10.1088/1757-899X/390/1/012063

unlubricated (dry) conditions. Experiments were designed based on RSM based box-behnken design method at at different parametric settings and illustrated in Table 1. The upper limit of a factor was coded as (+1), and the lower limit was coded as (−1). The selected process parameters burnishing speed (350 to 550 rpm), burnishing feed rate (0.06 to 0.12 mm/rev) and burnishing depth (0.1 to 0.2 mm) were considered based on the available machine capacity. A precise surface roughness tester (Mitutoyo- Surftest SJ-210 - Series 178) was used to measure surface roughness of the burnished workpiece at five distinct locations. Totally, seventeen experimental runs are carried out according to the design matrix in a random order to avoid any systematic error creeping into the system.

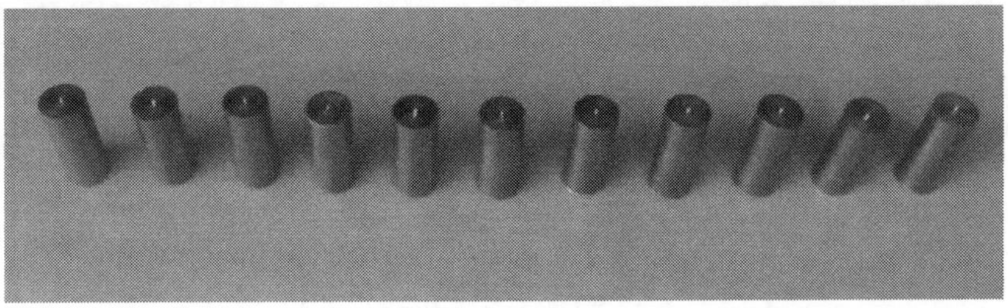

Figure 2. Sample of burnished workpieces

Table 2. Box-Behnken Design arrangement and responses

Exp. No	Process parameters			Response	
	Burnishing speed	Burnishing feed rate	Burnishing Depth	Exp. Surface Roughness	Pred. Surface Roughness
	rpm	mm/rev	mm	Micron	
1	450	0.06	0.20	0.284	0.283
2	350	0.09	0.10	0.505	0.510
3	450	0.09	0.15	0.505	0.510
4	450	0.09	0.15	0.515	0.514
5	450	0.09	0.15	0.510	0.511
6	550	0.06	0.15	0.291	0.291
7	550	0.09	0.20	0.505	0.505
8	450	0.09	0.15	0.502	0.504
9	350	0.09	0.20	0.529	0.530
10	450	0.12	0.20	0.730	0.730
11	450	0.06	0.10	0.252	0.258
12	350	0.06	0.15	0.313	0.321
13	550	0.12	0.15	0.855	0.854
14	550	0.09	0.10	0.553	0.551
15	350	0.12	0.15	0.769	0.768
16	450	0.12	0.10	0.791	0.792
17	450	0.09	0.15	0.505	0.505

3. Model development and ANOVA analysis

Table 2. summarizes various process parameter combinations for 17 experiments with measured surface roughness value.A nonlinear regression analysis was conducted using the experimental data collected as per BBD to establish input-output relationships of an roller burnishing process by the Design Expert software. The application of RSM is yielded the following second-order polynomial equation, which are empirical relations between surface roughness, and the process test variables (X_1 – burnishing speed, X_2 - burnishing feed, and X_3 – burnishing depth) in un-coded units as follows:

$$\text{SurfaceRoughness}=0.28705-0.00277359X_1+2.58253X_2+4.50086X_3+0.00890689X_1X_2-0.00360645X_1X_3-15.46230\,X_2X_3+0.00000291389X_1^2+22.69179\,X_2^2-5.40438\,X_3^2 \quad (1)$$

As shown in Table 2, the experimental surface roughness values fitted well with the predicted surface roughness values which were calculated using the regression model (equation 1). It demonstrated the feasibility to apply BBD to setting up the experimental optimization design for measuring the surface roughness values and establishing the regression equation model. The adequacy of the models so developed was tested using the analysis of variance technique (ANOVA)[14]. The ANOVA for the second-order equation is presented in Table 3 at 95% confidence level. Model summary statistics composed of the R-squared, the adjusted R-squared, the predicted R-squared, Adeq Precision and the predicted residual error sum of squares (PRESS).

Table 3. ANOVA Table for Surface Roughness

Source	Sum of Squares	df	Mean Square	F Value	*p*-value (Prob > F)
Model	0.516434	9	0.057382	890.1053	< 0.0001
X_1	0.00096	1	0.00096	14.89588	0.0062
X_2	0.5027	1	0.5027	7797.912	< 0.0001
X_3	0.000364	1	0.000364	5.651209	0.0491
X_1X_2	0.002856	1	0.002856	44.30205	0.0003
X_1X_3	0.001301	1	0.001301	20.17571	0.0028
X_2X_3	0.002152	1	0.002152	33.37796	0.0007
X_1^2	0.003575	1	0.003575	55.45645	0.0001
X_2^2	0.001756	1	0.001756	27.24131	0.0012
X_3^2	0.000769	1	0.000769	11.92279	0.0106
Residual	0.000451	7	6.45E-05		
Lack of Fit	0.000359	3	0.00012	5.172013	0.0732
Pure Error	9.25E-05	4	2.31E-05		
Cor Total	0.516885	16			
Std. Dev.	0.008029		R-Squared	0.999	
Mean	0.524272		Adj R-Squared	0.998	
C.V. %	1.53147		Pred R-Squared	0.988	
PRESS	0.005885		Adeq Precision	97.128	

The ANOVA of the quadratic regression model indicated that the model was highly significant, as the *F* value for the model was 890.1053. There was only a 0.01% chance that the "model *F* value" this large could occur because of noise. The *p* value Prob>*F* value of the model was <0.0001, which also confirmed that the model was highly significant. A lackof-fit value of 5.172013 implies that the lack of fit is not significant relative to the pure error, when p value is 0.0732>0.05 also supports the fitness of the model. The R^2 is a gage of the amount of deviation around the mean explained by the model and adjusted R^2 is the R-squared adjusted for the number of terms in the model proportionate to the number of points in the design matrix. Predicted R^2 is a gage of the amount of degree of diversity in new data explained by the model. It displays that the quadratic model with maximum adjusted R^2 and predicted R^2 values and minimum PRESS value (0.00588is the best model) recommend as a suitable case. A normal probability plot of the residuals is depicted in Fig. 5, which reveals that the residuals generally fall on a least-square line which is used to estimate the cumulative distribution function for the population. As evident from the figure, the errors are normally distributed and there are almost no serious violations of the assumptions that underlie the analysis. Thus, for advance analysis, this model was suggested.

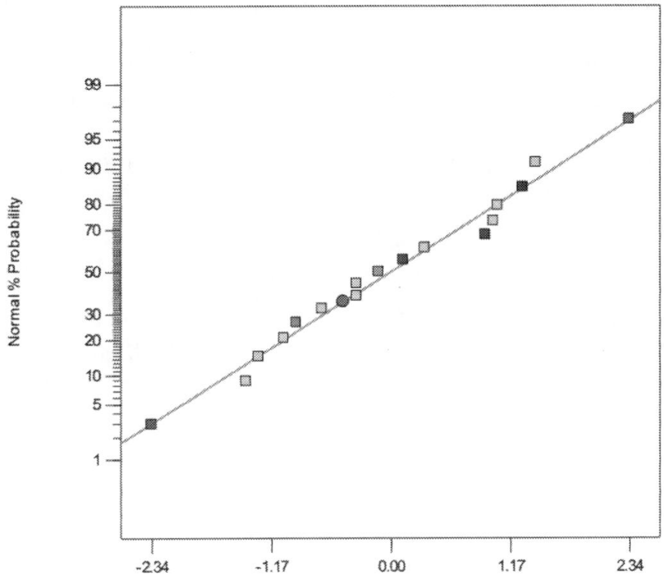

Figure 3. Normal probability plot

4. Optimization by lion optimization algorithm

The step by step evaluation for lion optimization algorithm [15] as follows:

1. Generate random sample of lions N_{pop}(N_{pop} is number of initial population).
2. Initiate prides and nomad lions
   i. Randomly select %N(percent of lions that are nomad) of initial population as nomad lion. Partition remained lions into P(P is number of prides) prides randomly , and formed each prides territory.
   ii. In each pride %S(Sex rate) of entire population are known as females and the rest as males. This rate in nomad lions is inversed.
3. For each pride do
   i. Some randomly selected female lion go hunting.

   ii. Each of remained female lion in pride go toward one of the best selected position from territory.
   iii. In pride, for each resident male; %R(Roaming percent) of territory randomly are selected and checked. %Ma(Mating probability) of females in pride mate with one or several resident male.(New cubs become mature).
   iv. Weakest male drive out from pride and become nomad.
4. For nomad do
   i. Nomad lion(both male and female) moving randomly in search space. %Ma(Mating probability) of nomad female mate with one of the best nomad male.(New cubs become mature).
   ii. Prides randomly attacked by nomad male.
5. For each pride do
   i. Some female with I rate ((Immigrate rate)) immigrate from pride and become nomad.
6. Do
   i. First, based on their fitness value each gender of the nomad lions are sorted. After that, the best females among them are selected and distributed to prides filling empty places of migrated females.
   ii. With respect to the maximum permitted number of each gender, nomad lions with the least fitness value will be removed.

If termination criterion is not satisfied, then go to step 3

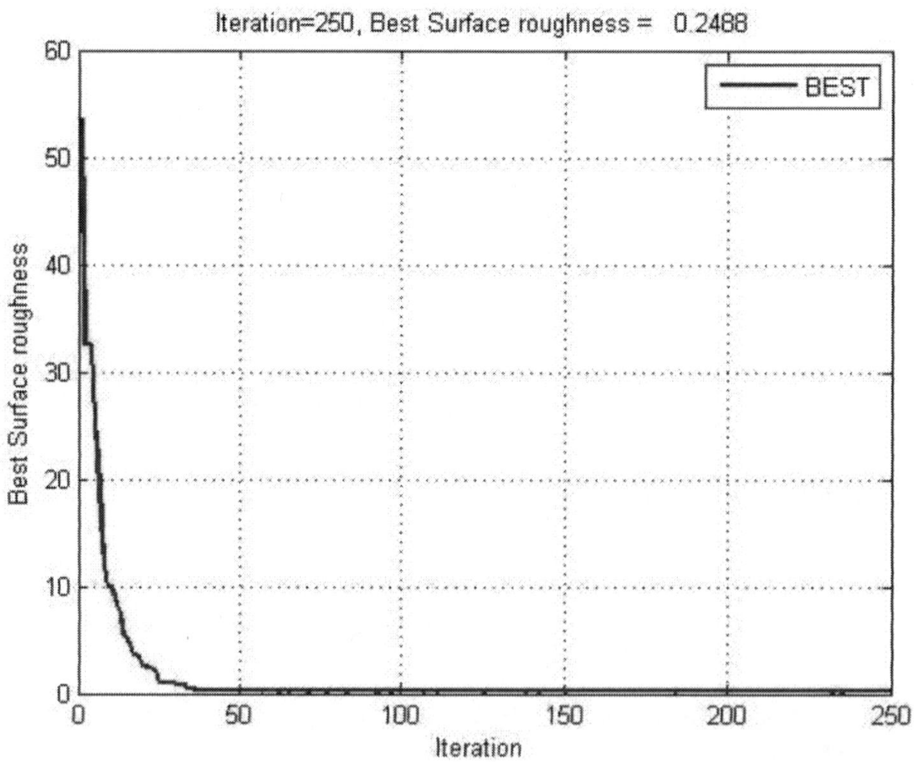

Figure 4. Convergence plot

At first, mathematical model of surface roughness is developed and considered as fitness function with subject to the constraints of upper and lower bounds of burnishing process parameters to minimize the surface roughness. The lion optimization algorithm is implemented and executed in matlab environment in several times to yield the best optimal process parameters by tuning the algorithm parameters. The convergence plot of surface roughness value over max number of generations using lion optimization algorithm is shown in Figure 4, which illustrates the algorithm, converges up to 200 generations. And, the best optimal is achieved at 40th iteration. Table.4 summarizes the comparison between the initial condition and optimal condition achieved by LO algorithm. The relative percentage of improvement is achieved 20.51%, and it is clear that the predictive capability of the lion optimization algorithm. After the selection of AWJ optimal process parameters, further experiments were carried out to verify the corresponding surface roughness predicted by algorithm. Table.5 shows that the percentage of error between the predicted and experimented values. From this analysis, it is observed that the calculated error(i.e. 3.225) is very small which confirms the excellent reproducibility of the experimental conditions.

Table 4. Comparison between the initial condition and optimal condition achieved by LO algorithm

Model Summary		Initial condition	Optimal condition	% of improvement
Process parameters	Burnishing speed (rpm)	350	458.11	
	Burnishing feed rate (mm/rev)	0.06	0.06	
	Burnishing depth(mm)	0.15	0.1	
Response		0.313	0.2488	20.51

Table 5. Experimental validation of optimal parameter settings

	Burnishing speed(rpm)	Burnishing feed rate(mm/rev)	Burnishing Depth(mm)	Surface roughness (Micron)
Predicted by algorithm	458.11	0.06	0.1	0.248
Experimental	458.11	0.06	0.1	0.256
% Error				3.225

5. Conclusions

The experimental work on burnishing of brass was performed successfully. Then, the Lion optimization algorithm was implemented to yield the best optimal values. The following conclusions are drawn as follows:

1. The Box-Behnken design approach is used for experimental design. The selected process parameters burnishing speed (350 to 550 rpm), burnishing feed rate (0.06 to 0.12 mm/rev) and burnishing depth (0.1 to 0.2 mm) were considered based on the available machine capacity to perform the experiments.
2. The predicted values match the experimental values reasonably well, with R^2 of 0.999 for surface roughness. Thus the developed model is more suitable for further analysis.
3. The group behaviour of lions is an effective model for constructing the lion optimization algorithm.

4. The optimal performance characteristics are observed to have minimum surface roughness (0.248 microns) when the process parameters are burnishing speed (458.11rpm), burnishing feed rate (0.01mm/rev) and burnishing depth (0.1 mm)
5. A significant improvement in surface roughness characteristics is observed at the optimum parameter settings in comparison to the initial settings.

References

[1] Kotiveerachari B and Murty R 1985 Study of some aspects of burnishing *International Journal of Production Research* 23 499-521.

[2] Sequera A, Fu C, Guo Y and Wei X 2014 Surface Integrity of Inconel 718 by Ball Burnishing *Journal of Materials Engineering and Performance* 23 3347-3353.

[3] Babu P, Ankamma K, Prasad T, Raju A and Prasad N 2012 Optimization of burnishing parameters and determination of select surface characteristics in engineering materials *Sadhana* 37 503-520.

[4] Okada M, Suenobu S, Watanabe K, Yamashita Y and Asakawa N 2015 Development and burnishing characteristics of roller burnishing method with rolling and sliding effects *Mechatronics* 29 110-118.

[5] Revankar G, Shetty R, Rao S and Gaitonde V 2014 Selection Of Optimal Process Parameters In Ball Burnishing Of Titanium Alloy *Machining Science and Technology* 18 464-483.

[6] Esme U 2010 Use of grey based taguchi method in ball burnishing process for the optimization of surface roughness and microhardness of AA 7075 aluminum alloy Materials and technology 44 (3), 129-135.

[7] Sagbas A 2011 Analysis and optimization of surface roughness in the ball burnishing process using response surface methodology and desirabilty function *Advances in Engineering Software* 42 992-998.

[8] Basak H and Goktas H 2009 Burnishing process on al-alloy and optimization of surface roughness and surface hardness by fuzzy logic *Materials & Design* 30 1275-1281.

[9] El-Tayeb N, Low K and Brevern P 2007 Influence of roller burnishing contact width and burnishing orientation on surface quality and tribological behaviour of Aluminium 6061 *Journal of Materials Processing Technology* 186 272-278.

[10] Gharbi F, Sghaier S, Al-Fadhalah K and Benameur T 2010 Effect of Ball Burnishing Process on the Surface Quality and Microstructure Properties of AISI 1010 Steel Plates *Journal of Materials Engineering and Performance* 20 903-910.

[11] Stalin John M and Vinayagam B 2011 Optimization of Ball Burnishing Process on Tool Steel (T215Cr12) in CNC Machining Centre Using Response Surface Methodology *Arabian Journal for Science and Engineering* 36 1407-1422.

[12] El-Taweel T and El-Axir M 2008 Analysis and optimization of the ball burnishing process through the Taguchi technique *The International Journal of Advanced Manufacturing Technology* 41 301-310.

[13] Tamilarasan A and Marimuthu K 2014 Multi-response optimisation of hard milling process parameters based on integrated Box-Behnken design with desirability function approach *International Journal of Machining and Machinability of Materials* 15 300.

[14] Tamilarasan A and Rajamani D 2017 Multi-response optimization of Nd:YAG laser cutting parameters of Ti-6Al-4V superalloy sheet *Journal of Mechanical Science and Technology* 31 813-821.

[15] Yazdani M and Jolai F 2016 Lion Optimization Algorithm (LOA): A nature-inspired metaheuristic algorithm *Journal of Computational Design and Engineering* 3 24-36.

IOP Conf. Series: Materials Science and Engineering **390** (2018) 012064 doi:10.1088/1757-899X/390/1/012064

The impact of workplace friendships on organizational commitment and intention to leave

A Akila[1] and R G Priyadarshini[2]

[1] Amrita School of Business, Amrita Vishwa Vidyapeetham, Coimbatore, India, akila.94@gmail.com
[2] Amrita School of Business, Amrita Vishwa Vidyapeetham, Coimbatore, India, rg_priyadarshini@cb.amrita.edu

Abstract- Friendships play a role in different aspects of our life. These friendships within organisations might hinder or facilitate both the organisational and employee functioning. There are various degrees of friendships found in organisations. The different degrees have different roles in functioning of the organisation. Therefore, this paper tries to find out the different degrees of friendship that exist in organisations and their effect on the employee's job commitment and intention to leave. The different levels of friendships studied in this paper are "work spouse, special peer, collegial peer and information peer". The study aims to find out the prominent types of friendships that exist in the workplace and to determine the influence of different types of friendships on the employee's job commitment and intention to leave. A questionnaire was constructed using a friendship scale, commitment scale and intention to leave scale. The samples for the study are professionals who are working in the service industries. Snowballing and convenient sampling technique was used to collect data.

1. Introduction

Friendships are very necessary to bring in a sense of happiness and well-being. Although friendships may mean different things to different people from various cultures and societies, there are some types of friendships that are prevalent only at a workplace. The social, personal and organisational benefits of friendships have been highlighted in research before. Study done by Isabella and Kram in 1985 studies the mentor alternatives of the role of peer relationships, Sias M and Cahill J (2014) examine friendships at work, Markiewicz, Devine and Kausilas (1999) studied the women and men and its impact on job satisfaction and resource implications in friendships. Workplace friendships enhance the job involvement and organisational involvement which in turn enhances the job performance (Ting SC, Ho MH, 2017). According to a study done by Gallup organisation, employees who have good friends at work are seen to be seven times more engaged in their work. There are some characteristics in a relationship like trust, mutual respect, open communication, and mindfulness that make it a healthy and good relationship. Some close workplace friendships may even be destructive or be more sinister than bringing any positive behaviour. These kinds of behaviours are unacceptable as it would lead to degradation of culture, growth and productivity of the organisation. Nevertheless, positive friendships help the employees to buy into the culture emotionally.

There are different degrees of friendships which can be found at workplace. "Work spouse is defined as a platonic, special friendship with a colleague in the work place which is characterized by a close bond, high disclosure, and high levels of support, mutual trust, honesty, loyalty, and respect. A Special friend is a person whom you consider to be a best friend. This person would still remain to be your best friend even if you don't work together. This person is considered to be much more than just a co-worker to you. A collegial peer is a person whom you consider as a "work buddy". This person is more than a mere acquaintance but you do not share every detail of your life with this person. You interact with this person fairly regularly on an equal basis as you consider this person a friend or colleague only. An Information peer is one whom you do not know that well or whom you feel is very close to. You would consider this person to only be an acquaintance more than a friend. Though you interact with this person on a regular basis, you would probably not continue your relationship with this person if you did not work here.

Organisational Commitment is the degree of connection an employee has with the organisation. On the whole, employees who are committed to the organisation feel connected with the goals of the organisation as well as have the feeling of fitting in. Therefore, this is one important aspect of which is important for organisational success. The three types of commitment which are distinctive according to the study by Allen and Meyer – "affective commitment" is affection for the job, "continuance commitment" is fear of loss, and "normative commitment" which is the obligation to stay.

Intention to leave a job talks about whether the employee would like to continue his/her job at the organisation or quit and move to another organisation. This may be impacted by many factors such as job stress, burnout, person organisation fit, job satisfaction and so on. The relationships employees have at their workplace may also determine their intention to leave the job.
This study tries to study the different degrees of friendship that are found in workplaces and the effect these friendships have on the organisational commitment and their intention to leave.

2. Literature Review

"Social exchange" theory can be considered as a kind of relationship between people that is predominantly transactional in nature (give and take). According to a theory (**Rook**, 1984), "we choose our social ties on the basis of their capacity to provide rewards relative to costs and as a function of the alternatives available."

Rawlins (1992) explains, "work affiliations range from patently 'friendly relations' to more involved bonds transcending the work setting (**Kurth**, 1970)". While some relationships which are "friendly relations" may primarily act to have advancement and growth one's career (**Maines**, 1981) and provide outlets for some informal, personalised interaction (**Wright**, 1978), and those which are closer friendships may be providing self-affirmation, emotional support, interdependence, and so on (**Wright**, 1985). Looking at the way "friends" grow into "close friends" over a period of time and then they become "best friends" provides a deeper understanding of these phenomena.

The importance of workplace social relationships has been recognized by scholars for nearly a century now. **Elton Mayo** (1945) brought out the workplace friendships to attention, he was the first to advocate the role of emotional factors that played an important role in determining employee behaviour and also in contending that the most dominant influence on productivity was the extent to which employees received social satisfaction in the workplace (**Mayo**, 1945).
"The impact of friendships at work", "Friendship opportunity" was developed by **Hackman and Lawler** (1971) as a dimension of perceived job characteristics. Its intent was to examine the degree to which a job allowed employees to talk to one another and establish relationships at work.

IOP Conf. Series: Materials Science and Engineering **390** (2018) 012064 doi:10.1088/1757-899X/390/1/012064

An essential factor that facilitates interpersonal relations is Self- Disclosure and it is capable of diminishing the likelihood of loneliness when "there is reciprocity of the sharing of information between the people in a relationship" as quoted by **Berg & Peplau**, 1982.
Covert cognitive and affective responses are both a part of the Friendship processes and overt behavioural events that occur when people interact. (**Adams & Blieszner**, 1994).
Kram and Isabella (1985) believe that co-worker friendships are very important ways to grow with their support and help. Another organisatonal relationship that has received research attention is the mentor-protégé relationship. (**Kram**, 1983). Organizational commitment could be defined as the employees' state of being committed to assist when it comes to the achievement of the organization's goals, and it also involves the employees' levels of identification, loyalty and involvement (**Caught & Shadur**, 2000). Therefore this commitment might be impacted by the friendships they have at workplace.
Intention to leave is one of the biggest predictors, sign of an employee who is going to leave the organisation **Irvine and Evans** (1995) have found positive relationship between intention to leave and turnover. This may be seen because they have the choice to leave if they are not satisfied with any aspects about their job or organisation.

3. Conceptual Model

From the above literature review, some of the different types of friendships are taken as variables and its impact on "organisational commitment" (OC) and "intention to leave" (ITL)

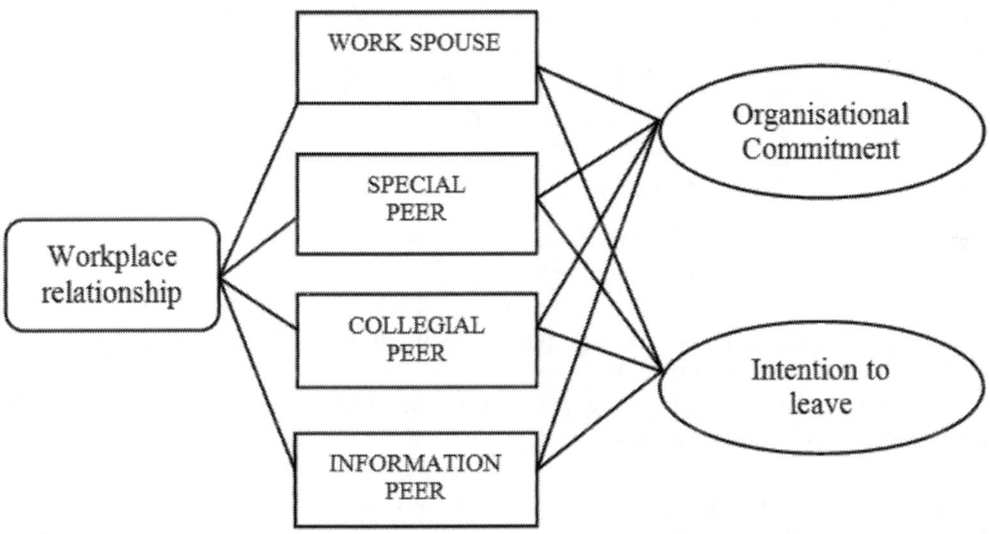

Figure 1. Model depicting workplace relationship

4. Methodology

The respondents for this study were employees from service sector where there would be a lot of team or group work involved. The size of the sample is 51 respondents. The sample consisted of 51.9% female (27) and 47.05% male (24) participants. Questionnaire was used to collect data for the research. Snowballing and convenient sampling technique was used to collect data. Friendship scale which was used in the study was Nielsen (2000) which measures the ability and prevalence of friendship. The friendship scale measures 6 different aspects of friendship – Stimulating Companionship (F1), Help (F2), Intimacy (F3), Reliable Alliance (F4), Self-validation (F5), Emotional Security (F6). To measure the organisational commitment of the employees, Meyer and

Allen (1991) Commitment scale was used. This measures 3 types of commitment – Affective Commitment (AffC), Continuance Commitment (ContC), Normative Commitment (NormC). Intention to leave is measured with Morrison (2004) scale which concentrates in 3 important precursors of turnover.

5. Research Questions

1. What are the prominent types of friendships that exist at workplace?
2. How do different types of friendships influence the employee's job commitment and intention to leave?

6. Analysis

The obtained responses were quantitatively analysed. The sample had 21 respondents (41.17%) below the age of 25, 26 respondents (50.98%) from 25 to 35 ages, between the ages 36 to 45 there are 2 respondents (3.92%) and greater than 46 has 2 respondents (3.92%). There were 3 respondents (5.88%) who has worked for lesser than 6 months in the organisation, 6 respondents (11.76%) in 6 months to 1 year category, 22 respondent in 1 year to 2 years, 4 respondents (7.84%) with 2 years to 3 years of experience in the organisation, 17 responses (33.33%) who have greater than 3 years of experience. Descriptive statistics was used to know how many respondents out of the whole sample had each type of friendship at their workplace.

Table 1. Shows the responses of participants indicating the preferred type of friendship

Type of friendship	Number of respondents
Work Spouse	35
Special Peer	50
Collegial Peer	49
Information Peer	50

It was also observed that the respondents on an average have around 1 - 2 work spouses, 2 - 3 special friends, 4 collegial peers and more than 5 information peers. Close to 30% of the respondents have more than 10 information peers at their workplace. This shows that the most predominantly found friendship at workplace is information peer.

The second preferred type of friendship is special peer (SP) as the number of special friend each respondent has is lesser than the information peers.

The average organisational commitment of the sample was found to be 3.04 on scale of 5. The average affective commitment of the sample was found to be 3.07, continuance commitment was 3.06 and normative commitment was 3. All three types of commitment are almost at the same level. However affective commitment seems to be the highest among the three, therefore, that shows that the individuals are involved emotionally with the organisation's goals and they feel responsible personally for the success of the organisation.
The average score of intention to leave was 2.97 on a scale of 5. So it is observed that the intention to leave is on the lower side of the scale.

There seems to be no significant correlation between "work spouse" type of friendship and organisational commitment. No significant correlation is found between every type of friendship (special friend, collegial peer and information peer) and organisational commitment. Similarly there appears to be no correlation between intention to leave and every type of friendship (special friend, collegial peer and information peer). As each type of friendship does not affect the organisational commitment or intention to leave, we can come to a conclusion that other factors such as working conditions, job satisfaction, benefits, work stress levels may play a significant role on them. There also seems to be no significant relationship between the "organisational commitment" and the "intention to leave" of the respondent.

There is a significant negative correlation at the 0.01 level between intention to leave (ITL) and reliable alliance variable of collegial peer (-.400). This shows that some aspects of friendships do have a critical role in intention to leave. Therefore if the employees do not have a sense of reliable alliance in their workplace, they might have a drive to leave the organisation. Since this relationship is seen with collegial peers, it proves that the employees do not have the need to feel very close to their peers on a personal level, but as acquaintances.

Similarly there seems to be negative relation (-.310) between "help" variable of collegial peer and intention to leave. This shows that the feeling of being helped by the peers play a role in intention to leave. Collegial peer seems to be more predominant gender wise; that is women seem to prefer this kind of friendship at their workplace. Work spouse type of friendship is found more in respondents who are below the age of 35, whereas it is not prevalent in respondents who are above the age of 45. This shows that the younger generation look towards friendships at work place for emotional bond, support and respect, whereas the older generation look at their colleagues as mere acquaintance.

7. Results and Discussion

From this study we can conclude that the prominent types of friendships that are observed at workplace are information peers and special friend. The least preferred type of friendship is work spouse, although this gives a lot of support both emotionally and growth wise.
There seems to be not much impact of different types of friendships on the organisational commitment and intention to leave. Only few aspects in different friendships seem to correlate with "organisational commitment and intention to leave".

8. Limitations

The limitation of the study is that the sample did not include equal number of employees from the older generation. For further studies researchers could study the impact of friendship in different fields of each industry by taking the employees hierarchy as well in to consideration

9. Conclusion

The different types of friendships do not affect the degree of organisational commitment and intention to leave. They might be individuals who are not relationship focused or their jobs might not pertain to any interdependent nature. Their organisational commitment maybe affected by other variables such as job satisfaction, job stress, person-organisation fit which have not been measured in the study. According to Pillemer, J., & Rothbard, N. (2018) as companies become more socially bound and connected, leaders and employees should be able to manage the tensions and challenges which arise from friendships inside these companies, in order make the best use of the benefits from these personal bonds.Therefore, this can be a scope for further study that can be done in this area.

IOP Conf. Series: Materials Science and Engineering **390** (2018) 012064 doi:10.1088/1757-899X/390/1/012064

10. References

[1] Adams, R. G., & Blieszner, R. (1994) *Journal of Social and Personal Relationships* **11(2)**, 163 – 184
[2] Berg, J. H., & Peplau, L. A. (1982). Loneliness: The relationship of self-disclosure and androgyny. *Personality and Social Psychology Bulletin*, **8(4)**, 624-630.
[3] Caught, K. & Shadur. (2000). The measurement artifact in the organizational commitment questionnaire. Psychological Reports, 87, 777-788.
[4] Hackman, J. R., & Lawler, E. E. (1971). Employee reactions to job characteristics. Journal of applied psychology, 55(3), 259.
[5] Irvine D M., & Evans, M. G. (1995) *Nursing Research,* **44**, 246- 253.
[6] Kram, K. E. (1983). Phases of the mentor relationship. *Academy of Management journal*, **26(4)**, 608-625.
[7] Kram, K. E., & Isabella, L. A. (1985). Mentoring alternatives: The role of peer relationships in career development. *Academy of management Journal*, **28(1)**, 110-132.
[8] Kurth, S. B. (1970). Friendships and friendly relations. *Social relationships*, 136-170.
[9] Maines, D. R. (1981). The organizational and career contexts of friendship among postdoctoral students. Research in the Interweave of Social Roles.
[10] Mayo, Elton 1945 The Social Problems of an In- dustrial Civilization. Boston: Division of Research, Graduate School of Business Administration, Harvard University.
[11] Pillemer, J., & Rothbard, N. (2018). Friends Without Benefits: Understanding the Dark Sides of Workplace Friendship. Academy of Management Review, amr-2016.
[12] Rawlins, W. K. (1992). Friendship matters: Communication, dialectics, and the lifecourse. Hawthorne, NY: Aldine
[13] Rook, K. S. (1984). The negative side of social interaction: impact on psychological well-being. *Journal of personality and social psychology*, **46(5)**, 1097.
[14] Ting, S. C., & Ho, M. H. (2017). The Influence of Workplace Friendship, Job Involvement, and Organizational Identification on Job Performance: Administrative Staffs of Private Science and Technology Universities in South Taiwan as an Example. International Journal of Humanities and Social Sciences, 9(6), 46-57.
[15] Wright, P. H. (1978). Toward a theory of friendship based on a conception of self. Human Communication Research, 4(3), 196-207
[16] Wright, P. H. (1985). The Acquaintance Description Form. In S. Duck, & D. Perlman (Eds.), Understanding personal relationships: An interdisciplinary approach (pp. 39-62). Newbury Park, CA: Sage

Natural fibre composites-A Review

V.M.Manickavasagam*, B.Vijaya Ramnath, C.Elanchezhian, V.Aravinthan, A.Vignesh.
Department of Mechanical Engineering
Sri Sairam Engineering College, West Tambaram, Chennai-44.

***Corresponding author mail ID: manickavasagam.mech@sairam.edu.in**

Abstract: Composite materials are playing a major role in engineering applications. They are used in automobiles, marine, aerospace, sports and various other fields. Nowadays, reinforcing is done with natural fiber in place of synthetic fiber. Natural fibers are cheap and abundantly available. They are nontoxic and can be treated chemically in order to enhance the mechanical properties. Such composite materials are 'Go by Green" and a lot of research is focused on such natural fiber based composites. A few of such research works are reviewed in this paper.

Keywords: Natural fiber, Composites, Banana, Jute.

1.0 Review on natural fiber composite

Nowadays, a natural fiber composite materials are replacing conventional materials because of their properties like high Strength to Weight ratio, high specific strength, high density and their bio degradability. This Section reviews some of the research carried out using natural fibers as reinforcement to fabricate composites. It reviews effect of mechanical behavior of composites while using various resigns and fibers.

2.0 Review on sisal fiber

Maya Jacob et al. [1] considered two natural fibers namely sisal fiber and oil palm fiber along with natural rubber composite. They studied the mechanical property of natural rubber composite when it was reinforced with sisal and palm fibers. The composite was fabricated using two row mill. Two different types of composites were made, one using untreated fiber and another with treated fiber. The tensile and tear strength were calculated by using Universal testing machine and tearing strength tester. They concluded that addition of sisal and oil palm fibre to the natural rubber composite gives high tensile and tear strength. They also found that the mechanical properties were improved much when compared to rubber composite was reinforced with chemically treated fibers than untreated fibers.Leandro Joséda Silva et al. [2] studied the addition and the effect of silica micro particles on polymeric composites. In this complex process, most of the mechanical properties were improved. They also concluded that the addition of maleic anhydride and silica nanoparticles improved the flexural strength and elastic modulus.

2.1Epoxy hybrid composites

Yuvaraj et al. [3] analysed Flexural and water absorption behaviour of epoxy hybrid composites.. Epoxy is used as a resin and hand layup methods were used to make up the composites. It seems that there is a slight increase in flexural property of composites with increase in fibre loading and also there is an decrease in water absorption property with increase in glass fibre content in composites. This results in better mechanical properties for glass fiber- sisal fiber construction than dispersed construction.Yuvaraj et al. [4] investigated Mechanical behaviour of Sisal epoxy Hybrid composites. It seems that the fiber is fabricated by hand layup process. It shows that the breaking load during double shear test, hardness, inter delamination tests are increased. The results prove that the composite having equal ratio of glass and sisal have improved mechanical properties than composite having minimum sisal fibre percentage

2.2 Twisted fiber composites

Vijaya Ramnath et al. [5] investigated twisted Natural Fiber Hybrid Composite on its mechanical behaviour and Fabrication is done by using Vacuum assisted compression moulding technique. To enhance the strength and stiffness of laminates two fibres are sandwiched between the layers of glass fibres. For the improvement of mechanical properties fibres are also alkalized. It shows that due to the presence of twisted fibres there is an significant improvement in its mechanical properties.

3.0 Review on banana fiber

Amira et al. [6] studied the effects of fiber configuration on mechanical properties of banana fiber. Four types of banana fibers were considered in this process and they are unreinforced PP, PP/Raw banana fiber, PP/Banana yarn and PP/Banana mat. The process used in preparation of composite was compression molding method. Tensile test, flexural test and other microstructural analysis were carried out . It was observed that the fiber produces high mechanical strength, and also banana composites high tensile and flexural strength compared to other fibres. It can be seen that the banana fiber yarn provides the best mechanical properties in PP composite.William Jordana et al. [7] proved that properties of Banana fibers reinforced polymer composites were improved by treating the fibres chemically. The two treatments were carried out such as Peroxide treatment and Permanganate treatment. For peroxide treatment, dicumyl peroxide was used and for permanganate treatment potassium permanganate was used. Then mechanical testing was done followed by computation of Single fiber results. Both treatments increased their interfacial bonding of banana pseudo stem to LPDE matrix. Flexural properties and tensile properties indicated that there was an increase in strength and stiffness with increased fiber volume fraction. Benitez et al. [8] showed that in order to improve the adherence property of banana fiber to plastic matrices, fibers must be treated chemically or physically. The two chemicals namely, Sodium hydroxide and Maleic anhydride were used to treat the fibres at different temperatures and pressures. The leaves from inside and outside part of the pseudo-stem of not being exposed to environment were used. The fibres were subjected to injection moulding process after the treatment. FTIR analysis has been done to check whether there are any changes in fiber surface due to different chemical treatments. FTIR results seem to prove that banana fiber is similar to pure cellulose. It was shown that canary banana tree has higher tensile ,strength and elastic modulus than other types of fibers.

3.1Treatment of banana fiber

Ramesh et al. [9] used different fiber volume fractions and used hand layup process by applying pressure at room temperature for fabricating banana fiber reinforced composites. The processed samples were subjected to different tests like tensile, flexural and impact tests and various results were studied. In order to analyses the structure of fractured surfaces; Scanning Electron Microscope was used to study the crack analysis. Different materials including catalyst methyl ethyl ketone peroxide and acetone thinner were used. After the SEM analysis, crack formation in matrix layer is clearly observed.

3.2 Mechanical and thermal characteristics of Banana fiber

Satish et al. [10]studied Banana-Kenaf Hybrid Epoxy Composites in terms of their mechanical and thermal characteristics in effect of fiber orientation and stacking sequence. Here the composite is prepared using Hand Layup process with different fiber orientations and volume fractions. It seems that fibres produce better properties when they are arranged at 45 degree inclinations than the other orientations. It was also seen that when the thickness increases compressive strength also increases. Manickavasagam et al. [11] evaluated Abaca and Flax reinforced polymer composites for double shear and hardness in automotive applications. Hand layup method is used for the production of composite. It is found that Ultimate strength and hardness of abaca is greater than that of flax composites.Vijaya Ramnath et al. [12] experimented Mechanical properties of Banana Jute Hybrid Composite. Hand LayUp method is used for the Fabrication of composite. Tensile test, Flexural test, delamination test are carried out. To know fracture direction, matrix structure, fiber orientation failure morphology tests is done using Scanning Electron Microscope. It shows that when compared to mono composites, hybrid composites have better mechanical properties.

3.3 Dynamic mechanical analysis

Laly Pothan et al. [13] carried put dynamic mechanical analysis of banana fiber reinforced polyester composites. This was carried out with reference to effect of temperature and pressure. The dynamic properties included the intrinsic properties, morphology and the nature of interface. Important materials included methyl ethyl ketone peroxide and cobalt nalphthenate. These dynamic properties were depending mostly on volume fraction of the fiber. The dynamic analysis showed that incorporation of fiber is decreased and is below the glass transition temperature. Sapuan et al. [14] studied the mechanical properties of three different geometries of woven banana fibres were prepared and mechanical properties of those geometries were studied. The mould was prepared and the process was carried out in either compression or injection moulding process. When analyzed statistically using ANOVA, it showed that insignificant results were obtained in three samples which confirmed that under different tests composites provided a stable mechanical behaviour. These positive results in banana fibers was more useful for people near village sites as they use this banana fibres for producing household utilities. Pothan et al. [15] considered the polarity parameters of a chemically modified banana fiber reinforced polyester composites, in order to improve adhesion and there was a need of chemical modification in fibres. These modifications had been analyzed to investigate the interfacial properties. These were investigated by solvatochromy and zeta potential measurements. Here, cellulose fibers and silanes were used. It was shown that composites having better modulus and low damping can be used as a substitute for building material, which can be developed from banana fibers.

3.4 Mechanical properties of intra-layer abaca fiber

Vijaya Ramnath et al. [16] determined Mechanical properties of intra-layer Abaca-jute-glass fibre reinforced composites. The composites are manufactured and varied in three different orientations and proportions. It is observed that for determining the mechanical properties of composite fiber orientation plays a vital role. samples having higher abaca content exhibited better properties when compared to other samples. Vijaya Ramnath et al. [17] tested Manila Fibber Reinforced Composite is for its tensile properties. Glass fibres are used to provide rigidity and strength to composites. Hand layup method is used for production of composites.Fiber matrix interfacial bonding determines the strength of composites. Homogeneous distribution of fibres results in uniform tensile properties in the samples. Rajesh et al. [18] studied the free vibration behavior of banana fibers. Hydrophilic nature plays a very important role in natural fiber where it produces poor adhesion between fiber-matrix. The fibres were pre-treated to produce good results. This chemical treatment increased the free vibrational properties of composites. In this work, short and random fibers were used. When investigated for damping factor and frequency, it was found that 50%composite had maximum natural frequency, it was because of the stiffness and modulus were increased by the chemical treatment. Venkateshwaran et al. [19] studied the water absorption behavior of natural fiber based composites. First of all, weight percentage and fiber length were determined. Fracture behavior was studied by Morphological analysis using Scanning Electron Microscope. The samples were prepared in mold machine. Due to the poor interfacial bonding between matrix and fiber there was a marginal increase in mechanical properties. By chemical treatment, interfacial bonding was improved. Bhoopalan et al. [20] studied the thermal properties of banana fiber reinforced epoxy composites. The composites were prepared by molding technique by preparing them in different weight ratios. Tensile, flexural and water absorption tests were carried out. There was an increase in thermal properties and considerable decrease in moisture absorption property. There was a marginal increase in mechanical property which was due to interfacial bonding between fiber and matrix.

4.0 CONCLUSION

In this paper, the natural fibre composite properties are reviewed on Banana and jute based hybrid composite .It is found that the natural fiber composites possess higher properties as compared to traditional one. Also it is concluded that natural fiber composites can be used as replacement material for automotive, construction, marine and domestic applications.

REFERENCES

1. Maya Jacob ,Sabu Thomas, K.T.Varughese, Mechanical properties of sisal/oil palm hybrid fiber reinforced natural rubber composites, *Composites Science and Technology*, 64 (2004), 955-965.

2.Leandro Joseda Silva,Túlio Hallak Panzera[a],Vânia ReginaVelloso[a], Andre Luis Christoforo, FabrizioScarpa ,Hybrid polymeric composites reinforced with sisal fibres and silica microparticles, *Composites*, 43 (2012), 3436-3444.

3.G. Yuvaraj, Dr.B. Vijaya Ramnath, G. Hemanth raagavendiran, 4R. Karthikraja, Investigation Of Flexural And Water absorption behaviour of epoxy hybrid composites, *Advances in Natural and Applied Sciences*, 2016 Special,10, (7): pages 34-39

4. G. Yuvaraj B.Vijaya Ramnath A.Abinash B.srinivasan R.Vikas Nair, Investigation of Mechanical Behaviour of Sisal Epoxy Hybrid Composites, *Indian Journal of Science and Technology*, Vol 9(34), DOI: 10.17485/ijst/2016/v9i34/101004, September 2016

5. B.Vijaya Ramnath S.Rajesh C.Elanchezian A.santosh Shankar S.pithchai pandian S.vickneshwaran R.Sundar Rajan, Investigation on Mechanical Behaviour of Twisted Natural Fiber Hybrid Composite Fabricated by Vacuum Assisted Compression Molding Technique, *Fibers and Polymers* 2016, Vol.17, No.1, 80-87

6.N. Amira, Kamal Ariff Zainal Abidinb , Faizzaty Binti Md Shiric, Effects of Fibre Configuration on Mechanical Properties of Banana Fibre/PP/MAPP Natural Fibre Reinforced Polymer Composite, *Procedia Engineering* ,184 (2017) 573 – 580

7.William Jordana , Patrick Chestera, Improving the Properties of Banana Fiber Reinforced Polymeric Composites by Treating the Fibers, *Procedia Engineering*, 200 (2017) ,283–289

8.A.N. Benítez a, M.D. Monzón b , I. Angulo a , Z. Ortega a , P.M. Hernández b , M.D. Marrero b, Treatment of banana fiber for use in the reinforcement of polymeric matrices, *Measurement*, 46 (2013), 1065–1073

9.M. Ramesha, T.Sri Ananda Atreyaa , U. S. Aswina , H. Eashwara , C. Deepab, Processing and Mechanical Property Evaluation of Banana Fiber Reinforced Polymer Composites , *Procedia Engineering* ,97 (2014) ,563 – 572

10. P.Sathish · R. Kesavan1 · B. Vijaya Ramnath · C. Vishal, Effect of Fiber Orientation and Stacking Sequence on Mechanical and Thermal Characteristics of Banana-Kenaf Hybrid Epoxy Composite, *Silicon DOI* 10.1007/s12633-015-9314-7.

11. Manickavasagam V.M , Vijaya Ramnath B Elanchezhian C Sundarrajan R Vickneshwaran S, Santhosh Shankar A, Kaosik R, Santhosh Kumar K, Evaluation of the double shear and hardness of abaca and flax reinforced polymer composite for automotive applications, *Applied Mechanics and Materials* Vols 766-767 (2015) pp 85-89.

12. Vijaya Ramnath, R. Sharavanan1, M. Chandrasekaran, C. Elanchezhian, R. Sathyanarayanan, R. Niranjan Raja, and S. Junaid Kokan, Experimental Determination of Mechanical Properties of Banana Jute Hybrid Composite, *Fibers and Polymers* 2015, Vol.16, No.1, 164-172.

13.Laly A. Pothan , Zachariah Oommen, Sabu Thomas ,Dynamic mechanical analysis of banana fiber reinforced polyester composite , *Composites Science and Technology*, 63 (2003) , 283–293

14 S.M. Sapuan,A.Leenie,M.Harimi,Y.K.Beng, Mechanical properties of woven banana fibre reinforced epoxy composites ,*Materials & design* ,27 (2006) ,689-693

15.Laly A.Pothan,Sabu Thomas ,Polarity parameters and dynamic mechanical behaviour of chemically modified banana fiber reinforced polyester composites ,*Composites science and Technology*, 63 (2003) ,1231-1240

16. B. Vijaya Ramnath a, V.M. Manickavasagam b, C. Elanchezhian a, C. Vinodh Krishna , S. Karthik, Determination of mechanical properties of intra-layer abaca–jute–glass fiber reinforced composite, *Materials and Design*(2014)643-652

17. Vijaya Ramnath B,, Manickavasagam V.M b, Elanchezhian C , Rajendra Prasad A , Kavin C, Karthik Subramanian B, Rahul V, Investigation of Tensile Properties of Manila Fibre Reinforced Composite, *Applied Mechanics and Materials* Vols 766-767 (2015) pp 96-99.

18.M Rajesh , Jeyaraj Pitchaiman, N Rajini , Free Vibration Characteristics of Banana/Sisal Natural Fibers Reinforced Hybrid Polymer Composite Beam, *Procedia Engineering* ,144 (2016), 1055 – 1059

19.N.Venkateshwaran,A.ElayaPerumal,A.Alavudeen,M.Thiruchitrambalam ,Mechanical and water absorption behaviour of banana/sisal reinforced hybrid composites ,*materials and design* ,32 (2011) ,4017-4021

20.M. Boopalan, M. Niranjanaa, M.J. Umapathy ,Study on the mechanical properties and thermal properties of jute and banana fiber reinforced epoxy hybrid composites, *Composites part B:Engineering,* 51 (2013), 54–57.

Process Parameter Optimization in TIG Welding of AISI 4340 Low Alloy Steel Welds by Genetic Algorithm

D.Vijayan[1*], V. Seshagiri Rao[2]
[1]Assistant Professor, Department of Mechanical Engineering, Sri Chandrasekharendra Saraswathi Viswa Mahavidyalaya, Enathur, Kanchipuram, Tamilnadu, India.
[2]Professor, St. Joseph's College of Engineering, Chennai, Tamilnadu, India.

[1*]Corresponding author: vijaiand2012@gmail.com

Abstract Weld quality is an essential requirement to the fabrication and construction industries. In this present investigation, an attempt has been made to optimize the process parameters of Tungsten Inert Gas (TIG) welding of AISI 4340 low alloy steel. Taguchi L_9 orthogonal array is used to design the experiment and response surface methodology was used to develop mathematical model and the process parameters were optimized using Genetic Algorithm (GA). The important process parameters namely welding current, voltage, welding speed, and gas flow rate are considered. The results indicate that the current and voltage have a maximum influence on improving Ultimate Tensile Strength (UTS) in TIG-welded joint.

1. Introduction

Tungsten Inert Gas (TIG) welding is one of the usual arc welding processes where arc produces between non-consumable tungsten electrode and workpiece and metal fusion happens. TIG welding is broadly utilized for joining alloy steels, stainless steels in numerous fabrication industries due to high caliber and great appearance. Be that as it may, one potential issue related to traditional TIG is the constraint on workpiece thickness that can be welded in single pass thus offers limited efficiency. And, TIG welding should have the following characteristics, (1) a straight polarity direct current, (2) high frequency arc starting, and (3) power supply reduction device (1). Increment in current may not be shrewd dependably as it will prompt increment in bead width with relatively low pick up in infiltration. Nickel-Chromium-Molybdenum steels are broadly utilized for joining as a part of various applications including marine, automotive and machine instruments, pressure vessels, transport, mining and so on. Impressive measures of spotlight on accomplishing high caliber welded joints of such materials for similar and dissimilar joining are being investigated at quicker production rates with higher amount of infiltration. Exertion has been made by a few researchers to modify the TIG welding process by utilizing fluxes to increase the weld penetration. The depth of penetration amid TIG welding could be altogether expanded laying an activated flux layer on workpiece surface before welding, accordingly eliminates the need of edge preparation, reduces number of passes required and improves efficiency. The procedure is instituted as flux activated TIG welding. Numerous analysts directed experimental studies to assess the impact of various oxide, halide or mixture of fluxes on increase in DOP and relating decrease in width or width-to-penetration (WP) on various high nickel-chromium-molybdenum steels at various instances (2-4). Be that as it may, the welding innovation has not been exceptionally fruitful to join nickel-chromium-molybdenum steels in commercial ventures in this way, and related studies are likewise to a great degree constrained. So, it gets to be important to explore the relevance of welding innovation in joining little scale parts.

2. Materials and Methods

The base materials used for the present investigation were AISI 4340 (ASTM A29), produced by M/s ALOKINGOTS India Pvt. Ltd, Mumbai, India is a medium carbon, low alloy steel. The chemical compositions of base metal as received is C – 0.39, Mn – 0.72, Cr – 0.7, Ni – 1.65, Si – 0.15, Mo – 0.2, P – 0.035 and S – 0.04. All AISI 4340 steel plates were sized into 150mm (L) × 75mm (W) × 5mm (t) were butt welded at different process parameters as per the design matrix. Figure.1 demonstrate the AISI 4340 steel plates to be welded. Before welding, acetone was used to eradicate the oils that remained on the surface of the specimens. Extensive welding trials were conducted in autogenous mode with ESAB make A-TIG power source having capacity of 250A with 25% duty cycle to examine the performance and appearance of the weld.

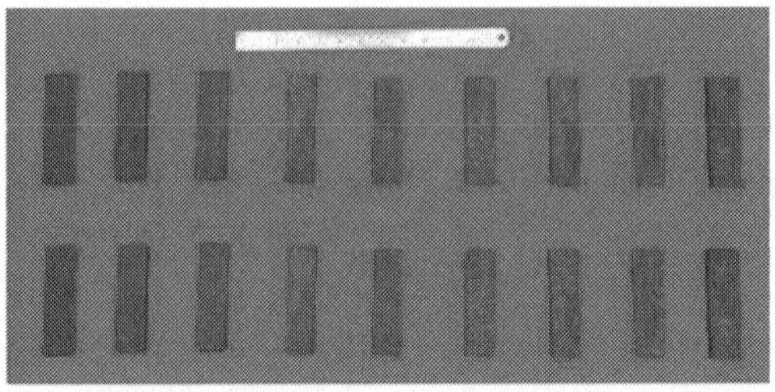

Figure 1. Test specimens of AISI 4340 – medium carbon, low alloy steels

Then, all the steel plates were welded by TIG welding as per the variables designed by orthogonal array using Taguchi technique. A 3 mm diameter and 2.2 % thoriated tungsten electrode with an air-cooled torch was used. The welding variables and their corresponding ranges are presented in Table 1. All the samples were welded under full penetration. The electrode angle was kept as 19 – 22° pure argon was used as a shielding gas during welding. The flux paste was applied later welding, acetone evaporated, parting a layer of flux on the surface. The torch was moved along the center line of the weld specimens.

Table.1 – Welding variables and their range

Process Factors	Current (A)	Voltage (V)	Welding Speed (WS) (mm/sec)	Gas Flow Rate (GFR) (lpm)
Levels	70 - 80	12 - 14	55 - 65	8-12

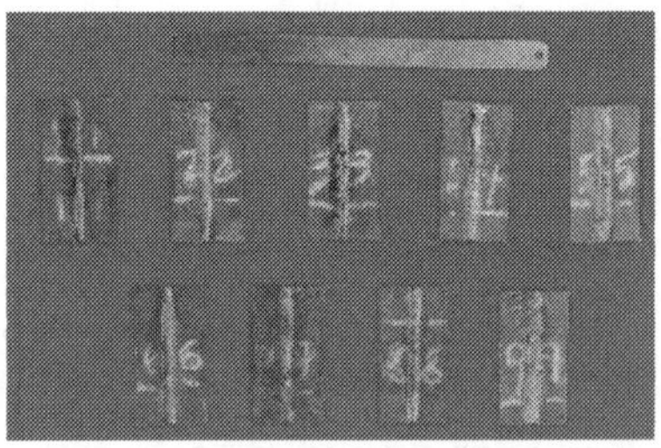

Figure 2. Fabricated AISI 4340 low alloy steel

The fabricated AISI 4340 steel plates after welding are presented in Figure 2 and their corresponding design matrix is presented in Table 2. After welding, specimens were collected from the center of the welded plate and subjected to Ultimate Tensile Strength (UTS). Subsequently, for conducting the hardness test, samples were allowed for grinding and polishing followed by etching in a solution of 35% concentrated HCL at 60 - 80°C for 12–15 minutes to produce a bright surface. Scanning electron microscope (SEM) (Make: FEI Quanta FEG 200 high resolution – Micro lab, Chennai, Tamilnadu, India) is used to develop the microstructure of the welded samples. Further, hardness variation in HAZ, NZ and TMAZ on the effect of TIG welding process parameters the specimens were subjected to vickers hardness and Izod impact test. Micro-hardness measurements were taken across the specimen at 300 gf load and 15 s dwell time on Vickers micro-hardness tester (Make: Fuel Instruments and Engineers (FIE), Mettax laboratory, Chennai, Tamilnadu, India). Indentations were typically performed at 3 mm spacing.

IOP Conf. Series: Materials Science and Engineering **390** (2018) 012066 doi:10.1088/1757-899X/390/1/012066

Table 2. Experimental results

Std	Current (A)	Volatge (V)	Welding speed (WS) (mm/min)	Gas Flow Rate (GFR) (lpm)	Ultimate tensile strength (UTS)
1	70	12	55	8	972
2	70	13	60	10	961
3	70	14	65	12	951
4	75	12	60	12	970
5	75	13	65	8	933
6	75	14	55	10	943
7	80	12	65	10	938
8	80	13	55	12	952
9	80	14	60	8	915

3. Optimization of welding parameters

3.1. Response surface methodology

Response Surface Methodology (RSM) is collection of mathematical and statistical techniques for process development optimization(5). RSM is often accomplished directly for the objectives of quality improvement including reduction of variability, process enhancement and product performance (6). The RSM is widely used for setting the parameter and optimization such as chemical, electronics, manufacturing and metal cutting industries. Lot of process parameters may affect the output quality responses in the TIG welding process. Therefore, optimization and finding several independent variables affecting the TIG welding process is quite complex and time-consuming process. However, using design of experiments and applying regression analysis, process modelling of the favorite response to numerous independent input variables can be gained. The response surface can be expressed if all the variables are assumed to be measurable,

$$y = f(x_1 + x_2 + ... + x_k) + \in$$

In many engineering analysis, there is a relationship between an output variable on interest 'y' and set of controllable variables $(x_1 + x_2 + ... + x_k)$, $\in$ is a noise or experimental error. Usually in most of the engineering and non-engineering RSM problems the relationship between the input and output variable are unknown. Thus, finding the suitable approximation for the true functional relationship between 'y' and set of the independent variables is the first step in RSM. Usually the second order model is utilized in RSM,

$$y = \beta_0 + \sum_{i=1}^{k} \beta_i x_i + \sum_{i=1}^{k} \beta_{ii} x_i^2 + \sum_{i} \sum_{j} \beta_{ij} x_i x_j + \in$$

The β coefficients can be calculated by means of using least square method in above model. The second-order model is normally used when the response function is not known or nonlinear (7).

3.2 Genetic algorithm

Genetic algorithm (GA) is used for stochastic problems, which is introduced by Holland (8) based on mechanics of a biological evolution process. Usually, a decision variable of a coded gene is represented by a real number, a bit or a string in genetic algorithm. And a corresponding gene of each parameter creates a chromosome which represents a binary of strings or an array of a problem that can be varied depending on a treated problem (9). A set of chromosomes is called population. Among solutions, one population is preferred through a selection operator such as roulette wheel or tournament selection, where the probability is to be selected relative to individual fitness. The selected set of individuals is used further to create a new population through genetic operators called mutation and crossover. Crossover chooses genes from parent chromosomes and builds a new offspring. Besides, the mutation is preventing all the solutions of the population get trapped into a local optimum. This procedure is repeated under some situations are satisfied, until goodness of best solution is attained. The GA parameters used in the present investigation is presented in Table 3.

Table 3. GA parameters

Parameters	Setting value
Population size	50
Crossover rate	0.8
Mutation rate	0.7
Selection operator	Adaptive feasible
Fitness operator	Tensile strength

4. Results and Discussion

4.1. Mathematical design of the problem

The maximization of the present fitness function values is subjected to the boundaries of weld parameters. The weld parameters and their range are shown in Table 4 is selected to show the limitations of the optimization solution and is given as follows,

Table 4. Weld parameters and their ranges

Parameters	Range
Current	$70 \leq A \leq 80$
Voltage	$12 \leq V \leq 14$
Welding speed	$55 \leq WS \leq 65$
Gas flow rate	$8 \leq GFR \leq 12$

In order to apply GA several parameters to be considered while initiating the problem toward the objective function of the problem. The MATLAB optimization toolbox is used to maximize the TIG welded AISI4340 steels. The optimal values for the maximization of UTS of TIG welded parameters obtained from GA is Current (A) = 70A, Voltage (V) = 12V, Welding speed = 55mm/min and Gas flow rate = 11.807 lpm and it is indicated that the optimal solution is obtained in 52nd iteration. The best fitness value obtained from GA for maximizing the UTS of TIG welded AISI 4340 is 988.816 MPa.

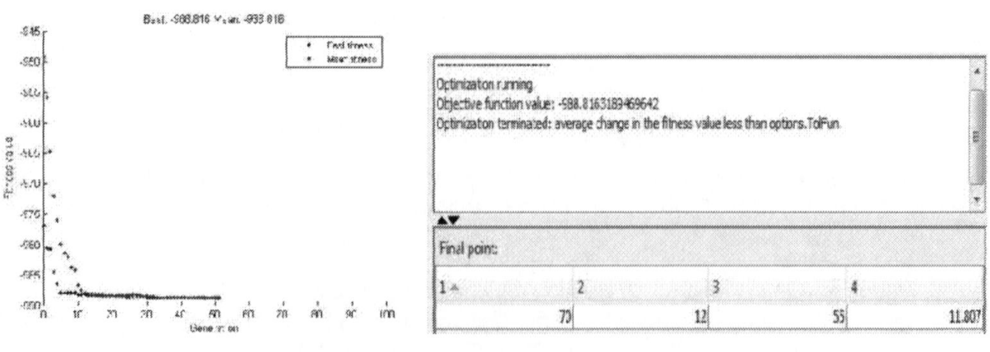

Figure 3. GA Convergence plot

Fitness value of the convergence result obtained using GA is graphically presented in Figure 3 that shows a dotted line signifies that there is no infringement of constrains in each generation. The maximum tensile strength by GA optimization was observed to be low current (70 A), low voltage (12 V), welding speed (55 mm/s), and the high gas flow rate (11.80 lpm). The optimal values of the process parameters obtained by the GA is 988.816 MPa. Based on Taguchi L9 orthogonal array nine experimental data was assessed from the experiments. Regression analysis was carried out model for UTS is,

$$UTS = +1345.5 - 2.63\times Current - 11.83\times Voltage - 1.50\times Welding\ speed + 4.41\times Gas\ flow\ rate$$

The analysis of variance (ANOVA) test was employed to determine the percentage contribution of the parameters at 95% confidence level and to investigate which process factor significantly affected the quality characteristic of TIG welding. The result of ANOVA is presented in Table 5. The significance of each process factor was tested using F – test. In general, if F of experimental runs are higher than the table 10.128 (from the Standard Fishers table), which means that the varying the process factor makes a big change on the process performance. In addition, the percentage contribution (P) indicates the significance of process factors for the response. For a factor with a high values of P, a small variation will have a great influence on the performance (10). The graphical representation of percentage contribution is presented in Figure 4, it observed that, current (A) was found to be a major factor affecting the welding process significantly the mean average of UTS of 38.64%. The voltage (V), gas flowrate (GFR) and welding speed (F) have an effect on the UTS of 31.21%, 17.39% and 12.54% respectively. And, the predicted and adjusted coefficient of determination (R-Squared) are 99.78% and 99.55% respectively. The "predicted R-Squared" of 99.78% is in reasonable agreement with the "Adjusted R-Squared" of 99.55%.

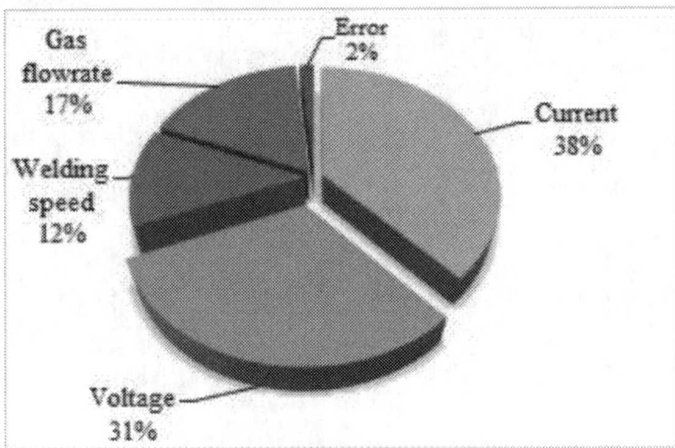

Figure 4 Contribution plot

Table 5. ANOVA results

Source	df	Adj SS	Adj MS	F value	p value	
Regression	4	2686.00	671.50	447.67	< 0.0001	Significant
A – Current	1	1040.17	1040.17	693.44	0.003	
B – Voltage	1	840.17	840.17	560.11	<0.0001	
C – Welding speed	1	337.50	337.50	225.00	<0.0001	
D – Gasflow rate	1	468.17	468.17	312.11	0.0002	
Error	4	6.00	1.50			
Total	8	5378.01				
S = 1.22	R-Sq = 99.78%		R-Squ (adj) = 99.55		R-Sq (Pre) = 98.95%	

4.2. *Ultimate tensile strength*

A SEM image of base materials were taken in order to understand primary phases of the materials and to ascertain morphological changes in various phases of materials after TIG welding. The developed SEM images of base materials as presented in Figure 5(a) and Figure 5(b). The temperature of weld metal with a much lower amount of hot molten mass comparatively to surrounding temperatures were always low. Hence an air-cooling effect was experienced by the weld metal produces irregular harder

IOP Conf. Series: Materials Science and Engineering **390** (2018) 012066 doi:10.1088/1757-899X/390/1/012066

(martensite) phases owing to insufficient solidification time. Moreover, a couple of micro holes have been observed on the weld zone due to rapid solidification of molten metals as shown in Figure 5(a). This phenomenon is observed in all samples being welded irrespective to the process parameters. Besides that, from over all micro structures a clear elongated grain of irregular harder martensitic interfaces was seen in all the fabricated weld joints as shown in Figure 5(b) And in the HAZ region, exhibited fine acicular ferrite and coarse harder martenstic structure was seen, hence a lower hardness was observed. The readings were taken in Vickers hardness scale is plotted. Figure 6 shows the hardness distribution at the cross section of the welded joints of AISI 4340 steel plates. The weld zone from -5mm to +5mm showed lower hardness value when compared to base materials of AISI 4340 steel plates. Due to more heat capacity of the argon gas flame quickly softening the metals thus leads the formation of micro holes on the weld zone as shown in Figure 5(a). This tendency in nature, decreasing the hardness on the weld zone and HAZ. However, the obtained hardness plot is mostly owing for ductile nature of the joints as evidenced from the fracture locations of welded plates. At weld nugget zone, the weld metal hardness is achieved was maximum due to the formation of irregular and harder martensitic structures. The HAZ region exhibited a mix of coarse and fine microstructures. And due to full penetration welding, cooling rate is slow hence there is no noticeable changes in the micro structures were observed in the metallurgical properties (4).

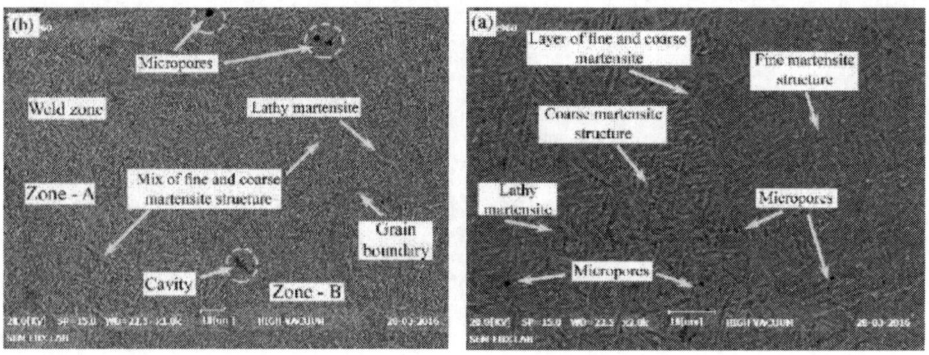

Figure 5. (a) Micro holes in TIG weld zone, (b) Harder martenstic interface

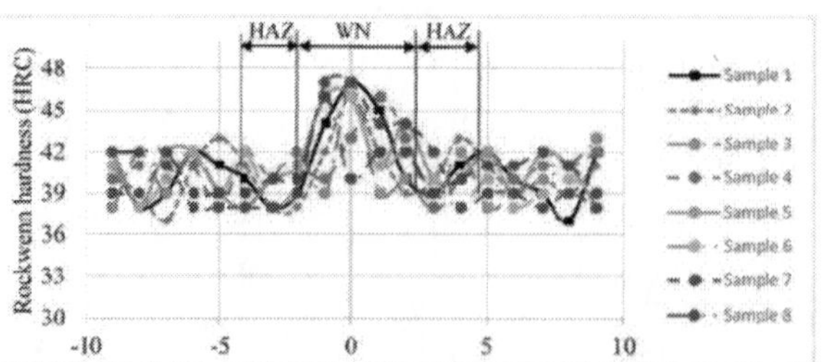

Figure 6. Rockwell hardness distribution of fabricated TIG welded samples

The main effects plot for UTS with individual effect of current, welding speed, voltage, and gas flow rate are presented in Figure 7. It indiactes lower current, welding speed, and voltage maximizing tensile strength because of lower heat input. The tensile strength increases with increase in gas flow rate. The reason is that shielding gases are used to prevent atmospheric contamination of the weld metal. A too low heat input result more weld defects thus tensile strength reduced in the fabricated weld joints. On the other hand increasing gas flow rate increasing tensile strength of the weld joint. Since, shielding gas has low ionization potential and is heavier than air, subsequently it giving exceptional shielding of the molten weld pool which improves the mechanical properties of the weld joint.

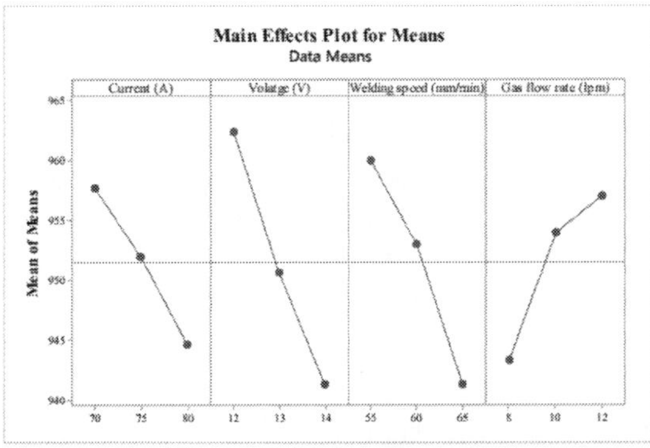

Figure 7. Main effect plot

The interaction effect of welding speed versus current on UTS of TIG welding process parameters is presented in Figure 8. It can be seen that increasing welding current increases UTS from 927 MPa to 954 MPa. Increasing welding current increases weld penetaration therefore heat input increases subsequently. Moreover, increasing weld penetration improves fluid flow in the weld pool, which is driven by the electromagnetic force, surface tension gradient, buoyancy force, and impinging force. Thus UTS is increases with increasing welding current of the fabricated weld joints.

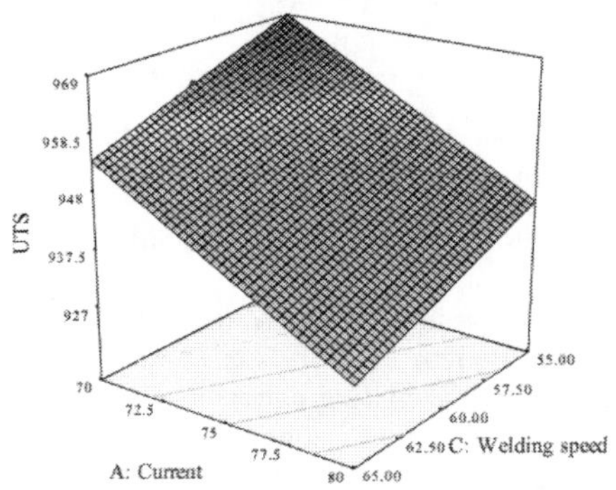

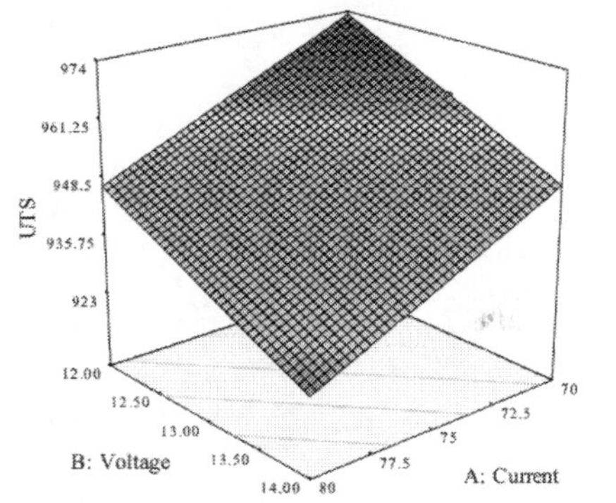

Figure 8. Welding speed, Current vs UTS **Figure 9. Welding speed, Gas flow rate vs UTS**

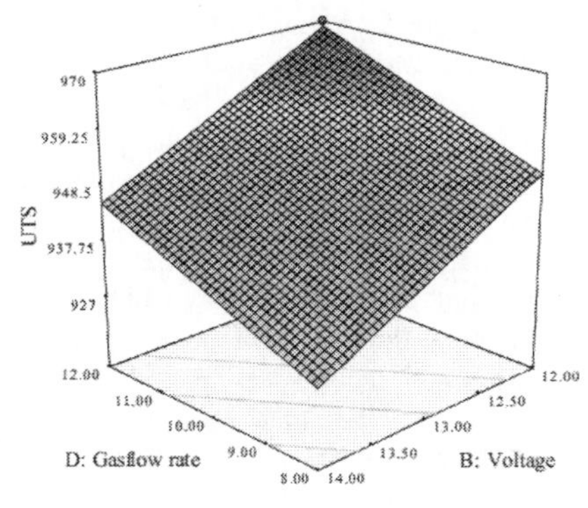

Figure 10. Voltage, Gas flow rate vs UTS

Figure 9. shows the welding current, voltage vs UTS. It is observed that the voltage is directly responsible to the UTS. For voltage, F value is 694.44 in the ANOVA which is the most significant term and consequential term in the TIG welding. In the surface plot, it can be observed that the incrementing current will decrease the UTS values from 961 MPa to 943 MPa. A change in microstructure directly affects the mechanical properties of the weld. Consequently, when the heat input increases highly, the hardness of the welded joint decreases slightly due to microstructure coarsening. Figure 10. shows surface plot of welding speed, gas flow rate vs. on UTS. When the voltage varies from 12 to 13.5 V, the UTS extremely decreases from 970 MPa to 950 MPa. In addition to that, an increasing gas flow rate from 8 to 12 lpm, the UTS increases slightly from 927 MPa to 948.5 MPa. Gases with low ionization potential encourage the ignition of the electric arc, and those with low thermal conductivity tend to build the arc stability. The corresponding F value is 560.11, which implies that voltage is the most critical term in the conducted A-TIG welding process.

5. Conclusion

In this present investigation, an attempt has been made to optimize the process parameters of A-TIG welding of AISI 4340 steel. Taguchi L_9 orthogonal array has been utilized in the present investigation to optimize the process parameters of TIG welding process. The mathematical model has been developed to understand the relationship between the ultimate tensile strength and the individual factors considered in the experiment namely welding current, voltage, welding speed, and gas flow rate were used and the ultimate tensile strength of AISI 4340 steel has predicted at 95 % confidence level. ANOVA has been employed to determine the percentage contributions of the process parameters. Optimization was carried out with the response surface method based genetic algorithm. Confirmation test reveals that GA can be utlized effectively to evaluate input process parameters to get the required ultimate tensile strength.

References

1. Zhu Y-P, Li C-Y, Zhang L-Y. Corrosion resistance and antibacterial activity of different zones in TA2 weldment by TIG welding. Rare Metals. 2014.
2. Bhattacharya A. Revisiting Arc, Metal Flow Behavior in Flux Activated Tungsten Inert Gas Welding. Materials and Manufacturing Processes. 2015;31(3):343-51.
3. Singh R, Slathia RS. Effect of Cryogenic Treatment on Sensitization of 304 Stainless Steel in TIG Welding. Journal of The Institution of Engineers (India): Series C. 2015;97(2):127-30.
4. Joseph J, Muthukumaran S. Optimization of activated TIG welding parameters for improving weld joint strength of AISI 4135 PM steel by genetic algorithm and simulated annealing. The International Journal of Advanced Manufacturing Technology. 2015.
5. Vijayan D, Rao V. Friction Stir Welding of Age-Hardenable Aluminum Alloys: A Parametric Approach Using RSM Based GRA Coupled With PCA. Journal of The Institution of Engineers (India): Series C. 2014;95(2):127-41.
6. Vijayan D, Seshagiri Rao V. Parametric optimization of friction stir welding process of age hardenable aluminum alloys–ANFIS modeling. Journal of Central South University. 2016;23(8):1847-57.
7. Palanikumar K. Application of Taguchi and response surface methodologies for surface roughness in machining glass fiber reinforced plastics by PCD tooling. The International Journal of Advanced Manufacturing Technology. 2006;36(1-2):19-27.
8. Holland JH. Adaptation in natural and artificial systems: an introductory analysis with applications to biology, control, and artificial intelligence: U Michigan Press; 1975.
9. D.Vijayan VSR. Tensile Properties Improvement on Friction Stir Welded Age-Hardenable Aluminum Alloys: An Evolutionary Approach using RSM based GA and SA. Revista Tecnica De La Facultad De Ingenieria Universidad Del Zulia; 2016. p. 55 - 70.
10. Souissi N, Souissi S, Lecompte J-P, Amar MB, Bradai C, Halouani F. Improvement of ductility for squeeze cast 2017 A wrought aluminum alloy using the Taguchi method. The International Journal of Advanced Manufacturing Technology. 2015;78(9-12):2069-77.

IOP Conf. Series: Materials Science and Engineering **390** (2018) 012067 doi:10.1088/1757-899X/390/1/012067

Sandwich and Natural fiber composites - A review

G Ramakrishnan *, B. Vijaya Ramnath, C.Vignesh, L.Vignesh, T.Perunddevan and M VidhyaRajan
Department of Mechanical Engineering
Sri Sairam Engineering College, West Tambaram, Chennai-44.

*** Corresponding author mail ID: ramakrishnan.mech@sairam.edu.in**

Abstract. Sandwich composite is a special form of a laminated composite comprising a combination of different materials that are bonded to each other so as to utilize the properties of each separate component to the structural advantage of the whole assembly. Usually a sandwich structure consists of two relatively thin, stiff and strong faces separated by a relatively thick lightweight core, for instance, honeycomb, balsa or foam cores. The purpose of sandwich structure is to achieve a stiff and simultaneously light component. A natural fiber composite also plays a vital role in automotive and construction filed due to their bio degradable behaviour and ease of manufacturing. This paper reviews sandwich structures, various tests performed on it and their properties along with some studies of mechanical behaviour of natural fiber composites. Major applications of the sandwich specimens are also highlighted in this work.

1.0 Introduction

A Composite material is a type of material made from two or more constituent materials, when combined, produces a material with characteristics superior to the individual components. The New Materials are added to improve characteristics such as Specific Strength, Stiffness, Hardness etc.A Sandwich-Structured Composite is a special class of Composite materials that is fabricated by attaching two thin but stiff skins to a lightweight but thick core.The important characteristics of sandwich composites are high stiffness to weight ratio and high bending strength to weight ratio.

2.0 Types of sandwich composites

1.FOAM INSULATED CONCRETE SANDWICH COMPOSITE.

2.BALSA WOOD SANDWICH COMPOSITE

3.ALUMINIUM HONEY COMB SANDWICH COMPOSITE.

2.1 Foam insulated concrete sandwich composite

FengtaoBaiand James S. Davidson. [1] in their paper presents the theoretical development that defines the small displacement behavior of foam insulated concrete sandwich panels. Composite theories presented by other researchers are first thoroughly reviewed and scrutinized in the context of their use for precast concrete beams. A more rigorous Discrete Model that incorporates the complex shear deformation behaviors and independent flexural resistance of the concrete wythes is then derived. Experimental data from full-scale precast sandwich panel tests are used to validate the developed methodology. Finally, it is demonstrated that this study provides a rigorous analysis methodology for foam insulated concrete sandwich structures. The mechanical behaviors of metallic sandwich panels, nailed timber systems, composite bridge beams and ICSPś are similar in some aspects. For example, they all consist of a relatively deformable middle layer that results in large overall shear deformation. However, there are also significant differences. For example, metal faced sandwich structures, due to the slenderness of the face plates, are more susceptible to face buckling and other local failure modes. For nailed timber and composite bridge beams, components are connected tightly through shear connectors, leaving little room for shear deformation. Actually, those two-layer composite structures could have fairly high composite action that in some circumstances may be considered as fully composite. Whereas metallic sandwich panels and ICSPs have a thick middle insulation bearing considerable amount of shear deformation and therefore are usually Composite material. Granholm [2] published his theory and work in the field of nailed timber structures. Granholm's theory focused on the equilibrium of axial force within individual layer and overall bending moment of the whole cross-section. Zhiqiang Fan et al. [3] conducted blast resistance of metallic sandwich panels subjected to proximity underwater explosion with honey comb structured aluminum-5052 alloy as core and concluded that specimens with thicker face-sheets and denser cores produced smaller deflections. Their sandwich specimens found applications in naval engineering.

2.2 Balsa wood sandwich composite
Grenestedt and Burak Bekisli [4] analyzed a new arrangement of balsa blocks in a sandwich core. This new core, high consisted of an assembly of oddly shaped balsa blocks, allowed the grains in various balsa blocks to be oriented to the normal direction of the core. The effective out-of-plane shear stiffness was isotropic in nature. Numerical analyses predicted superior shear stiffness. The shear stiffness depends on the grain orientation as well as boundary conditions. Experiments proved the new balsa core to be 70% stiffer than end grain balsa. Burak Bekisli and Grenestedt et al. [5] analyzed a new balsa core structure that better utilizes the anisotropy of balsa. Improved shear stiffness was predicted with this new core. Manufacturing methods for this core were developed and more extensive experiments were performed to verify the improvement in shear stiffness as well as to evaluate the shear strength. The increased stiffness and strength were obtained at the cost of complexity in terms of manufacturing and increased scrap. Apart from the CNC milling, the manufacturing methods produced balsa blocks that were not geometrically perfect. This led to gap between the balsa blocks, which filled with resin during the assembly and therefore increased the weight. However, it is believed that the wire saw manufacturing process easily be automated and tuned to produce almost perfect balsa blocks, which would lead to new cores with increased mechanical properties without a prohibitive cost or weight penalty.

Cesim Atas and Cenk Sevim [6] studied the damage process of the sandwich composites and analyzed the cross-examining load–deflection curves, energy profile diagrams and the damaged specimens. The primary damage modes observed are; fiber fractures at upper and lower skins, delaminations between adjacent glass–epoxy layers, core shear fractures, and face/core debonding. Experimental investigation on drop weight impact response of sandwich composite panels with PVC foam core and balsa wood core. Load– deflection curves of the both composites were of a mountain-like shape with two peaks under higher impact energies. It is noted that the second peak value for foam core sandwich samples is higher than first peak value generally while those of the balsa core samples are of the nearly the same values. On the other hand, in two-peaked curves, it is also noted that the first peak load value of foam core samples were smaller than that of balsa core samples as in the one-peaked curves. On the contrary, the second peak load value of foam core samples is found to be higher generally than that of balsa core ones. It is also seen from the Load–Deflection curves that the foam core sandwich is of lower bending stiffness. Balsa core samples are stiffer than PVC ones. It results in higher contact force but smaller deflection of the balsa wood sandwiches for smaller impact energy levels, before failure. This case implies a higher deformation capability of PVC foam samples, resulting in a larger delamination area formed between layers of top face-sheets due to stiffness mismatching. However, for balsa core sandwich, due to poor interface between balsa wood with the glass/epoxy layer compared with PVC foam core; the debonding between core and face-sheets becomes the dominant damage mode along with the fiber bending fractures, as the impact energy increases. As the impact energy is decreased, the Hardness increases for foam core samples compared to balsa wood composites. Vincent Legrand et al. [7] investigated the internal stresses induced by moisture absorption in sandwich composite material. Moisture absorption effect was analyzed for balsa core specimens and sandwich structures composed of E-glass/polyester face sheets bonded to a balsa core. A new experimental technique for measuring hygroscopic expansion has been developed and tested for the balsa core. The obtained results show that the diffusion of water in the balsa core conducts the diffusivity process in the entire sandwich structure. Therefore, the calculated diffusion parameters of the single balsa are higher than those of the balsa embedded in a sandwich structure.

Kepler [8] studied the concept based on utilization of the strongly orthotropic properties of the balsa wood, applying an appropriate transverse layup sequence. Comparing the standard balsa core systems, a substantial increase in the shear stiffness is demonstrated, whereas the transverse stiffness is reduced. The exposure to excess moisture degraded the mechanical properties of the balsa wood. Michael Osei-Antwi et al. [9] studied an experimental to evaluate the effects of parameters such as shear plane, density and adhesive joints on the shear stiffness and strength of balsa wood panels as well as the variation of ductility with respect to the shear planes. The shear planes exerted a significant effect on shear stiffness and strength. The strength of specimens with the shear plane transverse to the flat grain was reduced because of a change in the failure mode. The shear planes exerted a significant effect on shear stiffness and strength. Highest values were obtained for the shear plane (parallel to end grain), intermediate values for the plane (parallel to flat grain) and lowest values for the plane (transverse to flat grain). Shear stiffness and strength increased with increasing density of the balsa. The thin adhesive joints in the balsa panels between the lumber blocks increased the shear stiffness and strength with one exception. the strength of specimens was reduced because of a change in the failure mode. Julia de Castro et al. [10] investigated the fracture in the complex balsa cores of fiber-reinforced polymer (FRP) sandwich beams. The balsa layers were cut from panels which consisted of balsa blocks adhesively bonded together. Failure in the beams was initiated by cracks propagating through the balsa core thickness The effects of varying block densities and orientations and adhesive bonding between the blocks on the crack propagation and failure mode were investigated. Failure in the beams was initiated by cracks through the balsa core thickness. Cracks initiated and propagated in the low-density blocks due to their low fracture toughness. Cracks were not able to propagate through the transverse adhesive joints between blocks if the bonding was good and thus the fracture toughness was high.

2.3 Aluminium honeycomb sandwich composites

Yahaya et al. [11] studied the resistance of sandwich panels with different aluminum honey comb cores subjected to impact from foam projectiles. They determined that increasing the foil thickness and reducing the cell size led to a decrease in the back-face deflection, but increases in the overall honeycomb sandwich panels' weight. A comparison was also made that the honeycomb sandwich panels out perform both the air sandwich panels and the monolithic plates within an impulse range of 2.25 $kNsm^2$ ~ 4.70 $kNsm^2$. Crupi et al. [12] analyzed the static and low-velocity impact response of two typologies of aluminum honeycomb sandwich structures with different cell size. The failure mode and damage of the honeycomb panels have been investigated using the 3D Computed Tomography. The experiments concluded that the energy absorption capability of sandwich is strongly influenced by the size of the honeycomb cells. Also they proposed that their specimen would find applications in the transportation industry. Shiqiang Li et al. [13] investigated the response of metallic sandwich panels with stepwise graded aluminum honeycomb cores under blast loading. Their study using LS-DYNA software showed that under the same charge condition, the plastic energy dissipated by the core layers and the contact force attenuation were larger than those of the ungraded ones, especially for the graded panels with relative density descending core arrangement. Foo et al. [14] analyzed the low-velocity impact failure of aluminum honeycomb sandwich panels and inferred that the energy absorbed during impact is independent of the core density. However, a foil with higher density and greater thickness will result in a more damage tolerant core. Smaller cell sizes will also improve the tolerance of the core to impact damage. Xin Li et al. [15] examined the blast-resistance of square sandwich panels with hexagon aluminum honeycomb cores. The major conclusion of the tests is that when the scaled distance decreases, peak pressure loaded on the face sheet enhanced obviously, thus pitting and fragment easily occurred. However, with the increasing of scaled distance, large plastic deformation is prone to occur. Shanshan Shi et al. [16] inspected the flexural strength and energy absorption of carbon-fiber–aluminum-honeycomb composite sandwich reinforced by aluminum grid.

They concluded that the honeycomb filled orthogrid sandwich panels possess higher specific properties which are crucial for aerospace applications.

3.0 Natural fiber composites and their mechanical characteristics

Raman Bharath et al. [17] reviewed the Kenaf fibre reinforced composites. The various properties and the applications of kenaf fibre have been elaborately studied and has concluded that the kenaf composite has many advantages including waste disposal and non-toxic. Vijaya Ramnath et al. [18] studied the mechanical properties of composites with the alternation in their composition that affects the mechanical properties due to the changes in fibre orientation. It has been found that the samples with high abaca content has better properties of tension, flexural and impact loadings. Sathish et al. [19] used the abaca and raffia fibres to fabricate the hybrid composites with epoxy resin and woven glass fibres. The abaca composite has maximum displacement as compared to others while the abaca and raffia hybrid composite has better de-lamination strength. Vijaya Ramnath et al. [20,21] evaluated the mechanical properties of epoxy composites with abaca-jute-glass fibre reinforcement and found that the properties of tensile and shear strength are improved in abaca-jute hybrid composite when compared to abaca fibre alone. But in the case of flexural and impact strength abaca composite turned out to be a better result. This is because the strength increases as interfacial adhesion in case of abaca increases. They also investigated the mechanical properties like double shear and hardness of abaca-raffia hybrid composite. It has been concluded that the composite with GFRP, Abaca and Raffia has more hardness due to the reason that Raffia with abaca makes it harder for the indenter to indent inside the specimen. Hence, this fibre composite can suitably replace the present materials in automobiles.

Yuvaraj et al. [22] studied the flexural and water absorption property of sisal and glass fibre reinforced hybrid composites with different ratios of fibre in random orientation with epoxy composite by hand layup method. It has been found that there is a rise in the flexural property and decrease in the water absorption capacity with the fibre content of the composite being increased. IndaraSoto et al. [23] used sisal fiber reinforced hollow concrete blocks and the mechanical behaviour was studied. Sisal fibre was cut into the required length and then it was mixed in the concrete mixture (cement, sand, coarse aggregates) with respect to concrete volume. Concrete blocks with and without sisal fibres were made. Three types of specimens (blocks, prisms, wallets) were made. The compression and tensile strength on blocks, prisms and wallets were tested. The compression test was done using servo hydraulic controlled device. Tensile strength was carried out in Universal testing machine as per standards. The mechanical properties remain the same in both the cases (with and without sisal) but addition of sisal increased the block ductility.

Partha haldar et al. [24] evaluated two different types of composites were manufactured with combining layup technique and compression moulding. One composite was made with addition of aluminium powder to sisal-epoxy composite and another one was made without the addition of aluminium powder and the testing was done. Micro-hardness test was done using micro hardness tester. Tensile test was done using Instron machine. Impact test was done using impact tester. Moisture absorption test was also conducted. It was found that sisal-epoxy composite with aluminium powder gave more density, high tensile strength, high elastic modulus and high impact strength. The sisal-epoxy composite without aluminium powder showed high moisture absorption and thickness swelling. Kejariwala et al. [25] studied in flammability and moisture absorption behaviour of sisal-polyester composite. Different specimens of different thickness were made for different volume fraction of sisal fibre and the fabrication was done using hot compression moulding and the tests were conducted as per standards. Flammability was tested as per the ASTM standard and the moisture absorption test was also tested as per ASTM standard. It was found that the rate of burning decreased when the fibre volume fraction and thickness increased. The moisture absorption increased with the increase in fiber volume fraction and thickness

Vijaya Ramnath et al. [26] fabricated and studied the tensile and impact properties of natual fibre composites with abaca and raffia fibres in different compositions. The ultimate tensile strength of the Glass fibre reinforced plastics with abaca and raffia composite is very much higher when compared to the single composite. Also it has been found that the abaca glass fibre has more energy absorbing capacity than the raffia. Srinivasan et al. [27] evaluated the flexural property of Kenaf and Flax hybrid composite along with the mono fibre composite. The effect of hybridization on various aspects has been studied and the crack propagation and shearing styles of the composites are discussed elaborately which are due to the effect of hand layup process. It leads to breakage early when compared to other processes. Sathish et al. [28] conducted the experimental investigation with abaca and raffia hybrid composites on the flexural properties. It has been found that the raffia fibre composite has better strength when compared to abaca composite. The flexural test which has been carried out has better results for raffia composite.

4.0 Conclusion

This paper reviewed on sandwich as well as natural fiber composite and work done by various researchers. It is found that sandwich composites finds its applications in Marines and Aerospace. Moreover they also concluded that natural fiber composites which are biodegradable finds its applications in Domestic and Automotive field.

References

[1] FengtaoBaiand James S Davidson 2015 *Analysis of partially composite foam insulated concrete sandwich structures,* Engineering Structures 91 pp 197–209

[2] Granholm H 1949 *On composite beams and columns with particular regard to nailed timber structures,* Transaction of Chalmer University of Technical, No. 88. Gothenburg

[3] Zhiqiang Fan, Yingbin Liu and Peng Xu. 2016 *Blast resistance of metallic sandwich panels subjected to proximity underwater explosion.* International journal of Impact Engineering 93 pp 128–135.

[4] Joachim L Grenestedt and Burak Bekisli *2003 Analyses and preliminary tests of a balsa sandwich core with improved shear properties,* International Journal of Mechanical Sciences 45 pp 1327–46

[5] Burak Bekisli and Joachim L Grenestedt 2004 *Experimental evaluation of a balsa sandwich core with improved shear properties,* Composites Science and Technology 64 pp 667–674

[6] Cesim Atas and Cenk Sevim 2010 *On the impact response of sandwich composites with cores of balsa wood and PVC foam,* Composite Structures 93 pp 40–48

[7] Vincent Legrand, Luan TranVan, Frédéric Jacquemin and Pascal Casari 2015 *Moisture-uptake induced internal stresses in balsa core sandwich composite plate:* Modeling and experimental, Composite Structures 119 pp 355–364

[8] J.A. Kepler 2011 *Simple stiffness tailoring of balsa sandwich core material,* Composites Science and Technology 71 pp 46–51

[9] Michael Osei-Antwi, Julia de Castro, Anastasios P Vassilopoulos and Thomas Keller 2013 *Shear mechanical characterization of balsa wood as core material of composite sandwich panels,* Construction and Building Materials 41 pp 231–238

[10] Michael Osei-Antwi, Julia de Castro, Anastasios P Vassilopoulos and Thomas Keller 2014 // *Fracture in complex balsa cores of fiber-reinforced polymer sandwich Structures*, Construction and Building Materials 71 pp 194–201

[11] Yahaya M A , Ruan D, Lu G and Dargusch M S. 2015 *Collapse modes in aluminum honeycomb sandwich panels under bending and impact loading.* International Journal of Impact Engineering 75 .

[12] V. Crupi, G. Epasto, E. Guglielmino. 2012 *Collapse modes in aluminum honeycomb sandwich panels under bending and impact loading.* International Journal of Impact Engineering 43

[13] Shiqiang Li, Xin Li, Zhihua Wanga, Guiying Wuc, Guoxing Lu, Longmao Zhao.2016 *Finite element analysis of sandwich panels with stepwise graded aluminum honeycomb cores under blast loading.* Composites: Part A 80 pp 1–12

[14] C.C. Foo, L.K. Seah, G.B. Chai.2008 *Low-velocity impact failure of aluminum honeycomb sandwich panels.* Composite Structures 85 pp 20–28

[15] Xin Li, Peiwen Zhang, Zhihua Wang, Guiying Wu and Longmao Zhao.2014 *Dynamic behavior of aluminum honeycomb sandwich panels under air blast: Experiment and numerical analysis.* Composite Structures 108 pp 1001–08

[16] Shanshan Shi, ZhiSun, XiaozhiHu and HaoranChen. 2016 *Flexural strength and energy absorption of carbon-fiber–aluminum-honeycomb composite sandwich reinforced by aluminum grid.* Thin-Walled Structures 84 pp 416–422

[17] Raman Bharath V R, Vijaya Ramnath B, Manoharan N. July 2016 *Kenaf Fibre Reinforced Composites: A Review.* ARPN Journal of Engineering and Applied Sciences. Vol.10 NO.13.

[18] Vijaya Ramnath B, Manickavasagam V M, Elanchezhian C, Vinodh Krishna C, Karthik S and Saravanan K 2014 *Determination of mechanical properties of intra-layer abaca-jute-glass fibre reinforced composite* Materials and Design 60 pp 643-652.

[19] Sathish P, Kesavan R, Vijaya Ramnath B. July 2015 *Experimental Investigation on delamination behavior of abaca based hybrid composites.* ARPN Journal of Engineering and Applied Sciences. Vol.10.No.13

[20] Vijaya Ramnath B, Junaid Kokan S, Niranjan Raja R, Sathya Narayanan R, Elanchezhian C, Rajendra Prasad A, Manickavasagam V M. 2013 *Evaluation of mechanical properties of abaca-jute-glass-fibre reinforced epoxy composite* Materials and Design 51 pp 357-366.

[21] Vijaya Ramnath B, Elanchezhian C, Manickavasagam V M, Gowri Prasad S, Aravind Swamy S and Keshav Raj R. 2016 *Experimental Investigation on shear and hardness of abaca based hybrid composites.* MATECWeb of Conferences 74,

[22] Yuvaraj G, Vijaya Ramnath B, Hemanth Raagavendiran G, Karthikraja R. 2016 *Investigation of flexural and Water absorption Behaviour of epoxy Hybrid composites.* Advances in Natural and Applied Sciences Special 10(7): pp 34-39.

[23] .IndaraSotoIzquierdo, OrietaSotoIzquierdo, MarciAntonioRamalho, AlbertoTaliercio, 2017 *Sisal fiber reinforced hollow concrete blocks for structural applications*: Testing and modeling, Construction and Building Materials 151,pp 98-112.

[24].ParthaHaldar, NipuModak, GoutamSutradhar, 2017 *Comparative Evaluation of Mechanical Properties of Sisal-Epoxy Composites With and Without Addition of Aluminium Powder*, Materials today, 3397-3406.

[25].Roshan K. Kejariwala, , Keerthi Gowda B S,2017 *Flammability and moisture absorption behavior of sisal-polyester composites*, Materialstoday, 8040-8044.

[26] Vijaya Ramnath B, Manickavasagam V M, Elanchezhian C, Karthick R, Bharath U, Maruthi Kishore R A and Rahul R. 2015 *Investigation of tensile and impact behavior of abaca-raffia hybrid composites.* APRN Journal of Engineering and Applied Sciences Vol.No.13

[27] Srinivasan V S, Rajendra Boopathy S, Vijaya Ramnath B 2015 *Investigation of Flexural property of Kenaf –Flax hybrid composite.* ARPN Journal of Engineering and Applied Sciences. Vol.10. NO.13.

[28] Sathish P, Kesavan R, Vijaya Ramnath B. 2015 *Experimental investigation on Flexural property of abaca and raffia hybrid composites.* ARPN Journal of Engineering and Applied Sciences. Vol.10. NO.13.

Preparation and Analysis of the Thermal properties of Engine oil Reinforced with Multi-walled Carbon Nanotubes

S D Sathishkumar[1] and T Rajmohan[2]

[1,2]Department of Mechanical Engineering, Sri Chandrasekharendra Saraswathi Viswa Maha Vidyalaya Deemed to be University, Enathur, Kanchipuram-631561, India

E-mail: [1]sathishsd.2@gmail.com and [2]rajmohanscsvmv@yahoo.com

Abstract. In the present study, the engine oil reinforced with multi-walled carbon nanotubes is used to improve the thermal characteristics of the SAE 20W40 engine oil. Multi-walled carbon nanotube (MWCNT) is used as the Nano particle and SAE 20W40 engine oil is chosen as the base fluid. Viscosity, flash point, fire point, pour point and thermal conductivity are the quality parameters effective in analysing the thermal performance of engine oil were also studied. Bath type Ultra sonication method is used for dispersing the Nano particle in the engine oil at two concentrations of 0.2 and 0.4 wt. % respectively. The experimental results reveal that the thermal properties of the engine oil with MWCNTs is enhanced significantly as compared to that of the base oil.

Keywords: Nanoparticle, Engine oil, Viscosity, Ultra sonication, MWCNT.

1. Introduction

Li et al. [1] made an experimental investigation to study the effect of Nano particle sizes on the thermal properties of Al_2O_3/DiW Nano fluids at different filler fractions. They have found that the thermal conductivity of Nano fluids is enhanced significantly due to the particle size (36nm) of Nano material at 35°C by varying the concentration up to 6.0 vol. %. He et al. [2] investigated the thermal conductivity characteristics of TiO_2-water Nano fluid of varying sizes of Nano particle at different filler fractions. They have used Nano particles with diameters of 95nm, 145nm and 210nm. They have found that the thermal conductivity is enhanced by 5% at different concentrations for the Nano particle size of 95nm. It is observed that the thermal conductivity is increased when the filler fraction increases. Nguyen et al. [3] have investigated the effect of particle size on viscosity characteristics of Al_2O_3 aqueous based Nano fluids. They have found that the effects of particle size on viscosity are more significant for high particles size concentration. Taha-Tijerina et al. [4] have investigated the effect of Nano structure properties on viscosity characteristics of Nano fluids. They have used h-BN-mineral oil and grapheme-mineral oil as Nano fluids. They have found that at low filler fraction (<0.10 wt. %) the viscosity of the Nano fluids decreases greatly with temperature. Choi et al. [5] have found that the thermal conductivity of MWCNTs/engine oil Nano fluid increased significantly by 160% at 1.0 vol. % filler fraction of nanotubes. Liu et al. [6] investigated the effect of filler fraction on the thermal conductivities of Nano fluids. They have used Nano fluids containing MWCNTs of 10-50nm in diameter dispersed in ethylene glycol and synthetic engine oil. They have found that at 1.0 vol. % thermal conductivity enhancement is more for EG/MWCNT as compared to synthetic oil based Nano fluids. Hong et al. [7] have investigated the effect of concentration of Nano particle on the thermal properties of Nano fluids. They have prepared Nano fluids by mixing CNTs in Poly-olefin oils. They have found that the enhancement in thermal conductivity is increased as the filler fraction increases. Murshed et al. [8] have investigated the effect of Nano particle morphology on the thermal properties of Nano fluids. They have used water reinforced with TiO_2 rod type and spherical Nano particles as Nano fluids. The particle size of TiO_2 Nano particle is 10nm in diameter and 40nm in length. The particle size of the TiO_2 spherical shape Nano particle is 15nm. They have found that the thermal conductivity of Nano fluids is enhanced by 30% and 33% respectively at 5.0 vol. % filler fraction as compared to the base fluid. Hong et al. [9] have investigated the morphological characteristics of Nano fluids at varying concentrations of the Nano particle. They have used Ethylene glycol reinforced with Fe Nano particles as the Nano fluids. The size of the Nano particle used in this study is 10nm. They have found that the thermal conductivity is enhanced by 18% by using Nano fluid at a concentration of 0.55 vol. %. Beck et al. [10] have made an experimental study to investigate the effect of Nano particle size on the thermal conductivity of Nano fluids. They have used water based ceria (CeO_2) Nano fluids in their study, with particle sizes of 12nm and 74nm respectively. They have found that the effective thermal conductivity is increased with an increase of Nano particle size. From the literature survey it is clear that the inclusion of Nano particles in the preparation of Nano fluids has improved the thermal properties of the base fluids significantly. Therefore, the present work deals with the preparation of Nano fluids and the analysis of the thermal properties of the engine oil reinforced with multi-walled carbon nanotubes.

2. Experimental procedure

2.1. Materials used

Multi-walled carbon nanotubes is selected as the Nano material in this research work. Figure 1 shows the SEM image of the MWCNTs. The SEM image explains the Nano structures of the carbon nanotubes very clearly.

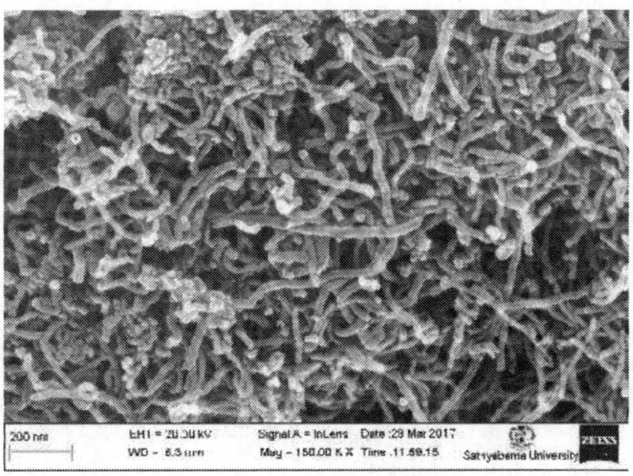

Fig. 1. SEM image of MWCNTs.

The SAE 20W40 engine oil is selected as the base oil. Figure 2 shows the details of the engine oil. The specifications of engine oil used are shown in Table 1. Bath type ultra sonicator is used for uniform mixing of carbon nanotubes with the engine oil.

Table 1. Specification of SAE 20W40 engine oil

Characteristic Test	Observation
Viscosity @ 100° C	14.36
Viscosity @ 40° C	127.47
Pour point (Deg. C)	-10
Flash Point (°C)	238
Fire Point (°C)	244
Thermal Conductivity @@40° C	0.142

2.2 Preparation of Nano lubricants

The Nano fluids were prepared by dispersing MWCNTs into the SAE20W40 oil base fluid in two different concentrations of 0.2 and 0.4 wt. % respectively. The mixture was then sonicated for 10min at 100% amplitude using a 130W, 20KHz ultrasonic processor (Nanotechnology research centre, Sathyabama University, Chennai) and it was followed by 20min stirring using a magnetic stirrer. The process was repeated until the total mixing time was 2 hours. The specification of MWCNT is presented in Table 2. The bath type ultra-sonication apparatus is shown in figure 3.

Table 2. Specifications of the multi-walled carbon nanotube

MWCNT	Description
Available Form	Black Powder
Diameter	Av. Outer Dia : 20nm
Length	Av 20 micrometer
Nanotubes Purity	>98%
Metal Particles	<1%
Amorphous Carbon	<1%
Specific Surface Area 2	220 m2/g
Bulk Density	0.20g/cm3

Figure 2. SAE 20W40 Engine oil

Figure 3. Bath type Ultra sonication apparatus

2.3 Testing the characteristics of lubricant with carbon nanotube

The properties of Nano fluids are tested in Delta laboratory, Chennai, India. The thermal conductivity of the base oil and Nano lubricants of all samples was measured on the basis standard of DM&AL/256 at a temperature of 40°C. The thermal properties of the engine oil with and without CNTs are calculated on the basis of established standards ASTM D 445, ASTM D 92 and ASTM D 97 respectively. The properties of Nano fluids is shown in Table 3.

Table 3. Properties of Nano fluids

Thermal properties	**Observation**	
	Sample 1 (0.2 wt. %)	**Sample 2 (0.4 wt. %)**
Viscosity @ 100° C	15.21	15.99
Viscosity @ 40° C	133.85	145.17
Pour point (Deg. C)	-12	-16
Flash Point (°C)	243	253
Fire Point (°C)	251	267
Thermal Conductivity @@40° C	0.151	0.157

3. Results and Discussions

3.1 Thermal Conductivity

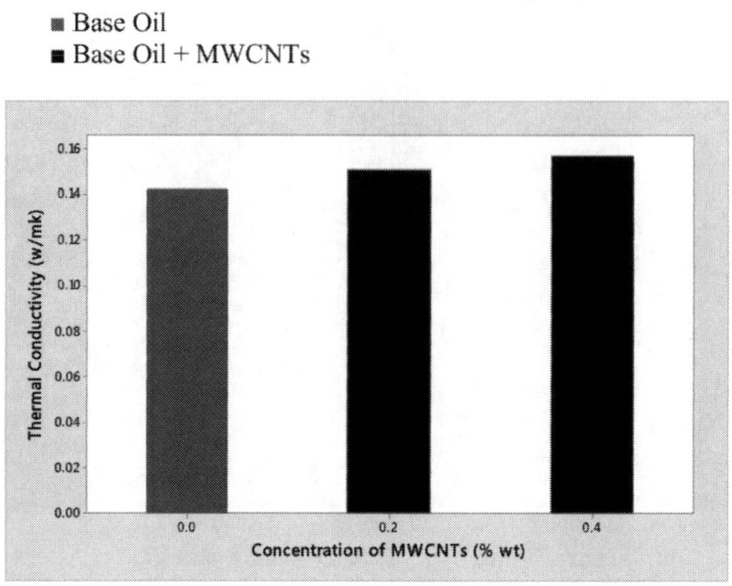

Fig. 4. Thermal conductivity of Nano-lubricants and base lubricant.

The results of thermal conductivity measurements of base lubricants containing MWCNTs are shown in Figure 4. The experimental results reveal that the thermal conductivity of Nano-lubricants is enhanced significantly as compared to the base oil. This is mainly due to the higher thermal conductivity of the MWCNTs with respect to engine oil. The thermal conductivity is enhanced by 6.3% and 10.5% at 0.2 and 0.4 wt. % concentrations of Nano lubricants with respect to the engine oil respectively.

3.2 Viscosity

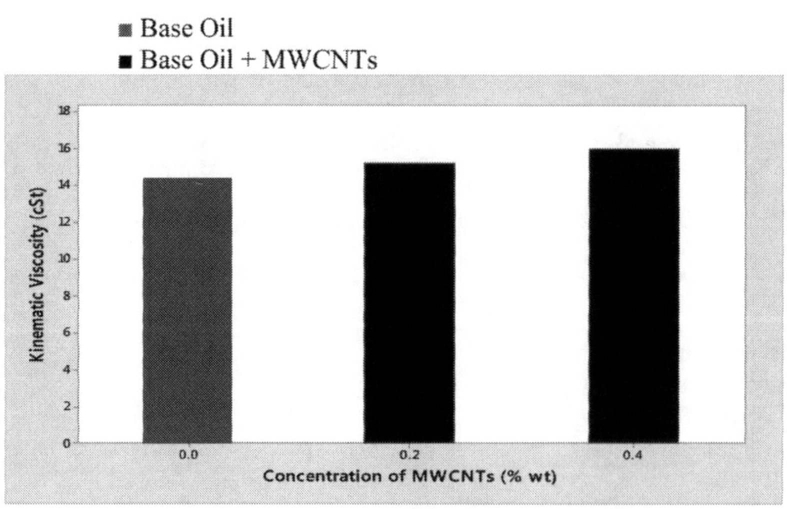

Fig. 5. Kinematic viscosity at 100°C.

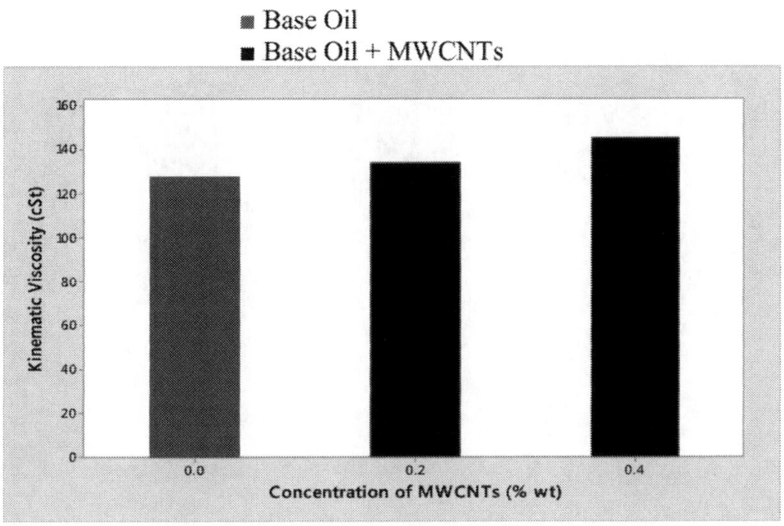

Fig. 6. Kinematic viscosity at 40°C.

The kinematic viscosities of Nano lubricants at 40°C and 100°C were measured at 0.2 and 0.4 wt. % respectively. The results of viscosity measurements of engine oil containing MWCNTs are shown in Figure 5 and Figure 6. The experimental results indicate that the viscosity of all the samples is reduced by increasing the temperature. Also it is observed that the amount of viscosity had an increasing trend with increasing the concentration of carbon nanotubes.

3.3 Pour point, Flash point and Fire point

The oil's pour point is the temperature at which the rate of flow will stop. The changes in pour point of engine oil with and without CNTs at two different concentrations are shown in Figure 7. The results indicate that the pour point of all the samples has improved with increasing the concentration of carbon nanotubes. Flash point is the temperature of the oil at which its vapour will ignite. The comparison of engine oil with and without CNTs at varying concentrations in terms of flash point is shown in Figure 8. The results indicate that the flash point of all the samples has an increasing trend with increasing the concentration of carbon nanotubes. The changes in the fire point values of Nano lubricants at varying concentrations is shown in Figure 9. The results reveal that the fire point of all the samples showing an increasing trend with increasing the concentration of carbon nanotubes. It was observed that flash point and fire point of Nano-lubricants had their best amounts at 0.4 wt. % concentration.

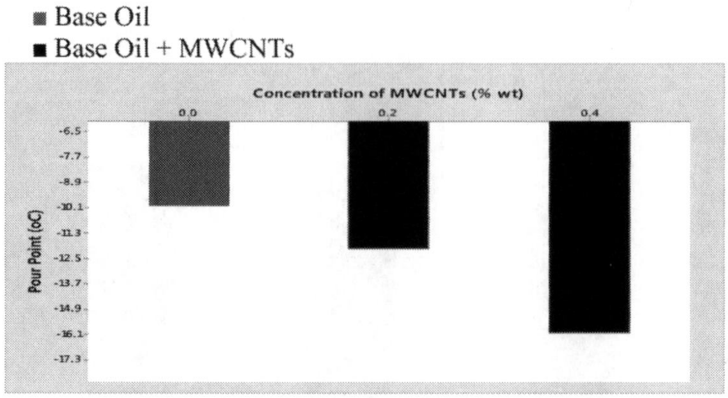

Fig. 7. Pour point of lubricants.

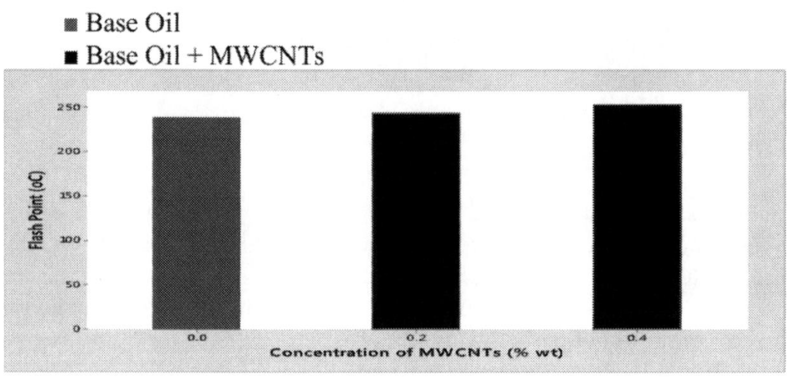

Fig. 8. Flash point of lubricants.

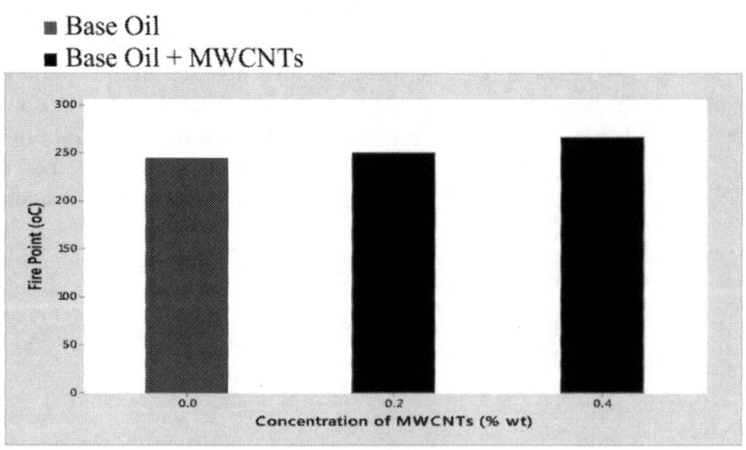

Fig. 9. Fire point of lubricants.

4. Conclusion

In this research work, the effects of adding MWCNTs in engine oil at two different concentrations was analysed. The results reveal that the concentration of MWCNTs in the engine oil had a significant improvement in the thermal properties of the base oil. The flash point and fire point of Nano lubricants is enhanced by about 6.3% and 9.4% respectively at the concentration of 0.4 wt. %. The results indicate that the heat harvesting capability of Nano lubricant is increased. It is concluded that the addition of Nano particle plays a vital role in enhancing the thermal performance of the engine oil.

References

[1] Li C H and Peterson G P 2007 *Journal of Applied Physics* **101**(4) 044312-044312-5
[2] He Y, Jin Y, Chen H, Ding Y, Cang D and Lu H 2007 *International Journal of Heat and Mass Transfer* **50**(11-12) 2272-2281
[3] Nguyen C T, Desgranges F, Roy G, Galanis N, Mare T, Boucher S and Angue Mintsa 2007 *International Journal of Heat and Fluid Flow* **28**(6) 1492-1506
[4] Taha-Tijerina J, Narayanan T N, Gao G, Rohde M, Tsentalovich D A, Pasquali M and Ajayan P M 2012 *ACS Nano* **6**(2) 1214-1220
[5] Choi S U S, Zhang Z G, Yu W, Lockwood F E and Grulke E A 2001 *Applied Physics Letters* **79**(14) 2252-2254
[6] Liu M S, Ching-Cheng Lin M, Huang I T and Wang C -C 2005 *International Communications in Heat and Mass Transfer* **32**(9) 1202-1210
[7] Hong H, Wensel J and Roy W 2007 *ECS Transactions* **2**(12) 133-138
[8] Murshed S M S, Leong K C and Yang C 2005 *International Journal of Thermal Sciences* **44**(4) 367-373
[9] Hong T K, Yang H S and Choi C J 2005 *Journal of Applied Physics* **97**(6) 064311
[10] Beck M P, Yuan Y, Warrier P and Teja A S 2010 *Journal of Applied Physics* **107**(6) 066101-066101-3

A REVIEW ONSANDWICH AND NATURAL FIBER COMPOSITES

B Vijaya Ramnath, C Elanchezhian, P S Shreyas, S Shriramkumar

N Ramanan, R Dinesh Riju

Department of Mechanical Engineering, Sri Sairam Engineering College, West Tambaram, Chennai-600 044, India

Abstract: Sandwich composite is a particular form of laminated composite in which a combination of two different and stiff materials is bonded together. In these composites, properties of the sandwich specimens are enhanced by its individual constituents. Sandwich composites usually consist of two thin, stiff materials separated by thick lightweight cores. The cores that are generally used are honeycomb and foam cores. In this paper research work is carried out on the sandwich composites. The various tests are performed on the fabricated models and major applicationsare also discussed below.

1.0INTRODUCTION:

A Composite material is a material that is made up of two or more constituent materials. The individual materials are bonded to produce a composite with characteristics superior to its constituents .In composites the individual properties of materials are retained. ASandwich-Structured Composite is a special class of Composite materials that is fabricated by attaching two thin but stiff skins to a lightweight but thick core. The sandwich composites possess properties like high strength and high stiffness to weight ratio.

2.0 TYPES OF SANDWICH COMPOSITES

1. Steel–concrete–steel (SCS) sandwich composite.
2. Glass fibre reinforced sandwich composite.

3.0 DESCRIPTION:

Here, various types of sandwich composites, based on steel concrete and glass fiber reinforced are discussed

3.1 STEEL–CONCRETE–STEEL (SCS) SANDWICH COMPOSITE:

Zhen-Yu Huang et al. [1] investigated the ultimate strength behavior of SCS sandwich shell experimentally and analytically. Two pilot quasi-static tests on the lightweight SCS sandwich composite shells subject to patch loading are carried out.Tests show that the punching shear resistance depends on the control perimeter of punched concrete frustum and shear connectors. The membrane action of the outer steel plates provides post-hardening strength.Meng Chu et al. [2] investigates the structural performance of steel-concrete-steel (SCS) sandwich members comprising a concrete core sandwiched in between two steel plates which are interconnected by channel steel connectors with large interval. Angle steel is adopted to serve as the connection between channel steel and surface plates. The behavior of a series of 8 large-scale, simply supported tests was investigated, covering a wide range of applied shear span to depth ratios from 1.5 to 5.5. Other parameters include the interval of channel steel and axial tension force. Although critical inclined cracks appeared in most of the beams, the observed modes of failure were all ductile failure characterized by yielding of the tension steel plate. Unlike the negligible dowel action of longitudinal reinforcement in reinforced concrete members, the dowel action of steel plate in SCS member seems to play a decisive role.C. Abraham and K.C.G.Ong. [3] investigated the performance of steel–concrete composite panels, which consist of fiber-reinforced high-strength concrete (FRHSC) on the compression face and a specially configured steel sandwich as tension reinforcement. The performance of the composite panel is compared to similar ordinary reinforced concrete panels. The latter is reinforced with the maximum reinforcement ratio specified by the codes. The static response of the composite panel is obtained numerically using finite element modeling and experimentally under third point bending tests.M. Leekitwattana et al. [4] suggested a steel-concrete-steel (SCS) sandwich beam represents a special form of sandwich structure. It consists of steel face plates and concrete core which are connected together by mean of a series of shear connectors. The state-of-the-art construction forms of SCS sandwich structures are double-skin sandwich construction (DSC), Bi-Steel sandwich construction (Bi-Steel), and alternative SCS sandwich construction. They are different only due to the pattern of their shear Connectors. O. Dogan and T.M.Roberts. [5] Investigated Double skin composite construction consists of a layer of a plain concrete, sandwiched between two layers of relatively thin steel plate, connected to the concrete by welded stud shear connectors. This const0r0uction 0acts in a similar way to doubly reinforced concrete elements but the flexibility of connection between the steel plates and concrete gives rise to interface slip and additional overall element deflection. This results in a strong and efficient structure with certain potential advantages over conventional forms of construction. Steel–concrete composite systems generally consist of steel plate, concrete and reinforcement. Shear connectors are usually utilized to develop the composite action between steel and concrete. In steel-concrete composite members, the natural bonding, friction, and mechanical interlocking actions of shear connectors have a significant influence on the degree of interaction.

3.2 GLASS FIBRE REINFORCED SANDWICH COMPOSITE

V. Crupi et al. [6] investigated the collapse of GFR-AFS occurred for the debonding of the aluminum faces from the core and is strongly influenced by the rigidity of glass fiber skin material, whereas the aluminum foam sandwiches collapsed for the foam crushing and their energy absorbing capacity and it is found that the light-weight aluminum sandwiches have good properties of energy dissipation and the amount of energy absorption under bending and impact tests can be improved reinforcing them by means of GFRP outer skins. Nicolas J. Lombardi and Judy Liu [7] studied GFRP-steel hybrid parametric to evaluate improvements on the GFRP honeycomb deck panel stiffness and concluded that in one direction, the hybrid concept results in a significant increase in both flexural and shear stiffness. In the other direction beam samples, the hybrid and Kansas Structural Composites, Inc. flexural stiffness results were about the same, apparently due to changes in the KSCI baseline beam. Amir Shahdin et al. [8] studied the entangled sandwich specimens having glass fiber cores and glass woven fabric as skin materials. The tested glass fiber entangled sandwich beams possess low compressive and shear modulus as compared to honeycomb and foam sandwich beams of the same specifications. The vibration tests show that the entangled sandwich beams possess higher damping ratios and low vibratory levels as compared to honeycomb and foam sandwich beams. The test results also proved that entangled sandwich specimens with shorter glass fiber lengths have high structural strength but on the other hand low damping ratios and higher vibratory levels when compared to entangled sandwich specimens with longer glass fiber length in the core. Tine Lanssens [9] analyzed the computational and experimental study on the mechanical behavior of a 3-dimensional GFRP sandwich panel with through thickness fiber insertions. Samples of the GFRP skin and glass fiber insertions were cut out from larger panels and are tested to determine basic material properties of the components of the sandwich panels. The results show that increasing the density of fiber insertions makes the panel stiffer. An alternating pattern for the insertions also increased the overall panel stiffness. Varying the overall panel thickness between a minimum and a maximum thickness indicted that an optimal overall thickness could be established .Mechanical tests were performed on the sandwich panels and the components of sandwich panel: skin and fiber insertion. Experimental results on the components were used to create a nonlinear finite element model that can predict the large-scale mechanical deformation of the sandwich panels. The model can predict force–displacement behavior of the panel accurately under a variety of experimental conditions. J. ARBAOUI et al. [10] The first, tow face sheet of composite glass fibers (T800/M300)/polyester resin. The second part is a honeycomb polypropylene core. It is found that decrease in shear modulus with increasing core thickness and increases with number of layer increasing. The effect of variations in core thickness and intermediate layers from single to quadruple on mechanical properties of a polypropylene honeycomb core was evaluated and results show that mechanical properties increase with increasing core thickness and intermediate layers.

3.3OTHER SANDWICH COMPOSITES

Jiayi Liu et al. [11] investigated the effect of temperature on the bending properties and failure mechanism of pyramidal truss structured carbon FRP core sandwich structures with carbon epoxy prepegs and found that bending failure load decreased by 92.27% when the temperature increased from 293K to 473K, which was mainly due to the degradation of the matrix properties and fiber-matrix interface properties. In addition to the decreasing of the bending failure load, the failure mode of composite sandwich structure also changed with temperature.M. BotshekananDehkordi& S.M.R. Khalili [12] performed frequency analysis of sandwich plate with active SMA hybrid composite face-sheets and temperature dependent flexible core bonded using epoxy resin. They proposed the result that using the pre-strained SMA wires the stiffness of the sandwich plate can be improved and therefore has a positive effect on the natural frequencies of the sandwich plate.P. Nagasankar et al. [13] assayed the role of different fiber orientations and thicknesses of the skins and the core on the transverse shear damping of polypropylene honeycomb sandwich structures. They inferred that increasing the thickness of PPHC core increases the damping loss factor since it undergoes greater transverse shear deformation. Also they found that 0° fiber oriented sandwich with large thickness is considered as the best choice since large damping loss factor, natural frequency and stiffness are obtained.SinduSatasivam and Yu Bai[14] investigated mechanical performance of bolted modular GFRP composite sandwich structures using standard and blind bolts. These sandwich specimens are used in beam and slab applications. The major conclusions made are: (i). The bending stiffness of all the sandwich specimens with bolts as shear connections decreased as the load increased due to the decrease in composite action between the panels and web sections with increasing load. (ii).As the bolt spacing increased, both the stiffness and degree of composite action were reduced. (iii). Standard through-bolts were shown to cause premature local buckling at large bolt spacing, and this premature buckling was avoided with the use of blind bolts. This is because blind bolts couple the upper or lower panels with the adjacent flange of the core profile. (vi). All-composite GFRP sandwich systems with bolted shear connections had superior performance to a one-way spanning reinforced concrete slab of similar dimensions, including a significant reduction in self-weight and an improvement of load-carrying capacity. In addition, the use of bolted shear connections allows for easy assembly and disassembly.M. Samadpour et al. [15] conducted vibration analysis of thermally buckled SMA hybrid composite sandwich plate with the temperature dependent laminated composite face sheets and inferred that during austenite phase transformation, the conspicuous tensile recovery stress is generated so that increasing the pre-strain leads to raising recovery stress intensely. This increasing of pre-strain leads to natural frequency decline in the post-buckling domain, which is due to lower deflection in this domain.A. Anjang et al. [16] assayed post-fire mechanical properties of sandwich composite structures and their experiment reveals that the core plays an important role in insulating the back skin, and this has significant influence on the post-fire properties. There was a large reduction to the compression stiffness and strength following exposure to the thermal flux, even for short times (e.g. 60 s) followed by a rapid reduction in the tension modulus and strength within the first 200 s, which is the time taken to decompose the front skin exposed directly to the thermal flux.Guo-dong Xu et al. [17] presented a new method for forming the graded corrugated truss core sandwich structures based on an auto-cutting and mould-press process and analyzed the bending stiffness, strength and failure mechanism of the specimens. Finally, they found that the strength of the graded corrugated truss core composite sandwich beam with strong struts in the middle region is much higher than the one with weak struts. The main failure mode of the graded corrugated truss core composite sandwich beam with weak struts in the middle is core member delamination induced by the indentation failure.

4.0NATURAL FIBRE COMPOSITES

VijayaRamnath et al. [18] studied the property comparison of abaca-jute hybrid composite and abaca composite and discovered that while abaca-jute hybrid composite had better tensile and shear properties, abaca had better flexural and impact properties.Srinivasan et al. [19] investigated the mechanical and thermal properties of banana-flax based natural fibre composite and discovered that hybrid composite has far better properties than single fibre glass reinforced composite when tested for flexural and impact loads.Manickavasagam et al. [20] observed that fibre orientation plays an important role in determining the mechanical properties of the composite by testing an intra layer abaca jute glass fibre reinforced composite. VijayaRamnath et al. [21, 22, 23] discovered that hybrid natural composites have excellent tensile and flexural properties investigating the properties of a jute flax based glass fibre reinforced composite. While testing of the mechanical properties of banana jute hybrid composite discovered that, overall, the hybrid composites had better mechanical properties than mono composites containing the same materials [22]. It was observed that observed a significant improvement in mechanical properties of composites due to the presence of twisted fibre and this also reveals the influence of fibre orientation on mechanical properties [23].Navin Chand et al [23] investigated the effect of sisal fibre orientation on electrical properties of sisal fibre reinforced epoxy composites and discovered that there was a significant effect on the properties observed. Lucas et al [24] investigated the proconcentration of copper in tobacco leaves samples using a minicolumn of sisal fibre loaded with alizarin blue and discovered that the achieved concentrations varied from 0.15 to 0.52 μg g−1.Rashnal Hossain et al [25] studied the tensile behavior of environment friendly jute epoxy laminated composite and discovered that the properties are very defect sensitive. Belaadi et al [26] researched the fatigue in sisal fibre reinforced polyester composites and discovered that failure is reached within a few cycles for high loading levels while it is only partial for lower loading levels. Yan Li et al [27] gave us a succinct and thorough review on sisal fibre and its composites as well as recent developments in the field.Dwivedi et al [28] investigated the influence of MA-g-PP on abrasive wear behavior of chopped sisal fibre reinforced polypropylene composites and discovered that the sample having 2% by weight of MA-g-PP offered maximum wear resistance.

5.0 CONCLUSION

The purpose of this review was to study about the various majorsandwich structures, tests performed on them and their properties. It is clear from the review that various types of sandwich composites have been studied and analyzed on the basis of the tests conducted upon them.

6.0 REFERENCES

[1] Zhen-YuHuang, Jun-YanWang, J.Y.RichardLiew andP W Marshall Lightweightsteel–concrete–steelsandwichcompositeshellsubject topunchingshear. *Ocean Engineering***102**(2015)146–161

[2] Meng Chu, Xiaobing Song and HonghuiGeStructural performance of steel-concrete-steel sandwich composite beams with channel steel Transactions, SMiRT-22 San Francisco, California, USA - August 18-23, 2013 Division X (include assigned division number from I to X)

[3] C. Abraham & K. C. G. Ong Performance of fiber-reinforced high-strength concrete with steel sandwich composite panels under static loading, *Computational Methods and Experimental Measurements* **XVII** 495

[4] M. Leekitwattana, S. W. Boyd and R. A. Shenoi An alternative design of steel-concrete-steel sandwich beam, *9th International Conference on Sandwich Structures ICSS***9**

[5] O. Dogan and T. M. Roberts, Comparing experimental deformations of steel concrete- steel sandwich beams with full and partial interaction theories, *International Journal of the Physical Sciences* **Vol. 5(10),** pp. 1544-1557, 2010

[6] V. Crupi , E. Kara , G. Epasto , E. Guglielmino , H. Aykul,Prediction model for the impact response of glass fiber reinforced aluminum foam sandwiches, *International Journal of Impact Engineering* **77** (2015) 97-107

[7] Nicolas J. Lombardi, Judy Liu ,Glass fiber-reinforced polymer/steel hybrid honeycomb sandwich concept for bridge deck applications, *Composite Structures* **93** (2011) 1275–1283

[8] Amir Shahdin, Laurent Mezeix, Christophe Bouvet, Joseph Morlier,and YvesGourinat, Fabrication and mechanical testing of glass fiber entangled sandwich beams: A comparison with honeycomb and foam sandwich beams, *Composite Structures***90** (2009) 404–412

[9] Tine Lanssens , CharlotTanghe , NimaRahbar , Pinar Okumus , Steven Van Dessel, Tahar El-Korchi,Mechanical behavior of a glass fiber-reinforced polymer sandwich panel with through-thickness fiber insertions,*Construction and Building Materials***64** (2014) 473–479

[10] J. arbaoui Y, Schmitt, J.-L. Pierrot, F.-X. Royer, Effect of core thickness and intermediate layers on mechanical properties of polypropylene honeycomb multi-layer sandwich structures, Volume 59 2014 Issue 1DOI: 10.2478/amm-2014-0002

[11] Jiayi Liu, Linling Xiang, Tao Kan, The effect of temperature on the bending properties and failure mechanism of composite truss core sandwich structures,*Composites: Part A***79** (2015) 146–154.

[12] M. Botshekanan Dehkordi, S.M.R. Khalili. Frequency analysis of sandwich plate with active SMA hybrid composite face-sheets and temperature dependent flexible core. *Composite Structures***123** (2015) 408–419.

[13] P. Nagasankar, S. BalasivanandhaPrabu, R. Velmurugan,Role of different fiber orientations and thicknesses of the skins and the core on the transverse shear damping of polypropylene honeycomb sandwich structures,*Mechanics of Materials* **91** (2015) 252–261.

[14] SinduSatasivam, Yu Bai, Mechanical performance of bolted modular GFRP composite sandwich structures using standard and blind bolts,*Composite Structures***117** (2014) 59–70.

[15] M. Samadpour, M. Sadighi, M. Shakeri and H.A. Zamani, Vibration analysis of thermally buckled SMA hybrid composite sandwich plate. *Composite Structures***119** (2015) 251–263.

[16] A. Anjang, V.S. Chevali, B.Y. Lattimer, S.W. Case, S. Feih and A.P. Mouritz, Post-fire mechanical properties of sandwich composite structures,*Composite Structures***132** (2015) 1019–1028.

[17] Guo-dong Xu, Fan Yang, Tao Zeng, Su Cheng and Zhi-hai Wang, Bending behavior of graded corrugated truss core composite sandwich beams. *Composite Structures***138** (2016) 342–351.

[18]B. Vijaya Ramnath, S. JunaidKokan, R. NiranjanRaja , R. Sathyanarayanan, C. Elanchezhian, A. Rajendra Prasad and V.M. Manickavasagam, Evaluation of mechanical properties of abaca–jute–glass fibre reinforced epoxy composite. *Materials and Design***51** (2013) 357–366.

[19] V.S. Srinivasan, S.RajendraBoopathy, D. Sangeetha and B. VijayaRamnath, Evaluation of mechanical and thermal properties of banana–flax basednatural fibre composite. *Materials and Design***60** (2014) 620–627.

[20] B. VijayaRamnath, V.M. Manickavasagam, C. Elanchezhian, C. Vinodh Krishna, S. Karthik and K. Saravanan, Determination of mechanical properties of intra-layer abaca–jute–glassfiber reinforced composite, *Materials and Design***60** (2014) 643–652.

[21] B. Vijaya Ramnath, C. Elanchezhian, P. V. Nirmal, G. Prem Kumar, V. Santhosh Kumar, S. Karthick, S. Rajesh and K. Suresh, Experimental Investigation of Mechanical behavior of Jute-Flax Based Glass Fiber Reinforced Composite,*Fibers and Polymers*, **Vol.15**, No.6, (2014) 1251-1262.

[22] B. Vijaya Ramnath, R. Sharavanan, M. Chandrasekaran, C. Elanchezhian, R. Sathyanarayanan,R. Niranjan Raja and S. JunaidKokan,Experimental Determination of Mechanical Properties ofBanana Jute Hybrid Composite,*Fibers and Polymers*, **Vol.16**, No.1, (2015), 164-172.

[23] B. Vijaya Ramnath, S. Rajesh, C. Elanchezhian, A. Santosh Shankar,S. PithchaiPandian, S. Vickneshwaran, and R. SundarRajan, Investigation on Mechanical Behaviour of Twisted Natural FiberHybridComposite Fabricated by Vacuum Assisted Compression Molding Technique. *Fibers and Polymers*, **Vol.17**, No.1, (2016), 80-87

[24] Navin Chand and Deepak Jain, Effect of sisal fibre orientation on electrical properties of sisal fibre reinforced epoxy composites. *Composites: Part A* **36** (2005), 594–602

[25]Fábio de S. Dias a ,Josemário S. Bonsucessoa, Lucas C. Oliveiraa and Walter N.L. dos Santos, Preconcentration and determination of copper in tobacco leaves samples by using a minicolumn of sisal fiber (Agave sisalana) loaded with Alizarin fluorine blue by FAAS,*Talanta***89** (2012) , 276-279

[26]Md. Rashnal Hossain, Md. Aminul Islam, Aart Van Vuurea and IgnaasVerpoest, Tensile behavior of environment friendly jute epoxy laminated composite. *Procedia Engineering* **56**(2013), 782 – 788

[27] A. Belaadi, A. Bezazi, M. Maache and F. Scarpa, Fatigue in sisal fiber reinforced polyester composites: hysteresis and energy dissipation,*Procedia Engineering***74**(2014) 325 – 328

[28]Yan Li, Yiu-Wing Mai and Lin Ye, Sisal fibre and its composites: a review of recent developments, *Composites Science and Technology***60** (2000) ,2037-2055

[29] U.K. Dwivedi andNavin Chand, Influence of MA-g-PP on abrasive wear behaviour of chopped sisal fibre reinforced polypropylene composites,*Journal of Materials Processing Technology***209** (2009), 5371–5375

Tribological Analysis of Biodegradable Refrigeration oil for Vapour Compression Refrigeration System using Four Ball Tribometer

P. Vithya[1*], G. Sriram[2], S. Arumugam[3], B.Viswa Harinath[4], A. Saijagadeeswar[5]

[1-5] Department of Mechanical Engineering, Sri Chandrasekharendra Saraswathi Viswa Mahavidyalaya, Kanchipuram, Tamilnadu - 631561, India.

Email: vithyagautham@gmail.com

Abstract. The purpose of this investigation is to accomplish the experimental study of rapeseed oil-based TMP (trimethylolpropane) ester as biodegradable refrigeration oil and correlate with synthetic refrigeration oil using the four ball tribometer. The biodegradable refrigeration oil is produced from rapeseed oil, which is biodegradable and has high lubricity properties such as a higher flash point temperature and VI (viscosity index). The friction and wear experiments of biodegradable refrigeration oil were performed using ASTM D 5183 and ASTM D2783. The surface morphology analysis of tested ball materials was also carried out using scanning electron microscope. The results show that on the similar experimental conditions, biodegradable refrigeration oil exhibited good friction reduction behaviour and wear reduction behaviour up to 50kg as compared to synthetic refrigeration oil. The friction coefficient has minimum value biodegradable refrigeration oil for 10kg and 50kg, maximum for 100 kg. The last non-seizure load and the weld load for biodegradable refrigeration oil are 40 kg and 126 kg respectively and last non-seizure load and the weld load for synthetic refrigeration oil are 50 kg and 160 kg. However, biodegradable refrigeration oil is eco friendly, can be encouraged for an alternative for commercial refrigeration lubricant.

1. Introduction

The Numerous researches have been investigated on the effect of synthetic refrigeration oil in the vapour compression refrigeration system .Several techniques have been tried out for increasing the co- efficient of performance (COP) of the vapour compression refrigeration system, as stated in the literature [1-5]. In general, majority of the study use the nano particle as additive in the refrigeration oil to improve co- efficient of performance of the refrigeration system by means of improving the tribological properties of the lubricant. The main purposes of the refrigeration oil are lubrication, removal of heat and for sealing. Refrigeration oil cooling is necessitating in reciprocating compressor as the piston compresses the refrigerant gas.

Most of the lubricant oil produced from the oil seed crop in worldwide. Brassica napus are commonly called as rapeseed or mustard, and third most predominant origin of vegetable oil, rating after only soybean and palm. Bio lubricants are potential alternative lubricants for the reason that of their low toxicity, excellent lubricating properties, high viscosity index, increased service life, more load carrying capacities and low emissions [6].

Yadav et al [7] examined the extreme pressure and anti-wear properties of fresh and used SAE15W40 and SAE 20W50 grade engine lubricating oil and stated that anti-wear and anti-friction properties in terms of wear scar diameter for Diesel engine D1 oil and petrol engine P1 sample lubricated steel-chromium alloy ball have less wear scar diameter and shallow groves as compare to Diesel engine car D2 engine oil samples and Petrol engine P2 engine oil samples. Dongare et al [8] investigated the tribological properties of various lubricating oils SAE20, SAE30, SAE40, SAE68, SAE90, SAE120, & SAE140 with and without of different anti-wear and extreme pressure additives (MOLYVAN A & VANLUBE 73). The results concluded that wear scar diameter, weld Load and even temperature increases with the increase of pressure or applied load.

Habibullah et al [9] analysed the tribological properties of Calophyllum inophyllum (CI) based trimethylolpropane (TMP) ester as lubricant using four ball tribometer and depict the similarities with commercial lubricant and paraffin mineral oil on energy economy feature. The results demonstrate that the biodegradable lubricant possess the better friction behaviour with less energy consumption. The present study synthesised the bio degradable refrigeration oil from rapeseed oil through transesterification with trimethylolpropane for hermetically sealed reciprocating compressors. Further, tribological performance of the bio degradable refrigeration oil has been investigated using four ball tribometer for observing the compatibility for use as refrigeration oil.

2. Experimental

2.1. Synthesis of pentaerythritol ester based Biodegradable Refrigeration oil

The synthesis of bio degradable refrigeration oil for the vapour compression refrigeration system of refrigeration compressors was carried out by the transesterification process of rapeseed oil. The concise procedure of rapeseed oil transesterification process is adopted from Hiekel et al [10]. The synthesis of raw rapeseed oil was carried in two stages. The first stage is the process of using methanol in the presence of sodium hydroxide to produce formulation of rapeseed oil methyl ester that is raw rapeseed oil (300g, 0.33mol) and sodium hydroxide (54g, 1.35 ml, in 675ml of water) was mechanically stirred for 1 hours maintaining the temperature at 80^0 - 90^0 C. The reaction is examined by thin layer chromatography until the complete disappearance of triglyceride. The second stage is the reaction of methyl ester with trimethylolpropane using sodium methoxide as catalyst to yield palm or rapeseed oil base trimethylolpropane esters. TMP (44.09g, 0.302mol) was first mixed with rapeseed fatty acid (300g, 1.06 Mol), after that the xylene was introduced as a catalyst.

Table.1. Thermo-Physical Properties of synthetic refrigeration oil and bio degradable refrigeration oil

Test parameters	Test method	Units	Synthetic refrigeration oil	Bio degradable refrigeration
Viscosity at 40ºC	ASTMD445	cSt	68	38.5
Viscosity at 100ºC	ASTMD445	cSt	9.3	10.15
Viscosity index	ASTMD445	-	114	266
Flash point	ASTM D92	^{0}C	220	240
Pour point	ASTM D97	^{0}C	-39	-6
Total Acid Number	ASTMD974	mgkoh/g	0.30	1.2

This course of reaction mixture is stirred at 135^0 -140^0C in the presence of para- Tolune Sulphonic acid (3.0g, weight% based on the fatty acid mixture) as catalyst in the reaction equipment. The mixture was further distilled at 110^0C-115^0C under the reduced pressure of 2-3mm of Hg to remove the xylene catalyst.

2.2. Tribological investigation

The tribological behaviour of biodegradable refrigeration oil and synthetic refrigeration were further experimentally investigated by the Four ball Tribometer [supplied by M/s DUCOM Instruments Pvt. Ltd, (TR 500), India]. The anti-wear and extreme pressure of biodegradable refrigeration oil was performed using ASTM D 5183 and ASTM D2783. The four-ball wear tribometer consists of three balls which are held stationary in a ball pot and a fourth ball fixed in a rotating spindle as shown in Fig.1. The test region is a top ball revolving in the hollow space of three similar balls in contact and press together in a cup below surrounded with the test lubricant. The tribological study was conceded out with ball material as steel, AISI 52-100, 12.7 mm in diameter with hardness of 64-66 Rc. The test balls, oil cup and holding parts were properly cleaned with acetone before starting of each experiment. The lubricant required for each test was 10 ml. The test conditions were the load is varied from 10 kg to 100 kg, operating temperature is at room temperature, rotational speed of 1200 rpm and operation time is 10minutes.

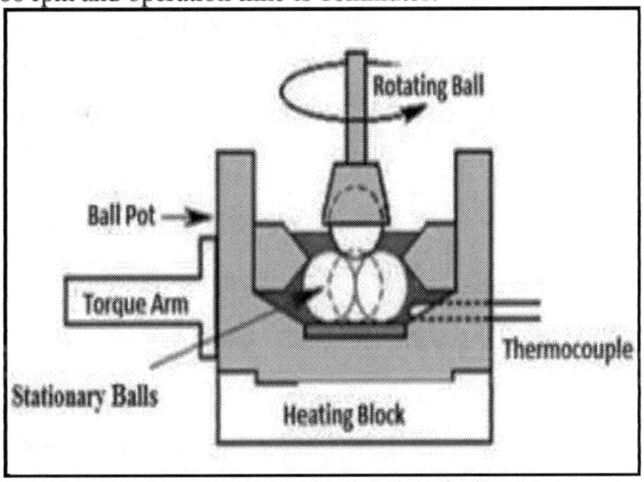

Fig.1 Schematic diagram of Four ball tribometer

The friction coefficient is a rate that illustrates the correlation between the friction force between two objects and the normal reaction between the objects that are concerned. When two objects are in moving contact, the torque generated by the frictional force is the friction torque. Wear is associated to interactions between surfaces and the removal of material on a surface and deformation as an effect of mechanical action of the adjacent surface. The friction coefficient and friction torque are analysed and calculated by the tribological study of four ball tribometer. The wear scar diameter (WSD) was measured and analyzed by image acquisition system. Scanning Electron Microscope (SEM) was used to evaluate the wear characteristics of the ball surfaces under 100X magnification.

3. Results and Discussion

3.1. Fourier Transform Infrared (FTIR) Analysis

The Fourier transform infrared spectroscopy (Bruker Alpha T, Germany) technique was used to identify the molecular bonding and functioning groups by means of supplying an infrared absorption spectrum for bio lubricant and synthetic refrigeration oil. Fig.1(a) and 1(b) show the clear view of FTIR for the synthetic refrigeration oil and bio degradable refrigeration oil respectively. The structure of the trimethylpropane tri ester from the esterification of rapeseed oil with TMP was confirmed by the disappearance of the peak of the ester of C=O stretching vibration at 1742.92 cm^{-1} in the moiety of glyceride. The appearance of broadband at 3300-2500 cm^{-1} can be seen in bio lubricant linked to the O-H stretching vibration of carboxylic acid and the presence of broadband at 3468.19 cm^{-1}. The stretching C-O-C peak appeared at 1162.71 cm^{-1}.

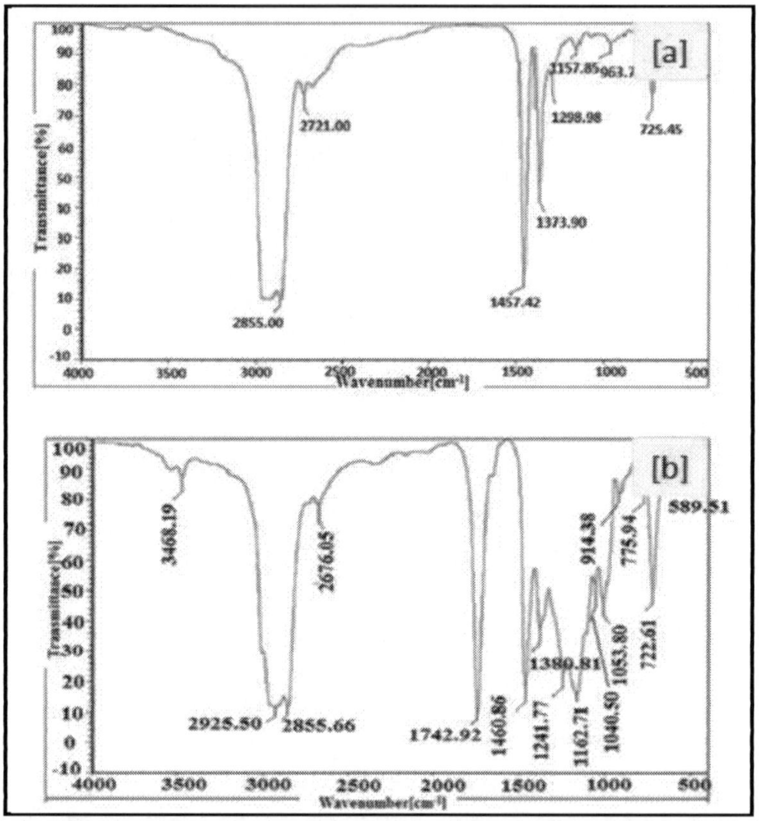

Fig.2. FTIR for (a) synthetic refrigeration oil(b) bio degradable refrigeration oil

The hydroxyl group peak appeared at 3468.19 cm^{-1} was too low and negligible showing the evidence for the completion of esterification reactions. The absorption bands at 2925.50-2855.66cm^{-1} and 1460.86-1380.81cm^{-1} are owing to the C-H stretching, demonstrates the alkane functional group. The peak occurring at 722.61 cm^{-1} appears appropriate in the presence of long chains of alkyls. The results are well agreed with the Yanxia Wu et al [11].

3.2. Tribological investigation of biodegradable refrigeration oil

3.2.1 Friction Behaviour

When two objects are in moving contact, the torque generated by the frictional force is the friction torque. Fig.3 (a-c) shows the variant of friction torque for synthetic refrigeration oil and bio degradable refrigeration oil at load of 10kg, 50kg and 100 kg. The friction profiles of biodegradable refrigeration oil having an uneven or irregular surface as compared to friction profiles of synthetic refrigeration oil for 100kg. The friction torque of bio degradable refrigeration oil at the load of 10kg, 50kg and 100 kg are 0.015Nm, 0.20Nm and 1.75 Nm respectively, whereas for the synthetic refrigeration oil at load of 10kg, 50kg and 100 kg are 0.078Nm, 1.45Nm and 0.50 Nm respectively.

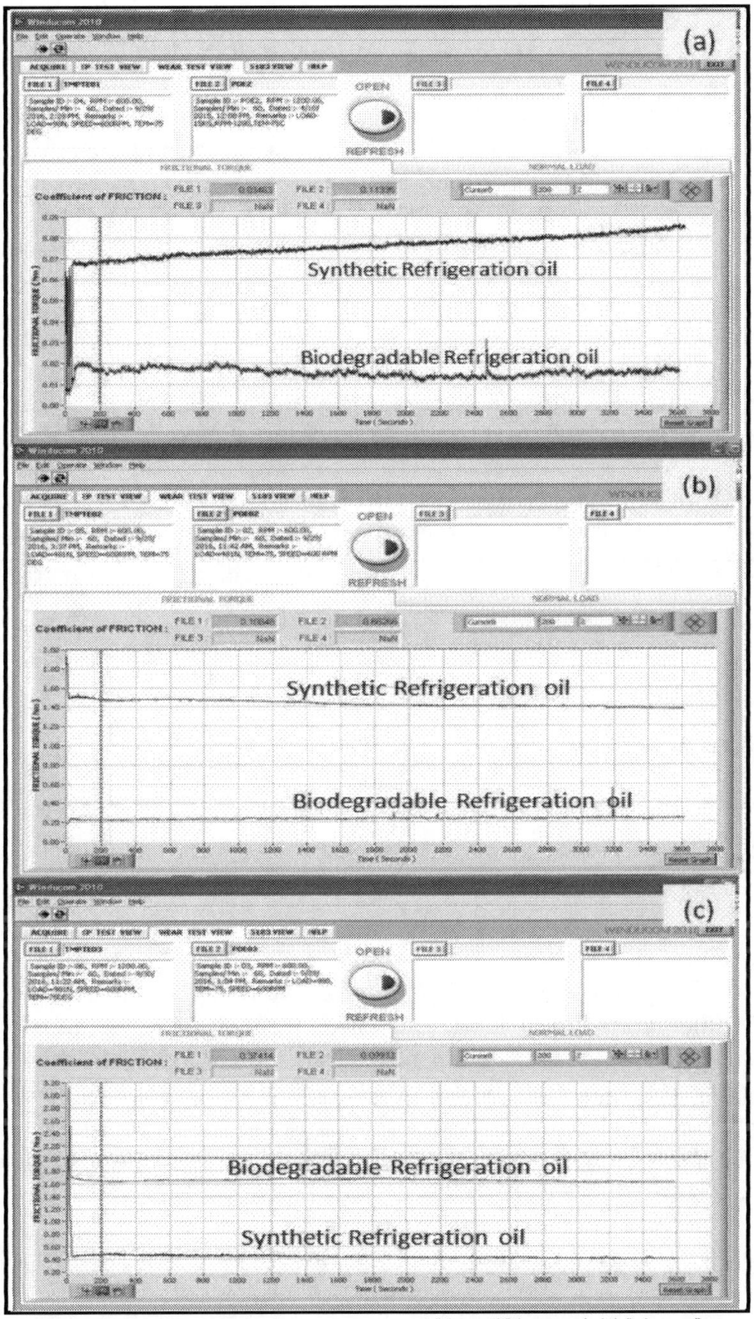

Fig.3. (a, b, c) Friction torque at 10kg, 50kg and 100 kg of synthetic refrigeration oil and bio degradable refrigeration oil

It is apparent from Fig. 3 (a-c), the bio degradable refrigeration oil exerts lower friction torque when compared to the synthetic refrigeration oil up to 50kg. This is due to the cause of presence of maximum portion of oxygen double bond of TMP ester in its chemical chain [12]. This is originated by the material metal-to-metal contact expected to abrasion process under higher loads. However, bio degradable refrigeration oil has superior friction behaviour in addition to capability to keep hold of its lubrication characteristics. Further increase in load more than 50 kg tends to increase the friction torque in bio degradable refrigeration oil.

3.2.2 Coefficient of Friction

The tribological studies clearly show the assessment of the lubrication characteristics of the biodegradable refrigeration oil from the base stocks of rapeseed oil. Fig.3 (a-c) shows the variation in the coefficient of friction of biodegradable refrigeration oil and synthetic refrigeration oil at load of 10kg, 50kg and 100 kg of 40 kg. The coefficient of friction of bio degradable refrigeration oil at load of 10kg, 50kg and 100 kg are 0.0346, 0.1088 and 0.3741 respectively, whereas the coefficient of friction for the synthetic refrigeration oil at load of 10kg, 50kg and 100 kg are 0.11335, 0.6623 and 0.09933 respectively. Table.2. depicts the COF of synthetic refrigeration oil and bio degradable refrigeration oil. The biodegradable refrigeration oil possesses good friction reduction behaviour when correlated with synthetic refrigeration oil up to 50 kg. The results entail that the biodegradable refrigeration oil has the maximum capability to preserve its properties devoid of the collapse of the lubricating layer. The COF is increased when the load increased in general. The high stress concentration will be produced in localized region due to the increase of load. That lead to the localized plastic deformation and followed by the abrupt propagation of crack [13].

Table.2. COF of synthetic refrigeration oil and bio degradable refrigeration oil

Lubricant	Load (kg)	Co-efficient of friction
Synthetic Refrigeration oil	10	0.0134
	50	0.6626
	100	0.09933
Biodegradable Refrigeration oil	10	0.0346
	50	0.1088
	100	0.3741

3.3 Wear Behaviour

Fig.4 (a-f) shows the optical micrographs of wear scar on the ball surface at 10kg, 50kg and 100 kg of 40 kg under the influence of the synthetic refrigeration oil and bio degradable refrigeration oil. It is apparent from the optical micrographs of wear scar on the ball surface at 10kg, 50kg and 100 kg, the bio degradable refrigeration oil displays lower wear scar diameter with the synthetic refrigeration. The biodegradable refrigeration oil had the lower mean wear scar diameter (WSD) than the synthetic refrigeration oil. Fig.4 (a-f) shows the difference in the mean wear scar diameter (WSD) for biodegradable refrigeration oil at load of 10kg, 50kg and 100 kg with the synthetic refrigeration oil. The severe wear is indicated generally by the large wear scar diameter [14]. Table 3 depicts the WSD of synthetic refrigeration oil and biodegradable refrigeration oil at load of 10kg, 50kg and 100 kg.

Table.3. Wear scar diameter of synthetic refrigeration oil and bio degradable refrigeration oil

Lubricant	Load (kg)	Wear scar (mm)			Average wear scar (mm)
		Ball 1	Ball 2	Ball 3	
Synthetic Refrigeration oil	10	572.3	499.5	546.7	539.5
	50	979.4	979.5	979.5	979.4
	100	1136.7	1105.3	1083.6	1108.5
Biodegradable Refrigeration oil	10	407.1	373.6	395.3	392
	50	770.4	810.3	847.6	809.4
	100	1233.1	1235.5	1235.1	1234.4

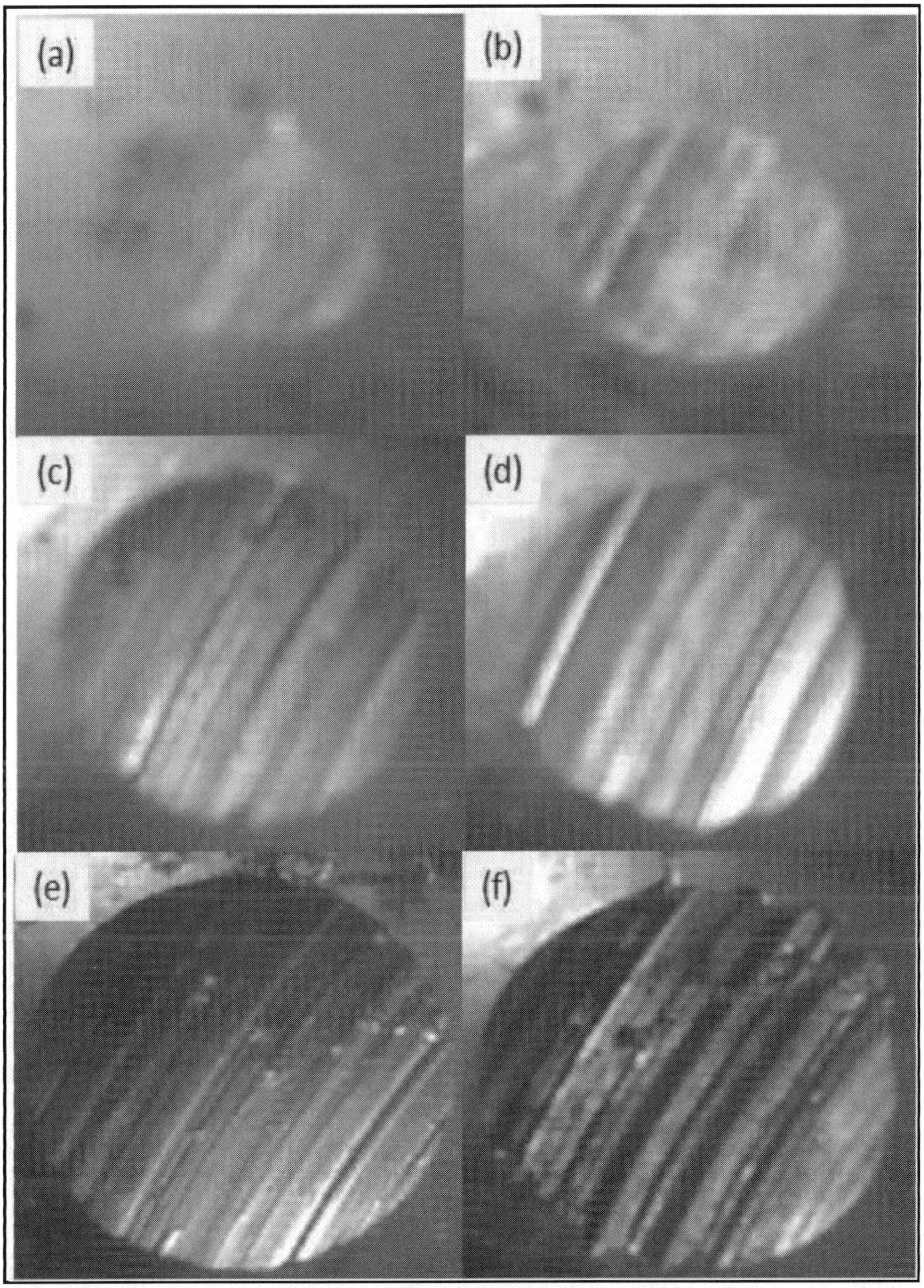

Fig.4. Optical micrographs of wear scar diameter at 10kg,50kg and 100 kg of (a, c, e) synthetic refrigeration oil, (b d, f) bio degradable refrigeration oil

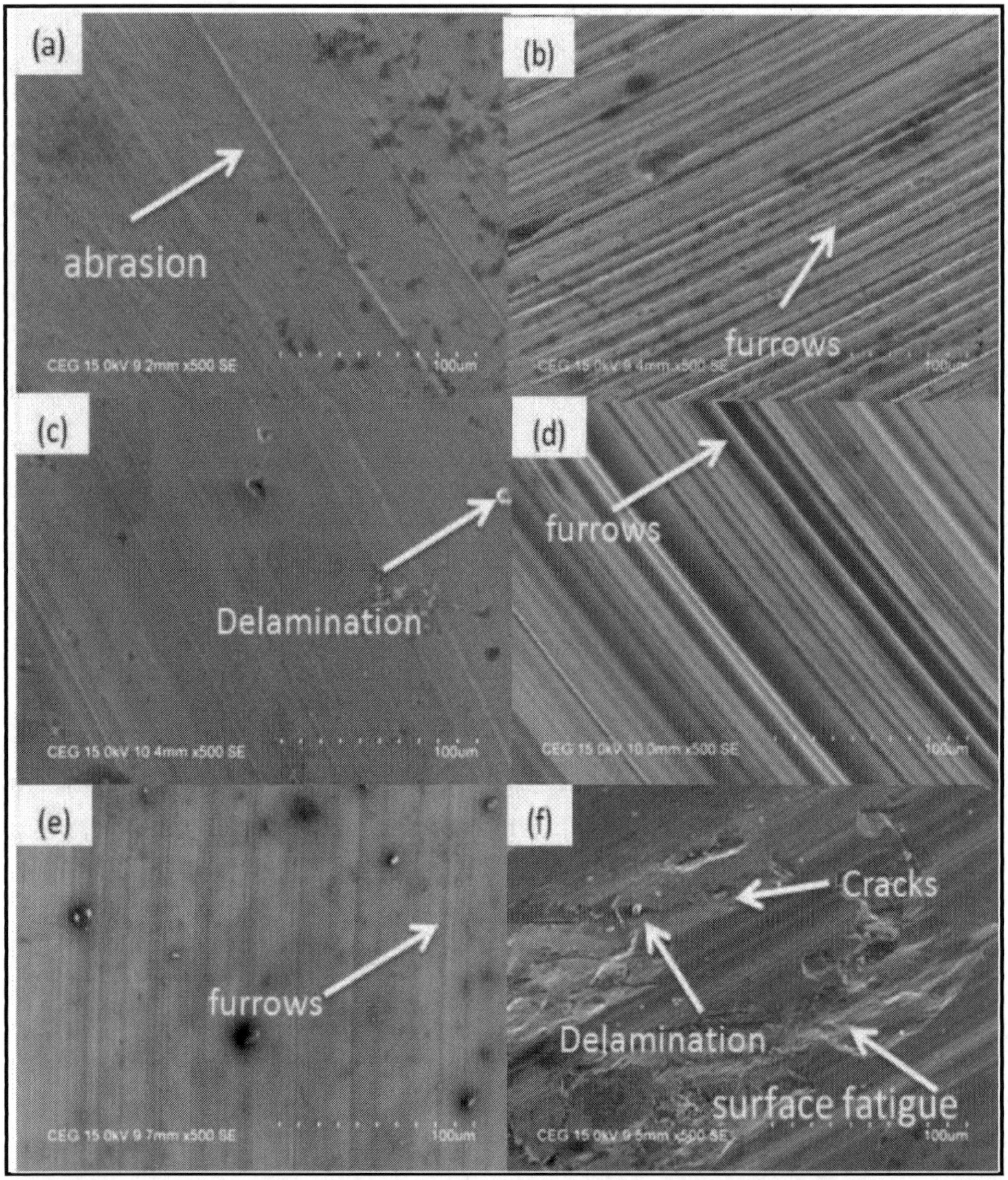

Fig.5. SEM images of tha ball surface of (a, c, e) synthetic refrigeration oil at 10kg,50kg and 100 kg, (b d, f) bio degradable refrigeration oil at 10kg,50kg and 100 kg

Scanning Electron Microscope (SEM) was used to evaluate the wear characteristics of the ball surfaces 100X magnification. The abrasive and adhesive type of wear mechanism was observed on the boundary lubrication regime. From the Fig. 5 (a-f), it is clear that the biodegradable refrigeration acquire smooth surface with inferior wear behaviour when the load increased. This is because of improved surface adsorption rate on metal surface. The abrasive lines on the wear surface on the ball lubricated with biodegradable refrigeration oil are confirmation of abrasive wear action as shown in Fig.5 (a-f). This will originate the shearing of physical bonding and chemical bonding due to aggressive mechanical shearing and straight metal-to-metal [15]. The biodegradable refrigeration oil the inferior tribological properties in terms of wear when evaluate the synthetic refrigeration oil for higher loads. The furrows and wear debris formed in the case of bio degradable refrigeration oil seen on the ball surface shown in Fig.5 (b, d, f). When the two surfaces contacted because of the high-pressure load which cause the plastic deformation of the surfaces and the severe wear occurred at higher loads [16]. This is due to that the bio degradable refrigeration oil does not contain any additives, however, the synthetic refrigeration oil was prepared with the various additives to better functioning in the compressor. Further this study has to be extended to improve the wear behaviour.

3.4 Extreme Pressure Analysis

The extreme pressure of biodegradable refrigeration oil was analysed using ASTM D2783. The intention of the extreme pressure test has to determination of the load carrying capacity of the biodegradable refrigeration oil. The extreme pressure test was conducted at a fixed speed of 1770 ± 60 rpm and 10 sec, increasing the loads until welding of the balls occurs. The test run is made first with a starting load of 40 kg and consequent test runs were made by increasing the loads until welding of the four balls occurs.

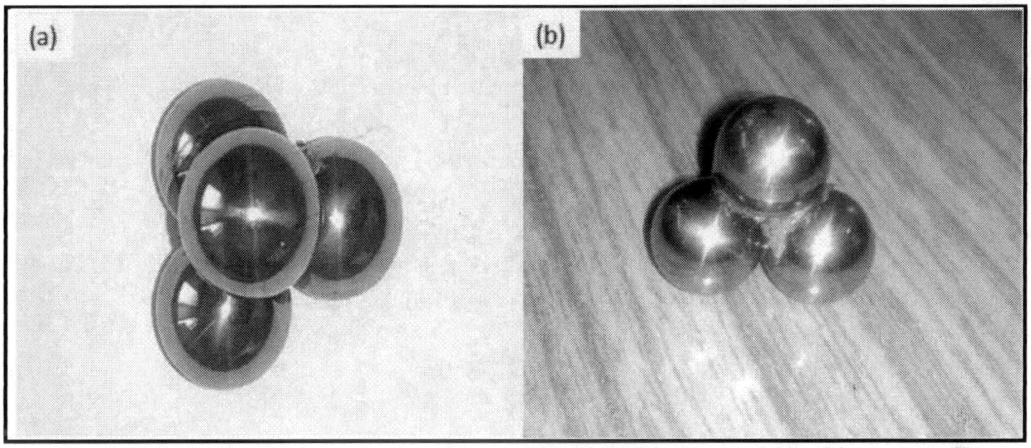

Fig.6 Weld load for (a) synthetic refrigeration oil (b) biodegradable refrigeration oil

The experiment was performed same for synthetic refrigeration oil. Fig. 6(a,b) depicts the weld load for synthetic refrigeration oil and biodegradable refrigeration oil. The last Non-seizure load for the biodegradable refrigeration oil and synthetic refrigeration oil are 40 kg and 50 kg respectively. The Weld load obtained for the biodegradable refrigeration oil and the synthetic refrigeration oil are 126 kg and 160 kg respectively.

4. Conclusions

The tribological behaviour of biodegradable refrigeration oil was performed using ASTM D 5183 and ASTM D2783 and following conclusions can be found from the investigational results.

- The biodegradable refrigeration oil demonstrates superior lubricity properties such as a higher flash point temperature and VI (viscosity index).
- The bio degradable refrigeration oil exerts lower friction torque when compared to the synthetic refrigeration oil up to 50kg.
- The biodegradable refrigeration oil possesses good friction reduction behaviour when correlated with synthetic refrigeration oil up to 50 kg.
- The deep furrows and cracks found in the bio degradable refrigeration oil exhibit inferior wear behaviour on higher loads.
- The last non seizure and weld load for bio degradable refrigeration oil are 40 kg and 126 kg and for synthetic refrigeration oil are 50 kg and 160 kg.

References

[1] Saidur R, Kazi S N, Hossain M S, Rahman M M, Mohammed H A, Renew. Sustainable Energy Rev 15, (2011) 310–323.
[2] Lee K, Hwang Y, Cheong S, Kwon L, Kim S, Lee J, Current Applied Physics 9,(2009) 128–31.
[3] Krishna Sabareesh R, Gobinath N, Sajith V, Das S, Sobhan C B, Int. J. Refrigeration 35,7(2012) 1989-1996.
[4] Tao Jia, Ruixiang Wang, Rongji Xu, Int. J. Refrigeration 45, (2014)120-127.
[5] Bobbo S, Fedele L, Fabrizio M, Barison S, Battiston S, Pagura C, Int. J. Refrigeration 33, 6(2010) 1180-1186.
[6] Mobarak H M, Niza Mohamad E, Masjuki H H, Kalam M A, Al ahmud K A H, Habibullah M, Ashraful M A, Renewable and sustainable energy resources, 33 (2014) 34-43.
[7] Gautam Yadav, Sudhir Tiwari, Jain M L, Materials Today: Proceedings 5 (2018) 248–253.
[8] Dongare A D, Gite A J, IOSR Journal of Engineering, 04, 08 (2014) 10-27.
[9] Habibullah M, Masjuki H H, Kalam M A, Lubricants, Vol. 58,6, (2015) 1002-1011.
[10] Ebtisam K. Heikal, M.S. Elmelawy, Salah A. Khalil, N.M. Elbasuny, Egyptian J. Petroleum 03, (2016)1-13
[11] Yanxia Wu, Weimin Li, Xiaobo Wang, Lubrication Science, (2014) 1-11.
[12] Yashvir Singh, Int. J. Mechanical, Aerospace, Industrial, Mechatronics and Mfg.Engg.10,7, (2016) 1268-1274.
[13] AL-Bukhaiti M A, Al-Hatab K A, Gabbert U, Emirates Journal for Engineering Research ,16 (2011) 9-22.
[14] Farooq M, Ramli A, Gul S, Muhammad N, J. Applied Sciences, (2011) 1381-1385.
[15] Syahrullail S, Wira J Y, Wan Nik, W, Fawwaz W, Appl, Mech. Materials. 315, (2013)936–940
[16] Iman Golshokouh, Syahrullail S, Farid Nasir Ani, Masjuki H H, Journal of Oil Palm Research Vol. 26,1 (2014) 25- 36.

Optimization of process parameters in Electro Discharge Machining of AISI 304 stainless steel

Asit Behera[1], Debashrita Panda[2] and Mayank Bhadauriya[2]
[1]Assistant Professor, SME, KIIT, Bhubaneswar, India
[2]Research Scholar, SME, KIIT, Bhubaneswar, India

Email: asit.nitrkl@gmail.com

Abstract. Due to the profound use of AISI 304 stainless steel in household appliances optimization of process parameters in Electro Discharge Machining of AISI 304 is mostly needed for quality improvement. In this experiment AISI 304 stainless steel has been taken for machining with EDM oil dielectric. Three controllable parameters like peak current (I_p), pulse on time (T_{on}) and Flushing pressure (F_p) at three levels were taken for optimization. Material Removal Rate (MRR) was taken as response parameter and was measured for various combinations of controllable parameters. Taguchi optimization technique with L_9 orthogonal array was used to optimize the specified controllable parameters. It was found that peak current of 15Amp, pulse on time of 200ms and Flushing pressure of 9Kgf/cm^2 bears the optimal quality characteristics. At these levels Material Removal Rate was found to be 16.250mm^3/min.

Keywords: EDM, Taguchi methodology, MRR

1. Introduction

Electric Discharge Machining is one advanced manufacturing process. It is the removal of segments of electrically conductive metals by electrical pulses between two terminals (electrode and work piece) within the dielectric. It is utilized for materials that are too hard for ordinary machining process such as Tungsten and Titanium. As there is no physical contact between the tool and the job, thus the job is not subjected to any mechanical forces during machining. Thus, thin and delicate work pieces can be machined properly. As AISI 304 steel have profound application in automobile industry and food handling equipment so it's desired for the final product to have greater quality and machining process should have higher productivity. Therefore we are intended to optimize the material removal rate of the product in the EDM operation.

P. Malhotra and A. Kohli [1] studied the effect of various process parameters like pulse on time, current, pulse off time and flushing pressure on tool wear rate, MRR and surface finish of the H-11 steel using one variable at a time approach. S. Chandramouli et al. [2] optimized the EDM process separately for MRR, TWR and surface finish. Zhang et al. [3] tried various dielectrics at same condition and analyzed various craters of different sizes and shapes. They discovered that liquid dielectrics possess greater removal efficiency. Wang et al. [4] made a comparative analysis between different dielectrics such as kerosene, distilled water and compound dielectric on Titanium alloy in order to evaluate MRR, SR and EWR. Guu et al. [5] took pulse current and gap voltage as process parameters and MRR, surface finish and de-lamination factor as responses. He studied characteristics of carbon fiber reinforced carbon composite machined in EDM. Thomas et al. [6] studied the effect of process parameters and developed the model for material removal rate through Response Surface Methodology while machining on EN31 tool steel. The model has been verified by experimental results and is observed adequate. M.J. Mohd et al. [7] optimized surface roughness using Response Surface Methodology. S.K Singh and N. Kumar [8] optimized the EDM process parameter to get better surface finish on the Titanium alloys.

The objective of this paper is to machine AISI 304 steel by EDM by taking various parameters. Three process parameters viz. peak current (I_p), pulse on time (T_{on}) and flushing pressure (F_p) were taken at three stages. Machining took place according to L9 orthogonal array and for each run Material removal rate was calculated. Then the three controllable parameters were optimized using MINITAB.

2. Material Selection and experimental procedure

In this experiment, AISI 304 stainless steel has been used as the work material and Cu electrode is used as the tool for the EDM process. Each work-piece material was of 25mm diameter and 10mm of width. Nine pieces of specified dimensioned work-pieces were prepared for the operation. The composition of AISI 304 stainless steel has given in table 1. In this experiment, paraffin based EDM oil was used.

Table 1. Composition of AISI 304 stainless steel

Grade	C	Si	P	S	Mn	Cr	Ni	N
304	0.08	0.75	0.045	0.030	2.0	20.0	10.5	0.10

Before machining the initial weight of the work-piece was measured by weighting machine. The specified dimension of the stainless steel was fixed by a fixture inside the EDM tank which was filled by EDM oil. Various machining parameters were set on the machine. As the machining started the time was recorded by means of a stopwatch until the process finished. Again the final weight of the machined work-piece was measured. Finally the material removal rate (MRR) was measured using the equation 1. The controlling parameters were set according to orthogonal array design and MRR was measured for each run.

$$MRR = \frac{(Wi-Wf)}{\rho t} \qquad [1]$$

Where,
W_i = Weight of the job(before machining)
W_f = Weight of the job(after machining)
t = Machining time
ρ = AISI 304 stainless steel work piece density = 8000 kg/m3

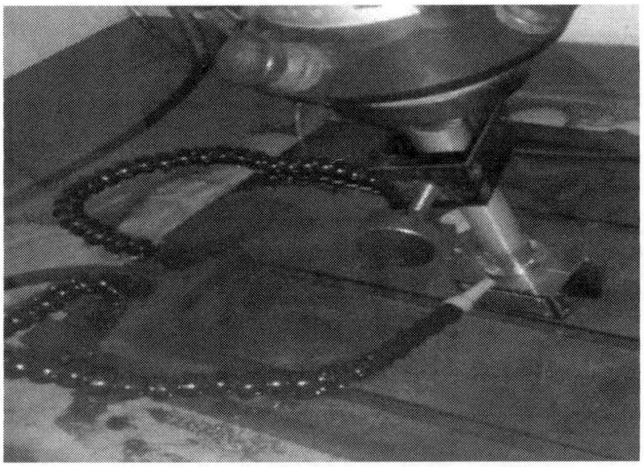

Figure 1. Machining in EDM

Figure 2. Machined Product

According to the orthogonal design experiments were carried out in a random order so as to avoid systematic error. Three laser parameters at three levels were taken into consideration for finding effective MRR for the machining. Levels of various laser parameters are given in table 2.

Table 2. Process parameters and their limits

Parameter	Unit	Level 1	Level 2	Level 3
Peak Current (I_P)	Ampere	5	10	15
Pulse on time (T_{on})	millisecond	50	100	150
Flushing Pressure (F_p)	Kgf/cm^2	6	9	12

3. Result and discussion

The orthogonal array with input parameters and responses were shown in table 3. Taguchi method uses S/N ratio which reduces variability and can be calculated as,

$$\text{S/N ratio} = -10\log10 \sum_{i=1}^{n}((1/Yi^2)/\text{n})) \qquad \text{(Larger is better)}$$

Where, Y= Response
n= no. of runs

Table 3. S/N Ratios as calculated in the Minitab

Obs. no.	I_p	T_{on}	F_p	MRR	S/N Ratio
1	5	100	6	3.030	9.628
2	5	150	9	3.232	10.189
3	5	200	12	4.440	12.947
4	10	100	9	5.000	13.947
5	10	150	12	5.537	14.587
6	10	200	6	7.031	16.940
7	15	100	12	9.375	19.439
8	15	150	6	11.805	21.441
9	15	200	9	16.250	24.217

The main effect plot for S/N ratio for all the three parameters were obtained in the MINITAB software during the analysis. In the machining process the last observation yields the highest MRR, having corresponding input variables values viz. 15 amp., 200 ms and 9 kgf/cm^2. For this observation the value of MRR is 16.250mm3/min.

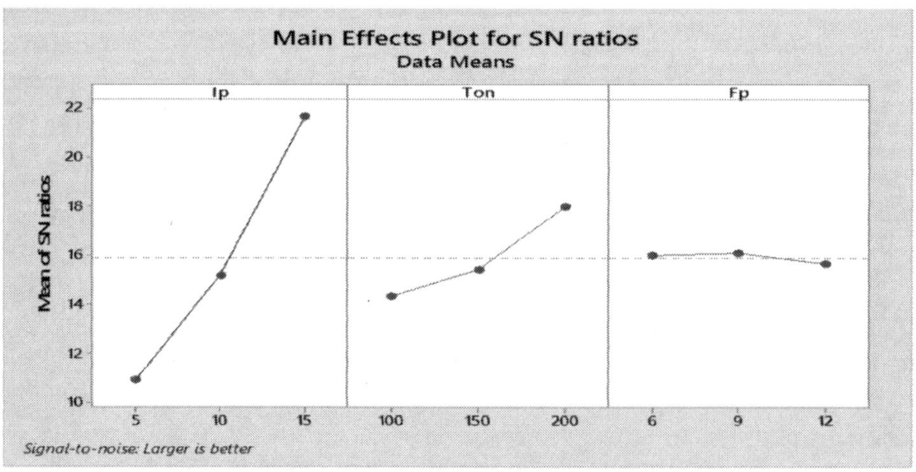

Figure 3. Main effect plot for S/N ratio

It is clear from figure 3 that peak current at 3rd level, pulse on time at 3rd level and Flushing pressure at 2nd level have the highest S/N ratio. So I_P at 15 amp. T_{on} at 200 ms and F_p at 12 Kgf/cm^2 bears the optimal solution to get the desired response.

We can observe from Figure 3 that the MRR increases with the increase in the peak current. This MRR increases such rapidly due to the high concentration of electrical energy in the space between the job and copper electrode, thus resulting in sudden melting and then vaporization of metal. MRR also increases as the pulse on time increases. However, it is also clear from the figure that material removal rate decreases with the increase in flushing pressure. One possible cause behind this decrement is that we took the value of flushing so high that jet stream was unable to properly remove the material.

The linear model of the different S/N ratio is developed by regression analysis which is a statistical measure for predicting the relationship among variables. In this analysis we got the following important values.

1. Standard deviation (S) = 0.6093
2. R-Sq = 99.6%
3. R-Sq(adj) = 98.5%

As R-Sq is close to 100%, this shows the reliability of our experiment with accurate results. The calculated value of S/N ratio from Taguchi analysis is found to be 24.217 which is the highest among any of the calculated values in the table. Hence the corresponding input factors are the most optimized controllable parameters. These factors at their optimized levels are also predicted by the Taguchi analysis on the Minitab software, which is depicted in Table 4.

Table 4. Predicted S/N Ratio

I_P	T_{ON}	F_P	SNRA
15	200	9	24.218

By taking the optimized controllable parameters The predicted value of the S/N ratio by Taguchi analysis is 24.218 which is greater than all individual S/N ratio thus satisfying the experiment.

4. Conclusion

Analyzing results, we concluded that the peak current is the highest influencing factor in determining MRR. MRR is directly proportional to the peak current and pulse on time but depends inversely on the flushing pressure. It was found that peak current of 15 Amp, pulse on time of 200 ms and Flushing pressure of 9Kgf/cm^2 bears the optimal quality characteristics. At these levels Material Removal Rate was found to be 16.250 mm^3/min.

References

[1] Malhotra P and Kohli A 2013 *International Journal for Science and Emerging Technologies with Latest Trends* **12(1)** 15-20
[2] Chandramouli S, Balraj US and Eswaraiah K 2014 *International Journal of Advanced Mechanical Engineering* **4** 425-434
[3] Zhang Y, Li J and Yu Z 2014 *J Mater Process Tech* **214(5)** 1052-1061.
[4] Wang X, Liu Z, Xue R, Tian Z and Huang Y 2014 *IJAMT 2014* **72(5)** 979-987.
[5] Guu YH, Hocheng H, Tai NH and Liu SY 2001 *J. Mater. Sci.* **36** 2037–2043.

[6] Thomas D, Kumar R, Singh GK, Sinha P and Mishra S 2015 *Materials Today: Proceedings* **2(4-5)** 1642–1648
[7] Mohd MJ, Sheikh K, Singh B and Malhotra N 2012 *International Journal of Engineering, Science and Technology* **4(3)** 45-52
[8] Singh SK and Kumar N 2013 *International Journal of Emerging Science and Engineering* **1** 2319–6378
[9] Ho KH and Newman ST 2003, *International Journal of Machine Tools & Manufacture* **43** 1287–1300.
[10] Durairaj M, Sudharsun D and Swamynathan N 2013 *procedia Engineering*, **64** 868-877.
[11] Puertas, Luis CJ and Alvarez L 2004, *Journal of Materials Processing Technology*, **153–154**, 1026–1032.
[12] Gary F and Benedict 1987 *Marcel Decker*, New York
[13] Tarang YS and Yang WH 1998 *Mater. Manuf. Process.* **13 (3)** 455–467.

IOP Conf. Series: Materials Science and Engineering **390** (2018) 012072 doi:10.1088/1757-899X/390/1/012072

Effect of welding parameters on mechanical properties of dissimilar Friction Stir Processed AA 8011 and AA 5083-H321 aluminium alloys

K Palani[1], C Elanchezhian[2], K H V Saiprakash[1], K Sreekanth[1], Dayanand[1], Keshav kumar[1], Deepak kumar[1]

[1] Department of Mechanical Engineering, SCSVMV University, Kanchipuram, 631 561, Tamil Nadu, India
[2] Department of Mechanical Engineering, Sri Sairam Engineering College, Chennai, 600 044, Tamil Nadu, India

E-Mail: saiprakash500@gmail.com, palani.k@kanchiuniv.ac.in

Abstract. The weight reduction and fuel economy are gaining importance in marine, defence and transportation industries to get the quality of the welded joints. The problems in fusion welding of aluminium alloys such as solidification shrinkage, hot cracking and residual stresses which degrade the strength of the welded joints. The innovative green joining technology, friction stir welding (FSW) process was implemented in addition to nanoparticles such as Al_2O_3 and SiC in similar and dissimilar aluminium alloys. In this work, the study was done on the effects of Friction Stir Processing (FSP) parameters on microhardness and tensile properties of AA5083-H321 and AA8011-H24 dissimilar aluminium alloy joints. The tensile test was conducted to evaluate the tensile properties of the welded joints. The Vickers microhardness test was carried out to examine the quality of the welded joints. It was observed that the similar AA8011 aluminium alloy combination with nanoparticle Al_2O_3 has maximum tensile properties and microhardness compared to other combination of joints.

1. Introduction

The friction stir welding process is the modern and eco-friendly solid-state joining process which is used to join relatively the light-weight materials, especially Aluminium and its alloys based on its industrial applications, such as shipbuilding, aerospace and automobile [1]. This process is ideally well suited for filling the voids, welding of dissimilar aluminium alloys because it does not involve melting, the weld solidification cracking of the welded joints. Further improvement of the weld quality of the joints, the Friction stir processing was applied in addition with nanoparticles which will overcome the variety of problems as in fusion welding of aluminium alloys such as porosity, the formation of brittle structures, and liquation cracking in the thermo-mechanically affected zone (TMAZ) [2].

The tool geometry plays a critical role in material flow and plastic deformation, it governs the development of heat due to the friction between the friction stir welding tool and the butt surfaces of the base materials [3]. The newly developed non-consumable tool consists of a different shoulder and pin geometry which is used to produce the dissimilar joints using the friction stir welding process [4]. The function of tool shoulder is to provide the latent heat by the application of a tool rotation speed and transverse speed over the surface of the base material being welded and the large amount of compressive force which makes the recrystallization process between the atoms in the welded joints [5-8]. The function of the tool pin is to move the highly plasticized material from the tool pin in the vertical direction. The different profiles of pin shape which plasticizes the material due to friction and the compressive force of the tool into the work materials [9]. The process parameters, such as rotational speed, welding speed and others are used for aluminium alloys and their combination with or without nanoparticles for generating the oxide layer breaking and mixing of the combinations [10]. The tool geometry plays a critical role in material flow and in turn, governs the transverse rate. The tool shoulder minimizes the formation of voids in the aluminium alloys [11, 12]. The friction stir processing (FSP) improves the aluminium alloys and their combinations toughness and flexibility in the specific area of the microstructure of the aluminium alloys with the formation of fine grain [13, 14]. The very limited earlier discussions have been made on the effects of process parameters on the strength properties of the dissimilar friction stir welded AA5083-H321 with AA 8011-H24 aluminium alloys. During this investigation, the focus was made on the effects of the process parameters on the microstructure and microhardness of the dissimilar friction stir welded/processed AA 5083-H321 with AA 8011-H24 aluminium alloys.

2. Experimentation

The cold worked non-heat treatable of AA 8011-H24 [15] and AA 5083-H321 [16] aluminium alloys (purchased from Virwadia steels, Chennai). The plates were cut into the dimensions of 150 x 150 x 6 mm by Power Hacksaw cutting machine. The mechanical properties and chemical compositions of the AA 8011-H24 and AA 5083-H321 aluminium alloys are specified in Table 1 and Table 2 respectively. The cut specimens for friction stir welding process are cleaned by an ultrasonic sonicator bath at a temperature at 40° C for 15 minutes for each specimen to remove the foreign particles and also cleaned with acetone. The computerized friction stir welding machine was used to produce the dissimilar friction stir butt welded joints of AA8011-H24 and AA5083-H321 aluminium alloys.

Table 1 Mechanical properties of AA8011-H24 and AA5083-H321 aluminium alloys

Material	Tensile strength (MPa)	Yield strength (MPa)	Elongation at Break (%)	Density kg/m^3
AA 8011-H24	110	140	5.6	2869
AA 5083- H321	317	228	16	2660

Table 2 Chemical composition of AA8011-H24 and AA5083-H321 aluminium alloys

Material	Fe	Si	Mn	Mg	Zn	Ti	Cu	Al
AA 8011	0.7	0.5	0.5	0.3	0.1	0.02	0.13	Balance
AA 5083	0.4	0.4	0.4-1	4.9	0.25	0.15	0.25	Balance

In this work, the D2 steel tool with hexagonal pin shape, having the other dimensions of 18 mm shoulder diameter, 6 mm pin diameter having 2 mm side length and 5.7 mm pin length was used and shown in Figure 1. The FSW and tool parameters play a key role in deciding the quality of the weld strength and the details of the FSW parameters are tabulated in Table 3.

Table 3 Welding parameters used for friction stir welding of butt joints

Sl. No.	Process Parameters	Details
1	Rotational speed, r/min	1200
2	Welding speed, mm/min	50
3	Plunge depth, mm	2
4	Tool Pin Profile	Hexagonal pin

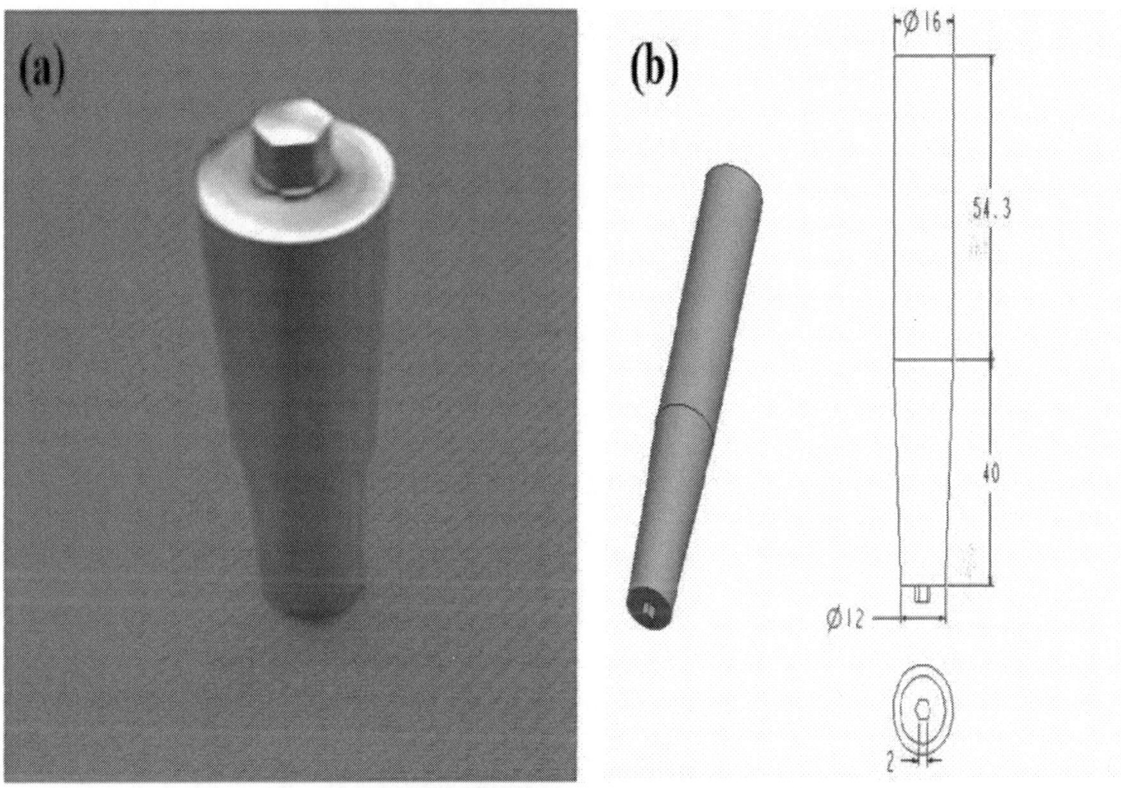

Figure 1 FSW Tool geometry

Ultrasonic cleaned specimens are cut into 10 x 10 x 6 on either side equal to the weldment for examining the microstructural properties. These specimens are moulded by the Bakelite granules and etched with Keller's etchant made in a composition of 5 cm^3 HNO_3 conc., 3 cm^3 HCL conc., 2 cm^3 HF conc., 190 cm^3 H_2O for 20 min and immediately these specimens are dipped in the water. This process increases the visualization of microstructure at the weldment. The microstructure analysis was made and analyzed on the etched specimens by Stereomicroscope (Magnification: 2 mm) and microhardness is also investigated by Vickers microhardness testing machine (Model: Wolpert, HV 0.5Kg) at Omega Inspection and Analytical Laboratory. The microstructures were observed by the scanning electron microscope with different magnifications (500 KX) and with EDS at PSG Tech College of Engineering, Coimbatore. Bothe macrostructure and hardness specimens were prepared as per its standards ASME E340 and IS: 1501 respectively.

3. Results and Discussions

3.1 Analysis on the Macrostructure of the FSW/FSPed Butt Joints

The friction stir welding joints are successfully produced. The obtained joints show no porosity with some noticeable defects.

Table 4 Macrostructure of the FSW/FSPed Butt Joints

Sl.No.	Specimen Details	Macrostructures	Inferences
		AA 8011 – AA 8011 Aluminium alloys	
1.	Without Nanoparticles	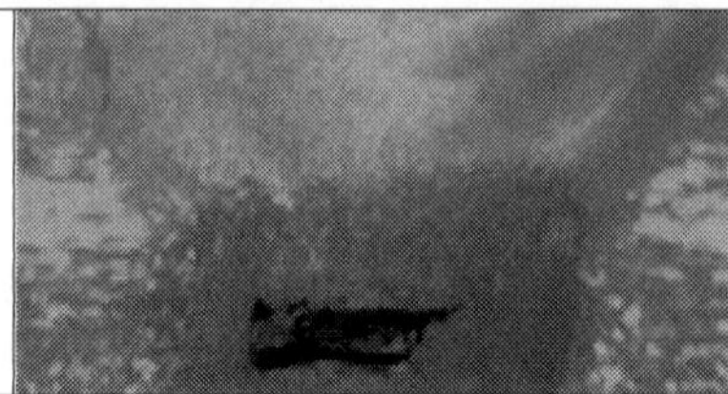	The tunnel defect is found at the advancing side due to higher welding speed and pin length.
2.	With Al_2O_3 Nanoparticles	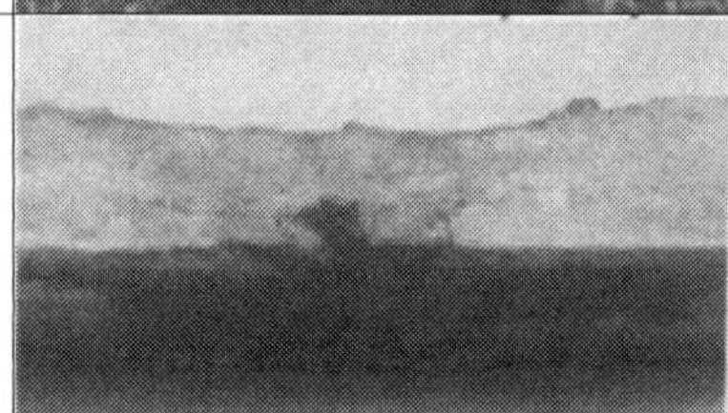	The weldment is good and there is a proper mixing of nano particles.
3.	With SiC Nanoparticles		The weldment is good and there is a proper mixing of nano particles
		AA 5083 – AA 5083 Aluminium alloys	
1.	Without Nanoparticles	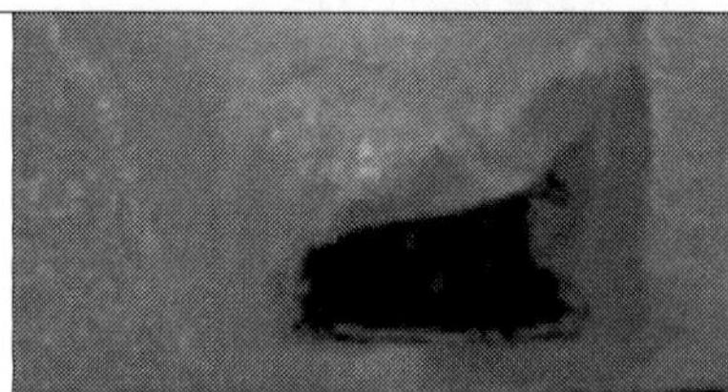	The tunnel defect is found at the advancing side due to higher welding speed and pin length.
2.	With Al_2O_3 Nanoparticles		Cracks and Tunnel defect is formed due to improper mixing of nano particles due to higher welding speed
3.	With SiC Nanoparticles		The defect-free welded joints are observed.

AA 8011 – AA 5083 Aluminium alloys			
1.	Without application of Nanoparticles	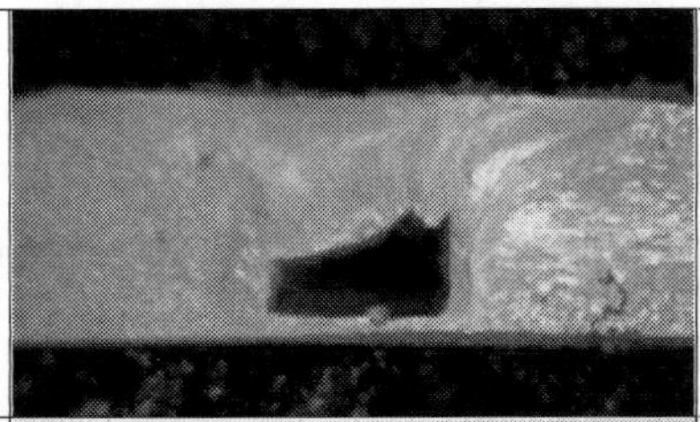	The cavity is formed at the lower part of the welded zone nearer the retreating side.
2.	With Al_2O_3 Nanoparticles	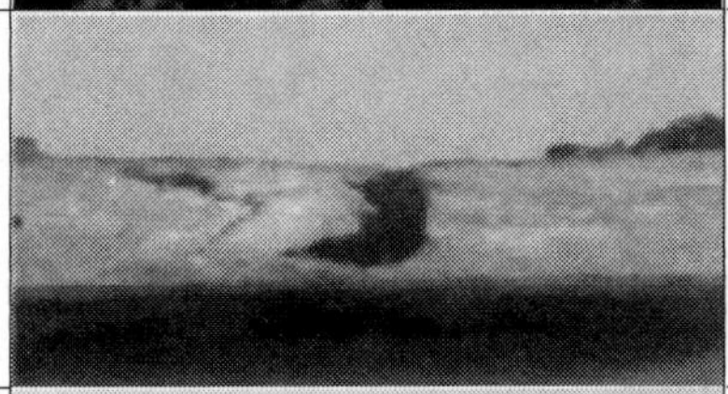	The cavity is formed as the material is dissimilar in the welding zone.
3.	With SiC Nanoparticles	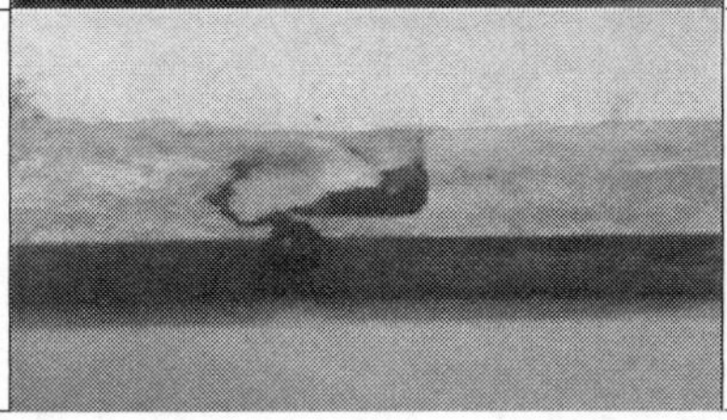	The tunnel defect is found at the advancing side due to the higher welding speed.

From the Table 4, it is observed that the formation of tunnel defect observed in the root of the tool due to higher welding speed [17] and lower penetration into the AA8011-AA8011 aluminium alloys without addition of nanoparticles, whereas in addition with nanoparticles, such as Al_2O_3, and SiC in friction stir processed aluminium alloy of AA8011-AA8011 showed the defect-free structure in the stir zone due to the proper dynamic crystallization and material flow in the welded joints. The AA5083-AA5083 aluminium alloy With SiC Nan particles showed the sound defect-free weld while comparing to other welds of Similar AA 5083 aluminium alloy without nano addition and with Al_2O_3. Furthermore, the cavity and tunnel defect were found for all conditions of dissimilar aluminium alloy combinations [18] with and without nanoparticles in stir zone.

3.2 Analysis of the Micro Hardness of the FSW/FSPed Butt Joints

The similar friction stir welding of aluminium alloys showed the same base metal hardness in both advancing and retreating side. From the Figures 2 (b) and 2 (c),) the maximum hardness was obtained in friction stir processed aluminium alloys in addition to Al_2O_3 nanoparticles. The maximum hardness was achieved in the AA8011-H24 and AA5083-H321 aluminium alloys with Al_2O_3 addition and stated that the proper mixing of the nanoparticles in the aluminium alloys with the help of heat developed by the tool on the butt surfaces compared to the hardness of the other butt joints. Furthermore, in thermomechanically affected zone (TMAZ), the hardness value initially decreased than the hardness of the base aluminium alloys due to plastic deformation of material then it is drastically increasing in the stir zone similar results obtained by Ghosh et al. [19].

IOP Conf. Series: Materials Science and Engineering **390** (2018) 012072 doi:10.1088/1757-899X/390/1/012072

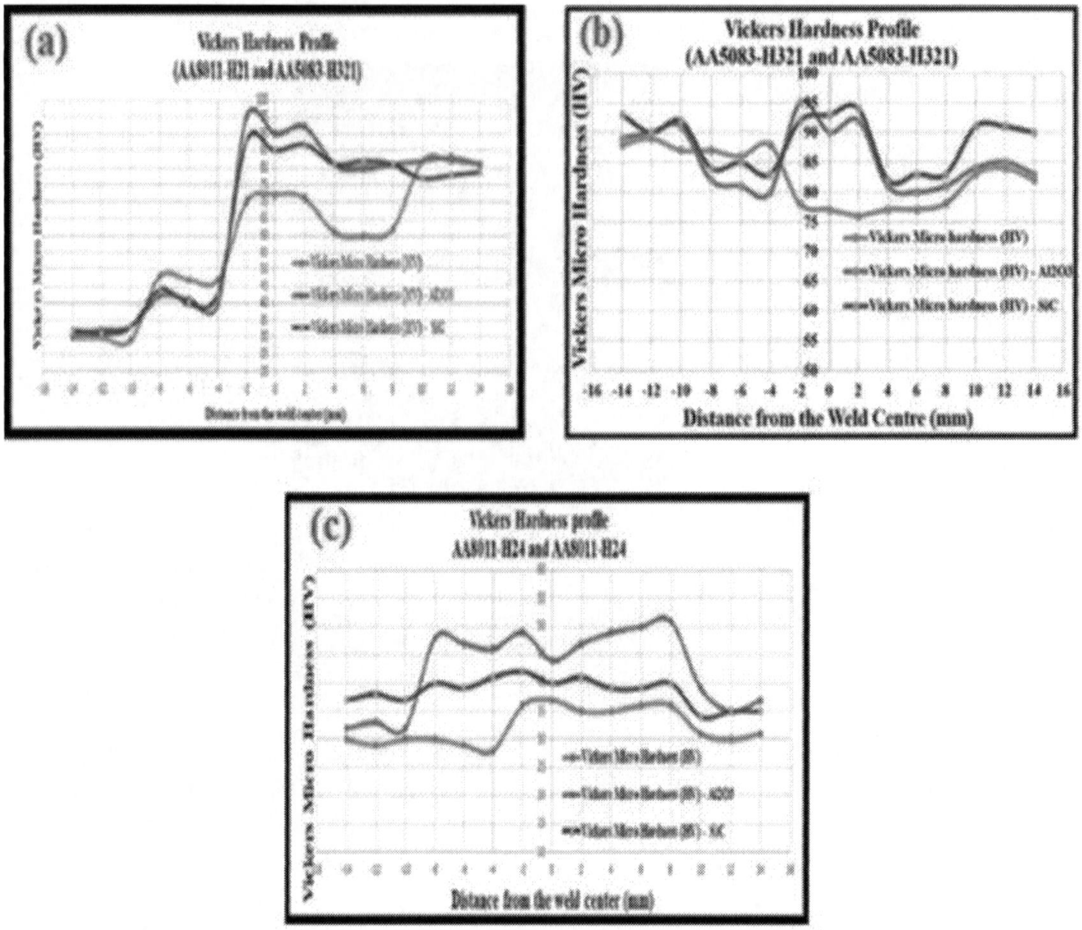

Figure 2 Micro Hardness of test specimens for different material combinations

3.3 Analysis of the Microstructure of the FSW/FSPed Butt Joints

The microstructure of the welded joints showed the quality of the weld based on the improvement of the strength properties and sound defect free joints. The microstructure analysis was carried for the dissimilar aluminium alloys of AA5083 H321-AA5083 H321 in addition to SiC Nano Particles shown in Figure 3. From the Figure, it was observed that the better weld appearance and joint strength was examined with nanoparticle SiC in AA5083 H321-AA5083 H321 aluminium alloys due to sufficient heat generation between the alloy materials and tool pin surfaces in stir zone whereas the improper diffusion of nanoparticles with aluminium alloys were observed in the thermo-mechanically affected zone [20] and heat affected zone which degrades the joint strength properties.

4. Conclusions

The effects of the process parameters on the microstructure and microhardness of the dissimilar friction stir welded/processed AA5083-H321 with AA 8011-H24 aluminium alloys were investigated and the following conclusions were derived:

- The macrostructure of both similar and dissimilar FSW/FSP of AA5083-H321 with AA 8011-H24 aluminium alloys in addition to nanoparticles Al_2O_3 and SiC showed the sound defect free joints compared to other joints.
- The maximum hardness was observed at the weld nugget zone in addition to nanoparticles Al_2O_3 in dissimilar AA8011-H24 and AA5083-H321 welded joints by proper material mixing and plastic deformation and fine grain formation occurred by the maximum frictional heat at the stir zone whereas the variations of hardness occurred at the thermomechanically affected zone which is lower than the hardness at the stir zone.
- The better microstructural results were observed in friction stir processing joints relate to the friction stir welded joints and the addition of nanoparticles improved the surface and joint properties.

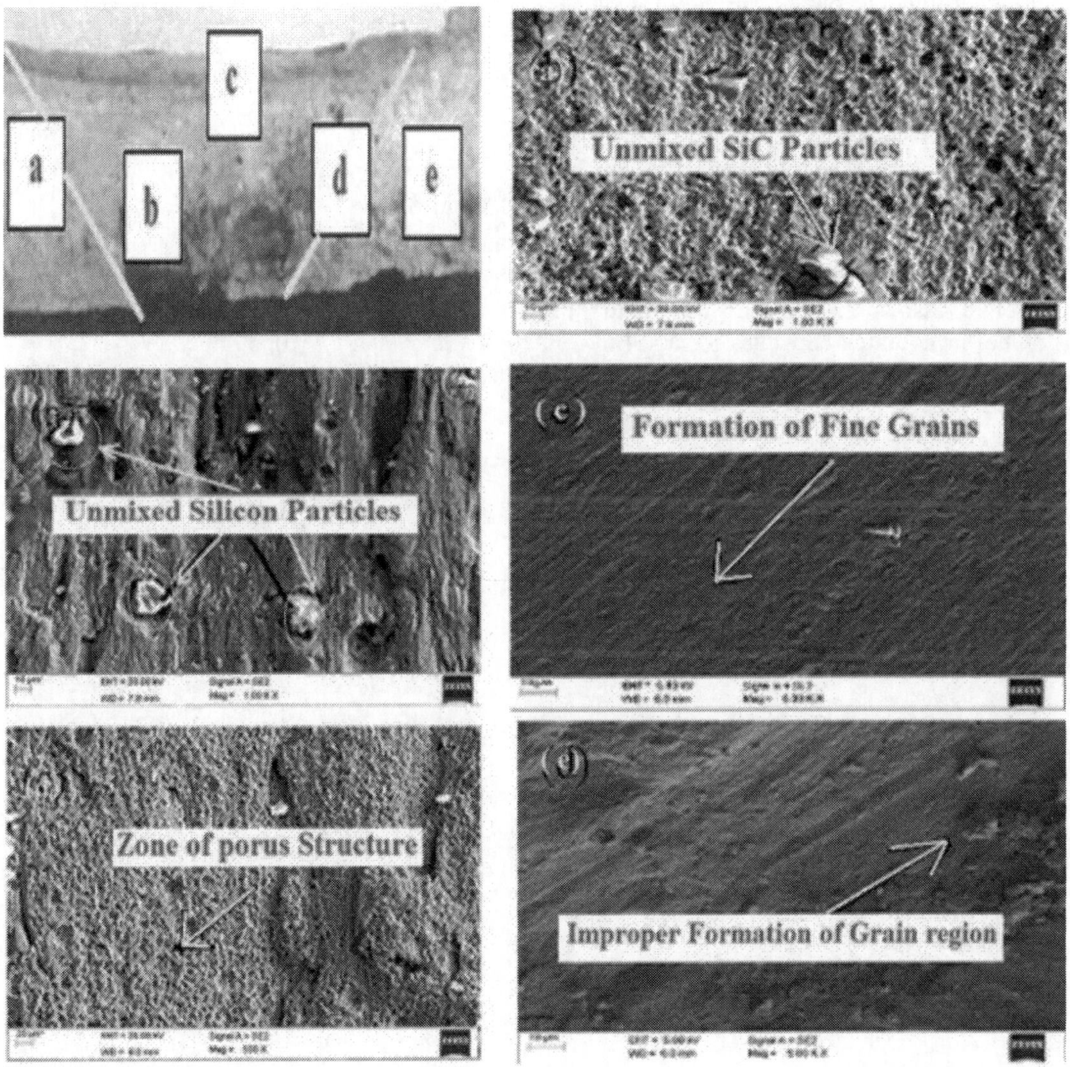

Figure 3 Microstructural analysis of the FSW/FSPed joints of the AA5083-H321 - AA5083-H321 with SiC (a) TMAZ at advancing side (b) TMAZ at retreating side (c) Nugget zone (d) HAZ at advancing side (e) HAZ at retreating side

5. References

[1] Thomas WM and Nicholas ED 1997 *Mater. Des.* **18** 269-73
[2] Koilraj M, Sundareswaran V, Vijayan S and Koteswara Rao SR 2012 Mater. Des. **42** 1-7
[3] Mishra RS and Mab ZY 2005 Mater. Sci. Engg. R **50** 1-78
[4] Vilaca P Santos, Gois A, Quintino AL 2005 *weld. Jour.* **49** 1681-05
[5] Palani K and Elanchezhian C 2015 *Appl. Mech. Mater* **766** 921-27
[6] Mastanaiah P, Sharma A, Reddy GM 2016 *Trans. Ind. Inst. Met.***69** (7) 1397-1415
[7] Palani K and Elanchezhian C 2015 *Appl. Mech. Mater* **813** 446-450
[8] Saravanan V, Rajakumar S and Banerjee 2016 *Int. J. Adv.Manuf. Tech.* **87** (9-12) 3637-45
[9] Madhusudhan R, Sarkar MM and Ramanaiah N 2013 *Int J. of Mech. Prod. Engg.* **4** (2) 204-208
[10] Palani K and Elanchezhian C 2015 *Appl. Mech. Mater* **813** 451-55
[11] Vijayan S, Raju R and Rao SRK *Mater. Manuf. Proc* **25** 1206-12
[12] Tanabe H and Watanabe T 2008 *weld. Int.* **22** (9) 588-96
[13] Cao X and Jahazi M 2011 *Mater. Des.* **30** (6) 2033-42
[14] Amini S, Amiri MR and Barani A 2015 J. Adv. *Manuf. Tech.* **76** (1-4) 255-61
[15] Palani K and Elanchezhian C 2015 Appl. Mech. Mater 813 446-450
[16] Palani K, Elanchezhian C, Ramnath BV and Bhaskar GB 2015 *Adv. Sci. Eng. Medic.* **10** 1-6
[17] Elangovan K, Balasubramanian V and Valliappan M 2009 Mat. M*anuf. Proc.* **15** (2) 321-30
[18] Şefika Kasman 2013 Int. J. Adv. Manuf. Tech **68** 795–804
[19] Ghosh M, Kumar K, Kailas SV and Ray AK 2010 *Mater. Des* **31** (6), 3033-37
[20] Palani K, Elanchezhian C, Ramnath BV, Bhaskar GB, Jagadeesh JS and Kumar GM 2017 *Int. J. Res. App. Sci. Eng. Tech* **5** 437–442

IOP Conf. Series: Materials Science and Engineering **390** (2018) 012073 doi:10.1088/1757-899X/390/1/012073

LIMITATIONS OF PROBABLE VEHICLE HEADLIGHT TECHNOLOGIES – A CRITICAL REVIEW

Chitradeep Sarma, Ankita Gupta, Abhinav Singh, Indrani Bhattacharjee, Abhilash Mohanta

S Tripathy, D Singhal

School of Mechanical Engineering, Kalinga Institute of Industrial Technology, Bhubaneswar, Odisha, India

Email- 1502183@kiit.ac.in

ABSTRACT. Day in and out there is an increase in the number of vehicles on the road. People are starting to prefer travelling at night times in order to avoid the heavy traffic. But while driving at night, a clear perception of the road and upcoming obstacles in not possible. Drivers generally switch on to high-intensity headlights while driving at night times which result in discomfort for the drivers driving from the opposite direction. People experience temporary blindness due to sudden glare of the high beam lights of the upcoming vehicle which in medical terms is referred as Troxler effect. In this paper, we have studied the possible methods which have been released by some developers but are limited to academic models. We have basically summarized all those methods and determined the reasons behind their failure to come into practical use. These technologies have certain loopholes which limit these to academic prototypes only. In addition to this, we have presented a new direction to work on this issue through the implementation of Internet of Things (IoT). This technology paves new roadways to share large information using cloud databases.

KEYWORDS
Photopic, Scotopic, Troxler Effect, LDR, Fuzzy logic, Pulse Width Modulation, Wireless Modulation Sensor.

1.INTRODUCTION

Light is an electromagnetic radiation. The word usually refers to visible light which is detected by naked human eyes. Light indeed is responsible for the sense of vision.401-701nm(nanometer) holds the range of visible wavelengths surrounded by infrared (the longer wavelength) and ultrasonic (the shorter wavelength). The word usually refers to visible light which is noticeable to the human eye and it is responsible for what we see. Visible light is generally characterized as having the wavelength in the range of 401-701 nm(nanometer). Between the infrared (Longer wavelength) and the ultraviolet (Shorter wavelength). Light can be produced by nature or by people. `Artificial` light is normally created by lighting frameworks that change electrical energy into light. The human eyes are versatile to a specific range of vision. There are two visions to be specific Photopic and Scotopic vision [3]. Human eyes vary with various conditions of light surrounding it. In a bright environment, human eye opposes up to 3cd/m2, called photopic view. During dull and dim surroundings our eye changes to Scotopic vision, ranging from 31-45μcd/m2. It takes generally about 4 secs for human eyes to change from photopic to scotopic and vice-versa. This is also an additional case of TROXLER effect. With the brightness increase the strain, that expands the reaction time delaying the reaction. The need of headlights is inevitable for the night journey. The same headlight which helps the driver for better vision during night travel is also responsible for numerous accidents.Usually, the control is under the driver to switch in between high beam to low beam as per requirement.In Dark conditions, in absence of any other alternate source of light, high beam is preferred, to track the road or any obstacle ahead. Whereas in busy roads or heavy traffic low beam is preferred. In case of two-way traffic, the vehicles ply on both lanes. As the bright light from the vehicles of the opposite side falls on to the eyes of the driver of another side, results in glaring effect for sufficient amount of time. Even fraction of diversion might be the main cause of a fatal accident. Here, in the paper, we have discussed some most common concepts for changing the intensity of vehicle headlight which includes the use of LDR, Fuzzy logic sensor, Wireless Sensor Network, IR transmitter-receiver, Camera-based and PWM. But as of today these technologies have not been used in the practical scenario and are restricted within academic prototype. So this paper reflects the loopholes in these above-mentioned concepts and finally, we provide a different idea of these concepts.

2. LITERATURE REVIEW

Several authors and developers have researched and worked on different vehicle headlight technologies. These technologies or methods have certain limitations due to which they have been restricted to the theoretical domain. We studied these methods and segregated them and analyzed their drawbacks. Our critical review has been briefed on the following points:

2.1. LDR BASED INTENSITY CONTROL

In the LDR (Light Dependent Resistor) based headlight technology, LDR is implemented to change the intensity of light of the vehicle headlight. The resistance value of LDR varies according to the intensity of the incident light upon it. BJT and MOSFET are triggered by the gate current and are controlled by the implementation of the potential divider which consists of resistors. The switching device used in this method is Relay which has two contacts. Among the two contacts, high intensity LED is closed contact, whereas open contact is used to connect to low intensity LED [1] [2] [8].
LDR resistance varies over a wide range. They have a wide tolerance and their value changes with temperature change and with age. Sunlight and humidity destroy an LDR in a couple of years. Photo-diodes and photo-transistors are much better. Moreover, in this case, the light coming from the upcoming vehicle must fall on the LDR such that the LDR can work. This implies the LDR must be placed in the front. As a result, the LDR might experience Environmental effects and reduce its life [22]. In addition to this, the distance factor of the upcoming vehicle is not considered, which will also have its effect on the intensity of light. It hence is clear that use of LDR is not possible with this presented approach and has practical limits for use.

2.2 FUZZY LOGIC BASED INTENSITY CONTROL

PIC controller along with the implementation of fuzzy logic is used to control the intensity. The distance between the two vehicles, the Light intensity from the coming vehicle is received as inputs

by the fuzzy sensor. That checks whether it is confined within the tolerable limit. The information is conveyed to the micro-controller which converts it to the ambient light intensity. This software could be easily developed through MATLAB fuzzy logic toolbar, comprising fuzzy rules and values and act as the interference platform for the sensor and controller. The controller completes the process of fuzzification and defuzzification, the sensor output passed the information whether it is within or beyond the limit of tolerance [1] [24].
But the use of this concept is being limited as there are thousands of different fuzzy system configuration may arise conjunction disjunction, implication and diffusion choices, certain stem from the root interference means of fuzziness. The problem of dimension makes it an impossible practical implementation to set up rule base with more than three variables (if not only the light and distance but also considering the environmental aspect and other obstacles are detected, which is common the practical scenario). Application of learning controller from the input is still restricted to an academic prototype. Moreover, the recognition of gradual colour change is quite difficult and the technology used to acquire upcoming vehicle distance and light intensity is not clearly stated.

2.3 WIRELESS SENSOR NETWORK

A network that consists spatial distribution of autonomous devices comprising a collection of sensor nodes, that communicates and records surrounding conditions. Environmental monitoring, robotic control, smart grid, industrial automation and energy control systems etc are the fields of wireless sensors application [25]. The following system has an autonomous gateway which provides wireless connectivity to the wired databases and sensor nodes. Some of the distinguished parameters that can be monitored through WSN are humidity, temperature, pressure and sound intensity. With the help of this particular technology, a system can be established to control the headlight intensity. The process includes the sensing of the intensity of headlight by the sensor node which is then passed digitally to the wired or wireless media, to control the light strength [1]. But obviously, this technology and system working in the entire process have some flaws as it is still not in practical use. With intensity variable changing in different areas the sensor would find it difficult to capture the particular intensity of light. Fog, mist, humidity and rain decreases the intensity strength which almost makes the system absurd to sense the light. Alongside, with wireless media interference is always possible with other frequencies which will make it difficult to transfer the data efficiently and within a stipulated amount of time.

2.4 IR TRANSMITTER-RECEIVER

In this method, Infrared radiation is transmitted, which is being emitted by the LED, connected to the IR transmitter. The photodiode which is attached to the IR system receives the signal at receiving end. IR radiation which is not detected by human eye hence it is the preferable electro spectrum radiation used in wireless communication. The power supply is provided to this transmitting-receiving circuit. The IR rays which were emitted are reflected the receiver. Then the receiver receives and the device is turned on. The comparator gets the input voltage from the receiver, comparator thereby quickly compares the predefined threshold value with the received voltage. Comparator thereafter triggers the device to either switch on or off. Transmitter or MOSFET can act as a triggering device and Relay plays the switching device. Triggering device turns on followed by the switching device, then the LED shifted to low intensity. Again as the transistor is switched the high intensity LED too gets switched. Henceforth, the controller is developed through IR [1] [6].
But the uses of Infrared rays are being not practical in case of vehicles. The infrared signal sometimes faces an obstacle and gets blocked. Being a form of light, infrared only moves in straight lines. Therefore, the transmitters receivers have to be closely aligned to communicate directly inside of each other. The weather can also cause an effect, interference from sunlight, hail, storm, rain, dust

IOP Conf. Series: Materials Science and Engineering **390** (2018) 012073 doi:10.1088/1757-899X/390/1/012073

particles and pollution. The transmission range is quite short (within 10 meters). As a result, the performance also drops if the distance between the transmitter and the receiver is away from the range of IR device. Moreover, the speed at which transmission takes is quite slow [23].

2.5 CAMERA BASED INTENSITY CONTROL

A monochromatic or polychromatic optical device such as a still camera that captures still frames and easily transmits it or sends and record it via an electronic sensor. Generally, 4.3mm focal length is used due to its compatibility due to its human vision. Through various image processing techniques, the information of any vehicle approaching as well as vehicle coming from opposite direction as obtained. Hence the extraction is featured. After that, the analysis of the level of intensity is detected and henceforth the intensity of the vehicle headlights switches from high to low beam when required. High beams are preferred during the night only in absence of vehicles surrounding it and further switching it into low intensity with the increasing number of vehicles [1] [11]. As high-intensity headlights cause difficulties in the vision of the drivers and in certain cases it might be the cause of fatal accidents.

This method may not be reliable as the weather is prone to change and the diversity of reasons that come in handy place to place as the vehicle travels and climatic conditions that would affect the frames captured by the optical sensors. Fog mist dust particles etc. contribute to it. Rain, snow may ruin the frames and might not match causing the deviation. Resulting error in the detection of sensing and prediction of the light coming from ahead. Besides frames are varied with every light conditions from dawn to dusk. As it is a camera-based image processing unit it might mismatch with the database. The database which is predefined. Failure of matching of data in real-time environment conducts unwanted intensity of light. This might again arise a fatal effect while driving on highways.

2.6 PULSE WIDTH MODULATION

Pulse Width Modulation is the methods that deal with controlling simple circuits in the association of the microcontroller. The signal is digital in between processor to the controller. Noise is not an influential factor except when it can be strong enough to manipulate the binary logic. It has noise immunity over analogue controlling systems. By taking the output from the controller the intensity of the emitting headlight can be controlled. The duty cycles of PWM develops the power and glows the headlight, for instance, the duty cycle of the PWM considered 75% then the brightness would appear $3/4^{th}$. Another important factor is Pulse Width [1]. There will be a gradient of ON and OFF of the flashing of the headlights to change its brightness, then to have a constant focus light beam. The mean setback lies on the microcontroller and its action to control and execute faster calculations to perform this operation [26].

PWM signals work in the 2kHz to 16 kHz, but sometimes it causes damage in DC and AC motors due to pulsating current (PWM). Among several effects, ball bearings are sensitive, and sometimes can be damaged due to the balls themselves being charged and discharged inside the rings (mainly due to EMF). Also, due to rise and fall of the current in PWM dimming Electromagnetic Interference (EMI) happens. This system also requires semiconductor device with low turn on and high turn off times but are very expensive.

3. CONCLUSION

In this paper, we emphasized on all the possible technology and methods that have been introduced to control the intensity of vehicle headlight for the purpose of reducing the discomfort of sudden glare at night time. We have studied and presented all the limitations of these methods and the reasons why these have not been implemented in the real scenario and are limited to academic prototype only. Our motivation was to direct the enthusiastic developer towards better methods, leaving the presented methods. Use of Internet of Things(IoT) can be a different direction for solving this issue. IOT can be implemented to distribute the information of headlight beam intensity of a car to other cars within a limited range. This information will be notified to both the drivers when both are within a particular range, to change their beam intensity as per the situation. If the driver does not change the intensity, after a minimum distance between the two cars, the low beam will be triggered automatically. Although this is a rough idea, a certain direction is being provided to work on this issue through the implementation of IoT.

References

[1] Devashree Chilla, Manasi Joshi, Sanjyot Kajale, Seema Deoghar- "**Headlight Intensity Control Methods- A Review**", *IJIRCCE*, Volume **4**, Issue 2, February 2016.

[2] Muralikrishnan.R- "**Automatic Headlight Dimmer a Prototype for Vehicles**", *IJRET*, B.E, Electrical and Electronics Engg, Sri Venkateswara College of Engineering, Tamil Nadu, India.

[3] AlsamMusthafa R, Bala Krishnan T, Seetha Raman N, Shankar M, Asst.prof Ms Swathi R – "**Automatic Headlight Beam Controller**", *IJTRD*, Dept.of Instrumentation and Control Engg., Sri Krishna College of Technology, Coimbatore, India.

[4] Mr Vishal G. Nandokar, Prof. K. Sujatha, Prof. HarshvardhanUpadhyay- "**Advanced Head-Light Controlling System for Vehicles**", *IJEEE*, Shree Ramchandra College of Engineering, Lonikand, Pune, India.

[5] G.M. Pushpanjali, P.S. Mali, R.R. Naman-"**Automatic Headlight Dipper with Respect to Upcoming Vehicles Response**", *ICRIET*, Department of Physics, Sangameshwar College, Solapur, Maharastra, India.

[6] A.S.M Asaduzzaman, Mohammad Mahmudul Islam, Shuva Paul, MdFarhatAlam, MdMashuker Rahman- "**Automatic High Beam Controller for Vehicle**", *IJSER*, Vol **4**, Issue 3, March-2013.

[7] Tejas Vijay Narkar-"**Automatic Dipper Light Control for Vehicles**", *IJRET*, B.E., Instrumentation Engineering, VPM's Maharshi Parshuram College of Engg., Maharastra, India

[8] Prof. BalajiWattamwar, Raghavendra Rao, Pratik Vyas, MithileshTipkari-"**Automatic Headlight Beam Shifter for Controlling High Beam to Low Beam of Vehicles**", *IJIRCCE*, Vol.**4**, Issue 4, April 2016, Dept. of E&TC, Maharashtra Institute of Technology, Aurangabad, India.

[9] Karthik M.M-"**Automated Headlight Intensity Control and Obstacle Alerting System**", *IRJET*, Vol.**3**, Issue 06, June-2016, M.Tech, ECE, Srinivas Institute of Technology, Valachil, Mangaluru, Karnataka, India.

[10] DarkoJuric, Steve Lancaric-"**A Method for On-road Night-time Vehicle Headlight Detection and Tracking**", Dept. of Electronic Systems and Information Processing, University of Zagreb, Croatia.

[11] Feng LUO, Fengjian HU-"**A comprehensive Survey of Vision-Based Vehicle Intelligent Front Light System**",*IJSSIS*, Clean Energy Automotive Engg. Centre, School of Automotive Studies, Tongji University, Shanghai, China.

[12] Sushil Kumar Choudhary, Rajiv Suman, Sonali, Honey Banga-"**Electronic Head Lamp Glare Management System for Automobile Application**",*IJRAT*, Vol.**2**, No.5, May 2014, Dept. of Industrial & Production Engg, Dept. of Mechanical Engg., College of Technology, G.B Pant University of Agriculture & Technology, Pantnagar, India.

IOP Conf. Series: Materials Science and Engineering **390** (2018) 012073 doi:10.1088/1757-899X/390/1/012073

[13] Priyanka Dubal, Mr Nanaware J.D- "**Design of Adaptive Headlights for Automobiles**",*IJRITCC*, Vol.**3**, Issue 3, College of Engg. And Polytechnic, Satara, Maharashtra, India

[14] O.Akinsanmi,A.D.ganjang,H.U.Ezea "**Design and Development of an Automatic Automobile Headlight Switching System**",ISSN:2394-3661, Volume-**2**, Issue-8, August 2015

[15] Akogbe A. Michael,Folorunso T. Abiodun and Omeligbe O. Samuel -"**Development of a Microcontroller based light dependent resistor (LDR) vehicle tracking system**", Department of Computer Engineering, Federal University of Technology Minna,Nigeria,ISSN:2319-7900

[16] Robert Tamburo,ErikoNurvitadhi ,Abhishek Chugh, Mei Chen,AnthonyRowe,TakeoKanade,Srinivasa G. Narasimhan -"**Programmable Automotive Headlights**",Carnegie Mellon University, Pittsburgh, PA USA.

[17] Antonio Lopez. JorgHilgenstock, Andreas Busse, Ramon Baldrich, Felipe Lumbreras and Joan Serrat-"**Nighttime Vehicle Detection for Intelligent Headlight Control**", Computer Vision Center and Computer Science Dept., Auton. Univ. of Barcekona

[18] Kalayani Gaikwad, Ramesh Mali-"**A Survey o n Automobile Headlights Adaaptive Control Based on Can/Lin Network**", Dept. of Elecronic and Telecommunication, MAEER`S MIT College of Engineering, Kothrude, Pune, Maharastra , India

[19] S. Kanagamaliga, Dr. S. Vasuki, A. Kanimozhidevi, S. Priyadharshani, S. Rajeswari-"**Tracking And Counting The Vehicle in Night Scenes**", Department of Electronic and Communication Engineering, Velammal College of Engineering and Technology, Madurai

[20] Susan T. Chrylser, Paul J. Carlson and H. Gene Hawkins-"**Headlamp Illuminations Provided to Sign Positions by Passengers Vehicles**", Texas Transportation Institute, The Texas A&M University, *Report-0-1796-3*, October 2003

[21] http://www.ni.com/white-paper/7142/en/.html

[22] http://www.edaboard.com/showthread.php?t=303331.html

[23] http://www.rfwireless-world.com/Terminology/Advantages-and-Disadvantages-of-Infrared-Sensor.html

[24] http://whatis.techtarget.com/definition/fuzzy-logic

[25] http://www.elprocus.com/archiecture-of-wireless-sensor-network-and-applications/

[26] http://www.learn.sparkfun.com/tutorials/pulse-width-modulation

IOP Conf. Series: Materials Science and Engineering **390** (2018) 012074 doi:10.1088/1757-899X/390/1/012074

Tribological investigation of cashew nut shell oil as lubricant additive

G.Venkatakoteswararao [1*], S. Baskar[2], S. Arumugam [3]

[1, 3] Department of Mechanical Engineering, Sri Chandrasekharendra Saraswathi Viswa Mahavidyalaya, Enathur, Kanchipuram - 631561, Tamil Nadu, India.

[2] Department of Mechanical Engineering, Adhi College of Engineering and Technology

Email - Venkatakoteswararao02@gmail.com

Abstract. The ecological and harmfulness issues of conservative lubricants as well as their increasing cost lead to converted interest in the growth of environmentally friendly oils as lubricants and manufacturing fluids. This investigation examines friction and wear behaviour of cashew nut shell methyl ester contaminated bio-lubricant using four-ball tribometer. The cashew nut shell methyl ester(CSME) was assorted with conservative lubricant in dissimilar fractions of 15% and 30% by volume. The investigation results shown that 30% addition of cashew nut shell methyl ester shows less coefficient of friction and wear scar diameter also reduced. the results show 30% addition of cashew nut shell methyl ester in bio-lubricant produces better presentation and antiwear features.

1. Introduction

Ecological oils are flattering a vital alternative to conservative oils as an outcome of the improved responsiveness of eco-friendly pollution. Vegetal oil, including animal fat, was used as a lubricant thousands of years ago vegetal oils were used in the structure of tributes. Therefore, the use of vegetal oils as a lubricant in the manufacturing sector is not a new idea. For the last three decades, the lubrication industry has been trying to formulate environmentally friendly lubricants with practical features equal to those of mineral oil. [1-2]. The benefits of choosing vegetal oil rather than oils from other sources are that they are ecological, less toxic compared to petroleum-based oils, easily produced, and renewable. Vegetable oils have lubricating abilities that are better than those of currently used mineral or synthetic oils because of a large amount of unsaturated and polar ester groups composition resulting in the development of thick films among the metal to metal interactions. [3-5] There are certain limitations to bio-based lubricants which include greater corrosion due to the lower thermal stability that donates to the sticking effect. Oxidation stability and low pour point can be adapted by partial use of additives. To make bio-based lubricants maintainable, there is a need to improve the available range of viscosity. do so, viscosity transformers can also be used which are ecologically friendly. the friction and wear behaviour of Pongamia biodiesel-based bio-lubricant were investigated under varying dissimilar load to determine their potential as an alternative to mineral oil. This example was selected as it is comparable to edible oils that are also used in food applications, and also because it can be planted on marginal or semi-desiccated lands. There are some misinterpretations in the comparison of the fuels' lubrication properties and viscosity.[6-10] The lubrication of the fuel is not directly provided by the viscosity of the fuel but by other components in the fuel that prevent wear on contacting metal surfaces to make vegetable oil-based lubricants sustainable, there is a need to improve their narrow range of viscosities Viscosity is one of the significant factors in defining coefficient of friction between the sliding surfaces, as it acts as a protective film between the surfaces in contact to protect them from wear. To do so, viscosity Carried to your modifiers can also be used which are globally friendly. [11-15]There are some negatives of vegetable oil-based lubricants that they have lower thermal/oxidative stability, higher flash point and high-temperature operability important to the higher coefficient of friction to overcome these boundaries, several investigators have been carried out Oxidation stability and low pour point can be adapted by partially adding additives and using N-Phenyl-alpha-naphthylamine (Am2) as an antioxidant to improve oxidation stability Moreover, transesterification or epoxidation.

2. Experimental development:

2.1Lubricant preparations

The different percentages of the cashew nut shell methyl ester (CSME) are with lubricant oil SAE 20W40.it was varied from 15% and 30%by volume mixed with cashew nut shell methyl ester. And the lubricant was considered as SAE 20W40 with comparison of above two samples. The blending percentages is assorted with conservative bio lubricant by a standardized mixture equipment.

2.2Apparatus

A four-ball Tribometer TR-30L-IAS, a versatile apparatus supplied by Ducom Instruments Bangalore, used to conduct the experiments the use of a four-ball wear machine is a reputable method for examining lubricant characteristics. This device uses four balls: three on the bottom and one on top. The upper ball is seized in a collet at the lower end of the vertical spindle, which is driven by the motor. The bottom three balls are held definitely in a ball pot containing the lubricant being tested and are pressed against the upper rotating ball. The important components are the oil cup assembly, collet, locknut adaptor, and standard steel ball bearings. The components' surfaces were cleaned with acetone before the tests. The balls are made up of AISI E-52100 chrome alloy steel with a diameter of 12.7 mm, extra polished to grade 25 and hardness of 64 HRC. In each test, the new balls are used. The ball pot is supported above the loading lever on a thrust bearing and plunger with a load cell which is fixed to a loading lever to measure the normal load. The frictional torque exerted on the three balls is measured by a frictional load cell. A RTD type temperature sensor is mounted at the bottom part of ball pot between the two heaters. The test lubricant is heated to 75 °C by a heater fixed in a four ball tribometer with a speed of about 1200 rpm and normal load of 148 N. The specific CCD microscope is used for capturing the images the wear scar of three fixed balls in the ball pot. The specific image capture software is used for measuring the wear scar diameter and wear scar image on the ball surface. The test was conducted in accordance with ASTM D 4172.

Figure 1 Four ball wear tester

3.Results and Discussion

The coefficient of friction of various concentrations of cashew nut shell oil with biolubricants and SAE20W40 are shown in Fig.2. The values of cashew nut shell methyl ester (CSME) are 0.065010 and 0.8992 respectively. Whereas for the SAE20W40 is 0.10152. The coefficient of friction is lowered by about 11% and 35% as compared with SAE20W40.the values wear scare diameter of cashew nut shell methyl ester (CSME) are 0.55mm and 0.508 mm and SAE20W40 is 0.932mm shown in fig: 3.the wear scare diameter is lowered by about 38.2% and 42.4% as compared with SAE20W40.The lubrication properties shows better results in CSME 30% compared to SAE 20W40 and CSME 15%. shown in table1

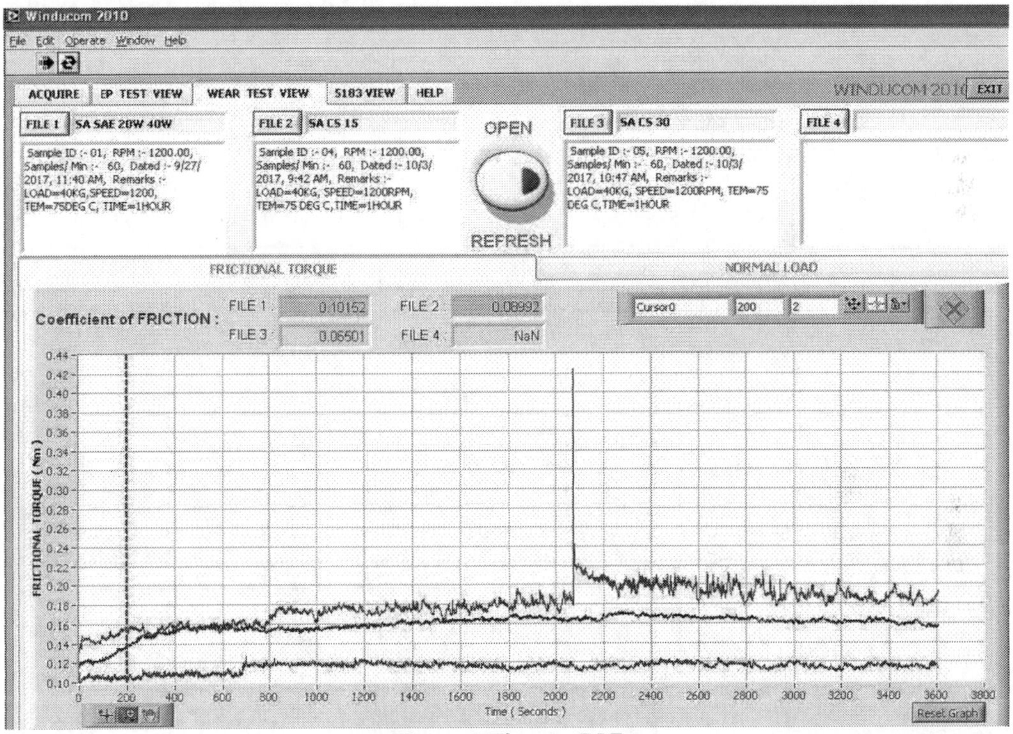

Figure 2 Time vs COF

Table 1: Properties of Lubricants

Properties	SAE20W40	CSME 15% WITH SAE20W40	CSME30% WITH SAE20W40
Flash point (°C) [ASTM D92]	200	214.27	190.2
Viscosity @100°C (cSt) [ASTM D445]	121	74.79	42.61
Viscosity @100°C (cSt) [ASTM D445]	15.2	15.48	12.56
Viscosity index [ASTM D2270]	120	126	158

The wear scar morphology was recorded using confocal microscope. The most important capability of confocal microscopy is the ability to acquire high quality metallographic images with the capability to reconstruct the sectional plane information into three-dimensional model. SEM images have taken in mag500x. from the SEM images observed minimum wear occurred in cashew nut shell methyl ester CSME 30% compared to other two lubricants. From the EDAX results the carbon and silicon present are lesser in the CSME30% compared two other two lubricants. Chromium and iron values are higher in the CSME30% compared two other two lubricants. Are shown in fig.4

Table 2. Elemental Analysis of SAE20W40 and its Blends

	Lubricants		
Elements	**SAE20W40**	**CSME 15% WITH SAE20W40**	**CSME 30% WITH SAE20W40**
Carbon	2.46	3.39	1.32
Silicon	0.12	0.17	0.09
Chromium	1.42	1.31	1.46
Iron	96	95.14	97.13

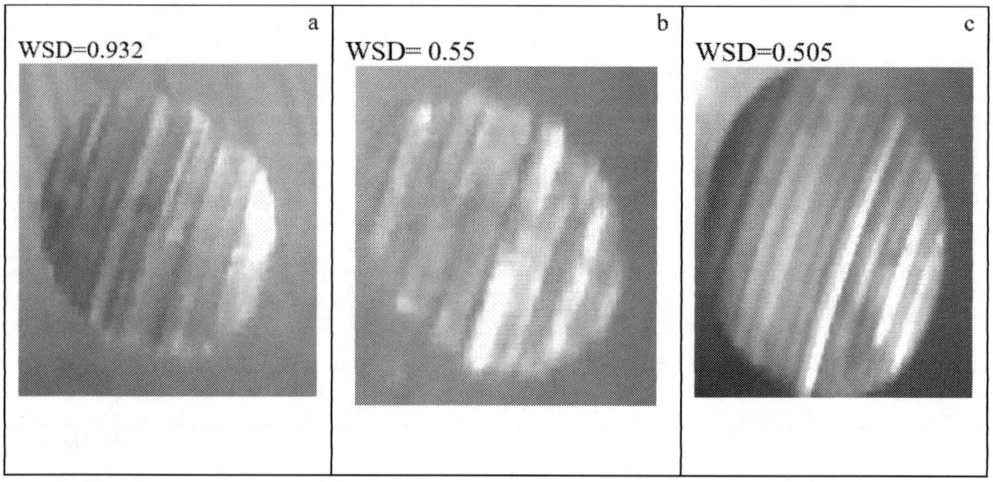

Figure 3 wear scar diameter images a) SAE20W40, b) CSME 15% WITH SAE20W40 c) CSME 30% WITH SAE20W40

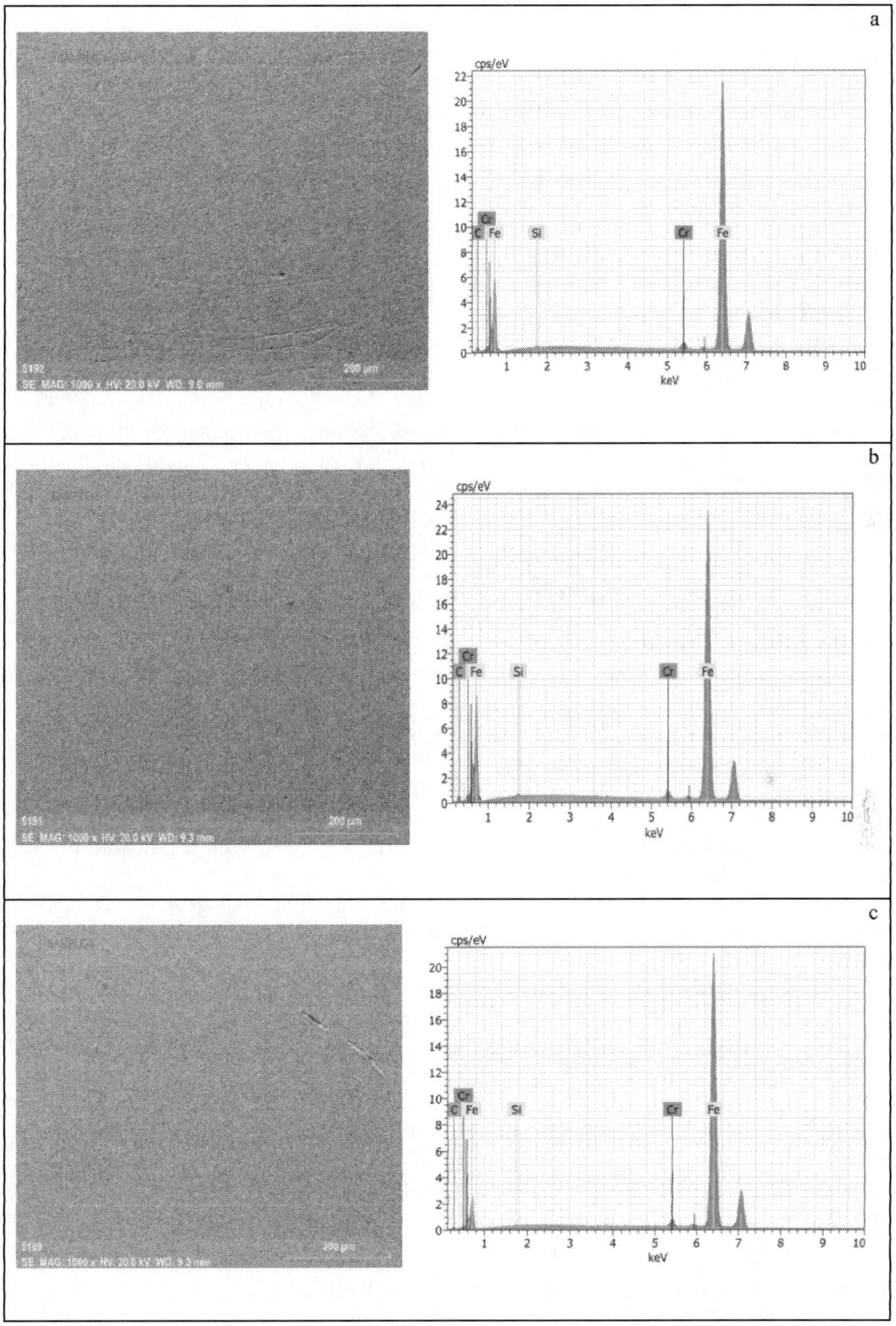

Figure 4 SEM And Edax images a) SAE20W40, b) CSME 15% WITH SAE20W40 c) CSME 30% WITH SAE20W40

Conclusions

The tribological investigation of diffèrent Lubricants was tested using Pin on Disc Tribometer under various conditions and the results were compared with the SAE20W40 Lubricant. The following conclusions were made from the experimental study:

- The properties of the viscosity and viscosity index of the three samples shows better results in the CSME 30% compared two other two lubricants.
- The coefficient friction wear also shows better in the CSME 30%with SAE20W40 compared two other two lubricants
- SEM and EDAX results also concluded that better results shows in CSME 30%with SAE 20W40

References

[1] Nosonovsky, M. (2000), "Oil as a Lubricant in the Acient Middle East," *Tribology Online*, 2(2), pp 44–49.

[2] Alla, M. P. and Richard, J. A. (2004), "Oxidation Stability and Tribological Behavior of Vegetable Oil Hydraulic Fluids," *Tribology Transactions*, 47(2), pp 182–187.

[3] Yashvir Singh (2015), Tribological behavior as lubricant additive and physiochemical characterization of Jatropha oil blends *Friction* 3(4): 320–332

[4] S sempebwa, J.C.& Carpenter, D.O. (2009), The generation, use and disposal of waste crank case oil in developing countries: a case for kampala district, Uganda. *J. Hazard Mater*. 161: 835-841.

[5] Maleque, M. A., Masjuki, H. H., Haseeb, A.S.M.A. (2000.), Effect of mechanical factors on tribological properties of palm oil methyl ester blended lubricant. *Wear*. 239: 117-125.

[6] Hwang H S, Erhan S Z. (2002), Lubricant Base Stocks from Modified Soybean Oil. *Champaign (USA): AOCS Press,*

[7] Saliha N, Salimona J, Yousifb J. (2011), the physicochemical and tribological properties of oleic acid based triester biolubricants. *Ind Crop Prod* 34: 1089–1096

[8] Yashvir Singh et al (2017), Tribological characterization of Pongamia pinnata oil blended bio-lubricant ISSN: 1759-7269 (Print) 1759-7277

[9] Ponnekanti, N., and Kaul, S. (2012), Development of eco-friendly/biodegradable lubricants: an overview. *Renew. Sust. Energy Rev*. 16:764–774.

[10] Quinchia, L., Delgado, M. A., Valencia, C., Franco, J., and Gallegos, C. (2010), Viscosity modification of different vegetable oils with EVA copolymer for lubricant applications. *Ind. Crop. Prod*. 32:607–612.

[11] Grigg, H. C. (1994), "Reformulated Diesel Fuels and Fuel Injection Equipment, Lucan Powertrain Systems," *New Fuels and Vehicles for Cleaner Air Conference*, Phoenix, AZ, January 11

[12] Rani S, Joy M L, Nair K P. (2015), Evaluation of physiochemical and tribological properties of rice bran oil–biodegradable and potential base stoke for industrial lubricants. *Ind Crop Prod* 65: 328–333

[13] Hwang H-S, Erhan S. (2001), Modification of epoxidized soybean oil for lubricant formulations with improved oxidative stability and low pour point. *J Am Oil Chem Soc* 78: 1179–1184

[14] He Z, Lu J, Zeng X, Shao H, Ren T, Liu W. (2004). Study of the tribological behaviors of S, P-containing triazine derivatives as additives in rapeseed oil. *Wear* 257: 389–394

[15] Wu Y, Li W, Zhang M, Wang X. (2013), Improvement of oxidative stability of trimethylolpropane trioleate lubricant. *Thermochim Acta* 559: 112–118.

[16] S. Baskar et al. (2015), Experimental Analysis on Tribological Behavior of Nano Based Bio-Lubricants using Four Ball Tribometer, *Tribology in Industry* Vol. 37, No. 4 449-454.

[17] G. Venkatakoteswararao et al (2018), Experimental Investigation on Tribological Behaviour of Lubricating Oil with the Addition of CuO and TiO2 Nano Additives *Adv. Sci. Eng. Med*. 10, 390–394.

IOP Conf. Series: Materials Science and Engineering **390** (2018) 012075 doi:10.1088/1757-899X/390/1/012075

Hardness and impact behaviour of aluminium metal matrix composite

C. Parswajinan[1, a*], **B. Vijaya Ramnath**[2, b], **S. Abishek**[3, c], **B. Niharishsagar**[4, d], **G.Sridhar**[5,e]

[1]Assistant Professor Department of Mechanical Engineering, Sri Sai Ram Engineering College, Chennai-600044, Tamil Nadu, India

[2]Professor, Department of Mechanical Engineering, Sri Sai Ram Engineering College, Chennai – 600 044, India

[3,4,5]Department of Mechanical Engineering, Sri Sai Ram Engineering College, Chennai - 600 044, India

[a*]parswajinan@gmail.com, [b]vijayaramnathb@gmail.com

Abstract - This research paper deals with the fabrication procedure and the testing process of metal matrix composite material comprising of Aluminium and Silicon Carbide in addition with Carbon Nanotubes. Aluminium composites are preferred in manufacturing of various industrial machine components and automobile parts due to their favourable properties like low density which thereby results in their light weight, high strength and superior malleability comparatively to iron carbon alloys like steel and cast iron. When Silicon Carbide is added it helps in furthering the thermal conductivity and melting point properties of Aluminium in addition to improving its hardness.The addition of Carbon Nanotubes potentially helps in further improving the tensile strength of the metal matrix composite. In this study four samples of different compositions of Silicon Carbide are prepared and studied. The fabrication process is done by Stir casting carries out by adding the required additives into the molten mixture of aluminium followed by continuous stirring. After solidification for a period of time the samples are cut according to standard dimensions for carrying out the testing procedures. The various test procedures like hardness and impact tests were conducted to study the mechanical properties of obtained metal matrix composite. The various results inferred from the tests are analysed and subjected to comparison with each other in this study.

1. Introduction

Conventional materials have drawbacks in attaining good combination of strength, stiffness, hardness and density. To come across these disadvantages composites are more promising materials of recent history. Distinct properties like high strength, damping capacity, specific modulus also good wear resistance are present in metal matrix composites when compared to un-reinforced alloys. They were made by addition of reinforced material into metal matrix. The reinforcement layer is coated to prevent any chemical attack on the matrix. For example, carbon fibres are used in aluminium metal matrix to produce composites with low density and high strength.

The strength can be further increased by adding an appropriate material. High homogeneity is required to reach optimum mechanical properties for the composite material [1]. Metal-matrix composites of discrete-SiC-reinforced aluminium alloy is contrived by the method of melt-stirring. The microstructure and mechanical properties were investigated. The rise in regularity of particle scattering is the main cause for improvement in dependability resulting in hot extrusion by awell-shaped die. Young's modulus and 0.2% offset yield strength of the extruded MMCs were compared with theoretical values calculated by the Eshelby method [2].Mechanical properties of the CNT/Al composite with the increase in Young's modulus clearly the shear lag model seems to be applicable, since the other two models are not associated with Young's modulus, but an increase in yield strength could imply the additional applicability of Orowan looping and thermal mismatch models to the composite system [3].The tensile behaviour of a 20% SiC particle reinforced Al-Si composite with different degrees of clustering showed that an increase in particle clustering yielded a higher work hardening rate, with reduction in ductility [4].When the percentage of the reinforcements in MMC increases, the mechanical properties such as tensile strength as well the impact strength increases the reinforcements obstructs the deformation in the composite [5].The relation between fabrication situations and identical scattering of particles is examined for Al composites with SiC*p*-reinforced. By evaluating experimental results methodically, connection is among the micro-structure, the particle character, the mechanical properties and process structures for abest melt-stirring with the processing of MMCs [6]. Development of carbide on the surface of CNTs and its effects on the mechanical properties of composites needs to be assessed as the load transfer ability and strength of CNTs can be improved. Further studies are required in assessing the various health and environmental hazards imposed by the CNT reinforced composites [7]. The impact behaviour of aluminium was significantly reduced by the presence of SiC particles. On the other hand, the reduction in impact strength due to aging was of no importance [8]. Al MMC has been fabricated by reinforcing with titanium di-boride (TiB_2) as reinforcements by stir casting technique and the mechanical properties have been found and have been reported that the properties have been improved on adding the reinforcements [9]. Aluminium with CNT MMC are fabricated and their mechanical behaviour are studied [14,15,16]. Review on AMMC were made and found suitability of MMC for automotive applications [13,17]. Mechanical behaviours of AMMC were identified and it was concluded that reinforcement increases mechanical properties. [10,11,12,18]

2. Experimental Procedure

2.1. Stir casting process

Stir casting is a good economical process for producing composite material and best suited for Aluminium-Silicon carbide combination. The process is simple where molten aluminium is stirred continuously with addition of reinforcements in appropriate proportions. During the process particles tend to form agglomerates that can only dissolved by vigorous stirring at high temperature. This process is chosen mainly due to both flexibility and feasibility that it offers as shown in Figure 1. It is very economical, simple, effective and adaptive. The chance of controlling the matrix structure is easy by using different stirrer blades and cooling times. The whirlpool technique provides high strength and homogenous set of aluminium composites.

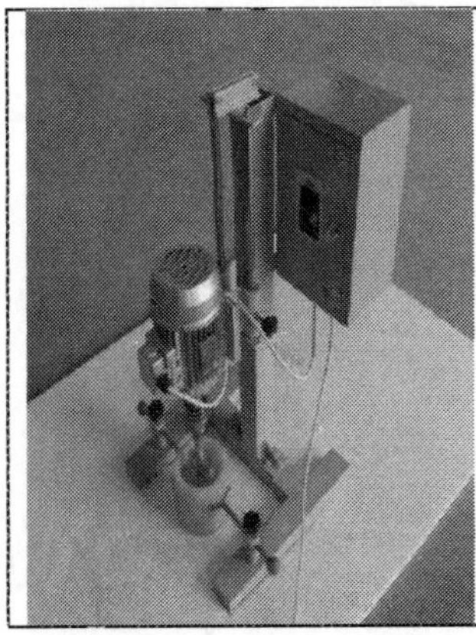

Figure 1. Stir casting process.

2.2 Fabrication process

The foremost part of stir casting is preheating the materials to be casted. In this case the aluminium rod weighing 1.5kg approximately is taken in a graphite crucible and the reinforcements in suitable proportions are mixed in another crucible. Both of them are heated separately for around 3-4hrs at different temperatures ~900°C and ~400°C respectively. After the preheating process both the crucible are taken out and the reinforcement materials are added to the molten aluminium. De-gasser and Dust remover are added to remove any gasses and impurities present in the mixture. This whole mixture in the crucible is again left to the furnace for further heating do that the aluminium alloy completely fuses with the powders of CNT and silicon carbide. The furnace is then taken out after 2hrs of heating is introduced to the stirring machine where vigorous stirring is done by the rotating blades of the motor at 550 rpm. Constant stirring is done for about 6-10 min the matrix structure began to form while stirring and the mixture begins to solidify, ensuring that the mixture is thoroughly mixed it is then poured in preheated die to the temperature of 550°C to take the shape of the die.

3. Results and Discussion

The tested result for compression and hardness tests performed on the metal matrix composites are as follows.

Table 1. Experimental Values of Aluminium MMC

	Samples	Composition	Impact load (Joules)	Hardness in HBW
1	Sample 1	Al+2%SiC	8	55.4
2	Sample 2	Al+2%SiC+0.2%CNT	6	55.2
3	Sample 3	Al+3%SiC+0.2%CNT	10	47.5
4	Sample 4	Al+4%SiC+0.2%CNT	8	54.9

IOP Conf. Series: Materials Science and Engineering **390** (2018) 012075 doi:10.1088/1757-899X/390/1/012075

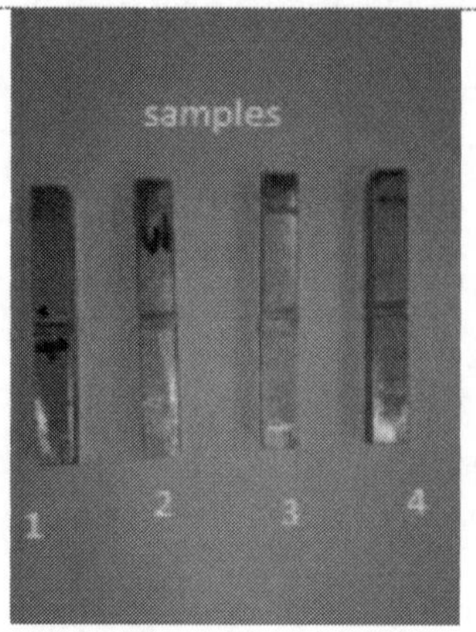

Figure 2. Impact load test result sample.

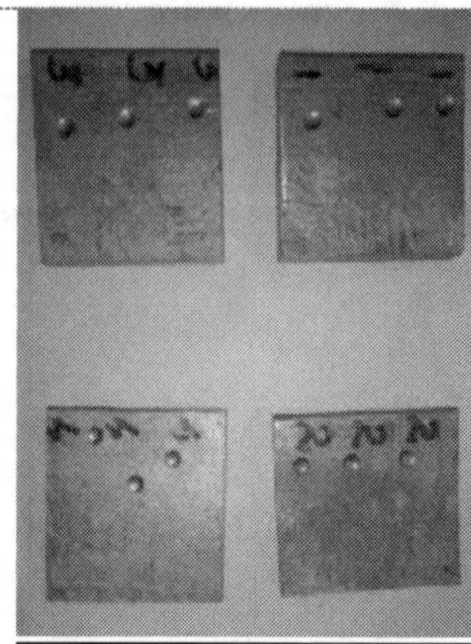

Figure 3. Hardness test result sample.

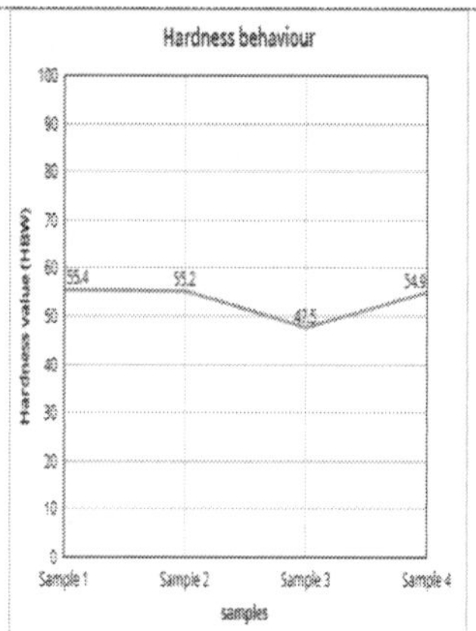

Figure 4. Graphical representation of hardness behaviour.

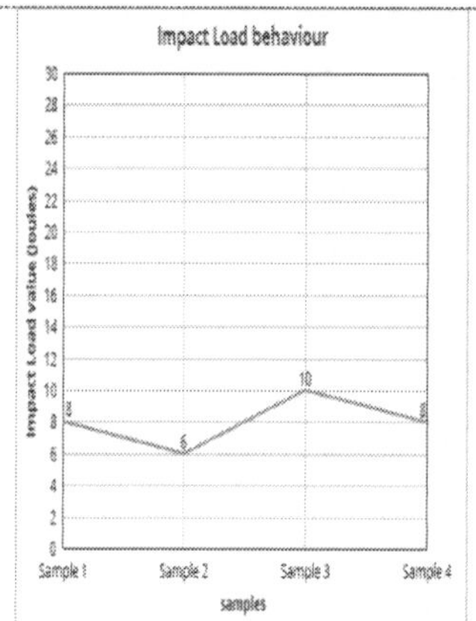

Figure 5. Graphical representation of impact load behaviour.

4. Conclusion

It is found that the composition Al+3%SiC+0.2%CNT withstands more impact load compared to other compositions but at the same time there's a huge drop in the hardness of that material which may be due to its plasticity. There's a stable increase in hardness upon the addition of silicon carbide but if the same is added more, hardness decreases gradually after attaining a peak value. Discussing the load withstanding capability, it's also starts to decrease after attaining a limiting value. Thus we infer with optimum addition of silicon carbide and carbon nanotubes in their corresponding ratios to the base metal aluminium yields a metal matrix composite that have a superior balance between high hardness and prevail against exceptional impact load.

References

[1] Brian Ralph, H.C. Yuen, W.B. Lee, The processing of metal matrix composites — an overview, Journal of Materials Processing Technology, Volume 63, Issues 1–3, January 1997, Pages 339-353, ISSN 0924-0136.

[2] Y.H. Seo, C.G. Kang, Effects of hot extrusion through a curved die on the mechanical properties of SiCp/Al composites fabricated by melt-stirring, Composites Science and Technology, Volume 59, Issue 5, April 1999, Pages 643-654, ISSN 0266-3538,

[3] George, R., et al. "Strengthening in carbon nanotube/aluminium (CNT/Al) composites." Scripta Materialia 53.10 (2005): 1159-1163.

[4] Deng X, Chawla N. Modelling the effect of particle clustering on the mechanical behaviour of SiC particle reinforced Al matrix composites. J Mater Sci 2006; 41:5731–4.

[5] J. Jeykrishnan, B. VijayaRamnath, X. Hervin Savariraj, R. David Prakash, V. R. Dhinesh Rajan and D. Dinesh Kumar. Investigation on Tensile and Impact Behaviour of Aluminum Base Silicon Carbide Metal Matrix Composites. Indian Journal of Science and Technology, Vol 9(37), DOI: 10.17485/ijst/2016/v9i37/101979 , October 2016

[6] Young HS, Chung GK. The effect of applied pressure on particle-dispersion characteristics and mechanical properties in melt-stirring squeeze-cast SiC/Al composites. J Mater Process Technol 1995;55:370–9.

[7] Ramnath, B. V., Parswajinan, C., Elanchezhian, C., Pragadeesh, S. V., Ramkishore, P. R., Sabarish, V. (2014). A Review on CNT Reinforced Aluminium and Magnesium Matrix Composites. In Applied Mechanics and Materials (Vol. 591, pp. 120-123). Trans Tech Publications.

[8] Ozden S, Ekici R, Nair F. Investigation of impact behaviour of aluminium based SiC particle reinforced metal– matrix composites. Composites: Part A, Applied Science and Manufacturing. 2007 Feb; 38(2):484-94.

[9] J.Jeykrishnan, S.Jayanthi Nathan, M.R.Karthik. Fabrication and characterization of aluminium titanium di-boride metal matrix composites using stir casting technique. International journal of mechanical engineering and technology. Vol. 8(4). 2017.

[10] B.Vijaya Ramnath, C. Elanchezhian, M.Jaivignesh, S. Rajesh, C. Parswajinan, A. Siddique Ahmed Ghias, "Evaluation of Mechanical Properties of Aluminium Alloy-Alumina-Boron Carbide Metal Matrix Composites", Materials and design, Volume 58, June 2014, Pages 332-338.

[11] Siddique Ahmed Ghias B.Vijaya Ramnath, 'Investigation of Tensile Property of Aluminium SiC Metal Matrix ', International Journal Applied Mechanics and Materials, Volume 766-767, 2015, Pages 252-256.

[12] P.Subramaniya Reddy, R.Kesavan, and B.Vijaya Ramnath, "Investigation of Mechanical Properties of Aluminium6061-Silicon Carbide, Boron Carbide Metal Matrix Composite, DOI: 10.1007/s12633-016-9479-8 2016, Silicon

[13] C.Elanchezhian, B.VijayaRamnath, G.Ramakrishnan, K.N.SripadaRaghavendra, MithunMuralidharan, V.Kishore, "Review on metal matrix composites for marine

applications". Accepted for publication in Elsevier – Materials Today Proceedings, 5(1), PP.1211-1218

[14] Parswajinan C, Vijaya Ramnath B, Syed Mohammed Buhari A. S. A., Muthukumaaran C and Riyaz Ahmed A, "Investigations of mechanical properties of Aluminium Matrix Composite reinforced with B4C and MWCNT", Advanced Science, Engineering and Medicine, Vol. 10, 2018, 1–3

[15] C. Parswajinan, B.Vijaya Ramnath, N. Sai Prasath, S. Kuralamudhan and S. Balaji, "Mechanical investigation of Aluminum hybrid Composite Reinforced with CNT and Aluminum oxide (Al2O3)", Advanced Science, Engineering and Medicine, Vol. 10, 2018, 1–3,

[16] C. Parswajinan, B. Vijaya Ramnath, M. Vetrivel, M. Aswin, and B. Mohana Vignesh, "Mechanical Behavior of Aluminium Metal Matrix Composite", Advanced Science, Engineering and Medicine, Vol. 10, 2018, 1–3

[17] G. Ramakrishnan , Vijaya Ramnath B, Naveen E, Gowtham S, Arun Kumar A, Muthuvel M.S and Akil R," A review on aluminium metal matrix composites", Advanced Science, Engineering and Medicine Volume 10, 2018, Pages 1-4

[18] Shreyas P, Shriramkumar , Sruthisagar R , Vijaya Ramnath B and Elanchezhian C ,"Studies on Mechanical Behaviour of Aluminium-Silicon Carbide-Copper Metal Matrix Composite", MATEC Web of Conferences 166, 01001 (2018).

IOP Conf. Series: Materials Science and Engineering **390** (2018) 012076 doi:10.1088/1757-899X/390/1/012076

Optimization of Drilling Parameters of Carbon Fiber Composites Using RSM based Desirability Function

D.Vijayan[1*], P. Abhishek[2], Y.G. Manoj Kumar[2], P. Balaji[2], P. Siva kumar Reddy[2]

[1]Assistant Professor, Department of Mechanical Engineering, Sri Chandrasekharendra Saraswathi Mahavidyalaya, Kanchipuram, Tamilnadu, India.

[2]Student, Department of Mechanical Engineering, Sri Chandrasekharendra Saraswathi Mahavidyalaya, Kanchipuram, Tamilnadu, India.

Corresponding author: [1*]vijaiand2012@gmail.com

Abstract. This paper deals with the drilling of Carbon Fiber Reinforced Polymer (CFRP) composite filled with Carbon Nano Tube (CNT) using Response Surface Method (RSM) based desirability function. In drilling of hybrid metal matrix CFRP composites, excess thrust force and burr height degrades the performance of composite. Therefore, thrust force and burr height of the composite to be minimized in order to improve the performance of the hybrid metal matrix composite. Hence, thrust force and burr height are factors being considered in the present investigation are the major responses to be minimized using RSM based Desirability function. Four important input factors are considered such as drilling speed, feed rate, percentage of CNT and cone angle of drill bit to analyze the performance of drilling process. The results showed that, feed rate is highly influential parameter which influences thrust force and burr height in hybrid metal matrix composites. During drilling operation, due to mutual rubbing of CNT abrasive particles causes extensive surface damages like voids, cracks and fiber pullouts. ANOVA result indicates experimental data are well correlated at 95 % confidence interval and this technique can be very much useful and reliable for predicting the drilling parameters of hybrid CFRP metal matrix composite.

1. Introduction

Carbon fiber reinforced plastics (CFRPs) have been extensively used in various engineering applications due to their outstanding mechanical properties (1, 2). Therefore, it enables light weight products manufacturing and extend their service-life structures. In the past decades metallic materials have been widely replaced by CFRP. In view of growing engineering applications with composites requires a systematic study for their machining characteristics was envisaged. The efficient and economical machining parameters required for obtaining desired profile and dimensions of the composite. However, it is well known that the many factors affecting the mechanical properties of the CFRP during loading. Especially during manufacturing processes, the factors such as temperature, pressure and processing time may lead cracking, fiber delamination's and breakages in the composites that affects the mechanical properties of fabricated composite (3). Hence composites can be considerably modified by resins, adding nano filler and method of fiber layups, etc. Therefore, numerous surface treatments have been developed to improve interfacial adhesion characteristics of the polymer composites namely electrochemical method, plasma treatment and high energy irradiation and oxidation method (4). Although homogeneous scattered nano particles are of great importance to enhance the surface strength and hardness of the composite. Thus, increase of addition amount of nano particles may lead poor surface finish, and increase thrust force which cause excess tool wear, exhibit numerous pits and burrs (5). Therefore, by selecting appropriate machining parameters can minimize thrust force and reduces burr and pits during composite machining. Therefore, in view of above a detailed systematic study was conducted using RSM based Desirability function. The mathematical model was developed using RSM for modeling and analysis performance in the drilling of hybrid metal matrix composites.

2. Experimental setup

2.1. Materials and methods

Carbon fibers of single ply lamina having a diameter of 5-10μm was fabricated with a thickness of 5 mm sheet for conducting a drilling process. 60% resin and 40% hardener are mixed together in a required proportion, and the gel time is measured. In addition to that 5% of carbon nano tubes having the sizes of 50 – 80nm were added in a liquid epoxy resin based on weight fraction. The mixture stirred at room temperature for 48 hours under a seal condition to make homogeneous CNT filled epoxy solution. Carbon fiber reinforced polymers (CFRP) composites is made into laminas by 300 mm x 300 mm with 0, 90, 45 degrees of angle. Weights are added on the setup to remove the excess amount of resin and hardener mixture. The fabricated CNT filled CFRP composite plate as presented in Figure 1.

Figure 1. Fabricated CNT filled CFRP composite

2.2. Experimental design

The experiment was conducted with 29 runs as per the RSM design matrix. The design matrix was selected as per the number of input factors planned for the drilling experiments. The input parameters and their corresponding range chosen in the experiment are presented in Table 1.

Table 1. Input parameters and their range

PARAMETERS	RANGE
Drilling speed (N)	500-1100 (rpm)
Feed rate (f)	0.4-0.8 (mm/min)
% of CNT	3-5
Point angle (Θ)	110-120

3. Experimental procedure

Drilling experiment was conducted using a vertical CNC machining center. Each CFRP sample to be drilled was cut into the sizes of 100mm x 100mm. The solid carbide drill was used to conduct each run involved in the design matrix. The Point angle of the carbide drill tool 110°, 115° and 120° respectively was maintained according to the sequence of the experimental run in the design matrix. Two responses namely thrust force on z – axis and burr height were considered as a response in the present investigation. The schematic representation of drilling process is presented in Figure 2 and the design matrix and their corresponding response obtained is presented in Table 2.

Figure 2. Schematic representation of drilling process

Table 2. Design matrix and their corresponding responses

Std	Run	Rotational speed (N)	Feed rate (mm/min)	Percentage of CNT %)	Point angle (Θ)	Thrust force (N)	Burr height (mm)
1	10	500	4	4	115	15.75	0.02
2	9	1100	4	4	115	58.5	0.36
3	25	500	8	4	115	15.2	0.09
4	24	1100	8	4	115	104.33	0.3
5	11	800	6	3	110	33.17	0.12
6	18	800	6	5	110	22.14	0.06
7	8	800	6	3	120	18.16	0.26
8	4	800	6	5	120	45.85	0.2
9	17	500	6	4	110	12.36	0.02
10	19	1100	6	4	110	80.14	0.29
11	7	500	6	4	120	16.86	0.14
12	5	1100	6	4	120	95.5	0.41
13	21	800	4	3	115	15.4	0.12
14	14	800	8	3	115	35.68	0.21
15	2	800	4	5	115	23.18	0.16
16	26	800	8	5	115	44.68	0.05
17	12	500	6	3	115	14.33	0.08
18	20	1100	6	3	115	65.67	0.38
19	3	500	6	5	115	12.17	0.06
20	22	1100	6	5	115	90.3	0.29
21	29	800	4	4	110	21.17	0.09
22	27	800	8	4	110	42.94	0.09
23	6	800	4	4	120	22.63	0.22
24	1	800	8	4	120	46.75	0.21
25	15	800	6	4	115	42.48	0.26
26	16	800	6	4	115	42.18	0.27
27	13	800	6	4	115	41.63	0.29
28	23	800	6	4	115	41.84	0.28
29	28	800	6	4	115	44.94	0.28

4. Modeling the process parameters using RSM

Response Surface Methodology (RSM) is a statistical technique used for modeling, optimization, and to analyze the problems (6-10). It is also used to establish the relationship between the input drilling process parameters and the required responses. Usually, the goal is to estimate the values of input process parameters in the drilling process at which the responses are optimized. The optimum may be either minimum or maximum depends upon the user interest in the experiment. Hence RSM is widely used in many application areas in order to get an optimum values on their corresponding processes (11-13). The thrust force and burr height of the CRFP composites are related to rotational speed (N), feed rate (f), percentage of CNT (% CNT), and Point angle (Θ). The surface is represented by the following,

Thrust force (T) = f(N, f, %CNT)

Burr height (B_H) = f(N, f, %CNT)

The second order polynomial response surface model used to evaluate the parametric effects of the drilling process parameters are as follows,

$$Y=\beta_o+\sum_{i=1}^{k}\beta_i x_i+\sum_{i=1}^{k}\beta_{ii}^2 x_i^2+\sum_{i<j}\beta_{ij}x_i x_j+\in$$

Where β_o is the average response, β_i, β_{ii} and β_{ij} are the coefficients that depend on the major and interaction effects of the parameters and $\in$ is the statistical error (14). A second order response surface for thrust force and burr height can be represented as below,

Thrust force (T) =+42.61+33.98*A+11.08*B+4.66*C+2.82 * D+11.60 * A * B+6.70 * A * C+2.72 * A * D+0.30 * B * C+0.59 * B * D+9.68* C * D+11.63 * A^2 -5.23 * B^2-8.42 * C^2-3.80 * D^2

Burr Height (BH) = +0.28 +0.13* A -1.667E-003 * B -0.029 * C +0.064 * D -0.033 * A * B -0.018 * A * C -0.050 * B * C -2.500E-003* B * D + -9.250E-003 * A^2 -0.074 * B^2 -0.066 * C^2 -0.051 * D^2

ANOVA is applied for estimating the significance of the model at 95% significance level. The main objective of conducting ANOVA is to estimate the interaction and significant effects on thrust force and burr height of drilling process. In ANOVA, adequacy of the developed model was conformed with R-Sq values. In the present investigation the R-Sq was obtained 0.9974 and 0.9953 for thrust force and

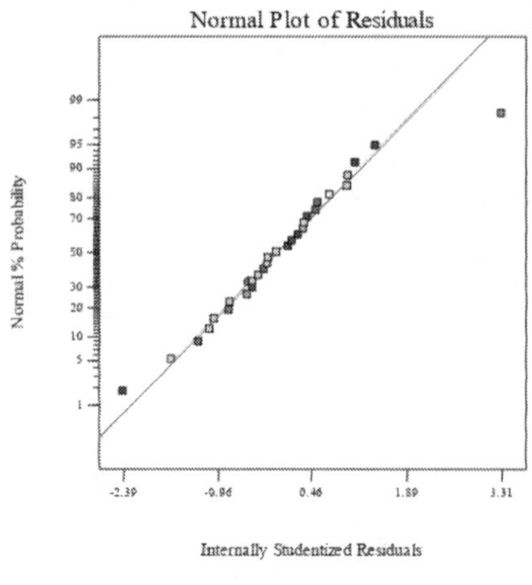

Figure 3. Experimental vs. predicted values of thrust force

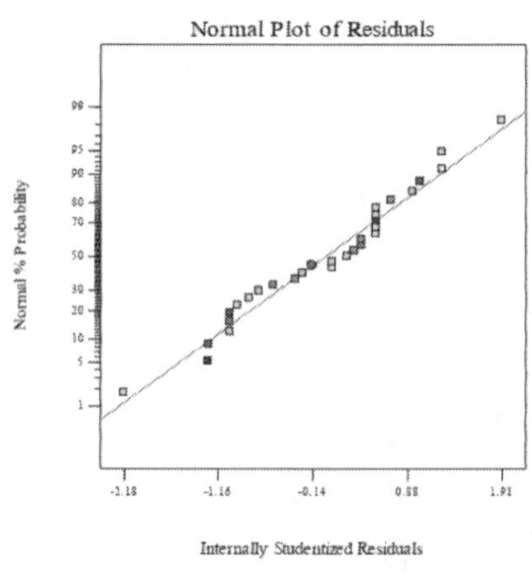

Figure 4. Experimental vs. predicted values of Burr height

burr height respectively. The larger R-Sq values is desirable. Therefore, the present R-Sq value is indicates high correlation between experimental and predicted values. Also, the aforementioned experimental and predicted values are plotted in Figure 3 and Figure 4. Hence the predicted mathematical model is very useful and effectively predict the thrust force and burr height of CFRP composite filled with CNT nano particles in drilling process within the limits of factors studied.

5. Optimization using RSM based desirability function

Derringer and Suich (1980) proposed a multi-response optimization technique namely desirability function. The overall desirability function of the multi-response system can be estimated by combining the individual desirability function. The desirability function can be represented as $D = d_1^{w1}.d_2^{w2}...d_n^{wn}$ where w_j ($0<w_j<1$) is the weight value given for the importance of jth response variable and $\sum_{j=1}^{n} w_j = 1$. The parameter settings with maximum overall desirability is considered for optimum parameter combination. In the present investigation the goal is to minimize the thrust force and burr height of the drilling process of CFRP composite. The optimization analysis was performed using Design Expert software. The goal sets and targets of input parameters namely N, f, % of CNT and Θ and the response are presented in Table 3.

Table 3. Goals, Targets of input parameters of drilling process

Drilling parameters	Goal	Lower limit	Upper limit	Lower weight	Upper weight	Importance
Drilling Speed (N)	is in range	500	1100	1	1	3
Feed rate (f)	is in range	4	8	1	1	3
% of CNT	is in range	3	5	1	1	3
Point angle (Θ)	is in range	110	120	1	1	3
Thrust force (T)	minimize	12.17	104.33	1	1	3
Burr height (h)	minimize	0.02	0.41	1	1	3

The solution with high desirability is preferred. Therefore, the best three combination of obtained desirability values and their corresponding suggested input process parameters are presented in Table 4.

Table 4. Desirability values

Number	Drilling Speed (N)	Feed rate (f)	% of CNT	Point angle (Θ)	Thrust force (T)	Burr height (h)	Desirability	
1	511.56	4.11	4.74	113.48	7.55313	0.00867	1	Selected
2	538.44	4.57	4.99	111.1	7.0906	0.0308	1	
3	551.72	8	4.55	112.58	11.7932	0.01169	1	
4	502.25	4	4.98	116.68	7.85585	0.01972	1	

6. Results and discussion

The contour plots for drilling speed, feed rate, % of CNT and Point angle on thrust force and burr height are presented in Figure 5. Increasing feed rate increasing thrust force of drilling process and vice versa. Therefore, the increase in thrust force bends the materials larger to the extent interfacial bond cracking progresses, resulting larger burr formation. The increase in weight percentage of CNT increases thrust force. The presence of addition of CNT in the metal matrix composited increases tensile strength, hardness and heat resistance of the composite (15, 16). This can be attributed increasing brittleness of the composite that reduce built-up edge in the drilling of hybrid composites. Increasing point angle of drill increases thrust force of drilling process whereas the burr height of the composite is decreases. As the drill advances it forms a pivot point thereby it decrease the burr height of the metal matrix composite. This results are in good agreement with Basavarajappa et al (17).

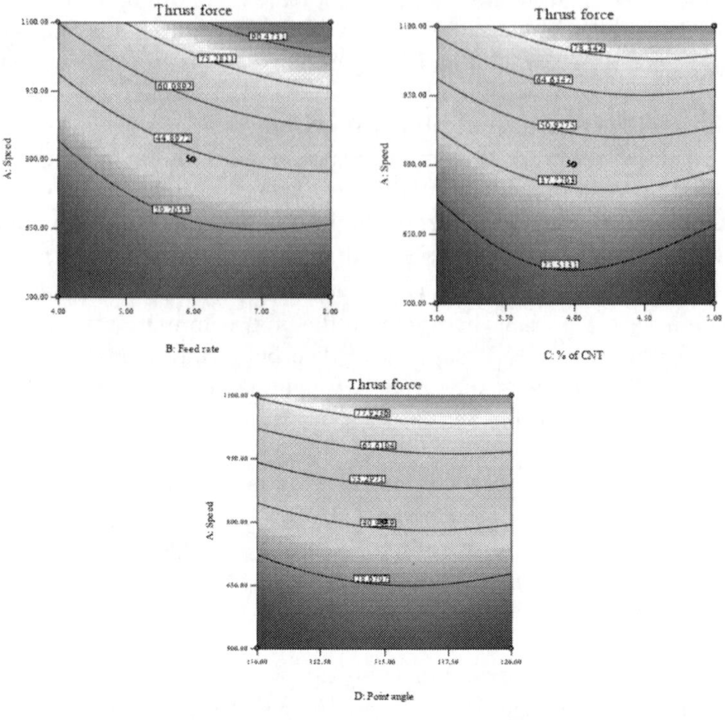

Figure 5. Contour plots for Thrust force

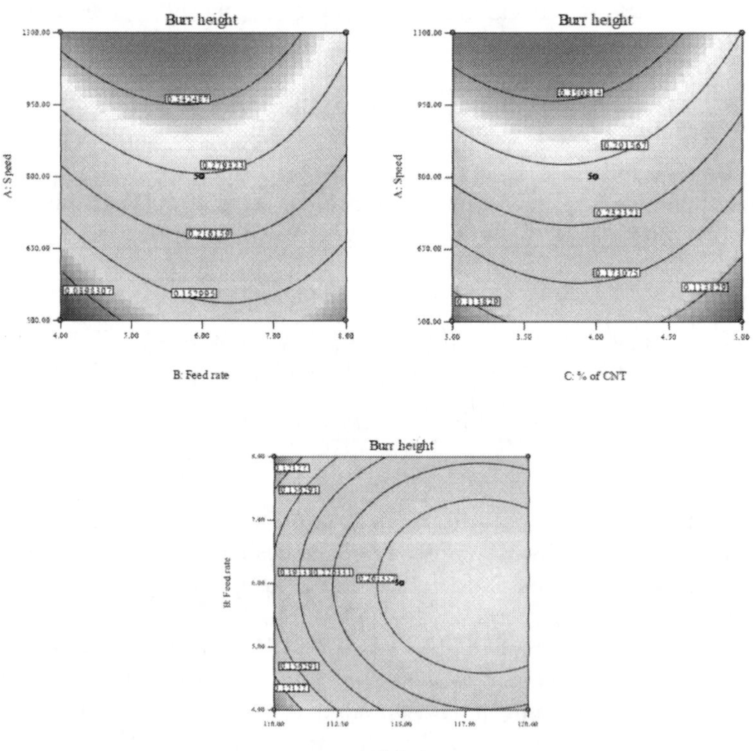

Figure 6. Contour plots for burr height

Low feed rate encourages less thrust force therefore exist poor plastic deformation on the metal matrix composite. Due to low interfacial strength CNT particles smears before bending the material during the plastic deformation in the drilling process. Increasing feed rate increase temperature at the tool-work surface therefore softening plastic deformation. Hence, burr height predominantly increases with increase in spindle speed. SEM was used analyze the machined surface. During drilling operation, due to mutual rubbing of CNT abrasive particles causes extensive surface damages like voids, cracks and fiber pullouts.

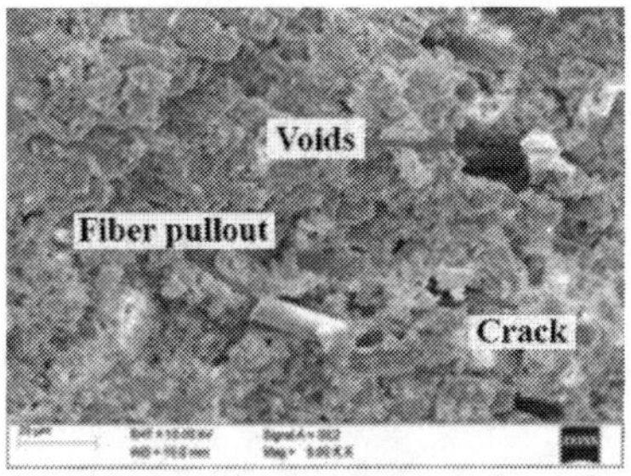

Figure 7. SEM image of machined surface

7. Conclusions

The mathematical models were developed for estimating the thrust force and burr height in drilling of hybrid carbon fiber composites. Response surface modeling-based desirability-based optimization was carried out to minimize the thrust force and burr height in drilling process. Based on the experimentation the following conclusions can be drawn,

1. Response surface method-based desirability is very useful in predicting the thrust force and burr height of the hybrid metal matrix composites.
2. The result indicates feed rate is highly influential parameter which influences thrust force and burr height in hybrid metal matrix composites.
3. SEM analysis of the drillied hole surface indicates higher feed rate and spindle speed increase voids, cracks and fiber pullouts.
4. The ANOVA result indicates that the experimental data are well correlated with the predicted data at 95 % confidence interval; hence, this technique can be very much useful for a reliable prediction of different performances.

8. References

1. Xu B, Du Y, Wang P, Yan L, Sun B, Du B, et al. Microstructure, surface emissivity and ablation resistance of multilayer coating for lightweight and porous carbon–bonded carbon fiber composites. Journal of Alloys and Compounds. 2016;685:799-805.
2. Carvalho O, Buciumeanu M, Soares D, Silva FS, Miranda G. Evaluation of CNT Dispersion Methodology Effect on Mechanical Properties of an AlSi Composite. Journal of Materials Engineering and Performance. 2015;24(6):2535-45.
3. Ma Y, Ueda M, Yokozeki T, Sugahara T, Yang Y, Hamada H. A comparative study of the mechanical properties and failure behavior of carbon fiber/epoxy and carbon fiber/polyamide 6 unidirectional composites. Composite Structures. 2017;160:89-99.
4. Jiang B, Zhang T, Zhao L, Huang Y. Interfacially reinforced carbon fiber composites by grafting modified methylsilicone resin. Composites Science and Technology. 2017;140:39-45.
5. Gopalakrishnan S, Murugan N. Prediction of tensile strength of friction stir welded aluminium matrix TiCp particulate reinforced composite. Materials & Design. 2011;32(1):462-7.
6. Rajakumar S, Balasubramanian V. Multi-Response Optimization of Friction-Stir-Welded AA1100 Aluminum Alloy Joints. Journal of Materials Engineering and Performance. 2011.

7. Rajakumar S, Balasubramanian V. Diffusion bonding of titanium and AA 7075 aluminum alloy dissimilar joints—process modeling and optimization using desirability approach. The International Journal of Advanced Manufacturing Technology. 2016.
8. Vijayan D, Rao V. Friction Stir Welding of Age-Hardenable Aluminum Alloys: A Parametric Approach Using RSM Based GRA Coupled With PCA. Journal of The Institution of Engineers (India): Series C. 2014;95(2):127-41.
9. D Vijayan VSr. Tensile Properties Improvement on Friction Stir Welded Age-Hardenable Aluminum Alloys: An Evolutionary Approach using RSM based GA and SA. Revista Tecnica De La Facultad De Ingenieria Universidad Del Zulia. 2016;39(1):55 - 70.
10. Vijayan D, Seshagiri Rao V. Parametric optimization of friction stir welding process of age hardenable aluminum alloys−ANFIS modeling. Journal of Central South University. 2016;23(8):1847-57.
11. Mm IW, Azmi AI, Lee CC, Mansor AF. Kerf taper and delamination damage minimization of FRP hybrid composites under abrasive water-jet machining. The International Journal of Advanced Manufacturing Technology. 2016;94(5-8):1727-44.
12. Wang H, Qin X, Wu D, Song A. Optimization of Cutting Parameters in Helical Milling of Carbon Fiber Reinforced Polymer. Transactions of Tianjin University. 2017;24(1):91-100.
13. Vijayan D, Seshagiri Rao V. A parametric optimization of FSW process using RSM based grey relational analysis approach. International Review of Mechanical Engineering (IREME). 2014;8(2):328-37.
14. Palanikumar K, Muthukrishnan N, Hariprasad KS. Surface roughness parameters optimization in machining A356/SiC/20p metal matrix composites by PCD tool using response surface methodology and desirability function. Machining Science and Technology. 2008;12(4):529-45.
15. Palanikumar K, Karthikeyan R. Assessment of factors influencing surface roughness on the machining of Al/SiC particulate composites. Materials & Design. 2007;28(5):1584-91.
16. Rajmohan T, Palanikumar K. Modeling and analysis of performances in drilling hybrid metal matrix composites using D-optimal design. The International Journal of Advanced Manufacturing Technology. 2012;64(9-12):1249-61.
17. Basavarajappa S, Chandramohan G, Davim JP, Prabu M, Mukund K, Ashwin M, et al. Drilling of hybrid aluminium matrix composites. The International Journal of Advanced Manufacturing Technology. 2006;35(11-12):1244-50.

IOP Conf. Series: Materials Science and Engineering **390** (2018) 012077 doi:10.1088/1757-899X/390/1/012077

Dufour and Soret effects on MHD convection of Oldroyd-B liquid over stretching surface with chemical reaction and radiation using Cattaneo-Christov heat flux

K. Loganathan[b,#], **S. Sivasankaran**[a,*], **M. Bhuvaneswari**[a], **S. Rajan**[b]
[a]Department of Mathematics, King Abdulaziz University, Jeddah, Saudi Arabia
[b]Department of Mathematics, Erode Arts & Science College, Erode, Tamilnadu, India
[#]Email: loganathankaruppusamy304@gmail.com, [*]Email: sd.siva@yahoo.com

Abstract: The main purpose of this investigation is to present the effects of Dufour and Soret on MHD convection of an Oldroyd-B liquid over a stretchy sheet with Cattanao-Christov heat flux model in the appearance of radiation and chemical reaction. The governing systems of non-linear PDE's are reconstructed into non-linear ODE's. The resulting intricate non-linear boundary value problem is solutes through homotopy analysis method. The influence of various physical parameters on velocity, concentration, temperature profiles and dimensionless version of local skin friction, local mass and heat transfer rates are discussed through graphs.

1. Introduction

The study of heat transfer for an incompressible liquid past a stretchy surface has attracted many investigators by interest of its industrial applications. Such as petroleum industry, geothermal problems, cooling of nuclear reactors. Hayat et al.[1] provides MHD 2D steady flow of an Oldroyd-B liquid with Cattanao-Christov model. Similar studies in this region was found in Eswaramoorthi et al.[2] and Hayat et al. [3]. Bhuvaneshwari et al. [4] examined the heat transfer aspects of the free convection stream of a heat generating liquid towards a semi-infinite inclined surface impacted in a porous medium. Imtiaza et al. [5] explored the 2D flow of third grade liquid over a linear stretchy sheet with chemical reaction. Dufour and Soret effects on the mixed convection flow of an Oldroyd-B liquid was analyzed by Ashraf et al.[6]. Jagan et al.[7] investigate the non-linear thermal radiation effect on three-dimensional MHD convective flow of nanoliquid over a non-linear stretchy sheet with porous medium. Heat transfer and convective flow with chemical reaction is applied in many technical and industrial progress. Few more studies in this area are seen in Karthikeyan et al.[8], Ruhaila et al.[9] and Sivasankaran et al.[10].

Inspirited by the above applications and surveys explained, the purpose of this present work is to explore the effects of chemical reaction and radiation on 2D magneto-hydrodynamic flow of an Oldroyd-B liquid with Cattanao-Christov model in appearance of Dufour and Soret effects.

2. Mathematical formulation of the problem

Consider the MHD flow of an Oldroyd-B liquid over a linear stretchy sheet. The velocity components (u_1, v_1) are taken in the (x, y)-directions. The velocity of sheet is assumed as $U_W = ax$ where $a > 0$ is the stretching rate. The two temperatures on and apart from the surface are expressed by $T_W$and $T_\infty$with $T_W > T_\infty$. The magnetic field of strength B_m is applied upright to the stretchy surface. The electric and induced magnetic fields are neglected. We obtain the following governing equations

$$\frac{\partial u_1}{\partial x} + \frac{\partial v_1}{\partial y} = 0 \tag{1}$$

$$u_1\frac{\partial u_1}{\partial x} + v_1\frac{\partial u_1}{\partial y} + A_1\left(u_1^2\frac{\partial^2 u_1}{\partial x^2} + v_1^2\frac{\partial^2 u_1}{\partial x^2} + 2u_1v_1\frac{\partial^2 u_1}{\partial x\partial y}\right) = \mu\frac{\partial^2 u_1}{\partial y^2}$$

$$-\mu A_2\left(u_1\frac{\partial^3 u_1}{\partial x\partial y^2} + v_1\frac{\partial^3 u_1}{\partial y^3} - \frac{\partial u_1}{\partial x}\frac{\partial^2 u_1}{\partial y^2} - \frac{\partial u_1}{\partial y}\frac{\partial^2 v_1}{\partial y^2}\right) - \frac{\sigma B_m^2}{\rho}\left(u_1 + A_1\mu\frac{\partial u_1}{\partial y}\right) \tag{2}$$

$$\rho c_p\left(u_1\frac{\partial T}{\partial x} + v_1\frac{\partial T}{\partial y}\right) = -\nabla.q \tag{3}$$

$$u_1\frac{\partial c}{\partial x}+v_1\frac{\partial c}{\partial y}=D_m\frac{\partial^2 c}{\partial y^2}-k_m(c-c_\infty)+\frac{D_m k_T}{T_m}\frac{\partial^2 T}{\partial y^2} \tag{4}$$

Then, the boundary conditions are

$$u_1 = u_{1_w},\ v_1=0,\ c=c_w,\ T=T_w \text{ at } y=0$$

$$u_1\to 0, v_1\to 0,\ c\to c_\infty,\ T\to T_\infty \text{ as } y\to\infty \tag{5}$$

where $\mu,\ \rho,\ A_1,\ A_2, \sigma,\ c,\ c_p,\ c_w, c_\infty, D_m$ and k_T are the kinematic viscosity, liquid density, relaxation time, retardation time, electrical conductivity, concentration, specific heat, wall concentration, ambient liquid concentration, diffusion coefficient and first order chemical reaction parameters respectively. Using Cattaneo-Christov heat flux theory, we obtain the following energy equation

$$u_1\frac{\partial T}{\partial x}+v_1\frac{\partial T}{\partial y}+\lambda\left(u_1^2\frac{\partial^2 T}{\partial x^2}+v_1^2\frac{\partial^2 T}{\partial y^2}+\left(u_1\frac{\partial u_1}{\partial x}+v_1\frac{\partial u_1}{\partial y}\right)\frac{\partial T}{\partial x}+2u_1v_1\frac{\partial T^2}{\partial x\partial y}\right)$$

$$+\left(u_1\frac{\partial u_1}{\partial x}+v_1\frac{\partial u_1}{\partial y}\right)=\frac{k}{\rho c_p}\frac{\partial^2 T}{\partial y^2}-\frac{1}{\rho c_p}\frac{\partial q_r}{\partial y}+\frac{D_m k_T}{c_s c_p}\frac{\partial^2 c}{\partial y^2} \tag{6}$$

Consider the following similarity transformations

$$\eta=\sqrt{\frac{a}{\mu}}y,\ u_1=axf'(\eta),\ v_1=-\sqrt{a\mu}f(\eta),\ \theta(\eta)=\frac{T-T_\infty}{T_w-T_\infty},\phi(\eta)=\frac{c-c_\infty}{c_w-c_\infty} \tag{7}$$

Apply equation (7) in equations (2), (4) and (6) we have

$$f'''+ff''-f'^2+\alpha_1(2ff'f''-f^2f''')+\beta_1(f''^2-ff^{iv})-M(f'-\alpha_1ff'')=0 \tag{8}$$

$$\left(1+\frac{4}{3}Rd\right)\theta''+Prf\theta'-Pr\gamma(f^2\theta''+ff'\theta')+PrD_f\phi''=0 \tag{9}$$

$$\frac{1}{Sc}\phi''+f\phi'-Cr\phi+Sr\theta''=0 \tag{10}$$

with boundary conditions

$$f(\eta)=0, f'(\eta)=1, (\eta)=1,\ \phi(\eta)=1 \text{ at } y=0$$

$$f'(\eta)\to 0, \theta(\eta)\to 0, \phi(\eta)\to 0 \text{ as } y\to\infty \tag{11}$$

Here we declare the non-dimensional variables with respect to $\alpha_1=A_1a$ and $\beta_1=A_2a$ are the relaxation time &retardation time parameters respectively, $M=\sigma B_0^2/\rho a$ the Hartmann number, $Pr=\rho C_p/k$ the Prandtl number, $Rd=(4\sigma^*T_\infty^3)/(kk^*)$ the radiation parameter, $\gamma=\lambda a$ the non-dimensional thermal relaxation time, $D_f=\frac{D_m k_T}{\mu c_s c_p}\frac{(c_w-c_\infty)}{(T_w-T_\infty)}$ the Dufour number, $Cr=\frac{k_m}{a}$ the chemical reaction parameter, $Sc=\mu/D_m$ the Schmidt number, $Sr=\frac{D_m k_T}{\mu T_m}\frac{(T_w-T_\infty)}{(c_w-c_\infty)}$ the Soret number.

The dimensionless forms of skin friction, heat transfer rate and local mass transfer rate are represented below

$$Re^{\frac{1}{2}}Cf_x=\frac{1+\alpha_1}{1+\beta_1}f''(0),\ Re^{-\frac{1}{2}}Nu_x=-\left(1+\frac{4}{3}Rd\right)\theta'(0),\ Re^{-\frac{1}{2}}Sh_x=-\phi'(0)$$

3. HAM Solutions

The initial guesses for homotopy solutions are defined as $f_0 = 1 + e^{-\eta}$, $\theta_0 = e^{-\eta}$ and $\phi_0 = e^{-\eta}$ the auxiliary linear operators L_f, L_θ and L_ϕ are derived as $L_f = f'''(\eta) - f'(\eta)$, $L_\theta = \theta''(\eta) - \theta(\eta)$ and $L_\phi = \phi''(\eta) - \phi(\eta)$ with satisfying the following properties $L_f[E_1 + E_2 e^{\eta} + E_3 e^{-\eta}] = 0$, $L_\theta[E_4 e^{\eta} + E_5 e^{-\eta}] = 0$ and $L_\phi[E_6 e^{\eta} + E_7 e^{-\eta}] = 0$, where $E_j (j = 1 - 7)$ denote the arbitrary conditions. $f_m(\eta) = f_m^*(\eta) + E_1 + E_2 e^{\eta} + E_3 e^{-\eta}$, $\theta_m(\eta) = \theta_m^*(\eta) + E_4 e^{\eta} + E_5 e^{-\eta}$ and $\phi_m(\eta) = \phi_m^*(\eta) + E_6 e^{\eta} + E_7 e^{-\eta}$. Where $f_m^*(\eta)$, $\theta_m^*(\eta)$ and $\phi_m^*(\eta)$ are the appropriate solutions.

The auxiliary parameters h_f, h_θ and h_ϕ play an vital part for convergence series solutions. The h-profiles are drawn at 15th order of approximations to accomplish valid ranges of parameters(see Figure.1). The admissible values of h_f, h_θ and h_ϕ are $-1.3 \le h_f \le -0.3$, $-1.0 \le h_\theta \le -0.3$, $-1.2 \le h_\phi \le -0.3$. More over the solutions converge in the total range of η. When $h_f = h_\theta = h_\phi = -0.7$. Table.1 displays the order of approximations of HAM solutions.

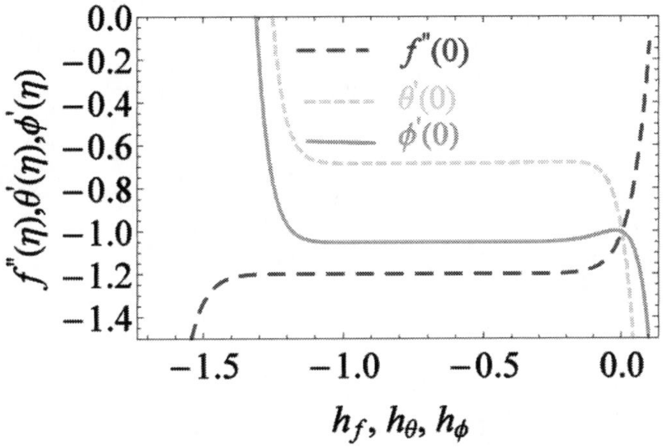

Figure.1. h-curves for h_f, h_θ *and* h_ϕ

4. Results and discussion

The efficient numerical calculations are execute for various values of α_1, β_1, Rd, Cr, D_f and Sr on velocity, temperature and concentration profiles with fixed values of $M = 0.5$, $\alpha_1 = 0.1$, $\beta_1 = 0.2$, $\gamma = 0.5$, $Pr = 1.0$, $Sc = 0.9$, $Rd = 0.3$, $Cr = 1.0$, $D_f = 0.5$ and $Sr = 0.3$. Figure 2 show that relaxation time(α_1) and retardation time(β_1) have inverse effects on the velocity distribution $f'(\eta)$. Figure 3(a) depicts that extensive values of radiation constant(Rd) on temperature component the thermal and momentum boundary layer thickness rises. The influence of Dufour number(D_f) on temperature distribution is plotted in Figure 3(b) it is noted that an enlargement in Dufour number the temperature enhances. Figure 4(a) illustrates the influence of chemical reaction(Cr) on concentration profiles. For larger values of chemical reaction reduces in species the concentration. Figure 4(b) demonstrates the effects of Soret number(Sr) on the concentration distribution. From this figure it is identify that an enhances in the Soret number the concentration of the liquid and solutal boundary layer thickness enhances. From Figure 5(a)&5(b) sketched for relaxation time (α_1), retardation time $(\beta_1)$with magnetic field parameter(M) on skin friction co-efficient. It is observed that α_1 and β_1 has opposite effects on local skin friction. Figure 6(a) is for the influence of radiation constant(Rd) and non thermal relaxation time constant(γ). Larger values of Rd lead to enhance the Nusselt number. Similar behavior is observed for γ. Figure 6(b) is plotted for the influence of Dufour number (D_f) and non thermal relaxation time(γ). Here Nusselt number diminishes via Dufour number. Characteristics of M and Sr with Cr on Sherwood number are displayed in Figure 7(a) & 7(b). The local mass transfer rate diminishes by uplifting the values of M, Sr and Cr.

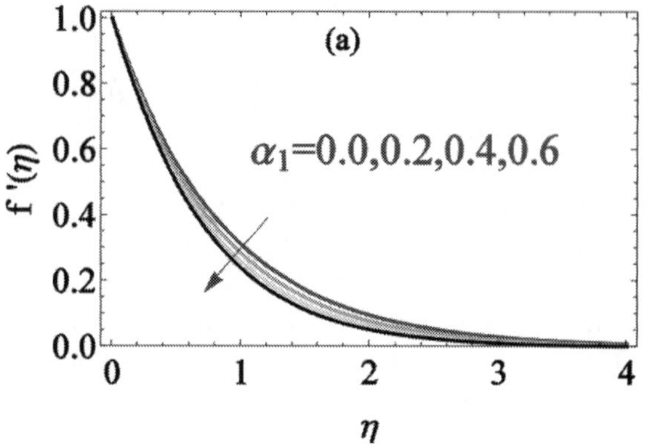

Figure.2. Influence of α_1 and β_1 on velocity profile $f'(\eta)$.

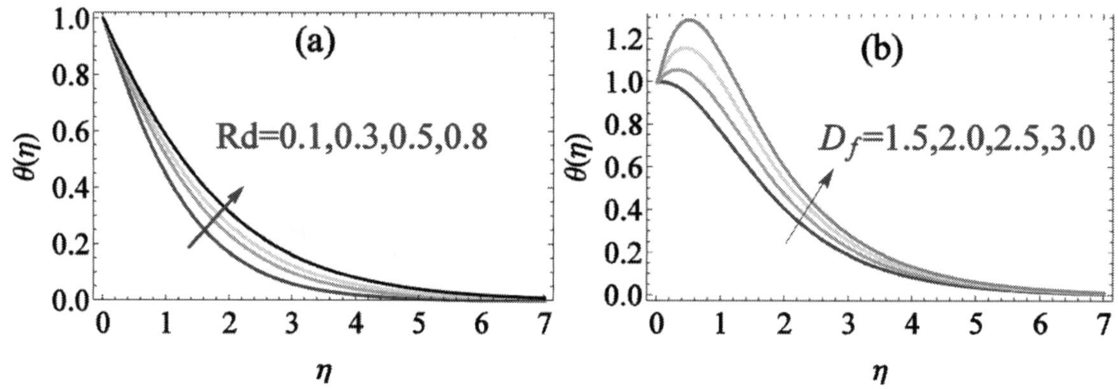

Figure.3. Influence of Rd and D_f on temperature profile $\theta(\eta)$.

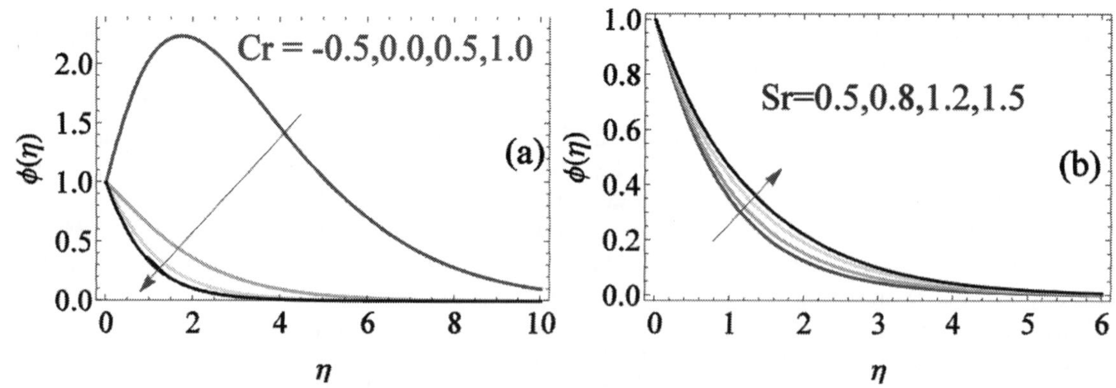

Figure.4. Influence of Cr and Sr on concentration profile $\phi(\eta)$

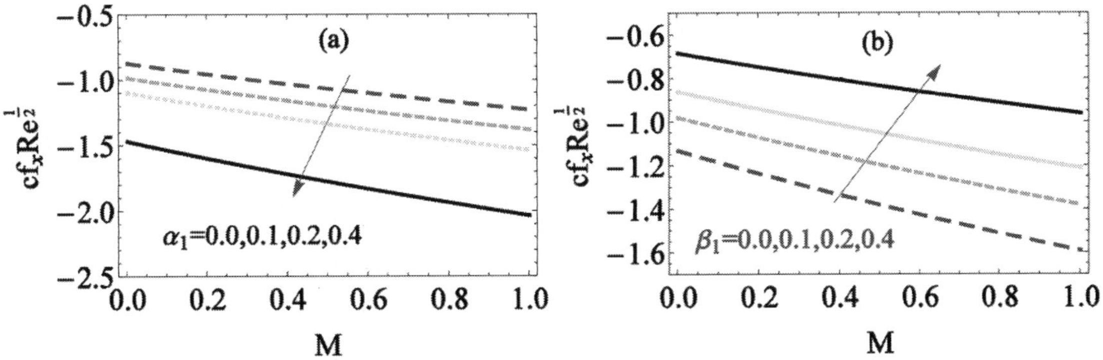

Figure.5. Influence of α_1 and β_1 on skin friction coefficient $Re^{\frac{1}{2}}Cf_x$.

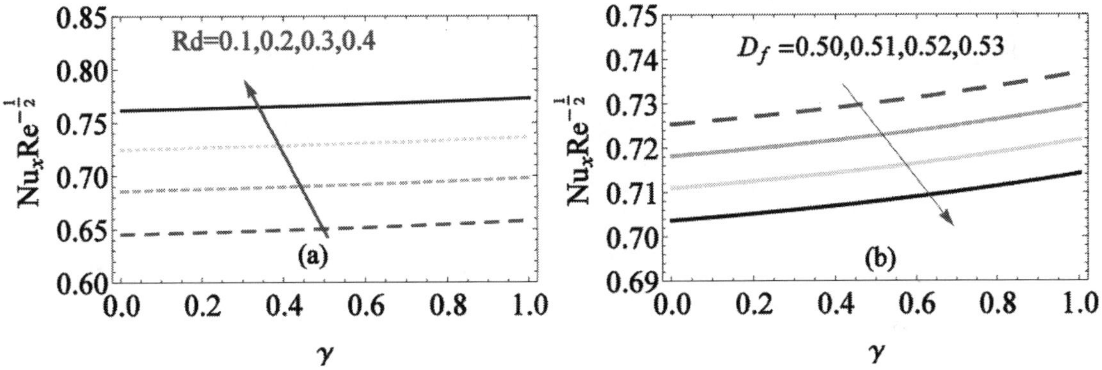

Figure.6. Influence of Rd and D_f on local Nusselt number $Re^{-\frac{1}{2}}Nu_x$.

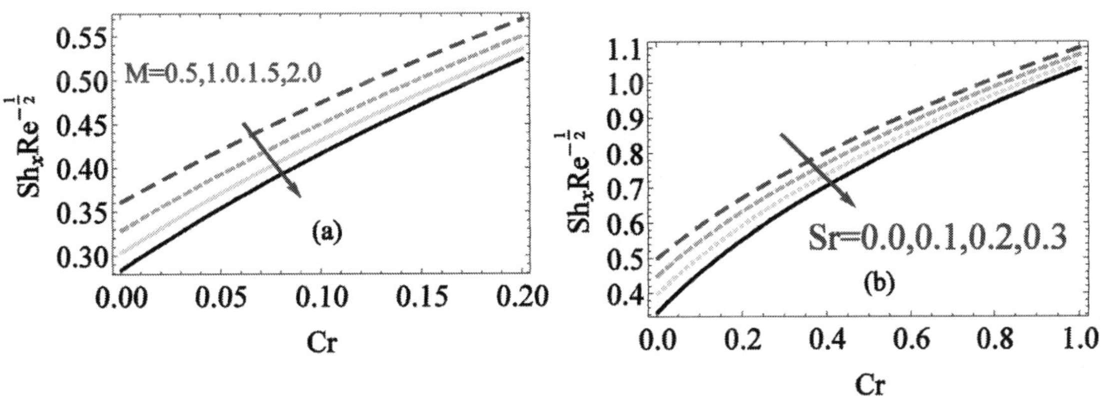

Figure.7. Influence of M and Sr on local Sherwood number $Re^{-\frac{1}{2}}Sh_x$.

Table:1 Order of approximations when $M = 0.5, \beta_1 = 0.2, \alpha_1 = 0.1, \gamma = 0.5, Pr = 1.0, Sc = 0.9, Rd = 0.3, Cr = 1.0, D_f = 0.5, Sr = 0.3$ and h = -0.7

Order of approximation	$-f''(0)$	$-\theta'(0)$	$-\phi'(0)$
1	1.6333	0.6266	0.9727
5	1.1944	0.5253	1.0595
10	1.1944	0.5217	1.0625
16	1.1944	0.5216	1.0625
20	1.1944	0.5216	1.0625
30	1.1944	0.5216	1.0625
40	1.1944	0.5216	1.0625
50	1.1944	0.5216	1.0625

5. Final report

The magneto-convection of an Oldroyd-B liquid under the effects of Cattaneo-Christov heat flux and chemical reactions in attendance of Dufour and Soret effects are investigated. The thermal and momentum boundary layer thickness enhances with upgrading radiation constant. The distribution of temperature of the liquid enhances with Dufour number increases. The concentration diminishes for the larger chemical reaction constant. Finally, the solutal boundary layer thickness enhances with Soret number.

6.References

[1] T.Hayat,M. Imtiaz, A.Alsaedi,S. Almezal, Journal of magnetic materials, 401(1), (2016), 296-303.

[2] S Eswaramoorthi M Bhuvaneswari, S Sivasankaran, H Niranjan and S Rajan 2017 IOP Conf. Ser.: Mater. Sci. Eng. 263 062009.

[3] T. Hayat, Z. Hussain, A. Alsaedi, Results in physics, (2016) 1161-1167.

[4] M. Bhuvaneswari, S. Sivasankaran and M. Ferdows, (2009) Nonlinear Analysis: Hybrid Systems (3), (2009), 536–542.

[5] M. Imtiaza, A. Alsaedi, A. Shafiq, T. Hayat, Journal of molecular liquids, (229), (2017) 501-507.

[6] B. Ashraf, T. Hayat, A. Alsaedi, S,A. Shehzad, Resltults in physics, (2016), 917-924.

[7] K.Jagan, S. Sivasankaran, M. Bhüvaneswari, S. Rajan, International journal of pure and applied mathematics, 117(13), (2017), 43-51.

[8] Karthikeyan S, Bhuvaneswari M, Sivasankaran S and Rajan S, Journal of Applied Fluid Mechanics, (9), (2016), 1447-1455.

[9] Ruhaila M K, Sivasankaran S, Bhuvaneswari M and SiriZ, Journal of Applied Fluid Mechanics,(9), (2016). 379-388.

[10] Sivasankaran S, Niranjan H and Bhuvaneswari M, International Journal of Numerical Methods and Heat Fluid Flow (27), (2017), 454-470.

Optimization in tribological behaviour of Al-nanoTiO_2 powder metallurgy composites using response surface method

D Balaji[1], M Vetrivel [1], M Ekambaram [1], B Sai Ganesh[1], M Sai Raghava Datta[1], S Manideep Kumar[1], V Sai Darshan[1] and G Vignesh[1]

[1] Department of mechanical engineering, SCSVMV, Enathur 631 561.Tamil Nadu, India.
balajibala81@gmail.com, km_vetrivel@yahoo.com , e4ekam@gmail.com

Abstract. The Al- nano TiO_2 composite processed powder metallurgy and wear behaviour was experimentally investigated on aluminium with combined nano TiO_2 weight percent reinforcement of 4%,7% and 10% at different sliding velocity of 1.5 m/s,2.0 m/s,2.5 m/s and different normal loads of 10N,30N,50N. The effect of specific wear rate and co efficient of friction optimised by response surface method. The results predicted by the developed model are concurring well within the measured values. The correlation co-efficient of model, regarding the specific wear rate and co efficient of friction, were about 0.96 and 0.94 respective with conform the degree of accuracy of the mathematical model. The optimal condition attain the minimum specific wear rate and co-efficient of friction at sliding velocity 2.04 m/s and Normal load 13.5 N with Al- TiO_2 10% weight. The predicted value of specific wear rate $1.08e^{-7}$ and co efficient of friction 0.31 respectively.

1. Introduction

Powder Metallurgy (PM) techniques are used in manufacture of new composite material component in aerospace and automobile industry due to its superior characteristics. Aluminium metal matrix composites (AMCs) are quite attractive and various studies were reported on wear behavior of Aluminium composite reinforced with different materials [1-3]. Powder Metallurgy routed technique in manufacture of metal matrix nano composite will produce uniformity in the reinforced distribution[4,5].

Aluminium based hybrid AMCs along with TiO_2 reinforced with various particle of SiC,SiO_2 and Gr will improve the mechanical and wear resistance[6,7]. Addition of metal oxide TiO_2 with Aluminium alloy possesses better wear resistant property [8-11]. The porosity and tensile strength of Al with nano TiO_2 (size 50nm) will provide superior wear resistance[12].

1.1. Response Surface Methodology

As mentioned, RSM can be an excellent approach to study a process response and to figure out the best correlation among the parameters of a process. This is done via developed models based on the statistical methods to investigate the relation between the inputs and outputs of any process.

The mathematical model has been developed based on the experimental results, with an objective to determine the specific wear rate and coefficient of friction; also to study the inter relationships between the test variables of Al – nano TiO2 composite.

The basic concept of response method [13] is to construct an approximation of this implicit limit state by an approximate simple and explicit polynomial function. A response surface equation is usually formulated as

$$Y = \beta_0 + \sum_{i=1}^{m} \beta_i x_i + \sum_{i=1}^{m} \beta_{ii} x_i^2 + \sum_{i=1}^{m} \sum_{i<j}^{m} \beta_{ij} x_i x_j + \varepsilon \qquad (1)$$

Where β_0, β_i, β_{ii}, and β_{ij}, represent regression coefficients; x_i (i = 1,2,…,m) are design variables, ε is the random error, Y is the response and m is the total number of design variables.

Design Expert V8.0.6.0 software was used for the regression graphical analysis of the results and also to obtain the coefficient regression equations. To analyses statistically the model, the Analysis of the Variance (ANOVA) was used. Further the results have been used for the prediction of the outputs for new conditions. The Fisher's F-test (overall model significance), its probability p(F), determination coefficient R^2 (used to measure the fitness of regression model) were also studied by ANOVA [9].

1.2. Multiple response optimizations

In most of the modern technological situations, more than one response variable is pertinent to the success of an industrial process or system. In this study, the influence of sliding velocity, sliding velocity and weight percent reinforcement of nano TiO_2 on the specific wear rate and
Co efficient of friction is investigated and further the optimal input parameter is investigated and further the optimal minimizing specific wear rate and minimizing co-efficient of friction. A desirability function based simultaneous optimization technique is used in this work [14].

2. Materials and Experimental Details

2.1. Materials

The materials used in present work are Aluminium fine powder with size range 130 – 180 micrometer, the purity 98% as base material bought from Loba Chemi, India. The reinforced material TiO_2 powder of size 5 nm, the purity 99.89 % bought from US Research Nano Materials Inc, USA.

2.2. Pin Specimen preparation

The Blending process is done by high energy planetary ball milling machine using Al powder with 4 wt%, 7 wt%, 10 wt% of TiO_2 nano powder. The Blended powder is compacted in a Compacting machine to attain 30mm height with applied pressure of 8.5MPa. The compacted specimens are sintered using sintering furnace for 90min at 580°C.

2.3. Wear test methodology

Wear teat conducted using Pin on Disc tribometer on Al- TiO_2 P/M pin specimen as per ASTM G99 standard and recorded the output value of wear rates, frictional force, coefficient of friction based on the design variable mention in Table 1.

Table 1. Experimental factor and their levels

Symbol	Parameter	Unit	Coded values and the corresponding values of parameters		
			-1	0	+1
A	Load	N	10	30	50
B	Sliding Velocity	m/s	1.5	2.0	2.5
C	TiO_2 Weight Fraction	%	4	7	10

Sliding distance 1000 m

Specific wear rate (SWR) are calculated by wear volume per unit distance and load and Co efficient of friction (COF).

3. Results and Discussion

3.1. RSM Modeling

The experimental design and corresponding results are presented in Table 2. The following model in Eq.(2) and (3), were respectively developed model for the Specific wear rate and Co efficient of friction using regression coefficient summarized in Table 2. Following the calculation of the regression coefficient, the model for prediction of Specific wear rate and Co efficient of friction was determined as follows:

$$SWR = 2.11e^{-6} + 4.64e^{-7}A + 2.31e^{-6}B - 1.04e^{-6}C + 2.37e^{-7}A^2 + 9.68e^{-7}B^2$$
$$- 4.94e^{-8}C^2 + 4.3e^{-7}AB - 3.39e^{-7}AC + -9.06e^{-7}BC \quad (2)$$

$$COF = 0.4989 + 0.0390A + 0.1192B - 0.0683C + 0.0119A^2 + 0.0257B^2$$
$$- 0.0199C^2 + 0.0137AB - 0.0055AC - 0.0159BC \quad (3)$$

Where A, B and C are coded values pertaining to the temperature, strain rate and solid lubricant respectively.

The significance of each coefficient in the regression in Equation (2) and (3) are determined by t-values and p-values listed in Table 2. Higher t-value and lower p - value regarding a coefficient indicate that the coefficient is more significant. In general, any term having the value of $p < 0.05$ is significant.

Table 2
Analysis of the design results for Specific wear rate and Co efficient of friction

	Specific wear rate (SWR)			Co efficient of friction(COF)		
	Regression Coefficient	t-value	p-value	Regression Coefficient	t-value	p-value
Constant	2.11E-06	8.091	< 0.0001	0.49898	24.775	< 0.0001
A	4.64E-07	3.822	0.0014	0.03903	4.165	0.0006
B	2.31E-06	19.923	**< 0.0001**	0.11924	13.344	**< 0.0001**
C	-1.04E-06	-8.589	**< 0.0001**	-0.06837	-7.295	**< 0.0001**
A^2	4.13E-07	2.998	0.0081	0.01372	1.290	0.2144
B^2	-3.39E-07	-2.285	0.0354	-0.00550	-0.481	0.6367
C^2	-9.06E-07	-6.572	< 0.0001	-0.01597	-1.501	0.1517
AB	2.37E-07	1.132	0.2734	0.01189	0.735	0.4723
AC	9.68E-07	5.339	**< 0.0001**	0.02571	1.838	0.0837
BC	-4.94E-08	-0.236	0.8165	-0.01994	-1.233	0.2343

Referring to the value of specific wear rate presented in Table 3, the sliding velocity and the weight percentage of reinforcement that have p - value less than 0.05 indicates that they are significance with respect to specific wear rate, the interaction among the load and weight percentage of reinforcement parameters was significant. In terms of co efficient of friction, sliding velocity and the weight percentage of reinforcement parameters affects considerably the co efficient of friction values. Regarding the variables interaction, no interaction among the parameters was significant for co efficient of friction as well.

Table 3. Response for Specific wear rate and Co efficient of friction observed experimental and predicted

Exp. No	Process parameters levels			Experimental result		Predicted value	
	Load N	Sliding Velocity m/s	TiO_2 Weight Fraction %	specific wear rate m^3/Nm	coefficient of friction	specific wear rate m^3/Nm	coefficient of friction
1	10	2.0	4	2.30E-06	0.495	2.33E-06	0.534
2	30	2.5	10	3.47E-06	0.519	3.39E-06	0.561
3	30	2.5	7	4.97E-06	0.613	5.39E-06	0.629
4	30	1.5	4	1.01E-06	0.440	6.90E-07	0.448
5	50	1.5	10	7.25E-07	0.361	5.75E-07	0.351
6	50	2.5	7	6.91E-06	0.722	6.50E-06	0.667
7	30	2.5	4	6.90E-06	0.721	7.29E-06	0.696
8	50	2.0	10	1.46E-06	0.474	1.25E-06	0.475
9	30	2.0	7	2.10E-06	0.491	1.94E-06	0.505
10	30	2.0	10	1.55E-06	0.450	9.19E-07	0.437
11	10	1.5	7	8.74E-07	0.391	9.87E-07	0.343
12	50	1.5	4	1.16E-06	0.473	1.25E-06	0.486
13	50	2.5	4	9.77E-06	0.848	8.74E-06	0.734
14	50	2.0	4	3.12E-06	0.529	3.88E-06	0.610
15	50	2.5	10	3.39E-06	0.569	4.16E-06	0.599
16	10	2.5	4	6.37E-06	0.651	6.31E-06	0.658
17	10	2.0	7	1.94E-06	0.486	1.75E-06	0.467
18	50	1.5	7	1.01E-06	0.440	9.65E-07	0.419
19	30	2.0	4	2.40E-06	0.503	2.87E-06	0.572
20	10	1.5	4	9.32E-07	0.426	6.00E-07	0.410
21	10	2.0	10	1.35E-06	0.432	1.06E-06	0.399
22	10	1.5	10	5.52E-07	0.250	1.28E-06	0.275
23	10	2.5	10	3.29E-06	0.510	3.09E-06	0.523
24	30	1.5	10	6.24E-07	0.312	6.88E-07	0.313
25	30	1.5	7	8.87E-07	0.421	7.39E-07	0.381
26	50	2.0	7	2.40E-06	0.503	2.61E-06	0.543
27	10	2.5	7	4.54E-06	0.594	4.75E-06	0.591

The comparison of the results predicted by the model with those measured by experimental testing is presented in Fig.1 for SWR and COF. From the figure, it has been seen that most of the points are close to the center line and hence this empirical model with excellent agreement between this predicted and measured values implies the high accuracy of the model.

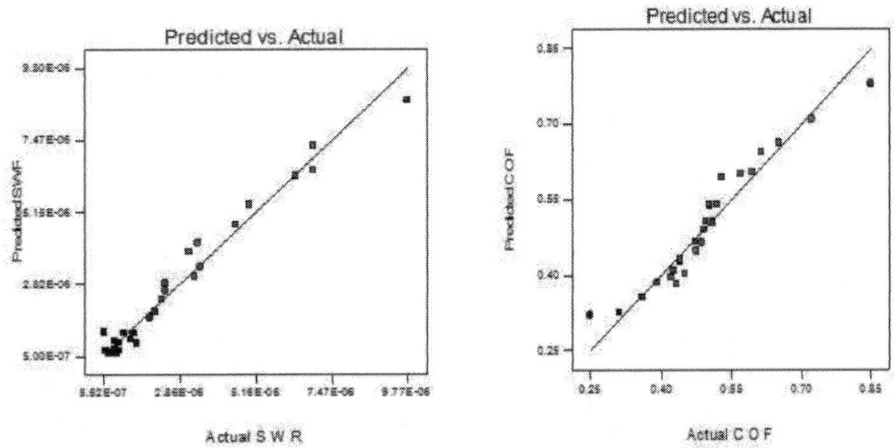

Figure 1. Experimental and predicted value of SWR and COF

Table 4. ANOVA based on RSM for models

Source	Specific wear rate (SWR)				
	SS	DF	MS	F-value	p >F
Model	1.41e^{-10}	9	1.56e^{-11}	59.31	<0.0001
Residual	4.48e^{-12}	17	2.64e^{-13}		
Total	1.45e^{-10}	26			
R^2			**0.9617**		
	Co efficient of friction (COF)				
Model	0.39925	9	0.04436	162.72	<0.0001
Residual	0.02668	17	0.00157		
Total	0.4259	26			
Total			**0.9374**		

The results of the analysis of variance (ANOVA) have been shown in Table 4. ANOVA is applied to analyze statistically the performance of the models developed by RSM. The ANOVA, performed on the RSM regression, indicates that p is greater than F (p>F) demonstrating that the models are significant and there is only a 0.01% chance that a model F-value this large could occur due to noise. Another way to check the accuracy of a model is normally by calculating the R^2 coefficient. The R^2 values were 0.9617 and 0.9374 for SWR and COF, respectively. This implies that about experimental results regarding the SWR and COF measurements are compatible with the ones predicted by the developed models. This again confirms the accuracy of models.

IOP Conf. Series: Materials Science and Engineering **390** (2018) 012078 doi:10.1088/1757-899X/390/1/012078

3.2. Contour plats

After verification of the model, the contour plots can be significant approaches to study the effect of interaction among the variables on SWR and COF of investigated AMCs. The 2D contour plots represent graphically for the response surface of SWR and COF the different interaction among load (A), sliding velocity (B) at different TiO_2 wt % reinforcement (C) are shown in Figure2 and Figure3.

From the Figure 2, it is clear that as specific wear rate increases with increase in sliding velocity and percentage of reinforcement. It is more prominent that the combination of load and the percentage of reinforcement increases, specific wear rate decreases.

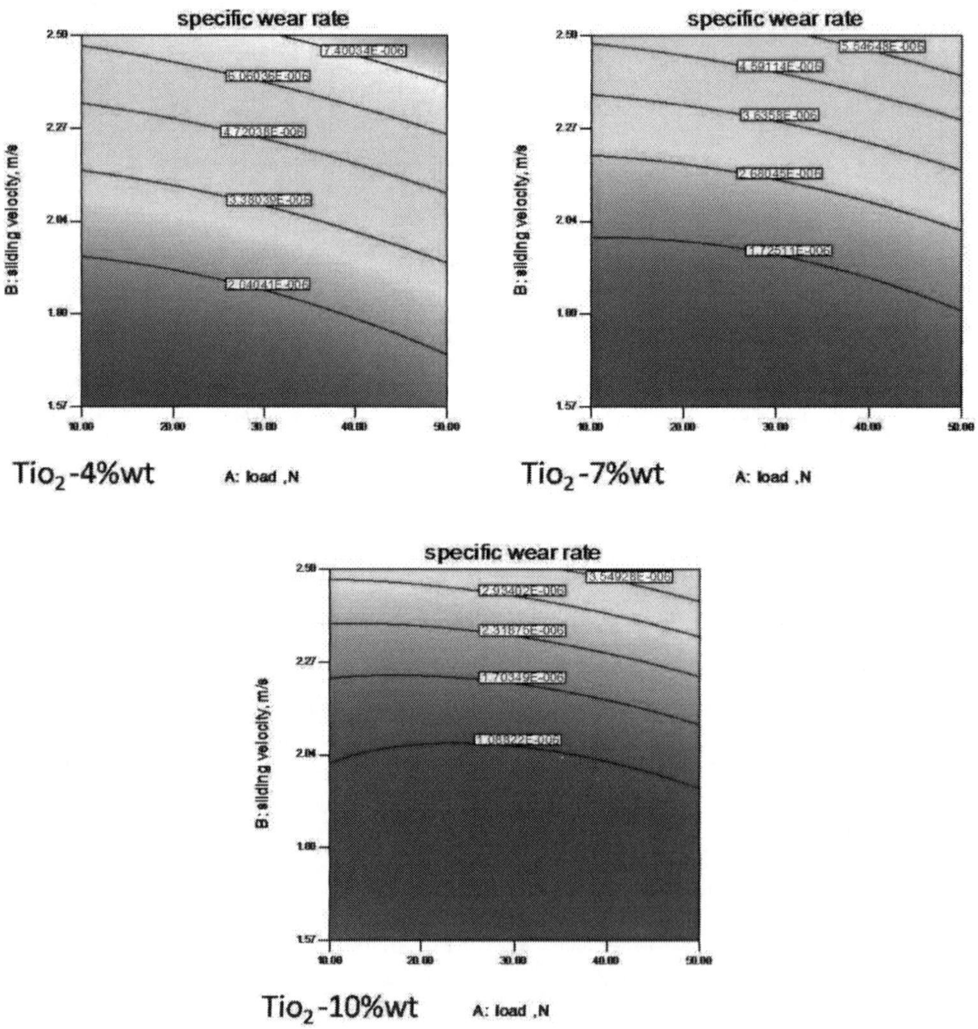

IOP Conf. Series: Materials Science and Engineering **390** (2018) 012078 doi:10.1088/1757-899X/390/1/012078

Figure 2. Contour plots of the SWR with different interaction among load (A), sliding velocity (B) at different TiO_2 wt % reinforcement (C)

From the Figure 3, it is clear that as coefficient of friction increases with increase in sliding velocity and load. It is more significant that as the percentage of reinforcement increases, coefficient of friction decreases.

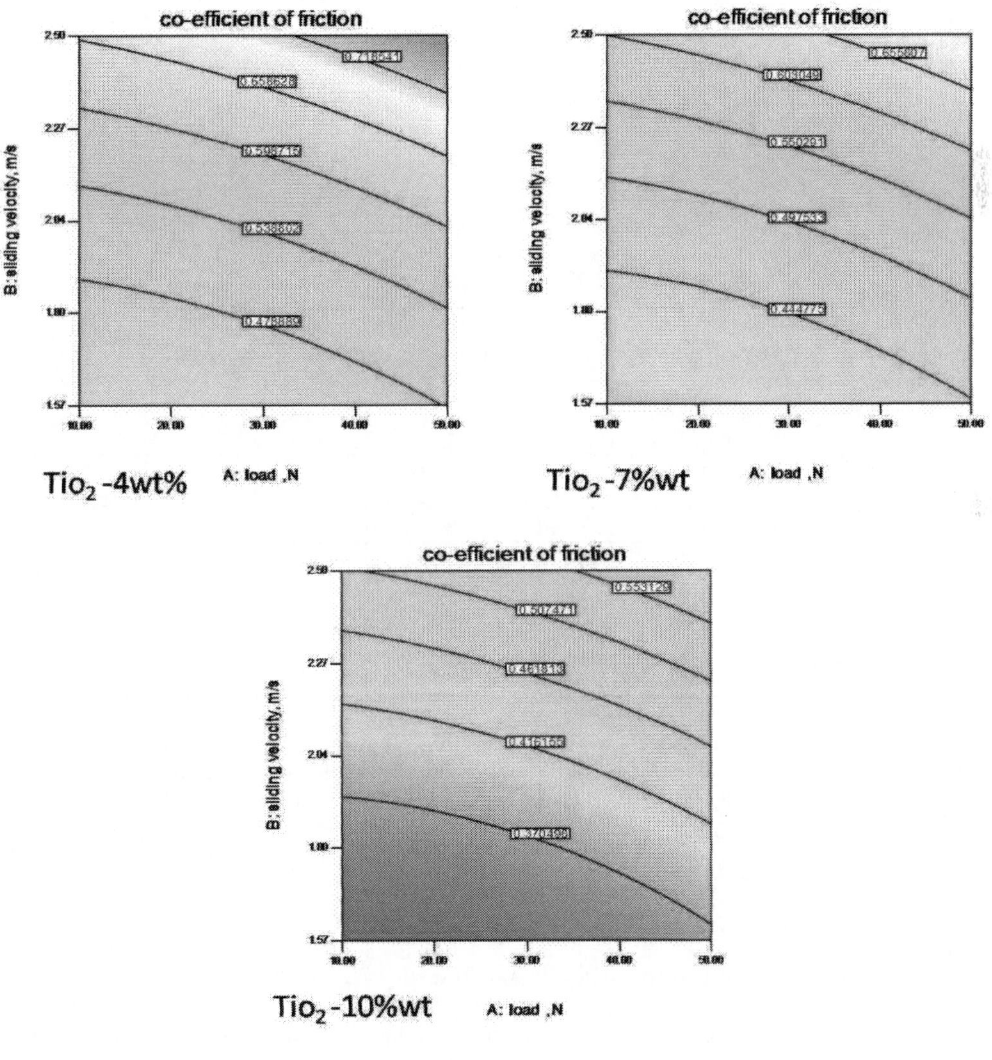

Figure 3. Contour plots of COF with different interaction among load (A), sliding velocity (B) at different TiO_2 wt % reinforcement (C)

3.3. Multiple response optimizations

The optimization analysis was carried out using Design-Expert software. The goal set, lower limits used, upper limits used, weights used, and importance of the factors given are presented in Table 5. In desirability-based approach, different best solutions were obtained.

Table 5. Goals set and limits used for optimization

Name	Goal	Lower Limit	Upper Limit	Lower Weight	Upper Weight	Importance
Load (N)	is in range	10	50	1	1	3
Sliding Velocity(m/s)	is in range	1.5	2.51.0	1	1	3
TiO_2 Weight Fraction (wt %)	is in range	4	10	1	1	3
Specific wear rate (mm^3/Nm)	minimize	$5.52e^{-7}$	$9.77e^{-6}$	1	1	3
Co efficient of friction	minimize	0.250	0.848	1	1	3

The solutions with high desirability are preferred. The best three solutions obtained for the multi-response optimization is presented in Table 6.

Table 6. Best solution for optimization

Number	Load N	sliding velocity m/s	TiO_2 weight percentage	Specific wear rate mm^3/Nm	Co efficient of friction	Desirability
1	13.49	2.04	10	1.08E-06	0.391	**0.849**
2	13.75	2.04	10	1.07E-06	0.391	0.849
3	13.20	2.04	10	1.08E-06	0.391	0.849

From the Table 6, the developed models were used for multiple response optimizations by desirability function approach to obtain maximum strength coefficient and minimum strain hardening exponent. The desirability is found to be 0.849 as acceptable and excellent significant based on Harrington's rating system[14]. The optimized Specific wear rate $1.08e^{-6}$ mm^3/Nm and Co efficient of friction 0.3910 for load 13.49 N and sliding velocity 2.04 m/s , when the TiO_2 weight percentage 10(wt %) are Selected from Table 6

4. Conclusion

Response surface model and experimental design were used to model the tribological behaviour of Al-nano TiO_2 powder metallurgy composite response of AMCs and predict the optimum Wear behaviour, which is characterized by a significant effect of load, sliding velocity and reinforcement weight percentage condition. According to the analysis of variance (ANOVA), the accuracy of developed model conformed by R2 and p-value. The predicted and measured values have excellent agreement. Further, optimum combination of process parameters to attain the minimizing the specific wear rate and minimizing co efficient of friction can be predicted by the statistically developed models.

References

[1] Ravindran P, Manisekhar K, Narayanasamy P, Selvakumar N and Narayanasamy R 2012 *Mat.and Des.* **39** 42-54.
[2] Wang Y Q, Afsar A M, Jang J H, Han K S and Song J I 2010 *Wear* **268** 863-870.
[3] Mazahery A and Shabani M O 2011 *Matr.Sci.Compos.* **45** 2579-2586.
[4] Sivashankaran S, Sivaprasad K, Narayasamy R and Iyer V K 2010 *Powder Tech.* **201** 70-82.
[5] Vaezi M R 2008 *Matr.Process.Technol.* **205** 332-337.
[6] Bindumadhavan P N, Chia T K, Chandrasekaran M, Wah H K, Lam L N and Prabhakar O 2001 *Matr.and Sci.EngA* **315** 217-226.
[7] Manindar Singh, Khushdeep Goyal and Deepak kumar Goyal 2015 *Uni.J.Mech.Engg.* **3** 142-146.
[8] Suresha S and Sridhara P K, 2010 *Comp. Sci. & Tech.* **70** 1652-1659.
[9] Suresha S and Sridhara P K, 2010 *Mat.and Des.* **31** 1804-1812.
[10] Manickam Ravichandran and Abdullah Naveen Sait 2014 *J.Matr.Res.* **29** 1480-1487.
[11] Padmavathy K R and Ramakrishnan R 2017 *Int.J.Engg.Tech.* **9** 3368 – 3373.
[12] Amal E. Nassar and Eman E. Nassar 2017 *J.King Saud Univ.-Engg.Sci.* **29** 295-299.
[13] Vetrivel M, Senthilvelan S 2012 *J.Archives Des Sci.* **65** 12.
[14] Harington , J 1965 *Industrial Quality Control* ,**21**, 494-498.

A Review on Recent Trends in Surface Coatings

Chintada.Vinod Babu[1], Meghana Yaradi[2], Dharma Teja Thammineni[3]
[1]Asst.Professor, Department of Mechanical Engineering, GMRIT, Rajam.
[2, 3] 3rdB.Tech student, Department of Mechanical Engineering, GMRIT, Rajam.
Email: vinodbabu.chintada@gmail.com, meghanayaradi006@gmail.com, dharmateja15798@gmail.com

Abstract: Corrosion of the components is a predominant factor which controls the life of metallic parts. Metals parts of various structures are often subjected to damage due to corrosion driven failures causing loss of functionality and dimensions. In order to reduce corrosion, there are two ways one is introducing new materials and the other is improving the existing materials by adding alloying materials or by surface coating techniques. Surface coatings are a feasible way to enhance the corrosion resistance of metallic materials, thus expanding the life span of the metal components. Aim of this paper is to draw comparison between various coating techniques and analyze which surface coating technique and coating materials improves the corrosion resistance of the materials. Further, in this review paper coat materials on the base material by mostly cold spray and plasma spray techniques. So, these two surface coating modification techniques are discussed in detail.

Keywords: Corrosion, wear, coatings.

1. Introduction:

Corrosion is a natural phenomenon in which gradual degradation of the material takes place by chemical or electrochemical reaction. In general, marine corrosion is classified into 5 types as galvanic corrosion, crevice corrosion, stress corrosion, intergranular corrosion and flow accelerated corrosion.[1].

1.1. Economic impact of corrosion globally:

According to NACE's(National Association of Corrosion Engineers), IMPACT (International Measures of Prevention, Application and Economics of corrosion Technologies). By 2013, the global cost of corrosion is estimated to be almost US $3trillion which is equal to 34% of global GDP. This indicates picture perfectly the entire scenario of the corrosion impact on world. So there is a need to enhance the corrosion resistance of the material[2].Due to cost constraints new material usage is difficult. So we opted to surface coatings. Out of all surface coatings Thermal spray is an age old process still it is dominant technique because of the emerging new techniques along with the feasibility in the choice of the both substrate and the coating material[3].

2. Discussions:

2.1. Zn alloy coatings on the mild steel:

The Zn-Al and Zn-Sn is coated on the mild steel using plasma spraying technique. The coating materials are mixed in following proportions Zn-Al(75/25), Zn-Al(50/50), Zn-Al(25/75) and Zn-Sn(75/25), Zn-Sn(50/50), Zn-Sn(25/75).Corrosion tests are performed in 3.65% NaCl solution at room temperature. From this the below results are obtained shown in table-1.The relation between the corrosion potential and corrosion densities are shown by potentiodynamic polarization curve known as TAFEL plot.

From the slope of the tafel plot, it is observed that lower current densities and higher polarization resistance provides highest corrosion resistance [4-5]. Among Zn-Sn and Zn-Al alloy compositions the Zn-Sn(50/50) has better corrosion resistance than Zn-Al(25/75) [4,5].

Table 1.Corrosion current densities of Zn-Sn coatings in 3.5wt% NaCl [*Ref 4-5*]

Substrate	Coating material	Technique	Current density i_{corr}(A/cm^2)	Polarization resistance (R_p)
Mild steel	Zn-Sn(25/75)	Plasma spraying	5.24×10^{-4}	43.49
Mild steel	Zn-Sn(50/50)	Plasma spraying	4.44×10^{-5}	354.71
Mild steel	Zn-Sn(75/25)	Plasma spraying	9.29×10^{-4}	66.288
Mild steel	Zn-Sn(25/75)	Plasma spraying	5.08×10^{-4}	144.68
Mild steel	Zn-Sn(50/50)	Plasma spraying	5.66×10^{-4}	92.67
Mild steel	Zn-Sn(75/25)	Plasma spraying	6.67×10^{-4}	60.482
Mild steel	uncoated	-	2.39×10^{-2}	1.968

Now the Znand its alloy coating takes place by using an unbalanced magnetron sputtering in presence of Argon gas at the base pressure of 1.4×10^{-3}pa and thickness of 2μm.. The samples are tested in 3.5 wt% NaCl solution by spraying it on the samples. The pure Zn film has corrosion current density of 3.918×10^{-2} A/cm^2 while the 87.8Zn-12.2Mg sample has corrosion current density of 2.586×10^{-3} A/ cm^2 .The preferred composition for obtaining better corrosion resistance is 87.8Zn-12.2 wt% Mgbeyond the 15.8% Mg the corrosion resistance falls down[6]. In the preferred composition of the Mg content in the Zn composition coating shows better corrosion resistance than the pure Zn coating [6].

The Zn-5wt%Al is coated on the mild steel substrate by using sputter deposition.Due to the different temperatures employed for the substrates, the thickness is coated on the sample is differed as at low temperature 3μm and at high temperature 30μm, and the microstructure also varies.As at high temperature an open woolen-like microstructure is shown and at lower temperature close and compact microstructure is observed.The samples are tested in 3.5wt% NaCl to check the corrosion resistance.It shows that the corrosion resistance is mainly depends on the number of Zn atoms present on the surface of the substrate [7]. The mild steel is coated with the CrAlSiN by varying the Si content from 0 to11.5wt%. The CrAl interlayer layer is deposited with a thickness of 100nm in presence of the sputter and reactive gas- Ar and nitrogen with a flow rate of 20 and 10 sccm. The Si is added with varying the compositionand deposited with a thickness of 1.2μm. The samples are treated in the 3.5wt% NaCl. The Si content with 8.7wt% possesses high hardness and corrosion resistance. With further increasing the Si content in the coating the hardness and corrosion properties are decreased [8].

2.2. Aluminum alloy coating on the magnesium substrate:

By using thermal spray method the Al/SiC matrix composite is coated on the Mg substrate. The Sic particle distance is average 50μm and standoff distance is 15-20cm.Compared to mono layer multi-layer coatings provides a better corrosion resistance due to low porosity. In this technique sand off distance influence the corrosion resistance of the coating. It was confirmed by 98% improvement in corrosion resistance of the coating in 3.6 wt% NaCl solution. by reducing the standoff distance from 20cm to 15cm[9].By using the magnetron sputtering coating method the coating materials aluminum, aluminum nitride, diamond-like carbon are coated on Mg alloy.The coating deposited as three layers. The Ar flux maintained at 40 sccm in chamber. At first, the pure aluminum is coated on the Mg substrate at sputter current of 2 A in presence of base pressure of 2.2×10^{-3} Torr and the chamber temperature is 35°C.This acts like an intermediate layer between the substrate and AlNcoating. In the later stage, AlN is deposited on the surface and in the final stage, the Diamond-like carbon(DLC) coating is applied on the substrate with an sputter current of 0.2 A. Now it was tested in 3.5 wt% NaCl solution by immersing the coated and uncoated samples for 6hr. The results obtained from the immersion test are for uncoated sample$E_{corr} = -1.57V \& i_{corr} = 2.25 \times 10^{-5} Acm^{-2}$andfor coated sample is $E_{corr} = -1.48V \& i_{corr} = 1.28 \times 10^{-6} A\, cm^{-2}$.The multi-layer coated Mg substrate possesses good corrosion resistance[10]. Now the AlN coatings are applied on the Mg alloys like AZ31and AZ91 by DC magnetron sputtering. The coatings are applied on the substrate with bias voltage of -60V, pressure 0.4pa, magnetron current 5A and standoff distance 90mm. Al interlayer is used for more adhesion and hard coating, so AlN is applied as an interlayer. The AlN coated AZ91 alloy showed the i_{corr}=3mA/cm^{-2} and E_{corr}=-725mV almost [11]. The thickness of Al layer deposited on the substrate is 250nm. The i_{corr} values of Al coated AZ31 substrate material is $7.155\mu Acm^{-2}$. The corrosion formation on the surface is mainly due to the dissolution of the OH^- ions in 3.5wt% NaCl by situating the Cl^- ion on the coating surface [12].

2.3. Ti alloy coatings on Al substrate:

Ti-6Al-4V is coated on the 6061-7651 aluminum alloys by using cold spray technique. In this techniqueDe-Laval nozzle (i.e. converging and diverging) type of spray gun is used for impinging Ti-6Al-4V on Al substrate by maintaining the following parameters. Supersonic speeds are achieved by accelerating particles with expansion ratio of 5.6 and standoff distance is 30mm. The feed rates of the Ti-6Al-4V are 40g/min through 15 passes and 2 mm thickness coating is achieved. In this method there is 56% reduction of corrosion current density(i_{corr}) of the coated sample with uncoated one [13]. Now the Al 7075-T6 disk shaped substrate is coated with TiN by using Dc magnetron sputtering. As the Ti is deposited as an intermediate layer for increasing adhesiveness of the coating.The samples are tested by exposing it to 5 wt% NaCl for 72 hr. The E_{corr} for 25wt% N in TiN is 1.6V. Bias voltage influence the corrosion resistance of the coating bias voltage value-150 gives the better corrosion resistance of the coating in 3.5% NaCl solution[14]. Al-W is coated on 7075 alloy by using sputter coating technique.Al-W coatings corrosion resistance is depends on W concentration. DarjaKekMerl confirmed that 3.5% of W in Al-W coating proved a better corrosion resistance when compred with 0.6% and 11%.The concentration of W gives better corrosion resistance up to some limit he also confirmed that if W concentration greater than or equal to 11% shows lower corrosion resistance when compared with uncoated sample in 5%NaCl solution it was confirmed by oxide layer formed in corrosion test[15].The $N_2$with different compositions(0.035wt%, 0.13wt%, 1wt%) is deposited by magnetron sputtering on AA7075 Al alloy. The Al-0.43wt%N compositionshown good corrosion potential of -0.35V when compared to bare substrate has -0.75V [16].The CrN and Ni/Cr-N is deposited on the AA7075 alloy with N flow rate is $9.5cm^3$. The thickness of the coating formed is 2.5μm for CrN and 3μm for Ni/Cr-N. The samples are tested in 0.1M NaCl. The corrosion properties on the CrN on AA7075 are not good as the steel coated with CrN. The observed values of the Ni/Cr-N on AA7075 alloy is E_{corr}= -0.645V and i_{corr}= $0.41\mu Acm^{-2}$ and for CrN, the i_{corr}=0.99μA/cm^2 and E_{corr}=-0.679V [17].

2.4. Coatings on the 316L stainless steel:

TiO_2 and $Al_2O_3$are deposited on the 316L stainless steel by arc plasma process method. Corrosion resistance is observed by the bond strength parameter. The bond strength between 316L steel and the TiO_2 is 40.607N/mm^2 and for $Al_2O_3$26.639N/mm^2. The bond strength of TiO_2 is 1.5 times than that of the Al_2O_3 [18,19]. The bond strength Al_2O_3 coating reduced due to the steady state crystal transition of α-$Al_2O_3$to the metastable phase of Y-Al_2O_3 shows erosion with much pores and cracks results lower corrosion resistance of $Al_2O_3$when compared with TiO_2[19]. By using HVOF(High velocity Oxygen-Fuel) technique 316L stainless steel is coated on ZE41 Mg alloy. Usually the thickness of the coating material is in between the range of 42.418μm. Spraying distanc for the 3 specimens is 1 layer is deposited from 200mm for SS-1 , for SS-2 is 400mm for 3 layers and 4 layers at 300mm distance for SS-3 Table-4 Shows corrosion current densities of ZE41 Mg coating on stainless steel substrate [20].

Table 2. Corrosion current densities of ZE41 Mg coating on stainless steel substrate [*Ref20*]

Substrate	Coating material	i_{corr}(μA/cm^2)
ZE41 Mg alloy	Uncoated	17.9
ZE41 Mg alloy	SS-1	2368.8
ZE41 Mg alloy	SS-2	3.19
ZE41 Mg alloy	SS-3	4.62

316L stainless steel is coated with Ti by using the sputtering method with vacuum pressure of 5×10^{-5}mbar by maintaining the thermal time plasma oxidation of 15,45,75min.The working gas is O_2 as 25% and H_2 of 75% with a pressure of 2.6mbar used after coating; the samples are cooled in nitrogen atmosphere. Coating formed at 45 minutes of oxidation time gives better wear resistance under load 1N 100rpm speed. Corrosion and hardness of the TiO_2coating depends on nitrating range of $7.9 \leq N \leq 32$.High hardness and corrosion resistance obtained at 28%-30%N and 23%. At 1N load TiO_2-45 showed a wear rate of $0.024\times10^{-3}mm^3/Nm$ [21].

The CrSiN coatings are deposited on the 316L stainless steel by using magnetron sputtering.The thickness of the deposited Cr interlayer is 0.4μm and for CrSiCN is 2μm.The corrosion tested are conducted in 3.5wt% NaCl solution, wear on ball-on-disk. With the Si content of 25-30wt% have better corrosion resistance than the CrN, because the Si settles over the cracks and pores formed on the surface of coating [22]. The 25wt% CrSiCN coating posses $E_{corr} = -0.19V$ &$i_{corr} = 5.2 \times 10^{-9}Acm^{-2}$ [22].

2.5. Other coating techniques:

Deposition of TiN and TiAlN on the 316L stainless steel by PEALD(Plasma Enhanced Atomic Layer Deposition) reveals that the corrosion resistance of bare substrate is low followed by the TiN when compared TiAlN because of the high corrosion current density. The corrosion current density for bare sample is 1.5μ Acm^{-2}, for Tin coated is 1.35μ Acm^{-2} , and for TiAlN coated sample is 0.12μAcm^{-2}[23]. Ni coated at different temperature on low carbon steel by using HVAF technique. Ni coated at 1520°C reduces the i_{corr}value by 33% when coated at 1490°C.It is concluded that high in-flight particle temperature leads to better corrosion.By using this technique we can develop Cr_2C_3-NiCr and Ni/Cr_2C_3-NiCr coatings on low carbon steel when compare the two coating second one give the better corrosion resistance this was confirmed by 77% reduction of i_{corr}value. These two coatings also provide better corrosion résistance when compared with Ni coating. The corrosion resistance of the samples are shown as Substrate <Ni(2) < Ni(1) <Cr_2C_3-NiCr< Ni/Cr_2C_3-NiCr resistance [24].

3. Conclusions:

From the above study, the conclusions drawn are as follows:

- As the number of layers of the coating material increased, porosity decreases which leads to better corrosion resistance.
- Variation in the weight percentage of the composition determines the corrosion current densities and polarization resistance.
- Inflight temperature is another vital parameter that governs the corrosion resistance of the material.

- Stand-off distance also decides the thickness of the coating material which ultimately leads to the enhancement of the corrosion resistance. Decrease in the standoff (i.e spraying distance) increases corrosion potential.

References:

[1] G. H. Koch, J. A. Beavers, And W. E. Berry, Marine Corrosion, Treatise On Materials Science And Technology, Vol. 28
[2] NACE International's IMPACT Breaks New Ground in the Study of Corrosion Management, 2017
[3] NuriaEspallargas, Future Development of Thermal Spray Coatings Types, Designs, Manufactureand Applications
[4] O.P. Oladijo, M.H. Mathabatha, A.P.I. Popoola, T.P. Ntsoane, Characterization and corrosion behaviour of plasma sprayed Zn-Snalloycoating on mild steel, 2017
[5] M.H. Mathabatha, A.P.I. Popoola, O.P. Oladijo, Residual stresses and corrosion performance of plasma sprayed zinc-based alloy coating on mild steel substrate, 2016
[6] JoungHyun La, SangYul Lee, SeokJun Hong, Synthesis of Zn–Mg coatings using unbalanced magnetron sputteringand theirs corrosion resistance, Surface & Coatings Technology 259 (2014) 56–61
[7] M.A. Baker 1,W. Gissler *, S. Klose 2, M. Trampert, F. Weber, Morphologies and corrosion properties of PVD Zn–Al coatings, Surface and Coatings Technology 125
[8] C.H. Lin, J.G. Duh, Electrochemical impedance spectroscopy (EIS) study on corrosion performance of CrAlSiN coated steels in 3.5 wt.%NaCl solution, Surface & Coatings Technology 204 (2009) 784–787
[9] M. Campo , M. Carboneras, M.D. López, B. Torres, P. Rodrigo, E. Otero, J. Rams, Corrosion resistance of thermally sprayed Al and Al/SiC coatings on Mg, Surface & Coatings Technology 203 (2009) 3224–3230
[10] Guosong Wu, Wei Dai, He Zheng, Aiying Wang, Improving wear resistance and corrosion resistance of AZ31 magnesium alloy byDLC/AlN/Al coating, Surface & Coatings Technology 205 (2010) 2067–2073
[11] HikmetAltun*, Sadri Sen, The effect of DC magnetron sputtering AlN coatings on the corrosionbehaviour of magnesium alloys, Surface & Coatings Technology 197 (2005) 193– 200
[12] Guosong Wu ⁎, XiaoqinZeng, Guangyin Yuan, Growth and corrosion of aluminum PVD-coating on AZ31 magnesium alloy, Materials Letters 62 (2008) 4325–4327
[13] Wen Sun, Adrian Wei Yee Tan, IulianMarinescu, Wei QuanToh, Erjia Liu, Adhesion, Tribological and Corrosion Properties of Cold-Sprayed CoCrMo andTi6Al4V Coatings on 6061-T651 Al Alloy, 2017
[14] M. Diesselberg, H-R.Stock, P. Mayr, Corrosion protection of magnetron sputtered TiN coatings deposited on high strength aluminium alloys, Surface and Coatings Technology 177 –178 (2004) 399–403
[15] DarjaKekMerl, Peter Panjan, JanezKovac, Corrosion and surface study of sputtered Al–W coatings with a rangeof tungsten contents, Corrosion Science 69 (2013) 359–368
[16] H. Schafer, H.R. Stock, Improving the corrosion protection ofaluminium alloys using reactivemagnetron sputtering, Corrosion Science 47 (2005) 953–964
[17] DarjaKek Merl1, Ingrid Milo{ev1, Peter Panjan1, Franc Zupani~2, morphology and corrosion properties pvdcr-n coatings deposited on aluminium alloys, issn 1580-2949, mtaec9, 45(6)593(2011)

[18] Shinichiro Adachi ⇑, Nobuhiro Ueda, Surface hardness improvement of plasma-sprayed AISI 316L stainlesssteel coating by low-temperature plasma carburizing, Advanced Powder Technology 24 (2013) 818–823

[19] Yuzhen Wang a, Fen Gao a, Jianqiao Yang b, Yitong Zhu a,Changqing Fang Shuzhong Wang b, Gaoyang Zhao a, Comparative study on corrosion characteristics ofAl2O3/316L and TiO2/316L stainless steel insupercritical water, i n t e rna t i o n a l journa l o f hydrogen energy xxx (2 0 1 7) 1 e7

[20] S. García-Rodríguez, A.J. López⁎, B. Torres, J. Rams, 316L stainless steel coatings on ZE41 magnesium alloy using HVOFthermal spray for corrosion protection, Surface & Coatings Technology 287 (2016) 9–1

[21] M. Alvarez-Vera a,n, H.M.Hdz-García a, R.Muñoz-Arroyo a, J.C.Diaz-Guillen a, A.I.Mtz-Enriquez, J.L.Acevedo-Dávila, TribologicalstudyofathinTiO2 nanolayercoatingon316Lsteel, Wear376-377(2017)1702–1706

[22] Peter A. Dearnley⁎, Giles Aldrich-Smith, Corrosion–wear mechanisms of hard coatedaustenitic 316L stainless steels, Wear 256 (2004) 491–499

[23] A. Fossati *, F. Borgioli, E. Galvanetto, T. Bacci, Corrosion resistance propertiesof glow-discharge nitrided AISI 316Laustenitic stainless steel in NaCl solutions, Corrosion Science 48 (2006) 1513–1527

[24] ZhiweiWu,abcFei Zhou,*abQiangMa,abQianzhiWang,abZhifengZhoudand Lawrence Kwok-Yan Li, Tribological and electrochemical properties ofCr–Si–C–N coatings in artificial seawater, RSC Adv., 2016, 6, 76724

Analysis on Tribological Characteristics of Waste Ayurvedic Oil Biodiesel Blends using Four Ball Tribometer

R. Balakumar [a],*, G. Sriram [b], S. Arumugam [c], A. Shree Navin [d]

[a, b, c] Department of Mechanical Engineering, [d] Master in AeroMat-Innovation,

[a, b, c] Sri Chandrasekharendra Saraswathi Viswa Mahavidyalaya,

Enathur, Kanchipuram – 631 561. Tamilnadu, India.

[d] Institute Mines Telecom Mines Albi, France.

[a,] *balavhd@gmail.com, [b]drg.sriram@gmail.com, [c]aru_amace@yahoo.co.in, [d]shreenavin.a92@gmail.com

Abstract. Tribology related problems in an engine give challenges to the automobile sector. To minimize the issues raised by diesel fuel, the biodiesel gains a vertical attention by the researchers. Especially, the waste oils will lead a high pollutant material to the environment and it needs a proper procedure to disposal. In this present study it is aimed to investigate the friction and wear characteristics of Waste Ayurvedic Oil (WAO) biodiesel blends by varying the applied load in a four ball wear test Tribotester. The investigations of fuels were WAO biodiesel blends (B10, B15 and B20) and diesel fuel was analysed with test conditions viz: load 40Kg and 80Kg, constant temperature 75°C, duration: 300 sec, sliding speed: 1200rpm for all the tested fuels. For the given experimental conditions more wear scar diameter were observed for diesel than WAO biodiesel blends for both the loads. The coefficient of friction and frictional torque were absorbed less with WAO15 for both the loads than other blended fuels and diesel fuels. The worn surfaces of the steel balls were examined by a Scanning Electron Microscope (SEM). The morphology image results showed that minimal friction, wear and smoother surface were absorbed for all the biodiesel concentrations than the diesel fuel.
Keywords: WAO Biodiesel, Wear, Friction, load variation, Four Ball Tribometer.

1. Introduction

The ever growing environmental issues and fastest depletion of petroleum based fossil fuels leads the researchers to work out in the way of alternate fuels [1, 2]. In this scenario the energy supply by fuel to the vehicles is not only the important point as well as it should lead less emission and give good lubricity without changing the atmospheric condition [3]. Many researchers investigated the biodiesel in terms of conversation and Tribology from neat vegetable oils, waste cooking oil, waste transformer oil and animal fats [4-8]. Mostly the biodiesels are made from edible vegetable oil sources in all over the world. There is equal weightage and importance given for edible oils for human nutrition's purpose. The vegetable oils used in engines will be good for present research but this may lead more important than petroleum products in near coming years. So, it is necessary to replace the input oil source with low cost or price less oil sources viz waste oils like cooking oil and animal fats etc [4, 5].

There are many studies were undertaken to investigate the production of biodiesel from waste oils and animal fats [6]. In the engines the temperature inside the cylinder will be more and it will affect the lubricity of the fuel. This may lead more friction in the mating parts inside the engine [7]. In-fact the commercially available petroleum based fossil fuel supplies poor lubricity. So, from the above issues it leads to search an alternate fuel for diesel engines called biodiesel. The biodiesel is a renewable and it can be generated from many vegetable oil sources, animal facts and waste oils [8, 9]. Nabi et al. [10] investigated the waste transformer oil (WTO) as a biodiesel blends (B10, B15 B20) in a single cylinder diesel engine. From the test results the WTO biodiesel blend (B20) produce high brake thermal efficiency, less fuel consumption and reduced exhaust emission than other biodiesel blends and diesel fuel. Biodiesel supplies some good advantages than diesel fuel like reduction in emission, biodegradability, higher flash point, good cetane number etc [11].

From all the above, the impact of friction and wear by the diesel or the biodiesel will affect the engine component life cycle. The wear and friction properties of the alternate fuel as well as the diesel were studied by many researchers using four ball wear test tribometer, High Frequency Reciprocating Rig and pin on disc tribometer. Fazal et al. [12] investigated the friction and wear characteristics of palm oil biodiesel using four ball wear tester for the different biodiesel blends viz. B10, B20, B50, B100 with a input parameters like speed: 600, 900, 1200,1500 rpm, load: 40Kg, temperature: 75°C. From the investigation results the increase of biodiesel blends and speed will decrease the wear and friction characteristics. If the temperature increases then the wear and friction will increase but at the same time the biodiesel blends will reduce the wear and friction. Kumar N and Chauhan S [13] analyzed the jatropha oil methyl ester (B20, B40 and B100) in a four ball wear tester with a speed: 1500 rpm, duration: 1 hour, temperature: 45, 60, 75°C and load: 40, 50, 60 Kg. There was a positive impact were observed by reduction in load, temperature and increase in biodiesel blends. Basically the biodiesels have high viscosity this will form a stable film between the mating parts and it will reduce the wear and friction between the liner and ring in the engine [14].

My previous study investigates the tribological characteristics of Waste Ayurvedic Oil (WAO) biodiesel blends (B10, B15, and B20) along with diesel as a lubricity enhancer using Four Ball Wear Tribometer. The co-efficient of friction and wear scar diameter were observed less for the blend WAO (B15) than other blended fuel and diesel fuel. From the SEM results the abrasive and adhesive were absorbed less for B15 than other fuels [15]. From the many number of researches conducted on the lubricity of various biodiesel and its blends none of them touched in the area of Waste Ayurvedic oil biodiesel blends by varying the loads (40kg and 80kg) in a four ball wear tester.

2. Materials and methods

2.1 Synthesis of Waste Ayurvedic Oil Biodiesel production and blend preparation

After one use of ayurvedic oil becomes waste and it is highly difficult to dispose. The waste ayurvedic oil (after the treatment) was procured from Sri Jayendra Saraswathi Ayurveda College and Hospital, Chennai, India. The chemical used for transesterification process viz. methanol, sodium hydroxide, hydrochloric acid and anhydrous sodium sulphate was purchased from M/S Ganapathy Trading Company, Chennai. The oil contains more number of unwanted impurities and it need to be filtered. The figure 1shows the WAO filtration process.

Figure 1 WAO filtration setup

The waste ayurvedic oil was converted as waste ayurvedic oil methyl ester via transesterification process. The 400ml of methoxide (methanol and NaOH) and 1400ml of WAO both were mixed with constant rotation and heated in the mantle for one hour at a temperature of 60°C. Then the mixture was poured in a separating funnel to separate the glycerol and methyl ester. Then the methyl ester was separated and heated above 60°C to remove excess methanol. Finally the bubble was wash was carried out to remove the left out catalyst and glycerol. At final the blends were prepared by mixing with the diesel fuel i.e. 90% of diesel fuel with 10% WAO Methyl ester referred as WAO10, 85% of diesel fuel with 15% WAO Methyl ester referred as WAO15 and 80% of diesel fuel with 20% WAO Methyl ester referred as WAO20. The properties of biodiesel and diesel were considered from balakumar et al. [15].

2.2 Tribological test on four ball wear Tribotester

To analyse the wear and frictional characteristics of WAO biodiesel blends the four ball wear tester was used. The stainless steel balls were used for this analysis. Before testing the four balls need to be cleaned thoroughly by using toluene and wiped with wet tissues. Three balls were tightly fixed in the steel cup at the bottom and fourth ball is kept at the top of rotating spindle using chuck. The test was conducted as per the standard ASTM D4172 for wear and ASTM D5183 for frictional characteristics.

The lower three balls were filled with test fuel. The analysis was carried under a test conditions viz. load: 40kg and 80kg, constant temperature: 75°C, duration: 300 seconds, sliding speed: 1200 rpm (constant speed). The computer was interconnected with the four ball machine to analyse the wear and friction. After the test the wear scar image and wear scar diameter was measured using the optical microscope for the lower three balls. The upper ball will not be subject to wear scar image analysis. The change in the wear scar diameter depends on the applied load and the friction.

3. Results and discussion

3.1 Wear scar diameter analysis

The figure 2 (a - h) shows wear scar diameter (WSD) of different WAO biodiesel blended fuel and diesel fuel for both applied loads. The oxidation rate during the experiment will be effect on the wear scar diameter. The WSD was observed less for all WAO biodiesel blends than the diesel fuel for both the loads.

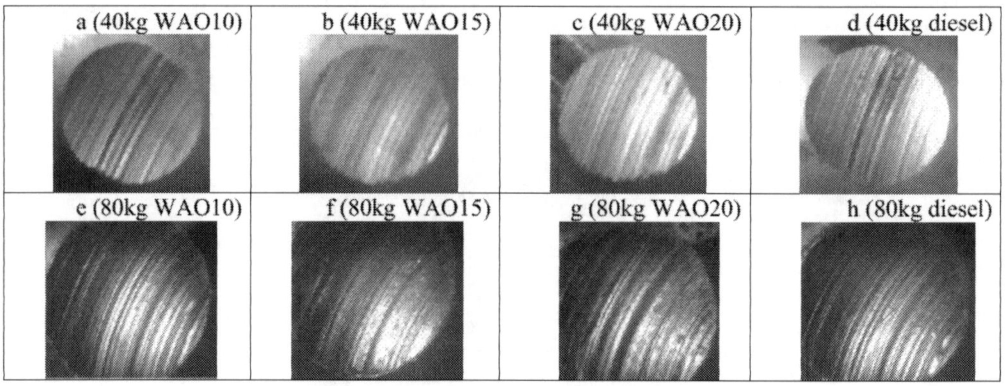

Figure 2 Optical microscope image of the tested steel ball

The WSD were noted less for WAO15 for both loads than other WAO blended fuel and diesel fuel. In general, if the applied load increases then the pressure applied on the balls will tends to increase the WSD. For the 40kg applied load the WSD value for WAO15 is 0.392mm for other bended biodiesel fuel and diesel WSD varies between 0.439 to 0.491mm. Similarly, for higher load the least WSD was absorbed for WAO15 as 2.458mm which is better than other blended fuels and diesel fuel. Similarly largest WSD was observed for 80kg applied load which is higher than the 40kg load. For the applied load WAO15shows less wear scar diameter than the other blended fuel and diesel fuel. Marginally the scar diameter was measured for WAO15 is 2.458 mm. At the same time the WSD value for WAO20, WAO10 and diesel were measured as 2.490, 2.564 and 2.682 mm which is higher than WAO15. The test results were consistent with the results by Syahrullail et al. [16].

3.2 Wear and friction characteristics

The figure 3 (a & b) shows the time and frictional torque characteristics for WAO biodiesel blends and diesel fuel for 40kg and 80kg. After few minutes the frictional torque reached the stable condition for all the tested fuel and at all the test conditions. The frictional torque was observed less for all the WAO biodiesel blends than the diesel fuel.

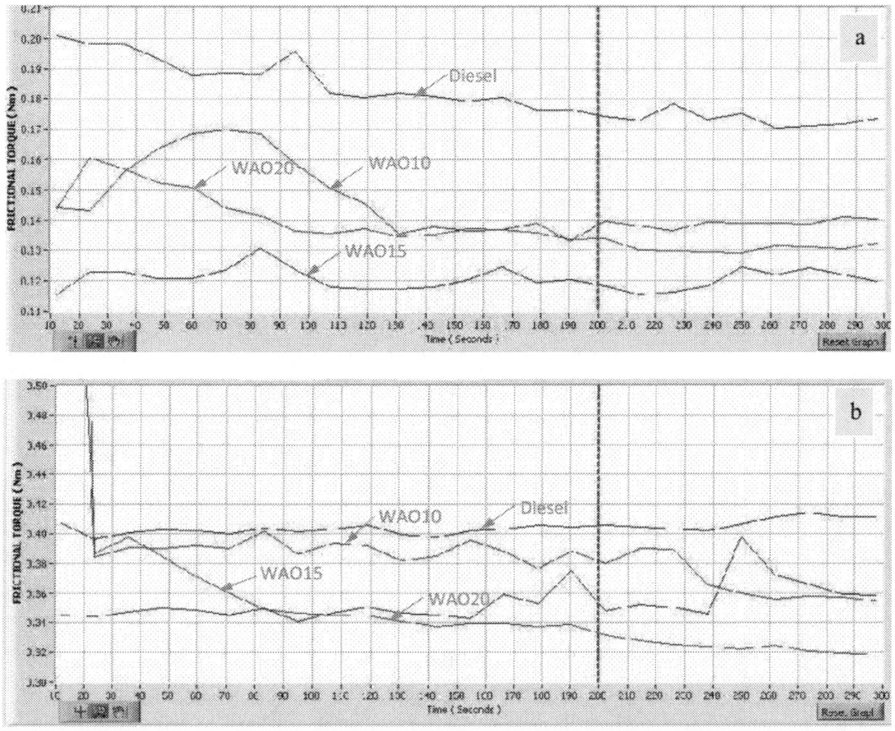

Figure 3 Frictional torque versus time for (a) 40kg, (b) 80kg.

The figure depicts that the frictional coefficient was decreasing with the increase in time for all the tested fuels. For 40kg applied load the WAO15 marginal reduction in frictional torque was observed i.e. 0.117Nm to 0.13Nm. However the frictional torque for WAO20 and WAO15 shows slightly higher than the WAO15 and lower than diesel fuel the frictional torque value varies between 0.13Nm to 0.17Nm. At the same time the diesel fuel shows higher value of frictional torque than the biodiesel blends which falls between 0.17Nm to 0.2Nm.

Similarly for higher load i.e. 80kg applied load also the WAO biodiesel blends shows reduction in frictional torque than the diesel fuel. Generally, for higher load the frictional torque will be high because the more pressure over the ball will be high. More fluctuations were observed at the beginning for all the fuels used. After 30 seconds all the blends the WAO biodiesel blends and diesel fuel become stable conditions. For the WAO15 the frictional torque fluctuates between 0.32 to 0.34Nm. At the same time higher level of frictional torque were observed for diesel fuel which lies between 0.40 to 0.42Nm. The other WAO biodiesel blends shows less frictional torque than diesel fuel and higher than WAO15. The test results consistent with the results from Habibullah et al. [17].

3.3 Coefficient of Friction

The figure 4 shows the graph between the different fuels and coefficient of friction (COF) value for 40kg and 80kg applied load. From the figure it is found that high coefficient of friction was observed at higher load.

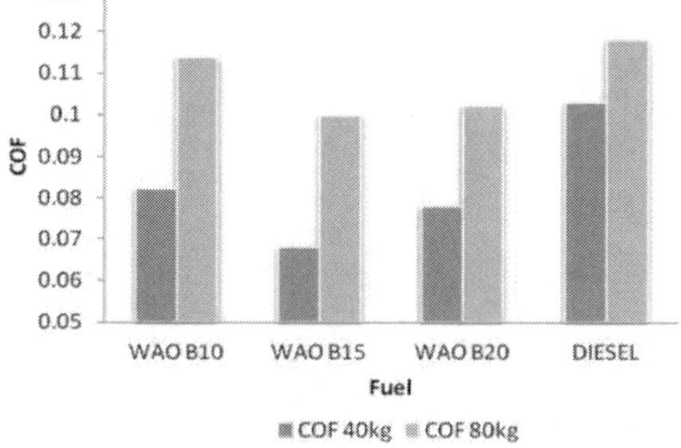

Figure 4 Co-efficient of friction with respect to fuel for 40kg and 80kg.

The variation frictional coefficient was minimal for all the fuels. From the graph it is found that for 40kg load the coefficient of friction were fluctuates between 0.06817 to 0.08244 for WAO biodiesel blends and for the diesel 0.10312 was absorbed which is higher than biodiesel blends. Similarly for 80kg the WAO biodiesel blended fuels the coefficient of friction varies between 0.09995 to 0.11386 and for the diesel fuel it is found higher than biodiesel fuel. The results were consistent with the results from wain et al [18].

3.4 SEM analysis of steel ball worn surfaces

Figure 5 (a-h) shows the SEM micrographs of worn surfaces of the steel balls via WAO10, B15, B20 and diesel fuel for the loads of 40kg and 80kg. From the analysis of worn surfaces of tested steel balls it shows that the sizes of the particles removed from cavities of the tested worn balls are much lesser for WAO15 followed by WAO10, B20 and diesel fuel.

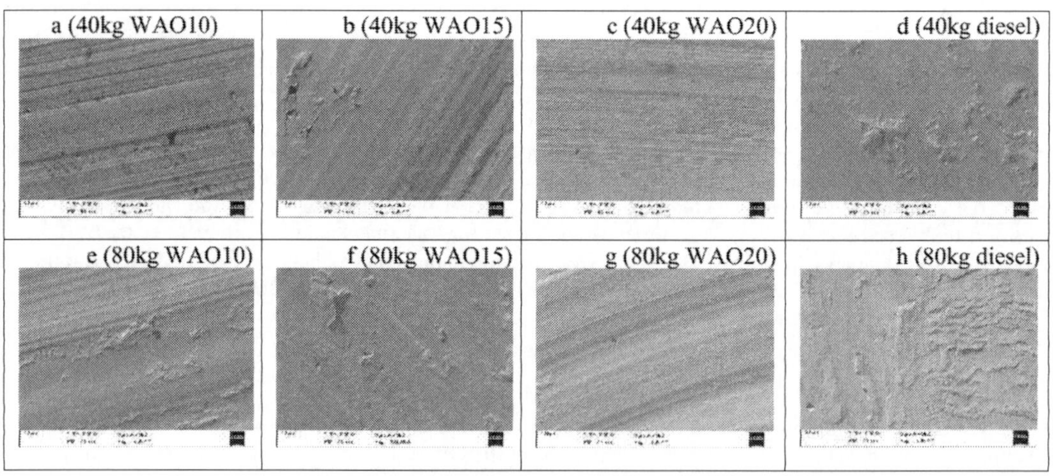

Figure 5 SEM image of tested steel ball

The deformation of the wear surface of steel balls was severe for diesel fuel. Further the wear scar was less for all the biodiesel blends than the diesel fuel. The surface morphology demonstrates that the edge of the surface at the diesel fuel is plastically elongated which gives cracks and wear debris. From the definition removal of wear debris more than 20μm is called as adhesive wear and less than 20μm is called as abrasive wear. From the image morphology the diesel fuel falls in the category of adhesive wear and the WAO biodiesel blends were falls in the abrasive wear. From the image the layer of metals from the worn surfaces was excluded out in the direction of sliding of the test balls. Although the extrusion of metal is less for WAO15 than other blended fuels and diesel fuel. The wear debris and the deformation were less for WAO15 followed by WAO10, B20 and diesel fuel. Especially in diesel fuel very micro level cracks were occurred than blended biodiesel. No severe wear and damages can be seen in WAO15 than other blended fuel and diesel fuel. Some regular, thin and radial scratches were observed on the tested surface which results abrasive wear. Finally the morphology shows that less wear was observed with all the biodiesel concentrations than the diesel fuel. The test results consistent with the results from Habibullah et al. [17].

Conclusion

From the investigation results the following were concluded

- The lubricity of the WAO biodiesel blends gives good lubricity which leads for reduction in wear and friction for both the applied loads.
- The WSD increases with increase of load which leads increase of wear. The WSD for WAO15 biodiesel blend at 40kg and 80kg load was 9.82% and 22.4% better than the diesel fuel. The WSD was considerable better for WAO15 than other blended biodiesel and diesel fuel.
- The lubricity in terms of wear and friction were decreased for WAO15 than other blended fuel and diesel fuel. The frictional torque value fluctuates between 0.117Nm to 0.13Nm for WAO15 which is marginally lesser than other blended fuels and diesel fuels.
- The coefficient of friction for WAO15 at 40kg and 80kg applied load was 3.5% and 1.8% better than diesel fuel. The increase in applied load will act more pressure on the balls which leads the increase in friction.
- The surface morphology of the tested steel balls revealed that more wear debris was found for diesel fuel which falls in adhesive wear. But for all the concentrations of biodiesel less wear were absorbed and falls in abrasive wear.
- From the investigation results it is suggested that WAO15 exhibits better lubricating performance in terms of friction and wear for both the applied loads and hence it can be used as a engine fuel which leads the engine life.

Reference

[1] Restrepo-Flórez, J.-M., Bassi, A., Rehmann, L., Thompson, M.R., (2014). Investigation of bio film formation on polyethylene in a diesel/biodiesel fuel storage environment. *Fuel* 128,240–247.

[2] Silitonga, A.S., Masjuki, H.H., Mahlia, T.M.I., Ong, H.C., Chong, W.T., Boosroh, M.H., (2013). Overview properties of biodiesel diesel blends from edible and non-edible feedstock. *Renewable Sustain. Energy Rev.* 22 (0), 346–360

[3] Tung SC, McMillan ML. (2004) Automotive tribology overview of current advances and challenges for the future. *Tribol Int*;37:517–36.

[4] Girotto, F., Alibardi, L., Cossu, R., (2015). Food waste generation and industrial uses: a review. *Waste Manage*. 45, 32–41.

[5] R. Balakumar, G. Sriram, and S. Arumugam Waste Ayurvedic Oil as an Engine Fuel—A Review. (2018) *Adv. Sci. Eng. Med.* 10, 249–254

[6] Bhatti, H.N., Hanif, M.A., Qasim, M., Rehman, A., (2008). Biodiesel production from waste tallow. *Fuel* 87, 2961–2966.

[7] Sarvi A, Fogelholm CJ, Zevenhoven R. (2008) Emissions from large-scale medium speed diesel engines: 2 – Influence of fuel type and operating mode. *Fuel Process Technol*; 89:520–7.

[8] Demirbas A. (2009) Biodiesel from waste cooking oil via base-catalytic and supercritical methanol transesterification. *Energy Convers Manage*;50: 923–7.

[9] Ushakov S, Valland H, Aesoy V. (2013) Combustion and emissions characteristics of fish oil fuel in a heavy-duty diesel engine. *Energy Convers Manage*; 65: 228–38.

[10] Md uran abi, Md shamin akhter, Md atiqur rahman, (2013) aste transformer oil as an alternative fuel for diesel engine. *Procedia Engineering*: 56: 401-406.

[11] Ozsezen AN, Canakci M (2011). Determination of performance and combustion characteristics of a diesel engine fueled with canola and waste palm oil methyl esters. *Energy Convers Manage*; 52:108–16.

[12] Fazal, M.A., Haseeb, A.S.M.A., Masjuki, H.H., (2013). Investigation of friction and wear characteristics of palm biodiesel. *Energy Convers. Manage*. 67 (0), 251–256.
[13] Kumar, N., Chauhan, S., (2014). Analysis of tribological performance of biodiesel. *Proc. Inst. Mech. Eng., Part J: J. Eng. Tribol.*, http://dx.doi.org/10.1177/ 1350650114532452.
[14] Syahrullail, S., Zubil, B.M., Azwadi, C.S.N., Ridzuan, M.J.M., (2011). Experimental Evaluation of Palm Oil as Lubricant in Cold Forward Extrusion Process, *International Journal of Mechanical Sciences*, 53, pp.549-555.
[15] R. Balakumar, G. Sriram, S. Arumugam, (2018) Investigation of Friction and Wear Characteristics of Waste Ayurvedic Oil Biodiesel Blends, *Advanced Science, Engineering and Medicine* Vol. 10, 1–4.
[16] Syahrullail, S., Wira, J.Y., Wan Nik, W., Fawwaz, W., (2013). Friction characteristics of RBD. Palm olein using four-ball tribotester. *Appl, Mech. Mater*. 315, 936–940.
[17] M. Habibullah, H.H. Masjuki, M.A. Kalam, N.W.M. Zulkifli, B.M. Masum, A. Arslan, M. Gulzar (2015). Friction and wear characteristics of Calophyllum inophyllum biodiesel. *Ind. Crops and Products* 76: 188–197
[18] Wain KS, Perez JM, Chapman E, Boehman AL (2005). Alternative and low sulfer fuel options: boundary lubrication performance and potential problems. *Tribol Int*; 38:313–9.

Effect of second order slip and non-linear thermal radiation on mixed convection flow of MHD Jeffrey nanofluid with double stratification under convective boundary condition

K. Jagan[1,3], S. Sivasankaran[2,*], M. Bhuvaneswari[2] and S. Rajan[3]

[1]Department of Mathematics, Nandha Engineering College, Erode, Tamilnadu, India.
[2]Department of Mathematics, King Abdulaziz University, Jeddah 21589, Saudi Arabia.
[3]Department of Mathematics, Erode Arts & Science College, Erode, Tamilnadu, India.

Email : sd.siva@yahoo.com

Abstract. In this article, the effect of non-linear thermal radiation and second order slip velocity in MHD convective Jeffrey nanofluid flow over a stretching surface with convective boundary condition is investigated in the presence of thermal and concentration stratification. The ordinary differential equations are obtained by using suitable transformations in the governing equations. Homotopy analysis method is used to solve the ordinary differential equations. The velocity, temperature and concentration profile is computed for different pertinent parameters involved in the study. Also computation is done for skin friction coefficient and local Nusselt number and the results are discussed.

1. Introduction

The effect of thermal radiation on convective MHD viscoelastic fluid flow over a stretching sheet was analyzed by Eswaramoorthi et al. [1]. Stratification of the fluid is a deposition or formation of layers. The mixed convective micropolar fluid flow over a continuously moving vertical surface engrossed in a stratified thermally and solutally medium was discussed by Rashad et al. [2]. Hayat et al. [3] investigated about the unsteady nanofluid flow with magnetohydrodynamics and double stratification. The MHD nanofluid flow past a permeable stretching surface with second order velocity slip was examined by Zhu et al. [4]. The thermal radiation effect on magnetohydrodynamics stagnation point Jeffrey fluid flow of towards a stretching sheet was analyzed by Das et al. [5]. The thermal radiation effect on mixed convection Jeffrey nanofluid flow with thermal and solutal stratified medium was discussed by Abbasi et al. [6]. In this paper, the effect of second order slip and non-linear thermal radiation on magnetohydrodynamics mixed convection flow of Jeffrey nanofluid towards a stretching sheet with double stratification under convective boundary condition is investigated which is the extension of the paper Abbasi et al. [6].

2. Mathematical Formulation

The steady, incompressible and laminar Jeffrey nanofluid flow towards stretching surface with non-linear thermal radiation and convective boundary condition is considered. The stretching sheet velocity is taken as $u_{1,w} = a_1 x$ where a_1 are positive constants. The fluid flow is considered along y-axis with second order slip and magnetic field normal to the fluid flow. The thermophoresis and Brownian motion are considered due to the presence of nanoparticles. The ambient temperature and concentration are assumed as $T_\infty = T_0 + d_1x$ and $C_\infty = C_0 + d_3x$. The fluid temperature and concentration are considered as $T_w = T_0 + d_2x$ and $C_w = C_0 + d_4x$. The thermal and concentration stratification effects are also considered. The governing equations for this flow analysis can be constructed as

$$\frac{\partial u_1}{\partial x}+\frac{\partial v_1}{\partial y}=0, \tag{1}$$

$$u_1\frac{\partial u_1}{\partial x}+v_1\frac{\partial u_1}{\partial y}=\nu\,\frac{1}{\lambda_1+1}\frac{\partial^2 u_1}{\partial y^2}+\nu\,\frac{\lambda_2}{\lambda_1+1}\left[\frac{\partial u_1}{\partial y}\frac{\partial^2 u_1}{\partial x\partial y}+u_1\frac{\partial^3 u_1}{\partial x\partial y^2}-\frac{\partial u_1}{\partial x}\frac{\partial^2 u_1}{\partial y^2}+v_1\frac{\partial^3 u_1}{\partial y^3}\right]$$
$$-\frac{\sigma B_0^2}{\rho}u_1+[\beta_T(T-T_\infty)+\beta_C(C-C_\infty)]\,g, \tag{2}$$

$$u_1\frac{\partial T}{\partial x}+v_1\frac{\partial T}{\partial y}=\alpha\,\frac{\partial^2 T}{\partial y^2}+D_B\,\tau\left(\frac{\partial T}{\partial y}\frac{\partial C}{\partial y}\right)+\tau\frac{D_T}{T_\infty}\left(\frac{\partial T}{\partial y}\right)^2-\frac{1}{(\rho c_p)}\frac{\partial q_r}{\partial y}, \tag{3}$$

$$u_1\frac{\partial C}{\partial x}+v_1\frac{\partial C}{\partial y}=D_B\left(\frac{\partial^2 C}{\partial y^2}\right)+\frac{D_T}{T_\infty}\left(\frac{\partial^2 T}{\partial y^2}\right), \tag{4}$$

where the boundary conditions are

$$u_1=u_{1,w}+u_{1,Slip}=a_1x+A\frac{\partial u_1}{\partial y}+B\frac{\partial^2 u_1}{\partial y^2}, v_1=0, -k_T\frac{\partial T}{\partial y}=h_1(T_w-T), C=C_w \text{ at y=0},$$
$$u_1\to 0,\ T\to T_\infty,\ C\to C_\infty \text{ as } y\to\infty, \tag{5}$$

where A & B, α, β_C, β_T, c_p, D_B, D_T, g, k_T, λ_1 & λ_2 , ν, q_r, ρ & σ and τ are first and second order slip velocity factor, thermal diffusivity, concentration expansion co-efficient, thermal expansion co-efficient, specific heat, Brownian motion co-efficient, thermophoresis co-efficient, acceleration due to gravity, fluid thermal conductivity, ratio of relaxation to retardation & retardation time parameter, kinematic viscosity, radiative heat flux, fluid density and electrical conductivity and ratio between the effective nanoparticle materials and fluid heat capacity. The similarity transformations are

$$\eta=y\sqrt{\frac{a_1}{\nu}},\ u_1=a_1x\,s'(\eta),\ v_1=-\sqrt{a_1\nu}s(\eta), \theta(\eta)=\frac{T-T_\infty}{T_w-T_0}, \varphi(\eta)=\frac{C-C_\infty}{C_w-C_0} \tag{6}$$

Substituting (**6**) in (**1**), (**1**) is satisfied identically. Substituting (**6**) in (2) to (**5**), we get

$$s'''-(\lambda_1+1)\,[s'^2-s\,s"]-\beta\left[s\,s^{iv}-s"^2\right]+(\lambda_1+1)[Ri\,(\theta+N\,\varphi)-M^2\,s']=0, \tag{7}$$

$$\left(1+\frac{4}{3}Rd\right)\theta"+\frac{4}{3}Rd[(\theta_w)^3(3\theta^2(\theta')^2+\theta^3\theta")+3(\theta_w)^2(2\theta(\theta')^2+\theta^2\theta")]$$
$$+4Rd[\theta_w((\theta')^2+\theta\theta")]+Pr[s\,\theta'+Nb\,\theta'\,\varphi'+Nt\,(\theta')^2+s'\,\theta-S_T\,s']=0, \tag{8}$$

$$\varphi"+\text{Pr Le s}\,\varphi'+\frac{\text{Nt}}{\text{Nb}}\theta"-\text{Pr}\,Le\,s'\,\varphi-\text{Pr}\,Le\,S_C\,s'=0, \tag{9}$$

$$s(0)=0, s'(0)=1+\epsilon\,s"(0)+\epsilon_2\,s'''(0),\ \theta'(0)=-\alpha_T(1-S_T-\theta(0)),$$
$$\varphi(0)=1-S_C,\ s'(\eta)=0, \theta(\eta)=0, \varphi(\eta)=0, as\,\eta\to\infty, \tag{10}$$

where α_T, N, β, ϵ & ϵ_2, Le & Pr, M^2, Nd & Nt, Rd, Re_x, Ri, S_C, S_T and θ_w are thermal Biot number, buoyancy ratio parameter, Deborah number, first & second order slip velocity parameter, Lewis & Prandtl number, Hartmann number, Brownian motion and thermophoresis parameters, non-linear thermal radiation, local Reynolds number, Richardson number, nanoparticle volume fraction stratification parameter, thermal stratification parameter and temperature ratio parameter are defined as

$$\alpha_T = \frac{h_1\sqrt{v/a_1}}{k}, N = \frac{Gr^*}{Gr}, \beta = \lambda_2 a_1, \epsilon = A\sqrt{\frac{a_1}{v}}, \epsilon_2 = B\left(\frac{a_1}{v}\right), Le = \frac{\alpha}{D_B},$$
$$M^2 = \frac{\sigma B_0^2}{\rho a_1}, Nb = \tau\, v^{-1} D_B(C_w - C_\infty), Nt = \tau\, v^{-1} D_T\left(\frac{T_w - T_\infty}{T_\infty}\right), \mathrm{Pr} = \frac{\mu c_p}{k},$$
$$Rd = \frac{4\sigma^* T_\infty^3}{kk^*}, Re_x = \frac{u_w(x)\, x}{v}, \mathrm{Ri} = \frac{Gr}{Re_x^2}, \mathrm{S_C} = \frac{\mathrm{d}_3}{\mathrm{d}_4}, \mathrm{S_T} = \frac{\mathrm{d}_1}{\mathrm{d}_2}, \theta_\mathrm{w} = \frac{T_w - T_0}{T_\infty}.$$

The skin-friction co-efficient & local Nusselt number are

$$Re_x^{1/2} C_{f_x} = \left(\frac{2}{1+\lambda_1}\right)(1+\beta)\, s''(0), Re_x^{-1/2} Nu = -\left(1 + \frac{4}{3} Rd(\theta_w)^3\right)\theta'(0).$$

3. Method of the solution

The equations (7) to (10) are solved using HAM by choosing the initial approximation and auxiliary linear operators (refer Jagan et al. [7]) as

$$s_0(\eta) = \eta\, e^{-\eta} + \left(\frac{3*\epsilon_2 - 2*\epsilon}{\epsilon_2 - 1 - \epsilon}\right) * e^{-\eta} - \left(\frac{3*\epsilon_2 - 2*\epsilon}{\epsilon_2 - 1 - \epsilon}\right),$$
$$\theta_0(\eta) = -\alpha_T\left[\mathrm{S_T} * \eta - \left(\frac{1}{\alpha_T + 1}\right)\exp(-\eta)\right], \varphi_0(\eta) = -(\mathrm{S_C} - 1)\exp(-\eta).$$
$$L_s(s) = \frac{d^3 s}{d\eta^3} - \frac{ds}{d\eta}, \; L_\theta(\theta) = \frac{d^2\theta}{d\eta^2} - \theta, L_\varphi(\varphi) = \frac{d^2\varphi}{d\eta^2} - \varphi,$$

which satisfies the property

$$L_s[Q_1 + Q_2\exp(-\eta) + Q_3\exp(\eta)] = 0, \; L_\theta[Q_4\exp(-\eta) + Q_5\exp(\eta)] = 0,$$
$$L_\varphi[Q_6\exp(-\eta) + Q_7\exp(\eta)] = 0$$

where $Q_1, Q_2, ..., Q_7$ are the arbitrary constants. The h-curve is plotted for $\lambda_1 = 0.3$, $\beta = 0.2$, M = 0.7, Ri = 0.3, N = 0.3, Pr = 1, Le = 1, Nt = Nb = 0.2, Rd = 0.4, S_T= 0.1, S_C = 0.2, $\theta_w = 0.1$, $\alpha_T = 0.1$, $\epsilon = 0.2$ and $\epsilon_2 = 0.3$. The range for admissible values of h_s, h_θ and h_φ are $-1.4 \le h_s \le -0.2$, $-1.4 \le h_\theta \le 0$ and $-1.2 \le h_\varphi \le -0.1$ (see Figure1).

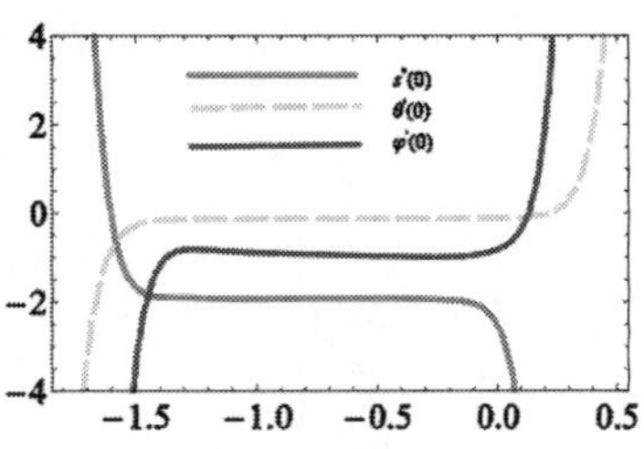

Figure 1: h-curve

4. Results and Discussion

Here the pertinent parameters are discussed with their different combinations which are involved in this study. From Figure 2, while increasing ϵ, the velocity profile gets decreased. It is also evident that the velocity profile enlarges with raise in second order slip parameter (ϵ_2) for $0 \le \eta < 1.675$ whereas it decreases with increase in ϵ_2 for $1.675 < \eta \le 5$, but, it remains same (0.12) for all the values of ϵ_2 at $\eta = 1.675$. The velocity profile increases with increase in ratio of relaxation to retardation time (λ_1)

for $0 \leq \eta < 0.7558$ whereas it decreases with increase in ϵ_2 for $0.7558 < \eta \leq 5$. But it remains same (0.50) for all the values of λ_1 at η = 0.7558. Also the velocity profile diminishes first and then rises with enlargement in β. In Figure 3, the temperature profile enlarge with rise in radiation parameter (Rd) and thermal Biot number (α_T) but decreases with increase in S_T. The concentration profile diminishes with rise in S_C. From Figure 4, the skin friction co-efficient enlarge with rise in ϵ and β whereas it diminishes with increase in ϵ_2 and M. It is concluded from Figure 5 that when ϵ_2 and α_T are increased, the local Nusselt number increases. Also the local Nusselt number diminishes with rise in ϵ and Rd.

5. Conclusion

The study of mixed convection MHD Jeffrey nanofluid flow with non-linear radiation and second order slip with thermal and solutal stratification under convective boundary condition is presented. The thickness of thermal boundary layer enlarges with rise in thermal radiation parameter. When thermal radiation parameter value is raised, the heat transfer rate diminishes where as local mass transfer rate enlarges. The skin friction enhances (diminishes) when first (second) order slip velocity is raised. While rising the second (first) order slip velocity, the heat & mass transfer rate enhances (decreases). The thickness of thermal boundary layer and local Nusselt number enlarges with rise in thermal Biot number. When the thermal (solutal) stratification is increased, the thickness of thermal (solutal) boundary layer diminishes.

References

[1] Eswaramoorthi S, Bhuvaneswari M, Sivasankaran S and Rajan S 2015 *Procedia Engineering* **127** 916-923.
[2] Rashad A.M, Abbasbandy S and Chamkha A.J 2014 *Journal of the Taiwan Institute of Chemical Engineers* **45** 2163-2169.
[3] Hayat T, Imtiaz M and Alsaedi A 2016 International *Journal of Heat and Mass Transfer* **92** 100-109.
[4] Zhu J, Zheng L and Zhang X 2015 Applied *Mathematics and Mechanics* **36** 1131-1146.
[5] Das K, Acharya N and Kundu P.K 2015 *Alexandria Engineering Journal* **54** 815-821.
[6] Abbasi F.M, Shehzad S.A, Hayat T and Alhuthali M.S 2016 *Journal of Hydrodynamics, Ser. B* **28** 840-849.
[7] Jagan K, Sivasankaran S, Bhuvaneswari M and Rajan S 2017 *International Journal of Pure and Applied Mathematics* **117(13)** 43-51.

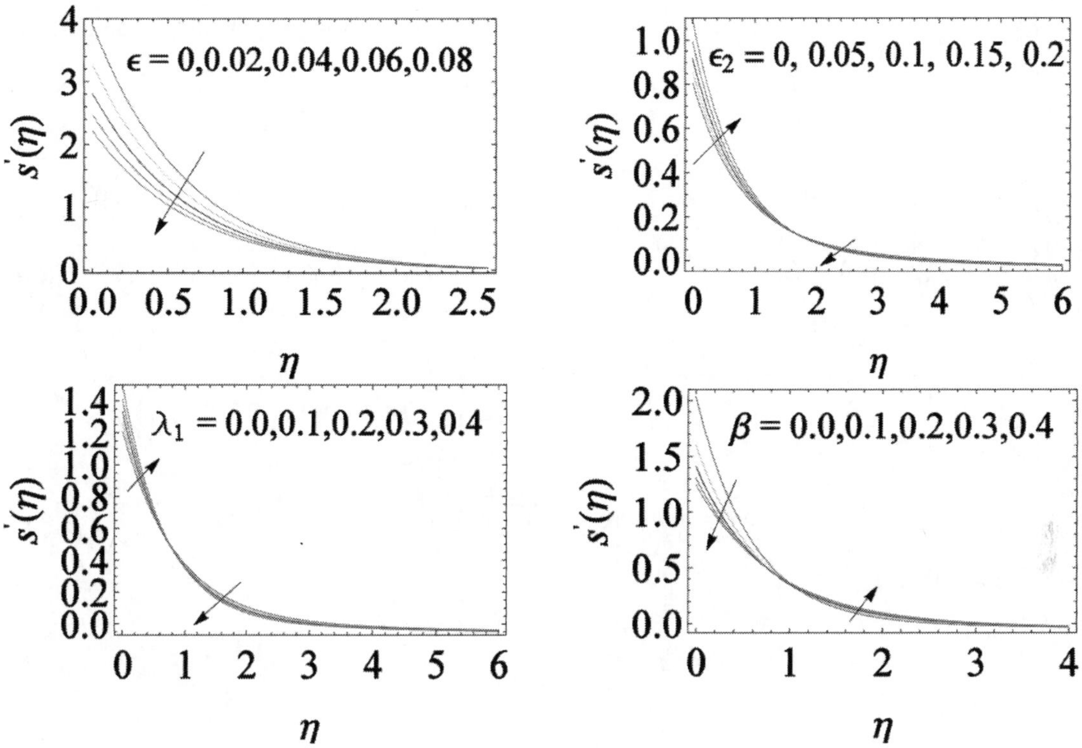

Figure 2: Influence of ϵ, ϵ_2, λ_1 and β on $s'(\eta)$

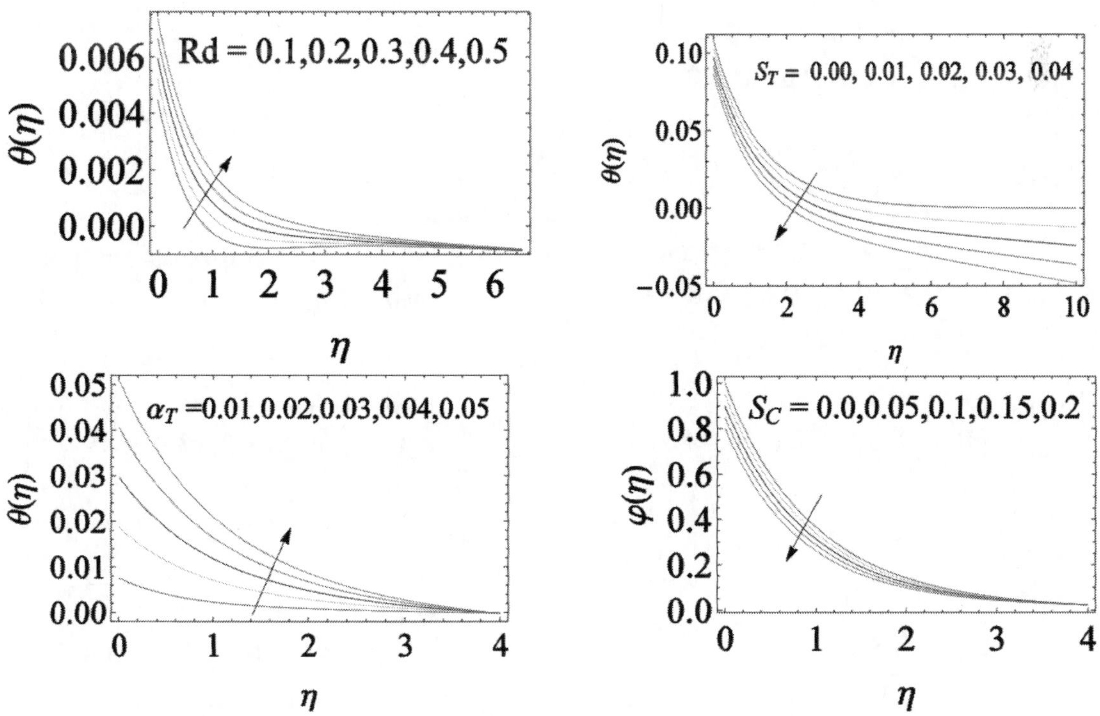

Figure 3: Influence of Rd, S_T and α_T on $\theta(\eta)$ and S_C on $\varphi(\eta)$

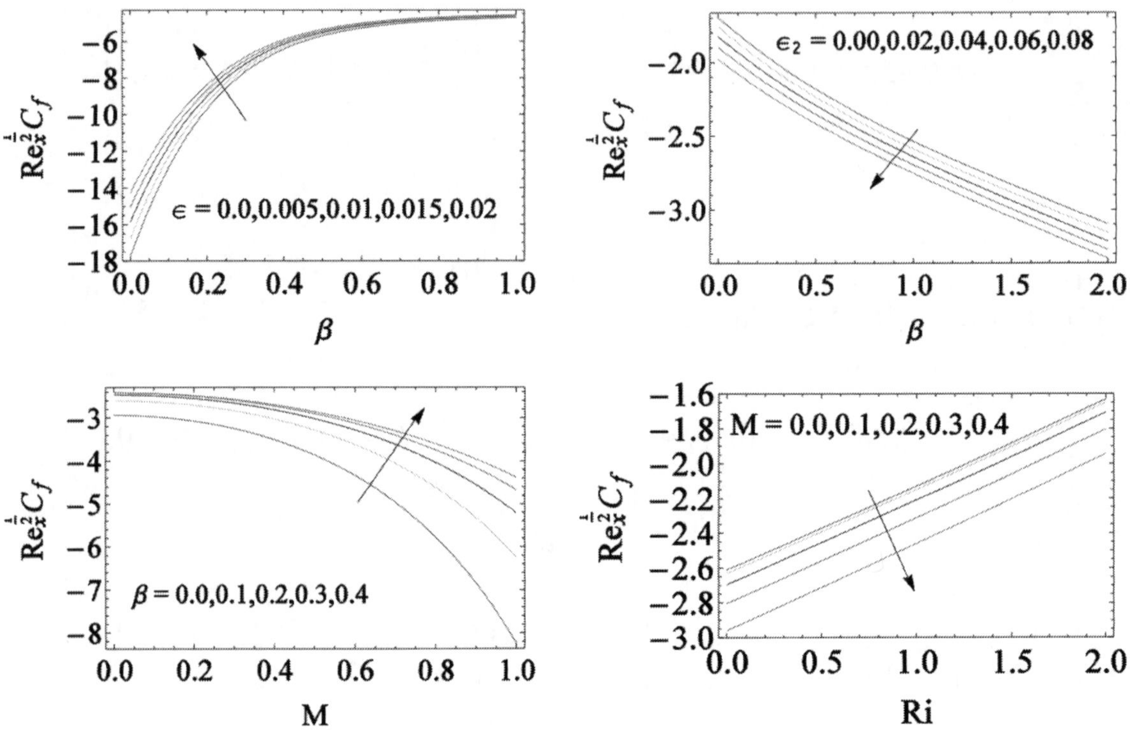

Figure 4: Influence of ϵ, ϵ_2, β and M on $Re_x^{1/2}C_f$

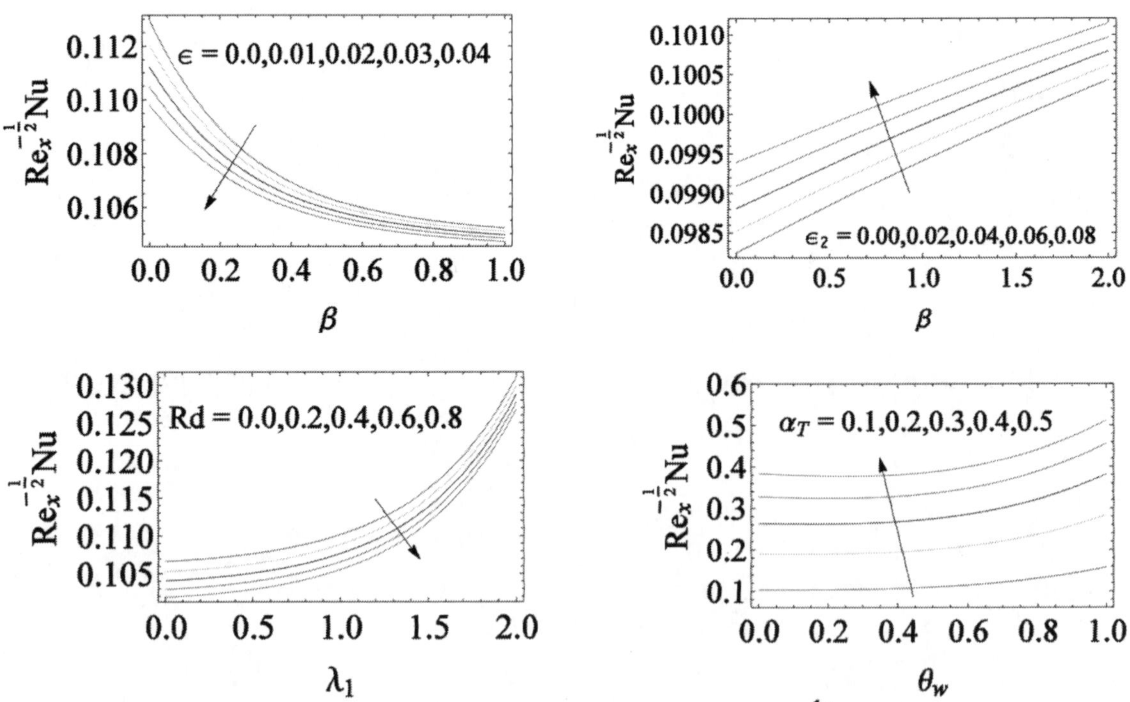

Figure 5: Influence of ϵ, ϵ_2, Rd and α_T on $Re_x^{-1/2}Nu$

Power sector reality and a way to improve service quality with community involvement.

K V James[1] and RG Priyadarshini[2]

[1]Amrita School of Business, Amrita Viswa Vidyapeetham, Coimbatore, Tamilnadu, India, jameskv09@gmail.com

[2]Amrita School of Business, Amrita Viswa Vidyapeetham, Coimbatore, Tamilnadu, India, rg_priyadarshini@cb.amrita.edu

Abstract. This paper looks into a brief view on Indian power sector including recent advancements and a study on its customer services in Kerala region. The customer service quality was measured from the customer perspective. Primary data were collected by questionnaire method from three main regions like Southern, Central and Northern areas of the state. A sample of 282 results was analyzed from different category of consumers across the state for this study. Moreover many numbers of consumers were interviewed from different parts of the state based on typical questions to assess their expectations and actual services obtained. The study found that the service quality does not match with the expectations of the consumers. Based on the study author underlines, considering the peculiarity of the state, the involvement of the community can improve employee outcome and the service expectations of consumers.

1. Introduction

1.1 *Indian electricity scenario*: In a view of providing reliable and affordable service to the people of the country, Indian power industry was operated under the ownership and control of the Government. The case was the same in other parts of the world. The quality of service was adversely affected by numerous problems in the power sector. The power sector under government control became financially unstable and unsustainable.

1.2 *Power sector in India*: At the time of independence the power generation was very less in India. As on 31st March 1951, the total installed power generation capacity was 1713 MW. After the successive five year plans, the installed power generation capacity rose to 94,300MW by 1995-1996. This was enhanced to 2, 72,687 MW as on March, 2015 and became 3,06,358MW as on 30/09/2016.

The split up details are as follows:

As on 30/09/2016

TYPE	INSTALLED CAPACITY MW
COAL	1,87,258.88
GAS	25,057.13
DIESEL	918.89
NUCLEAR	5774.00
HYDRO	43,112.43
RENEWABLE	44,236.92
TOTAL	3,06,358.25 MW

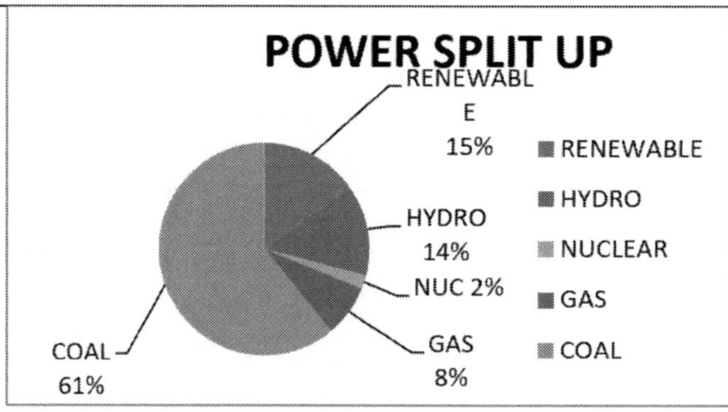

Figure 1. Power split up as on 30-09-2016

1.3 Power generation types and influences

The major portion of India's Power generation comes from thermal power stations, the fuel used is coal. CO_2 emission is calculated as 10% from the coal power stations. Huge amounts are required to be spent for limiting emission. More Hydro power stations are not coming up due to environmental concerns even though, country has a huge potential. Till nineties, Hydro Power had a considerable share in the power generation.

Renewable energy sources: Renewable energy resources are the power of the future, particularly from solar and wind. There are reports that scientists have cleared critical hurdles in manufacturing solar cells so that they can be printable as easy and inexpensive as printing a newspaper. Because of the advancements taking place in the manufacturing of renewable sources and their cost coming down, no new coal plants will be necessary after 2025 and by 2050, the power generation from coal plants may come to a halt –report says. This can reduce pollution and carbon dioxide emissions by about 10% after 2030, a big achievement. There are also reports that researches are taking place in mining Moon dust which is rich in Helium 3 is an abundant source of power that can meet the energy requirements of the whole world

1.4 *Access to electricity and its performance in India*

Due to different reasons like excessive political interferences, inefficient management and other operational problems, Utilities are still in poor performance .Consumer is suffered from unreliable and poor quality of service also with higher tariff. Hence this paper intends to have a study on service quality at Power Utility Services from consumer perspective focusing on the state of Kerala. Also considering the peculiarity of the state, study undertakes the community involvement can improve the service quality.

2. Service quality

Employee attitudes, Service behaviors affect service quality (Hartline and Farell 1996). Employees must be able to recognize customer needs or employees attempts to make this behavior suitable to customer needs; the service quality will be favorable from customer's perspective (Betner ET al.1990). The security feeling of the customer and the perception of employee integrity and competence in their mind can be improved if the employees behave in an assuring manner with the customers.(Johnston,1995).

When an employee underestimates the expectations of the consumer or fails to understand the expectations and needs of the consumers, there emerges a 'service gap'. If the quality of service extended to the customer is less than what is expected there will be customer dissatisfaction and a service quality gap exists (Parasuraman, Zeithamal &Berry 1990).

<u>**A SERVICE QUALITY MODEL IS SHOWN BELOW:**</u>

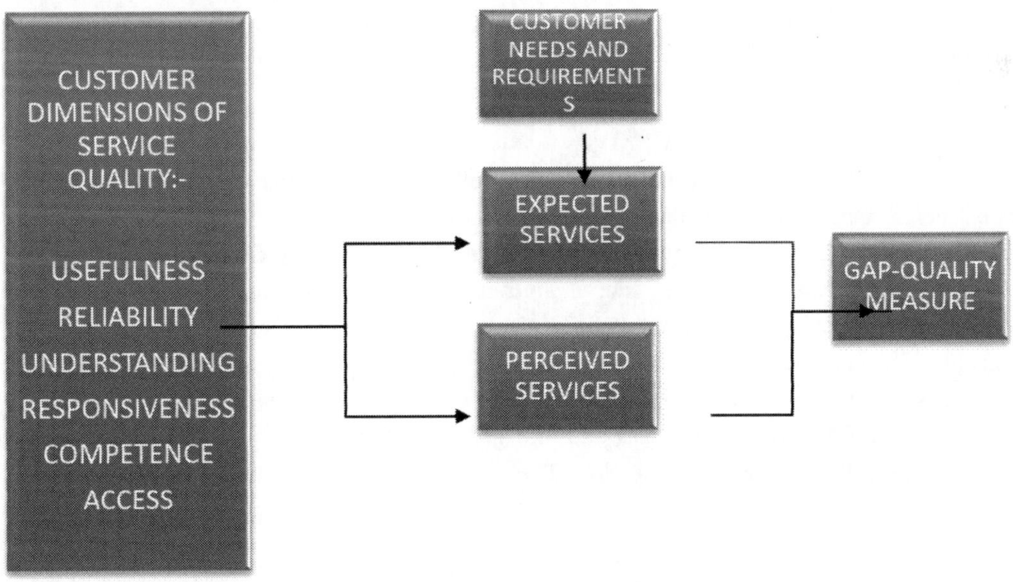

Figure 2 Service Quality Model

(ADAPTED FROM ZEITHAMAL, PARASURAMAN A ND BERRY, 1990

3. Objectives of the study: The major objective of the research is to measure the present level of service quality from consumer perspective and improvement in consumer satisfaction with the involvement of the community at power utility services.

4. Research methodology

4.1 *Research design*: The study is limited to the investigation of the power sector services in the state of Kerala alone. A detailed research was done to identify the different factors that influence the consumers mind in the power sector. Then a questionnaire was developed giving emphasis to various factors like human aspects of service delivery including Reliability, Responsiveness, Technical Capability of employees, Safety and Security aspects, interests of the employees to involve the community etc. Tangibles like cleanliness of the facility, neatness of the employee dress code etc., Social responsibility of the employees and also non- human aspects and advancements of easiness. Also how the service quality is reflected among various categories of consumers, their gender, different age groups and consumers at different income level. Primary data was collected from consumers of three different regions -Southern, Central and the Northern part of the state.

The sampling frame consisted of all categories of Electricity consumers in the Kerala region. Primary data were collected from 282 consumers from different parts of the state and utilized for the purpose of this study.

4.2. *Data analysis tools*

For the purpose of collecting primary data for this study, structured questionnaire on 5 point liker scale was designed and validated. The statistical technique like Pivot table was used for the analysis and interpretation of data.

4.3. *Analysis of data*

Data was loaded in Excel and Pivot table was developed for the study analysis. Questions H1 to H15 was loaded for the analysis of human aspects of Service delivery like Responsibility, Responsiveness, Technical Capability, Safety and Security aspects, employee interest to involve community etc.T1, T2, T3, T4 for the analysis of tangibles like cleanliness of the surroundings, Housekeeping, wearing uniforms and other safety measures, professional functioning etc. S1, S2and S3 were loaded for the measurement of social responsibility.N1 and N2 for the non- human aspects and measures of advancement on easiness and simplicity. The row values were filtered with consumer type C1, C2, C3, and C4, different age groups Y1, Y2, and Y3, gender F and M and also with different income level to know their influence on service quality. The results are tabulated.

Table 1. Analysis of Human Aspects with different factors

	FACTORS	OVERALL RATING	CONSUMER TYPES				GENDER		INCOME GROUPS					AGE GROUP		
			C1	C2	C3	C4	M	F	<2.5L	2.5-5L	5-10L	10-15L	>15L	Y1	Y2	Y3
1	HUMAN ASPECT OF SERVICE DELIVERY	2.93	3.39	2.74	3.13	2.46	3.27	3.49	3.04	3.38	3.43	3.05	3.05	3.3	3.49	3.40
A	RELIABIL-ITY	3.40					3.30	3.55	3.05	3.34	3.52	2.95	3	3.3	3.52	3.53
B	TECH-CAPABIL-ITY	3.52					3.46	3.60	3.08	3.57	3.60	3.24	3.39	3.46	3.68	3.52
C	SAFETY - SECURITY	3.185					3.11	3.29	3.02	3.17	3.22	2.70	2.72	3.17	3.31	3.12
D	COMMUNITY INVOLVEMENT	3.14					3.05	3.26	3.03	3.10	3.03	2.70	2.67	3.14	3.23	3.06

4.4. *Data analysis and discussion*

Data furnished in the Pivot table generally shows that the service quality from consumer perspective in the region of Kerala is just below or just above the average (3) in the liker scale. It is not good and not poor, but in between. Let us have a look at different categories of factors. In the Human aspects of service delivery , the overall value rating is 2.93 means just below average. There is a reliability factor of 3.40 means between average and good. Technical Capability of employees, 3.52 is also above average and near to good. Safety and Security aspects 3.185 mean just above average. Employee interest to involve the community for better performance was found as 3.14 i.e., just above average. Now if we look at these aspects to different category of consumers like Domestic, Commercial, Agricultural and Industrial consumers, the value rating has no much change i.e. just below and just above the average. That means, whatever be the type of consumers the human aspects don't change much. Similarly the gender aspect, the consumers at different income groups, consumers in different age groups etc. has no much influence on the Human aspects of service quality i.e. around three i.e. Just below or just above 3 which is the average. The service quality on tangibles like Cleanliness, employee professionalism, working in uniform etc., again shows just above and just below average. Again if we look at the influence of category of consumers, age groups, gender and different income levels, there is not much change in the value rating i.e. just below or just above the average.

Social responsibility of the employees was also seen as a rating of in and around three i.e. average. The type of consumers, age group, gender and income level do not find much difference on the social responsibility of the employees i.e. around means average.

Service quality regards to office advancements and simplification of the procedures, the value rating is just below and just above three i.e. Average. If we look at across consumer types, age group, gender and income level, there is not much influence, value being in and around 3 i.e. average.

5. Community involvement

From the Analysis we could see that the consumer satisfaction level is only average across various types of consumers. This can very well be improved with the involvement of the community with the organization. When we consider with power sector activities, whether it is new additions or maintaining and up keeping the existing system, all are directly linked with the community. It is possible that the responsibility can be shared within groups and everyone can be involved for contributing their strength to the organization. The problems and disputes that arise out can also be subjected for discussion and awareness and can be settled for common benefits within the permissible organizational regulations. Taking that the leaders in the community are open minded with their ideas and hence implemented with positive involvement.

How the community involvement become beneficial. Let us consider one example.

Suppose the organization gives an electric connection to a new consumer in a locality. The probable issues can be , in marking the location of electric poles through roads or private properties and drawing lines, in providing stays or supports, cutting and removing tree branches and over hangings- in all such cases the community support and leadership can solve the work execution smooth and delays can be avoided keeping harmony among people.

6. Community Involvement Ideas for Business Organizations

When the organizational employees have good relations with the community and showing interest in involving community for various activities, people will often show loyalty to the organization. The ideas like *formation of an Energy Conservation Cell, Forming Environment protection Cells, Sponsor a Sports Team, Support a local charity, Host Community Meetings, Create a Scholarship, Become A Local Expert* for supporting community members when they need assistance etc. are some of the ideas for community involvement in business organizations. Long lasting community involvement programs will definitely improve the service quality of the employees, values in the society and satisfaction to consumers. Various activities demonstrating commitment to the community improve the perceptions of all stakeholders of the organization including its employees. This is necessary for the long term success of the Firm.

7. Conclusion

Power sector plays a vital role in the development of the nation and the standard of living of the people. Service quality of power utilities has a direct impact on the day to day life of the consumers. The study shows the service quality is at the average level at all areas of function including safety and security matters. Also it was seen that factors like age group, gender, income level and consumer types have no much influence on service quality. Community involvement with the organizational activities will definitely improve the service quality and the level of consumer satisfaction.

8. Implications

This study is significant for policy makers and the management for designing, training and developing plans and adapting measures to improve service quality to a higher level.

References

[1] *GOI, Ministry of power, Central electricity authority report,* 2011, New Delhi

[2] Sharma and Anupriya, 2016, *Journal of economic policy and research,* Financing of power sector in India: issues and challenges, **11**, Hyderabad, 34-51.

[3] *Central electricity authority,* General electricity statistics (various issues)

[4] *GOI, Planning commission,* Annual report on the working of state electricity boards and electricity departments (various issues1990-2002), New Delhi

[5] *GOI, 2003,* The electricity act 2003, New Delhi

[6] *GOI, Ministry of power, 2002,* Gokak committee report on power sector reforms, New Delhi

[7] Saravanan R and Rao KSP, June 2007, *Total quality management,* **18**, 435-449

[8] Andrew M Farrell, Anne L Souchon and Geoffrey R Durden, 2001, *Journal of marketing management, 17*, 577-593

[9] Satapathy and Suchismita, 2014, *Emerald publishing*, An analysis for service quality enhancement in electricity utility sector of India by SEM, **20**, 964-986

[10] Mahesh Subramony, 2017, *Academy of management perspectives,* **31**, 28-43

[11] Carol C, Bienstock, Carol W and De Moran Ville, 2006, *Service marketing,* **28**, 103-118

[12] AllenKlose and Todd Finkle, 1995, *Psychology and Marketing,* **20**

[13] Jennifer Rowley, 1998, *Total quality management,* **9**, 321-333

[14] Satapathy S, Mahapatra SS, Biswas A, Mishra PD and Patel SK, 2014, *Emerald Publishing,* **21**, 634-650

[15] Snoj and Boris, 1994, *Systems Research,* **11**, 155-165

[16] Susan J Devlin and Dong HK, 1994, *Marketing Research,* **6**, Issue 1

Effect of Biodegradable metal cutting fluids in machining applications - A Review

K Venkatesh[1*], G Sriram[2], S Sai Raj Pavan[3] and S Suresh[4]

[1]Associate Professor, Dept. of Mechanical Engg., Sri Chandrasekharendra Saraswathi Viswa Mahavidyalaya, Enathur, Kanchipuram – 631561, Tamilnadu, India.

[2]Professor and Dean (E&T), Dept. of Mechanical Engg., Sri Chandrasekharendra Saraswathi Viswa Mahavidyalaya, Enathur, Kanchipuram – 631561, Tamilnadu, India.

[3, 4] B.E. Final Year Student, Dept. of Mechanical Engg., Sri Chandrasekharendra Saraswathi Viswa Mahavidyalaya, Enathur, Kanchipuram – 631561, Tamilnadu, India.

Corresconding Author Email [1*] kvenkateshme@gmail.com

Abstract. The vegetable based metal cutting fluids are widely used in machining applications because of its effectiveness in reducing the cutting temperature, cutting force as well in getting better the tool life and surface integrity. Both the cultivation and production of edible and non-edible oils are comparatively uncomplicated and viable. The sustainability and biodegradability are the two major attributes which expands the usage of edible and non-edible oils in the present scenario while comparing with various technical aspects and feasibilities from the different research work so far reported. This review article is mainly exploring the advantages and usage of vegetable based cutting fluids for various machining applications for yielding better machinability criterion.

Keywords: Biodegradable, metal cutting fluids, edible, non-edible, machinability

1. Introduction

Machining is a mechanical process that involves in the removal of work material into a desired shape with high electrical energy consumption due to presence of relative motion between tool and work material. During the machining process, mild to severe plastic deformation may occur and consequently 99% of the energy is transformed into thermal form and conducted into the workpiece [1]. This review article has elaborating the availability and usage of many edible and non-edible based metal cutting fluids in machining applications. At present in addition to traditional flood cooling methods, several alternative cooling methods like cryogenic cooling, minimum quantity lubrication (MQL) etc., are being used [2]. Rahim et al introduced the supercritical carbon dioxide technique to reduce the heat generation at the tool-chip interface zone [3].The prime aim of machining process is to obtain the high quality components at minimal investment. Moreover vegetable based fluids are also being employed in power transformers as insulating and cooling medium [4].

2. General Cutting fluids

Practically metal cutting fluids are unavoidable due to its plenty of advantages in machining process like enhancing the costlier cutting tool life, minimising the occurrence of thermal damage over metal component surface, assisting in clearing the accumulated chips from cutting zone and helping a lot in obtaining the good surface texture.

2.1. Kinds of Cutting fluid

2.1.1. Straight oils: Straight oils are often used in machining where they function effectively in direct form. This oil consists of mineral or petroleum base and may also contain some polar lubricants similar to vegetable oils, fats and esters.

2.1.2. Synthetic Fluids: A synthetic fluid consists of alkaline organic and in-organic compounds alongside additives to prevent corrosion. They function well in diluted form and offer best cooling performance.

2.1.3. Soluble oil: These are formed emulsions by mixing with water. The resulting concentration of emulsion and a base mineral oil to form a stable emulsion. These soluble oils also perform well in their diluted form and provide satisfactory lubrication in addition to heat transfer performance.

2.1.4. Semi synthetic fluids: These are basically combination of soluble and synthetic fluids. In addition their better performance, cost of this kind of fluids is considerably low [5].

2.2. Uses of vegetable based cutting fluid

It is feasible technically to improve the product quality at minimised cutting tool cost, and essentially presenting safe environment through the application of vegetable based cutting fluids. For achieving energy independence and clean surroundings then it is required to use massively available vegetable based cutting oils. Other merits of this kind of oil are:

- diminished frictional coefficient
- High flash point
- Less smoke and fire risk
- Easy clean up
- Safe to operators and environment

3. Edible cutting fluids

The elaborate development of metal working fluids and further utilisation of cutting fluids have resulted in significant industrial revolution. During 18th century, mineral based cutting oils are widely employed in machining applications. However mineral oils are non-renewable whereas edible oils are renewable and offer several advantages for the sustainable manufacturing theme. Edible oils are generally considered and used as food substance other than dairy products which are useful for human consumption. Plant derived edible oil consists of carboxylic acids with long hydro carbon chain. The most commonly used edible oils are coconut oil, corn oil, cotton seed oil, olive oil, palm oil, peanut oil, rapeseed oil, sunflower oil etc. Generally emulsifier has vital role in dissolving the oil in normal water to formulate steady oil in water emulsion. The increased use of petroleum based cutting fluids causing environmental imbalance is a major negative aspect to eco-system. To overcome this adverse effects of using conventional cutting fluids, the production and usage of edible and non-edible based cutting oils have to be motivated which the directed to achieve sustainability machining process [1].

4. Non-edible cutting fluids

Non-edible oils are not suitable for human and animal consumption due to their toxic composition nature. The recent research articles highlight the use of non-edible oils in machining than the edible oils by considering economical and technical features.

4.1. Available Non-edible oils:

- Jatropha cruces
- Calophylluminophyllum
- Mahua
- Pongame oil tree
- Neem oil
- Rubber tree
- Croton megalocarpus

Moreover these non-edible crops have to be planted and cultivated massively at unproductive land areas as well in poverty suffering areas and also in forest as physical boundaries to other useful type of crops.

5. Physical properties of edible and non-edible oils

Normally edible oils are suggested for human intake however non-edible based oil has been completely restricted from consuming due to the presence of toxins. Moreover this is a vital reason why non edible oils are preferred for production of metal cutting fluids. The production of non-edible based oil is comparatively cheaper than the production of edible oils. Since edible oils are human consumable they are not found ideal to use as metal cutting fluids. Various research studies have highlights the eligibility criteria for the use of non-edible oils [6] in industrial applications. Table 1 presents some of the basic physical properties of edible and non-edible oils.

Table 1. Properties of edible and non-edible oil

Properties	Sunflower oil	Rapeseed oil	Soybean oil	Jatropha oil	Neem oil	Castor oil
Kinematic viscosity@40° C	40.05	45.6	32.93	47.48	68.03	220.6
Kinematic viscosity@100° C	8.65	10.07	8.08	8.04	10.14	19.72
Viscosity index	206	216	219	208	135	220
Pour point	-12	-12	-9	0	9	-27
Flash point	252	240	240	240	-	250

6. Experimental Outcomes through the usage of Biodegradable metal cutting fluids

Talib et al. [7] performed some experimental trials to enhance thermal oxidative stability and cooling cum lubrication behaviour of crude jatrohpha. Crude Jatropha Oil (CJO) has been currently acknowledged as an environmentally benign metal working fluid (MWF). The authors have found improvement on the properties of CJO by adding molar ratios of jatropha methyl ester (JME) to trimethylolpropane (TMP) (MJO, MJO3, MJO5) through by addition of various concentrations of Hexagonal Boron Nitrate (HBN) particles ranging between 0.05 to 0.5 wt %. This research article evaluated the machinability aspects such as cutting force, cutting temperature, chip thickness, tool chip contact length, and specific energy with the employment of minimum quantity lubrication (MQL) setup and reported that by adding 0.05 wt% of HBN particles enhanced the overall machining performances. Carbon chain could affect the lubrication performance; CJO containing a high percentage of unsaturated carbon chain could be inferred to use as metal working fluid.

Rao et al. [8] used the minimum quantity lubrication (MQL) technique in their research work. Since the convectional metal working fluids has facing numerous limits in ecological system and disposal of used matters safely hence the concept of MQL is becoming significant. Because of the better properties of vegetable based cutting fluids, enhanced performance during machining process can be yielded compared to mineral oils. Further inclusion of Nano particles in both vegetable based and mineral based oils has been tried by many researches to augment the overall performance. Future research may be directed to improve the individual performance of fluids through viable alternate measures.

Abdalla et al. [9] developed novel sustainable neat oil for machining applications especially with non-edible based one. Further with this oil, they have conducted some experimental trials with the stainless steel and titanium as work materials. The experimental results have revealed that for reducing the cutting force and temperature, vegetable based cutting oil should be preferred.

6.1 Turning Process

Elmunafi et al. [10] used castor oil with a MQL setup having a flow rate of 50 ml/hr along with 5 bar inlet air pressure to turn hardened stainless steel using coated carbide cutting tool. Finally authors have assessed some of the vital machinability aspects: tool life, surface roughness and cutting force were improved considerably. Adekunle et al. [11] made a study on chip morphology and behaviour of cutting tool during turning of AISI 301 with different biodegradable based cutting oils. Elmunafi et al. [12] carried out experimental work with the use of castor oil as cutting fluid. They performed trials under dry condition and also with MQL technique using wiper geometry - coated carbide cutting tool. The outcome shows that the use of MQL seems better to turn the hardened stainless steel using coated tool. The process is limited due to formation of elevated cutting temperature at high cutting speeds and observed that Air - oil mist get evaporated. They also observed that through supply of smaller amount of coolant oil (50 ml/hr) yielded a better tool life.

Dhar et al. [13] observed that the drop in cutting temperature, moderate tool wear and improved surface finish while turning AISI 1060 work material with MQL system filled with vegetable based oil compared with flood and dry machining conditions. Kuram et al. [14] used the canola and sunflower oil based cutting fluids in a MQL system and found good effect on process outcomes like tool wear and surface finish. Ojolo et al. [15] machined the aluminium component with ground nut, coconut oil, palm and kernel vegetable based cutting fluid in a turning process and observed the considerable changes in tool wear. Paul et al. [16] observed that the lower temperature was recorded while using neem oil based cutting fluid compared to conventional soluble oil and other techniques. Yakubu et al. [17] performed machining experiment on aluminium - manganese alloy with neem seed oil based cutting fluid, conventional soluble oil and also at dry condition so as to evaluate surface roughness and tool wear. Experimental results were also compared and presented in the article.

6.2 Drilling Process

Kelly et al. [18] employed MQL setup and used vegetable based oil as coolant medium during drilling of cast aluminium silicon alloy. The machinability characteristics like cutting torque, surface roughness and tool wear were studied under MQL and dry environment. It is observed that vegetable based coolant oil with a MQL technique minimises the torque and enhances the surface finish however found increasing at higher feed rate and cutting speed levels. Costa et al. [19] observed that burr height was smaller at dry machining environment and same was found larger at MQL machining situation.

Rahim et al [20] performed high speed drilling on Ti-6-Al-4V work material with palm oil and synthetic ester based coolant oils in a MQL setup separately and noticed that palm oil reduced the tool wear effectively. Belluco et al. [21] carried out drilling experiments on AISI 316L material with

IOP Conf. Series: Materials Science and Engineering **390** (2018) 012083 doi:10.1088/1757-899X/390/1/012083

vegetable based cutting fluid and noticed that the good results are obtained in the process outcomes like tool wear, tool life, cutting force and chip formation. Kuram et al. [22] noticed better improvement on surface finish and declined thrust force with minimised frictional effect while machining with sunflower oil based cutting fluid.

6.3 Milling Process

Kuram et al. [23] noticed that usage of sunflower and canola vegetable based cutting oil helps in obtaining best surface finish, improved tool life and reduced worn out of cutting edges during milling of aluminium work pieces while compared with commercial synthetic oil. Sales et al. [24] studied the effect of different feed rates on tool wear, surface roughness and chip formation aspects during the milling process with MQL setup. It is observed that friction decreases with the increase of feed rate and also there is reduction of tool wear, surface roughness, burr length etc.

6.4 Grinding Process

Kalitha et al. [25] conducted tribological study on soybean oil mixed with MoS_2 nano particles to provide nano lubrication for machining of cast iron using MQL technique and used vitrified bonded aluminium oxide abrasives as grinding wheel material. It is observed that a notable reduction in the coefficient of friction and grinding forces while using soybean oil as a base fluid. This reduction in grinding forces is reduced due to formation of Mo-S-P tribofilm. Guo et al. [26] carried out grinding experiments on Ni alloy to assess the lubrication performance of castor oil while mixed with different vegetable oil. In this study, the castor oil was mixed in 1:1 ratio with vegetables oils. Ultimately they observed that soybean and castor oil mixture showed better performance and reduced the grinding force and specific normal grinding force by 27.03% and 23.15 % respectively while comparing with pure castor oil.

Conclusion

The application of traditional cutting fluids in machining process repeatedly overtime induces chemical changes of fluids due to the environmental effects, contamination from the metal chips and tramp oil. These effects can be easily overcome by the suitable selection and use of edible and non-edible based cutting fluids in order to yield better overall performance and machinability criterion. The various process input parameters like cutting velocity, feed rate, depth of cut, tool geometry and others have to be optimized for getting better overall process outcomes with the applications of these vegetable based cutting fluids. The process of cultivation and production of Edible and non-edible oils are found trouble-free and economical and the process wastage obtained during the extraction of vegetable oil can be easily degradable. Therefore overall machinability aspects of any kind of machining applications can be enhanced by using the edible and non-edible cutting fluids appropriately.

References

[1] Anuj Kumar Sharma, Arun Kumar Tiwari and Amit Rai Dixit 2016 J Clean Prod **127** 01-18.

[2] Wenfeng Ding, Liangchi Zhang, Zheng Li, Yejun Zhu, Honghua Su, and Jiuhua Xu 2017 Int J Adv Manuf Technol **88** 2939-68.

[3] Rahim E A, Rahim A A, Ibrahim M R and Mohid Z 2016 Procedia CIRP, **40** 638– 42.

[4] Oommen T V 2002 Electrical Insulation Magazine, IEEE **18** 6 - 11.

[5] UNIST Australia Fluid application solutions for metal cutting https://unist.com.au/mql/

[6] Emel Kuram, Babur Ozcelik and Erhan Demirbas 2013 Green Manufacturing Processes and Systems, Materials Forming, Machining and Tribology Springer-Verlag Berlin Heidelberg.

[7] Talib N, Sasahara H, and Rahim E A 2017 Int J Adv Manuf Technol **92** 371–391.

[8] Nageswara Rao D, Srikant R R, and Srinivasa Rao Ch, 2007 ABCM **XXIX**, (4) 396-400.

[9] Abdalla H S, Baines W, Mclntre G, and slade C 2006 Int J Adv Manuf Technol **34** 21-33.

[10] Mohamed Handawi Saad Elmunafi, Noordin M Y, and Kurniawan D 2015 Procedia

IOP Conf. Series: Materials Science and Engineering **390** (2018) 012083 doi:10.1088/1757-899X/390/1/012083

Manufacturing **2** 563 – 567.
[11] Adekunle A S, Adedayo S M, Ohijeagbon I O and Olusegun H D 2015 JPE **18** (1) 18-22.
[12] Mohamed Handawi Saad Elmunafi, Kurniawan D and Noordin M Y 2015 Procedia CIRP **26** 408 – 411.
[13] Dhar N R, Ahmed M T and Khoda A K M B 2007 ICME07- AM-37:1-5, 2007.
[14] Kuram E, Ozcelick B, Demirads E, Sik E and Cetin M H 2011 TribolInt **44** 1864-1871.
[15] Ojolo S J, Amuda M O H, Ogunmola O Y and Ononiwu C U 2008 Revista Material **13(4)** 650-663.
[16] Paul S and Pal P K 2011 Mach Technol Mater **11** 24-28.
[17] Yakubu S O and Bello Y IJIERT (2015) FEB.
[18] Kelly J F and Cotterell M G 2002 J Mater process technol **120** 327-334.
[19] Costa E S, Silva M B and Machado A R 2009 J BrazsocMechSci Eng. **31(1)** 57-36.
[20] Rahim E A and Sasahara H 2011 Proc. Inst. Mech. Eng. B **225** (9) 1505-1511
[21] Belluco W and De Chiffre L 2004 J Mater process Technol **148** 171-176.
[22] Kuram E, Ozcelick B, Demirads E and Sik E 2011 Mater manuf processes **26(9)** 1136-1146.
[23] Kuram E, Ozcelick B, Demirads E, Sik E and Askin S 2010 Proceedings of the world congress on engineering,London 1292-1296.
[24] Sales W, Becker M, Barcellos C S, Jr JL, Bonney J and Ezugwu E O 2009 IndLubrTribol **61(2)** 84-90.
[25] Kalita P, Malshe Ajay P, Wenping Jiang, and Shih Albert J 2010 Trans NAMRI/SME **38** 137-144.
[26] Guo S, Li C, Zhang Y, Wang Y,Li B ,Yang M, Zhang X and Liu G 2017 J clean prod **140** 1060-1076.

Cross diffusion effects on combined bioconvection of nanofluid in a flat channel along with microorganisms

S P Geetha[1], S Sivasankaran[2,*], M Bhuvaneswari[2] and S Rajan[3]

[1] Department of Mathematics, Vellalar College for Women, Erode-638 012, Tamilnadu, India

[2] Department of Mathematics, King Abdulaziz University, Jeddah 21589, Saudi Arabia

[3] Department of Mathematics, Erode Arts and Science College, Erode-638 009, Tamilnadu, India

*e-mail: sd.siva@yahoo.com

Abstract. This paper investigates the cross diffusion effects on fully developed mixed bioconvection of gyrotactic microorganisms through horizontal channel filled with nanoliquid under the influence of Soret and Dufour effects. The governing nonlinear partial differential equations are converted into set of nonlinear ordinary differential equations using appropriate similarity variables. The equations together with the boundary conditions are solved by using homotopy analysis method (HAM). The effects of thermophoresis, Brownian motion, Soret and Dufour parameters on mass and heat transfer are analysed graphically. It was found that Soret and Dufour effects was highly influenced in the temperature and the movement of microorganisms.

Keywords: Bioconvection, Soret effect, Dufour effect, Nanofluid, Channel, Microorganisms.

1. Introduction

A great amount of studies on convection of nanoliquid has been reported in recent years due to their energy transport control potential in different energy systems [1, 2, 3]. Bioconvection pattern happen due to up-swimming of microorganisms which are a slightly denser than water in suspensions. The movable microorganisms are commonly larger than water and they swim skyward when stimuli such as light and chemical attraction exist, which move under the variable density concept and hydro-dynamic inconstancy. Combinations of nanoliquid and microorganisms in convective flows is influenced by many researchers due to its significant in microfluidic devices. The nanofluid bioconvection of gyrotactic microorganisms was examined by several authors [4, 5, 6]. Geetha et al. [6] explored the bioconvection of nanoliquid in the presence of Joule heating in a channel. Soret and Dufour effects along with various effects on convection was discussed by [7, 8, 9, 10, 11].

The present investigation extends the work of Xu and Pop[12] proposed by Kuznetsov and Nield by considering the reactions of Dufour and Soret over doubly diffusive combined convection with gyrotactic microorganisms.

2. Mathematical Modelling

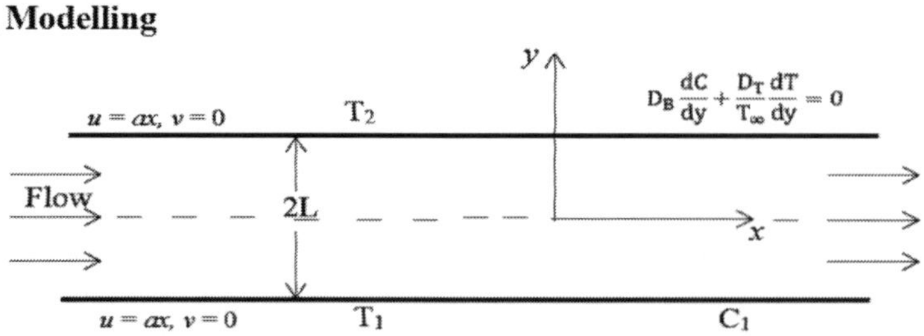

Figure 1. Physical configuration and co-ordinate systems.

Consider the combined bio-convection of nanofluid in a horizontal channel of length $2L$ filled along gyrotactic microorganisms as shown in Figure 1. The axes of the horizontal and vertical directions to the channel walls are along the axes of x and y respectively. Let the velocity of upper and lower walls are $u_1 = ax$. Let T_1 and T_2 are the constant temperatures lie on the top and the bottom of the walls. Suppose N_1 and N_2 are constant densities of the movable microorganisms. Two different solutal concentrations are taken on the walls. Let us consider on the upper wall, the nanoparticle volume fraction satisfies the passively controlled model, but on the lower wall, it is constant. The equations of continuity, momentum, heat energy, nanoliquid concentration, solutal concentration and the microorganisms density are:

$$u_{1x} + v_{1y} = 0, \tag{1}$$

$$u_1\zeta_x + v_1\zeta_y = \nu(\zeta_{xx} + \zeta_{yy}), \tag{2}$$

$$u_1T_x + v_1T_y = \alpha(T_{xx} + T_{yy}) + \frac{Dk}{c_s c_p}(\Gamma_{xx} + \Gamma_{yy}) + \tau\left\{D_B(C_xT_x + C_yT_y) + \frac{D_T}{T_0}\left[(T_x)^2 + (T_y)^2\right]\right\}, \tag{3}$$

$$u_1C_x + v_1C_y = D_B(C_{xx} + C_{yy}) + \frac{D_T}{T_0}(T_{xx} + T_{yy}), \tag{4}$$

$$u_1\Gamma_x + v_1\Gamma_y = D(\Gamma_{xx} + \Gamma_{yy}) + \frac{Dk}{T_m}(T_{xx} + T_{yy}), \tag{5}$$

$$u_1N_x + v_1N_y + (N\hat{v})_y = D_n N_{yy}. \tag{6}$$

where p and T be the pressure and heat energy, C is the nanoparticle volume fraction, Γ - the solutal concentration, $\hat{v} = (u_1, v_1)$ being the velocity components of nanoliquid in the Cartesian coordinates, ρ, μ and α are respectively density, viscosity and thermal diffusivity of the nanofluid, c_s be the concentration susceptibility and c_p is the specific heat. The ratio of the efficient heat energy measure of the nanoparticle to the basic liquid is represented by $\tau = (\rho c_p)_p/(\rho c_p)_f$. D_B and D_T are called the two diffusion coefficients.

The conditions at wall and centre of the channel are as follows:

$$u_1 = ax,\quad v_1 = 0,\quad T = T_2,\quad D_B\frac{dC}{dy} + \frac{D_T}{T_\infty}\frac{dT}{dy} = 0,\quad \Gamma = \Gamma_1,\quad N = N_2 \quad \text{as} \quad y = L;$$

$$u_{1y} = 0,\ v_1 = 0 \ \text{as} \ y = 0; \quad u_1 = ax,\ v_1 = 0,\ T = T_1,\ C = C_1,\ \Gamma = \Gamma_2,\ N = N_1 \ \text{as} \ y = -L.$$

By using the successive similarity transformations

$$\psi(x,y) = axLf_1(\eta),\quad \eta = \frac{y}{L},\quad \theta(\eta) = \frac{T - T_0}{T_2 - T_0},\quad \phi(\eta) = \frac{C - C_0}{C_0},\quad \gamma(\eta) = \frac{\Gamma - \Gamma_0}{\Gamma_2 - \Gamma_0},\quad S(\eta) = \frac{N}{N_2}.$$

in Eqs. $(1)-(6)$, continuity equation (1) is obviously true and the simplified form of other equations are given by

$$f_1'''' + Re(f_1 f_1''' - f_1' f_1'') = 0, \tag{7}$$

$$\theta'' + (RePr)f_1\theta' + Nb\theta'\phi' + Nt(\theta')^2 + Df\gamma'' = 0, \tag{8}$$

$$\phi'' + \frac{Nt}{Nb}\theta'' + (ReLe)f_1\phi' = 0, \tag{9}$$

$$\gamma'' + ReSc_\gamma f_1\gamma' + SrSc_\gamma\theta'' = 0, \tag{10}$$

$$s'' - Pe_b(\phi's' + s\phi'') + (ReSc)f_1s' = 0, \tag{11}$$

along with the following conditions at boundary

$$\theta = \delta_\theta,\quad \phi = \delta_\phi,\quad \gamma = \delta_\gamma,\quad s = \delta_s \quad \text{as} \quad \eta = -1;\quad f_1 = 0,\quad f_1'' = 0 \quad \text{as} \quad \eta = 0;$$

$$f_1 = 0, \quad f_1' = 1, \quad \theta = 1, \quad Nb\phi' + Nt\theta' = 0, \quad \gamma = 1, \quad s = 1 \quad \text{as} \quad \eta = +1.$$

The dimensionless numbers are given by

$$Re = \frac{aL^2}{\nu}, \quad Pr = \frac{\nu}{\alpha}, \quad Nb = \frac{\tau D_B C_0}{\alpha}, \quad Nt = \tau\left(\frac{D_T}{T_0}\right)\frac{T_2 - T_0}{\alpha}, \quad Le = \frac{\nu}{D_B}, \quad Pe_b = \frac{bW_c}{D_n},$$

$$Sc = \frac{\nu}{D_n}, \quad Df = \frac{Dk(\Gamma_2 - \Gamma_1)}{\alpha c_s c_p (T_2 - T_0)}, \quad Sc_\gamma = \frac{\nu}{D}, \quad Sr = \frac{Dk(T_2 - T_0)}{\nu T_m (\Gamma_2 - \Gamma_0)}, \quad \delta_\theta = \frac{T_1 - T_0}{T_2 - T_0},$$

$$\delta_\phi = \frac{C_1 - C_0}{C_0}, \quad \delta_\gamma = \frac{\Gamma_1 - \Gamma_0}{\Gamma_2 - \Gamma_0}, \quad \delta_s = \frac{N_1}{N_2},$$

where Re-Reynolds number, Pr-Prandtl number, Nb-Brownian motion parameter, Nt-thermophoresis parameter, Le-Lewis number, Df-Dufour number, Sr-Soret number, Sc-Schmidt number, Sc_γ-Schmidt number for mass transfer and Pe_b-Peclet number.

In convection flow problems, very important to aware the amount of heat transfer between the solid wall and the liquid. To calculate the heat and mass transfer rates over the channel, it is essential to give description of the local Nusselt number and the local Sherwood number along the boundary. The definition of the local Nusselt number and local Sherwood number are listed by

$$Nu = \frac{q_w x}{k(T_2 - T_0)}, \quad \text{where the heat flux} \quad q_w = -k\left(\frac{\partial T}{\partial y}\right)_{y=0},$$

$$Sh = \frac{j_w x}{D(\Gamma_2 - \Gamma_0)}, \quad \text{where the mass flux} \quad j_w = -D\left(\frac{\partial \Gamma}{\partial y}\right)_{y=0}.$$

From the above equation, the dimensionless form of the Nusselt and the Sherwood numbers of this model are given by $Nu = -\theta'(\eta)$ and $Sh = -\gamma'(\eta)$ at $\eta = 0$.

3. Results and analysis

The set of non-linear differential equations $(7)-(11)$ are evaluated by using HAM technique subject to the values of parameters $h_f = h_\phi = -1.5, h_\theta = h_\gamma = -0.2, h_s = -0.5$. The fixed values of other parameters are taken as $Pr = Sc = Le = Pe_b = Sc_\gamma = Sr = Df = Nb = Nt = 1$, $\delta_\theta = \delta_\phi = \delta_\gamma = 0.5, \delta_s = 1$. The Soret number ($Sr$) and Dufour number ($Df$) lie between 0 to 4, $-10 \le Re \le 10$ for Reynolds number, Brownian motion and Thermophoresis parameters lie between 0 to 2.

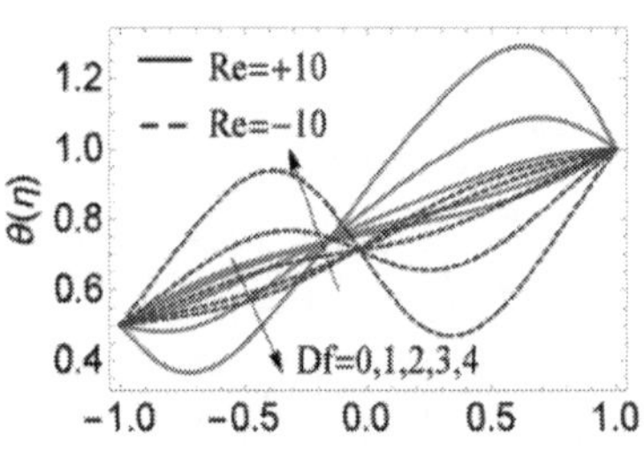

Figure 2. The temperature for different entries of Df with $Re = \pm 10$, $Sr = 1$.

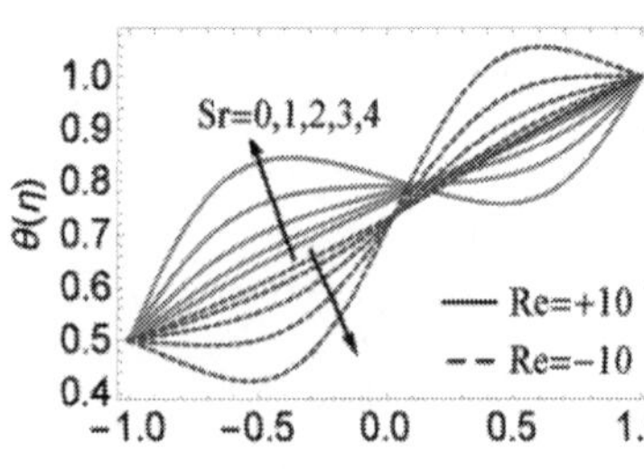

Figure 3. The temperature for various entries of Sr with $Re = \pm 10$, $Df = 1$.

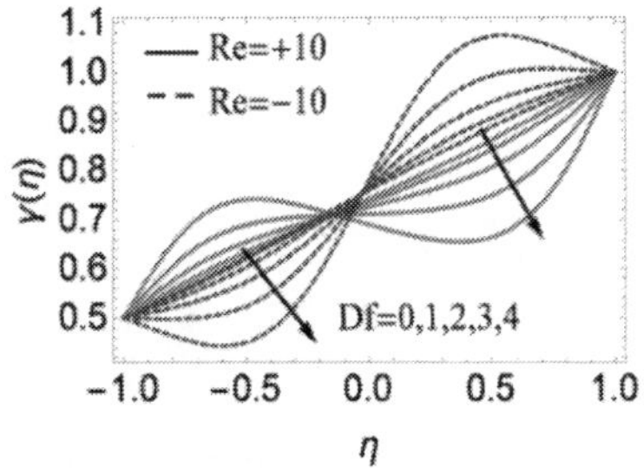

Figure 4. The concentration for various entries of Df including $Re = \pm 10$, $Sr = 1$.

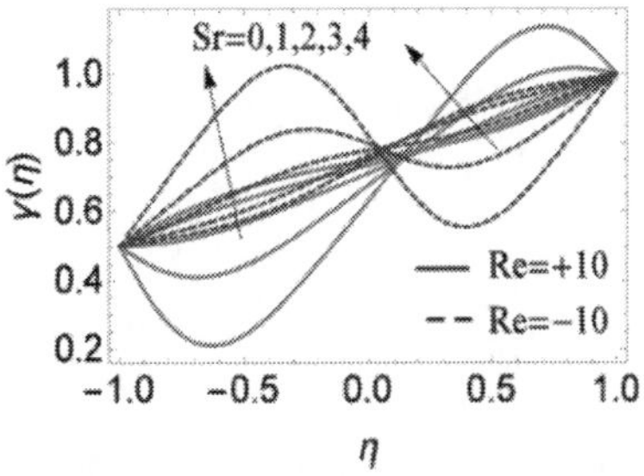

Figure 5. The concentration for various entries of Sr including $Re = \pm 10$, $Df = 1$.

Fig 2-3 analyse the temperature profile under Sr and Df respectively. The thermal energy of the liquid enhances with the rising values of Dufour and Soret numbers. These two effects provide opposite influence on the nanoliquid temperature. This is because of the Soret effect reflects reverse effect of Dufour number. Figures 4-5 discuss the impact of Sr and Df for the solute concentration. It is seen that the variation of Sr and Df can change concentration to a large extent. The solutal concentration get opposite behaviour in the upper and lower parts of the channel. The Soret number more pronounces on solutal concentration than the Dufour number. Figures 6-7 deal the effect of microorganisms density under Sr and Df. It is detected that $S(\eta)$ enhances with Df in the lower portion of the channel and it raises with Sr values in the upper portion of the channel. The reverse trend is discovered on growing the values of Df and Sr in the upper and lower portions of channel respectively.

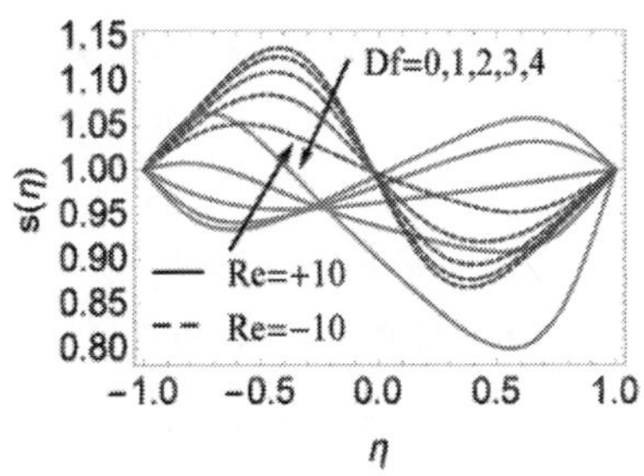

Figure 6. The microorganisms density for various Df values with $Re = \pm 10$, $Sr = 1$.

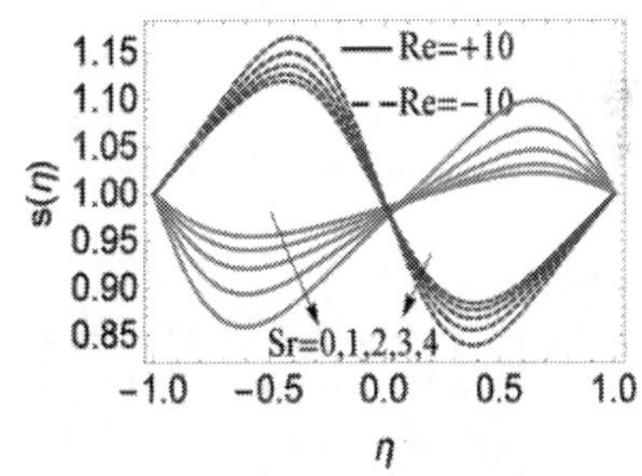

Figure 7. The microorganisms density for various Sr values with $Re = \pm 10$, $Df = 1$.

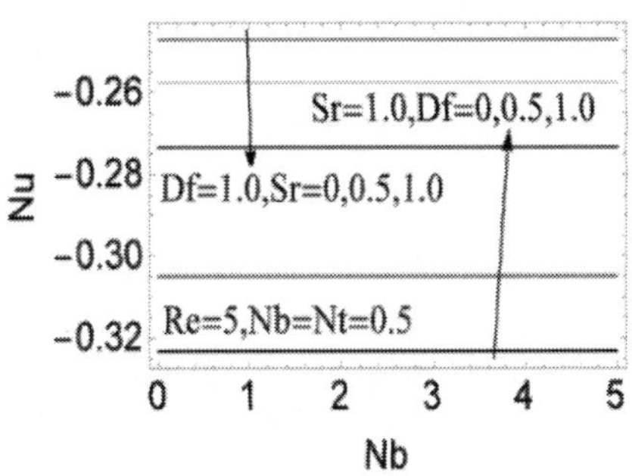

Figure 8. Nusselt number versus Nb.

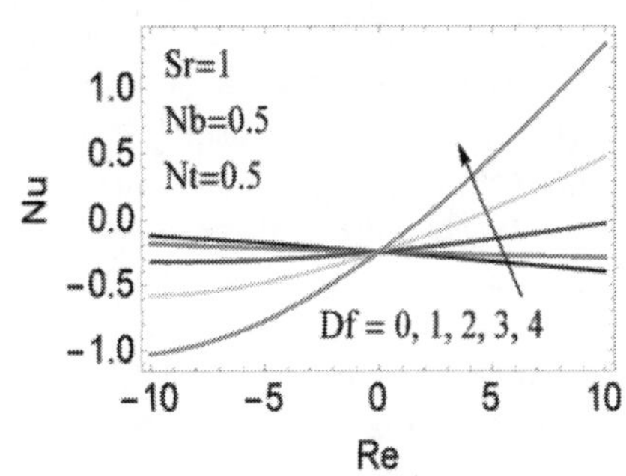

Figure 9. Nusselt number versus Re.

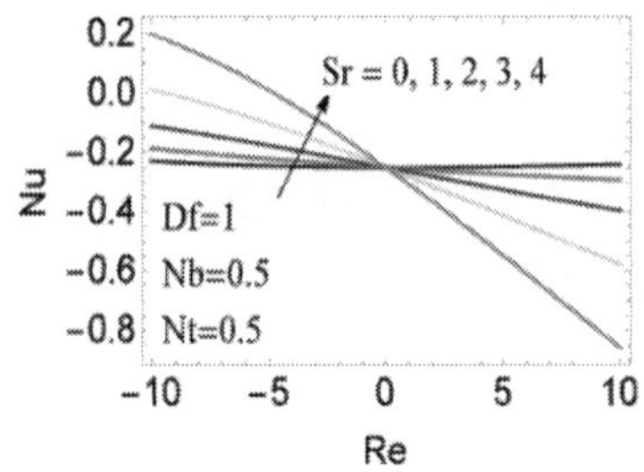

Figure 10. Nusselt number versus Re.

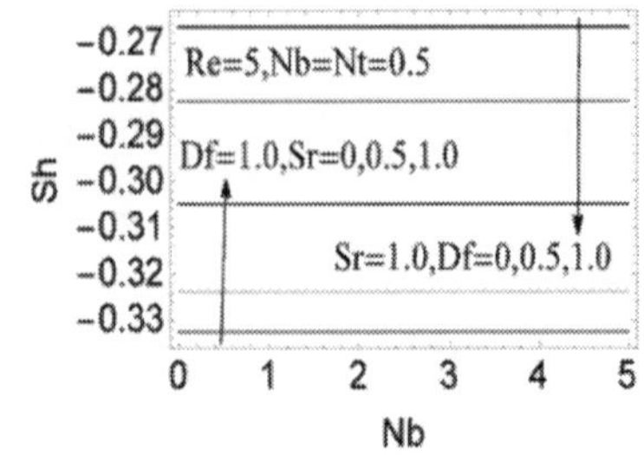

Figure 11. Sherwood number versus Nb.

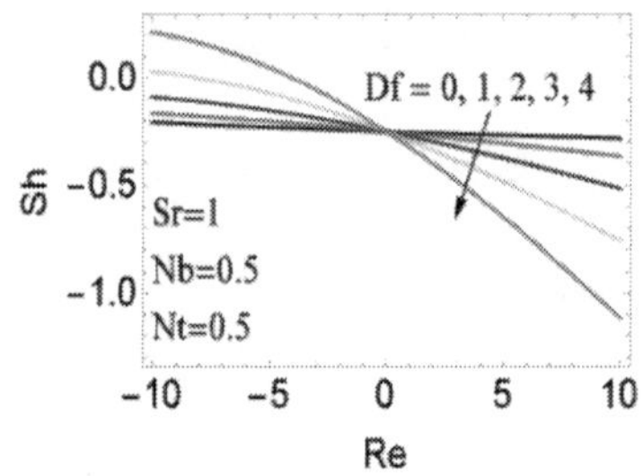

Figure 12. Sherwood number versus Re.

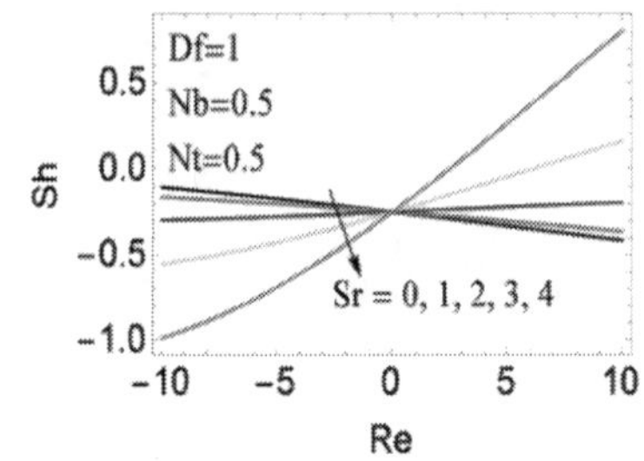

Figure 13. Sherwood number versus Re.

The effect of energy transport inside the channel is demonstrated by Nusselt number and it is pictured in the Figures 8-10 for different entries of Sr and Df along with Reynolds number. The energy transport enhances on raising the values of Soret number and it diminishes by growing the values of Dufour number. Figures 11-13 demonstrated the influence of solutal transport inside the channel for several combinations of Sr, Df and Re. The Dufour number reduces the mass transfer rate inside the channel whereas the mass transport raises when increasing the values of Sr. The local Sherwood number behaves in the opposite manner with respect to Nu in front of Soret and Dufour number, Brownian number and Reynolds number.

4. Conclusion

In this paper, mixed bioconvection of nanofluid with microorganisms in the horizontal channel in front of Brownian motion, diffusion-thermo and thermo-diffusion reactions are examined. The important concepts are noted in the following lines:

- The temperature and solute concentration enhance for raising the values of Dufour and Soret numbers.
- The local Nusselt number enhances whereas the local Sherwood number diminishes on raising the values of Dufour number. The rate of heat transport diminishes whereas the mass transport rate enhances when the Soret number raises.
- The density of microorganisms enhances with Soret and Dufour numbers.
- In overall, it is discovered that the physical quantities behaves opposite manner in the upper and lower portion of the channel.

References

[1] Sivasankaran S, Aasaithambi T and Rajan S 2010 Numerical analysis on convection heat transfer using different nanofluids in a cavity with linearly varying wall temperature *Maejo Int. J. Sci. Tech.* **4(03)** 468 - 482

[2] Sivasankaran S and Pan K L 2014 Natural Convection of Nanofluids in a Cavity with Non-uniform Temperature Distributions on Side Walls *Numerical Heat Transfer* A **65** 247268

[3] Rashad A M, Sivasankaran S, Mansour M A and Bhuvaneswari M 2017 Magneto-convection

of nanofluids in a lid-driven trapezoidal cavity with internal heat generation and discrete heating *Numerical Heat Transfer* A **71(12)** 1223-1234

[4] Xun S, Zhao J, Zheng L and Zhang X 2017 Bioconvection in rotating system immersed in nanofluid with temperature dependent viscosity and thermal conductivity *Int. J. Heat and Mass Trans.* **111** 100106

[5] Shaw S, Sibanda P, Sutradhar A and Murthy P V S N 2014 Magnetohydrodynamics and Soret Effects on Bioconvection in a Porous Medium Saturated With a Nanofluid Containing Gyrotactic Microorganisms *J. Heat Trans.* **136** 052601-10

[6] Geetha S P, Bhuvaneswari M, Sivasankaran S and Rajan R 2017 MHD bioconvection of nanofluid in a channel with Joule heating and passive control *Int. J. Pure and Applied Math.* **117(12)** 213-221

[7] Karthikeyan S, Bhuvaneswari M, Sivasankaran S and Rajan R 2016 Soret and Dufour Effects on MHD Mixed Convection Heat and Mass Transfer of a Stagnation Point Flow towards a vertical plate in a Porous medium with Chemical Reaction, Radiation and Heat Generation *J. Applied Fluid Mech.* **9(3)** 1447-55

[8] Eswaramoorthi S, Bhuvaneswari M, Sivasankaran S and Rajan S 2016 Soret and Dufour effects on viscoelastic boundary layer flow over a stretching surface with convective boundary condition with radiation and chemical reaction *Scientia Iranica* B **23(6)** 2575-86

[9] Abreu C R A, Alfradique M F and Silva Telles A 2006 Boundary layer flows with Dufour and Soret effects: I: Forced and natural convection *Chemical Engg. Sci.* **61** 4282-89

[10] Postelnicu A 2007 Influence of chemical reaction on heat and mass transfer by natural convection from vertical surfaces in porous media considering Soret and Dufour effects *Heat Mass Trans.* **43** 595602

[11] Bhuvaneswari M Sivasankaran S and Kim YJ 2011 Numerical Study on Double diffusive mixed convection in a two-sided lid driven cavity with Soret effect *Numerical Heat Transfer* A, **59** 543-560

[12] Xu H and Pop L 2014 Fully developed mixed convection flow in a horizontal channel filled by a nanofluid containing both nanoparticles and gyrotactic microorganisms *European Journal of Mechanics* B/*Fluids* **46** 37-45

Tribological Characteristics of Aluminium Metal Matrix with nano BN Powder Metallurgy Composite.

M Ekambaram [1], M Vetrivel [1], D.Balaji [1], Abdul Shahid Afrid [2], Bhuma Naveenkumar[2], Daruga Raja Manikanta[2], Dara Amruthraj[2], Kalidasu Jayaram Krishna[2]

[1] Department of Mechanical Engineering, SCSVMV, Enathur 631 561.Tamil Nadu, India
[2] Department of Mechanical Engineering, Ramachandra college of Engineering, Eluru, Andhra Pradesh.

e4ekam@gmail.com, km_vetrivel@yahoo.com , balajibala81@gmail.com

Abstract. The tribological characteristics of Aluminium metal matrix with different nano BN powder compositions were investigated using pin on disc tribometer. The pure aluminium, Al +1 wt% BN, Al + 3wt% BN, and Al + 5wt% BN compositions are prepared as 36 specimens using Powder metallurgy technique. In each composition 9 specimens undergone testing with three different load conditions and three different sliding velocities applied at constant sliding distance of 1000m. The result shows that when there is increasing the weight percentage of BN nano particles in aluminium metal matrix there is reduction of coefficient of friction by increasing the load and sliding velocities. Similarly the specific wear rate also gets reduced by increasing the load and sliding velocity.

1. Introduction

The most widely used metal in Metal matrix composites is Aluminium due its wide applications in the industries [1, 2]. There is a good improvement in the tribological properties of aluminium metal matrix reinforced by different nano materials [3-7]. The specific wear rate of Aluminium alloy composites decreases with increase in volume fraction and size of reinforcement. Load and sliding distance show a positive influence on wear implying increase of wear with increase of either load or sliding distance or both. [8]. The sliding speed variations will also causes changes in the wear behavior of aluminium metal matrix composites [9]. Boron Nitride is the solid lubricant material with low density and high thermal conductivity and high electrical resistivity. It also possesses the better wettability and also suggested BN might be an optimum reinforcement for an Al matrix composite [9-16]. Adel Mahmood Hassan et al. have reported decrease in hardness with increase in % reinforcement of Gr due to increased porosity [22].The coefficient of friction is decreased with increasing wt.% of MoS2 and increasing the due to load. At a higher load and sliding speed the tribolayer that is formed increases, due to squeezing out of solid lubricant from the hybrid composites [23].

IOP Conf. Series: Materials Science and Engineering **390** (2018) 012085 doi:10.1088/1757-899X/390/1/012085

2. Materials

The materials used in present work are 98% pure Aluminium and 99.8% pure Boron nitride nano powder. Boron Nitride Nano particles size limited to 70-80 nm with hexagonal crystal structure. Its density is 2.29 g /cm^3. Aluminium Powder with size limited to 130 – 180 micrometer, density 2.70 g /cm^3 and bears a melting point of 660.32°C.The Aluminium fine powder purchased from LOBA CHEMIE PVT.LTD., MUMBAI, INDIA. The Boron Nitride nano powder purchased from US Research Nanomaterials, Inc, Houston, USA.

3. Experimental Procedure

3.1. Fabrication of specimens

The Blending process is done by planetary ball milling machine using Al powder with 1 wt%, 3 wt%, 5 wt% of BN nano powder. The Blended powder is compacted in a Compacting machine to attain 30mm height. The compacted specimens are sintered using Sintering machine for 2hrs at 600°C.

3.2. Wear testing experimentation.

Each specimen prepared as per ASTM G99 standards and undergone wears testing using Pin on Disc tribometer. There are 3 different load conditions such as 5N, 7N, 10N were applied for testing. There are 3 different sliding velocities conditions such as 1.5m/s, 2 m/s, 2.5m/s were also applied for testing. Table.1 and Table.2 represented below shows the testing results obtained. The wear rates, frictional force, coefficient of friction are calculated for each specimen using Pin on Disc tribometer.

4. Results and Discussions

4.1. Wear testing results.

Table 1 represents the Specific wear rate of all the specimens, hence wear volume per unit distance can be calculated. The formula for Specific wear rate is given below in Equation (1).

Specific wear rate = [illegible] m^3/Nm - Equation (1)

Table 1. Specific Wear Rate calculation for all specimens.

Sl. No	Sliding Distance (m)	Sliding Velocities (m/s)	Load (N)	Specific Wear Rate (SWR) In terms of E^{-9} mm^3/Nm			
				100% Pure Al	99 wt%Al + 1wt%BN	97 wt% Al + 3wt%BN	95 wt% Al + 5wt%BN
1	1000	1.5	5	1.0322	0.9084	8.4592	11.1094
2	1000	1.5	7	9.8238	8.4363	10.5602	17.0982
3	1000	1.5	10	10.0963	9.9551	10.7387	19.1312
4	1000	2	5	7.4948	4.3245	9.3451	19.1756
5	1000	2	7	7.5354	3.9867	11.6831	20.9876
6	1000	2	10	10.0758	8.5645	14.3412	24.1023
7	1000	2.5	5	10.2008	8.5213	11.5106	18.3417
8	1000	2.5	7	10.395	8.7623	14.1214	22.5106
9	1000	2.5	10	10.5537	9.1234	15.0987	24.3234

IOP Conf. Series: Materials Science and Engineering **390** (2018) 012085 doi:10.1088/1757-899X/390/1/012085

Coefficient of Friction = $\frac{(\quad)}{\text{[illegible]} \; (\;)}$ - Equation (2)

Table 2. Co-Efficient of friction calculation for all specimens.

S.No	Sliding Distance (m)	Sliding Velocities (m/s)	Load (N)	Co-efficient of friction			
				100% Pure Al	99 wt%Al + 1wt%BN	97 wt% Al + 3wt%BN	95 wt% Al + 5wt%BN
1	1000	1.5	5	0.71	0.46	0.18	0.1
2	1000	1.5	7	0.71	0.51	0.08	0.03
3	1000	1.5	10	0.72	0.54	0.03	0.01
4	1000	2	5	0.76	0.69	0.14	0.08
5	1000	2	7	0.56	0.53	0.05	0.02
6	1000	2	10	0.54	0.51	0.01	0.01
7	1000	2.5	5	0.6	0.43	0.09	0.04
8	1000	2.5	7	0.47	0.34	0.02	0.01
9	1000	2.5	10	0.57	0.29	0.01	0.01

4.2 Specific wear rate versus load:

The test results shows that when 1.5m/s, 2m/s, 2.5m/s sliding velocities were applied, the specific wear rate of Al + 1wt% BN is decreasing when compared with other compositions. When there is increase in Normal load also causes increase in specific wear rate. When there is increase in BN nano particles composition the specific wear rate also gets increased. The Figures 1 (a)-(c) represents the specific wear rate versus normal load at different sliding velocity conditions.

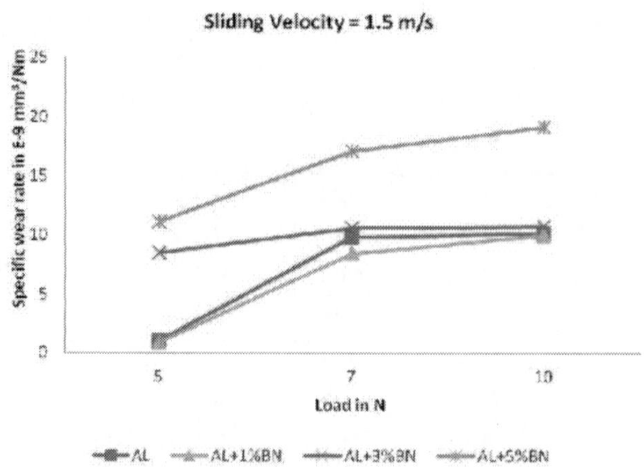

Figure 1 (a) Specific wear rate versus Load at 1.5 m/s Sliding velocity condition,

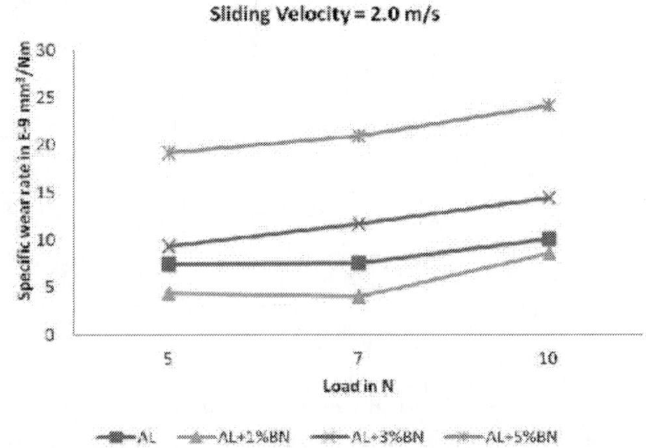

Figure 1 (b) Specific wear rate versus Load at 2 m /s Sliding velocity conditions,

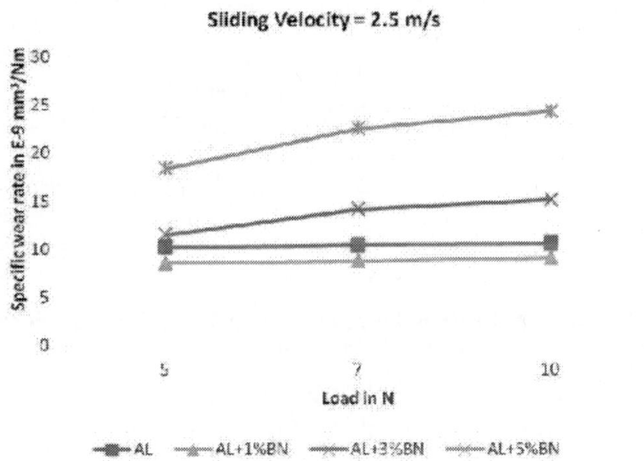

Figure 1 (c) Specific wear rate versus Load at 2.5 m /s Sliding velocity conditions,

4.3 Coefficient of friction versus load:

The test results shows that when 1.5m/s, 2m/s, 2.5m/s sliding velocities were applied, the coefficient of friction is decreasing when BN nano particle compositions are increased with increasing load and sliding velocities.. The Figures 2 (a)-(c) represents the coefficient of friction versus normal load at different sliding velocity conditions.

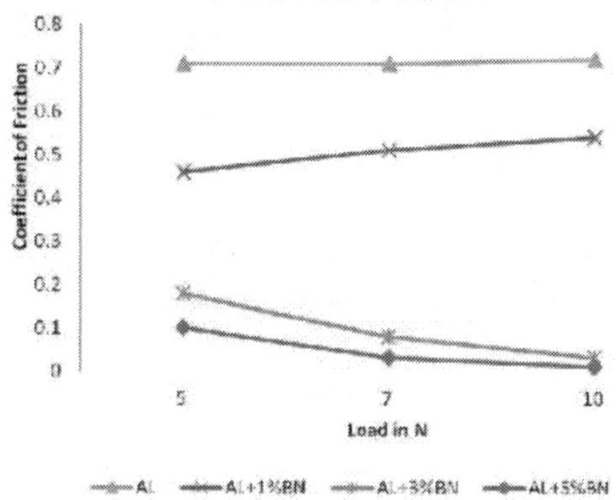

Figure 2 (a) Coefficient of Friction versus Load at 1.5 m /s Sliding velocity conditions,

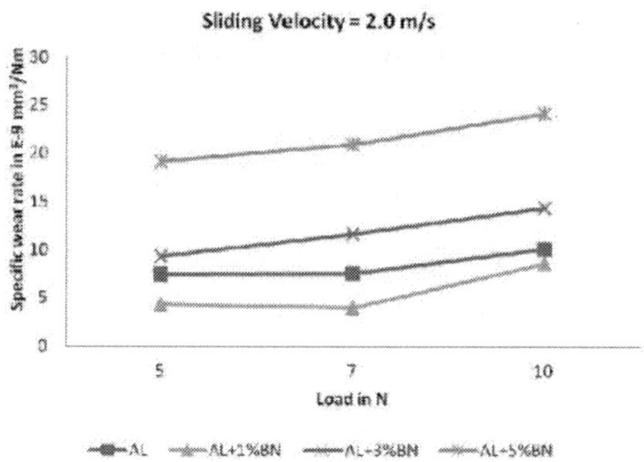

Figure 2 (b) Coefficient of Friction versus Load at 2 m /s Sliding velocity conditions,

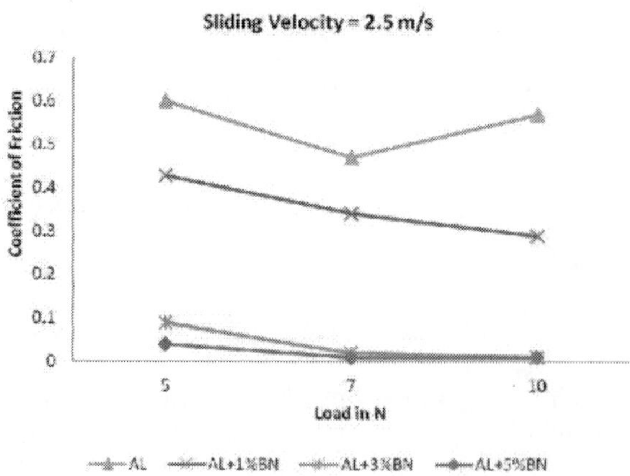

Figure 2 (c) Coefficient of Friction versus Load at 2.5 m /s Sliding velocity conditions,

5. Conclusion

The results showed that Al+1 wt% BN has better wear resistive ability compared with other compositions used in present work. Further reduction of wt % may increase the wear resistive ability and reduction in specific wear rate. The increase in sliding velocity and load causes decrease in coefficient of friction.

References

[1] Casati R, Vedani M, 2014 *Metal matrix composites reinforced by nano-particles - a review, Metals* **4** 65–83.

[2] Veeresh Kumar G.B, Rao C.S.P, Selvaraj N, 2O11 *Mechanical and tribological behavior of particulate reinforced aluminum metal matrix composites – a review, J. Miner. Mater. Charact. Eng.* **10** 59–91.

[3] Deuis, R.L., Subramanian, C., Yellup, J.M.1997: *Dry sliding wear of aluminium composites—a review. Compos. Sci. Technol.* **57**, 415–435

[4] Hosking FM, Portillo FF, Wunderlin R, Mehrabian R.1982 *Composites of aluminium alloys: fabrication and wear behaviour. Journal of Materials Science.* **17**:477-98

[5] Hutchings IM. 1992 London *Tribology: Friction and wear of engineering materials.* **1st ed**..

[6] Rohatgi PK, Liu Y, Ray S. 2004 *Friction and wear of metal–matrix composites, ASM hand book*, **vol. 18**; p. 801–11

[7] Mohammad Sharifi.E, Karimzadeh.F, 2011 *Wear behavior of aluminum matrix hybrid nano composites fabricated by powder metallurgy, Wear* **271** 1072–1079

[8] Garcia-Cordovilla C, Narciso J, Louis E. 1996 *Abrasive wear resistance of aluminium alloy/ceramic particulate composites.* Wear;**192**:170

[9] Basavarajappa S, Chandramohan G, Mahadevan Arjun, Tangavelu Mukundan, Subramanian R, Gopalakrishnan P. 2007 *Influence of sliding speed on the dry sliding wear behaviour and the subsurface deformation on hybrid metal matrix composite. Wear*;**262**:1007–12.

[10] Xue XM, Wang JT, Quan MX. 1991 *Mater Sci Engng A*;**132**:277

[11] Shen L, Tan BJ, Willis WS, Galasso FS, Suib SL.1994 *J Am Ceram Soc*;**77**:1011.

[12] Fujii H, Nakae H, Okada K. 1993 *Acta Metall Mater*;**41**:2963

[13] Xue XM, Wang JT, Quan MX. 1991 *J Mater Sci*;**26**:6391

[14] Xue XM, Wang JT, Zhao FM. 1992 *Mater Sci Lett*;**11**:199.

[15] Kobashi M, Choh T. 1997 *J Mater Sci*;**32**:6283

[16] Nicholas MG, Mortimer DA, Jones LM, Crispin RM, 1990 *J Mater Sci*;**25**:2679.

[17] Chiaramonte FP, Rosenthal BN. 1991 *J Am Ceram Soc*;**74**:658

[18] Suresh S, Moorthi S V N, Vettivel S C, Selvakumar N, Jinu G R, 2014 *Effect of graphite addition on mechanical behavior of Al6061/TiB2 hybrid composite using acoustic emission,Mater.Sci.Eng.***A612** 16–27.

[19] Chi H, Jianga L, Chen G, Kang P, Lin X, . Wu G, 2015 *Dry sliding friction and wear behavior of (TiB2+h-BN)/2024Alcomposites,Mater.Des.***87** 960–968.

[20] Amal E Nassar,Eman E Nassar 2015 *Journal of King saud University- Engineering science.*Volume **29**, 295-299

[21] Hassan Adel Mahmood, Tashtoush Ghassan Mousa, Al-Khalil Jafar Ahmed. 2007 *J Compos Mater*; **41**:453–65.

[22] Pitchayyapillai G, Seenikannan P, Raja K, and Chandrasekaran K 2016 *Advances in Materials Science and Engineering* **2016**, 1-9.

IOP Conf. Series: Materials Science and Engineering **390** (2018) 012086 doi:10.1088/1757-899X/390/1/012086

Free convection of nanoliquids in an enclosure with sinusoidal heating

S. Sivasankaran[1,*] T. Aasaithambi[2], M. Bhuvaneswari[1], S. Rajan[3]
[1]Department of Mathematics, King Abdulaziz University, Jeddah, Saudi Arabia
[2]Department of Mathematics, Sri Shanmugha College of Engineering and Technology, Pullipalayam, Sankari, Tamilnadu, India
[3]Department of Mathematics, Erode Arts and Science College, Erode, Tamilnadu, India
[*]Corresponding author email: sd.siva@yahoo.com; sdsiva@gmail.com.

Abstract. The goal of the current numerical research is to explore the convection of different nanoliquids in a square cavity. The temperature at left wall varies sinusoidally whereas the temperature at right wall is kept as constant. The horizontal walls are taken as adiabatic. The finite volume method is utilized to discretize the governing equations and the solutions are found iteratively for diverse combinations of relevant parameters involved in the study. It is established that the energy transfer enhances with raising the nanoliquid volume fraction. The growth in the averaged energy transport strongly depends on the nanoparticle chosen.

1. Introduction

Convective cooling augmentation method has been a key problem in many industrial and technological applications because of thermal controlling of devices. In recent years, nanoliquids are considered in heat transfer research to analyze the number of factors in different ways. Ho *et al.* [1] numerically examined the properties of nanoliquids on convection in the enclosure. Sivasankaran et al. [2] numerically examined the free convection of nanoliquids in an enclosed space with linearly heated wall. Ho et al. [3] experimentally investigated the free convection of nanoliquid in a vertical enclosure. The thermophoresis, sedimentation, and Brownian motion effects on Rayleigh–Bénard free convection of nanoliquids in an enclosed space were examined by Ho et al. [4].

The maximum number of the reports on convection of nanoliquids in cavities have been considered either isoflux and/or isothermal thermal conditions in boundaries. But, these temperature conditions at boundary are not appropriate in numerous applications. Therefore, we want the awareness of the outcome of non-isothermal wall temperature to analyze several applications. The non-uniform heating effect on both walls in a cavity was examined by Sivasankaran et al. [5]. The MHD convection in cavities with non-uniform boundary heating was examined by several authors [6-7]. Convection in a porous enclosed space with sinusoidal boundary heating on side walls was reported in Ref. [8-9]. The aspect ratio and inclination effects on free convection with sinusoidal surface condition was explored by Cheong et al. [10]. Sivasankaran and Pan [11] numerically examined the free convection of nanoliquids with sinusoidal heating walls. The non-uniform temperature on sidewall(s) in a cavity under different situations is examined by several authors in different studies [12-16].

No work has been executed on convection flow of nanoliquids for non-uniform boundary conditions with different nanoliquids. Hence, the present report aims to explore the effects of free convection characteristics with sinusoidally varying wall temperature utilizing different nanoliquids.

2. Mathematical Modelling

A two-dimensional enclosed space of height L filled with water-based nanoliquids containing various nano-particles (Al_2O_3, Cu, Ag and TiO_2) is taken for examination as displayed in the Figure 1. The vertical left wall is sinusoidally heated and the vertical right wall is cooled at a uniform temperature. The bottom & top boundaries are adiabatic. The nanoliquid is a liquid-solid blend with uniform shape, size, and volume fraction of nano-particles distributed within base liquid water. The flow is presumed to be incompressible and laminar. It is presumed that both water and nano-particles are in thermally equilibrium. The buoyancy term based on Boussinesq approximation is added and other thermo-physical properties are expected to be constant. It is presumed that no chemical reaction between the nanoparticles and water. The mathematical model for the above said geometrical and physical conditions can be written as follows.

IOP Conf. Series: Materials Science and Engineering **390** (2018) 012086 doi:10.1088/1757-899X/390/1/012086

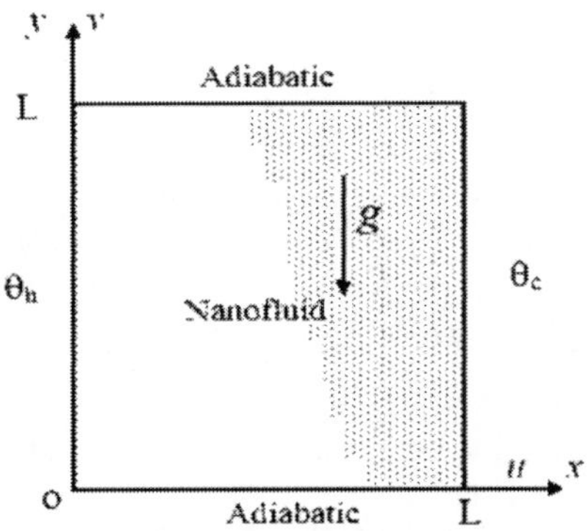

Fig. 1. Physical configuration

$$u_x + v_y = 0 \tag{1}$$

$$u_t + uu_x + vu_x = -(1/\rho_{nf})p_x + (\mu_{nf}/\rho_{nf})[\nabla^2 u] \tag{2}$$

$$v_t + uv_x + vv_x = -(1/\rho_{nf})p_y + (\mu_{nf}/\rho_{nf})[\nabla^2 v] + (g(\rho\beta)_{nf}/\rho_{nf})(\theta - \theta_c) \tag{3}$$

$$\theta_t + u\theta_x + v\theta_y = \alpha_{nf}[\nabla^2 \theta] \tag{4}$$

The no-slip velocity conditions are taken on all walls. The temperature on left and right walls are taken as θ = sin(y/L) and $\theta = 0$. The thermo-physical properties of the nanoliquid calculated using several formulae are as follows.

$$\rho_{nf} = \phi\rho_p + (1-\phi)\rho_f$$

$$(\rho\beta)_{nf} = \phi(\rho\beta)_p + (1-\phi)(\rho\beta)_f$$

$$(\rho c_p)_{nf} = \phi(\rho c_p)_p + (1-\phi)(\rho c_p)_f$$

$$k_{nf} = k_f\left[\frac{2\phi(k^*_{pf}-1) + k^*_{pf} + 2}{k^*_{pf} - \phi(k^*_{pf}-1) + 2}\right] \quad \text{with} \quad k^*_{pf} = k_p / k_f$$

$$\mu_{nf} = \mu_f(1-\phi)^{-2.5}$$

The properties of the water and various nano-particles are found in (Ho et al., 2008). The dimensionless variables are derived as follows.

$$(U,V) = (u,v)L/\alpha_{nf},\ (X,Y) = (x,y)/L,\ T = \frac{\theta - \theta_c}{(\theta_h - \theta_c)},\ F_0 = \frac{t\alpha_{nf}}{L^2},\ \text{and}\ P = \frac{pL^2}{\rho_{nf}\alpha^2_{nf}}.$$

Using this non-dimensional variables, the equations are non-dimensionalized and they are given as follows:

$$\frac{\partial U}{\partial X} + \frac{\partial V}{\partial Y} = 0 \tag{5}$$

$$\frac{\partial U}{\partial F_O} + U\frac{\partial U}{\partial X} + V\frac{\partial U}{\partial Y} = -\frac{\partial P}{\partial X} + \Pr_f\left(\frac{C^*_{p,nf}\mu^*_{nf}}{k^*_{nf}}\right)\left[\frac{\partial^2 U}{\partial X^2} + \frac{\partial^2 U}{\partial Y^2}\right] \tag{6}$$

$$\frac{\partial V}{\partial F_O} + U\frac{\partial V}{\partial X} + V\frac{\partial V}{\partial Y} = -\frac{\partial P}{\partial Y} + \Pr_f\left(\frac{C^*_{p,nf}\mu^*_{nf}}{k^*_{nf}}\right)\left[\frac{\partial^2 V}{\partial X^2} + \frac{\partial^2 V}{\partial Y^2}\right] + Ra_f \Pr_f \beta^*_{nf}\left(\frac{\rho^*_{nf}C^*_{p,nf}}{k^*_{nf}}\right)^2 T \tag{7}$$

$$\frac{\partial T}{\partial F_O}+U\frac{\partial T}{\partial X}+V\frac{\partial T}{\partial Y}=\left[\frac{\partial}{\partial X}\left(k_{nf}^{*}\frac{\partial T}{\partial X}\right)+\frac{\partial}{\partial Y}\left(k_{nf}^{*}\frac{\partial T}{\partial Y}\right)\right] \tag{8}$$

The surface conditions in the non-dimensional form are as follows. No slip condition for velocity on side walls. The left and right walls temperature are taken as T = sin(Y) and T=0.

The physical properties ratios appear in equations (2-4) are as follows. $k_{nf}^{*}=k_{nf}/k_{f}$, $\mu_{nf}^{*}=\mu_{nf}/\mu_{f}$, $Cp_{nf}^{*}=Cp_{nf}/Cp_{f}$, $\rho_{nf}^{*}=\rho_{nf}/\rho_{f}$, and $\beta_{nf}^{*}=\beta_{nf}/\beta_{f}$, where the subscripts *nf* and *f* denote the nanoliquid and water. The non-dimensional numbers performed in the above equations are, $Ra_{f}=\frac{g\beta_{f}(\theta_{h}-\theta_{c})L^{3}}{\nu_{f}\alpha_{f}}$, Rayleigh number, and $\Pr_{f}=\frac{\nu_{f}}{\alpha_{f}}$, Prandtl number (Pr=6.7). The stream function is defined by $U=\frac{\partial\Psi}{\partial Y}$ and $V=-\frac{\partial\Psi}{\partial X}$.

The energy transfer rate at the heated surface is obtained by the Nusselt number, which is estimated as $Nu_{h}=\frac{h_{n}L}{k_{f}}=-k_{nf}^{*}\left.\frac{\partial T}{\partial Y}\right|_{X=0}$. The averaged Nusselt number along the heated surface is acquired as $\overline{Nu}=\int_{0}^{1}Nu_{h}dY$. Moreover, it is also significant to enumerate the energy transport efficacy of using the nanoliquid to that of the water. The averaged energy transfer coefficient ratio of nanoliquid to that of the water, ε_h, is computed as $\varepsilon_{h}=\overline{h_{n}}/\overline{h_{f}}$. The solution of the dimensionless governing systems are done by finite volume method. The detailed solution methodology is found in Ref. Sivasankaran et al. (2010).

3. Results and Discussion

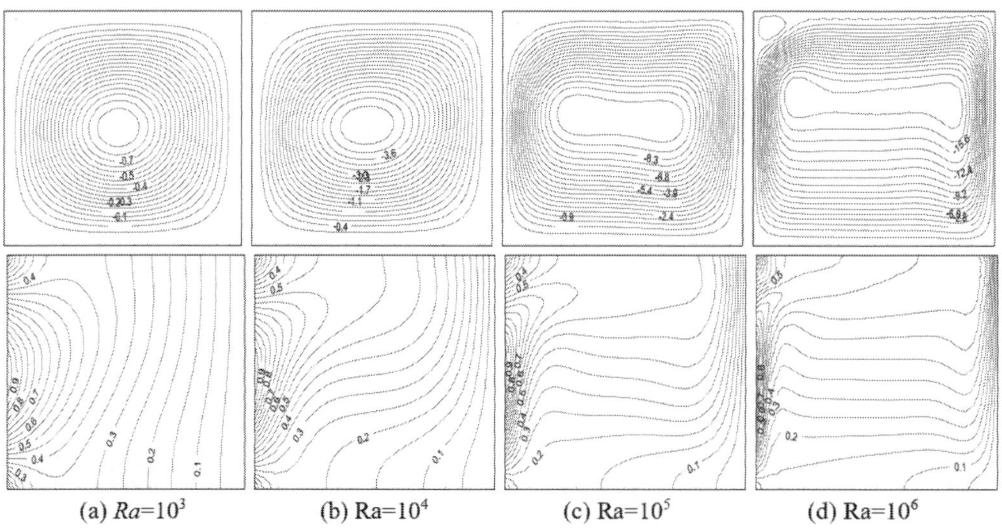

(a) $Ra=10^3$ (b) $Ra=10^4$ (c) $Ra=10^5$ (d) $Ra=10^6$

Fig. 2. Streamlines (up) and isotherms (down) of Al_2O_3-nanofluid with ϕ=4% for various Ra.

The calculations are done for the Rayleigh numbers from 10^3 to 10^6, and the volume fraction ϕ of nano-particles from 0% to 4%. Figure 2 displays liquid stream and temperature characteristics of Al_2O_3-nanofluid for different Rayleigh numbers. The stream pattern is involves a cell occupying the complete enclosure. The core area of the cell is at central part of the cavity for the weak buoyancy force ($Ra=10^3$). The isotherms are spread equally inside the cavity. Here energy transport is by means of conduction dominated mode. When increasing Ra, the core area of the cell elongated horizontally ($Ra \geq 10^5$). The minor cell exists in the high values of Rayleigh number at the left-top corner due to non-uniform heating. The isotherms illustrate the vertical stratification of temperature in the cavity. The isotherms are crowded beside the thermal walls and forming the thermal boundary layers. It clearly indicates that convection mode of heat transfer is dominated here.

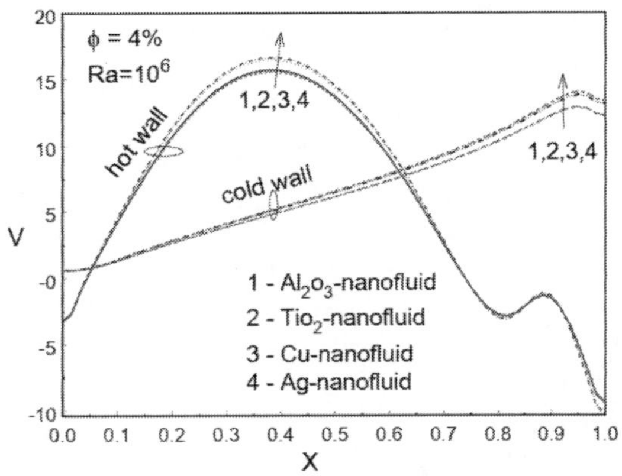

Fig. 3. Local Nusselt number for various nanofluids with $Ra=10^6$ and $\phi=4\%$.

The effect of local energy transfer rate among the different nanofluid is displayed in Figure 3. It is witnessed that the higher energy transfer is gained when using Ag-nanoliquid. A further inspection of the profiles exposes that the highest value of local energy transfer rate is achieved at Y=0.35 for $Ra=10^6$, that is ¼ from the bottom of the warmer wall. It is witnessed that the highest local energy transport rate is achieved at the middle part of left wall for sinusoidally varying wall temperature and at the top of the wall for isothermal surface temperature. The wavy shape of local Nusselt number profile indicates the direct effect of non-uniform heating of sidewall on energy transfer.

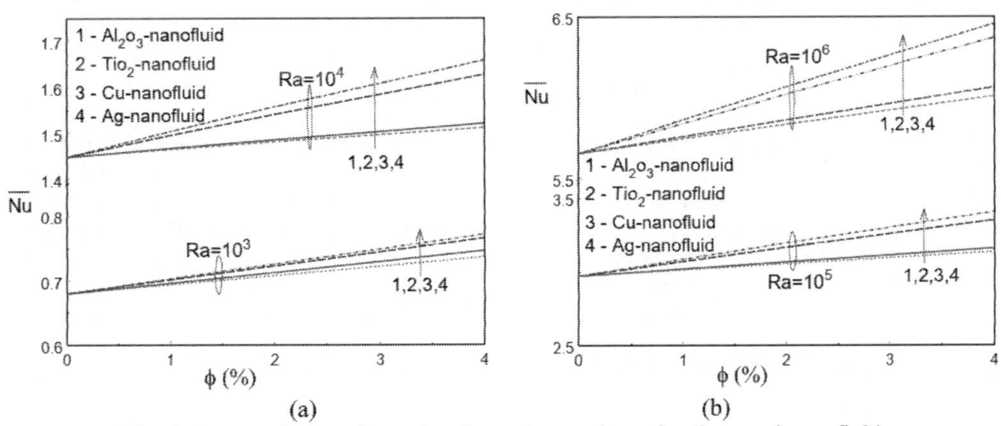

Fig. 4. Averaged Nusselt number for various volume fractions and nanofluids.

Figure 4 provides the effect of total energy transport rate through the enclosure among the various nanoliquids. The favorable effects of the nano-particle fraction on the averaged Nusselt number trend can be perceived clearly with raising the particle fraction. Averaged Nusselt number enhances on raising the values of Ra. It is established that the maximum energy transfer rate is witnessed for using Ag-nanoparticle comparing among other nanoparticles, like Cu, Al_2O_3 and TiO_2. When raising the nanoliquid volume fraction, the variation in the energy transport rate amongst the various nanoliquids is increased. The change amongst several nano-particles plays a key factor on the convective energy transport rate, which is evidently shown in the Figure 4.

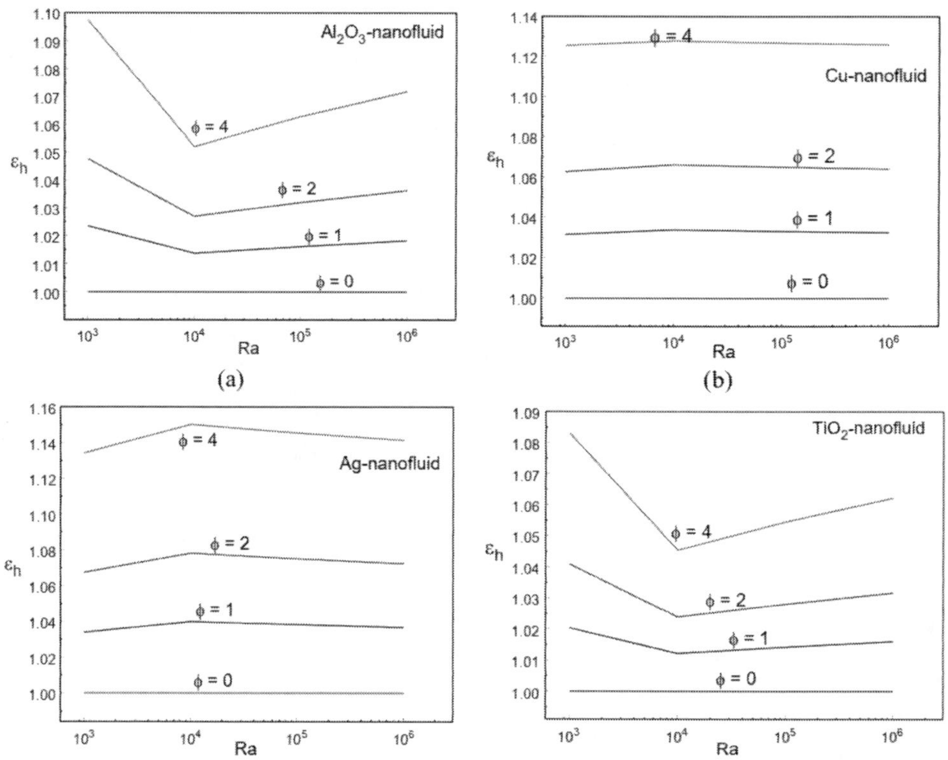

Fig. 5. Heat transfer ratio for different nanofluids.

Figure 5 demonstrated the energy transfer efficacy of the nanoliquid for various volume fraction of the nanoparticle versus Ra. The heat transfer ratio of oxide-nanoliquids (Al_2O_3 & TiO_2) gets minimum when $Ra=10^4$. However, the reverse tendency is detected for using metal nano-particles (Cu and Ag), that is, maximum energy transport rate is achieved at $Ra=10^4$. It is also acquired that energy transport coefficient ratio is higher than one always. It is concluded from these figures that nanoliquids with Al_2O_3 and TiO_2-particle are having similar behaviour while Cu and Ag-nanoliquids are so. The maximum value of the heat transport ratio for the nanoliquids with the oxide particle is attained for $Ra=10^3$ whereas highest value of the heat transfer coefficient ratio for Ag- and Cu-nanoliquids is obtained for $Ra=10^4$. Table 1 shows a significant variance on the averaged Nusselt number for various nano-particles.

4. Final Remarks

The study numerically inspects the heat transfer augmentation of nanoliquids of four different nanoparticles (Al_2O_3, Cu, Ag and TiO_2) in a rectangular cavity with sinusoidally changing wall temperature. The energy transfer ability of base liquid can be improved when immersing the nano-particles in base fluid and the effect is noticeable as the particle volume fraction increases. However, the growth in averaged Nusselt number is powerfully depending on the nano-particle selected. The heat transfer ratio of Al_2O_3- and TiO_2-nanoliquids is minimum when $Ra=10^4$ while it is high for Cu- and Ag-nanofluid. The nanoliquids with oxide-particles having the similar behaviour on the flow and heat transfer while the nanoliquids with metal particles are having similar trend. A significant variance on the averaged Nusselt number is detected for various nano-particles. The outcomes evidently demonstrate that the type of nanoparticle considered is vital factor on the convective cooling applications.

Table 1. Heat transfer enhancement for different values of ϕ (%) and different nanofluids

		Enhancement (%) (A-B)/B×100			
Ra	ϕ (%)	Al_2O_3	TiO_2	Cu	Ag
10^3	1	2.3578345	2.0362916	3.1498564	3.3799499
	2	4.7674217	4.0930196	6.2907442	6.7438742
	4	9.7377374	8.3093928	12.567080	13.425409
10^4	1	1.3770573	1.2174648	3.3874794	4.0003232
	2	2.7013461	2.3919569	6.6296668	7.8183877
	4	5.2004845	4.5435240	12.788564	15.026077
10^5	1	1.6224753	1.4230049	3.3109584	3.8243867
	2	3.2103653	2.8061427	6.5233759	7.5138307
	4	6.2830292	5.4566521	12.689394	14.556959
10^6	1	1.8318063	1.5984382	3.2496021	3.6731750
	2	3.6393797	3.1680655	6.4284496	7.2449994
	4	7.1787521	6.2168752	12.606351	14.142465

References

[1]. Ho C.J., Chen M.W., and Li, Z.W. 2008. *Int. J. Heat Mass Transfer*, 51, 4506-4516.
[2]. Sivasankaran, S., Aasaithambi, T., and Rajan, S. 2010. *Maejo Int. J. Sci. Tech*. 4(3), 468-482.
[3]. Ho, C.J., Liu, W.K., Chang, Y.S., Lin, C.C. 2010. *Int. J. Thermal Sciences*, 49, 1345-1353.
[4]. Ho C.J., Chen D., Yan W.M., Mahian O. 2014. *Int. Commun. Heat Mass Transfer* 57, 22-26.
[5]. Sivasankaran S., Sivakumar V., Prakash P. 2010. *Int. J. Heat Mass Transfer*, 53, 4304-4315.
[6]. Bhuvaneswari M, Sivasankaran S, Kim YJ, 2011, *Numerical Heat Transfer A*, 59, 167-184.
[7]. Sivasankaran S., Malleswaran A., Lee J., Sundar P. 2011. *Int. J. Heat Mass Transfer*, 54, 512-525.
[8]. Sivasankaran S., Bhuvaneswari M. 2013. *Numerical Heat Transfer A*, 63(1), 14-30.
[9]. Sivasankaran S., Pan K.L. 2012. *Numerical Heat Transfer A*, 61(2), 101-121.
[10]. Cheong H.T., Zailan Siri, Sivasankaran S. 2013, Int. Commun. Heat Mass Transfer, 45, 75-85.
[11]. Sivasankaran S., Pan K.L. 2014. Numerical Heat Transfer A, 65, 247-268.
[12]. Janagi, K., Sivasankaran, S., Bhuvaneswari, M., Eswaramurthi, M. 2017. *Int. J. Numerical Methods Heat Fluid Flow*, 27(4), 1000-1014.
[13]. Sivakumar, V., Sivasankaran, S. 2014. J. Applied Mech. Tech. Physics, 55(4), 634-649.
[14]. Cheong H.T., Sivasankaran S., Bhuvaneswari M, 2017. *Int. J. Numerical Methods Heat Fluid Flow*, 27 (2), 287-309.
[15]. Sivasankaran, S., Ananthan, S.S., Abdul Hakeem, A.K. 2016. Scientia Iranica -B Mech Engg, 23(3), 1027-1036.
[16]. Sivasankaran S., Cheong H.T., Bhuvaneswari M., Ganesan, P. 2016. Numerical Heat Transfer A, 69(6), 630-642.

Multi Objective Process Parameters Optimization of Friction Stir Welding using NSGA – II

D.Vijayan[1*], P. Abhishek[2]
[1]Assistant Professor, Department of Mechanical Engineering, Sri Chandrasekharendra Saraswathi Viswa Mahavidyalaya (SCSVMV), Enathur, Kanchipuram, Tamilnadu, India.
[2]Student, Department of Mechanical Engineering, Sri Chandrasekharendra Saraswathi Viswa Mahavidyalaya (SCSVMV), Enathur, Kanchipuram, Tamilnadu, India.

[1*]Corresponding author: vijaiand2012@gmail.com

Abstract

Age hardenable aluminum alloys such as AA2024 and AA6061 are extensively usd in automotive, aircraft and marine industries because of its high strength to weight ratio and good ductility. When compared to fusion welding techniques Friction Stir Welding (FSW) is a promising technique to improve the quality of the weld joint. FSW process parameters such as tool rotational speed (N), welding speed (F), axial load (P) and tool pin profiles (Cylindrical, Square and Taper) have crucial effects on the mechanical properties of FS welded joints significantly. The mathematical model was developed for the responses namely Ultimate Tensile Strength (UTS) and Tensile Elongation (TE) using Response Surface Method (RSM) and the developed mathematical model was effectively utilized to optimize the FS welded process parameters using non-dominated sorting genetic algorithm-II (NSGA-II). The maximized tensile properties of FS welded age hardenable aluminum alloys were obtained nearly 143.3 N/mm^2 with its corresponding tensile elongation of 12.53%. It is concluded that, the NSGA – II is very useful in optimizing the process parameters of FS Welding and improving the quality of the weld joint.

Keyword: Friction stir welding, NSGA, tensile, elongation, micro structure, response surface

1. Introduction

In the present scenario light weight structure applications are more demanding in nuclear, aircraft and marine industries to improve fuel economy and payload capability. Friction stir welding (FSW) is a modern solid-state welding is used for joining many light-weight alloys such as aluminum alloys, magnesium-based alloys. In FSW, a rotating tool comprising of a pin and a shoulder is squeezed against the synchronized ends of two metal plates to be welded, while traversing along the weld line. A significant advantage in FSW is, heat transfer between the plates is accomplished by the deformation of the two plates and the tool therefore no melting occurs therefore weld line is made in solid state. In recent years joining of age hardenable aluminum alloys is an intriguing topic in the welding research. Due to inherent weld qualities such as porosity and sensitive to cracking, age hardenable aluminum alloys are not suitable to weld using conventional welding technique hence; these materials are categorized by difficult to weld metals. Owing to its excellent mechanical and metallurgical properties such as fine recrystalized grains, and reduction or absence of superficial and internal flaw on the weld zone etc., Weld quality is affected by many process parameters in FSW that leads imperfections like voids, pinholes during the welding process (1). It is difficult to select appropriate process parameters even a skilled labor to obtain optimum weld quality. Therefore, it is necessary to obtain a mathematical relationship between the process parameters to optimize the weld quality. In the past 5 years, evolutionary algorithms such as artificial neural network (2), genetic algorithm (3-5), simulated annealing algorithm (6), particle swarm optimization, and Non Dominated Sorting Genetic Algorithm (NSGA – II) have become popular in finding the suitable process parameters, especially if the process is too complex.

Among the various popular evolutionary algorithms, Non-Dominated Sorting Genetic Algorithm - II (NSGA - II) is extensively adopted in all manufacturing processes such as machining, forming and welding as an effective approach to improve the quality of the product. NSGA - II searches a whole design space for global optimum operation by various design points using random genetic operations (7). The conventional NSGA has been generally criticized for its lack of elitism and need for specifying the sharing parameters. But the modified NSGA – II which has a better sorting algorithm and no sharing parameter needs to be specify sharing parameter. Despite the fact that the NSGA - II offers numerous advantages on manufacturing processes, very limited numbers of investigations were carried out so far on FSW of age hardenable alloys using NSGA - II. For instance, Bandaru et al. (8) proposed an investigation to minimize the temperature difference between the leading edge of the FSW tool probe and the work piece material using NSGA - II. Akabri et al (9) optimized the process parameters of A356 composite fabricated by FSP using TOPSIS and NSGA – II. Residual stress across the weld position on age hardenable aluminum alloys was estimated by Hidalgo et al (10) using NSGA – II algorithm. Though numerous attempts have been made to optimize the friction stir welding process parameters for improving the ultimate tensile strength and tensile elongation, an attempt of NSGA – II to find the optimal friction stir welding process parameters for improving the ultimate tensile strength and tensile elongation of AA2024 – AA6061 age hardenable aluminum alloys are very rare. Hence, in the present investigation the aluminum alloys AA2024 – AA6061 friction stir weld joints are considered to optimize the process parameters using NSGA – II.

2. Experimental setup

The schematic arrangement of FSW setup of joining age hardenable aluminum alloys and the fabricated FSW tools are presented in Figure 1. Age hardenable aluminum alloys of AA2024 and AA6061 were cut from 6.25mm sheet with dimensions of 300 mm x 150 mm by computerized friction stir welding machine to fabricate the dissimilar weld joints of AA2024-AA6061 aluminum alloys.

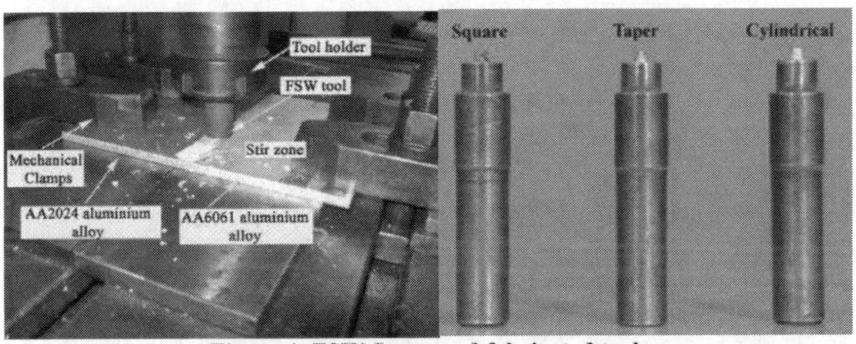

Figure 1. FSW Setup and fabricated tools

Four important FS welding process parameters namely tool rotational speed (N), welding speed (F), axial force (P) and pin profiles (PP) such as cylindrical (1), Square (2) and taper (3) is considered. A mechanical clamping device was used to hold the plates in position. H13 hot work steel was used to fabricate the rotating tools. Due to pin rotation the friction created heat in the pieces and turned the intended area for welding into plastic state. The use of protection gas during welding was not required, since there was no melting during this process (11). The FS welding process parameters and their range is presented in Table 1. 3 levels, 4 factors RSM based BBD design was used to design the experiment. Therefore, totally 31 experiments were conducted as per the design matrix.

Table 1. FS welding parameters and their range

FSW Parameters	Range
Tool rotational speed, N (rpm)	1700 – 1900
Welding speed, F (mm/min)	30 – 90
Axial force, P (kN)	3 – 9
Pin Profiles (PP)	Cylindrical (1), Square (2) and Taper (3)

All the fabricated specimens were subjected to ultimate tensile strength and tensile elongation as per ASTM E384. The tensile specimens were cut longitudinal direction along the stirred line. The fabricated FS welded samples as per the design matrix is presented in Figure 2. Scanning electron microscope (SEM) was used to analyze the microstructural properties of FS welds. Few defects were absorbed on the stir zone and are presented visually in Figure 3. Two responses namely ultimate tensile strength (UTS) and tensile elongation (TE) are considered for optimization. The design matrix and their corresponding results were presented in Table 2. When comparing to other pin profiles, square pins playing a significant role in yielding the maximum tensile strength and tensile elongation of AA2024 and AA6061 age hardenable aluminum alloys.

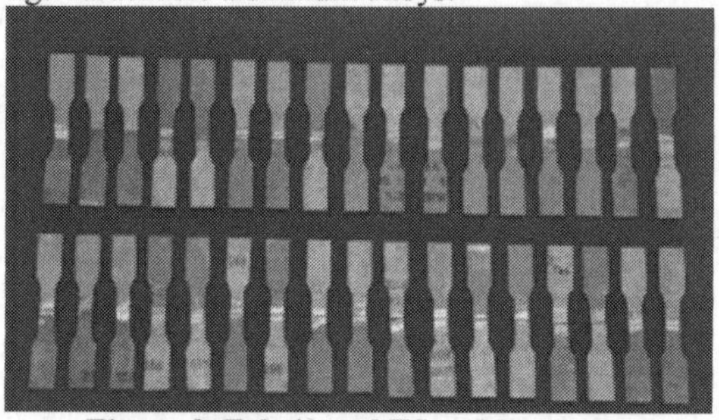

Figure 2. Fabricated FS weld samples

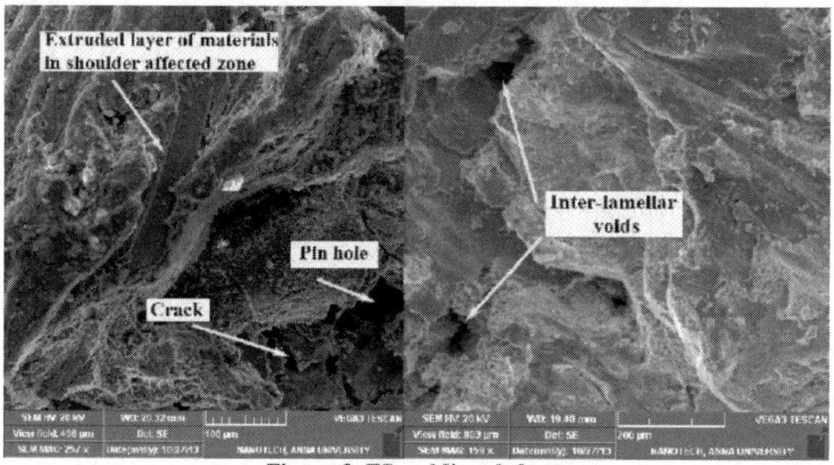

Figure 3. FS welding defects

Table 2. Design matrix and results

Std	Run	A:Rotational speed (RPM)	B:Welding speed (mm/min)	C:Axial load (kN)	D:Pin Profile	Ultimate Tensile Strength (UTS) (MPa)	Tensile Elongation (TE) (%)
1	7	1500	30	6	2	108.31	10.97
2	8	1900	30	6	2	109.33	9.23
3	11	1500	90	6	2	109.32	11.45
4	5	1900	90	6	2	101.06	9.45
5	25	1700	60	3	1	122.25	12.38
6	3	1700	60	9	1	127.09	10.16
7	4	1700	60	3	3	117.64	11.37

8	22	1700	60	9	3	124.31	9.72
9	18	1500	60	6	1	122.34	11.26
10	28	1900	60	6	1	119.06	10.54
11	6	1500	60	6	3	120.21	11.1
12	30	1900	60	6	3	115.37	9.11
13	21	1700	30	3	2	107.34	11.56
14	14	1700	90	3	2	106.13	11.77
15	23	1700	30	9	2	116.31	10.03
16	17	1700	90	9	2	112.31	10.49
17	20	1500	60	3	2	118.52	12.28
18	29	1900	60	3	2	115.98	10.12
19	19	1500	60	9	2	122.66	10.26
20	27	1900	60	9	2	124.22	9.37
21	31	1700	30	6	1	115.32	10.67
22	16	1700	90	6	1	104.34	11.67
23	26	1700	30	6	3	110.27	9.92
24	13	1700	90	6	3	108.31	10.89
25	24	1700	60	6	2	138.4	12.11
26	15	1700	60	6	2	143.64	12.25
27	2	1700	60	6	2	139.21	12.12
28	12	1700	60	6	2	139.28	12.48
29	1	1700	60	6	2	139.91	12.45
30	10	1700	60	6	2	141.34	12.38
31	9	1700	60	6	2	143.3	12.53

3. Multi objective optimization using NSGA

The main objective of the present investigation is to maximize the tensile properties such as ultimate tensile strength and tensile elongation by optimizing the FS welding process parameters. Therefore, it can be considered as multi-objective optimization problem. The Matlab code was developed using the version R2010b. The developed mathematic model of multi-objective problem can be developed as below:

Ultimate tensile strength (UTS) = -(-902.25563+1.03486*X(1)+3.33452*X(2)+12.33048*X(3)+32.16226*X(4)-0.000386*X(1)*X(2)+0.075167*X(2)*X(4)-0.000299*X(1)^2-0.024151*X(2)^2-0.93717*X(3)^2-9.46577*X(4)^2);
Tensile Elongation (TE) = -(-73.10948+0.096367*X(1)+0.11732*X(2)-0.31669*X(3)+5.21411*X(4)+0.000529167*X(1)*X(3)-0.0015875*X(1)*X(4)-0.000029507*X(1)^2-0.000903*X(2)^2-0.070450*X(3)^2 -0.72405*X(4)^2);

Subject to,
1700rpm $\leq$ Rotational speed (N) $\leq$ 1900
30 $\leq$ Welding speed (F) $\leq$ 90
3 $\leq$ Axial force (P) $\leq$ 9
1 $\leq$ Tool pin $\leq$ 3 where 1 = Cylindrical pin, 2 = Square Pin and 3 = Taper Pin

Since there are trade-offs among these two objectives, the optimization problem in equation (1) generally has a set of optimum solutions. These solutions are optimum in the Pareto sense, that is, there is no optimum better than other designs in all objectives. NSGA-II is implemented to solve the

FS welding process parameters optimization problem and obtain the Pareto optimal front. The settings for the NSGA-II in this optimization problem are presented in Table 3.

Table 3. NSGA input parameters

Parameters	Value
Population size	100
Mutation fraction	0.2
Maximum iterations	500
Pareto fraction	0.4
Elite count	2
Function tolerance	10E-06
Crossover function	0.8
Scaling fitness function	Rank

4. Optimization results

Figure.4 show plots the obtained Pareto-optimal front for Ultimate Tensile Strength and Tensile Elongation. Each point in Figure illustrates a specific optimal solution, whose corresponding process parameters can be adopted according to the requirements of the design makers. The obtained NSGA solutions are presented in Table. It can be concluded from Figure that all optima are non-inferior with each other because there are trade-offs among the ultimate tensile strength and tensile elongation, e.g., the maximum ultimate tensile strength obtained is nearly 143.3 N/mm^2 with its corresponding tensile elongation of 12.53%.

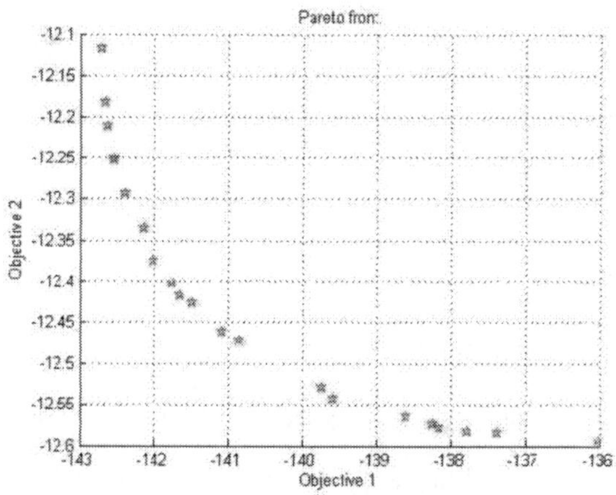

Figure 4. Pareto Optimal plot

Table 4. Pareto Optimal solutions

Run	Ultimate tensile strength (Mpa)	Tensile elongation (%)	Rotational speed (rpm)	Welding speed (mm/min)	Axial load (kN)	Pin profile
1	-133.47	-12.82	1626.34	65.20	3.86	1.82
2	-134.37	-12.82	1630.98	63.85	3.93	1.82
3	-142.79	-12.18	1692.15	59.25	6.59	1.94
4	-133.47	-12.82	1626.34	65.20	3.86	1.82
5	-139.58	-12.71	1643.75	61.74	5.04	1.84
6	-141.67	-12.54	1666.21	61.36	5.70	1.89

7	**-142.70**	**-12.30**	**1688.76**	**59.27**	**6.30**	**1.90**
8	**-142.27**	**-12.41**	**1669.84**	**61.27**	**6.15**	**1.87**
9	**-142.40**	**-12.38**	**1663.52**	**59.87**	**6.26**	**1.88**
10	-136.47	-12.80	1639.58	64.16	4.34	1.83
11	**-142.04**	**-12.49**	**1664.70**	**59.62**	**5.85**	**1.90**
12	-136.98	-12.79	1636.73	63.66	4.46	1.82
13	-139.18	-12.72	1655.52	62.39	4.82	1.92
14	-140.36	-12.67	1652.99	61.05	5.21	1.84
15	-135.98	-12.80	1634.61	64.29	4.25	1.88
16	-141.75	-12.52	1672.45	61.62	5.73	1.87
17	-139.70	-12.70	1640.17	62.38	5.13	1.89
18	**-142.62**	**-12.31**	**1683.17**	**60.24**	**6.36**	**1.87**
19	-138.63	-12.74	1659.14	62.31	4.66	1.86
20	**-142.18**	**-12.44**	**1668.58**	**61.32**	**6.06**	**1.89**
21	-138.98	-12.73	1645.32	63.63	4.95	1.84

Six solutions in the Pareto-optimal front were selected randomly (based on obtained maximum tensile strength and 'tensile elongation) (run no: 7,8,9,11,18, & 20) to verify the effectiveness of the optimal results. The fabricated weld samples based on confirmation experiments is presented in Figure 5. The confirmation experiment results are compared with optimization results. The obtained confirmation results and optimization results are plotted in Table 5.

Table 5. Confirmation experiment results

Experimenal run	Rotational speed (N), rpm	Welding speed (F), mm/min	Axial force (F), kN	Pin profile	NSGA results		Confirmation results	
					UTS (N/mm^2)	TE (%)	UTS (N/mm^2)	TE (%)
7	1688.76	59.27	6.3	1.9	142.7	12.3	143.14	12.54
8	1669.84	61.27	6.15	1.87	142.27	12.41	141.45	11.58
9	1663.52	59.87	6.26	1.88	142.4	12.38	141.35	12.11
11	1664.7	59.62	5.85	1.9	142.04	12.49	142.65	12.14
18	1683.17	60.24	6.36	1.87	142.62	12.31	143.65	12.65
20	1668.58	61.32	6.06	1.89	142.18	12.44	140.15	11.64

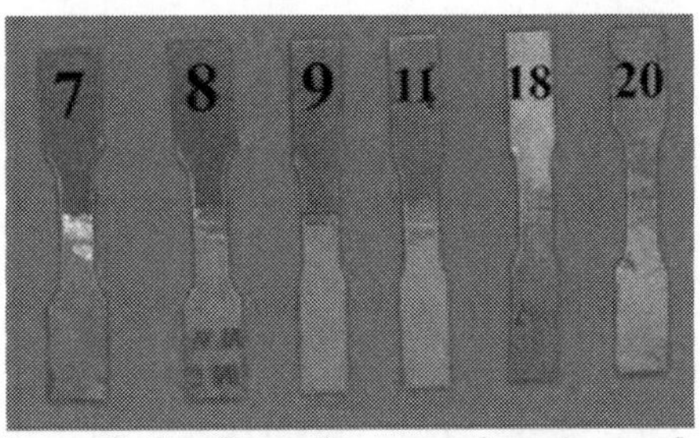

Figure 5. Confirmation experiment samples

The confirmation experiments indicated that the prediction accuracy of proposed NSGA – II optimization approach can meet the requirement of FS welding of AA2024 and AA6061 age hardenalbe aluminum alloys.

5. Conclusion

Non-Sorted Genetic Algorithm (NSGA – II) optimization technique is proposed to obtain the optimal process parameters in FS welding of AA2024 and AA6061 age hardenalbe aluminum alloys. Following conclusions can be drawn from the above investigation:

1. The tensile properties of AA 2024 and AA6061 age hardenable aluminum alloys have greatly increased by the optimizing the FS welding process parameters namely tool rotational speed (N), welding speed (F), axial force (P) and pin profiles (PP) such as cylindrical, square and taper.
2. Square pin profile playing a significant role and yielded a maximum tensile strength and tensile elongation when fabricated AA2024 and AA6061 age hardenable aluminum alloys.
3. Confirmation experiments is in good agreement with optimization results. NSGA-II was proved to be feasible and would be useful to guide FS welding of age hardenable aluminum alloys.

6. References

1. Vijayan D, Seshagiri Rao V. A parametric optimization of FSW process using RSM based grey relational analysis approach. International Review of Mechanical Engineering (IREME). 2014;8(2):328-37.
2. Maleki E, editor Artificial neural networks application for modeling of friction stir welding effects on mechanical properties of 7075-T6 aluminum alloy. IOP Conference Series: Materials Science and Engineering; 2015: IOP Publishing.
3. Kamal Babu K, Panneerselvam K, Sathiya P, Noorul Haq A, Sundarrajan S, Mastanaiah P, et al. Parameter optimization of friction stir welding of cryorolled AA2219 alloy using artificial neural network modeling with genetic algorithm. The International Journal of Advanced Manufacturing Technology. 2017.
4. D.Vijayan VSR. Tensile Properties Improvement on Friction Stir Welded Age-Hardenable Aluminum Alloys: An Evolutionary Approach using RSM based GA and SA. Revista Tecnica De La Facultad De Ingenieria Universidad Del Zulia; 2016. p. 55 - 70.
5. Vijayan D, Seshagiri Rao V. Parametric optimization of friction stir welding process of age hardenable aluminum alloys–ANFIS modeling. Journal of Central South University. 2016;23(8):1847-57.
6. Babajanzade Roshan S, Behboodi Jooibari M, Teimouri R, Asgharzadeh-Ahmadi G, Falahati-Naghibi M, Sohrabpoor H. Optimization of friction stir welding process of AA7075 aluminum alloy to achieve desirable mechanical properties using ANFIS models and simulated annealing algorithm. The International Journal of Advanced Manufacturing Technology. 2013;69(5-8):1803-18.
7. Yang Y, Cao L, Zhou Q, Wang C, Wu Q, Jiang P. Multi-objective process parameters optimization of Laser-magnetic hybrid welding combining Kriging and NSGA-II. Robotics and Computer-Integrated Manufacturing. 2018;49:253-62.
8. Bandaru S, Tutum CC, Deb K, Hattel JH. Higher-level innovization: A case study from Friction Stir Welding process optimization. 2011:2782-9.
9. Akbari M, Shojaeefard MH, Asadi P, Khalkhali A. Hybrid multi-objective optimization of microstructural and mechanical properties of B 4 C/A356 composites fabricated by FSP using TOPSIS and modified NSGA-II. Transactions of Nonferrous Metals Society of China. 2017;27(11):2317-33.
10. Ignacio Hidalgo J, Fernández R, Manuel Colmenar J, Cioffi F, Risco-Martín JL, González-Doncel G. Using evolutionary algorithms to determine the residual stress profile across welds of age-hardenable aluminum alloys. Applied Soft Computing. 2016;40:429-38.
11. Jiang P, Wang C, Zhou Q, Shao X, Shu L, Li X. Optimization of laser welding process parameters of stainless steel 316L using FEM, Kriging and NSGA-II. Advances in Engineering Software. 2016;99:147-60.

IOP Conf. Series: Materials Science and Engineering **390** (2018) 012088 doi:10.1088/1757-899X/390/1/012088

Cross diffusion, radiation and chemical reaction effects on MHD combined convective flow towards a stagnation-point upon vertical plate with heat generation

S. Karthikeyan[1], M. Bhuvaneswari[2], S. Sivasankaran[2,#], S. Rajan[1]
[1]Department of Mathematics, Erode Arts and Science College, Erode, India
[2]Department of Mathematics, King Abdulaziz University, Jeddah, Saudi Arabia
#Email: sd.siva@yahoo.com (corresponding author)

Abstract. The impacts of cross diffusion on stagnation point MHD combined convective flow approaching a plate which is upright in a porous surrounding subject to chemical reaction, heat generation and radiation is explored. Relations of similarity are employed for the conversion of flow relations as ordinary differential equations upon the application of relations of similarity, shooting method and Runge-Kutta algorithm are applied to find the solution. An analysis is made upon the graphical depictions on velocity of the liquid, its temperature and its concentration with respect to some physical entities and conclusions drawn. It is noticed that velocity shoots up across the boundary layer with a rise in heat generation, Dufour number and Biot number whereas it reduces for strong chemical reaction. For ascending values of Dufour number, chemical reaction parameter and Biot number, the temperature soars up and as a result, thickness of thermal boundary layer inflates.

Keywords: MHD; Radiation; combined convection; stagnation-point; porous medium; Soret/Dufour effect.

1 Introduction

The case of mass and energy transport of an electrically conducting liquid flowing on a porous plate upon a magnetic domain has attracted numerous investigators in view of reactors in nuclear power stations, the controlling of boundary layer in aerodynamics and plasma studies. Dufour and Soret effects on MHD mass energy transfer with chemical reaction and convective condition on boundary over a plate placed upright were reported in [1]. Radiation effects on the flowing of MHD electrically conducting liquid over a porous surrounding semi-infinite vertical plate with unsteady suction were dealt in [2]. Lie group analysis was employed in [3] to investigate the normal convection past a semi-infinite inclined plate. Many analytical as well as numerical considerations were performed on the flow upon the point of stagnation in a liquid saturated porous surrounding. In [4], the transfer of mass and heat by normal convection in the flow of a liquid upon the point of stagnating with Dufour and Soret effects, blowing/suction using model of Darcy-Boussinesq, was investigated. MHD combined convective flowing on an upright surface in porous neighborhood having convective condition on boundary was considered in [5]. A study on the similarity solution of the flow on stagnating-point in a stretching sheet with slip effects was carried out by the authors in [6]. This paper explores the behavior of the flow on MHD combined convection stagnating-point towards porous surrounding plate placed upright with mass and energy transfer influenced by radiation, heat generation, Soret and Dufour effects subject to convective boundary condition.

2 Flow Analysis

Two dimensional steady MHD flow of an incompressible electrically conducting viscous liquid in the vicinity of a point of stagnation at the surface $y = 0$ with the flow region as $y > 0$ is considered. By setting the origin fixed, we apply two forces of equal magnitude and opposite in direction along x-axis. The liquid arriving from y-axis make an impact on the wall at $y = 0$ and as a result, there are two streams leaving in upper and lower directions. In nearby surrounding of a stagnating point, flow velocity is taken as $U_\infty = lx$, where l is positive. A magnetic domain B_0, a constant, is employed in y direction. The electric field as well as the magnetic field is negligible because of polarization. The concentration and the temperature of the liquid are C_∞ and T_∞ and that of the liquid at stretching wall are C_w and T_w respectively. We assume that the plate gets heated due to convection by a hot liquid having T_f as temperature. Further, we assume that the electrical and viscous dissipation are negligibly small. The equations pertaining to this MHD stagnating-point flow of mass and energy transfer upon a hot vertical plate are

$$\frac{\partial \hat{u}}{\partial x}+\frac{\partial \hat{v}}{\partial y}=0 \tag{1}$$

$$\hat{u}\frac{\partial \hat{u}}{\partial x}+\hat{v}\frac{\partial \hat{u}}{\partial y}=\nu\frac{\partial^2 \hat{u}}{\partial y^2}+(\hat{T}-\hat{T}_\infty)g\beta+(\hat{C}-\hat{C}_\infty)g\beta^* + \left(\frac{\sigma_e B_0^2}{\rho}+\frac{\nu}{\tilde{K}}\right)(U_\infty-\hat{u})+\frac{dU_\infty}{dx}U_\infty \tag{2}$$

$$\hat{u}\frac{\partial \hat{T}}{\partial x}+\hat{v}\frac{\partial \hat{T}}{\partial y}=\alpha\frac{\partial^2 \hat{T}}{\partial y^2}-\frac{\alpha}{k}\frac{\partial q_r}{\partial y}+Q(\hat{T}-\hat{T}_\infty)+\frac{D_m K_T}{c_s c_p}\frac{\partial^2 \hat{C}}{\partial y^2} \tag{3}$$

$$\hat{u}\frac{\partial \hat{C}}{\partial x}+\hat{v}\frac{\partial \hat{C}}{\partial y}=D_m\frac{\partial^2 \hat{C}}{\partial y^2}+\frac{D_m K_T}{\hat{T}_m}\frac{\partial^2 \hat{T}}{\partial y^2}-K_1\hat{C} \tag{4}$$

with $\hat{u}=0,\ \hat{v}=0,\ k\frac{\partial \hat{T}}{\partial y}=h(\hat{T}-\hat{T}_f),\ \hat{C}=C_w$ when $y=0$

$$\hat{u}\to\hat{U}_\infty=lx,\ \hat{T}\to\hat{T}_\infty,\ \hat{C}\to\hat{C}_\infty \text{ when } y\to\infty \tag{5}$$

where, with usual notations, $q_r=-4\sigma^*/3K'\frac{\partial \hat{T}^4}{\partial y}$. On the assumption that the temperature variations are too minimal, we can take $\hat{T}^4$ as

$$\hat{T}^4\approx 4\hat{T}_\infty^3\hat{T}-3\hat{T}_\infty^4. \tag{6}$$

The dimensionless quantities are

$$Gr=\frac{\beta g(\hat{T}_w-\hat{T}_\infty)x^3}{\upsilon^2},\ Gc=\frac{\beta g(\hat{C}_w-\hat{C}_\infty)x^3}{\upsilon^2},\ S=\frac{Q\upsilon}{\alpha l},\ R=\frac{4\sigma^*\hat{T}_\infty^3}{kK'},$$
$$Df=\frac{K_T D_m(\hat{C}_w-\hat{C}_\infty)}{c_p\alpha c_s(\hat{T}_w-\hat{T}_\infty)},\ K=\frac{\upsilon}{l\tilde{K}},\ \Pr=\frac{\upsilon}{\alpha},\ Sc=\frac{\upsilon}{D_m},\ M=\frac{\sigma_e B_0^2}{l\rho},\ \upsilon=\frac{\mu}{\rho}, \tag{7}$$
$$Sr=\frac{K_T(\hat{T}_w-\hat{T}_\infty)}{\hat{T}_m(\hat{C}_w-\hat{C}_\infty)},\ \mathrm{Re}_x=\frac{U_\infty x}{\upsilon},\ Ri_T=\frac{Gr}{\mathrm{Re}_x^2},\ Ri_C=\frac{Gc}{\mathrm{Re}_x^2}$$

We introduce the following similarity relations

$$\eta=y\sqrt{\frac{c}{\hat{v}}},\ \psi(x,y)=\sqrt{\hat{v}l}x\hat{f}(\eta),\ \theta(\eta)=\frac{\hat{T}-\hat{T}_\infty}{\hat{T}_f-\hat{T}_\infty},\ \phi(\eta)=\frac{\hat{C}-\hat{C}_\infty}{\hat{C}_w-\hat{C}_\infty} \tag{8}$$

where $\psi(x,y)$ is given by $\hat{u}=\frac{\partial\psi}{\partial y}$ and $\hat{v}=\frac{-\partial\psi}{\partial x}$ so that Eq.(1) satisfied identically. Eqs.(2)-(5), on the application of Eqs.(6)-(8), yield the following equivalent set of equations.

$$\hat{f}\hat{f}''+\hat{f}'''-\hat{f}'^2+Ri_T\theta+Ri_C\phi+(M+K)(1-\hat{f}')=-1, \tag{9}$$

$$(1+4R/3)\theta''+\Pr f\theta'+S\theta+Df\phi''=0, \tag{10}$$

$$\phi''+ScSr\theta''+Scf\phi'-\frac{K_1}{C}Sc\phi=0, \tag{11}$$

subject to $\hat{f}'=0,\ \hat{f}=0,\ \theta'=B_i(\theta-1),\ \phi=1,$ when $\eta=0,$

$$\hat{f}'=1,\ \theta=0,\ \phi=0 \text{ when } \eta\to\infty \tag{12}$$

Here $B_i = \frac{h}{k}\sqrt{\frac{\nu}{c}}$, the convective heat transfer parameter.

Solutions of Eqs.(8)-(11) are arrived by the application of shooting method combined with Runge-Kutta fourth-order algorithm. Localized Nusselt and Sherwood numbers and then the coefficient of skin-friction are given by

$$Nu = \frac{xq_w}{(\hat{T}_w - \hat{T}_\infty)k},\ Sh = \frac{xq_m}{(\hat{C}_w - \hat{C}_\infty)D_m},\ \hat{C}_f = \frac{2\tau_w}{\rho U_\infty^2}, \tag{13}$$

Here

$$\tau_w = \mu \left.\frac{\partial \hat{u}}{\partial y}\right|_{(y=0)},\ q_w = -k\left.\frac{\partial \hat{T}}{\partial y}\right|_{(y=0)} = -\left.\frac{4\sigma^*}{3K'}\frac{\partial \hat{T}^4}{\partial y}\right|_{(y=0)},\ q_m = -D\left.\frac{\partial \hat{C}}{\partial y}\right|_{(y=0)} \tag{14}$$

The relations for Localized Nusselt and Sherwood numbers and then the coefficient of skin-friction are derived as

$$\mathrm{Re}_x^{-1/2} Nu = -(1+4R/3)\theta'(0),\ \mathrm{Re}_x^{-1/2} Sh = -\phi'(0),\ \mathrm{Re}_x^{1/2} C_f = \hat{f}''(0) \tag{15}$$

3 Results and Discussion

Eqs. (9)-(11) with its associated conditions on the boundary (12) were dealt with shooting method along with fourth order Runge–Kutta algorithm for the solution. By applying shooting method, we first break down the non-linear differential equations of higher order into first order linear simultaneous differential equations and further we transform these into initial value problem. The resulting equation is evaluated by incorporating Runge-Kutta algorithm of order four. The numerical iteration procedure was performed repeatedly until the level of tolerance of error is attained.

Table 1. Computations of $-\theta'(0)$, $-\phi'(0)$ and $f''(0)$ for varying governing parameters when $Sc = 0.5$, Ri_T=1, Ri_c=0.5 and $Pr = 1.0$.

K	M	S	R	Cr	Sr	Df	$f''(0)$	$-\theta'(0)$	$-\phi'(0)$
0.1	0.1	0.1	0.1	0.1	0.2	0.2	2.246971	0.0978520	0.5451982
0.5							2.319502	0.1008649	0.5458624
1.0							2.415747	0.1037379	0.5465281
1.0	0	0.1	0.1	0.1	0.2	0.2	2.396287	0.1032492	0.5463949
	4						3.245195	0.1212068	0.5505797
	8						4.999324	0.09244849	0.5652972
1.0	0.1	0.1	0.1	0.1	0.2	0.2	2.415747	0.10373790	0.5465281
							2.968114	0.09464353	0.5712622
							3.507120	0.08798717	0.5951335
1.0	0.1	0.1	0.1	0.1	0.2	0.2	2.632905	0.1021410	0.5552750
							3.054221	0.1009250	0.5715686
							3.462389	0.1021722	0.5863139
1.0	0.1	-0.5	0.1	0.1	0.2	0.2	2.336319	0.5544120	0.4726206
		0.0					2.399971	0.1909342	0.5327260
		0.5					2.491805	-0.3105458	0.6091604
1.0	0.1	0.1	0	0.1	0.2	0.2	2.416824	0.0901831	0.5486991
			0.5				2.411396	0.1356563	0.5411857
			1.0				2.410166	0.3011181	0.5600268
1.0	0.1	0.1	0.1	-1.0	0.2	0.2	2.448736	0.1746870	0.3498008
				0			2.418123	0.1094058	0.5075595
				1.0			2.398602	0.05824485	0.8481207
1.0	0.1	0.1	0.1	0.1	0	0.2	2.415747	0.1037379	0.5465281
					0.5		2.415123	0.09949238	0.5708835
					1.0		2.414324	0.09286911	0.6136006
1.0	0.1	0.1	0.1	0.1	0.2	0	2.407324	0.1385104	0.5409370
						0.5	2.428890	0.04820966	0.5555990
						1.0	2.452206	-0.05507862	0.5728061

IOP Conf. Series: Materials Science and Engineering **390** (2018) 012088 doi:10.1088/1757-899X/390/1/012088

The coefficient of skin-friction, Nusselt and the Sherwood numbers respectively vary with $f''(0)$, $-\theta'(0)$and $-\phi'(0)$. These are computed for some values of the parameters *K, M,* Ri_T, Ri_C , *S, R, Cr, Sr* and *Df* when Pr = 1.0, *Sc* = 0.5and their numerical values are presented in Table 1. We observe that, at the plate surface, the skin-friction become high when there is a rise in *K, M,* Ri_T, Ri_C, *S* and *Df* and lowers for raising values of *R, Cr* and *Sr*. It can thus be understood that the influence of magnetic field, buoyancy forces, permeability of porous medium, thermal radiation and Dufour effect tend to accelerate skin-friction. The rise in the localized Nusselt number is seen when *K* and *R* rise. But it reduces for soaring values of Ri_T, *S, Cr, Sr* and *Df.* Further, we notice that, except for the thermal radiation parameter *R*, the localized Sherwood number goes up for raising values of all other parameters.

3.1 Velocity Profiles

Figures 1(a-d) illustrate the consequences of physical parameters *S, Df, Cr* and B_i upon the velocity profile. Generally, at the plate surface, the velocity is zero and it accelerates to the free stream velocity which is away from the surface of the plate for all parameters. A rise in the liquid velocity in the case of increasing *S* values as depicted in Fig. 1(a). The action of Dufour number *Df* on velocity is shown in Fig. 1(b).From this plot, it is observed that a rise in the Dufour number *Df* enhances the velocity inside the boundary layer. From Fig. 1(c), we observe an opposite trend in the velocity profile. Here we see that the velocity diminishes as the chemical reaction parameter *Cr* goes up. The velocity profiles with respect to the Biot number B_i are plotted in Fig.1(d). Further, when Biot number takes high value, the velocity rises.

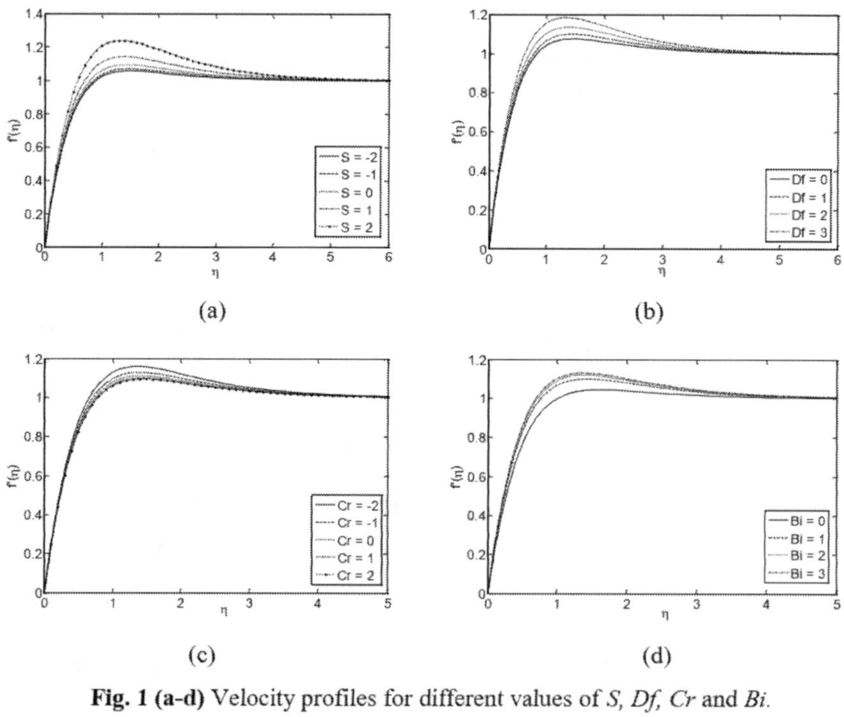

Fig. 1 (a-d) Velocity profiles for different values of *S, Df, Cr* and *Bi.*

3.2 Temperature Profiles

Variations on temperature by the influence of *Df, Sr, Cr* and *Bi* are represented in Figures 2 (a-d). The impact of Dufour number *Df* is depicted in Fig. 2(a). Temperature inside the boundary layer soars up for increasing Dufour number *Df*. Profiles of temperature for varyingSoret number *Sr* is illustrated in Fig. 2(b). We see that there is a significant rise in temperature as the Soret number becomes larger. Fig. 2(c) represents the variations on temperature due to the influence of chemical reaction *Cr*. It is noted that the temperature shoots up for higher values of *Cr*. The temperature depictions for varying Biot number B_i are illustrated in Fig. 2(d). It is observed that thermal boundary layer inflates and the temperature of the plate surface rises as the parameter B_i takes high values.

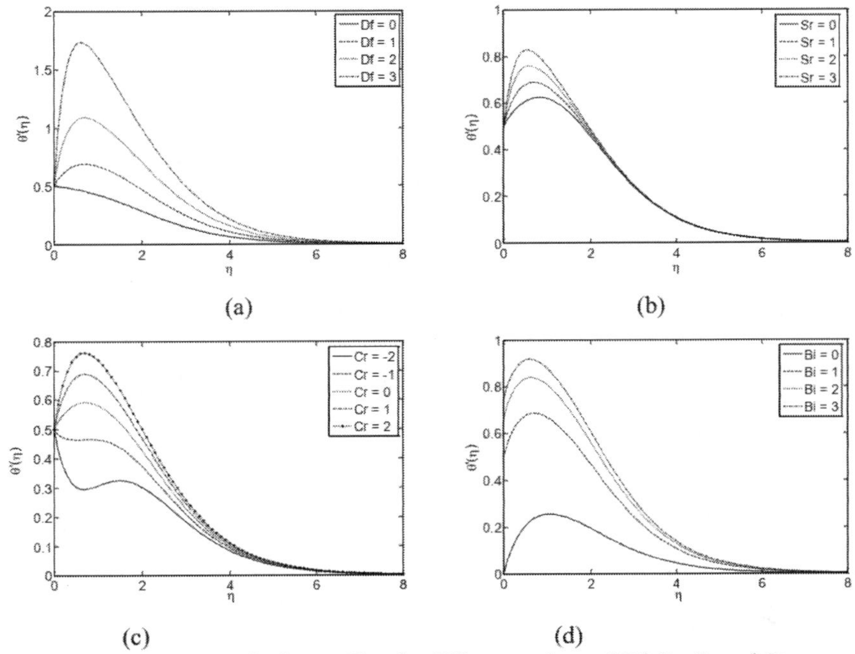

Fig. 2 (a-d) Velocity profiles for different values of *Df, Sr, Cr* and *Bi*.

3.3 Concentration Profiles

The graphs plotted in the Figs. 3 (a-d) depict the variations of concentration profiles with respect to the physical parameters *S, Df, Sr* and *Cr*. For the case of different values of the heat source parameter *S*, the profiles of concentration are depicted in Fig. 3(a). We notice that the concentration declines for soaring values of *S*. We notice a similar trend in the concentration profile for the cases of *Df, Sr* and *Cr* which are respectively plotted in Figures 3(b-d). Ascending values of these parameters diminish the species concentration at the boundary layer.

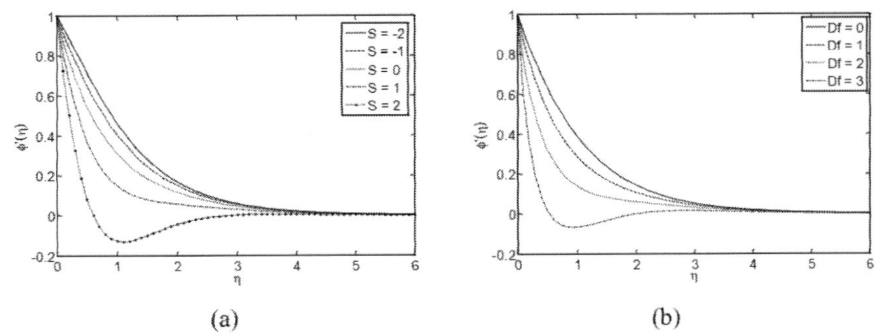

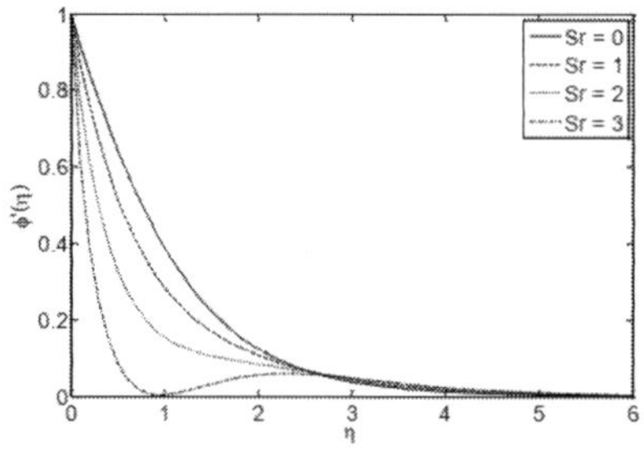

(c)

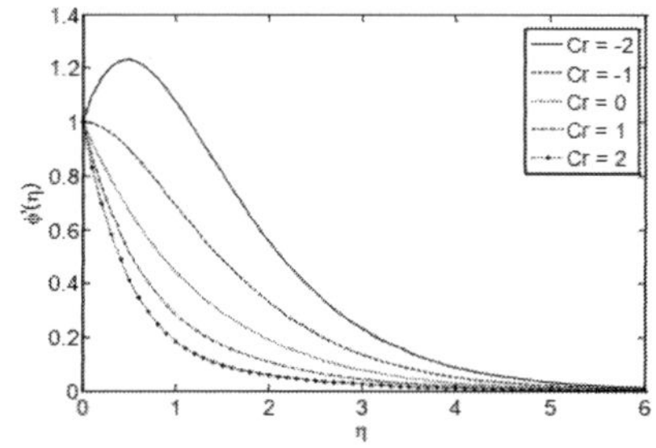

(d)

Fig. 3 (a-d) Velocity profiles for different values of *S, Df, Sr* and *Cr*.

4 Conclusions

Flow of MHD combined convection stagnating-point upon a plate placed upright in a porous surrounding with mass and heat transfer in the influence of internal heat generation, radiation, Dufour and Soret effects subject to convective boundary condition is studied. Shooting method with Runge-Kutta algorithm is applied for finding the solution. Conclusions of the study are as follows.

The velocity shoots up across the boundary layer with a rise in heat generation, Dufour number and Biot number whereas it reduces for strong chemical reaction. For ascending values of Dufour number, chemical reaction parameter and Biot number, the temperature soars up and as a result, thickness of thermal boundary layer inflates. A fall in liquid temperature in the boundary layer occurs for a rise in Soret number. The concentration of species diminishes when the values of *S, Df, Sr* and *Cr* ascend.

The coefficient of skin-friction and the localized Sherwood number rise whereas the localized Nusselt number diminishes when Dufour number and internal heat generation rise. When the radiation becomes high, the localized Nusselt and Sherwood numbers rise whereas the skin-friction coefficient diminishes. The coefficient of skin-friction and the localized Sherwood number rise with accelerating magnetic parameter.

References

[1] Gangadhar K 2013 *Journal of Applied Fluid Mechanics* **6(1)** 95-105
[2] Karthikeyan S, Bhuvaneswari M, Rajan S, Sivasankaran S 2013 *Applied Mathematics and Computational Intelligence* **2(1)** 75-83
[3] Bhuvaneswari M, Sivasankaran S, Kim Y J 2012 *Journal of Porous Media* **15(12)** 1155-1164
[4] Postelnicu A 2010 *Heat Mass Transfer* **46** 831-840
[5] Makinde O D, Aziz A 2010 *International Journal of Thermal Sciences* **49** 1813-1820
[6] Zaimi K, Ishak A 2016 *Mathematics* **4(27)** 1-8

IOP Conf. Series: Materials Science and Engineering **390** (2018) 012089 doi:10.1088/1757-899X/390/1/012089

Investigation on Nano Particulate Aerosol at Idling Conditions of Vehicles

Rahul Sunny[1], **D Senthil Kumar**[2*]

[1,2]Dept of Mechanical Engineering, Amrita School of Engineering, Coimbatore, Amrita Vishwa Vidyapeetham, India

*Corresponding Author

Email: d_senthilkumar@cb.amrita.edu

Abstract. In general, Engine will emit more particulate tailpipe emissions during idling and acceleration conditions. This is a common phenomenon in location such as Toll Plazas which has one of the highest number of vehicle density in urban areas. Therefore the study of aerosol particulates characteristics like average particle concentration and average particle diameter in air is important. The objective of this work is to describe the process of measurement and analysis of ambient Nano Particulate Matter at a Toll Plaza in a Tier II city. A nano particulate counter was used to measure particulate samples and it was mounted at a position that best replicates human exposure to the pollutants. The measurement was carried out for duration of 20 consecutive days and the time of the measurement was planned so as to cover peak traffic time experienced at the Toll Plaza. Particulate number concentration was measured at varying vehicle densities and also at different wind speeds. The results revealed that high concentration was observed for an increase in vehicle concentration as anticipated. Also high number concentration was observed for smaller particulate size (diameter 20 to 80 μm). The relation with number concentration and wind speed was inconclusive.

Introduction

Over two million deaths happen each year due to respiratory system damage caused by air pollution [2]. Among these death due fine particulate matter is 2.1 million [5]. Particle matter in the air can be roughly classified into coarse (diameter ~ 10μm), fine (diameter ~2.5μm) and ultrafine particles (diameter ≤0.1μm). The characteristics of the particle varies drastically with their size. Large particles having diameter around 10μm, stays suspended in air for a few minutes to few hours and has a range of 1 to 10km before settling, while smaller particles having diameter around 2.5μm will stay suspended in air for days to weeks and has a range of 100 to1000km [3, 5]. The sources of particulate matter varies from combustion of oil, gasoline, coal, transformation products of SO_2, NO_x, metal oxides, dust, fly ash, sea salt and various organic parts [3].

A major contributor to this particulate matter air pollution is from vehicles as tail pipe emission due to the combustion. An ideal engine condition that facilitates particulate formation is vehicle idling. At idle condition the manifold throttle is only partly open, this creates additional load on engine due to formation of low pressure zone on the underside of the throttle. In addition the volumetric efficiency deteriorates due to blockage of free flow. But during cruising conditions the throttle is at wide open position and thus engine is at high efficiency. Therefore it is logical that high air pollution is observed in areas with high idling vehicle density such as in traffic junctions and toll plazas.

A Toll Plaza was selected within the city limit. The selected Toll Plaza has an average traffic flow of 25000 vehicles on weekdays with an average flow of 58 vehicles per minute. The majority of these are heavy duty vehicles as shown in figure 1. Figure 2 shows the selected Toll Plaza and the surrounding areas. It is to be noted that a hospital and a school were in the exposure range of the particles emitted from this Toll Booth.

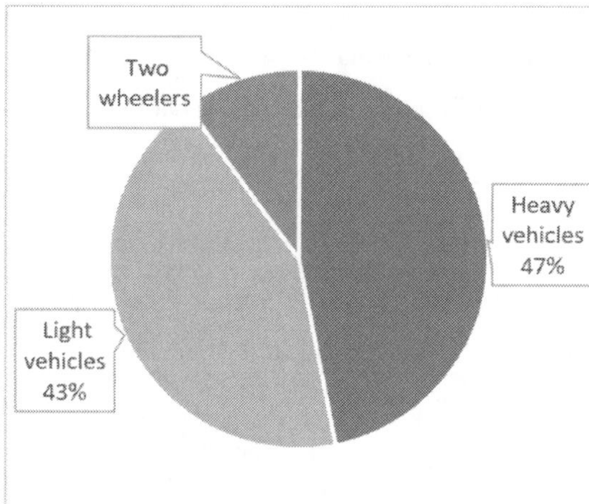

Figure 1. Classification according to vehicle type

Figure 2. Map showing the relevant Toll Plaza and the surrounding areas

Experimental Setup

The Nano Particulate counting system used is a 'Diffusion Size Classifier Mini' or 'DiSC Mini' from 'Matter Aerosol', Switzerland. It has a time resolution of 1Hz and gives the particle count in real time. It has an average particle size range of 10 to 300nm and a concentration range of 1000 to 1000000 particles per cm^3. It further has the advantage of being very portable thus allowing the measuring point to be precisely selected.

In order to measure metrological parameters such as Wind Speed, a three cup hand-held anemometer (DHA 111) with a Sampling Time of 2 seconds and accuracy of +/- 3% was used. A standard Thermometer and Hygrometer was used to measure ambient temperature and humidity.

The Plaza had four lanes. Lane 1 catered to traffic towards west, Lane 3 & Lane 4 catered to traffic toward the opposite (East) direction. Lane 2 was kept blocked and was opened only to let emergency vehicles pass through and also when congestion increases in a particular lane. The experiment with all the instruments were mounted on the slightly elevated island between Lane 1 and Lane 2 in front of the Operator's Booth.

The Particle Counter was placed at a height of 3 meter from the road surface. The anemometer was placed at the same height and two feet away from the counter to get an accurate wind speed without obstructing inflow to the counter.

Methodology

The measurement was taken for a duration of six hours per day for a total of 20 days. The measurement time was designed to cover the peak evening traffic. The time for measurement was taken as 2 to 8 pm.

Results and Discussions

Table 1. Average Temperature and Relative Humidity values for each hour

Time Duration	Average Temperature (°C)	Average Relative Humidity (%)
2pm-3pm	33.5	46.8
3pm-4pm	31.5	46.7
4pm-5pm	30.4	484
5pm-6pm	30.7	49.3
6pm-7pm	29.1	58.2
7pm-8pm	27.7	67.4

Table 2. Particle characteristic data along with vehicle data

S. No.	Time	No. of Vehicles			Total Vehicles	Average Particle Concentration (Pt/ccm)	Average Particle Diameter (nm)	Avg LDSA (μm2/ccm)
		Heavy Vehicles	Four Wheelers	Two Wheelers				
1	3 - 3.15pm	38	42	8	88	56433.6	72.2	218.91
2	3.30 - 3.40pm	41	46	12	99	53429.8	68.3	355.8
3	3.40-3.55pm	42	44	4	90	56776	65.25	273.19
4	4.15-4.30pm	39	30	8	77	41373.6	67.7	254.8
5	4.45-5.00pm	56	73	9	138	62945.9	61.8	425.7
6	5.30-5.55pm	58	50	27	135	48236.5	54.6	276.1
7	6.10-6.30pm	51	34	11	96	49227.2	75.9	260.3
8	6.50-7.15pm	69	42	4	115	58822.9	78.1	352.27
9	7.30-7.45pm	59	34	7	100	63183.1	96.2	497.0
Average		50	44	10	104	54492.1	71.1	323.8

Average Ambient Temperature and Relative Humidity values are shown in Table 1 on an hourly basis. This data is measured since Ambient Temperature and Relative Humidity will influence behavior of suspended particle matter. Geometric mean values of Particle Concentration, Particle Diameter, LDSA for an average day is given in Table 2. It also gives the time duration of measurement and the Type and Number of vehicles plying during the specified time period. Table 3 gives the wind speed and the corresponding Particle Concentration. Wind Speed readings above 0.1 m/s is only taken and the values are arranged in the ascending order

Table 3. Wind speed data

S No.	Wind Speed (m/s)	Avg Particle Concentration (pt/ccm)
1	0.1	49456
2	0.2	51654
3	0.3	37594
4	0.4	7474
5	0.5	25017
6	0.6	274184
7	0.7	12550
8	0.8	89741
9	0.9	24958
10	1	36070
11	1.1	71328
12	1.2	51159
13	1.3	11005

S No.	Wind Speed (m/s)	Avg Particle Concentration (pt/ccm)
14	1.4	18512
15	1.5	13715
16	1.7	20817
17	1.8	45912
18	1.9	10277
19	2	175278
20	2.1	38885
21	2.2	10575
22	2.3	27706
23	2.4	26877
24	2.6	6401
25	3.2	8785

IOP Conf. Series: Materials Science and Engineering **390** (2018) 012089 doi:10.1088/1757-899X/390/1/012089

4.1 Assessing the Potential Impact of the Traffic on the Particulate Level

To assess the impact of vehicle concentration at toll booth on the particle concentration. A graph was plotted between the concentration level and the number of vehicles simultaneously present at the booth. It is seen in the figure 3 that there is a definite increase in Particle Number Concentration (PNC) with increase in the simultaneous number of vehicles. This confirms the fact that vehicles are the major contributing factor to the pollution at the toll booth.

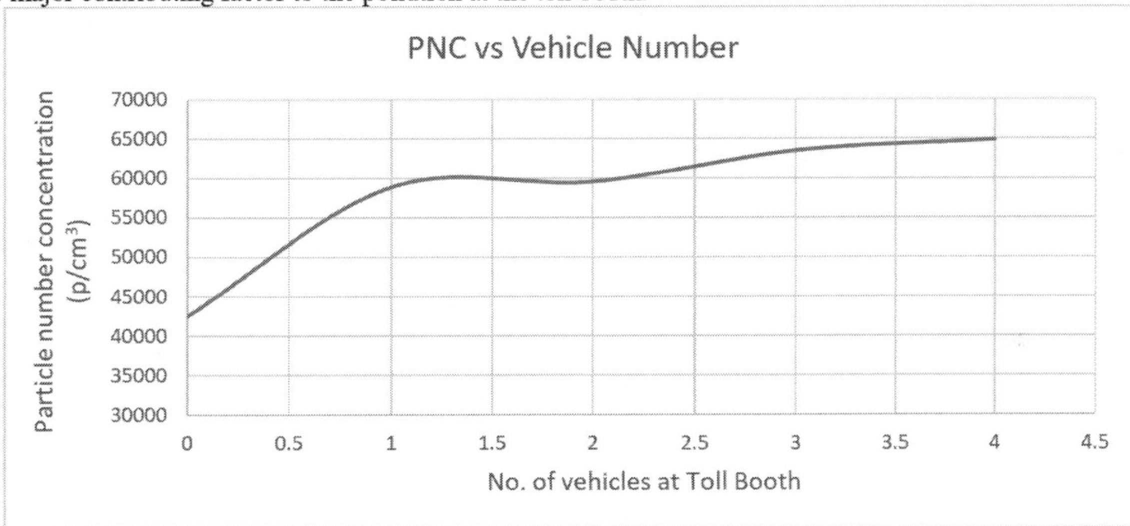

Figure 3. Variation PNC with no. of vehicle

4.2 Assessing Relationship between Particulate Size and Concentration

Figure 4 shows the size of particles plotted against the particle number concentration (PNC). The entire size range is split into two, the lower half and the higher half. It is seen that 91% of the particles lies in the lower half range (i.e. 0 to 150 nm range).

Also it is seen that the average diameter of the particles is 73nm. This diameter is associated with a PNC of 54702 particles /cm^3. This shows the presence of increasingly smaller sized particles in any measured particulate sample. This indicates more focus should be on Ultra-Fine particles. More accurate measuring techniques such as Particle Counting is suggested to be adopted.

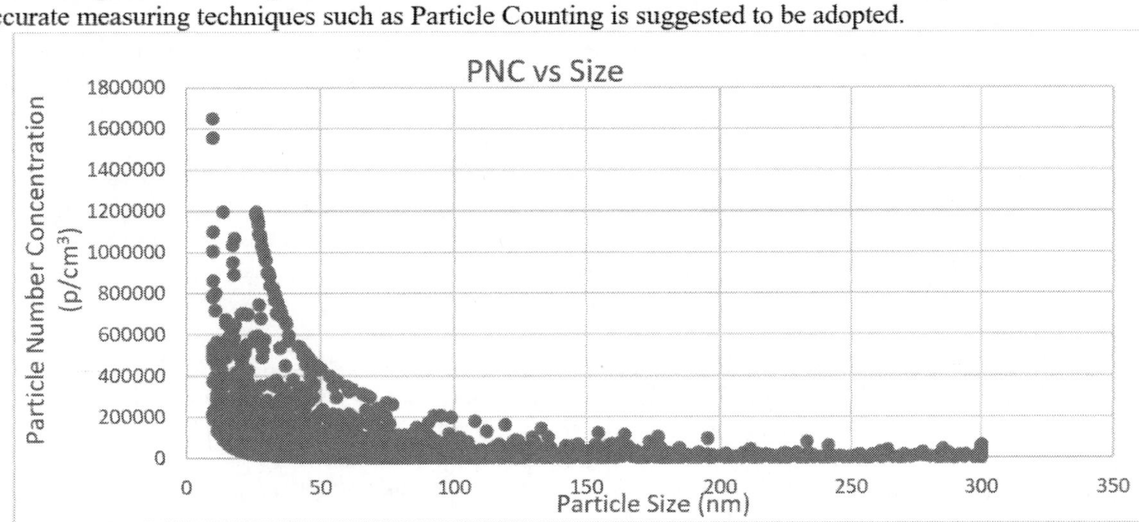

Figure 4. Variation of PNC along with particle size

4.3 Assessing the Potential Impact of Wind Speed in Particulate Level

It was *a priori* hypothesis that with increase in wind speed and relative increase in PNC would also be seen. But it is observed in figure 5 that PNC shows no strict trend along with the rising wind speed.

This may be due to the fact that the immediate surroundings of the toll plaza is an open space present with trees. Thus the higher order of wind speed brings with it a mixture of the vehicle pollutants as well fresh air from the surroundings.

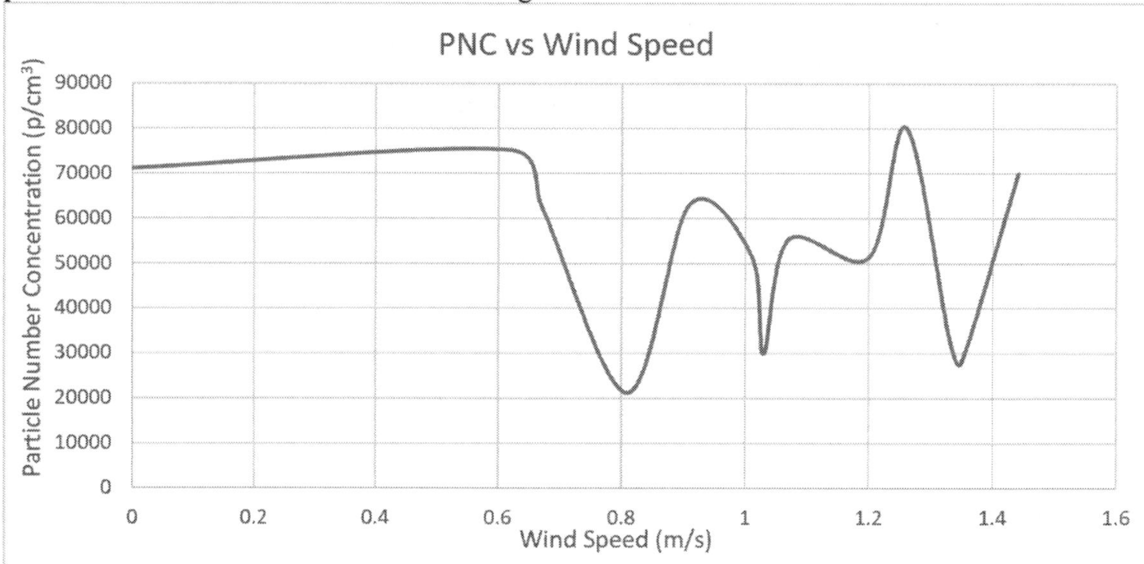

Figure 5. Variation of PNC along with wind speed

Conclusion and Future Work

a) PNC is seen to increase in relation to vehicle concentration. This confirms the fact that vehicles are a major contributing factor to the nano particulate matter.

b) In terms of relation with size, it is seen that 91% of the total number of particulates are seen to lie in the lower half of the total diameter range (i.e. 91% of particles has a diameter in the range of 0-150nm for a sample with a range of 0-300nm). This show the presence of increasingly smaller particles in the measured sample. This indicates to the importance of measurement of smaller particulates given the fact that as size decreases the danger to humans also increases. In order to get a better understanding of the level of pollutants, particle counting is advised over conventional mass measurement.

c) The relation with wind speed and PNC was studied. No systematic trend was observed. It is suggested to incorporate wind direction data during measurement. To further understand dynamics of such particles with respect to Brownian motion and turbulent diffusion, a controlled environment study such as a wind tunnel is suggested.

References

[1] Pant P, Guttikunda S and Peltier R 2016 *Environ. Res.* **147** 480–96

[2] Shah A S, Langrish J P, Nair H, McAllister D A, Hunter A L and Donaldson K 2013 *Lancet* **382** 39–48

[3] Srimuruganandam B and Nagendra S 2012 *Sci Total Environ.* **433** 8–19

[4] Brunekreef B 2003 *Health Aspects of Air Pollution with Particulate Matter, Ozone and Nitrogen Dioxide* (WHO Working Group, Germany)

[5] Kim Ki, Kabir E and Kabir S 2015 *Environ. Int.* **74** 136-143

[6] Ministy of Environment, Forest and Climate Change 2016 *Report on Ambient Air Quality Data during Odd and Even Period 15th to 30th April* (Central Pollution Control Board, Delhi)
[7] Thakur A 2017 *Orientjchem* **2** 33
[8] Atkinson R W, Barratt, Armstrong B, Anderson H R, Beevers S D, Mudway I S, Green D, Derwent R, Wilkinson P, Tonne C and Kelly F 2009 *Atmos. Environ* **43** 5493
[9] Clancy L, Goodman P, Sinclair H and Dockery D W 2002 *Lancet* **360** 10-14
[10] Cupples J 2008 *Atmos. Environ* **43** 207-217
[11] Gautam 1998 *Indian Standard Methods for Measurement of Air Pollution* (The Gazette of India, New Delhi)
[12] Lin H, Zhang Y, Liu T, Xiao J, Xu Y, Xu X, Qian Z, Tong S, Luo Y, Zeng W and Ma W 2014 *Atmos. Environ* **91** 24-31
[13] Sengupta B 2003 *Guidelines for Ambient Air Quality Monitoring* (Central Pollution Control Board, New Delhi)
[14] Schneider I, Teixeira E, Oliveira L and Wiegand F 2015 *Atmos Pollut Res* **6** 877-885
[15] Kokila M and Geetha P 2016 **9** *J. Chem.* 515-518
[16] Mehel A and Murzyn F 2015 *Atmos Pollut Res* **6** 612-617
[17] Matter Aerosol, DiSC Mini User Manual, an instrument for measuring nano particles, Wohlen, Switzerland, see https://www.testo.com/en-US/testo-discmini/p/133

Acknowledgements
The authors would like to thank Amrita Vishwa Vidyapeetham and Amrita Automotive Research and Technology Center (AARTC) for their support and encouragement. The authors would also like to thank Operations Manager of L&T Toll Plaza for his support given during the course of the experiment. We would also like to express deep gratitude to the entire Staff and Students of Amrita Vishwa Vidyapeetham for providing a productive and encouraging environment.

IOP Conf. Series: Materials Science and Engineering **390** (2018) 012090 doi:10.1088/1757-899X/390/1/012090

Mechanical behaviour of sisal – glass fiber reinforced hybrid Nano composites

K. Mohan[1*], T. Rajmohan[2]

[1-2]Department of Mechanical Engineering, Sri Chandrasekharendra Saraswathi Viswa Mahavidhyalaya, Kancheepuram, Tamilnadu - 631561, India.

Email: mohank84@gmail.com

Abstract.Natural fiber reinforced composites are playing a vital role in recent years due to its biodegradable, recycling and eco-friendly in nature. In this present work biodegradable hybrid composites is developed with sisal fiber and commercially available glass fiber as reinforcing fibers and nano (Multi walled carbon nanotube) incorporated epoxy used as a matrix. Using compression molding technique, composites were fabricated. Ultrasonic probe sonicator is used for incorporate the MWCNT into epoxy resin by varying their Wt% as 0%, 0.5% and 1% respectively. The composites were subjected to various mechanical testing like tensile, compressive and flexural. The Importance of this work can be predominantly affected by factors which include variation in nano additives and extreme fiber layers. The results indicated that, mechanical properties are enhanced by increasing the wt% of MWCNT. The microstructure of the developed composites were analyzed using scanning electron microscope (SEM) and the EDS analysis were used to confirm the presence of carbon nanotube and glass fiber in the developed composites.

1. Introduction

Depa ray et al, [1] Investigated the flexural strength of jute fiber vinylestar composites. The jute fibers are treated with NAOH (5%) solution at 30°c for 0, 2,4,6,8 hours respectively. Flexural strength and modulus improved by 20% and 23% for jute fiber reinforced composite at 35wt% for 4 hours treating with NAOH solution.HariSankar et al, [2] investigated the mechanical properties of shot carbon fiber reinforced composites filled with an nano clay. They have concluded, while increasing the wt% of nano clay indicates asignificant improvement in the impact strength of the composites. Beyond the optimum level of nano clay contents shows the negative sign in the ultimate strength. From the TGA analysis it was observed thermal stability of the composite is increased at 6Wt% of nano clay.
MD.FarukHossen et al, [3] investigated thetensile properties of chemically treated jute fiber reinforced polyethylene/clay nano composites. Treated fiber composite which is having 15% fiber content exhibiting the maximum tensile strength and modulus rather than raw fiber composites. On the other hand the nano clay filled composites shows the improvement in the mechanical properties over the composites without nano clay. Sathish et al, [4] investigated the mechanical and thermal properties of banana, kenaf and glass fiber reinforced composites with different volume fraction and fiber

IOP Conf. Series: Materials Science and Engineering **390** (2018) 012090 doi:10.1088/1757-899X/390/1/012090

orientation. The hybrid composite which is arranged in 45° orientation shows the enhanced mechanical properties.
Dramatic increase in lignin mechanical properties regardless of the presence of NCC after the thermal treatment. The strength and modulus of electro spun fiber mats are increased 400 and 143% respectively after the oxidation process and the carbonization dramatically improved the mechanical properties concluded by Mijung Cho et al [5]. Balajiet al, [6]examined the mechanical properties and water absorption of cardanol bio-composites reinforced with bagasse fiber. The highest tensile strength and flexural strength is achieved 15wt% and 20 mm fiber length of bagasse fiber reinforced cardanol composite. Further increasing with the wt% of fibers the tensile and flexural strength is decreased.
The glass fiber reinforced composite was prepared filled with various composition of cloisite clay particles (1, 3, 5, and 7%). The 5% Nano clay reinforcement exhibits the enhanced tensile and flexural strength along with their modulus. Impact strength also improved. Further increasing the Nano clay the mechanical properties are decreased have been reported by jeyakumar et al [7]. Lee et al, [8] investigated structural and mechanical properties of the composites filled with acid treated and un treated MWCNT with various compositions of MDI. The treated MWCNT composites shows the improved mechanical properties and better adhesion between the MWCNT and PU matrix. Li-Xiu Gong et al, [9]investigated the fracture behavior and interfacial strength between fiber and epoxy resin modified by silica nano particles. Addition of nano particles exhibiting improved in tensile, fracture toughness and elastic modulus of the composite. Silica content composites shows good interfacial strength between fibers and epoxy resin.

Islam et al, [10] have prepared nano composites using coir and wood fiber with MMT nano clay. They observed the tensile strength is reduced due to the similar cellulose content presence in the fibers. By adding the MMT, the composites shows the better tensile strength and highest rate of water absorption.Jute, banana and flax composites were prepared by varying their stacking sequence and wt% of MWCNT. Extreme layer of jute fiber reinforced composites with 1wt% of MWCNT exhibiting the maximum mechanical properties rather than others reported by Mohan et al, [11]. Rajmohan et al [12] investigated glass banana reinforced composites by incorporating the various wt% of MWCNT. Mechanical properties were enhanced while increasing the wt% of MWCNT. Vetivera fiber subjected to various surface chemical treatment then it is fabricated and tested. Benzoyl chloride treated composites exhibiting superior mechanical strength than the other treatments reported by Vinayagamoorthy [13]. In the review of natural fiber laminates, various fabrication methods, different surface treatments of fibers and various composition of fibers effect the mechanical properties of the composites [14]. CFRP composites filled with fly ash were subjected to analyse the surface roughness during drilling process. It's observed, highly influencing parameter is a feed rate which influences on the surface roughness [15]. Recent years many researchers have focused on machining of natural fibers reinforced composites. From the literature, it's observed very limited work has been carried out to find the characteristics of MWCNT filled composites.

2. Materials and methods

Sisal fiber is obtained from Eco craft enterprises, Osur. From M/s. Sun Tech fibers, Chennai. Glass fiber, Epoxy resin Araldite LY556 and Hardener Aradur HY951 are obtained. Multi Wall Carbon Nano Tube (MWCNT) is purchased from M/S US Research Nano materials Inc, USA. The modified nano additive resin is prepared by using ultrasonic probe sonicator. Initially in the separate beaker, the respective wt% of MWCNT are added with the epoxy resin and allowed to vibrate 3hours. Then the same blend is allowed to mechanical stirrer for the next two hours for homogeneous mixing of the MWCNT into the epoxy resin. Using compression molding technique the composites were fabricated in the size of 30X30 cm. During fabrication process the glass fibers were kept in the extreme layers, in-between sisal and glass fibers were arranged alternatively. The composites has been cured 24 hours in compression molding machine at the pressure rate of 350 psi. Then the specimens are prepared as per the ASTM standards.

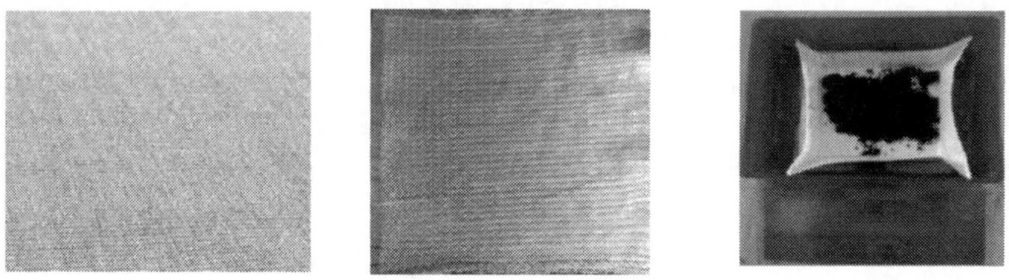

a) Sisal fiber b) Glass fiber c) MWCNT

Fig.1 Materials used for fabricating the composites

Table 1. Composition of composites

S.No	Particulars in Wt%.	G/S 0% CNT	G/S 0.5% CNT	G/S 1% CNT
1	Glass Fiber	25	25	25
2	Sisal Fiber	15	15	15
3	Epoxy resin	60	59.5	59
4	MWCNT	0	0.5	1

3. Results and Discussion

3.1 Tensile properties

The fabricated composites are cutted and sized into standard as per ASTM D638 for tensile test. The tensile test values are shown in the fig2. The maximum tensile strength is achieved 234.5Mpa for glass / Sisal composite which is having 1wt% of MWCNT followed by 198.6Mpa and 155.7 Mpa for 0.5Wt% and 0Wt% of MWCNT composite respectively. The results indicates while increasing the Wt% of MWCNT which increasing the tensile strength of the composites due to the high tensile and toughness properties of MWCNT. Similar result reported by Raj Mohan et al, [12].

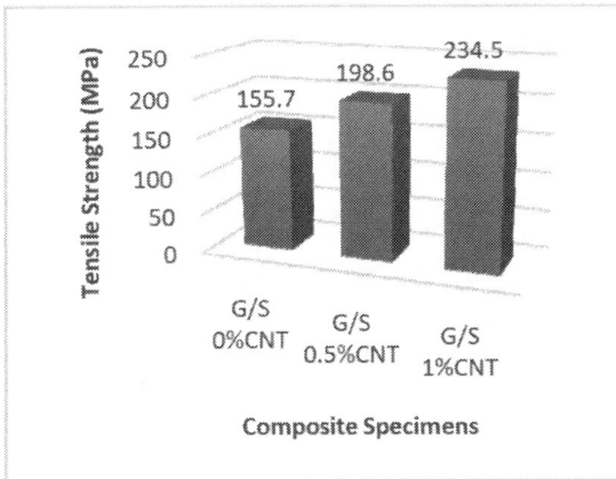

Fig 2. Effect of MWCNT on tensile

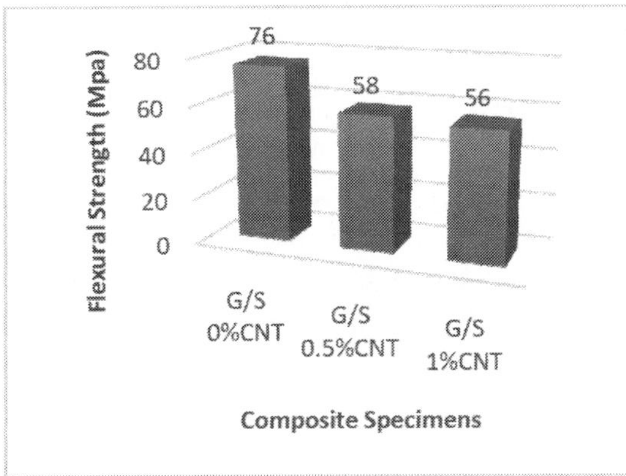

Fig 3. Effect of MWCNT on flexural

3.2 Flexural properties

Flexural specimens were prepared according to the standard as per ASTM D790. The test has been carried out using 3point flexural testing method. The obtained results are revealed in the fig 3. The maximum flexural value is 76 Mpa for 0% incorporated MWCNT followed by 58 Mpa and 56 Mpa for 0.5% and 1% MWCNT incorporated composites. From the results it is concluded incorporating the MWCNT is not improving the flexural properties of the composites. This fall is due to thepresence of higher hemicellulose and lignin content in the sisal fiber which is not making a proper adhesion with the matrix. Hence it is confirmed from the SEM analysis in fig 5c. Treated sisal fiber used as a reinforcement which improves the flexural strength and modulus of the composite rather than untreated sisal fiber reported by jhon et al, [16]Increasing the volume of sisal fiber also lead to decrease the flexural properties reported by venkatareddy et al, [17].

3.3 Impact properties

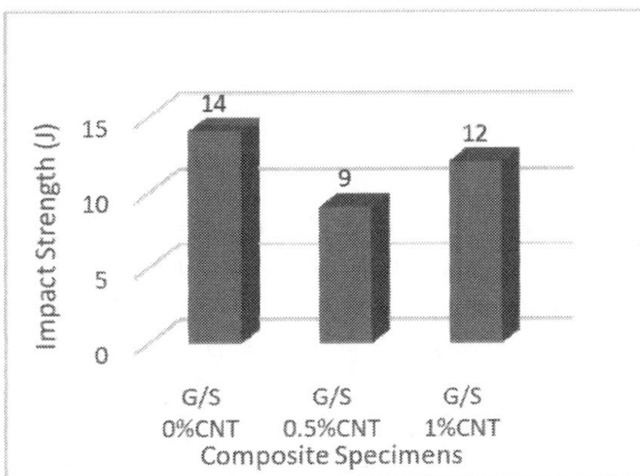

Fig 4. Effect of MWCNT on Impact

The impact test of the different composite specimens has been carried out using Charpy impact test method. The specimens are prepared as per ASTM A370 standard. During the test, the energy loss is found and the obtained results are plotted in the fig 4.G/S with 0% MWCNT composites exhibiting the maximum value compare to other two specimens which are incorporated with MWCNT 0.5% and 1%

respectively. The results indicates, initially the value is decreased then it is increased while increasing the wt% of MWCNT. The reduction of impact value due to the fiber breakage and delaminating of the composites. From the SEM images the breakages were observed and the similar results have been reported by Ramesh et al, [18]

4. Microstructure Analysis

The microstructure, fractured surface and interfacial properties of the sisal glass fiber reinforced filled with MWCNT composites were analysed using scanning electron microscope (SEM). The Fig.5 (a) shows the more elongation of fibers during tensile testing which indicates the improvement in the tensile properties of the composites also indicating the reinforcements are well bonded with the epoxy matrix.

a) Pulled out fiber during tensile test b) Blending of MWCNT & matrix

Fig. 5. SEM images of MWCNT filled glass – sisal composites

Proper blending of epoxy resin and MWCNT are shown in the fig 5(b) which indicates the homogeneous mixing of MWCNT into the epoxy resin. Mechanical properties enhanced by proper blending of nano particles with the resin.

5. Conclusion

The sisal and glass fiber reinforced composite filled with an MWCNT were fabricated using compression molding technique. The effects of MWCNT on mechanical properties were analysed and the following conclusions are drawn.

- The tensile properties of developed composites increased while increasing the wt% of MWCNT. The tensile property of the composites increases up to 33.6% with addition of 1wt% of MWCNT.
- Flexural strength of the composites considerably reduced while increasing the wt% of MWCNT. The composite which is having 0%MWCNT show the maximum flexural strength of the composite.
- Impact strength of the composite initially decreases when increasing the wt% of MWCNT at 0.5%. Further increasing with the wt% of MWCNT shows the significant improvement in the impact strength.

- SEM analysis shows the pulled out fiber during tensile and blending of MWCNT with the epoxy resin also it's confirmed the presence of MWCNT in the composite through EDS analysis.

References

[1] Dipa ray B, Sarkar K,Rana A K, and Bose N R 2001 *Mater. Sci.***24** 129.
[2] HariSankar P, Mohana Reddy Y R, and Hemachandra ReddyK 2015 *Fib. Polym.***16**443.
[3] FarukHossenMd,SininHamdanMd, Rezaur Rahman, Md. Mizanur Rahman, FuiKiewLiew and Josephine Changhui Lai 2015. *Fib. Polym.***16**475.
[4] Sathish P, Kesavan R,VijayaRamnath B and Vishal C 2017*J. Silic.* **9** 577.
[5] Mijung Cho, Muzaffer A. Karaaslan, Scott Rennecka and Frank K 2017*J. Mater. Sci.* **52** 9602.
[6] Balaji A,Karthikeyan B, Swaminathan J andSundar Raj C 2017 *Fib.Polym.* **18** 1193.
[7] Jeyakumar R, SampathP S, Ramamoorthi R and RamakrishnanT 2017*J.Adv. Manuf.Technol.***93** 527.
[8] Sang Do Lee, Oh Jin Kwon, ByoungChul Chun, Jae Whan Cho and Jong Shin Park2009*Fib. Polym.***10** 71.
[9] Li Xiu Gong, Li Li Hu, Jing Zang, Yong Bing Pei, Li Zhao and LongCheng Tang 2015*Fib. Polym.* **16** 2056.
[10] Islam S, AhmadM B, Hasan M, Aziz S A, Jawaid M, Haafiz M K M, Zakaria S A H 2015*J.prop.Bio.Res.***10** 1394.
[11] Mohan K, and RajmohanT 2017*J.Nat.fib.***14** 864.
[12] Rajmohan T, Mohan K and PalanikumarK 2015*Appl.Mech.Mater.***766-767** 193.
[13] Vinayagamoorhy R 2017 *J. Nat.fib.***10**108.
[14] Vinayagamoorthy R 2017 *J.Reinf.Plast.Compos.***36** 1577.
[15] Rajmohan T, Mohan K, VenkatramanMahalingam,Satendra Kumar Bajpai 2015 *Appl.Mech.Mater.***813-814**505.
[16] John K andVenkatanaidu S 2004 *J.Reinf.Plast.Compos.***23** 15.
[17] Venkata ReddyG, Venkata Naidu S and Shobha Rani T 2009*J.Reinf.Plast.Compos.***28** 16.
[18] RameshM, PalanikumarK and Hemachandra Reddy 2013*Compos.Part B***48** 9.

A study on factors affecting succession planning in small and medium scale Indian family business

Akash S Bathija[1] and R G Priyadarshini[2]

[1] Amrita School of Business, Coimbatore, Amrita Vishwa Vidyapeetham, India, akashbathija@gmail.com

[2] Amrita School of Business, Coimbatore, Amrita Vishwa Vidyapeetham, India, rg_priyadarshini@cb.amrita.edu

Abstract: Today, various factors determine the successor in a family business. Succession planning in family business is a tough and a critical challenge. Succession planning in family business is influenced by various factors. This paper attempts to study the effect of various personal and strategic factors on small and medium scale Indian family business in the succession planning process. It is important to understand the effect of these factors while selecting the successor as it may eventually determine the success or failure of the family business. The study uses a case research method which focuses on the companies from the manufacturing sector and the service sector in the small and medium enterprise category. The study focuses on cases from the succession planning of various companies and it helps to understand how succession planning in those firms is affected by various factors. Data is collected from the head of the family business in both the manufacturing and the service sector. The sample consists of detailed discussion with the owners of eight family businesses to understand various personal and strategic factors affecting succession planning across India.

Keywords: Succession planning, Family business, Small and medium scale enterprises, Successor, Personal factors, Strategic factors.

1. Introduction

In India, the major part of businesses is in the control of the families. From various small stores to large conglomerates we have the business that is run by the family in India. Many of the biggest known companies in India that include the Ambani, Birla and Tata are all family governed business. Entrepreneurship has been encouraged a lot and this has resulted in the formation of a lot of family business in India. . Family businesses help to promote the economic and social development and also provide for the creation of wealth around the world [1]. With so much importance given to family business in India, it has become very important to carefully select the successor to ensure the growth and success in the family business. Family businesses understand the importance of succession planning to ensure continuity of their business [2]. For most family businesses, determining who would be the successor is the toughest and most critical challenge they face. For each family, there may be different needs due to the size of the family, size of the business or gender preferences that determines the successor of the business. [11]

2. Literature Review

2.1. Family business

A Family business is the company or firm where the family has ownership of the major part of the stock and has complete control over all aspects in the firm. [3].

Family business can be defined as a firm which is owned, controlled and operated by the members of a particular family, there will be non-family members also employed into the business. It is a company in which there is strong influence of the family in the day to day running of the business [4].

2.2. Succession Planning

Succession planning in family business is a tough and a critical challenge that a firm may face while transferring the business successfully to the next generation. Today, succession in family business is an issue that is given maximum importance by family businesses [5]. Considering the various missions of the family businesses possibly succession planning and advancing the business to the following generations is the most significant one [6], and so it can be believed that proper succession planning is a requirement for a business to survive or sustain [7], [8], [9], [10].

2.3. Indian Family business

In India, the system of joint families has been prevalent and this has also ensured that the family business concept has been prevalent and successful, especially among the people from the 'Vaishya' community who have traditionally been in business. The sociological factors that are prevalent in India replicate the bond between families and businesses. Working together in a family business also deepens the relationship between families and the business is usually divided among the businessman's children, who in turn diversify and expand the family business. [1].

2.4. Small and Medium scale enterprises

The Ministry of Micro Small and Medium Enterprises of India(MSME) has different definitions for small and medium enterprises based on the sector they operate in. The below table explains the differentiation.

Sector	Category	Minimum Investment in Plant & Machinery(INR)	Maximum Investment in Plant & Machinery(INR)
Manufacturing	Small	>25 Lacs	<=5 Crores
	Medium	>5 Crores	<=10 Crores
Service	Small	>10 Lacs	<=2 Crores
	Medium	>2 Crores	<=10 Crores

2.5. Influencing factors of succession planning

Quite often it is observed that son is the legal heir of family businesses. Even in the practical world while you see that female successors possess more financial resources, the business owners by default chose male heirs for their businesses [11]. The Key personal attributes and deliverables are assumed to be present in the male heirs than female heirs. Negligence of knowledge has been emphasised. The key to a successful succession plan is for the successors to be equipped with sufficient knowledge to run the business. [12]. If a family business founder follows a religion, it will affect both the individual's life and the firm's behaviour. Using nationwide family firm survey data from Chinese family firms, it was understood that the family firm founders religiosity is conducive to their succession intention. [13]. The transfer of the family business to the successors an important strategic challenge for the family. The tax effect makes the process of family succession difficult with an important strategic decision to make keeping the tax laws in consideration during family succession [14]

3. Need for study

This study will help us understand the various personal factors affecting succession planning in small and medium scale family business.

This study will also help us understand the various strategic factors affecting succession planning in small and medium scale family business.

4. Research question

What are the various factors that affect the succession planning decision of selecting the successor in small and medium scale family business?

5. Objective

To study and understand the various factors influencing succession planning in small and medium scale Indian family business.

6. Methodology

An amalgamation of Primary research and Secondary research has been used in this research. As the word denotes – Secondary research is done through secondary (non-direct) resources. This involves literature gathering from various sources like the internet, books, case studies and understanding the concept of Family Business in India. Primary research involves research from Primary (direct) resources. Multiple interviews were conducted with the owners of various family businesses and the insights obtained were recorded as cases. The study of these cases helped provide an in-depth analysis of the subject in a real-life scenario.

Case study is a particular approach for qualitative research that provides for an in-depth analysis of an existing occurrence within its real-life perspective [15].

6.1 Sample Details:

The sample includes eight Indian family businesses from both the manufacturing sector and the service sector that are into business for more than a decade.

7. Analysis

7.1 Case 1

This organization was incorporated in the year 1994 and is engaged in construction, property and real estate development business. This organization has successfully completed many big commercial projects including housing projects. This organization had diversified its business into manufacturing of plastic components like bottles and jars in the year 2008. This organization has its headquarters in central Coimbatore.

In the case of selecting the successor of the business while succession planning, the CEO feels that would certainly looks for skills in the prospective successor that would fill the gap of skills that the organization may lack. The CEO also feels that the preparedness of the prospective successor to handle management tasks is a key factor and also the prospective successor must be knowledgeable and well qualified. The CEO would not consider gender and religious beliefs while choosing the prospective successor. The CEO would not have a particular time period for the successor to learn the business but he feels that it would be a natural thing to have the successor train for a duration so that the DNA of the organization will get imbibed in to the prospective successor. The CEO tells that neither the economic conditions nor the industry type will influence him on his succession planning decision.

7.2 Case 2

This company was founded in 1991 and was involved in the manufacturing of food products. Initially the company was involved in the manufacturing of waffles and then diversified into various varieties of biscuits and snacks with major sales coming in the form of the different flavours of biscuits. This organisation has its headquarters in Pollachi.

While selecting the successor of the business, the CEO would want the prospective successor to have the necessary skills and be capable enough to manage such a big company and be involved in all the departments of the company with abilities to manage all the departments with efficiency. The CEO in this company feels that practical knowledge and understanding of the market is more important than qualifications. The CEO also feels that he would prefer the prospective successor to be male. The CEO feels that his prospective successor should take a sufficient time of three years to learn the business well. In addition to these factors, the prospective successor's personality and attitude would be a crucial factor. The CEO would like the prospective successor to be motivated and achieve big goals in life.

7.3 Case 3

This company was established in the year 1989 and is engaged in the manufacturing of silk sarees. This company has its headquarters in Salem. The company initially started as a retail outlet dealing with silk sarees and slowly expanded its business into manufacturing their own products. The company today has a portfolio of several silk products.
While selecting the successor of the business, the CEO would look for skills in designing, creativity and ability to develop unique products that would stand out. The CEO feels that good management abilities are a must to succeed in a business and that his successor should be well qualified. In this business it was necessary for the prospective successor to have a well-defined role since it is a textile mill. [16] [17] The CEO states that the gender of the prospective successor would influence his decision as he wants a male to succeed him but religious and family beliefs would not be considered by him. The CEO would want a learning period of two to three years for the prospective successor to learn the business. The CEO believes that the size of the business would influence his decision as with larger size comes more responsibilities. The CEO also believes that in this competitive industry someone with a creative mindset would be preferred. Also according to the CEO the decision making ability of the prospective successor would be a determinant factor.

7.4 Case 4

This business was established in 1995 and it is engaged in the trading of various hardware products and power tools for individuals and industries. This firm has its headquarters in Coimbatore. This company initially started trading with only basic tools and then expanded its business slowly into a large portfolio of power tools and hardware products.
While selecting the successor, the CEO states the he would look at the leadership skills in his prospective successor as he feels that the leadership of the prospective successor can determine the outcome of the business. Also good managerial abilities would be required. Good knowledge about the industry and the products is required in the prospective successor. The CEO also tells that his family beliefs may be a major factor in determining the prospective successor of the business. The CEO would want the prospective successor to learn for a time period of two years to understand how the market works and how to deal with its competitors.

7.5 Case 5

This company was founded in 1994 and is engaged in the trading and distribution of electrical components to local retailers and neighbouring states. This company has its headquarters in Coimbatore. This firm initially started with the trading of wires and pipes and then slowly grew to become one of the leading suppliers of electrical components.
While being asked about choosing the prospective successor, the CEO tells that the business generally stays in the family and is handed over to the children when they grow up. Customer relationship is a major element in their business and the prospective successor must be good at it. The most viable candidate would be the one who has the most experience . However others with less experience would be involved in the start up of a smaller business or would be engaged n various other supervision activities. The CEO feels that knowledge would be pivotal and would also grow with experience. Qualifications would not be given too much importance. The gender o religious beliefs would have no impact on his decision to choose the prospective successor. The CEO believes that the type of industries and the economic condition do not matter as for the CEO only experience in the firm matters.

7.6 Case 6

This company was started in 1997 as an event management company and it was engaged in providing services of managing events for various occasions. This company has its headquarters in Bangalore and has successfully conducted a large number of events ranging from small gatherings to large industrial events.

While selecting the successor, the CEO feels that in this industry, the prospective successor must have excellent customer relationship skills as that is the most important part in this business. The CEO states that more than qualifications, knowledge matters the most. The CEO feels that a time period of three to four years would be a suitable time period to learn the business. The CEO believes in gender equality and the gender would not influence his decision in the succession planning decision. The CEO wants the prospective successor to be capable enough to adapt to the changing economic conditions. In addition to all these factors, the CEO wants the prospective successor to be dedicated and to be passionate and interested in this field.

7.7 Case 7

This organization was incorporated in 1985 and is engaged in the insurance and mutual funds advisories. The firm has its headquarters in Chennai. This company was initially started as a financial advisory company and then later expanded into Insurance and mutual funds advisors.

While selecting the prospective successors, the CEO feels that skills like sales, customer relationship management, time management and communication skills are very important. The CEO would look for good managerial qualities in the prospective successor as he feels that it would be necessary to assign work to multiple employees and make sure it doesn't affect the client satisfaction. For the CEO, qualifications do not matter but industry related knowledge is required. Gender and religious beliefs do not influence the decision of the CEO because he feels that gender and religious beliefs do not run the business. The CEO would want his prospective successor to learn the business for a time period of two years. When asked about the influence of the type of industry and the economic condition in the industry, the CEO said that the person who is better at understanding macroeconomic trends and predicting future returns would be more likely to suggest better products for the clients and thus be a suitable successor.

7.8 Case 8

This company was founded in the year 1992 and is involved in the manufacturing and distribution of glass and plywood. This company has its headquarters in Coimbatore. Initially the company was started as a glass trading organization. Today it has expanded into the manufacturing and distribution of glass and plywood.

While selecting the successor, the CEO feels that the prospective successor should have good managerial skills and leadership skills as both go hand in hand. The CEO feels the successor should be well qualified and knowledge is something that can always be gained. The CEO would want the prospective successor to learn the business for three to four years before the prospective successor would take over the business. Gender and religious beliefs do not have an influence on the CEO. The CEO tells that the type of industry and the economic condition would have an influence on his decision to choose the prospective successor as understanding the products and competitor in this industry is vital for the sustenance of the business.

8. Summary and Conclusion

To summarize, it can be seen from the above analysis of the various factors affecting succession planning that include personal factors like the skill set of the prospective successor, the management abilities, the gender, knowledge, qualifications, religious and family beliefs, and strategic factors like the economic conditions, the type of industry and the size of the business all have a vital role to play in determining who would succeed the business. These factors also give us insights as to what the owners expect in their successors.

Family businesses play a pivotal role in our country as the majority of the businesses are conducted by families. The major challenge that they face is to choose a suitable successor to ensure the growth and sustenance of the business. From the above analysis of the various Indian family businesses it can be seen that people take both personal factors as well as strategic factors into consideration as they need to choose a successor that would ensure the growth and sustenance of the business. Also from the analysis it can be seen that gender is not an important factor for most of the firms. But most of the firms have given importance to the skill set, knowledge and the experience of the prospective successor. Additionally leadership and understanding of the business has also been a decisive factor for the prospective successor. The religiosity of families influenced the decision in choosing the prospective successor. [13] Also as seen in many cases above the economic condition and the type of industry are also factors that would influence the decision of succession planning in the small and medium scale Indian family business.

9. Implications for future research

Further studies can be done taking the prospective successor's view of succession planning and the factors that influence the prospective successor to succeed the business.

10. References

[1] Panjwani, Ashok., Aggarwal Vijay and Dhameja Nand, Family Business: Yesterday, Today, Tomorrow, *Indian Journal of Industrial Relations*, Vol. **44**, No. 2 (Oct., 2008), pp. 272-291.

[2] Brockhaus, R. H. (2004). Family business succession: suggestions for future research, *Family BusinessReview*, **17** (2), 165–177.

[3] E. Tsang (2002), Learning from overseas venturing experience. The case of Chinese family businesses, *Journal of Business Venturing*, **17** (1) (2002), pp. 21-40

[4] Dutta, Sudipta (1997), Family Business in India, *New Delhi: Response Books*, 1997, pp. 267

[5] Bird, B., Welsch, H., Astrachan, J. H., & Pistrui, D. (2002). Family business research: the evolution of an academic field. Family Business Review, **15**(4), 337–350.

[6] Davis, S. M. (1968). Entrepreneurial succession, *Administrative Science Quarterly*, **13**(3), 402–416.

[7] Cabrera-Suarez, K., De Saa-Perez, P., & Garcia-Almeida, D. (2001). The succession process from a resource and knowledge based view of the family firm, *Family Business Review*, **14**(1), 37–46.

[8] Shepherd, D. A., & Zacharakis, A. (2000). Structuring family business succession: an analysis of the future leader's decision making, *Entrepreneurship Theory and Practice*, **24**, 25–39. Summer.

[9] Davis, P. S., & Harveston, P. D. (1998). The influence of family on the family business succession process: a multigenerational perspective, *Entrepreneurship Theory and Practice*, **22**(3), 31–49.

[10] Barnes, L. B. (1988). Incongruent hierarchies: daughters and younger sons in company CEOs, *Family Business Review*, **1**(1), 9–21.

[11] Jan-Philipp Ahrens , Andreas Landmann, Michael Woywodea, (2015) Gender preferences in the CEO successions of family firms, *Journal of Family Business Strategy* **6** (2015) 86–103

[12] Kalin P (2016), Knowledge Creation Aiding Family Business Succession Plan, *International Journal of Business & Management Science* **6**(1), p.63-84

[13] Na Shen, Jun Sub (2016), Religion and succession intention - Evidence from Chinese family firms, *Journal of Corporate Finance* **45** (2017) 150–161

[14] Eva Vincencováa, Roman Horákc (2015) , The Tax Effects of the Family Business Succession, *Procedia Economics and Finance* **34** (2015), 303 – 310

[15] Yin, K Robert (2003), Applications of case study research, *Sage Publications* **4** (2009), pp. 240

[16] R. G Priyadarshini(2005)" A study on Organizational Effectiveness in Textiles mills at Coimbatore *"Management and labour studies,***30**

[17] R. G Priyadarshini(2009)"The Importance Of Role Efficacy and Self Efficacy In Organizations and its Relationship With HR Practices *"Management and labour studies,***34**

IOP Conf. Series: Materials Science and Engineering **390** (2018) 012092 doi:10.1088/1757-899X/390/1/012092

The effect of corrosion inhibitor on corrosion of automotive materials in Biodegradable engine oil

R.Ellappan[1*], S.Arumugam[1]**

[1]Department of Mechanical Engineering, Sri Chandrasekharendra Saraswathi Viswa Mahavidyalaya, Kanchipuram - 631561, Tamilnadu, India.

*ellsraaj@gmail.com, **aru_amace@yahoo.co.in

Abstract - Bio-degradability is one of the major issue in the disposal of used synthetic engine lubricants, by considering this the development of vegetable based engine lubricants shows more attention, the corrosion of engine parts owing to its usage under bio-lubricant environment can be suspected due to its hygroscopic nature. This work aims to analyse the effect of blend of poly-ethylene glycol and benzotriazole as corrosion inhibitor on ferrous and non-ferrous metals in chemically modified rapeseed oil based bio- degradable engine oil as per ASTM D6594. Scanning Electron Microscope (SEM) analysis shows that the pitting corrosion over the surfaces of lead and copper is found more comparatively with other metals in the bio lubricant environment without corrosion inhibitor, the metals immersed in the bio-lubricant with inhibitor provides the protective oxide film layer over the metallic surfaces restrict the corrosion rate.

1. Introduction

Development of alternate fuels to alter the place of fossil fuel due to the higher consumption rate leads to search suitable substitute fuels [1]. Similarly focus towards eco-friendly bio-lubricant from canola oil is renewable and bio-degradable and nontoxic , therefore it can be considered as a replacement for synthetic lubricants [2]. Bio lubricants meet out the tribological properties compared with commercially available synthetic lubricants [3]. Arumugam et al., reveals that the bio-diesel contaminated bio lubricant shows better performance in terms of wear and frictional force and friction, over mineral lubricant under similar operating conditions [4]. Bio-diesel reacts with metals resulting in corrosion due to hygroscopic nature and oxygen moieties present in bio diesel [5]. Metal samples were prepared and cleaned as per ASTM standard G01 [6]. A test rig was developed as per ASTM D6594, to conduct laboratory corrosion test at high temperature environment for synthetic lubricant [7].

Katherine et al., reported that the effect of synthetic lubricant under high temperature environment suspected to corrosion over ferrous and non-ferrous metals [8]. Haseeb et al., concluded that the copper bronze shows higher corrosion among non ferrous metals [9]. Singh et al., describes about the cause for the corrosion is due to the auto oxidation and moisture presence in vegetable oils [10]. Fazal et al., express that the static immersion test on bio-diesel at 800c for 1200 hours, the copper and aluminium are susceptible to corrosion [11]. Haseeb et al., justifies that the increasing TAN number in oil leads to corrosion [12]. Krazan et al., investigated tribological properties of formulated rapeseed oil and compared with synthetic oil [13]. Arumugam et al., reported that rapeseed based bio-lubricant is very much comparable with synthetic lubricants in all respects [14]. Arumugam et al., concluded that bio-lubricants are more corrosive than synthetic lubricants by conducting on test as per ASTM D6594 [17]. Fazal, et al., and Matjaz Finsger et al., justifies that the effect of corrosion inhibitors like TBA, BTA over cast iron and low carbon steel on corrosion in palm biodiesel, shows better performance in reading the acidity of the bio diesel and creates protective layer over the metal surface which prevents the corrosion [15,16].

2. Materials and methods

2.1 Selection of materials and preparation of sample

Material selection is the key parameter to predict the corrosion probability on both ferrous and non-ferrous material which is used in diesel engines are cylinder liner and journal bearing. Ferrous metals such as CI and Mild steel and nonferrous metals such as copper, tin, lead and zinc were considered for testing. Metal samples of CI, mild steel and copper were cut from round bars and rest of Tin, lead and zinc samples has been copper from sheet.

All samples were polished by using 600 – 1200 grade silicon carbide emery sheets. Before immersion the specimen were cleaned to remove oiliness with acetone followed by ultrasonic cleanic in an absolute ethanol bath as per ASTM G1 STD [6]

2.2 Selection of corrosion inhibitors

2.2.1 Ferrous Metal and Its Inhibitor (Tert Butyl amine)

Fazal et al. [15] investigated the effects off TBA as inhibitor on cast iron and low carbon steel in palm biodiesel (B100) and concluded TBA exhibits best inhibition properties. At 300 K inhibitor efficiency at 250 ppm is 86.5% for cast iron and 86.71% for low carbon steel. At concentration of 100ppm the inhibitor efficiency is 47% for both cast iron and low carbon steel. The high inhibitor efficiency could be attributed to the dominant physical adsorption of N-containing compound which creates protective layer over the metal surface which prevents corrosion. The protective coating compounds indicates it is form of $Fe(NO_3)_3.9H_2O$. Figure 1 shows the structure of $Fe(NO_3)_3.9H_2O$.

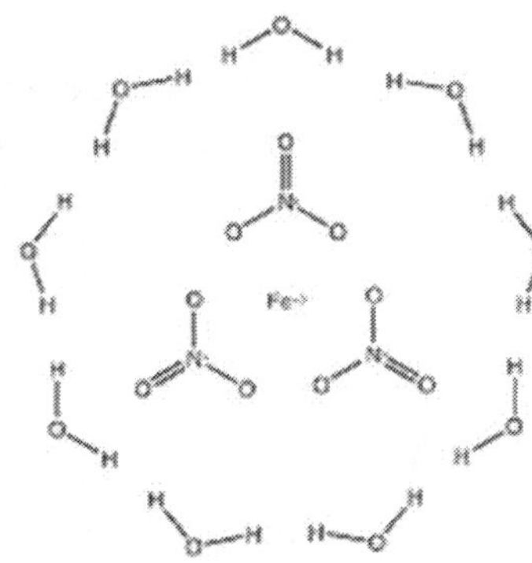

Figure 1 Structure of $Fe(NO_3)_3.9H_2O$

The leading and prominent group for corrosion inhibition is amine group. A peak of N-containing compound $Fe(NO3)_3.9H_2O$ at around 40 degree is formed on the metal surface may help from an N-rich layer which contributes in reducing corrosion attack. The adsorption molecules on metal surface via hetero cyclic moiety N atoms exhibits a stronger interaction compared with that via other hetero atoms. A nearly uniform region enriched with N and probably composed of Fe (NO3)3.9H2O is observed via XRD the main constituents within the layer are N , which proves the assumption of TBA adsorption on LCS surface forms protective layer to prevent further interaction between the metal and the bio diesel. The thickness of compound Fe (NO3)3.9H2O on CI and LCS are found out to be 19.97 μm and 7.78 μm respectively.

2.2.2 Non-Ferrous Metal and Its Inhibitor (Benzotriazole)

It is a heterocyclic organic compound produced by reaction of ortho-phenylenediamine with sodium and acetic acid giving a white powder at room temperature. It is sufficiently soluble in water solutions and used as corrosion inhibitor. Figure 2 (a, b, c) shows the structure of Benzotriazozle, Structure of BTAH and Cu-BTA complex chemisorbed on the Cu surface.

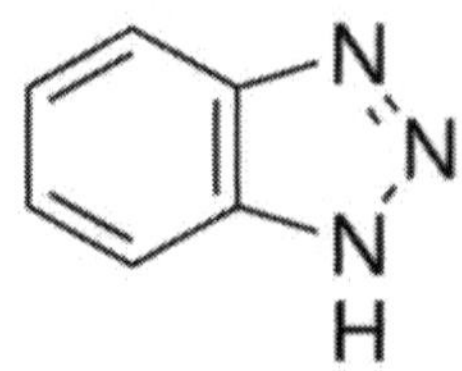

Figure 2 (a) Structure of Benzotriazole

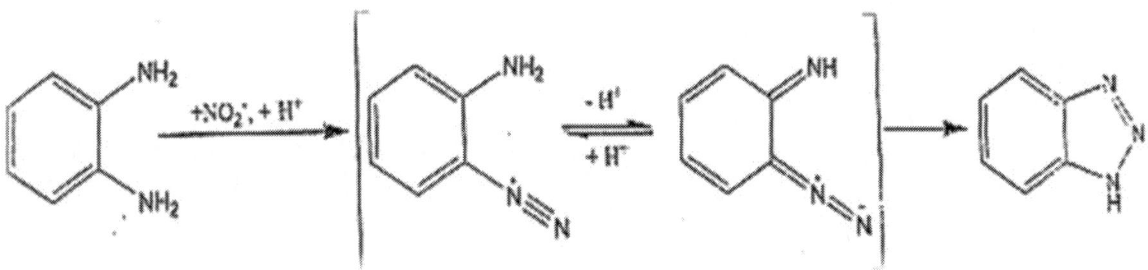

Figure 2 (b) Structure of BTAH

BTAH on reaction with Cu in 10% H_2SO_4 solutions under dynamic conditions and at low flow rate (Reynolds no. < 500) there is maximum IE at 10mM.For a medium of 3% NaCl and (3-5.3) pH limit 20mM concentration of BTAH the H_2 and air punged test gives an effectiveness of (63-95%).

Naphthatriazole is the one of the best inhibitors used for Corrosion of non-ferrous metals. The advantage is that effectiveness increases when compared to BTAH. But it is very expensive.

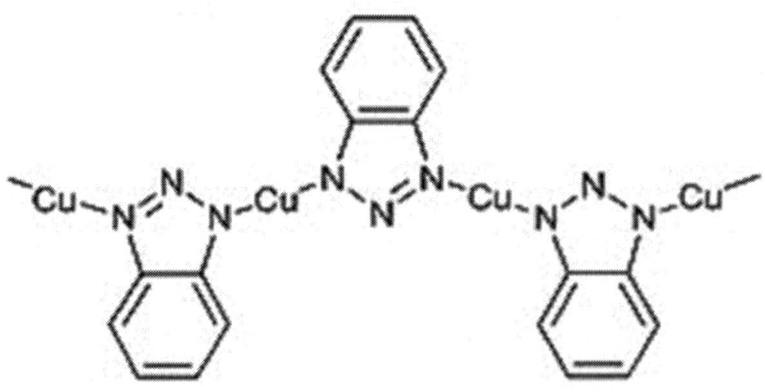

Figure 2 (c) Cu-BTA complex chemisorbed on the Cu surface

Protective coating is the film or multilayer and not monolayer formed by adsorption or chemisorption due to gas molecules present nearer to cu atoms. The above diagram denotes the co-ordination of (C-N) atoms where the lone pair of electron (N) is at the center. In the (N-H) bond the removal of H atom is replaced to form a C-BTA structure as shown. It's known as polymeric chain. The thickness of the film is around (50 A0) and the concentration increased if time of immersion is high. If the film is thin then effectiveness of BTAH is high. The film formed because of adsorption of BTAH on the surface of Cu.

2.3 High temperature corrosion bench test (HTCBT)

Fig 3 (a-c). Shows Laboratory test set up, as per ASTM D 6594 corrosion test was conducted about 160 hours at 1350c. [7] The samples immersed in both the lubricant with and without inhibitor were cleaned .the corrosion products on the metal surfaces were removed by dipping the metal coupons about twice for 5 – 10 sec in 10 – 15 % Hydrochloric acid with Clarke solution. The table 1 shows the corrosion rate of the bio-degradable engine oil with and without inhibitor. Metal samples were then cleaned by acetone and dried and instantly weighed in weigh balance with 5 decimal accuracy to find out the weight loss for evaluating corrosion rate. Rapeseed based bio lubricant with 10% diluted rapeseed methyl ester as immersion medium. Table 1 shows the values of corrosion rates.

Figure 3 photographic view of a) Test samples b) Samples immersed in without inhibitor c) Samples immersed in with inhibitor

2.4 Corrosion Rate

The rate of corrosion is the speed at which any metal deteriorates in a specific environment. Degradation of metals was investigated by measurement of corrosion rate. It is expressed in mpy (mils per year). The corrosion rate can be calculated using the equation given below,

$$\text{Corrosion rate (mpy)} = \frac{w \times 534}{D \times t \times A}$$

Where,

w – Weight loss in the metal in mg

D – Density of the metal in g/cm^3

A – Exposed surface area of the metal in inch^2

T – Time of exposure of the metal in lubricant in h = 168h

Table 1 Corrosion rate

S.no	Metal	Density at 20°C (g/cm^3)	Bio Lubricant Without Inhibitor		Bio Lubricant With Inhibitor	
			Weight lost (mg)	Corrosion rate (mpy)	Weight lost (mg)	Corrosion rate (mpy)
1	Cast Iron	7.272	3.04	0.9415	0.2	0.0531
2	Mild Steel	7.833	3.58	1.16	0.5	0.1447
3	Copper	8.954	4.03	1.27	2.31	0.4127
4	Zinc	7.144	3.04	0.9584	1.1	0.3739
5	Tin	7.304	2.05	0.6249	1.37	0.35074
6	Lead	11.393	5.02	7.171	3.95	0.98145

IOP Conf. Series: Materials Science and Engineering **390** (2018) 012092 doi:10.1088/1757-899X/390/1/012092

3. Result and Discussion

3.1 Visual examinations for the specimens used in the corrosion studies

The Figure 4 (a-c) demonstrates the specimens of the test samples before and after testing with and without inhibitor.

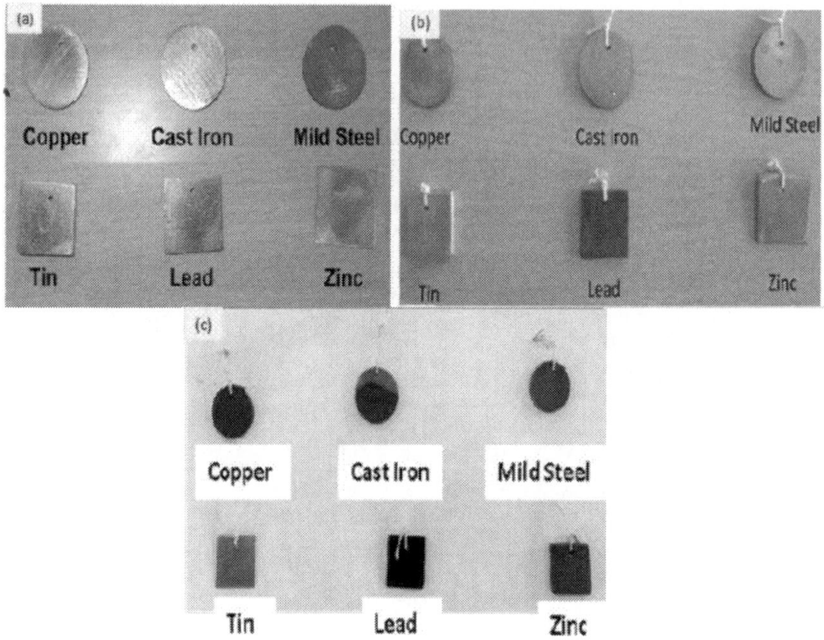

Figure 4 photographic view of a) Specimens before testing b) Specimens after testing in bio lubricant with inhibitor c) Specimens after testing in bio lubricant without inhibitor

3.2 Investigation of Surface Morphology Using SEM

Scanning Electron Microscopy (SEM) is a method for high-resolution imaging of surfaces. It is preferred for getting the information about qualitative differences in the surfaces before and after testing in bio and reference lubricants. The SEM images of different metals before and after testing in bio and reference lubricants are shown and discussed below. The figure 5 (a-i) shows the SEM images of different material used in this investigation.

3.3 Cast Iron

The SEM images taken for Cast iron shows no noticeable amount of corrosion even though some deposits can be seen in case specimen is exposed to bio lubricant with or without inhibitor. Hence it can be said that pitting corrosion may have started to take place in the specimens.

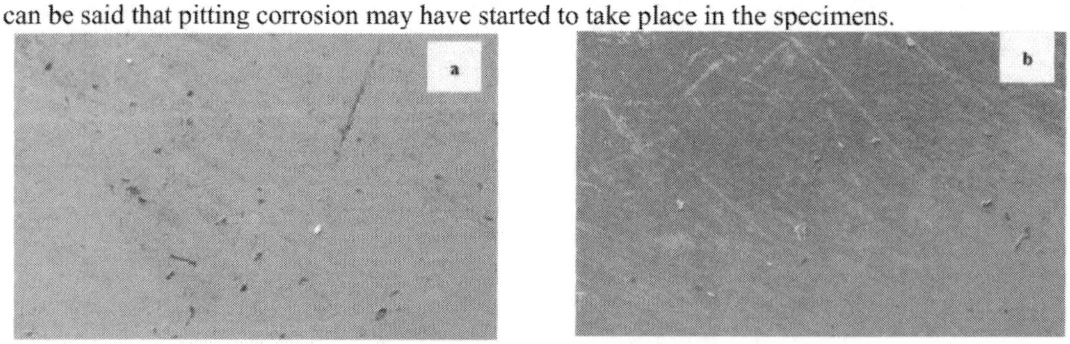

Figure 5 SEM images of (a) CI with inhibitor (b) CI without inhibitor

IOP Conf. Series: Materials Science and Engineering **390** (2018) 012092 doi:10.1088/1757-899X/390/1/012092

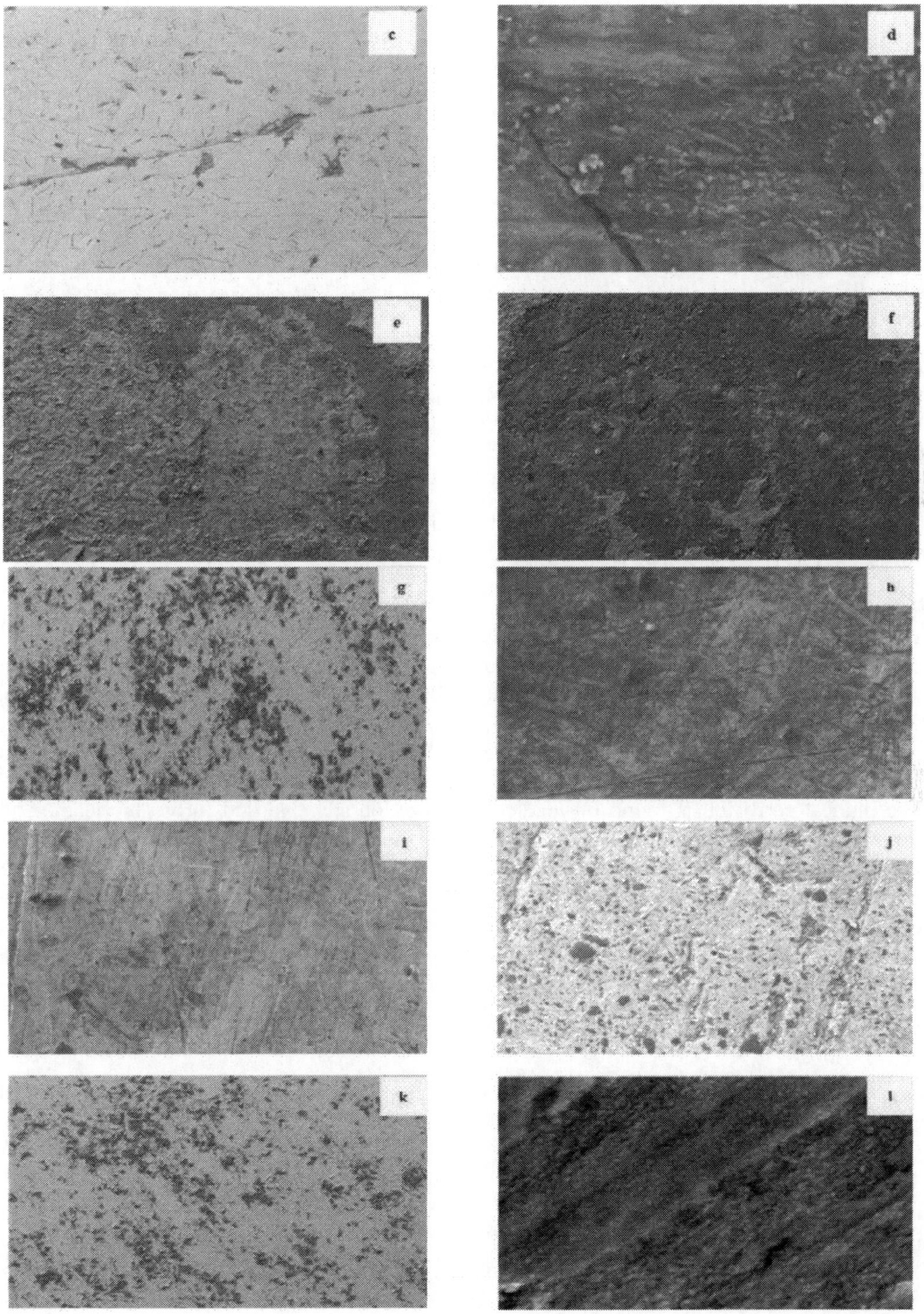

Figure 5 SEM images of (a) CI with inhibitor (b) CI without inhibitor (c) Mild steel with inhibitor (d) Mild steel without inhibitor (e) Copper with inhibitor (f) Copper without inhibitor (g) Zinc with inhibitor (h) Zinc without inhibitor (i) Tin with inhibitor (j) Tin without inhibitor (k) Lead with inhibitor (l) Lead without inhibitor

3.4 Mild Steel
There no noticeable amount of corrosion is found on mild steel except some white dots, which are non-metallic deposits in case specimen exposed to bio lubricant with inhibitor and some shaded regions which may denote compounds formed out of corrosion on specimen exposed to bio lubricant without lubricant.

3.5 Copper
In case of copper exposed to bio lubricant with inhibitor number of pits can be found throughout the surface which shows that pitting corrosion has taken place. Whereas the specimen exposed to bio lubricant without inhibitor is not corroded as much as the one exposed to bio lubricant with inhibitor but it also shows some very small pits which may be the starting of pitting corrosion.

3.6 Zinc
Zinc specimen exposed to bio lubricant with inhibitor shows some deposits over the surface which may later lead to corrosion and the specimen exposed to without inhibitor has white spots that indicates non-metallic deposits throughout the surface. Otherwise there is no noticeable amount of corrosion is found.

3.7 Tin
Tin exposed to both bio lubricants doesn't show any noticeable corrosion. It can be justified with the help of corrosion rate calculation in which tin shows least corrosion rate for both bio lubricants.

3.8 Lead
Among all the specimens Lead shows the maximum corrosion effect specifically on the specimen exposed to bio lubricant without inhibitor. Very large pits and debris are found over the specimen exposed to bio oil. Similarly specimen exposed to reference oil also shows number of small pits through the surface but no debris found. From the SEM images it can be clearly said that the corrosion is more in case of all specimens which are exposed to bio lubricant as compared to the specimens exposed to synthetic lubricants. In specimens exposed to bio oil deposits are found frequently over the surface. The highest corrosion is found on lead followed by copper exposed to bio lubricant. There was no noticeable corrosion in case other specimens. Tin is the least corroded metal among all other.

The density of the each metal at 20°C is taken from the Heat transfer data book and the weight loss is calculated from the weight of the specimens measured before and after testing, which is converted into corrosion rate by substituting the values in the above equation and the results are tabulated in Table From the above table 1 it can be found that for the specimens tested in bio lubricant without inhibitor, the corrosion rate increases in the order given below,

$$Pb > Cu > MS > Zn > CI > Sn$$

Whereas for the specimens tested in bio lubricant with inhibitor, the corrosion rate increases in the order given below,

$$Pb > Cu > Zn > Sn > MS > CI$$

4. CONCLUSIONS
The degradation of different metals which comprises in the engine lubrication system was investigated with High Temperature Corrosion Bench Test Rig. The following points has been concluded from the above study,

- All analysis carried on specimens and oils as well as corrosion rate calculations and properties measured before and after testing prove that the bio lubricant without inhibitor is more corrosive than the bio lubricant with inhibitor in majority of cases.

IOP Conf. Series: Materials Science and Engineering **390** (2018) 012092 doi:10.1088/1757-899X/390/1/012092

- Upon exposure to different metals the corrosivity of the lubricants increases. Among the metals tested Lead, Copper and Tin are more prone to corrosion.
- Ferrous compounds such as Cast Iron and Mild Steel doesn't show significant corrosion as compared to other metals.
- Cast iron is found to be more resistant to the corrosion in case of exposure to both bio and synthetic lubricants.
- Further investigation has to be done in order to understand the corrosion mechanism and different types of corrosion taking place.
- Suitable additives can be added to reduce the corrosion effect on the metals. The corrosion inhibitor, mixture of PEG 400 and Benzotriazole shows a downfall in corrosion rates for Cast iron, Mild steel, Zinc and Lead, however the corrosion rate for Tin and Copper have increased.

5. REFERENCES

[1] Singh. B, John Korstad, Y.C. Sharma, 2012 *Renewable and Sustainable Energy Reviews 16*, 3401–3408.

[2] Rajesh V. Sharma, Ajay K. Dalai, 2013,Applied *Catalysis B: Environmental*, Volumes 142–143, 604-614.

[3] I.M. Atadashi, M.K. Aroua, A. Abdul Aziz 2012, *A review, Renewable and Sustainable Energy Reviews*, Volume 14, Issue 7,1999-2008.

[4] Krzan B, Vizintin J 2003, *Tribology International*,36 (11): 827-833.

[5] Arumugam, S., and G. Sriram,2012 ,*Tribology Transactions* 55.4: 438-445.

[6] ASTM G:1.*Standard Practice for Preparing, Cleaning, and Evaluation Corrosion Test Specimens*

[7] ASTM D: 6594. *Standard Test Method for Evaluation of Corrosiveness of Diesel Engine Oil at 135°C.*

[8] Katherine M. Richard 2010, *SAE International Journal of Fuels Lubr*. 2(2):66-71.

[9] Haseeb. A.S.M.A, H.H. Masjuki, L.J. Ann, M.A. Fazal 2010, *Fuel processing Technology* 91, 329-334.

[10] Singh. B, John Korstad, Y.C. Sharma, 2012, *Renewable and Sustainable Energy Reviews* 16, 3401–3408.

[11] Fazal. M.A, A.S.M.A. Haseeb, H.H. Masjuki, 2012, *Energy* 40, 2012; 76 – 83.

[12] Haseeb. A.S.M.A, H.H. Masjuki, L.J. Ann, M.A. Fazal, 2010, *Fuel processing Technology* 91, 329-334.

[13] Krzan B, Vizintin J.2003, *Tribology International*. 36 (11): 827-833.

[14] S. Arumugam, G. Sriram, R. Ellappan, 2014, *Energy*, 72, , 618-627.

[15] Fazal, M.A.AU - Sazzad, B.S.AU - Haseeb, A.S.M.A.AU - Masjuki, H.H.Jo, 2016 *Energy Conversion and Management*, 122- 290 - 297.

[16] Matjaz Finsgar, Ingrid Milosev, 2010,*Corrosion Science*, 52, 9, 2737-2749.

[17] S. Arumugam, R. Ellappan, S. Sangavi, G. Sriram and P. Ramakrishna. 2018, *Arab J Sci Eng* 43: 1345.

Basic Problems faced by the Small Scale Industry Sector

A Bharadwaj, A K Pati, H Srivastava, A Panda, S Tripathy

School of Mechanical Engineering, KIIT Deemed to be University, Bhubaneswar-751024

stripathyfme@kiit.ac.in

Abstract. Industries contribute cordially towards massive developments in the country but however it's the Small Scale Industries which has a very low contribution towards this development. Most of the small scale industries are still struggling make the best out of them but due to numerous problems and issues their effort lies incomplete. In this paper, a short analysis about SSI sectors based on a case study on a Small scale industry conducted by us and an analysis of reviews about small scale enterprises and its development and growth by various authors. Mainly factors like lack of a proper work environment, technical and financial support from the government, high competition against the large scale units in the market etc. are the root reasons behind their halt in growth and development. Thus looking upon all the basic problems an industry faces and looking upon our result a list of feasible suggestions can be considered that could benefit the operations and working of these industries.

1. Introduction

Usually small scale industries are situated in semi-urban or rural areas. They do not have a large number of machineries nor they use highly advanced technologies. The main purpose of these industries is to use the locally available raw materials efficiently as well as it acts as a platform to create employment opportunities for the local people as they do not necessacarily need skilled workers. They have a limited boundary for their markets. As compared to large scale industries, the SSI sectors cannot enjoy the benefits and that's also due to their size and nature. These industries have importantly contributed to the economy but due to lack of support they face problems in their operations.

According to the amount of capital invested in plant and machinery the ministry of MSME has further divided these small scale into micro, small and medium enterprises-

Description	Investment in plant and machinery
Micro Enterprises	Up to 25 lakhs
Small Enterprises	Between 25 lakhs- 5 crore
Medium Enterprises	Between 5 crore to 10 crore

Table 1. Different categories of industries with their investment in plant and machineries (MSME report)

Few SSI sectors are provided with support for technical up-gradation. There were certain provisions by the government for the SSI sectors like reserving some particular products for manufacturing. This was the benefit the small scale units had over the large scale enterprises who were refrained from producing the items that was kept for the Small scale sectors. But due to the gradual rise in the economy and globalization, this provision which these sectors had have been affluently vulnerable. Thus these sectors are now facing the race against large scale industries in India as well as foreign countries.

2. Objectives

- To study and examine the working of a small scale industry.
- To analyze the problems faced in different aspects of manufacturing, by the industry

3. Literature Review

Jos Frijins et al.(1999)[6], has emphasized on safer and clean manufacturing methods by various technological and operational barriers. The research discusses feasible ways to prevent the deterioration of small scale industries in developing nations, giving importance to strategies towards clean manufacturing and production. The author refers to his project on controlling pollution in a small scale metal firm in Nairobi, Kenya. The study showed that cleaner production initiatives is a significant approach towards controlling pollution in an economic manner, but small-scale firms require external assistance to work onto these initiatives to minimize wastes.

Kateshiya Sumeet(2017)[2], has studied and examined the growth rate of SSI sectors and further researched about the problems of a Small Scale Industry based on his case study in Jamnagar city in context of Brass Part units. His study and analysis have resulted towards the need to utilize modern technology in production methods, suitable infrastructure and the role of the government in supporting these units.

Prof M. Chandraiah et al.(2014)[1], has focused the importance of MSMEs towards the nation's development. The study shows the purpose of MSMEs, the operations and results of the MSMEs in the Indian economy, its policies and the issues it encountered. On a conclusion it can be noted from the analysis that MSMEs sector faced Key Challenges like as Lack of access to markets at a global level, Limited access to equity capital, Storage issues, package and designing issues and marketing, etc.

Harri Ramcharran(2016)[3], has studied about the performance of small and medium sized industries by empirically estimating the efficiency of bank loans and its rates to the SME sector of India. It was done using data from 1979 to 2013. The results have important policy implications regarding the allocation of capital and risks associated since bank loans are the main source behind financing these industries. The results show the rise in productivity (i.e. output elasticity) of bank credit from 0.76 to 1.23; the productivity of labor turned out to be negative, but increased from −1.57 to −0.628. The sector's efficiency had improved from returns to scale of −0.89 to 0.607 which was mainly due to the increasing productivity of bank credit.

Sangita G.Patil et al.(2014)[4], has studied and analyzed on the problems faced by SSI's. According to the author small scale industries are the roots behind the entire economy and if they are developed then the other economy related problems can be resolved automatically. The author as a result urges towards supporting of MSMEs, their education and empowerment so as to achieve development.

Nafis Khuriyati et al.(2015)[5], has studied on the Cleaner Production Strategies for enhancing the Performance of a Small Scale Biscuit Industry through an analysis based on a case study on the Cracker Industry in Yogyakarta Special Province. Upon the analysis it was found that the End of pipe treatment strategy which has been known until that day is in fact has lost its utility due to its high cost. And Good Housekeeping techniques have been one of the cheap and easy techniques in clean production strategies.

4. Research Methodology

- The study and analysis is based on the information collected by us from the case study we conducted on an industry.

- Some information and sample data were gathered from related research manuscripts and online sources which complimented the further analysis of the problems faced by the SSI. Limitations

Although we managed to collect necessary information but there were few operative data to which we were unauthorized to access, due to official constraints of the firm.

4.1. About the Industry
The Industry is located in the Mancheswar Industrial Estate of Bhubaneswar city. It is a small scale manufacturer of a wide range of polymer products in Odisha. An Iso-9001:2008 certified venture, it was established in 2005. With all the passing years of production and growth the company achieved to expand its venture as a Polymer firm (Iso-9001:2015). The company has a two-level organizational structure with the owner being the sole proprietor. The industry was set up within an area of about 6500 sq. feet belonging to the owner himself, which comprises of their factory and storage space. With the strength of around thirty employees (out of which eleven are female) the company manages to earn about 10-15% profit annually, annual turnover being about ₹ 5 crore. They produce around 400 variety of products like PET water and juice Bottles, PVC Garden pipes, HDPE Jars, Electrical gang box, Plastic Buckets and mugs, Electrical fittings, Flip top caps etc and the products manufactured by the Polymer firm includes plug boards, electrical sockets, switches etc. Among different processes of manufacturing the products, a significant factor that benefits the overall revenue of the firm is the recycling of waste products. The products that are manufactured normally give out wastes from the processed raw materials at the end, which are recycled back using additives in a hopper and are processed similarly to make new products. The machineries used in the firm are modern injector molding machines for molding buckets, mugs, containers etc, another machine is used called PVC Garden pipe plant to produce the PVC TR pipes and regarding recycling the recycling PIECCO scrap grinder machine is used. The machine specifications are as follows:

Machine	SPECIFICATIONS
Injector Moulding-(200/125) - Based on tonnage of raw materials.	415 V, 3-Φ 50 c/s, 24VD, Rated current- 60 A. Power consumption- (80 kW/ 60-70 kW) Servo injection moulding using hydraulic clamping.
PVC Garden pipe plant	
Plastic scrap grinder machine	Motor-5HP Output- 100to110 kg/hr

Table2. Machine Specifications of the machineries used by the industry

The industry has a well connected web of clients throughout the city as well as in different parts of Odisha and nearby states. The industry also produces varieties of any single product as per the specifications required by the clients. Thus, it establishes a healthy Manufacturer-client relationship.

4.2. Different aspects of a manufacturing industry

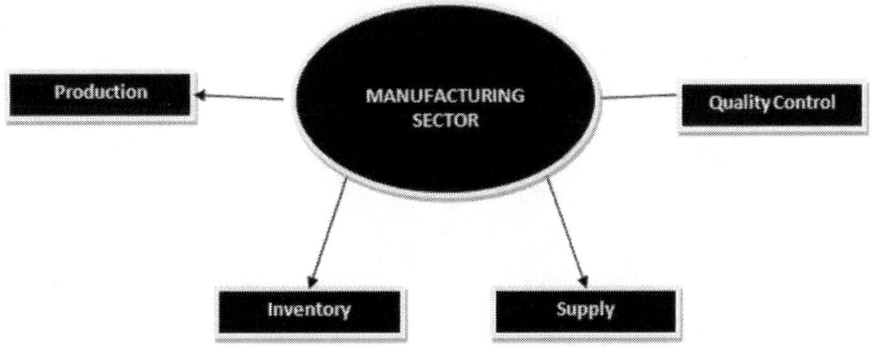

Fig 1. A schematic representation of the different aspect of manufacturing is as given-

4.3. Analysis of the problems faced by the SSI

PRODUCTION:	INVENTORY:	SUPPLY CHAIN:	QUALITY CONTROL:
• Problems due to sudden absence of the regular work force. • Machinery faults and errors. • Longer idle gaps • No fixed production rate	• No proper protocol to manage inventory. • Inadequate space for inventory and storage.	• Transportation issues • Limited market coverage and fluctuating demand	• Absence of proper quality control processes • Usage of polymers and chemicals which are less hazardous to the environment.

Table 3. Problems faced by the Industry with respect to the general problems faced by the SSI sectors.

In the manufacturing system of the industry, a lot of issues have been observed(some prominent ones mentioned in the chart above). The highlighted issues as in Table 2 with respect to different aspects have more likely occurred and are the key points which need to be focused. In comparison to other SSI sectors, the industry primarily face the following problems:-

4.3.1. Production Problems

- At times Unavailability of adequate workforce affects the production rate. The number of regular workers is very few and thus strict rules cannot be imposed on the workers against the risk of losing them.
- In case of absence of regular workers the machines cannot be operated with full efficiency. But usually it's the machine failure and error which causes the halt to machining processes. The machines used are quite modern and electronic, which requires high maintenance. But they don't have their own maintenance team so they have to send the parts workshops far away which lead to ideal time.
- Due to fluctuating demands by the clients, production quantity fluctuates due to which the company sometimes faces a longer idle time. These idle gaps are usually utilized to clean

up the machines work area or carryout other product manufacturing. These are instances when the demand highly falls and the production processes halt for the while.
- Being a SSI, there's no standard production rate. Amount of raw materials varies blindly meeting the random fluctuations of the demand. And hence, the output quantity of the product also varies.

4.3.2. Inventory Problems

- According to their policies the industry have no standardized inventory control. They don't have a fixed rate of production as to manage their Ordering quantity or Re-order level. They have a good friendly relationship with the dealer who unquestioningly provides them with required raw materials.
- No separate spaces for inventory and storage. The workspace is about 6500 sq. feet which restricts the proprietor to spend separately for inventory or storage. The backyard of their factory is used as inventory/storage.

4.3.3. Supply Chain Problems

- After the manufacturing of the products they need to be transported to the market. And concerning the supply (in) of raw materials to the supply (out) of products, its transportation that controls the process. The company has 1 vehicle of their own and borrows 3-4 vehicles from sources in order to carry out the processes. Having fewer vehicles consumes a lot of time regarding process completion.
- The market coverage of SSIs is limited so they fall behind in the competitive market. The net supply chain is also very complicated due to fluctuating demands and production which hampers the linear profit and growth of the firm.

5. Discussion
The analysis shows the prominent problems or constraints faced by the industry regarding its proper functioning and efficiency. Given below are a couple of suggestions based on reviews by prominent researchers and merging our opinions with their work which can preferably help towards the betterment of the industry-
- An improved working environment which can motivate the workers on giving more effort and punctuality towards their work. Implementation of proper safeguarding and cleaning of the workplace by improving the efforts of the workers during work hours and cleaning the waste scrap on the production equipments prior to washing improves performance and safety to the environment.[6]
- A well organized production system and procedure including legal protocols and proper project planning which can help in cordially governing the system and can also reduce unwanted idle time. This includes proper inventory control and quality control techniques.
- The authorities should focus more on enhancing the skills and knowledge of the existing workforce, such as good managing and housekeeping skills so as to compete with the high standards of the large scale units.[5],[4]
- The proprietor should pay more attention on promoting and marketing its products which can earn business bonds with other industries especially large scale industries. Thus in a way it can establish a strong web of corporate relationships and a healthy supply chain.[1]
- Favorable and clean manufacturing and production measures which are economical and which can be very feasible to the SSI Sectors. Government and large institutions must

undertake the initiative to provide technical and financial initiatives for better enhancement. . It requires effective organization of the firms. The SSI's must be given the opportunity and additional support to waste minimization methods.[6],[2]

- Apart from government bodies and large scale industries it's the banks that can provide the industries with capital, resulting in improvement in the scale efficiency of the Small Scale and middle scale sectors of India. It is mainly due to the increasing productivity of bank loans. Other studies discussed before also report a positive impact of external financing on growth of the industries.[3]
- Machineries must be renewed in order to avoid the inferior performance and quality of the output, comparing to the large scale firms. Moreover, the small-scale enterprises should efficiently care about the fluctuating preferences and of the people. Accordingly, modernization and rationalization is a must in these industries.

5.1. Other Problems

Besides the above discussed problems small scale industries face a lot of other problems like lack of financial assistance, from well established large-scale industries, poor project planning, etc. These problems prevent the SSIs to reach a prestigious stage.

6. Conclusion

And we could thus suggest the best optimal solutions and alternatives keeping in mind the methods and approaches adopted by other researchers that can help the sector to grow healthily. In developing nations like India quite a proportion of the industries are SSI and are still emerging. In order to succeed and make their best they must work on the constraints which act as hurdles in their paths. In the case study conducted upon the industry, the problems were identified and analyzed. It can be seen that in comparison to various other industries the problems that persisted in the industry were nearly in the same page. In India it's either the government or the large scale sectors that can help the sector to grow. With additional support and initiatives possible solutions can be expected. And the extent of succeeding the nation's growth with the SSI's depend on how fast and up to how deep are these initiatives implemented.

References

[1]. Prof M. Chandraiah, R. Vani-"The Prospects and Problems of MSMEs sector in India an Analytical study"- IJBMI, Vol.3, Issue 8, Aug-14, ISSN (Online): 2319 – 8028, ISSN (Print): 2319 – 801X

[2]. Kateshiya Sumeet-" PROBLEMS OF SMALL SCALE INDUSTRY IN JAMNAGAR CITY IN CONTEXT OF BRASS PART UNITS"- Abhinav Publication, Vol. 6, Issue 8 (Aug-17) UGC APPROVED Online ISSN-2277-1166

[3]. Harri Ramcharran-"Bank lending to small business in India: Analyzing productivity and efficiency"- QREF, June-16, Elsevier

[4]. Sangita G.Patil, Dr. P.T.Chaudhari-" Problems of Small Scale Industries in India"- IJEMR, Vol. 4, Issue-2, April-14, ISSN No. 2250-0758.

[5]. Nafis Khuriyati, Wagiman, Denok Kumalasari-" Cleaner Production Strategy for Improving Environmental Performance of Small Scale Cracker Industry"- ICoA-2014, Elsevier B.V. (CC BY-NC-ND license

[6]. Jos Frijns, Bas Van Vliet-"Small-Scale Industry and Cleaner Production Strategies"- IHUDS, Vol.27, Issue- 6, 1999 Elsevier Science Ltd.

[7]. www.accountlearning.com (https://accountlearning.com/what-are-the-problems-faced-by-small-scale-industries-in-india/)

[8]. www.msme.org.in

Influence of nano based metal working fluids on machining process – A Review

K Venkatesh[1*], G Sriram[2] A Uday Kiran Reddy[3] and S V R Nikhil[4]

[1]Associate Professor, Dept. of Mechanical Engg., Sri Chandrasekharendra Saraswathi Viswa Mahavidyalaya, Kanchipuram - 631561, Tamilnadu, India.

[2]Professor and Dean (E&T), Dept. of Mechanical Engg., Sri Chandrasekharendra Saraswathi Viswa Mahavidyalaya, Kanchipuram - 631561, Tamilnadu, India.

[3, 4] Final Year Student, Dept. of Mechanical Engg., Sri Chandrasekharendra Saraswathi Viswa Mahavidyalaya, Kanchipuram - 631561, Tamilnadu, India.

Corresconding Author Email [1*] kvenkateshme@gmail.com

Abstract. Tool wear is natural phenomenon, and this leads to tool failure which is caused mainly due to various aspects like the surface of the work piece, material of the work piece to be machined and other. The application of the metal working fluids (MWF's) (or generally coolant) is considered to reduce the friction between the work and tool and to provide lubrication as well as cooling. In a machining process, the use of coolant is essential to dissipate the heat generated and to enhance various properties like productivity, tool life, machinability, etc. There are many types of oil based MWF's are available like mineral oil, semi-synthetic oil, synthetic oil based MWF's. But lately vegetable oil based MWF's are being used because of its eco-friendly nature. Recently Nanoparticles are used to enhance the properties of metal working fluids. In this paper there is brief explanation on preparation of metal working fluids with enrichment of Nanoparticles.

Keywords: Tool wear, Eco-friendly, Metal working Fluids, Nano Additives

1. Introduction

Machining is a process of material removal in which the loss of material is caused by affecting a relative motion between tool and work piece. Due to removal of material in the form of chips, new surfaces are cleaved from the work piece accompanied by a large consumption of energy. The mechanical energy necessary for the machining operation is transformed into heat, leading to conditions of high pressure, high temperatures and severe thermal and frictional conditions, consequently making metal cutting process more and more inefficient in terms of tool life, dimensional accuracy and material removal rate.

IOP Conf. Series: Materials Science and Engineering **390** (2018) 012094 doi:10.1088/1757-899X/390/1/012094

To avoid difficulties in machining an effective coolant system is to be provided using a cutting fluid. The important functions of cutting fluids are cooling the tool and workpiece, lubricating chip, tool and workpiece and improve surface finish [1]. The main requirements of a cutting fluid are 1. Long life, 2.Should not form oxides, 3. Good thermal conductivity, 4.Less wetability, 5.Less odour, 6. Should not be flammable, 7.Should not affect the health of workers [1].

In manufacturing industry, many types of lubricants are used as a baseoil for metal working fluids. The fluids which are mainly used as base oil are water, distilled water, mineral oil, synthetic oil, vegetable oil, etc. Base oils cannot be used directly, since pure or raw base oil may cause corrosion, and some can be chemically imbalanced and cannot have a long-life period. So, additives are added to the base oil for better performance. Based on the substance that is used as base oil, additives changes. Additives are chemical substances that are used to enhance the properties of the base oil and improve the performance by taking out negative properties form the base oil. Additives may function as anti-wear, anti-corrosive, anti-oxidant and neutralizing agents and some can be anti-bacterial (mainly for vegetable oils). Simply additives are added to make base oil chemically stable and to become more user friendly to use.

Recently vegetable oils are widely used because of its high flash points and biodegradability when compared to the mineral oils[2], but vegetable oils contain unsaturated fatty acids which poses some problems in terms of corrosiveness and life of the coolant oil, because of high temperature raise in machining process [2].So, vegetable oils are hydrolysed to get ester oil compound which has more stability due to the presence of ester compound[3]. Additives are to this ester compound of their respective vegetable base oil. Lawal et al. [4] intensive research work on the formulation of vegetable based metal working fluids using appropriate additives like Zinc diamydithiocarbamate (ZDDC), Zinc dialkyldithio phosphate (ZDDP), etc. Lawal et al. also reported about various emulsifiers and surfactants used in preparation of the vegetable based oils, and then they compared the properties like surface roughness, tool wear between rapeseed, coconut, sunflower oil based metal working fluids and every result is compared with the commercial vegetable based oil and commercial mineral oil. And the paper concluded that vegetable based oil gave better performance than mineral oil and it is reported that surface roughness and tool wear is improved when coconut oil is used compared to the soluble oil. Commercial vegetable based cutting fluids are used extensively, because of its good properties and high efficient performance.

In this study it is discussed about process of preparation of MWF's with enrichment of nanoparticles.

2. Performance of MWF's with enrichment of nanoparticles

Nowadays nanoparticles are extensively used in many fields, in the field of manufacturing industry it is used to enhance the properties of MWF's. The most commonly used Nano particles are

1. Molybdenum disulphide (MoS_2)
2. Multi walled carbon Nano tubes (MWCNT)
3. Single walled carbon Nano tubes (SWCNT)
4. Aluminium oxide (Al_2O_3)
5. Titanium dioxide (TiO_2)
6. Copper oxide (CuO)
7. Silicon dioxide(SiO_2)
8. Zirconium dioxide (ZrO_2)
9. Hexagonal boron nitride (hBN)
10. Graphite Nanoplatelets (GnP)
11. Nano boric acid (NBA)
12. Nano diamond (ND)

Based on the literature study of Rabesh et al. [5], Nanofluids shown better rheological, tribological properties and thermal conductivity is increased. It is also reported that surface finish is increased because of Nanoparticles. Rabesh et al. suggested of hybrid Nanofluids which contains two or more Nanoparticles. Since Nanoparticles give a prominent change and increase in the properties of the metal working fluid, so it is used widely in all machining applications as coolant oil such as turning, milling, drilling and grinding in both CNC and lathe operations. The following study is based on the various works on different "machining applications" with different Nanoparticles and various base oils.

2.1. Turning process

Srikant et al. [6] experimented on mixing of copper oxide (CuO) in water which acts as base oil. It is reported that the CuO particles are mixed in proportions 0.5-8 vol% with 0.5 as equal interval. Finally, it is concluded that water based cutting fluid with CuO particles given good results in maintaining temperature which in turn results in increase of tool life.

Krishna et al. [7] experimented on mixing of Nano Boric acid in proportions 0.25, 0.5-8 wt% to SAE-40 oil and coconut oil, and it is reported that 0.5% of Nano Boric Acid reduced overall temperature compared to the other experiments carried out on different wt%, Taguchi method is used for design of experiments. It is also reported that the Nanoparticles have enhanced the oil properties greatly and given better surface finish and decreased great amount of temperature.

Amrita et al. [8] experimented on mixing of Nano Graphite (size < 80nm) in coconut oil in proportions 0.1, 0.3, 0.5 wt%. It is reported that they used MQL (minimal quantity lubrication) technique at different flow rates 10 and 15 ml/min and obtained good results in reducing temperature and cutting forces at the flow rate of 15 ml/min. They used Taguchi loss method for design of experiments for setting optimal parameters.

Sayuti et al. [9] experimented on mixing of silicon dioxide (SiO_2) which has size of 5-15nm in mineral oil. The proportions added were 0.2, 0.5 and 1 wt% and in this study Sayuti et al. shown better results in good surface finish and least tool wear at 0.5 wt% of Nanoparticle.

Gupta et al. [10] experimented on mixing of Al_2O_3, MoS_2 and Graphite in vegetable oil separately. All the Nanoparticles are mixed at 3 wt%. It is reported that comparing Al_2O_3, MoS_2 and graphite based vegetable oil due to good thermal conductivity graphite based vegetable oil given better performance than the other two in reducing overall temperature, but MoS_2 gave better surface finish.

Raju et al. [11] experimented on mixing of Multiwalled carbon Nanotubes (MWCNT) (size 10-20nm) in "distilled water" at a proportion of 0.2 vol%. In this paper it is reported that with the use of distilled water based fluid mixed with 0.2 vol% MWCNT cutting forces and surface roughness reduces 5-8% and 9-22%, respectively when compared to dry and base fluid.

Pavani et al. [12] experimented on mixing of Nano Boric Acid in "soybean oil" and "coconut oil". It is mentioned that the mixing is done with the use of "mechanical stirrer". Taguchi method is used to design the experiment. It was found that 2% wt of Nano Boric Acid is optimum for soybean oil and 3% wt for coconut oil. In this paper the comparison is made between tip temperature of the tungsten carbide (WC) tools, surface roughness and cutting forces. It is concluded that the vegetable based oils with enrichment of Nano Boric Acid given good results. Cutting temperatures reduced and surface finish increased to a maximum level compared with dry machining. It is also concluded that the study encourages to the use of vegetable based oil.

Su et al. [13] experimented on mixing of Nano Graphite (size 35nm) in "LB2000 vegetable based commercial cutting fluid" and "PriEco6000 unsaturated polyol ester oil". The proportions are 0.1 and 0.5%. Nanofluids are formulated in two-step method. Ultrasonication is used for mixing and it is done at 40KHz, 100W. It is reported that irrespective of cutting oil both graphite-based Nanofluids reduced cutting forces and a good amount of temperature. But LB2000 given maximum reduction in temperature when compared to dry cutting.

Amrita et al. [14] experimented on mixing of Functionalized Nano Graphite (FNG, size 80nm), Nano Boric Acid (NBA, size 100nm) and $nMoS_2$ (size 100nm) in water-miscible oil, at a definite proportion of 0.3 wt%. In the series of experiments $nMoS_2$ Nanoparticles at 0.3 wt% recorded reduction in cutting forces and recorded good wear and surface properties. But in terms of heat dissipation NBA at 0.3 wt% shown good improvement when compared other coolant conditions.

2.2. Drilling process

Chatha et al. [15] experimented on Soyabean oil as a base oil and taking Al_2O_3 1.5vol% in Drilling process, the size of the Nanoparticle is taken approximately 20nm. In this study, it has given difference between dry and wet drilling conditions by using MQL technique. In this, he has found out that by the application of the Nano particles no. of drilled holes increases which means that the life of the tool has been increased when compared with the dry conditions.

Chai et al. [16] experimented by taking hydrogenated oil as a base oil and Multi walled carbon Nano particles (MWCNT) in different proportions like 22, 50, 100ppm.This means that he has taken a very smaller ratio of MWCNT. The size of MWCNT is in between 10-12nm. The process used for experimentation is drilling process. As the volume percentage is very low then we can tell the quantity in terms of parts per million(ppm). It has been observed that the thermal conductivity of the base oil (hydrogenated oil) is increased with addition of MWCNT.

Nam et al. [17] experimented by taking vegetable oil and paraffin oils as base fluids and Nano diamond (ND) as Nano particle in Drilling process. Size of particle is 30nm. ND is taken in 1,2vol%. With the addition of 1% in paraffin oil gives more improvement in reducing thrust forces, tool wear than by adding 2% in vegetable oil. Nano diamond plays an important role in reducing the tool wear, but the only limitation is its high cost.

2.3. Grinding process

Prabhu and Vinayagam. [18] experimented by taking SAE20W40 oil and MWCNT and CNT (Carbon Nano tubes) in Grinding process. The size is 10-12nm for MWCNT and 1-2nm for CNT. Initially they have taken 0.2vol% of MWCNT and observed that there is slight improvement in surface quality from micro level to Nano level. After that, by taking 2vol% of MWCNT, there is an increase in flash and fire point of the Nano cutting fluid. By taking CNT it us observed that there is an improvement in the surface roughness value of the workpiece.

Kalita et al. [19] experimented Soyabean oil as a base oil and MoS_2 with 5, 20wt% in Grinding process. The size of Nanoparticle is less than100nm. By using MQL technique the cutting fluid produces lesser coefficient of friction and improves the grinding forces without effecting the material removal rate, and wheel wear reduces.

Mao et al. [20] experimented by taking deionized water and canola oil and their respective Nano particles Al_2O_3 and MoS_2 in Grinding process. Size of Al_2O_3 is 40, 80nm and size of MoS_2 is 70nm. When Al_2O_3 in added in deionized water and MoS_2 in canola oil it is observed that higher concentration of Nanoparticles reduces grinding forces and temperature and it depends upon the oil that will be chosen.

3. Conclusion

As discussed above Nanoparticles increase the properties of metal working fluids, based on the above study the Nanoparticles used for the enrichment of metal working fluids for machining are Multiwalled Carbon Nano Tubes (MWCNT), Molybdenum disulfide (MoS_2, $nMoS_2$), Nano Graphite, Aluminium oxide (Al_2O_3), Nano Boric Acid, Silicon dioxide (SiO_2), Copper oxide (CuO). The base oils used are distilled water, water, vegetable oils like soybean oil, coconut oil, commercial vegetable oils, mineral oil, SAE-40 oil, ester oil.

1. From the above study it is noted that Nanoparticles improve the properties of metal working fluids. Irrespective of base oil, Nanoparticle enriched cutting oils gave better performance than dry cutting in cutting forces and surface finish.
2. Although, water, distilled water, mineral oil, SAE-40 oil performed well and gave better performance it is recommended to use vegetable oil or commercial vegetable oil, because of their biodegradability and higher flash point.
3. Nanoparticles improve the thermal conductivity of the metal working fluids, which results in reducing the overall temperature of machining process.
4. Thermal conductivity of MWCNT is the highest at 6000 W/mK and Copper oxide has the least at 19.6 W/mK.

References

[1] R K Rajput The text book of manufacturing technology (2014)
[2] R R Srikant and P N Rao, J P Davim (ed.) *Sustainable Machining, Materials Forming, Machining and Tribology* (2017), DOI 10.1007/978-3-319-51961-6_2
[3] Nabel A Negm, Mohamed M El-Sukkary, Galal H Sayed, Manal G Mohamed, Mahmoud Bekheit, *J Surfact Deterg* (2016) 19:455–466
[4] S A Lawal, I A Choudhury & Y Nukman, *Int J Adv Manuf Technol* (2013) 67:1765–1776
[5] Rabesh Kumar Singh, Amit Rai Dixit, Amitava Mandal, Anuj Kumar Sharma, *J Braz Soc Mech Sci Eng* (2017) 39:4677–4717
[6] Srikant R R, Rao D N, Subrahmanyam M S, Vamsi Krishna P, *P I Mech Eng J-J Eng Tribol* (2010) 223:221
[7] Krishna P V, Srikant R R, Rao D N. *Int Tool Manuf J Mach* (2010) 50:911–916.[8]. Amrita M, Srikant R R, Sitaramaraju AV, Prasad M M S, Vamsi Krishna P, *P I Mech Eng J-J Eng Tribol* (2013) 227(12):1334–1346
[8] Amrita M, Srikant RR, Sitaramaraju AV, Prasad MMS, Vamsi Krishna P,*P I Mech Eng J-J EngTribol*(2013) 227(12):1334–1346
[9] Sayuti M, Sarhan A A D, Salem F, *J Clean Prod* (2014) 67:265–276
[10] Gupta M K, Sood P K, Sharma V S, *J Clean Prod* (2016) 135:1276–1288
[11] Raju R A, Andhare A, Sahu N K, *Mater Manuf Process* (2017) 32:1–7
[12] P N L Pavani, R Pola Rao and S. Srikiran, *J Mech Sci & Tech* 29 (11) (2015) 4877~4883
[13] Yu Su, Le Gon Bi Li, Zhiqiang Liu & Dandan Chen, *Int J Adv Manuf Technol* (2016) 83:2083–2089
[14] Amrita M, Shariq S A, Manoj Gopal Charan, *12th Global Cong Manuf Manage Proc Eng* (2014) 97:115–124
[15] Chatha SS, Pal A, Singh T (2016) J Clean Prod 137:537–545
[16] Chai YH, Yusup S, Chok VS, Arpin MT, Irawan S (2016) Appl Therm Eng 107:1019–102
[17] Nam JS, Lee PH, Lee SW (2011) Int J Mach Tool Manuf 51(7–8):649–652

[18] Prabhu S, Vinayagam BK (2012) Int J Adv Manuf Technol 60(1):149–160
[19] Kalita P, Malshe Ajay P, Wenping Jiang, Shih Albert J (2010) Trans NAMRI/ SME 38:137–144
[20] Mao C, Zhang J, Huang Y, Zou H, Huang X, Zhou Z (2013) Mater Manuf Process 28(4):436442

Free Convection of Water near its Density Maximum in a Heat Generating Porous Cavity with Sinusoidal Heating

K. Janagi[1], *S. Sivasankaran*[2], *M. Bhuvaneswari*[3] *and M. Eswaramurthi*[4]

[1]Department of Mathematics, Sri Shanmugha College of Engineering and Technology, Pullipalayam, Sankari, Tamil Nadu, India
[2]Department of Mathematics, King Abdulaziz University, Jeddah, Saudi Arabia,
[4]Department of Mathematics, Velalar College of Engineering and Technology, Erode, Tamil Nadu, India
[2]Email: sd.siva@yahoo.com (corresponding author)

Abstract. The aim of the present numerical investigation is to understand the unsteady free convective flow and heat transfer of cold water near its density maximum temperature in a two-dimensional square porous cavity with internal heat generation. The present study uses the Brinkmann–Forchheimer extended Darcy model for porous medium to study the effects of density inversion parameter, heat generation parameter, Darcy number and porosity. The finite volume method is used to solve the governing equations. The nonlinear behavior of local heat transfer is observed due to sinusoidal heating. The heat transfer enhances for heat absorption case than that of heat generation case in the presence of density maximum.

Keywords: *Heat transfer; Porous medium; Sinusoidal temperature; Density inversion.*

1. Introduction

Free convection in porous media plays a vital role in many engineering applications such as grain storage installations, solar energy system, oil recovery, underground water flow, heat exchangers, planar reactors and industrial furnaces. In general, density of the fluid decreases when temperature increases. This behavior is reversed in pure water when it attains its maximum density of 999.972[kgm^{-3}] around$4°C$ at a pressure of one atmosphere. The effect of density maximum on free convection has been studied by many authors on the basis of different phenomena. Sivasankaran and Ho [1, 2] numerically investigated the effect of density maximum and observed that the temperature of maximum density leaves strong effects on fluid flow and heat transfer. Varol *et al.* [3] numerically analyzed the free convection of cold water near $4°$C in a thick bottom walled porous cavity.

The numerical study on convective flow in cavities with sinusoidal temperature distribution is made by several authors. Sivasankaran *et al.* [4] numerically investigated the effect of non-uniform heating on both side walls on mixed convection in a lid-driven cavity. Bhuvaneswari *et al.* [5] numerically examined the effect of magneto-convective flow with non-uniform heating on side walls. It is observed that the heat transfer rate increases in the case of non-uniform heating of side walls. Sivasankaran *et al.* [6] numerically investigated the effects of sinusoidal boundary condition on mixed convection in a square cavity in the presence of magnetic field. Sivasankaran and Bhuvaneswari [7] numerically deliberate the convection in a porous cavity with sinusoidal heating on both sidewalls.

Since no work is reported on convection of cold water in enclosure in the presence of density maximum and heat generation, the present study involves on the solution of this problem numerically.

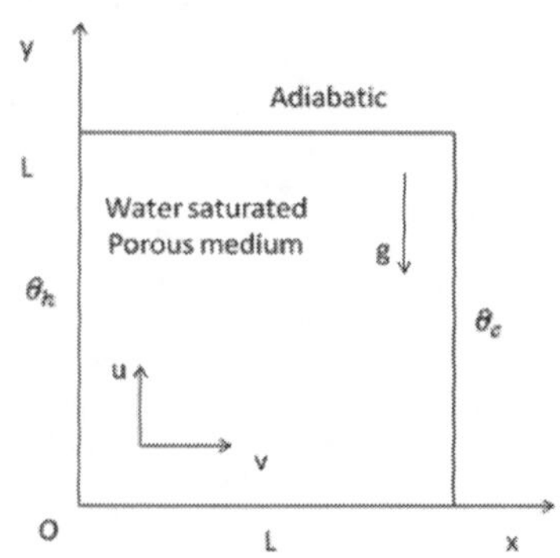

Figure 1. Physical configuration

2. Mathematical formulation

In the present study, cold water saturated square porous cavity of size L with internal heat generation is considered as shown in Figure 1. The right side wall is sustained with fixed temperature θ_c and left side wall of the cavity is maintained with sinusoidal temperature. The horizontal walls act in adiabatic condition. The gravity functions in downward direction. The velocity components u and v are considered in x and y directions respectively. A homogeneous, isotropic porous medium is considered in this study. The porous medium is modeled with Brinkman-Forchheimer extended Darcy model and it is assumed thermodynamic equilibrium with the water. Except density, all the thermal properties of the water are considered as constant and the density of the water follows the relation described by Sivasankaran and Ho [2]. The Boussinesq approximation is valid and it is further assumed that the viscous dissipation is negligible.

The equations governed the system are as follows:

$$\frac{\partial u}{\partial x}+\frac{\partial v}{\partial y}=0 \tag{1}$$

$$\frac{1}{\varepsilon}\frac{\partial u}{\partial t}+\frac{1}{\varepsilon^2}\left[u\frac{\partial u}{\partial x}+v\frac{\partial u}{\partial y}\right]=-\frac{1}{\rho_0}\frac{\partial p}{\partial x}+\frac{\nu}{\varepsilon}\left[\frac{\partial^2 u}{\partial x^2}+\frac{\partial^2 u}{\partial y^2}\right]-\frac{\nu}{K}u-\frac{Fc}{\sqrt{K}}u\sqrt{u^2+v^2} \tag{2}$$

$$\frac{1}{\varepsilon}\frac{\partial v}{\partial t}+\frac{1}{\varepsilon^2}\left[u\frac{\partial v}{\partial x}+v\frac{\partial v}{\partial y}\right]=-\frac{1}{\rho_0}\frac{\partial p}{\partial y}+\frac{\nu}{\varepsilon}\left[\frac{\partial^2 v}{\partial x^2}+\frac{\partial^2 v}{\partial y^2}\right]-\frac{\nu}{K}v-\frac{Fc}{\sqrt{K}}v\sqrt{u^2+v^2}+g\beta|\theta-\theta_m|^b \tag{3}$$

$$\sigma\frac{\partial \theta}{\partial t}+u\frac{\partial \theta}{\partial x}+v\frac{\partial \theta}{\partial y}=\alpha\left[\frac{\partial^2 \theta}{\partial x^2}+\frac{\partial^2 \theta}{\partial y^2}\right]+\frac{q}{\rho C_p} \tag{4}$$

The following are the initial and boundary conditions for the problem:

$$\begin{array}{llll} t=0, & u=v=0, & \theta=\theta_c, & 0\le (x,y)\le L \\ t>0, & u=v=0, & \dfrac{\partial\theta}{\partial y}=0, & y=0\,\&\,L \\ & u=v=0, & \theta=\theta(y)=\sin\pi y, & x=0 \\ & u=v=0, & \theta=\theta_c, & x=L \end{array} \tag{5}$$

The non-dimensional equations are obtained using the following dimensionless variables: $X=\frac{x}{L}$, $Y=\frac{y}{L}$, $U=\frac{uL}{\nu}$, $V=\frac{vL}{\nu}$, $T=\frac{\theta-\theta_c}{\theta_h-\theta_c}$, $\tau=\frac{t\nu}{L^2}$, $\zeta=\frac{\omega}{\nu/L^2}$ and $\Psi=\frac{\psi}{\nu}$. Then we get,

$$\frac{1}{\varepsilon}\frac{\partial \zeta}{\partial \tau}+\frac{1}{\varepsilon^2}\left[U\frac{\partial U}{\partial X}+V\frac{\partial U}{\partial Y}\right]=\frac{1}{\varepsilon}\nabla^2\zeta-\frac{1}{Da}\zeta-\frac{Fc}{\sqrt{Da}}\zeta|V|-\frac{Fc}{\sqrt{Da}}\left[U\frac{\partial|V|}{\partial Y}-V\frac{\partial|V|}{\partial X}\right]+\frac{Ra}{\Pr}\frac{\partial}{\partial X}|T-T_m|^b \tag{6}$$

$$\frac{\partial^2 \psi}{\partial X^2}+\frac{\partial^2 \psi}{\partial Y^2}=-\zeta \tag{7}$$

$$\frac{\partial T}{\partial \tau}+U\frac{\partial T}{\partial X}+V\frac{\partial T}{\partial Y}=\frac{1}{\Pr}\nabla^2 T+Hg \tag{8}$$

$U=\frac{\partial \psi}{\partial Y}$, $V=-\frac{\partial \psi}{\partial X}$ and $\zeta=\frac{\partial V}{\partial X}-\frac{\partial U}{\partial Y}$, $Fc=\frac{1.75}{\sqrt{150\varepsilon^{3/2}}}$, and $|V|=\sqrt{U^2+V^2}$.

The present problem is governed by the following non-dimensional parameters: Prandtl number, $\Pr = \frac{\nu}{\alpha}$; Darcy number $Da = \frac{K}{L^2}$; Rayleigh number, $Ra = \frac{g\beta|\theta_h - \theta_c|^b L^3}{\nu\alpha}$; density inversion parameter, $T_m = \frac{\theta_m - \theta_c}{\theta_h - \theta_c}$ and $Hg = \frac{qL^2}{\rho C_p \nu(\theta_h - \theta_c)}$, the heat generation parameter.

The dimensionless form for the initial and boundary conditions are as follows:

$$\tau = 0, \quad U = V = \psi = 0, \quad \zeta = T = 0, \quad 0 \le X \le 1, \quad 0 \le Y \le 1$$

$$\tau > 0, \quad U = V = \psi = 0, \quad \zeta = \frac{\partial^2 \psi}{\partial Y^2}, \frac{\partial T}{\partial Y} = 0, \quad Y = 0 \,\&\, 1$$

$$\tau > 0, \quad U = V = \psi = 0, \quad \zeta = \frac{\partial^2 \psi}{\partial X^2}, T = \sin \pi Y, \quad X = 0 \tag{9}$$

$$\tau > 0, \quad U = V = \psi = 0, \quad \zeta = \frac{\partial^2 \psi}{\partial X^2}, T = 0, \quad X = 1$$

The local heat transfer across the hot wall is calculated by the local Nusselt number $Nu = \left(\frac{\partial T}{\partial Y}\right)_{Y=0}$ and the average Nusselt number is given by $Nu = \int_0^1 Nu dX$. The non-dimensional governing equations are solved using finite volume method. The detailed method of solution and code validation can be found in Janagi et al. [8].

3. Results and discussion

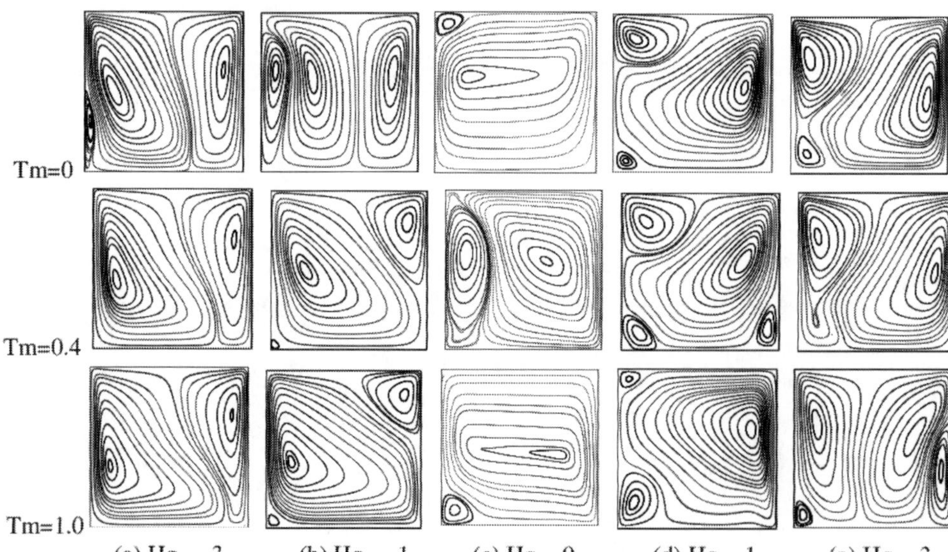

Figure 2. Streamlines for various density inversion parameter and heat generation parameter with Ra=10^6, Da=10^{-3}, and ε=0.6.

In the present investigation, the controlling parameters are the heat generation parameter ($-3 \le Hg \le 3$), Rayleigh number ($10^3 \le Ra \le 10^6$), Darcy number ($10^{-1} \le Da \le 10^{-5}$), porosity ($0.4 \le \varepsilon \le 0.8$) and density inversion parameter ($0 \le T_m \le 1$). Fig. 2(a)-(e) presents the flow for different values of density inversion parameter and heat generation parameter with $Ra = 10^6$, and

$Da = 10^{-3}$. When $T_m = 0$ and Hg=-3, there appears a bicellular pattern along with a small eddy near the bottom left corner of the cavity. Though there is no density inversion inside the cavity, bicellular pattern exists due to heat absorption. As T_m raises, the cell near the hot wall strengthens and cell near the cold wall is pushed towards the cold wall. Also, it is observed that the small eddy disappears as shown in Fig. 2(a).There appears a tricellular pattern for $T_m = 0$ on increasing Hg, see Fig. 2(b). As T_m raises, the cell near the hot wall strengthens and suppressing the secondary cell as described in the previous case. A single largest cell is formed when $T_m = 0$ in the absence of heat generation parameter. This is because the density maximum plane exists along the cold wall and hence the water rises along the hot wall and falls along the cold wall. As T_m increases, a hot cell is suppressed by the formation of secondary cell due to density inversion and the secondary cell enlarges in size to occupy the entire cavity. With increase in heat generation parameter and when $T_m = 0$, two small secondary cells formed near the hot wall. The secondary cells grow in size as T_m raises. When Hg=3, T_m=0, the small eddy grows in size and rotates with more dense near the top-left corner of the cavity. As T_m raises, the bi-cellular pattern is formed. In addition, the multicellular pattern is formed when $T_m = 1$ and Hg=3. When Hg=3, the counterclockwise rotating primary cell pushes the secondary cell towards the right wall as T_m raises. Further increase in T_m, the tri-cellular pattern is formed.

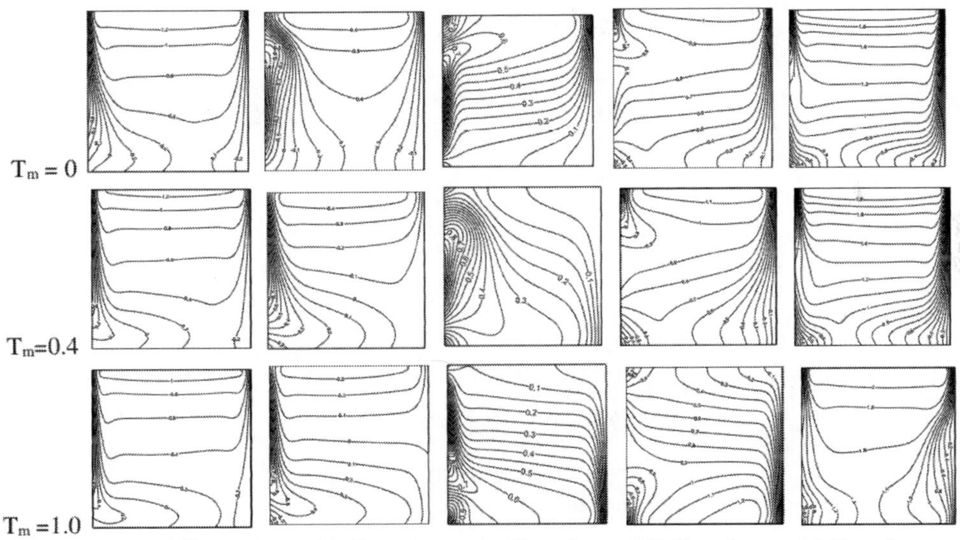

Figure 3. Isotherms for different density inversion parameter and heat generation parameter with Ra=10^6, Da=10^{-3}, and ε=0.6.

The isotherms are depicted for different density inversion parameter and heat generation parameter with $Ra = 10^6$, $Da = 10^{-3}$ and $\varepsilon = 0.6$ in the fig. 3(a) – 3(e). When there is no heat generation or absorption (Hg = 0), $T_m = 0$, thermal boundary layer forms near the left-bottom wall and right-top wall of the cavity, which shows the strong convective motion inside the cavity. When Hg=−3, there is no considerable change in the convective heat transfer as T_m raises. The strong thermal boundary layer exists along left wall for all values of T_m in the case of heat absorption. But, we found the reverse phenomena in the case of heat generation, that is, the strong boundary layers are formed along right wall in the case of heat generation, which is evidently seen in Fig. 3(a,b,d,e). The vertical temperature stratification occurs in the upper portion of the cavity in the case of strong heat absorption and heat generation. Fig. 4 shows the local Nusselt number for different values of heat generation parameter. The local Nusselt number profiles clearly indicate that the local heat transfer is directly affected by sinusoidal heating. It is observed from Fig. 4(a)-4(c) that the local heat transfer is very high in the upper portion of the cavity. Also local heat transfer rate behaves nonlinearly with the height of the wall (*Y*).

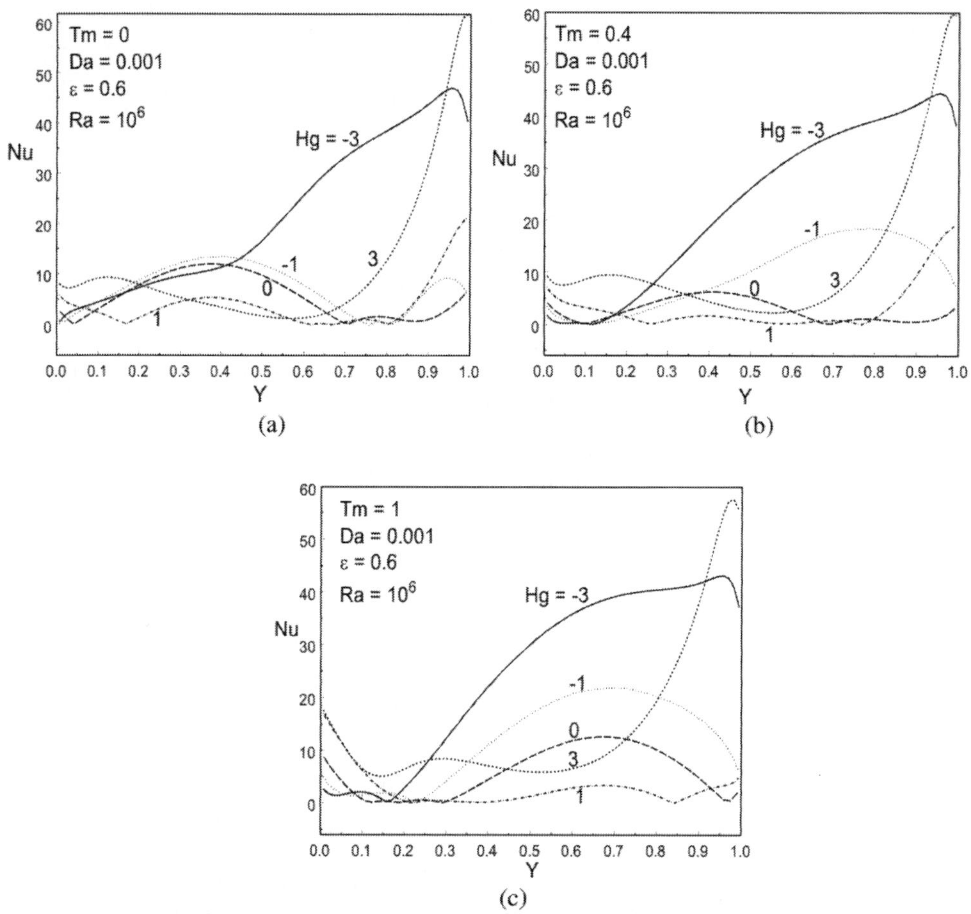

(c)

Figure 4 Local Nusselt number for different heat generation parameters.

Fig. 5(a) presents average Nusselt number for different values of the density inversion parameter and heat generation parameter. The average heat transfer rate raises on increasing density inversion parameter except Hg = 1. The heat transfer rate is higher for heat absorption case than that of heat generation case. This is because the heat energy is transformed into the cavity rapidly due to heat absorption. Fig. 5(b) exemplifies the average Nusselt number against the Darcy number for different values of the density inversion parameter and heat generation parameter. It is observed that higher heat transfer rate is found at Hg = -1 than Hg = 1 for all values of T_m and Da. In the presence of heat absorption, the average heat transfer rate decreases on increasing the values of the Darcy numbers for all values of T_m. In the case of heat generation, average Nusselt number decreases slightly and then it raises on increasing the values of the Darcy numbers. The average Nusselt number raises on increasing the values of T_m in the case of heat absorption. Also the average heat transfer rate decreases on increasing the value of T_m for low values of Da $(10^{-4}, 10^{-5})$ in the case of heat generation. The opposite behavior is observed for higher values of Da $(10^{-1}, 10^{-2})$ in the case of heat generation. It is interesting to note that two different behaviors on heat transfer rate are observed in the presence of heat generation and heat absorption, respectively.

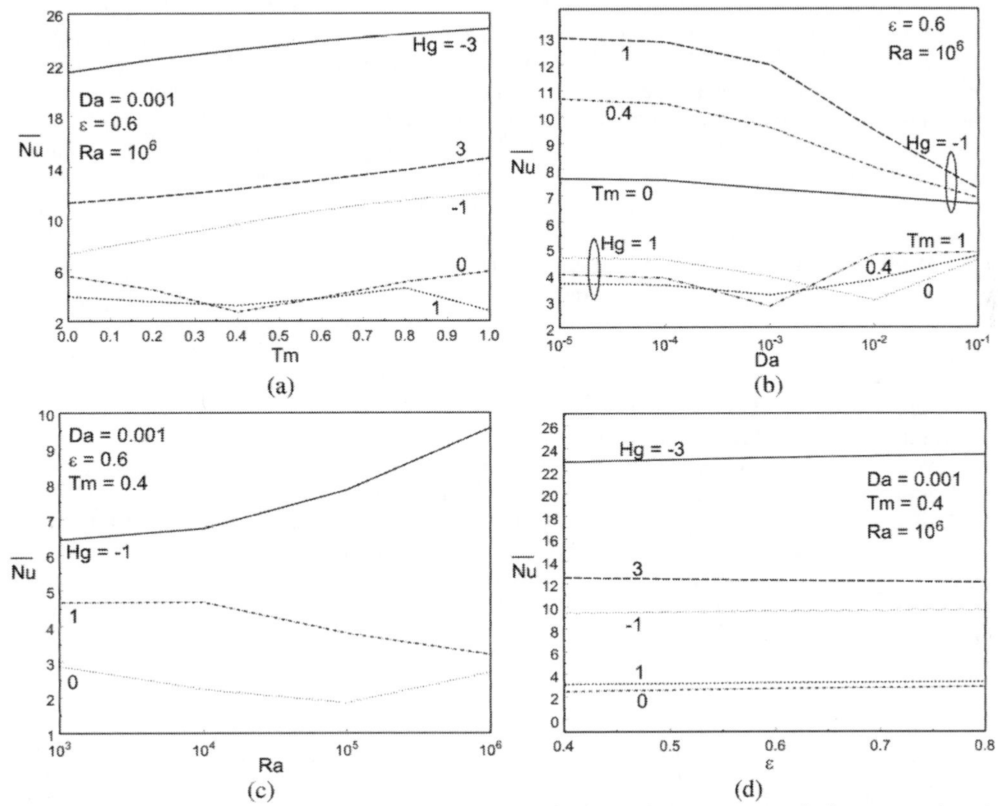

Figure 5. Average Nusselt number for (a) different density inversion parameter, (b) Darcy number, (c) Rayleigh number and (d) porosity.

Fig. 5(c) indicates the average Nusselt number for different values of heat generation parameter and Rayleigh number. It is found that the average heat transfer rate decreases on increasing the value of Ra for Hg = 1 in the presence of density maximum inside the enclosure. The average heat transfer rate decreases first $(Ra \leq 10^5)$ and then raises in the absence of heat generation/absorption. Here we observed a contradict result from the universal fact that average Nusselt number raises with Ra. This is because of the density maximum plane exists inside the enclosure and non-uniform heating. Also the average heat transfer rate raises with Ra in the presence of heat absorption. Further scrutinizing the curves, it is observed that heat transfer rate is high in the presence of heat absorption/generation with comparing to the non-heat generating case. Fig. 5(d) shows the effect of porosity on average Nusselt number for different values of the heat generating parameter. It is clearly observed that heat absorption/generation enhances the heat transfer rate comparing the non-heat generating (Hg = 0) case. There is no significant change on average heat transfer rate when increasing the porosity for given values of Hg. Comparing heat absorption and heat generation cases, the enhancement of heat transfer is observed for heat absorption case. This is because of the fluid inside the cavity absorbs the heat energy, so the heat is transferred quickly from the vertical walls.

4. Conclusions

The effects of heat generation/absorption and sinusoidal heating on free convection in a square porous cavity are studied numerically in the presence of density maximum of water. It is detected that the strong thermal boundary layers are formed along the left wall in the case of heat absorption and along the right wall in the case of heat generation. The multi-cellular flow patterns are formed inside the enclosure due to sinusoidal heating and density inversion effect. The local heat transfer behaves nonlinearly and it is directly affected by the sinusoidal heating. The average heat transfer rate decreases first and then enhances on increasing the values of the Rayleigh number in the absence of heat generation/absorption, which contradicts the universal fact that average Nusselt number enhances with Ra. There is no significant effect on average Nusselt number when the porosity raises in the presence of density maximum for different values of heat generation parameter. The average heat transfer rate diminishes on rising Da in the presence of heat absorption. The average heat transfer rate decreases first and increases on increasing Da in the presence of heat generation. The heat transfer enhances for heat absorption case than that of heat generation case for all values of Ra , Da , ε in the presence of density maximum.

References

[1]. Sivasankaran S and Ho C J 2008, *Numer. Heat Trans. A,* **53** 507-23.
[2]. Sivasankaran S and Ho C J 2008, *Int. J. Thermal Sci.* **47** 1184-94.
[3]. Varol Y Oztop H F Mobedi M Pop I 2010, *Int. J. Heat Mass Trans.* **53** 1691-98.
[4]. Sivasankaran S Sivakumar V Prakash P 2010, *Int. J. Heat Mass Trans.* **53** 4304-15.
[5]. Bhuvaneswari M Sivasankaran S Kim Y J 2011, *Numer. Heat Trans. A* **59** 167-84.
[6]. Sivasankaran S Malleswaran A Lee J Sundar P 2011, *Int. J. Heat and Mass Trans.* **54** 512-25.
[7]. Sivasankaran S and Bhuvaneswari M 2013, *Numer. Heat Trans. A* **63** 14-30.
[8]. Janagi K Sivasankaran S Bhuvaneswari M Eswaramurthi M 2017, Int. J. Numer. Methods for Heat & Fluid Flow, **27**(4) 1000-14.

Effects of multiple slip on MHD combined convective flow of viscoelastic nanofluid over a stretchy sheet with heat absorption

S. Eswaramoorthi[1,a], S. Sivasankaran[2,b], M. Bhuvaneswari[2] and S. Rajan[3]

[1]Department of Mathematics, Dr. N.G.P. Arts and Science College, Coimbatore, India.

[2]Department of Mathematics, King Abdulaziz University, Jeddah, Saudi Arabia.

[3]Department of Mathematics, Erode Arts and Science College, Erode, India.

[a]eswaran.bharathiar@gmail.com, [b]sd.siva@yahoo.com

Abstract We explore the impact of MHD combined convective flow of viscoelastic nanofluid over a stretchy sheet with heat absorption & multiple slip conditions. The similarity variables are employed to convert the PDE's into dimensionless ODE's and they are solved using homotopy analysis method (HAM). The impact of governing parameters on the velocity, temperature & nanoparticle concentration are analyzed with the help of tables and graphs.

Keywords: Viscoelastic nanofluid, magnetic field, heat absorption, slip boundary conditions.

1. Introduction

The convective heat transfer of nanofluids are imperative in huge number of industrial and engineering applications, like, biological sensors, advanced nuclear systems, electronic devices, hybrid-powered engines. Many authors are interested to study the nanofluids. Ahmed et al. [1] investigate the nanofluid flow over a stretching tube. Some significant studies on these directions are [2]-[4]. In petroleum production, fabrication of adhesive tapes, drawing of paper films, etc, the non-Newtonian materials are essential. Due to these applications, more researchers are examining the behavior of non-Newtonian materials. Viscoelastic fluid is a subclass of non-Newtonian fluid and having both viscous and elastic behavior. Several studies in this literature, like, Cortell [5], Abel and Mahesha [6], Eswaramoorthi et al.([7]-[9]) are examined the viscoelastic fluid model without nanofluids. Shit et al.[10] examined the heat transfer characteristic of a viscoelastic nanofluid. Some recent investigations of viscoelastic nanofluid are shown in the Refs. [11]-[15].

The primary objective is to investigate the impact of MHD combined convective flow of viscoelastic nanofluid over a stretchy sheet. The effects of heat generation and multiple slip conditions are additionally taken into our study.

2. Mathematical Formulation:

A steady 2D flow of a MHD viscoelastic nanofluid over a stretchy sheet. An extrinsic constant magnetic field (B_m) is applied and the induced magnetic field is omitted because of small values of magnetic Reynolds number. The flow direction is considered in $x-$ axis which is taken along the sheet and $y-$ axis is perpendicular to it. Heat absorption/generation property of the fluid is additionally considered. Under the above prescriptive assumptions, the governing boundary layer equations are considered as follows,

$$u_x + u_y = 0 \tag{1}$$

$$uu_x + vu_y = \nu u_{yy} + k_0\left(uu_{xyy} + u_x u_{yy} + u_y v_{yy} + vu_{yyy}\right)$$
$$+g\left[\beta_T(T - T_\infty) + \beta_C((C - C_\infty))\right] - \frac{\sigma B_m^2}{\rho}u \quad (2)$$
$$uT_x + vT_y = \alpha_m T_{yy} + \tau\left[D_B C_y T_y + \frac{D_T}{T_\infty}{T_y}^2\right] + \frac{Q}{\rho C_p}(T - T_\infty) \quad (3)$$
$$uC_z + vC_y = D_e C_{zz} + \frac{D_T}{T_\infty}T_{yy} \quad (4)$$

where $(u, v), \nu, k_0, g, \beta_T \,\&\, \beta_C, \sigma, \rho, T, D_T \,\&\, D_B, \alpha_m \,\&\, D_e, C_p, Q$ and C are the velocities, kinematic viscosity, viscoelastic parameter, acceleration due to gravity, thermal & concentration expansion coefficients, electrical conductivity, density, temperature, thermophoretic diffusion & Brownian motion coefficients, thermal & mass diffusivities, specific heat, internal heat absorption(< 0)/generation(> 0) and concentration.

The boundary conditions can be derived as,

$$u = U_z(x) = ax + L_1 u_y,\ v = 0,\ T = T_z + L_2 T_y,\ C = C_z + L_3 C_z \text{ at } y = 0$$
$$u \to 0, u_y \to 0, T \to T_\infty, C \to C_\infty \text{ as } y \to \infty \quad (5)$$

where a > 0 is constant. The non-dimensional variables are

$$\eta = \sqrt{\frac{a}{\nu}}y,\ u = axf'\ v = \sqrt{a\nu}f,\ \theta = \frac{T-T_\infty}{T_z-T_\infty},\ \phi = \frac{C-C_\infty}{C_z-C_\infty}, \quad (6)$$

Reduced form of the Equations (2) - (4) are

$$f''' + ff'' - f'^2 + K\left(2f'f''' - f''^2 - ff^{iv}\right) - Mf' + Ri(\theta + N\phi) = 0 \quad (7)$$
$$\left(\frac{1}{Pr}\right)\theta'' + f\theta' + Hg\theta + Nb\theta'\phi' + Nt\theta'^2 = 0 \quad (8)$$
$$\phi'' + PrLef\phi' + \frac{Nt}{Nb}\theta'' = 0 \quad (9)$$

with boundary conditions:

$$f(0) = 0, f'(0) = 1 + \lambda_1 f''(0), \theta(0) = 1 + \lambda_2\theta'(0), \phi(0) = 1 + \lambda_3\phi'(0),$$
$$f'(\infty) = 0, f''(\infty) = 0, \theta(\infty) = 0, \phi(\infty) = 0.$$

where $K, M, N, Ri, Nt, Pr, Hg, Nb, Le, \lambda_1, \lambda_2$ and λ_3 are the viscoelastic parameter, magnetic field parameter, buoyancy ratio parameter, Richardson number, thermophoresis parameters, Prandtl number, heat absorption/generation, Brownian motion, Lewis number, corresponding slip parameters.

The reduced skin friction coefficient and local Nusselt number are given by

$$\frac{1}{2}C_f\sqrt{Re} = (1 + 3K)f''(0), \frac{Nu}{\sqrt{Re}} = -\theta'(0)$$

3. HAM Solution:

The initial guess of the HAM are $f_0 = \frac{1-e^{-\eta}}{1+\lambda_1}$; $\theta_0 = \frac{e^{-\eta}}{1+\lambda_1}$; $\phi_0 = \frac{e^{-\eta}}{1+\lambda_1}$ and the linear operators are $L_f = \frac{d^3 f}{d\eta^3} - \frac{df}{d\eta}$; $L_\theta = \frac{d^2\theta}{d\eta^2} - \theta$; $L_\phi = \frac{d^2\phi}{d\eta^2} - \phi$ with $L_f[D_1 + D_2 e^\eta + D_3 e^{-\eta}] = 0$; $L_f[D_4 e^\eta + D_5 e^{-\eta}] = 0$; $L_f[D_6 e^\eta + D_7 e^{-\eta}] = 0$; where $D_j (j = 1 - 7)$ denote the constants. The general solution of equations (7)-(9) are $f_m(\eta) = f_m^+(\eta) + D_1 + D_2 e^\eta + D_3 e^{-\eta}$; $\theta_m(\eta) = \theta_m^+(\eta) + D_4 e^\eta + D_5 e^{-\eta}$; $\phi_m(\eta) = \phi_m^+(\eta) + D_6 e^\eta + D_7 e^{-\eta}$, where $f_m^+(\eta), \theta_m^+(\eta)$ and $\phi_m^+(\eta)$ are the special solutions.

The HAM solutions having the auxiliary parameters $\hbar_f, \hbar_\theta$ and $\hbar_\phi$ and these are controlling the solution convergences. Theses curves are displayed in Figure 1 and noticed that the limit ranges are $-1.4 \le \hbar_f \le -0.1$ and $-1.6 \le \hbar_\theta, \hbar_\phi \le -0.2$. We choose the $\hbar$ value in the total region is -1. Table 1 shows the different order of approximation of HAM solutions and noted that 15th order is suitable for the velocity, temperature and nanoparticle concentration.

4. Correlation Analysis:

The correlation equations of $\frac{1}{2}C_f\sqrt{Re}$ and $\frac{Nu}{\sqrt{Re}}$ are

$$\frac{1}{2}C_f\sqrt{Re} = -0.434534 - 1.242040K - 0.141117M + 0.092979Ri + 0.106843Hg$$
$$-0.001819Nt + 0.056822Nb$$
$$\frac{Nu}{\sqrt{Re}} = 0.2817453 - 0.017885K - 0.011085M + 0.012414Ri - 0.530821Hg$$
$$-0.067075Nt - 0.080416Nb$$

This equation is valid for $0 \le K \le 0.3, 0 \le M \le 1.5, 0 \le Ri \le 1 -0.3 \le Hg \le 0.3, 0.0 \le Nt \le 1.5$, $0.1 \le Nb \le 1.3$ and $0.5 \le \lambda(\lambda_1 = \lambda_2 = \lambda_3) \le 2.5$ with the maximum error is 0.09.

5. Results and Discussion:

In this section, we explore the impact of pertinent physical parameters, like, viscoelastic parameter (K), magnetic field parameter (M), Richardson number (Ri), Brownian motion parameter (Nb), buoyancy ratio parameter (N), heat absorption/generation parameter (Hg), thermophoresis parameter (Nt), velocity slip parameter (λ_1), temperature slip parameter (λ_2) and concentration slip parameter (λ_3) with constant values of Prandtl and Lewis numbers. Figures 2(a-d) portray the impact of M, Ri, N and λ_1 of $f'(\eta)$. It is seen that the fluid velocity enhances with large values of Ri and N whereas it declining with escalating the values on M and λ_1. The variations of λ_2, Hg, Nb and Nt on $\theta(\eta)$ are presented in Figures 3(a-b) and observed that the fluid temperature boosted when rising the values of Hg, Nb and Nt and it suppresses with increasing the values of λ_2. Figures 4(a-b) exhibit the behavior of M, λ_3, Nb and Nt on $\phi(\eta)$. We concluded from these figures that the nanoparticle concentration growing when strengthening the values of M and Nt and it found the opposite behavior for increasing the values of Nb and λ_3.

6. Conclusions:

The mixed convective flow of a MHD viscoelastic nanofluid over a stretchy sheet with heat absorption and slip conditions are analyzed here. The present investigation have indicated the followings

- The fluid velocity enhances with large values of Ri and N and it declining with escalating the values on M and λ_1.
- The fluid temperature boosted when rising the values of Hg, Nb and Nt and it suppresses with increasing the values of λ_2.
- The nanoparticle concentration growing when strengthening the values of M and Nt and it found the opposite behavior for increasing the values of Nb and λ_3.

Table 1. Different order of approximations of HAM solutions

Order	$-f''(0)$	$-\theta'(0)$	$-\phi'(0)$	Order	$-f''(0)$	$-\theta'(0)$	$-\phi'(0)$
1	0.39375	0.48750	0.39881	25	0.42531	0.48336	0.37534
5	0.42453	0.48397	0.37542	30	0.42531	0.48336	0.37534
10	0.42534	0.48339	0.37521	35	0.42531	0.48336	0.37534
15	0.42531	0.48336	0.37534	40	0.42531	0.48336	0.37534
20	0.42531	0.48336	0.37534	45	0.42531	0.48336	0.37534

References:

[1] Ahmed S.E, Hussein A.K, Mohammed H.A and Sivasankaran S 2014 *Appl. Math. Comput.* 238 149-162.
[2] Bachok N, Ishak A and Pop I 2010 *Int. J. Thermal Sci.* 49 1663-1668.
[3] Khan W.A and Pop I 2010 *Int. J. Heat Mass Transf.* 53 2477-2483.
[4] Hassani M, Tabar M.M, Nemati H, Domairry G and Noori F 2011 *Int. J. Thermal Sci.* 50 2256-2263
[5] Cortell R. 2006 *Physics Lett.* A 357(4-5), 298-305.
[6] Abel M.S and Mahesha N 2008 *Appl. Math. Model.* 32(10) 1965-1983.
[7] Eswaramoorthi S, Bhuvaneswari M, Sivasankaran S and Rajan S 2015 *Procedia Engin.* 127 916-923.
[8] Eswaramoorthi S, Bhuvaneswari M, Sivasankaran S and Rajan S 2016 *Scientia Iranica Transaction B, Mech. Engin.* 23(6) 2575-86.
[9] Eswaramoorthi S, Bhuvaneswari M, Sivasankaran S, Niranjan H and Rajan S 2017 *IOP Conf. Series: Mater. Sci. Engin.* 263 1-8.
[10] Shit G.C, Haldar S.K, Ghosh S.K 2016 *Int. J. Appl. Math. Comput.* 2 593-608.
[11] Sheu L.J 2011 *Trans. Porous Media* 88, 461-477.
[12] Goyal M and Bhargava R 2014 *Appl. Nanosci.* 4 761-767.
[13] Ramzan M and Yousaf F 2015 *AIP Advances* 5 057132-1-057132-15.
[14] Ramzan M, Inam S and Shehzad S.A 2016 *Alexandria Engin. J.* 55(1) 311-319 .
[15] Seth G.S, Mishra M.K and Chamkha A.J 2016 *J. Nanofluids* 5(4) 511-521.

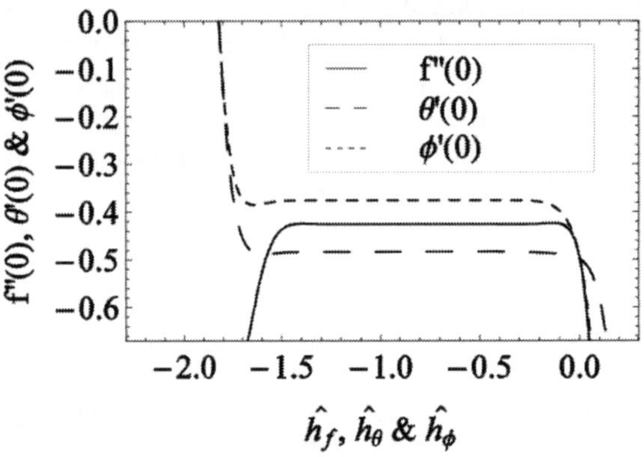

Figure 1: h curves of $f''(0), \theta'(0)$ and $\phi'(0)$.

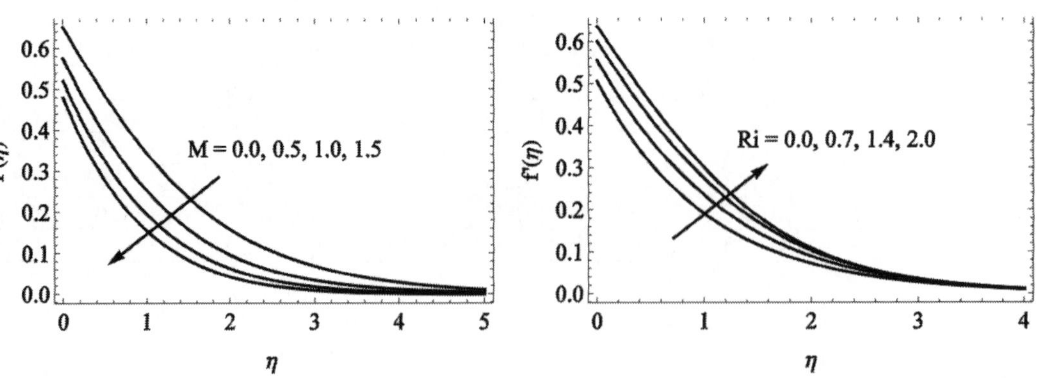

IOP Conf. Series: Materials Science and Engineering **390** (2018) 012096 doi:10.1088/1757-899X/390/1/012096

(a) (b)

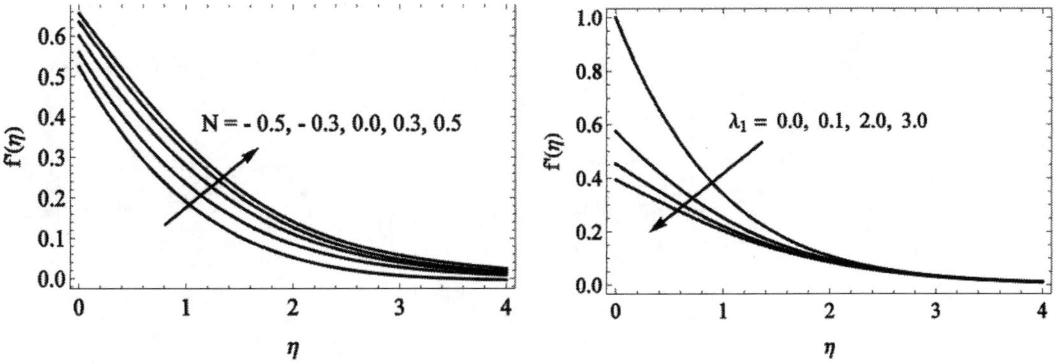

(c) (d)

Figure 2: Velocity for various values of M, Ri, N & λ_1.

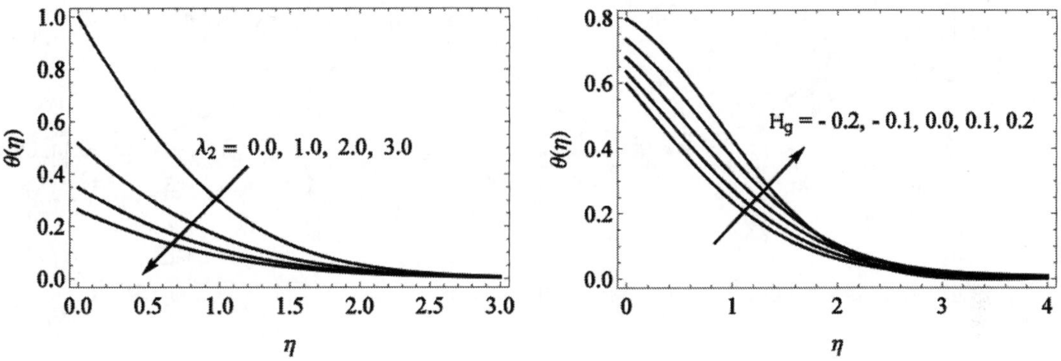

(a) (b)

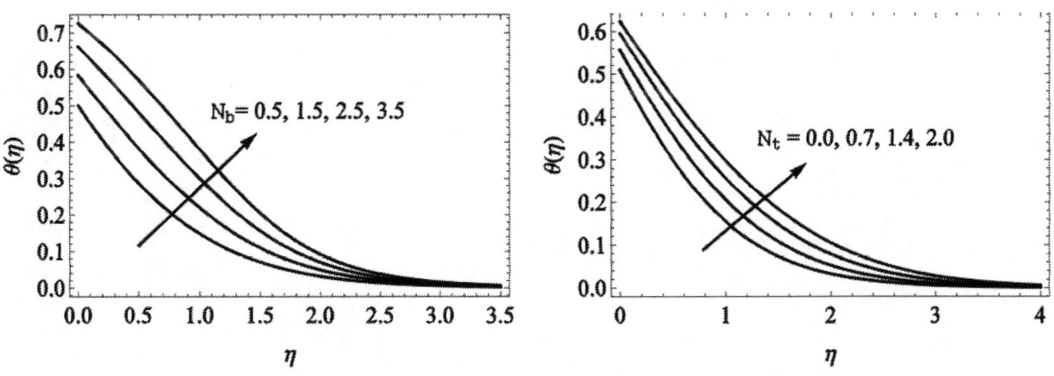

(c) (d)

Figure 3: Temperature for various values of λ_2, Hg, Nb & Nt

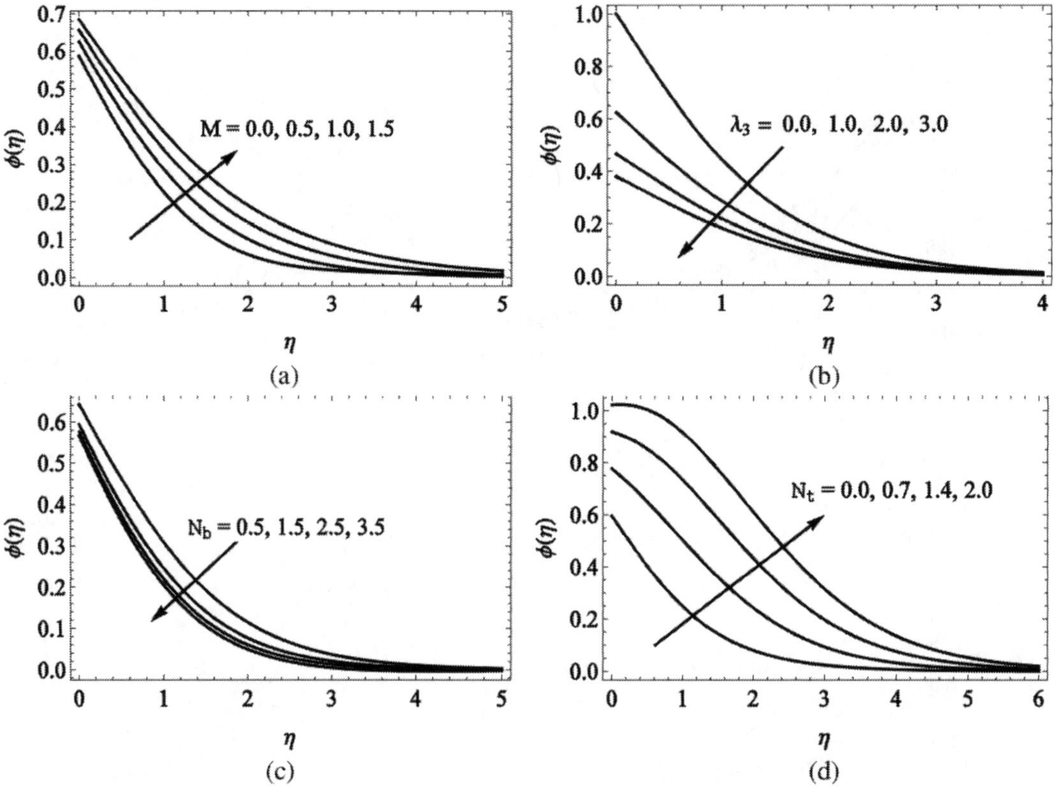

Figure 4: Nanoparticle concentration profile for different values of M, λ_3, Nb & Nt.

IOP Conf. Series: Materials Science and Engineering **390** (2018) 012097 doi:10.1088/1757-899X/390/1/012097

An overview on performance of metal working fluids with enrichment of nanoparticles for machining process

K Venkatesh[1*], G Sriram[2] A Uday Kiran Reddy[3] and S V R Nikhil[3]

[1]Associate Professor, Dept. of Mechanical Engg., Sri Chandrasekharendra Saraswathi Viswa Mahavidyalaya, Kanchipuram - 631561, Tamilnadu, India.

[2]Professor and Dean (E&T), Dept. of Mechanical Engg., Sri Chandrasekharendra Saraswathi Viswa Mahavidyalaya, Kanchipuram - 631561, Tamilnadu, India.

[3]Final Year Student, Dept. of Mechanical Engg., Sri Chandrasekharendra Saraswathi Viswa Mahavidyalaya, Kanchipuram - 631561, Tamilnadu, India.

Corresconding Author Email [1*] kvenkateshme@gmail.com

Abstract. Tool wear is natural phenomenon, and this leads to tool failure which is caused mainly due to various aspects like the surface of the work piece, material of the work piece to be machined and other. The application of the metal working fluids (MWF's) (or generally coolant) is considered to reduce the friction between the work and tool and to provide lubrication as well as cooling. In a machining process, the use of coolant is essential to dissipate the heat generated and to enhance various properties like productivity, tool life, machinability, etc. There are many types of oil based MWF's are available like mineral oil, semi-synthetic oil, synthetic oil based MWF's. But lately vegetable oil based MWF's are being used because of its eco-friendly nature. Recently Nanoparticles are used to enhance the properties of metal working fluids. In this paper there is brief explanation on preparation of metal working fluids with enrichment of Nanoparticles.

Keywords: Tool wear, Eco-friendly, Metal working Fluids, Nano Additives

1. Introduction

Machining is a process of material removal in which the loss of material is caused by affecting a relative motion between tool and work piece. Due to removal of material in the form of chips, new surfaces are cleaved from the work piece accompanied by a large consumption of energy. The mechanical energy necessary for the machining operation is transformed into heat, leading to conditions of high pressure, high temperatures and severe thermal and frictional conditions, consequently making metal cutting process more and more inefficient in terms of tool life, dimensional accuracy and material removal rate.

IOP Conf. Series: Materials Science and Engineering **390** (2018) 012097 doi:10.1088/1757-899X/390/1/012097

To avoid difficulties in machining an effective coolant system is to be provided using a cutting fluid. The important functions of cutting fluids are cooling the tool and workpiece, lubricating chip, tool and workpiece and improve surface finish [1]. The main requirements of a cutting fluid are 1. Long life, 2.Should not form oxides, 3. Good thermal conductivity, 4.Less wetability, 5.Less odour, 6. Should not be flammable, 7.Should not affect the health of workers [1].

In manufacturing industry, many types of lubricants are used as a baseoil for metal working fluids. The fluids which are mainly used as base oil are water, distilled water, mineral oil, synthetic oil, vegetable oil, etc. Base oils cannot be used directly, since pure or raw base oil may cause corrosion, and some can be chemically imbalanced and cannot have a long-life period. So, additives are added to the base oil for better performance. Based on the substance that is used as base oil, additives changes. Additives are chemical substances that are used to enhance the properties of the base oil and improve the performance by taking out negative properties form the base oil. Additives may function as anti-wear, anti-corrosive, anti-oxidant and neutralizing agents and some can be anti-bacterial (mainly for vegetable oils). Simply additives are added to make base oil chemically stable and to become more user friendly to use.

Recently vegetable oils are widely used because of its high flash points and biodegradability when compared to the mineral oils[2], but vegetable oils contain unsaturated fatty acids which poses some problems in terms of corrosiveness and life of the coolant oil, because of high temperature raise in machining process [2].So, vegetable oils are hydrolysed to get ester oil compound which has more stability due to the presence of ester compound[3]. Additives are to this ester compound of their respective vegetable base oil. Lawal et al. [4] intensive research work on the formulation of vegetable based metal working fluids using appropriate additives like Zinc diamydithiocarbamate (ZDDC), Zinc dialkyldithio phosphate (ZDDP), etc. Lawal et al. also reported about various emulsifiers and surfactants used in preparation of the vegetable based oils, and then they compared the properties like surface roughness, tool wear between rapeseed, coconut, sunflower oil based metal working fluids and every result is compared with the commercial vegetable based oil and commercial mineral oil. And the paper concluded that vegetable based oil gave better performance than mineral oil and it is reported that surface roughness and tool wear is improved when coconut oil is used compared to the soluble oil. Commercial vegetable based cutting fluids are used extensively, because of its good properties and high efficient performance.

In this study it is discussed about process of preparation of MWF's with enrichment of nanoparticles.

2. Performance of MWF's with enrichment of nanoparticles

Nowadays nanoparticles are extensively used in many fields, in the field of manufacturing industry it is used to enhance the properties of MWF's. The most commonly used Nano particles are

1. Molybdenum disulphide (MoS_2)
2. Multi walled carbon Nano tubes (MWCNT)
3. Single walled carbon Nano tubes (SWCNT)
4. Aluminium oxide (Al_2O_3)
5. Titanium dioxide (TiO_2)
6. Copper oxide (CuO)
7. Silicon dioxide(SiO_2)
8. Zirconium dioxide (ZrO_2)
9. Hexagonal boron nitride (hBN)
10. Graphite Nanoplatelets (GnP)
11. Nano boric acid (NBA)
12. Nano diamond (ND)

IOP Conf. Series: Materials Science and Engineering **390** (2018) 012097 doi:10.1088/1757-899X/390/1/012097

Based on the literature study of Rabesh et al. [5], Nanofluids shown better rheological, tribological properties and thermal conductivity is increased. It is also reported that surface finish is increased because of Nanoparticles. Rabesh et al. suggested of hybrid Nanofluids which contains two or more Nanoparticles. Since Nanoparticles give a prominent change and increase in the properties of the metal working fluid, so it is used widely in all machining applications as coolant oil such as turning, milling, drilling and grinding in both CNC and lathe operations. The following study is based on the various works on different "machining applications" with different Nanoparticles and various base oils.

2.1. Turning process

Srikant et al. [6] experimented on mixing of copper oxide (CuO) in water which acts as base oil. It is reported that the CuO particles are mixed in proportions 0.5-8 vol% with 0.5 as equal interval. Finally, it is concluded that water based cutting fluid with CuO particles given good results in maintaining temperature which in turn results in increase of tool life.

Krishna et al. [7] experimented on mixing of Nano Boric acid in proportions 0.25, 0.5-8 wt% to SAE-40 oil and coconut oil, and it is reported that 0.5% of Nano Boric Acid reduced overall temperature compared to the other experiments carried out on different wt%, Taguchi method is used for design of experiments. It is also reported that the Nanoparticles have enhanced the oil properties greatly and given better surface finish and decreased great amount of temperature.

Amrita et al. [8] experimented on mixing of Nano Graphite (size < 80nm) in coconut oil in proportions 0.1, 0.3, 0.5 wt%. It is reported that they used MQL (minimal quantity lubrication) technique at different flow rates 10 and 15 ml/min and obtained good results in reducing temperature and cutting forces at the flow rate of 15 ml/min. They used Taguchi loss method for design of experiments for setting optimal parameters.

Sayuti et al. [9] experimented on mixing of silicon dioxide (SiO_2) which has size of 5-15nm in mineral oil. The proportions added were 0.2, 0.5 and 1 wt% and in this study Sayuti et al. shown better results in good surface finish and least tool wear at 0.5 wt% of Nanoparticle.

Gupta et al. [10] experimented on mixing of Al_2O_3, MoS_2 and Graphite in vegetable oil separately. All the Nanoparticles are mixed at 3 wt%. It is reported that comparing Al_2O_3, MoS_2 and graphite based vegetable oil due to good thermal conductivity graphite based vegetable oil given better performance than the other two in reducing overall temperature, but MoS_2 gave better surface finish.

Raju et al. [11] experimented on mixing of Multiwalled carbon Nanotubes (MWCNT) (size 10-20nm) in "distilled water" at a proportion of 0.2 vol%. In this paper it is reported that with the use of distilled water based fluid mixed with 0.2 vol% MWCNT cutting forces and surface roughness reduces 5-8% and 9-22%, respectively when compared to dry and base fluid.

Pavani et al. [12] experimented on mixing of Nano Boric Acid in "soybean oil" and "coconut oil". It is mentioned that the mixing is done with the use of "mechanical stirrer". Taguchi method is used to design the experiment. It was found that 2% wt of Nano Boric Acid is optimum for soybean oil and 3% wt for coconut oil. In this paper the comparison is made between tip temperature of the tungsten carbide (WC) tools, surface roughness and cutting forces. It is concluded that the vegetable based oils with enrichment of Nano Boric Acid given good results. Cutting temperatures reduced and surface finish increased to a maximum level compared with dry machining. It is also concluded that the study encourages to the use of vegetable based oil.

Su et al. [13] experimented on mixing of Nano Graphite (size 35nm) in "LB2000 vegetable based commercial cutting fluid" and "PriEco6000 unsaturated polyol ester oil". The proportions are 0.1 and 0.5%. Nanofluids are formulated in two-step method. Ultrasonication is used for mixing and it is done at 40KHz, 100W. It is reported that irrespective of cutting oil both graphite-based Nanofluids reduced cutting forces and a good amount of temperature. But LB2000 given maximum reduction in temperature when compared to dry cutting.

Amrita et al. [14] experimented on mixing of Functionalized Nano Graphite (FNG, size 80nm), Nano Boric Acid (NBA, size 100nm) and $nMoS_2$ (size 100nm) in water-miscible oil, at a definite proportion of 0.3 wt%. In the series of experiments $nMoS_2$ Nanoparticles at 0.3 wt% recorded reduction in cutting forces and recorded good wear and surface properties. But in terms of heat dissipation NBA at 0.3 wt% shown good improvement when compared other coolant conditions.

2.2. Drilling process

Chatha et al. [15] experimented on Soyabean oil as a base oil and taking Al_2O_3 1.5vol% in Drilling process, the size of the Nanoparticle is taken approximately 20nm. In this study, it has given difference between dry and wet drilling conditions by using MQL technique. In this, he has found out that by the application of the Nano particles no. of drilled holes increases which means that the life of the tool has been increased when compared with the dry conditions.

Chai et al. [16] experimented by taking hydrogenated oil as a base oil and Multi walled carbon Nano particles (MWCNT) in different proportions like 22, 50, 100ppm.This means that he has taken a very smaller ratio of MWCNT. The size of MWCNT is in between 10-12nm. The process used for experimentation is drilling process. As the volume percentage is very low then we can tell the quantity in terms of parts per million(ppm). It has been observed that the thermal conductivity of the base oil (hydrogenated oil) is increased with addition of MWCNT.

Nam et al. [17] experimented by taking vegetable oil and paraffin oils as base fluids and Nano diamond (ND) as Nano particle in Drilling process. Size of particle is 30nm. ND is taken in 1,2vol%. With the addition of 1% in paraffin oil gives more improvement in reducing thrust forces, tool wear than by adding 2% in vegetable oil. Nano diamond plays an important role in reducing the tool wear, but the only limitation is its high cost.

2.3. Grinding process

Prabhu and Vinayagam. [18] experimented by taking SAE20W40 oil and MWCNT and CNT (Carbon Nano tubes) in Grinding process. The size is 10-12nm for MWCNT and 1-2nm for CNT. Initially they have taken 0.2vol% of MWCNT and observed that there is slight improvement in surface quality from micro level to Nano level. After that, by taking 2vol% of MWCNT, there is an increase in flash and fire point of the Nano cutting fluid. By taking CNT it us observed that there is an improvement in the surface roughness value of the workpiece.

Kalita et al. [19] experimented Soyabean oil as a base oil and MoS_2 with 5, 20wt% in Grinding process. The size of Nanoparticle is less than100nm. By using MQL technique the cutting fluid produces lesser coefficient of friction and improves the grinding forces without effecting the material removal rate, and wheel wear reduces.

Mao et al. [20] experimented by taking deionized water and canola oil and their respective Nano particles Al_2O_3 and MoS_2 in Grinding process. Size of Al_2O_3 is 40, 80nm and size of MoS_2 is 70nm. When Al_2O_3 in added in deionized water and MoS_2 in canola oil it is observed that higher concentration of Nanoparticles reduces grinding forces and temperature and it depends upon the oil that will be chosen.

3. Conclusion

As discussed above Nanoparticles increase the properties of metal working fluids, based on the above study the Nanoparticles used for the enrichment of metal working fluids for machining are Multiwalled Carbon Nano Tubes (MWCNT), Molybdenum disulfide (MoS_2, $nMoS_2$), Nano Graphite, Aluminium oxide (Al_2O_3), Nano Boric Acid, Silicon dioxide (SiO_2), Copper oxide (CuO). The base oils used are distilled water, water, vegetable oils like soybean oil, coconut oil, commercial vegetable oils, mineral oil, SAE-40 oil, ester oil.

1. From the above study it is noted that Nanoparticles improve the properties of metal working fluids. Irrespective of base oil, Nanoparticle enriched cutting oils gave better performance than dry cutting in cutting forces and surface finish.
2. Although, water, distilled water, mineral oil, SAE-40 oil performed well and gave better performance it is recommended to use vegetable oil or commercial vegetable oil, because of their biodegradability and higher flash point.
3. Nanoparticles improve the thermal conductivity of the metal working fluids, which results in reducing the overall temperature of machining process.
4. Thermal conductivity of MWCNT is the highest at 6000 W/mK and Copper oxide has the least at 19.6 W/mK.

References

[1] R K Rajput The text book of manufacturing technology (2014)
[2] R R Srikant and P N Rao, J P Davim (ed.) *Sustainable Machining, Materials Forming, Machining and Tribology* (2017), DOI 10.1007/978-3-319-51961-6_2
[3] Nabel A Negm, Mohamed M El-Sukkary, Galal H Sayed, Manal G Mohamed, Mahmoud Bekheit, *J Surfact Deterg* (2016) 19:455–466
[4] S A Lawal, I A Choudhury & Y Nukman, *Int J Adv Manuf Technol* (2013) 67:1765–1776
[5] Rabesh Kumar Singh, Amit Rai Dixit, Amitava Mandal, Anuj Kumar Sharma, *J Braz Soc Mech Sci Eng* (2017) 39:4677–4717
[6] Srikant R R, Rao D N, Subrahmanyam M S, Vamsi Krishna P, *P I Mech Eng J-J Eng Tribol* (2010) 223:221
[7] Krishna P V, Srikant R R, Rao D N. *Int Tool Manuf J Mach* (2010) 50:911–916.[8]. Amrita M, Srikant R R, Sitaramaraju AV, Prasad M M S, Vamsi Krishna P, *P I Mech Eng J-J Eng Tribol* (2013) 227(12):1334–1346
[8] Amrita M, Srikant RR, Sitaramaraju AV, Prasad MMS, Vamsi Krishna P,*P I Mech Eng J-J EngTribol*(2013) 227(12):1334–1346
[9] Sayuti M, Sarhan A A D, Salem F, *J Clean Prod* (2014) 67:265–276
[10] Gupta M K, Sood P K, Sharma V S, *J Clean Prod* (2016) 135:1276–1288
[11] Raju R A, Andhare A, Sahu N K, *Mater Manuf Process* (2017) 32:1–7
[12] P N L Pavani, R Pola Rao and S. Srikiran, *J Mech Sci & Tech* 29 (11) (2015) 4877~4883
[13] Yu Su, Le Gon Bi Li, Zhiqiang Liu & Dandan Chen, *Int J Adv Manuf Technol* (2016) 83:2083–2089
[14] Amrita M, Shariq S A, Manoj Gopal Charan, *12th Global Cong Manuf Manage Proc Eng* (2014) 97:115–124
[15] Chatha SS, Pal A, Singh T (2016) J Clean Prod 137:537–545
[16] Chai YH, Yusup S, Chok VS, Arpin MT, Irawan S (2016) Appl Therm Eng 107:1019–102
[17] Nam JS, Lee PH, Lee SW (2011) Int J Mach Tool Manuf 51(7–8):649–652

[18] Prabhu S, Vinayagam BK (2012) Int J Adv Manuf Technol 60(1):149–160
[19] Kalita P, Malshe Ajay P, Wenping Jiang, Shih Albert J (2010) Trans NAMRI/ SME 38:137–144
[20] Mao C, Zhang J, Huang Y, Zou H, Huang X, Zhou Z (2013) Mater Manuf Process 28(4):436442

Joint two-echelon inventory model for optimality of cycle time and inventory decisions with finite production rate under credit period

Seelam Krugon[1], **Dega Nagaraju**[2,*]

[1]Department of Mechanical Engineering, Bapatla Engineering College, Bapatla-522101, A.P., India.
[2]Center for Innovative Manufacturing Research (CIMR), Vellore Institute of Technology (VIT), Vellore-632014, Tamil Nadu, India.

E-mail:seelamkrugon@gmail.com
[*]**Corresponding Author:**deganagarajulc@gmail.com

Abstract. In the universal competition of business environment, to face the global challenges around the business, it has turn into more traditional of giving trade credit by the vendor to the buyer as part of improving the sales revenue. The buyer need not to pay the vendor immediately whenever buyer received the goods, if a vendor offers credit period to the buyer. In this scenario focusing on the entire supply chain (SC) performance in terms of cycle time, inventory levels, number of shipment and the overall variable cost of the entire SC is essential. Mathematical models are developed for the joint two-echelon inventory system by assuming that production rate is finite. The present work aims to illustrate the optimality of cycle time, levels of inventory, number of shipments and the overall variable cost of the SC under the phenomena of credit period. Finally, numerical examples are solved to illustrate the theoretical results and also the sensitivity analysis is carried out.

1. Introduction

The goal of SCM is to increase the differential or aggressive benefit of the channels as a whole rather than to enlarge the improvement of any sole firm. Client value is produced through cooperation and teamwork to improve effectiveness (lower costs) and market value (added benefits) in ways that are most precious to key customers.

In this prospective, Goyal [1] is the pioneer of developing integrated inventory model. Saha and Goyal [2] proposed SC cooperation contracts where the demand is expressed in terms of retailer's price. They have expressed three coordinating contracts for coordination in two-echelon SC like combined repayment contract, price discount contract in wholesale and contract with cost sharing. Gallego et al [3] had presented a paper on pricing decision under coordination and stock replenishment policy for single merchant and single or extra geologically discrete buyers. In addition to the above, they derived the price-sensitive buyer order to analyze a SC with issues in coordination. A two-level SC model developed by Rahdar et al [4] with coordination methods for items of deterioration type. The authors considered a SC, comprising of a vendor and several buyers and made an assumption that the inventory items deteriorate over time and its inventory level reduces. Li [5] had made an attempt on a method of analytical for price analysis in multi-stage SC. A mathematical model for cost distribution analysis is proposed under stochastic environment for multi-stage SC network.

In this study, we made an attempt to develop a coordinated inventory model with a finite production rate for a two-echelon SC. Further, it is imagined that the supplier adopts credit period strategy to the buyer. Besides, the overall variable cost of SC is assumed to be a convex function of the cycle time. In addition, it is assumed that no shortages are allowed. The objective of this present research is to show the optimal cycle time, inventory levels, and overall variable cost and number of shipments from the vendor to the buyer in order to minimize the total variable cost. The current model is illustrated with a numerical examples and sensitivity analysis is done with reference to the parameters of the model.

This research article is described as follows. Section 2 deals with the mathematical model development consisting of the assumptions and features, notations used. A numerical illustration is described in Section 3. Section 4 deals with the summary of the findings and further extension of the current work.

2. Mathematical Model Development

For the development of the present model, one product is shipped from one vendor to one buyer is considered for joint system.

2.1 Notations

λ	Buyer demand rate per annum (units/year)
C	Production rate (units)
ε_B	Ordering price of Buyer (in Rs. per order)
ε_v	Cost at the manufacturer per setup (in Rs. per batch)
β_B	Price of shipment incurred to buyer to receive goods from vendor (in Rs. per delivery)
β_v	Cost of shipment incurred to vendor to transport goods to buyer (in Rs. per shipment)
U_B	Buyer's cost per unit (in Rs./unit)
U_v	Vendor price per unit (in Rs./unit)
S_B	Buyer's selling price per unit (in Rs./unit)
W_e	Earned rate of interest (in Rs. per Re per year)
W_P	Paid rate of interest (in Rs./Re./year)
t	Allowable credit period
q	Replenishment quantity (in units)
η	Number of shipments (+ve integer, decision variable)
M	Cycle time length at buyer (in years, decision variable)
π_B	Overall variable cost of the buyer (in Rs.)
π_v	Overall variable cost of the vendor (in Rs.)
π_S	Overall variable cost of the SC (in Rs.)

2.2 Features and assumptions

a) Deterministic rate of demand
b) Replenishment rate is instantaneous
c) Vendor 's inventory level is numeral multiple of buyer's inventory level
d) Trade credit is provided by vendor to the buyer
e) Shortages are not considered
f) Finite production rate

2.3 Model Formulation

2.3.1 Buyer Node

Case I: ($M \geq t$)

Annual variable cost of the buyer = Annual ordering cost + Annual carrying cost + Annual transportation cost + Interest amount paid per annum – Interest amount earned per annum.

$$\pi_B(\eta, M) = \frac{\varepsilon_B}{M} + \frac{\lambda M U_B W_p}{2\eta} + \frac{\beta_B \delta}{M} + \frac{U_B \lambda (M-t)^2 W_p}{2M} - \frac{S_B \lambda t^2 W_e}{2M} \tag{1}$$

Case II: ($M < t$)

Annual variable cost of the buyer = Annual ordering cost + Annual carrying cost + Annual transportation cost – Interest amount earned per annum.

$$\pi_B(\eta, M) = \frac{\varepsilon_B}{M} + \frac{\beta_B \delta}{M} + \frac{\lambda M U_B W_p}{2\eta} - S_B \lambda W_e \left(t - \frac{M}{2}\right) \tag{2}$$

2.3.2 Vendor Node

Annual variable cost of the vendor = Annual ordering/setup cost + Annual transportation cost+ Annual carrying cost

$$\pi_v(\eta, M) = \frac{\varepsilon_v}{M} + \frac{\beta_v \eta}{M} + \frac{\lambda M C_v W_p}{2}\left(1 - \frac{\lambda}{C} - \frac{1}{\eta} + \frac{2\lambda}{\eta C}\right) \tag{3}$$

2.3.3 Joint Supply Chain

Vendor &buyer jointly decided a combined policy for optimal inventory decision.

Case I: ($M \geq t$)

Annual variable cost of the SC = Annual variable cost of the buyer + annual variable cost of the vendor- Interest amount earned per annum. Upon simplification, the annual variable cost of the SC is expresses as:

$$\pi_S(\eta, M) = \frac{1}{M}(\varepsilon_B + \beta_B \eta + \varepsilon_v + \beta_v \eta) + \lambda W_p \left(\frac{U_B(M-t)^2}{2M} + \frac{M U_v}{2}\left(1 - \frac{\lambda}{P} - \frac{1}{\eta} + \frac{2\lambda}{\mu C}\right) + \frac{M U_B}{2\eta}\right) - \frac{S_B \lambda t^2 W_e}{2M} \tag{4}$$

Optimality Criteria:

For known value of η , π_S is convex in terms of M . The proof is as shown below.

$$\frac{\partial}{\partial M}(\pi_S(\eta, M)) = 0$$

$$\lambda W_p \left(U_B + U_v\left(1 - \frac{\lambda}{P} - \frac{1}{\eta} + \frac{2\lambda}{\eta C}\right) + \frac{U_B}{\eta}\right) = \frac{2}{M^2}\left\{(\varepsilon_B + \beta_B \eta + \varepsilon_v + \beta_v \eta) + \left(\frac{\lambda t^2}{2}\right)(U_B W_p - S_B W_e)\right\} \tag{5}$$

$$\frac{\partial^2}{\partial M^2}(\psi_S(\eta, M)) = \frac{2}{M^3}(\varepsilon_B + \beta_B \eta + \varepsilon_v + \beta_v \eta) + \frac{\lambda t 2}{M^3}(U_B W_p - S_B W_e) \tag{6}$$

Since $\frac{\partial^2}{\partial M^2}(\pi_S(\eta, M)) > 0$ for all values of cycle time, rate of shipment and other model parameters, the cycle time becomes optimal. Optimal cycle time is derived from equation (5) as shown in equation (7).

$$M^* = \left(\left(2\left(\varepsilon_B + \beta_B + \frac{\varepsilon_v}{\eta} + \beta_v\right) + 0.5\lambda t^2 (U_B W_P - S_B W_e)\right) \Big/ \left(\lambda W_P \left(\left(U_B + U_v\left(1 - \frac{\lambda}{C} - \frac{1}{\eta} + \frac{2\lambda}{\eta P}\right)\right) + \frac{U_B}{\eta}\right)\right)\right)^{0.5} \tag{7}$$

In the same way, for the known value of M , π_S is convex in terms of η . The inequality conditions for optimal η are derived and are expressed as shown below.

$$\pi_S(\eta^*) \leq \pi_S(\eta^* - 1) \quad and \quad \pi_S(\eta^*) \leq \pi_S(\eta^* + 1)$$

$$\eta^*(\eta^* - 1) \leq \left(\left(\lambda M^2 W_P \left(U_B + U_v\left(\frac{2\lambda}{C} - 1\right)\right)\right) \Big/ (2(\beta_B + \beta_v))\right) \leq \eta^*(\eta^* + 1) \tag{8}$$

Case II: ($M < t$)

Annual variable cost of the SC = Annual variable cost of the buyer + annual variable cost of the vendor- Interest amount earned per annum. Upon simplification, the annual variable cost of the SC is expresses as:

$$\pi_S(\eta, M) = \frac{1}{M}(\varepsilon_B + \beta_B \eta + \varepsilon_v + \beta_v \eta) + \frac{\lambda M W_p}{2}\left(\frac{U_B}{\eta} + U_v\left(1 - \frac{\lambda}{C} - \frac{1}{\eta} + \frac{2\lambda}{\eta P}\right)\right) - S_B \lambda W_e \left(t - \frac{M}{2}\right) \tag{9}$$

For known value of η , π_S is convex in terms of M . The proof is as shown below.

$$\frac{\partial}{\partial M}(\pi_S(\eta, M)) = 0$$

$$\frac{1}{M^2}(\varepsilon_B + \beta_B \eta + \varepsilon_v + \beta_v \eta) = \frac{\lambda W_P}{2}\left(\frac{U_B}{\mu} + U_v\left(1 - \frac{\lambda}{C} - \frac{1}{\eta} + \frac{2\lambda}{\eta C}\right)\right) + \left(\frac{S_B \lambda W_e}{2}\right) \tag{10}$$

$$\frac{\partial^2}{\partial M^2}(\pi_S(\eta, M)) = \frac{2}{M^3}(\varepsilon_B + \beta_B \eta + \varepsilon_v + \beta_v \eta)$$

Since $\frac{\partial^2}{\partial M^2}\left(\pi_S(\eta,M)\right)>0$ for all values of cycle time, rate of shipment and other model parameters, the cycle time becomes optimal. Optimal cycle time is derived from equation (10) as shown in equation (11).

$$M^*=\left(\left(2\left(\varepsilon_B+\beta_B+\frac{\varepsilon_v}{\eta}+\beta_v\right)\right)\Big/\left(\lambda W_p\left(\frac{U_B}{\eta}+U_v\left(1-\frac{\lambda}{C}-\frac{1}{\eta}+\frac{2\lambda}{\eta C}\right)\right)+S_B\lambda W_e\right)\right)^{0.5} \quad (11)$$

In the same way, for the known value of M, π_S is convex in terms of η. The inequality conditions for optimal η are derived and are expressed as shown below.

$\pi_S(\eta^*)\le\pi_S(\eta^*-1)$ *and* $\pi_S(\eta^*)\le\pi_S(\eta^*+1)$

$$\eta^*(\eta^*-1)\le\left(\left(\lambda M^2 I_P\left(U_B+U_v\left(\frac{2\lambda}{C}-1\right)\right)\right)\Big/\left(2(\beta_B+\beta_v)\right)\right)\le\eta^*(\eta^*+1) \quad (12)$$

Case III: ($t=M$)

If credit period and cycle time both are equal, then overall variable cost of the total SC is obtained as:

$$\pi_S(\eta,M)=\frac{1}{M}(\varepsilon_B+\beta_B\eta+\varepsilon_v+\beta_v\eta)+\frac{\lambda M W_p}{2}\left(\frac{U_B}{\eta}+U_v\left(1-\frac{\lambda}{C}-\frac{1}{\eta}+\frac{2\lambda}{\eta C}\right)\right)-\left(\frac{S_B\lambda W_e M}{2}\right) \quad (13)$$

For known value of η, π_S is convex in terms of M. In the same way, for the known value of M, π_S is convex in terms of η. By using the similar procedure followed in Case I and Case II, the optimality criteria is derived for Case III.

3. Numerical Investigation

Numerical data is set for illustrating the model and is solved using MATLAB program and the results are shown in Table 1.The values of various parameters for present inventory system are: ε_v = INR 1600/setup, ε_B = INR 400/order, U_v = INR 100/unit, U_B = INR 140 /unit, S_B = INR 160 /unit, λ = 1800 units /year, C= 2000 units / year W_P = 0.18 INR / INR /year, W_e= 0.12 INR/ INR/year, β_B = INR 750/shipment, β_v = INR 250/shipment.

Table 1: Optimal Values of Decision Variables and Objective Function

Description	$t=10$	$t=20$	$t=30$	$t=40$	$t=50$	$t=60$
M^* (in Years)	0.1925	0.1943	0.1973	0.2015	0.2067	0.2129
q^* (in Units)	365.5	349.74	355.14	362.7	372.06	383.22
μ^* (Integer)	1.00	1.00	1.00	1.00	1.00	1.00
Q^* (in Units)	365.5	349.74	355.14	362.7	372.06	383.22
π_B^* (inRs.)	14136.02	12679.32	11440.55	10404.71	9553.14	8865.37
π_v^* (inRs.)	11253.71	11168.96	11033.31	10854.27	10640.91	10402.71
π_S^* (inRs.)	25389.73	23848.28	22473.86	21258.98	20194.05	19268.08

The optimal values of decision variables and objective function are shown in Table 1. From table 1, it is shown that under trade credit scenario, the buyer cost and vendor cost are decreased, with increase in the trade credit. The overall variable cost of the SC can be reduced with the effect of trade credit. The shipment quantities are not influenced with increment and decrement of trade credit. Table 1, also demonstrates the optimal cycle time, overall variable cost, replenishment quantities and rate of shipments from the vendor to the buyer. The cost of SC is decreased with increasing trade credit. The overall buyer cost is reduced with increasing trade credit. Similarly, the cost of vendor is also reduced by increasing trade credit. Finally, it is evident that the overall variable SC cost is reduced with increasing trade credit. From Figure1, the credit period is directly proportional to the cycle time. As the trade credit increases, the cycle time also increases. From Figure 2, the buyer replenishment quantity is suddenly decreased to certain period and then suddenly increases as credit period goes on increases. From Figure 3, as long as the credit period is increased or decreased, there will be no effect on the shipment quantity

Table 2: SENSITIVITY ANALYSIS

Para-meter	(in %)	$t=20$					$t=50$					$t=80$					$t=110$				
		M^*	μ^*	π_B^*	π_v^*	π_S^*	M^*	μ^*	π_B^*	π_v^*	π_S^*	M^*	μ^*	π_B^*	π_v^*	π_S^*	M^*	μ^*	π_B^*	π_v^*	π_S^*
$\frac{\lambda}{C}$	−40	0.3045	1	8942.61	9315.98	18258.59	0.3071	1	6898.85	9263.42	16162.26	0.3120	1	5039.31	9169.91	14209.22	0.3189	1	3350.81	9042.13	12392.94
	−20	0.3451	2	10023.10	11218.86	21241.97	0.3481	2	7258.97	11160.96	18419.92	0.3536	2	4708.32	11058.13	15766.44	0.3615	2	2355.84	10918.00	13273.83
	+20	0.2820	2	11751.33	13733.39	25484.73	0.2857	2	7690.48	13628.16	21318.64	0.2924	2	4012.05	13445.30	17457.35	0.3013	2	662.72	13217.57	13880.29
	+40	0.2612	2	12447.64	14830.07	27277.71	0.2652	2	7752.99	14698.17	22451.17	0.3348	3	4612.76	13371.72	17984.48	0.3450	3	581.89	13088.75	13670.64
ε_v	−40	0.2831	2	10349.91	10981.73	21331.64	0.2868	2	6957.87	10900.83	17858.70	0.2934	2	3878.52	10760.31	14638.83	0.3025	2	1072.66	10582.57	11655.24
	−20	0.2963	2	10652.70	11783.55	22436.25	0.2997	2	7247.22	11702.66	18949.88	0.3061	2	4145.03	11561.29	15706.32	0.3152	2	1317.57	11373.29	12690.86
	+20	0.3209	2	11252.90	13257.48	24510.38	0.3241	2	7824.07	13177.65	21001.71	0.3300	2	4681.78	13036.81	17718.59	0.3384	2	1802.42	12846.80	14649.22
	+40	0.3325	2	11548.58	13941.34	25489.91	0.3356	2	8109.53	13862.31	21971.84	0.3413	2	4949.57	13722.39	18671.96	0.3495	2	2047.02	13532.56	15579.59
ε_B	−40	0.3026	2	10274.88	12695.73	22970.61	0.3060	2	6869.05	12609.11	19478.15	0.3122	2	3766.63	12457.20	16223.82	0.3212	2	939.64	12254.11	13193.75
	−20	0.3057	2	10617.24	12616.39	23233.63	0.3090	2	7205.354	12532.91	19738.27	0.3153	2	4092.39	12386.41	16478.80	0.3241	2	1251.44	12190.28	13441.71
	+20	0.3119	2	11285.55	12466.23	23751.78	0.3151	2	7862.284	12388.58	20250.86	0.3212	2	4729.45	12252.08	16981.53	0.3299	2	1862.09	12068.92	13931.01
	+40	0.3148	2	11611.94	12395.12	24007.06	0.3181	2	8183.314	12320.17	20503.48	0.3242	2	5041.11	12188.32	17229.43	0.3328	2	2161.26	12011.20	14172.46
U_v	−40	0.3214	2	11266.58	11206.86	22473.44	0.3249	2	7843.79	11120.42	18964.20	0.3313	2	4711.59	10967.91	15679.50	0.3404	2	1844.68	10762.06	12606.73
	−20	0.3149	2	11104.46	11884.36	22988.82	0.3183	2	7684.20	11800.97	19485.17	0.3246	2	4556.52	11654.17	16210.68	0.3335	2	1695.80	11456.70	13152.50
	+20	0.3030	2	10814.22	13175.33	23989.55	0.3063	2	7399.13	13097.60	20496.73	0.3123	2	4280.69	12961.39	17242.08	0.3209	2	1432.70	12779.51	14212.21
	+40	0.2976	2	10683.96	13792.04	24476.00	0.3008	2	7271.51	13716.95	20988.46	0.3067	2	4157.78	13585.67	17743.44	0.3151	2	1316.28	13411.06	14727.34
U_B	−40	-	-	-	-	-	-	-	-	-	-	-	-	-	-	-	-	-	-	-	-
	−20	0.3367	2	9827.74	11935.09	21762.83	0.3372	2	6856.59	11923.79	18780.38	0.3383	2	3930.34	11903.01	15833.35	0.3399	2	1048.30	11873.10	12921.40
	+20	0.2870	2	11936.12	13125.20	25061.32	0.2927	2	8119.01	12961.74	21080.75	0.3030	2	4862.47	12686.19	17548.65	0.3174	2	2083.00	12338.58	14421.58
	+40	0.2694	2	12809.69	13689.04	26498.73	0.2771	2	8633.23	13431.68	22064.91	0.2909	2	5292.52	13013.92	18306.45	0.3099	2	2619.50	12514.00	15133.51
S_B	−40	0.3096	2	11042.96	12520.02	23562.99	0.3172	2	8076.84	12342.71	20419.54	0.3308	2	5731.64	12051.52	17783.16	0.3497	2	3896.74	11698.11	15594.84
	−20	0.3092	2	10998.56	12529.96	23528.51	0.3147	2	7808.31	12400.22	20208.53	0.3246	2	5082.33	12179.98	17262.31	0.3385	2	2757.62	11899.48	14657.10
	+20	0.3084	2	10909.46	12549.97	23459.43	0.3096	2	7260.89	12520.49	19781.37	0.3118	2	3722.81	12466.83	16189.64	0.3151	2	291.50	12390.90	12682.40
	+40	-	-	-	-	-	-	-	-	-	-	-	-	-	-	-	-	-	-	-	-
W_P	−40	-	-	-	-	-	-	-	-	-	-	-	-	-	-	-	-	-	-	-	-
	−20	0.3447	2	9983.19	11227.85	21211.04	0.3453	2	7011.56	11216.06	18227.63	0.3464	2	4084.44	11194.37	15278.81	0.3480	2	1201.13	11163.13	12364.26
	+20	0.2824	2	11800.05	13722.47	25522.52	0.2880	2	7987.20	13563.87	21551.07	0.2981	2	4738.02	13297.44	18035.46	0.3122	2	1968.25	12963.28	14931.52
	+40	0.2619	2	12552.70	14806.57	27359.27	0.2694	2	8387.48	14562.56	22950.04	0.2828	2	5065.66	14169.87	19235.53	0.2998	2	2382.78	13739.35	16122.13
W_e	−40	0.3096	2	11042.96	12520.02	23562.99	0.3172	2	8076.84	12342.71	20419.54	0.3308	2	5731.64	12051.52	17783.16	0.3497	2	3896.74	11698.11	15594.84
	−20	0.3092	2	10998.56	12529.96	23528.51	0.3147	2	7808.31	12400.22	20208.53	0.3246	2	5082.33	12179.98	17262.31	0.3385	2	2757.62	11899.48	14657.10
	+20	0.3084	2	10909.46	12549.97	23459.43	0.3096	2	7260.89	12520.49	19781.37	0.3118	2	3722.81	12466.83	16189.64	0.3151	2	291.50	12390.90	12682.40
	+40	-	-	-	-	-	-	-	-	-	-	-	-	-	-	-	-	-	-	-	-

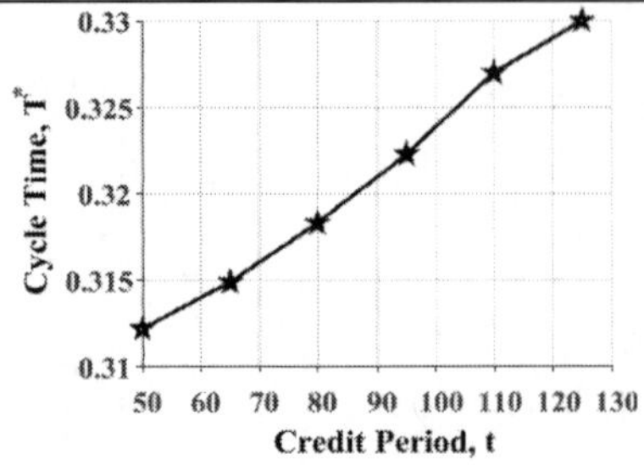

Fig. 1 Cycle Time vs Credit Period

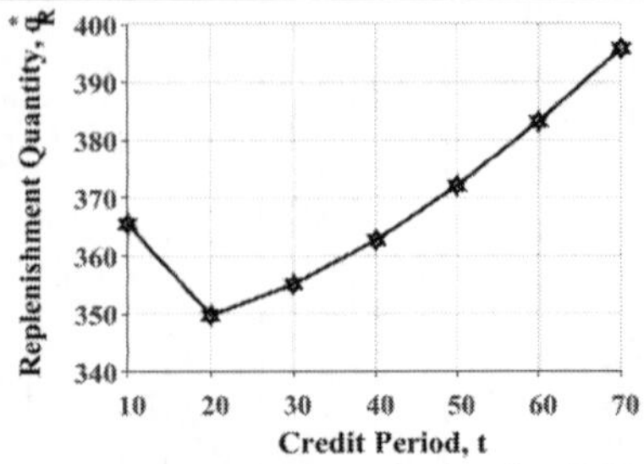

Fig. 2 Replenishment Quantity vs Credit Period

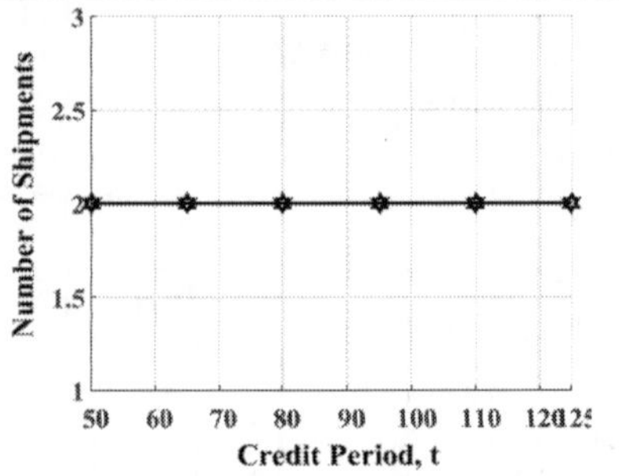

Fig. 3 Shipment Frequency vs Credit Period

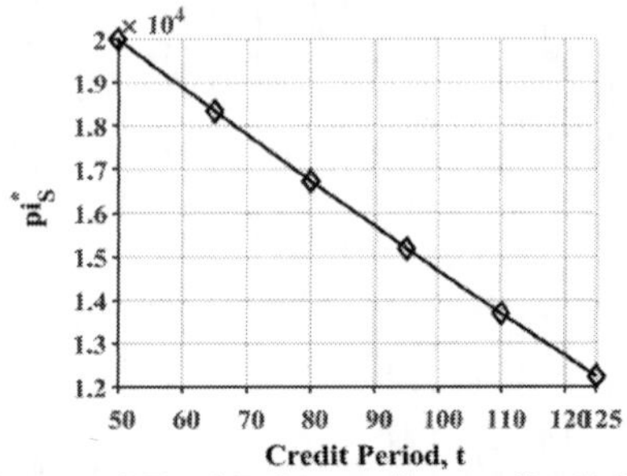

Fig. 4 Annual Total Relevant Cost vs Credit Period

Otherwise, simply the rate of shipment is remains constant with respect to credit period. Constant rate of shipment can be observed by increasing the credit period in this scenario. It is also observed that from Figure 4, the overall variable cost is reduced with increasing credit period. Further, the sensitivity analysis is carried out to analyze the variation in decision variables and objective function with respect to variation in model parameters. The results are tabulated in Table 2 and it is observed that variation in model parameters significantly influence the optimality of decision variables and objective function.

4. Conclusions

In this particular work, a mathematical model is developed for a two-echelon inventory system with a one kind of product supplying from one-vendor to the one-buyer under finite production rate with trade credit. In the development of the model, inventory related costs like ordering/setup costs, costs of carrying and transportation are considered. The optimality condition is established for the decision variables and objective function. MATLAB computer program is developed to solve the model, based upon on the optimality condition. From research findings, it is concluded that cycle time increases with increasing trade credit and no change in the shipment quantity. Also, it is concluded that the overall variable cost of the SC decreases with increasing credit period.

References

[1] Goyal SK. An Integrated inventory model for a single-supplier single-customer problem. International Journal of Production Research.1976, **15(1)**, 107-111.
[2] Saha S, Goyal, SK. Supply chain coordination contracts with inventory level and retail price dependent demand. International Journal of Production Economics. 2015,**161**, 140-152.
[3] BoyacıT, Gallego G, Coordinating pricing and inventory replenishment policies for one wholesaler and one or more geographically dispersed buyers. International Journal of Production Economics. 2002, **77(2)**95-111.
[4] Rahdar M, Nookabadi, AS. Coordination mechanism for a deteriorating item in a two-level supply chain system. Applied Mathematical Modelling.2014, **38(11)**, 2884-2900.
[5] LiC.An analytical method for cost analysis in multi-stage supply chains: A stochastic network model approach. Applied Mathematical Modelling. 2014, **38(11)**, 2819-2836.

Optimization of machining parameters using Entropy-VIKOR method in drilling of A356-TiB_2/TiC in-situ composites

Ismail.Kakaravada[1], A. Mahamani[2] and V.Pandurangadu[3]

[1,3]Department of Mechanical Engineering, Jawaharlal Nehru Technological University (A) College of Engineering, Anantapuramu,India.
[2]Department of Mechanical Engineering, Sri Venkateswara College of Engineering & Technology, Chittoor, AndhraPradesh, India

[1]Corresponding author: ismailmtech23@gmail.com

Abstract. This study presents the multi-performance analysis in drilling of A356-TiB_2/TiC in-situ composite produced through stir casting route. A novel Entropy-VIKOR approach is adopted for the optimization of parameters in drilling like spindle speed, drill diameter and % of reinforcement over the MRR, delamination and surface roughness. Experiments were conducted on vertical machining centre by using of carbide drills followed by Taguchi L_{27} orthogonal array. The Entropy weight measurement method was used to evaluate individual weights of each response. For the multi responses optimization VIKOR indexing method was adopted. In this paper, the multi-objective responses are converted as a single VIKOR index, and the ANOVA was performed for VIKOR index to determine the influence parameter for drilling. Finally, confirmation experiment was performed and validated using the regression equation to recognize the effectiveness of proposed method. Result of ANOVA indicates that reinforcement ratio has strongest influence on VIKOR index.

1. Introduction

Aluminium particulate metal matrix composite has wide spread attention to the researchers because of these composites exhibits the promising mechanical and metallurgical properties. As a result, many current applications for AMMCs are in aerospace, automobile components [1]. In in-situ method of fabrication the reinforcement material was synthesized by using inorganic salts through an exothermic reaction which creates the amalgamation of liquid matrix metal results the thermo-dynamically stabled, uniformly distributed ceramic particles are generated during the fabrication [2]. Drilling of composites has received wide-range of attention because of poor machinability. The abrasive natured ceramic reinforcement present in composites causes the deterioration of machining responses in all manufacturing processes [3].Taguchi orthogonal array was followed to conduct the drilling experiments by considering cutting velocity and cutting time on drilling of composites with 5 mm drill against responses like tool wear, surface roughness and specific cutting pressure are correlated by regression equation [4]. It also observed that when feed rate and point angle increases the delamination increases [5].The multi-objective characteristic optimization was conducted by using grey relational analysis to optimize the machining responses by considering the input parameters as speed, feed rate and drill point angle [6]. The present work deals the optimisation of machining parameters to optimize the material removal rate, delamination and surface roughness in drilling of A356-TiB_2/TiC composites. Taguchi L_{27} array was followed to conduct experimentation. The Entropy -VIKOR was performed to combine multiple responses into one desired value that is VIKOR Index. Therefore, the trail with Minimum VIKOR Index is identified as optimal machining parameters. Analysis of variance (ANOVA) also performed to identify the significant influential parameter to affect the combine responses.

2. Experimental procedure

2.1 Work material

In the present work A356 alloy was selected for drilling study, the chemical composition of A356 alloy is Si -6.5%, Fe-0.15%, Cu-0.03%, Mn-0.10%, Mg-0.4%, Zn-0.07%, Ni-0.05 and Ti-0.1%. A356-TiB_2/TiC composites fabricated with various volume fractions through mixed salt reaction system. A measured quantity of halide salts namely potassium hexa-fluro-titanate (K_2TiF_6), potassium hexa-fluro-borate (KBF_4) and graphite(C) are used to synthesis the composites.TiB_2 and TiC reinforcements are formed by exothermic chemical reaction. The energy dispersive X-Ray (Hitachi S-300 H a model) spectra of A356-TiB_2 / TiC composite are recorded by lithium drift silicon analyzer with operating voltage 20 KV and displayed in the Figure 2. This spectrum confirms the various elements presence in composites like titanium, carbon, copper and aluminium. Microstructure analysis was conducted by using scanning electron microscope (JEOL 6360 LV model) and presented in Figure 1. The Figure1 depicts the presence of homogenously distributed and cluster free TiB_2 and TiC reinforcements present in the fabricated composites. L_{27} orthogonal array was used to conduct the experiments by considering cutting speed, diameter of drill, and % of reinforcement as parameters. Further, the Entropy-VIKOR Index was adopted to optimize the multi responses

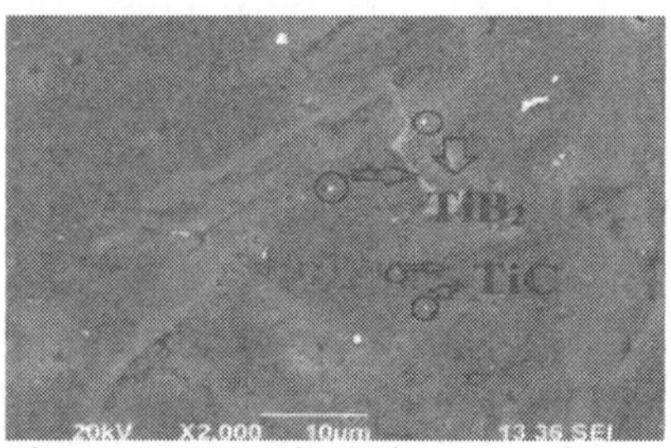

Figure1. A356- TiB_2/ TiC composites

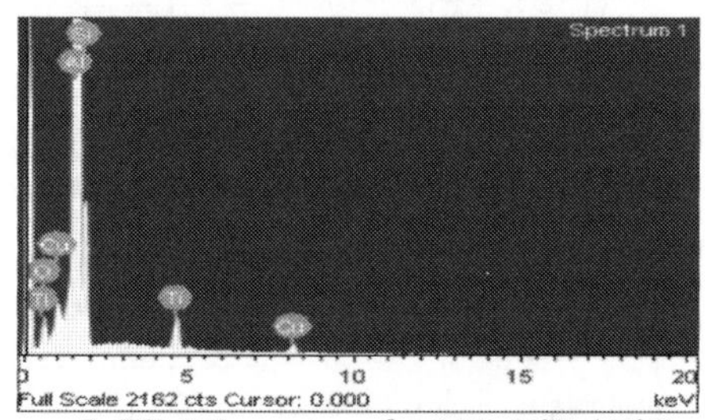

Figure 2. EDAX spectra of A356- TiB_2/ TiC

2.2 Experimental work

The drilling experiments were carried out on CNC Vertical Machining Center (VMC) as shown in Figure 3 .For the experimentation samples were prepared in the form of 150×50×10 mm by cutting in milling machine. Experiments were performed by using orthogonal L_{27} array considered machining parameter namely drill diameter, spindle speed and %of reinforcement of material and their levels shown in Table 1.The carbide drill bits are used for conducting experiments. The surface roughness was measured by using of SJ210 stylus type surface roughness tester (Mitutoya Make, 0.001μm). Electronic Top Loading Balance (Shimadzu) was used to measure the work piece weights before after machining processes to evaluate MRR. The delamination was measured by using Metallurgical microspore (Metzer Optical Instruments Private Limited, Mathura, India). The mathematical relation shown in below is used to evaluate the MRR and delamination.

MRR= $\frac{Wb-Wa}{t}$ mg/minWhere Wb= weight of work piece before machining (mg) Wa= weight of work piece after machining (mg) and t= drilling time.

Delamination factor = $\frac{Dmax}{D}$ (mm) where D_{max}= Measured Diameter (mm) D = Diameter of drill in (mm)

Figure 3. Drilling on Vertical Machining Center

Table 1. Drilling parameters and their levels

Factor	Processes parameter	Unit	Level 1	Level 2	Level 3
A	Spindle speed	rpm	795	1250	1980
B	Drill diameter	mm	2.5	5	10
D	% of Reinforcement	%	2.5	5	7.5

2.3 Entropy weight measurement

The entropy weight method the number of alternatives and various criterions considered to evaluate multi parameter optimization by considering a relative decision matrix. If numbers of alternatives are spindle speed, drill diameter and % of reinforcement considers as 'M' and numbers of criteria are assumed MRR, delamination and surface roughness considers as 'N' then decision matrix having an order of M×N.

$$D = \begin{bmatrix} X_{11} & X_{12} & X_{1n} \\ X_{21} & X_{22} & X_{23} \\ X_{m1} & X_{m2} & X_{mn} \end{bmatrix} \qquad r_{ij} = \frac{x_{ij}-min(x)_{ij}}{\max(x_{ij})-\min(x_{ij})} \quad (1) \qquad r_{ij} = \frac{\max(x_{ij})-x_{ij}}{\max(x_{ij})-\min(x_{ij})} \quad (2)$$

Where i=1,2,3,....m & j=1,2,3....*n in equations*

The decision matrix is normalized by maximum values (i.e. beneficial attribute) and minimum values (non-beneficial attribute). The normalized values calculated by following mathematical equations 1 and 2 and values are tabulated in Table 4

2.4 Entropy weight generation model

In this stage the various criterion are resolute by the entropy method projection values (T_{ij}) are evaluated by equation (4).and the projection of alternative is used to determine the entropy index (e_{ij}) using equation (5). The projection and entropy indexed values tabulate in Table 4

$$T_{ij} = \frac{r_{ij}}{\sum_{i=1}^{m} r_{ij}} \quad (4) \qquad e_{ij} = -\frac{1}{lnm}\sum_{i=1}^{n} T_{ij} lnT_{ij} \quad (5)$$

2.5 VIKOR optimization method

The VIKOR method is applicable for multi parameter optimization to find a final solution. In this method statistical procedure to adopt to find the solution close to ideal positive solution (f^*) and the ideal negative solution (f^-) were identified by following equations 6 and 7.

$$f^* = f^- = \{(\min f_{ij} /j\epsilon J)OR \left(\max f_{ij}\right)/j \epsilon j')\} \ (6) \qquad f^- = \{(\min f_{ij} /j\epsilon J)OR \left(\max f_{ij}\right)/j \epsilon j')\} \ (7)$$

Where J= {j=1, 2...n}, f_{ij} if desired responses is large. J'= {1, 2…n}, f_{ij} if desired responses is small. For the calculation of utility and regret measure for each response calculated by the following equation 8 and 9 respectively and Tabulated in Table 4

$$s_i = \sum_{j=1}^{n} w_j \frac{(f^*-f_{ij})}{(f_j^*-f_j^-)'} \quad (8) \qquad R_i = max_j \left[w_j \frac{(f^*-f_{ij})}{(f_j^*-f_j^-)'}\right]' (9)$$

Where $s_i$utility measure, R_i is the regret measure w_j is the corresponding weight of the jth attribute. The individual weight calculated by entropy weight method and finally the VIKOR index calculated by equation 10 and presented in Table 4 and to determine rank of VIKOR indexed values the least index value treated as first rank.

$$Q_i = v\left[\frac{(s_i-s^*)}{(s^- -s^*)}\right] + (1-v)\left[\frac{(R_i-R^*)}{(R^- -R^*)}\right] \quad (10)$$

Where, Q_i is the VIKOR index of the ith result of alternatives and v is the higher weight of the group utility. In general the value of v considers as 0.5. In the above equation $s^{-}= max_i\, s_i$, $s^{*}= min_i\, s_i$, $R^{-} = max_i\, R_i$, And $R^{*} = min_i\, R_i$ are the maximum and minimum values of utility measure and regret measure. The following condition must be satisfied to compromise the solution of the consider weight of given alternative. The alternative A^1 is considered lower VIKOR index values by measure Q (minimum) and A^2 is the second lowest VIKOR index in list.

a) Acceptable advantage is $Q(A^2)$-$Q(A^1) \geq DQ = \frac{1}{(m-1)}$ where m is the number of experiments

b) Acceptable decision making stability

The alternative A^1 is the lowest VIKOR index ranked by S or/ and R the solution is stable in decision making. This could be the strategy of maximum of utility value v is more than 0.5 needed or equal to 0.5 or with veto less than 0.5 if one condition not satisfied the following must be followed.

I. The alternatives A^1 *and* A^2 if only second condition is not satisfied.

II. The first condition not satisfied A^n is calculated by the following relation $Q(A^n)$-$Q(A^1) <$ DQ for maximum of n

From the above equation Q value is minimum, which results the least rank which compromise the better results for multi optimization problem.

3. Results and discussion

This study reveals the combination of Entropy weight measurement and VIKOR method to evaluate the decision matrix D $m \times n$ is represent by the experimental results as shown in Table 4. For the calculation of Entropy weight of each response equation 1 and 2 are used. For the material removal rate beneficial attribute, and the surface roughness and delamination non beneficial attribute equation was used. After calculation the entropy method projection values (T_{ij}) are calculated by equation (4). The projection of alternative is used to determine the entropy index (e_{ij}) using equation (5). The projection and entropy indexed values tabulated in Table 4. Finally the VIKOR method was adopted for multiresponse optimization. From the individual matrix is analyze f^{*} and f^{-} possible best solution values (14.735, 1.3389, 8.948) and worst solution values (2.9278, 1.3046, 2.471) were analyzed by equation 6 and 7.

The values of utility measure and regret measure is by equation 8 and 9, respectively. The maximum and minimum utility and regret measure are (0.9909, 0.0246 and 0.07029, 0.0156) respectively. The lowest VIKOR index calculated using equation 10 and lowest index values ranked as 1 as shown in Table 4. The validation of experiment calculated acceptable advantage equation is satisfied with the value (0.0253) is less than DQ value (0.146). Therefore condition is acceptable.

3.1 optimal level selection and analysis of variance

To find the optimum level machining parameter the main effect parameter are evaluated on the basis of each level of VIKOR index as shown in Table 3. From the table 3 it is found that the 7.5 % of reinforcement, spindle speed at 1980 rpm and diameter of drill is 10 mm the optimal parameter for the multi character machining. Also table 3 shows the difference value of maximum and minimum values and their rank among the processes parameter for the selective model. The results of ANOVA test for VIKOR index shown in Table 2. From the table it has been observed that contribution each parameter on output response. And finally the interaction process parameters are significant at 95% confidence.

Table 2 ANOVA for VIKOR Index

Source	DF	Adj SS	Adj MS	F-Value	contribution
% of Reinforcement	2	0.9826	0.491310	90.09	**61.72**
Spindle speed	2	0.3699	0.184972	33.92	**23.23**
drill Diameter	2	0.1304	0.065196	11.95	**8.19**
Error	20	0.1091	0.005454		0.685
Total	26	1.5920			100

Table 3 Main effect factor on VIKOR level

Parameter	Level 1	Level 2	Level 3	Max-Min	Rank
% of Reinforcement	0.7246	0.3695	0.2840	**0.4406**	1
Spindle speed	0.6133	0.4352	0.3297	**0.2836**	2
Drill diameter	0.5262	0.4884	0.3636	**0.1626**	3

IOP Conf. Series: Materials Science and Engineering **390** (2018) 012099 doi:10.1088/1757-899X/390/1/012099

Table 4. Taguchi L_{27} orthogonal array with machining parameters, experimental results, projected values, individual weighted Entropy Index values, utility, regret measures values and VIKOR index of alternative with their rank

	Machining Parameters			Experimental results			Projection values (Tij)			Entropy Index (Tij)ln(Tij)						
Trail. No	% of Rein-force ment	Speed	Drill diameter	MRR	Delamination	surface roughness	C-1	C-2	C-3	C-1	C-2	C-3	Utility Measure (Sij)	Regret Measure (Rij)	VIKOR index (Qij)	Rank order
A-1	2.5	795	2.5	3.6884	1.03461	2.471	0.0160	0.0302	0.0149	-0.0661	-0.1058	-0.0628	0.9909	0.7029	0.8067	27
A-2	2.5	795	5	7.1948	1.09311	3.774	0.0312	0.0319	0.0228	-0.1082	-0.1100	-0.0863	0.7776	0.5615	0.7867	25
A-3	2.5	795	10	11.7424	1.13239	3.668	0.0509	0.0331	0.0222	-0.1516	-0.1128	-0.0845	0.7148	0.5730	0.7626	23
A-4	2.5	1250	2.5	4.8414	1.13311	3.229	0.0210	0.0331	0.0195	-0.0811	-0.1138	-0.0769	0.8443	0.6207	0.8642	26
A-5	2.5	1250	5	8.3518	1.19161	3.531	0.0362	0.0348	0.0213	-0.1202	-0.1169	-0.0822	0.7396	0.5879	0.7862	24
A-6	2.5	1250	10	12.8940	1.23240	4.426	0.0559	0.0360	0.0268	-0.1612	-0.1197	-0.0970	0.5674	0.4907	0.6265	19
A-7	2.5	1980	2.5	6.6778	1.22944	3.979	0.0289	0.0359	0.0241	-0.1025	-0.1195	-0.0897	0.6915	0.5393	0.7260	22
A-8	2.5	1980	5	10.1881	1.28794	4.281	0.0442	0.0376	0.0259	-0.1378	-0.1234	-0.0947	0.5869	0.5065	0.6480	21
A-9	2.5	1980	10	14.7358	1.22878	5.176	0.0639	0.0359	0.0313	-0.1758	-0.1194	-0.1085	0.4659	0.4093	0.5147	15
A-10	5	795	2.5	3.2814	1.24478	5.522	0.0143	0.0363	0.0334	-0.0605	-0.1205	-0.1136	0.5566	0.3718	0.5343	17
A-11	5	795	5	6.7918	1.11328	4.824	0.0295	0.0325	0.0292	-0.1038	-0.1114	-0.1032	0.6581	0.4475	0.6420	20
A-12	5	795	10	11.3394	1.14411	5.719	0.0492	0.0334	0.0346	-0.1481	-0.1136	-0.1164	0.4910	0.3504	0.4848	14
A-13	5	1250	2.5	4.4384	1.14328	6.279	0.0193	0.0334	0.0380	-0.0760	-0.1135	-0.1243	0.5128	0.2897	0.4519	13
A-14	5	1250	5	7.9488	1.20178	6.582	0.0345	0.0351	0.0398	-0.1161	-0.1176	-0.1284	0.4081	0.2567	0.3738	12
A-15	5	1250	10	12.4964	1.24261	7.476	0.0542	0.0363	0.0452	-0.1580	-0.1204	-0.1401	0.2359	0.1597	0.2141	6
A-16	5	1980	2.5	6.2748	1.23961	7.029	0.0273	0.0362	0.0425	-0.0980	-0.1202	-0.1343	0.3601	0.2082	0.3136	10
A-17	5	1980	5	9.7851	1.29811	7.332	0.0425	0.0379	0.0444	-0.1341	-0.1241	-0.1383	0.2553	0.1753	0.2355	7
A-18	5	1980	10	14.3328	1.33894	8.226	0.0622	0.0391	0.0498	-0.1727	-0.1268	-0.1494	0.0831	0.0783	0.0758	1
A-19	7.5	795	2.5	2.9278	1.14428	5.243	0.0128	0.0334	0.0317	-0.0554	-0.1136	-0.1095	0.6427	0.4021	0.6009	18
A-20	7.5	795	5	6.4381	1.17278	5.546	0.0278	0.0342	0.0335	-0.0999	-0.1156	-0.1140	0.5534	0.3692	0.5308	17
A-21	7.5	795	10	10.9858	1.21361	6.44	0.0477	0.0354	0.0390	-0.1450	-0.1184	-0.1265	0.3812	0.2722	0.3711	11
A-22	7.5	1250	2.5	4.0848	1.31278	7.541	0.0178	0.0383	0.0456	-0.0714	-0.1251	-0.1409	0.2930	0.1527	0.2385	8
A-23	7.5	1250	5	7.59511	1.27128	7.303	0.0329	0.0371	0.0442	-0.1124	-0.1223	-0.1379	0.2983	0.1785	0.26010	9
A-24	7.5	1250	10	12.1428	1.31211	8.1918	0.0526	0.0383	0.0496	-0.1550	-0.1250	-0.1490	0.1267	0.0820	0.1011	2
A-25	7.5	1980	2.5	5.9211	1.30911	7.751	0.0256	0.0382	0.0469	-0.0940	-0.1248	-0.1436	0.2502	0.1299	0.1998	5
A-26	7.5	1980	5	9.4314	1.32761	8.053	0.0409	0.0388	0.0487	-0.1307	-0.1261	-0.1473	0.1661	0.0971	0.1324	4
A-27	7.5	1980	10	13.9791	1.30844	8.948	0.0606	0.0382	0.0542	-0.1699	-0.1248	-0.1580	0.0246	0.0156	0.1214	3

Conformation test

To verify the model the conformation experiments is analyzed. The predicted value of conformation experiment is obtained by regression equation. From the confirmation test the percentage of error with is experiments value and evaluated by regression equation with correspond processes parameter of lowest VIKOR index. The results are shown in Table 5.

Regression equation for VIKOR index = 0.4594 + 0.2652 % of reinforcement_2.5 0.0899 % of reinforcement_5.0 - 0.1754 % of reinforcement_7.5 + 0.1539 speed_795 - 0.0242 speed_125 0.1297 speed_1980+0.0668 dia drill_2.5 + 0.0290 dia drill_5.0 - 0.0958 dia drill_10.0

Table 5 Results of conformation experiment

Numerical Index	Lowest main effect level of VIKOR Index	Predicted Value	Experimental results	% of Error
VIKOR index	Spindle speed=1980 Drill Diameter=10 % of reinforcement=5	0.0732	0.0758	0.0026

Conclusions

In this study, the machinability of A356-TiB_2/TiC composites produced by K_2TiF_6-KBF_4-Graphite reaction was investigated and the following conclusions are drawn.

A356-TiB_2/TiC composites are successfully produced by K_2TiF_6-KBF_4-Graphite reaction with different reinforcement ratio and characterization report from scanning electronic microscopy and EDAX analysis confirms presence of TiB_2 and TiC reinforcements and their uniform distribution. From the analysis of results exhibits the % of reinforcement is most influential machining parameter followed by spindle speed and diameter of drill.

The Entropy-VIKOR index analysis of the experimental results of material removal rate, delamination and surface roughness can convert the multi characteristics performance optimization of the single performance characteristic is called composite VIKOR index.

From the ANOVA of VIKOR index reveals the moderate reinforcement (5%), higher spindle speed (1980 rpm) and larger diameter of drill (10 mm) is the optimum parameters for machinability of A356-TiB_2/TiC composites. Finally it has been concluded that the VIKOR index of predicted and experiment values are attained with less error value. As a result, the optimization of multiple performance characteristics can be greatly simplified.

References

[1] Seeman M , Ganesan G and Karthikeyan R 2010 *I.J.Adv.Manf.Tech.* **48**, 613-624
[2] Tham L M , Gupta M and Cheng L 2001 *Acta Materialia* **49**, 3243-3253
[3] Mahamani A, Karthik A, Karthikeyan P and Praveen Kumar 2010 *Fron.Auto.Mech.Engg (FAME)*,50-53
[4] Jahanmir S, Ramulu M and Koshy P 2000 *Marcel dekker .Inc*, Newyork
[5] Mahamani A 2014 *Proce.Mater.Sci*,**6** 950-960
[6] Nihant Tosun 2006 *I.J.Adv.Manf.Tech* **28**, 450-455

IOP Conf. Series: Materials Science and Engineering **390** (2018) 012100 doi:10.1088/1757-899X/390/1/012100

Mono and multi-response optimization of 3D printer parameters to attain improved hardness and surface roughness

A Mahamani[1], P Kumar[2], K Ismail[3], S Jawahar[4], T Chiranjeevi Reddy[5], T Vijayasai[6], T Venkata Phaneendra[7], V Uday sankar[8] and V Gopi chandu[9]

[2,3] Department of Mechanical Engineering
Jawaharlal Nehru Technological University, Anantapur-515002, AP, India.
[1,4,5,6,7,8,9]Department of Mechanical Engineering
Sri Venkateswara College of Engineering & Technology (Autonomous), Chittoor-517127, AP, India.

Abstract. 3D printing refers to process used to create a 3D object in which the material is joined or solidified with materials being added together. 3D printing Prototyping and tooling to direct part manufacturing in industrial sectors such as architectural, medical, dental, aerospace, automotive, furniture and jewellery are the potential applications of the additive manufacturing. This study describes certain aspects that must be controlled in and around an entry-level rapid prototyping 3-dimensional technology platform to investigate printing quality and optimization of the process. PLA (Poly Lactic Acid) material is to be used for making components. Printing speed, Layer thickness and Fill density are selected as 3D printing process parameters whereas component hardness and surface roughness are selected as responses. L9 layout is too followed for experimentation. Taguchi method is to be employed for the optimisation. Result of the study shows that, the longer thickness and printing speed are most influential parameters in developing surface roughness and component hardness respectively.

1. Introduction:

3D printing is a process of making physical object from a three dimensional digital model, typically by laying down many successive thin layers of material. It brings digital object into a physical form by a adding layer by layer of a material.3D printing is also known as additive manufacturing, refers to processes used to create a 3-dimensional object in which material is joined or solidified under computer control to create an object with materials being added together. Stereolithography, fused deposition

IOP Conf. Series: Materials Science and Engineering **390** (2018) 012100 doi:10.1088/1757-899X/390/1/012100

modeling is the most common types that are used for 3D printing. Taguchi methods are statistical method is also called robust design method developed by Genichi Taguchi to improve the quality of manufactured goods, and also applied to engineering sectors. Yang [1] attempted to optimize the lemon juice gel based 3D printed food material to attain better rheological and mechanical properties. Nozzle height, nozzle diameter, extrusion rate and nozzle movement speed are sleeted as 3D printing process parameter. Analysis of the 3D printed component showed that good resolution, unwrinkled surface texture, lesser point defects and compressed deformation free surfaces. Lavi and Toyserkani [2] proposed a hybrid approach that coupled with material jetting and extrusion processes to make a 3 D printed silicone component. They employed statistical optimization to improve the 3D printing resolution and improve the surface quality of 3D printed features. It is clear from the analysis speed of jetting print head and velocity of the extrusion print head plays a decisive role to improve the responses. Lanaro et al [3] developed a complex chocolate object by 3D printing technique. The fineness of the extruded chocolate is assessed by achieving the span large detachment without collapsing. It is seen from the results the speed of moving nozzle, rate of extrusion rates and rate of cooling offer noteworthy influence on fineness of the extruded chocolate. The results show that the addition of pure cork and cork waste can be processed with polymers such as HDPE, having adequate physical and mechanical properties. Brites et al.,[4] developed a cork waste reinforced HDPE by 3D printing technique. Physical and mechanical properties of the material is investigated under different volume fraction of cork waste. Analysis of various properties shows that 3D printing parameters and cork waste content yields considerable improvement in mechanical properties. Manti had et al [5] designed a 3D printed chocolate with various different support structures including cross support, parallel support and without support. Fineness of the chocolate is assessed by completeness of the dimensions, weight and physical properties of the component. Result shows that, the breaking strength of the 3D printed chocolate improved by providing support structure. 3D printed chocolate made with cross support has superior breaking strength than other support structures. For the further contribution, an effort has been made to optimize of printing speed, layer thickness and fill density to accomplish the improved component hardness and surface roughness of gear component which is made by PLA.

2. Experimental work:

A spur gear with 65mm outer diameter, 52mm inner diameter and 6mm thickness was designed by using Solid works software. Designed model converted into dot STL file and then it was converted to G code. Developed G code was imported to into a computer controlled 3D Printer. During the process the PLA material heated up to 185^{0}C and the molten PLA forms the object. L_9 experimental layout is generated by Minitab 18 software. Printing speed, layer thickness and fill density is preferred as process parameter whereas hardness and surface roughness of the 3D printed component is considered as responses. The specification of the 3D printer used for the experimental work is exhibited in the Table 1. Surface roughness and hardness of the 3D printed component for different levels of the *3D* printer parameter is exposed in Table 2. The photographic view of 3D printed component is illustrated in Figure 1.

Table 1 Specification of 3D printer*:*

S No	Parameters	Properties
1	Build Volume	Up to 25X21X21 cm
2	Layer Height	0.05-0.35 mm
3	Printing surface	Heated Printed Bed
4	No of Extruders	02
5	Supported materials	Nylon and Polycarbonate

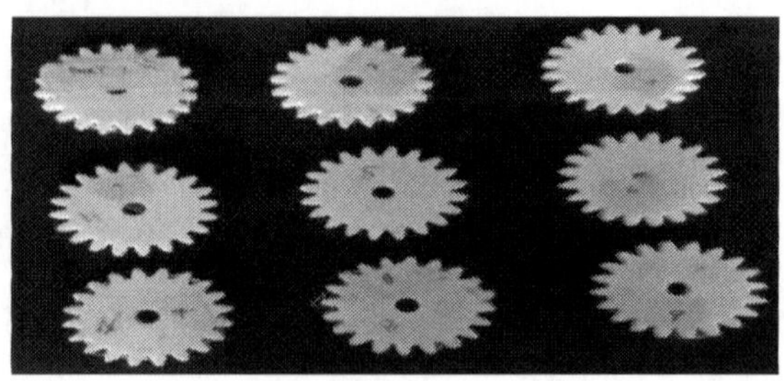

IOP Conf. Series: Materials Science and Engineering **390** (2018) 012100 doi:10.1088/1757-899X/390/1/012100

Figure 1. 3D printed component

Table 2 L9 experimental layout

Component Number	Speed (mm/sec)	Layer thickness (mm) (%)	Fill Density	Surface roughness	Hardness
1	30	0.1	20	2.04	37.64
2	30	0.2	50	17.19	46.94
3	30	0.3	100	6.99	34.65
4	40	0.1	50	3.14	46.96
5	40	0.2	100	4.96	44.36
6	40	0.3	20	21.62	34.16
7	50	0.1	100	4.93	37.16
8	50	0.2	20	26.30	30.81

3. Analysis of results

3.1 Mono response optimization

Response graph for surface roughness and hardness are generated by Minitab software 18 and presented in Figure 2 and 3. From the Figure 2, it is clearly understood that the printing speed 30mm/sec to 40mm/sec the surface roughness increases, again the printing speed is 40mm/sec to 50mm/sec the surface roughness is drastically increased. The layer thickness 0.1mm to 0.2mm the surface roughness is drastically increased, again increase the layer thickness 0.2mm to 0.3mm the surface roughness is increased. If the fill density increase 20% to 50% the surface roughness decreases, again increase the fill density 50% to 100% the surface roughness drastically decreases. Among the above results, the minimum surface roughness occurs in 30mm/sec of printing speed, 0.1mm of layer thickness 20% of fill density.

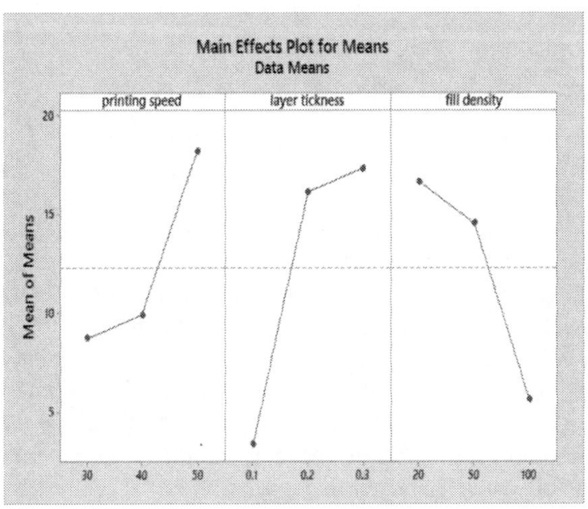

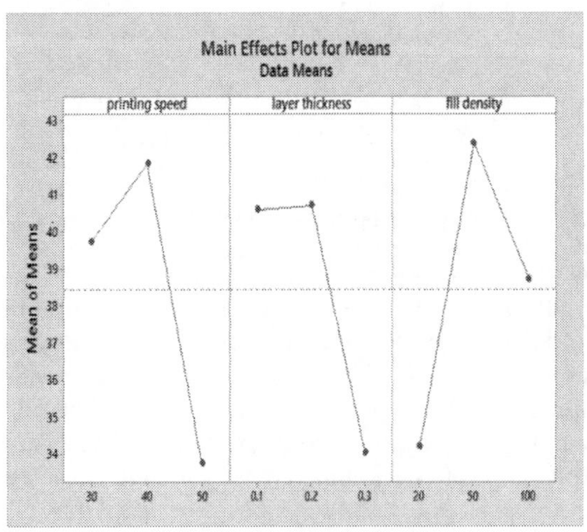

Figure 2 Response graph for surface roughness *Figure 3 Response graph for hardness*

From the Figure 3, it is clearly seen that by increasing the printing speed 30mm/sec to 40mm/sec the hardness value is increased, again increasing printing speed from 40mm/sec to 50mm/sec the hardness value is drastically decreased. If increase the layer thickness from 0.1mm to 0.2mm the hardness value is slight increases, again the increasing the layer thickness 0.2mm to 0.3mm the hardness value is drastically decreased. If increase the fill density is increased to 20% to 50% the hardness value is drastically increased if again increase the fill density from 50% to 100% the hardness value is decreased. Among the above result, the maximum hardness occurs 40mm/sec of printing speed, 0.1 mm layer thickness 50% of fill density.

Analysis of the variance for the surface roughness and hardness is performed and presented in Table 3 and 4. The Table 3 clearly brings out that layer thickness has the strongest influence on the surface roughness than the fill density and printing speed. The percentage of contribution of the layer thickness was recorded as 47.62.Further, the percentage contribution of the fill density and printing speed are recorded as 27.27 and 21.14 respectively. Table 4 illustrates that printing speed is a most influential parameter on the hardness of the component and the percentage of contribution was estimated about 35.18. The percentage of contribution of fill density and layer thickness was estimated as 33.73 and 29.18 correspondingly.

Table 3 ANOVA for surface roughness:

S No	Source	DF	Adj SS	Adj MS	% of Contribution
1	Printing Speed	2	159.63	79.82	21.14
2	Layer thickness	2	359.5	179.75	47.62
3	Fill density	2	205.87	102.93	27.27
4	Error	2	29.88	14.94	3.95
5	Total	8	754.88		100

Table 4 ANOVA for hardness:

S No	Source	DF	Adj SS	Adj MS	% of Contribution
1	Printing Speed	2	105.399	52.7	35.18143
2	Layer thickness	2	87.449	43.724	29.18985
3	Fill density	2	101.054	50.527	33.7311
4	Error	2	5.685	2.842	1.897612
5	Total	8	299.587		100

3.2 Multi-response optimization

Desirability function analysis is deployed to renovate the multi-response distinctiveness into single-response distinctiveness. Therefore, optimization of the intricate natured multi-response characteristics can be transformed into optimization of a solo response characteristic called as composite desirability. The single response optimization is aimed minimize the surface roughness and maximize the component hardness. The current work deals the estimation of composite desirability for surface roughness and hardness. Step by step procedure followed to compute the composite desirability is discussed by Sait et al [6]. Composite desirability value for each trail is estimated and tabulated in Table 5. Response graph for the composite desirability values is generated by Minitab software illustrated in Figure 4.

IOP Conf. Series: Materials Science and Engineering **390** (2018) 012100 doi:10.1088/1757-899X/390/1/012100

Table 5 Composite desirability for the L9 layout

S No	Printing Speed (mm/sec)	Layer thickness (mm)	Fill Density (%)	Surface Roughness (μm)	Hardness	Individual desirability for surface roughness	Individual desirability for hardness	Composite desirability	Rank
1	30	0.1	20	2.04	37.64	1	0.42	0.650	3
2	30	0.2	50	17.19	46.94	0.37	0.99	0.6126	4
3	30	0.3	100	6.99	34.65	0.79	0.23	0.435	6
4	40	0.1	50	3.14	46.96	0.953	1	0.977	1
5	40	0.2	100	4.96	44.36	0.87	0.83	0.859	2
6	40	0.3	20	21.62	34.16	0.19	0.02	0.064	8
7	50	0.1	100	4.93	37.16	0.8866	0.394	0.588	5
8	50	0.2	20	26.30	30.81	0	0	0	9
9	50	0.3	50	23.37	33.29	0.12	0.15	0.136	7

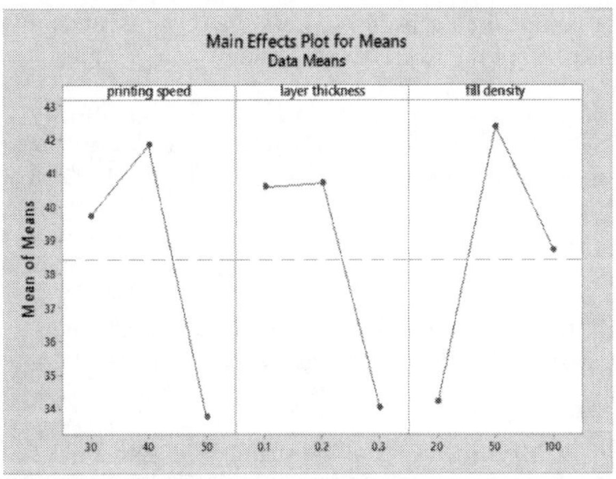

Figure 4 Response graph for composite desirability

From the Figure 4, it is clearly implicit that the increase in printing speed from 30mm/sec to 40mm/sec the desirability is increased, the further increase in printing speed from 40mm/sec to 50mm/sec the desirability drastically decreases. By increasing the layer thickness from 0.1mm to 0.2 mm the desirability value decreases, whereas the increase in layer thickness from 0.2mm to 0.3mm the desirability drastically decreases. By increasing the fill density from 20% to 50% the desirability radically increases while increasing the fill density from 50% to 100% the desirability increases. Among the 9 components, the component produced by the 4th trail has the highest value of composite desirability. Hence, the optimum parameters to attain the improved surface roughness and hardness are at 40mm/sec printing speed, 0.1mm layer thickness, and 50 % fill density. Analysis of variance also performed for the composite desirability value and tabulated in Table 6. Layer thickness has the strongest influence on the composite desirability than other parameters. The percentage contribution for the Layer thickness and fill density and printing speed is estimated as 42.84, 27.53 and 27.1 respectively.

Table 6 ANOVA for composite desirability

S No	Source	DF	Adj SS	Adj MS	% of Contribution
1	Printing Speed	2	0.26348	0.13174	27.10086
2	Layer thickness	2	0.41654	0.20827	42.84421
3	Fill density	2	0.26773	0.13386	27.53698
4	Error	2	0.02449	0.01224	2.517949
5	Total	8	0.97223		100

4 Conclusion

Optimization of printing speed, layer thickness and fill density to achieve the improved component hardness and surface roughness of gear component was carried out and the following conclusions are drawn. The minimum surface roughness of the component was achieved at 30mm/sec printing speed, 0.1mm of layer thickness and 20% of fill density. The maximum hardness was attained 40mm/sec printing speed, 0.1mm of layer thickness and 50% of fill density. Maximum composite desirability accomplished at 40mm/sec printing speed, 0.1mm layer thickness and 50% of fill density. Layer thickness is significant parameters for surface roughness of the component; fill density is the most influential parameter for hardness and layer thickness also most considerable for composite desirability. Less than 5 % error in all analysis variance estimation reflects the statistical significance of the experimental value.

Acknowledgment
The Authors acknowledge the Science and Engineering Research Board, DST, New Delhi for the provision of Microhardness tester and surface roughness tester under Young Scientist Scheme (Ref. No. SB/FTP/ETA-0190/2014).

References

[1] Yang F, Zhang M, Bhandari B and Liu Y 2017 Investigation on lemon juice gel as food material for 3D printing and optimization of printing parameters *LWT Food Sci and Techno* doi: 10.1016/j, Lwt.2017.08.054.

[2] Lavi F and Toyserkani E 2018 A hybrid additive manufacturing method for the fabrication of silicone bio-structures: 3D printing optimization and surface characterization *Mater & Des* ***138**, 46-61*

[3] Lanaro M, David P, Forrestal, Scheurer S, Damien J, Liao S, Sean K, Powell, Maria and Woodruff 2017 3D printing complex chocolate objects: Platform design, optimization, and evaluation *J food Eng.* doi:10.1016/j.jfoodeng.2017.06.029.

[4] Brites F, Malca C, Gaspar F, Horta J F, Franco MC, Biscaia S and Mateus A 2017 Cork plastic composite optimization for 3D printing applications In *conf sustain and intellig manuf* doi: 10.1016/j.promfg.2017.08.020.

[5] Mantihal S, Sanjeev Prakash, F, Godoi C and Bhandari B 2017 Optimization of chocolate 3D printing by correlating thermal and flow properties with 3D structure modeling *Inn food sci emerg techno* Doi: 10.1016/j.ifset.2017.09.012.

[6] Naveen Sait, A., Aravindan, S. and Noorul Haq, A. 2009 Optimisation of machining parameters

IOP Conf. Series: Materials Science and Engineering **390** (2018) 012100 doi:10.1088/1757-899X/390/1/012100

of glass-fiber-reinforced plastic (GFRP) pipes by desirability function analysis using Taguchi technique Int J Adv Manuf Technol **43** 581-589.

IOP Conf. Series: Materials Science and Engineering **390** (2018) 012101 doi:10.1088/1757-899X/390/1/012101

A comparative study on development of small and medium enterprises (SMEs) in kanchipuram district

V. Venkatasubramanian[1*], Dr. K.P.V. Ramanakumar[a],
[1*]Department of Mechanical Engineering, Sri Chandrasekharendra Saraswathi Viswa Mahavidyalaya, Enathur, Kanchipuram – 631561, Tamilnadu, India.

a[*]Department of Management Studies, Sri Chandrasekharendra Saraswathi Viswa Mahavidyalaya, Enathur, Kanchipuram – 631561, Tamilnadu, India.

[*1]Corresponding Author: venmoon@gmail.com

Abstract. This research paper explains the role of small- and medium-sized enterprises in the financial scheme of diverse countries, through a picky focal point on Kanchipurm District. The function of the research is to maintain and build up the growth of small- and medium-sized businesses in the country by formative the efficiency of programs held by the government to support better SMEs activity, as well as presenting successful examples from international practices in order to adopt them to Kanchipurm District conditions. In the present period, the development of the entrepreneurship culture and attractive the scope of small- and medium-sized business are significant for the outlook economic growth, employment and income of the country's population.

1. Introduction

Developed countries' experiences show that, in the last years, the role of SMEs in the economy has strengthened and the majority of enterprises have been ranked in the SMEs category – more than 95% (Ayyagari et al, 2011). Especially after the recent economic crisis beginning in 2008, experience has shown that SMEs have been able to respond to the challenges of the crisis more quickly and flexibly than large corporations and, indeed, they succeeded to survive and thrive. With a view to this observation, many developed countries have devised various kinds of non-financial and monetary incentive mechanisms aimed at enhancing the weight of SMEs in different sectors. The research is structured as follows. In the 1th chapter, literature review, definition of small and medium entrepreneurship have been explained and readers could get information about different kind of models related SMEs and their realization mechanism. In the following chapter, the level of development of small and medium sized entrepreneurship in different regions and countries such as United State, European Union, Asia region, Georgia, Russia, Turkey, and its role in economic development are investigated. In the next chapter, the realistic situation of SMEs in the economy of Azerbaijan, SWOT analysis of SME, the possible constrains for the SMEs development and growth as well as the role of the government policies for improving SMEs have been analyzed. In conclusion, in order to development of the small and medium-sized entrepreneurs several suggestions have been offered.

2. Literature Review

In a world of globalization, transitioning and developing countries and their enterprises face major challenges in trying to take advantage of trade and investment opportunities. While governments make policies in trade and investment areas, it is enterprises that trade and invest and so, their role in the economy is pivotal. SMEs are significant players in transitioning and developing countries. These firms typically account for more than 90% of all firms outside the agricultural sector, constitute a major source of employment and generate substantial domestic and export earnings (OECD, 2009). As such, SME development emerges as a key instrument in further poverty reduction efforts.

2.1. Small and Medium Enterprise in India As we all know that the industrial sector of any country particularly small sector are the back bone of any developing country but India showed little concern for industrialization especially for small scale industry until 1970s. But after that period, government policy of industrialization were promoted through the restricted foreign trade regime which discriminated the large firm from the small firm and the large scale sector obtain better place from the small scale sector. Since large firms were better placed to obtain import permits for capital equipment, components, and raw materials and were also better able to obtain tariff rebates intended to alleviate some of the harmful effects of high protection. Since the inception of the industry policy resolution the role of cottage and small scale industries were stressed for balanced growth and individual development i.e., for utilization of local resources and providing employment opportunity. So there were lots many efforts put up by the government by providing tax rebates and subsidies, and fix some items in the reserve category in which there exclusive production were made from the small scale sector. But after 1977 Industrial resolution the policy of government had emphasized on development of SSI in rural area and small town through the promotion of khadi and village industries and handloom sector. These industrial initiatives not only increased the items in reservation category for production in small scale sector, and but also established district industries center (DIC) to serve as focal point for development of small and cottage industries, emphasized on product standardization, quality control, market survey etc. These changes had shifted the emphasis of industrial resolution towards raising investment limits in these small scale industries. But after the 1991 industrial policy resolution the process of globalization had enhanced the relevance of Special economic zones (SEZs) which have become an important contact in the export led industrialization strategy playing a crucial role in promoting the manufacturing sector including providing an enabling international climate for small and medium enterprises (SMEs) and offer platform for attracting export oriented. Therefore there is the need to improve the operation of these firms through improving their supply chain and promoting them globally

SMEs are perceived as vital to socio-economic growth, including Gross Domestic Product (GDP) growth, and sustainable development in emerging economies for many reasons. These include such aspects as their familiarity with local communities, their perceived capacity to create employment for low-skilled workers due to their labour-intensive nature and their "innovation and sustainable initiatives due to inherent flexibility and risk-taking ability". [1] In South Africa, "at least 90% of the manufacturing sector is small and medium, in terms of enterprise size...[These enterprises] contribute about half of the employment; half of the gross domestic product and 1 out of 5 units exported is produced in the small and medium sector in South Africa"[2]. Hence, South African Manufacturing SMEs perform a vital role in the economy and form critical links in large Supply Chains. Finch [3] posed the question of whether large companies increase their exposure to risk by incorporating small to medium enterprises (SMEs) in "business critical positions" in their supply chains. He concluded that "inter-organizational networking" did elevate risk

IOP Conf. Series: Materials Science and Engineering **390** (2018) 012101 doi:10.1088/1757-899X/390/1/012101

exposure and this was exacerbated by the presence of SMEs as supply chain partners. This is because risk events impact SMEs more severely due to their "size and limited resources" (human, structural and information). This consequently highlights the importance of understanding risk management and business continuity planning across the whole supply chain. The importance of SMEs in the supply chains of large companies, including Multi-national Corporations, cannot, however, be under-estimated [4]. Benefits, such as, productivity improvement, greater innovative capacity, increased access to talent and local markets can be derived from partnering with SMEs. This is because smaller businesses have lower overheads, specialized knowledge, innovative products and adaptability to changing conditions [4], [1]. Additionally, SME's that manage their own risks and assist larger supply chain partners in mitigating their risks are sought after in supply networks [4], thus, increasing the competitiveness and sustainability of SMEs.

2.2. Kanchipuram Kanchipuram district is one of the industrially fastest growing districts in the state of Tamilnadu. The electronic and software industry is one of the fastest growing sectors in the district. Special industrial parks for software industries assisted the growth of this sector. Software industry in the district grows at over 40- 50%. Auto components and automobile industry grows at 15-20%. The recent development in this sector has created a strong base for engineering industry. Major car manufacturers located in the district like Ford, Hyundai, Nissan and BMW have spurred the development of auto component and engineering sectors, which have seen a growth trend of over 15 %. Garment industry has faced a slight set back but at present picket up growth momentum. The sound agro base in the district has helped in the development of rice milling and Oil extraction units. Its bovine population helped in the growth of dairy and dairy products. In fact Kanchipuram district has attained self-sufficiency in milk availability and started supplying to nearby districts also. The growth of traditional industries has also assisted in export of traditional items like silk sarees/products. Leather & Leather products grow at 10%.

Table 1. Potential for new Manufacturing MSMEs and Block-wise Potential for MSMEs in Kanchipuram

s.no	Name of the Block	Potential Activities
1	Kancheepuram	Engineering Auto Components Rice Milling, Silk Weaving, Garments
2	Walajabad	Engineering, Rice Milling, Oil Crushing, Readymade Garments
3	Sriperumbudur	Automobile Ancillaries, Plastic items Electronics, Engineering Industries
4	Kundrathur	Bio tech, Electronics Electronic Industries Software
5	St. Thomas mount	Computer Software Engineering Components Multiple products, Electronics, paints
6	Kattankulathur	Auto Ancillaries, Engineering, Wood Works, Electrical Electronic Works
7	Thiruporur	Pharmaceuticals, Rice Milling, Oil Crushing Software, Engineering
8	Thirukalukundram	Rice Milling, Food Products Fish net, Engineering Agricultural based products
9	Uthiramerur	Agricultural based products garments, Engineering
10	Acharapakkam	Weaving, Rice Milling, PVC Plastic Pipe, Engineering
11	Lathur	Cashewnut, Fish Food, Rice Milling, Oil Crushing, Bricks
12	Chithamur	Fish Net, Cashew Nut, Rice Milling Bricks

Table 2. Year Wise Trend of Units Registered in Kanchipuram District

Year	Micro	Small	Medium	Total
2007-2008	1377	213	6	1,596
2008-2009	1832	339	11	2,182
2009-2010	2264	632	63	2,959
2010-2011	3693	712	52	4,457
2011-2012	3956	1065	254	5,275
2012-2013	5656	2503	610	8,769
2013-2014	7073	4824	848	12,745
2014-2015	9117	6483	909	16,509

Table 3. Details of Udyog Aadhaar Memorandum (UAM) Filed/Registered

Particular	Micro	Small	Medium	Total
Total Units registered under UAM	6882	1036	22	7940
Manufacturing Units	2114	400	18	2532
Service units	4766	636	4	5406

Table 4. Details of Existing Micro & Small Enterprises in Kanchipuram District

Type of industries	Numbers	Investment in Lakhs	Employment
Agro based	931	5586	3736
Soda water	164	820	492
Cotton textile	1500	8624	5920
Woollen, silk & artificial Thread based clothes	1043	4172	5020
Jute and jute based	876	2678	1232
Readymade garments	4502	20646	27012
Wood / wooden based furniture	1008	3062	3624
Paper and paper products	785	3925	3630
Leather based	914	2940	3125
Chemical / chemical based	700	2316	2012
Rubber, plastics and petrol based	674	1930	1876
Mineral based	1419	7095	3860
Metal based	1450	7452	3928
Engineering based	1879	5217	4212
Electrical machinery and transport equipment	522	1502	1420
Repairing and services	1690	1020	3978
Others	340	1020	1216

3. Research Design and Methodology

3.1 Scope of the Study The present study has collected the Comparative Study on Small Medium Large Scale Strategies. This study opens beautiful vistas over on Small Medium Large scale Strategies preference and awareness of various major objectives. This also paves the way to SMEs region attract the outsider state investors and have continued need for foreign as well as government assistance for improvement of their industrial base approval in an intensified manner. It also focuses exact problems related with scale manufacturers and retailers supply Chain Strategies in Kanchipuram Distract

3.2 Nature of the Study The researcher was adopted by descriptive and nature of the study. In this study to examine various direct as well as indirect factors that lead to comparative and Competitive Study on Small Medium Large Strategies. This study is restricted only to dyadic chain i.e., Manufacturers and retailers in business to business orientation, there by exclusive of small and medium enterprises. The considered work will endow with useful insights to small and medium enterprises of Kanchipuram District, policy makers and promotion researchers, as it will help top management in framing such policies that can increase Manufacturers and Retailers Supply Chain Strategies in business to business orientation

3.3 Pilot Study A beginning inquiry is undertaken by contacting 75 small medium large scale to identify the important variables about characteristic features of strategy, problem of Manufacturers, factors of comparative, instrument and the changes in return of manufactures. The purpose of the pilot study is to test the quality of the items in the questionnaire and to confirm the feasibility of the study. This preliminary investigation is conducted in different parts of small and medium enterprises of Kanchipuram District. The study used principal component analysis; alpha method and T-square test were applied. It is found that the alpha value is 0.912 and T-square value is 422.31 which are statistically significant at 5 per cent level. It is ascertained that the items in Liker's five point scale of the questionnaire are highly dependable and the samples satisfy the ordinary distribution reasonably. So, the items in the questionnaire can be used further in this study.

3.4 Research Gap The small scale industrial sector which plays an important role in the Indian economy in terms of employment and growth has recorded a high rate of growth since independence inspire of stiff competition from the large sector and so encouraging support. Moreover the fourth census (2015-2016) of MSME the most of industries registered were of manufacturing, assembling, processing, repairing and maintenance in nature. The census of MSME also shows that 90.08% of enterprises registered were proprietary enterprises, which means they may carry out the concept of supply chain management directly or indirectly in their firms. Hence this sector has been acknowledged as a growing as well as most important one for the local, regional and national economy of the country. The focus of the study therefore is to find out comparative strategies the customer service measures including service strategy, service interface, service quality and overall size of the firm in supply chain management of SMEs in business to business orientation. Moreover supply chain management in small and medium enterprise helps to improve product quality, timely delivery, and integration with logistics functions to deliver improved customer service.

3.5 the Objectives of the Study

1. To study comparative study on small medium large scale strategies
2. To examine the role of service quality in attractive customer service in supply chain management

3.6 Formulation of Hypotheses

1. Ho =Association between comparative study on small medium large scale strategies
2. Ho =Association between the impact of overall size of firms and the customer service strategy

IOP Conf. Series: Materials Science and Engineering **390** (2018) 012101 doi:10.1088/1757-899X/390/1/012101

4. Field Work and Collection of Data

This study was conducted by district industrial center in Kanchipuram distract and other SMEs and government employees the study identified various items to measure the supply customer service in supply chain management that includes the service quality, service interface, service strategy in terms of retailer and customer orientation for business to business chain. The researcher used structured interview schedule for the purpose of personal interviews. Interviews are conducted in different manufacturers, retailers and customers in their convenience. Before the interview, proper understanding is conventional. The data collected are recorded by the researcher in the interview schedule. The schedules thus filled up are systematically checked to ensure accuracy, reliability and totality. On an average, each interview took about an hour. The data thus collected were categorized and posted in the master table for further processing. The generation of items for customer service, its influence on customer and supplier are discussed as under: Basically, the study was based on the customer service construct in the supply chain management of the SMEs, so single instrument was constructed with business to business orientation in supply chain management. In the supply chain management of the small firm which has only single owner/manager who carried out the function of the customer for their different supplier and supplier for their customer in business to business orientation had been consulted for the response through convenient sampling technique.

4.1 Collection of Data The data collected by the researcher primary as well as secondary sources relevant for gathering necessary information pertaining to the research problem in hand have been used in the study. Primary data based on the first hand information has been collected from the bank customers through self personalized and well structured schedule. Secondary data has been generated from books, journals, annual reports, magazine, news paper and Internet has also been used to substantiate primary information.

4.2 Sample Size

4.2.1 Data Collection Forms The pretesting of the instrument designed was done before going for the final construct. Since the pretesting was done on only 50 firms of small and medium enterprises on the basis of the convenient sampling technique. The single construct was used in the study which consisted of two orientations one in terms of suppliers and other in terms of customers in a single questionnaire with business to business orientation. The total number of the items included in the pretesting instrument consists of 150 items in which the only 67 items were retained with reliable value of Cronbach Alpha and KMO (Kaiser-Meyer-Olkin) measure of sample adequacy. The overall value of these measure lies above the 0.5 for the cronbach alpha and KMO which is acceptable in nature (Malhotra, 2000). The final outcome of this instrument would be laid down for the study in order to achieve the objectives of this study. The respondents chosen from an age group ranging from more than 20 years to 65 years in Kanchipuram district including different strata of like Small medium large scale strategies. Initially 720 questionnaires were distributed to the Manufacturers and Retailers spread over in Kanchipuram district. But only 615 respondents returned the filled up questionnaires out of 720. In which 600 of them are found usable. Hence, the accurate sample of the study is 600.

5. Framework Analysis

The sources of data are primary as well as secondary. The data collected from small medium large scale manufacturers and retailer's survey constitutes primary and information gathered through books, journals, magazines, reports, dairies are considered as the secondary source. The data collected from both the sources is scrutinized, edited and tabulated. The data is analyzed using statistical package for social sciences (SPSS). The following statistical tools are used in the study univariate, bivariate and multivariate, Kruskal Wallis test, One-way analysis of variance, Factor analysis, value of Cronbach Alpha and KMO (Kaiser-Meyer-Olkin) Multiple discriminate analyses, multiple regression analysis, Non-parametric chi-square analysis, and percentage analysis have been employed. The reliability and validity of the final instrument were carried out after the collection of the final data in the small and medium sized enterprises. The final questionnaire consists of 12 blocks for the manufacturers and retailers service make in terms of supply chain strategies of the firm. These includes manufacturing units constitutes for supply chain management segmentation, 5406 items for service strategy for the service quality were lastly dropped in the final run of the model through confirmatory factor analysis (CFA) and lastly the customer service construct consist of overall size of the firm with 5 items. The normality of these items was tested through the Kolmogorov-Smirnov (K-S) test for all dependent and independent variables were conducted.

6. Limitations of the Study

The study is limited to Kanchipuram district only. Therefore the conclusion may not be comprehensive for the other parts of the district. The study is appropriate to the comparative and competitive study on small medium large scale manufacturers and retailers supply chain strategies. Public limited companies remain excluded. The limitations associated with the statistical tools are related for the tools in work in this study. Period of the study commenced from 2015 to 2016.

7. CONCLUSION

The study is appropriate to the comparative and competitive study on small medium large scale manufacturers and retailers supply chain strategies is an essential condition in logistics activities and is affected by various environmental factors shaping today's marketplace. Logistics manufacturers and retailers supply chain service has its roots in the logistics discipline, and logisticians can use and learn from marketing techniques and methodologies to investigate customer service. A strategy for logistics customer service requires a basic trade-off between costs incurred and enhanced profit received. Each industrial sector will also have its own unique needs and issues that further complicate such considerations. However, while the importance of individual customer service elements varies among firms, there is a common set of elements offered higher than that should provide a useful starting point for most firms. A global point of view focuses on seeking common market demands universal rather than cutting up world markets and treating them as separate entities with very diverse product needs.

Reference

[1] Beyene, A., Africa Development Vol. XXVII, No.3 & 4. 2002, 27(3) : p.130-156
[2] Blinov, A.O. and Shapkin I.N., 1993.
[3] Caloghirou, Y., Aimilia, P., Yiannis, S., and Lefteris, P., European Management Journal. 2004 22(2) p.231-243
[4] Castrogiovanni, G. J. Journal of Management. 1996 22(6), p. 801–823.
[5] Clark III, M. and Moutray, C., journal of public procurement, volume 4, issue 3, 2004, 450-470
[6] Duarte, N.,. ersa04p39, European Regional Science Association.. 2004
[7] EBRD (2014), BEEPS V (2011-2014),
[8] Commission of European Communities , 1999.Sixth periodic report on the social and economic situation and development of regions in the European Union. Brussels: Commission of European Communities.

IOP Conf. Series: Materials Science and Engineering **390** (2018) 012102 doi:10.1088/1757-899X/390/1/012102

Solar powered Hybrid vehicle

K. Mohan[1*], S.Sankaranarayanan[2], Shyam Sundar Devi Prasad[3], V.Sivasubramaniam[4] and V. Sairam[5]

[1-5]Department of Mechanical Engineering,
Sri Chandrasekharendra Saraswathi Viswa Mahavidyalaya,
Kanchipuram, Tamilnadu - 631561, India.

Email: mohank84@gmail.com

Abstract.In the current scenario, global warming is a threat to the society. One of the major reasons is the release of carbon-di-oxide from an automobile exhaust due to the combustion of fossil fuels which pollutes the environment. One of the optimistic solutions for this problem is to use of hybrid vehicles. Generally, Hybrid vehicle involves a combination of transmission system driven through electrical, solar as well as internal combustion (IC) engine. This work involves hybridization with solar power and conventional power IC engine. Hence it is called a Hybrid Solar Vehicle (HSV). It can be driven both on internal combustion engines as well as on solar energy assisted with electrical motor. In real life applications using solar vehicle produces zero emissions. At present, hybrid electric vehicles are being developed and launched into the market. For long distance travelling its necessary of periodic charging of their batteries, so these vehicles are depends the electrical sources also leads to increase the cost of electricity. These kinds of problems will be solved by using hybrid solar vehicle also HSV supporting to the green environment.

1. Introduction

Hybrid-electric vehicle use IC engines and electric motors, to reduce pollution and achieve better fuel economy. G.J. Offer Et.al [1] investigated performance of battery electric vehicles (BEV) hydrogen fuel cell electric vehicles (FCEV) hydrogen fuel cell plug-in hybrid vehicles (FCHEV) and reported that BEV and FCHEV are relatively insensitive to electricity costs but the FCHEV and FCV are sensitive to hydrogen cost. Saeid Bashash Et.al [2] studied the problem of optimizing the charge pattern of a plug-in hybrid electric vehicle (PHEV) and reported optimizes PHEV charge patterns for both battery longevity and energy cost, for the first time.

Andrew Burke Et.al [3] concluded the power capability of ultra-capacitors and lithium batteries for electric and hybrid vehicle applications and reported for plug-in hybrid and battery electric vehicle applications, the maximum useable power density for a lithium-ion battery can be higher than that corresponding to 95% efficiency. Matthieu Dubarry Et.al [4] investigated using in situ electrochemical inference approach on commercial lithium-ion cells based on composite positive electrode for plug-in hybrid electric vehicle applications and reported loss of lithium inventory (LLI) may occur as a result of parasitic reactions to form SEI layer on the electrode surfaces.

Jenn-Jiang Hwang Et.al [5] studied on the power management system in a fuel cell hybrid vehicle the dynamic performance and power distribution for a fuel cell hybrid electric vehicle (HEV). and reported the power provided by the Li-ion battery is more than that of the fuel cell. S. Kermani Et.al [6] concluded Predictive energy management for hybrid vehicle the powertrain operating point in order to minimize the fuel consumption, and reported guarantying bounds on the state of charge error while providing satisfactory fuel consumption. Bor Yann Liaw Et.al [7] investigated driving cycle to understand the battery performance in real-life electric hybrid vehicle and reported comprehensive driving cycle analysis could be achieved with unique fuzzy logic Pattern recognition (FL-PR) technique and use this FL-PR technique to analyse duty Cycles and identify useful stress factors to define vehicle usage profile.

Georgios Fontaras Et.al [8] studied fuel economy and pollutant emissions measurements of two hybrid electric production vehicles namely Prius II and a Honda Civic and reported that both vehicles Present improved energy efficiency and pollutant emissions compared to conventional cars. The fuel economy benefit of the two HEVs peaked under urban driving conditions where reductions of 60% and 40% were observed, respectively. Karan C. Prajapati Et al [9] investigated about the type of hybrid vehicles and have given us information about the different types of hybrids in their paper and concluded that there are six types of hybrids which include the plug ins and self-charging types. Robert F Nelson Et.al [10] studied the operation of batteries in hybrid electric vehicles. His work reviews the specifications and operational requirements imposed on batteries due to the projected architectures for HEVs.
Anuragh.M.Lullhe Et al [11] investigated about the drives used in electric and hybrid electrical vehicles. He stated that the two main types of drives are AC and DC, of which the most widely used, are 3phase Induction motors, Switched reluctance motors, and Brushless DC motors. Teresa Donateo Et.al [12] in her paper of intelligent usage of IC engines in hybrid vehicles stated that focuses on different approaches presented in literature on the usage of information about traffic and weather conditions for the optimal energy management of hybrid electric vehicles. Ivan Arsie Et al [13] have given us information on the possibilities hybridizing the currently used Fuel engines with renewable sources of energy. This paper also focusses on the main problems related to the development of these vehicles, with specific attention on photovoltaic panels. AsllanHajderi Et al [14] have investigated about construction of hybrid vehicles and efficiency of their use in transport, to reduce environmental pollution in urban areas, compared to vehicles with engine. Jinmingliu Et al [15] investigated about the Toyota Hybrid System (THS) is used in Toyota Prius and compared the two major types of hybrids with the split power hybrids.

2. Materials and methods
The detail specifications of the accessories like engine, motor, solar panels, and batteries which are used in this work are discussed below.

2.1 Internal combustion Engine
A four-stroke 150cc engine was assembled in the chassis which produces power of 13.8Bhpand torque of 13.6NM.

2.2 Motor
Motor and its Controller are the two most critical components for HSV. As the selected motor must be able to produce enough torque and power to pull the load. Three types of motors: AC motor, DC motor and Brushless DC motor were taken into consideration. According to calculations and references we concluded by choosing 1.3hp, 48V BLDC motor.

2.3 Motor controller
A motor controller is a component which is used to start and stop the motor also which is used control and varying the speed of the motor drive. The direction of motor rotation also possible to change using this controller during vehicle reverses movement/drive.

2.4 Braking system
Braking system is a key feature in any vehicle. Two types of braking system namely disc brake and drum brake. In this work disc brake is mounted on the chassis for considering the safety precautions.

2.5 Battery
Three types of battery namely lead acid battery, Nickel metal hydride battery and Li-ion battery were examined. Li-ion battery is widely used in the application of hybrid vehicle. In this work Li-ion batteries each of 34 Ah are preferred to store the energy and which is used to drive the motor. Li-ion batteries of 20Amps are used as they will last for about 800 full charge cycles before any replacement required.

2.6 Solar panel
According to the design calculation and torque requirements, 200 watts solar panel has been selected in this work. Four panels of 50 watts capacity each were used. All the materials were effectively utilised to avoid energy losses.

2.7 Solar charge controller
Solar charge controller controls the voltage received from the panel to the battery. It is generally used to convert non-linear energy form into linear energy form. It is used to prevent the short circuiting of the components.

3. Results and Discussion

3.1 Basics of Hybridization
A hybrid vehicle has multiple distinct energy sources which could be separately or simultaneously operated to drive the vehicle. In the solar vehicles, the drive obtained from electric motor through the power stored from the batteries by solar energy.

3.2 Hybrid Solar Vehicle (HSV)
The most commonly adapted hybrid vehicle has sources of a solar power assisted with electric motor and an I.C. engine. In a HSV, the I.C. engine cooperates with an electric motor which has optimal usage of the engine. Especially driving in city traffic involves frequent starts and stops of the vehicle and during idling, the engine consumes more fuel without producing useful work thus it leads to higher fuel consumption, less efficiency and unnecessary emission from the exhaust. The HSV solves those problem and also no exhaust emission from the vehicle. The parallel power train runs on both, combustion engine and electric motor. The primary drive of the vehicle from solar powered and secondary drive by the combustion. These vehicles gain maximum efficiency in city during heavy traffic conditions without producing any emissions which supports to the green environment.

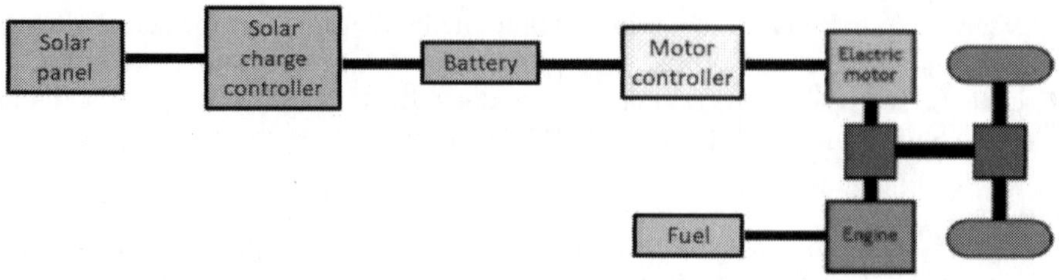

Figure.1. Schematic layout of HSV

3.3 Charging of Batteries

The power supply to the electric motor comes from batteries .When driving a vehicle on electric motor, the stored energy in the batteries gets used up quickly. Batteries of a HSV are charged by solar power. The solar energy is an efficient way to save more energy. Charging the battery using solar power will be a challenge but can turn out to be a good scope for further development as solar power is freely available.

There are some of the important results inferred from the study.
1. Results were positive for cold rolled steel:

Table 1. Material properties

PROPERTY	HOT ROLLED STEEL	COLDROLLED STEEL	ALUMINIUM
Density	7.77g/cc	7.8g/cc	2.7g/cc
Bulk modulus	140Gpa	140Gpa	76Gpa
Shear modulus	80Gpa	70Gpa	26Gpa
Elongation in brake	15%	20%	17%
Reduction of area	40%	50%	55%

Cold rolled steel has the highest elongation in brake when compared to the other materials. It also has a considerable reduction of area of 50%. It is also commercially available in the market. Considering all these reasons, we used cold rolled steel and fabricated the chassis frame.

- Calculation of the required torque for the selection of engine
  Max. Load needs to be pulled= 600 kg
  Width of the frame = 150cm
  W.K.T Torque = load * perpendicular distance
  = 600* 150
  = 90000 kg-cm= 9.0Nm.
- So we selected a 150cc engine which has a maximum torque of 13.4 Nm and a maximum output power of 13.8 Bhp. Since the required torque is less than the design torque, the design is safe
- We have selected 48V, 1000 Watts BLDC motor and used in this vehicle. The output torque of the motor with respect to speed and current are plotted and shown in the following figure 2 and 3.

Table 2. Motor output torque with respect to speed

SPEED (KM/HR)	SPEED (RPM)	TORQUE (NM)
10	208.97	3.58
20	417.94	1.79
30	626.91	1.19
40	835.88	0.897

Calculations for obtaining speed and torque:

Speed in rpm= (speed in km/hr*1000)/ (2*pi* radius*60)

Torque = power/ (speed in rpm)

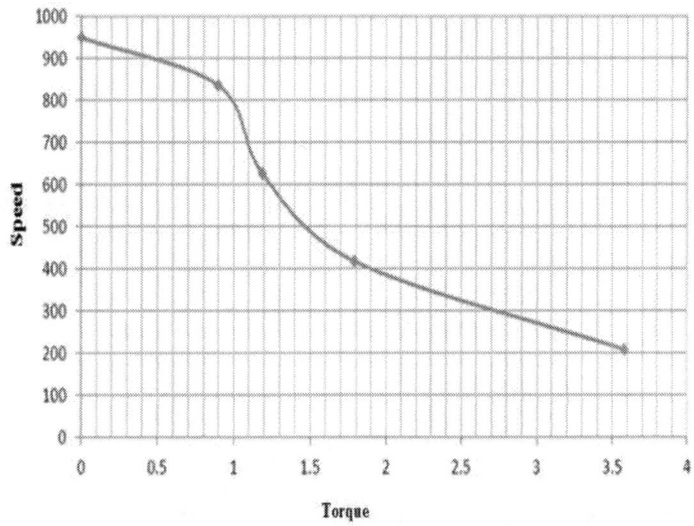

Figure.2.Mechanical charateristic of motor

From the above obtained value in the graphs and calculations, a motor of 1000watts-48V was selected. Fig. 2 shows the torque versus speed of the motor which indicates the mechanical characteristics. Selected motor gives the maximum output torque of 14.2 Nm thus more than satisfying than the required torque of 9.0 Nm. The design is safe as the required torque is less than the maximum output torque of the motor. Fig3 depicts current versus speed, known as electrical characteristics which indicate, at the maximum load conditions while increasing the speed of the vehicle the required input current increases. The required input current for the motor, calculated as per design is only 20Amps. The selected motor can produce an output of 14.3Nm when the maximum input current of 31Amps is provided.

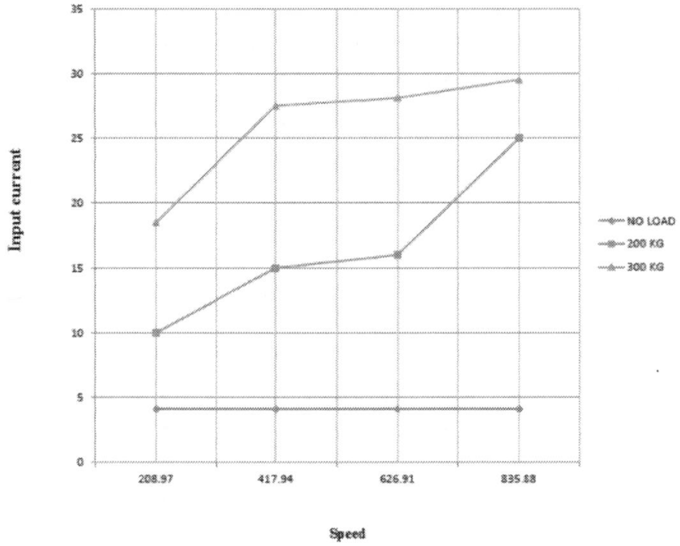

Figure.3 Electrical characteristics of motor

4. Conclusion

In this work solar powered hybrid vehicle has been successfully fabricated built in with internal combustion engine. Generally, the electric hybrid vehicles are disadvantageous during long distance travels also required periodic plug in of their batteries. These kinds of problems have been solved using hybrid solar vehicle which gives a car, self-charging potential from the solar panels. The efficiency of the solar panels which is used is 15-20%, but there are various means to increase the efficiency of the panels by changing their silicon materials. The future of energy sector lies solely on alternative energy resources. The cost of HSVs is more than the conventional cars but they are more efficient and cause less exhaust emissions. This challenge now can turn out to be a good scope for further development of a pollution free vehicle.

5. References

[1] Offer GJ, Howey D, Contestabile M, Clague R, and Brandon NP 2010 *J. Energ.Poli.* **38** 24.
[2] Saeid B, Scott J. Moura, Joel C. Forman, Hosam K. Fathy 2010 *J. Power Sources* **196** 8.
[3] Andrew Burke, Marshall Miller 2010 *J. Power Sources*, **196** 514.
[4] MatthieuDubarrya, Cyril Truchota, Bor Yann Liawa,, Kevin Geringb, SergiySazhinb, David Jamisonb,ChristopherMichelbacherb 2011, *J. Power Sources*,**196** 10336.
[5] Jenn-Jiang Hwang, Yu-Jie Chen, Jenn-Kun Kuo 2012 , *Int. J. of hydrogen energy*, **37** 4476.
[6] Kermani.S,Delprat.S, Guerra, Trigui.R 2012, Jeanneret Control Engineering Practice, **20** 408.
[7] Bor Yann Liaw,MatthieuDubarry 2007 *J. Power Sources*, **174** 76.
[8] GeorgiosFontaras, PanayotisPistikopoulos, Zissis Samaras 2008 Atmospheric Environment **42** 4023.
[9] Karan.C.Prajapati,Ravi Patel, Rachit 2014, **3** 1076.
[10] Robert F Nelson 2000 *J Power Source.* **91** 26.
[11] Anuragh.M.Lulhe and Tanuja.M.Date 2015 IEEE Xplore digital library.
[12] TeresaDonateo 2012, *INTECH*, Chapter 6.
[13] Ivan Arsie, Gianfranco Rizzo, GiovanniePetrone, Giovannie Spagnuolo 2006 *ICAT06*, **17**.
[14] AsllanHajderi, Interdisplinary 2016 *J. Research and development*. **4**.
[15] Jinming Liu and Huei Peng 2006 *IEEE transactions on control system tech*. **16** 1242.
[16] Rushikesh TrusharSoni 2015, *IOSR-JMCE*. **12** 11.

IOP Conf. Series: Materials Science and Engineering **390** (2018) 012103 doi:10.1088/1757-899X/390/1/012103

A review on friction stir welding of aluminium metal matrix composites

B Vijaya Ramnath[1], S Abhishek Subramanian[2] , R Rakesh[3], S Sharun Krishnan[4] and A L Ashwin Ramanathan[5]

[1] Professor, Department of Mechanical Engineering, Sri Sai Ram Engineering College, Chennai-44.
[2,3,4,5]Department of Mechanical Engineering, Sri Sai Ram Engineering College, Chennai-44.

Abstract. In the emerging world of technology, metal matrix composites have been playing the vital role in various applications to reduce the weight, increase the physical properties like hardness, tensile strength etc. Among these Aluminium metal matrix composites are the ones where the research is concentrated since Iron and steel can be replaced by Aluminium because of their properties. Thus, the usage of such composites in daily life is difficult because of the tough joining combining process like welding, casting etc. In this paper we are going to review the effects of such Aluminium metal matrix composites when joined using Friction Stir Welding and the related effects on the materials.

1.0 Introduction

In this review paper we have stated the effects of Friction Stir Welding on different composites of Aluminium as base material and formulating the results from the resulting material. The different tests have been made on the composite materials by viewing the different physical properties of it. The different advantages of using FSW is also seen in the following review paper.

2.0 Friction stir welding in aluminium-SiC metal matrix composite

Paola Bassani et al [2007] studied the effect of laser welding of the MMC using a CO2 and diode laser source. He has found out the most efficient method of welding from these two processes. The CO2 method of welding is done under deep penetration conditions and the diode laser uses a conduction welding method. From the results he has found that the CO2 welding process leads to the formation of Al4C3 which is found to decrease the toughness of the bead, while the laser welding process seems to give less formation of Al4C3 which is good and gives a good quality bead compared to C02 method of welding. Also the laser diode has emerged to be the best method in terms of high efficiency and high radiation wavelength absorption rate by the MMC's when compared with CO2 welding process. Rotundo et al [2010] has done his research on the effect of using Linear Friction Welding to make defect free joints in 2124Al with reinforcing material as SiCp of 25 vol%. The MMC joints were tested then at microstructural level and mechanically characterized on the basis of hardness, tensile, fatigue without and weld heat treatment process after LFW. The results showed that the joints produced were defect free with a uniform distribution of particles in the centre zone and the plasticity nature in the aluminium metal matrix obtained is reasonably good as expected. The decrease in hardness at the welded zone was 10% of hardness of the base material. The efficiency of the metal matrix joint produced by FSW was found to be more than 80%. Similarly, the S-N probability curves were also obtained with reasonable results. The only defect occurring was the fracture occurring at the Thermo-Mechanically Affected Zone(TMAZ). Won-Bae Lee et al [2006] has studied the effect of FSW on AZ91 Mg alloy with the reinforcing particles as SiC and has found the micro-structural property and the wear produced. The AZ91/SiC/10p microstructures initially were made up of irregularly distributed β-phases(Al12Mg17) and a collective mass group of SiC particles. The joint consisted of evenly distributed particles of SiC in a recrystallized manner and resulted in β-phase dissolution. Finally, the results showed that the composites hardness and wear resistance increased in the welded zone due to microstructural modification when compared to the base metal. Wei Wang et al [2009] investigated the effect of making Al metal matrix composites with bulk dispersal of SiCp reinforced in it. The SiCp were well uniformly distributed over a range of 5mm × 2mm on the cross section of the joint. About 1.5% of SiCp was found in the reinforced region which was not restricted to 100μm magnitude on the upper surface of the matrix. The hardness at the micro level reached upto 10% higher steady state than the base metal at a depth of 1.0mm down the upper surface. Zadi et al [2013] used the method of Friction Stir Process(FSP) to redefine the micro level structure of sintered Al-SiC composites with the concentration of the particles ranging over 4 to 16 vol%. Initially two SiC particles of sizes (490N and 800 grades) were checked for. The results showed that for the composites containing 4 and 8 vol% of 490N grade SiC gained an increase in hardness from 130 HV and 145 HV to 171 HV and 177 HV respectively. This was due to the uniform distribution of the SiC particles during FSP. While, the composites which contained 16 vol% of SiC gave bad results due to the formation of residual pores and lack of mass grouping. Therefore, the composites containing 4 and 8 vol% of SiC showed better inter particle mean spacing than the use of 16 vol% of Sic particles. Devaraju Aruvi et al [2013] examined the impact of tool rotational speed on mechanical and wear properties of aluminium alloy based composites using Friction Stir Processing and the resulting surface was studied using an optical microscope for finding the extent of dispersion of reinforced particles. The results showed that the micro level structures of hybrid composites with the acting reinforced particles namely SiC, Gr and Al2O3 were found to be evenly distributed in the small zone. The micro level hardness was found to be decreased by increasing the rotational speed and showed greater value in the Al-SiC composite and Al-Al2O3 composite surface due to the action of SiC and Al2O3 particles respectively. Also, the Aluminium and SiC/Gr (reinforcing materials) hybrid composites showed high wear resistance as the SiC particles worked as a load bearing element and Gr particles worked as a lubricant.

3.0 Friction stir welding in aluminium-Al_2O_3 metal matrix composite

Ceschini et al [2007] investigated the effect of Friction stir welding of Aluminium alloy (AA6061) reinforced with 20% volume fraction of Al2O3.The tests are carried out without any heat treatment after welding process. The results show that there is a reduction in amount of reinforcement particles up to 60% near the weld zone. It also contributes to the refinement of grains in the Friction stir welded sample. The microhardness, proof strength and ultimate tensile strength is also decreasing gradually of 60%,43% and 28% respectively from the base material to that of the welded region due to over raging of Aluminium whereas Failure of elongation increased by 64% and low cycle fatigue life was less than that of the base material. Ceschini et al investigated the effect of friction stir welding of Aluminium alloy (AA7005) reinforced with 10% of alumina particles by employing a tool rotating speed of 600 rpm and a welding speed of 250 mm/min. The optical and scanningelectron microscopy observations performed on the different zones of FSW joints cross-section revealed the different structures of the nugget, the thermo-mechanical affected zone and the heat affected zones. There occurs a strong grain refinement produced by the dynamic recrystallization. Fatigue tests were carried out under pulsating, tension loading using a resonant electro-mechanical testing machine. All the mechanical tests were performed up to failure which occurred at the interface with the welded area. The strong deformation of the material during the process leads to a strong grain size reduction. A fatigue life of 2 · 107 cycles was recorded for a stress amplitude of 120 MPa. Pirondi and Collini [2009] carried out analysis on fatigue crack propagation resistance of particulate metal-matrix composites butt joints obtained by friction stir welding. Two different alloy namely al 6061 and al7005 reinforced with 20% and 10% al2o3 respectively were under test. The results showed that the fracture toughness of the FSW joint is about 25% lower than the parent material in the case of W6A20A, while it is 10–20% higher in the case of W7A10A and crack propagation rate is lower than in the parent material in the case of W6A20A, while it is higher in the case of W7A10A. The comparison between parent material and joint showed that the welding process affects fracture toughness and fatigue crack growth rate differently depending on the material. Marzoli et al [2006] established the friction stir welding (FSW) process parameters envelope for an AA 6061 alloy reinforced with 20% of Al2O3 particles and he also explained the difficulties in joining MMCs. Microstructure has been observed with optical microscope, and images have been analysed with an image analysis software. Microhardness and tensile tests have been also carried out. The tools stirring effect has a substantial influence on the reinforcement particles distribution and shape. It has been found that the particles which were larger in size with edges being sharp would break way, making them rounded. The tensile testing of welded specimens revealed joint efficiencies of over 80% for the YS and of slightly more than 70% for the UTS. Failure in tensile testing always took place outside the stir zone.

4.0 FRICTION STIR WELDING IN ALUMINIUM ALLOYS

Fratini et al [2010] has experimented the Friction Stir Welding of titanium alloys which in his case is the more challenging as the material has intricate thermos-mechanical and thermos-chemical characteristics. The main aim is to increase the mechanical performance of the welded regions. The FSW of titanium blanks was performed and the mechanical and metallurgical tests were carried out showing the properties of FSW of titanium alloys. Results show that the microhardness level increase with decreasing average grain size. Tracie Prater [2014] has presented his research paper on the extent of usability of MMC's in the aerospace industries. He has extended his research on the effect of tool wear in FSW. He has explained that the wear process in FSW is due to a shear instead of a drag. His results show that he best way to tackle wear is to use a tool material which is harder than the reinforcement particles and use reinforcement particles which comes at the top order in the FEPA scale. He also states that diamond coatings are the best alternative as a wear resistant tool material. Minak et al [2010] has mainly concentrated on the fatigue factor of FSW joints on an aluminium based composite (AA6061/22 vol.% Al2O3p). The composite specimens obtained had an adequate increase in micro hardness property and showed an superior joint efficiency when compared to the tensile strength of the material. Although the analysis of fatigue resistance was very poor for the FSW specimens obtained with different process parameters. The failure mostly occurred in the stirred zone on the application of high loads and fine reinforcement particles in the fractured zone were visible. Heurtier et al [2006] has investigated the thermomechanical model in three dimension for FSW process. He has presented his views on the temperature estimations and microhardness levels in different weld zones. Results show that the increase in velocity of the tool smoothens the hardness profile close to the bulk zone and it is mainly due to the decrease in the temperature near the weld zone. Meshram et al [2007] has pointed out the necessity of joining dissimilar metals in the industries and has given a solution for it by using FSW process. The commonly used metal combinations are Cu-Ti, Fe-Ti, Fe-Cu, Fe-Ni. His research points out that he has taken the pure form of each dissimilar metal and has joined them using FSW process. His results prove that an increased time of interaction led to the reduction in strength in insoluble system and gain in strength in soluble system. Darras et al [2007] has experimented the FSW of commercial AZ31 magnesium alloy as most of the research papers have been concentrated on aluminium alloy. The effect of FSW on magnesium alloys is dynamically reported and it is shown that microstructure properties of magnesium alloys have tend to be increased and homogeneous grain structure and recrystallized structure is also obtained in the FSW weld zone. Adem Kurt et al [2011] has studied the effect of reinforcing SiC particles into pure form of aluminium. The composite samples were then tested at different tool rotating speeds and traverse rates. The microstructural results showed that by increasing the effect of tool rotating speed and traverse rate gave a even distribution of SiC particles in the base material and also the hardness property was three times more than the base parent metal. Results also showed an increase in bending strength of the metal matrix composite which was the ultimate aim of this study. Dora Siva Prasad et al [2014] utilised the method of double stir casting in his experimentation. The aluminium base material is reinforced with different proportions of rice husk ash namely 2,4,6 and 8 wt% and SiC particles in a similar proportion. The mechanical properties such as hardness, density, porosity of the resulting composite has been viewed in detail. For this purpose, SEM observations have been made and finally it was seen that hardness, porosity, ultimate tensile strength and yield strength increased significantly with decrease in density of the composite material. The precipitation kinetics of the resulting material was also found to be greatly increased. Kunal Yeole et al [2014] has used the technique of including a layer of zinc oxide nanoparticles on the upper surface of fly ash by a unique chemical method. The zinc oxide is known to be corrosive in nature. With the differing concentrations of fly ash being 5, 10, 15(w/w) % of zinc oxide, the results were rather pleasing with the increase in crystallinity and reduction in size of the particle, specific gravity and oil absorption rates which is due to evenly distribution of zinc oxide particles on fly ash. The main use of this nanocomposite is it features as an anticorroding coater by serving as an additive. Dora Siva Prasad and Chintada Shoba [2014] has experimented the wear resistant property of hybrid aluminium composite with the main reinforcement particles as rice husk ash and SiC particles with a proportion of 8%. The test carried out was pin-on desk wear test and SEM observations are made out to study the given characteristics. Results made out were quite positive with the increase in wear resistance of the hybrid composite which was the sole purpose of the experiment. Storjohann et al() made out an investigation by comparing the method of fusion welding of Al-MMC's and FSW matrix composite. Both showed their own positive results as fusion welding tends to increase the hardness of the Al base material due to formation of aluminium carbide(Al4C3) needle like structures in the welded region while the later method proved to produce an evenly distributed microstructure with an even hardness nature which was not seen in fusion welding of Al-MMC's. Yahya Bozkurt et al [2011] has presented his research topic on the friction stir welding of AA2124 consisting of 25% of SiC particles in hat treated T4 aluminium MMC plates and more importantly at low welding parameters. The composite was also tested for heat dissipation from a specific distance of 15mm from the weld portion. The results showed an increase in Ultimate tensile strength which resulted in a improved joint efficiency of approximately 73%. Gan et al [2010] in his research paper has provided a basic view of the use of friction stir welding process in the various manufacturing industries and aerospace industries. He has pointed out that FSP is very useful when it comes to increasing the microstructural characteristics of the base material by reinforcing it with particles. His results show that the FSP method increase the hardness and other challenges met are in increasing the fatigue factor and joint strength. Some parameter studies are done by varying process parameters [23,24,25].

5.0 Conclusion

The various physical properties such as hardness, toughness have been found to be increased and
The density and strength to wear ratio has found to be decreased and there is an uniform grain distribution of the particles in the welded zone. Therefore, Friction Stir Welding process is more efficient than the other methods of preparing composites.

References

[1] Paola Bassani, Edoardo Capello, Daniele Colombo, Barbara Previtali, Maurizio Vedani Effect of process parameters on bead properties of A359/SiC MMCs welded by laser, Composites: Part A 38 (2007) 1089–1098

[2] F. Rotundo, L. Ceschini, A. Morri, T-S. Jun c, A.M. Korsunsky, Mechanical and microstructural characterization of 2124Al/25 vol.%SiCp joints obtained by linear friction welding (LFW), Composites: Part A 41 (2010) 1028–1037

[3] Won-Bae Lee, Chang-Yong Lee, Myoung-Kyun Kim, Jung-Il Yoon, Young-Jig Kim, Yun-Mo Yoen c, Seung-Boo Jung, Microstructures and wear property of friction stir welded AZ91 Mg/SiC particle reinforced composite, Composites Science and Technology 66 (2006) 1513–51520

[4] Wei Wang, Qing-yu Shi, Peng Liu, Hong-ke Li, Ting Li, A novel way to produce bulk SiCp reinforced aluminium metal matrix composites by friction stir processing, journal of materialsprocessing technology 209(2009) 2099–2103

[5] H. Izadi, A. Nolting, C. Munro, D.P. Bishop, K.P. Plucknett, A.P. Gerlich, Friction stir processing of Al/SiC composites fabricated by powder metallurgy, Journal of Materials Processing Technology 213 (2013) 1900– 1907Reference lists

[6] Devaraju Aruria, Kumar Adepu, Kumaraswamy Adepu, Kotiveerachari Bazavada, Wear and mechanical properties of 6061-T6 aluminum alloy surface hybrid composites [(SiC + Gr) and (SiC + Al2O3)] fabricated by friction stir processing mater res technol.2013;2(4):362–369

[7] L. Fratini, F. Micari, G. Buffa, V.F. Ruisi, A new fixture for FSW processes of titanium alloys, CIRP Annals - Manufacturing Technology 59 (2010) 271–274

[8] Tracie Prater, Friction Stir Welding of Metal Matrix Composites for use in aerospace structures, Acta Astronautica 93 (2014) 366–373

[9] G. Minak, L. Ceschini, I. Boromei, M. Ponte, Fatigue properties of friction stir welded particulate reinforced aluminium matrix composites, International Journal of Fatigue 32 (2010) 218–226

[10] P. Heurtier, M.J. Jones, C. Desrayaud, J.H. Driver, F. Montheillet, D. Allehaux, Mechanical and thermal modelling of Friction Stir Welding, Journal of Materials Processing Technology 171 (2006) 348–357

[11] S.D. Meshram, T. Mohandas, G. Madhusudhan Reddy, Friction welding of dissimilar pure metals, Journal of Materials Processing Technology 184 (2007) 330–337

[12] B.M. Darras, M.K. Khraisheh, F.K. Abu-Farha, M.A. Omar, Friction stir processing of commercial AZ31 magnesium alloy, Journal of Materials Processing Technology 191 (2007) 77–81

[13] Adem Kurt, Ilyas Uygur, Eren Cete, Surface modification of aluminium by friction stir processing, Journal of Materials Processing Technology 211 (2011) 313–317

[14] Dora Siva Prasada, Chintada Shobab, Nallu Ramanaiahc, Investigations on mechanical properties of aluminum hybrid composites, j mater res technol. 2014;3(1):79–85

[15] Kunal Yeole, Pravin Kadam, Shashank Mhaske, Synthesis and characterization of fly ash-zinc oxide nanocomposite, j mater res technol. 2014;3(2):186–190

[16] Dora Siva Prasad, Chintada Shoba, Hybrid composites – a better choice for high wear resistant materials, j mater res technol. 2014;3(2):172–178

[17] Yong X. Gan, Daniel Solomonand Michael Reinbolt, Friction Stir Processing of Particle Reinforced Composite Materials, Materials 2010, 3, 329-350; doi:10.3390/ma3010329

[18] L. Ceschini, I. Boromei, G. Minak, A. Morri, F. Tarterini, Microstructure, tensile and fatigue properties of AA6061/20 vol.%Al2O3p friction stir welded joints, Composites: Part A 38 (2007) 1200–1210

[19] L. Ceschini, I. Boromei, G. Minak, A. Morri, F. Tarterini, Effect of friction stir welding on microstructure, tensile and fatigue properties of the AA7005/10 vol.%Al2O3p composite, Composites Science and Technology 67 (2007) 605–615

[20] P. Cavaliere, G.L. Rossi, R. Di Sante, M. Moretti, Thermoelasticity for the evaluation of fatigue behaviour of 7005/Al2O3/10p metal matrix composite sheets joined by FSW, International Journal of Fatigue 30 (2008) 198–206

[21] A. Pirondi, L. Collini, Analysis of crack propagation resistance of Al–Al2O3 particulate-reinforced composite friction stir welded butt joints, International Journal of Fatigue 31 (2009) 111–121

[22] L.M. Marzoli, A.v. Strombeck, J.F. Dos Santos, C. Gambaro, L.M. Volpone, Friction stir welding of an AA6061/Al2O3/20p reinforced alloy, Composites Science and Technology 66 (2006) 363–371.

[23] B.Vijaya Ramnath, P Venkatesan, S Sathish, T Vignesh, Parameter optimization of friction stir welding of AA8011-6062 using mathematical method C Elanchezhian, Procedia Engineering 97, 775-782.

[24] B.Vijaya Ramnath, C Elanchezhian, S Rajesh, SJ Prakash, BM Kumaar , Design and Development of Milling Fixture for Friction Stir Welding, Materials Today: Proceedings 5 (1), 1832-1838

[25] C Elanchezhian, B Vijaya Ramnath, K Pazhanivel, A Vedhapuri, B Mano , Comparative Study and Analysis of Friction Stir Welding with Plasma ARC Welding , Applied Mechanics and Materials 766, 695-700

IOP Conf. Series: Materials Science and Engineering **390** (2018) 012104 doi:10.1088/1757-899X/390/1/012104

Automated line integration from through hole technology to integrated circuit testing

Roshan Abraham Joseph[1], G. Sakthivel[2]

School of mechanical and building sciences, Vellore Institute of Technology, Chennai

Emil ID: roshanj94@gmail.com

Abstract: Through Hole Technology (THT) is a method in electronic printed circuit board (PCB) mounting technology by which components are mounted on the PCB board and it is later sent for Integrated circuit testing (ICT). The in-process material movement is the real culprit for decreased production rates. Here we discuss the various identified reasons for the same and calculated effective possible maximum production rates and improvise the process with the help of an automated material transfer system and validate the same using production simulation software called Tecnomatix. Each of the system i.e. the manual and the automated system was studied, and appropriate corrective decision was taken to improvise the overall productivity.

1. Introduction

In any manufacturing unit, production process and its routing is very much essential to be updated in accordance with the upgrowing demands of the product. Here, two production lines are taken into consideration i.e. THT and ICT. THT line deals with huge amounts of PCBs and the same must be achieved in testing of the PCBs. This testing is being handled by ICT. Currently the manual material handling is leading to huge accumulation of PCBs and that further affects as an implication for slower production rate in THT line. The proposed automated material handling system deals with autonomous entry and exit of PCB bins from station to station thereby reducing lead time and a direct improvement in productivity by effective production process. Highlighted the new built-in platforms and remote controls for factory automation [1], Application of robotic process automation integrating platforms for factory process optimisation [2], Incorporating active energy drives with variable frequency drives is the key to new levels of control and precise automation [3], Advantages of using parallelism with crossover methods for production [4], Coining two major production concepts of capacity planning and time management along with modern computing platforms [5], Critical path analysis for batch production is the key way to improvise productivity [6], Usage of flexible manufacturing systems along with advanced special purpose automated machines for better production [7], Experimental analysis and study about modern controls like model based analysis, intelligent analysis, etc. for factory control and automation [8], Strategic planning and control for effective control in factory operations [9], Self-adaptive factory systems being discussed for advance agility of factory production system [10].

2. Experimental setup

2.1 Layout of THT to ICT line

There are two lines for THT and one line for ICT consisting 4 machines. THT line 1 has three working stations and THT line 2 has four working stations. Material movement takes place from THT line 1 and 2 to the four machines of ICT as per the testing required by the product. Out of these four machines, two are multimode machines and two are fly machines.

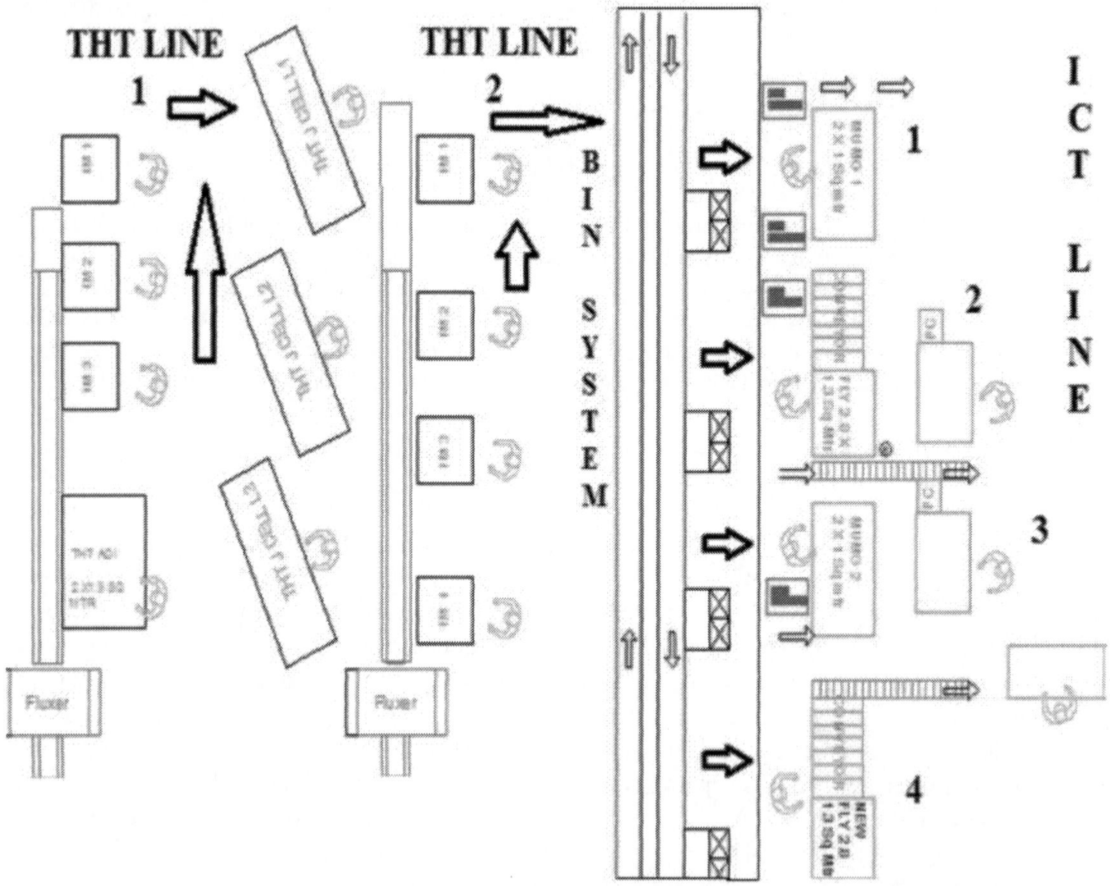

Figure 1. Layout of THT to ICT line

Figure 2. Bin trolley system for material transfer from THT to ICT line

2.2 Simulation using tecnomatix

The above THT line was taken into consideration and the mounting time cycle for each type and batch of PCB was taken into consideration. Similar was done at the ICT end with respect to corresponding PCB testing requirement.

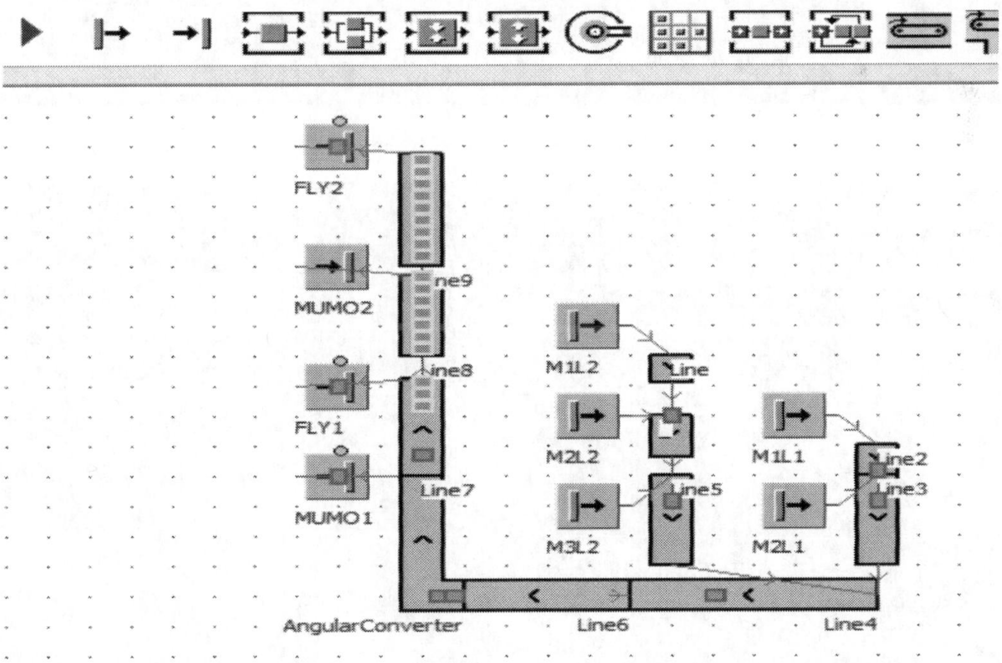

Figure 3. Tecnomatix simulation of automated line from THT to ICT for manual transfer

2.3 Automated material movement from THT to ICT

The transfer of materials from both THT lines to ICT (Integrated Circuit Testing) is done manually which leads to huge accumultion of bins containing PCBs and loss of space. Need for a automated system was required for better movement of PCBs and save floor space. The flowing system was proposed after analysing, designing, modifying,etc, the area and process that was taken into consideration. Movement is achiveded by belt & chain conveyor system for horizontal and vertical movement respectively.

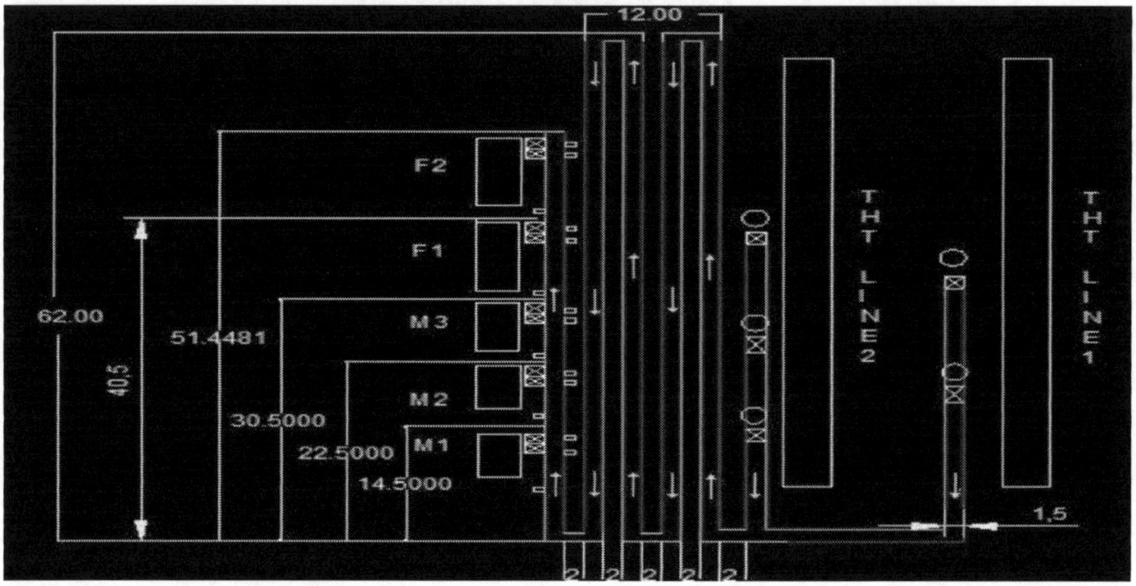

Figure 4. Top view 2D layout of automated material transfer system

Figure 5. Top view of model of automated material transfer system

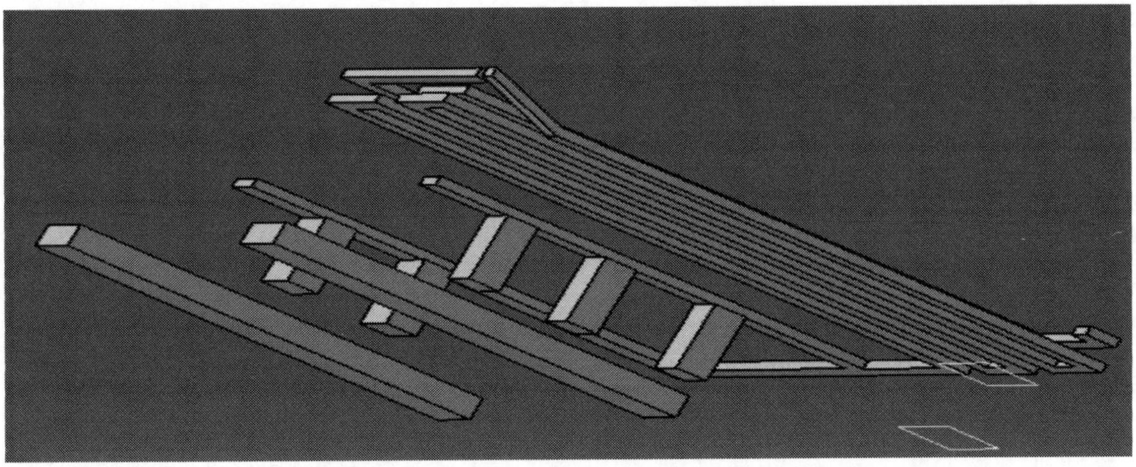

Figure 6. Bottom view model of automated material transfer system

Design Calculations

Minimum overhead distance of conveyor belt : $10\ ft.$

Maximum overhead distance of conveyor belt : $12\ ft.$

Dimension of cassette : $1(L) * 0.83(B) * 0.15(H)\ ft.$

Width of conveyor : $1.5\ ft.$

Rack storage dimension : $1.1(L) * 1(B) * 7.5(H)\ ft.$ --{$2.5\ ft.$ left for mounting of rack on floor}

Spacing between two conveyor belts : $2\ ft.$

Number of cassettes per rack:

$0.15 ft.$(height of cassette) $+ 0.05 ft.$ (clearance between 2 cassettes) = $0.2\ ft.$

Therefore, $7.5/0.2 = 37.5$ cassettes ~~ 36 cassettes (rounding off considering additional clearance)

Total rack capacity:

Total no. of racks : 15 (2 on THT Line1 and 3 on THT Line2 and 2 each on each ICT)

Therefore, total capacity of storage : $15 * 36 = 540$

Overhead Conveyor design:

Total considered area for conveyor travel: $62 * 12\ ft.$

No. of cassettes per line of conveyor :

$1ft. + 0.5ft.$ clearance between 2 cassettes $= 1.5ft.$

No. of cassette per line $= 62/1.5 = 41.33$ ~~ 40 cassettes (rounding off for additional clearance)

Total no. of lines $= 4$

Therefore, $4 * 40 = 160$ cassettes

Total capacity of conveyor system:

540 (rack storage) $+ 160$ (conveyor line) $= 700\ cassettes$

Drop points at ICT:

(Considering 2 feet gap between each ICT)

(Length of each MUMO machine : $6ft.$

Length of each FLY machine : $8.4ft$)

Mumo (ICT)1 : $12\ to\ 14.5\ ft.$ (from start point of measurement)

Mumo (ICT)2 : $20\ to\ 22.5\ ft.$

Mumo (ICT)3 : $28\ to\ 30.5\ ft.$

Fly (ICT)1 : $38\ to\ 40.5\ ft.$

Fly(ICT)2 : $49\ to\ 51.5\ ft.$

Loop System design for return flow of cassettes to ICT:

Take off point to $12ft.$ elevation : $54\ ft.$ ($3\ ft.$ after F2)

Drop point from $12ft.$ to $10ft.$: $3ft.$ from start of delivery line (delivery line is the line at ICT drop points)

Currently, the area is being simulated manually with a conveyor prototype for better judgement and precise achievement of design and results.

Figure 7. Prototype of conveyor developed on ground level for concept realization

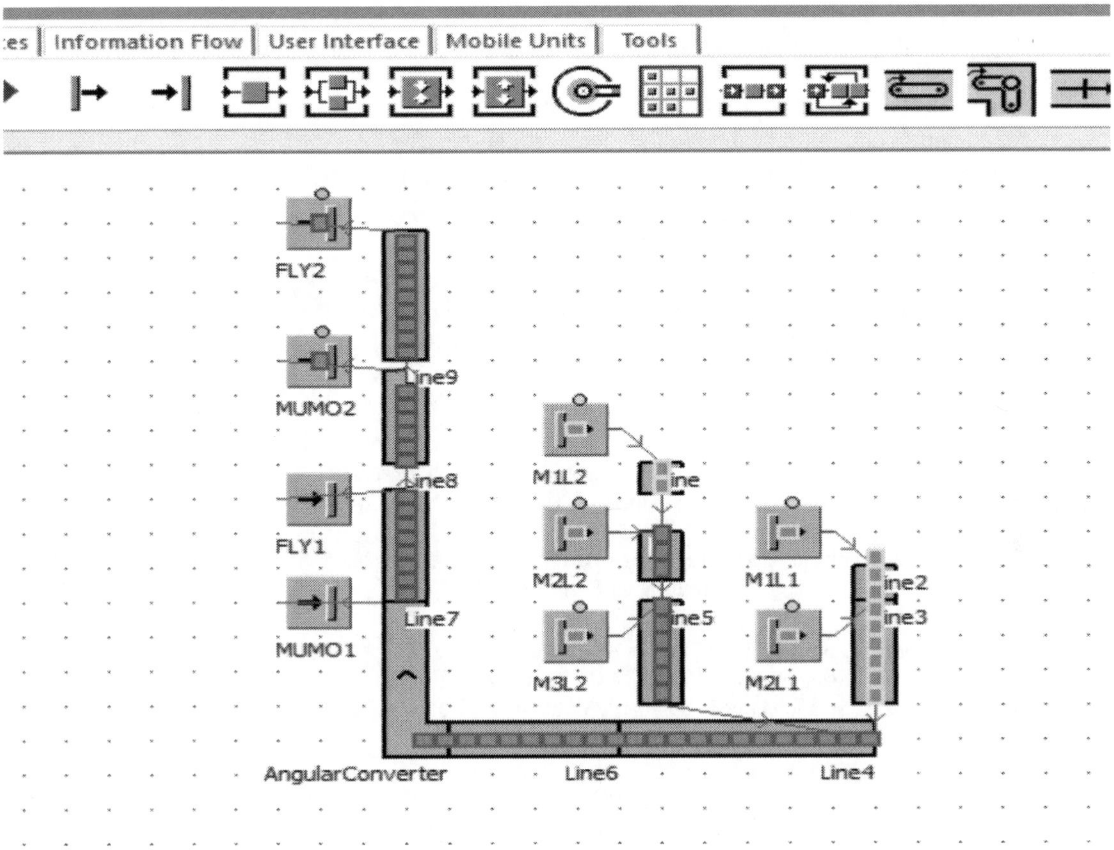

Figure 8. Tecnomatix simulation of automated material handling system highlighting greater density in material flow

3. Results and discussions

Analysing and understanding the two production run techniques, both were simulated in tecnomatix for their productivity/ throughput. It was found out that with an automated system the productivity increased considerably. Waiting wastages, ambiguity, space utilisation, manpower, etc. was taken into consideration and a clear-cut optimisation was achieved by incorporating automated material handling conveyor system.

Table 1. Productivity Chart comparing manual and automated process

Process	Output
Manual material transfer	1500
Automated material transfer	2400
Output improvement per day	900
Output improvement per week	5400
Output improvement per month	21600

4. Conclusion

An automated material handling system was being developed keeping the main focus on its throughput. Detailed study was carried out on the following for the development of automated system:

i. Process involved in manual transfer of material.
ii. Tecnomatix simulation analysing the ideal throughput.
iii. Designing of an automated system for improving the productivity.
iv. Validating the automated process by tecnomatix.

There was a huge benefit in the automated material handling system by overcoming many manual shortcomings and unnecessary waiting from one production line to another.

References

[1] Prof. Jaikaran Singh1, Prof. Mukesh Tiwari2, Mr. Manish Shrivastava3, August 2013,International Journal of Engineering Trends and Technology (IJETT) – Volume 4 Issue 8.

[2] Lacity, M., Willcocks,L., and Craig, A. (2015), "Robotic Process Automation at Telefónica O2," The Outsourcing Unit Working Research Paper Series

[3] Anuradha Tomar; Yog Raj Sood; Devesh Singh Tomar, Year: 2015,An Efficient Scheme of Automation and Control for Conventional Cable Manufacturing Industry . ,IEEE Transactions on Industrial Informatics , IEEE Journals & Magazines., , Volume: 11, Issue: 1 ,Pages: 132 - 140 ,Cited by: Papers (1) .

[4] T. Freiheit; Y. Koren; S. J. Hu, Year: 2004,Productivity of parallel production lines with unreliable Machines and material handling , IEEE Transactions on Automation Science and Engineering , IEEE Journals & Magazines, , Volume: 1, Issue: 1 ,Pages: 98 - 103 ,Cited by: Papers (7) .

[5] W.H.M. Zijm*, R. Buitenhek , (1996) , on January 9,1995.Capacity planning and lead time management, , Department [Mechanical Engineering, University of Twente, Int. J. Production Economics 46 47 1,165 179, P.O. Box ~17, 7500 AE Enschede, Netherlands.

[6] W. H. M. Zijm - Submitted by H. J. J. Kals (1),Year 1998,The Integration *of* Process Planning and Shop Floor Scheduling in Small Batch Part Manufacturing, , University of Twente, Dept. of Mechanical Engineering, Enschede, Netherlands Received .

[7] Arvind Kaushal, Ajay Vardhan, R. S. Rajput (April 2016), Flexible Manufacturing System A Modern Approach To Manufacturing Technology International Refereed Journal of Engineering and Science (IRJES) ISSN (Online) 2319-183X,(Print) 2319 -1821 Volume 5, Issue 4,PP.16 -23.

[8] MariagraziaDotoli,AlexanderFay,Year 2015, A Survey on Advanced Control Approaches in Factory Automation, IFAC-PapersOnLine on sciencedirect,Volume 48, Issue 3, 2015, Pages 394-399

[9] Jack R.Meredith,Year 1987, Strategic control of factory automation, Long Range Planning on Sciencedirect,Volume 20, Issue 6, December 1987, Pages 106-112.

[10] E.CarpanzanoF.Jovane, Year 2007, Advanced Automation Solutions for Future Adaptive Factories, CIRP Annals Volume 56, Issue 1, 2007, Pages 435-438.

Dry Sliding Wear behaviour of cast Al/Al_20_3/Gr hybrid nano-composite using response surface methodology

I. Manivannan[1], S. Ranganathan[2], S. Gopalakannan[3], S. Suresh[4]

[1]Department of Mechanical Engineering, Motilal Nehru Government polytechnic college, Puducherry, 605008, India.Tel.: +91 9443483684.
[2]Department of Mechanical Engineering, Saveetha School of Engineering, Saveetha University, Chennai-602 105, Tamilnadu, India. Tel.: +91 9787415094.
[3]Department of Mechanical Engineering, Adhiparasakthi Engineering College, Melmaruvathur 603 319,TamilNadu, India. Tel.: +91 9944949026.
[4]Department of Mechanical Engineering, University College of Engineering, Nagercoil, Tamilnadu, India, Pin: 629 004. Tel.: +91 9443494576.

E-mail: i.manivannan72@gmail.com.,ranganathan@saveetha.com

Abstract. The newly engineered metal matrix nanocomposite (MMNC) of Al 6061 reinforced with 1.2 wt % Al203 and 0.5 wt % Gr (Al-1.2Al203-0.5Gr) hybrid nanocomposite was prepared by ultrasonic assisted stir casting method. A three-level central composite design test was developed using response surface methodology; parameters such as load, velocity and sliding distance were varied in the range of 10-30 N, 0.4-1.1 m/s and 500–1500 m, respectively. Dry sliding wear tests were performed as per the experimental design using a pin-on disc setup at room temperature. Analysis of variance (Anova) was applied to investigate the influence of process parameters and their interactions viz., Load, velocity, sliding distance on wear, and coefficient of friction (COF). The aim of the study was to identify the process parameters that significantly affect the output features. Further, a mathematical model has been formulated by applying response surface method in order to estimate the tribology characteristics such as wear and COF of the hybrid nanocomposites.

1. Introduction

The Al/Al203/Gr hybrid nanocomposites find wide application in advanced engineered materials due to its tribological applications. When the reinforcement particles are of micro size they improve the strength of the materials, but the ductility of the micro composites weakens when the percentage of reinforcement increases. For less quantity of particulates the mechanical and tribological properties of nano particle dispersion strengthened metal matrix composites are far superior to those of their micro particle composites. Aluminum composites reinforced with nano-particles, strengthen the metal matrix and maintain good ductility compared to micro particles [1-4]. The scope of Al-Al203-Gr nano-composites is of great interest for tribological applications in automotive, aerospace industries including engine valves, cylinder liners. The usage of aluminum hybrid self-lubricating nanocomposite components leads to reduce the oil consumption, power loss and maintenance cost [5]. The present work is envisaged to develop a mathematical model and analyze the effects of tribology parameters on the performance characteristics of MMNC using response surface methodology (RSM)[6-10]. Accordingly, the quantitative mathematical models have been carried out to study influence of process parameters viz., load, velocity and sliding distance and their interactions on wear and coefficient of friction (COF) by using RSM.

2. Experimental Details

2.1 Ultrasonic assisted stir casting method.

The material used in the present investigation for the preparation ofAl-1.2Al_20_3-0.5Gr hybrid nanocompositeconsists of Aluminium 6061 as the base matrix alloy. Its chemical composition (%)

isMg-0.5, Fe- 0.17, Si- 0.2, Cu-0.2, Cr-0.3, Ag-0.4 and Al6061 remaining balance. Nano-Al_2O_3 particles supplied by US Research Nano materials, Inc. USA, with an average size of 50 *nm* and graphite ≤ 20 *µm* supplied by Sigma Aldrich are used for hybrid nano-composite fabrication.The hybrid nanocomposites were cast using ultrasonic assisted stir casting method as it ensures uniform distribution of the reinforcement[9].

2.2 Design of Experiments.

In the present study the experiments were designed based on central composite design (CCD) of response surface method.The factorial portion of CCD is a full factorialdesign with all combination of the factors at two levels (high, +1, and low, -1). The "face-centered CCD" involves 20 experimental observations at three independent input variables [4]. The experimental design layout that was adopted in this study in the actual form is shown in Table 1.

Table. 1 Design matrix and experimental results

Ex.No	Load(N)	Sliding speed(*m/sec*)	Sliding distance(*m*)	Specific wear rate(SWR)10^{-7}*g/Nm*	Coefficient of Friction(*COF*)
1	10	0.4	500	3.41	0.293
8	30	1.1	1500	1.06	0.211
12	20	1.33	1000	0.86	0.243
5	10	0.4	1500	2.75	0.273
15	20	0.75	1000	1.55	0.252
17	20	0.75	1000	1.55	0.253
16	20	0.75	1000	1.54	0.254
4	30	1.1	500	0.72	0.233
9	10	0.75	1000	2.11	0.272
20	20	0.75	1000	1.55	0.252
10	40	0.75	1000	0.85	0.221
19	20	0.75	1000	1.55	0.252
3	10	1.1	500	1.42	0.276
7	10	1.1	1500	1.08	0.257
18	20	0.75	1000	1.56	0.252
13	20	0.75	200	1.74	0.265
14	20	0.75	1900	1.93	0.241
2	30	0.4	500	1.82	0.243
11	20	0.16	1000	2.05	0.252
6	30	0.4	1500	1.43	0.224

3. Experimental Procedure.

The wear behaviour of the Al-1.2Al_2O_3-0.5Gr hybrid nano-composites samples under dry sliding condition were investigated using a pin-on-disc wear test machine (TR-20-PHM-M1 DUCOM), according to ASTM G99 standard. The wear tests were performed using cast hybrid nanocomposite pins on a hard steel disc of 62HRC under dry sliding condition. The conventional grinding and polishing techniques were used for surface preparation of cylindrical pin and disc surface. The sample of cylindrical pin and disc surface were polished down to 0.15µm with emery papers of 1000 and 1200 grid size. The applied load for the nano-composite pins were of 10–30 N. The sliding distance of 500, 1000 and 1500 m was employed, with sliding speed of 0.4, 0.75, 1.1 m/s. The dimensions of the wear test pins are of diameter 10 *mm* and height 20 *mm*. The sliding path on the disc surface was 25 mm diameter. After each test acetone is used to clean the pin and disc to remove the traces. An electronic balance of 0.1 *mg* accuracy was used to find the weight of pins. By weighing the pin before and after each wear test, loss of wear of the pin was calculated. The computer system connected with the pin-on-disc wear test machine was used to record the friction co-efficient between the specimen and rotating disc.

4. Results and Discussion

Mathematical Model for Specific Wear and Friction. The fit summary results of quadratic and linear models are given in ANOVA Table 2.

Table 2 The ANOVA table for the fitted models

Source	Sum of squares	df	Mean square	F value	Prob>F	
For SWR						
Model	9.705	9	1.078	5.697	0.005	significant
Residual	1.892	10	0.189			
Lack of Fit 1.891	5	0.378	2768.757	1.345e-08	Not significant	
Pure Error	0.0006	5	0.00013			
Cor Total	11.598	19				
Std. Dev.	0.435			R^2	0.836	
Mean	1.7805			Adjusted R^2	0.689	
C.V. %	24.434			Predicted R^2	0.237	
				Adeq Precision	8.687	
For COF						
Model	0.0056	6	0.0009	42.359	8.5648e-08	significant
Residual	0.0002	13	2.212e-05			
Lack of Fit	0.0002	8	3.534e-05	36.565	0.0005	Not significant
Pure Error	4.83e-06	5	9.66e-07			
Cor Total	0.005	19				
Std. Dev.	0.0047			R^2	0.951	
Mean	0.27985			Adjusted R^2	0.928	
C.V. %	1.680			Predicted R^2	0.797	
				Adeq Precision	25.40	

When the R^2 approaches unity, the better the response model fits the actual data [6]. It exists the less the difference between the predicted and actual data. Further the value of adequate precision (AP) in this model, which compares the range of the predicted value at the design point to the average prediction error, is well above 4. The values obtained are as follows: $R^2 = 0.8368$ and AP= 8.6872 for specific wear rate; $R^2 = 0.9513$ and AP= 25.4047 for COF. The response equations for Specific wear rateand COF are:

IOP Conf. Series: Materials Science and Engineering **390** (2018) 012105 doi:10.1088/1757-899X/390/1/012105

Specific wear rate:

$SWR=1.83481 + -0.401196 * A + -0.660605 * B + -0.209617 * C + 0.2425 * AB + 0.06 * AC + 0.075 * BC + -0.151433 * A^2 + 0.0253437 * B^2 + 0.0465569 * C^2$

Coefficient of friction:

$COF = 0.27985 + -0.0176567 * A + -0.00534526 * B + -0.00674647 * C + -0.00375 * AB + -0.00275 * AC + -0.00475 * BC$

4.1 Effect of dry sliding parameters on wear

It can be noted from the fig.1.that the wear decreases with increase in sliding speed for Al-1.2Al_2O_3-0.5Gr hybrid nanocomposites. The decrease of wear with increase of sliding speed may be because of the increased effectiveness of the mechanical mixed layer (MML) in Al-1.2Al_2O_3-0.5Gr hybrid nanocomposites. That is, at high sliding speed may result in quick formation of MML and there by the wear gets reduced. It can be noticed that the wear decreases with increase in load. The decrease of wear with increase of load in Al-1.2Al_2O_3-0.5Gr hybrid composites nanocomposites may be due to the increased effect of the MML. It can be observed from the fig.2. that the wear increases with sliding distance due to the weakened tribolayer. Similar results have been reported on wear rates of composite, which showed a drastic increase with the sliding distance and shortly after the removal of tribolayer the composite seized against the steel counterface [10].

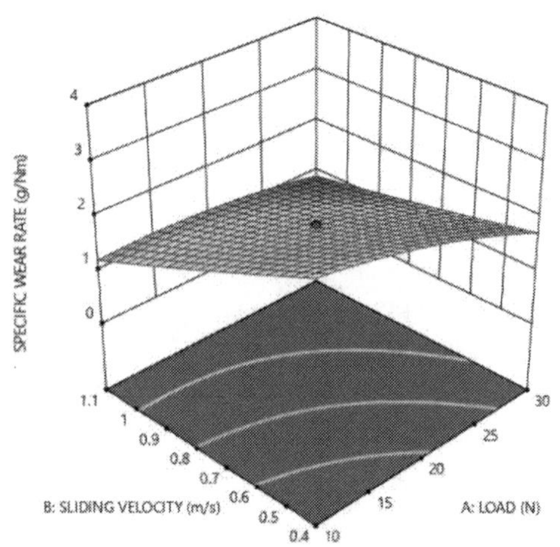

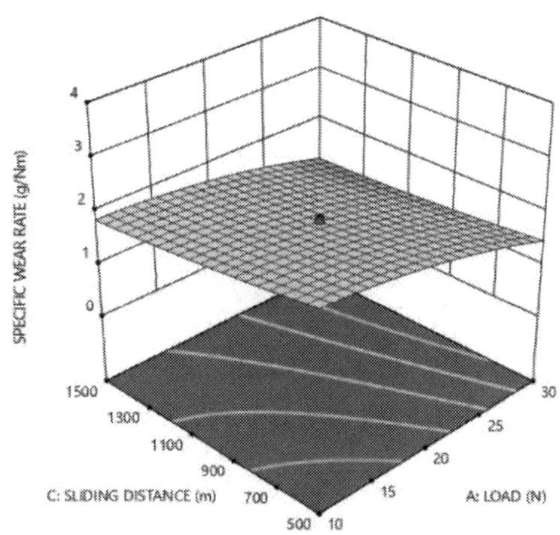

Figure.1 SWR versus sliding velocity and load **Figure.2** SWR versus sliding distance and load

The Al-1.2Al_2O_3-0.5Gr hybrid nanocomposites exhibit improved wear properties owing to the synchronized effect of enhanced strength and formation of a tribolayer predominantly consisting of Al_2O_3 and graphite particulates[7,8].

4.2 Effect of dry sliding parameters on friction

Friction coefficient is influenced by load and sliding speed. The surface plot of the effect of load and sliding speed on friction coefficient of Al-1.2Al_2O_3-0.5Gr hybrid nanocomposites is shown in Fig. 3&4. Sliding distance is not affecting the friction coefficient of composites. As the sliding

IOP Conf. Series: Materials Science and Engineering **390** (2018) 012105 doi:10.1088/1757-899X/390/1/012105

between the composite and the disc surface is separated by the tribolayer, friction coefficient is independent of the disc surface. This is true as long as a stable tribolayer exists at contact surfaces even for longer sliding distances. Thus, a stable tribolayer may exist for sliding distances considered in the present investigation, as a result of which the friction coefficient is not influenced by the sliding distance. Friction may get reduced with increase in sliding speed due to the quick formation of tribolayer at contact surfaces resulting in decrease of friction coefficient.

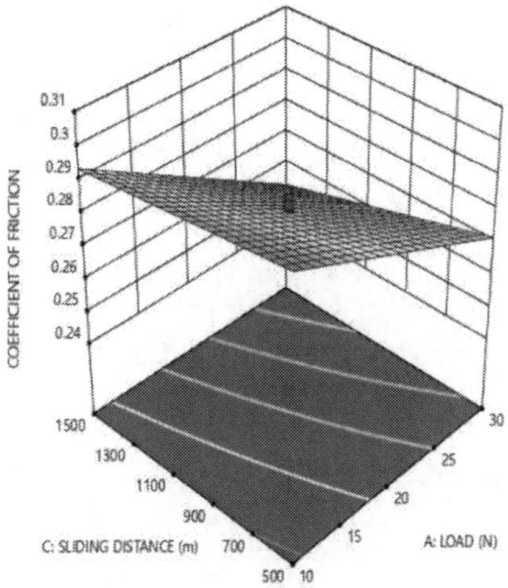

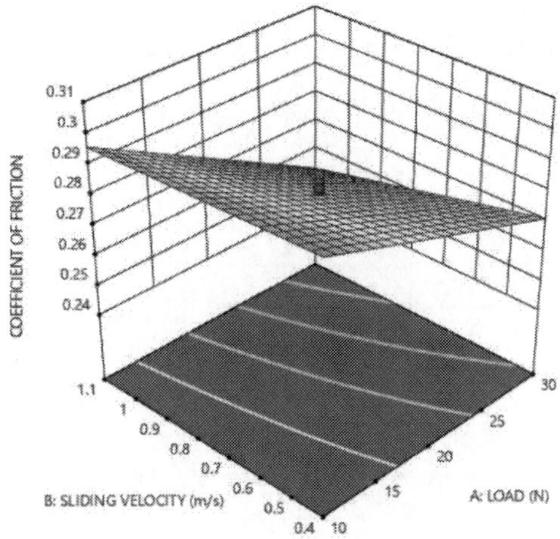

Figure. 3 COF versus sliding distance and load **Figure. 4** COF versus sliding velocity and load

The confirmation experiment was performed by conducting tests with specific dependent variables. In this procedure, the process variables were assigned intermediate values in order to carry out the conformity test runs. It was observed that the percentage errors between the predicted and the experimental values are within the acceptable range. Hence, these models can be used for prediction of wear rate and coefficient of friction at 95% level of confidence [10,11].

4.3 Multiobjective optimization

It is a challenging task in selecting the optimal combination for wear and coefficient of friction due to the presence of a number of process variables. Grey relational analysis was employed to optimise the tribological parameters having multiple responses. An integration of artificial neural networks with genetic algorithms is proposed to optimise the multiobjectives of material selection. Taguchi's signal to noise ratio was used for multiobjective optimisation using grey relational analysis. The desirability function approach is one of the most widely used methods in industry for the optimisation of multiple response processes. An application of combined central composite design and desirability approach to improve the multiple performance characteristics of wear of aluminium metal matrix composite was reported by Ravikumar and Sreebalaji [11]. Design Expert software was used to evaluate the desirability value. Wear rate and coefficient of friction were optimized using a set of data derived from RSM. The optimality solution is to evaluate the input process parameters in minimizing the wear rate and coefficient of friction. The range, goals and optimum values of input parameters and the output characteristics, namely, wear rate and coefficient of friction, are given in Table 3.

Table 3 Range of parameters and response for desirability

Sl.no.	Process parameter	Goal	Lower limit	Upper limit	optimum value
1	Load/N	is in range	10	30	30
2	Sliding speed/ms^{-1}	is in range	0.4	1.1	1.1
3	Sliding distance/m	is in range	500	1000	1400
4	Specific wear rate$10^{-7}$$_{(g/Nm)}$	minimise	0.63	3.97	0.85
5	Coefficient of friction	minimise	0.241	0.306	0.241

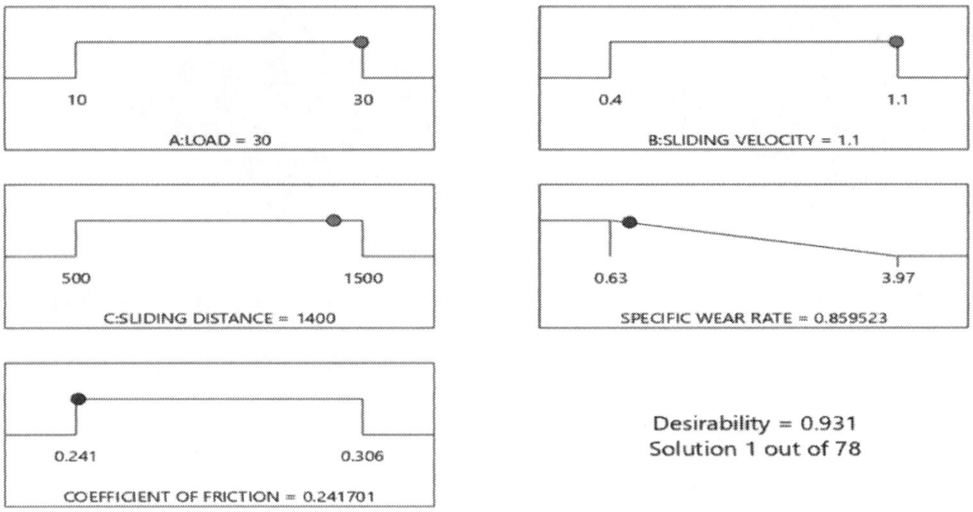

Figure.5 Ramp function graph of desirability

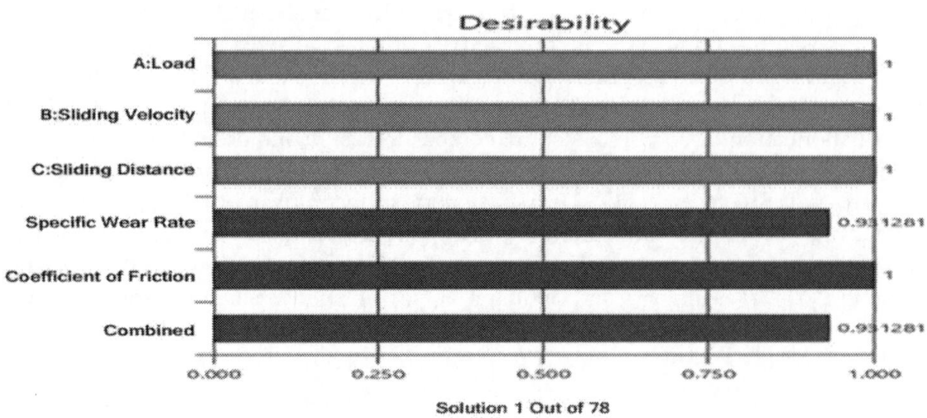

Figure. 6 Bar graph of desirability

IOP Conf. Series: Materials Science and Engineering **390** (2018) 012105 doi:10.1088/1757-899X/390/1/012105

The set of conditions with highest desirability value is selected as optimum condition for the responses. The optimal set of conditions with higher desirability function is given in Table 3. The ramp function graph and bar graph of desirability are shown in Figs. 5 and 6. Factor setting or response prediction for a particular characteristic is indicated by the dot on each ramp. The height of the dot shows the desirability of the response [11]. The overall desirability function of the responses, namely, the wear and coefficient of friction, is shown in the bar graph. The desirability value varies from 0 to 1 depending upon the closeness of the response towards the output. The near optimal region had an overall desirability value of 0.885, indicating the closeness of the target. The optimum parameter set for the current study is as follows: load, 30 N; sliding speed, 1.1 m/ s; sliding distance,1400m and with specific wear rate 0.859 10^{-7} g/Nm; coefficient of friction, 0.241.

5. Conclusion

The influence of process parameters on abrasive wear and coefficient of friction of Al-1.2Al_20_3-0.5Gr hybrid nanocomposites was investigated. A mathematical model was developed to predict the wear rate and coefficient of friction of Al-1.2Al_20_3-0.5Gr hybrid nanocomposite material incorporating the effects of applied load, sliding distance and sliding speed. The adequacy of the proposed model was then investigated using ANOVA. Confirmation tests were carried out to study that the accuracy of the developed model and the deviation between the actual value and the predicted value are within the acceptable limit. Coefficient of friction decreased with increase in sliding speed and load. The parameters were optimised using desirability based multi response optimisation technique to minimise the wear rate and coefficient of friction. Increase of speed reduces wear by supporting mechanically mixed tribolayer and increase of load decreases wear by the role of MML layer.

References

[1] Suresh S, Moorthi N SV, Vettivel S C, Selvakumar N, and Jinu G R, 2014, *Mater. Sci. Eng. A*, **612**, 16-27.

[2] Cao G, Choi H, Konishi H, Kou S, Lakes R, and Li X, 2008 *J. Mater. Sci*, **43,** 5521-5526.

[3] Baradeswaran A, and ElayaPerumal A, 2014, *Composites: Part B*, **56,** 464-471.

[4] Mosleh-Shirazi S, Akhlaghi F, and Li D.Y, 2016,*Tribol. Int*,**103,** 620-628.

[5] Mahdavi S, and Akhlaghi F, 2011, *Tribol. Lett*, **44**, 1.

[6] Kumar R,Dhiman S, 2013,*Mater. Des*, **50**, 351-359.

[7] Ravindran P, Manisekar K, Narayanasamy P, Selvakumar N, Narayanasamy R, 2012, *Mater. Des*, **39**, 42-54.

[8] Radhika N, Raghu R, 2015, *Tribol. Lett*, **59**:2.

[9] Gopalakannan S, Senthilvelan T, 2013, Int J AdvManufTechnol, **67**:485-493.

[10] Kumar S, Balasubramanian V, 2010, *Tribol. Int*, **43**, 414-422.

[11] Ravikumar K, Sreebalaji VS, 2015, *Tribology - Materials, Surfaces & Interfaces*,**9**,128-136.

IOP Conf. Series: Materials Science and Engineering **390** (2018) 012106 doi:10.1088/1757-899X/390/1/012106

Investigation of growth of oxides, amines and halogen compounds in engine oil using FT-IR spectroscopy

Pradeep Kewat

Department of Mining Machinery engineering
Indian Institute of technology(Indian School of Mines), Dhanbad, INDIA

pradeep_kewat@mme.ism.ac.in

Abstract. This study focuses on chemical degradation of engine oil chemical degdration causes formation of undesired chemicals which are harmful to the components os the engine. FTIR spectrum indicates the presence of some of the chemicals like oxisdes, Alkanes, Carbonyl Compounds, Amines, Carbodiimides and Halogen Compounds. The growth of the peaks in FTIR spectrum indicate the formation of the undesired compounds.

1. Introduction.

The functions of engine oil are to keep the engine lubricated and to remove heat, wear debris and combustion products. Wear debris are produced due to wear of the engine various components[1]. Wear of piston rings and cylinder liner results in the increased clearance between them thus passage for the combustion products to interact with the oil[2]–[5]. Interaction of engine oil with wear debris and combustion products under high-temperature condition results in the formation of several undesired chemical compounds[1], [4], [6], [7]. Formation and growth of these compounds are the indications of the degradation of the engine oil and frequent replacement of the engine oil is then required [8], [9]. some of the modern engine oil contains antioxidant and other additives which neutralize the chemical changes occurring in the engine oil and reduces friction and wear thus increased service life is achieved. Formation of oxides in a reaction is the primary indication of decomposition of that chemical. Decomposition of the engine oil results in the formation various undesired chemicals including oxides, amines and halogen compounds. To investigate the formation and growth of these compounds Fourier Transform Infrared Spectroscopy is used in this study. Fourier Transformation infrared spectroscopy is based on the absorption of a specific wavenumber of infrared light by a specific compound or chemical bond this detail is given in table 1 [5], [10]–[12]. A chemical compound or bond absorbs a specific wavenumber thus the absorption of a particular wavelength indicates the presence of that chemical or bond and Intensity of the absorption represents quantity [5], [10], [13], [14]. Wavenumber is an inverse of a wavelength so the graphs were plotted between the inverse of the wavenumber and absorbance. In the experiment, infrared light of the wavenumber 400 to 4000 were allowed to pass through each sample and absorption intensity for each wavenumber is recorded. Some chemicals or chemical bonds absorb numerous distinct wavenumbers. The highest peak of absorption is taken into consideration for the study.

Table 1. Characteristics groups their absorption wavenumber.

Characteristics group	**Absorption Wavenumber**	**Characteristics group**	**Absorption Wavenumber**
Alkanes	2850-2960/1350-1470	Carbodiimides	1580-1370
Alkenes	3020-3080/675-1000	Nitro Compound	1590-1570
Alkynes	3300	Thiocarbonyl Group	975
Acetals	1150	Halogen Compound	1380-1220 /760-640
Ethers	1180-970	Silicon Compound	980-910
Carbonyl Compound	3200-2500	Soot	2000
Acid Chlorides	1790	Oxidation (Carbon)	1700
Anhydrides	1800	Nitrarion	1630
Amides	1640-1590	Sulphation	1150
Amines	830-770 / 3180-2000	Diesel	800
Nitriles	2300	Petrol	750
Anti-wear additive	960	Antifreeze glycol	880

This study is conducted on engine oil of a mine excavator having a 13 litre six-cylinder diesel engine which produces max 265 kW and maximum torque 1800 Nm, with 42 litre capacity of engine oil in crank-case, Volvo Drain Specification-3 (VDS-3) recommends the oil to be replaced after every 500 hours of operation is based on the recommendation from American Petroleum Institute for Compression Ignition four stroke diesel engine refer table 2. With the recommended engine oil fuel is also recommended to have sulphur content less than 3000ppm because Higher sulphur content shortens service interval.

Table.2: Specification of oil used

Part no	VOE 11708321
Viscosity Grade	SAE 15W-40
Quality Level	VDS-3, API CI-4
Kinematic viscosity @100°C	14.432
Viscosity index	135
Flash point	220°C

2. Experimental Setup

Fourier Transform Infrared Spectroscopy (FTIR) The FTIR (Perkin Elmer Spectrum100 Series), Schematic diagram shown in figure 1, it consists of light source, interferometer, sample container, detector, interferogram and computer. Infrared energy is emitted from the glowing source detected by detector. The beam enters into interferometer where the spectral encoding is carried out, Signal then come out of the interferometer and enters into the sample where it transmitted through the sample, depending upon the type of analysis being accomplished. Here frequency of energy absorbed which depends on the characteristic of the sample. Then beam reaches the detector for final measurements. The detectors are specially designed to measure the interferogram signal and further this signal sent to the computer where the Fourier transformation takes place in the form of spectral for interpretation and any further manipulation[11].

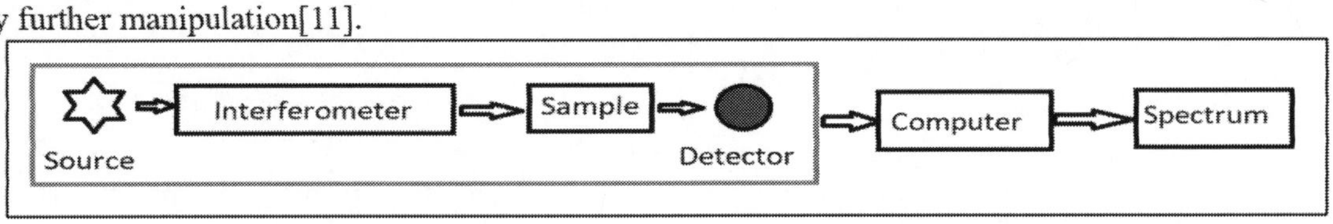

Figure 1. Schematic diagram of FTIR.

IOP Conf. Series: Materials Science and Engineering **390** (2018) 012106 doi:10.1088/1757-899X/390/1/012106

3. Results and discussion

Figures 1 illustrate percentage absorption of infrared light with wavenumbers, in these figures sharp peaks are observed at wavenumbers are 1226, 1376, 1457, 1704, 2360, 2850, 2877 and 2919. Out of which 1376, 1457, 2850,2877 and 2919 are fundamental peaks of the lubricant, fundamental peaks belong to basic constituents of the engine oil.

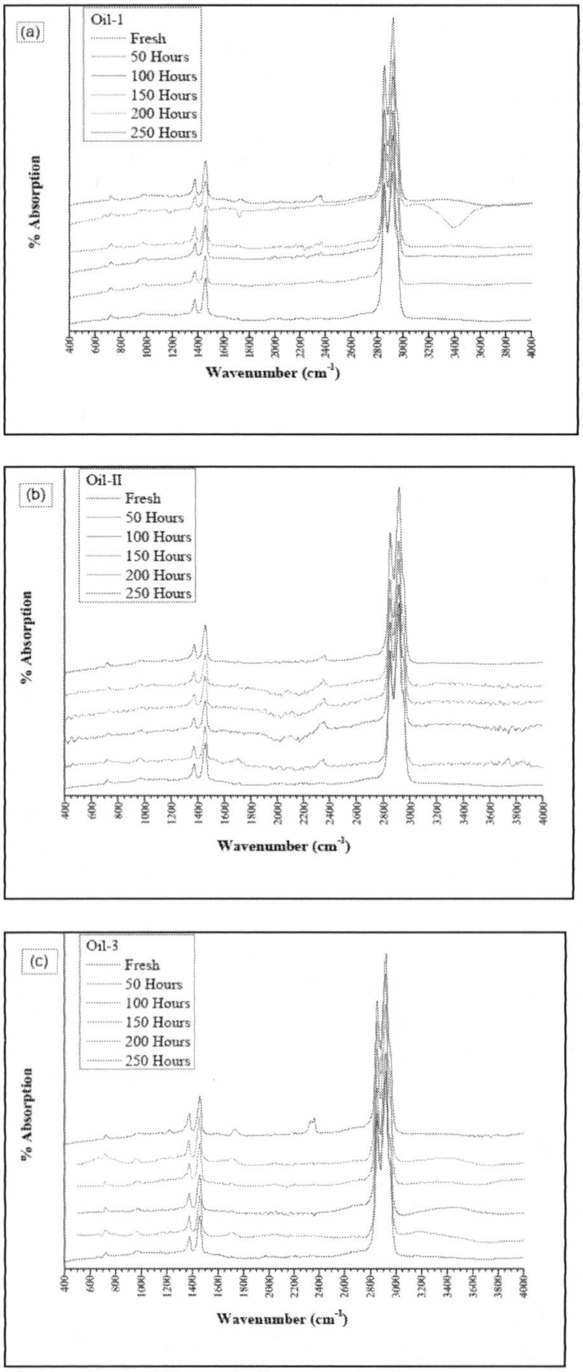

Figure 2. FTIR variation of % Absorbance with wavenumber (a) Oil-I, (b) Oil-II, (c) Oil-III.

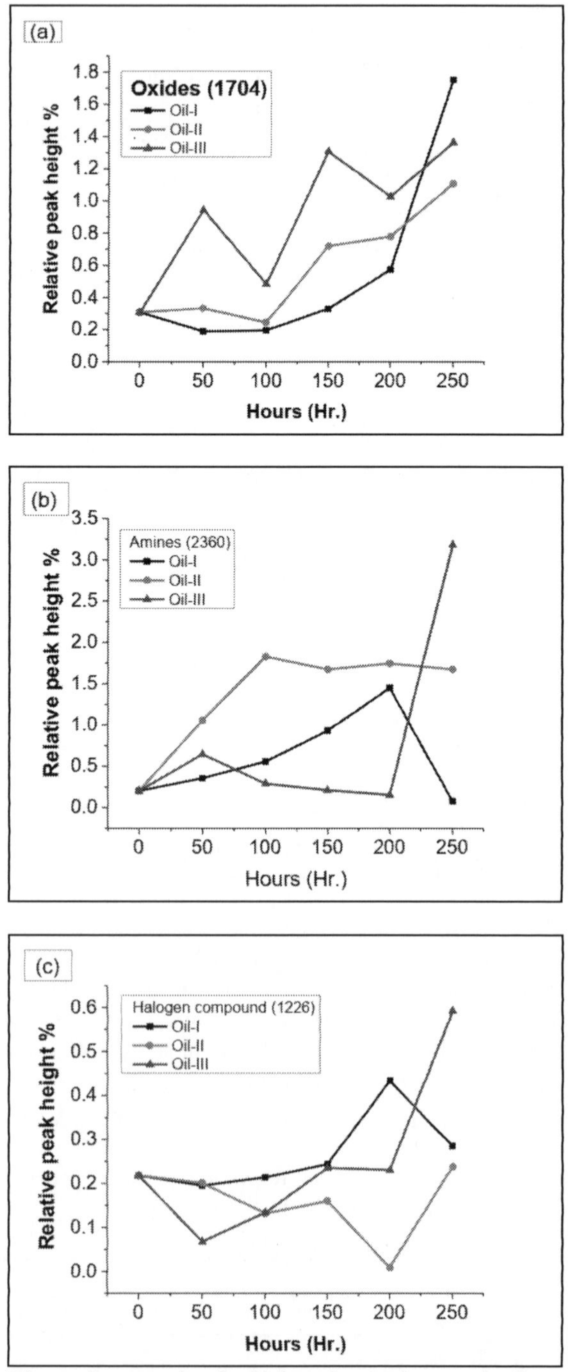

Figure 3. growth of relative absorption peaks for (a)oxides, (b) Amines (c) Halogen compound.

Basic constituents of oil are alkanes, carbonyl compounds, amines, carbodiimides, and halogen compounds refer table 1. Wavenumbers 1226, 1704 and 2360 do not belongs to the basic composition their presence indicates the growth of undesired chemicals and degradation of the engine oil.

3.1 Oxides

Oxidation is one of the most important and critical criterion of the evaluation of the engine oil, the oxidation of engine oil refers to formation of acid in the engine oil the amount of oxidation measured by neutralizing the acid formed in the engine oil and measuring the amount of potassium hydroxide to neutralize the acid formed in the engine oil called total acid number. In an engine oil low total acid number implies a better performance of the engine oil and lower corrosion tendency of the engine oil. In Fourier transformation spectroscopy wavenumber 1704 is the indication of the oxidation of the engine oil, higher peak indicates a higher amount of oxidation. Figure 2(a) illustrates oxidation peaks of Oils, I, II and III for 250 running hours. And side by side comparison of three oils is made, Oil-I, II and Oil-III are showing clear chronological growth of the oxidation in the engine oils. However, Oil II and III initial 50 hours oxidation took place followed by activation of anti-oxidants, which is further followed by oxidation and activation of anti-oxidants. Oil I has gone through the activation of anti-oxidants with starting of the evaluation period causing loss of anti-oxidants in the beginning and rapid oxidation is observed after 200 hours. This may be because of different operating conditions such as mining environment temperature, the load on the machine, and antioxidants present in the engine oil and their activation. Oxidation graphs for oil I, II, III agree to the growth of oxidation in the engine oil with time.

3.2 Amines

Amines are the basic constituents of the engine oil the wavenumbers 2850, 2877 and 2919 belongs to amines, the presence of peak at wavenumber at 2360 also belongs to amines (see table-1). However, the peak at wavenumber 2360 does not belong to the fundamental peak of the engine oil it is important to study this wavenumber due to changes occurring in the absorption peak is rapid. Figure 2(b) belongs to wavenumber 2360, growth of amines in Oil I is observed during initial 200 hours after that fall in concentration of amines is noticed, rapid growth of amines took place in Oil II which is stabilized after 100 hours. However, growth of concentration in amines in Oil III has grown for initial 50 hours followed by gradual fall up-to 200 hours then sudden increase in the concentration of amines is observed.

3.3 Halogen compounds

Halogen compounds are excited by wavenumbers 1380-1220 and 760-640 refer table-1. Noticeable changes are observed in Fourier transform spectroscopy in this range of wavenumbers are at wavenumber 1226 and 1376. Wavenumber 1376 belongs to the fundamental peak of the engine oil but peak corresponding to wavenumber 1226 is not a fundamental peak thus peak at wavenumber 1226 is the undesired growth of wavenumber 1226 is focused in this study. Figure 2(c) illustrates side by side comparison of the variation of the wavenumber 1226. Oil-I has stable peak for 150 hours followed by rise at 200 hours and fall at 250 hours the final amount of halogen compound was stable. Oil II has gone through fall in the halogen peak for 150 hours followed by fall at 200 hours and rise at 250 hours the final peak height has less significant changes. Oil III has gone through fall in halogen compounds for initial 50 hours of operation followed by gradual increase for 200 hours and rapid increase after 200 hours.

4. Conclusion

Activation of anti-oxidants in Oil-I has resulted in rapid oxidation of the oil after 200 hours and fall in Amines and Halogen compounds is observer against its's trend after 200 hours. Oxidation of Oil II and III is neutralized periodically resulted in less end oxidation against Oil-I, Oil-I has suffered 28.56 % more oxidation relative to Oil II and 58.30 % more oxidation relative to Oil III. Formation of Amines is stabilized for Oil II after 150 hours. However, rapid fall is observed in Oil I and rapid increase in observed in Oil III. Rapid formation of halogen compound in Oil II and Oil III after 200 hours is observed unlike Oil I. Oil I and Oil II attended close values those that of beginning but Oil III has a growth of 171.55% in halogen compound relative to initial value.

References

[1] A. K. Hasannuddin *et al.*, "Performance, emissions and lubricant oil analysis of diesel engine running on emulsion fuel," *Energy Convers. Manag.*, vol. 117, pp. 548–557, 2016.
[2] V. U. Garud, "Integrated wireless online oil Condition Monitoring system for IC Engine," *IEEE Digit. Explor. 978-1-4673-9939-5/16.*, pp. 209–214, 2016.
[3] S. R. Chowdhury, R. Kumar, R. Kaur, A. Sharma, and A. P. Bhondekar, "Quality Assessment of Engine Oil : An Impedance Spectroscopy Based Approach," pp. 608–612, 2016.
[4] R. Taheri Ghahrizjani, H. Sadeghi, and A. Mazaheri, "A Novel Method for onLine Monitoring Engine Oil Quality Based on Tapered Optical Fiber Sensor," *IEEE Sens. J.*, vol. 16, no. 10, pp. 3551–3555, 2016.
[5] S. Zzeyani, M. Mikou, J. Naja, and A. Elachhab, "Spectroscopic analysis of synthetic lubricating oil," *Tribol. Int.*, vol. 114, no. April, pp. 27–32, 2017.
[6] M. J. Souza De Carvalho, P. Rudolf Seidl, C. R. Pereira Belchior, and J. Ricardo Sodr, "Lubricant viscosity and viscosity improver additive effects on diesel fuel economy," *Tribol. Int.*, vol. 43, no. 12, pp. 2298–2302, 2010.
[7] S. George, S. Balla, V. Gautam, and M. Gautam, "Effect of diesel soot on lubricant oil viscosity," *Tribol. Int.*, vol. 40, no. 5, pp. 809–818, 2007.
[8] V. Macián, B. Tormos, P. Olmeda, and L. Montoro, "Analytical approach to wear rate determination for internal combustion engine condition monitoring based on oil analysis," *Tribol. Int.*, vol. 36, no. 10, pp. 771–776, 2003.
[9] A. D. M. Rosman, A. H. A. Razak, M. F. Azmi, S. A. M. Al Junid, A. K. Halim, and M. F. M. Idros, "Capability of Indium Tin Oxide as a sensing mechanism for engine oil degradation monitoring," *Proc. - 6th IEEE Int. Conf. Control Syst. Comput. Eng. ICCSCE 2016*, no. November, pp. 546–550, 2017.
[10] C. Zhao, Y. Jiao, Y. K. Chen, and G. Ren, "The Tribological Properties of Zinc Borate Ultrafine Powder as a Lubricant Additive in Sunflower Oil," *Tribol. Trans.*, vol. 57, no. 3, pp. 425–434, 2014.
[11] R. K. Upadhyay, L. A. Kumaraswamidhas, and M. M. Engineering, "Condition Based Monitoring of Ch 4 15W40 Engine Oil By Ftir , and Analytical Ferrography," vol. 3, no. 1, pp. 1–7, 2013.
[12] M. C. Garry and J. Bowman, "FT-IR Analysis of Used Lubricating Oils – General Considerations," *Thermo Sci.*, 2007.
[13] M. F. Sgroi *et al.*, "Engine bench and road testing of an engine oil containing MoS2particles as nano-additive for friction reduction," *Tribol. Int.*, vol. 105, no. July 2016, pp. 317–325, 2017.
[14] L. Xiang, C. Gao, Y. Wang, Z. Pan, and D. Hu, "Tribological and tribochemical properties of magnetite nanoflakes as additives in oil lubricants," *Particuology*, vol. 17, pp. 136–144, 2014.

Optimization of WEDM Process Parameters on Machining of AZ91 Magnesium alloy using MOORA method

A.Muniappan*,Mantri Sriram, C.Thiagarajan ,G.Bharathi Raja & T.Shaafi
Department of Mechanical Engineering, Saveetha School of Engineering,
Saveetha Institute of Medical and Technical Sciences, Chennai, Tamil Nadu, India

*E-Mail: inspire.munish@gmail.com

Abstract. Selection of optimum machining parameters in wire electrical discharge machining (WEDM) operations leads to good functional attributes for the machined components and increased productivity. In this present experimental work, the effect and advancement of machining control parameters over kerf and cutting speed in wirecut electrical discharge machining operations were analyzed. Multi objective optimization by ratio analysis (MOORA) applied for optimization of kerf width (KW) and cutting speed (CS). Magnesium AZ91 alloy is selected as work material of dimension 100 x 100 x 10 mm. The analyses were outlined with Taguchi L27 design matrix. WEDM parameters resemble Pulse on time, Pulse off time, current, gap voltage, wire speed & wire tension are considered. The effect of the cutting control parameters on the kerf width and cutting speed is directed by using examination of fluctuation. The parameters corresponding to experiment run number 2 are pulse on duration 106 units (Level 1), pulse off duration40 units (Level 1), peak pulse current 150 units (Level 2), gap voltage30 units (Level 2), wire feed 6 units (Level 2) & wire tension 8 units (Level 2) are the best combination to achieve better CS and KW.

Keywords: MOORA, WEDM, Taguchi method, AZ91, kerf width & cutting speed.

1. Introduction

Wire EDM has all the earmarks of being exceptionally encouraging strategy among all non-conventional machining strategies [1]. The industrial growing sectors utilizing Wire EDM technology comes under five main different categories: tool and die, power generation, aerospace, automotive, oil and gas industries. WEDM maintains the largest percentage of share, as much as 50% and ECM about 15% lagging behind laser processes which are 20%.In many application areas which offers several benefits that include better precision, higher cutting rate & controlled material removal and also different types of materials that can be processed. The main aim of Wire EDM users& manufactures is to obtain a better steadiness of the machining process and increased productivity. More colorful materials are produced, and more mind boggling shapes are introduced and regular machining activities achieve their restrictions; the expanded utilization of Wire EDM in assembling keeps on developing at a quickened rate. Wire EDM makers and client accentuate on accomplishment of higher machining profitability with a coveted exactness and surface wrap up. Magnesium composites are combinations of magnesium alloys with other metals, often aluminum, manganese, copper, zinc, silicon rare earths metals and zirconium [2]. Magnesium compounds have hexagonal cross section structure which influences the essential properties of this composite. That is why, alloys of magnesium are mostly used as cast alloys & wrought alloys which has been more extensive. Cast magnesium alloys are utilized for a little parts of present day auto parts, and magnesium square; incredible magnesium is likewise utilized for camera bodies and in focal points segments.

The selection of proper machining variables leads to accuracy, quality and functional attributes of machined component [3]. Kerf width and cutting speed are important objective responses in WEDM operations to obtain high machining performances [4]. Proper selection of control variables reduces the kerf width and increases cutting speed [5]. The optimal machining variables setting plays a significant role to improve the performance of any machining operation. In real time manufacturing, the decision making process is more difficult due to various interests and values of different decision makers. There is a need for simple, systematic and logical procedure to solve decision-making problems effectively. The MOORA method is one of the Multi criteria Decision Making methods which use statistical procedure for the selection of the best alternative from the given alternatives. This method generates most suitable alternatives by considering both beneficial and non beneficial alternatives and eliminates unsuitable alternative from the given alternatives for strengthening the existing selection procedure. The MOORA method involves lesser computations, comprehensiveness and robustness which can solve multiple number of criteria simultaneously [6]. The overall material grades of magnesium is available, in the research work the grade used to do the research is identified to be AZ91, which had most application in aircraft industry. The material magnesium AZ91 is selected for the process where these metals have good mechanical strength and are conductive. This research work also aims at investigating the main effects of the different Wire EDM control parameters like as wire speed, wire feed, Pulse on and Pulse off time on KW by using the technique named Taguchi. MINITAB version 18.0 software was used so as near optimize the operating variables of Wire EDM for magnesium AZE91.

2. Materials &method

AZ91 Magnesium alloys

The mass fabrication of lightweight metal matrix castings for moderate strength applications requires the make use of of high volume die-casting techniques. The predominant magnesium alloy used for both die as well as sand casting is AZ91, containing approximately 9% aluminum and 1% zinc, which possess excellent casting properties and reasonable strength.

Machining control variables

Machining process parameters in WEDM, Pulse on time, Pulse off time, current, gap voltage, wire speed and wire tension were considered as input parameters. Responses CS and KW were measured after machining for investigation. The scopes of these procedure parameters were chosen on the origin of the pilot tests. The different parameters level and its assignments are exhibited in Table2.

Table 1. Control Variables

Symbol	Process parameter	Level 1	Level 2	Level 3
A	Pulse on time	106	116	126
B	Pulse off time	40	50	60
C	Pulse current	70	150	230
D	Gap Voltage	20	30	40
E	Wire feed	4	6	8
F	Wire tension	4	8	12

Experimental Design by Taguchi technique

Taguchi approach is a proficient apparatus for the summarize of a great assembling framework. It also effectively solves many difficult problems in manufacturing [7].It is a strategy in view of OA tests, which give much decreased change to the try different things with ideal setting of process control parameters. The six control variables, that is, Pulse on time (A), Pulse off time (B), current (C),Gap voltage (D), wire feed (E) and wire tension(F) with three levels were chosen in this examination. The tests were finished by Table 3. This table just speaks to specific level of the different elements of the procedure at which the examinations would be directed. Kerf width ought to be as least as conceivable in the WEDM procedure. Kerf width is an imperative component of the laser cutting procedure that gives the benefit of this innovation contrasted with different strategies for form cutting.

Experimental Set Up

Analyses were led on Electronica Sprint cut CNC wire cut electrical discharge machine to think about the cutting speed and kerf width influenced by the machining control parameters at various levels. WEDM is a start disintegration process. The flashes are produced between the work material and the wire terminal. The dielectric liquid is ceaselessly encouraged into the cutting zone with required weight. The material is getting expelled by a progression of discrete sparkles occurring at the zone to be machined through electro-warm system. Test set up of the wire electrical release machine is appeared in Figure 1.

Amid machining process little gap kept up between the work and wire material. The wire is held by a stick direct at the upper & lower parts of the work piece. The work example measure utilized as a part of this examination is 100 x 100 x 10 mm rectangular plate. Zinc covered metal brass wire of 0.25 mm diameter was considered for part of this investigation. Deionized water was utilized dielectric liquid at room temperature. In the wake of machining, the examples were cleaned with acid after machining. The kerf was measured utilizing profile projector measuring framework. The kerf esteems were measured at six spots spread over the entire length of cut. The kerf esteems utilized a part of this examination are the numerical normal of three estimations produced using the example in each cut. The cutting speed has been observed from machine display board.

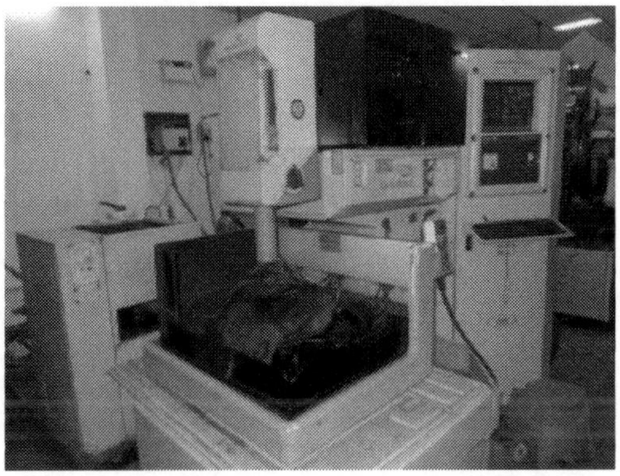

Fig. 1 WEDM experimental set up (Electronica Sprint cut)

Methodology

MOORA is the one of the MCDM techniques used to choose best options among a given number of choices [9]. This issue includes different goals or criteria and furthermore struggle with each other. They are advantageous (amplification) furthermore, non-valuable (minimization) targets. MOORA technique considers both advantageous and non gainful targets for illuminating and positioning ideal options all the while.

Step 1: Description of problem and objectives.
In the present analysis, cutting speed& kerf width are selected as non-beneficial attribute.
Step 2: Construction of decision matrix.
The decision matrix which is used to represent the experimental results with respect to various output parameters. The matrix D_{mxn} is represented as

$$D_{mxn} = \begin{bmatrix} X_{11} & X_{12} & \cdots & \cdots & X_{1n} \\ X_{21} & X_{22} & \cdots & \cdots & X_{2n} \\ \cdots & \cdots & \cdots & \cdots & \cdots \\ \cdots & \cdots & \cdots & \cdots & \cdots \\ X_{m1} & X_{m2} & \cdots & \cdots & X_{mn} \end{bmatrix} \quad (4)$$

where Xi j indicates the experimental result of ith alternative on jth attribute, m indicates the number of experiments, and n refers to the number of output parameters. The decision matrix is given in Table 4.

Step 3: Normalization of input data.

Generally, normalization is needed, as the variety and unit of output value may be different from others. The meaning of normalization is converting the original score into a comparable score. The output values presented in the decision matrix are normalized with the help of equation (3). The value of normalized decision matrix N27_3 is presented in Table 4. The expression used to determine the normalized decision matrix (Ni j) is given by

$$N_{ij} = \frac{x_{ij}}{\sqrt{\sum x_{ij^2}}} \quad \text{for} \quad i = 1, m : j = 1, \ldots, n, \tag{5}$$

where xi j and Ni j are original and normalized score of decision matrix, respectively.

Step 4: Determination of solution.

The standardized scores are included the instance of valuable or amplification objective and subtracted on account of non-helpful or minimization objective. At that point the multi-target streamlining moves toward becoming

where g speaks to the quantity of properties for maximization,(n-g) speaks to the quantity of traits for minimization, and Yi is the standardized evaluation estimation of ith elective regarding all characteristics. Table 4 demonstrates the standardized appraisal estimation of chose yield parameters

$$Y_i = \sum_{j=1}^{g} N_{ij} - \sum_{j=g+1}^{n} N_{ij}, \tag{6}$$

Step 5: The attributes being considered are more important than others in practical situations.

To identify the important attribute, it must be multiplied with its relative importance. If relative importance is considered, then the equation is modified as

$$Y_i = \sum_{j=1}^{g} W_j N_{ij} - \sum_{j=g+1}^{n} W_j N_{ij} \quad (j = 1, 2, \ldots, n), \tag{7}$$

where Wj represents weight of j_{th} attribute and it is calculated by entropy measurement method.

3. Result and discussion

The experimental values cutting speed and kerf width are tabulated in table 2.Table 3 shows the square of Xi, normalized matrix for Xi and rank calculated by MOORA method. The normalized CS & KW are multiplied with their corresponding weights. These values are listed in the table 6. The parameters, higher the better (maximum) and smaller the better (minimum) respectively added for rank calculation. The cutting speed observed in the experiment is in the range of from 1.92 to 7.52 mm/min. Similarly the kerf width observed in the experiment is in the range of from 0.32 to 0.374 mm. From these observations, experiment number 2 has the best rank. The parameters corresponding to experiment run number 2 are pulse on time 106 units (Level1), pulse off time 40units (Level1), peak current 150 units (Level 2), gap set voltage 30 units (Level 2), wire feed 6 units (Level2) and wire tension 8 units (Level2).

Table. 2 Cutting speed& Kerf width

Ex.No.	Pulse on Time	Pulse off Time	Current	Gap voltage	Wire Feed	Wire Tension	CS	KW
1	106	40	70	20	4	4	6.32	0.323
2	106	40	150	30	6	8	7.52	0.32
3	106	40	230	40	8	12	6.89	0.321
4	106	50	70	30	6	12	3.47	0.33
5	106	50	150	40	8	4	3.49	0.332
6	106	50	230	20	4	8	5.51	0.337

7	106	60	70	40	8	8	1.92	0.327
8	106	60	150	20	4	12	3.06	0.33
9	106	60	230	30	6	4	2.83	0.341
10	116	40	70	30	8	8	6.12	0.346
11	116	40	150	40	4	12	6.25	0.341
12	116	40	230	20	6	4	5.98	0.358
13	116	50	70	40	4	4	4.03	0.374
14	116	50	150	20	6	8	6.01	0.371
15	116	50	230	30	8	12	5.98	0.365
16	116	60	70	20	6	12	3.08	0.337
17	116	60	150	30	8	4	3.79	0.348
18	116	60	230	40	4	8	3.68	0.358
19	126	40	70	40	6	12	5.78	0.356
20	126	40	150	20	8	4	5.92	0.351
21	126	40	230	30	4	8	5.97	0.352
22	126	50	70	20	8	8	5.17	0.347
23	126	50	150	30	4	12	5.99	0.364
24	126	50	230	40	6	4	5.97	0.364
25	126	60	70	30	4	4	3.14	0.361
26	126	60	150	40	6	8	4.92	0.372
27	126	60	230	20	8	12	5.98	0.349

Table. 3 Normalized decision matrix & MOORA rank

Ex.No.	Square of Xi		Normalized Xi		Wj*Xi		∑max-∑min	Rank
	CS	KW	CS	KW	CS	KW		
1	39.9424	0.104329	0.2340	0.1788	0.11699	0.08942	0.02757	3
2	56.5504	0.1024	0.2784	0.1772	0.13921	0.08859	0.05062	1
3	47.4721	0.103041	0.2551	0.1777	0.12754	0.08887	0.03868	2
4	12.0409	0.1089	0.1285	0.1827	0.06423	0.09136	-0.02712	19
5	12.1801	0.110224	0.1292	0.1838	0.06460	0.09191	-0.02731	20
6	30.3601	0.113569	0.2040	0.1866	0.10200	0.09329	0.00870	13
7	3.6864	0.106929	0.0711	0.1811	0.03554	0.09053	-0.05498	27
8	9.3636	0.1089	0.1133	0.1827	0.05664	0.09136	-0.03471	23
9	8.0089	0.116281	0.1048	0.1888	0.05239	0.09440	-0.04202	26
10	37.4544	0.119716	0.2266	0.1916	0.11329	0.09579	0.01750	5
11	39.0625	0.116281	0.2314	0.1888	0.11570	0.09440	0.02129	4
12	35.7604	0.128164	0.2214	0.1982	0.11070	0.09911	0.01159	9
13	16.2409	0.139876	0.1492	0.2071	0.07460	0.10354	-0.02894	21
14	36.1201	0.137641	0.2225	0.2054	0.11125	0.10271	0.00855	14
15	35.7604	0.133225	0.2214	0.2021	0.11070	0.10105	0.00965	12
16	9.4864	0.113569	0.1140	0.1866	0.05701	0.09329	-0.03628	24
17	14.3641	0.121104	0.1403	0.1927	0.07016	0.09634	-0.02618	18
18	13.5424	0.128164	0.1362	0.1982	0.06812	0.09911	-0.03099	22

19	33.4084	0.126736	0.2140	0.1971	0.10700	0.09855	0.00844	15
20	35.0464	0.123201	0.2192	0.1943	0.10959	0.09717	0.01242	8
21	35.6409	0.123904	0.2210	0.1949	0.11051	0.09745	0.01307	7
22	26.7289	0.120409	0.1914	0.1921	0.09570	0.09606	-0.00036	16
23	35.8801	0.132496	0.2218	0.2015	0.11088	0.10077	0.01011	10
24	35.6409	0.132496	0.2210	0.2015	0.11051	0.10077	0.00974	11
25	9.8596	0.130321	0.1163	0.1999	0.05813	0.09994	-0.04181	25
26	24.2064	0.138384	0.1822	0.2060	0.09108	0.10298	-0.01191	17
27	35.7604	0.121801	0.2214	0.1932	0.11070	0.09662	0.01408	6

Conclusions

This paper has presented application of MOORA method for the selection of optimal control parameters in machining of AZ91 magnesium alloy using WEDM. The MOORA method, indeed, considers all the attributes along with their relative importance which enables this method to provide not only a better but also an accurate evaluation of the alternatives. As compared to many other MODM methods, this method is found to be simple, logical and robust. This method can simultaneously consider any number of quantitative and qualitative selection of attributes and is able to provide optimum solution quickly as computational time necessary to obtain the solution is reasonably low. MOORA method is employed to select the optimum machining parameters in WEDM machining of AZ91 Magnesium alloy with zinc coated brass wire electrode. From these observations, experiment number 2 has the best rank. The parameters corresponding to experiment run number 2 are pulse on time 106 units (Level1), pulse off time 40 units (Level 1), peak current 150 units (Level 2), gap set voltage 30 units (Level 2), wire feed 6 units (Level2) and wire tension 8 units (Level 2).The optimum results are adopted in validation study and the results based on WEDM process responses can be effectively improved.

References

[1] Chakraborty, S., (2010). Applications of the MOORA method for decision making in manufacturing environment, International Journal of Advanced Manufacturing Technology, 54(9•12), 1155•1166, d oi: 10.1007/s00170•010•2972•0.

[2] Snoeys R, Staelens F, Dekeyser W. "Current trends in non – conventional material removal process" Annals of CIRP, Vol. 35/2 (1986), pp. 467-480.

[3] Tatar, C., Ozdemir, N. (2010). "Investigation of thermal conductivity and microstructure of the a-Al_2O_3 particulate reinforced aluminium composites (Al/Al_2O_3-MMC) by powder metallurgy method". Physica B Condensed Matter. Vol.405, pp.896-899.

[4] Brauers, W. K. M., (2008). Multi•objective contractor's ranking by applying the MOORA method, J ournal of Business Economics and Management, 4,pp245–255.

[5] Brauers, W. K. M., and Zavadskas, E. K., (2009). Robustness of the multiobjective MOORA method with a test for the facilities sector, Technological and Economic Development of Economy: Baltic Jou rnal on Sustainability, 15(2), pp352–375.

[6] Gadakh, V.S., "Application of MOORA method for parametric optimization of milling process", International Journal of Applied Engineering Research, Dindigul, Vol. 1, No. 4, pp. 743–758, 2011.

[7] Rajesh, S., Rajakarunakaran, S., Suthakarapandian, R. and Pitchipoo, P., "MOORA-Based tribological studies on red mud reinforced aluminum metal matrix composites", Advances in Tribology, 2013.

[8] Brauers, W. K. M., Zavadskas, E. K., Peldschus, F., and Turskis, Z., (2008). Multiobjective de cision•making for road design, Transport,23(3),pp183–193.

[9] Jangra, K., Grover, S. and Aggarwal, A., "Optimization of multi machining characteristics in WEDM of WC5.3% Co composite using integrated approach of Taguchi, GRA and entropy method", Frontiers of Mechanical Engineering, Vol. 7, No. 3, pp. 288–299, 2012.

[10] Muniappan .A, Thiagarajan .C and Somasundaram .S (2017) Parametric Optimization of kerf width and surface roughness in Wire Electrical Discharge Machining (WEDM) of hybrid Aluminium (Al6061/SiC/Graphite) composite using Taguchi-based Gray Relational Analysis" International Journal of Mechanical & Mechatronics Engineering IJMME-IJENS Vol:16 No:05, pp. 95-103.

Influence of Friction Stir Processing Parameters on Tensile properties and Microstructure of Dissimilar AA 8011-H24 and AA 6061-T6 aluminum alloy joints in Nugget Zone

K Palani[1], C Elanchezhian[2], Karanam Avinash[1], C Karthik[1], Karra Chaitanya[1], Karthik Sivanur[1], K Yugandhar Reddy[1]

[1] Department of Mechanical Engineering, Sri Chandrasekharendra Saraswathi Viswa Mahavidhyalaya, Enathur, Kanchipuram, 631 561, Tamilnadu, India
[2] Department of Mechanical Engineering, Sri Sairam Engineering College, West Tambaram, Chennai, 600 044, Tamil Nadu, India

E-Mail: palani.k@kanchiuniv.ac.in, avinashkaranam09@gmail.com.

Abstract. The recently developed innovative joining process, Friction stir welding (FSW) is connected to make the best quality of strength and hardness of the welded joints. The lightweight nonferrous materials are focused on aircraft hydraulic systems, fuel tank and transportation ventures due to its high strength to weight ratio and corrosion resistance. The joining of aluminum alloys is extremely troublesome in the conventional fusion welding process because of its cementing shrinkage, hot breaking, scatter, fumes and residual stresses. The friction stir processing is connected to enhance the mechanical properties of the friction stir welded joints by the addition of Nanoparticles viz., SiC and Al_2O_3. The investigation was made on the microstructural analysis of the similar and dissimilar friction stir welded and processed AA 8011-H24 and AA 6061-T6 aluminum alloy joints by Scanning Electron Microscope and the presence of alloying elements in the welded region was found by Electron Dispersive Spectroscopy.

1. Introduction

The Welding Institute (TWI), Cambridge, UK in 1991 introduced the new material joining technique of Friction stir welding (FSW) which is very efficient, environmentally friendly, and flexible process to produce the material joining [1]. Compared to the fusion welding process, the outstanding advantage of the FSW process is that complete melting does not occur near the weld zone. This includes different process parameters like rotational speed, weld speed, plunge depth and tool parameters like tool pin diameter, shoulder diameter, pin profile etc. This process is applied to produce the good mechanical properties of the joints of the different alloy materials in a solid-state due to the formation of intermetallic compounds. The process parameters mainly affect the properties like heat dissipation, plastic flow and grain structure of the weld region, whereas the tool pin parameters affect the localized heating and material flow [2]. The requirements of lightweight materials like aluminium alloys are increasing in various fields like automation industry, aerospace, chemical and marine industrial applications due to its properties of high strength, low cost, low melting point and better metallurgical properties [3]. Welding of these aluminium alloys by Friction Stir welding process has an advantage over the fusion welding to minimize the defects such as porosity, slag formation, solidification shrinkage [4]. The characteristics of both similar and dissimilar friction stir welded aluminium alloys has shown the involvement of recrystallization phenomena as the basic cause to pave for the occurrence plastic deformation [5].

IOP Conf. Series: Materials Science and Engineering **390** (2018) 012108 doi:10.1088/1757-899X/390/1/012108

In the joining of dissimilar materials, a mixture of processes occurring in the stirred zone is expected to generate even more complex microstructure. Hence, the dissimilar welding of aluminium alloys is believed to improve the specific characteristics of weld joint viz., hardness, tensile strength [6]. The different tempered AA6013 aluminium alloys, namely T4 and T6 was successfully welded with friction stir welding process by Heinz B et al. [7] and also found the elongated needle-like structure and fine grains in the thermomechanically affected zone (TMAZ) and stir zone (SZ) due to dynamic recrystallization of the materials. The microstructural changes in dissimilar welds due to various process parameters have been widely studied to understand the behavior of AA 6061 aluminium alloy [8]. It is reported that the development of fine equiaxed grains and the occurrence of residual stress were found in the stir zone of the friction stir welding AA 6061 aluminium alloy joints Muthukumaran S et al. [9]. The material flow and plastic deformation which was controlled by the geometry of the tool via generated frictional between the FSW tool and the butt surfaces of the base materials [10]. The newly designed non-consumable tool which has more hardness than the base alloy materials [11, 12] and it consists of a different pin diameter, shoulder diameter and pin length which are used to produce butt joints. A large amount of frictional force developed by the tool shoulder which makes the recrystallization of the atoms in the nugget zone [13].

Hou JC et al. [14] made an investigation on the effects of the tool rotational speed with self-reacting friction stir welding (FSW) tool of the butt joint properties of aluminium alloys. They reported that the defect-free sound welds were observed for the lower tool rotational speeds, whereas the voids at the root of the tool were found for the higher rotational speeds with an increase in grain size at the weld nugget zone. The interface of the two alloying elements has the lower hardness relate to the stir zone and base metal zone. The plasticized materials were formed due to the plunging force and rotational speed of the tool pin and these materials were allowed to move towards the axis of the tool [15]. The formation of the oxide layers in the welded joints of aluminium alloys based on the proper selection of input parameters, viz., tool rotational speed, welding speed etc., were studied and breaking of these layers were investigated in addition to Nanoparticles [16]. The material flow depends on the tool geometry, such as pin shape, tool pin diameter and tool shoulder diameter and which is decided by the transverse speed of the FSW tool. The formation of voids in the aluminium alloys minimized by the tool shoulder [17]. The authors [18] have made a joining of AA6061-T6 aluminium alloys using friction stir welding and studied the behavior of the material flow and heat transfer of the aluminium alloys with the cylindrical concave tool. The friction stir processing (FSP) improves the aluminium alloys and their combinations toughness and flexibility in the specific area of the microstructure of the aluminium alloys with the formation of fine grains [19].

It is evident that very limited work carried out earlier on the friction stir processing of different grades of aluminium alloys. In this investigation, an attempt has been done to study the influence of FSP parameters on the tensile properties and microstructure of dissimilar AA8011-H24 and AA6061-T6 aluminum alloy joints in Nugget Zone.

Table 1 Chemical compositions (wt. %) and Mechanical properties of the AA 6061-T6 and AA 8011-H24 aluminium alloys

Alloy	Mg	Mn	Cu	Fe	Si	Zn	Ti	Al	Tensile strength Mpa	Density kg/m^3
AA 6061-T6	1	0.15	0.27	0.7	0.6	0.6	0.15	Balance	310	2700
AA 8011-H24	0.1	0.1	0.1	0.75	0.7	0.1	0.05	Balance	140	2689

2. Experimentation work

2.1 Selection of Workpiece Materials

The precipitation hardened, solutionized and artificially aged AA 6061-T6 aluminium alloy containing magnesium and silicon as major alloying elements which are placed on the advancing side, whereas the AA 8011-H24 aluminium alloy (purchased from Ultimate Enterprises, Chennai) [20] is placed on the retreating side. The chemical composition of the alloy elements in (Wt. %) and mechanical properties of the aluminium alloys are given in Table 1. The aluminium plates were cut to the dimensions of 150 mm x 150 mm x 6 mm by power hacksaw and machined by grinding. Furthermore, the cut plates were cleaned in an ultrasonic sonicator bath and cleaned with acetone.

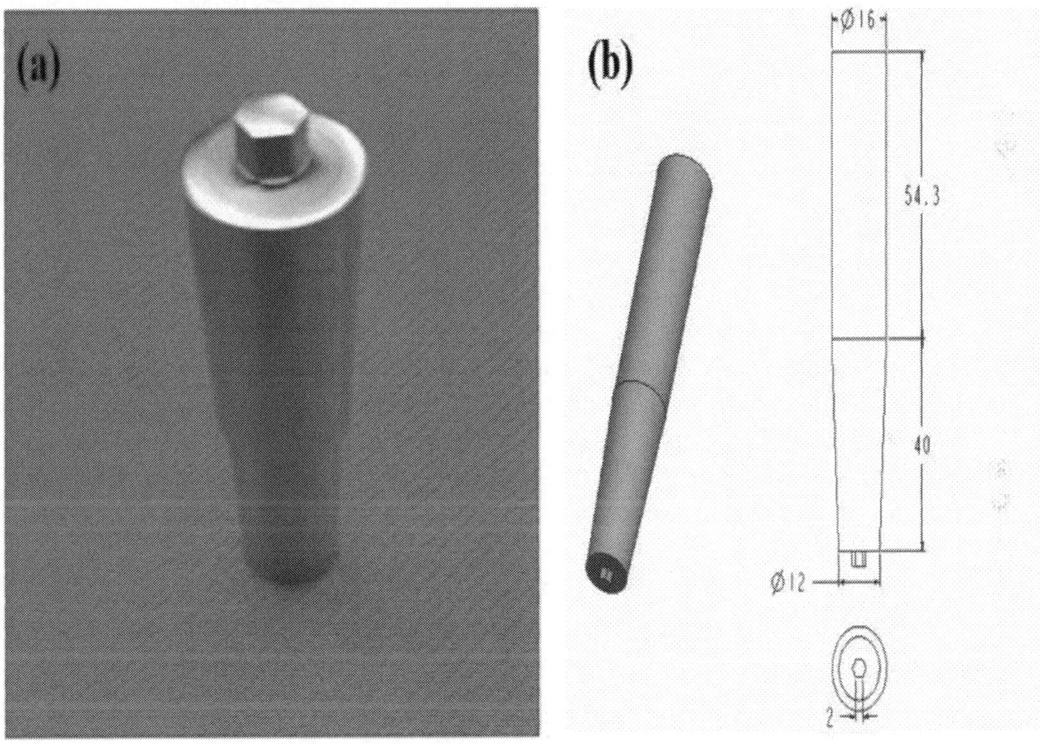

Figure 1 FSW Hexagon pin tool

2.2 Selection of Tool materials

The tool material, D2 grade steel (High Carbon-High Chromium steel) tool (Procured from SRK Diamond tools, Chennai) was selected for friction stir welding process to join the AA 8011-H24 and AA 6061-T6 aluminium alloys. The non-consumable FSW tool was designed based on the requirements with the dimensions of shoulder diameter 18 mm, shank diameter 16 mm, pin length of 5.7 mm, pin diameter of 6 mm and pin side length of 2 mm with hexagonal pin profile shown in Figure 1 which is used for joining of these alloys.

Table 2 Different combinations of aluminium alloys

Sl.No.	Material Combinations of Aluminium alloys	Status of Nano particles used
1	AA 8011 - AA 8011	Without Nano Particles
2	AA 8011 - AA 6061	Without Nano Particles
3	AA 6061 - AA 6061	Without Nano Particles
4	AA 8011 - AA 8011	With Al_2O_3 Nano Particles
5	AA 8011 - AA 6061	With Al_2O_3 Nano Particles
6	AA 6061 - AA 6061	With Al_2O_3 Nano Particles
7	AA 8011 - AA 8011	With SiC Nano Particles
8	AA 8011 - AA 6061	With SiC Nano Particles
9	AA 6061 - AA 6061	With SiC Nano Particles

In this investigation, the nine different aluminium alloy material combinations shown in Table 2 was friction stir welding and processing in addition to two different nanoparticles, such as SiC and Al2O3. The process parameters are the key factors in deciding the quality of the welded joints which is obtained in a similar way [21]. The FSW parameters were considered, such as 1200 RPM tool rotational speed, 50 mm/min welding speed and plunge depth of 2 mm in joining of two different grades of aluminium alloy in butt joint configuration. The microstructure of the friction stir welded joints were examined by preparing the specimens based on the ASTM E340-15 standard after dipped into the Keller's reagent for 20 seconds and immediately washed through water [22] and it reveals the good microstructure at the welded region. The Optical microscope, Scanning Electron Microscope (SEM) with Electron Dispersive Spectroscopy (EDS) respectively were used to find the macrostructure, microstructure analysis of the dissimilar welded joints. The tensile test was conducted on the welded specimen which was prepared as per ASME Sec IX standards.

3. Results and Discussions

The following results have been observed from the testing done for the weld specimens with different specifications.

3.1 Macrostructure of the FSW/FSPed Joints in the Nugget Zone

The defect-free sound weld was found to improve the joint strength by the macrostructure examination of the welded joints. Table 3 shows the macrostructure of the friction stir welding and processing joints of the different grades of aluminium alloys in the weld nugget zone. The appearance of tunnel defects was observed in the both similar and dissimilar AA 8011 and AA 6061 aluminium alloys without nanoparticles due to lack of material mixing in the root of the tool inside the aluminium alloys in the stir zone. The defect-free friction stir processed joints were observed on the dissimilar aluminium alloys due to the proper mixing of material and plastic deformation in the stir zone of both FSW/FSP welded joints and similar to Threadgill PL et al. [23].

IOP Conf. Series: Materials Science and Engineering **390** (2018) 012108 doi:10.1088/1757-899X/390/1/012108

Table 3 Macrostructure of the FSW/FSPed Joints in the Nugget Zone

Material Combinations	Macrostructure	Name of the Defects	Probable reason
(Without Nano Particles)			
AA8011 - AA 8011	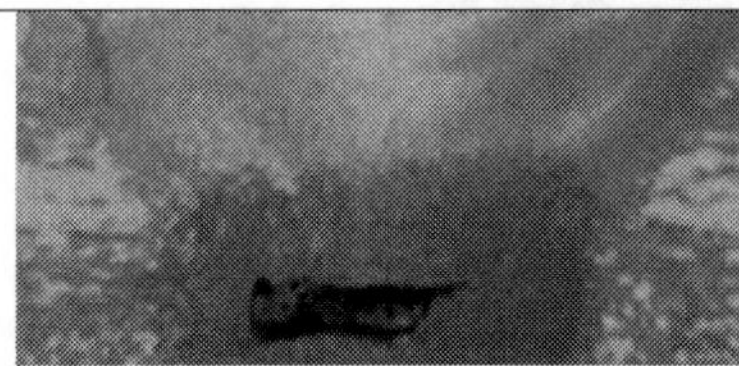	Tunnel Defect	Tunnel defect is identified due to improper heat generation.
AA6061 - AA6061		Tunnel defect	Cavities were observed in this joint due to uneven mixing of nano particles.
AA8011 - AA6061	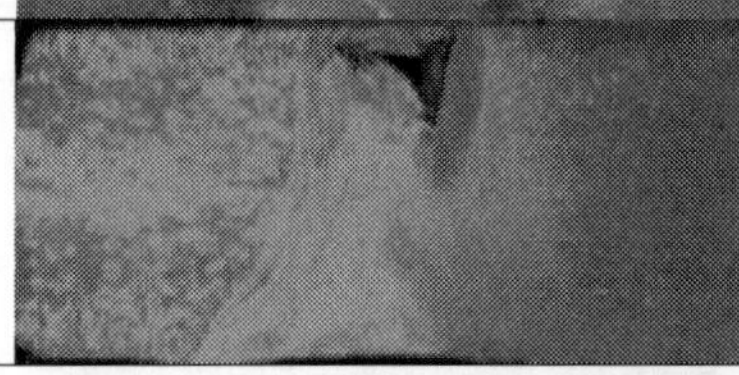	Tunnel defect	Tunnel defect is observed because of less hardness material at retreating side.
(With Al_2O_3Nano Particles)			
AA8011 - AA 8011	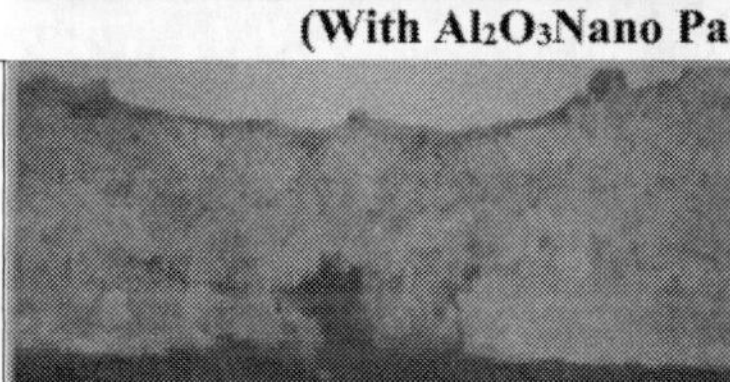	No defects	No defects are appeared due to proper mixing of materials.
AA6061 - AA6061	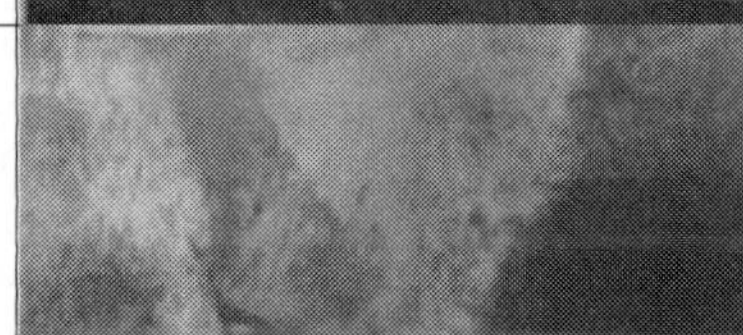	No defects	Mixing is done properly and no defects are found.
AA8011 - AA6061	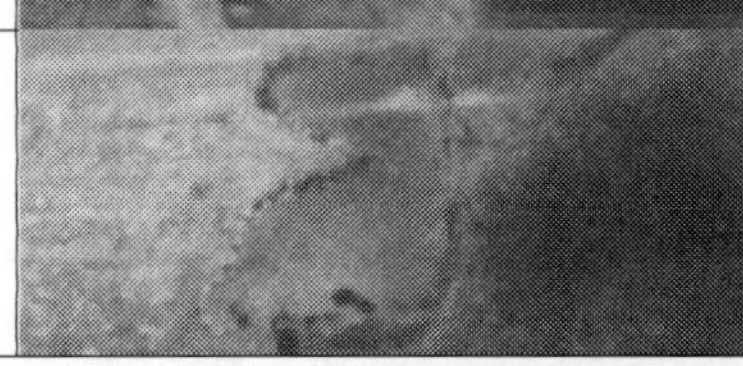	No defects	Good structure is observed due to proper mixing of nanoparticles at the stir zone.
(With SiC Nano Particles)			
AA8011 - AA 8011	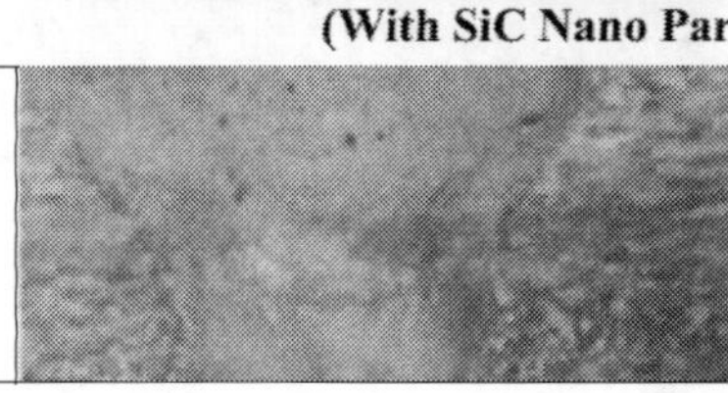	No defects	No defects are observed due to sufficient amount of heat generation.

<table>
<tr><td>AA8011 - AA6061</td><td>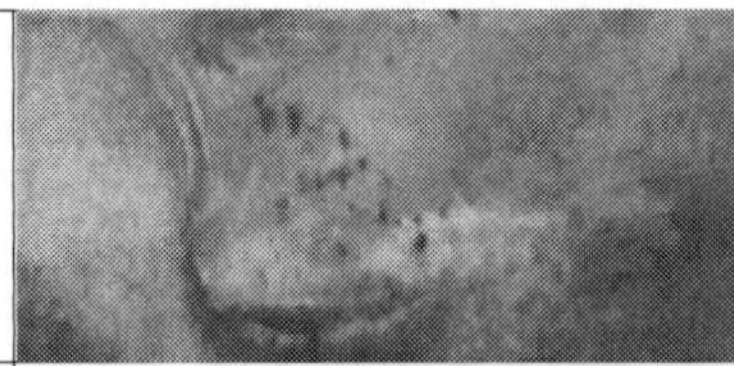</td><td>No defects</td><td>Due to sufficient material flow and plastic deformation by the generated heat, there is no defects are observed</td></tr>
<tr><td>AA6061 - AA6061</td><td>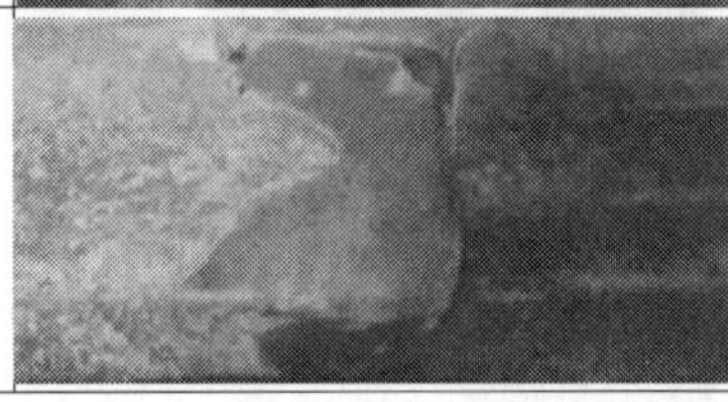</td><td>No defects</td><td>Due to the proper selection of nanoparticles and process parameters good structure is obtained.</td></tr>
</table>

Table 4 SEM and EDS Images of the FSW/FSPed Joints in the Nugget Zone

<table>
<tr><th>Material Combinations</th><th>SEM Microstructure</th><th>EDS Image</th><th>Inferences</th></tr>
<tr><td colspan="4">(Without Nano Particles)</td></tr>
<tr><td>AA8011-AA 8011</td><td>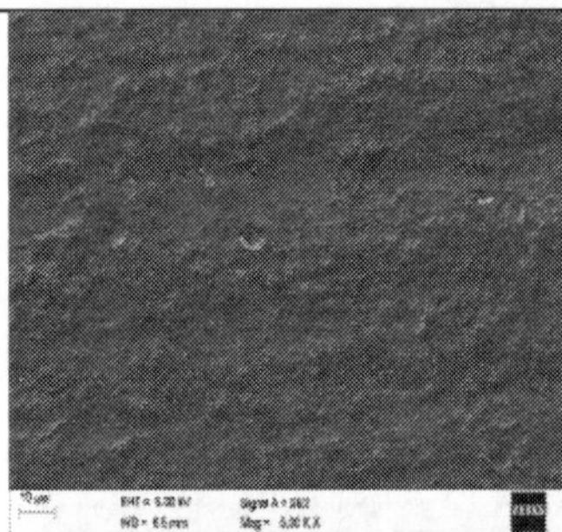</td><td>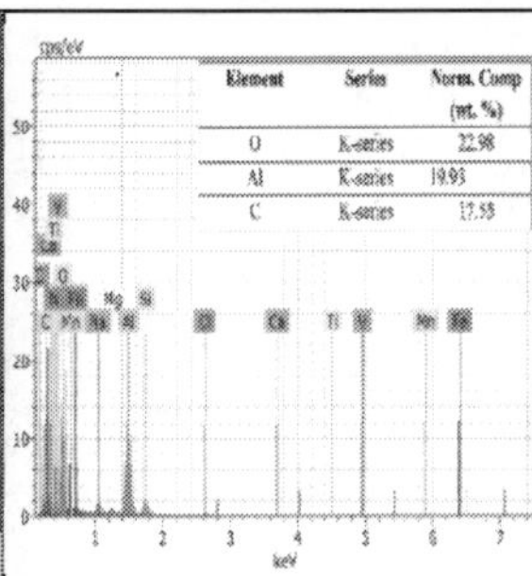
</td><td>Proper mixing has been done and intermetallic components were not found.</td></tr>
<tr><td>AA8011-AA6061</td><td>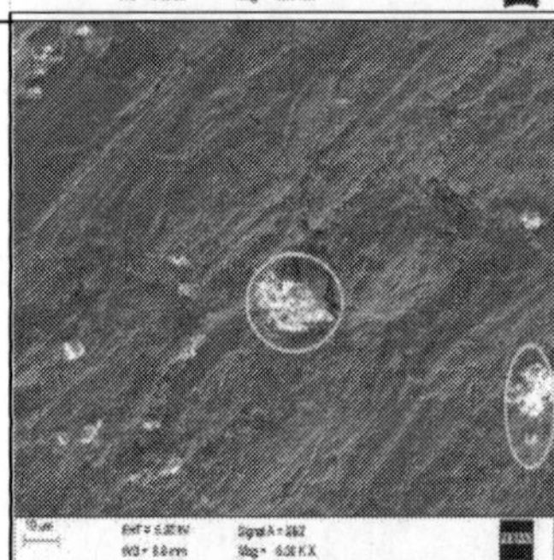</td><td>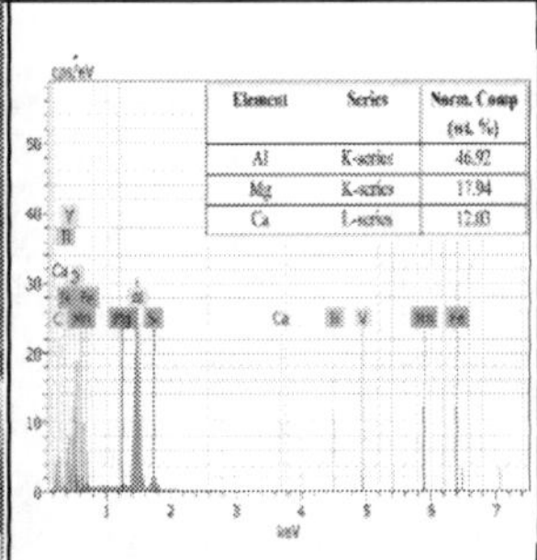
</td><td>The surface seems to be clear with evenly distributed grains but with the presence undissolved intermetallic components.</td></tr>
<tr><td>AA6061 - AA6061</td><td></td><td>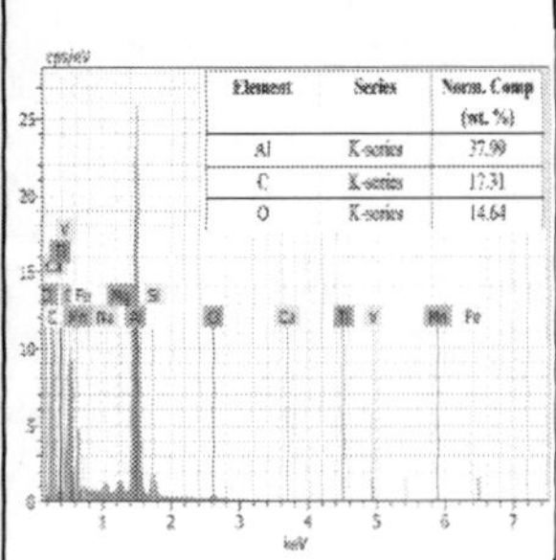
</td><td>The surface seems to be clear with evenly distributed grains but with the presence of little amount unprocessed intermetallic components here and there.</td></tr>
</table>

IOP Conf. Series: Materials Science and Engineering **390** (2018) 012108 doi:10.1088/1757-899X/390/1/012108

(With Al_2O_3 Nano Particles)			
AA8011-AA 8011	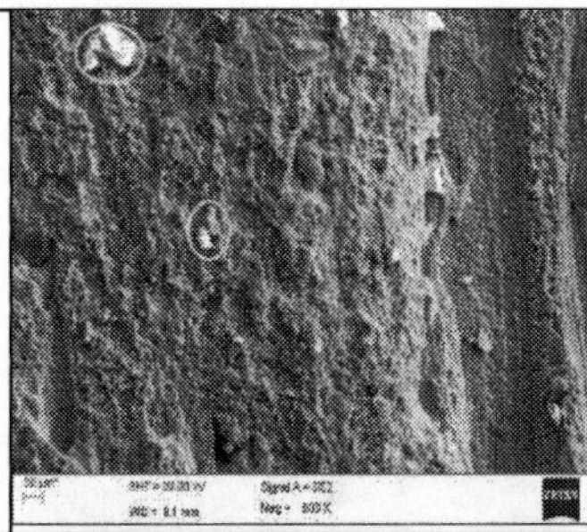	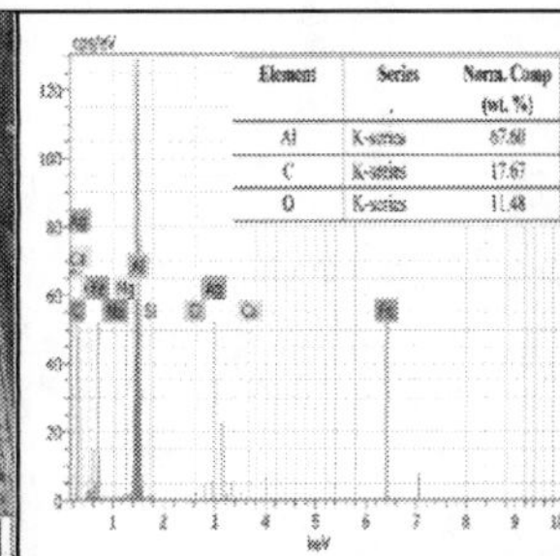	The surface is rough and intermetallic compounds are found due to improper mixing.
AA8011-AA6061	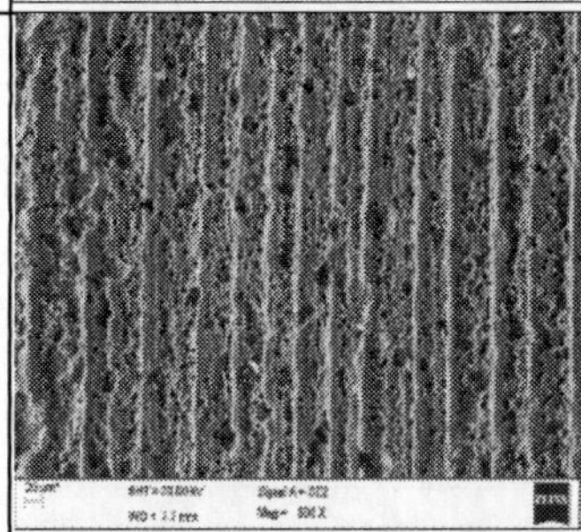	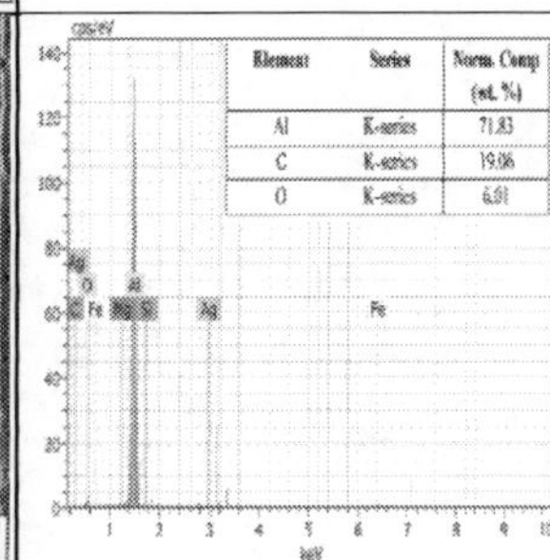	The surface is very clear with no presence of intermetallic components. But we can observe voids due to improper mixing or insufficient rotational speed.
AA6061-AA6061	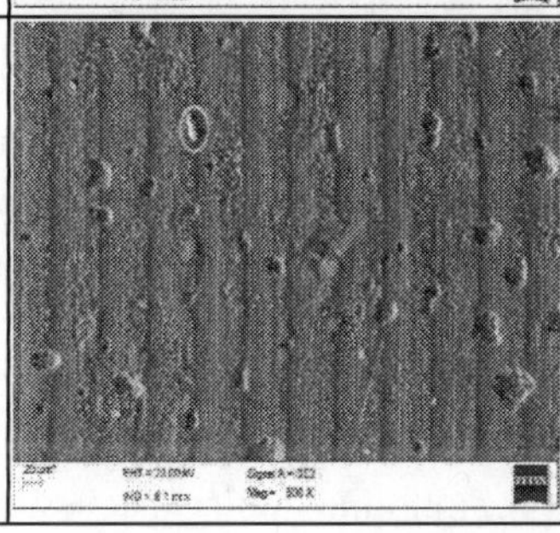	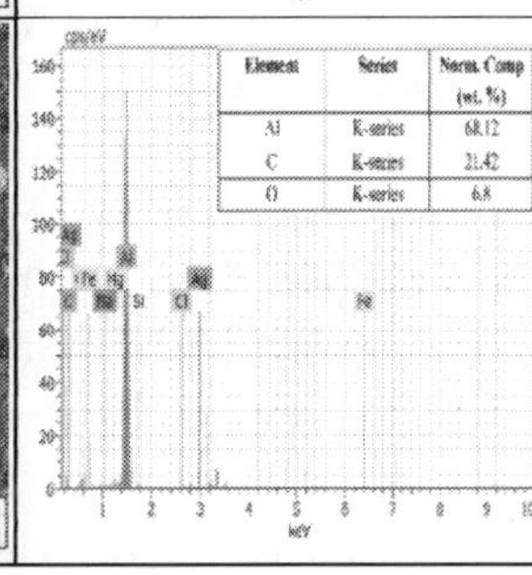	Grain distribution is properly done but the presence of Pits due to the mixing of Al_2O_3 nanoparticles has been improperly done which lead to the formation of intermetallic components.
(With SiC Nano Particles)			
AA8011-AA 8011	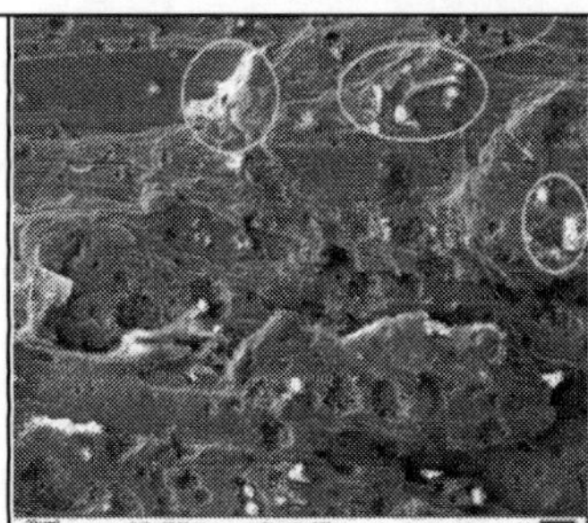	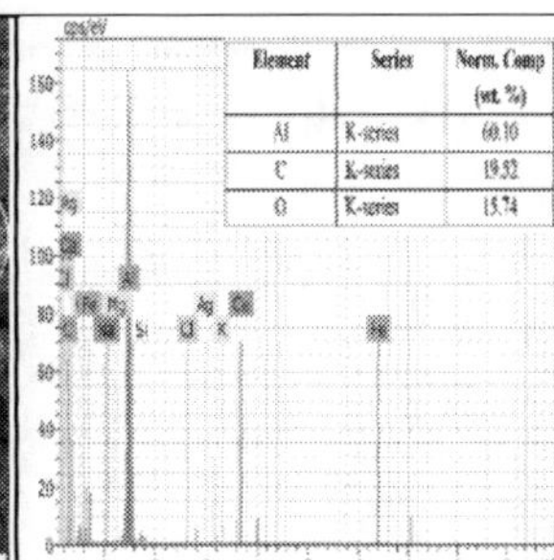	The surface seems to be rough and with the presence of intermetallic components which were not melted by insufficient heat generation.

AA8011-AA6061	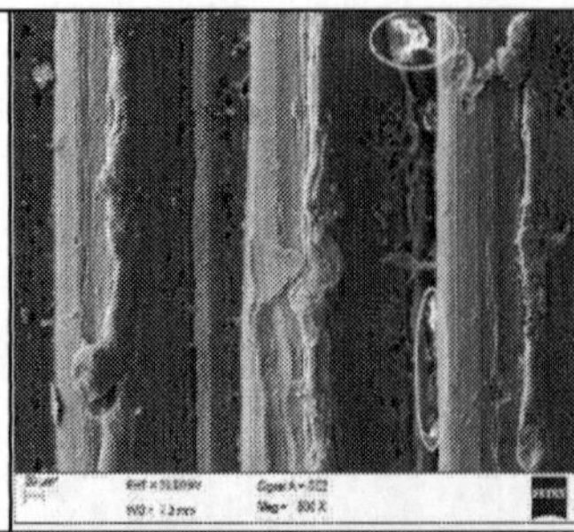	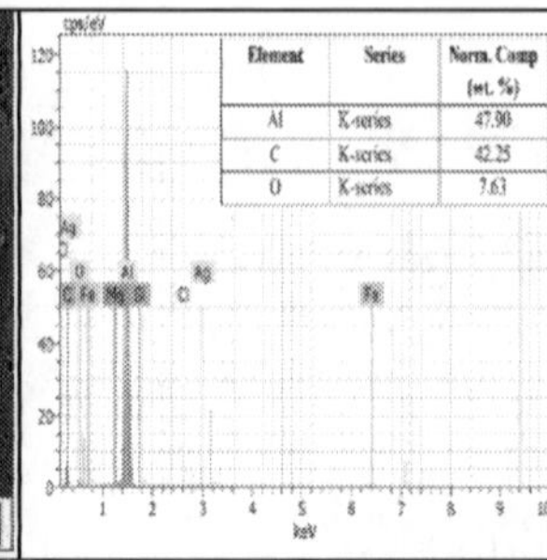	The surface shows some cracks in the material due to improper mixing of nanoparticles. This occurs if the welding speed is high or heat generated is not sufficient.
AA6061-AA6061	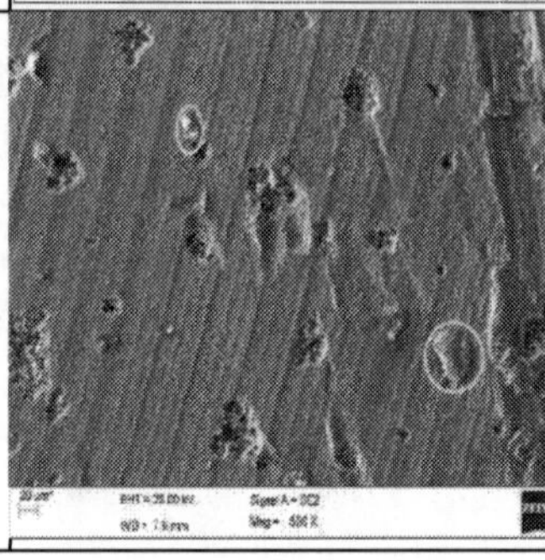	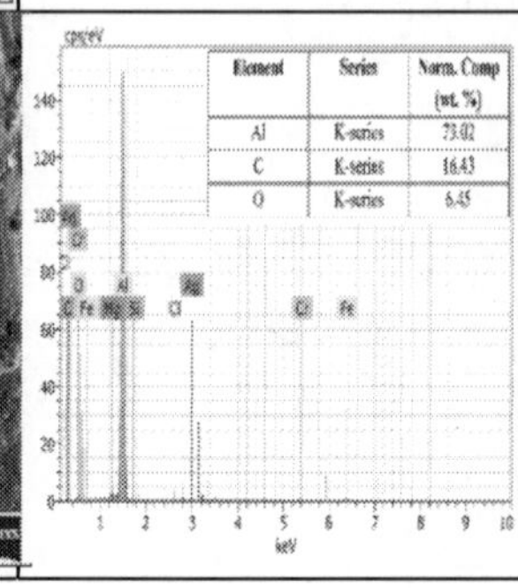	This surface and grain distribution are very good but the presence of pits due to the mixing of SiC nano particles

In both similar and dissimilar AA8011-H24 and AA6061-T6 aluminium alloys, the surface appearance is clear and there is proper distribution of grains without intermetallic components is presented in Table 4. The usage of Al_2o_3 nanoparticles in similar welding, especially in AA6061-AA6061-T6 aluminium alloys, has led to defect-free joints and clear surface without any intermetallic components and in the dissimilar welding joint was formed without any defects but the surface showed some voids and intermetallic components in the specimen. In the specimens where SiC was used as nanoparticles in similar welding the specimen (AA6061-AA6061-T6) gave better weld surface and joint when compared to (AA8011-AA8011-H24) and in dissimilar welding the intermetallic components were formed which leads to improper mixing of base metals and formation of tiny cracks in the weld region and similar discussion was made on the dissimilar aluminium alloys by Rodrigues DM et al. [24].

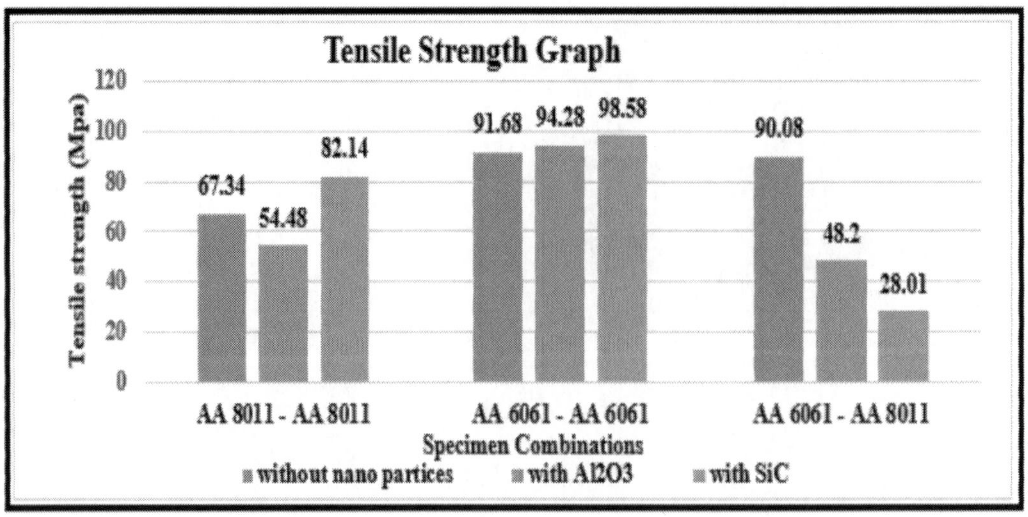

Figure 2 Tensile strength graph for different combinations

3.2 Analysis of the Tensile Strength of the FSW/FSP joints in Nugget zone

The tensile strength analysis of a specimen gives the value of axial pull up to which the material or substance can withstand without breaking. In this work, the Friction Stir Welded specimens were cut as per the ASME standard dimensions and tensile strength analysis has been made. The results are plotted on a graph for different combinations and is presented in Figure 2. The tensile strength graph is drawn based on the values obtained from the tensile test on the welded joints. From the graph, the similar AA 6061 with AA6061 aluminium alloys with and without nanoparticles showed the maximum tensile strength compared to the other material combinations of the welded joints. Furthermore, the similar AA6061 aluminium alloys combinations in addition to SiC nanoparticles gave better tensile strength, viz., 98.58 Mpa in the welded region when compared to other combinations of the welded joints due to the proper diffusion of the SiC nanoparticles in the stir zone of the similar friction stir welded AA 6061 aluminium alloys.

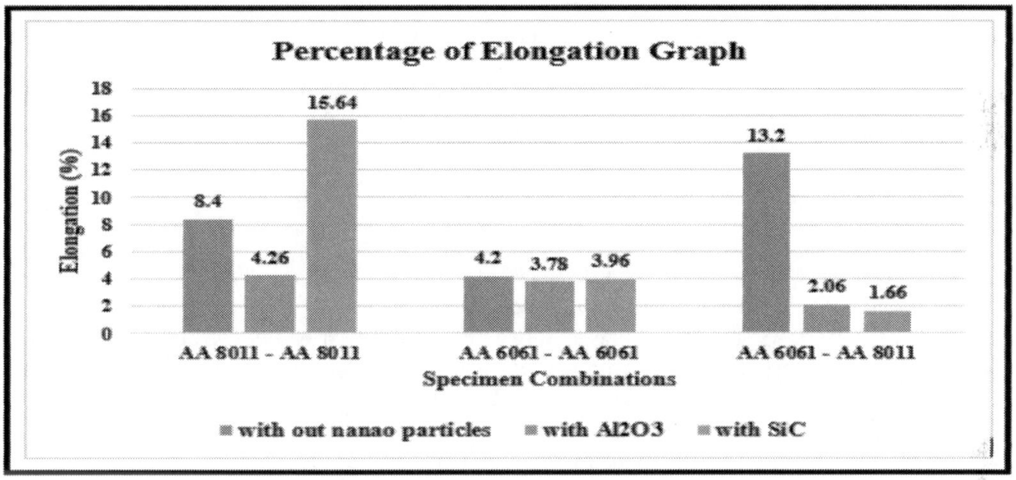

Figure 3 Percentage of Elongation of different specimens

The proper dynamic recrystallization of the materials was appeared in the weld nugget zone by the developed heat in the joints compared to other combinations of the materials with and without nanoparticles [25]. The tensile strength of 90.08 Mpa was found in dissimilar friction stir welding of AA 8011 with AA 6061 aluminium alloys without nano additions whereas the lower strength was observed with nanoparticles due to the improper material flow and the formation of the intermetallic compounds.

3.3 Analysis of the Percentage of Elongation of the FSW/FSP joints in Nugget zone

The percentage elongation showed the strength properties of the dissimilar friction stir welded different grades of aluminium alloys. Figure 3 shows the percentage of elongation of both the similar and dissimilar AA 6061 and AA 8011 aluminium alloys with and without the addition of the nanoparticles. The elongation of 8.4 % was achieved for the similar AA 8011 aluminium alloys in the stir zone, whereas these alloys with the SiC nanoparticles addition increases the elongation properties and showed 15.64 % compared to other combination of the welded joints. Furthermore, the minimum elongation was obtained in the hot worked aluminium alloy combinations with and without nanoparticles, while the dissimilar AA6061 and AA 8011 alloys without nanoparticles have the maximum hardness of 13.2 % compared to the addition of the nanoparticles.

4. Conclusion

The study was done successfully on the influence of FSP parameters on the tensile properties and microstructure of dissimilar AA8011-H24 and AA6061-T6 aluminum alloy butt joints in Nugget Zone and the following conclusions can be drawn:

- The microstructure analysis showed that both similar and dissimilar FSP joints in addition to Al_2O_3 revealed the better microstructure in the stir zone relate to the SiC nanoparticles in stir zone, whereas the SiC nanoparticles processed welded joints gave the better results than the others in the thermomechanically affected zone at Advancing side.
- The tensile strength of 98.58 % and 90.08 % was found at the weld nugget zone of the similar AA6061 aluminium alloys in addition to SiC nanoparticles and dissimilar friction stir welding of AA 8011 with AA 6061 aluminium alloys without nano additions respectively.
- The similar friction stir welded AA 8011aluminium alloys in addition to SiC nanoparticles showed the better percentage of elongation compared to other combinations of the welded joints.

5. References

[1] Mishra RS and Ma ZY 2005 *Mater. Sci. Eng. R.* **50** 1-78
[2] Ramulu PJ, Babu AS, Narayan RG, Prasad SD and Rao PS 2013 *Int. Con. Des. Manu.* **64** 862-67
[3] Palani K and Elanchezhian C 2015 *Appl. Mech. Mater.* **814** 451-55
[4] Chandu KVP, Rao AS and Subrahmanyam BV 2014 *Int. J.Res. Mech. Eng.Tech.* **4** 119-22
[5] Murr LE, Li Y, Flores RD and McClure JC 1998 *J. Mater. Proc. Manu. Sci.* **7** (10) 145-61
[6] Ouyang JH, Kovacevic R 2002 *J. Mat. Eng. Perf.* **11** 51-63
[7] Heinz B and Skrotzki B 2002 *Metal. Mater. Trans.* B **33B** 489-498
[8] Ghosh M, Kailas SV, Kumar K and Ray AK 2010 *Mater. Des.* **31** 30-3
[9] Muthukumaran S and Mukherjee SK 2008 *Int. J. Adv. Manuf. Technol.* **38** 68–73
[10] Palani K and Elanchezhian C 2015 *Appl. Mech. Mater.* **766** 921-27
[11] Girish G and Deepandurai K 2016 *Int. J. Sci. Tech. Eng.* **2** (10) 872-79
[12] Vilaca P, Gois AC and Quintino AL 2005 Weld. Wor. **49** 1681-05
[13] Palani K, Elanchezhian C, Bhaskar GB and Vijaya Ramnath B 2018 Adv. Sci. Eng. Medic. **10** 1-6
[14] Hou JC, Liu HJ and Zhao YQ 2014 *Int. J. Adv. Manuf. Tech.* **73** 1073-79
[15] Tanabe H and Watanabe T 2008 *weld. Int.* **22** (9) 588-96
[16] Madhusudhan R, Sarkar MM and Ramanaiah N 2013 *Int. J. of Mech. Prod. Engg.* **4** (2) 204-208
[17] Ouyang JH and Kovacevic R 2002 *J. Mater. Eng. Perf.* **11** 51-63
[18] Atharifar H, Lin D and Kovacevic R *J. Mater. Eng. Perf.* **18** 339-350
[19] Cao X and Jahazi M 2011 *Mater. Des.* **30** (6) 2033-42
[20] Palani K, Elanchezhian C, Vijaya Ramnath V, Bhaskar GB, Sai Jagadeesh J and Manoj Kumar G 2017 *Int. J. Res. Appl. Sci. Engg. Technol.* **5** 437-442
[21] Amini S, Amiri MR and Barani A 2015 *J. Adv. Manuf. Tech.* **76** (1-4) 255-61
[22] Palani K, Elanchezhian C 2015 *Appl. Mecha. Mater.* **813** 446-450
[23] Threadgill PL, Leonard AJ, Shercliff HR and Withers PJ 2013 *J. Int. Mater. Rev.* **54** (2) 49-93
[24] Rodrigues DM, Loureiro A, Leal RM, Chaparro BM, Vilaca P 2008 *Mater. Des.* 30 1913-21
[25] Tavares SMO, Dos Santos JF, De Castro PMST 2013 *Theo. Appl. Frac. Mech.* 65 8–13

Parametric optimization of WEDM control variables on magnesium AZ91 alloy by TOPSIS method

A.Muniappan[a]*, Thunga Bharath[a], S.Madhu[b], P.Arunagiri[a]& G.Md Shaqib[a]
[a]Department of Mechanical Engineering, Saveetha School of Engineering, Saveetha Institute of Medical and Technical Sciences, Chennai, Tamil Nadu, India
[b]Department of Automobile Engineering, Saveetha School of Engineering, SaveethaInstitute of Medical and Technical Sciences, Chennai, Tamil Nadu, India

*E-Mail: inspire.munish@gmail.com

Abstract. Multi-Criterion Decision Making (MCDM) strategies have gotten much consideration from scientists and professionals in assessing, evaluating and positioning choices over assorted manufacturing industry. Technique for Order Preference by Similarity to Ideal Solution (TOPSIS) continues on working attractively crosswise over various application territories. This paper proposes the TOPSIS-based Taguchi enhancement way to deal with decide the ideal WEDM process parameter for machining of magnesium AZ91 alloy by zinc coated wire electrode. This experiment was outlined with L27 orthogonal cluster. The test ponders were directed under shifting Pulse on time, Pulse off time, current, gap voltage, speed and tension of wire. The impact of the machining variables over kerf width (KW) and cutting speed (CS) is dictated by utilizing examination of fluctuation. The parameters corresponding to experiment run number 2 are Pulse on time 106 units (Level 1), Pulse off time 60 units (Level 3), peak current 70 units (Level 1), gap voltage 40 units (Level 3), wire feed 4 units (Level 1) & wire tension 4 units (Level 1) are the best combination to achieve better cutting speed& kerf width

Keywords: TOPSIS, WEDM, Taguchi method, AZ91, cutting speed & Kerf width criterion

1. Introduction

Manufacturing sector is growing rapidly by accommodating technology modernization. The mechanism for machining hard reinforced materials, intricate shapes & different contours which are very difficult for machining by conventional techniques which created many unconventional methods. CNC EDM machine was developed in the year 1969. WEDM which involves moving very thin electrode continuously. Electrode wire materials like pure brass, diffused coated and zinc coated wire of diameter ranges from 0.05-0.35 are widely applied in industry. The gap generally between work and wire electrode is generally ranges from 25 to 50 microns and is continuously maintained by a modern computer program controlled coordinating system. To achieve smooth surface quality on the tool and in the component, optimum process variables setting is a very important. Machining control variables are optimized by using different methods for improving the better quality. Taguchi method which is widely applied mainly experimental design in manufacturing application. It allows optimization of few parameters in machining by turning, milling, Electro Discharge Machining, wire cut EDM, welding, grinding etc. The optimization is achieved with lesser number of tests by this overall cost and time is saved.

2. Materials& method

Table 1. Composition of AZ91 alloy

Al	Zn	Mn	Fe	Si	Cu	Zr	Mg
8.84	0.61	0.18	0.02	0.02	0.005	0.002	Balance

2.1. Machining control variables

Six information process variables Wire EDM, to be specific, Pulse on time, Pulse off time, current, gap voltage, the wire drum speed & tension were considered their consequences for kerf & CS while machining the AZ91 alloy. The scopes of these procedure parameters were considered on the premise of the pilot tests. The levels of different variables & their assignments are exhibited in Table 2.

Table 2. Control variables levels.

Symbol	Process variables	Level -1	Level- 2	Level- 3
A	Pulse on time	106	116	126
B	Pulse off time	40	50	60
C	Pulse current	70	150	230
D	Gap Voltage	20	30	40
E	Wire speed	4	6	8
F	Wire tension	4	8	12

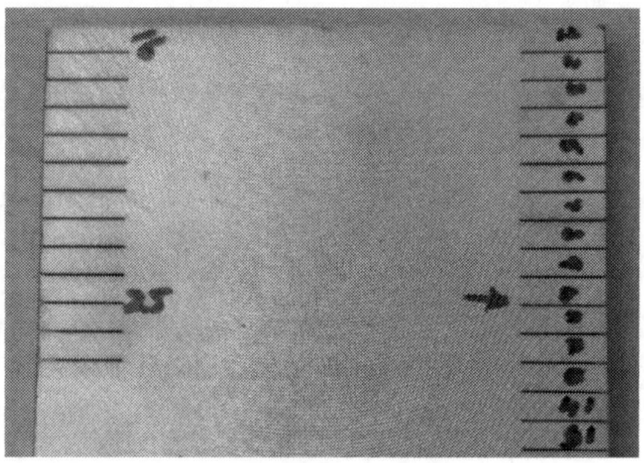

Fig. 1 Kerf width for AZ91 magnesium alloy.

2.2. Taguchi Method of experimental design

Taguchi strategy is a proficient apparatus for the outline of a great assembling framework. It also effectively solves some more complex problems in manufacturing (Roy 1990).It is a strategy in view of OA tests, which give much decreased change to the try different things with ideal setting of process control parameters. The six control parameters, that is, Pulse on time (A), Pulse off time (B), current (C),Gap voltage (D), wire drum speed (E) and wire tension(F) at three levels were picked in this examination. The tests were finished by Table 3. This table just speaks to specific level of the different elements of the procedure at which the examinations would be directed. Kerf width is considered to be as least as conceivable in the WEDM procedure. Kerf width is an imperative component of laser cutting procedure that gives the benefit of this innovation contrasted with different strategies for form cutting.

2.3. Experimental Set Up

Analyses were led on Electronica Sprint cut CNC wire cut electrical discharge machine to think about the surface finish quality and KW mostly influenced by the machining control parameters at various levels. WEDM is a start disintegration process. The flashes are produced between the work & the wire terminal. Dielectric liquid is ceaselessly encouraged into the machining zone with required weight. The material is getting expelled by a progression of discrete sparkles occurring at the zone to be machined through electro-warm system. Test set up of the wire electrical release machine is appeared in Figure 2. Amid machining process little hole kept up between the work and wire material. The machined small minute particles were flushed & carried away by the persistent stream of dielectric medium. The electrode is held by a stick direct at the upper & lower parts of the work piece.

The work example measure utilized a part of this examination is 100 x 100 x 10 mm rectangular plate. Zinc covered metal cathode wire of 0.25 mm diameter was utilized to a part of this investigation. Deionized water was preferred as dielectric liquid at room temperature. In the wake of machining, the examples were cleaned with acid after machining. The kerf was measured utilizing profile projector measuring framework. The kerf esteems were measured at six spots spread over the entire length of cut. The kerf esteems considered as a part of this examination are the numerical normal of three estimations produced using the example in each cut.

2.4. Methodology

Procedure for Order Preference by Similarity to Ideal Solution Method (TOPSIS) is one of the MCDM which is utilized to comprehend the sort of basic leadership issue. This technique depends on the idea that the picked option ought to have the most limited Euclidean separation from the perfect arrangement, and the most remote from the negative perfect arrangement. The perfect arrangement is a speculative answer for which all ascribe esteems compare to the most extreme trait esteems in the database containing the fantastic arrangements; the negative perfect arrangement is the theoretical answer for which all credit esteems relate to the base property estimations in the database. TOPSIS along these lines gives an answer that is not just nearest to the speculatively best, that is additionally the most distant from the theoretically most noticeably awful.

Step 1: Initial step to be followed in TOPSIS strategy is the standardization of execution of various model and the standardized grid is appeared in Table 4.This stage give way to contrasting distinctive foundation by changing over different properties measurement into non dimensional property. Standardize scores or information as takes after:

$Rij = xij/ (\Sigma x^2 ij)$ for i= 1…m; j= 1… n;

Step2:Apportioning weights for the whole model which are considered for streamlining. The weights considered for this exploration were: surface unpleasantness =0.522, kerf width =0.478. The aggregate of weight ought to be equivalent to one.

Step 3: Develop the weighted standardized choice grid as in Table 5. Assume we have weights for every basis wj for j = 1… n.

On duplicating every section of standardized choice lattice by its particular weight the component acquired is:

Vij =WjRij

Step4: The next step is determination of ideal and negative ideal solution Ideal solution.

Positive ideal solution

$A^+ = \{ V1^+,...,Vn^+\}$ where $Vi^+ =\{ \max (Vij)$ if $j \in J$; $\min (Vij)$ if $j \in J'\}$

Negative ideal solution

$A^- = \{ V1^-,...,Vn^-\}$ where $Vi^- =\{ \min (Vij)$ if $j \in J$; $\max (Vij)$ if $j \in J'\}$

Step 5: Separation measure determination is the fifth step in TOPSIS method. The value obtained is given below:The separation from the ideal alternative is:

$$S_i^+= [\Sigma (V_j^+- V_{ij})^2]^{½} \qquad i = 1, …, m$$

Similarly, the separation from the negative ideal alternative as shown below Table 6:

$$S_i^-= [\Sigma (V_j' - V_{ij})^2]^{½} \qquad i = 1, …, m$$

Step6: The relative closeness of a particular alternative

$P_i = S_i^- / (S_i^+ +S_i^-)$, $\qquad 0<P_i< 1$

Select the option with P_i closest to 1

3. Result and discussion

Table 4 shows the square of Xi and normalized matrix for Xi calculated by TOPSIS method. The normalized CS & KW are multiplied with their corresponding weights given in table 5. The parameters, higher the better (maximum) and smaller the better (minimum) respectively added for rank calculation. The surface roughness observed in the experiment is in the range of from 2.252 to 4.256. Similarly, the kerf width observed in the experiment is in the range of from 292 to 327. From these observations, experiment number 4 has the best rank. The parameters corresponding to experiment run number 2 are pulse on time 106 units (Level1), pulse off time 60 units(Level 3), peak current 70 units (Level 1), gap set voltage 40 units (Level 3), wire feed 4 units (Level1) and wire tension 4 units (Level1).Figure 2 shows the effect of process parameters on closeness values of coefficient.

Table.3 Cutting speed & Kerf width

Ex.No.	Pulse on Time	Pulse off Time	Current	Gap voltage	Wire Feed	Wire Tension	CS	KW
1	106	40	70	20	4	4	6.32	0.323
2	106	40	150	30	6	8	7.52	0.32
3	106	40	230	40	8	12	6.89	0.321
4	106	50	70	30	6	12	3.47	0.33
5	106	50	150	40	8	4	3.49	0.332
6	106	50	230	20	4	8	5.51	0.337
7	106	60	70	40	8	8	1.92	0.327
8	106	60	150	20	4	12	3.06	0.33
9	106	60	230	30	6	4	2.83	0.341
10	116	40	70	30	8	8	6.12	0.346
11	116	40	150	40	4	12	6.25	0.341
12	116	40	230	20	6	4	5.98	0.358
13	116	50	70	40	4	4	4.03	0.374
14	116	50	150	20	6	8	6.01	0.371
15	116	50	230	30	8	12	5.98	0.365
16	116	60	70	20	6	12	3.08	0.337
17	116	60	150	30	8	4	3.79	0.348
18	116	60	230	40	4	8	3.68	0.358
19	126	40	70	40	6	12	5.78	0.356
20	126	40	150	20	8	4	5.92	0.351
21	126	40	230	30	4	8	5.97	0.352
22	126	50	70	20	8	8	5.17	0.347
23	126	50	150	30	4	12	5.99	0.364
24	126	50	230	40	6	4	5.97	0.364
25	126	60	70	30	4	4	3.14	0.361
26	126	60	150	40	6	8	4.92	0.372
27	126	60	230	20	8	12	5.98	0.349

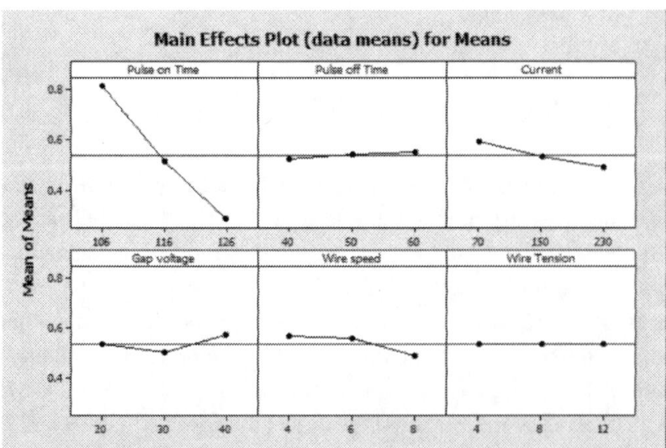

Fig. 2 Effect of process parameters on closeness co efficient

Table.4 Normalized Cutting speed& Kerf width

Ex.No.	Pulse on Time	Pulse off Time	Peak Current	Gap set voltage	Wire Feed	Wire Tension	CS	KW
1	108	40	90	10	3	4	0.11699	0.08942
2	108	40	90	10	4	8	0.13921	0.08859
3	108	40	90	10	5	12	0.12754	0.08887
4	108	50	160	30	3	4	0.06423	0.09136
5	108	50	160	30	4	8	0.0646	0.09191
6	108	50	160	30	5	12	0.102	0.09329
7	108	60	230	50	3	4	0.03554	0.09053
8	108	60	230	50	4	8	0.05664	0.09136
9	108	60	230	50	5	12	0.05239	0.0944
10	117	40	160	50	3	8	0.11329	0.09579
11	117	40	160	50	4	12	0.1157	0.0944
12	117	40	160	50	5	4	0.1107	0.09911
13	117	50	230	10	3	8	0.0746	0.10354
14	117	50	230	10	4	12	0.11125	0.10271
15	117	50	230	10	5	4	0.1107	0.10105
16	117	60	90	30	3	8	0.05701	0.09329
17	117	60	90	30	4	12	0.07016	0.09634
18	117	60	90	30	5	4	0.06812	0.09911
19	126	40	230	30	3	12	0.107	0.09855
20	126	40	230	30	4	4	0.10959	0.09717
21	126	40	230	30	5	8	0.11051	0.09745
22	126	50	90	50	3	12	0.0957	0.09606
23	126	50	90	50	4	4	0.11088	0.10077
24	126	50	90	50	5	8	0.11051	0.10077
25	126	60	160	10	3	12	0.05813	0.09994
26	126	60	160	10	4	4	0.09108	0.10298
27	126	60	160	10	5	8	0.1107	0.09662

Table 5 Closeness coefficient values and ranking of alternatives

Ex.no.	S^+	S^-	P_i	Rank
1	0.022235	0.082665	0.788032	3
2	0	0.104742	1	1
3	0.011673	0.093162	0.888651	2
4	0.075031	0.031168	0.293489	22
5	0.074684	0.031301	0.295333	21
6	0.037506	0.067246	0.641956	15
7	0.103688	0.01301	0.111484	27
8	0.082616	0.024363	0.227736	23
9	0.087014	0.019169	0.18053	26
10	0.026901	0.078135	0.743886	5

11	0.024217	0.080679	0.769132	4
12	0.030389	0.07529	0.712442	8
13	0.066317	0.03906	0.370669	18
14	0.031323	0.075715	0.707364	10
15	0.031114	0.075201	0.707343	11
16	0.082334	0.023791	0.22418	24
17	0.069484	0.035361	0.337269	19
18	0.071864	0.03288	0.313906	20
19	0.033715	0.071634	0.67997	14
20	0.030838	0.074323	0.706758	12
21	0.030036	0.075217	0.714627	6
22	0.044147	0.060623	0.578633	16
23	0.030837	0.075391	0.709707	9
24	0.031178	0.075021	0.706422	13
25	0.081871	0.022875	0.218387	25
26	0.050235	0.055543	0.525089	17
27	0.029619	0.075478	0.718173	7

Conclusions

In this experimental study, the TOPSIS method is applied for the estimation of optimum machining parameters to minimize cutting speed and kerf width. The conclusions drawn from this study are as follows:

1. TOPSIS method is employed to select the optimum machining parameters in WEDM machining of AZ91 Magnesium alloy with zinc coated brass wire electrode.
2. From these observations, experiment number 4 has the best rank. The parameters corresponding to experiment run number 2 are pulse on time 106 units (Level1), pulse off time 60 units(Level 3), peak current 70 units (Level 1), gap set voltage 40 units (Level 3), wire feed 4 units (Level 1) and wire tension 4 units (Level1).
3. The optimum results are adopted in validation study and the results based on WEDM process responses can be effectively improved.
4. As compared to many other MADAM methods, this method is found to be simple, logical and robust. This method can simultaneously consider any number of quantitative and qualitative selections of attributes and is able to provide optimum solution quickly as computational time required to obtain the solution is reasonably low. This technique can be effectively considered to take care of a few multi objective optimization problems with different environment.

REFERENCES

[1] Tatar, C., Ozdemir, N. (2010). "Investigation of thermal conductivity and microstructure of the a-Al_2O_3 particulate reinforced aluminium composites (Al/Al_2O_3-MMC) by powder metallurgy method". Physica B Condensed Matter. Vol.405, pp.896-899.

[2] Manna, A., Bains, H.S., Mahapatra. P.B. (2011)."Experimental study on fabrication of Al–Al_2O_3/Grp metal matrix composites". Journal of Composite Materials, Vol.45, No.19, pp.2003-2010.

[3] Tosun N., Cogun C. and Tosun G. 2004. A study on kerf and material removal rate in wire electrical discharge machining based on Taguchi method. Journal of Materials Processing Technology. 152,316-322.

[4] Hung, N.P., Loh, N.L., Xu, Z.M . (1996). "Cumulative tool wear in machining metal matrix composites part II: machinability". Journal of Material Processing Technology, Vol.58, pp.114–120.

[5] Singaravel Balasubramaniyan &Thangiah Selvaraj (2017), Application of integrated Taguchi and TOPSIS method for optimization of process parameters for dimensional accuracy in turning of EN25 steel, Journal of the Chinese Institute of Engineers, pp. 1-8.

[6] R. Manivannan & M. Pradeep Kumar (2016), Multi-attribute decision making of cryogenically cooled Micro-EDM drilling process parameters using TOPSIS method, Materials and Manufacturing Processes, pp. 1-31

[7] Arun Kumar Srirangan &Paulraj Sathiya (2017), Optimization of Process Parameters for Gas Tungsten Arc Welding of Incoloy 800HT Using TOPSIS, Materials Today: Proceedings 4 (2017) 2031–2039

[8] K.D. Mohapatraa and S.K. Sahoo (2018), A multi objective optimization of gear cutting in WEDM of Inconel 718 using TOPSIS method, Decision Science Letters, pp.157-170.

[9] Arun Kumar Parida and Bharat Chandra Routara (2014), Multiresponse Optimization of Process Parameters in Turning of GFRP Using TOPSIS Method, International Scholarly Research Notices, Hindawi Publishing Corporation, pp. 1-10.

[10] Gadakh, V. S (2012), parametric optimization of wire electrical discharge machining using TOPSIS method, Advances in Production Engineering & Management, pp. 157-164.

[11] P.C.Padhi, VikasUpadhyay, Raj Kumar (2016), Optimization of Process Parameters in WEDM of EN-31 Alloy Steel using Taguchi Technique and TOPSIS, Advanced Materials Manufacturing & Characterization, Vol 6 Issue 1, pp. 39-45.

[12] Muniappan .A, Thiagarajan .C and Somasundaram .S (2017) Parametric Optimization of kerf width and surface roughness in Wire Electrical Discharge Machining (WEDM) of hybrid Aluminium (Al6061/SiC/Graphite) composite using Taguchi-based Gray Relational Analysis" International Journal of Mechanical & Mechatronics Engineering IJMME-IJENS Vol:16 No:05, pp. 95-103.

MECHNANICAL PROPERTIES OF ALOE VERA-JUTE-GLASS FIBER REINFORCED POLYMER MATRIX COMPOSITES

AshwinSailesh, P.VigneshPothiraj, B.Ranjith, A.Subashchandar, S.Ganesh
Assistant Professor, Department of Mechanical Engineering, Sri Sai Ram Institute of Technology, Chennai, Tamil Nadu, India.
Undergraduate Students, Department of Mechanical Engineering, Sri Sai Ram Institute of Technology, Chennai, Tamil Nadu, India
E-mail: ashwinmech9191@gmail.com

Abstract. In this current experimental work, composite material is fabricated using a combination of natural fibers such as aloe vera, jute and glass fiber reinforced in a Epoxy Resin (Polymer Matrix). The composite samples were fabricated by compression molding Technique. The fabricated composite samples are tested for its mechanical properties such as tensile, flexural and impact properties. The composite sample dimensions and the testing of the samples are carried out as per ASTM standards.

Keywords: Aloe Vera fiber, Jute fiber, Glass fiber, Epoxy Resin

1. Introduction

Over the last two to three decades composite materials have been used immensely in many engineering fields. Composite materials provide high stiffness, high strength to weight ratio. Modern composite materials are used in many applications ranging low sophisticated to highly sophisticated applications. Polymer Matrix Composites are widely used in automotive industry especially in areas such as automobile window panel, vehicle doors etc., where the conventional materials have been replaced by composites. Natural fiber composites are attractive and most currently used materials, due to the cost effectiveness of the raw materials. Also, due to the easy availability of natural fibers many researchers are keen in conducting research in the field of composites. Composite material made up of combining a natural fiber with a synthetic fiber such as glass fiber is finding increased applications. Generally, a composite material is a material which is made up of two or more constituent materials. These constituent materials are physically and chemically distinct in nature. Polymer matrix composite is one type of composite material in which a polymer acts as a matrix. Epoxy resin is one of the important polymer matrix.

PongsathornKongkeaw et al studied and investigated the effect of fiber length on the tensile properties of epoxy resin composites reinforced by the fibers of bamboo. The composite samples were fabricated by manual layup method. The samples fabricated with five different fiber lengths. The tensile strength of the composite samples was cut according to ASTM standard. The samples were tested for its tensile strength using Universal Testing Machine and the fiber – matrix interaction was observed SEM micrographs. The result concluded that the tensile strength of the composite samples increases with the increase in the fiber length without affecting the elongation at break of the composite. Also, the presence of voids, fiber length and the interfacial adhesion between fiber – matrix can affect the mechanical properties of the fabricated composites [1].

Soma Dalbehera et al fabricated a hybrid composite using jute fiber and glass fiber as reinforcing materials and epoxy resin as the matrix material. Hand Layup technique was adopted for the fabrication of the composite samples. Stacking sequence of the fibers was followed. The effect of stacking sequence on the tensile, flexural properties of the composites was investigated and it was concluded that the combined effect of natural and synthetic fibers into the polymer matrix enhanced the mechanical properties of the composite material [2].

TemesgenBerhanu et al fabricated Jute-Polypropylene composites by compression molding technique. The weight percentage of the fiber reinforcement was varied. The effect of weight percentage of jute reinforcement on the mechanical properties of the composite material was investigated and was reported that the mechanical properties of the composites substantially improved with the addition of 40% of jute fiber reinforcement [3].

Rajesh Ghosh et al investigated the effect of fiber volume fraction on the tensile strength of banana fiber reinforced vinyl ester resin composites. The composite fabricated were tested for its tensile strength and the samples were prepared cut according to ASTM standard. They concluded that with the increase in the fiber volume fraction the tensile strength of the fabricated composite material increases after an initial dip in the value of tensile strength [4].

Yamini et al investigated the impact behavior of natural based composites using coir and Aloe Vera. The composite materials were fabricated by hand layup technique. Charpy impact test was carried out in a un-notched specimen. The moisture absorption test was also carried out. It was reported that the impact value of the aloe vera fiber was comparatively lesser due to the fact that it has higher bond strength at the interface between the fiber and the matrix. Also, the moisture absorption test shows that the coir samples absorb more moisture than aloe vera due to the presence of micro-voids [5].

SudeepDeshpande et al investigated the mechanical properties such as ultimate tensile strength, flexural strength, inter laminar shear strength, impact strength etc of E-Glass fiber/jute fiber reinforced epoxy composites with bone and coconut powder as filler materials. The composite samples were fabricated using hand layup technique. It was reported that flexural strength, hardness, inter laminar shear strength were superior for the composites filled with 15% volume coconut powder. Also, the composite materials with 15% volume bone powder exhibited superior impact strength [6].

2. EXPERIMENTAL DETAILS

2.1. Materials Used

In the present investigation Aloe Vera, Jute and E – Glass fiber are used as reinforcing material and Epoxy resin (LY556) and a suitable hardener (HY951) are used as the raw materials.

2.2. Method of Fabrication

The composite material is fabricated in the form of a plate of dimensions 27cm x 27cm by compression molding technique. Stacking sequence of the fiber is followed in the fabrication of the composite material. The stacking is G-A-J-A-G. This means the composite sample consist of five layers of fibers in which the Glass fiber are placed in the top and bottom layer. The individual composite samples for the mechanical testing are cut from the prepared composite plate. The dimensions of the composite samples are as per ASTM standards.

2.3. Preparation of Samples

The composite sample for the Tensile, Flexural and Impact (Izod) test is cut as per ASTM D3039, ASTM D790 and ASTM D256 respectively.

IOP Conf. Series: Materials Science and Engineering **390** (2018) 012110 doi:10.1088/1757-899X/390/1/012110

3. RESULTS AND DISCUSSIONS

The prepared composite samples are tested for its mechanical properties such as tensile strength, flexural strength and impact strength. The tensile test is carried out using a Universal Testing Machine (UTM). The three point flexure test is carried out to determine the flexural strength of the fabricated composite sample. Impact test is carried out using the Izod impact testing machine

The tensile test specimen loaded in the corresponding fixture in a UTM is shown in figure 1.

Figure 1. Tensile test specimen loaded in the UTM

Similarly the flexural test sample loaded in the three point flexure apparatus is shown in figure 2.

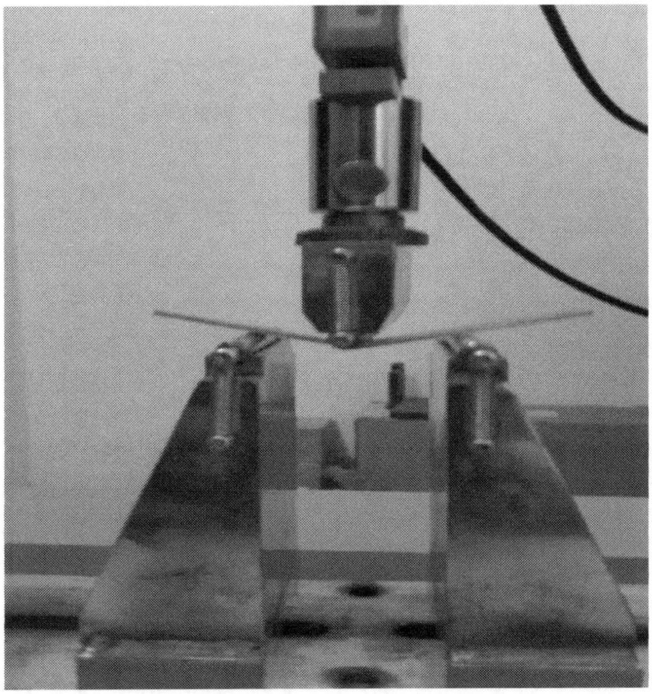

Figure 2. Flexural Test Specimen loaded in a three point flexure attachment

A representative plot of the Stress-Stain for the Tensile Test obtained from the UTM is shown in figure 3.

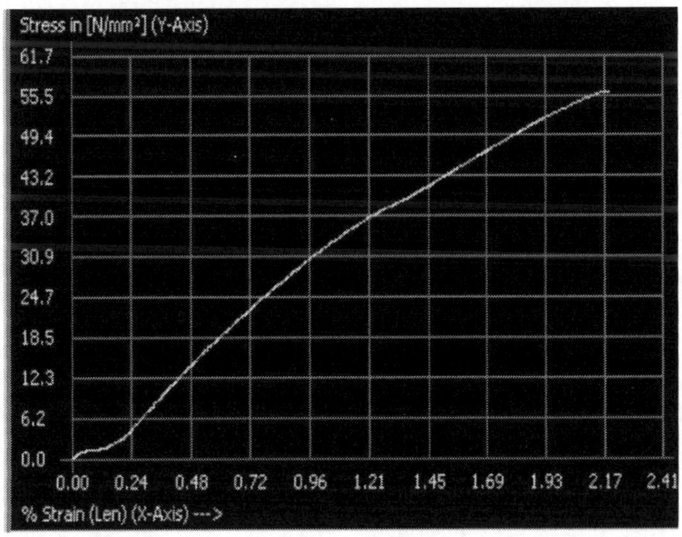

Figure 3. Representative plot of the Stress-Strain graph for the fabricated composite material obtained from UTM

A sample of the Load-Length plot for the Flexural Test obtained from the UTM is shown in figure 4.

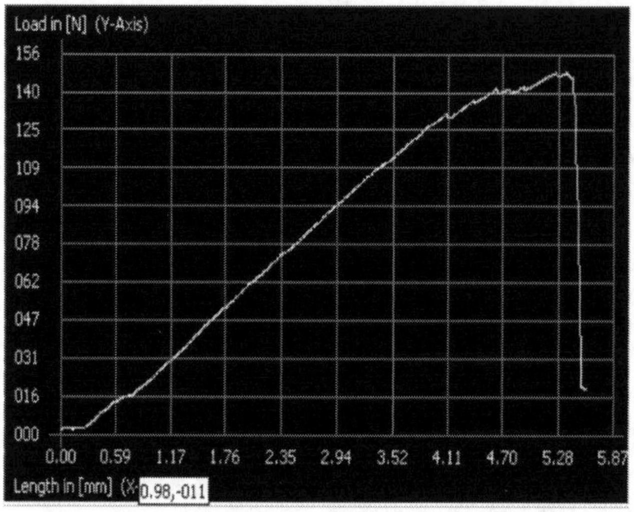

Figure 4. Representative plot of the Load-Length for the fabricated composite material obtained from UTM

The overall experimental result of the tested composite samples on an average of five samples for each test is shown in table 1.

Table 1 Experimental result of the tested composite samples

Tensile Strength (MPa)	Flexural Strength (MPa)	Impact Strength (J/mm^2)
60.017	107.896	0.026

4. CONCLUSION

In this present experimental work, Aloe Vera, Jute, Glass fiber reinforced Epoxy composites is fabricated by compression molding method. The fabricated composites are tested for its mechanical properties such as tensile strength, flexural strength and impact strength. ASTM standards are followed for the sample preparation as well as for the testing procedure. Based on the various mechanical testing carried out, it is concluded that the tensile strength, flexural strength and the impact strength of the Aloe Vera-Jute-Glass fiber reinforced epoxy composites is 60.017 MPa, 107.896 MPa and 0.026 J/mm^2 respectively.

5. REFERENCES

[1] PongsathornKongkeaw, WimNhuapeng, WandeeThamajaree (2011), "The effect of fiber length on the tensile properties of Epoxy resin composites reinforced by the fibers of bamboo", Journal of Microscopy Society of Thailand, Vol.4, No.1, pp.46 – 48.

[2] Soma Dalbehera, S.K.Acharya (2014), "Study on mechanical properties of natural fiber reinforced woven jute-glass hybrid epoxy composites" Advances in Polymer Science and Technology, Vol.4, No.1, pp.1-6

[3] TemesgenBerhanu, Pradeep Kumar, Inderdeep Singh (2014), "Mechanical behavior of Jute fiber reinforced Polypropylene composites", AIMTDR 2014.

[4] Rajesh Ghosh, Reena.G, Rama Krishna.A, LakshmipathiRaju.Bh, "Effect of Fiber Volume Fraction on the Tensile Strength of Banana Fiber Reinforced Vinyl Ester Resin Composites", International Journal of Advanced Engineering Sciences and Technologies, Vol.4, No.1, pp. 89-91

[5] S.Yamini, K.ShanmugaSundaram (2015), "Study on the mechanical behavior of natural based composite using coir and aloe vera, IJETCSE, Vol.13, No.4, pp.278-280.

[6] SudeepDeshpande, T.Rangaswamy (2014), "Effect on fillers on E-Glass/Jute fiber reinforced epoxy composites", International Journal of Engineering Research and Applications, Vol.4, No.8, pp.118-123.

Performance Estimation of Vapour Compression Refrigeration System using Real Gas Model

P. Vithya[1*], G. Sriram[2], S. Arumugam[3], V. Adithiya[4], D. Anand Raj[5], V. Aswin[6], R.Balaji[7]

[1-7] Department of Mechanical Engineering, Sri Chandrasekharendra Saraswathi Viswa Mahavidyalaya, Kanchipuram, Tamilnadu - 631561, India.

Email: vithyagautham@gmail.com

Abstract. The performance of a vapour compression refrigeration system depends greatly on the spatial arrangement of the components, properties of the refrigerant and lubricant. Numerous models have been developed by using these properties to predict the performance of the vapour compression refrigeration system. The usual ideal gas model does not yield accurate results due to its limited range. In this study, an analytical real gas model has been developed for heat transfer in the evaporator and work output rate of compressor to achieve the coefficient of performance (COP) of the system. This real gas model is validated against the investigational data and the end result obtained was consistent with an error band of 10% for exhaust temperature and 7% for COP. This model can be used to predict the COP of any vapour compression refrigeration system, given that the thermodynamic properties of the working fluid and required dimensions of the system are known. However, it cannot be extended to compressor oils due to it does not differentiate between necessary properties like viscosity of refrigerator oil.

1. Introduction

The utilization of energy by the refrigerators is rising progressively due to their rapid usage. These ever-increasing energy requisite forces heighten the shortage in the energy and environment pollution. The previous researchers found the numerous models to determine the performance of a hermetically sealed compressor under dissimilar conditions. The refrigerant and lubricants plays a vital role in the working and performance of a compressor. To attain a high COP can be attained, high values of latent heat, liquid thermal conductivity in vapour density and low values of liquid viscosity and molecular weight are required. The experimental studies reveal usage of nanoparticles in lubricant has reduced power input to the compressor [1].

Due to rising environmental pollution concerns, bio-lubricants are projected as substitute to petroleum oil based lubricants. Vegetable oils have been considered for use in industries. But at high temperature and pressure, their performance is poor. However, plant oils are able of given that extremely low friction and wear coefficients under highly controlled test conditions. High lubricity, less lubricant consumption, energy efficiency, public health and cost effectiveness are some of the advantages of bio lubricants [2]. Previous studies show that the optimization methods provide the superior statistical prediction efficiency [3, 4]. Jatinder Gill et al [5] examined the vapour compression refrigeration system performance with the mixture of R134a and LPG refrigerant on various evaporator temperatures. The adaptive neuro-fuzzy inference system (ANFIS) model was s to predict the COP of the system. The superior statistical efficiency results with less than 0.3% percentage error was found in the evaluation of statistical analysis to mathematical using ANFIS models. Wang et al [6] presented the lubricant effect on the heat transfer characteristics on conventional refrigerants and natural refrigerant R-744. The results concluded that parameters like heat flux, saturation temperatures, vapour quality, oil concentrations and geometric configurations and affect the heat transfer coefficient. The present work the real gas mathematical model was developed to investigate the behaviour of heat exchangers in steady state vapour compression refrigeration system is validated against published data.

2. Experimental

2.1 Description of physical system

The test rig for conducting a performance test is newly developed as per the refrigeration test rig standards. Fig.1 shows the schematic diagram of simple vapour compression refrigeration system. The model of testing hermetically sealed reciprocating compressor, AQAW77X thermally protected for R134a was procured from shangzhure refrigeration Co. Ltd, the power input of the compressor has ½ HP and single phase. The condenser was air cooled and the capillary tube expansion valve was used for expansion process, the evaporator cabin dimensions are height 0.65m, length 6.54m and breadth 0.42m. The thermocouples used were calibrated T type for observing the temperature of the refrigerant.

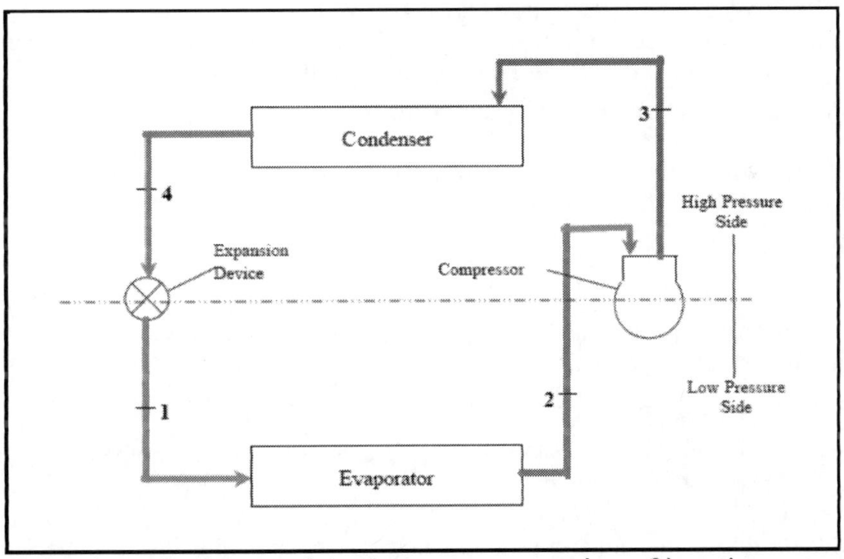

Fig.1 Schematic diagram of simple vapour compression refrigeration system

The temperatures were measured with thermocouples, read by 7-segment display of 4 digits. The pressures were measured using pressure gauges (absolute). The temperature and pressures were monitored and were recorded in steady state at the four thermodynamically significant points. Using a p-h chart, the enthalpy at those points were evaluated. From these enthalpies, the theoretical work input and evaporative heat were estimated. Also, the theoretical coefficient of performance was calculated. This experimental result was compared with the results obtained from the model.

2.2 Model Derivation and Evaluation

A mathematical model is used to predict the behaviour of the vapour compression refrigeration system. The execution of simulation program is based upon real gas mathematical models of the components of the refrigeration circuit including the compressor, heat exchangers, thermostatic expansion valve.

2.2.1 Heat Transfer in Constant Area Steady Flow

The following assumptions are made about the heat transfer in steady state condition. There is no heat transfer along the axis of the pipe. The flux, measured as energy per unit surface area per unit time, is constant. Pressure along the flow is constant. The surface tension energy is negligible. For such a pipe of constant area of cross-section, under steady state condition, the energy balance equation can be easily constructed from scratch. Since the fluid is a mixture of gas and liquid, the total enthalpy *dh* is the sum of changes in enthalpies in gas and liquid, weighted by a parameter called dryness fraction *x*.

Under constant pressure *dp=0,* and hence, the changes in temperatures are the only driving force for the changes in enthalpy. Thus, we write $(dhg) = cp,\ dT$ and $(dhf) = c,\ dT$ whereas, the dryness fraction *x* is defined as,

$$x = \frac{v}{v_g} = \frac{ZRT}{p} \times \frac{p}{Z_s R T_s} = \frac{ZT}{Z_s T_s} \tag{1}$$

Thus, the change in dryness fraction dx is expanded using logarithm in the following manner.

$$dx = \left(\frac{ZT}{Z_s T_s}\right) \times \left[\frac{dZ}{Z} + \frac{dT}{T} - \frac{dZ_s}{Z_s} - \frac{dT_s}{T_s}\right] \tag{2}$$

But, since, Ts = f(p) and Zs = g(Ts , p), dTs = 0 ; dZs = 0 since dp = 0. Hence

$$(dx)_p = \left(\frac{ZT}{Z_s T_s}\right)\left[\frac{dZ}{Z} + \frac{dT}{T}\right] = \left(\frac{ZT}{Z_s T_s}\right)\left[\frac{dZ}{dT}\frac{1}{Z} + \frac{1}{T}\right] dT \tag{3}$$

The total enthalpy difference (dh)p, however, is defined in terms of the above components, The isobar or constant pressure work done can be evaluated in the following way as

$$\delta w = pdv = p\left(\frac{\partial v}{\partial T}\right)_p dT = R\left[\left(\frac{\partial Z}{\partial T}\right)_p T + Z\right] dT \tag{4}$$

Velocity of the fluid flow, for a constant mass flux G is u = Gv. The infinitesimal change in velocity can be written in terms of the chosen variables

$u = Gv$ Implies $du = Gdv$ (or)

$$udu = G^2\left(\frac{ZRT}{p}\right)\left(\frac{\partial v}{\partial T}\right)_p dT = G^2\left(\frac{ZRT}{p}\right)\left(\frac{R}{p}\left[\left(\frac{\partial Z}{\partial T}\right)_p T + Z\right]\right) dT \tag{5}$$

Now, the heat balance equation can be written as $(dh_p) = (dh + udu + dw)_p$

Substituting for dh, udu & dw in above,

$$\delta p = \left(c_{pf} + c_{pg}\frac{ZT}{Z_s T_s} + \frac{\partial (ZT)}{\partial T}\left[\frac{\Delta h}{Z_s T_s} + (ZT)\left(\frac{G^2 R^2}{p^2}\right) + R\right]\right) dT \tag{6}$$

Substituting for constant flux and making change in heat as the subject in the above equation

$$\delta p = \frac{\dot{Q}\Pi}{GA} dl = \left(c_{pf} + c_{pg}\frac{ZT}{Z_s T_s} + \frac{\partial (ZT)}{\partial T}\left[\frac{\Delta h}{Z_s T_s} + (ZT)\frac{G^2 R^2}{p^2} + R\right]\right) dT \tag{7}$$

In other way, the above equation can be rearranged to to be written as

$$\frac{dT}{dl} = \frac{\dot{Q}\Pi}{GA}\left(c_{pf} + c_{pg}\frac{ZT}{Z_s T_s} + \frac{\partial (ZT)}{\partial T}\left[\frac{\Delta h}{Z_s T_s} + (ZT)\frac{G^2 R^2}{p^2} + R\right]\right)^{-1} \tag{8}$$

Alternatively, (7) can be integrated to obtain length – Temperature relationship, implicitly in terms of temperature. Thus, length *l* = f (T) only. However, this form does not allow closure in terms of length.

$$l = \left(\frac{GA}{\dot{Q}\Pi}\right)\int\left(c_{pf} + c_{pg}\frac{ZT}{Z_s T_s} + \frac{\partial (ZT)}{\partial T}\left[\frac{\Delta h}{Z_s T_s} + (ZT)\frac{G^2 R^2}{p^2} + R\right]\right) dT \tag{9}$$

Integrating the above, as an indefinite integral and introducing a constant of integration,

$$l = \left(\frac{GA}{\dot{Q}\Pi}\right)\left(c_{pf}T + \frac{\Delta h}{Z_s T_s} + R\right)(ZT) + \left(\frac{G^2 R^2}{2p^2}\right)(ZT)^2 + \frac{c_{pg}}{Z_s T_s}\int (ZT) dT + k \tag{10}$$

To calculate the integral in (10), we use integration by parts and simplify it.

$$l = \left(\frac{GA}{\dot{Q}\Pi}\right)\left(c_{pf}T + \frac{\Delta h}{Z_s T_s} + R\right)(ZT) + \left(\frac{G^2 R^2}{2p^2}\right)(ZT)^2 + \frac{c_{pg}}{Z_s T_s}[T\int Z dT + \int\int Z dT dT \tag{11}$$

However, it is enough to integrate the differential equation dh with respect to the temperature distribution to compute the change in heat along the duct. Thus,

$$\Delta q = \int_{h_0}^{h_L} dh = \int_{T_0}^{T_L} \frac{dh}{dT} dT = \int_0^L \frac{dh}{dT}\frac{dT}{dl} dl \tag{12}$$

The enthalpy change as written as

$$(dh)_p = c_{pf} dT + c_{pfg}\left(\frac{ZT}{Z_s T_s}\right) dT + \Delta h\left(\frac{ZT}{Z_s T_s}\right)\left[\left(\frac{\partial Z}{\partial T}\right)_p \times \frac{1}{Z} + \frac{1}{T}\right] dT \tag{13}$$

The third term in above equation can be simplified by taking LCM and observing that it is a term that can be simplified in a calculus method.

$$(dh)_p = c_{pf} dT + c_{pfg}\left(\frac{ZT}{Z_s T_s}\right) dT + \Delta h\left(\frac{1}{Z_s T_s}\right)\left[\left(\frac{\partial Z}{\partial T}\right)_p \times T + Z\right] dT$$

$$(dh)_p = \left(c_{pf} + ZT\left(\frac{c_{pfg}}{Z_s T_s}\right) + \left(\frac{\Delta h}{Z_s T_s}\right)\left[\frac{\partial}{\partial T}(ZT)_p\right]\right) dT \tag{14}$$

From, (12), we write

$$\Delta q = \int_T \left(c_{pf} + \frac{c_{pfg}}{Z_s T_s}(ZT) + \frac{\Delta h}{Z_s T_s}\left(\frac{\partial}{\partial T}(ZT)_p\right)\right) dT$$

$$= q_0 + c_{pf} T + \frac{Cpfg}{Z_s T_s}\int_T (ZT) dT + \frac{\Delta h}{Z_s T_s}(ZT) \tag{15}$$

The integral in (15) cannot be directly evaluated and in order to facilitate numerical integration, it is expanded using a mathematical technique called integration by parts. Also, as the expression thus obtained only qualifies for heat per unit mass while it is of general interest to obtain the rate of heat transfer to be computed. Thus, we multiply the above expression with rate of mass flow to obtain the gross heat transfer rate. Thus,

$$\dot{Q} = \dot{m}(q_o + c_{pf} T + \frac{c_{pfg}}{Z_s T_s}[T \int_T Z dT - \int_T \int_T Z dT dT + \frac{\Delta h}{Z_s T_s}(ZT) \tag{16}$$

The above equation gives insight on the nature of heat transfer. The integrals involved with the compression factor can be found numerically or analytically based on the model for compression factor used.

2.2.2 Work Transfer through Reciprocating Compressor

Compressor is the essential component of the vapour compression system. It creates the pressure difference that drives the working fluid through the system. In order to estimate the work transfer rate in the compressor, the compressor is a reciprocating compressor with a single slider crank mechanism is to be assumed. The single Slider-Crank Mechanism is shown in Fig.2.

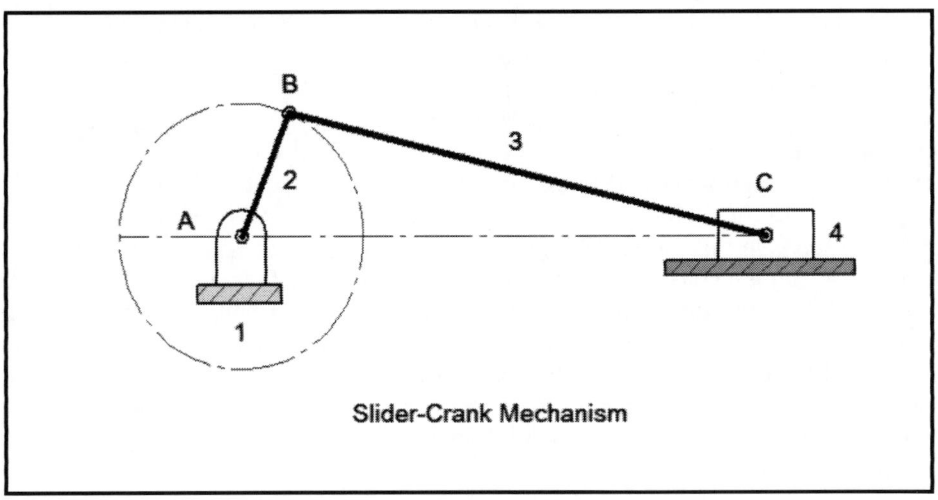

Fig.2 Single Slider-Crank Mechanism

Let the crank be of radius r and connecting rod of length c, and are mass less for the sake of idealization. The mass of the piston be m. Then, the position of the piston from the origin is,

$$y = r\left(\sqrt{n^2 - sin^2\theta} - cos\theta\right) \quad (17)$$

where, n is the ratio of the lengths of connecting rod to that of the crank rod. As the position is known from the above equation, it is easy to compute the first and second derivatives of the above form, to obtain ẏ,&ÿ respectively.

$$\dot{y} = -r\dot{\theta}\left(sin\theta + \frac{sin2\theta}{\sqrt{n^2-\sin^2\theta}}\right) \text{ and } \ddot{y} = r\,\dot{\theta}^2\left(cos\theta + \frac{sin2\theta}{\sqrt{n^2-\sin^2\theta}} + \frac{cos2\theta}{(n^2-\sin^2\theta)^{\frac{3}{2}}}\right) \quad (18)$$

With the above definitions intact, we may now define the pressure exerted, using Newton's second law of motion.

$$F = m \times \ddot{y} = pA \quad (19)$$

In other way, we can rewrite the above equation as,

$$e = y - (c - r) \quad (20)$$

The instantaneous volume of the cylinder is realized with transforming the slider position into piston's displacement from the cylinder BDC.

$$\forall = Ae = A(y - c + r) \quad (21)$$

The instantaneous volume is then calculated as the product of y and the area of cross section A.

The thermodynamic work done is computed as, $\delta W = pdA$

From the above equation, $d\forall = Ady = A\frac{dy}{dt}dt = A\dot{y}dt$

From the equation for Newton's second law,

$$p = \frac{\dot{y}\dot{m}}{A} \Rightarrow \delta W = \frac{m}{A}\dot{y}y\,\ddot{A}dt = m\ddot{y}\dot{y}dt$$

In other way, we can rewrite the above equation as

$$\frac{dW}{dt} = W = m\dot{y}\ddot{y} \quad (22)$$

However, it is evident from the above equation that the rate of work transfer is dependent on the first and second derivatives (velocity and the acceleration) of the piston, and is time dependent.

2.2.3 Algorithm to Calculate the Coefficient of Performance

To calculate the coefficient of performance of the system in consideration, the heat transfer rate and the work transfer rate are the required data. However, due to the complex calculations involved, despite the software used, it is a complex task to be done, and revert to define the steps involved in the calculation. The function files TGrad.m, heat.m, xdot.m and xddot.m were written in Octave and Matlab are usable in format. The first two m-files also use another m-file named compFactor.m, that computes the compressibility factor at a given temperature and pressure. The inbuilt wrapper for lsode, a FORTRAN package traditionally used to handle to solve initial value problem of ODE is used in this paper. Though, the inbuilt function ode45 can also be used to obtain the same solution. Using the predefined values and the accessory m-files, with the given initial value, the lsode computes the temperature distribution for a given range of values. Since the nature of the solution is not known initially, a very small step size of 0.01 is assumed. Using the temperature distribution, the m-file heat.m is used to estimate the heat transfer rate which is saved to the internal memory. Again, the m-files xdot.m and xddot.m are used to compute velocity and acceleration of the piston that in turn are used to estimate the work transfer rate. These results are used to calculate the coefficient of performance of the system.

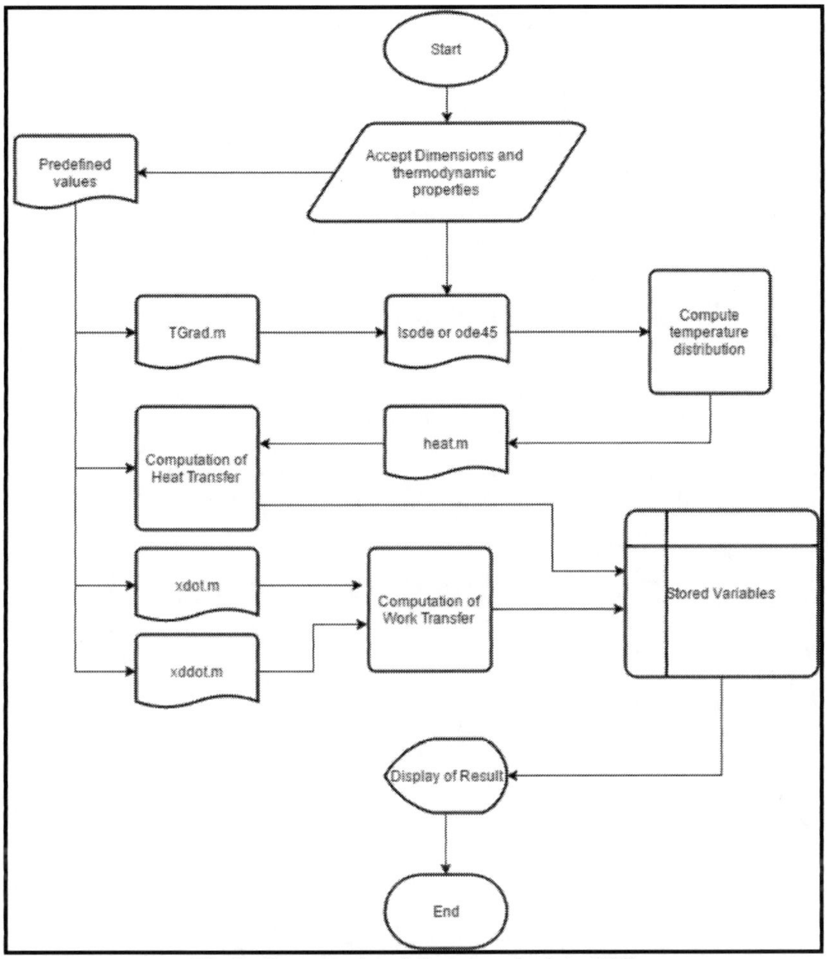

Fig.3. Computational flow to calculate COP

3. Results and Discussion

In this study, the steady-state behaviour and simulation of the refrigeration system using R134a and synthetic refrigeration oil was presented. Throughout the process, the heat and work changes of the system were analyzed.

3.1 Heat Transfer and Work Transfer Rate

In this study, experimental data were collected for the performance analysis of R134a synthetic refrigeration oil (ISO grade POE68). The results obtained from experimentation and model is given in the Tables.2 and 3. From the Fig.4,for this particular case, it can be shown that the temperature gradient does not depend upon the length, which is obvious from the definition of the gradient, in equation [22], apparent that the gradient changes direction of propagation around some temperature between 300 K and 400 K. The pressure of the evaporator is altered according to the difference.

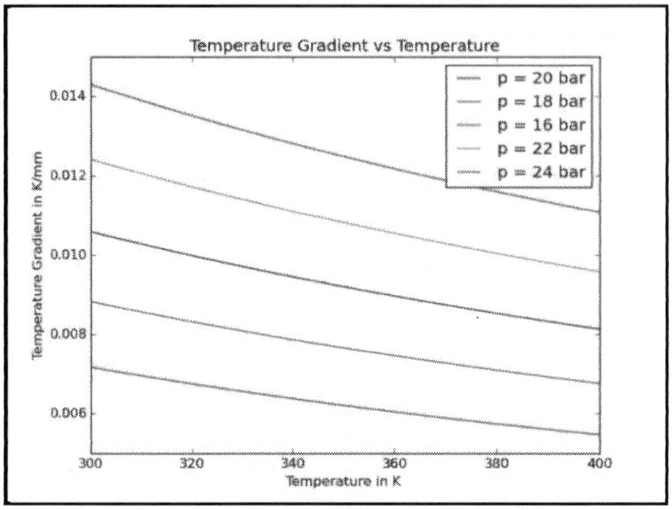

Fig.4 Temperature gradient vs Temperature

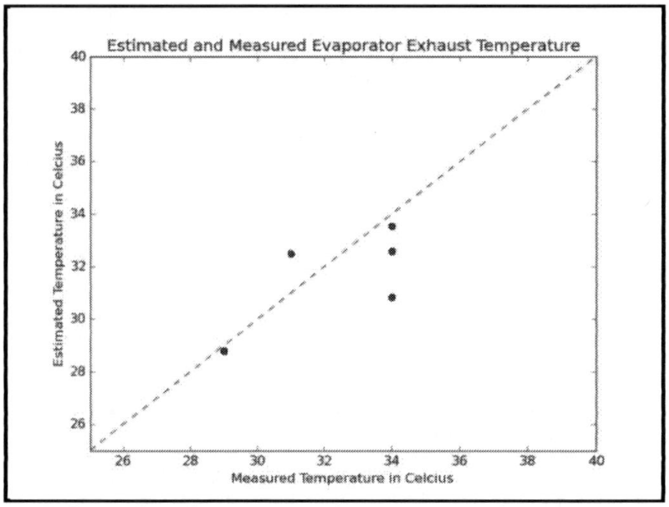

Fig.5 Estimated against measured evaporator exhaust temperature

Table.1 Temperatures and Pressures at Points of Interests

	Evaporator Suction		Compressor Suction		Compressor Exhaust	
S.No	Temperature (^0C)	Pressure psi	Temperature (^0C)	Pressure psi	Temperature (^0C)	Pressure psi
1	21	21	31	11	51	125
2	19	19	29	15	49	156
3	19	19	29	17	49	127
4	22	18	34	10	57	140
5	23	20	34	11	57	150
6	23	20	34	14	55	140

Table.2 Predicted and Measured Evaporator Temperatures

S.No	Evaporator Suction Temperature (°C)	Evaporator Suction Pressure psi	Estimated Temperature at Evaporator Exhaust (°C)	Measured Evaporator Temperature (°C)	Error (%)
1	21	21	32.498	31	-4.832
2	19	19	28.785	29	0.741
3	19	19	28.785	29	0.741
4	22	18	30.840	34	1.294
5	23	20	33.541	34	1.350
6	23	20	32.572	34	4.200

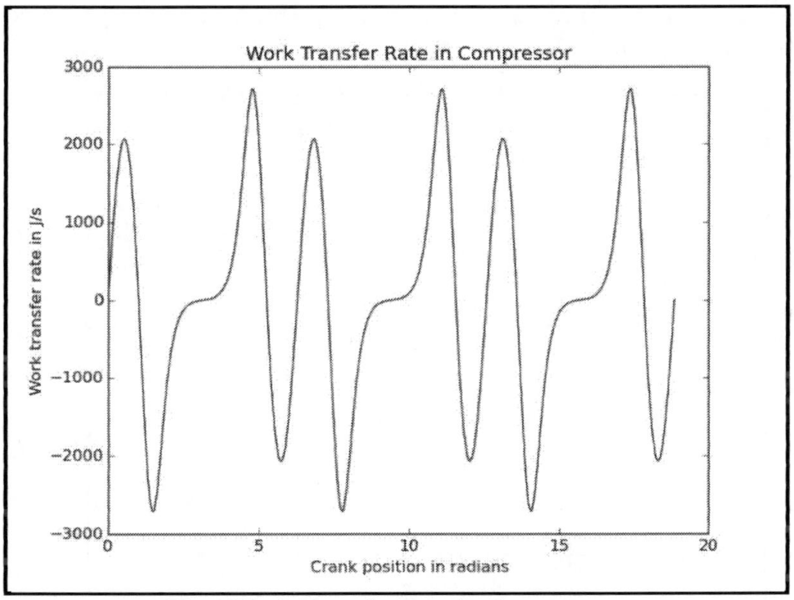

Fig.6 Work Transfer Rate in Compressor

Table.3 Heat and Work Calculated from Experimental Data

S.No	Heat Transfer (kJkg)	Work Transfer (kJkg)	Coefficient of Performance
1	9.811	4.594	2.135
2	9.047	4.586	1.972
3	8.802	4.695	1.874
4	11.248	6.211	1.810
5	10.534	5.042	2.091
6	10.196	4.571	2.230

The results from Fig.4 and Fig.5 show that the condenser parameters largely influence the discharge pressure and the compressor mass flow rate, while the evaporator parameters mostly have an effect on the refrigerant mass flow rate and the heat transfer rates which was also concluded by Harms [7].It is to be noted that the work transfer rate is very dynamic and the frequency is too high such that its peak value can be considered to be its steady state value. Due its complex form it is to be estimated only from graphical data. In this case, the work transfer rate is determined to be 2698 J/s from the Fig.6. Table.3 shows the heat and work transfer calculated from the experimental data.Fig.6 depicts the work transfer in the compressor that will influence by the pressure drop between condenser and evaporator. The work of compression decreases due to decreased condenser heat capacity increases for same cooling capacity of the system. The error occurred in the case of evaporator exhaust temperature estimation it is ±10%.

3.2 Coefficient of Performance

The variation in coefficient of performance with respect to the rate of heat transfer in (kW) is calculated from the experimental results and validated against the mathematical model. The results be evidence and as the same trend provided Akintunde et al [8]. Table.4 shows the predicted COP against calculated COP and the data for validation are compared are shown.

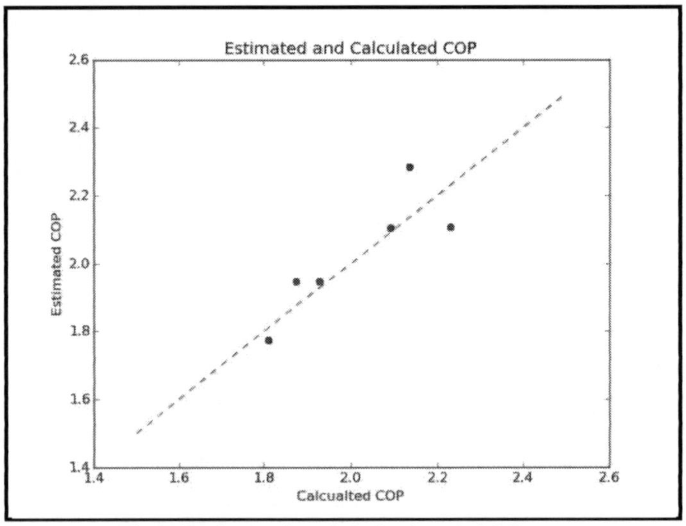

Fig.7 Estimated and Calculated COP

Table 4. Predicted COP against Calculated COP

S.No	Heat Transfer Rate (J/s)	Work Transfer Rate (J/s)	Coefficient of Performance Estimated	Coefficient of Performance Observed	Error (%)
1	6160.9	2698	2.284	2.135	-6.978
2	5255.6	2698	1.948	1.928	-1.037
3	5255.6	2698	1.948	1.875	-3.893
4	4782.5	2698	1.773	1.810	2.044
5	5678.5	2698	2.105	2.091	-0.669
6	5678.5	2698	2.108	2.230	5.470

The compressor work consumption reduction thus increases the COP. It is observable that the error in estimation of COP is ±7% and is well within the acceptable range when compared with the results of by Shurooq Talib Remedhan Al-Hemeri. Fig.7 depicts the estimated and calculated COP. The COP of the system increases due to an increase in the refrigeration capacity resulting from increases in the evaporator efficiency [10].

4. Conclusions

A mathematical model has been developed, analytic in nature in order to predict the evaporator exhaust temperature, the work transfer rate and COP of the system, given the evaporator suction temperature. The thus developed model when compared with experimental results, the model yields consistent results with an error band of ±10% and ±7% for temperature exhaust temperature and COP respectively. This model can be applied to any vapour compression refrigeration System to estimate its performance index. However, this model does not differentiate between the various parameters like volumetric efficiency and Viscosity of the refrigeration oil in the compressor for work transfer, and is based solely on the piston design and its angular velocity. Therefore, this model cannot be used to predict the changes when the compressor oil is changed.

References

[1] Saidur R, Kazi S N, Hossain M S, Rahman M M, Mohammed H A, Renew. Sustainable Energy Rev 15, (2011) 310–323.
[2] Mobarak H M, Niza Mohamad E, Masjuki H H, Kalam M A, Al ahmud K A H, Habibullah M, Ashraful M A, Renewable and sustainable energy resources, 33 (2014) 34-43.
[3] Domanski P A, Yashar D, Kim M, Int. Journal of Refrigeration 28,6 (2005) 820-827.
[4] Ding G L, Wu Z G, Wang K, Fukaya M, Int. J. Refrigeration 30 (3) (2007) 418-432.
[5] Jatinder Gill, Jagdev Singh, Int. J. Refrigeration 16 (2017) 436-446.
[6] Chi-Chuan Wang, Armin Hafner, Cheng-Shu Kuo, Wen-Der Hsieh, Renew. Sustainable Energy Rev 16 (2012) 5071-5086.
Baskar S, Sriram G, Arumugam S, Tribology Transactions Vol.60,3,(2017) 428-436.
[7] Harms T M, Braun J E, Groll E A, HVAC&R Research Journal 10,1(2004) 5–20.
[8] Akintunde M, Falade T, Bolaji B, Journal of sustainable energy and environment 2 (2015) 61-64.
[9] Shurooq Talib Remedhan Al-Hemeri, The Eighth Jordan International Chemical Engineering Conference (2017) 1-13.
[10] Yeunyongkul P, Sakulchangsatjatai P, Terdtoon P, Energy research Journal 2 (2012) 104-110.

Machining of Ti-6Al-4V ELI Alloy: A brief review

Anurag, R kumar*, S Roy, K K Joshi, A K Sahoo and R K Das

School of Mechanical Engineering, Kalinga Institute of Industrial Technology, Deemed to be University, Bhubaneswar-751024, India

E-mail: ramanujkumar22@gmail.com

Abstract: Ti-6Al-4V ELI has similar properties as of Ti-6Al-4V (Grade 5) except it has higher ductility as well as improved fracture toughness. It has the ability to withstand high temperature without losing properties, highly corrosion endurance and high strength to weight ratio. So it's widely applicable in biomedical implants, aerospace industry, military application, automobile component and high pressure cryogenic vessels. During the machining of the titanium alloys, a lot of machining difficulty occurs due to its hardness, chemical reaction take place between tool and machine workpiece, high amount of heat generated at cutting zone. Cutting speed and cutting feed are predominant factors for cutting tool wear as well as the finished quality of the surface. Mainly flank wear (adhesion), notch wear and crater wear (diffusion) are identified during the machining. Cutting temperature indirectly influenced the surface quality of machined surface and tool wear. The high cutting speed with low feed rate is a most favourable combination for cutting force. For cutting insert, performance tool coating plays a significant role and coated inserts performed better than the uncoated tool at various cutting conditions. Machining under minimum quality lubrication is more favourable than flooded and dry scenario.

1. Introduction

Now a day's due to advancement in cutting technology specially in the cutting tool, hard turning even at high speed successfully replaced the EDM and Grinding machining to improve the productivity. Ti-6Al-4V ELI (Grade23) comes under $\alpha + \beta$ categories of titanium alloys and ELI stands for an extra-low interstitial. Ti-6Al-4V ELI has similar properties as of Ti-6Al-4V (Grade 5) except it has higher ductility as well as improved fracture toughness. It has the ability to withstand at high temperature without losing properties, highly corrosion endurance and high strength to weight ratio. So it's widely applicable in biomedical implants, aerospace industry, military application, automobile component (at high stress), and high pressure cryogenic vessels.

Table 1. Chemical content of Ti-6Al-4V ELI [2-14]

Composition	C	Si	Fe	Al	N	V	S	O	H	Ti
Wt %	0.003 - 0.11	< 0.03	0.10 - 0.22	5.85 - 6.13	0.006 - 0.02	3.8 - 4.1	0.003	0.08 - 0.13	0.001 - 0.003	Balance

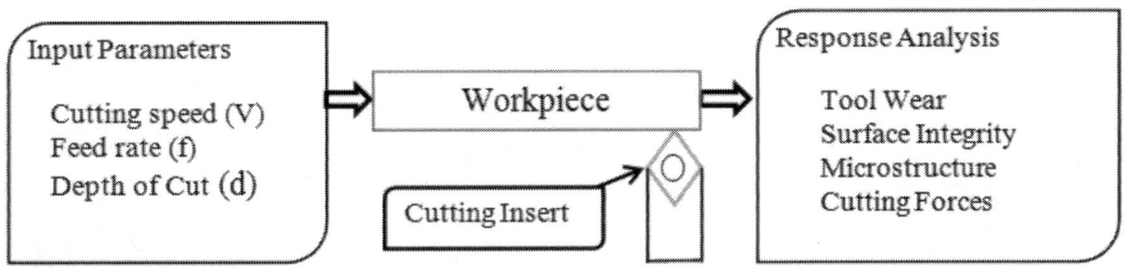

Figure 1. Schematic view of machining aspects

Table 2. Physical and mechanical properties of Ti-6Al-4V ELI alloy

Density (ρ)	Hardness HRC	Ultimate Tensile strength	Yield Tensile strength	Elongation of break	Modulus of elasticity	Poisson ratio
4.43 g/cc	35	860 MPa	790 MPa	15%	133.5 GPa	0.342

During the machining of the titanium alloys, a lot of machining difficulty occurs due to its hardness, chemical reaction take place between tool and machine workpiece, high amount of heat generated at cutting zone [1]. Figure 1 shows the schematic view of machining aspects and chemical content and physical and mechanical properties of Ti-6Al-4V ELI alloy shown in Table 1and Table 2.

2. Application of machining methods

Hard turning is successfully replaced the EDM and Grinding for machining the very high hardness materials. During machining of this type of titanium alloy difficult to machine because of the high amount of heat produced at the cutting region which affected the machining process. However, selection of suitable cutting parameters and favourable cooling environments are the key aspects which affect the entire cutting process. Very few amounts of machining investigations have been found on machining Ti-6Al-4V ELI till date. Some studies were made considering both turning and milling cutting operation and its cutting details with responses studied has been arranged in Table 3.

Table 3. Machining parameters for Turning and Milling process

Reference	Author's Name	Year	Machining Process	Process parameters			Response
				(V) (m/min)	(F) (mm/rev)	(D) (mm)	
1	Sulaiman	2014	Turning	120, 170, 220	0.2	0.4	Tool wear
2	Sulaiman et al.	2013	Turning	120, 170, 220	0.1, 0.15, 0.2	0.4, 0.5, 0.6	Optimized cutting parameters
3	Shin et al.	2013	Turning	110	0.20	1.0	Cutting Force, Tool life, and surface roughness
4	Ghani, Haron	2015	Turning	120-200	0.1-0.2	0.4-0.6	Wear mechanism

5	Karkaols et al	2016	Milling	75,100, 125	0.2, 0.25, 0.30	0.3, 0.6, 0.9	Surface Roughness
6-9	Ibrahim et al.	2013, 2011, 2009	Turning	55, 75, 95	0.15,0.25, 0.35	0.10, 0.15, 0.20	Microstructure alteration, surface roughness and cutting tool life
10	Gusri et al.	2010	Turning	55, 75, 95	0.15,0.20, 0.35	0.10, 0.15, 0.20	Surface Quality
11	Dillibabu et al.	2013	Turning	55, 75, 95	0.15,0.20, 0.35	0.10, 0.15, 0.20	Effect of cutting Parameters
12	Haron et al.	2016	Turning	120, 170, 220	0.1, 0.15, 0.2	0.4, 0.5, 0.6	Cutting Tool Performance
13	Sargade et al.	2016	Turning	50, 65, 80	0.08,0.15, 0.2	0.5	Surface roughness and Cutting force
14	Sharif	2014	Milling	200, 250, 300	0.03, 0.06	Radial0.5 Axial 1.5	Surface roughness and cutting Forces

3. Application of Cutting tools and cooling techniques

The performance of the cutting insert also depends on the types of the coating applied to the tools, thickness of the coating, working temperature. When the working on the titanium alloy the tool wear gradually increased after the damage of the coating surface. Various researchers suggested that use only coated insert for machining of titanium alloy [1, 4]. Basically, two types of coating (CVD & PVD) used during the machining of hard materials. When machining hard materials high range temperature occurred at the cutting region this excessive temperature causes softening of materials and required low cutting force to cut the materials but the increment in the temperature causes of tool wear and also affected the machining performance [2]. Also applied cooling technique affected the machining performance.

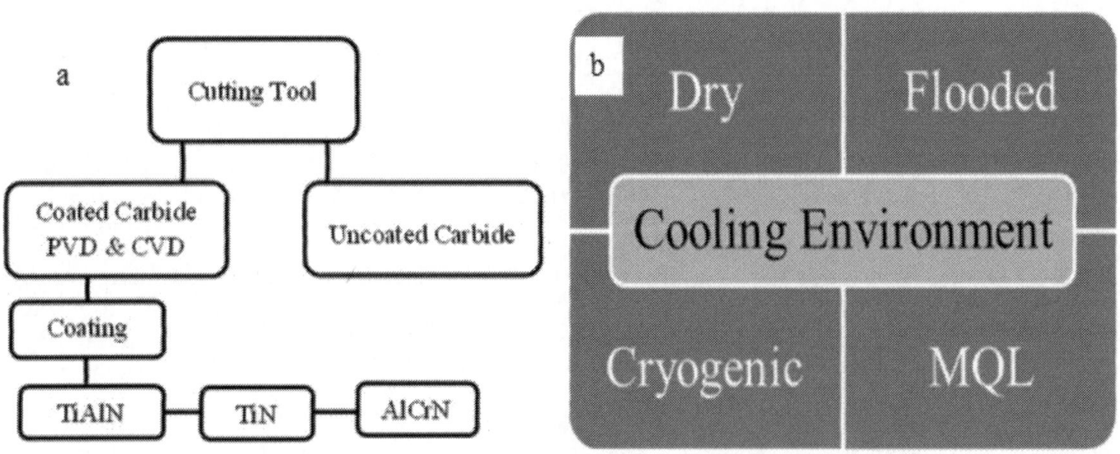

Figure 2. (a) Types of tool coating and (b) cooling condition

Table 4. Cooling Conditions and finding

Reference	Author's Name	Year	Machining Process	Cooling Condition	Finding
1	Sulaiman	2014	Turning	Flooded	Tool wear rapidly increase with cutting speed
2	Sulaiman et al.	2013	Turning	Dry	Low tool wear at high turning speed with low feed rate
3	Shin et al.	2013	Turning	Dry	Wear and force rapidly increase with an increase in cutting time.
4	Ghani and Haron	2015	Turning	Dry	Adhesion and diffusion.
5	Karkaols et al	2016	Milling	Dry	To reduce the cutting force and satisfied surface finish working at high turning speed with low feed rate.
6-9	Ibrahim et al.	2013, 2011, 2009	Turning	Dry	Cutting temperature has mainly affected the microstructure of workpiece surface. Surface integrity affected by tool wear.
10	Gusri et al.	2010	Turning	Dry	Surface quality highly correlated with feed rate
11	Dillibabu et al.	2013	Turning	Dry and flooded	Nose radius predominant factor for surface roughness
12	Haron et al.	2016	Turning	Flooded, MQL	MQL performed better than the flooded cooling condition.
13	Sargade et al.	2016	Turning	Dry	Feed rate is predominant factor for Surface roughness and force.
14	Sharif	2014	Milling	Dry	Feed rate affected surface roughness and cutting force decrease with increase in speed

Because when working in dry environment more heat engendered at the cutting area which causes of tool wear to eliminate this problem using a suitable cooling condition which is harmless for the environment and employed health such as MQL, Cryogenic etc. cooling condition [3, 5]. Sulaiman et al.[2] predicted the optimized cutting variables during turning of the Ti-6Al-4V ELI under wet cooling environment using RSM technique. Found that the speed and axil feed predominant factor for the tool life and roughness of the machined surface. According to the ISO 3685 performance of the cutting tool for Ti-6Al-4V ELI is acceptable. RSM give the optimized values of the input variables ($v = 220$ m/min, $f = 0.1$ mm/rev, $d = 0.4$ mm).Haron et al.[12] used a water based mineral oil in both MQL and flooded cooling environment and the result shows that in MQL cooling technique measured low wear at tool edge as compared to the flooded cooling condition. Figure 2(a, b) and Table 4 shows the types of cooling condition and tool coating used during machining Ti-6Al-4V ELI by various researchers.

4. Analysis of Tool Wear

During the machining friction between the tool and workpiece causes the garb in the tool this wear cannot eliminate but it can control by choosing appropriate cutting parameters and cooling condition.

A high amount of cost spent on the tooling section so it's very important to investigate the cause of tool failure and wear mechanism during the machining processes. When machining at excessive cutting speed more heat appeared at cutting region due to low thermal conductivity causes of the rise in the temperature at a cutting zone which affected the performance of the cutting insert and the quality of the finished machined surface [2, 3]. Tool wear is affected the overall machining performance. During cutting operation tool wear appeared due to the friction between the tool and workpiece. Wear in cutting inserts depends upon the tool parameters, types of coating and cutting condition. At elevated temperature, a chemical reaction takes place between the tool and workpiece causes the wear in tools also chips affected the rake surface of the tools. Mainly flank wear (adhesion), notch wear and crater wear (diffusion) produce during the machining of Ti-6Al-4V ELI alloy [4, 5]. Sulaiman et al.[1] analysed the tool wear rate for an uncoated tool insert during high speed turning of Ti-6Al-4V ELI alloy in a dry environment. The rate of wear significantly increase from low to high cutting speed and the tool life is less at high cutting speed. The performance of the uncoated carbide insert is low in the dry cooling condition. Shin et al.[3] used Back propagation neural networks technique for detecting the wear at cutting edge and the effect of the input parameters when turning of Ti-6Al-4V ELI in dry condition used CVD coated cemented carbide tool insert. Found that the surface roughness has closed relationship with tool wear.

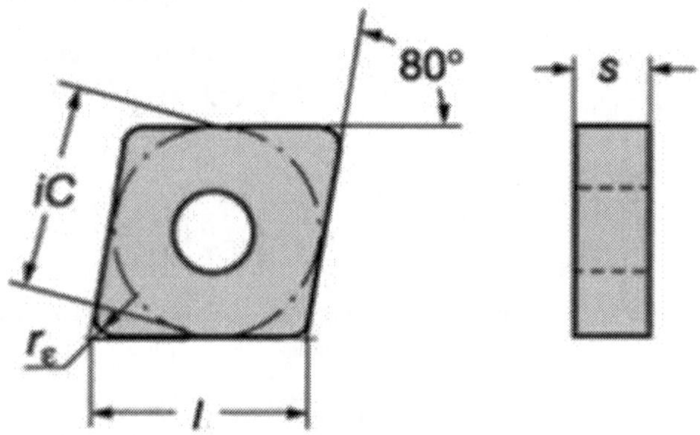

Figure 3. Schematic of the geometry of the carbide insert used in this study [1, 2]

Ghani and Haron [4] investigated the wear mechanism during the turning of Ti-6Al-4V ELI with coated (TiAlN/ AlCrN) and uncoated tungsten carbide tools under MQL cooling condition. Found that the flank wear at the flank face, notch wear and crater wear at rake surface cutting insert occur. Excessive temperature and high stress generated at the cutting area which cause of tool wear out. Adhesion and diffusion found that the predominant wear mechanism also chemical reaction takes place between the tool and workpiece materials causes of wear mechanism. Fig. 3 showed the geometry of an uncoated carbide cutting insert. Where s = 0.13 mm, l = 12 mm, iC = 12.7 mm and rє = 0.8 mm.

5. Analysis of Surface Roughness

Titanium alloy used in biomedical implants and biomedical implants where required satisfied surface roughness and integrity. Surface quality easily affected by the input parameters when machining on titanium alloy specially feed rate is the predominant factor for surface roughness. Due to the high hardness of Ti-6Al-4V ELI alloy and low thermal conductivity, a high amount of heat generated at the upper surface of the workpiece causes increased in the tool wear which damage the microstructure and surface integrity [5-12]. Karkaols et al.[5] predicted the surface roughness of Ti-6Al-4V ELI alloy in a milling operation. Aftermaths obtained that the feed rate is the predominant variable for affected the roughness of the finished surface of the workpiece. RSM method used to optimizes the input and output response and then compared the RSM and ANN models for more accurate prediction. Ibrahim et al.[6-9] investigated the microstructure alterations and surface integrity during turning of Ti-6Al-4V

ELI alloy in a dry cooling environment with coated carbide cutting insert. Found that when the temperature increases at cutting region it's damaged the upper machined workpiece surface and affected the surface microstructure. Surface roughness increased with increase in the temperature. Gusri et al.[10] investigated the surface quality of Ti-6al-4V ELI alloy using CVD coated insert in the absence of cooling environment. Found that the roughness of the machined surface is more influenced by the feed. The changing in the microstructure and texture of the surface affected the surface quality. Also, a white layer with 2µm thickness found during machining at high speed with high feed rate. Dillibabu et al. [11] investigated the influence of the process variables on quality of machined surface during machining of Ti-6Al-4V ELI alloy with coated (PVD- TiAlN) coated carbide inserts in the dry and wet environment. According to the ANOVA analysed data showed that the insert nose radius is predominant factor for surface quality than the other process variable and applied the cooling condition. Ibrahim et al. [8] investigated the surface integrity using coated carbide inserts during machining of Ti-6Al-4V ELI in the dry cooling environment. Aftermath showed that the feed rate is the dominated factor for surface quality. Surface roughness curves subsist of three aspects at first stage give better surface quality after that regularly decrease in the surface roughness. Also found a white layer at finished surface when machining at (v = 95m/min, f = 0.35 mm/rev, d = 0.10 mm). There was no physical damage (tears and crack) of the coated carbide tools. Ibrahim et al.[9] predicted the surface integrity and the tool (PVD coated) insert during the working of Ti-6Al-4V ELI used dry cooling environment. The result showed that the tool life high at low cutting parameters values as compare to the high parameters values. Tool wear increased rapidly, gradually and extremely at the initial, second and final stage. Surface quality is better at the initial stage as compared to the second and final cutting stage. Because when the change in the nose radius increment in the tool wear and affected the roughness of the machined surface. Haron et al.[12] investigate the performance of the uncoated carbide and quality of the machined surface of the Ti-6Al-4V ELI alloy in wet and MQL coolant condition during high speed turning operation. The result showed that in MQL cooling condition better tool life (25% and surface roughness (30%) more than the wet (flooded) cooling condition.

6. Analysis of Cutting Forces

Calculating the cutting force is very complexity during cutting operation because during machining the input parameters affected the cutting force specially feed rate is the predominant factor for cutting force in turning operation. Various researchers suggested that low cutting forces required when machining with high cutting speed and low axial feed. Sargade et al. [13] investigated the roughness of the machined surface and cutting force when turning operation of Ti-6Al-4V ELI with PVD coated carbide insert in absence of coolant. Found that specially feed rate is predominant variable for surface roughness and cutting force. ANOVA and RSM technique used to predict the effect of input parameters and optimization for minimizing surface roughness and cutting force. Sharif et al.[14] investigated the effect on upper finished surface roughness and cutting forces during milling machining on Ti-6Al-4V ELI under dry condition using PVD coated (TiAlN+TiN) and uncoated carbide tools. The result showed that the cutting force is low at the increment in the cutting speed with the low feed rate for both the coated and uncoated cutting insert. Coated tool show the better performance than the uncoated carbide tool at each cutting condition.

7. Conclusion

Machining of Ti-6Al-4V ELI comes under alloy which is difficult to cut. However, selection of process parameters, cutting tools, and cooling process plays a major role in machining. This raised in temperature affected the whole machining performance. The effect of the cutting parameters shown as

- Generation of higher ranges of cutting temperature may cause the poor surface quality as well as declines the dimensional accuracy and tool life. Also, higher heat accelerates the work softening effects and causes the variation in the microstructure of the finished surface. This change in the microstructure affected the surface quality. The cutting feed was predominant factor for surface roughness.

- Abrasion and diffusion are prime mechanisms involved in coated carbide tool failure. Damage to upper layers of the coating is identified at higher speeds. Cutting speed is predominant factor for tool wear. Tool wear can be controlled by the applied appropriate cooling condition for better cutting performance.
- The high cutting force required to cut the hard material. At high temperature, work softening appeared which reduces the shear strength of materials thus machining is easier but at the same time, this elevated temperature also accelerates the tool wear thus surface quality. The high cutting speed with low feed rate is a most favorable combination for cutting force.

8. Future research

Very few conventional machining work reported on Ti-6Al-4V ELI however further detail investigations are needed to study its machinability. Some points may be considered in future as follows:

- Applications of different kinds of cooling/lubrication techniques like minimum quantity lubrication, high pressure cooling, spray cooling, cryogenic cooling and nano fluids can be considered.
- Investigations on residual stresses, white layer formation, microstructure etc. can be considered for the betterment of overall machining process.
- Economic aspects can be considered

Reference

[1] Sulaiman M A, Haron C C H, Ghani J A and Kasim M S 2014 Effect of High-speed Parameters on Uncoated Carbide Tool in Finish Turning Titanium Ti-6Al-4V ELI *Sains Malaysiana* **43,1** pp 111–116

[2] Sulaiman M A, Haron C, Ghani J A and Kasim M S 2013 Optimization of Turning Parameters for Titanium Alloy Ti-6Al-4V ELI Using the Response Surface Method (RSM) *Journal of Advanced Manufacturing Technology* **7,2** pp 11-28

[3] Shin H G, Yoo S H, Park S W and Hong D P 2013 A Study on the Cutting Characteristics and Detection of the Abnormal Tool State in Turning of Ti-6Al-4V ELI *Applied Mechanics and Materials* **433-435** pp 2025-2030

[4] Ghani J A and Haron C C H 2015 Wear mechanism of coated and uncoated carbide cutting tool in machining process *Journal of Materials Research* pp 1-7

[5] Karkalos N E, Galanis N I and Markopoulos A P 2016 Surface roughness prediction for the milling of Ti–6Al–4V ELI alloy with the use of statistical and soft computing techniques *Measurement* **90** pp 25–35

[6] Ibharim G A, Arinal H, Zulhanif and Haron C H C 2013 Microstructure Alterations of Ti-6Al-4V ELI during Turning by Using Tungsten Carbide Inserts under Dry Cutting Condition *International Journal on Engineering and Technology Development* **1,2** pp 37-40

[7] Ibharim G A, Haron C H C and Ghani J A 2011 Evaluation of PVD-Inserts Performance and Surface Integrity When Turning Ti-6Al-4v ELI under Dry Machining *Advanced Materials Research* **264-265** pp 1050-1055

[8] Ibharim G.A, Haron C and Ghani J 2009 The effect of Dry machining on Surface Integrity of Titanium alloy Ti-6Al-4V ELI *Journal of Applied Sciences* **9,1** pp 12-127

[9] Ibharim G, Haron C H C and Ghani J A 2009 Surface Integrity of Ti-6Al-4V ELI When Machined using coated carbide tools under dry cutting condition *International Journal of Mechanical and Materials Engineering* **4,2** pp 191-196

[10] Gusri A I, Hassan C C H, Jaharah A G, Yasir A, Zaid Y and Yanuar B 2011 Surface Quality of Ti-6%Al-4%V ELI When Machined Using CVD-Carbide Tools at High Cutting Speed

International conference on Advances in Materials and Processing Technologies **1315,1** pp 1107-1112

[11] Dillibabu R, Sivasakthivel K and Kumar S 2013 Optimization of process parameters in dry and wet machining of Ti-6AL-4V ELI using Taguchi method *International Journal of Design and Manufacturing Technology* **4,3** pp 15-21

[12] Haron C H C, Sulaiman M A, Ghani J A, Kasim M S and Mohamad E 2016 Performance of carbide tool in high speed turning of Ti-6Al-4V ELI under conventional coolant and minimal quantity lubrication *ARPN Journal of Engineering and Applied Sciences* **11,7** pp 4817-4821

[13] Sargade V G, Nipanikar S R and Meshram S M 2016 Analysis of surface roughness and cutting force during turning of Ti6Al4V ELI in dry environment *International Journal of Industrial Engineering Computations* **7** pp 257–266

[14] Sharif S, Safari H, Izman S and Kurniawan D 2014 Effect of High Speed Dry End Milling on Surface Roughness and Cutting Forces of Ti–6Al–4V ELI *Applied Mechanics and Materials* **493** pp 546-551

IOP Conf. Series: Materials Science and Engineering **390** (2018) 012113 doi:10.1088/1757-899X/390/1/012113

Analysis of Chip Reduction Coefficient in Turning of Ti-6Al-4V ELI

Anurag, R kumar*, K K Joshi and R K Das

School of Mechanical Engineering, Kalinga Institute of Industrial Technology, Deemed to be University, Bhubaneswar-751024, India

E-mail: ramanujkumar22@gmail.com

Abstract: Titanium alloys are generally used in aerospace industries due to its ability to retain its high strength at elevated temperatures. Present paper deals with Taguchi optimization approach applied to optimize process parameter in turning of Ti-6Al-4V ELI (grade 23) titanium alloy workpiece with coated carbide tool under dry condition. The trails carried out by employing L9 orthogonal array. Linear regression modelling was adopted to develop a model for prediction of experimental results. The predicated results lay proximal to the experimental value. Main effects plot and ANOVA are used to analyse the effect of input variables namely speed (v), feed (f) and depth of cut (d) on chip reduction coefficient. The optimal parameter setting obtained for chip reduction coefficient is *d3-f1-v1* i.e (d = 0.4 mm; f = 0.04 mm/rev; v = 46 m/min).

1. Introduction

In recent decades titanium alloys have proven to be a novel material for industrial applications. Titanium alloy mostly used in aerospace industries and biomedical implants due to its ability to hold its high strength at elevated temperature, excessive resistance to corrosion, low density, and low thermal conductivity. But machining of this alloy recessed high cutting speed but it generates high amount heat at the cutting point, which alters the tool life, roughness of the machined surface. To analysis, the machinability aspects of titanium alloy during turning process are essential to study the chip morphology. Chip formation depends on the material mechanical and thermal properties. At high cutting speed, continuous chip converts into serrated chip formation by the thermal effect which affected the cutting force and wears in the cutting tool. Gao and Zhang [1] investigated the effect of the cutting condition on serration found in the chip during turning process and found that when the rake angle is a more than the decrease in the size of sawtooth but the increment in the total number of sawtooth. Hua and Shivpuri [2] analysed the segmentation of chip while the turning of titanium alloy. Stress state is changed at the cutting zone is change when the increments in the cutting speed cause in the shifting in the fracture mechanism from cutting insert edge to the rest surface of the chip. This changing in the fracture is main causes of converting of discontinuous chips into the serrated chip. Sutter and List [3] probed the changing in the mechanism of chip formation during the turning of the Ti-6Al-4v alloy at elevated rotational speed. It was observed that the rotational speed is a predominant aspect for formatting the formation of the chip also the angle of shearing and length of crack affected the chip formation. Dargusch et al. [4] deliberated the influence of the wear of cutting insert on chip formation during turning of Ti-6Al-4V alloy under dry cooling condition. When the tool worn increased then the friction between the cutting insert and chip also increased which causes of increased in twining deformation and chip formation through both tensile and compressive. Bai et al. [5] researched the formation chip throughout orthogonal cutting of Ti-6Al-4V alloy and observed that the contact length of tool-chip, cutting forces, coefficient of friction and shear strain at the shear region is reduced with in the decrease in the speed and chip-segmentation rise with an increment in rotational speed with a reduction in the feed. Wu and To [6] investigated the effect of the serrated formation of the chip during machining of titanium alloy using Johnson-Cook (JC) model and FEM model. Found that especially rake angle affected the serrated chip formation its decrease with large rake angle. Sun et al. [7] researched the machining of

Ti-6Al-4V alloy using Cryogenic compressed air cooling condition found that at a low speed more friction between tool and workpiece causes a reduction in the thickness of chip and at low speed, with low feed, the chip segment convert irregular to the regular segment. Aramcharoen [8] studied the influence of tool wear and chip formation in turning of titanium alloy using cryogenic cooling condition. When friction is reduced between tool and chip the produced chip shape is the helical and small radius of curvature. Sahoo and Sahoo [9] examined the effect of the multilayer coated cutting tool during hard turning and optimized the optimal cutting condition for the roughness of the machined workpiece surface using RSM and grey relation approach. Kumar and Kulkarni [10] investigated the surface finish throughout the turning of titanium alloy. Taguchi method reduced the total number of the experiment. Found that the cutting edge radius is specially affected the surface finish. Sulaiman et al. [11] predicted the impact of cutting variables on the machining performance during machining of Ti-6Al-4V ELI using RSM technique. Found out the optimization condition for tool life and surface roughness at elevated cutting speed with low feed or depth of cut. Ramana [12] investigated the optimal process parameters for the roughness of the machined surface using ANOVA technique during machining of the Ti-6Al-4V alloy under various lubrication conditions. Observed that axial feed rate is the major factor that influences the finish of turned surface in all cooling condition. Gupta et al. [13] examined the turning of titanium alloy (grade-2) utilizing RSM approach in MQL cooling condition. Predicted the optimized cutting condition for low surface roughness, low cutting force and tool worn machining at low cutting speed with low feed with high side cutting edge angle. Bandaoalli et al. [14] investigated the surface roughness during the milling of titanium alloy (grade-5) using ANN, GMDH and MRA technique. Result showed that the depth of cut and axial feed rate is predominant factor for the imprecision of machined surface but when the feed and depth of cut are constant than roughness declines with increment in the rotational speed. Kumar et al. [15] found that the turning speed and radial depth of cut were the predominant factor for the chip reduction coefficient also found that high CRC value measured while using uncoated carbide tool as compare to coated carbide tool. It was observed from the literature analysis that the most of the research papers deal with Taguchi methodology for optimization of the separate performance characters.

2. Experimental Methods

Ti-6Al-4V ELI (grade-23) titanium alloy workpiece having length 100 mm and diameter 35 mm used as the machined material. The hardness of Ti-6Al-4V ELI alloy is (35±1) at normal surrounding temperature. A WIDIA made coated carbide (CNMG 120408) cutting insert manipulated for the machining. HMT NH22 high-speed lathe machine used for present tuning process. All the experiment performed under dry cooling environment.

Figure 1. (a) Experimental set up (b) Chip thickness measuring instrument (Digital Vernier calliper)

The chip thickness measured by digimatic caliper manufactured by Mitutoyo Company (Made in Japan) having measuring range between (0-300mm / 0-12inch) and minimum indication value (0.01mm / 0.0005inch). The L9 orthogonal array was used as the design of the experiment. The range of cutting speed is 46- 102 m/min, the axial cutting feed is 0.04-0.2 mm/rev and depth of cut is 0.2-0.4 mm and the schematic diagram of turning process shown in the Fig. 1.

3. Analysis of Chip reduction coefficient

Generally, in the turning of any materials continuous, discontinuous and continuous with built-up-edge chips produce [16]. In the turning process, the materials ahead the cutting insert is compressed all sided and deformation takes place in the form of a chip. The chip thickness (a_2) is always larger than the thickness of uncut chip (a_1). The ratio of the chip thickness to the thickness of uncut chip known as the chip reduction coefficient (ξ) shown in the equation in (1) [16].

$$\xi = \frac{a2}{a1} \tag{1}$$

Where a_1 = feed . Sinϕ, and $\phi = 95^0$

If the value of (ξ) is more than more force and energy required to cut the materials. So it's very important to reduce the (ξ) without influence the productivity. So chip reduction coefficient is important machinability aspects in machining area. From the main effect plot for ξ, it was clearly noted that cutting speed and feed rate are inversely proportional to chip reduction coefficient. With the increment in speed and feed the chip reduction coefficient decreases. Taking into account the trend of the depth of cut, all the values of ξ generated lie very proximal or lie on the mean line without any sign of variation. So the depth of cut is not relevant for chip reduction coefficient shown in Fig. 2.

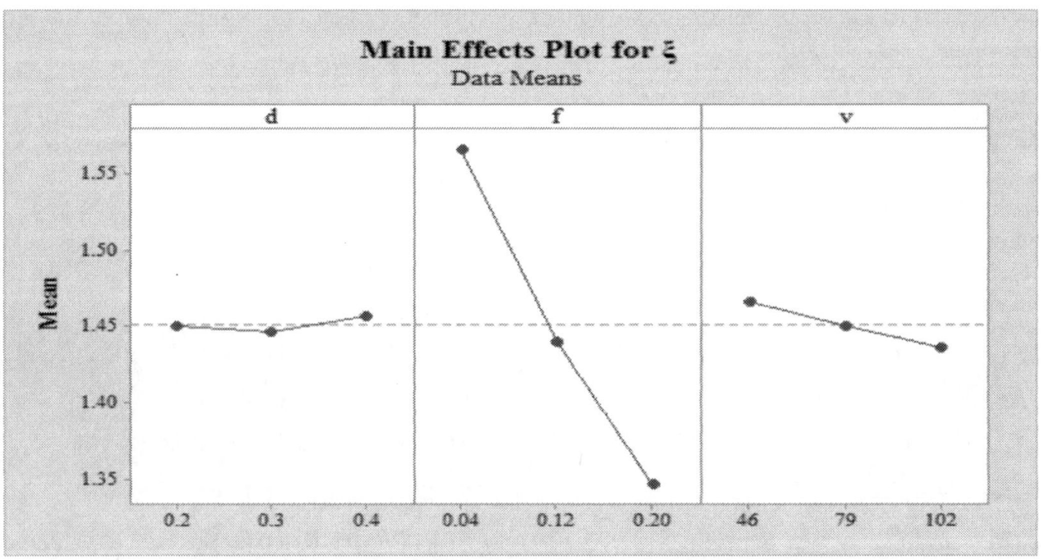

Figure 2. Main effect plot for chip reduction coefficient

4. Linear Regression Modelling

To relate a mathematical relationship between cutting parameters and responses, linear regression analyses were performed and further the following equations were developed:

ξ = 1.64649 + 0.0333333 d + 1.375 f - 0.000533586 v
R-Sq = 99.10% R-Sq (pred) = 96.98% R-Sq (adj) = 98.57%

From the linear regression model it was observed that R-Sq value (99.10%) is very close to R-Sq (adj) value (98.57%) and both the values were close to the R-Sq (pred) value which signifies the validation of the model. In this model R-Sq value approaches to unity (100%), the desired output response model is fitted closely and effectively with the actual experimental result data. The graph plotted between experimental observation and chip reduction coefficient signifies that experimental data and predicted data lie close to each other without any significant deviation shown in Fig. 3. Especially at trail no. 3, 6 and 7 the experimental and predicted value is the most proximally placed. This signifies that the model used for prediction is significant. Fig. 4 shows the various residual plots for chip reduction coefficient were plotted using the MINITAB-17 software. The normal probability plot shows that the residual value lay very close to the mean line, which indicates the validation of experimental values. Overall residuals plot confirms the significant establishment of the model to predict the results correctly.

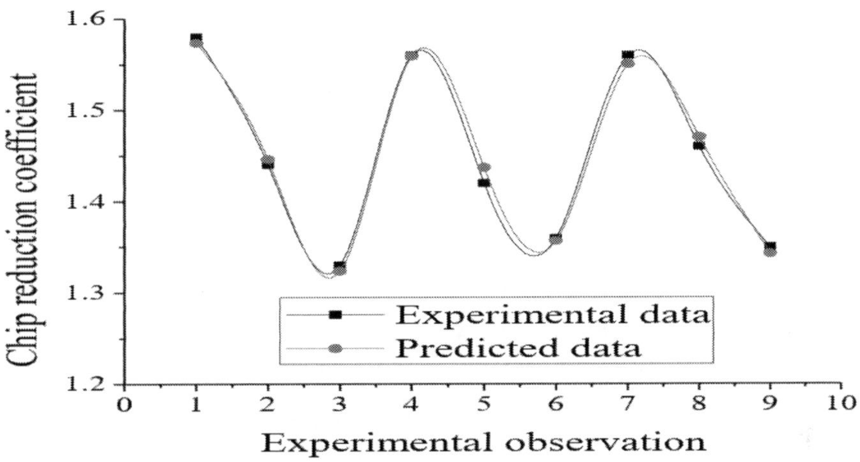

Figure 3. Compression graph of chip reduction coefficient between experimental & predicted data

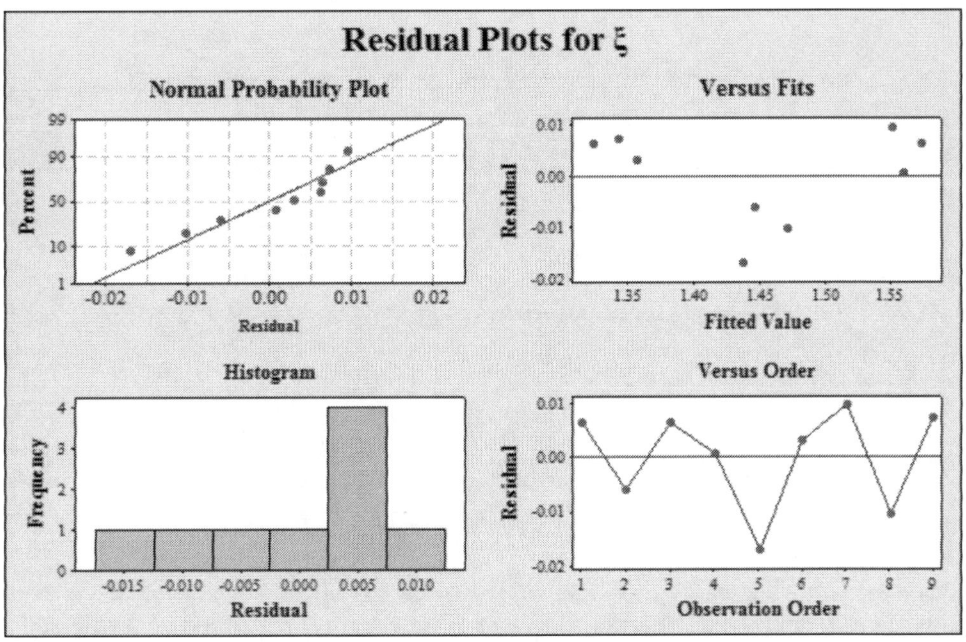

Figure 4. Residual plots for chip reduction coefficient

ANOVA method is used to determine the influence of given input parameter from a series of experimental results for a machining operation and also used to decode the experimental data. ANOVA is also used to determine the percentage contribution of the individual parameter, which is an advantage over Taguchi method. Here from Table 1, the P value (0.000015) for regression is below the standard value of 0.05 (95 % of confidence level) which ensures that the established model is valid. The contribution of feed on chip reduction coefficient is highest among all input variables which confirm that the feed is the most dominating term associated with chip reduction coefficient.

Table 1. Model ANOVA

Source	DF	Seq SS	Adj SS	Adj MS	F	P	Contribution (%)
Regression	3	0.0740202	0.0740202	0.0246734	184.490	0.000015	
D	1	0.0000607	0.0000667	0.0000667	0.498	0.511696	0.090
f	1	0.0726000	0.0726000	0.0726000	542.850	0.000003	97.20
V	1	0.0013535	0.0013535	0.0013535	10.121	0.024502	1.81
Error	5	0.0006687	0.0006687	0.0001337			0.90
Total	8	0.0746889					

5. Taguchi Optimization

The Taguchi design is power full design mostly acknowledged and used in engineering investigation and optimization. In Taguchi method used an orthogonal array to reduce the total number of experiment also eliminates the influence of uncontainable aspects.

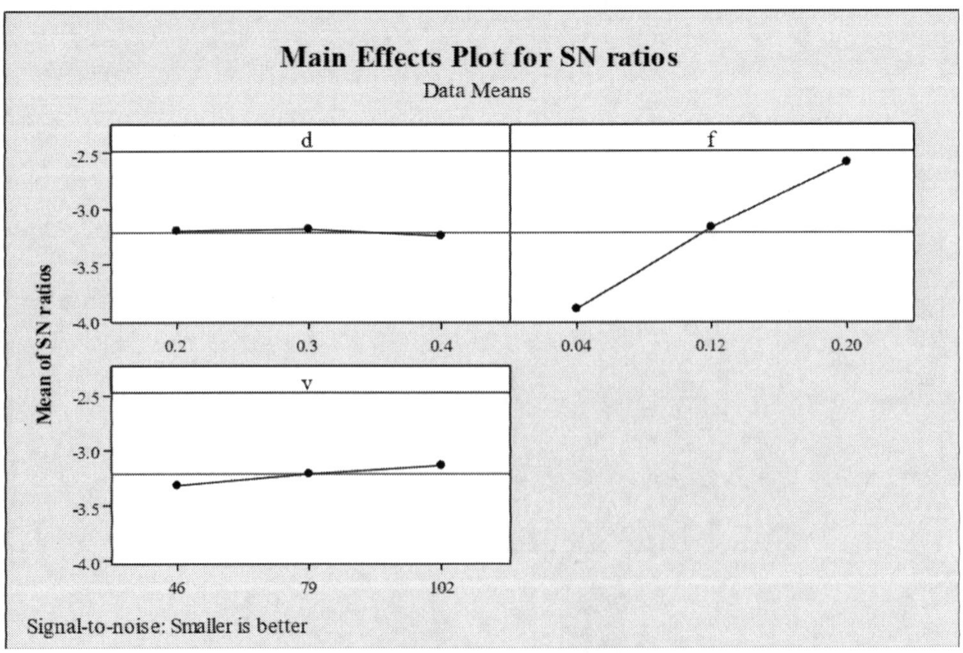

Figure 5. Main effect plots for S/N ratio

The advantaged of using the Taguchi method lessen the total number of test, less the experimental time, reduction in the production cost, and main advantage of this method is simple and precision [Manivel and Gandhinathan]. Taguchi (1990) design of the experiment is joined with the orthogonal array to analysis the complete parametrs within a lesser number of investigation tests. For investigation of S/N ratio Taguchi cites three types of performance structures (1) Lower the better,

(2) Higher the better, and (3) Nominal the better. In this work lower the better condition chosen to attain the optimal parameters during the turning process. From the main effect plot for S/N ratio the optimal cutting parameter combination is found to be d3-f1-v1 i.e (d = 0.4 mm; f = 0.04 mm/rev; v = 46 m/min). From the main effect plots for S/N ratio it was found that depth of has no significant effect on chip reduction coefficient as all its values lay proximal to the mean line whereas feed and speed are dominant as displayed in Fig. 5.

6. Conclusion

The following conclusions were drawn on the basis of an experimental investigation carried out in this present study.

- In metal cutting operation, Taguchi design of experiments was very efficiently used in optimization of input cutting parameters.
- The optimal parameter setting for better chip reduction coefficient is found to be *d3-f1-v1*, which denotes to a depth of cut (0.4 mm), feed (0.04 m/rev) and speed (46 m/min) and corresponding chip reduction coefficient is found to be 1.52.
- The regression model developed for prediction stands valid and the predicted value was found proximal to the experimental values.
- In this model R-Sq value approaches to unity (100%), the desired output response model is fitted closely and effectively with the actual experimental result data.
- From this present experimental investigation, it was concluded that with increment in feed and speed the chip reduction coefficient decreases, while the depth of cut has no significant effects.

Reference

[1] Gao C and Zhang L 2013 Effect of cutting conditions on the serrated chip formation in high-speed cutting *Machining Science and Technology* **17** pp 26–40
[2] Hua J and Shivpuri R 2014 Prediction of chip morphology and segmentation during the machining of titanium alloys *Journal of Materials Processing Technology* **150** pp 124–133
[3] Sutter G and List G 2013 Very high speed cutting of Ti–6Al–4V titanium alloy – change in morphology and mechanism of chip formation *International Journal of Machine Tools & Manufacture* **66** pp 37–43
[4] Daragusch M, Sun S, Kim J, Li T, Trimby P and Cairney J 2018 Effect of tool wear evolution on chip formation during dry machining of Ti-6Al-4V alloy *International Journal of Machine Tools and Manufacture* **126** pp 13–17
[5] Bai W, Sun R, Roy A and Silberschmidt V 2017 Improved analytical prediction of chip formation in orthogonal cutting of titanium alloy Ti6Al4V *International Journal of Mechanical Sciences* **133** pp 357–367
[6] Wu H and To S 2015 Serrated chip formation and their adiabatic analysis by using the constitutive model of titanium alloy in high speed cutting *Journal of Alloys and Compounds* **629** pp 368–373
[7] Sun S, Brandt M and Dargusch M 2010 Machining Ti–6Al–4V alloy with cryogenic compressed air cooling *International Journal of Machine Tools & Manufacture* **50** pp 933–942
[8] Aramcharoen A 2016 Influence of cryogenic cooling on tool wear and chip formation in turning of titanium alloy *Procedia CIRP* **46** pp 83 – 86
[9] Sahoo A K and Sahoo B 2013 Performance studies of multilayer hard surface coatings (TiN/ TiCN/Al2O3/TiN) of indexable carbide inserts in hard machining: Part-II (RSM, grey relational and techno economical approach) *Measurement* **46** pp 2868–2884
[10] Kumar S M and Kulkarni S K 2017 Analysis of Hard Machining of Titanium Alloy by Taguchi Method *Materials Today: Proceedings* **4** pp 10729–10738
[11] Sulaiman M A, Haron C C H, Ghani J A and Kasim M 2013 Optimization of Turning

Parameters for Titanium Alloy Ti-6Al-4V ELI Using the Response Surface Method (RSM) *Journal of Advanced Manufacturing Technology* **7,2** pp 11-28

[12] Ramana M V 2017 Optimization and Influence of Process Parameters on Surface Roughness in Turning of Titanium Alloy under Different Lubricant Conditions *Materials Today: Proceedings* **4** pp 8328–8335

[13] Gupta M, Sood P, Singh G and Sharma V 2017 Sustainable Machining of Aerospace Material - Ti (grade-2) alloy: Modelling and Optimization *Journal of Cleaner Production* pp 1-17

[14] Bandapali C, Sutaria B, Bhatt D and Singh K 2017 Experimental Investigation and Estimation of Surface Roughness using ANN, GMDH & MRA models in High Speed Micro End Milling of Titanium Alloy (Grade-5) *Materials Today: Proceedings* **4** pp 1019–1028

[15] Kumar R, Sahoo A k, Mishra P C and Das R K 2018 Comparative investigation towards machinability improvement in hard turning using coated and uncoated carbide inserts: part I experimental investigation *Advance in Manufacturing* **doi.org/10.1007/s40436-018-0215-z**

[16] Chattopadhyay AB 2017 Machining and Machine Tools 2_{nd} edition, *ISBN:* **978-81-265-6474-3**

IOP Conf. Series: Materials Science and Engineering **390** (2018) 012114 doi:10.1088/1757-899X/390/1/012114

To study the effect of inventory dependent consumption parameter for constant and time dependent holding cost

Pankaj S. Ardak,
P.R.Pote College of Engineering and Management, Amravati
pankajardak@gmail.com

Abstract. Economic production quantity models are effectively used to control inventory. In the present work EPQ model has been developed to study the effect of demand parameters on holding cost and total inventory. Demand used is inventory dependent during inventory buildup time and constant demand has been used during inventory depletion period. Rate of deterioration is considered to be constant. A numerical example has been included to validate the model. The sensitivity analysis shows that, higher inventory can attract customer to buy more. The result of models indicates that, the holding cost and total inventory is sensitive to change in inventory dependent demand parameter. Increase in demand parameter decreases the holding cost and total inventory.

1. Introduction

Every organization in many sectors of the economy has adopted some type of inventory system. Inventory is one of the most visible and tangible aspects of doing business. Unless inventories are controlled, they may be unreliable, inefficient and costly. The Economic Production quantity (EPQ model has been widely used for more than decades as an important tool to control the inventory. The EPQ is powerful to help practitioners and engineers to make a decision regarding inventory control. The complexity of a resulting EPQ model depends on assumptions one makes about various parameters of the inventory system.

The effect of an imperfect production process, by assuming different rates of deterioration during the production process, on the optimal cycle time had been studied by (Rosenblatt and Lee, 1986). The Economic order quantity (EOQ) model with power form of stock dependent demand for deteriorating items (Teng.*et.al*., 2005), investigated that the shape parameter of the demand and selling price is highly sensitive to optimal solution. The impact of random machine failure on the EPQ model (Teng, *et.al.* 2005) reveals the demand function and purchase cost is positive and fluctuating with time. The production process can shift at random from in control to out of control state. In such situation, the items produced out of control state deteriorate at higher rate than those at the normal rate, hence get consumed by demand under Last in First out (LIFO) policy (Garry and Dahl, 2006). Beerhouse and Babar, developed (EPQ) model by considering both the depreciation cost of stored items and process quality cost. Depreciation cost and process quality cost were assumed to be a continuous function of holding time and of production run length. Yuan et.al (2011) derived the optimal manufacturing batch size and number of shipment with scrap using mathematical modeling and algebraic approach. Ata (2011) discussed EPQ model for multi products single machine with discrete delivery. Naja and Naiki focused on EPQ model with a production capacity limitation and a random defective production rate. Gede and Hui(2010) analyzed an EPQ model for deteriorating items with stochastic machine unavailability and price- dependent demand. Jinn(2007), used time varying demand and cost to analyze EPQ model and characterize the influences of both demand and cost over the length of production run time and the economic production quantity. Hezari (2008) developed EPQ model by considering imperfect and defective items. Holding cost for defective items has not been considered in his study.David et. al (2009) considered partial backordering with constant demand to study inventory model. Jiang Tao (2009) considered a multi-item inventory system, where the vendor provides delay in payment to the retailer . Yao (2007) established a model for deteriorating items under delay in payment, in which the demand is a negative exponential function of price. Jia (2009) reported a supply chain system with trade credit. Liao established an EPQ model for deteriorating items

with delay in payment. Ardak *et. al.* (2017) developed EPQ model by considering mixed demand pattern.. Ardak and Borade (2017) considered time dependent holding cost to develop EPQ model.
Most of the researchers developed EPQ models by considering various demand patterns. The present study discusses the effect of inventory dependent parameter on holding cost and total inventory with the help of EPQ model. In the present study, it is assumed that the demand remains as inventory dependent during inventory buildup time and constant during inventory depletion period. This paper has five sections. Research motivation and literature is narrated in section 1. Section 2 contains notations and assumption. Methodology used to develop the model is discussed in section 3. Numerical and sensitivity analysis is discussed in section 4 and finally concluded in section 5.

2. Assumptions and Notation

2.1 Assumptions:-

The following assumptions are made in development of the model.

a) The production rate is known and treated as constant.
b) The production rate is greater than the demand.
c) The demand is inventory level dependent in up time and constant in down time.
d) Deterioration of the items is varying.
e) Inventory holding cost is known and termed as constant and time dependent.
f) Deterioration of the items start as it enters into inventory.
g) Shortages are not allowed.
h) Every produced items needs inspection.

2.2 Notation:-

D – Basic demand,
H – Holding cost per unit,
I_1 – Inventory level during production up time,
I_2 – Inventory level during production down time,
P – Rate of production,
T – Production cycle time,
T_1– Production up time. Maximum inventory buildup in this time,
T_2– Production down time. ,
Ci – Inspection cost per unit,
I – Total Inventory.
θ – Basic rate of Deterioration,
α – Inventory dependent consumption rate parameter,

3. Mathematical Model.

This model deals with deteriorating items those consider the effect of different demand patterns for different time periods. Inventories are build up gradually during production up time, new production run will start after complete consumption of the buildup inventory. To keep the production process to its initial working condition, setup is essential. At the beginning, it is assumed that the inventory is zero.

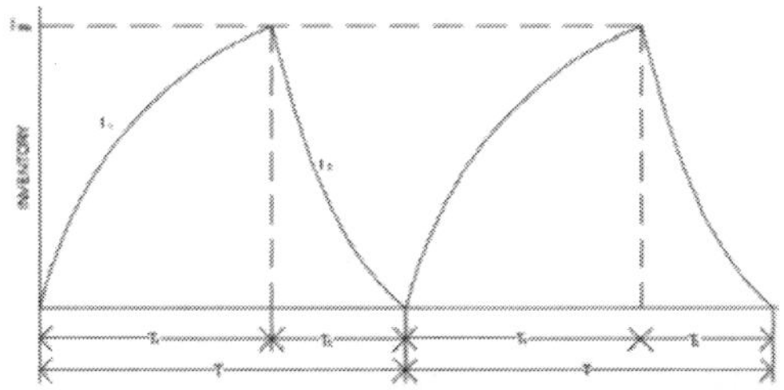

Fig 1. Inventory System

During the time period $(0, T_1)$, as the production rate is greater than the demand rate, inventory is gradually buildup at a rate of (P-D). Here the demand rate is inventory dependent. This rate is offset by a constant deterioration rate. At time T_1 the inventory will be maximum. At this stage, production is terminated and on hand inventory used to meet the demand and to offset the loss due to deterioration. In this period, demand remains constant. The deterioration of the product will start soon it enters into the inventory. Production system can be described by the following differential equations.
According to the assumptions, over time span $[0, T_1]$, deterioration rate is constant and demand rate is inventory dependent which makes variation of the inventory level with respect to time for the reference time governed by

$$\frac{dI_1(t)}{dt} = P - D - \alpha I_1(t) - \theta I_1(t) \qquad 0 \le t \le T_1 \qquad 1$$

In the time interval $(0,T_2)$, the system is affected by the combined effect of constant demand and deterioration. Hence, the change in inventory level is governed by the following differential equation

$$\frac{dI_2(t)}{dt} = -D - \theta I_2(t) \qquad 0 \le t \le T_2 \qquad 2$$

Above equations can be solved by using initial boundary conditions like at $t = 0$, $I_1(t) = 0$ and at $t = T_2$, $I_2(t) = 0$. And by using boundary condition $I_1(T_1) = I_2(0)$, approximate value of production down time can be given as,

$$T_2 \approx \frac{P-D}{D}\left[T_1 - \frac{(\alpha+\theta)T_1^2}{2}\right] \qquad 3.$$

Total inventory is given by sum of inventory in up time and inventory in down time.

$$TI = \int_0^{T_1} I_1(t)dt + \int_0^{T_2} I_2(t)dt \qquad 4.$$

By solving Equation no. 4,use Taylor's series approximation and after some simplification, the approximate total inventory can be expressed as,

$$TI \approx \frac{(P-D)T_1^2}{2} + \frac{DT_2^2}{2} \qquad 5.$$

Constant Holding cost is given by

$$HC = H\,(TI) \qquad 6.$$

Time dependent Inventory holding cost is given by

$$IH = \int_0^{T_1} H(t) I_1(t) dt + \int_0^{T2} H(t) I_2(t) dt$$

$$IH = (P - D) T_1^2 \left[\frac{a}{2} + \frac{b}{\alpha + \theta} \right] + DT_2^2 \left[\frac{a}{2} + \frac{b}{\theta} \right] \quad 7.$$

Total cost = Set up cost + Holding cost + Inspection cost + Deteriorating cost.

$TC = A + H\,(TI) + IC + C_d\,(TD)$

As total cost per unit time is a function of production up time, Optimum value of production up time can be derived by solving following equation.

$$\frac{dTCT}{dT_1} = 0 \quad 8.$$

4. Numerical experiment and sensitivity analysis

Numerical example and sensitivity analysis has been carried out to validate the theoretical aspects. The numerical data is adopted from work published elsewhere (Ardak and Borade)except some additional data like deterioration cost, set up cost and inspection cost. Let A= Rs.30 per production cycle, C_d = Rs.5 per unit per unit time, C_i = Rs 2 per unit per unit time, H = Rs.2 per unit per unit time, P = 2500 units per unit time, D = 1200 units per unit time, $\alpha = 0.5$, $\theta = 0.1$
The optimum value of T_1 can be found, as the total cost function is convex (fig. 2 and 3). The optimum value of T_1 when holding cost considered to be constant is 0.073 and 0.033 when holding cost is linear function of time.

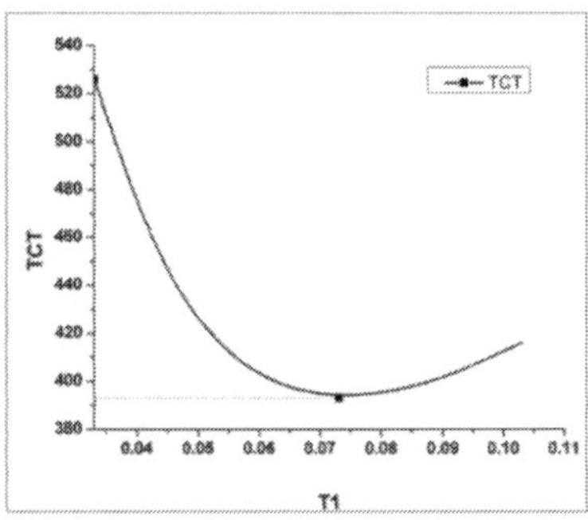

Fig. 2 T_1 v/s TCT for constant holding cost

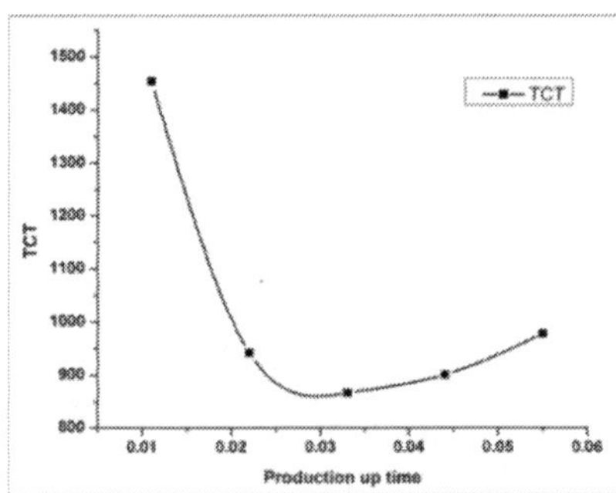

Fig. 3. T_1 v/s TCT for time dependent holding cost

Table 1 Values of constant HC, and TI for different values of α

α	HC	TI
0.3	106.3	53.15
0.4	102.11	51.055

IOP Conf. Series: Materials Science and Engineering **390** (2018) 012114 doi:10.1088/1757-899X/390/1/012114

0.5	99.32	49.66
0.6	96.55	48.27
0.7	92.58	46.29

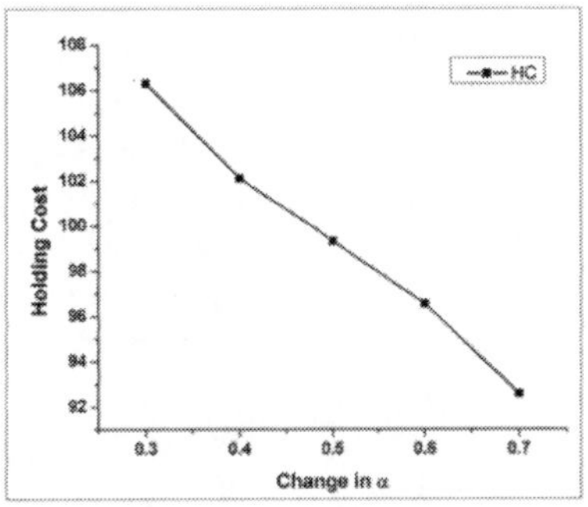

Fig.4. Demand parameters α v/s Holding Cost

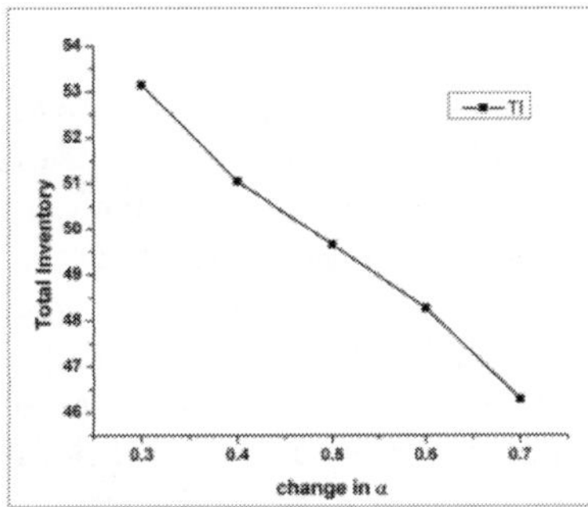

Fig 5. Demand parameters α v/s Total inventory

Table 2.Values of time dependent HC and TI for different values of α

α	HC	TI
0.3	29.100	1.97
0.4	28.520	1.62
0.5	28.470	1.46
0.6	27.960	1.29
0.7	27.590	1.14

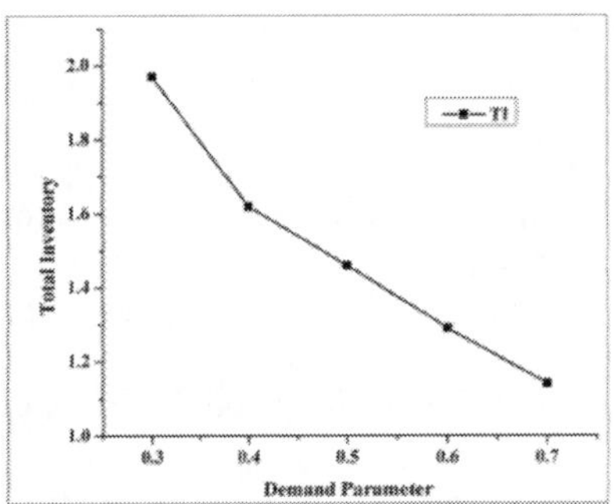

Fig.6. Demand parameters α v/s time dependent Holding Cost

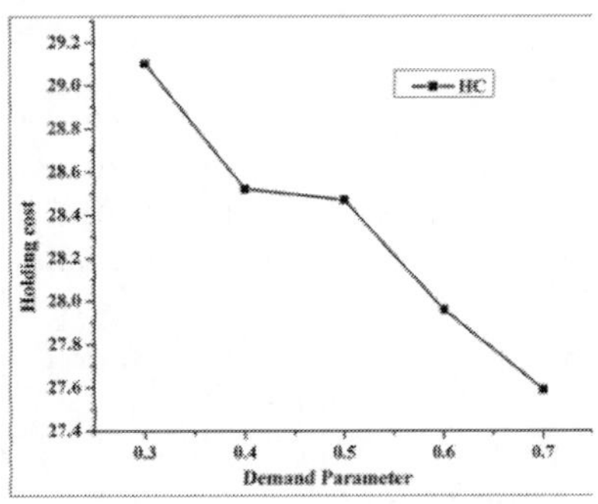

Fig 7. Demand parameters α v/s Total inventory

Figure 4,5,6 and 7 shows the relation between inventory dependent demand parameter, total cost and holding cost. Demand used for this work is inventory dependent for production up time. Curve in Fig.6

and 7 indicates that total inventory and holding cost both are decreasing function of 'α'. Inventory dependent consumption rate decreases the total inventory and holding cost, subsequently may lead in decreasing total cost. At higher inventory level, demand will be more and hence can reduce total cost. Decrease in holding cost implies that inventory dependent demand will be profitable. Thus selection of optimum value of 'α' attracts the attention of inventory manage.

5. Conclusion

The research findings of the present work will be useful to practitioners in realizing the importance of inventory dependent demand and time dependent holding cost on EPQ. During inventory buildup time, demand used was inventory dependent and during inventory depletion time it was constant. This approach has not been considered in previous research. The research indicates that the annual total cost function possesses convexities. It was found that holding cost and total inventory is sensitive to change in parameter 'α'. Increase in demand parameter 'α' decreases the holding cost and total inventory. This attracts the attention of inventory managers to estimate the accurate value of the inventory dependent consumption rate parameter. This indicates that higher inventory stock can attract the customer to buy more.

References

Ardak P.S. Borade A.B. Reneta S.B.(2017), 'An EPQ Model For Deteriorating Items with mix demand pattern', *International Journal of Mechanical Engineering and Technology*, Volume 8, Issue 6, June 2017, pp. 59–69 .

Ardak P. S. Borade A. B.(2017), 'An EPQ Model with time dependent holding cost and varying deterioration rate',*International Journal of Mechanical Engineering and Technology,* Volume 8, Issue 8, August 2017, pp 958-966

Ata ,A. T., Gede ,A. W., Hui, M. W.,Jahangir ,B.(2011) "Multi products single machine economic production quantity model with multiple batch size'. *International Journal of Industrial Engineering Computations,* 2, 213–224,.

Behrouz ,A. N., Babak(2009.), A. 'EPQ model with depreciation cost and process quality cost as continuous functions of time'. *International Journal of Industrial Engineering* 5, (8) , pp 77-89

David W. Pentico, Matthew J. Drake, Carl Toews,(2009.) 'The deterministic EPQ with partial backordering:Anewapproach', *International journal of Management Sciences, Omega* 37,624 – 636.

Gede, A. W., Hui, M. W.(2010) 'Production Inventory Models for Deteriorating Items with Stochastic Machine Unavailability Time, Lost Sales and Price Dependent Demand'. *Jurnal Teknik Industri*, 12, (2), 61-68.

Gary ,C.L,, Dah,C.G.,(2006) 'On a production inventory system of deteriorating items subject to random machine breakdown with a fixed repair time', *Mathematical and Computer modeling*, 43, pp 920-932.

Jinn,T. T., Liang,Y. O., Chun,T. C.(2007.) 'Deterministic economic production quantity models With time-varying demand and cost'.*Applied Mathematical Modelling* ,29, 987–1003,

Jia Tao, Xu Yu,(2009.) 'Study on the optimal trade credit period in the supply chain when end demand is stock-dependent', *Operations research and management science*, vol. 18, pp. 8-14,

Jui-Jung Liao,(2007) 'On an EPQ model for deteriorating items under permissible delay in payments', *Applied Mathematical Modeling*, vol. 31, pp.393-403,.

Jinn ,T.T., Liang,Y. O., Mei, C. C., (2005) 'A EOQ model for deteriorating items with power form stock dependent demand, *Information and Management science*, 16, Number 1 pp 1-16.

Jinn,T. T., Liang, Y.O., Chun,T.C., (2005) , Deterministic economic production quantity models with time varying demand and cost, *Applied Mathematical Modeling*, 26,pp 987-1003.

Mo Jiangtao, Chen Guimei, Mao Hong, Fan Ting,(2011) Optimal Ordering Policy for Multi-item with Stock-dependent Demand Rate under Delay in Payment, IEEE,464-468,.

P.S.Ardak, A.B.Borade, 'An EPQ Model with Varying Rate of Deterioration and Mixed Demand Pattern.', *International Journal of Mechanical and Production Engineering Research and Development*, Vol. 7, Issue 5, Dec 2017, 423-432
Rosenblatt, M., and Lee,H.L., (1986),Economic Production Cycles with Imperfect Production Processes , *IJE Transaction*,.pp 48-55.
S. R. Hejazi, J. C. Tsou, M. Rasti Barzoki(2008) 'Optimal lot size of EPQ model considering imperfect and defective products' , *Journal of Industrial Engineering International* , , Vol. 4, No. 7, 59-68,

Study of a peptide with a drug using ultrasonic velocity and molecular modelling techniques

R Padmavathy [1*], K Dhanalakshmi [1] and N Radha [2]

[1]PG & Research Department of Physics, Seethalakshmi Ramaswami College, Tiruchirappalli- 620002, Tamil Nadu, India.
[2]PG & Research Department of Chemistry, Seethalakshmi Ramaswami College, Tiruchirappalli- 620002, Tamil Nadu, India.
*Correspondence for Author, Principal & Head, PG & Research Department of Physics, Seethalakshmi Ramaswami College, Tiruchirappalli- 620 002, Tamil Nadu, India.
*Email: shreepadmaram@gmail.com ; Contact number: +91 9791595230

Abstract: Peptides are amazing diverse and multi-faceted class of compounds constructed from relatively simple building blocks, the amino acids. The ultrasonic velocity (U), density (ρ) and viscosity (η) measurements have been carried out for peptide with fixed concentration of the sulfadrug in non – aqueous solution of ternary mixture at 308.15 K. Internal pressure and free volume are thermodynamic parameters computed in the present study. Experimental data have been used to estimate the internal pressure (π_i) and free volume (V_f).The results are discussed in terms of structure making/breaking effects of drug in non-aqueous solution. The Insilco studies indicate the binding positions of the peptide and sulfa drug to active sites on three different types of proteins.

Keywords: Internal pressure, free volume, docking study

1.Introduction

In the recent years ultrasonic velocity measurements are helpful to interpret solute-solvent, ion-solvent, solvent-solvent interaction in binary/ ternary mixtures, aqueous and non-aqueous medium [1-7]. Sulfonamides also known as sulfa drugs, represent a kind of typical antibiotics and have been widely used in human and vertinary medicine to treat and prevent infections of bacterial diseases [8]. Peptides have gained many applications in medicine and biotechnology and other fields. Glycyl Glycine is the simplest dipeptide used in biochemical research and in the preparation of biodegradable polymers. Sulfadiazine is a sulfonamide antibiotic and it is well known as one as of the World Health Organisations list of essential medicines. It eliminates bacteria that cause infections by stopping the production of folate inside the bacterial cell, and is commonly used to treat urinary tract infections (UTIS). Therefore it is important to analyse the characterization, antibiotic testing and drug patent of the title compound for future studies. Furthermore, researches are also very helpful for the theoretical calculation of the experimental evidences since computational methods are reliable to characterize the molecule because of their efficiency and exactness with regard to the estimation of molecular properties.

Aamer Saeed et.al,[9] reported Sulfonamide-Linked ciprofloxacin, Sulfadiazine and Amantadine derivatives as a novel class of inhibitors of Jack Bean Urease; Synthesis, Kinetic Mechanism and Molecular Docking. Kuppamuthu Sarojini et.al [10] reported Molecular Docking studies of some sulphonamide derivatives as PBP-2X inhibitors as antimicrobial agents. Leonardo et.al [11] reported Molecular Docking and structure-based Drug design Strategies. Jayakumar et.al [12] reported Thermodynamic studies of molecular interactions in aqueous Glycine –A Biological molecule. Santosh et.al [13] reported Volumetric and ultrasonic investigation of Glycylglycine in aqueous $FeCl_3$ solutions. Padmavathy et.al [14] reported An investigation of acoustic, thermodynamic and antimicrobial activity study of a drug in a peptide. Raj Muhamed et.al [15] reported Synthesis, spectroscopic (FT-IR, FT-Raman, NMR, UV-Visible), NLO, NBO,

HOMO-LUMO, Fukui function and molecular docking study of (E)-1-(5-bromo-2-hydroxybenzylidene) semicarbazide.

Literature survey reveals that best of our knowledge, there is no ultrasonic and molecular docking studies for title compound peptide with drug in non-aqueous medium. In the present investigation, the polar protic solvent is taken along with a peptide and drug, the solutions were prepared at different molalities with peptide and a fixed concentration of drug. The thermodynamic parameters such as internal pressure are free volume were computed. The results arrived from these parameters were correlated with ultrasonic and docking studies. Docking investigation has been performed to find out the Bonded residues, hydrogen bond, Bond distance (Å), Estimated Inhibition constant (µm) and Binding energy (kcal/mol) of the molecule.

2. Materials and Methods

2.1Experimental Details

Ultrasonic velocity was measured using digital ultrasonic interferometer of fixed frequency 2MHz (Model F-81 Mittal enterprises, New Delhi) with an accuracy of ± 0.2 m/s. The density of the non-aqueous peptide with drug solutions were measured using Anton Paar DMA 4100 Digital densitometer with an accuracy of ± 0.0001 gm/cc. The viscosities of the solutions were measured using Cannon Fenske viscometer (±0.1 %) error with the experimental solution was immersed in a temperature controlled water bath. The time of flow was measured using a stop watch with an accuracy of 0.1 sec.

2.2 Computation

The measured data were used to calculate the thermodynamic, solvation and electrochemical parameters using the standard relations,

Internal Pressure $\prod_i = b\,RT\,(K\eta/U)^{1/2}\,\rho^{2/3}/M_{eff}^{7/6}$ (10^9) Pa ----- 1

Free volume $V_f = [M_{eff}\,U/K\,\eta]^{3/2}$ (10^{-9}) m^3 ------ 2

3. Results and discussion

3.1 Ultrasonic study

3.1.1 Internal pressure and Free volume

The internal pressure of Glycyl-L-glycine + Sulfadiazine + Formamide is found to increase with increasing concentration. A dip is observed at 0.025 and 0.01 molality of the solutions which are graphically represented in figure (1). Internal pressure is a measure of cohesive energy of the system [16]. Hence the variations may be due to the structure making nature of the peptide and drug in the solvent. It implies that there is a strong peptide - amide - drug interaction taking place in the solution. Free volume is the free space available for the molecule to move· Thus free volume is an inverse function of internal pressure [17]. In this study the free volume is found to decrease with increasing concentration. It is identified from internal pressure and free volume of Glycyl-L-glycine+ Sulfadiazine+ Formamide mixture that there is a strong intermolecular interaction exist between the molecules present in the ternary mixture.

Table 1. The computed values of internal pressure and free volume of non- aqueous ternary solutions, at different concentrations are shown in the table

Molality (m)	**Glycyl-L-glycine+ Sulfadiazine + Formamide (308.15 K)**	
	Internal pressure (10^9) Pa	Free volume (10^{-9})m^3
0.001	1.3404	34.9
0.005	1.3531	33.9
0.01	1.3822	31.8
0.025	1.3117	37.2
0.05	1.3899	31.1

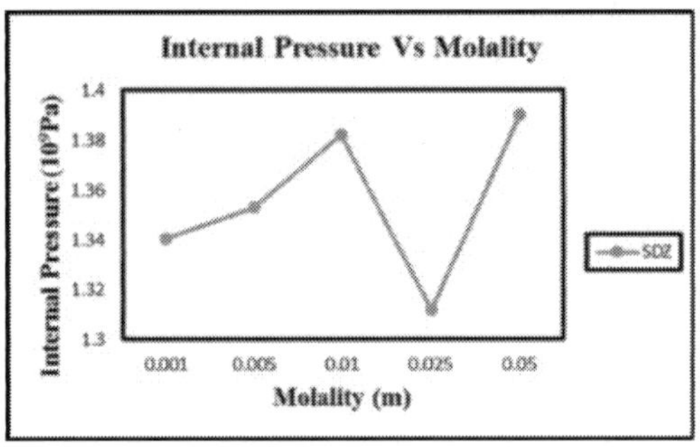

Figure 1. Internal pressure

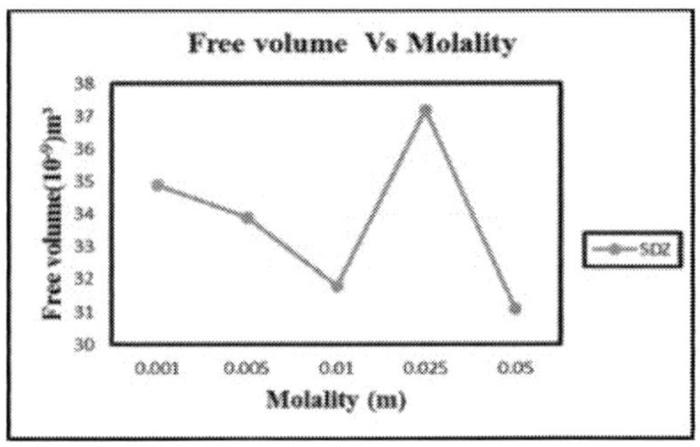

Figure 2. Free volume

3.2 Docking study

The Glycyl – L –glycine + Sulfadiazine + formamide entity is selected to be docked into the active site of 4HOE, 5IGW, 5IFP protein. The ligand is docked into the functional site of respective protein and docking energy is examined to achieve a minimum value. The Auto Dock results indicate the binding position of the peptide together with a rough estimate of interaction. The number of hydrogen bond formation in this molecular docking study is 4 with 4HOE, 6 with 51GW and 2 with 51FP (Table 2). Peptides are structurally related to proteins. Glycyl-L-glycine is a dipeptide with NH_2 group at one terminal and COOH group at another end. This dipeptide is considered as a ligand and used for docking analysis on antimicrobial protein targets. The binding sites on the protein surface is recognized by the ligand. The bond distance reduces with bond residues ASP170-1.9 Å in 4HOE, ASP55 – 1.8 Å & ARG 53 – 1.8 Å in 51GW and GLU200 – 1.7 (Å) in 51FP protein targets.

The autodock results indicate the binding position of the Peptide together with Sulfadiazine (Figure 3-5). The molecular docking binding energies and inhibition constants are also listed in the table (4). Among them 51GW is found to exhibit the lowest free energy of -6.25 kcal/mol. Hence it can be considered most docked inhibitors interacted well with the ligands within the 51GW. The term active sites in 4HOE protein clearly visible in Figure 3. In Figure 4 the six H- bonding sites are shown for the protein

51GW. Figure 5 reveals that large number binding sites eventhough only two hydrogen bonds are found between active sites.

Table 2. Hydrogen bonding and molecular docking with antimicrobial protein targets

Protein (PDB ID)	Bonded residues	No. of hydrogen bond	Bond distance (Å)	Estimated Inhibition Constant (µm)	Binding energy (kcal/mol)	Reference RMSD (Å)
4HOE	ASP 21 ASP 21 LYS 37 ASP 170	4	2.4 2.3 2.2 1.9	28.20	-2.11	38.78
5IGW	ASP 55 ASP 55 LYS 87 GLU 54 GLU 54 ARG 53	6	1.8 2.1 2.4 2.1 2.2 1.8	26.12	-6.25	43.42
5IFP	TYR 260 GLU 200	2	2.3 1.7	46.80	-5.87	60.440

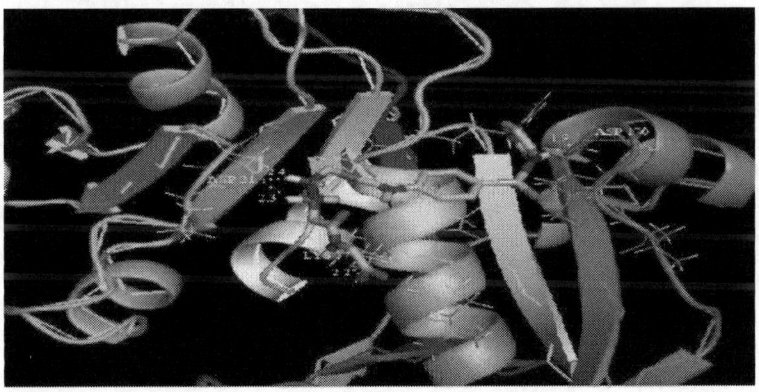

Figure 3. Docking and Hydrogen bonding interaction of Glycyl – L –glycine+ Sulfadiazine +Formamide with 4HOE Protein structure

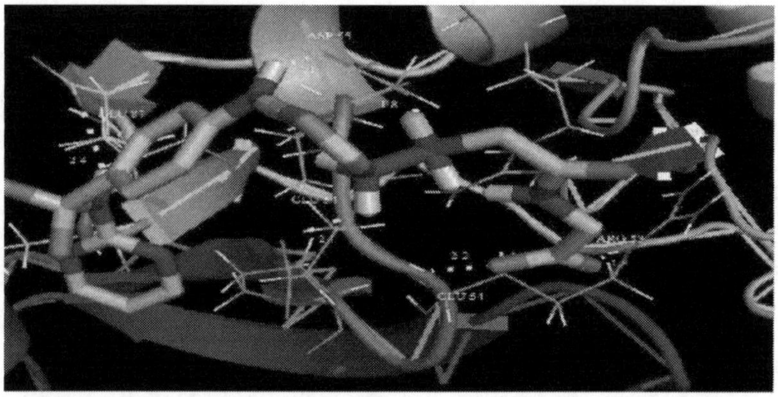

Figure 4. Docking and Hydrogen bonding interaction of Glycyl – L – glycine+ Sulfadiazine +Formamide with 51GW Protein structure

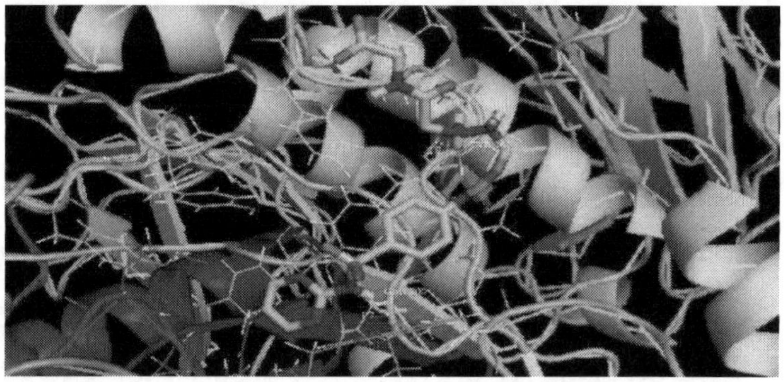

Figure 5. Docking and Hydrogen bonding interaction of Glycyl – L – glycine+ Sulfadiazine +Formamide with 51FP Protein structure

4. Conclusion

Ultrasonic method is a powerful probe for characterizing the physico-chemical properties and existence of molecular interaction in the liquid mixtures. Sulfadiazine is a sulfonamide antibiotic and it is well known as one of the World Health Organization's List of Essential Medicines. Internal pressure is a significant factor which various due to all the interactions present between a peptide-drug and solvent. Analysis of the thermodynamical properties, suggest that there is a strong peptide - drug - amide interaction in the solutions. The above results are confirmed by ultrasonic and docking studies. The molecular dock output shows the lowest binding energy for 51GW there six hydrogen bonding interaction and most docked inhibitors interacted with the ligand within the 51GW binding site.

References

[1] Chapke U D, Meshram B P, Agarwal P S, Berad B N, 2013 Ultrasonic velocity, density and viscosity measurements of some drug in aqueous 70% DMF at 300.15K, *International Journal of Emerging Technologies in Computational and Applied Sciences,* 13-147, 269-275.

[2] Sonar A N, Pawar N S, Khairnar M D, 2011 ultrasonic studies on molecular interaction of substituted heterocyclic compounds in acetone-water mixture at 303K, International Journal of Applied Biology and Pharmaceutical technology, 2(3) 291-295.
[3] Sonar A N, Chaudhari R T, Pawar N S, Nehete J P, 2014 Studies of acoustic, thermodynamic and additive properties of substituted heterocyclic drugs in Ethyl Alcohol, *International Journal of Experimental Pharmacology*,4(1) 51-54.
[4] Rajesh R, Raj Muhamed R, Justin Adaikala Baskar, Kannappan V, 2016 Ultrasonic and spectral studies on charge transfer complexes of anisole and certain aromatic amines, *Chemical Physics*, 478, 34-44.
[5] Dakhole S R, Wagh M S, Agrawal P S, Paliwal L J, 2015 Molecular interactions studies in Al_2O_3 Nanoparticles, *International Journal of Advances in Science Engineering and Technology*, 1,64-77.
[6] Thennarasu J, Meenakshi G, 2011 Evaluation of ultrasonic velocity and percentage deviation of ternary mixture at different temperatures, *RASAYAN J.Chem*, 4(4), 904-909.
[7] Colthup N B, Daly L H, Wiberly S E, 1990 In Introduction to Infrared and Raman Spectroscopy , *3rd ed, , Academic Press , New York*,340.
[8] Colthup N B, Daly L H, Wiberly S E, 1990 In Introduction to Infrared and Raman Spectroscopy , *3rd ed, Academic Press : New York*,340.
[9] Pervaiz Ali Channar, Aamer Saeed, Fernando Albericio, Fayaz Ali Larik, Qamar Abbas, Mubashir Hassan, Hussain Raza, Sung-Yum Seo, 2017 Sulfonamide-Linked ciprofloxacin, Sulfadiazine and Amantadine derivatives as a novel class of inhibitors of Jack Bean Urease; Synthesis, Kinetic Mechanism and Molecular Docking, *Molecules*, 22, 1-20.
[10] Kuppamuthu Sarojini, Krishnan H, 2014 Molecular Docking studies of some sulphonamide derivatives as PBP-2X inhibitors as antimicrobial agents, *Romanian J.Biophys*, 24(3) 175-184.
[11] Leonardo Ferreira G, Ricardo dos Santos N, Glaucius Oliva, Adriano D, Andricopulo, 2015 Molecular Docking and structure-based Drug design Strategies. Molecules, 20, 13384-13421.
[12] Jayakumar S, Preetha Mary George, Shubhashree N S, Divya P, Anees Ahmed M, 2015 Thermodynamic studies of molecular interactions in aqueous Glycine –A Biological molecule. *International Journal of Innovative Research in Science, Engineering and Technology*, 4(1), 119-122.
[13] Santosh M S, Krishna Bhat D, Aarti S, Bhat, 2009 Volumetric and ultrasonic investigation of Glycylglycine in aqueous $FeCl_3$ solutions, J.Solution Chem, 39,1763-1773.
[14]. Padmavathy R, Dhanalakshmi K, Jasmine Vasantha Rani E, Radha N,2017 An investigation of acoustic, thermodynamic and antimicrobial activity study of a drug in a peptide, International journal of Phamaceutical Sciences and Research, 8(8) 3477-3483.
[15] Raja M, Raj Muhamed R, Muthu S, Suresh M, 2017 Synthesis, spectroscopic (FT-IR, FT-Raman, NMR, UV-Visible), NLO, NBO, HOMO-LUMO, Fukui function and molecular docking study of (E)-1-(5-bromo-2-hydroxybenzylidene) semicarbazide, *Journal of Molecular Structure*, 1141, 284-298.
[16] Thirumaran S, Karthikeyan N, 2011 Thermo-Acoustical and excess thermodynamic studies of ternary liquid mixtures of substituted benzenes in aqueous mixed solvent systems at 303.15, 308.15 and 313.15K, *Int. J. Chem. Rese* vol.**3**, 83-98.
[17] Santhi N, Madumitha J, 2014 Molecular interaction studies in binary liquid mixture through ultrasonic measurements at 303.15K, *Inte. J. Adva Chem,* **2**(1), 12-16.

IOP Conf. Series: Materials Science and Engineering **390** (2018) 012116 doi:10.1088/1757-899X/390/1/012116

A comparative study of experimental and theoretical refractive index of binary liquid mixtures using mathematical methods

A Gayathri, T Venugopal, R Padmanaban, K Venkatramanan and R Vijayalakshmi

Faculty of Science, Sri Chandrasekharendra Saraswathi Viswa Mahavidyalaya, Enathur, Kanchipuram – 631561, India

E-mail :kv@kanchiuniv.ac.in

Abstract. A comparative study of refractive index of binary liquid mixtures such as Cyclohexane+O-xylene, PEG200+Water, PPG400+toluene having industrial applications are analysed at temperature 303K for different concentrations [0 to 1%]. In the present study experimental values are taken from literature and there values are compared with theoretical values obtained by various mathematical equations such as Lorentz – Lorenz relation, Gladstone- Dale equation, Weiner's relation, Heller's (H) equation, Arago-Biot equation. Thus the present study reveals the nature of interaction between component molecules in the mixtures and enables us to identify a suitable mathematical model for predicting the refractive index of various binary liquid mixtures.

1. Introduction

The general applicability of refractive index in chemical analysis and industry are revealed in literature survey. Many researchers have been studied the theoretical refractive index measurements by the refractive index mixing rules [1-4]. The generally used theoretical rules are Arago-biot, GladstoneDale, Lorentz-Lorentz, Weiner, Heller, Newton and Eyring-John. The validity of the mixing rules has been tested by several researchers and pointed out the importance of mixing rules [1-5]. In the present study, we report the experimental and theoretical refractive index of binary mixtures such as Cyclohexane+O-xylene, PEG200+Water, PPG400+toluene and from these the validity of the mixing rules have been analyzed.

2. Methods and Materials

The binary mixtures are prepared by dissolving Cyclohexane+O-xylene, PEG200+water, PPG400+toluene to get desired concentrations. Magnetic stirrer [REMI make] was used for this purpose at a rate of 1000 rpm. Refractive index values are measured using thermostatically controlled Abbe Refractometer having accuracy less than 0.0001 units. Density values are measured using specific gravity method. The mass of the liquid is measured using a K-ROY make electronic balance, with an accuracy of ±0.001gm.

3. Results and Discussions

In the present study, refractive index is determined for aqueous mixtures of Cyclohexane+O-xylene, PEG200+water, PPG400+toluene for different concentrations (0 to 1%) at a temperature 303K. The experimental refractive index values of these mixtures are compared with theoretical refractive indices by mathematical models like Lorentz-Lorenz equation, Gladstone-Dale equation, Newton's equation, Arago-Biot(AB) equation, Heller equation and Weiner equation which are listed below.

Lorentz-Lorentz (L-L)

Lorentz-Lorenz (L-L) relation for refractive index is based on the change in the molecular interaction with volume fraction:

$$[(n_m^2 - 1)/(n_m^2 + 2)](1/\rho_m) = \phi_1(n_1^2 - 1/n_1^2 + 2) + \phi_2(n_2^2 - 1/n_2^2 + 2)$$

where, $\phi_1 = w_1/\rho_1$ and $\phi_2 = w_2/\rho_2$, w_1 and w_2 are the weight fractions of the pure components 1 and 2 respectively, and, ρ_m, ρ_1 and ρ_2 are the density of the mixture, and densities of the pure components 1 and 2 respectively and n_m, n_1 and n_2 are the refractive index of the mixture, refractive indices of the pure components 1 and 2 respectively.

Galdstone-Dale relation (G-D)

The Gladstone-Dale relation is used for optical analysis (the determination of composition from optical measurements), or to calculate the density of a liquid for use in fluid dynamics. The relation has also been used to calculate refractive index. Gladstone-Dale (G-D) equation for predicting the refractive index of a binary liquid mixture is as follows

$$(n_m - 1) = \phi_1(n_1 - 1) + \phi_2(n_2 - 1)$$

Where, n_m, n_1 and n_2 are the refractive index of the mixture, refractive indices of the pure components 1 and 2 respectively.

Weiner's relation (W)

Weiner's (W) relation may be represented as

$$(n_m^2 - n_1^2)/(n_m^2 + 2n_1^2) = [(n_2^2 - n_1^2)/(n_2^2 + 2n_1^2)]\phi_2$$

where, $\phi_2 = w_2/\rho_2$ and, w_2 is the weight fraction of the pure component and ρ_2 is the density of the pure component 2 and n_m, n_1 and n_2 are the refractive index of the mixture, refractive indices of the pure components 1 and 2 respectively.

Heller's equation (H)

Heller's (H) equation is given by

$$(n_m - n_1)/n_1 = 3/2[(m^2 - 1)/(m^2 + 2)]\phi_2$$

where, $m = n_2/n_1$, $\phi_2 = w_2/\rho_2$, w_2 is the weight fraction of the pure component 2 and ρ_2 is the density of the pure component 2 and n_m, n_1 and n_2 are the refractive index of the mixture, refractive indices of the pure components 1 and 2 respectively.

Arago-Biot (A-B)

Arago-Biot (A-B), assuming volume additively, proposed the following relation for refractive index of binary liquid mixtures

$$n_m = \phi_1 n_1 + \phi_2 n_2$$

where, $\phi_1 = w_1/\rho_1$ and $\phi_2 = w_2/\rho_2$, w_1 and w_2 are the weight fractions of the pure components 1 and 2 respectively, and, ρ_m, ρ_1 and ρ_2 are the density of the mixture, and densities of the pure components 1 and 2 respectively and n_m, n_1 and n_2 are the refractive index of the mixture, refractive indices of the pure components 1 and 2 respectively.

Newton (N)

Newton (N) gave the following equation

$$(n_m^2 - 1) = \phi_1(n_1^2 - 1) + \phi_2(n_2^2 - 1)$$

where, $\phi_1 = w_1/\rho_1$ and $\phi_2 = w_2/\rho_2$, w_1 and w_2 are the weight fractions of the pure components 1 and 2 respectively, and, ρ_m, ρ_1 and ρ_2 are the density of the mixture, and densities of the pure components 1 and 2 respectively and n_m, n_1 and n_2 are the refractive index of the mixture, refractive indices of the pure components 1 and 2 respectively.

Average percentage error

The percentage of deviation is calculated by the following equation

$$APE = 1/n\sum[(R.I._{exp} - R.I._{cal})/R.I._{exp}]*100$$

Where n is the number of data used, R.I.exp is the experimental refractive index and R.I.cal is the theoretical refractive index. The experimentally measured refractive index and the estimated refractive index from various mathematical models for the binary liquid mixture Cyclohexane+O-Xylene at different concentrations at 303K are presented in the Table 1.

Table 1. Experimental and mathematical values of Refractive index Cyclohexane + O-Xylene

Conc(%)	EXP	A-B	G-D	L-L	W	H
0.0000	1.4200	1.4200	1.4200	1.4200	1.4200	1.4200
0.0908	1.4333	1.4423	1.4409	1.4265	1.4287	1.4284
0.1835	1.4422	1.4604	1.4606	1.4335	1.4373	1.4368
0.2781	1.4495	1.4851	1.4793	1.4413	1.4458	1.4452
0.3747	1.4556	1.4976	1.4968	1.4499	1.4544	1.4536
0.4734	1.4606	1.5232	1.5127	1.4589	1.4631	1.4622
0.5742	1.4654	1.5345	1.5261	1.4651	1.4715	1.4705
0.6772	1.4702	1.5487	1.5387	1.4733	1.4799	1.4790
0.7824	1.4792	1.5509	1.5496	1.4787	1.4880	1.4872
0.8900	1.4919	1.5673	1.5598	1.4880	1.4966	1.4959
1.0000	1.4950	1.5780	1.5685	1.4950	1.5048	1.5043

Table 2. Average percentage values of Cyclohexane + O-Xylene

Conc(%)	EXP	%A-B	%G-D	%L-L	%W	%H
0.0908	1.4333	-0.6279	-0.5302	0.4744	0.32093	0.3418
0.1835	1.4422	-1.2619	-1.2758	0.60324	0.33975	0.3744
0.2781	1.4495	-2.4560	-2.0558	0.56571	0.2552	0.2966
0.3747	1.4556	-2.8854	-2.8304	0.39159	0.0824	0.1374
0.4734	1.4606	-4.2859	-3.5670	0.11639	-0.1711	-0.1095
0.5742	1.4654	-4.7154	-4.1422	0.02047	-0.4162	-0.3480
0.6772	1.4702	-5.3394	-4.6592	-0.2108	-0.6597	-0.5985
0.7824	1.4792	-4.8472	-4.7593	0.03380	-0.5949	-0.5408
0.89	1.4919	-5.0539	-4.5512	0.26141	-0.3150	-0.2681
1	1.495	-5.5518	-4.9163	0.0000	-0.6555	-0.6220
AVG		-3.3659	-3.0261	0.20510	-0.1649	-0.1215

From Table 1, it is observed that the experimental values of refractive index increases with increase in concentration. Arago-Biot relation, Weiner relations, Glad-stone relations shows good agreement with theoretical values of binary liquid mixture Cyclohexane + O-Xylene. The limitations and approximations incorporated in these theories are responsible for the deviations of theoretical values from experimental values.
For the binary liquid mixture Cyclohexane+O-Xylene at 303K the average percentage values are given in Table 2 Further from the table 2 the deviations are more in Lorentz-Lorenz relation. At all concentrations the values obtained from Glad-Stone relations and Arago-Biot relations are found to be nearer. Exactly **Lorentz-Lorenz** relations performs the best of all methods.
The experimentally measured refractive index and the estimated refractive index from various mathematical models for the binary liquid mixture PEG200+Water at different concentrations at 303K are presented in the Table 3.

Table 3. Experimental and mathematical values of Refractive index PEG 200 + Water

Conc(%)	EXP	A-B	G-D	L-L	W	H
2	1.341	1.3458	1.3416	1.3418	1.3397	1.3383
4	1.343	1.3454	1.3418	1.3419	1.3399	1.3385
6	1.346	1.3457	1.3419	1.342	1.3401	1.3387
8	1.348	1.3456	1.342	1.3421	1.3403	1.3389
10	1.350	1.3455	1.3422	1.3422	1.3405	1.3391

Table 4. Average percentage values of PEG 200 + Water

Conc(%)	%EXP	%A-B	%G-D	%L-L	%W	%H
2	1.341	-0.3579	-0.0444	-0.0597	0.0969	0.2013
4	1.343	-0.1787	0.0894	0.0819	0.2308	0.3350
6	1.346	0.0223	0.3046	0.2972	0.4383	0.5423
8	1.348	0.1780	0.4451	0.4377	0.5712	0.6751
10	1.350	0.3333	0.5778	0.5778	0.7037	0.8074
AVG		-0.0006	0.2744	0.2670	0.4082	0.5122

From Table 3, it is observed that the experimental values of refractive index vary with increase in concentration. In Gladstone values the refractive indices first decreases and suddenly increases and further decreases with increase in concentration. It shows that Arago-Biot relation, L-L relations, Glad-stone relations shows good agreement with theoretical values.
Average percentage values for binary liquid mixture PEG 200 + Water is given in Table 4 Further in the Table 4 the deviations are more in Heller's relation. At all concentrations the values obtained from L-L realtions and Arago-Biot relations are found to be nearer. Particularly Arago-Biot relations perform the best of all methods.
The experimentally measured refractive index and the estimated refractive index from the various mathematical models like Arago-Biot, Gladstone-Dale, Lorentz-lorenz, Weiner equation, Heller

equation for the binary liquid mixture of PPG 400 + toluene at five different concentrations at 303K are summarized in the Table 5.

Table 5. Experimental and mathematical values of Refractive index PPG 400 + toluene

Conc(%)	EXP	A-B	G-D	L-L	W	H
2	1.494	1.4935	1.4939	1.4942	1.4947	1.4944
4	1.493	1.4945	1.4937	1.4945	1.4944	1.4942
6	1.492	1.4955	1.4935	1.4947	1.4942	1.494
8	1.49	1.4965	1.4933	1.4949	1.494	1.4938
10	1.489	1.4975	1.4932	1.4951	1.4938	1.4935

From Table 5 it is observed that the experimental values of refractive index decreases with increase in concentration.Whereas the Arago-Biot relation increases with increase in concentration. This table reveals that Arago-Biot relation, L-L relations, Glad-stone relations shows good agreement with theoretical values. The limitations and approximations incorporated in these theories are responsible for the deviations of theoretical values from experimental values.

For the binary liquid mixture PPG 400 + Toluene at 303K the average percentage values are given in Table 6. Also from the Table 6 the deviations are more in Glad-Stone relation. At all concentrations the values obtained from L-L realtions and Arago-Biot relations are found to be nearer. Altogether Arago-Biot relations perform the best of all methods.

Table 6. Average percentage values of PPG 400 + toluene at 303K

Conc(%)	%EXP	%A-B	%G-D	%L-L	%W	%H
2	1.494	0.0335	0.0067	-0.0134	-0.0469	-0.0268
4	1.493	-0.1005	-0.0469	-0.1005	-0.0938	-0.0804
6	1.492	-0.2346	-0.1005	-0.1810	-0.1475	-0.1341
8	1.49	-0.4362	-0.2215	-0.3289	-0.2685	-0.2550
10	1.489	-0.5709	-0.2821	-0.4097	-0.3224	-0.3022
AVG		-0.2617	-0.1289	-0.2067	-0.1758	-0.1597

4. Conclusion

In this present study the experimental refractive index of binary liquid mixtures such as Cyclohexane + O-xylene, PEG200 + water, PPG400 + toluene are compared with their theoretical refractive index for different concentrations at a temperature 303K. From these it may be concluded that the binary mixtures PEG200 + water, PPG400 + toluene showed Arago-Biot relation as the best suitable method for calculating the refractive index in the liquid systems.

Acknowledgments

The authors acknowledge the financial support of SCSVMV, Enathur, Kanchipuram for carrying out this research work.

References

[1] Gayathri A and Venugopal T (2016) *International Journal of Mathematics Trends and Technology* **34** 126
[2] Narendra K, Krishnamurthy P and Srinivasu Ch (2011) *Asian Journal of applied sciences* **4** 535
[3] Dyaneshwar S and Wankhede (2011) *International Journal of Chemistry research* **2** 23
[4] Padmanaban R, Venkatramanan K, Girivel S, Kasthuri K, Usharani A, Gayathri A and Roy Vellaichamy (2017) *Springer Proceedings in Physics* **189** 709
[5] Azhagiri S, Jayakumar S, Padmanaban R, Gunasekaran S and Srinivasan S (2009) *Journal of Solution Chemistry* **38** 441

Institute of Physics
Dirac House, Temple Back
Bristol BS1 6BE UK

ISSN: 1757-8981
ISBN 978-1-5108-6821-2

International Scientific-Technical Conference on Innovative Engineering Technologies, Equipment and Materials (ISTC-IETEM 2017)

IOP Conference Series: Materials Science and Engineering
Volume 412

Kazan, Russia
6 – 8 December 2017

IOP
Institute of Physics

International Scientific-Technical Conference on Innovative Engineering Technologies, Equipment and Materials (ISTC-IETEM 2017)

IOP Conference Series: Materials Science and Engineering
Volume 412

Kazan, Russia
6 – 8 December 2017

IOP
Institute of Physics